# Studies in Logic

Volume 96

# BCK Algebras versus m-BCK Algebras

Foundations

Volume 89
Essays on Set Theory
Akihiro Kanamori

Volume 90
Model Theory for Beginners. 15 Lectures
Roman Kossak

Volume 91
A View of Connexive Logics
Nissim Francez

Volume 92
Aristotle's Syllogistic Underlying Logic: His Model with his Proofs of Soundness and Completeness
George Boger

Volume 93
Truth and Knowledge
Karl Schlechta

Volume 94
A Lambda Calculus Satellite
Henk Barendregt and Giulio Manzonetto

Volume 95
Transparent Intensional Logic. Selected Recent Essays
Marie Duží, Daniela Glavaničovà, Bjørn Jespersen and Miloš Kosterec, eds

Volume 96
BCK Algebras versus m-BCK Algebras. Foundations
Afrodita Iorgulescu

Studies in Logic Series Editor
Dov Gabbay dov.gabbay@kcl.ac.uk

# BCK Algebras versus m-BCK Algebras

## Foundations

Afrodita Iorgulescu

ISBN 978-1-84890-416-3

College Publications
Scientific Director: Dov Gabbay
Managing Director: Jane Spurr

http://www.collegepublications.co.uk

---

**To the memory of William W. McCune**

**To my dear Mother, on the occasion of her centenary**

# Preface

In my previous monograph [116], I have introduced, among many others, the *M, ME, ML* and *MEL algebras* (in the non-commutative case), thus completing the **old frame-work centered on BCK algebras** begun in papers [114] - see Figures 2.1, 2.4. A *MEL algebra* is an algebra $(A, \to, 1)$ of type $(2, 0)$ verifying the properties (M) $(1 \to x = x)$, (Ex) $(x \to (y \to z) = y \to (x \to z))$ and (La') $(x \leq 1)$, where $x \leq y \overset{def.}{\Longleftrightarrow} x \to y = 1$. I also established a general 'bridge theorem' ([116], Theorem 5.3.1) saying that, in the involutive case (i.e. in an algebra $(A, \to, 0, 1)$ with $x^- \overset{def.}{=} x \to 0$ verifying (DN) $((x^-)^- = x)$), the properties (M) and (Ex) are equivalent with the properties (Pcomm) $(x \odot y = y \odot x)$, (PU) $(1 \odot x = x)$ and (Pass) $(x \odot (y \odot z) = (x \odot y) \odot z)$, respectively, of an algebra $(A, \odot, ^-, 1)$ of type $(2, 1, 0)$ verifying (DN).

Then, searching around the *MV algebra* - an algebra $(A, \odot, ^-, 1)$ verifying the properties (PU), (Pcomm), (Pass), $x \odot 0 = 0$ $(0 \overset{def.}{=} 1^-)$, (DN) and ($\wedge_m$-comm) $((x^- \odot y)^- \odot y = (y^- \odot x)^- \odot x)$, where the binary relation $x \leq_m y \overset{def.}{\Longleftrightarrow} x \odot y^- = 0$ can be defined - on April 21st, 2019, the Easter Sunday, I had the 'revelation' that the property $x \odot 0 = 0$ means $x \leq_m 1$, hence it is the analogous property, called by me (m-La') ('m' coming from 'magma'), of the above property (La'); thus, I was able to define a new algebra $(A, \odot, ^-, 1)$, the *m-MEL algebra*, verifying (PU), (Pcomm), (Pass), (m-La') and, hence, I realized that **the MV algebra is just the involutive m-MEL algebra verifying ($\wedge_m$-comm).** So, April 21, 2019, was the starting day of this monograph.

I continued finding the properties (m-B), (m-BB), (m-Re), (m-K) etc., the analogous of the properties (B), (BB), (Re), (K) etc., respectively, from the 'world' of BCK algebras; finally, I introduced many other new algebras, including the *m-BCK algebra* (which is always involutive); the m-BCK algebras contain the MV algebras and the Boolean algebras and they are definitionally equivalent to the involutive BCK algebras (Corollary 17.1.4, in this monograph). Thus, I have obtained a **new frame-work, centered on m-BCK algebras** - see Figure 8.1. A resulting long paper of 102 pages was submitted on September 1, 2019, to Adrian Rezuş [118].

Then, I started to 'incorporate' in this new frame-work the quantum-structures: the bounded involutive lattices, the De Morgan algebras, the ortholattices and the quantum MV algebras, in the joint papers [122], [123], [124] with Michael Kinyon. Michael helped me before in my papers [114], so I asked again for his help in

July 2020; Michael proved theorems and found examples by using the computing program *Prover9/Mace4* (created by his friend William W. McCune). Then, on September 16, 2020, I downloaded from internet *Prover9/Mace4* and I started to work myself with this amazing computing program. Thus, I was able to quickly finish the papers [119], [120], [121], also related to quantum structures.

This monograph has 17 chapters, divided into three parts: Part I contains Chapters 1 - 4, Part II contains Chapters 5 - 15 and Part III contains 'bridge theorems' (Chapters 16, 17), as the following Figure shows. It gathers mainly the results from above mentioned papers, in Chapters 2, 3, 4, 5 (Section 5.2), 6 - 14, 17, in the *commutative case*. My very recent results, never submitted, were also included in the book: results on L algebras, in Section 3.5, and very important final results on quantum structures, with many examples, in Chapter 15; the new results related to quantum-B algebras, pseudo-BCI/pseudo-BCK algebras and prealgebras determined me to consider also the *non-commutative case*, in Chapter 1, Chapter 5 (Sections 5.1, 5.3 - 5.5) and Chapter 16.

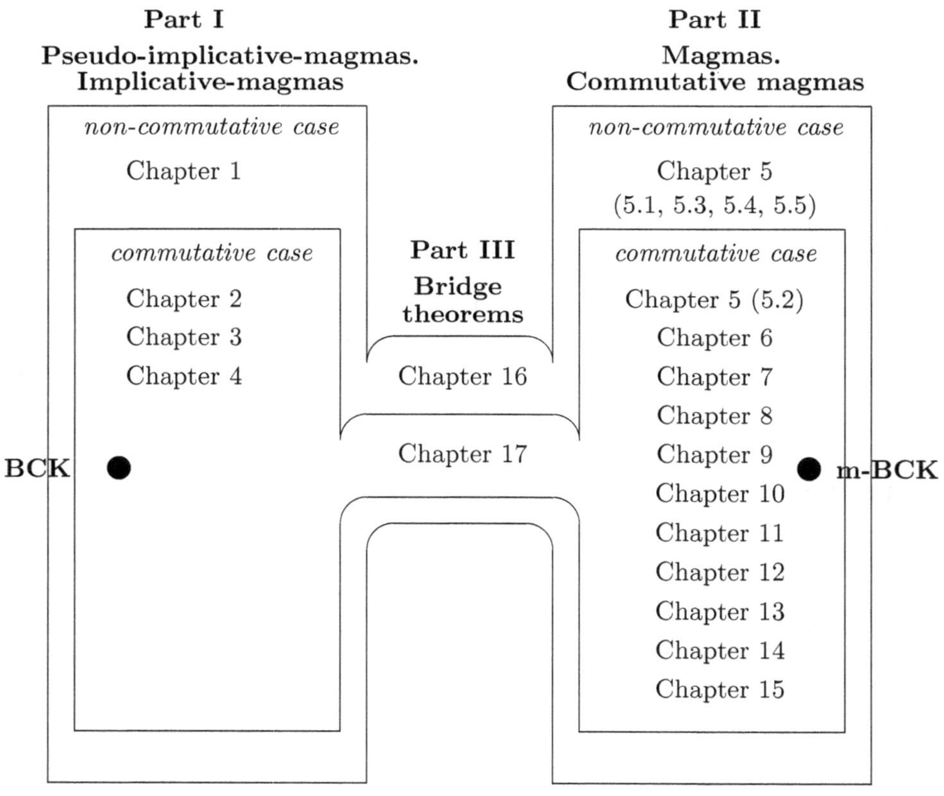


I due first many thanks to Adrian Rezuş, who so kindly accepted to publish my very long paper [118].

I due many, many thanks to Michael Kinyon for our very fruitful colaboration; without his kind help, my book would be poor in results and examples.

I due also thanks to Claudia Mureşan, who very kindly helped me to start working with *Prover9/Mace4*. I also due thanks to people I have never met, like Martin Lukac, Jane Spurr, Jair Minoro Abe, and to Arsham Borumand Saeid, for their help with the publication of the six papers.

I am specially grateful to my colleague George Georgescu, for his encouragement during this work.

But this monograph would not be written in so little time and with so many important results and examples without the help of this wonderful instrument which is *Prover9/Mace4*. Therefore, I dedicate this monograph to the memory of William W. McCune (1953 - 2011).

Finally, thanks to God, I have crossed long years of lockdown (including the pandemic years 2020, 2021) near my dear mother, born on February 1922, so now we can both celebrate healthy my book and her centenary. Therefore, I dedicate also this monograph to my mother, Elisabeta Iorgulescu, former teacher of mathematics, my first mentor.

Afrodita Iorgulescu
afrodita.iorgulescu@ase.ro

Bucharest, October 2022

# Contents

# Introduction

The *algebraists* work usually with the commutative groups defined additively and with the positive (right) cone of a partially-ordered commutative group $(G, \leq, +, -, 0)$, where there are essentially a sum $\oplus = +$ and an element 0. Sometimes, the negative (left) cone is needed also, where there are essentially a product $\odot = +$ and an element $1 = 0$. They work with algebras that have associated an (pre-order) order relation, which usually does not appear explicitely in the definitions. The presence of the (pre-order) order relation implies the presence of the (generalized) duality principle. Thus, each algebra has a dual one, the (pre-order) order relation has a dual one. We have given names to the pair of dual algebras [104], [116], [118]: "left" algebra and "right" algebra, names related to the left-continuity of a t-norm $\odot$ and to the right-continuity of a t-conorm $\oplus$, respectively. Hence, the algebraists usually work with *unital commutative right-magmas.*

By contrary, the *logicians* work with the logic of *truth*, where the *truth* is represented by 1, and there is essentially one implication (two, in the non-commutative case); we could name this logic "left-logic". One can imagine also a "right-logic", as a logic of *false*, where the *false* is represented by 0. Hence, the logicians usually work with commutative *left-algebras of logic* (or *algebras of commutative left-logic*). In this monograph, we shall call these 'implicational' or 'implicative' algebras: *unital left-implicative-magmas* (*unital left-pseudo-implicative-magmas*, in the non-commutative case).

Summarizing, for algebraists, the appropriate algebras are the *unital commutative magmas* and, among these, the appropriate algebras are the *right-algebras.* For logicians, by contrary, the appropriate algebras are the *unital implicative-magmas* and, among these, the appropriate algebras are the *left-algebras.* This explains why, for examples, the MV algebras were initially introduced as unital commutative right-magmas, while the Wajsberg algebras were initially introduced as left-algebras of logic (unital left-implicative-magmas).

In this monograph, since we are coming from (algebras of) logic side, **we shall work with left-algebras** in first place, therefore the unital commutative magmas will be defined multiplicatively, i.e. we shall work with *unital commutative left-magmas*, in first place.

Beside the classical and non-classical logics, there exist the quantum logics. Examples of algebraic structures connected with quantum logics (= quantum structures/algebras) are: the bounded involutive (involution) lattices, the De Morgan algebras, the ortholattices, the orthomodular lattices, the quantum MV algebras (a better name is perhaps *quantum-MV algebras*, because they are generalizations of MV algebras, and not particular cases of MV algebras, i.e. they are not MV algebras that are 'quantum'), etc.

*De Morgan algebras* (De Morgan lattices, in [38]) were introduced by Gr.C. Moisil [156] (see also Antonio Monteiro [157]).

*The ortholattice* is an important example of *sharp* structure (which satisfies the noncontradiction principle) from sharp quantum theory [38] (Birkhoff, 1967; Kalmbach, 1983).

*Orthomodular lattices* (particular ortholattices) generalize the Boolean algebras. They have arisen, cf. [166], "in the study of quantum logic, that is, the logic which supports quantum mechanics and which does not conform to classical logic. As noted by Birkhoff and von Neumann in 1936 [5], the calculus of propositions in quantum logic "is formally indistinguishable from the calculus of linear subspaces [of a Hilbert space] with respect to set products, linear sums and orthogonal complements" in the role of *and*, *or* and *not*, respectively. This has led to the study of the closed subspaces of a Hilbert space, which form an orthomodular lattice in contemporary terminology. As often happens in algebraic logic, the study of orthomodular lattices has tremendously developed, both for their interest in logic and for their own sake, see Kalmbach [133]".

*Quantum-MV algebras* (or QMV algebras) were introduced by Roberto Giuntini in [66] (see also [67], [68], [69], [70], [71], [72], [45]), as non-lattice theoretic generalizations of MV algebras and as non-idempotent generalizations of orthomodular lattices. Cf. [45], from an algebraic point of view, MV algebras and QMV algebras share a common set of axioms, which S. Gudder [76] has called *supplement algebra* (*S algebra*). An MV algebra is an S-algebra verifying the axiom (MV), while an QMV algebra is an S algebra verifying the axiom (QMV), which is weaker than (MV).

Thus, the commutative algebraic structures connected directly or indirectly with classical/non-classical logics belong to two parallel "worlds":

1. the "world" of *commutative left-algebras of logic* (called here *unital left-implicative-magmas*) $(A, \rightarrow, 1)$, where there are (essentially) one implication, $\rightarrow$, and an element 1 verifying the property (M) $(1 \rightarrow x = x)$, called *unit*. Here are some algebras belonging to this 'world':

*Hilbert algebras* were introduced in 1950, in a dual form, by Henkin [83], under the name "implicative model", as a model of positive implicative propositional calculus – an important fragment of classical propositional calculus introduced by Hilbert [86], [87].

*BCK algebras* and *BCI algebras* were introduced in 1966 by K. Iséki [126], as algebraic models of BCK logic and of BCI logic, respectively. Hilbert algebras are particular cases of BCK algebras [104]. Hundred of papers were written on BCK and BCI algebras, and the books [153] and [104] on BCK algebras and the book [91] on BCI algebras. Most of the commutative algebras of logic (such as Wajsberg algebras, implicative-Boolean algebras, Hilbert algebras, $R_0$ algebras, weak-$R_0$ algebras etc.) are particular cases of BCK algebras (see [104]; see also [169], [147]); and the BCK algebras are particular cases of BCI algebras.

*Wajsberg algebras* were introduced in 1984, by Font, Rodriguez and Torrens [56], but they were also considered earlier by Komori in [136], [138], under the name of *CN algebras*; they are a model of $\aleph_0$-valued Łukasiewicz logic, studied by Wajsberg in 1935 [207]. They are termwise equivalent to MV algebras [56].

Several generalizations of BCI or of BCK algebras were introduced in time, namely:

*BCH algebras* were introduced in 1983 by Q.P. Hu and X. Li [89]. There are many papers on BCH algebras since then, but the exact connection between BCH algebras and BCI algebras was presented in [114].

*BCC algebras*, also called *BIK$^+$ algebras*, were introduced in 1984 by Y. Komori [139], [140] (see [218]).

*BZ algebras*, also called *weak-BCC algebras*, were introduced in 1995 by X.H. Zhang and R. Ye [219].

*BH algebras* were introduced in 1998 by Y.B. Jun, E.H. Roh and H.S. Kim [130], as a generalization of BCH and BZ algebras.

*BE algebras* were introduced in 2006 by H.S. Kim and Y.H. Kim [134].

*L algebras* were introduced by Wolfgang Rump in 2008 [182].

*Implicative-Boolean algebras*, a term equivalent definition of Boolean algebras, were introduced in 2009 [105] and presented also in [63], motivated by the axioms system of the classical propositional logic:
(G1) $\varphi \to (\psi \to \varphi)$,
(G2) $(\varphi \to (\psi \to \chi)) \to ((\varphi \to \psi) \to (\varphi \to \chi))$,
(G3) $(\neg\psi \to \neg\varphi) \to (\varphi \to \psi)$.

*CI algebras* were introduced in 2010 by B.L. Meng [151], as a generalization of BE algebras.

*Pre-BCK algebras* were introduced in 2010 by D. Buşneag and S. Rudeanu [19].

In the papers [114] from 2013, 2016, we have introduced new generalizations of BCI or of BCK algebras (*RM, pre-BZ, RME (= CI), RME**, pre-BCI, aRM (= BH), BCH** algebras* and *RML, pre-BCC, aRML, BE**, aBE, aBE** algebras*, respectively, and many others) and, consequently, new generalizations of Hilbert algebras. Namely, we have found thirty one new distinct generalizations of BCI or of BCK algebras and twenty new distinct generalizations of Hilbert algebras. We have presented the hierarchies existing between all these algebras, old or new ones. We have presented proper examples for each old or new algebra.

In the monograph [116] from 2018, we have introduced the *M, ME, ML, MEL algebras* (in the non-commutative case), thus completing the **frame-work centered on BCK algebras** begun in papers [114] - see Figures 2.1, 2.4 - which has in the top of the hierarchy the *M algebras*. The algebras $(A, \to, 1)$, verifying the basic property (M): $1 \to x = x$, are called *left-M algebras* [116], [118]; among the M algebras with additional operations, there are the algebras $(A, \to, 0, 1)$ (where a negation can be defined by: $x^- = x \to 0$), or $(A, \to, ^-, 1)$, with $1^- = 0$, where 1 is the *last element*, verifying (or not) (Ex) (Exchange): $x \to (y \to z) = y \to (x \to z)$; an internal binary relation can be defined by: $x \leq y \overset{def.}{\Longleftrightarrow} x \to y = 1$ ($\leq$ can be a pre-order, an order, or even a lattice order).

We have also introduced the *implicative-group*, a term-equivalent definition of the group, as a particular BCI algebra, in the preprints [107], [108] and in the papers [109], [110] – in the non-commutative case. We have developed this notion in [116].

We have introduced the *implicative-ortholattices* in 2020 [118], as definitionally equivalent notions with the ortholattices.

Note that these algebras (unital left-implicative-magmas) are particular cases of *left-implicative-magmas*, where the existance of the unit 1 is not compulsory (i.e an element 1 may exist, but it does not verify the property (M)) - see for example the *d-algebras* [160].

More generally, we have the "world" of *non-commutative left-algebras of logic* (called here *unital left-pseudo-implicative-magmas*) $(A, \to, \leadsto, 1)$, where there are (essentially) two implications, $\to$ and $\leadsto$, and an element 1 verifying the property (pM) $(1 \to x = x = 1 \leadsto x)$, called *unit*. Here, there are two types of algebras: (i) algebras with a unique binary relation: $x \leq y \stackrel{def.}{\Longleftrightarrow} x \to y = 1 \Longleftrightarrow x \leadsto y = 1$, called *pseudo-algebras* in [116], as for examples the *pseudo-BCI algebras* and the *pseudo-BCK algebras*, recalled in Chapter 1, and (ii) algebras with two different binary relations: $x \ll y \stackrel{def.}{\Longleftrightarrow} x \to y = 1$ and $x \leq y \stackrel{def.}{\Longleftrightarrow} x \leadsto y = 1$, as for examples the *semi-BCI algebras* and the *semi-BCK algebras* [199], recalled in Chapter 1. Note that these non-commutative algebras are particular cases of *pseudo-implicative-magmas*, where the existance of the unit 1 is not compulsory - see, for examples, the *quantum B algebras*, recalled in Chapter 1, where an element 1 does not exist in general, and the *pseudo d-algebras* [131], where an element 1 exists, but not verifying (pM).

2. the "world" of *unital commutative left-magmas* $(A, \odot, 1)$, where there are (essentially) a product, $\odot$, and an element 1 verifying the properties (PU) $(1 \odot x = x = x \odot 1)$ and (Pcomm) $(x \odot y = y \odot x)$. Here are some algebras belonging to this 'world':

The earliest study of *groups* as such probably goes back to the work of Lagrange in the late 18th century. However, this work was somewhat isolated, and 1846 publications of Augustin Louis Cauchy and Galois are more commonly referred to as the beginning of group theory. *Lattice-ordered groups or l-groups* have been studied in different contexts, cf. [1]. The theory blossomed under the leadership of Paul Conrad in the 1960s, cf. [1].

*Boolean algebras* were introduced in 1847 by George Boole [9], [10].

*Residuated lattices*, the algebraic counterpart of logics without contraction rule, were introduced in 1924 by Krull [144].

*MV algebras* were introduced in 1958 by C.C. Chang [23], as a model of $\aleph_0$-valued Łukasiewicz logic. Chang's definition of MV algebras has 17 axioms. There is a huge literature concerning the MV algebras; we mention only a reference book, [24].

*Divisible residuated lattices* (or "divisible integral, residuated, commutative l-monoids" [88]) were introduced in 1965, in a dual, more general form, by Swamy

[202].

*BL algebras* were introduced in 1996, by Petr Hájek [77], [79], [78], as a common generalization of *MV algebras*, *Product algebras* and *Gödel algebras*, in connection with continuous t-norms on the real unit interval [0,1].

*MTL algebras* were introduced in 2001 [50], as algebraic model for the monoidal t-norm logic, a generalization of Hájek's Basic Logic.

*WNM*, *IMTL* and *NM algebras* are particular classes of MTL algebras [50], introduced in 2001 too.

In 2020 [118], we have introduced the *m-MEL, m-BE, m-BCK algebras* etc. ('m' coming from 'magma'), as the analogous of MEL, BE, BCK algebras etc. from the 'world' of algebras of logic. They are unital commutative left-magmas with additional operation, of the form $(A, \odot, ^-, 1)$, with $1^- = 0$, where 1 is the *last element*, verifying (or not) (Pass) (associativity of product) $(x\odot(y\odot z) = (x\odot y)\odot z)$; an internal binary relation can be defined by: $x \leq_m y \Longleftrightarrow x \odot y^- = 0$ ($\leq_m$ can be a pre-order, an order, or even a lattice order). Thus, we have obtained a **new frame-work, centered on m-BCK algebras** - see Figure 8.1. The m-BCK algebras (which are always involutive) contain the MV algebras, the IMTL and NM algebras, the Boolean algebras.

Then, we were able to 'incorporate' in the new frame-work all the quantum structures/algebras (i.e. putting them on the 'map'): first, the bounded involutive lattices, the De Morgan algebras and the ortholattices, in [122], [123]; then, the quantum-MV algebras, in [124], [119], [121]; then, the orthomodular lattices, in [120]. The connections between algebras of logic/algebras and quantum algebras, which were not very clear before, were thus clarified: we have proved that these quantum algebras belong, in fact, to the "world" of *involutive unital commutative left-magmas*.

Note that these algebras (unital commutative left-magmas) are particular cases of *commutative left-magmas*, where the unit 1 is no more compulsory (i.e. an element 1 may exist, but not verifying (PU)) - see for example the *commutative semigroup*.

More generally, we have the "world" of *unital left-magmas* $(A, \odot, 1)$, where there are (essentially) a product, $\odot$, and an element 1 verifying the property (PU) $(1 \odot x = x = x \odot 1)$ - see for example the *monoid*, which verifies (PU) and (Pass). Note that the unital left-magmas are particular cases of *left-magmas*, where the unit 1 is no more compolsory - see for examples the *semigroup*.

Between the two parallel " worlds" there are some connections, 'bridge theorems', as for examples: the equivalence between BCK(P') algebras and pocrims, in the non-involutive case, and the definitional equivalence between Wajsberg algebras and MV algebras, in the involutive case ($(x^-)^- = x$). In [118], two general 'bridge theorems' - recalled in Chapter 17 - connect the two 'worlds' in the involutive commutative case, by the inverse maps $\Phi$ ($x \odot y \stackrel{def.}{=} (x \to y^-)^-$) and $\Psi$ ($x \to y \stackrel{def.}{=} (x \odot y^-)^-$); recall, for examples, that $\leq \Longleftrightarrow \leq_m$, (M) $\Longleftrightarrow$ (PU) + (Pcomm), (Ex) $\Longleftrightarrow$ (Pass) etc. These theorems can be used to prove the *definitionally equivalence* (d.e.) between the analogous involutive (left-) algebras from the two "worlds" simply by choosing appropriate definitions of these algebras.

This monograph has 17 chapters, divided into three parts: Part I (Pseudo-implicative-magmas. Implicative-magmas) (including the 'world' of BCK algebras) contains Chapters 1 - 4; Part II (Magmas. Commutative magmas) (including the 'world' of m-BCK algebras) contains Chapters 5 - 15; Part III contains 'bridge theorems' (Chapters 16, 17). This book gathers mainly results from the papers [114], [118], [122], [123], [124], [119], [120], [121] and [116] in Chapters 2, 3, 4, 5 (Section 5.2), 6 - 14, 17, in the *commutative case*. There are also incorporated in the book some new results, never published: some new results on L algebras, in Section 3.5 (of Chapter 3), and very important final results on quantum structures, with many examples, in Chapter 15. The new results related to quantum-B algebras determined us to consider also the *non-commutative case*, in Chapter 1, Chapter 5 (Sections 5.1, 5.3 - 5.5) and Chapter 16.

The monograph is written in a **unifying way**, which consists in **fixing unique names for the defining properties**, making lists of these properties and then using them for defining the different algebras and for obtaining results.

The Bibliography has a minimum of titles, otherwise it would be huge.

I love the geographical (tourist) maps; in a book of geography, a map says more than thousand words. Similarly, I consider that in a book of mathematics, a "map", i.e. a Figure showing the connections between the elements of an algebra or between different algebras, says more that thousand words. Therefore, in this monograph there are many Figures.

**Part I (Pseudo-implicative-magmas. Implicative-magmas)** has four chapters:

In **Chapter 1 (Pseudo-BCI algebras and pseudo-BCK algebras vs. quantum-B algebras. Implicative-groups)**, we discuss about different notions related to quantum-B algebras (QB algebras), in order to better understand these 'algebras' introduced by W. Rump. Therefore, this chapter is the only one, in Part I, which deals with the non-commutative case. We introduce the notions of pseudo-implicative-magmas (pi-magmas) and implicative-magmas (i-magmas), the notions of pseudo-exchange (pe) and pseudo-residoid (pr). We recall the notions of pseudo-BCI algebra (pBCI algebra) and pseudo-BCK algebra (pBCK algebra) from [116]; we present 10 old or new equivalent definitions of pBCI and pBCK algebras, needed in the sequel and we recall another non-commutative generalization of BCI/BCK algebras, called semi-BCI/BCK algebras. We recall facts about QB algebras. We discuss about partially ordered pi-magmas (po-pi-magmas) and lattice ordered pi-magmas ($l$-pi-magmas), namely we introduce po-pe ($l$-pe), po-pr ($l$-pr), po-mpr ($l$-mpr) and po-ipr ($l$-ipr) and show their connections with QB algebras and pBCI algebras; this section is entirely new. We recall facts about implicative-groups from [116]. We discuss about the p-semisimple (or discrete) property (p-s) and about p-semisimple pseudo-structures and po-pi-magmas; this section is partially new. We discuss some logics, namely the pseudo-BCK→ logic, the pseudo-BCI→ (or QB) logic, the pseudo-BCI→' (or QB') logic, the pseudo-BCI→"(or QB") logic; this section is partially new. We introduce and study some prealgebras/prestructures, namely the pseudo-BCK→ prealgebras, the pseudo-BCI→ (or QB) prestructures, the pseudo-BCI→' (or QB') prealgebras, the pseudo-BCI→" (or QB") prealgebras; this section is entirely new.

In **Chapter 2 (M algebras)**, we make an overview in an unifying way of the most important 24 M algebras introduced/analysed in [114] and [116], with many clarifications and new results. We recall the list **A** of basic properties, the 24 M algebras and the "Big map" connecting them. We recall the list **B** of particular properties and some other M algebras. We study in some detail the BCK algebras and the Hilbert algebras and we establish connections between classes of BCK algebras. We discuss about (involutive) negations. The content is taken mainly from [118].

In **Chapter 3 (M algebras $(A, \rightarrow, 0, 1)$, or $(A, \rightarrow, ^-, 1)$, with $1^- = 0$)**, we analyse in some detail the M algebras *with last element,* 1, with some additional operations, of the form $(A, \rightarrow, 0, 1)$ or $(A, \rightarrow, ^-, 1)$ with $1^- = 0$, which determine Hierarchies $2^b$, $2'^b$. We present new results. We introduce the new notion of *implicative-ortholattice* as a term-equivalent definition of the ortholattice recalled and studied in Chapter 6. We establish connections mainly between BE algebras, BCK algebras, Wajsberg algebras, implicative-ortholattices and implicative-Boolean algebras. We present the list **C** of some properties of negation. We discuss about involutive MEL algebras. We discuss about involutive BCK algebras, namely about Wajsberg algebras, weak-$R_0$ algebras and $R_0$ algebras, implicative-Boolean algebras. We present new results on L algebras introduced by W. Rump and some examples. The content of this chapter is taken mainly from [118].

In **Chapter 4 (M algebras $(A, \rightarrow, 1)$, or $(A, \rightarrow, ^{-1}, 1)$, with $1^{-1} = 1$)**, we analyse in some detail the M algebras *without last element*, with some additional operations, of the form $(A, \rightarrow, 1)$ or $(A, \rightarrow, ^{-1}, 1)$ with $1^{-1} = 1$, which determine Hierarchies 1, 1'. We present new results and the connection with the implicative-group and the implicative-goop introduced in [116]. We present the list $\mathbf{C}_1$ of properties of the negation $^{-1}$ and some results. We discuss about the p-semisimple property (p-s) and the p-semisimple algebras. The content of this chapter is taken mainly from [118].

**Part II (Magmas. Commutative magmas)** has eleven chapters:

In **Chapter 5 (Preliminaries/Miscellany)**, we recall some notions needed in Part II (in commutative case) or related to notions from Chapter 1 (in non-commutative case). We recall the notions of magma and commutative magma and discuss about semigroups and monoids. We recall definitions and facts about lattices, namely Ore and Dedekind equivalent definitions of lattices. We discuss about partially ordered magmas (po-magmas) and about lattice ordered magmas ($l$-magmas), namely about po-s ($l$-s), po-m ($l$-m), po-mm ($l$-mm), po-im ($l$-im). We recall things about groups from [116]. We discuss about the m-p-semisimple property (m-p-s) and about m-p-semisimple po-magmas.

In **Chapter 6 (m-M algebras)**, we introduce 12 new algebras, called *m-M algebras* here for the first time, (including the m-BCK algebras) as particular cases of unital commutative magmas with some additional operation of the form $(A, \odot, ^-, 1)$ with $1^- = 0$, which determine Hierarchies m-2, m-2'. We present some particular properties and properties of the negation. We study in some details the involutive m-MEL and m-BE algebras and the connections between m-M algebras and MV algebras. Most of the results from this chapter are taken mainly from [118] and [124].

In **Chapter 7 ($m_1$-M algebras)**, we introduce 10 new algebras as particular cases of unital commutative magmas with some additional operation of the form $(A, \cdot, ^{-1}, 1)$ with $1^{-1} = 1$, called *$m_1$-M algebras* here for the first time, which determine Hierarchies $m_1$-1, $m_1$-1'. We present the properties of the inverse (negation) $^{-1}$ and we study the $m_1$-p-semisimple algebras. We make the connections with the commutative group, moon [116], goop [116]. The content of this chapter is taken mainly from [118].

In **Chapter 8 (The "Big m-map" - final connections between the algebras defined in Chapters 6, 7)**, we establish the global hierarchies of the new algebras, containing Hierarchies $m_1$-1 and m-2, $m_1$-1' and m-2'. The content of this chapter is taken mainly from [118].

In **Chapter 9 (Putting bounded involutive lattices, De Morgan algebras, ortholattices and Boolean algebras on the "map")**, we continue the results from Chapter 6. We extend the initial "Big m-map" from Chapter 8 with a "Little m-map" and study the involutive case. We redefine the bounded involutive lattices (**BIL**) and the De Morgan algebras (**De Morgan**) as involutive m-MEL algebras. Finally, we put **BIL** and **De Morgan**, and their subclasses, on the "involutive Little m-map". We redefine also the ortholattices (**OL**) and the Boolean algebras (**Boole**) as involutive m-BE algebras. Finally, we put **OL**, and its subclass, and **Boole** on the "involutive Little m-map". We present 18 examples of the involved algebras. The content of this chapter is taken from [122] and [118].

In **Chapter 10 (Two generalizations of bounded involutive lattices and of ortholattices)**, we continue the results from Chapter 9. We introduce and study two dual independent absorbtion laws: (m-Wabs-i) and (m-Vabs-i) and two generalizations of bounded lattices (**BL**): bounded softlattices (**BSL**) and bounded widelattices (**BWL**). We then introduce and study two corresponding generalizations of **BIL**: bounded involutive softlattices (**BISL**) and bounded involutive widelattices (**BIWL**). Finally, we put **BIL**, **BISL**, **BIWL**, and their subclasses, on the "involutive Little m-map". We then introduce and study two corresponding generalizations of **OL**: orthosoftlattices (**OSL**) and orthowidelattices (**OWL**). We prove that transitive and antisymmetric orthowidelattices are MV algebras; thus, a proper subclass of MV algebras (**taOWL**) is obtained; a generalization of ($\wedge_m$-comm) property, called ($\Delta_m$), is introduced on this occasion. We put **OL**, **OSL**, **OWL**, and their subclasses, on the "involutive Little m-map". We present 11 examples of the various algebras discussed herein. The content of this chapter is taken from [123].

In **Chapter 11 (Putting quantum MV algebras on the "map")**, we redefine the quantum MV algebras introduced by R. Giuntini as involutive m-BE algebras verifying (Pqmv). We dig around the structure of quantum-MV (QMV) algebras and obtain a decomposition of (Pqmv) into only two properties: ($\Delta_m$) and

(Pom). We shall thus introduce three generalizations of QMV algebras (**QMV**): the pre-MV algebras (**PreMV**), the metha-MV algebras (**MMV**) and the orthomodular algebras (**OM**). We shall also introduce and study the transitive QMV algebras (**tQMV**), the transitive PreMV algebras (**tPreMV**), the transitive MMV algebras (**tMMV**) and the transitive OM algebras (**tOM**). We prove that MV algebras coincide with the antisymmetric QMV algebras (**aQMV**) - but also with **aPreMV** and with **aMMV**. Consequently, MV algebras and QMV algebras, and also tQMV algebras, will be put on the same "map" (involutive "Big m-map"). The *taOM algebra*, a proper generalization of MV algebra inside the class of m-BCK algebras, is put in evidence. The content of this chapter is taken from [124].

In **Chapter 12 (Orthomodular algebras)**, we continue the results from Chapter 11: we analyse in some details the orthomodular (OM) algebras, with a special insight on taOM algebras. We prove that almost all the properties of QMV algebras are also verified by OM algebras; we put OM algebras on the "map". We mainly prove that any m-BCK algebra verifies the property (trans) (the binary relation $\leq_m^M$ is transitive) and we introduce and analyse the so called *trans algebras.* We prove the definitional equivalence between involutive residuated lattices and m-BCK lattices, thus putting IMTL, NM, MV and ${}_{(WNM)}$MV algebras on the same "map". We present 15 examples of the involved algebras. The content of this chapter is taken from [119].

In **Chapter 13 (Two generalizations of orthomodular lattices)**, we continue the results from Chapters 10 - 12. We study in some details the orthomodular lattices (**OML**). We prove that transitive OMLs coincide with Boolean algebras. We introduce and study two generalizations of **OML**: the orthomodular softlattices (**OMSL**) and the orthomodular widelattices (**OMWL**). We prove that the **OMSL** coincide with **OML** and that **OMSL** $\subset$ **OMWL**. We prove that OMWLs are a (proper) subclass of QMV algebras. Hence, the transitive OMWLs are a proper subclass of transitive QMV algebras. We present 23 examples of the various algebras discussed herein. The content of this chapter is taken mainly from [120].

In **Chapter 14 (The properties (m-Pabs-i) and ($\mathbf{WNM}_m$))**, we continue the results from Chapters 11 - 13. We introduce the properties ($\text{WNM}_m$) and ($\text{aWNM}_m$) in connection with the property (WNM) introduced in [50] and recalled in Chapter 12, and we prove that, in an MV algebra, we have:

(m-Pabs-i) $\Longleftrightarrow$ ($\text{aWNM}_m$) $\Longleftrightarrow$ ($\text{WNM}_m$) (= (WNM)).

We generalize the above result, proving that, in a transitive QMV algebra, we have:

(m-Pabs-i) $\Longleftrightarrow$ ($\text{aWNM}_m$) $\Longleftrightarrow$ ($\text{WNM}_m$).

The content of this chapter is taken from [121].

In **Chapter 15 ((Involutive) m-BCK algebras and involutive m-pre-BCK algebras)**, we continue the results from the previous chapters culminating here with the build of transitive quantum algebras. We prove that any IMTL algebra verifies ($\text{prel}_m$) and any NM algebra verifies ($\text{prel}_m$) and ($\text{WNM}_m$). We introduce, on any involutive m-pre-BCK algebra, a congruence relation, $\|_Q$, based on the pre-order $\leq_m$, and the corresponding quotient algebra, that is an m-BCK algebra. We introduce and study two new quantum structures/algebras, the tQIMTL and tQNM algebras, as non-lattice generalizations of (transitive) IMTL and NM algebras, respectively. We give a general method, the "method of Q-parallel rows/columns", and a particular method, the "method of identic rows/columns", to obtain involutive m-pre-BCK algebras, hence transitive quantum algebras, from a given finite m-BCK algebra. We present 29 types of examples; we build transitive quantum algebras: involutive m-pre-BCK algebras/lattices, tQIMTL, tQNM, tPreMV, tMMV algebras, tOMWLs = $_{(WNM_m)}$tQMV algebras, tOWLs, tOSLs, tOLs, tQMV algebras. Conclusions end the chapter. The content of this chapter is entirely new.

**Part III (Bridge theorems between the two 'worlds')** has two chapters:

In **Chapater 16 (Connections in the non-commutative case)**, we present bridge theorems in the non-commutative case. We present bridge theorems connecting po($l$)-pi-magmas (Chapter 1) and po($l$)-magmas (Chapter 5) and also i-groups (Chapter 1) and groups (Chapter 5). The content of this chapter is taken mostly from [116].

In **Chapter 17 (Connections between M and m-M, m$_1$-M algebras, in the involutive case)**, we present bridge theorems in the commutative case. We establish two general theorems connecting the two 'worlds', in the involutive case: one connecting bounded M algebras *with last element* with m-M algebras, the other connecting M algebras *without last element* with m$_1$-M algebras. These theorems can be used to prove the *definitionally equivalence* (d.e.) between the analogous involutive algebras from the two 'worlds', simply by choosing appropriate definitions of the algebras. The content of this chapter is taken mainly from [118].

# Part I

# Pseudo-implicative-magmas. Implicative-magmas

# Chapter 1

# Pseudo-BCI algebras and pseudo-BCK algebras vs. quantum-B algebras. Implicative-groups

In this chapter, we discuss about different notions related to quantum-B algebras, in order to better understand the quantum-B algebras introduced by W. Rump. Therefore, this chapter is the only one, in Part I, which deals with the non-commutative case.

In Section 1.1, we introduce the notions of pseudo-implicative-magmas (pi-magmas) and implicative-magmas (i-magmas) - in the 'world' of algebras of logic, as the analogous of magmas and commutative magmas - in the 'world' of algebras (magmas). We also introduce here the notions of pseudo-exchange (pe) and pseudo-residoid (pr), as the analogous of the notions of semigroups and monoids - see Section 5.1 from Part II. We recall the notions of pseudo-algebra and pseudo-structure (of logic), the list **pA** of basic properties and the pseudo-M algebras from [116].

In Section 1.2, we present 10 old or new equivalent definitions of pBCI/pBCK algebras needed in the sequel. We make the connection with another non-commutative generalizations of BCI/BCK algebras, namely the semi-BCI/semi-BCK algebras introduced in [199].

In Section 1.3, we recall facts about quantum-B algebras (QB algebras). We present three equivalent definitions, two old ones and one new one. We recall the locally unital quantum-B algebras and the unital, the maximal and the integral quantum-B algebras.

In Section 1.4, we discuss about the partially ordered pi-magmas (po-pi-magmas) and the lattice ordered pi-magmas ($l$-pi-magmas), namely we introduce po-pe ($l$-pe), po-pr ($l$-pr), po-mpr ($l$-mpr) and po-ipr ($l$-ipr) - the analogous of the po($l$)-

magmas discussed in Section 5.3 from Part II - and show their connections with QB algebras and pBCI algebras. Here also we study po-pi-magmas with (pP') and with (pRP') and their connections. This section is entirely new, never published before.

In Section 1.5, we recall facts about implicative-groups and we present some examples - in connection with Section 5.4 from Part II.

In Section 1.6, we discuss about the p-semisimple (or discrete) property (p-s) and about p-semisimple pseudo-structures and po-pi-magmas. This section is partially new.

In Section 1.7, we discuss some logics. We also discuss about the connections between these logics and their algebras (of logic). This section is partially new.

In Section 1.8, in connection with the previous section, we introduce and study some prealgebras/prestructures. We also study the connections between these prealgebras/prestructures and their algebras/structures. We present some examples. This section is entirely new, never published before.

## 1.1 Pseudo-implicative-magmas

We introduce here the following definitions, by analogy with magmas, unital magmas and commutative magmas, unital commutative magmas, respectively (see Part II).

**Definitions 1.1.1**

- A *pseudo-implicative-magma*, or a *pseudo-i-magma*, or a *pi-magma* for short, is an algebra $(A, \Rightarrow, \approx>)$ of type $(2, 2)$.

- A *unital pi-magma* is an algebra $(A, \Rightarrow, \approx>, u)$ of type $(2, 2, 0)$ verifying the axiom: for all $x \in A$,

(pM) $\quad u \Rightarrow x = x = u \approx> x.$

We have a *left-notation* and a *right-notation* for pi-magmas. A pi-magma in left-notation (right-notation) will be called a *left-pi-magma* (*right-pi-magma*, respectively). Thus, we have, for example:

**Definitions 1.1.2**

- A *unital left-pi-magma* is an algebra $(A^L, \rightarrow=\rightarrow^L, \rightsquigarrow=\rightsquigarrow^L, 1)$ of type $(2, 2, 0)$ verifying the axiom: for all $x, y \in A^L$,

(pM) $\quad 1 \rightarrow x = x = 1 \rightsquigarrow x,$

i.e. 1 is the *unit* element of $A^L$.

- A *unital right-pi-magma* is an algebra $(A^R, \rightarrow^R, \rightsquigarrow^R, 0)$ of type $(2, 2, 0)$ verifying the axiom: for all $x, y \in A^R$,

($pM^R$) $\quad 0 \rightarrow^R x = x = 0 \rightsquigarrow^R x,$

i.e. 0 is the *unit* element of $A^R$.

Denote by **pi-magma** the class of all left-pi-magmas and by **upi-magma** the class of all unital left-pi-magmas. We have: **upi-magma** $\subset$ **pi-magma**.

**Definitions 1.1.3**

- A *commutative pi-magma*, or an *implicative-magma*, or an *i-magma* for short, is a pi-magma $(A, \Rightarrow, \approx>)$ verifying $\Rightarrow = \approx>$, i.e. is an algebra $(A, \Rightarrow)$ of type (2).
- A *unital i-magma* is an algebra $(A, \Rightarrow u)$ of type $(2, 0)$ verifying the axiom: for all $x \in A$,

(M) $\qquad u \Rightarrow x = x.$

Denote by **i-magma** the class of all left-i-magmas and by **ui-magma** the class of all unital left-i-magmas. We have: **ui-magma** $\subset$ **i-magma**.

### 1.1.1 Pseudo-exchanges. Pseudo-residoids

We shall introduce, in the 'world' of pi-magmas, the new notions of *pseudo-exchange* and *pseudo-residoid* as the analogous of the notions of *semigroup* and *monoid*, respectively, from the 'world' of magmas.

**Definitions 1.1.4** (In left-notation)

(1) A *left-pseudo-exchange*, or *left-p-exchange* or *left-pe* for short, is a left-pi-magma $\mathcal{A}^L = (A^L, \rightarrow, \rightsquigarrow)$ verifying the axiom: for all $x, y, z \in A^L$,

(pEx) (pseudo-Exchange) $\qquad x \rightarrow (y \rightsquigarrow z) = y \rightsquigarrow (x \rightarrow z).$

(2) A *left-pseudo-residoid*, or *left-p-residoid* or *left-pr* for short, is a unital left-pi-magma $\mathcal{A}^L = (A^L, \rightarrow, \rightsquigarrow, 1)$ verifying the axiom (pEx) (beside (pM)), i.e. is a unital left-p-exchange.

Denote by **p-exchange** the class of all left-p-exchanges and by **p-residoid** the class of all left-p-residoids. Hence, we have:

**p-exchange** = **pi-magma** + (pEx) and **p-residoid** = **upi-magma** + (pEx),

(1.1) $\qquad \mathbf{p-residoid} \subset \mathbf{p-exchange} \subset \mathbf{pi-magma}.$

In commutative case, we have the new notions of *exchange* and *residoid* as the analogous of the notions of *commutative semigroup* and *commutative monoid*, respectively, from the 'world' of commutative magmas.

**Definitions 1.1.5** (In left-notation)

(1') A *left-exchange* is a left-i-magma $\mathcal{A}^L = (A^L, \rightarrow)$ verifying the axiom: for all $x, y, z \in A^L$,

(Ex) (Exchange) $\qquad x \rightarrow (y \rightarrow z) = y \rightarrow (x \rightarrow z).$

(2') A *left-residoid* is a unital left-i-magma $\mathcal{A}^L = (A^L, \rightarrow, 1)$ verifying the axiom (Ex) (beside (M)), i.e. is a unital left-exchange.

Denote by **exchange** the class of all left-exchanges and by **residoid** the class of all left-residoids. Hence, we have:

**exchange** = **i-magma** + (Ex) and **residoid** = **ui-magma** + (Ex),

(1.2) $\qquad \mathbf{residoid} \subset \mathbf{exchange} \subset \mathbf{i-magma}.$

### 1.1.2 Pseudo-algebras, pseudo-structures

We recall from [116] the general equivalent notions of *pseudo-algebra* and *pseudo-structure* of logic, the non-commutative notions corresponding to *algebra* and *structure* of logic, respectively, in the commutative case. Following the context, we shall use one or the other equivalent notion.

**Definitions 1.1.6** ("L" comes from "Left" )

(i) A *left-pseudo-algebra of logic*, or a *left-pseudo-algebra* for short, is an algebra $\mathcal{A}^L = (A^L, \to=\to^L, \rightsquigarrow=\rightsquigarrow^L, 1)$ of type $(2,2,0)$ verifying the property: for all $x, y \in A^L$,

$$(IdEqR) \quad x \to y = 1 \Longleftrightarrow x \rightsquigarrow y = 1.$$

An *internal* binary relation $\leq$ can then be defined on $A^L$ by: for all $x, y \in A^L$,

$$(pdfrelR) \quad x \leq y \stackrel{def.}{\Longleftrightarrow} x \to y = 1 \; (\stackrel{(IdEqR)}{\Longleftrightarrow} x \rightsquigarrow y = 1).$$

(i') Any algebra $\mathcal{A}'^L = (A^L, \sigma)$ whose signature $\sigma$ contains $\to, \rightsquigarrow$, 1 and (IdEqR) is verified is also called *left-pseudo-algebra*.

(i") Any algebra $\mathcal{A}''^L = (A^L, \tau)$ which is term equivalent to a left-pseudo-algebra $\mathcal{A}'^L = (A^L, \sigma)$ is also called *left-pseudo-algebra*.

Equivalently,

(j) a *left-pseudo-structure of logic*, or a *left-pseudo-structure* for short, is a structure $\mathcal{A}^{\mathcal{L}} = (A^L, \leq, \to, \rightsquigarrow, 1)$, with a binary relation $\leq$ on $A^L$, two binary operations on $A^L$, $\to$ and $\rightsquigarrow$, and an element $1 \in A^L$ such that the above property (IdEqR) holds and $\leq$ and $\to$, 1 are connected by:

$$(pEqrelR) \quad x \leq y \Longleftrightarrow x \to y = 1 \; (\stackrel{(IdEqR)}{\Leftrightarrow} x \rightsquigarrow y = 1),$$

equivalence which says that $\leq$ is an *internal* binary relation.

(j') Any structure $\mathcal{A}'^L = (A^L, \leq, \sigma)$ whose signature $\sigma$ contains $\to, \rightsquigarrow$, 1 and (IdEqR) and (pEqrelR) are verified is also called *left-pseudo-structure*.

(j") Any structure $\mathcal{A}''^L = (A^L, \leq, \tau)$ which is term equivalent to a left-pseudo-structure $\mathcal{A}'^L = (A^L, \leq, \sigma)$ is also called *left-pseudo-structure*.

Denote by **pseudo-algebra** the class of all left-pseudo-algebras. We have: **pseudo-algebra** $\subset$ **upi-magma**.

The dual definitions of *right-pseudo-algebra* and *right-pseudo-structure* are obvious and are omitted.

**Remark 1.1.7** *In Definitions 1.1.6, the equivalence*

$$x \leq y \Longleftrightarrow x \to y = 1 \; (\stackrel{(IdEqR)}{\Longleftrightarrow} x \rightsquigarrow y = 1)$$

*is used either*
*- as the definition (pdfrelR) of the binary relation $\leq$, in the algebra $(A^L, \to, \rightsquigarrow, 1)$, or*
*- as the connection (pEqrelR) between the binary relation $\leq$ and the operations $\to, \rightsquigarrow, 1$, in the structure $(A^L, \leq, \to, \rightsquigarrow, 1)$.*

• The left-pseudo-algebra (left-pseudo-structure) is *commutative* if $\to\ =\ \rightsquigarrow$, and in this case: the unique binary operation is denoted by $\to$, (IdEqR) becomes superfluous, (pdfrelR) becomes (dfrelR), (pEqrelR) becomes (EqrelR) and we say that we have a *left-algebra* (*left-structure*, respectively) (of logic) (see Chapter 2).

**Remark 1.1.8** *The pseudo-algebras/pseudo-structures are unital pi-magmas, but there are unital pi-magmas which are not pseudo-algebras/pseudo-structures, i.e. where the two binary operations (implications) and* 1 *determine two different binary relations,* $\ll$ *and* $\leq$ *- see the semi-BCI and the semi-BCK algebras recalled in the next section.*

### 1.1.3 List pA of basic properties and pseudo-M algebras

Let $\mathcal{A}^L = (A^L, \to, \rightsquigarrow, 1)$ be a left-pseudo-algebra or, equivalently,
let $\mathcal{A}^L = (A^L, \leq, \to, \rightsquigarrow, 1)$ be a left-pseudo-structure.

Consider the following list **pA** of basic properties that can be verified by $\mathcal{A}^L$: for all $x, y, z \in A^L$,

(An) (Antisymmetry) $x \to y = 1$ and $y \to x = 1 \Longrightarrow x = y$,
(An') (Antisymmetry) $x \leq y$ and $y \leq x \Longrightarrow x = y$;

(pB) = (pB$_\to$) + (pB$_\rightsquigarrow$), where:
(pB$_\to$) $(z \to x) \to ((y \to z) \to (y \to x)) = 1$,
(pB$_\rightsquigarrow$) $(z \rightsquigarrow x) \to ((y \rightsquigarrow z) \rightsquigarrow (y \rightsquigarrow x)) = 1$,
(pB') = (pB$_\to$') + (pB$_\rightsquigarrow$'), where
(pB$_\to$') $z \to x \leq (y \to z) \to (y \to x)$,
(pB$_\rightsquigarrow$') $z \rightsquigarrow x \leq (y \rightsquigarrow z) \rightsquigarrow (y \rightsquigarrow x)$;

(pBB) = (pBB$_1$) + (pBB$_2$), where:
(pBB$_1$) $(y \to z) \to ((z \to x) \rightsquigarrow (y \to x)) = 1$,
(pBB$_2$) $(y \rightsquigarrow z) \to ((z \rightsquigarrow x) \to (y \rightsquigarrow x)) = 1$,
(pBB') = (pBB$_1$') + (pBB$_2$'), where:
(pBB$_1$') $y \to z \leq (z \to x) \rightsquigarrow (y \to x)$,
(pBB$_2$') $y \rightsquigarrow z \leq (z \rightsquigarrow x) \to (y \rightsquigarrow x)$;

(p*) = (p*$_\to$) + (p*$_\rightsquigarrow$), where:
(p*$_\to$) $x \to y = 1 \Longrightarrow (z \to x) \to (z \to y) = 1$,
(p*$_\rightsquigarrow$) $x \to y = 1 \Longrightarrow (z \rightsquigarrow x) \to (z \rightsquigarrow y) = 1$,
(p*') = (p*$_\to$') + (p*$_\rightsquigarrow$'), where:
(p*$_\to$') $x \leq y \Longrightarrow z \to x \leq z \to y$,
(p*$_\rightsquigarrow$') $x \leq y \Longrightarrow z \rightsquigarrow x \leq z \rightsquigarrow y$);

(p**) $x \to y = 1 \Longrightarrow ((y \to z) \to (x \to z) = 1,\ (y \rightsquigarrow z) \to (x \rightsquigarrow z) = 1)$;
(p**') $x \leq y \Longrightarrow (y \to z \leq x \to z,\ y \rightsquigarrow z \leq x \rightsquigarrow z)$;

(pC) =(pC$_1$) + (pC$_2$), where:

(pC$_1$) $(x \to (y \rightsquigarrow z)) \to (y \rightsquigarrow (x \to z)) = 1$,
(pC$_2$) $(x \rightsquigarrow (y \to z)) \to (y \to (x \rightsquigarrow z)) = 1$,
(pC') =(pC$_1$') + (pC$_2$'), where:
(pC$_1$') $x \to (y \rightsquigarrow z) \leq y \rightsquigarrow (x \to z)$,
(pC$_2$') $x \rightsquigarrow (y \to z) \leq y \to (x \rightsquigarrow z)$;

(pD) $y \to ((y \to x) \rightsquigarrow x) = 1, \ y \to ((y \rightsquigarrow x) \to x) = 1$,
(pD') $y \leq (y \to x) \rightsquigarrow x, \ y \leq (y \rightsquigarrow x) \to x$;

(pEx) $x \to (y \rightsquigarrow z) = y \rightsquigarrow (x \to z)$;

(Fi) (First element) $0 \to x = 1$,
(Fi') (First element) $0 \leq x$;

(pH) $((x \to y) \rightsquigarrow y) \to y = x \to y, \quad ((x \rightsquigarrow y) \to y) \rightsquigarrow y = x \rightsquigarrow y$;

(pK) = (pK$_\to$) + (pK$_\rightsquigarrow$), where:
(pK$_\to$) $x \to (y \to x) = 1$,
(pK$_\rightsquigarrow$) $x \to (y \rightsquigarrow x) = 1$,
(pK') = (pK$_\to$') + (pK$_\rightsquigarrow$'), where:
(pK$_\to$') $x \leq y \to x$,
(pK$_\rightsquigarrow$') $x \leq y \rightsquigarrow x$;

(pKLa) $x \rightsquigarrow (y \to x) = y \to 1, \ x \to (y \rightsquigarrow x) = y \rightsquigarrow 1$;

(La) (Last element) $x \to 1 = 1$,
(La') (Last element) $x \leq 1$;

(pM) $1 \to x = x = 1 \rightsquigarrow x$;

(N) (Maximal element) $1 \to x = 1 \Longrightarrow x = 1$ (*modus ponens*),
(N') (Maximal element) $1 \leq x \Longrightarrow x = 1$ (*modus ponens*);

(Re) (Reflexivity) $x \to x = 1$,
(Re') (Reflexivity) $x \leq x$;

(Tr) (Transitivity) $x \to y = 1$ and $y \to z = 1 \Longrightarrow x \to z = 1$,
(Tr') (Transitivity) $x \leq y$ and $y \leq z \Longrightarrow x \leq z$;

(pEq#) $x \rightsquigarrow (y \to z) = 1 \Longleftrightarrow y \to (x \rightsquigarrow z) = 1$,
(pEq#') $x \leq y \to z \Longleftrightarrow y \leq x \rightsquigarrow z$;

(pEqrel) $x \leq y \iff 1 \leq x \to y \iff 1 \leq x \rightsquigarrow y$,

where (pdfrelR) $\Longrightarrow$ ((X') $\stackrel{def.}{\Leftrightarrow}$ (X)), for any property (X) of a left-pseudo-algebra

$\mathcal{A}^L$, while (pEqrelR) $\Longrightarrow$ ((X') $\Leftrightarrow$ (X)), for any property (X') of a left-pseudo-structure $\mathcal{A}^L$.

**Conventions.** In order to simplify the writting, if the algebra $(A^L, \to, \leadsto, 1)$ verifies the property (Re), for example, then we shall no more mention that the associated structure $(A^L, \leq, \to, \leadsto, 1)$ verifies the associated property (Re'). Also, we shall freely use (Re) or (Re') in a proof.

### 1.1.3.1 Pseudo-M algebras

*Pseudo-BCK algebras* were introduced by G. Georgescu and A. Iorgulescu in 2001 [62], as a non-commutative generalization of Iseki's BCK algebras. *Pseudo-BCI algebras* were introduced by W.A. Dudek and Y.B. Jun in 2008 [44], as a non-commutative generalization of Iseki's BCI algebras. *Pseudo-BE algebras* were introduced by R.A. Borzooei et al. in 2013 [11], as a non-commutative generalization of BE algebras. *Pseudo-BCH algebras* were introduced by A. Walendziak in 2015 [209], as a non-commutative generalization of BCH algebras. *Pseudo-CI algebras* were introduced by A. Rezaei et al. in 2016 [180], as a non-commutative generalization of RME = CI algebras. The other non-commutative generalizations were introduced in [116].

**Definitions 1.1.9** [116] (The dual ones are omitted)

• Algebras **without last element**:

- **without pseudo-exchange property**:

A left-pseudo-algebra $(A^L, \to, \leadsto, 1)$ is called:
- *left-pseudo-M algebra*, or *left-pM algebra* for short, if it verifies the axiom (pM);
- *left-pseudo-RM algebra*, or *left-pRM algebra* for short, if it verifies (pM), (Re);
- *left-pseudo-pre-BZ algebra*, or *left-pre-pBZ algebra* for short, if it verifies (pM), (Re), (pB);
- *left-pseudo-aRM algebra*, or *left-paRM = pBH algebra* for short, if it verifies (pM), (Re), (An);
- *left-pseudo-BZ algebra*, or *left-pBZ algebra* for short, if it verifies (pM), (Re), (An), (pB).

-**with pseudo-exchange property**:

A left-pseudo-algebra $(A^L, \to, \leadsto, 1)$ is called:
- *left-pseudo-ME algebra*, or *left-pME algebra* for short, if it verifies (pM), (pEx);
- *left-pseudo-RME algebra*, or *left-pRME = pCI algebra* for short, if it verifies (pM), (pEx), (Re);
- *left-pseudo-pre-BCI algebra*, or *left-pre-pBCI algebra* for short, if it verifies (pM), (pEx), (Re), (pB);
- *left-pseudo-BCH algebra*, or *left-pBCH algebra* for short, if it verifies (Re), (pEx), (An), hence (pM);
- *left-pseudo-BCI algebra*, or *left-pBCI algebra* for short, if it verifies (pBB), (pD), (Re), (An).

• Algebras **with last element**:

- **without pseudo-exchange property**:

A left-pseudo-algebra $(A^L, \rightarrow, \rightsquigarrow, 1)$ is called:
- *left-pseudo-ML algebra*, or *left-pML algebra* for short, if it verifies (pM) and (La);
- *left-pseudo-RML algebra*, or *left-pRML algebra* for short, if it verifies (pM), (Re) and (La);
- *left-pseudo-pre-BCC algebra*, or *left-pre-pBCC algebra* for short, if it verifies (pM), (Re), (pB) and (La);
- *left-pseudo-aRML algebra*, or *left-paRML algebra* for short, if it verifies (pM), (Re), (An) and (La). - *left-pseudo-BCC algebra*, or *left-pBCC algebra* for short, if it verifies (pM), (Re), (pB), (An) and (La).

**-with pseudo-exchange property**:

A left-pseudo-algebra $(A^L, \rightarrow, \rightsquigarrow, 1)$ is called:
- *left-pseudo-MEL algebra*, or *left-pMEL algebra* for short, if it verifies (pM), (pEx) and (La);
- *left-pseudo-BE algebra*, or *left-pBE algebra* for short, if it verifies (pM), (pEx), (Re) and (La);
- *left-pseudo-pre-BCK algebra*, or *left-pre-pBCK algebra* for short, if it verifies (pM), (pEx), (Re), (pB) and (La);
- *left-pseudo-aBE algebra*, or *left-paBE algebra* for short, if it verifies (pM), (pEx), (Re), (An) and (La);
- *left-pseudo-BCK algebra*, or *left-pBCK algebra* for short, if it verifies (pBB), (pD), (Re), (An). and (La).

We have denoted [116] by **pM**, ... **pBCK** the corresponding classes of left-pM algebras.

**Remarks 1.1.10**

*(1) Not all unital left-pi-magmas are left-pseudo-algebras (Definition 1.1.6), because (IdEqR) $(x \rightarrow y = 1 \iff x \rightsquigarrow y = 1)$ does not always hold.*

*(2) Consequently, a unital left-pi-magma differs from a left-pM algebra (Definitions 1.1.9), because a left-pM algebra is a left-pseudo-algebra verifying (pM), i.e. the property (IdEqR) also holds. Hence, we have:* $\mathbf{pM} \subset \mathbf{upi-magma}$.

*(2') In the commutative case, unital i-magmas coincide with M algebras:* **M** = **ui-magma**.

*(3) A left-p-residoid differs from a left-pME algebra (Definitions 1.1.9), because a left-pME algebra is a left-pseudo-algebra verifying (pEx) and (pM), i.e. the property (IdEqR) also holds. Hence, we have:* $\mathbf{pME} \subset \mathbf{p-residoid}$.

*(3') In the commutative case, residoids coincide with ME algebras (see Figure 2.1):* **ME** = **residoid**.

**Proposition 1.1.11** *(See [116])*

*We have (following the numbering from Propositions 2.1.3, 2.1.4):*
*(pA00) (pM) + (pEqrelR) $\Longrightarrow$ (N'),*
*(pA1) (La') + (An') $\Longrightarrow$ (N'),*
*(pA2) (pK') + (An') + (pEqrelR) $\Longrightarrow$ (N'),*
*(pA3) (pC') + (An') $\Longrightarrow$ (pEx),*
*(pA4) (Re) + (pEx) + (pEqrelR) $\Longrightarrow$ (pD'),*

(pA4') (pD') + (Re) + (An') + (pEqrelR) $\Longrightarrow$ (N');
(pA5) (Re) + (An') + (pEx) + (pEqrelR) $\Longrightarrow$ (pM),
(pA6) (pK') + (Re) + (pEqrelR) $\Longrightarrow$ (La'),
(pA7) (pK') + (N') + (pEqrelR) $\Longrightarrow$ (La'),
(pA7') (pK) + (pM) + (pEqrelR) $\Longrightarrow$ (La'),
(pA8) (La) + (pEx) + (Re) $\Longrightarrow$ (pK),
(pA8') (pC) + (Re) + (An) +(IdEqR) $\Longrightarrow$ (pKLa),
(pA8") (pKLa) $\Longrightarrow$ ((pK) $\Leftrightarrow$ (La));
(pA9) (pM) + (La) + (IdEqR) + (pB) $\Longrightarrow$ (pK),
(pA9') (pM) + (La') + (p**') $\Longrightarrow$ (pK');
(pA10) (pEx) + (pB') + (pEqrelR) $\Longrightarrow$ (pBB'),
(pA10') (pEx) + (pBB') + (pEqrelR) $\Longrightarrow$ (pB'),
(pA10") (pEx) + (pEqrelR) $\Longrightarrow$ ((pB') $\Leftrightarrow$ (pBB'));
(pA10$_\rightarrow$) (pB$_\rightarrow$) + (pC$_2$) + (N) + (IdEqR) $\Longrightarrow$ (pBB$_1$),
(pA10$_\rightsquigarrow$) (pB$_\rightsquigarrow$) + (pC$_1$) + (N) + (IdEqR) $\Longrightarrow$ (pBB$_2$),
(pA10'$_\rightarrow$) (pBB$_1$) + (pC$_2$) + (N) + (IdEqR) $\Longrightarrow$ (pB$_\rightarrow$),
(pA10'$_\rightsquigarrow$) (pBB$_2$) + (pC$_1$) + (N) + (IdEqR) $\Longrightarrow$ (pB$_\rightsquigarrow$);
(pA11) (Re) + (pEx) + (p*') + (pEqrelR) $\Longrightarrow$ (pBB'),
(pA11') (pEx) + (p*') + (pD') $\Longrightarrow$ (pBB'),
(pA12') (pM) + (pB') + (pEqrelR) $\Longrightarrow$ (p*'),
(pA13) (N') + (p*') + (pEqrelR) $\Longrightarrow$ (Tr'),
(pA14$_\rightarrow$) (pB$_\rightarrow$) + (N) $\Longrightarrow$ (Tr),
(pA14$_\rightarrow$') (pB$_\rightarrow$) + (pM) $\Longrightarrow$ (Tr),
(pA14$_\rightsquigarrow$) (pB$_\rightsquigarrow$) + (N) $\Longrightarrow$ (Tr),
(pA14$_\rightsquigarrow$') (pB$_\rightsquigarrow$) + (pM) $\Longrightarrow$ (Tr);
(pA15) (N') + (pBB') + (pEqrelR) $\Longrightarrow$ (p**');
(pA16) (N') + (p**') + (pEqrelR) $\Longrightarrow$ (Tr'),
(pA17) (N) + (pBB$_1$) + (IdEqR) $\Longrightarrow$ (Tr),
(pA17') (pM) + (pBB$_1$) $\Longrightarrow$ (Tr);
(pA18) (pBB') + (pM) + (pEqrelR) $\Longrightarrow$ (Re');
(pA18') (pBB') + (pM) $\Longrightarrow$ (pD');
(pA19) (pM) + (pB$_\rightarrow$) $\Longrightarrow$ (Re),
(pA19') (pM) + (pB$_\rightsquigarrow$) $\Longrightarrow$ (Re),
(pA20) (pBB') + (pD) + (N') + (pEqrelR) $\Longrightarrow$ (pC'),
(pA21") (pBB') + (pM) + (An') + (pEqrelR) $\Longrightarrow$ (pEx),
(pA22) (pK) + (pEx) + (pM) $\Longrightarrow$ (Re),
(pA23) (pK') + (pC') + (An') + (pEqrelR) $\Longrightarrow$ (Re'),
(pA23') (pK) + (pC) + (N) + (IdEqR) $\Longrightarrow$ (Re);
(pA24) (Re) + (pEx) + (Tr) + (pEqrelR) $\Longrightarrow$ (p**);

(pA33) (pD') + (p**') + (An') $\Longrightarrow$ (pH),
(pA34) (pEqrel) + (N') + (Re') $\Longrightarrow$ (pEqrelR).

**Proof.** (pA00): If $1 \leq x$, i.e. $1 \to x = 1 = 1 \rightsquigarrow x$, by (pEqrelR), then $x = 1$, by (pM).

(pA1): Suppose $1 \leq x$; by (La'), we also have $x \leq 1$; hence, $x = 1$, by (An').

(pA2): Suppose $1 \leq x$, i.e. $1 \to x = 1$, by (pEqrelR); by (pK'), $x \leq 1 \to x = 1$; then, $x = 1$, by (An'). Thus, (N') holds.

(pA3): By (pC'), $x \to (y \rightsquigarrow z) \leq y \rightsquigarrow (x \to z)$ and $y \rightsquigarrow (x \to z) \leq x \to (y \rightsquigarrow z)$; then, by (An'), $x \to (y \rightsquigarrow z) = y \rightsquigarrow (x \to z)$, i.e. (pEx) holds.

(pA4): $y \to [(y \to x) \rightsquigarrow x] \overset{(pEx)}{=} (y \to x) \rightsquigarrow (y \to x) = 1$, by (Re) and (pEqrelR), hence $y \leq (y \to x) \rightsquigarrow x$, by (pEqrelR). Similarly,
$y \rightsquigarrow [(y \rightsquigarrow x) \to x] \overset{(pEx)}{=} (y \rightsquigarrow x) \to (y \rightsquigarrow x) = 1$, hence $y \leq (y \rightsquigarrow x) \to x$.

(pA4'): Suppose $1 \leq x$, i.e. $1 \to x = 1$, by (pEqrelR); by (pD'),
$x \leq (x \rightsquigarrow x) \to x \overset{(Re)+(pEqrelR)}{=} 1 \to x = 1$; hence, $x = 1$, by (An').

(pA5): By (pA4), (Re) + (pEx) + (pEqrelR) $\Longrightarrow$ (pD') and,
by (pA4'), (pD') + (Re) + (An') + (pEqrelR) $\Longrightarrow$ (N').
By (Re) and (pEqrelR), $x \rightsquigarrow (1 \to x) \overset{(pEx)}{=} 1 \to (x \rightsquigarrow x) = 1 \to 1 = 1$; hence, $x \leq 1 \to x$.
On the other hand, $(1 \to x) \rightsquigarrow x = 1$; indeed, by (pD'), $1 \leq (1 \to x) \rightsquigarrow x$, hence, by (N'), $(1 \to x) \rightsquigarrow x = 1$; hence, $1 \to x \leq x$. And $x \leq 1 \to x$ and $1 \to x \leq x$ imply $1 \to x = x$, by (An').
Similarly, $1 \rightsquigarrow x = x$. Thus, (pM) holds.

(pA6): Take $y = x$ in (pK'); we obtain: $x \leq x \to x = 1$, by (Re) and (pEqrelR); thus, (La') holds.

(pA7): By (pK'), for all $x$, $1 \leq x \to 1$; then, by (N'), $x \to 1 = 1$, i.e. $x \leq 1$, by (pEqrelR); thus, (La') holds.

(pA7'): By above (pA00), (pM) + (pEqrelR) $\Longrightarrow$ (N'), and by above (pA7), (pK') + (N') + (pEqrelR) $\Longrightarrow$ (La'); thus, (La') holds.

(pA8): In (pEx) ($x \to (y \rightsquigarrow z) = y \rightsquigarrow (x \to z)$), take $z = x$; we obtain: $x \to (y \rightsquigarrow x) = y \rightsquigarrow (x \to x) = y \rightsquigarrow 1 = 1$, by (Re), (La). Similarly, in (pEx) ($x \rightsquigarrow (y \to z) = y \to (x \rightsquigarrow z)$), take $z = x$; we obtain: $x \rightsquigarrow (y \to x) = y \to (x \rightsquigarrow x) = y \to 1 = 1$. Thus, (pK) holds.

(pA8'): By (pC$_1$), $(y \to (x \rightsquigarrow z)) \to (x \rightsquigarrow (y \to z)) = 1$;
for $z := x$, we obtain, by (Re) and (IdEqR), (a) $(y \to 1) \to (x \rightsquigarrow (y \to x)) = 1$;
by (pC$_2$), $x \rightsquigarrow (y \to z)) \to (y \to (x \rightsquigarrow z)) = 1$;
for $z := x$, we obtain, by (Re) and (IdEqR), (b) $(x \rightsquigarrow (y \to x)) \to (y \to 1) = 1$;
from (a), (b) and (An), $x \rightsquigarrow (y \to x) = y \to 1$.

By (pC$_1$), $(x \to (y \rightsquigarrow z)) \to (y \rightsquigarrow (x \to z)) = 1$;
for $z := x$, we obtain, by (Re), (a') $(x \to (y \rightsquigarrow x)) \to (y \rightsquigarrow 1) = 1$;
by (pC$_2$), $(y \rightsquigarrow (x \to z)) \to (x \to (y \rightsquigarrow z)) = 1$;
for $z := x$, we obtain, by (Re), (b') $(y \rightsquigarrow 1) \to (x \to (y \rightsquigarrow x)) = 1$;
from (a'), (b') and (An), $x \to (y \rightsquigarrow x) = y \rightsquigarrow 1$. Thus, (pKLa) holds.

(pA8"): Obviously.

(pA9): In (pB$_{\to}$) ($(z \to x) \to ((y \to z) \to (y \to x)) = 1$), take $z := 1$ to obtain: $(1 \to x) \to ((y \to 1) \to (y \to x)) = 1$, hence, by (pM) and (La), $x \to (y \to x) = 1$.
In (pB$_{\rightsquigarrow}$) ($(z \rightsquigarrow x) \to ((y \rightsquigarrow z) \rightsquigarrow (y \rightsquigarrow x)) = 1$), take $z := 1$ to obtain:
$(1 \rightsquigarrow x) \to ((y \rightsquigarrow 1) \rightsquigarrow (y \rightsquigarrow x)) = 1$, hence, by (pM), (La) and (IdEqR), $x \to (1 \rightsquigarrow (y \rightsquigarrow x)) = 1$, hence, by (pM) again, $x \to (y \rightsquigarrow x) = 1$. Thus, (pK) holds.

(pA9'): Since $x \leq 1$, for all $x$, by (La'), it follows, by (p**'), that: $1 \to z \leq x \to z$ and $1 \rightsquigarrow z \leq x \rightsquigarrow z$; then, by (pM), we obtain: $z \leq x \to z$ and $z \leq x \rightsquigarrow z$, i.e. (pK') holds.

(pA10): By (pB'),
$z \to x \leq (y \to z) \to (y \to x)$ and $z \rightsquigarrow x \leq (y \rightsquigarrow z) \rightsquigarrow (y \rightsquigarrow x)$;
hence, by (pEqrelR),
$(z \to x) \rightsquigarrow ((y \to z) \to (y \to x)) = 1$, $(z \rightsquigarrow x) \to ((y \rightsquigarrow z) \rightsquigarrow (y \rightsquigarrow x)) = 1$;
then, by (pEx), we obtain:
$(y \to z) \to ((z \to x) \rightsquigarrow (y \to x)) = 1$, $(y \rightsquigarrow z) \rightsquigarrow ((z \rightsquigarrow x) \to (y \rightsquigarrow x)) = 1$,
i.e., by (pEqrelR),
$y \to z \leq (z \to x) \rightsquigarrow (y \to x)$ and $y \rightsquigarrow z \leq (z \rightsquigarrow x) \to (y \rightsquigarrow x)$; thus, (pBB') holds.

(pA10'): By (pBB'),
$y \to z \leq (z \to x) \rightsquigarrow (y \to x)$ and $y \rightsquigarrow z \leq (z \rightsquigarrow x) \to (y \rightsquigarrow x)$;
hence, by (pEqrelR),
$(y \to z) \to ((z \to x) \rightsquigarrow (y \to x)) = 1$, $(y \rightsquigarrow z) \rightsquigarrow ((z \rightsquigarrow x) \to (y \rightsquigarrow x)) = 1$;
then, by (pEx), we obtain:
$(z \to x) \rightsquigarrow ((y \to z) \to (y \to x)) = 1$, $(z \rightsquigarrow x) \to ((y \rightsquigarrow z) \rightsquigarrow (y \rightsquigarrow x)) = 1$, i.e., by (pEqrelR),
$z \to x \leq (y \to z) \to (y \to x)$ and $z \rightsquigarrow x \leq (y \rightsquigarrow z) \rightsquigarrow (y \rightsquigarrow x)$; thus, (pB') holds.

(pA10"): By (pA10) and (pA10').

(pA10$_{\to}$): By (pB$_{\to}$), $(y \to z) \to ((x \to y) \to (x \to z)) = 1$, hence by (IdEqR), $(y \to z) \rightsquigarrow ((x \to y) \to (x \to z)) = 1$;
by (pC$_2$), $((y \to z) \rightsquigarrow ((x \to y) \to (x \to z))) \to ((x \to y) \to ((y \to z) \rightsquigarrow (x \to z))) = 1$;
by (N), $(x \to y) \to ((y \to z) \rightsquigarrow (x \to z)) = 1$, i.e. (pBB$_1$ holds.

(pA10$_{\rightsquigarrow}$): By (pB$_{\rightsquigarrow}$), $(y \rightsquigarrow z) \to ((x \rightsquigarrow y) \rightsquigarrow (x \rightsquigarrow z)) = 1$;
by (pC$_1$), $((y \rightsquigarrow z) \to ((x \rightsquigarrow y) \rightsquigarrow (x \rightsquigarrow z))) \to ((x \rightsquigarrow y) \rightsquigarrow ((y \rightsquigarrow z) \to (x \rightsquigarrow z))) = 1$;
by (N), $(x \rightsquigarrow y) \rightsquigarrow ((y \rightsquigarrow z) \to (x \rightsquigarrow z)) = 1$, hence
by (IdEqR), $(x \rightsquigarrow y) \to ((y \rightsquigarrow z) \to (x \rightsquigarrow z)) = 1$, i.e. (pBB$_2$ holds.

(pA10'$_{\to}$): By (pBB$_1$), $(x \to y) \to ((y \to z) \rightsquigarrow (x \to z)) = 1$ (pB$_{\to}$);
by (pC$_1$), $((x \to y) \to ((y \to z) \rightsquigarrow (x \to z))) \to ((y \to z) \rightsquigarrow ((x \to y) \to (y \to z)))$;
by (N), $(y \to z) \rightsquigarrow ((x \to y) \to (x \to z)) = 1$, hence
by (IdEqR), $(y \to z) \to ((x \to y) \to (x \to z)) = 1$, i.e. (pB$_{\to}$) holds.

(pA10'$_{\rightsquigarrow}$): By (pBB$_2$, $(x \rightsquigarrow y) \to ((y \rightsquigarrow z) \to (x \rightsquigarrow z)) = 1$, hence
by (IdEqR), $(x \rightsquigarrow y) \rightsquigarrow ((y \rightsquigarrow z) \to (x \rightsquigarrow z)) = 1$;
by (pC$_2$), $((x \rightsquigarrow y) \rightsquigarrow ((y \rightsquigarrow z) \to (x \rightsquigarrow z))) \to ((y \rightsquigarrow z) \to ((x \rightsquigarrow y) \rightsquigarrow (x \rightsquigarrow z))) = 1$; by (N), $(y \rightsquigarrow z) \to ((x \rightsquigarrow y) \rightsquigarrow (x \rightsquigarrow z))) = 1$, i.e. (pB$_{\rightsquigarrow}$) holds.

(pA11): By (pA4), (Re) + (pEx) + (pEqrelR) $\Longrightarrow$ (pD'), thus (pD') holds. Hence, $z \leq (z \to x) \rightsquigarrow x$ and $z \leq (z \rightsquigarrow x) \to x$. Then, by (p*'), we obtain:
$y \to z \leq y \to [(z \to x) \rightsquigarrow x] \overset{(pEx)}{=} (z \to x) \rightsquigarrow (y \to x)$ and
$y \rightsquigarrow z \leq y \rightsquigarrow [(z \rightsquigarrow x) \to x] \overset{(pEx)}{=} (z \rightsquigarrow x) \to (y \rightsquigarrow x)$; thus, (pBB') holds.

(pA11'): By (pD'), $y \leq (y \to z) \rightsquigarrow z$; then,
by (p*$_{\to}$'), $x \to y \leq x \to ((y \to z) \rightsquigarrow z) \overset{(pEx)}{=} (y \to z) \rightsquigarrow (x \to z)$.

Similarly, by (pD'), $y \leq (y \rightsquigarrow z) \to z$; then,
by (p$*_{\rightsquigarrow}$'), $x \rightsquigarrow y \leq x \rightsquigarrow ((y \rightsquigarrow z) \to z) \overset{(pEx)}{=} (y \rightsquigarrow z) \to (x \rightsquigarrow z)$.
Thus, (pBB') holds.

(pA12'): By (pA00), (pM) + (pEqrelR) $\Longrightarrow$ (N'). Suppose $y \leq z$, i.e. $y \to z = 1 = y \rightsquigarrow z$, by (pEqrelR). Since, by (pB'),
$y \to z \leq (x \to y) \to (x \to z)$ and $y \rightsquigarrow z \leq (x \rightsquigarrow y) \rightsquigarrow (x \rightsquigarrow z)$,
then, by (N'), $(x \to y) \to (x \to z) = 1$ and $(x \rightsquigarrow y) \rightsquigarrow (x \rightsquigarrow z) = 1$, i.e. $x \to y \leq x \to z$ and $x \rightsquigarrow y \leq x \rightsquigarrow z$; thus, (p*') holds.

(pA13): Suppose $x \leq y$ and $y \leq z$, i.e. $x \to y = 1$, by (pEqrelR); then, by (p*'), $x \to y \leq x \to z$, hence $1 \leq x \to z$; then, by (N'), $x \to z = 1$, i.e. $x \leq z$, by (pEqrelR) again. Thus, (Tr') holds.

(pA14$_{\to}$): Suppose $x \to y = 1$ and $y \to z = 1$;
by (pB$_{\to}$), $(y \to z) \to ((x \to y) \to (x \to z)) = 1$, hence $1 \to (1 \to (x \to z)) = 1$;
then, by (N), $1 \to (x \to z) = 1$, hence by (N) again, $x \to z = 1$; thus, (Tr) holds.

(pA14$_{\to}$'): Suppose $x \to y = 1$ and $y \to z = 1$;
by (pB$_{\to}$), $(y \to z) \to ((x \to y) \to (x \to z)) = 1$, hence $1 \to (1 \to (x \to z)) = 1$;
then, by (pM) twice, $x \to z = 1$; thus, (Tr) holds.

(pA14$_{\rightsquigarrow}$): Suppose $x \to y = 1$ and $y \to z = 1$;
by (pB$_{\rightsquigarrow}$), $(y \rightsquigarrow z) \to ((x \rightsquigarrow y) \rightsquigarrow (x \rightsquigarrow z)) = 1$, hence $1 \to (1 \rightsquigarrow (x \rightsquigarrow z)) = 1$;
then, by (N), $1 \rightsquigarrow (x \to z) = 1$, hence by (IdEqR), $1 \to (x \to z) = 1$, hence, (N) again, $x \to z = 1$; thus, (Tr) holds.

(pA14$_{\rightsquigarrow}$'): Suppose $x \to y = 1$ and $y \to z = 1$;
by (pB$_{\rightsquigarrow}$), $(y \rightsquigarrow z) \to ((x \rightsquigarrow y) \rightsquigarrow (x \rightsquigarrow z)) = 1$, hence $1 \to (1 \rightsquigarrow (x \rightsquigarrow z)) = 1$;
then, by (pM) twice, $x \to z = 1$; thus, (Tr) holds.

(pA15): By (pBB'),
$y \to z \leq (z \to x) \rightsquigarrow (y \to x)$ and $y \rightsquigarrow z \leq (z \rightsquigarrow x) \to (y \rightsquigarrow x)$.
Suppose $y \leq z$, i.e. $y \to z = 1 = y \rightsquigarrow z$, by (pEqrelR); then,
$1 \leq (z \to x) \rightsquigarrow (y \to x)$ and $1 \leq (z \rightsquigarrow x) \to (y \rightsquigarrow x)$; hence, by (N'),
$1 = (z \to x) \rightsquigarrow (y \to x)$ and $1 = (z \rightsquigarrow x) \to (y \rightsquigarrow x)$, i.e. by (pEqrelR),
$z \to x \leq y \to x$ and $z \rightsquigarrow x \leq y \rightsquigarrow x$; thus, (p**') holds.

(pA16): Suppose $x \leq y$ and $y \leq z$, i.e. $y \to z = 1$, by (pEqrelR); then, by (p**'), $y \to z \leq x \to z$, hence $1 \leq x \to z$; then, by (N'), $x \to z = 1$, i.e. $x \leq z$, by (pEqrelR) again. Thus, (Tr') holds.

(pA17): Suppose $y \to z = 1$ and $z \to x = 1$;
then, by (pBB$_1$) ($(y \to z) \to ((z \to x) \rightsquigarrow (y \to x)) = 1$), we obtain:
$1 \to (1 \rightsquigarrow (y \to x)) = 1$, hence, by (N), $1 \rightsquigarrow (y \to x) = 1$; then, by (IdEqR), $1 \to (y \to x) = 1$, hence, by (N) again, $y \to x = 1$; thus, (Tr) holds.

(pA17'): Suppose $y \to z = 1$ and $z \to x = 1$; then, by (pBB$_1$), we obtain:
$1 \to (1 \rightsquigarrow (y \to x)) = 1$, hence, by (pM), $y \to x = 1$; thus, (Tr) holds.

(pA18): First, by (pA00), (pM) + (pEqrelR) $\Longrightarrow$ (N'). Then, take $y = z = 1$ in (pBB'): $y \to z \leq (z \to x) \rightsquigarrow (y \to x)$ and $y \rightsquigarrow z \leq (z \rightsquigarrow x) \to (y \rightsquigarrow x)$;
by (pM), we then obtain:
$1 \leq (1 \to x) \rightsquigarrow (1 \to x) = x \rightsquigarrow x$ and $1 \leq (1 \rightsquigarrow x) \to (1 \rightsquigarrow x) = x \to x$;
hence, by (N'), $x \rightsquigarrow x = 1$ and $x \to x = 1$, i.e. $x \leq x$. Thus, (Re') holds.

(pA18'): In (pBB'), take $y = 1$; we obtain:

$z \overset{(pM)}{=} 1 \to z \leq (z \to x) \rightsquigarrow (1 \to x) \overset{(pM)}{=} (z \to x) \rightsquigarrow x$ and
$z \overset{(pM)}{=} 1 \rightsquigarrow z \leq (z \rightsquigarrow x) \to (1 \rightsquigarrow x) \overset{(pM)}{=} (z \rightsquigarrow x) \to x$. Thus, (pD') holds.

(pA19): In ($pB_{\to}$) $((z \to x) \to ((y \to z) \to (y \to x)) = 1)$, take $y = z := 1$, to obtain: $(1 \to x) \to ((1 \to 1) \to (1 \to x)) = 1$, hence, by (pM), $x \to x = 1$; thus, (Re) holds.

(pA19'): In ($pB_{\rightsquigarrow}$) $((z \rightsquigarrow x) \to ((y \rightsquigarrow z) \rightsquigarrow (y \rightsquigarrow x)) = 1)$, take $y = z := 1$, to obtain: $(1 \rightsquigarrow x) \to ((1 \rightsquigarrow 1) \rightsquigarrow (1 \rightsquigarrow x)) = 1$, hence, by (pM), $x \to x = 1$.

(pA20): First, by (pA15), (pBB') + (N') + (pEqrelR) imply (p**').

Then, by (pBB'): $(Y \to Z \leq (Z \to X) \rightsquigarrow (Y \to X))$,
for $X = u \rightsquigarrow x$, $Y = y$, $Z = z \rightsquigarrow x$, we obtain:
$y \to (z \rightsquigarrow x) \leq ((z \rightsquigarrow x) \to (u \rightsquigarrow x)) \rightsquigarrow (y \to (u \rightsquigarrow x))$.
Then, by (p**'), we obtain:
$V_1 \overset{notation}{=} [((z \rightsquigarrow x) \to (u \rightsquigarrow x)) \rightsquigarrow (y \to (u \rightsquigarrow x))] \to$
$[(u \rightsquigarrow z) \rightsquigarrow (y \to (u \rightsquigarrow x))] \leq$
$(y \to (z \rightsquigarrow x)) \to [(u \rightsquigarrow z) \rightsquigarrow (y \to (u \rightsquigarrow x))] \overset{notation}{=} W_1$.
But, the left side $V_1 = 1$; indeed, by (pBB'), we have:
$u \rightsquigarrow z \leq (z \rightsquigarrow x) \to (u \rightsquigarrow x)$;
then, by (p**'), we obtain:
$[(z \rightsquigarrow x) \to (u \rightsquigarrow x)) \rightsquigarrow (y \to (u \rightsquigarrow x)) \leq (u \rightsquigarrow z) \rightsquigarrow (y \to (u \rightsquigarrow x))$, i.e. $V_1 = 1$.
Then, by (N'), $W_1 = 1$, i.e.
$y \to (z \rightsquigarrow x) \leq (u \rightsquigarrow z) \rightsquigarrow (y \to (u \rightsquigarrow x))$,
which for $z = y \to x$ and $u = z$ gives:
$y \to ((y \to x) \rightsquigarrow x) \leq (z \rightsquigarrow (y \to x)) \rightsquigarrow (y \to (z \rightsquigarrow x))$;
but, by (pD), the left side $y \to ((y \to x) \rightsquigarrow x) = 1$; hence, by (N'),
$(z \rightsquigarrow (y \to x)) \rightsquigarrow (y \to (z \rightsquigarrow x)) = 1$,
i.e. $z \rightsquigarrow (y \to x) \leq y \to (z \rightsquigarrow x)$, by (pEqrelR).

Similarly, by (pBB') again: $(Y \rightsquigarrow Z \leq (Z \rightsquigarrow X) \to (Y \rightsquigarrow X))$,
for $X = u \to x$, $Y = y$, $Z = z \to x$, we obtain:
$y \rightsquigarrow (z \to x) \leq ((z \to x) \rightsquigarrow (u \to x)) \to (y \rightsquigarrow (u \to x))$.
Then, by (p**'), we obtain:
$V_2 \overset{notation}{=} [((z \to x) \rightsquigarrow (u \to x)) \to (y \rightsquigarrow (u \to x))] \rightsquigarrow$
$[(u \to z) \to (y \rightsquigarrow (u \to x))] \leq$
$(y \rightsquigarrow (z \to x)) \rightsquigarrow [(u \to z) \to (y \rightsquigarrow (u \to x))] \overset{notation}{=} W_2$.
But, the left side $V_2 = 1$; indeed, by (pBB'), we have:
$u \to z \leq (z \to x) \rightsquigarrow (u \to x)$;
then, by (p**'), we obtain:
$[(z \to x) \rightsquigarrow (u \to x)] \to (y \rightsquigarrow (u \to x)) \leq (u \to z) \to (y \rightsquigarrow (u \to x))$, i.e. $V_2 = 1$.
Then, by (N'), $W_2 = 1$, i.e.
$y \rightsquigarrow (z \to x) \leq (u \to z) \to (y \rightsquigarrow (u \to x))$,
which for $z = y \rightsquigarrow x$ and $u = z$ gives:
$y \rightsquigarrow ((y \rightsquigarrow x) \to x) \leq (z \to (y \rightsquigarrow x)) \to (y \rightsquigarrow (z \to x))$;
but, by (pD), the left side $y \rightsquigarrow ((y \rightsquigarrow x) \to x) = 1$; hence, by (N'),
$(z \to (y \rightsquigarrow x)) \to (y \rightsquigarrow (z \to x)) = 1$, i.e $z \to (y \rightsquigarrow x) \leq y \rightsquigarrow (z \to x)$; hence, $y \to (z \rightsquigarrow x) \leq z \rightsquigarrow (y \to x)$. Thus, (pC') holds.

(pA21"): By (pA00), (pM) + (pEqrelR) $\Longrightarrow$ (N'),
by (pA18'), (pBB') + (pM) $\Longrightarrow$ (pD'),
by (pA20), (pBB') + (pD) + (N') $\Longrightarrow$ (pC') and
by (pA3), (pC') + (An') $\Longrightarrow$ (pEx). Thus, (pEx) holds.

(pA22): $x \to x \stackrel{(pM)}{=} 1 \rightsquigarrow (x \to x) \stackrel{(pEx)}{=} x \to (1 \rightsquigarrow x) \stackrel{(pK)}{=} 1$; thus, (Re) holds.

(pA23): By above (pA2), (pK') + (An') + (pEqrelR) $\Longrightarrow$ (N'); by above (pA3), (pC') + (An') $\Longrightarrow$ (pEx); by (pK'), $1 \leq [(x \to x) \to 1]$; by (N'), $(x \to x) \to 1 = 1$, i.e. $x \to x \leq 1$, by (pEqrelR).

On the other hand, $1 \rightsquigarrow (x \to x) \stackrel{(pEx)}{=} x \to (1 \rightsquigarrow x) = 1$, by (pK) and (pEqrelR); hence, $1 \leq x \to x$.
Now, by (An'), $x \to x = 1$, i.e. $x \leq x$, by (pEqrelR). Thus, (Re') holds.

(pA23'): By (pK), $x \to ((x \to (x \to x)) \rightsquigarrow x) = 1$;
by (pC), $(x \to ((x \to (x \to x)) \rightsquigarrow x)) \to ((x \to (x \to x)) \rightsquigarrow (x \to x)) = 1$;
by (N), $(x \to (x \to x)) \rightsquigarrow (x \to x) = 1$;
but, by (pK) again, $x \to (x \to x) = 1$;
then, by (N) again and by (IdEqR), it follows that $x \to x = 1$, i.e. (Re) holds.

(pA24): Supp. (Tr) holds, i.e. $X \to Y = 1$, $Y \to Z = 1$ imply $X \to Z = 1$.
We must prove that (p**) holds, i.e.
$y \to z = 1$ implies $(z \to x) \rightsquigarrow (y \to x) = 1$ and $(z \rightsquigarrow x) \to (y \rightsquigarrow x) = 1$.

Suppose $y \to z = 1$; we prove that $A \stackrel{notation}{=} (z \to x) \rightsquigarrow (y \to x) = 1$.
Indeed, $A \stackrel{(pEx)}{=} y \to ((z \to x) \rightsquigarrow x)$. Take $X = y$, $Z = (z \to x) \rightsquigarrow x$ and $Y = z$; then we have $A = X \to Z$ and:
$X \to Y = y \to z = 1$, by hypothesis;
$Y \to Z = z \to [(z \to x) \rightsquigarrow x] \stackrel{(pEx)}{=} (z \to x) \rightsquigarrow (z \to x) \stackrel{(Re),(pEqrelR)}{=} 1.$
Hence, by (Tr), it follows that $X \to Z = 1$, i.e. $A = 1$.

Suppose $y \to z = 1$; we prove that $B \stackrel{notation}{=} (z \rightsquigarrow x) \to (y \rightsquigarrow x) = 1$.
Indeed, $B \stackrel{(pEx)}{=} y \rightsquigarrow ((z \rightsquigarrow x) \to x)$. Take $X = y$, $Z = (z \rightsquigarrow x) \to x$ and $Y = z$; then we have $B = X \to Z$ and:
$X \to Y = y \to z = 1$, by hypothesis;
$Y \rightsquigarrow Z = z \rightsquigarrow [(z \rightsquigarrow x) \to x] \stackrel{(pEx)}{=} (z \rightsquigarrow x) \to (z \rightsquigarrow x) \stackrel{(Re),(pEqrelR)}{=} 1$; hence, $Y \to Z = 1$, by (pEqrelR).
Hence, by (Tr), it follows that $X \to Z = 1$, i.e. $B = 1$.

(pA33): By (pD'), $x \leq (x \to y) \rightsquigarrow y$, which by (p**') gives $((x \to y) \rightsquigarrow y) \to y \leq x \to y$; but, by (pD') also, $x \to y \leq ((x \to y) \rightsquigarrow y) \to y$. Then, by (An'), we obtain $((x \to y) \rightsquigarrow y) \to y = x \to y$. Similarly, by (pD'), $x \leq (x \rightsquigarrow y) \to y$, which by (p**') gives $((x \rightsquigarrow y) \to y) \rightsquigarrow y \leq x \rightsquigarrow y$; but, by (pD') also, $x \rightsquigarrow y \leq ((x \rightsquigarrow y) \to y) \rightsquigarrow y$. Then, by (An'), we obtain $((x \rightsquigarrow y) \to y) \rightsquigarrow y = x \rightsquigarrow y$. Thus, (pH) holds.

(pA34): Suppose $x \leq y$; then, $1 \leq x \to y$, by (pEqrel). By (N'), it follows that $x \to y = 1$. Conversely, suppose $x \to y = 1$; by (Re'), $x \to y \leq x \to y$, hence $1 \leq x \to y$; then, by (pEqrel), $x \leq y$. Thus, (pEqrelR) holds. $\square$

**Theorem 1.1.12** *([116], Theorem 2.1.9)*
*Let $\mathcal{A}^L = (A^L, \leq, \to, \rightsquigarrow, 1)$ be a left-pseudo-structure such that ((IdEqR), (pEqrelR)) (Re), (pM), (pEx) hold. Then, we have:*

$$(pBB) \iff (pB) \iff (p*).$$

**Proof.** By Proposition 1.1.11 (10"), (pEx) + (pEqrelR) $\Longrightarrow$ ((pBB') $\Leftrightarrow$ (pB')).
By (pA11), (Re) + (pEx) + (pEqrelR) + (p*') $\Longrightarrow$ (pBB').
By (pA12'), (pM) + (pEqrelR) + (pB') $\Longrightarrow$ (p*'). Hence, we have:
$(p*) \Longrightarrow (pBB) \iff (pB) \Longrightarrow (p*)$. □

**Theorem 1.1.13**
*Let $\mathcal{A}^L = (A^L, \leq, \to, \rightsquigarrow, 1)$ be a left-pseudo-structure such that ((IdEqR), (pEqrelR)) (Re), (pM), (pEx) hold. Then, we have:*

$$(p**) \iff (Tr).$$

**Proof.** By (pA00), (pM) + (pEqrelR) $\Longrightarrow$ (N'); by (pA16), (N') + (pEqrelR) + (p**') $\Longrightarrow$ (Tr'). Conversely, by (pA24), (Re) + (pEx) + (Tr) $\Longrightarrow$ (p**). □

**Theorem 1.1.14** *([116], Theorem 2.1.10)*
*Let $\mathcal{A}^L = (A^L, \leq, \to, \rightsquigarrow, 1)$ be a left-pseudo-structure such that ((IdEqR), (pEqrelR)) (An), (pM), (pB) hold. Then, we have:*

$$(pEx) \iff (pBB).$$

**Proof.** By Proposition 1.1.11 (pA10), (pEx) + (pEqrelR) + (pB') $\Longrightarrow$ (pBB').
By (pA21"), (pM) + (pBB') + (An') + (pEqrelR) $\Longrightarrow$ (pEx). □

The next theorem was proved first in the commutative case by Michael Kinyon, by using the automated theorem proving tool *Prover9*.

**Theorem 1.1.15** *([116], Theorem 2.1.11)*
*In any left-pseudo-structure $(A^L, \leq, \to, \rightsquigarrow, 1)$, we have:*
*(i) (pM) + (pBB') + (pEqrelR) $\Longrightarrow$ (pB'),*
*(ii) (pM) + (pB') + (pEqrelR) $\Longrightarrow$ (p**').*

**Proof.** (i): By Proposition 1.1.11 (pA18'), (pM) + (pBB) $\Longrightarrow$ (pD).
Suppose that (pBB') holds, i.e.
(pBB$'_1$) $\quad x \to y \leq (y \to z) \rightsquigarrow (x \to z)$,
(pBB$'_2$) $\quad x \rightsquigarrow y \leq (y \rightsquigarrow z) \to (x \rightsquigarrow z)$.
First, in (pBB$'_1$), set $x = u$ and $y = (u \to v) \rightsquigarrow v$, to get:
$(u \to [(u \to v) \rightsquigarrow v]) \to [(((u \to v) \rightsquigarrow v) \to z) \rightsquigarrow (u \to z)] \overset{(pD)}{=}$
$1 \to [(((u \to v) \rightsquigarrow v) \to z) \rightsquigarrow (u \to z)] \overset{(pM)}{=} (((u \to v) \rightsquigarrow v) \to z) \rightsquigarrow (u \to z) = 1.$
After renaming variables, we get:
(a1) $\quad (((x \to y) \rightsquigarrow y) \to z) \rightsquigarrow (x \to z) = 1.$
Next, in (pBB$_2$'), set $x = u \rightsquigarrow v$ and $y = (v \rightsquigarrow w) \to (u \rightsquigarrow w)$, to get:

$((u \rightsquigarrow v) \rightsquigarrow [(v \rightsquigarrow w) \to (u \rightsquigarrow w)]) \to$
$[(((v \rightsquigarrow w) \to (u \rightsquigarrow w)) \rightsquigarrow z) \to ((u \rightsquigarrow v) \rightsquigarrow z)] \stackrel{(pBB_2')}{=}$
$1 \to [(((v \rightsquigarrow w) \to (u \rightsquigarrow w)) \rightsquigarrow z) \to ((u \rightsquigarrow v) \rightsquigarrow z)] \stackrel{(pM)}{=}$
$(((v \rightsquigarrow w) \to (u \rightsquigarrow w)) \rightsquigarrow z) \to ((u \rightsquigarrow v) \rightsquigarrow z) = 1.$
After renaming variables, we get:
(b1) $$(((x \rightsquigarrow y) \to (u \rightsquigarrow y)) \rightsquigarrow z) \to ((u \rightsquigarrow x) \rightsquigarrow z) = 1.$$
Taking $z = u \rightsquigarrow y$ in (b1), we get:
(c1) $$(((x \rightsquigarrow y) \to (u \rightsquigarrow y)) \rightsquigarrow (u \rightsquigarrow y)) \to ((u \rightsquigarrow x) \rightsquigarrow (u \rightsquigarrow y)) = 1.$$
Now, in (a1), set $x = v \rightsquigarrow w,\ y = t \rightsquigarrow w,\ z = (t \rightsquigarrow v) \rightsquigarrow (t \rightsquigarrow w)$ to get:
$[(((v \rightsquigarrow w) \to (t \rightsquigarrow w)) \rightsquigarrow (t \rightsquigarrow w)) \to ((t \rightsquigarrow v) \rightsquigarrow (t \rightsquigarrow w))] \rightsquigarrow$
$((v \rightsquigarrow w) \to ((t \rightsquigarrow v) \rightsquigarrow (t \rightsquigarrow w))) \stackrel{(c1)}{=}$
$1 \rightsquigarrow ((v \rightsquigarrow w) \to ((t \rightsquigarrow v) \rightsquigarrow (t \rightsquigarrow w))) \stackrel{(pM)}{=}$
$(v \rightsquigarrow w) \to ((t \rightsquigarrow v) \rightsquigarrow (t \rightsquigarrow w)) = 1,$
i.e. ($pB_{\rightsquigarrow}$') ($v \rightsquigarrow w \leq (t \rightsquigarrow v) \rightsquigarrow (t \rightsquigarrow w)$) holds.

Now, "dually", in ($pBB_2$'), set $x = u$ and $y = (u \rightsquigarrow v) \to v$, to get:
$(u \rightsquigarrow [(u \rightsquigarrow v) \to v]) \to [(((u \rightsquigarrow v) \to v) \rightsquigarrow z) \to (u \rightsquigarrow z)] \stackrel{(pD)}{=}$
$1 \to [(((u \rightsquigarrow v) \to v) \rightsquigarrow z) \to (u \rightsquigarrow z)] \stackrel{(pM)}{=} (((u \rightsquigarrow v) \to v) \rightsquigarrow z) \to (u \rightsquigarrow z) = 1.$
After renaming variables, we get:
(a2) $$(((x \rightsquigarrow y) \to y) \rightsquigarrow z) \to (x \rightsquigarrow z) = 1.$$
Next, in ($pBB_1$'), set $x = u \to v$ and $y = (v \to w) \rightsquigarrow (u \to w)$, to get:
$((u \to v) \to [(v \to w) \rightsquigarrow (u \to w)]) \to$
$[(((v \to w) \rightsquigarrow (u \to w)) \to z) \rightsquigarrow ((u \to v) \to z)] \stackrel{(pBB_1')}{=}$
$1 \to [(((v \to w) \rightsquigarrow (u \to w)) \to z) \rightsquigarrow ((u \to v) \to z)] \stackrel{(pM)}{=}$
$(((v \to w) \rightsquigarrow (u \to w)) \to z) \rightsquigarrow ((u \to v) \to z) = 1.$
After renaming variables, we get:
(b2) $$(((x \to y) \rightsquigarrow (u \to y)) \to z) \rightsquigarrow ((u \to x) \to z) = 1.$$
Taking $z = u \to y$ in (b2), we get:
(c2) $$(((x \to y) \rightsquigarrow (u \to y)) \to (u \to y)) \rightsquigarrow ((u \to x) \to (u \to y)) = 1.$$
Now, in (a2), set $x = v \to w,\ y = t \to w,\ z = (t \to v) \to (t \to w)$ to get:
$[(((v \to w) \rightsquigarrow (t \to w)) \to (t \to w)) \rightsquigarrow ((t \to v) \to (t \to w))] \to$
$((v \to w) \rightsquigarrow ((t \to v) \to (t \to w))) \stackrel{(c2)}{=}$
$1 \to ((v \to w) \rightsquigarrow ((t \to v) \to (t \to w))) \stackrel{(pM)}{=}$
$(v \to w) \rightsquigarrow ((t \to v) \to (t \to w)) = 1,$
i.e. ($pB_{\to}$') ($v \to w \leq (t \to v) \to (t \to w)$) holds.

(ii): Suppose (pB') is

$$y \to z \leq (x \to y) \to (x \to z),\ y \rightsquigarrow z \leq (x \rightsquigarrow y) \rightsquigarrow (x \rightsquigarrow z).$$

If $x \leq y$, i.e. $x \to y = 1 = x \rightsquigarrow y$, then we get, from (pB'):
$y \to z \leq 1 \to (x \to z) \stackrel{(pM)}{=} x \to z$ and $y \rightsquigarrow z \leq 1 \rightsquigarrow (x \rightsquigarrow z) \stackrel{(pM)}{=} x \rightsquigarrow z$, i.e. (p**') holds. □

By Theorem 3.1.9 (i) and (pA12'), we obtain immediately that:

**Corollary 1.1.16** *([116], Corollary 2.1.12)*

$$(pM) \ + \ (pBB') \ + \ (pEqrelR) \implies (p*').$$

Concluding, by above Theorem 3.1.9 and Proposition 1.1.11, we immediately obtain:

**Corollary 1.1.17** *([116], Corollary 2.1.13)*
*In any pseudo-structure $(A^L, \leq, \rightarrow, \rightsquigarrow, 1)$ verifying (pM), we have:*

$$(pBB') \implies (pB') \implies (p*'), (p**') \implies (Tr').$$

**Proposition 1.1.18** *We have:*
*(pA#1) (pEx) $\Longrightarrow$ (pEq#);*
*(pA#2) (pEq#) + ($p*_{\rightarrow}$) + (Re) + (An) $\Longrightarrow$ (Tr);*
*(pA#3$_{\rightarrow}$) (pEq#') $\Longrightarrow$ ((pB$_{\rightarrow}$') $\Leftrightarrow$ (pBB$_1$')),*
*(pA#3$_{\rightsquigarrow}$) (pEq#') $\Longrightarrow$ ((pB$_{\rightsquigarrow}$') $\Leftrightarrow$ (pBB$_2$')),*
*(pA#3) (pEq#') $\Longrightarrow$ ((pB') $\Leftrightarrow$ (pBB'));*
*(pA#4) (pEq#') + (Re') $\Longrightarrow$ (pD'),*
*(pA#5) (pEq#') + (pD') + (Tr') $\Longrightarrow$ (p**'),*
*(pA#5') (pEq#') + (Re') + (Tr') $\Longrightarrow$ (p**'),*
*(pA#6) (pEq#') + (pB') + (pD') + (Tr') $\Longrightarrow$ (pC'),*
*(pA#6') (pEq#') + (pB') + (Re') + (Tr') $\Longrightarrow$ (pC'),*
*(pA#7) (pEq#') + (pEx) + (pD') + (p*') $\Longrightarrow$ (pB'),*
*(pA#7') (pEq#') + (pEx) + (Re') + (p*') $\Longrightarrow$ (pB'),*
*(pA#8) (pEq#') + (pM) $\Longrightarrow$ (pEqrel),*
*(pA#9) (pEq#') + (pM) + (N') + (Re') $\Longrightarrow$ (pEqrelR),*
*(pA#9') (pEq#') + (pM) + (An') + (La') $\Longrightarrow$ (IdEqR) + (pEqrelR);*
*(pA#10) (pEq#') + ($p*_{\rightarrow}$') + (pD') + (Tr') $\Longrightarrow$ ($p*_{\rightsquigarrow}$'),*
*(pA#10') (pEq#') + ($p*_{\rightsquigarrow}$') + (pD') + (Tr') $\Longrightarrow$ ($p*_{\rightarrow}$').*

**Proof.** (pA#1): $x \rightarrow (y \rightsquigarrow z) = 1 \overset{(pEx)}{\Longleftrightarrow} y \rightsquigarrow (x \rightarrow z) = 1$.

(pA#2): (By *Prover9*) Suppose that $x \rightarrow y = 1$ and $y \rightarrow z = 1$; we must prove that $x \rightarrow z = 1$.
Since $y \rightarrow z = 1$, it follows, by ($p*_{\rightarrow}$), that $(x \rightarrow y) \rightarrow (x \rightarrow z) = 1$, i.e.

$$1 \rightarrow (x \rightarrow z) = 1. \tag{1.3}$$

Since $x \rightarrow y = 1$ and, by (Re), $x \rightsquigarrow x = 1$, it follows that $x \rightarrow y = x \rightsquigarrow x$; then, by (Re'), $x \rightarrow y \leq x \rightsquigarrow x$; hence, by (pEq#'), $x \leq (x \rightarrow y) \rightarrow x$, i.e. $x \rightarrow ((x \rightarrow y) \rightarrow x) = 1$, i.e.

$$x \rightarrow (1 \rightarrow x) = 1. \tag{1.4}$$

Take now $X := x \rightarrow z$ in (1.4), to obtain: $(x \rightarrow z) \rightarrow (1 \rightarrow (x \rightarrow z)) = 1$, hence, by (1.3),

$$(x \rightarrow z) \rightarrow 1 = 1. \tag{1.5}$$

Finally, from (1.3) and (1.5), by (An), we obtain $x \to z = 1$. Thus, (Tr) holds.

(pA#3$_\to$): (pB$_\to$'), i.e. $y \to z \leq (x \to y) \to (x \to z)$, is equivalent, by (pEq#'), with $x \to y \leq (y \to z) \rightsquigarrow (x \to z)$, which is (pBB$_1$').

(pA#3$_\rightsquigarrow$): (pB$_\rightsquigarrow$'), i.e. $y \rightsquigarrow z \leq (x \rightsquigarrow y) \rightsquigarrow (x \rightsquigarrow z)$, is equivalent, by (pEq#'), with $x \rightsquigarrow y \leq (y \rightsquigarrow z) \to (x \rightsquigarrow z)$, which is (pBB$_2$').

(pA#3): By (pA#3$_\to$) and (pA#3$_\rightsquigarrow$).

(pA#4): By (Re'), $x \to y \leq x \to y$; then, by (pEq#'), $x \leq (x \to y) \rightsquigarrow y$. By (Re'), $x \rightsquigarrow y \leq x \rightsquigarrow y$; then, by (pEq#'), $x \leq (x \rightsquigarrow y) \to y$. Thus, (pD') holds.

(pA#5): Let $x \leq y$; by (pD'), $y \leq (y \to z) \rightsquigarrow z$; then,
by (Tr'), $x \leq (y \to z) \rightsquigarrow z$; hence,
by (pEq#'), $y \to z \leq x \to z$; thus, (p**$_\to$') holds.
Let $x \leq y$; by (pD'), $y \leq (y \rightsquigarrow z) \to z$; then,
by (Tr'), $x \leq (y \rightsquigarrow z) \to z$; hence,
by (pEq#'), $y \rightsquigarrow z) \leq x \rightsquigarrow z$; thus, (p**$_\rightsquigarrow$') holds too.

(pA#5'): By (pA#4) and (pA#5).

(pA#6): [197] By (pD'), $y \leq (y \rightsquigarrow z) \to z$;
by (pB$_\to$'), $(y \rightsquigarrow z) \to z \leq (x \to (y \rightsquigarrow z)) \to (x \to z)$; then,
by (Tr'), $y \leq (x \to (y \rightsquigarrow z)) \to (x \to z)$; hence,
by (pEq#'), $x \to (y \rightsquigarrow z) \leq y \rightsquigarrow (x \to z)$, which is (pC$_1$').
By (pD'), $y \leq (y \rightsquigarrow z) \rightsquigarrow z$;
by (pB$_\rightsquigarrow$'), $(y \rightsquigarrow z) \rightsquigarrow z \leq (x \rightsquigarrow (y \rightsquigarrow z)) \to (x \rightsquigarrow z)$; then,
by (Tr'), $y \leq (x \rightsquigarrow (y \to z)) \rightsquigarrow (x \rightsquigarrow z)$; hence,
by (pEq#'), $x \rightsquigarrow (y \to z) \leq y \to (x \rightsquigarrow z)$, which is (pC$_2$').

(pA#6'): By (pA#4) and (pA#6).

(pA#7): [197] By (pD'), $y \leq (y \to z) \rightsquigarrow z$; hence,
by (p*$_\to$'), $x \to y \leq x \to ((y \to z) \rightsquigarrow z)$, i.e.,
by (pEx), $x \to y \leq (y \to z) \rightsquigarrow (x \to z)$, which is (pBB$_1$'); then,
by (pEq#'), we obtain $y \to z \leq (x \to y) \to (x \to z)$, which is (pB$_\to$').
By (pD'), $y \leq (y \rightsquigarrow z) \to z$; hence,
by (p*$_\rightsquigarrow$'), $x \rightsquigarrow y \leq x \rightsquigarrow ((y \rightsquigarrow z) \to z)$, i.e.,
by (pEx), $x \rightsquigarrow y \leq (y \rightsquigarrow z) \to (x \rightsquigarrow z)$, which is (pBB$_2$'); then,
by (pEq#'), we obtain $y \rightsquigarrow z \leq (x \rightsquigarrow y) \rightsquigarrow (x \rightsquigarrow z)$, which is (pB$_\rightsquigarrow$').

(pA#7'): By (pA#4) and (pA#7).

(pA#8): By (pM), $y = 1 \to y$, hence $x \leq y \Longleftrightarrow x \leq 1 \to y$; then, by p(Eq#'), $x \leq 1 \to y \Longleftrightarrow 1 \leq x \rightsquigarrow y$. Hence, $x \leq y \Longleftrightarrow 1 \leq x \rightsquigarrow y$.
By (pM), $y = 1 \rightsquigarrow y$, hence $x \leq y \Longleftrightarrow x \leq 1 \rightsquigarrow y$; then, by (pEq#'), $x \leq 1 \rightsquigarrow y \Longleftrightarrow 1 \leq x \to y$. Hence, $x \leq y \Longleftrightarrow 1 \leq x \to y$. Thus, (pEqrel) holds.

(pA#9): By (pA#8) and (pA34).

(pA#9'): By (An'), $x \leq 1$ and $1 \leq x$ imply $x = 1$, and by (La'), $1 \leq 1$; hence, we have: $x = 1 \Longleftrightarrow (x \leq 1$ *and* $1 \leq x)$; but, since (La') holds, it follows that we have: (a) $x = 1 \Longleftrightarrow 1 \leq x$. Then,
$x \to y = 1 \overset{(a)}{\Longleftrightarrow} 1 \leq x \to y \overset{(pEq\#')}{\Longleftrightarrow} x \leq 1 \rightsquigarrow y \overset{(pM)}{=} y$ and
$x \rightsquigarrow y = 1 \overset{(a)}{\Longleftrightarrow} 1 \leq x \rightsquigarrow y \overset{(pEq\#')}{\Longleftrightarrow} x \leq 1 \to y \overset{(pM)}{=} y$; thus, (IdEqR) and (pEqrelR) hold.

(pA#10): Let $y \leq z$; by (p*$_{\to}$') , $(x \rightsquigarrow y) \to y \leq (x \rightsquigarrow y) \to z$; but, $x \leq (x \rightsquigarrow y) \to y$, by (pD'); hence, by (Tr'), $x \leq (x \rightsquigarrow y) \to z$, which, by (pEq#'), gives $x \rightsquigarrow y \leq x \rightsquigarrow z$. Thus, (p*$_{\rightsquigarrow}$') holds.

(pA#10'): Let $y \leq z$; by (p*$_{\rightsquigarrow}$') , $(x \to y) \rightsquigarrow y \leq (x \to y) \rightsquigarrow z$; but, $x \leq (x \to y) \rightsquigarrow y$, by (pD'); hence, by (Tr'), $x \leq (x \to y) \rightsquigarrow z$, which, by (pEq#'), gives $x \to y \leq x \to z$. Thus, (p*$_{\to}$') holds. □

### 1.1.4 The "Big map": the global hierarchy of pseudo-M algebras

We recall now from [116] the following definition (see [114] for the commutative case).

**Definition 1.1.19** Let $\mathcal{A} = (A, \leq, \to, \rightsquigarrow, 1)$ be a left-(right-) pseudo-structure.

1) We shall say that $\mathcal{A}$ is *reflexive*, if $\leq$ is reflexive (i.e. it satisfies property (Re)).

2) We shall say that $\mathcal{A}$ is *antisymmetric*, if $\leq$ is antisymmetric (i.e. it satisfies property (An)).

3) We shall say that $\mathcal{A}$ is *transitive*, if $\leq$ is transitive (i.e. it satisfies property (Tr)).

4) We shall say that $\mathcal{A}$ is *pre-ordered*, if $\leq$ is a pre-order relation (i.e. it is reflexive and transitive).

5) We shall say that $\mathcal{A}$ is *ordered*, if $\leq$ is a partial order relation (i.e. it is reflexive, antisymmetrique and transitive).

6) We shall say that $\mathcal{A}$ is a *lattice*, if $\leq$ is a lattice order relation (i.e. it is a partial order such that there exist $\inf(x, y)$ and $\sup(x, y)$ for each $x, y \in A$); we shall use the notation $x \wedge y$ for $\inf(x, y)$ and $x \vee y$ for $\sup(x, y)$, with $x \leq y \Leftrightarrow x \wedge y = x \Leftrightarrow x \vee y = y$ (see Section 5.2).

- **Hasse diagrams and Hasse-type diagrams**

**Remark 1.1.20** ([111], Remark 3.11) (see also ([116], Remark 2.1.21) )

(1) If the pseudo-algebra (pseudo-structure) is ordered, then the usual *Hasse diagram* is used, where each element is represented by a bullet • and if $x \leq y$ and there is no $z$ such that $x < z < y$, then $x$ is reprezented below $y$ and a line will connect them.

(2) If the pseudo-algebra (pseudo-structure) is not ordered (i.e. it is neither reflexive nor antisymmetric nor transitive, or it is only reflexive, or reflexive and transitive, or reflexive and antisymmetric), then a *Hasse-type diagram* is used [111], where each element is represented by a circ $\circ$ and if $x \leq y$ and $y \leq x$ and $x \neq y$ (i.e. $x$ and $y$ have the *same 'height'*, or are *'parallel'*), then a horizontal line will connect them.

**Remark 1.1.21** [114] In the diagram of a hierarchy of classes of algebras, we shall represent:

- *reflexive* algebras by ○

- *antisymmetric* algebras by ○
- *transitive* algebras by •
- *reflexive* and *antisymmetric* algebras by ◎
- *reflexive* and *transitive* algebras by ⊙
- *ordered* algebras by ●

Then, the above subclasses of pseudo-M algebras are connected as in the "Big map" recalled from [116] in next Figure 1.1.

Pseudo-BCI algebras and pseudo-BCK algebras are discussed in some details in next section.

## 1.2 Pseudo-BCI and pseudo-BCK algebras

### 1.2.1 Ten equivalent definitions

**• We present first their definitions as left-pseudo-algebras (the dual ones are omitted)**

**Definitions 1.2.1** (Definition 1 of pBCI/pBCK algebras)
(1) A *left-pBCI algebra* is an algebra $\mathcal{A}^L = (A^L, \to, \rightsquigarrow, 1)$ verifying (pBB0, (pD), (Re), (An), (IdEqR).
(2) A *left-pBCK algebra* is a left-pBCI algebra verifying additionally (La) (Last element).

Note that left-pBCI and left-pBCK algebras are left-pseudo-algebras. Note that a binary relation $\leq$ can then be defined by: for all $x, y \in A^L$,

(pdfrelR) $$x \leq y \overset{def.}{\Longleftrightarrow} x \to y = 1 \ (\overset{(IdEqR)}{\Longleftrightarrow} x \rightsquigarrow y = 1).$$

**• We present their equivalent definitions as left-pseudo-structures**

**Definitions 1.2.2** (Definition 1' of pBCI/pBCK algebras)
(1) A *left-pseudo-BCI algebra*, or a *left-pBCI algebra* for short, is a structure $\mathcal{A}^L = (A^L, \leq, \to, \rightsquigarrow, 1)$, where $\leq$ is a binary relation on $A^L$, $(\to, \rightsquigarrow)$ is a pair of binary operations on $A^L$ (called *pseudo-implication*) and 1 is an element of $A^L$ verifying the axioms (pBB'), (pD'), (Re'), (An'), (IdEqR), (pEqrelR).
(2) A *left-pseudo-BCK algebra*, or a *left-pBCK algebra* for short, is a left-pBCI algebra verifying additionally (La') (Last element).

We have the following result (the dual ones are omitted), by Proposition 1.1.11.

**Proposition 1.2.3** *(See ([116], Proposition 2.3.4))*
*The following properties (including the defining ones) hold in a left-pBCI algebra (hence in a left-pBCK algebra): (pM), (N'), (Re'), (Tr'), (An'), (pEx), (pC'), (pEq#'), (pB'), (pBB'), (pD'), (p*'), (p**'), (pH).*

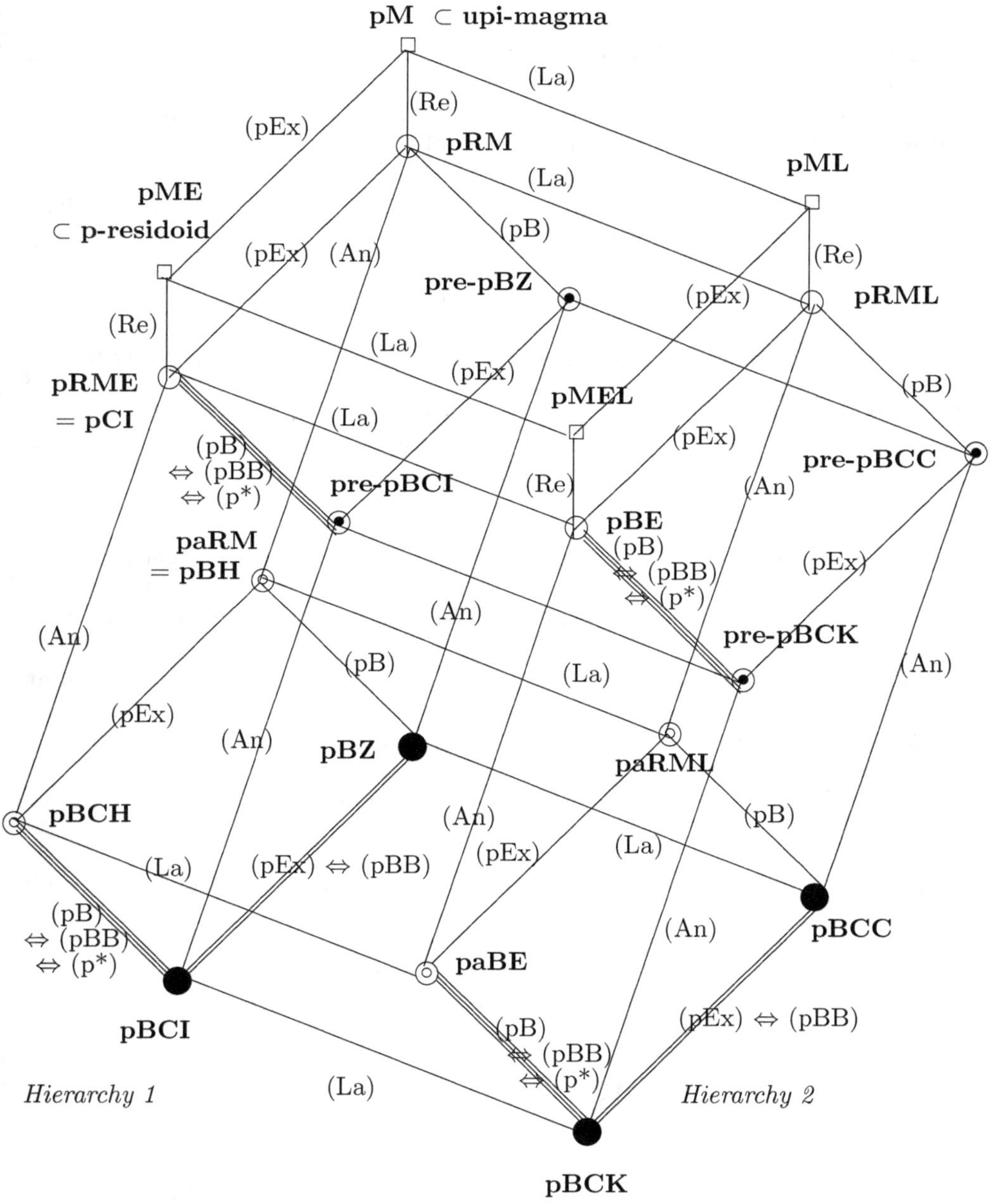


Figure 1.1: The "Big map": the global hierarchy of pM algebras

**Remark 1.2.4** *In a left-pBCI algebra (left-pBCK algebra)* $\mathcal{A}^L = (A^L, \leq, \rightarrow, \rightsquigarrow, 1)$, *the binary relation* $\leq$ *is a partial order relation, since the properties (Re'), (An'), (Tr') hold.*

Denote by **pBCI** the class of left-pBCI algebras and by **pBCK** the class of all left-pBCK algebras.

If the binary relation $\leq$ is a lattice order, then denote by **pBCI-L** the class of left-pBCI lattices and by **pBCK-L** the class of left-pBCK lattices.

Hence, a left-pBCK algebra is a left-pBCI algebra verifying (La') and, dually, a right-pBCK algebra is a right-pBCI algebra verifying (Fi'), i.e.:

$$\mathbf{pBCK} = \mathbf{pBCI} + \text{(La') and, dually, } \mathbf{pBCK}^R = \mathbf{pBCI}^R + \text{(Fi')}.$$

Note that we could say that a pBCK algebra is an "integral" pBCI algebra.

**Remark 1.2.5** *Since the property (La') holds in a left-pBCK algebra, but does not hold in a left-pBCI algebra, it follows, by Proposition 1.1.11 (pA6), that the property (pK') holds in a left-pBCK algebra, but does not hold in a left-pBCI algebra.*

**Remarks 1.2.6** *(See [114]) Consider the above property (N')* $(1 \leq x \implies x = 1)$.

*Note that this property says that* 1 **is the maximal element** *of the left-pBCI algebra. Hence,*

- *if* $\mathcal{A}^L$ *is a left-pBCI algebra, then* $(A^L, \leq)$ *is a* **poset with maximal element** 1, *while*
- *if* $\mathcal{A}^L$ *is a left-pBCK algebra, then* $(A^L, \leq, 1)$ *is a* **poset with greatest element** 1.

*Note also that, by (pA1), (La') + (An')* $\Longrightarrow$ *(N').*

**Theorem 1.2.7** *[48] Any pseudo-BCK algebra can be extended to a pseudo-BCI algebra.*

**Proof.** Let $(A, \leq, \rightarrow, \rightsquigarrow, 1)$ be a pseudo-BCK algebra and $\delta \notin A$. Define on $A' = A \cup \{\delta\}$ the operations: for all $x, y \in A'$,

$$x \rightarrow' y = \begin{cases} x \rightarrow y, & \text{if } x, y \in A, \\ \delta, & \text{if } x = \delta,\ y \in A \text{ or } x \in A,\ y = \delta, \\ 1, & \text{if } x = y = \delta, \end{cases}$$

$$x \rightsquigarrow' y = \begin{cases} x \rightsquigarrow y, & \text{if } x, y \in A, \\ \delta, & \text{if } x = \delta,\ y \in A \text{ or } x \in A,\ y = \delta, \\ 1, & \text{if } x = y = \delta. \end{cases}$$

Then, by routine calculations, $(A', \leq, \rightarrow', \rightsquigarrow', 1)$ is only a pseudo-BCI algebra, since $\delta \not\leq 1$. □

**Open problem 1.2.8** *Can any pseudo-BCI algebra be extended to a quantum-B algebra?*

In the rest of the section, we shall present other, old or new, equivalent definitions of pBCI (pBCK) algebras as **left-pseudo-structures**, in order to see their deep connections with the quantum-B algebras (presented in the next section), which are in fact structures (partially ordered pi-magmas).

- **Other nine equivalent definitions of pBCI/pBCK algbras**

To introduce the following first equivalent definition of pBCI (pBCK) algebras, coming from their position on the "map", we need the following result.

**Theorem 1.2.9**

*(1) Let $\mathcal{A}^L = (A^L, \leq, \rightarrow, \rightsquigarrow, 1)$ be a structure such that (pM), (Re'), (An'), (pEx), (pB'), (IdEqR) and (pEqrelR) hold. Then, $\mathcal{A}^L$ is a left-pBCI algebra (Definition 1').*

*(1') Conversely, every left-pBCI algebra verifies (pM), (Re'), (An'), (pEx), (pB'), (IdEqR), (pEqrelR).*

**Proof.** (1): We must prove, by Definition 1', that (pBB'), (pD'), (Re'), (An'), (IdEqR), (pEqrelR) hold. It remains to prove that (pBB') and (pD') hold. Indeed, by (pA10), (pEx) + (pB') + (pEqrelR) $\Longrightarrow$ (pBB') and
by (pA4), (Re) + (pEx) + (pEqrelR) $\Longrightarrow$ (pD').

(1'): We must prove that (pM), (pEx), (pB') hold. Indeed,
by (pA4'), (pD') + (Re) + (An') + (pEqrelR) $\Longrightarrow$ (N'),
by (pA20), (pBB') + (pD) + (N') $\Longrightarrow$ (pC'),
by (pA3), (pC') + (An') $\Longrightarrow$ (pEx);
by (pA10'), (pEx) + (pBB') + (pEqrelR) $\Longrightarrow$ (pB');
by (pA5), (Re) + (An') + (pEx) + (pEqrelR) $\Longrightarrow$ (pM). □

By above theorem, we obtain the following equivalent definition of left-pBCI algebras.

**Definition 1.2.10** (Definition 2' of pBCI algebras)

A *left-pBCI algebra* is a structure $\mathcal{A}^L = (A^L, \leq, \rightarrow, \rightsquigarrow, 1)$ such that the properties (pM), (Re'), (An'), (pEx), (pB'), (IdEqR), (pEqrelR) hold.

The following theorem is presented in [104] for pseudo-BCK algebras. We proved in [116] that it is valid for pseudo-BCI algebras too. It gives us an equivalent definition of left-pBCI algebras (and hence of left-pBCK algebras) (the right-case is omitted).

**Theorem 1.2.11** *([116], Theorem 2.3.7)*

*(1) Let $\mathcal{A}^L = (A^L, \leq, \rightarrow, \rightsquigarrow, 1)$ be a structure such that $(A^L, \leq)$ is a poset (i.e. (Re'), (An'), (Tr') hold) and the following properties hold: (pBB'), (pM), (IdEqR), (pEqrelR) (i.e. the pair $(\rightarrow, \rightsquigarrow)$ is a pseudo-residuum). Then, $\mathcal{A}^L$ is a left-pBCI algebra (Definition 1').*

*(1') Conversely, every left-pBCI algebra $(A^L, \leq, \rightarrow, \rightsquigarrow, 1)$ is a poset satisfying (pBB'), (pM), (IdEqR), (pEqrelR).*

**Proof.** (1): We have to prove that (pBB'), (pD'), (Re'), (An'), (IdEqR), (pEqrelR) hold. It remains to prove that (pD') holds; indeed, by (pA18'), (pBB') + (pM) $\Longrightarrow$ (pD'); thus, (pD') holds.

(1'): Suppose that (pBB'), (pD'), (Re'), (An'), (IdEqR), (pEqrelR) hold. It remains to prove that (pM) and (Tr') hold. Indeed,
by (pA4'), (pD') + (Re) + (An') + (pEqrelR) $\Longrightarrow$ (N'),
by (pA20), (pBB') + (pD) + (N') $\Longrightarrow$ (pC'),
by (pA3), (pC') + (An') $\Longrightarrow$ (pEx),
by (pA5), (Re) + (An') + (pEx) + (pEqrelR) $\Longrightarrow$ (pM); thus, (pM) holds.
by (pA10'), (pEx) + (pEqrelR) + (pBB') $\Longrightarrow$ (pB'),
by (pA12'), (pM) + (pEqrelR) + (pB') $\Longrightarrow$ (p*'),
by (pA13), (N') + (pEqrelR) + (p*') $\Longrightarrow$ (Tr'); thus, (Tr') holds. □

By above theorem, we obtain the following equivalent definitions (the dual case is omitted).

**Definition 1.2.12** (Definition 3' of pBCI algebras) [116]
A *left-pBCI algebra* is a structure $\mathcal{A}^L = (A^L, \leq, \rightarrow, \rightsquigarrow, 1)$ such that $(A^L, \leq)$ is a poset (with maximal element 1) and the properties (pBB'), (pM), (IdEqR), (pEqrelR) hold.

The following theorem gives us another equivalent definition of a left-pBCI algebra (and hence of a left-pBCK algebra) (the right-case is omitted).

**Theorem 1.2.13** *([116], Theorem 2.3.10)*
*(1) Let $\mathcal{A}^L = (A^L, \leq, \rightarrow, \rightsquigarrow, 1)$ be a structure such that $(A^L, \leq)$ is a poset (i.e. (Re'), (An'), (Tr') hold) and the following properties hold: (pM), (pEx), (p*'), (IdEqR), (pEqrelR). Then, $\mathcal{A}^L$ is a left-pBCI algebra (Definition 3').*
*(1') Conversely, every left-pBCI algebra $(A^L, \leq, \rightarrow, \rightsquigarrow, 1)$ is a poset satisfying (pM), (pEx), (p*'), (IdEqR), (pEqrelR).*

**Proof.** (1): We must prove that $(A^L, \leq)$ is a poset and (pBB'), (pM), (IdEqR), (pEqrelR) hold. It remains to prove that (pBB') holds. Indeed, by (pA11), (Re) + (pEx) + (pEqrelR) + (p*') $\Longrightarrow$ (pBB').

(1'): We know that $(A^L, \leq)$ is a poset and (pBB'), (pM), (IdEqR), (pEqrelR) hold. It remains to prove (pEx) and (p*') hold. Indeed,
by (pA21"), (pBB') + (pM) + (An') + (pEqrelR) $\Longrightarrow$ (pEx);
by (pA10'), (pEx) + (pEqrelR) + (pBB') $\Longrightarrow$ (pB'),
by (pA12'), (pM) + (pEqrelR) + (pB') $\Longrightarrow$ (p*'). □

By above Theorem 1.2.13, we obtain the following equivalent definitions (the dual case is omitted).

**Definition 1.2.14** (Definition 4' of pBCI algebras)
A *left-pBCI algebra* is a structure $\mathcal{A}^L = (A^L, \leq, \rightarrow, \rightsquigarrow, 1)$ such that $(A^L, \leq)$ is a poset (with maximal element 1) and the properties (pM), (pEx), (p*'), (IdEqR), (pEqrelR) hold.

To introduce the equivalent definition of pBCI algebras coming from the non-commutative BCI logic, we need the following result.

**Theorem 1.2.15** *([116], Theorem 2.3.12)*

*(1) Let $\mathcal{A}^L = (A^L, \leq, \rightarrow, \rightsquigarrow, 1)$ be a structure such that (pB'), (pC'), (Re'), (An'), (IdEqR), (pEqrelR) hold. Then, $\mathcal{A}^L$ is a left-pBCI algebra (Definition 1').*

*(1') Conversely, every left-pBCI algebra verifies (pB'), (pC'), (Re') and (An'), (IdEqR), (pEqrelR).*

**Proof.** (1): Following Definition 1' of left-pBCI algebras, we must prove that (pBB'), (pD'), (Re'), (An'), (IdEqR), (pEqrelR) hold. it remains to prove that (pBB') and (pD') hold. Indeed,
by (pA3), (pC') + (An') $\Longrightarrow$ (pEx),
by (pA10), (pEx) + (pB') + (pEqrelR) $\Longrightarrow$ (pBB') and
by (pA4), (Re') + (pEx) + (pEqrelR) $\Longrightarrow$ (pD').

(1'): It remains to prove that every left-pBCI verifies (pB'). Indeed, by Proposition 1.2.3, (pB') holds. □

By above theorem, we obtain the following equivalent definition of left-pBCI algebras coming from the non-commutative BCI logic, called *pBCI logic* (see Definitions 1.7.3). Note that the name "BCI" for pBCI (BCI) algebras comes from the names ("B", "C", "I") given to the logical axioms of BCI logic, where the name (Re) (coming from "Reflexivity") is used in the monograph, instead of (I) (coming from "Identity"), for the corresponding axiom.

**Definition 1.2.16** (Definition 5' of pBCI algebras) [116]

A *left-pBCI algebra* is a structure $\mathcal{A}^L = (A^L, \leq, \rightarrow, \rightsquigarrow, 1)$ such that the properties (pB'), (pC'), (Re'), (An'), (IdEqR), (pEqrelR) hold.

Since by (pA3), (pC') + (An') $\Longrightarrow$ (pEx), and since, by (pA10"), (pEx) + (pEqrelR) $\Longrightarrow$ ((pB') $\Leftrightarrow$ (pBB')), we obtain immediately the following equivalent definition of pBCI algebras.

**Definition 1.2.17** (Definition 6' of pBCI algebras)

A *left-pBCI algebra* is a structure $\mathcal{A}^L = (A^L, \leq, \rightarrow, \rightsquigarrow, 1)$ such that the properties (pBB'), (pC'), (Re'), (An'), (IdEqR), (pEqrelR) hold.

We shall prove now that, in Definition 5' of pBCI algebras, we can take only ($pB_{\rightarrow}$') instead of (pB'). A similar proof can made by taking ($pB_{\rightsquigarrow}$') instead of (pB').

**Theorem 1.2.18**

*(1) Let $\mathcal{A}^L = (A^L, \leq, \rightarrow, \rightsquigarrow, 1)$ be a structure such that ($pB_{\rightarrow}$'), (pC'), (Re'), (An'), (IdEqR), (pEqrelR) hold. Then, $\mathcal{A}^L$ is a left-pBCI algebra (Definition 5').*

*(1') Conversely, every left-pBCI algebra verifies ($pB_{\rightarrow}$'), (pC'), (Re'), (An'), (IdEqR), (pEqrelR).*

**Proof.** (1): (**By Prover9.**) By Definition 5 of pBCI algebras, it remains to prove that (pB$_{\rightsquigarrow}$') holds. Indeed,
by (pA3), (pC') + (An') $\Longrightarrow$ (pEx),
by (pA#1), (pEx) $\Longrightarrow$ (pEq#),
by (pA#4), (pEq#') + (Re') $\Longrightarrow$ (pD'),
by (pA4'), (pD') + (Re) + (An') + (pEqrelR) $\Longrightarrow$ (N') and
by (pA#3$_{\rightarrow}$), (pEq#') $\Longrightarrow$ ((pB$_{\rightarrow}$') $\Leftrightarrow$ (pBB$_1$')).

Now, we prove

$$x \to y \ \leq \ x \to ((y \rightsquigarrow z) \to z). \tag{1.6}$$

Indeed, we have:
by (pD), $y \to ((y \rightsquigarrow z) \to z) = 1$ and
by (pB$_{\rightarrow}$'), $y \to ((y \rightsquigarrow z) \to z) \ \leq \ (x \to y) \to (x \to ((y \rightsquigarrow z) \to z))$; hence,
by (N'), $(x \to y) \to (x \to ((y \rightsquigarrow z) \to z)) = 1$, i.e. (1.6) holds, by (pEqrelR).

Now, we prove

$$((x \rightsquigarrow y) \to y) \to z \ \leq \ x \to z. \tag{1.7}$$

Indeed, we have:
by (pD), $x \to ((x \rightsquigarrow y) \to y) = 1$ and
by (pBB$_1$'), $x \to ((x \rightsquigarrow y) \to y) \ \leq \ (((x \rightsquigarrow y) \to y) \to z) \rightsquigarrow (x \to z)$; hence,
by (N'), $(((x \rightsquigarrow y) \to y) \to z) \rightsquigarrow (x \to z)$, i.e. (1.7) holds, by (pEqrelR).

Now, we prove:

$$x \rightsquigarrow y \ \leq \ x \rightsquigarrow ((y \rightsquigarrow z) \to z). \tag{1.8}$$

Indeed,
by (1.6), taking $X := x \rightsquigarrow y$, we obtain:
$(x \rightsquigarrow y) \to y \ \leq \ (x \rightsquigarrow y) \to ((y \rightsquigarrow z) \to z)$, hence, by (pEqrelR),
(a) $\quad ((x \rightsquigarrow y) \to y) \ \to \ ((x \rightsquigarrow y) \to ((y \rightsquigarrow z) \to z)) = 1.$
By (1.7), taking $Z := (x \rightsquigarrow y) \to ((y \rightsquigarrow z) \to z)$, we obtain:
(b) $\quad ((x \rightsquigarrow y) \to y) \to [(x \rightsquigarrow y) \to ((y \rightsquigarrow z) \to z)] \ \leq$
$\quad x \to [(x \rightsquigarrow y) \to ((y \rightsquigarrow z) \to z)].$
Then, from (a) and (b), by (N'), we obtain
$x \to [(x \rightsquigarrow y) \to ((y \rightsquigarrow z) \to z)] = 1$
$\overset{(pEqrelR)}{\Longleftrightarrow} x \ \leq \ (x \rightsquigarrow y) \to ((y \rightsquigarrow z) \to z)$
$\overset{(pEq\#')}{\Longleftrightarrow} x \rightsquigarrow y \ \leq \ x \rightsquigarrow ((y \rightsquigarrow z) \to z)$, i.e. (1.8) holds.

Finally, (1.8), i.e. $x \rightsquigarrow y \ \leq \ x \rightsquigarrow ((y \rightsquigarrow z) \to z)$
$\overset{(pEx)}{\Longleftrightarrow} x \rightsquigarrow y \ \leq \ (y \rightsquigarrow z) \to (x \rightsquigarrow z)$, that is (pBB$_2$')
$\overset{(pEq\#'}{\Longleftrightarrow} y \rightsquigarrow z \ \leq \ (x \rightsquigarrow y) \rightsquigarrow (x \rightsquigarrow z)$, that is (pB$_{\rightsquigarrow}$'). Thus, $\mathcal{A}^L$ is a left-pBCI algebra.

(1'): Obviously. □

By above theorem, we obtain the following equivalent definition of left-pBCI algebras coming from the reduct of *pBCI logic* called in [193] *quantum-B logic* (*QB logic*) (see Definitions 1.7.4).

**Definition 1.2.19** (Definition 7' of pBCI algebras)
A *left-pBCI algebra* is a structure $\mathcal{A}^L = (A^L, \leq, \rightarrow, \rightsquigarrow, 1)$ such that the properties (pB$_{\rightarrow}$'), (pC'), (Re'), (An'), (IdEqR), (pEqrelR) hold.

To introduce the equivalent definition of pBCK algebras coming from the *pseudo-BCK logic* (see Definitions 1.7.3), we need the following result.

**Theorem 1.2.20** *([116], Theorem 2.3.14)*
*(1) Let $\mathcal{A}^L = (A^L, \leq, \rightarrow, \rightsquigarrow, 1)$ be a structure such that (pB'), (pC'), (pK'), (An'), (IdEqR), (pEqrelR) hold. Then, $\mathcal{A}^L$ is a left-pBCK algebra (Definition 1').*
*(1') Conversely, every left-pBCK algebra verifies (pB'), (pC'), (pK') and (An'), (IdEqR), (pEqrelR).*

**Proof.** (1): Following the Definition 1' of left-pBCK algebras, the properties (pBB'), (pD'), (Re'), (La'), (An'), (IdEqR), (pEqrelR) hold; hence, it remains to prove that (pBB'), (pD'), (Re'), (La') hold. Indeed, by Proposition 1.1.11:
by (pA3), (pC') + (An') $\Longrightarrow$ (pEx) and
by (pA10), (pEx) + (pB') + (pEqrelR) $\Longrightarrow$ (pBB'); thus, (pBB') holds;
by (pA2), (pK') + (An') + (pEqrelR) $\Longrightarrow$ (N') and
by (pA7), (pK') + (N') + (pEqrelR) $\Longrightarrow$ (La'); thus, (La') holds;
by (pA23), (pK') + (pC') + (An') + (pEqrelR) $\Longrightarrow$ (Re'); thus (Re') holds;
by (pA4), (Re) + (pEx) + (pEqrelR) $\Longrightarrow$ (pD'); thus, (pD') holds.
(1'): It remains to prove that (pB'), (pC') and (pK') hold. Indeed,
by Proposition 1.2.3, (pB') and (pC') hold, also (pEx) holds and
by (pA8), (La) + (pEx) + (Re) $\Longrightarrow$ (pK). □

By above theorem, we obtain the following equivalent definition of left-pBCK algebras coming from the *pseudo-BCK logic*. Note that the name "BCK" for pBCK (BCK) algebras comes from the names ("B", "C", "K") given to the logical axioms of the *BCK logic*.

**Definition 1.2.21** (Definition 5' of pBCK algebras) [116]
A *left-pBCK algebra* is a structure $\mathcal{A}^L = (A^L, \leq, \rightarrow, \rightsquigarrow, 1)$ such that the properties (pB'), (pC'), (pK'), (An'), (IdEqR), (pEqrelR) hold.

Also, since, by (pA3), (pC') + (An') $\Longrightarrow$ (pEx), and since, by (pA10"), (pEx) + (pEqrelR) $\Longrightarrow$ ((pB') $\Leftrightarrow$ (pBB')), we obtain immediately the following equivalent definition of pBCK algebras.

**Definition 1.2.22** (Definition 6' of pBCK algebras)
A *left-pBCK algebra* is a structure $\mathcal{A}^L = (A^L, \leq, \rightarrow, \rightsquigarrow, 1)$ such that the properties (pBB'), (pC'), (pK'), (An'), (IdEqR), (pEqrelR) hold.

We shall prove now that, in Definition 5' of pBCK algebras, we can take only (pB$_{\rightarrow}$') instead of (pB'). A similar proof can made by taking (pB$_{\rightsquigarrow}$') instead of (pB').

**Theorem 1.2.23**
*(1) Let $\mathcal{A}^L = (A^L, \leq, \rightarrow, \rightsquigarrow, 1)$ be a structure such that (pB$_\rightarrow$'), (pC'), (pK'), (An'), (IdEqR), (pEqrelR) hold. Then, $\mathcal{A}^L$ is a left-pBCK algebra (Definition 5').*
*(1') Conversely, every left-pBCK algebra verifies (pB$_\rightarrow$'), (pC'), (pK'), (An'), (IdEqR), (pEqrelR).*

**Proof.** (1): By (pA23), (pK') + (pC') + (An') + (pEqrelR) $\Longrightarrow$ (Re'). Then, apply Theorem 1.2.18 (1).
(1'): Obviously. □

By above theorem, we obtain the following equivalent definition of left-pBCK algebras.

**Definition 1.2.24** (Definition 7' of pBCK algebras)
A *left-pBCK algebra* is a structure $\mathcal{A}^L = (A^L, \leq, \rightarrow, \rightsquigarrow, 1)$ such that the properties (pB$_\rightarrow$'), (pC'), (pK'), (An'), (IdEqR), (pEqrelR) hold.

To introduce the equivalent definition of pBCI algebras coming from the Definition 1.3.1 of quantum-B algebras (Definition 1), we need the following result.

**Theorem 1.2.25**
*(1) Let $\mathcal{A}^L = (A^L, \leq, \rightarrow, \rightsquigarrow, 1)$ be a structure such that (pB'), (p*$_\rightarrow$'), (pEq#'), (Re'), (An'), (IdEqR) and (pEqrelR) hold. Then, $\mathcal{A}^L$ is a left-pBCI algebra (Definition 5').*
*(1') Conversely, every left-pBCI algebra verifies (pB'), (p*$_\rightarrow$'), (pEq#'), (Re'), (An'), (IdEqR), (pEqrelR).*

**Proof.** (1): We have to prove that (pB'), (pC'), (Re'), (An'), (IdEqR), (pEqrelR) hold. It remains to prove that (pC') holds. Indeed,
by (pA#4), (pEq#') + (Re') $\Longrightarrow$ (pD'),
by (pA#2), (pEq#) + (p*$_\rightarrow$) + (Re) + (An) $\Longrightarrow$ (Tr),
by (pA#6), (pEq#') + (pB') + (pD') + (Tr') $\Longrightarrow$ (pC').
(1'): It remains to prove that (p*$_\rightarrow$') and (pEq#') hold. Indeed,
by (pA3), (pC') + (An') $\Longrightarrow$ (pEx),
by (pA5), (Re) + (An') +(pEx) + (pEqrelR) $\Longrightarrow$ pM),
by (pA12'), (pM) + (pEqrelR) + (pB') $\Longrightarrow$ (p*').
By (pA#1), (pEx) $\Longrightarrow$ (pEq#). □

By above theorem, we obtain the following equivalent definition of left-pBCI algebras coming from the Definition 1 of quantum-B algebras.

**Definition 1.2.26** (Definition 8' of pBCI algebras)
A *left-pBCI algebra* is a structure $\mathcal{A}^L = (A^L, \leq, \rightarrow, \rightsquigarrow, 1)$ such that the properties (pB'), (p*$_\rightarrow$'), (pEq#'), (Re'), (An'), (IdEqR), (pEqrelR) hold.

To introduce the equivalent definition of pBCI algebras coming from the equivalent Definition 1.3.5 of quantum-B algebras (Definition 2), we need the following result.

**Theorem 1.2.27**

*(1) Let $\mathcal{A}^L = (A^L, \leq, \rightarrow, \rightsquigarrow, 1)$ be a structure such that (pEx), (p$*_{\rightarrow}$'), (pEq#'), (Re'), (An'), (IdEqR) and (pEqrelR) hold. Then, $\mathcal{A}^L$ is a left-pBCI algebra (Definition 7').*

*(1') Conversely, every left-pBCI algebra verifies (pEx), (p$*_{\rightarrow}$'), (pEq#'), (Re'), (An'), (IdEqR), (pEqrelR).*

**Proof.** (1): We must prove that (pB'), (p$*_{\rightarrow}$'), (pEq#'), (Re'), (An'), (IdEqR), (pEqrelR) hold. It remains to prove that (pB') holds. Indeed,
by (pA#4), (pEq#') + (Re') $\Longrightarrow$ (pD'),
by (pA#10), (pEq#') + (p$*_{\rightarrow}$') + (pD') + (Tr') $\Longrightarrow$ (p$*_{\rightsquigarrow}$'); thus, (p*') holdss;
by (pA#7), (pEq#') + (pEx) + (pD') + (p*') $\Longrightarrow$ (pB').

(1'): It remains to prove that (pEx) holds. Indeed,
by (pA#6'), (pEq#') + (pB') + (Re') + (Tr') $\Longrightarrow$ (pC'),
by (pA3), (pC') + (An') $\Longrightarrow$ (pEx). □

By above theorem, we obtain the following equivalent definition of left-pBCI algebras coming from Definition 2 of quantum-B algebras.

**Definition 1.2.28** (Definition 9' of pBCI algebras)

A *left-pBCI algebra* is a structure $\mathcal{A}^L = (A^L, \leq, \rightarrow, \rightsquigarrow, 1)$ such that the properties (pEx), (p$*_{\rightarrow}$'), (pEq#'), (Re'), (An') and (IdEqR), (pEqrelR) hold.

In order to better see the connections with the *maximal/integral quantum-B algebras* - presented in next section - we introduce a final equivalent definition of pBCI/pBCK alagebras.

**Theorem 1.2.29**

*(1) Let $\mathcal{A}^L = (A^L, \leq, \rightarrow, \rightsquigarrow, 1)$ be a structure such that $(A^L, \leq)$ is a poset (i.e. (Re'), (An'), (Tr') hold) and the following properties hold: (pEx), (p$*_{\rightarrow}$'), (pEq#'), (pM), (N') and (IdEqR), (pEqrelR). Then, $\mathcal{A}^L$ is a left-pBCI algebra (Definition 9').*

*(1') Conversely, every left-pBCI algebra verifies the above properties.*

**Proof.** (1): Obviously, by Definition 1.2.28 (Definition 9') of pBCI algebras.

(1'): By Definition 1.2.28, it remains to prove that (pM), (N'), (Tr') hold. Indeed,
by (pA5), (Re) + (An') + (pEx) + (pEqrelR) $\Longrightarrow$ (pM);
by (pA00), (pM) + (pEqrelR) $\Longrightarrow$ (N');
by (pA#2), (pEq#) + (p$*_{\rightarrow}$) + (Re) + (An) $\Longrightarrow$ (Tr). □

By above theorem, we obtain the following final equivalent definition of left-pBCI algebras.

**Definition 1.2.30** (Definition 10' of pBCI algebras)

A *left-pBCI algebra* is a structure $\mathcal{A}^L = (A^L, \leq, \rightarrow, \rightsquigarrow, 1)$ such that $(A^L, \leq)$ is a poset (i.e. (Re'), (An'), (Tr') hold) and the following properties hold: (pEx), (p$*_{\rightarrow}$'), (pEq#'), (pM), (N') and (IdEqR), (pEqrelR).

Consequently, we obtain immediately the following final equivalent definition of left-pBCK algebras.

**Definition 1.2.31** (Definition 10' of pBCK algebras)
A *left-pBCK algebra* is a structure $\mathcal{A}^L = (A^L, \leq, \rightarrow, \rightsquigarrow, 1)$ such that $(A^L, \leq, 1)$ is a poset with greatest element (i.e. (Re'), (An'), (Tr'), (La') hold) and the following properties hold: (pEx), (p*$_{\rightarrow}$'), (pEq#'), (pM) and (IdEqR), (pEqrelR).

• **Compatible deductive systems**

Recall the following definitions needed in Section 1.7.

**Definition 1.2.32**
(i) Let $\mathcal{A}^L = (A^L, \leq, \rightarrow, \rightsquigarrow, 1)$ be a left-pBCI algebra.
A *$(\rightarrow, \rightsquigarrow)$-deductive system*, or simply a *deductive system* when there is no danger of confusion, of $\mathcal{A}^L$ is a subset $S \subseteq A^L$ which satisfies:
(ds1) $1 \in S$,
(pds2) $x \in S$ and $x \rightarrow y \in S$ imply $y \in S$
(or $x \in S$ and $x \rightsquigarrow y \in S$ imply $y \in S$).
(i') Dually, let $\mathcal{A}^R = (A^R, \leq, \rightarrow^R, \rightsquigarrow^R, 0)$ be a right-pBCI algebra.
A *$(\rightarrow^R, \rightsquigarrow^R)$-deductive system* or simply a *deductive system*, when there is no danger of confusion, of $\mathcal{A}^R$ is a subset $S' \subseteq A^R$ which satisfies:
(ds1') $0 \in S'$,
(pds2') $x \in S'$ and $x \rightarrow^R y \in S'$ imply $y \in S'$
(or $x \in S'$ and $x \rightsquigarrow^R y \in S'$ imply $y \in S'$).

**Definition 1.2.33** (see ([145], Definition 2.2.1)
(1) Let $\mathcal{A}^L = (A^L, \leq, \rightarrow, \rightsquigarrow, 1)$ be a left-pBCI algebra.
We say that a $(\rightarrow, \rightsquigarrow)$-deductive system $S^L$ of $\mathcal{A}^L$ is *compatible*, if the following condition holds: for any $x, y \in A^L$,
(Co$^L$) $$x \rightarrow y \in S^L \Longleftrightarrow x \rightsquigarrow y \in S^L.$$
(1') Let $\mathcal{A}^R = (A^R, \leq, \rightarrow^R, \rightsquigarrow^R, 0)$ be a right-pBCI algebra.
We say that a $(\rightarrow^R, \rightsquigarrow^R)$-deductive system $S^R$ of $\mathcal{A}^R$ is *compatible*, if the following condition holds: for any $x, y \in A^R$,
(Co$^R$) $$x \rightarrow^R y \in S^R \Longleftrightarrow x \rightsquigarrow^R y \in S^R.$$

• **The commutative case**

A left-pBCI (left-pBCK) algebra $\mathcal{A}^L = (A^L, \leq, \rightarrow, \rightsquigarrow, 1)$ is *commutative* if $\rightarrow = \rightsquigarrow$. A commutative left-pBCI (left-pBCK) algebra is a left-BCI (left-BCK) algebra (see Chapter 2).

Dually, a right-pBCI (right-pBCK) algebra $\mathcal{A}^R = (A^R, \geq, \rightarrow^R, \rightsquigarrow^R, 0)$ is *commutative* if $\rightarrow^R = \rightsquigarrow^R$. A commutative right-pBCI (right-pBCK) algebra is a right-BCI (right-BCK) algebra.

### 1.2.2 Pseudo-Hilbert algebras do not exist

You can see, in Chapter 2, that Hilbert algebras are definitionally equivalent to positive-implicative BCK algebras (i.e. BCK algebras verifying (pimpl)) (Theorem 2.2.33), and recall that, in BCK algebras, (pimpl) $\Longleftrightarrow$ (pi) ([127], Theorem 8).

You can see also, in Remark 3.5.4, that the non-commutative generalizations of the properties (pimpl) and (pi) are the properties (ppimpl) and (ppi), respectively, where:
(ppimpl) = (pimpl$_{\rightarrow}$) + (pimpl$_{\rightsquigarrow}$), where
(pimpl$_{\rightarrow}$) $x \to (y \to z) = (x \to y) \to (x \to z)$ and
(pimpl$_{\rightsquigarrow}$) $x \rightsquigarrow (y \rightsquigarrow z) = (x \rightsquigarrow y) \rightsquigarrow (x \rightsquigarrow z)$;

(ppi) = (pi$_{\rightarrow}$) + (pi$_{\rightsquigarrow}$), where
(pi$_{\rightarrow}$) $y \to (y \to x) = y \to x$,
(pi$_{\rightsquigarrow}$) $y \rightsquigarrow (y \rightsquigarrow x) = y \rightsquigarrow x$
and that we have:
(pB3) (ppimpl) + (Re) + (pM) $\Longrightarrow$ (ppi).

We shall prove that pseudo-BCK algebras verifying (ppimpl) are in fact BCK algebras verifying (pimpl), i.e. Hilbert algebras; thus, non-commutative generalizations of Hilbert algebras do not exist. Moreover, we shall prove that there are pseudo-BCK algebras verifying (ppi) and not verifying (ppimpl).

**Proposition 1.2.34** *Let $\mathcal{A}^L = (A^L, \to, \rightsquigarrow, 1)$ be a left-pseudo-algebra verifying (pEx), (pD), (pK) and (An). If (ppimpl) holds, then $\to = \rightsquigarrow$.*

**Proof.** (Following the ideea of a *Prover9* proof.)
By (pD), $x \rightsquigarrow ((x \to y) \rightsquigarrow y) = 1$.
But, $x \rightsquigarrow ((x \to y) \rightsquigarrow y) \overset{(ppimpl)}{=} (x \rightsquigarrow (x \to y)) \rightsquigarrow (x \rightsquigarrow y)$.
It follows that $(x \rightsquigarrow (x \to y)) \rightsquigarrow (x \rightsquigarrow y) = 1$, hence, by (pEx),
$(x \to (x \rightsquigarrow y)) \rightsquigarrow (x \rightsquigarrow y) = 1$.
On the other hand, $(x \rightsquigarrow y) \to (x \to (x \rightsquigarrow y)) = 1$, by (pK).
Hence, by (An), we obtain

$$x \to (x \rightsquigarrow y) = x \rightsquigarrow y. \tag{1.9}$$

Similarly, by (pD), $x \to ((x \rightsquigarrow y) \to y) = 1$.
But, $x \to ((x \rightsquigarrow y) \to y) \overset{(ppimpl)}{=} (x \to (x \rightsquigarrow y)) \to (x \to y)$.
It follows that $(x \to (x \rightsquigarrow y)) \to (x \to y) = 1$.
On the other hand, $(x \to y) \to (x \to (x \rightsquigarrow y))$
$\overset{(pEx)}{=} (x \to y) \to (x \rightsquigarrow (x \to y)) \overset{(pK)}{=} 1$.
Hence, by (An), we obtain

$$x \to (x \rightsquigarrow y) = x \to y. \tag{1.10}$$

From (1.9) and (1.10), we obtain $x \to y = x \rightsquigarrow y$, for all $x, y \in A^L$, i.e. $\to = \rightsquigarrow$. □

**Corollary 1.2.35** *Let $\mathcal{A}^L = (A^L, \to, \rightsquigarrow, 1)$ be a left-pBCK algebra verifying (ppimpl). Then, $\to = \rightsquigarrow$, i.e. $\mathcal{A}^L$ is a left-BCK algebra verifying (pimpl), i.e. a Hilbert algebra.*

**Proof.** Any left-pBCK algebra verifies (pEx), (pD), (pK) and (An). Then, apply Proposition 1.2.34. □

**Example 1.2.36 Pseudo-BCK algebra verifying (ppi) and not verifying (ppimpl),**

The algebra $\mathcal{A}_4 = (A_4 = \{0, a, b, 1\}, \to, \rightsquigarrow, 1)$, with the tables of $\to$ and $\rightsquigarrow$ given below, is a pseudo-BCK algebra verifying (ppi) and not verifying (ppimpl) for $(0, b, a)$: $b = 0 \to a = 0 \to (b \to a) \neq (0 \to b) \to (0 \to a) = b \to b = 1$.

| $\to$ | 0 | a | b | 1 |
|---|---|---|---|---|
| 0 | 1 | b | b | 1 |
| a | 1 | 1 | 1 | 1 |
| b | 0 | a | 1 | 1 |
| 1 | 0 | a | b | 1 |

and

| $\rightsquigarrow$ | 0 | a | b | 1 |
|---|---|---|---|---|
| 0 | 1 | a | b | 1 |
| a | 1 | 1 | 1 | 1 |
| b | 0 | 0 | 1 | 1 |
| 1 | 0 | a | b | 1 |

### 1.2.3 Semi-BCI and semi-BCK algebras

Semi-BCI algebras and semi-BCK algebras were introduced in 2019 [199], as follows:

**Definitions 1.2.37** [199] (The dual ones are omitted)

(i) An algebra $\mathcal{A}^L = (A^L, \Rightarrow, \to, 1)$ of type $(2, 2, 0)$ is called a *left-semi-BCI algebra*, or a *left-SBCI algebra* for short, whenever for all $x, y, z \in A^L$,
(sb1) $x \Rightarrow (y \Rightarrow z) = y \Rightarrow (x \Rightarrow z)$ (it is property (Ex) for $\Rightarrow$),
(sb2) $x \to (y \to z) = y \to (x \to z)$ (it is property (Ex) for $\to$),
(sb3) $x \Rightarrow y \leq (z \Rightarrow x) \to (z \Rightarrow y)$,
(sb4) $1 \Rightarrow x = x$ (it is property (M) for $\Rightarrow$),
(sb5) if $x \ll y \leq z$ then $x \ll z$,
(sb6) if $x \leq y \ll z$ then $x \ll z$,
(sb7) if $x \leq y$ and $y \leq x$ then $x = y$ (it is property (An') for $\leq$),
where $x \ll y \overset{def.}{\Longleftrightarrow} x \Rightarrow y = 1$ and $x \leq y \overset{def.}{\Longleftrightarrow} x \to y = 1$.

(ii) A left-SBCI algebra $(A^L, \Rightarrow, \to, 1)$ is called a *left-semi-BCK algebra*, or a *left-SBCK algebra* for short, whenever for any $x \in A^L$,
(sbk) $x \ll 1$ (it is property (La') for $\ll$).

**Proposition 1.2.38** *[199] In a left-SBCI algebra $\mathcal{A}^L$, the following hold: for any $x, y, z \in A^L$,*
*(sb8) if $x \ll y$ and $y \ll z$ then $x \ll z$ (it is property (Tr') for $\ll$),*
*(sb9) if $x \ll y$ and $y \ll x$ then $x = y$ (it is property (An') for $\ll$),*
*(sb10) $(y \Rightarrow z) \leq (z \Rightarrow x) \to (y \Rightarrow x)$ (it is (pBB2') property),*
*(sb11) if $1 \leq x$, then $x = 1$ (it is property (N') for $\leq$),*
*(sb12) $x \to x = 1$ (it is property (Re) for $\leq$),*
*(sb13) if $x \ll y$ then $x \leq y$,*
*(sb14) $x \Rightarrow y \leq x \to y$,*
*(sb15) $x \to ((x \to y) \to y) = 1$ (it is property (D) for $\to$),*
*(sb16) if $x \ll y$ then $x \Rightarrow ((x \Rightarrow y) \Rightarrow y) = 1$,*

*(sb17) if* $x \ll y$ *then* $z \Rightarrow x \leq z \Rightarrow y$,
*(sb18) if* $x \ll y$ *then* $y \Rightarrow z \leq x \Rightarrow z$.

**Proposition 1.2.39** *In a left-SBCI algebra* $\mathcal{A}^L$*, the following hold: for any* $x \in A^L$,
*(sb19)* $1 \to x = x$ *(it is property (M) for* $\to$*),*
*(sb20) if* $x \ll 1$ *then* $x \leq 1$ *(i.e. if (La') holds for* $\ll$ *then (La') holds for* $\leq$ *too).*

**Proof.** (sb19): (By *Prover9*) First, we prove that $1 \to x \leq x$, i.e. that we have:

$$(1.11) \qquad (1 \to x) \to x = 1.$$

Indeed, in (sb2) $(x \to (y \to z) = y \to (x \to z))$, take $Y := x \to y$ and $Z := y$, to obtain:
(a) $x \to ((x \to y) \to y) = (x \to y) \to (x \to y) \stackrel{(sb12)}{=} 1$;
taking now $x := 1$ in (a), we obtain:
$1 \to ((1 \to y) \to y) = 1$, i.e. $1 \leq (1 \to y) \to y$;
then, by (sb11), we obtain $(1 \to y) \to y = 1$, thus (1.11) holds.
Now, we prove that $x \leq 1 \to x$, i.e. that we have:

$$(1.12) \qquad x \to (1 \to x) = 1.$$

Indeed, again in (sb2) $(x \to (y \to z) = y \to (x \to z))$, take $z := y$ to obtain:
(b) $x \to (y \to y) = y \to (x \to y)$, hence, by (sb12), to obtain:
(b') $x \to 1 = y \to (x \to y)$; taking now $x := 1$ in (b'), we obtain:
$1 \to 1 = y \to (1 \to y)$, hence, by (sb12), $y \to (1 \to y) = 1$, thus (1.12) holds.
Finally, from (1.11) and (1.12), by (sb7), we obtain that $1 \to x = x$.
(sb20): By (sb13). □

We conclude, from (sb4) and (sb19), that property (pM) holds for any left-SBCI/SBCK algebra, hence any left-SBCI/SBCK algebra is a *unital pseudo-i-magma*.

We also conclude [199], from (sb8), (sb9), that the relation "$\ll$" is transitive and antisymmetric, but it is not necessarily reflexive, whereas, by (sb7), (sb12), the relation "$\leq$" is antisymmetric and reflexive, but it is not necessarily transitive. The following proposition provides a condition for them to be partial orders: ([199], Proposition 5.2) The relation "$\ll$" coincides with "$\leq$" if and only if "$\ll$" is reflexive.

**Definition 1.2.40** We shall say that a left-SBCI/SBCK algebra $\mathcal{A}^L$ is *strong*, if the binary relations "$\ll$" and "$\leq$" coincide, i.e. if $x \Rightarrow y = 1 \Longleftrightarrow x \to y = 1$, for all $x, y \in A^L$, i.e. if (IdEqR) holds, where:

$$(IdEqR) \qquad x \Rightarrow y = 1 \Longleftrightarrow x \to y = 1.$$

In this case, we define: for all $x, y \in A^L$,

$$(pdfrelR) \qquad x \leq y \stackrel{def.}{\Longleftrightarrow} x \to y = 1 \; (\stackrel{(IdEqR)}{\Longleftrightarrow} x \Rightarrow y = 1).$$

We conclude, since (IdEqR) holds, that any strong left-SBCI/SBCK algebra is a *left-pseudo-algebra*, as any left-pBCI/pBCK algebra is.

Since, by ([199], Theorem 5.1), $(A, \to, 1)$ is a BCI algebra if and only if $(A, \to, \to, 1)$ is an SBCI algebra, it follows that the SBCI algebras are another non-commutative generalization of BCI algebras, beside the pseudo-BCI algebras.

By ([199], Proposition 6.2), there are strong SBCI algebras $(A, \Rightarrow, \to, 1)$ that are not pBCI algebras.

By ([199], Proposition 6.3), there are pBCI algebras $(A, \leq, \Rightarrow, \to, 1)$ which are not (strong) SBCI algebras.

Let $(A, \Rightarrow, \to, 1)$ be a SBCI algebra such that the relations "$\ll$" and "$\leq$" correspond to the operations "$\Rightarrow$" and "$\to$", respectively. By ([199], Proposition 6.4), if both $(A, \ll, \Rightarrow, \to, 1)$ and $(A, \leq, \Rightarrow, \to, 1)$ are pBCI algebras, then $(A, \to, 1)$ is a BCI algebra.

If **SBCI** denotes the class of all left-SBCI algebras and **SSBCI** denotes the class of all strong left-SBCI algebras, if **pBCI** denotes the class of all left-pBCI algebras and **BCI** denotes the class of all left-BCI algebras, then there are the following connections [199]:

$$\mathbf{BCI} = \mathbf{SBCI} \cap \mathbf{pBCI} = \mathbf{SSBCI} \cap \mathbf{pBCI}.$$

Note that both pBCI algebras and SBCI algebras are unital pseudo-i-magmas and that both pBCI algebras and SSBCI algebras are pseudo-algebras.

Resuming, there are the connections from Figure 1.2 (see [199]).

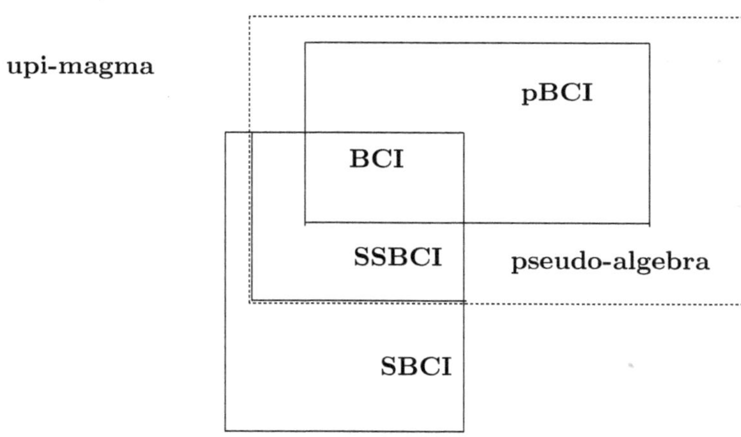


Figure 1.2: Resuming connections between **SBCI**, **SSBCI**, **BCI** and **pBCI**

Also, if **SBCK** denotes the class of all left-SBCK algebras and **SSBCK** denotes the class of all strong left-SBCK algebras, if **pBCK** denotes the class of all left-pBCK algebras and **BCK** denotes the class of all left-BCK algebras, then there are the following connections:

$$\mathbf{BCK} = \mathbf{SBCK} \cap \mathbf{pBCK} = \mathbf{SSBCK} \cap \mathbf{pBCK} \textit{ and}$$

**SBCK $\subset$ SBCI, SSBCK $\subset$ SSBCI, pBCK $\subset$ pBCI, BCK $\subset$ BCI.**

Consequently, both pBCK algebras and SBCK algebras are unital pseudo-i-magmas and both pBCK algebras and SSBCK algebras are pseudo-algebras. Hence, there are similar connections as those from Figure 1.2.

**Remarks 1.2.41**

*(1) Let $\mathcal{A}^L$ be a left-SSBCI algebra. Then, the unique binary relation $\leq$ is an order, i.e. (Re'), (An'), (Tr') hold. Since (pM) also holds, it follows that $\mathcal{A}^L$ is a transitive left-paRM = left-pBH algebra.*

*(2) Let $\mathcal{A}^L$ be a left-SSBCK algebra. Since the order relation $\leq$ satisfies also (La'), it follows that $\mathcal{A}^L$ is a transitive left-paRML algebra.*

The following examples were found by *Mace4 program.*

**Examples 1.2.42 Examples of SBCI algebras**

**1. Example of proper SBCI algebra** (i.e. not verifying $\ll = \leq$):
The algebra $(A_5 = \{0, 2, 3, 4, 1\}, \Rightarrow, \rightarrow, 1)$ is a proper SBCI algebra, where the tables of $\Rightarrow$ and $\rightarrow$ are the following:

| $\Rightarrow$ | 0 | 2 | 3 | 4 | 1 |
|---|---|---|---|---|---|
| 0 | 4 | 4 | 4 | 2 | 2 |
| 2 | 4 | 4 | 4 | 2 | 0 |
| 3 | 4 | 4 | 4 | 2 | 3 |
| 4 | 0 | 2 | 3 | 4 | 1 |
| 1 | 0 | 2 | 3 | 4 | 1 |

| $\rightarrow$ | 0 | 2 | 3 | 4 | 1 |
|---|---|---|---|---|---|
| 0 | 1 | 4 | 1 | 3 | 3 |
| 2 | 1 | 1 | 1 | 3 | 3 |
| 3 | 4 | 4 | 1 | 3 | 3 |
| 4 | 3 | 3 | 3 | 1 | 1 |
| 1 | 0 | 2 | 3 | 4 | 1 |

.

**2. Example of proper SSBCI algebra** (i.e. not being BCI algebra):
The algebra $(A_6 = \{0, 2, 3, 4, 5, 1\}, \Rightarrow, \rightarrow, 1)$ is a proper SSBCI algebra, where the tables of $\Rightarrow$ and $\rightarrow$ are the following:

| $\Rightarrow$ | 0 | 2 | 3 | 4 | 5 | 1 |
|---|---|---|---|---|---|---|
| 0 | 1 | 0 | 0 | 3 | 4 | 0 |
| 2 | 0 | 1 | 3 | 4 | 5 | 1 |
| 3 | 0 | 1 | 1 | 0 | 3 | 1 |
| 4 | 1 | 0 | 0 | 1 | 0 | 0 |
| 5 | 0 | 1 | 1 | 0 | 1 | 1 |
| 1 | 0 | 2 | 3 | 4 | 5 | 1 |

| $\rightarrow$ | 0 | 2 | 3 | 4 | 5 | 1 |
|---|---|---|---|---|---|---|
| 0 | 1 | 0 | 0 | 2 | 4 | 0 |
| 2 | 0 | 1 | 3 | 0 | 3 | 1 |
| 3 | 0 | 1 | 1 | 0 | 2 | 1 |
| 4 | 1 | 0 | 0 | 1 | 0 | 0 |
| 5 | 0 | 1 | 1 | 0 | 1 | 1 |
| 1 | 0 | 2 | 3 | 4 | 5 | 1 |

.

Note that this SSBCI algebra is not a pBCI algebra because (pEx) is not verified for $(3, 2, 5)$.

**Examples 1.2.43 Examples of SBCK algebras**

**1. Example of proper SBCK algebra** (i.e. not verifying $\ll = \leq$):
The algebra $(A_5 = \{0, 2, 3, 4, 1\}, \Rightarrow, \rightarrow, 1)$ is a proper SBCK algebra, where the tables of $\Rightarrow$ and $\rightarrow$ are the following:

| $\Rightarrow$ | 0 | 2 | 3 | 4 | 1 |
|---|---|---|---|---|---|
| 0 | 0 | 0 | 0 | 0 | 1 |
| 2 | 0 | 0 | 3 | 2 | 1 |
| 3 | 0 | 0 | 0 | 0 | 1 |
| 4 | 0 | 0 | 3 | 0 | 1 |
| 1 | 0 | 2 | 3 | 4 | 1 |

| $\rightarrow$ | 0 | 2 | 3 | 4 | 1 |
|---|---|---|---|---|---|
| 0 | 1 | 0 | 0 | 0 | 1 |
| 2 | 1 | 1 | 0 | 0 | 1 |
| 3 | 1 | 0 | 1 | 0 | 1 |
| 4 | 1 | 1 | 0 | 1 | 1 |
| 1 | 0 | 2 | 3 | 4 | 1 |

.

**2. Example of proper SSBCK algebra** (i.e. not being BCK algebra):
The algebra $(A_5 = \{0, 2, 3, 4, 1\}, \Rightarrow, \rightarrow, 1)$ is a proper SSBCK algebra, where the tables of $\Rightarrow$ and $\rightarrow$ are the following:

| $\Rightarrow$ | 0 | 2 | 3 | 4 | 1 |
|---|---|---|---|---|---|
| 0 | 1 | 2 | 1 | 4 | 1 |
| 2 | 1 | 1 | 1 | 2 | 1 |
| 3 | 0 | 2 | 1 | 4 | 1 |
| 4 | 1 | 1 | 1 | 1 | 1 |
| 1 | 0 | 2 | 3 | 4 | 1 |

| $\rightarrow$ | 0 | 2 | 3 | 4 | 1 |
|---|---|---|---|---|---|
| 0 | 1 | 3 | 1 | 3 | 1 |
| 2 | 1 | 1 | 1 | 3 | 1 |
| 3 | 0 | 0 | 1 | 3 | 1 |
| 4 | 1 | 1 | 1 | 1 | 1 |
| 1 | 0 | 2 | 3 | 4 | 1 |

.

Note that this SSBCK algebra is not a pBCK algebra because (pEx) is not verified for $(2, 3, 4)$.

# 1.3 Quantum-B algebras

Quantum-B algebras were introduced by W. Rump in [193], [197] as subreducts of quantales. We shall recall here the necessary results on quantum-B algebras from [193], [197] in order to show the deep connections existing between quantum-B algebras and pseudo-BCI, pseudo-BCK algebras, this being the goal of this chapter.

We begin noting that all three algebras have an *order relation*, $\leq$, but, while pseudo-BCK algebras and pseudo-BCI algebras are indeed algebras (algebras in the *universal algebras* sense concern sets equipped with operations (and no relations)), defined either as algebras (pseudo-algebras) or as structures (pseudo-structures) (because the order relation $\leq$ is *internal*, i.e. it can be defined by the operations of the algebra), the quantum-B algebras are 'fake algebras', they being structures (their order relation $\leq$ is an *external* one).

## 1.3.1 Three equivalent definitions of quantum-B algebras

**Definition 1.3.1** (Definition 1) [193], [197]
A *quantum-B algebra*, or a *QB algebra* for short, is a structure

$$\mathcal{X} = (X, \leq, \rightarrow, \rightsquigarrow)$$

such that $(X, \leq)$ is a poset (i.e. (Re'), (Tr'), (An') hold) and the following properties hold: for all $x, y, z \in X$,

(pB') $\quad y \rightarrow z \leq (x \rightarrow y) \rightarrow (x \rightarrow z), \quad y \rightsquigarrow z \leq (x \rightsquigarrow y) \rightsquigarrow (x \rightsquigarrow z),$
($\text{p*}_{\rightarrow}$') $\quad y \leq z \Longrightarrow x \rightarrow y \leq x \rightarrow z,$
(pEq#') $\quad x \leq y \rightarrow z \Longleftrightarrow y \leq x \rightsquigarrow z.$

We shall denote by **QB** the class of all QB algebras.

If the order $\leq$ is a lattice order, with lattice operations $\wedge$ and $\vee$, we shall denote the *quantum-B lattice* by $\mathcal{X} = (X, \wedge, \vee, \rightarrow, \rightsquigarrow)$ and their class by **QB-L**.

**Proposition 1.3.2** *[193], [197]*
*Let $\mathcal{X} = (X, \leq, \rightarrow, \rightsquigarrow)$ be a QB algebra. Then, the properties (pD'), ($\text{p*}_{\rightsquigarrow}$'), (p**'), (pH), (pC'), (pEx), (pBB') hold.*

**Proof.** We use Propositions 1.1.11 and 1.1.18.
By (pA#4), (pEq#') + (Re') $\Longrightarrow$ (pD');
by (pA#10), (pEq#') + (p$*_{\rightarrow}$') + (pD') + (Tr') $\Longrightarrow$ (p$*_{\rightsquigarrow}$'); thus, (p*') holds.
by (pA#5), (pEq#') + (pD') + (Tr') $\Longrightarrow$ (p**');
by (pA33), (pD') + (p**') + (An') $\Longrightarrow$ (pH);
by (pA#6), (pEq#') + (pB') + (pD') + (Tr') $\Longrightarrow$ (pC');
by (pA3), (pC') + (An') $\Longrightarrow$ (pEx);
by (pA#3), (pEq#') $\Longrightarrow$ ((pB') $\Leftrightarrow$ (pBB')). □

Note that , since by (pA#3), (pEq#') $\Longrightarrow$ ((pB') $\Leftrightarrow$ (pBB')), we obtain immediately the following equivalent definition.

**Definition 1.3.3** (Definition 1')
A *quantum-B algebra*, or a *QB algebra* for short, is a structure

$$\mathcal{X} = (X, \leq, \rightarrow, \rightsquigarrow)$$

such that $(X, \leq)$ is a poset (i.e. (Re'), (Tr'), (An') hold) and the following properties hold: for all $x, y, z \in X$,
(pBB') $\quad y \rightarrow z \leq (z \rightarrow x) \rightsquigarrow (y \rightarrow x), \quad y \rightsquigarrow z \leq (z \rightsquigarrow x) \rightarrow (y \rightsquigarrow x)$,
(p$*_{\rightarrow}$') $\quad y \leq z \Longrightarrow x \rightarrow y \leq x \rightarrow z$,
(pEq#') $\quad x \leq y \rightarrow z \Longleftrightarrow y \leq x \rightsquigarrow z$.

**Theorem 1.3.4**
*(1) Let $\mathcal{X} = (X, \leq, \rightarrow, \rightsquigarrow)$ be a structure such that $(X, \leq)$ is a poset and the following properties hold: (pEx), (p$*_{\rightarrow}$'), (pEq#'). Then, $\mathcal{X}$ is a QB algebra.*
*(1') Conversely, every QB algebra $\mathcal{X} = (X, \leq, \rightarrow, \rightsquigarrow)$ verifies the above properties.*

**Proof.** (1): By (pA#4), (pEq#') + (Re') $\Longrightarrow$ (pD');
by (pA#10), (pEq#') + (p$*_{\rightarrow}$') + (pD') + (Tr') $\Longrightarrow$ (p$*_{\rightsquigarrow}$');
by (pA#7'), (pEq#') + (pEx) + (p*') + (Re') $\Longrightarrow$ (pB'). Thus, (pB') holds and, hence, $\mathcal{X}$ is a QB algebra.
(1'): Every QB algebra has the property (pEx), by Proposition 1.3.2. □

By Theorem 1.3.4, the following equivalent definition of QB algebras follows [193], [197].

**Definition 1.3.5** (Definition 2) [193], [197]
A *quantum-B algebra*, or a *QB algebra* for short, is a structure

$$\mathcal{X} = (X, \leq, \rightarrow, \rightsquigarrow)$$

such that $(X, \leq)$ is a poset (i.e. (Re'), (Tr'), (An') hold) and the following properties hold: for all $x, y, z \in X$,
(pEx) $\quad x \rightarrow (y \rightsquigarrow z) = y \rightsquigarrow (x \rightarrow z)$,
(p$*_{\rightarrow}$') $\quad y \leq z \Longrightarrow x \rightarrow y \leq x \rightarrow z$,
(pEq#') $\quad x \leq y \rightarrow z \Longleftrightarrow y \leq x \rightsquigarrow z$.

### 1.3.2 Locally unital quantum-B algebras

**Definitions 1.3.6** [193] Let $\mathcal{X} = (X, \leq, \rightarrow, \rightsquigarrow)$ be a QB algebra.

(i) An element $e \in X$ verifying $e \rightarrow e = e = e \rightsquigarrow e$ is called a *local unit.*

(ii) $\mathcal{X}$ is called with *enough local units*, if for all $x \in X$,

$$(x \rightarrow x) \rightsquigarrow x = x = (x \rightsquigarrow x) \rightarrow x. \tag{1.13}$$

(iii) $\mathcal{X}$ is *locally unital*, if is with enough local units (i.e. (1.13) holds) and for all $x, y \in X$,

$$x \rightarrow x \leq y \Longrightarrow y \rightarrow y \leq y. \tag{1.14}$$

**Proposition 1.3.7** *[193] Let $\mathcal{X} = (X, \leq, \rightarrow, \rightsquigarrow)$ be a QB algebra with enough local units (i.e. (1.13) holds). Then, for all $x \in X$,*

$$e \rightarrow e = e \Longleftrightarrow e \rightsquigarrow e = e, \tag{1.15}$$

$$x \rightarrow x \leq x \Longleftrightarrow x \rightsquigarrow x \leq x. \tag{1.16}$$

**Proof.** (1.15): If $e \rightarrow e = e$, then, by (1.13), $(e \rightarrow e) \rightsquigarrow e = e$, which implies $e \rightsquigarrow e = e$. Conversely, if $e \rightsquigarrow e = e$, then, by (1.13), $(e \rightsquigarrow e) \rightarrow e = e$, which implies $e \rightarrow e = e$.

(1.16): If $x \rightarrow x \leq x$, then, by (p**'), $x \rightsquigarrow x \leq (x \rightarrow x) \rightsquigarrow x \overset{(1.13)}{=} x$. Conversely, if $x \rightsquigarrow x \leq x$, then, by (p**'), $x \rightarrow x \leq (x \rightsquigarrow x) \rightarrow x \overset{(1.13)}{=} x$. □

**Definition 1.3.8** Let $\mathcal{X} = (X, \leq, \rightarrow, \rightsquigarrow)$ be a QB algebra with enough local units (i.e. (1.13) holds). An element $x \in X$ is *positive*, if $x \rightarrow x \leq x$ ($\overset{(1.16)}{\Longleftrightarrow} x \rightsquigarrow x \leq x$).

Denote by $X^+$ the set of all positive elements of $X$, i.e.

$$X^+ = \{x \in X \mid x \rightarrow x \leq x \ (\overset{(1.16)}{\Longleftrightarrow} x \rightsquigarrow x \leq x)\}.$$

**Proposition 1.3.9** *Let $\mathcal{X} = (X, \leq, \rightarrow, \rightsquigarrow)$ be a QB algebra with enough local units (i.e. (1.13) holds). Then, we have: for all $x, y \in X$,*

$$(IdEq_{X^+}) \quad x \rightarrow y \in X^+ \Longleftrightarrow x \rightsquigarrow y \in X^+ \quad \text{and} \tag{1.17}$$

$$(pEqrel_{X^+}) \quad x \leq y \Longleftrightarrow x \rightarrow y \in X^+ \ (\overset{(IdEq_{X^+})}{\Longleftrightarrow} x \rightsquigarrow y \in X^+). \tag{1.18}$$

**Proof.** Suppose $x \leq y$; then, by (p**'), $y \rightarrow y \leq x \rightarrow y$ and $(x \rightarrow y) \rightsquigarrow y \leq (y \rightarrow y) \rightsquigarrow y \overset{(1.13)}{=} y$, and, by (p*$_{\rightarrow}$'), $x \rightarrow ((x \rightarrow y) \rightsquigarrow y) \leq x \rightarrow y$. Then, $(x \rightarrow y) \rightsquigarrow (x \rightarrow y) \overset{(pEx)}{=} x \rightarrow ((x \rightarrow y) \rightsquigarrow y) \leq x \rightarrow y$, i.e. $x \rightarrow y \in X^+$. Cnversely, suppose $x \rightarrow y \in X^+$, i.e. $(x \rightarrow y) \rightarrow (x \rightarrow y) \leq x \rightarrow y$; by (pB'), $y \rightarrow y \leq (x \rightarrow y) \rightarrow (x \rightarrow y)$; then, by (Tr'), $y \rightarrow y \leq x \rightarrow y$, and by (p**'),

$(x \to y) \rightsquigarrow y \leq (y \to y) \rightsquigarrow y \overset{(1.13)}{=} y$; but, by (pD'), $x \leq (x \to y) \rightsquigarrow y$; then, by (Tr') again, $x \leq y$. Thus, $x \leq y \iff x \to y \in X^+$. Similarly, $x \leq y \iff x \rightsquigarrow y \in X^+$. Hence, (1.17) and (1.18) hold. □

We shall denote by $(X, \leq, \to, \rightsquigarrow, X^+)$ a locally unital QB algebra. We shall denote by **luQB** the subclass of all locally unital QB algebras.

Let us consider the set $\Delta$ of all elements $x \to x$, $x \rightsquigarrow x$, for all $x \in X$, i.e.

$$\Delta = \{x \to x,\ x \rightsquigarrow x \mid x \in X\}.$$

**Proposition 1.3.10** *Let $\mathcal{X} = (X, \leq, \to, \rightsquigarrow, X^+)$ be a locally unital QB algebra (i.e. (1.13) and (1.14) hold). Then,*

*(a) $X^+$ is an upper set (i.e. $x \in X^+$ and $x \leq y$ imply $y \in X^+$).*

*(b) $\Delta \subseteq X^+$.*

**Proof.** (a): Let $x \in X^+$, i.e. $x \to x \leq x$, and let $x \leq y$; then, $x \to x \leq y$, by (Tr'); then, by (1.14), $y \to y \leq y$, i.e. $y \in X^+$.

(b): Let $y \in \Delta$, i.e. there exists $x_0 \in X$ such that $y = x_0 \to x_0$ (or $y = x_0 \rightsquigarrow x_0$). Then, by (Re'), $x_0 \to x_0 \leq y$, hence, by (1.14), $y \to y \leq y$, i.e. $y \in X^+$. □

**Corollary 1.3.11** *(See [8], page 179)*

*Let $\mathcal{X} = (X, \leq, \to, \rightsquigarrow, X^+)$ be a locally unital QB algebra (i.e. (1.13) and (1.14) hold). Then,*

$$(x \to x) \to (x \to x) \ = \ x \to x \quad \textit{and} \quad (x \rightsquigarrow x) \rightsquigarrow (x \rightsquigarrow x) \ = \ x \rightsquigarrow x,$$

*i.e. $x \to x$ and $x \rightsquigarrow x$ are local units.*

**Proof.** By definitions and by Proposition 1.3.10 (b), we have:
$(x \to x) \to (x \to x) \ \leq \ x \to x$ and $(x \rightsquigarrow x) \rightsquigarrow (x \rightsquigarrow x) \ \leq \ x \rightsquigarrow x$.
By (pB'), we have:
$x \to x \leq (x \to x) \to (x \to x)$ and $x \rightsquigarrow x \leq (x \rightsquigarrow x) \rightsquigarrow (x \rightsquigarrow x)$.
Then, by (An'), we obtain:
$(x \to x) \to (x \to x) \ = \ x \to x$ and $(x \rightsquigarrow x) \rightsquigarrow (x \rightsquigarrow x) \ = \ x \rightsquigarrow x$. □

### 1.3.3 Unital (maximal, integral) quantum-B algebras

**Definitions 1.3.12** Let $\mathcal{X} = (X, \leq, \to, \rightsquigarrow)$ be a QB algebra.

(i) We say that $\mathcal{X}$ is *unital*, if there exists an element $1 \in X$, called *unit*, verifying:
(pM) for all $x \in X$, $\quad 1 \to x = x = 1 \rightsquigarrow x$.

(ii) We say that $\mathcal{X}$ is *maximal*, if it is unital and the unit 1 is the maximal element of the poset $(X, \leq)$, i.e.
(N') for all $x \in X$, $\quad 1 \leq x$ implies $x = 1$.

(iii) We say that $\mathcal{X}$ is *integral*, if it is unital and the unit 1 is the greatest element of the poset $(X, \leq)$, i.e.
(La') for all $x \in X$, $\quad x \leq 1$.

**Lemma 1.3.13** *[193] The unit, when exists, is unique.*

**Proof.** Suppose there are two units, 1 and $1'$, i.e. we have:
$1 \to 1 = 1 = 1 \rightsquigarrow 1$ and $\quad 1 \to 1' = 1' = 1 \rightsquigarrow 1'$,
$1' \to 1' = 1' = 1' \rightsquigarrow 1'$ and $\quad 1' \to 1 = 1 = 1' \rightsquigarrow 1$.
Then, $1' = 1 \to 1' \leq (1' \to 1) \rightsquigarrow (1 \to 1) = 1 \rightsquigarrow 1 = 1$
and $1 = 1' \to 1 \leq (1' \to 1') \rightsquigarrow (1 \to 1') = 1' \rightsquigarrow 1' = 1'$, by (pBB$_1$').
Hence, by (An'), $1 = 1'$. □

We shall denote by $(X, \leq, \to, \rightsquigarrow, 1)$ a unital (maximal, integral) QB algebra.

We shall denote by **uQB** (**mQB**, **iQB**) the class of all unital (maximal, integral, respectively) QB algebras and by **uQB-L** (**mQB-L**, **iQB-L**) their corresponding lattice ordered ones.

A fudamental result is the following.

**Proposition 1.3.14** *[193] Every unital QB algebra $\mathcal{X} = (X, \leq, \to, \rightsquigarrow, 1)$ is locally unital and $1 \in \Delta \subseteq X^+$.*

**Proof.** Let $\mathcal{X} = (X, \leq, \to, \rightsquigarrow, 1)$ be a unital QB algebra. Since $1 \to x = x = 1 \rightsquigarrow x$, by (pM), it follows, by (Re'), that $x \leq 1 \to x$ and $x \leq 1 \rightsquigarrow x$, which, by (pEq#'), give $1 \leq x \rightsquigarrow x$ and $1 \leq x \to x$, respectively.

Then, by (p**'), we obtain $(x \rightsquigarrow x) \to x \leq 1 \to x \overset{(pM)}{=} x$ and $(x \to x) \rightsquigarrow x \leq 1 \rightsquigarrow x \overset{(pM)}{=} x$, respectively; but, by (pD'), we have $x \leq (x \rightsquigarrow x) \to x$ and $x \leq (x \to x) \rightsquigarrow x$ . Then, by (An'), $(x \rightsquigarrow x) \to x = x$ and $(x \to x) \rightsquigarrow x = x$. Thus, (1.13) holds.

Suppose now that $x \to x \leq y$; since $1 \leq x \to x$, it follows, by (Tr'), that $1 \leq y$; then, $1 \leq y \overset{(1.13)}{=} (y \to y) \rightsquigarrow y$, which, by (pEq#'), gives $y \to y \leq 1 \to y \overset{(pM)}{=} y$. Thus, (1.14) holds.

Hence, $\mathcal{X}$ is locally unital. Since $1 \to 1 = 1 \rightsquigarrow 1 = 1$, by (pM), it follows that $1 \in \Delta \subseteq X^+$. □

**Proposition 1.3.15** *Let $\mathcal{X} = (X, \leq, \to, \rightsquigarrow, 1)$ be a unital QB algebra. Then,*

$$X^+ = \{x \in X \mid 1 \leq x\},$$

*i.e. we have:*

(pEqrel1) $\quad 1 \leq x \iff x \to x \leq x \quad (\overset{(1.16)}{\iff} x \rightsquigarrow x \leq x)$,
(IdEq) $\quad 1 \leq x \to y \iff 1 \leq x \rightsquigarrow y$ *and*
(pEqrel) $\quad x \leq y \iff 1 \leq x \to y \quad (\overset{(IdEq)}{\iff} 1 \leq x \rightsquigarrow y)$.

**Proof.** $x \to x \leq x \overset{(pM)}{\iff} x \to x \leq 1 \to x \overset{(pEq\#')}{\iff} 1 \leq (x \to x) \rightsquigarrow x \overset{(1.13)}{\iff} 1 \leq x$, hence $X^+ = \{x \in X \mid 1 \leq x\}$; then, by (IdEq$_{X^+}$), we obtain (IdEq) and, by (pEqrel$_{X^+}$), we obtain (pEqrel). □

**Proposition 1.3.16** *Let $\mathcal{X} = (X, \leq, \to, \rightsquigarrow, 1)$ be a maximal QB algebra. Then, $X^+ = \{1\}$, hence we have:*

(IdEqR) $\qquad x \to y = 1 \iff x \rightsquigarrow y = 1,$

(pEqrelR) $\qquad x \leq y \iff x \to y = 1 \quad (\stackrel{(IdEqR)}{\iff} \quad x \rightsquigarrow y = 1).$

**Proof.** Since $\mathcal{X}$ is unital, by Proposition 1.3.15, we have $X^+ = \{x \in X \mid 1 \leq x\}$. By (N'), for all $x \in X$, $1 \leq x$ implies $x = 1$. It follows that

$$X^+ = \{x \in X \mid x = 1\} = \{1\},$$

i.e. (IdEqR) and (pEqrel) become (IdEqR) and (pEqrelR), respectively. □

**Remark 1.3.17** *Note that (IdEqR) and (pEqrelR) say that the order relation $\leq$ can be defined by $\to, \rightsquigarrow$ and 1, i.e. the external order $\leq$ became an internal order. Hence, a maximal QB algebra (i.e. a QB verifying (pM) and (N')) is indeed an algebra, i.e. it can be defined, equivalently, as an algebra $(X, \to, \rightsquigarrow, 1)$ of type $(2, 2, 0)$, where we can define $\leq$ by:*

$$(pdfrelR) \qquad x \leq y \stackrel{def.}{\iff} x \to y = 1 \quad (\stackrel{(IdEqR)}{\iff} \quad x \rightsquigarrow y = 1).$$

**Proposition 1.3.18** *(See ([193], Proposition 6))*
*For a unital QB algebra $\mathcal{X} = (X, \leq, \to, \rightsquigarrow, 1)$, the following are equivalent:*
*(a) $\mathcal{X}$ is a maximal QB algebra,*
*(b) $\mathcal{X}$ is a pseudo-BCI algebra,*
*i.e. we have:*

$$\textbf{pBCI} \iff \textbf{mQB}.$$

**Proof.** By Definitions 1.3.12 and 1.2.30 (Definition 10' of pBCI algebras). □

**Proposition 1.3.19** *(See ([197], Corollary 2))*
*For a unital QB algebra $\mathcal{X} = (X, \leq, \to, \rightsquigarrow, 1)$, the following are equivalent:*
*(a) $\mathcal{X}$ is an integral QB algebra,*
*(b) $\mathcal{X}$ is a pseudo-BCK algebra,*
*i.e. we have:*

$$\textbf{pBCK} \iff \textbf{iQB}.$$

**Proof.** By Definitions 1.3.12 and 1.2.31 (Definition 10' of pBCK algebras). □

Hence, we have the following connections:

$$\textbf{pBCK} \iff \textbf{iQB} \subset \textbf{pBCI} \iff \textbf{mQB} \subset \textbf{uQB} \subset \textbf{luQB} \subset \textbf{QB}.$$

We shall point out, in next section, how the QB algebra is the great piece which missed from the puzzle showing the connections (presented in Part III) between partially ordered pseudo-i-magmas and partially ordered magmas (presented in Part II).

### 1.3.4 Commutative quantum-B algebras

We say that a QB algebra $\mathcal{X} = (X, \leq, \to, \leadsto)$ is *commutative*, if $\to\ =\ \leadsto$. A commutative QB algebra will be denoted by $\mathcal{X} = (X, \leq, \to)$. Hence, we have the following equivalent definitions.

**Definition 1.3.20** (Definition 1)
A *commutative quantum-B algebra*, or a *commutative QB algebra* for short, is a structure $\mathcal{X} = (X, \leq, \to)$ such that $(X, \leq)$ is a poset (i.e. (Re'), (Tr'), (An') hold) and the following properties hold: for all $x, y, z \in X$,
(B') $y \to z \leq (x \to y) \to (x \to z)$,
(*') $y \leq z \Longrightarrow x \to y \leq x \to z$,
(Eq#') $x \leq y \to z \Longleftrightarrow y \leq x \to z$.

**Definition 1.3.21** (Definition 2)
A *commutaive quantum-B algebra*, or a *commutative QB algebra* for short, is a structure $\mathcal{X} = (X, \leq, \to, )$ such that $(X, \leq)$ is a poset (i.e. (Re'), (Tr'), (An') hold) and the following properties hold: for all $x, y, z \in X$,
(Ex) $x \to (y \to z) = y \to (x \to z)$,
(*') $y \leq z \Longrightarrow x \to y \leq x \to z$,
(Eq#') $x \leq y \to z \Longleftrightarrow y \leq x \to z$.

We shall denote by **cQB** the class of all commutative QB algebras.
We shall denote by **lucQB** (**ucQB** (**mcQB**, **icQB**) the class of all locally unital (unital, maximal, integral, respectively) commutative QB algebras.

## 1.4 Partially (lattice) ordered pseudo-i-magmas

### 1.4.1 Partially (lattice) ordered pi-magmas: po-pe ($l$-pe), po-pr, po-mpr, po-ipr ($l$-pr, $l$-mpr, $l$-ipr)

**Definitions 1.4.1**
(i) A *partially ordered pi-magma*, or a *po-pi-magma* for short, is a structure $(A, \leq, \to, \leadsto)$, such that the reduct $(A, \to, \leadsto)$ is a pi-magma and $(A, \leq)$ is a poset (i.e. $\leq$ is a partial order on $A$). No compatibility between $\leq$ and $\to$, $\leadsto$ is assumed.
(ii) If the partial order $\leq$ is a lattice order, then we say that we have a *lattice ordered pi-magma*, or an *$l$-pi-magma* for short.

Since $\leq$ is an order relation, recall that we shall denote by (Re') ($x \leq x$, for all $x$) the reflexivity, by (Tr') ($x \leq y$ and $y \leq z$ imply $x \leq z$, for all $x, y, z$) the transitivity and by (An') ($x \leq y$ and $y \leq x$ imply $x = y$, for all $x, y$) the antisymmetry.
Note that if $\leq$ is a lattice order, with lattice operations $\wedge$ and $\vee$, then the $l$-pi-magma is an *algebra* (in the universal algebra sense).

We shall introduce the following po-pi-magmas: po-pes and po-prs, po-mprs, po-iprs, and their corresponding lattice ordered ones.

• **Po-pes ($l$-pes)**

**Definitions 1.4.2** (Definition of po-pe ($l$-pe))

(i) A *left-partially ordered p-exchange*, or a *left-po-pe* for short, is a structure

$$\mathcal{A}^L = (A^L, \leq^e, \to, \rightsquigarrow)$$

such that the reduct $(A^L, \to, \rightsquigarrow)$ is a p-exchange (i.e. (pEx) holds), $(A^L, \leq^e)$ is a poset ('L' comes from 'left' and 'e' means *external* partial order), and $\leq^e$ is compatible with $\to$, $\rightsquigarrow$, i.e. the following properties hold: for all $x, y, z \in A^L$,
(p*$_{\to}$') $\quad x \leq^e y \Longrightarrow z \to x \leq^e z \to y$,
(pEq#') $\quad x \leq^e y \to z \Longleftrightarrow y \leq^e x \rightsquigarrow z$.

(i') Dually, a *right-partially ordered p-exchange*, or a *right-po-pe* for short, is a structure

$$\mathcal{A}^R = (A^R, \geq^e, \to^R, \rightsquigarrow^R)$$

such that the reduct $(A^R, \to^R, \rightsquigarrow^R)$ is a p-exchange (i.e. (pEx$^R$) holds), $(A^R, \leq^e)$ is a poset ('R' comes from 'right' and 'e' means *external* partial order), (where $x \leq^e y \Longleftrightarrow y \geq^e x$, for all $x, y \in A^R$), and $\geq^e$ is compatible with $\to^R$, $\rightsquigarrow^R$, i.e. the following properties hold: for all $x, y, z \in A^R$,
(p*$_{\to^R}$') $\quad x \geq^e y \Longrightarrow z \to^R x \geq^e z \to^R y$,
(pEq#$^R$') $\quad x \geq^e y \to^R z \Longleftrightarrow y \geq^e x \rightsquigarrow^R z$.

(ii) If the partial external order $\leq^e$, or $\geq^e$, is a lattice order, then we say that we have a *left-lattice ordered p-exchange* (*left-$l$-pe*), denoted by $(A^L, \wedge, \vee, \to, \rightsquigarrow)$, or a *right-lattice ordered p-exchange* (*right-$l$-pe*), denoted by $(A^R, \vee, \wedge, \to^R, \rightsquigarrow^R)$, respectively.

Denote by **po-pe** (**$l$-pe**) the class of all left-po-pes (left-$l$-pes, respectively) and by **po-pe**$^R$ (**$l$-pe**$^R$) the class of all right-po-pes (right-$l$-pes, respectively).

**Remarks 1.4.3**

*(1) The left-po-pe was introduced in the literature under the name " quantum-B algebra" (Definition 1.3.5) (Definition 2), by W. Rump [193]; see also [197]. We have denoted, in the previous section, by* **QB** *the class of all quantum-B algebras and by* **QB-L** *the class of all quantum-B lattices. Hence, we have:*

$$\mathbf{po-pe} = \mathbf{QB} \quad \textit{and} \quad \mathbf{l-pe} = \mathbf{QB-L}. \tag{1.19}$$

*(2) A left-po-pe verifies also the property (p*$_{\rightsquigarrow}$'): for all $x, y, z \in A^L$,*
*(p*$_{\rightsquigarrow}$')* $\quad x \leq^e y \Longrightarrow z \rightsquigarrow x \leq^e z \rightsquigarrow y.$
*Indeed, see Proposition 1.3.2.*

*(3) A left-po-pe verifies also the property (p**'): for all $x, y, z \in A^L$,*
*(p**')* $\quad x \leq^e y \Longrightarrow (y \to z \leq^e x \to z,\ y \rightsquigarrow z \leq^e x \rightsquigarrow z).$
*Indeed, see Proposition 1.3.2.*

• **po-prs, po-mprs, po-iprs ($l$-prs, $l$-mprs, $l$-iprs)**

**Definitions 1.4.4** (Definition of po-pr ($l$-pr)) (The dual ones are omitted)

(i) A *left-partially ordered p-residoid*, or a *left-po-pr* for short, is a structure

$$\mathcal{A}^L = (A^L, \leq, \rightarrow, \rightsquigarrow, 1)$$

such that the reduct $(A^L, \rightarrow, \rightsquigarrow, 1)$ is a p-residoid (i.e. (pEx), (pM) hold), $(A^L, \leq)$ is a poset and the properties (p$*_{\rightarrow}$') , (pEq#') hold, i.e. $\mathcal{A}^L$ is a unital left-po-pe (= unital (left-) quantum-B algebra).

(ii) If the partial order $\leq$ is a lattice order, then we say that we have a *left-lattice ordered p-residoid* (*left-$l$-pr*).

Denote by **po-pr** ($l$**-pr**) the class of all left-po-prs (left-$l$-prs).

**Remarks 1.4.5**

*(1) The left-po-pr was introduced in the literature under the name "unital quantum-B algebra" by W. Rump [193]. We have denoted, in the previous section, by* **uQB** *the class of all unital quantum-B algebras and by* **uQB-L** *the class of all unital quantum-B lattices. Hence, we have:*

(1.20) $$\mathbf{po-pr} = \mathbf{uQB} \quad \textit{and} \quad \mathbf{l-pr} = \mathbf{uQB-L}.$$

*(2) A left-po-pr verifies also (p$*_{\rightsquigarrow}$') (hence (p$*$')) and (p$**$').*

*(3) If $x \rightarrow y = 1 \Longleftrightarrow x \rightsquigarrow y = 1$ and $x \leq y \Longleftrightarrow x \rightarrow y = 1$, i.e. if (IdEqR) and (pEqrelR) hold, then a left-po-pr is a left-pME algebra verifying (p$*$'), (p$**$'), (pEq#') and (Re'), (Tr'), (An'), hence is a left-pBCI algebra.*

We obviously have the inclusions (the dual ones are omitted):

(1.21) $$\mathbf{po-pr} \subset \mathbf{po-pe} \quad \textit{and} \quad \mathbf{l-pr} \subset \mathbf{l-pe}.$$

**Definitions 1.4.6** (Definition of po-mpr ($l$-mpr)) (The dual ones are omitted)

(i) A *left-partially ordered maximal p-residoid*, or a *left-po-mpr* for short, is a left-po-pr

$$\mathcal{A}^L = (A^L, \leq, \rightarrow, \rightsquigarrow, 1),$$

i.e. (pEx), (pM), (p$*_{\rightarrow}$'), (pEq#') hold, such that $(A^L, \leq, 1)$ is a poset with 1 as maximal element, i.e. the additional property (N') holds: for all $x \in A^L$,

(N') $\quad 1 \leq x$ implies $x = 1$;

*maximal* means that the maximal element of the poset coincides with the unit element of the p-residoid.

(ii) If the partial order $\leq$ is a lattice order, then we say that we have a *left-lattice ordered maximal p-residoid* (*left-$l$-mpr*).

Denote by **po-mpr** the class of all left-po-mprs. In lattice ordered case, denote by $l$**-mpr** the class of all left-$l$-mprs.

Hence, a left-po-mpr is a left-po-pr verifying (N') and, similarly, in lattice ordered case. We write:

**po-mpr** = **po-pr** + (N') and $l$**-mpr** = $l$**-pr** + (N').

**Remarks 1.4.7**

*(1) The left-po-mpr was introduced in the previous section under the name "maximal unital quantum-B algebra". We have denoted by* **mQB** *the class of all maximal unital quantum-B algebras and by* **mQB-L** *the class of all maximal unital quantum-B lattices. Hence, we have:*

$$(1.22) \qquad \mathbf{po-mpr} \;=\; \mathbf{mQB} \quad \textit{and} \quad \mathbf{l-mpr} \;=\; \mathbf{mQB-L}.$$

*(2) A left-po-mpr verifies also ($p*_{\rightsquigarrow}$') (hence (p*')) and (p**').*

Hence, we obviously have the inclusions:

$$(1.23) \qquad \mathbf{po-mpr} \subset \mathbf{po-pr} \quad \textit{and} \quad \mathbf{l-mpr} \subset \mathbf{l-pr}.$$

By Proposition 1.3.18, **pBCI** $\Longleftrightarrow$ **mQB**, hence we have:
(1.24)

$$\mathbf{po-mpr} = \mathbf{mQB} \Longleftrightarrow \mathbf{pBCI} \quad \textit{and} \quad \mathbf{l-mpr} = \mathbf{mQB-L} \Longleftrightarrow \mathbf{pBCI-L}.$$

**Definitions 1.4.8** (Definition of po-ipr ($l$-ipr)) (The dual ones are omitted)

(i) A *left-partially ordered integral p-residoid*, or a *left-po-ipr* for short, is a left-po-pr

$$\mathcal{A}^L = (A^L, \leq, \rightarrow, \rightsquigarrow, 1),$$

i.e. (pEx), (pM), ($p*_{\rightarrow}$'), (pEq#') hold, such that $(A^L, \leq, 1)$ is a poset with 1 as greatest element, i.e. the additional property (La') holds: for all $x \in A^L$,
(La') (Last element) $\quad x \leq 1$;
*integral* means that the greatest element of the poset coincides with the unit element of the p-residoid.

(ii) If the partial order $\leq$ is a lattice order, then we say that we have a *left-lattice ordered integral p-residoid* (*left-l-ipr*).

Denote by **po-ipr** the class of all left-po-iprs. In lattice ordered case, denote by $l$-**ipr** the class of all left-$l$-iprs.

A left-$l$-pr (left-$l$-mpr, left-$l$-ipr) will be denoted by $(A^L, \wedge, \vee, \rightarrow, \rightsquigarrow, 1)$. A right-$l$-pr (right-$l$-mpr, right-$l$-ipr) will be denoted by $(A^R, \vee, \wedge, \rightarrow^R, \rightsquigarrow^R, 0)$.

Hence, a left-po-ipr is a left-po-pr verifying (La') and, similarly, in lattice ordered case. We write:

**po-ipr** $=$ **po-pr** $+$ (La') and $l$-**ipr** $=$ $l$-**pr** $+$ (La').

**Remarks 1.4.9**

*(1) The left-po-ipr was introduced in the literature under the name "integral unital quantum-B algebra", by W. Rump [193]; see also [197]. We have denoted, in the previous section, by* **iQB** *the class of all integral unital quantum-B algebras and by* **iQB-L** *the class of all integral unital quantum-B lattices. Hence, we have:*

$$(1.25) \qquad \mathbf{po-ipr} \;=\; \mathbf{iQB} \quad \textit{and} \quad \mathbf{l-ipr} \;=\; \mathbf{iQB-L}.$$

*(2) A left-po-ipr verifies also ($p*_{\rightsquigarrow}$') (hence (p*')) and (p**').*

**Lemma 1.4.10** *Let $\mathcal{A}^L = (A^L, \leq, \to, \rightsquigarrow, 1)$ be a left-po-ipr. Then, $\mathcal{A}^L$ is a left-po-mpr.*

**Proof.** We have to prove that (La') implies (N'). Indeed, by (A1), (La') + (An') imply (N'). □

We then have, by above Lemma 1.4.10, the inclusions (the dual ones are omitted):

$$\textbf{po}-\textbf{ipr} \subset \textbf{po}-\textbf{mpr} \quad \textit{and} \quad \textbf{l}-\textbf{ipr} \subset \textbf{l}-\textbf{mpr}. \tag{1.26}$$

By Proposition 1.3.19, **pBCK** $\Longleftrightarrow$ **iQB**, hence we have:
(1.27)

$$\textbf{po}-\textbf{ipr} = \textbf{iQB} \Longleftrightarrow \textbf{pBCK} \quad \textit{and} \quad \textbf{l}-\textbf{ipr} = \textbf{iQB}-\textbf{L} \Longleftrightarrow \textbf{pBCK}-\textbf{L}.$$

Finally, by inclusions 1.21, 1.23, 1.26, we have (the dual cases are omitted):

$$\textbf{po}-\textbf{ipr} \subset \textbf{po}-\textbf{mpr} \subset \textbf{po}-\textbf{pr} \subset \textbf{po}-\textbf{pe} \tag{1.28}$$

$$\textit{and} \quad \textbf{l}-\textbf{ipr} \subset \textbf{l}-\textbf{mpr} \subset \textbf{l}-\textbf{pr} \subset \textbf{l}-\textbf{pe}.$$

More precisely, we have:

| | | | | | | |
|---|---|---|---|---|---|---|
| **po-ipr** | $\subset$ | **po-mpr** | $\subset$ | **po-pr** | $\subset$ | **po-pe** |
| = **iQB** | | = **mQB** | | = **uQB** | | = **QB** |
| $\Longleftrightarrow$ **pBCK** | | $\Longleftrightarrow$ **pBCI** | | | | |

and

| | | | | | | |
|---|---|---|---|---|---|---|
| ***l*-ipr** | $\subset$ | ***l*-mpr** | $\subset$ | ***l*-pr** | $\subset$ | ***l*-pe** |
| = **iQB-L** | | = **mQB-L** | | = **uQB-L** | | = **QB-L** . |
| $\Longleftrightarrow$ **pBCK=L** | | $\Longleftrightarrow$ **pBCI-L** | | | | |

## 1.4.2 Po($l$)-pi-magmas with pseudo-product/pseudo-sum

We shall study here particular left-po-pi-magmas (left-$l$-pi-magmas) which have also a *pseudo-product* $\odot$ (*product*, in the commutative case).

We introduce the following two general definitions:

**Definitions 1.4.11** (Definition of structure with (pP')/(pS'))

(i) Let $\mathcal{A}^L = (A^L, \leq, \to, \rightsquigarrow)$ be a structure such that the reduct $(A^L, \leq)$ is a poset and $\to$, $\rightsquigarrow$ are binary operations. We say that $\mathcal{A}^L$ is a *structure with pseudo-product*, or a *structure with (pP')*, if, for any $x, y \in A^L$, there exist the smallest elements (under $\leq$) of the sets $\{z \in A^L \mid x \leq y \to z\}$, $\{z \in A^L \mid y \leq x \rightsquigarrow z\}$, they are equal and they define a new operation, $\odot$, called *pseudo-product*, i.e. $\mathcal{A}^L$ satisfies the following condition:

(pP') for all $x, y \in A^L$, $\quad \exists\, x \odot y \overset{def.}{=} \min\{z \mid x \leq y \to z\} = \min\{z \mid y \leq x \rightsquigarrow z\}$.

(i') Dually, let $\mathcal{A}^R = (A^L, \geq, \to^R, \rightsquigarrow^R)$ be a structure such that the reduct $(A^R, \geq)$ is a poset and $\to^R$, $\rightsquigarrow^R$ are binary operations. We say that $\mathcal{A}^R$ is a *structure with pseudo-sum*, or a *structure with (pS')*, if, for any $x, y \in A^R$, there exist the greatest elements of the sets $\{z \in A^R \mid x \geq y \to^R z\}$, $\{z \in A^R \mid y \geq x \rightsquigarrow^R z\}$, they are equal and they define a new operation, $\oplus$, called *pseudo-sum*, i.e. $\mathcal{A}^R$ satisfies the following condition:

(pS') for all $x, y \in A^R$, $\exists\, x \oplus y \stackrel{def.}{=} \max\{z \mid x \geq y \to^R z\} = \max\{z \mid y \geq x \rightsquigarrow^R z\}$.

(ii) If the partial order $\leq$, or $\geq$, is a lattice order, then we say that we have a *lattice structure with (pP')/(pS')*.

**Remark 1.4.12** *(The dual one is omitted)*
*By Definitions 1.4.11 (i), we have $x \odot y \in \{z \mid x \leq y \to z\}$ and $x \odot y \in \{z \mid y \leq x \rightsquigarrow z\}$, hence we have the property:*
*(pPP') for all $x, y$,* $\quad x \leq y \to (x \odot y), \quad y \leq x \rightsquigarrow (x \odot y)$.

*Consequently, (pP') $\Longrightarrow$ (pPP').*

**Definitions 1.4.13** (Definition of structure with (pRP')/(coRS'))
(i) Let $\mathcal{A}^L = (A^L, \leq, \to, \rightsquigarrow, \odot)$ be a structure such that the reduct $(A^L, \leq)$ is a partially preordered set (i.e. only (Re'), (Tr') hold) and $\to$, $\rightsquigarrow$, $\odot$ are binary operations. We say that $\mathcal{A}^L$ is a *structure with pseudo-residuum and pseudo-product*, or a *structure with (pRP')*, if $\leq$, $\to$, $\rightsquigarrow$ and $\odot$ are connected by the property:

(pRP') for all $x, y, z \in A^L$, $\quad x \leq y \to z \Longleftrightarrow y \leq x \rightsquigarrow z \Longleftrightarrow x \odot y \leq z$.

(i') Dually, let $\mathcal{A}^R = (A^R, \geq, \to^R, \rightsquigarrow^R, \oplus)$ be a structure such that the reduct $(A^R, \geq)$ is a partially preordered set (i.e. only (Re'), (Tr') hold) and $\to^R$, $\rightsquigarrow^R$, $\oplus$ are binary operations. We say that $\mathcal{A}^R$ is a *structure with pseudo-coresiduum and pseudo-sum*, or a *structure with (pcoRS')*, if $\geq$, $\to^R$, $\rightsquigarrow^R$ and $\oplus$ are connected by the property:

(pcoRS') for all $x, y, z \in A^R$, $\quad x \geq y \to^R z \Longleftrightarrow y \geq x \rightsquigarrow^R z \Longleftrightarrow x \oplus y \geq z$.

(ii) If the partial order $\leq$, or $\geq$, is a lattice order, then we say that we have a *lattice structure with (pRP')/(pcoRS')*.

Note that condition (pRP') needs not $\leq$ be an order.

**Remark 1.4.14**

$$(pRP') \Longrightarrow (pEq\#') \qquad and \qquad (pcoRS') \Longrightarrow (pEq\#^{R'}).$$

**Remark 1.4.15** *Just as the equivalences*

$$x \leq y \Longleftrightarrow x \to y = 1 \Longleftrightarrow x \rightsquigarrow y = 1$$

*were used either:*
*- to define the binary relation $\leq$ in the algebra $(A^L, \to, \rightsquigarrow, 1)$, by*

*(pdfrelR)* $(x \leq y \stackrel{def.}{\Longleftrightarrow} x \to y = 1 \Longleftrightarrow x \rightsquigarrow y = 1)$, *or*
*- to connect* $\leq$ *and* $\to, \rightsquigarrow$, $1$ *in the structure* $(A^L, \leq, \to, \rightsquigarrow, 1)$, *by*
*(pEqrelR)* $(x \leq y \Longleftrightarrow x \to y = 1 \Longleftrightarrow x \rightsquigarrow y = 1)$,

*the same happens with the equalities*

$$x \odot y = \min\{z \mid x \leq y \to z\} = \min\{z \mid y \leq x \rightsquigarrow z\},$$

*that can be used either:*
*- to define a new binary operation* $\odot$ *in the structure* $(A^L, \leq, \to, \rightsquigarrow)$, *by (pP'), or*
*- to connect* $\odot$ *and* $\leq, \to, \rightsquigarrow$ *in the structure* $(A^L, \leq, \to, \rightsquigarrow, \odot)$, *by (pRP').*

We prove now the following important general lemma.

**Lemma 1.4.16** *Let* $\mathcal{A}^L = (A^L, \leq, \to, \rightsquigarrow)$ *be a structure with (pP') (i.e. (Re'), (Tr'), (An'), (pP') hold), where the pseudo-product is* $\odot$. *If the property (p*') also holds, then* $(A^L, \leq, \to, \rightsquigarrow, \odot)$ *is a structure with (pRP') (i.e. (Re'), (Tr'), (pRP') hold).*

**Proof.** We must prove that (pRP') holds. Indeed, if $x \odot y \leq z$, then, it follows by (p*'), that $y \to (x \odot y) \leq y \to z$; and, since we also have, by (pP'), that (pPP') holds, i.e. $x \leq y \to (x \odot y)$ holds, we obtain $x \leq y \to z$, by (Tr'). Conversely, if $x \leq y \to z$, then, by (pP'), $x \odot y \leq z$. Also, if $x \odot y \leq z$, then, it follows, by (p*'), that $x \rightsquigarrow (x \odot y) \leq x \rightsquigarrow z$, and, by (pPP'), $y \leq x \rightsquigarrow (x \odot y)$; then, $y \leq x \rightsquigarrow z$, by (Tr'). Conversely, if $y \leq x \rightsquigarrow z$, then, by (pP'), $x \odot y \leq z$. Thus, (pRP') holds. □

A basic result is the following general lemma.

**Lemma 1.4.17** *(See ([116], Lemma 2.6.3))*

*Let* $\mathcal{A}^L = (A^L, \leq, \to, \rightsquigarrow, \odot)$ *be a structure (or, Galois dually, let* $\mathcal{A}^L = (A^L, \leq, \odot, \to, \rightsquigarrow)$ *be a structure) such that the property (pRP') (=(pPR'), respectively - see Part II) holds.*

*(a) If the reduct* $(A^L, \leq)$ *is a partially preordered set (i.e. only (Re'), (Tr') hold), then the following properties hold: for all* $x, y, z \in A^L$,

*(pEq#')* $\quad x \leq y \to z \Longleftrightarrow y \leq x \rightsquigarrow z$,
*(pPP')* $\quad y \leq x \to (y \odot x), \quad x \leq y \rightsquigarrow (y \odot x)$,
*(pRR')* $\quad (y \to x) \odot y \leq x, \quad y \odot (y \rightsquigarrow x) \leq x$,
*(pCp')* $\quad x \leq y \Longrightarrow (x \odot z \leq y \odot z, \quad z \odot x \leq z \odot y)$,
*(p*')* $\quad x \leq y \Longrightarrow (z \to x \leq z \to y, \quad z \rightsquigarrow x \leq z \rightsquigarrow y)$,
*(p**')* $\quad x \leq y \Longrightarrow (y \to z \leq x \to z, \quad y \rightsquigarrow z \leq x \rightsquigarrow z)$.

*(b) If the reduct* $(A^L, \leq)$ *is a poset (i.e. (Re'), (Tr'), (An') hold), then the following properties hold too:*
*(pP') for all* $x, y \in A^L$, $\exists\ x \odot y \stackrel{def.}{=} \min\{z \mid x \leq y \to z\} = \min\{z \mid y \leq x \rightsquigarrow z\}$,
*(pR') for all* $y, z \in A^L$, $\exists\ y \to z \stackrel{def.}{=} \max\{x \mid x \odot y \leq z\}$, $y \rightsquigarrow z \stackrel{def.}{=} \max\{x \mid y \odot x \leq z\}$.

**Proof.** (a): (pEq#'): Obviously, because it is contained in (pRP').

(pPP'): $y \le x \to (y \odot x) \overset{(pRP')}{\Leftrightarrow} y \odot x \le y \odot x$, which is true by (Re').
$x \le y \rightsquigarrow (y \odot x) \overset{(pRP')}{\Leftrightarrow} y \odot x \le y \odot x$, which is true by (Re').

(pRR'): $(y \to x) \odot y \le x \overset{(pRP')}{\Leftrightarrow} y \to x \le y \to x$, which is true by (Re').
$y \odot (y \rightsquigarrow x) \le x \overset{(pRP')}{\Leftrightarrow} y \rightsquigarrow x \le y \rightsquigarrow x$, which is true by (Re').

(pCp'): Let $x \le y$; by (pPP'), $y \le z \to (y \odot z)$; it follows, by (Tr'), that $x \le z \to (y \odot z)$; hence, $x \odot z \le y \odot z$, by (pRP').
Let $x \le y$; by (pPP'), $y \le z \rightsquigarrow (z \odot y)$; it follows, by (Tr'), that $x \le z \rightsquigarrow (z \odot y)$; hence, $z \odot x \le z \odot y$, by (pRP').

(p*'): Let $x \le y$; by (pRR'), $(z \to x) \odot z \le x$; it follows, by (Tr'), that $(z \to x) \odot z \le y$; hence, $z \to x \le z \to y$, by (pRP').
Let $x \le y$; by (pRR'), $z \odot (z \rightsquigarrow x) \le x$; it follows, by (Tr'), that $z \odot (z \rightsquigarrow x) \le y$; hence, $z \rightsquigarrow x \le z \rightsquigarrow y$, by (pRP').

(p**'): Let $x \le y$; by (pCp'), $(y \to z) \odot x \le (y \to z) \odot y$, and, by (pRR'), $(y \to z) \odot y \le z$; then, by (Tr'), $(y \to z) \odot x \le z$, hence $y \to z \le x \to z$, by (pRP').
Let $x \le y$; by (pCp'), $x \odot (y \rightsquigarrow z) \le y \odot (y \rightsquigarrow z)$, and, by (pRR'), $y \odot (y \rightsquigarrow z) \le z$; then, by (Tr'), $x \odot (y \rightsquigarrow z) \le z$, hence $y \rightsquigarrow z \le x \rightsquigarrow z$, by (pRP').

(b): (pP'): By (pPP'), $x \le y \to (x \odot y)$, i.e. $x \odot y \in \{z \mid x \le y \to z\}$. If $z' \in \{z \mid x \le y \to z\}$, i.e. $x \le y \to z'$, then, by (pRP'), $x \odot y \le z'$. Thus, $\min\{z \mid x \le y \to z\}$ exists and equals $x \odot y$.
By (pPP') again, $y \le x \rightsquigarrow (x \odot y)$, i.e. $x \odot y \in \{z \mid y \le x \rightsquigarrow z\}$. If $z' \in \{z \mid y \le x \rightsquigarrow z\}$, i.e. $y \le x \rightsquigarrow z'$, then, by (pRP'), $x \odot y \le z'$. Thus, $\min\{z \mid y \le x \rightsquigarrow z\}$ exists and equals $x \odot y$. Thus, (pP') holds.

(pR'): By (pRR'), $(y \to z) \odot y \le z$, i.e. $y \to z \in \{x \mid x \odot y \le z\}$. If $x' \in \{x \mid x \odot y \le z\}$, i.e. $x' \odot y \le z$, then, by (pRP'), $x' \le y \to z$. Thus, $\max\{x \mid x \odot y \le z\}$ exists and equals $y \to z$.
By (pRR') again, $y \odot (y \rightsquigarrow z) \le z$, i.e. $y \rightsquigarrow z \in \{x \mid y \odot x \le z\}$. If $x' \in \{x \mid y \odot x \le z\}$, i.e. $y \odot x' \le z$, then, by (pRP'), $x' \le y \rightsquigarrow z$. Thus, $\max\{x \mid y \odot x \le z\}$ exists and equals $y \rightsquigarrow z$. Thus, (pR') holds. □

Now the following general theorem follows immediately:

**Theorem 1.4.18** *We have:*

$$(pP') + (p*') \iff (pRP').$$

**Proof.** $\Longrightarrow$: By Lemma 1.4.16.
$\Longleftarrow$: By Lemma 1.4.17, (pP') and (p*') hold. □

### 1.4.3 Po-pi-magmas ($l$-pi-magmas) with (pP')/(pS')

We introduce po-pes(pP') and po-prs(pP'), po-mprs(pP'), po-iprs(pP'), and their corresponding lattice ordered ones. We recall the definitions of pBCI(pP') and pBCK(pP') algebras (lattices).

• **po-pes(pP') (*l*-pes(pP'))**

By general Definitions 1.4.11, we obtain the following definitions (the dual ones are omitted).

**Definitions 1.4.19** (Definition of po-pe(pP') (*l*-pe(pP')))

(i) A *left-po-pe with pseudo-product*, or a *po-pe(pP')* for short, is a left-po-pe

$$\mathcal{A}^L = (A^L, \leq^e, \rightarrow, \rightsquigarrow)$$

verifying additionally:

(pP') for all $x, y \in A^L$, $\exists\, x \odot y \stackrel{def.}{=} \min\{z \mid x \leq^e y \rightarrow z\} = \min\{z \mid y \leq^e x \rightsquigarrow z\}$.

(ii) If the partial order $\leq^e$ is a lattice order, then we say that we have a *left-lattice ordered p-exchange with pseudo-product* (*l-pe(pP')*).

Denote by **po-pe(pP')** (***l*-pe(pP')**) the class of all po-pes(pP') (*l*-pes(pP')).

• **po-prs(pP'), po-mprs(pP'), po-iprs(pP')**
**(*l*-prs(pP'), *l*-mprs(pP'), *l*-iprs(pP'))**

**Definitions 1.4.20** (Definition of po-pr(pP'), po-mpr(pP'), po-ipr(pP') (*l*-pr(pP'), *l*-mpr(pP'), *l*-ipr(pP')))

(i) A *left-po-pr (left-po-mpr, left-po-ipr) with pseudo-product*, or a *po-pr(pP') (po-mpr(pP'), po-ipr(pP'))* for short, is a left-po-pr (left-po-mpr, left-po-ipr, respectively)

$$\mathcal{A}^L = (A^L, \leq, \rightarrow, \rightsquigarrow, 1)$$

verifying additionally:

(pP') for all $x, y \in A^L$, $\exists\, x \odot y \stackrel{def.}{=} \min\{z \mid x \leq y \rightarrow z\} = \min\{z \mid y \leq x \rightsquigarrow z\}$.

(ii) If the partial order $\leq$ is a lattice order, then we say that we have a *l-pr(pP') (l-mpr(pP'), l-ipr(pP'))* .

Denote by **po-pr(pP')** the class of all po-prs(pP').
Denote by **po-mpr(pP')** the class of all po-mprs(pP').
Denote by **po-ipr(pP')** the class of all po-iprs(pP').
Denote analogously the lattice ordered case.

By general Remark 1.4.12, (pP') implies (pPP') and, dually, (pS') implies (pSS').

By inclusions 1.28, we have the inclusions:

(1.29)

$$\mathbf{po-ipr(pP')} \subset \mathbf{po-mpr(pP')} \subset \mathbf{po-pr(pP')} \subset \mathbf{po-pe(pP')} \quad \textit{and}$$

$$\mathbf{l-ipr(pP')} \subset \mathbf{l-mpr(pP')} \subset \mathbf{l-pr(pP')} \subset \mathbf{l-pe(pP')}.$$

• **pBCI(pP') and pBCK(pP') algebras (lattices)**

**Definitions 1.4.21** (Definition of pBCI(pP') algebra (lattice))

(i) A *left-pBCI algebra with pseudo-product*, or a *pBCI(pP') algebra* for short, is a left-pBCI algebra $\mathcal{A}^L = (A^L, \leq, \rightarrow, \rightsquigarrow, 1)$ verifying additionally (pP').

(ii) If the partial order $\leq$ is a lattice order, then we say that we have a *pBCI(pP') lattice.*

Denote by **pBCI(pP')** the class of all pBCI(pP') algebras and by **pBCI(pP')-L** the class of all pBCI(pP') lattices.

**Definitions 1.4.22** (Definition of pBCK(pP') algebra (lattice))

(i) A *left-pBCK algebra with pseudo-product*, or a *pBCK(pP') algebra* for short, is left-pBCK algebra $\mathcal{A}^L = (A^L, \leq, \rightarrow, \rightsquigarrow, 1)$ verifying additionally (pP').

(ii) If the partial order $\leq$ is a lattice order, then we say that we have a *pBCK(pP') lattice.*

Denote by **pBCK(pP')** the class of all pBCK(pP') algebras and by **pBCK(pP')-L** the class of all pBCK(pP') lattices.

### 1.4.4 Po-pi-magmas ($l$-pi-magmas) with (pRP')/(pcoRS')

We introduce the po-pes(pRP') and po-prs(pRP'), po-mprs(pRP'), po-iprs(pRP'), and their corresponding lattice ordered ones. We recall the pBCI(pRP') and pBCK(pRP') algebras (lattices).

By general Definitions 1.4.13, we obtain the following definitions (the dual ones are omitted).

- **po-pes(pRP') ($l$-pes(pRP'))**

**Definitions 1.4.23** (Definition of po-pe(pRP') ($l$-pe(pRP')))

(i) A *left-po-pe with the property (pRP')*, or a *po-pe(pRP')* for short, is a structure

$$\mathcal{A}^L = (A^L, \leq^e, \rightarrow, \rightsquigarrow, \odot)$$

such that $(A^L, \leq^e)$ is a poset, $(A^L, \rightarrow, \rightsquigarrow)$ is a p-exchange (i.e. (pEx) holds) and the pseudo-product $\odot$ is connected to $\leq^e$, $\rightarrow$, $\rightsquigarrow$ by the following property:
(pRP') for all $x, y, z \in A^L$, $\quad x \leq^e y \rightarrow z \Longleftrightarrow y \leq^e x \rightsquigarrow z \Longleftrightarrow x \odot y \leq^e z$.

(ii) If the partial order $\leq^e$ is a lattice order, then we say that we have a *left-lattice ordered p-exchange with (pRP') (l-pe(pRP'))*.

Denote by **po-pe(pRP')** the class of all po-pes(pRP') and analogously in lattice ordered case.

- **po-prs(pRP'), po-mprs(pRP'), po-iprs(pRP') ($l$-prs(pRP'), $l$-mprs(pRP'), $l$-iprs(pRP'))**

**Definitions 1.4.24** (Definition of po-pr(pRP'), po-mpr(pRP'), po-ipr(pRP') ($l$-pr(pRP'), $l$-mpr(pRP'), $l$-ipr(pRP')))

(i) A *left-po-pr (left-po-mpr, left-po-ipr) with the property (pRP'),* or a *po-pr(pRP') (po-mpr(pRP'), po-ipr(pRP'))* for short, is a structure

$$\mathcal{A}^L = (A^L, \leq, \rightarrow, \rightsquigarrow, \odot, 1)$$

such that $(A^L, \leq)$ is a poset ($(A^L, \leq, 1)$ is a poset with maximal element 1, with greatest element 1, respectively), $(A^L, \rightarrow, \rightsquigarrow, 1)$ is a p-residoid and the pseudo-product $\odot$ is connected to $\leq$, $\rightarrow$, $\rightsquigarrow$ by the following property:
(pRP') for all $x, y, z \in A^L$, $\quad x \leq y \rightarrow z \Longleftrightarrow y \leq x \rightsquigarrow z \Longleftrightarrow x \odot y \leq z$.

(ii) If the partial order $\leq$ is a lattice order, then we say that we have a *l-pr(pRP') (l-mpr(pRP'), l-ipr(pRP'))*.

Denote by **po-pr(pRP')** the class of all po-prs(pRP').
Denote by **po-mpr(pRP')** the class of all po-mprs(pRP').
Denote by **po-ipr(pRP')** the class of all po-iprs(pRP').
Denote analogously in lattice ordered case.
By inclusions 1.28, we have the inclusions:
(1.30)

$$\mathbf{po-ipr(pRP')} \subset \mathbf{po-mpr(pRP')} \subset \mathbf{po-pr(pRP')} \subset \mathbf{po-pe(pRP')}$$

$$and \quad \mathbf{l-ipr(pRP')} \subset \mathbf{l-mpr(pRP')} \subset \mathbf{l-pr(pRP')} \subset \mathbf{l-pe(pRP')}.$$

**Proposition 1.4.25** *(The dual case is omitted)*
*Any po-pe(pRP'), po-pr(pRP'), po-mpr(pRP'), po-ipr(pRP') verify the properties (pEq#'), (pP'), (pR'), (pPP'), (pRR'), (pCp'), (p*'), (p**').*

**Proof.** Obviously, by Lemma 1.4.17. □

**Corollary 1.4.26** *We have:*
*(1) Any po-pe(pRP') is a left-po-pe.*
*(2) Any po-pr(pRP') is a left-po-pr.*
*(3) Any po-mpr(pRP') is a left-po-mpr.*
*(4) Any po-ipr(pRP') is a left-po-ipr.*

**Proof.** Since (pEq#') and (p*') hold, by Lemma 1.4.17. □

**Proposition 1.4.27** *Let* $\mathcal{A}^L = (A^L, \leq, \rightarrow, \rightsquigarrow, 1)$ *be a left-po-pr = unital QB algebra. Then, we have:*
*(IdEq)* $\quad 1 \leq x \rightarrow y \iff 1 \leq x \rightsquigarrow y$ *and*
*(pEqrel)* $\quad x \leq y \iff 1 \leq x \rightarrow y \quad (\overset{(IdEq)}{\iff} 1 \leq x \rightsquigarrow y)$.

**Proof.** Obviously, by (1.20) and Proposition 1.3.15. □

**Proposition 1.4.28** *Let* $\mathcal{A}^L = (A^L, \leq, \rightarrow, \rightsquigarrow, 1)$ *be a left-po-mpr = maximal QB algebra* $\iff$ *pBCI algebra. Then, we have:*
*(IdEqR)* $\quad x \rightarrow y = 1 \iff x \rightsquigarrow y = 1$ *and*
*(pEqrelR)* $\quad x \leq y \iff x \rightarrow y = 1 \quad (\overset{(IdEqR)}{\iff} x \rightsquigarrow y = 1)$.

**Proof.** Obviously, by (1.24) and Proposition 1.3.16. □

**Proposition 1.4.29** *Let $\mathcal{A}^L = (A^L, \leq, \rightarrow, \rightsquigarrow, 1)$ be a left-po-ipr = integral QB algebra $\Longleftrightarrow$ pBCK algebra. Then, we have (IdEqR) and (pEqrelR).*

**Proof.** Obviously, by (1.27), since any pBCK algebra is a pBCI algebra. □

- **pBCI(pRP') and pBCK(pRP') algebras (lattices)**

**Definitions 1.4.30** (Definition of pBCI(pRP') algebra (lattice))

(i) A *left-pBCI algebra with pseudo-residuum and pseudo-product*, or a *pBCI(pRP') algebra* for short, is a structure $\mathcal{A}^L = (A^L, \leq, \rightarrow, \rightsquigarrow, \odot, 1)$ such that the reduct $(A^L, \leq, \rightarrow, \rightsquigarrow, 1)$ is a left-pBCI algebra verifying additionally (pRP').

(ii) If the partial order $\leq$ is a lattice order, then we say that we have a *pBCI(pRP') lattice.*

Denote by **pBCI(pRP')** the class of all pBCI(pRP') algebras and by **pBCI(pRP')-L** the class of all pBCI(pRP') lattices.

**Definitions 1.4.31** (Definition of pBCK(pRP') algebra (lattice))

(i) A *left-pBCK algebra with pseudo-residuum and pseudo-product*, or a *pBCK(pRP') algbera* for short, is a structure $\mathcal{A}^L = (A^L, \leq, \rightarrow, \rightsquigarrow, \odot, 1)$ such that the reduct $(A^L, \leq, \rightarrow, \rightsquigarrow, 1)$ is a left-pBCK algebra verifying additionally (pRP').

(ii) If the partial order $\leq$ is a lattice order, then we say that we have a *pBCK(pRP') lattice.*

Denote by **pBCK(pRP')** the class of all pBCK(pRP') algebras and by **pBCK(pRP')-L** the class of all pBCK(pRP') lattices.

**Proposition 1.4.32** *We have:*

*(pRP') + (pM) + (An') + (La') $\Longrightarrow$ (IdEqR) + (pEqrelR).*

**Proof.** (pRP') contains (pEq#') and by (pA#9'), (pEq#') + (pM) + (An') + (La') $\Longrightarrow$ (IdEqR) + (pEqrelR). □

### 1.4.5 Connections (pP') - (pRP')

By the general Theorem 1.4.18, we obtain immediately the following theorem.

**Theorem 1.4.33** *(The dual case is omitted)*

*(1) $(A^L, \leq^e, \rightarrow, \rightsquigarrow)$ is a po-pe(pP') iff $(A^L, \leq^e, \rightarrow, \rightsquigarrow, \odot)$ is a po-pe(pRP').*

*(2) $(A^L, \leq, \rightarrow, \rightsquigarrow, 1)$ is a po-pr(pP') iff $(A^L, \leq, \rightarrow, \rightsquigarrow, \odot, 1)$ is a po-pr(pRP').*

*(3) $(A^L, \leq, \rightarrow, \rightsquigarrow, 1)$ is a po-mpr(pP') iff $(A^L, \leq, \rightarrow, \rightsquigarrow, \odot, 1)$ is a po-mpr(pRP').*

*(4) $(A^L, \leq, \rightarrow, \rightsquigarrow, 1)$ is a po-ipr(pP') iff $(A^L, \leq, \rightarrow, \rightsquigarrow, \odot, 1)$ is a po-ipr(pRP').*

By Theorem 1.4.33 and (1.19), (1.20), (1.24), (1.27), we have the equivalences:

**Corollary 1.4.34**

$$\mathbf{QB(pP')} = \mathbf{po-pe(pP')} \Longleftrightarrow \mathbf{po-pe(pRP')} = \mathbf{QB(pRP')},$$
$$\mathbf{uQB(pP')} = \mathbf{po-pr(pP')} \Longleftrightarrow \mathbf{po-pr(pRP')} = \mathbf{uQB(pRP')},$$
$$\mathbf{pBCI(pP')} \Longleftrightarrow \mathbf{mQB(pP')} = \mathbf{po-mpr(pP')} \Longleftrightarrow$$
$$\Longleftrightarrow \mathbf{po-mpr(pRP')} = \mathbf{mQB(pRP')} \Longleftrightarrow \mathbf{pBCI(pRP')},$$
$$\mathbf{pBCK(pP')} \Longleftrightarrow \mathbf{iQB(pP')} = \mathbf{po-ipr(pP')} \Longleftrightarrow$$
$$\Longleftrightarrow \mathbf{po-ipr(pRP')} = \mathbf{iQB(pRP')} \Longleftrightarrow \mathbf{pBCK(pRP')}$$

and also the similar equivalences, in the lattice ordered case:

**Corollary 1.4.35**

$$\mathbf{QB(pP')-L} = \mathbf{l-pe(pP')} \Longleftrightarrow \mathbf{l-pe(pRP')} = \mathbf{QB(pRP')-L},$$
$$\mathbf{uQB(pP')-L} = \mathbf{l-pr(pP')} \Longleftrightarrow \mathbf{l-pr(pRP')} = \mathbf{uQB(pRP')-L},$$
$$\mathbf{pBCI(pP')-L} = \mathbf{l-mpr(pP')} \Longleftrightarrow \mathbf{l-mpr(pRP')} = \mathbf{pBCI(pRP')-L},$$
$$\mathbf{pBCK(pP')-L} = \mathbf{l-ipr(pP')} \Longleftrightarrow \mathbf{l-ipr(pRP')} = \mathbf{pBCK(pRP')-L}.$$

**Proposition 1.4.36** *(The dual case is omitted)*
*Let $\mathcal{A}^L = (A^L, \leq=\leq^e, \to, \rightsquigarrow, \odot)$ be a po-pe(pRP') = QB(pRP'). Then, $(A^L, \odot)$ is a semigroup (i.e. (Pass) holds).*

**Proof.**
$$\begin{aligned}
(x \odot y) \odot z \leq a &\overset{(pRP')}{\Longleftrightarrow} x \odot y \leq z \to a \\
&\overset{(pRP')}{\Longleftrightarrow} x \leq y \to (z \to a) \\
&\overset{(pEq\#')}{\Longleftrightarrow} y \leq x \rightsquigarrow (z \to a) \\
&\overset{(pEx)}{\Longleftrightarrow} y \leq z \to (x \rightsquigarrow a) \\
&\overset{(pRP')}{\Longleftrightarrow} y \odot z \leq x \rightsquigarrow a \\
&\overset{(pEq\#')}{\Longleftrightarrow} x \leq (y \odot z) \to a \\
&\overset{(pRP')}{\Longleftrightarrow} x \odot (y \odot z) \leq a;
\end{aligned}$$

hence, $(x \odot y) \odot z = x \odot (y \odot z)$, by the reflexivity and antisymmetry of $\leq$. □

**Proposition 1.4.37** *(The dual case is omitted)*
*Let $\mathcal{A}^L = (A^L, \leq, \to, \rightsquigarrow, \odot, 1)$ be a po-pr(pRP') = uQB(pRP'). Then, $(A^L, \odot, 1)$ is a monoid (i.e. (Pass), (PU) hold).*

**Proof.** Let $\mathcal{A}^L = (A^L, \leq, \to, \rightsquigarrow, \odot, 1)$ be a po-pr(pRP'). Then, the reduct $(A^L, \leq, \to, \rightsquigarrow, \odot)$ is a po-pe(pRP'). Then, by Proposition 1.4.36, $(A^L, \odot)$ is a semigroup. To prove that $(A^L, \odot, 1)$ is a monoid, it remains to prove (PU). Indeed,
(PU): $1 \odot x \leq a \overset{(pRP')}{\Longleftrightarrow} 1 \leq x \to a \overset{(pEq\#')}{\Longleftrightarrow} x \leq 1 \rightsquigarrow a \overset{(pM)}{\Longleftrightarrow} x \leq a$; hence, $1 \odot x = x$, by the reflexivity and antisymmetry of $\leq$. Also, $x \odot 1 \leq a \overset{(pRP')}{\Longleftrightarrow} x \leq 1 \to a \overset{(pM)}{\Longleftrightarrow} x \leq a$; hence, $x \odot 1 = x$, by the reflexivity and antisymmetry of $\leq$. □

Connections between these notions and the corresponding one from the 'world' of magmas (presented in Part II) are presented in Part III (Section 16.1).

### 1.4.6 Examples

All the examples were found by *Mace4 program*; they are all lattices.

**Examples 1.4.38 Proper lo-pes = quantum-B lattices**, i.e. not locally unital

• **Example 1.**

The algebra $\mathcal{A}_3^L = (A_3 = \{0,1,2\}, \wedge, \vee, \rightarrow, \rightsquigarrow)$, with the following tables of $\rightarrow$, $\rightsquigarrow$, $\wedge$, $\vee$, is a proper (i.e. not locally unital, since $(0 \rightsquigarrow 0) \rightarrow 0 = 0 \rightarrow 0 = 1 \neq 0$) lattice ordered pseudo-exchange (= quantum-B lattice). Since $x \leq y \Longleftrightarrow x \wedge y = x$, it follows that the lattice ordered set $(A_3, \leq)$ is that from the Hasse diagram below.

Note that for all $x, y \in A_3$, $x \odot y = \min\{z \mid x \leq y \rightarrow z\}$ exists, so $\mathcal{A}_3^L$ is a lo-pe(pP), where the table of $\odot$ is given below.

| $\rightarrow$ | 0 | 1 | 2 |
|---|---|---|---|
| 0 | 1 | 1 | 2 |
| 1 | 2 | 1 | 2 |
| 2 | 1 | 1 | 1 |

,

| $\rightsquigarrow$ | 0 | 1 | 2 |
|---|---|---|---|
| 0 | 0 | 1 | 2 |
| 1 | 0 | 1 | 2 |
| 2 | 1 | 1 | 1 |

,

| $\wedge$ | 0 | 1 | 2 |
|---|---|---|---|
| 0 | 0 | 0 | 2 |
| 1 | 0 | 1 | 2 |
| 2 | 2 | 2 | 2 |

,

| $\vee$ | 0 | 1 | 2 |
|---|---|---|---|
| 0 | 0 | 1 | 0 |
| 1 | 1 | 1 | 1 |
| 2 | 0 | 1 | 2 |

.

| $\odot$ | 0 | 1 | 2 |
|---|---|---|---|
| 0 | 0 | 1 | 2 |
| 1 | 0 | 1 | 2 |
| 2 | 2 | 2 | 2 |

.

• **Example 2.**

The algebra $\mathcal{A}_6^L = (A_6 = \{0,1,2,3,4,5\}, \wedge, \vee, \rightarrow, \rightsquigarrow)$, with the following tables of $\rightarrow$, $\rightsquigarrow$, $\wedge$, is a proper (i.e. not locally unital, since $(3 \rightarrow 3) \rightsquigarrow 3 = 2 \rightsquigarrow 3 = 1 \neq 3$) lattice ordered pseudo-exchange (= quantum-B lattice). Since $x \leq y \Longleftrightarrow x \wedge y = x$, it follows that the lattice ordered set $(A_6, \leq)$ is that from the Hasse diagram below.

| $\rightarrow$ | 0 | 1 | 2 | 3 | 4 | 5 |
|---|---|---|---|---|---|---|
| 0 | 1 | 1 | 2 | 1 | 1 | 1 |
| 1 | 2 | 1 | 2 | 2 | 2 | 2 |
| 2 | 1 | 1 | 1 | 1 | 1 | 1 |
| 3 | 2 | 1 | 2 | 2 | 2 | 2 |
| 4 | 2 | 1 | 2 | 2 | 2 | 2 |
| 5 | 2 | 1 | 2 | 2 | 2 | 2 |

| $\rightsquigarrow$ | 0 | 1 | 2 | 3 | 4 | 5 |
|---|---|---|---|---|---|---|
| 0 | 0 | 1 | 2 | 0 | 0 | 0 |
| 1 | 0 | 1 | 2 | 0 | 0 | 0 |
| 2 | 1 | 1 | 1 | 1 | 1 | 1 |
| 3 | 0 | 1 | 2 | 0 | 0 | 0 |
| 4 | 0 | 1 | 2 | 0 | 0 | 0 |
| 5 | 0 | 1 | 2 | 0 | 0 | 0 |

| $\wedge$ | 0 | 1 | 2 | 3 | 4 | 5 |
|---|---|---|---|---|---|---|
| 0 | 0 | 0 | 2 | 0 | 0 | 0 |
| 1 | 0 | 1 | 2 | 3 | 4 | 5 |
| 2 | 2 | 2 | 2 | 2 | 2 | 2 |
| 3 | 0 | 3 | 2 | 3 | 0 | 0 |
| 4 | 0 | 4 | 2 | 0 | 4 | 4 |
| 5 | 0 | 5 | 2 | 0 | 4 | 5 |

.

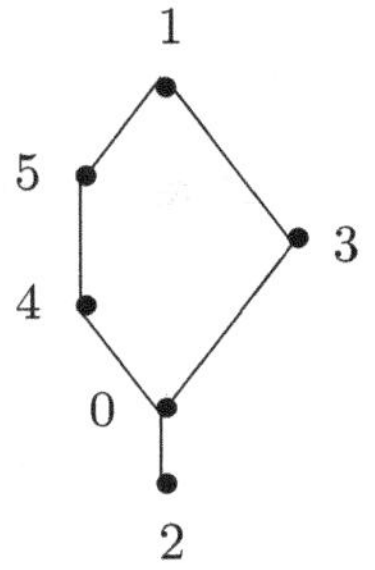


Note that for all $x, y \in A_6$, $x \odot y = \min\{z \mid x \leq y \rightarrow z\}$ exists, so $\mathcal{A}_6^L$ is a lo-pe(pP), where the table of $\odot$ is given below:

| $\odot$ | 0 | 1 | 2 | 3 | 4 | 5 |
|---|---|---|---|---|---|---|
| 0 | 0 | 1 | 2 | 1 | 1 | 1 |
| 1 | 0 | 1 | 2 | 1 | 1 | 1 |
| 2 | 2 | 2 | 2 | 2 | 2 | 2 |
| 3 | 0 | 1 | 2 | 1 | 1 | 1 |
| 4 | 0 | 1 | 2 | 1 | 1 | 1 |
| 5 | 0 | 1 | 2 | 1 | 1 | 1 |

.

**• Example 3.**

The algebra $\mathcal{A}_6^L = (A_6 = \{0, 1, 2, 3, 4, 5\}, \wedge, \vee, \to, \leadsto)$, with the following tables of $\to$, $\leadsto$, $\wedge$, is a proper (i.e. not locally unital, since $(3 \to 3) \leadsto 3 = 2 \leadsto 3 = 1 \neq 3$) lattice ordered pseudo-exchange (= quantum-B lattice). Since $x \leq y \Longleftrightarrow x \wedge y = x$, it follows that the lattice ordered set $(A_6, \leq)$ is that from the Hasse diagram below.

| $\to$ | 0 | 1 | 2 | 3 | 4 | 5 |
|---|---|---|---|---|---|---|
| 0 | 1 | 1 | 2 | 1 | 1 | 1 |
| 1 | 2 | 1 | 2 | 2 | 2 | 2 |
| 2 | 1 | 1 | 1 | 1 | 1 | 1 |
| 3 | 2 | 1 | 2 | 2 | 2 | 2 |
| 4 | 2 | 1 | 2 | 2 | 2 | 2 |
| 5 | 2 | 1 | 2 | 2 | 2 | 2 |

| $\leadsto$ | 0 | 1 | 2 | 3 | 4 | 5 |
|---|---|---|---|---|---|---|
| 0 | 0 | 1 | 2 | 0 | 0 | 0 |
| 1 | 0 | 1 | 2 | 0 | 0 | 0 |
| 2 | 1 | 1 | 1 | 1 | 1 | 1 |
| 3 | 0 | 1 | 2 | 0 | 0 | 0 |
| 4 | 0 | 1 | 2 | 0 | 0 | 0 |
| 5 | 0 | 1 | 2 | 0 | 0 | 0 |

| $\wedge$ | 0 | 1 | 2 | 3 | 4 | 5 |
|---|---|---|---|---|---|---|
| 0 | 0 | 0 | 2 | 2 | 2 | 2 |
| 1 | 0 | 1 | 2 | 3 | 4 | 5 |
| 2 | 2 | 2 | 2 | 2 | 2 | 2 |
| 3 | 2 | 3 | 2 | 3 | 2 | 2 |
| 4 | 2 | 4 | 2 | 2 | 4 | 2 |
| 5 | 2 | 5 | 2 | 2 | 2 | 5 |

.

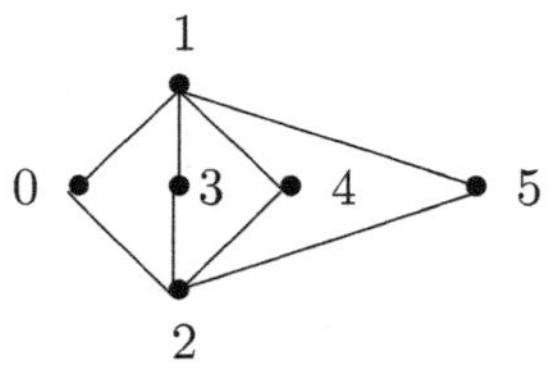


Note that $0 \odot 0 = \min\{z \mid 0 \leq 0 \to z\} = \min\{0, 1, 3, 4, 5\}$ does not exist, so $\mathcal{A}_6^L$ is not anymore a lo-pe(pP).

**Example 1.4.39 Proper locally-unital quantum-B lattice**, i.e. not unital

The algebra $\mathcal{A}_6^L = (A_6 = \{0, 1, 2, 3, 4, 5\}, \wedge, \vee, \to, \leadsto)$, with the following tables of $\to$, $\leadsto$, $\wedge$, is a proper (i.e. not unital) lattice ordered locally unital pseudo-exchange (= quantum-B lattice). Since $x \leq y \Longleftrightarrow x \wedge y = x$, it follows that the lattice ordered set $(A_6, \leq)$ is that from the Hasse diagram below.

| $\to$ | 0 | 1 | 2 | 3 | 4 | 5 |
|---|---|---|---|---|---|---|
| 0 | 1 | 3 | 3 | 3 | 4 | 3 |
| 1 | 3 | 1 | 3 | 3 | 4 | 5 |
| 2 | 0 | 3 | 2 | 3 | 4 | 3 |
| 3 | 4 | 4 | 4 | 4 | 4 | 4 |
| 4 | 3 | 3 | 3 | 3 | 4 | 3 |
| 5 | 3 | 3 | 3 | 3 | 4 | 1 |

| $\leadsto$ | 0 | 1 | 2 | 3 | 4 | 5 |
|---|---|---|---|---|---|---|
| 0 | 2 | 3 | 3 | 3 | 4 | 3 |
| 1 | 0 | 1 | 3 | 3 | 4 | 5 |
| 2 | 3 | 3 | 2 | 3 | 4 | 3 |
| 3 | 4 | 4 | 4 | 4 | 4 | 4 |
| 4 | 3 | 3 | 3 | 3 | 4 | 3 |
| 5 | 3 | 3 | 3 | 3 | 4 | 1 |

| $\wedge$ | 0 | 1 | 2 | 3 | 4 | 5 |
|---|---|---|---|---|---|---|
| 0 | 0 | 3 | 3 | 3 | 0 | 3 |
| 1 | 3 | 1 | 3 | 3 | 1 | 3 |
| 2 | 3 | 3 | 2 | 3 | 2 | 3 |
| 3 | 3 | 3 | 3 | 3 | 3 | 3 |
| 4 | 0 | 1 | 2 | 3 | 4 | 5 |
| 5 | 3 | 3 | 3 | 3 | 5 | 5 |

.

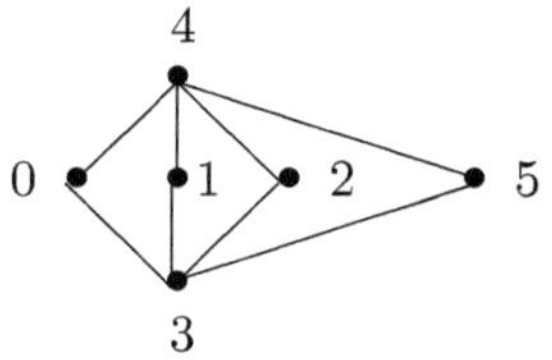


Note that for all $x, y \in A_6$, $x \odot y = \min\{z \mid x \leq y \to z\}$ exists, so $\mathcal{A}_6^L$ is a lo-locally unital (pP), where the table of $\odot$ is given below.

| $\odot$ | 0 | 1 | 2 | 3 | 4 | 5 |
|---|---|---|---|---|---|---|
| 0 | 4 | 4 | 0 | 3 | 4 | 4 |
| 1 | 0 | 1 | 4 | 3 | 4 | 5 |
| 2 | 4 | 4 | 2 | 3 | 4 | 4 |
| 3 | 3 | 3 | 3 | 3 | 3 | 3 |
| 4 | 4 | 4 | 4 | 3 | 4 | 4 |
| 5 | 4 | 5 | 4 | 3 | 4 | 4 |

.

**Examples 1.4.40 Proper unital lo-pes**, i.e. not maximal

• **Example 1.**

The algebra $\mathcal{A}_7^L = (A_7 = \{0,1,2,3,4,5,6\}, \wedge, \vee, \to, \rightsquigarrow)$, with the following tables of $\to$, $\rightsquigarrow$, $\wedge$, is a proper (i.e. not maximal) unital lattice ordered pseudo-exchange (= quantum-B lattice). Since $x \leq y \Longleftrightarrow x \wedge y = x$, it follows that the lattice ordered set $(A_7, \leq)$ is that from the Hasse diagram below.

| $\to$ | 0 | 1 | 2 | 3 | 4 | 5 | 6 |
|---|---|---|---|---|---|---|---|
| 0 | 0 | 2 | 2 | 2 | 2 | 5 | 2 |
| 1 | 0 | 1 | 2 | 3 | 4 | 5 | 6 |
| 2 | 0 | 0 | 0 | 0 | 0 | 0 | 0 |
| 3 | 0 | 1 | 2 | 1 | 1 | 0 | 4 |
| 4 | 0 | 1 | 2 | 3 | 1 | 0 | 6 |
| 5 | 0 | 2 | 2 | 2 | 2 | 0 | 2 |
| 6 | 0 | 1 | 2 | 1 | 1 | 0 | 1 |

| $\rightsquigarrow$ | 0 | 1 | 2 | 3 | 4 | 5 | 6 |
|---|---|---|---|---|---|---|---|
| 0 | 0 | 2 | 2 | 2 | 2 | 5 | 2 |
| 1 | 0 | 1 | 2 | 3 | 4 | 5 | 6 |
| 2 | 0 | 0 | 0 | 0 | 0 | 0 | 0 |
| 3 | 0 | 1 | 2 | 1 | 1 | 0 | 3 |
| 4 | 0 | 1 | 2 | 3 | 1 | 0 | 3 |
| 5 | 0 | 2 | 2 | 2 | 2 | 0 | 2 |
| 6 | 0 | 1 | 2 | 1 | 1 | 0 | 1 |

| $\wedge$ | 0 | 1 | 2 | 3 | 4 | 5 | 6 |
|---|---|---|---|---|---|---|---|
| 0 | 0 | 1 | 2 | 3 | 4 | 5 | 6 |
| 1 | 1 | 1 | 2 | 3 | 4 | 4 | 6 |
| 2 | 2 | 2 | 2 | 2 | 2 | 2 | 2 |
| 3 | 3 | 3 | 2 | 3 | 3 | 3 | 6 |
| 4 | 4 | 4 | 2 | 3 | 4 | 4 | 6 |
| 5 | 5 | 4 | 2 | 3 | 4 | 5 | 6 |
| 6 | 6 | 6 | 2 | 6 | 6 | 6 | 6 |

.

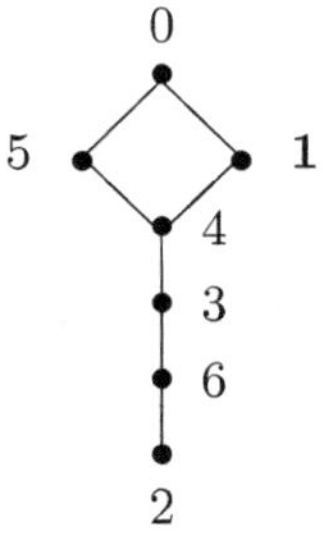


Note that for all $x, y \in A_7$, $x \odot y = \min\{z \mid x \leq y \to z\}$ exists, so $\mathcal{A}_7^L$ is an unital $l$o-pe(pP), where the table of $\odot$ is given below.

| $\odot$ | 0 | 1 | 2 | 3 | 4 | 5 | 6 |
|---|---|---|---|---|---|---|---|
| 0 | 0 | 0 | 2 | 5 | 5 | 5 | 5 |
| 1 | 0 | 1 | 2 | 3 | 4 | 5 | 6 |
| 2 | 2 | 2 | 2 | 2 | 2 | 2 | 2 |
| 3 | 5 | 3 | 2 | 6 | 3 | 5 | 6 |
| 4 | 5 | 4 | 2 | 6 | 4 | 5 | 6 |
| 5 | 5 | 5 | 2 | 5 | 5 | 5 | 5 |
| 6 | 5 | 6 | 2 | 6 | 6 | 5 | 6 |

.

• **Example 2.**

The algebra $\mathcal{A}_7^L = (A_7 = \{0,1,2,3,4,5,6\}, \wedge, \vee, \rightarrow, \rightsquigarrow)$, with the following tables of $\rightarrow$, $\rightsquigarrow$, $\wedge$, is a proper (i.e. not maximal) unital lattice ordered pseudo-exchange (= quantum-B lattice). Since $x \leq y \Longleftrightarrow x \wedge y = x$, it follows that the lattice ordered set $(A_7, \leq)$ is that from the Hasse diagram below.

| $\rightarrow$ | 0 | 1 | 2 | 3 | 4 | 5 | 6 |
|---|---|---|---|---|---|---|---|
| 0 | 0 | 2 | 2 | 2 | 2 | 2 | 2 |
| 1 | 0 | 1 | 2 | 3 | 4 | 5 | 6 |
| 2 | 0 | 0 | 0 | 0 | 0 | 0 | 0 |
| 3 | 0 | 4 | 2 | 4 | 4 | 5 | 6 |
| 4 | 0 | 2 | 2 | 2 | 4 | 5 | 6 |
| 5 | 0 | 2 | 2 | 2 | 2 | 4 | 2 |
| 6 | 0 | 2 | 2 | 2 | 4 | 5 | 4 |

| $\rightsquigarrow$ | 0 | 1 | 2 | 3 | 4 | 5 | 6 |
|---|---|---|---|---|---|---|---|
| 0 | 0 | 2 | 2 | 2 | 2 | 2 | 2 |
| 1 | 0 | 1 | 2 | 3 | 4 | 5 | 6 |
| 2 | 0 | 0 | 0 | 0 | 0 | 0 | 0 |
| 3 | 0 | 1 | 2 | 1 | 4 | 5 | 6 |
| 4 | 0 | 3 | 2 | 3 | 4 | 5 | 6 |
| 5 | 0 | 2 | 2 | 2 | 2 | 4 | 2 |
| 6 | 0 | 3 | 2 | 3 | 4 | 5 | 4 |

| $\wedge$ | 0 | 1 | 2 | 3 | 4 | 5 | 6 |
|---|---|---|---|---|---|---|---|
| 0 | 0 | 1 | 2 | 3 | 4 | 5 | 6 |
| 1 | 1 | 1 | 2 | 3 | 1 | 1 | 2 |
| 2 | 2 | 2 | 2 | 2 | 2 | 2 | 2 |
| 3 | 3 | 3 | 2 | 3 | 3 | 3 | 2 |
| 4 | 4 | 1 | 2 | 3 | 4 | 4 | 6 |
| 5 | 5 | 1 | 2 | 3 | 4 | 5 | 6 |
| 6 | 6 | 2 | 2 | 2 | 6 | 6 | 6 |

.

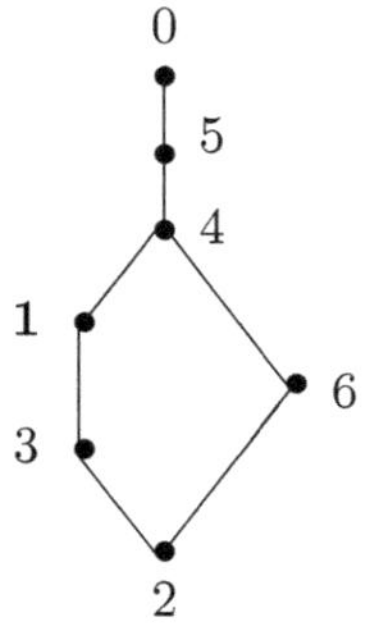


Note that $6 \odot 3 = \min\{z \mid 6 \leq 3 \rightarrow z\} = \min\{0,1,3,4,5,6\}$ does not exist.

## 1.5 Implicative-groups, po-($l$-)implicative-groups

The implicative-groups were introduced in [107], then developed in [110] and in the monograph [116], as a term-equivalent definition of the groups (see Section 16.2). In this section, we recall from [116] the definitions of the implicative-group, the po-implicative-group, the $l$-implicative-group and the intermediary notions of X-implicative-group, X-po-implicative-group, X-$l$-implicative-group, respectively, some results and the equivalences (EI1), (EI2), (EI3) between these notions, respectively.

The section has three subsections.

## 1.5.1 Implicative-groups

**• Basic properties. Connections**

Let $(A^L, \to=\to^L, \rightsquigarrow=\rightsquigarrow^L, 1)$ be a left-pseudo-algebra, as defined in the first section by Definition 1.1.6 (the dual case is omitted), where:
(IdEqR) $x \to y = 1 \Longleftrightarrow x \rightsquigarrow y = 1$.

We can then define on $A^L$ an *internal* (*natural*) binary relation $\leq$ by: for all $x, y \in A^L$,

$$(1.31) \qquad (pdfrelR) \qquad x \leq y \overset{def.}{\Longleftrightarrow} x \to y = 1 \; (\overset{(IdEqR)}{\Longleftrightarrow} x \rightsquigarrow y = 1).$$

Equivalently, let $(A^L, \leq, \to, \rightsquigarrow, 1)$ be a left-pseudo-structure, as defined in a previous section by Definition 1.1.6 (the dual case is omitted), where $\leq$ and $\to$, $\rightsquigarrow$, 1 are connected by:
(pEqrelR) $x \leq y \Longleftrightarrow x \to y = 1 \; (\overset{(IdEqR)}{\Longleftrightarrow} x \rightsquigarrow y = 1)$.

Consider the following new properties:
(pB=) $z \to x = (y \to z) \to (y \to x)$, $z \rightsquigarrow x = (y \rightsquigarrow z) \rightsquigarrow (y \rightsquigarrow x)$,
(pBB=) $y \to z = (z \to x) \rightsquigarrow (y \to x)$, $y \rightsquigarrow z = (z \rightsquigarrow x) \to (y \rightsquigarrow x)$;
(pD=) $(y \to x) \rightsquigarrow x = y = (y \rightsquigarrow x) \to x$,
(Id) $x \to 1 = x \rightsquigarrow 1$,
(pEqrelR=) $x = y \Longleftrightarrow x \to y = 1 \; (\overset{(IdEqR)}{\Longleftrightarrow} x \rightsquigarrow y = 1)$;
(pEq#$^=$) $x = y \to z \Longleftrightarrow y = x \rightsquigarrow z$,
(pRe) $x \to x = 1 = x \rightsquigarrow x$;
(p-s) (p-semisimple) $x \leq y \Longrightarrow x = y$ and, dually,
(p-s$^R$) (p-semisimple) $x \geq y \Longrightarrow x = y$;

and recall the old properties from list **pA**:
(pB') $z \to x \leq (y \to z) \to (y \to x)$, $z \rightsquigarrow x \leq (y \rightsquigarrow z) \rightsquigarrow (y \rightsquigarrow x)$,
(pBB') $y \to z \leq (z \to x) \rightsquigarrow (y \to x)$, $y \rightsquigarrow z \leq (z \rightsquigarrow x) \to (y \rightsquigarrow x)$,
(pC') $x \to (y \rightsquigarrow z) \leq y \rightsquigarrow (x \to z)$, $x \rightsquigarrow (y \to z) \leq y \to (x \rightsquigarrow z)$,
(pD') $y \leq (y \to x) \rightsquigarrow x$, $y \leq (y \rightsquigarrow x) \to x$,
(IdEqR) $x \to y = 1 \Longleftrightarrow x \rightsquigarrow y = 1$;
(pEqrelR) $x \leq y \Longleftrightarrow x \to y = 1 \; (\overset{(IdEqR)}{\Longleftrightarrow} x \rightsquigarrow y = 1)$;
(pEq#') $x \leq y \to z \Longleftrightarrow y \leq x \rightsquigarrow z$,
(pEx) $x \to (y \rightsquigarrow z) = y \rightsquigarrow (x \to z)$;
(pM) $1 \to x = x$, $1 \rightsquigarrow x = x$;
(An') $x \leq y$ and $y \leq x \Longrightarrow x = y$,
(Tr') $x \leq y$ and $y \leq z \Longrightarrow x \leq z$,
(p*') $x \leq y \Longrightarrow (z \to x \leq z \to y$ and $z \rightsquigarrow x \leq z \rightsquigarrow y)$,
(p**') $x \leq y \Longrightarrow (y \to z \leq x \to z$ and $y \rightsquigarrow z \leq x \rightsquigarrow z)$.

**Proposition 1.5.1** *(See Proposition 1.1.11) We have:*
*(0) (pD=) + (pM) + (pEqrelR) $\Longrightarrow$ (p-s);*
*(1) (pM) + (pBB=) $\Longrightarrow$ (pRe),*
*(1') (pEqrelR=) + (IdEqR) $\Longrightarrow$ (pRe),*
*(2) (pM) + (pBB=) $\Longrightarrow$ (pD=),*
*(3) (pEx) + (pRe) + (pEqrelR=) $\Longrightarrow$ (pD=);*
*(4) (pRe) + (pD=) $\Longrightarrow$ (pEqrelR=) + (IdEqR),*
*(5) (pBB=) + (pRe) + (pD=) + (pEqrelR) $\Longrightarrow$ (pC'),*
*(5') (pBB=) + (pRe) + (pD=) + (pEqrelR=) $\Longrightarrow$ (pEx),*
*(6) (pEx) $\Longrightarrow$ (pC'),*
*(6') (pC') + (p-s) $\Longrightarrow$ (pEx);*
*(7) (pBB=) + (pEx) + (pEqrelR=) + (IdEqR) $\Longrightarrow$ (pB=),*
*(7') (pB=) + (pEx) + (pEqrelR=) + (IdEqR) $\Longrightarrow$ (pBB=),*
*(7") (pEx) + (pEqrelR=) + (IdEqR) $\Longrightarrow$ ((pBB=) $\Leftrightarrow$ (pB=)),*
*(8) (pEx) + (pRe) $\Longrightarrow$ (Id),*
*(9) (pEqrelR=) + (IdEqR) + (pEx) $\Longrightarrow$ (pEq#$^=$),*
*(10) (pD=) + (pEqrelR=) + (IdEqR) $\Longrightarrow$ (pM),*
*(11) (pEx) + (pD=) $\Longrightarrow$ (pBB=);*
*(12) (pBB=) $\Longrightarrow$ (pBB'),*
*(12') (pBB') + (p-s) $\Longrightarrow$ (pBB=);*
*(13) (pB=) $\Longrightarrow$ (pB'),*
*(13') (pB') + (p-s) $\Longrightarrow$ (pB=);*
*(14) (pD=) $\Longrightarrow$ (pD'),*
*(14') (pD') + (p-s) $\Longrightarrow$ (pD=);*
*(15) (pEq#$^=$) $\Longrightarrow$ (pEq#'),*
*(15') (pEq#') + (p-s) $\Longrightarrow$ (pEq#$^=$);*
*(16) (pEqrelR=) $\Longrightarrow$ (pEqrelR),*
*(16') (pEqrelR) + (p-s) $\Longrightarrow$ (pEqrelR=);*
*(17) (pEqrelR=) $\Longrightarrow$ (p-s),*
*(18) (pEqrelR=) $\Longrightarrow$ (Tr'),*
*(19) (pEqrelR=) $\Longrightarrow$ (An'),*
*(20) (pEqrelR=) $\Longrightarrow$ (p*') + (p**').*

**Proof.** (0): If $x \leq y$, i.e. $x \to y = 1$, by (pEqrelR), then:
$x \overset{(pD=)}{=} (x \to y) \rightsquigarrow y = 1 \rightsquigarrow y \overset{(pM)}{=} y$. Thus, (p-s) holds.

(1): $x \to x \overset{(pM)}{=} (1 \rightsquigarrow x) \to (1 \rightsquigarrow x) \overset{(pBB=)}{=} 1 \rightsquigarrow 1 \overset{(pM)}{=} 1$ and
$x \rightsquigarrow x \overset{(pM)}{=} (1 \to x) \rightsquigarrow (1 \to x) \overset{(pBB=)}{=} 1 \to 1 \overset{(pM)}{=} 1$. Thus, (pRe) holds.

(1'): Obviously, $x = x$; hence, $x \to x = 1 = x \rightsquigarrow x$, by (pEqrelR=) + (IdEqR). Thus, (pRe) holds.

(2): $(y \to x) \rightsquigarrow x \overset{(pM)}{=} (y \to x) \rightsquigarrow (1 \to x) \overset{(pBB=)}{=} 1 \to y \overset{(pM)}{=} y$ and
$(y \rightsquigarrow x) \to x \overset{(pM)}{=} (y \rightsquigarrow x) \to (1 \rightsquigarrow x) \overset{(pBB=)}{=} 1 \rightsquigarrow y \overset{(pM)}{=} y$. Thus, (pD=) holds.

(3): $y \to ((y \to x) \rightsquigarrow x) \overset{(pEx)}{=} (y \to x) \rightsquigarrow (y \to x) \overset{(pRe)}{=} 1$; hence,
$y = (y \to x) \rightsquigarrow x$, by (pEqrelR=); $y \rightsquigarrow ((y \rightsquigarrow x) \to x) \overset{(pEx)}{=} (y \rightsquigarrow x) \to (y \rightsquigarrow$

$x) \overset{(pRe)}{=} 1$; hence, $y = (y \rightsquigarrow x) \to x$, by (pEqrelR=) again. Thus, (pD=) holds.

(4): If $x = y$, then $x \to y = y \to y \overset{(pRe)}{=} 1$ and $x \rightsquigarrow y = y \rightsquigarrow y = 1$; conversely, if $x \to y = 1$, then $x \overset{(pD=)}{=} (x \to y) \rightsquigarrow y = 1 \rightsquigarrow y \overset{(pRe)}{=} (y \to y) \rightsquigarrow y \overset{(pD=)}{=} y$; if $x \rightsquigarrow y = 1$, then $x = (x \rightsquigarrow y) \to y = 1 \to y \overset{(pRe)}{=} (y \rightsquigarrow y) \to y \overset{(pD=)}{=} y$. Thus, (pEqrelR=) + (IdEqR) hold.

(5): First, by (pBB=), we have: $Y \rightsquigarrow Z = (Z \rightsquigarrow X) \to (Y \rightsquigarrow X)$.
Take $X = u \to x$, $Y = y$, $Z = z \to x$; we obtain:
$y \rightsquigarrow (z \to x) = ((z \to x) \rightsquigarrow (u \to x)) \to (y \rightsquigarrow (u \to x))$
$\overset{(pBB=)}{=} (u \to z) \to (y \rightsquigarrow (u \to x))$.
Take now $z = y \rightsquigarrow x$ and $u = z$; we obtain:
$1 \overset{(pRe)}{=} y \rightsquigarrow y \overset{(pD=)}{=} y \rightsquigarrow [(y \rightsquigarrow x) \to x] = (z \to (y \rightsquigarrow x)) \to (y \rightsquigarrow (z \to x))$,
hence $z \to (y \rightsquigarrow x) \leq y \rightsquigarrow (z \to x)$, by (pEqrelR).

Then, by (pBB=) again, we have: $Y \to Z = (Z \to X) \rightsquigarrow (Y \to X)$.
Take $X = u \rightsquigarrow x$, $Y = y$, $Z = z \rightsquigarrow x$; we obtain:
$y \to (z \rightsquigarrow x) = ((z \rightsquigarrow x) \to (u \rightsquigarrow x)) \rightsquigarrow (y \to (u \rightsquigarrow x))$
$\overset{(pBB=)}{=} (u \rightsquigarrow z) \rightsquigarrow (y \to (u \rightsquigarrow x))$.
Take now $z = y \to x$ and $u = z$; we obtain:
$1 \overset{(pRe)}{=} y \to y \overset{(pD=)}{=} y \to [(y \to x) \rightsquigarrow x] = (z \rightsquigarrow (y \to x)) \rightsquigarrow (y \to (z \rightsquigarrow x))$,
hence $z \rightsquigarrow (y \to x) \leq y \to (z \rightsquigarrow x)$, by (pEqrelR). Thus, (pC') holds.

(5'): First, by (pBB=), we have: $Y \rightsquigarrow Z = (Z \rightsquigarrow X) \to (Y \rightsquigarrow X)$.
Take $X = u \to x$, $Y = y$, $Z = z \to x$; we obtain:
$y \rightsquigarrow (z \to x) = ((z \to x) \rightsquigarrow (u \to x)) \to (y \rightsquigarrow (u \to x))$
$\overset{(pBB=)}{=} (u \to z) \to (y \rightsquigarrow (u \to x))$.
Take now $z = y \rightsquigarrow x$ and $u = z$; we obtain:
$1 \overset{(pRe)}{=} y \rightsquigarrow y \overset{(pD=)}{=} y \rightsquigarrow [(y \rightsquigarrow x) \to x] = (z \to (y \rightsquigarrow x)) \to (y \rightsquigarrow (z \to x))$,
hence $z \to (y \rightsquigarrow x) = y \rightsquigarrow (z \to x)$, by (pEqrelR=).

(6): Obviously, since $= \Longrightarrow \leq$.

(6'): By (pC'), we have: $x \to (y \rightsquigarrow z) \leq y \rightsquigarrow (x \to z)$.
Hence, by (p-s) , we have: $x \to (y \rightsquigarrow z) = y \rightsquigarrow (x \to z)$. Thus, (pEx) holds.

(7): By (pBB=), we have:
$y \to z = (z \to x) \rightsquigarrow (y \to x)$ and $y \rightsquigarrow z = (z \rightsquigarrow x) \to (y \rightsquigarrow x)$;
then, by (pEqrelR=) and (IdEqR), we obtain:
$(y \to z) \to [(z \to x) \rightsquigarrow (y \to x)] = 1$ and $(y \rightsquigarrow z) \rightsquigarrow [(z \rightsquigarrow x) \to (y \rightsquigarrow x)] = 1$;
then, by (pEx), we obtain:
$(z \to x) \rightsquigarrow [(y \to z) \to (y \to x)] = 1$ and $(z \rightsquigarrow x) \to [(y \rightsquigarrow z) \rightsquigarrow (y \rightsquigarrow x)] = 1$;
then, by (pEqrelR=) and (IdEqR), we obtain:
$z \to x = (y \to z) \to (y \to x)$ and $z \rightsquigarrow x = (y \rightsquigarrow z) \rightsquigarrow (y \rightsquigarrow x)$. Thus, (pB=) holds.

(7'): By (pB=), we have:
$z \to x = (y \to z) \to (y \to x)$ and $z \rightsquigarrow x = (y \rightsquigarrow z) \rightsquigarrow (y \rightsquigarrow x)$;
then, by (pEqrelR=) and (IdEqR), we obtain:
$(z \to x) \rightsquigarrow [(y \to z) \to (y \to x)] = 1$ and $(z \rightsquigarrow x) \to [(y \rightsquigarrow z) \rightsquigarrow (y \rightsquigarrow x)] = 1$;

then, by (pEx), we obtain:
$(y \to z) \to [(z \to x) \rightsquigarrow (y \to x)] = 1$ and $(y \rightsquigarrow z) \rightsquigarrow [(z \rightsquigarrow x) \to (y \rightsquigarrow x)] = 1$;
then, by (pEqrelR=) and (IdEqR), we have:
$y \to z = (z \to x) \rightsquigarrow (y \to x)$ and $y \rightsquigarrow z = (z \rightsquigarrow x) \to (y \rightsquigarrow x)$; thus (pBB=) holds.

(7"): By (7) and (7').

(8): $x \to 1 \overset{(pRe)}{=} x \to (x \rightsquigarrow x) \overset{(pEx)}{=} x \rightsquigarrow (x \to x) \overset{(pRe)}{=} x \rightsquigarrow 1$. Thus, (Id) holds.

(9): $x = y \to z \overset{(pEqrelR=)+(IdEqR)}{\Longleftrightarrow} x \rightsquigarrow (y \to z) = 1 \overset{(pEx)}{\Longleftrightarrow} y \to (x \rightsquigarrow z) = 1 \overset{(pEqrelR=)}{\Longleftrightarrow} y = x \rightsquigarrow z$. Thus, (pEq$\#^=$) holds.

(10): $1 \to x = x \overset{(pEqrelR=)+(IdEqR)}{\Longleftrightarrow} (1 \to x) \rightsquigarrow x = 1$, that is true by (pD=), and
$1 \rightsquigarrow x = x \overset{(pEqrelR=)}{\Longleftrightarrow} (1 \rightsquigarrow x) \to x = 1$, that is true by (pD=) again. Thus, (pM) holds.

(11): $(z \to x) \rightsquigarrow (y \to x) \overset{(pEx)}{=} y \to ((z \to x) \rightsquigarrow x) \overset{(pD=)}{=} y \to z$ and
$(z \rightsquigarrow x) \to (y \rightsquigarrow x) \overset{(pEx)}{=} y \rightsquigarrow ((z \rightsquigarrow x) \to x) \overset{(pD=)}{=} y \rightsquigarrow z$. Thus, (pBB=) holds.

(12) - (20): Immediately, since $= \Longrightarrow \leq$. □

**Proposition 1.5.2**

*Let $(A^L, \to, \rightsquigarrow)$ be an algebra of type $(2,2)$. We have:*

*(pA#1=)* *(pEq$\#^=$) $\Longrightarrow$ (pD=),*
*(pA#2=)* *(pEq$\#^=$) $\Longrightarrow$ ((pB=) $\Leftrightarrow$ (pBB=)),*
*(pA#3=)* *(pEq$\#^=$) + (pD=) + (pBB=) $\Longrightarrow$ there exists $1 \in A^L$, such that (pRe) + (pM) hold (see ([193], Proposition 4)),*
*(pA#4=)* *(pEq$\#^=$) + (pD=) + (pB=) $\Longrightarrow$ (pEx) (see (pA#6) [197]).*

**Proof.** (pA#1=): $x \to y = x \to y \overset{(pEq\#^=)}{\Longleftrightarrow} x = (x \to y) \rightsquigarrow y$ and
$x \rightsquigarrow y = x \rightsquigarrow y \overset{(pEq\#^=)}{\Longleftrightarrow} x = (x \rightsquigarrow y) \to y$; thus, (pD=) holds.

(pA#2=): By (pEq$\#^=$),
(pB$_\to$=) $z \to x = (y \to z) \to (y \to x) \Longleftrightarrow y \to z = (z \to x) \rightsquigarrow (y \to x)$ (pBB$_1$=) and
(pB$_\rightsquigarrow$=) $z \rightsquigarrow x = (y \rightsquigarrow z) \rightsquigarrow (y \rightsquigarrow x) \Longleftrightarrow y \rightsquigarrow z = (z \rightsquigarrow x) \to (y \rightsquigarrow x)$ (pBB$_2$=).

(pA#3=): $x \overset{(pD=)}{=} (x \rightsquigarrow y) \to y \overset{(pBB=)}{=} (y \to y) \rightsquigarrow ((x \rightsquigarrow y) \to y)$
$= (y \to y) \rightsquigarrow x$; then, $x \to x = ((y \to y) \rightsquigarrow x) \to x \overset{(pD=)}{=} y \to y$.

Denote $u \overset{def.}{=} x \to x$, for all $x \in A^L$. We have: for all $x \in A^L$,

$$u \rightsquigarrow x = (x \to x) \rightsquigarrow x \overset{(pD=)}{=} x, \tag{1.32}$$

$$\textit{hence,} \qquad u \rightsquigarrow u = u. \tag{1.33}$$

Similarly, $x \overset{(pD=)}{=} (x \to y) \rightsquigarrow y \overset{(pBB=)}{=} (y \rightsquigarrow y) \to ((x \to y) \rightsquigarrow y)$
$= (y \rightsquigarrow y) \to x$; then, $x \rightsquigarrow x = ((y \rightsquigarrow y) \to x) \rightsquigarrow x \overset{(pD=)}{=} y \rightsquigarrow y$.

Denote $v \stackrel{def.}{=} x \rightsquigarrow x$, for all $x \in A^L$. We have: for all $x \in A^L$,

$$v \to x = (x \rightsquigarrow x) \to x \stackrel{(pD=)}{=} x, \tag{1.34}$$

$$hence, \qquad v \to v = v. \tag{1.35}$$

By (1.32), (1.34), we obtain: $u \rightsquigarrow v = v$ and $v \to u = u$, respectively.

We prove that $u = v$; indeed, $u \rightsquigarrow v = v \stackrel{(pEq\#^=)}{\Longleftrightarrow} u = v \to v \stackrel{(1.35)}{=} v$ (and also $v \to u = u \stackrel{(pEq\#^=)}{\Longleftrightarrow} v = u \rightsquigarrow u \stackrel{(1.33)}{=} u$).
Thus, $u = v$; we shall denote it by 1:

$$1 \stackrel{def.}{=} x \to x = x \rightsquigarrow x,$$

hence, (pRe) holds. By (1.32), (1.34), we have $1 \rightsquigarrow x = x$ and $1 \to x = x$, i.e. (pM) holds too.

(pA#4=): By (pD=), $y = (y \rightsquigarrow z) \to z$ and
by (pB=), in fact by (pB$_\to$=), $(y \rightsquigarrow z) \to z = (x \to (y \rightsquigarrow z)) \to (x \to z)$;
hence, $y = (x \to (y \rightsquigarrow z)) \to (x \to z)$ and then, by (pEq#$^=$),
$x \to (y \rightsquigarrow z) = y \rightsquigarrow (x \to z)$, i.e. (pEx) holds. An alternative proof can be made by (pB$_\rightsquigarrow$=). □

#### 1.5.1.1 The implicative-groups - a subclass of pBCI algebras

**Definition 1.5.3** (Definition 1 of implicative-groups)
An *implicative-group*, or an *i-group* for short, is an algebra

$$\mathcal{G} = (G, \to = \to^L, \rightsquigarrow = \rightsquigarrow^L, 1)$$

of type $(2, 2, 0)$ such that the following axioms hold: for all $x, y, z \in G$,
(pM) $\quad 1 \to x = x = 1 \rightsquigarrow x$,
(pBB=) $\quad y \to z = (z \to x) \rightsquigarrow (y \to x), \quad y \rightsquigarrow z = (z \rightsquigarrow x) \to (y \rightsquigarrow x)$.

Denote by **i-group** the class of all i-groups.

The i-group is said to be *commutative*, or *abelian*, if $x \to y = x \rightsquigarrow y$, for all $x, y \in G$.

**Proposition 1.5.4** *Let $(G, \to, \rightsquigarrow, 1)$ be an i-group. Then, the following properties hold: for all $x, y, z \in G$,*
*(pRe)* $\quad x \to x = 1 = x \rightsquigarrow x$,
*(pD=)* $\quad y = (y \to x) \rightsquigarrow x, \quad y = (y \rightsquigarrow x) \to x$,
*(IdEqR)* $\quad x \to y = 1 \Longleftrightarrow x \rightsquigarrow y = 1$,
*(pEqrelR=)* $\quad x = y \Longleftrightarrow x \to y = 1 \; (\stackrel{(IdEqR)}{\Longleftrightarrow} x \rightsquigarrow y = 1)$.

**Proof.** (pRe): By Proposition 1.5.1 (1), (pM) + (pBB=) $\Longrightarrow$ (pRe).
(pD=): By Proposition 1.5.1, (2), (pM) + (pBB=) $\Longrightarrow$ (pD=).
(IdEqR) and (pEqrelR=): By Proposition 1.5.1 (4), (pRe) + (pD=) $\Longrightarrow$ (pEqrelR=) + (IdEqR). □

The following theorem provides an equivalent definition of i-groups.

**Theorem 1.5.5**
*(1) Let $\mathcal{G} = (G, \to, \rightsquigarrow, 1)$ be an i-group. Then, the properties (pM), (pBB=), (IdEqR), (pEqrelR=) hold.*
*(1') Let $\mathcal{G} = (G, \to, \rightsquigarrow, 1)$ be an algebra of type $(2, 2, 0)$ verifying the properties (pM), (pBB=), (IdEqR), (pEqrelR=). Then, $\mathcal{G}$ is an i-group.*

**Proof.** (1): By Proposition 1.5.4, (IdEqR) and (pEqrelR=) hold; the other two properties hold by Definition 1 of i-groups.
(1'): Obviously, by Definition 1 of i-groups. □

It follows that we can define equivalently the i-groups as follows.

**Definition 1.5.6** (Definition 1' of i-groups) (See Definition 3' of pBCI algebras)
An *implicative-group*, or an *i-group* for short, is an algebra $\mathcal{G} = (G, \to, \rightsquigarrow, 1)$ of type $(2, 2, 0)$ such that the following axioms hold: (pM), (pBB=), (IdEqR), (pEqrelR=).

**Proposition 1.5.7** *Let $\mathcal{G} = (G, \to, \rightsquigarrow, 1)$ be an i-group. Then, the following properties hold: (p-s), (pM), (pBB=), (pB=), (pD=), (pEx), (pEq#$^=$), (pRe), (IdEqR), (pEqrelR=), (Id) and (pBB'), (pB'), (pC'), (pD'), (pEq#'), (pEqrelR), (An'), (Tr'), (p*'), (p**').*

**Proof.** By Proposition 1.5.1. □
We then have the following important remarks.

**Remarks 1.5.8**
*(1) The i-group, as defined by Definition 1', is a special left-pseudo-algebra, namely a left-pseudo-algebra $(G, \to, \rightsquigarrow, 1)$ for which the binary relation $\leq$ defined by (1.31) becomes the equality relation $=$ (i.e. the trivial partial order) (by (pEqrelR) and (pEqrelR=)), and so the property (p-s) holds.*
*(2) The i-group can be defined, equivalently, as a special left-pseudo-structure: $(G, \leq, \to, \rightsquigarrow, 1)$, where (pEqrelR) becomes (pEqrelR=), by Proposition 1.5.7; therefore, as left-pseudo-structure, the i-group $(G, =, \to, \rightsquigarrow, 1)$ is denoted also by $(G, \to, \rightsquigarrow, 1)$ .*

By above Proposition 1.5.7, we immediately obtain the following important result (the dual one is omitted).

**Corollary 1.5.9** *Any i-group is a left-pBCI algebra.*

- **Four other equivalent definitions of i-groups** [116]

The following theorem provides an equivalent definition of the i-group.

**Theorem 1.5.10**

*Let $\mathcal{G} = (G, \rightarrow, \rightsquigarrow, 1)$ be an algebra of type $(2, 2, 0)$. The following are equivalent:*

*(a) $\mathcal{G}$ is an i-group (i.e. (pM), (pBB=), (IdEqR), (pEqrelR=) hold, by Definition 1').*

*(b) (pD=), (pBB=), (IdEqR), (pEqrelR=) hold.*

**Proof.** (a) $\Longrightarrow$ (b): By Proposition 1.5.1 (2), (pM) + (pBB=) $\Longrightarrow$ (pD=).

(b) $\Longrightarrow$ (a): By Proposition 1.5.1 (10), (pD=) + (pEqrelR=) + (IdEqR) imply (pM). □

Hence, we have the following new equivalent definition of the i-group:

**Definition 1.5.11** (Definition 2 of i-groups) (See Definition 1' of pBCI algebras)

An *implicative-group*, or an *i-group* for short, is an algebra $\mathcal{G} = (G, \rightarrow, \rightsquigarrow, 1)$ of type $(2, 2, 0)$, such that the following axioms hold: (pBB=), (pD=), (IdEqR), (pEqrelR=).

To introduce another equivalent definition of i-groups, we need the following result.

**Theorem 1.5.12**

*(1) Let $(G, \rightarrow, \rightsquigarrow, 1)$ be an i-group. Then, the properties (pEx), (IdEqR), (pEqrelR=) hold.*

*(1') Let $\mathcal{G} = (G, \rightarrow, \rightsquigarrow, 1)$ be an algebra of type $(2, 2, 0)$ verifying the properties (pEx), (IdEqR), (pEqrelR=). Then, $\mathcal{G}$ is an i-group.*

**Proof.** (1): By Proposition 1.5.7.

(1'): We use Definition 1 of i-groups, hence we must prove that (pM) and (pBB=) hold. First, by Proposition 1.5.1 (1'), (pEqrelR=) + (IdEqR) implies (pRe). Then, by Proposition 1.5.1 (3), (pEx) + (pRe) + (pEqrelR=) $\Longrightarrow$ (pD=). By Proposition 1.5.1 (10), (pD=) + (pEqrelR=) + (IdEqR) $\Longrightarrow$ (pM), thus (pM) holds. Finally, by Proposition 1.5.1 (11), (pEx) + (pD=) $\Longrightarrow$ (pBB=). Thus, (pBB=) holds. □

By above Theorem 1.5.12, we obtain the following equivalent definition of i-groups.

**Definition 1.5.13** (Definition 3 of i-groups)

An *implicative-group*, or an *i-group* for short, is an algebra $\mathcal{G} = (G, \rightarrow, \rightsquigarrow, 1)$ of type $(2, 2, 0)$ verifying the following axioms: (pEx), (IdEqR), (pEqrelR=).

**Theorem 1.5.14**

*(1) Let $(G, \rightarrow, \rightsquigarrow, 1)$ be an i-group. Then, the properties (pEx), (pM), (pRe), (IdEqR), (pEqrelR=) hold.*

*(1') Let $\mathcal{G} = (G, \rightarrow, \rightsquigarrow, 1)$ be an algebra of type $(2, 2, 0)$ verifying the properties (pEx), (pM), (pRe), (IdEqR), (pEqrelR=). Then, $\mathcal{G}$ is an i-group.*

**Proof.** (1): Suppose (pEx), (IdEqR), (pEqrelR=) hold, by Definition 3. We must prove that (pM) and (pRe) hold. Indeed, by Proposition 1.5.1: by (1'), (IdEqR) + (pEqrelR=) $\Longrightarrow$ (pRe); by (3), (pEx) + (pRe) + (pEqrelR=) $\Longrightarrow$ (pD=) and by (10), (pD=) + (IdEqR) + (pEqrelR=) $\Longrightarrow$ (pM).

(1'): Obviously. □

By above Theorem 1.5.14, we obtain two other equivalent definitions of i-groups.

**Definition 1.5.15** (Definition 4 of i-groups) (See Definition 4' of pBCI algebras)

An *implicative-group*, or an *i-group* for short, is an algebra $\mathcal{G} = (G, \rightarrow, \rightsquigarrow, 1)$ of type $(2, 2, 0)$ verifying the following axioms: (pEx), (pM), (pRe), (IdEqR), (pEqrelR=).

**Remarks 1.5.16**

*(1) Definition 4 of i-groups shows the connection between pseudo-residoids (Definition 1.1.4) (the analogous of monoids) and i-groups (the term-equivalent notion of groups):*

**i-group** = **p-residoid** + *(pRe)* + *(IdEqR)* + *(pEqrelR=).*

*(2) Note the analogies with groups: note that (pM), (pEx), (pRe) correspond to (PU), (Pass), (pIv) =* $(m_1$*-pRe), respectively.*

*(3) Note that (pM), (pEx), (pRe) do not imply (IdEqR). Indeed, consider the algebra* $(A, \rightarrow, \rightsquigarrow, 1)$ *with the tables:*

| $\rightarrow$ | 0 | a | 1 |
|---|---|---|---|
| 0 | 1 | 1 | 1 |
| a | 1 | 1 | 1 |
| 1 | 0 | a | 1 |

| $\rightsquigarrow$ | 0 | a | 1 |
|---|---|---|---|
| 0 | 1 | 0 | 1 |
| a | 0 | 1 | 1 |
| 1 | 0 | a | 1 |

.

*It verifies (pM), (pEx), (pRe); then,* $0 \rightarrow a = 1$*, but* $0 \rightsquigarrow a = 0 \neq 1$.

*(4) Note that (pM), (pEx), (pRe) do not imply (pEqrelR=) too. Indeed, consider the algebra* $(A, \rightarrow = \rightsquigarrow, 1)$ *with the table:*

| $\rightarrow$ | 0 | 1 |
|---|---|---|
| 0 | 1 | 1 |
| 1 | 0 | 1 |

. *It verifies (pM), (pEx), (pRe); we have* $0 \rightarrow 1 = 1$*, but* $0 \neq 1$.

**Definition 1.5.17** (Definition 4' of i-groups)

An *implicative-group*, or an *i-group* for short, is an algebra $\mathcal{G} = (G, \rightarrow, \rightsquigarrow, 1)$ of type $(2, 2, 0)$ verifying the following axioms: (pEx), (pM), (IdEqR), (pEqrelR=).

**• Two other special equivalent definitions of i-groups coming from QB algebras**

The following equivalent definitions of i-groups coming from the definitions (2 and 1) of quantum-B algebras are very special, because they do not contain 1 anymore.

**Theorem 1.5.18** *(See Definition 2 of QB algebras)*
*(1) Let $(G, \rightarrow, \rightsquigarrow, 1)$ be an i-group. Then, the properties (pEx), (pEq#$^=$) hold.*
*(1') Let $\mathcal{G} = (G, \rightarrow, \rightsquigarrow)$ be an algebra of type $(2,2)$ verifying the properties (pEx), (pEq#$^=$). Then, $\mathcal{G}$ is an i-group.*

**Proof.** (1): Let $(G, \rightarrow, \rightsquigarrow, 1)$ be an i-group following Definition 3, i.e. (pEx), (IdEqR), (pEqrelR=) hold. Then, by Proposition 1.5.1 (9), (pEx) + (IdEqR), (pEqrelR=) $\Longrightarrow$ (pEx).
(1'): Let now $\mathcal{G} = (G, \rightarrow, \rightsquigarrow)$ be an algebra of type $(2,2)$ verifying the properties (pEx), (pEq#$^=$). Then, by (pA#1=), (pEq#$^=$) $\Longrightarrow$ (pD=);
by Proposition 1.5.1 (11), (pEx) + (pD=) $\Longrightarrow$ (pBB=);
by (pA#3=), (pEq#$^=$) + (pD=) + (pBB=) $\Longrightarrow \exists\, 1 \overset{def.}{=} x \rightarrow x = x \rightsquigarrow x$ such that (pRe), (pM) hold;
by Proposition 1.5.1 (4), (pRe) + (pD=) $\Longrightarrow$ (IdEqR) + (pEqrelR=);
then, $(G, \rightarrow, \rightsquigarrow, 1)$ is an i-group (Definition 3). □

By above Theorem 1.5.18, we obtain the following equivalent definition of i-groups.

**Definition 1.5.19** (Definition 5 of i-groups)
An *implicative-group*, or an *i-group* for short, is an algebra $\mathcal{G} = (G, \rightarrow, \rightsquigarrow)$ of type $(2,2)$ verifying the following axioms: (pEx), (pEq#$^=$).

**Theorem 1.5.20** *(See Definition 1 of QB algebras)*
*(1) Let $(G, \rightarrow, \rightsquigarrow)$ be an i-group. Then, the properties (pB=), (pEq#$^=$) hold.*
*(1') Let $\mathcal{G} = (G, \rightarrow, \rightsquigarrow)$ be an algebra of type $(2,2)$ verifying the properties (pB=), (pEq#$^=$). Then, $\mathcal{G}$ is an i-group.*

**Proof.** (1): ) Let $(G, \rightarrow, \rightsquigarrow)$ be an i-group following Definition 5, i.e. (pEx), (pEq#$^=$) hold. Then,
by (pA#1=), (pEq#$^=$) $\Longrightarrow$ (pD=);
by Proposition 1.5.1 (11), (pEx) + (pD=) $\Longrightarrow$ (pBB=);
by (pA#2=), (pEq#$^=$) + (pBB=) $\Longrightarrow$ (pB=).
(1'): Let $\mathcal{G} = (G, \rightarrow, \rightsquigarrow)$ be an algebra of type $(2,2)$ verifying the properties (pB=), (pEq#$^=$). Then,
by (pA#1=), (pEq#$^=$) $\Longrightarrow$ (pD=);
by (pA#4=), (pEq#$^=$) + (pB=) + (pD=) $\Longrightarrow$ (pEx). □

By above Theorem 1.5.20, we obtain the following equivalent definition of i-groups.

**Definition 1.5.21** (Definition 6 of i-groups)
An *implicative-group*, or an *i-group* for short, is an algebra $\mathcal{G} = (G, \rightarrow, \rightsquigarrow)$ of type $(2,2)$ verifying the following axioms: (pB=), (pEq#$^=$).

- **One involutive negation in i-groups:** $^{-1}$

Let $\mathcal{G} = (G, \rightarrow, \rightsquigarrow, 1)$ be an i-group. Define two *negations* on $G$ by: for all $x \in G$,

$$x^{-} \stackrel{def.}{=} x \rightarrow 1, \quad x^{\sim} \stackrel{def.}{=} x \rightsquigarrow 1.$$

But (Id) holds, by Proposition 1.5.7, hence $^{-} = {}^{\sim}$, i.e. there exists in fact only one negation (see the similar one negation in pBCI algebras in [116]), defined as follows: for all $x \in G$,

$$x^{-1} \stackrel{def.}{=} x \rightarrow 1 \stackrel{(Id)}{=} x \rightsquigarrow 1. \tag{1.36}$$

**Proposition 1.5.22** *Let $\mathcal{G}$ be an i-group. Then, we have: for all $x, y \in G$,*
*(N1)* $1^{-1} = 1$,
*(DN)* $(x^{-1})^{-1} = x$,
*(pNg)* $(x \rightarrow y)^{-1} = y \rightarrow x, \quad (x \rightsquigarrow y)^{-1} = y \rightsquigarrow x$,
*(pNg1)* $x \rightarrow y^{-1} = y \rightsquigarrow x^{-1}$,
*(pDNeg1)* $y \rightarrow z = z^{-1} \rightsquigarrow y^{-1}$ *and* $y \rightsquigarrow z = z^{-1} \rightarrow y^{-1}$,
*(pDNeg2)* $y^{-1} \rightarrow x = x^{-1} \rightsquigarrow y$,
*(pDNeg3)* $(x \rightarrow y^{-1})^{-1} = (y \rightsquigarrow x^{-1})^{-1}$.

**Proof.** (N1): Take $x = 1$ in (pM).
(DN): Take $x = 1$ in (pD=): $y = (y \rightarrow 1) \rightsquigarrow 1 = (y^{-1})^{-1}$.
(pNg): $(x \rightarrow y)^{-1} = (x \rightarrow y) \rightsquigarrow 1 \stackrel{(pRe)}{=} (x \rightarrow y) \rightsquigarrow (y \rightarrow y) \stackrel{(pBB=)}{=} y \rightarrow x$ and $(x \rightsquigarrow y)^{-1} = (x \rightsquigarrow y) \rightarrow 1 = (x \rightsquigarrow y) \rightarrow (y \rightsquigarrow y) = y \rightsquigarrow x$.
(pNg1): $y \rightsquigarrow x^{-1} \stackrel{def.}{=} y \rightsquigarrow (x \rightarrow 1) \stackrel{(pEx)}{=} x \rightarrow (y \rightsquigarrow 1) = x \rightarrow y^{-1}$.
(pDNeg1): Take $x = 1$ in (pBB=).
(pDNeg2): $x^{-1} \rightsquigarrow y \stackrel{(DN)}{=} x^{-1} \rightsquigarrow (y^{-1})^{-1} \stackrel{(pNg1)}{=} y^{-1} \rightarrow (x^{-1})^{-1} = y^{-1} \rightarrow x$.
(pDNeg3): By (pNg1). □

**Remark 1.5.23** *The negation $^{-1}$ is involutive, by (DN).*

Note that we can define equivalently the i-group as follows.

**Definition 1.5.24** (Definition 7 of i-groups) (See Definition 3 of i-groups)
An *implicative-group*, or an *i-group* for short, is an algebra $(G, \rightarrow, \rightsquigarrow, ^{-1}, 1)$ of type $(2, 2, 1, 0)$ verifying (IdEqR), (pEqrelR=), (pM) (not necessary), (pEx), (DN) (not necessary) and: (pNeg) $\quad x^{-1} = x \rightarrow 1 \ (\stackrel{(Id)}{=} x \rightsquigarrow 1)$.

**Remark 1.5.25** *Note that the equality:* $\quad x^{-1} = x \rightarrow 1 \ (\stackrel{(Id)}{=} x \rightsquigarrow 1)$
*can be used:*
*- either as the definition of the negation $^{-1}$ in the implicative-hub $(A, \rightarrow, \rightsquigarrow, 1)$ [116],*
*- or as the connection (pNeg) between the negation $^{-1}$ and $\rightarrow, \rightsquigarrow$, 1 in the implicative-group (Definition 7).*

**Proposition 1.5.26** *Let $(G, \rightarrow, \rightsquigarrow, 1)$ be an i-group. Then, we have: for all $x, y \in G$,*

$$(1.37) \qquad x \rightarrow y = (x^{-1} \rightsquigarrow y^{-1})^{-1}, \quad x \rightsquigarrow y = (x^{-1} \rightarrow y^{-1})^{-1} \quad (see\ (5.10)),$$

$$(1.38) \qquad (x \rightarrow y^{-1})^{-1} = y^{-1} \rightarrow x, \quad (y \rightsquigarrow x^{-1})^{-1} = x^{-1} \rightsquigarrow y.$$

**Proof.** (1.37): $x \rightarrow y \stackrel{(pNg)}{=} (y \rightarrow x)^{-1} \stackrel{(pDNeg1)}{=} (x^{-1} \rightsquigarrow y^{-1})^{-1}$
and $x \rightsquigarrow y = (y \rightsquigarrow x)^{-1} = (x^{-1} \rightarrow y^{-1})^{-1}$.
(1.38): by (pNg). □

**Remark 1.5.27** *Relations (1.37) say that, in the involutive algebra that is the i-group:*
*- the dual implications of $\rightarrow$, $\rightsquigarrow$ are $\rightsquigarrow$, $\rightarrow$, respectively, i.e.*

$$\rightarrow^R = \rightsquigarrow \quad and \quad \rightsquigarrow^R = \rightarrow;$$

*- the two implications $\rightarrow$ and $\rightsquigarrow$ are not independent, hence, we can define the i-group by using only $\rightarrow$ - open problem.*

- **New operation in i-groups: the pseudo-product $\cdot$**

The i-groups have a very interesting property.

**Proposition 1.5.28** *Let $(G, \rightarrow, \rightsquigarrow, 1)$ be an i-group and $x^{-1} \stackrel{def.}{=} x \rightarrow 1 = x \rightsquigarrow 1$, for all $x \in G$. For all $x \in G$, define the mapings $h_{x\rightarrow} : G \longrightarrow G$ and $h_{x\rightsquigarrow} : G \longrightarrow G$ by: for all $g \in G$,*

$$h_{x\rightarrow}(g) = x \rightarrow g, \quad h_{x\rightsquigarrow}(g) = x \rightsquigarrow g.$$

*Then, $h_{x\rightarrow}$ and $h_{x\rightsquigarrow}$ are bijective.*

**Proof.** Injectivity: Let $g_1,\ g_2 \in G$, $g_1 \neq g_2$. We must prove that $h_{x\rightarrow}(g_1) \neq h_{x\rightarrow}(g_2)$ and $h_{x\rightsquigarrow}(g_1) \neq h_{x\rightsquigarrow}(g_2)$.
Suppose, by absurdum hypothesis, that $h_{x\rightarrow}(g_1) = h_{x\rightarrow}(g_2)$, i.e. $x \rightarrow g_1 = x \rightsquigarrow g_2$; then, $g_1 \rightarrow g_2 \stackrel{(pB=)}{=} (x \rightarrow g_1) \rightarrow (x \rightarrow g_2) \stackrel{(pRe)}{=} 1$, hence, by (pEqrelR=), $g_1 = g_2$: contradiction.
Similarly, suppose, by absurdum hypothesis, that $h_{x\rightsquigarrow}(g_1) = h_{x\rightsquigarrow}(g_2)$, i.e. $x \rightsquigarrow g_1 = x \rightsquigarrow g_2$; then, $g_1 \rightsquigarrow g_2 \stackrel{(pB=)}{=} (x \rightsquigarrow g_1) \rightsquigarrow (x \rightsquigarrow g_2) \stackrel{(pRe)}{=} 1$, hence $g_1 = g_2$: contradiction.

Surjectivity of $h_{x\rightarrow}$: We must prove that for all $y \in G$, there exists $z \in G$, such that $y = h_{x\rightarrow}(z) = x \rightarrow z$. Indeed, let $y \in G$ and take $z = y^{-1} \rightsquigarrow x$; then, $h_{x\rightarrow}(z) = x \rightarrow z = x \rightarrow (y^{-1} \rightsquigarrow x) \stackrel{(pEx)}{=} y^{-1} \rightsquigarrow (x \rightarrow x) \stackrel{(pRe)}{=} y^{-1} \rightsquigarrow 1 = (y^{-1})^{-1} \stackrel{(DN)}{=} y$.

Surjectivity of $h_{x\rightsquigarrow}$: We must prove that for all $y \in G$, there exists $u \in G$, such that $y = h_{x\rightsquigarrow}(u) = x \rightsquigarrow u$. Indeed, let $y \in G$ and take $u = y^{-1} \rightarrow x$; then,

$h_{x \rightsquigarrow}(u) = x \rightsquigarrow u = x \rightsquigarrow (y^{-1} \to x) \overset{(pEx)}{=} y^{-1} \to (x \rightsquigarrow x) \overset{(pRe)}{=} y^{-1} \to 1 = (y^{-1})^{-1} \overset{(DN)}{=} y.$ □

We then have immediately the following corollary.

**Corollary 1.5.29** *Let $(G, \to, \rightsquigarrow, 1)$ be an i-group. For all $x, y \in G$,*
*- there exists a unique $z \in G$, $z = y^{-1} \rightsquigarrow x$, such that $y = h_{x\to}(z) = x \to z$ and*
*- there exists a unique $u \in G$, $u = y^{-1} \to x$, such that $y = h_{x\rightsquigarrow}(u) = x \rightsquigarrow u$.*

By above Corollary 1.5.29, we can introduce a binary operation $\cdot$ on $(G, \to, \rightsquigarrow, 1)$ as follows: for all $x, y \in G$,

$$y \cdot x \overset{def.}{=} z = y^{-1} \rightsquigarrow x, \quad x \cdot y \overset{def.}{=} u = y^{-1} \to x,$$

i.e.

$$x \cdot y \overset{def.}{=} y^{-1} \to x \overset{(pDNeg2)}{=} x^{-1} \rightsquigarrow y. \tag{1.39}$$

Note that $\cdot$ satisfies the following property: for all $x, y \in G$,

(dfpP=) $\exists\, y \cdot x = z$ such that $x \to z = y$ and $x \cdot y = u$ such that $x \rightsquigarrow u = y$.

**Remark 1.5.30** *Recall the dual properties from the previous section: for all $x, y$,*
*(dfpP') $\exists\, y \odot x = \min\{z \mid y \leq x \to z\}$ and $x \odot y = \min\{u \mid y \leq x \rightsquigarrow u\}$,*
*(dfpS') $\exists\, y \oplus x = \max\{z \mid y \geq x \to^R z\}$ and $x \oplus y = \max\{u \mid y \geq x \rightsquigarrow^R u\}$.*
*Recall also the property (p-s) $(x \leq y \Longrightarrow x = y)$. Then, note that:*
*(dfpP') + (p-s) $\Longrightarrow$ (dfpP=) and, dually, (dfpS') + (p-$s^R$) $\Longrightarrow$ (dfpP=).*

Since the i-group is, by (DN), an involutive algebra (where there is only one involutive negation), we then define the new operation $\cdot$ on $G$ alternatively by: for all $x, y \in G$,

$$x \cdot y \overset{def.}{=} (x \to y^{-1})^{-1} \overset{(pDNeg3)}{=} (y \rightsquigarrow x^{-1})^{-1}. \tag{1.40}$$

Indeed, it is the same operation as defined by (1.39), because, by (1.38), $(x \to y^{-1})^{-1} = y^{-1} \to x$ and $(y \rightsquigarrow x^{-1})^{-1} = x^{-1} \rightsquigarrow y$.

**Remark 1.5.31** *Note that the i-group may be defined equivalently as an algebra $(G, \to, \rightsquigarrow, \cdot, ^{-1}, 1)$, where $^{-1}$, $\cdot$ are expressed in terms of $\to$, $\rightsquigarrow$, 1 by (1.36), (1.40).*

**Remark 1.5.32** *Let $(G, \to, \rightsquigarrow, 1)$ be an i-group. We have the special properties: for all $x, y \in G$,*

$$x \to y = (x \cdot y^{-1})^{-1}, \quad x \rightsquigarrow y = (y^{-1} \cdot x)^{-1}, \tag{1.41}$$

$$x \to y = y \cdot x^{-1}, \quad x \rightsquigarrow y = x^{-1} \cdot y. \tag{1.42}$$

*Indeed, $(x \cdot y^{-1})^{-1} \overset{(1.40)}{=} ((x \to (y^{-1})^{-1})^{-1})^{-1} \overset{(DN)}{=} x \to y$ and $(y^{-1} \cdot x)^{-1} = ((x \rightsquigarrow (y^{-1})^{-1})^{-1})^{-1} = x \rightsquigarrow y$. Thus, (3.6) holds.*

*Also,* $y \cdot x^{-1} \overset{(1.40)}{=} (y \to (x^{-1})^{-1})^{-1} \overset{(DN)}{=} (y \to x)^{-1} \overset{(pNg)}{=} x \to y$ *and* $x^{-1} \cdot y \overset{(1.40)}{=} (y \leadsto (x^{-1})^{-1})^{-1} \overset{(DN)}{=} (y \leadsto x)^{-1} \overset{(pNg)}{=} x \leadsto y$. *Thus, (1.42) holds.*

**Theorem 1.5.33** *Let* $(G, \to, \leadsto, 1)$ *be an i-group and consider the operation* $\cdot$ *defined by (1.40). Then, the property* $(pGa^=)$ *holds, where for all* $x, y, z \in G$,

$(pGa^=)$ $\qquad x = y \to z \Longleftrightarrow y = x \leadsto z \Longleftrightarrow x \cdot y = z.$

**Proof.** The equivalence $x = y \to z \Leftrightarrow y = x \leadsto z$ is just the property (pEq#$^=$), that holds by Proposition 1.5.7. We shall prove that $x \cdot y = z \Longleftrightarrow x = y \to z$. Indeed, $x \cdot y = z \overset{(1.39)}{\Longleftrightarrow} x^{-1} \leadsto y = z \overset{(pEq\#^=)}{\Longleftrightarrow} z \to y = x^{-1} \Longleftrightarrow y \to z = x$: indeed, if $z \to y = x^{-1}$, then $(z \to y)^{-1} = (x^{-1})^{-1}$, i.e. $y \to z = x$, by (pNg), (DN), and if $y \to z = x$, then $(y \to z)^{-1} = x^{-1}$, i.e. $z \to y = x^{-1}$. □

**Remark 1.5.34**

$$(pGa^=) \Longrightarrow (pEq\#^=).$$

The following result, by its Corollary, will be useful in Part III to prove the equivalence between i-groups and groups.

**Proposition 1.5.35** *Let* $(G, \to, \leadsto, 1)$ *be an i-group. Then, the following properties hold: for all* $x, y, z \in G$,

(Pass) $\quad x \cdot (y \cdot z) = (x \cdot y) \cdot z$,
(PU) $\quad x \cdot 1 = x = 1 \cdot x$,
(pIv) $\quad x \cdot x^{-1} = 1 = x^{-1} \cdot x$,
($\alpha$) $\quad (x \to y) \cdot x = y = x \cdot (x \leadsto y)$,
($\beta$) $\quad x \to (y \cdot x) = y = x \leadsto (x \cdot y)$.

**Proof.** (Pass): $x \cdot (y \cdot z) \overset{(1.39)}{=} x \cdot (z^{-1} \to y) = x^{-1} \leadsto (z^{-1} \to y)$ and $(x \cdot y) \cdot z = (x^{-1} \leadsto y) \cdot z = z^{-1} \to (x^{-1} \leadsto y) \overset{(pEx)}{=} x^{-1} \leadsto (z^{-1} \to y)$; thus, (Pass) holds.

(PU): $x \cdot 1 \overset{(1.39)}{=} 1^{-1} \to x \overset{(N1)}{=} 1 \to x \overset{(pM)}{=} x$ and $1 \cdot x = 1^{-1} \leadsto x = 1 \leadsto x = x$.

(pIv): $x \cdot x^{-1} \overset{(1.39)}{=} x^{-1} \leadsto x^{-1} \overset{(pRe)}{=} 1$ and $x^{-1} \cdot x = x^{-1} \to x^{-1} = 1$.

($\alpha$): $(x \to y) \cdot x \overset{(1.39)}{=} (x \to y)^{-1} \leadsto x \overset{(pNg)}{=} (y \to x) \leadsto x \overset{(pD=)}{=} y$ and $x \cdot (x \leadsto y) \overset{(1.39)}{=} (x \leadsto y)^{-1} \to x \overset{(pNg)}{=} (y \leadsto x) \to x = y$. Thus, ($\alpha$) holds.

($\beta$): $x \to (y \cdot x) \overset{(1.42)}{=} (y \cdot x) \cdot x^{-1} \overset{(Pass)}{=} y \cdot (x \cdot x^{-1}) \overset{(pIv)}{=} y \cdot 1 \overset{(PU)}{=} y$ and $x \leadsto (x \cdot y) = x^{-1} \cdot (x \cdot y) = (x^{-1} \cdot x) \cdot y = 1 \cdot y = y$. Thus, ($\beta$) holds. □

**Corollary 1.5.36** *Let* $(G, \to, \leadsto, 1)$ *be an i-group. Then,* $(G, \cdot, ^{-1}, 1)$ *is a group.*

**Proof.** By Proposition 1.5.35. □

#### 1.5.1.2 The X-implicative-group - an intermediary notion

**Definition 1.5.37** (Definition 1)
An *X-implicative-group*, or an *X-i-group* for short, is an algebra

$$(G, \to, \rightsquigarrow, \cdot, 1)$$

of type $(2, 2, 2, 0)$ such that (pEx), (pM) hold and the pseudo-implication $(\to, \rightsquigarrow)$ and the operation $\cdot$ verify the following property: for all $x, y, z \in G$,
(pGa$^=$) $\qquad x = y \to z \Longleftrightarrow y = x \rightsquigarrow z \Longleftrightarrow x \cdot y = z.$

Denote by **X-i-group** the class of all X-i-groups.

First, we present the following general lemma:

**Lemma 1.5.38** *([116], Lemma 7.1.21) (See Lemma 1.4.17)*
*Let $\mathcal{A} = (A, \to, \rightsquigarrow, \cdot)$ (or $\mathcal{A} = (A, \cdot, \to, \rightsquigarrow)$) be an algebra of type $(2, 2, 2)$. Then, we have the following equivalence:*

$$(pGa^=) \iff (\alpha) + (\beta),$$

*where: for all $x, y \in A$,*

$$(\alpha) \qquad (x \to y) \cdot x = y = x \cdot (x \rightsquigarrow y),$$

$$(\beta) \qquad x \to (y \cdot x) = y = x \rightsquigarrow (x \cdot y).$$

**Proof.** (pGa$^=$) $\Longrightarrow$ $(\alpha) + (\beta)$:
$(\alpha)$: $(x \to y) \cdot x = y \overset{(pGa^=)}{\Longleftrightarrow} x \to y = x \to y$ and $x \cdot (x \rightsquigarrow y) = y \overset{(pGa^=)}{\Longleftrightarrow} x \rightsquigarrow y = x \rightsquigarrow y$, which are true.
$(\beta)$: $y = x \to (y \cdot x) \overset{(pGa^=)}{\Longleftrightarrow} y \cdot x = y \cdot x$ and $y = x \rightsquigarrow (x \cdot y) \overset{(pGa^=)}{\Longleftrightarrow} x \cdot y = x \cdot y$, which are true.
$(\alpha) + (\beta) \Longrightarrow$ (pGa$^=$):
Suppose $x \cdot y = z$; then $y \to z = y \to (x \cdot y) \overset{(\beta)}{=} x$ and $x \rightsquigarrow z = x \rightsquigarrow (x \cdot y) \overset{(\beta)}{=} y$.
Conversely, suppose $x = y \to z$; then $x \cdot y = (y \to z) \cdot y \overset{(\alpha)}{=} z$; suppose $y = x \rightsquigarrow z$; then $x \cdot y = x \cdot (x \rightsquigarrow z) \overset{(\alpha)}{=} z$. $\square$

**Remark 1.5.39** *The property (pGa$^=$) is the particular non-commutative Galois connection (pRP') (and also (pcoRS')) (therefore we called it (pGa$^=$)), where we have $=$ instead of $\leq$ (or $\geq$), i.e. (p-s) holds; also, $(\alpha)$ is (pRR') and $(\beta)$ is (pPP') (see Lemma 1.4.17). We have:*
*(pGa$^=$) $\Longrightarrow$ (pEq#$^=$); (pGa$^=$) $\Longrightarrow$ (pRP') + (pcoRS'),*
*(pRP') + (p-s) $\Longrightarrow$ (pGa$^=$), (pcoRS') + (p-s$^R$) $\Longrightarrow$ (pGa$^=$).*

**Proposition 1.5.40** *Let $(G, \to, \rightsquigarrow, \cdot, 1)$ be an X-i-group. Then,*

$$(pGa^=) \iff (\alpha) + (\beta)$$

*and the properties $(\alpha)$ and $(\beta)$ hold.*

**Proof.** By Lemma 1.5.38, $(pGa^{=}) \iff (\alpha) + (\beta)$. Since $(pGa^{=})$ holds, then $(\alpha)$ and $(\beta)$ hold too. □

**Proposition 1.5.41** *Let* $(G, \rightarrow, \rightsquigarrow, \cdot, 1)$ *be an X-i-group. Then we have, for all* $x, y, z \in G$,

*(pD=)* $\quad y = (y \rightarrow x) \rightsquigarrow x, \quad y = (y \rightsquigarrow x) \rightarrow x,$

*(pBB=)* $\quad y \rightarrow z = (z \rightarrow x) \rightsquigarrow (y \rightarrow x), \quad y \rightsquigarrow z = (z \rightsquigarrow x) \rightarrow (y \rightsquigarrow x),$

*(PU)* $\quad x \cdot 1 = x = 1 \cdot x,$

*(Pass)* $\quad x \cdot (y \cdot z) = (x \cdot y) \cdot z,$

*(IdEqR)* $\quad x \rightarrow y = 1 \iff x \rightsquigarrow y = 1,$

*(pEqrelR=)* $\quad x = y \iff x \rightarrow y = 1 \stackrel{(IdEqR)}{\iff} x \rightsquigarrow y = 1.$

**Proof.** (pD=): $y = (y \rightarrow x) \rightsquigarrow x \stackrel{(pGa^{=})}{\iff} y \rightarrow x = y \rightarrow x$, which is true and $y = (y \rightsquigarrow x) \rightarrow x \stackrel{(pGa^{=})}{\iff} y \rightsquigarrow x = y \rightsquigarrow x$, which is also true; thus, (pD=) holds.

(pBB=): $(z \rightarrow x) \rightsquigarrow (y \rightarrow x) \stackrel{(pEx)}{=} y \rightarrow ((z \rightarrow x) \rightsquigarrow x) \stackrel{(pD=)}{=} y \rightarrow z$ and $(z \rightsquigarrow x) \rightarrow (y \rightsquigarrow x) \stackrel{(pEx)}{=} y \rightsquigarrow ((z \rightsquigarrow x) \rightarrow x) \stackrel{(pD=)}{=} y \rightsquigarrow z$; thus, (pBB=) holds.

(PU): $x \cdot 1 = x \stackrel{(pGa^{=})}{\iff} x = 1 \rightarrow x$ and $1 \cdot x = x \stackrel{(pGa^{=})}{\iff} x = 1 \rightsquigarrow x$, which are true by (pM).

(Pass): $x \cdot (y \cdot z) = a \stackrel{(pGa^{=})}{\iff} y \cdot z = x \rightsquigarrow a \stackrel{(pGa^{=})}{\iff} y = z \rightarrow (x \rightsquigarrow a) \stackrel{(pEx)}{=} x \rightsquigarrow (z \rightarrow a) \stackrel{(pGa^{=})}{\iff} x \cdot y = z \rightarrow a \stackrel{(pGa^{=})}{\iff} (x \cdot y) \cdot z = a$.

(IdEqR) + (pEqrelR=): $1 = x \rightarrow y \stackrel{(pGa^{=})}{\iff} 1 \cdot x = y \stackrel{(PU)}{\iff} x = y$ and $1 = x \rightsquigarrow y \stackrel{(pGa^{=})}{\iff} x \cdot 1 = y \stackrel{(PU)}{\iff} x = y$; thus, (IdEqR) and (pEqrelR=) hold. □

Note that, by Proposition 1.5.41, any X-i-group verifies (IdEqR) and (pEqrelR=). It follows that we can define equivalently the X-i-groups as follows.

**Definition 1.5.42** (Definition 2 of X-i-groups)
An *X-implicative-group*, or an *X-i-group* for short, is an algebra

$$(G, \rightarrow, \rightsquigarrow, \cdot, 1)$$

of type $(2, 2, 2, 0)$ such that (pEx), (pM), (IdEqR), (pEqrelR=) hold and the pseudo-implication $(\rightarrow, \rightsquigarrow)$ and the operation $\cdot$ verify the following property: for all $x, y, z \in G$,
$(\text{pGa}^{=}) \quad x = y \rightarrow z \iff y = x \rightsquigarrow z \iff x \cdot y = z.$

We then immediately obtain the following important result, by using Definition 4' of pBCI algebras.

**Corollary 1.5.43** *(See Proposition 1.5.9)*
*Any X-i-group is a pBCI(pRP') algebra.*

A special equivalent definition of X-i-groups coming from QB algebras with (pRP') is obtained from the following Proposition.

**Proposition 1.5.44**
*(1) If $(G, \to, \rightsquigarrow, \cdot, 1)$ is an X-i-group, then the reduct $(X, \to, \rightsquigarrow, \cdot)$ verifies (pEx) and (pGa$^=$).*
*(2) Conversely, if an algebra $(G, \to, \rightsquigarrow, \cdot)$ of type $(2, 2, 2)$ verifies (pEx) and (pGa$^=$), then there exists $1 \in G$ such that $(G, \to, \rightsquigarrow, \cdot, 1)$ is an X-i-group.*

**Proof.** (1): Obviously, by Definition 1.
(2): Let $(G, \to, \rightsquigarrow, \cdot)$ be an algebra verifying (pEx) and (pGa$^=$); then, by Remark 1.5.39, (pGa$^=$) $\Longrightarrow$ (pEq#$^=$);
by (pA#1=), (pEq#$^=$) $\Longrightarrow$ (pD=);
by Proposition 1.5.1 (11), (pEx) + (pD=) $\Longrightarrow$ (pBB=);
by (pA#3=), (pEq#$^=$) + (pD=) + (pBB=) $\Longrightarrow$ $\exists 1 \in G$ such that (pRe) and (pM) hold. Then, the algebra $(G, \to, \rightsquigarrow, \cdot, 1)$ verifies (pEx), (pM) and (pGa$^=$), i.e. is an X-i-group. □

We then obtain:

**Definition 1.5.45** (Definition 3 of X-i-groups)
An *X-implicative-group*, or an *X-i-group* for short, is an algebra

$$(G, \to, \rightsquigarrow, \cdot)$$

of type $(2, 2, 2)$ such that (pEx) holds and the pseudo-implication $(\to, \rightsquigarrow)$ and the operation $\cdot$ verify the following property: for all $x, y, z \in G$,
(pGa$^=$) $\quad x = y \to z \Longleftrightarrow y = x \rightsquigarrow z \Longleftrightarrow x \cdot y = z.$

**Remark 1.5.46** *If $(G, \to, \rightsquigarrow, \cdot, 1)$ is an X-i-group, then $(G, \cdot, \to, \rightsquigarrow, 1)$ is an X-group, since (Pass), (PU) hold.*

### 1.5.1.3 The basic equivalence (EI1)

Here we present the equivalence (EI1) between the i-groups and the X-i-groups. We prove that the i-groups are termwise equivalent to the X-i-groups:

**Theorem 1.5.47**
*(1) Let $\mathcal{G} = (G, \to, \rightsquigarrow, 1)$ be an i-group.*
*Define $\pi(\mathcal{G}) \stackrel{def.}{=} (G, \to, \rightsquigarrow, \cdot, 1)$ by: $x \cdot y \stackrel{def.}{=} (x \to y^{-1})^{-1} \stackrel{(pNg1)}{=} (y \rightsquigarrow x^{-1})^{-1}$, where $x^{-1} \stackrel{def.}{=} x \to 1 \stackrel{(Id)}{=} x \rightsquigarrow 1$.*
*Then, $\pi(\mathcal{G})$ is an X-i-group.*
*(1') Conversely, let $\mathcal{G} = (G, \to, \rightsquigarrow, \cdot, 1)$ be an X-i-group.*
*Define $\pi^*(\mathcal{G}) \stackrel{def.}{=} (G, \to, \rightsquigarrow, 1)$.*
*Then, $\pi^*(\mathcal{G})$ is an i-group.*
*(2) The above defined mappings are mutually inverse.*

**Proof.** (1): By Definition 1, (pM) holds, by Proposition 1.5.7, (pEx) holds and by Theorem 1.5.33, (pGa$^=$) holds.
(1'): By Proposition 1.5.41, (pBB=) holds.

(2): If $(G, \to, \leadsto, 1) \stackrel{\pi}{\longrightarrow} (G, \to, \leadsto, \cdot, 1) \stackrel{\pi^*}{\longrightarrow} (G, \to, \leadsto, 1)$, then there is nothing to prove.
If $(G, \to, \leadsto, \cdot, 1) \stackrel{\pi^*}{\longrightarrow} (G, \to, \leadsto, 1) \stackrel{\pi}{\longrightarrow} (G, \to, \leadsto, \odot, 1)$, then we have to prove that $x \odot y = x \cdot y$, for all $x, y \in G$. Indeed,
$x \odot y = (x \to y^{-1})^{-1} = [x \to (y \to 1)] \leadsto 1 = x \cdot y \stackrel{(pGa^=)}{\Longleftrightarrow}$
$x = y \to ([x \to (y \to 1)] \leadsto 1)$, which is true, since
$y \to ([x \to (y \to 1)] \leadsto 1) = [x \to (y \to 1)] \leadsto (y \to 1) = x$, by (pEx) and (pD=). □

Hence, we have the basic equivalence:

(EI1) **i-group** $\Longleftrightarrow$ **X-i-group**.

## 1.5.2 Po-implicative-groups

### 1.5.2.1 The po-implicative-group

**Definition 1.5.48** A *partially-ordered implicative-group*, or a *po-implicative-group* or even a *po-i-group* for short, is a structure

$$\mathcal{G} = (G, \leq, \to, \leadsto, 1),$$

such that $(G, \to, \leadsto, 1)$ is an i-group, $(G, \leq)$ is a poset and $\leq$ is compatible with $\to, \leadsto$, i.e. we have: for all $x, y, z \in G$,
(p*') $x \leq y$ implies $z \to x \leq z \to y$ and $z \leadsto x \leq z \leadsto y$.

Note that in a po-i-group we can define a negation $^{-1}$ by (1.36) and a new operation $\cdot$ by (1.39).

**Remark 1.5.49** *Note that a po-i-group may be defined equivalently as a structure*

$$\mathcal{G} = (G, \leq, \to, \leadsto, \cdot, ^{-1}, 1),$$

*by Remark 1.5.31.*

The presence of the order relation implies the presence of the *Duality Principle*. It follows that there are two dual po-i-groups. If their support sets coincide ($G = G_1 = G_2$), we say that they are *self-dual*, i.e. $(G, \leq, \to, \leadsto, 1)$ is in the same time left-po-i-group and right-po-i-group; if their support sets differ ($G_1 \neq G_2$), then their unit elements differ and say that $1_1 \leq 1_2$ in the union set $G_1 \cup G_2$; we then call $\mathcal{G}_1$ as the *left-po-i-group* and $\mathcal{G}_2$ as the *right-po-i-group*.

Denote by **po-i-group** the class of all po-i-groups, namely:
- by **po-i-group** the class of all left-po-i-groups and
- by **po-i-group**$^R$ the class of all right-po-i-groups.

**Remark 1.5.50** *While a left-po-group is a left-po-m, a left-po-i-group is a left-po-pr and not a left-pBCI algebra, because the property (pEqrelR) is not verified.*

If the partial order relation $\leq$ is linear (total), then $\mathcal{G}$ is a linearly-ordered po-i-group.

**Proposition 1.5.51** *Let $(G, \leq, \rightarrow, \rightsquigarrow, 1)$ be a po-i-group. Then, the following properties hold: for all $x, y, a \in G$,*

(Neg2') $\quad x \leq y \Longrightarrow y^{-1} \leq x^{-1}$,
(pCp') $\quad x \leq y \Longrightarrow (a \cdot x \leq a \cdot y$ and $x \cdot a \leq y \cdot a)$,
(pGa$^=$) $\quad x \cdot y = z \Longleftrightarrow x = y \rightarrow z \Longleftrightarrow y = x \rightsquigarrow z$,
($\alpha$) $\quad (x \rightarrow y) \cdot x = y = x \cdot (x \rightsquigarrow y)$,
($\beta$) $\quad x \rightarrow (y \cdot x) = y = x \rightsquigarrow (x \cdot y)$;
(p*Eq) $\quad x \leq y \Longleftrightarrow z \rightarrow x \leq z \rightarrow y \Longleftrightarrow z \rightsquigarrow x \leq z \rightsquigarrow y$,
(pEq#$^{\leq}$) $\quad x \leq y \rightarrow z \Longleftrightarrow y \leq x \rightsquigarrow z$,
(pEq#$^{\geq}$) $\quad x \geq y \rightarrow z \Longleftrightarrow y \geq x \rightsquigarrow z$.

**Proof.** (Neg2'): Let $x \leq y$; then, by (p*'), (pRe), we have $1 = x \rightarrow x \leq x \rightarrow y$; then, by (p*') again and by (pEx), (pRe), we obtain $y^{-1} = y \rightsquigarrow 1 \leq y \rightsquigarrow (x \rightarrow y) = x \rightarrow (y \rightsquigarrow y) = x \rightarrow 1 = x^{-1}$.

(pCp'): Let $x \leq y$; then $y^{-1} \leq x^{-1}$, by (Neg2'); by (p*'), $a \rightarrow (y^{-1}) \leq a \rightarrow (x^{-1})$, hence by (Neg2')again, we obtain $a \cdot x = (a \rightarrow x^{-1})^{-1} \leq (a \rightarrow y^{-1})^{-1} = a \cdot y$; by (p*'), $a \rightsquigarrow (y^{-1}) \leq a \rightsquigarrow (x^{-1})$, hence by (Neg2')again, we obtain $x \cdot a = (a \rightsquigarrow x^{-1})^{-1} \leq (a \rightsquigarrow y^{-1})^{-1} = y \cdot a$.

(pGa$^=$): By Theorem 1.5.33.

($\alpha$) and ($\beta$): By Lemma 1.5.38, since (pGa$^=$) holds, it follows that ($\alpha$) and ($\beta$) hold.

(p*Eq): By (p*'), it is sufficient to prove that $z \rightarrow x \leq z \rightarrow y$ implies $x \leq y$ and that $z \rightsquigarrow x \leq z \rightsquigarrow y$ implies $x \leq y$. Indeed, $z \rightarrow x \leq z \rightarrow y$ implies, by above (pCp'), that $x \overset{(\alpha)}{=} (z \rightarrow x) \cdot z \leq (z \rightarrow y) \cdot z \overset{(\alpha)}{=} y$ and $z \rightsquigarrow x \leq z \rightsquigarrow y$ implies, by above (pCp'), that $x \overset{(\alpha)}{=} z \cdot (z \rightsquigarrow x) \leq z \cdot (z \rightsquigarrow y) \overset{(\alpha)}{=} y$.

(pEq#$^{\leq}$): If $x \leq y \rightarrow z$, then, by (Neg2'), $z \rightarrow y \overset{(pNg)}{=} (y \rightarrow z)^{-1} \leq x^{-1}$; then, by (pCp'), $y \overset{(\alpha)}{=} (z \rightarrow y) \cdot z \leq x^{-1} \cdot z \overset{(1.39)}{=} ((x^{-1})^{-1}) \rightsquigarrow z \overset{(DN)}{=} x \rightsquigarrow z$.
Similarly, if $y \leq x \rightsquigarrow z$, then, by (Neg2'), $z \rightsquigarrow x \overset{(pNg)}{=} (x \rightsquigarrow z)^{-1} \leq y^{-1}$; then, by (pCp'), $x \overset{(\alpha)}{=} z \cdot (z \rightsquigarrow x) \leq z \cdot (y^{-1}) \overset{(1.39)}{=} ((y^{-1})^{-1}) \rightarrow z \overset{(DN)}{=} y \rightarrow z$.

(pEq#$^{\geq}$): dually. □

**Remark 1.5.52** *(See Remark 1.5.16)*

*Since (pEq#$^{\leq}$) is just (pEq#'), then any po-i-group is a po-pr and we have:*

$$\textbf{po-i-group} = \textbf{po-pr} + (pRe) + (IdEqR) + (pEqrelR{=}).$$

**Theorem 1.5.53** *Let $(G, \leq, \rightarrow, \rightsquigarrow, 1)$ be a po-i-group. Then the properties (pGa$^{\leq}$) and (pGa$^{\geq}$) hold, where for all $x, y, z \in G$,*

(pGa$^{\leq}$) $\quad x \leq y \rightarrow z \Longleftrightarrow y \leq x \rightsquigarrow z \Longleftrightarrow x \cdot y \leq z$,
(pGa$^{\geq}$) $\quad x \geq y \rightarrow z \Longleftrightarrow y \geq x \rightsquigarrow z \Longleftrightarrow x \cdot y \geq z$

*and $\cdot$ is the operation defined by (1.40), via (1.36).*

**Proof.** (pGa$^{\leq}$): The equivalence $x \leq y \rightarrow z \Longleftrightarrow y \leq x \rightsquigarrow z$ is just (pEq#$^{\leq}$) from above Proposition 1.5.51. We prove now that $x \cdot y \leq z \Longleftrightarrow x \leq y \rightarrow z$. Indeed,

if $x \cdot y \leq z$, then, by (pCp'), $x \stackrel{(PU)}{=} x \cdot 1 \stackrel{(pIv)}{=} x \cdot (y \cdot (y^{-1})) \stackrel{(Pass)}{=} (x \cdot y) \cdot (y^{-1}) \leq z \cdot (y^{-1}) \stackrel{(1.39)}{=} ((y^{-1})^{-1}) \to z \stackrel{(DN)}{=} y \to z$.
Conversely, if $x \leq y \to z$, then, by (pCp'), $x \cdot y \leq (y \to z) \cdot y \stackrel{(\alpha)}{=} z$.
(pGa$^{\geq}$): dually. □

**Remark 1.5.54** *Note that:*
*- (pGa$^{\leq}$) is the analogous of the (pRP')=(pPR'),*
*- (pGa$^{\geq}$) is the analogous of the (pcoRS')=(pScoR'),*
*and that both are non-commutative Galois connections.*

**Corollary 1.5.55** *Let $(G, \leq, \to, \rightsquigarrow, 1)$ be a po-i-group. Then, $(G, \leq, \cdot, ^{-1}, 1)$ is a po-group.*

**Proof.** By Corollary 1.5.36 and Proposition 1.5.51. □

### 1.5.2.2 The negative and the positive cones of po-i-groups

Let $\mathcal{G} = (G, \leq, \to, \rightsquigarrow, 1)$ be a po-i-group (left-po-i-group or right-po-i-group).
Define the *negative cone* of $\mathcal{G}$ as follows: $G^- \stackrel{def.}{=} \{x \in G \mid x \leq 1\}$.
Define the *positive cone* of $\mathcal{G}$ as follows: $G^+ \stackrel{def.}{=} \{x \in G \mid x \geq 1\}$.
Then, we have:

**Lemma 1.5.56** *Let $\mathcal{G} = (G, \leq, \to, \rightsquigarrow, 1)$ be a po-i-group.*
*(1) For all $x \in G$, we have:*
*- if $x \in G^-$, then $x^{-1} \in G^+$ and*
*- if $x \in G^+$, then $x^{-1} \in G^-$.*
*(2) $G^-$, $G^+$ are not closed under $\to$, $\rightsquigarrow$.*

**Proof.** (1): Let $x \in G^-$, i.e. $x \leq 1$; then, $x \to x \leq x \to 1$ and $x \rightsquigarrow x \leq x \rightsquigarrow 1$, by (p*'); hence, $1 \leq x \to 1$ and $1 \leq x \rightsquigarrow 1$, by (pRe), hence $x^{-1} = x \to 1 = x \rightsquigarrow 1 \in G^+$. Similarly, if $x \in G^+$, then $x^{-1} = x \to 1 = x \rightsquigarrow 1 \in G^-$.

(2): Let $x, y \in G^-$, i.e. $x \leq 1$ and $y \leq 1$; then, $x \to y \stackrel{(p*')}{\leq} x \to 1$ and $x \rightsquigarrow y \stackrel{(p*')}{\leq} x \rightsquigarrow 1$; but, $x \to 1$, $x \rightsquigarrow 1 \in G^+$, by above (1); hence, $G^-$ is not closed under $\to, \rightsquigarrow$. Similarly, $G^+$ is not closed under $\to$, $\rightsquigarrow$. □

### 1.5.2.3 The X-po-implicative-group - an intermediary notion

**Definition 1.5.57** We shall name *X-po-implicative-group*, or *X-po-i-group* for short, a structure

$$(G, \leq, \to, \rightsquigarrow, \cdot, 1)$$

such that the following properties (pM) and (pEx) hold: for all $x, y, z \in G$,
(pM) $\quad 1 \to x = x = 1 \rightsquigarrow x$,
(pEx) $\quad x \to (y \rightsquigarrow z) = y \rightsquigarrow (x \to z)$,

$(G, \leq)$ is a poset and the pseudo-implication $(\rightarrow, \rightsquigarrow)$ and $\cdot$ verify the two properties: for all $x, y, z \in G$,
(pGa$^{\leq}$) $\quad x \leq y \rightarrow z \Longleftrightarrow y \leq x \rightsquigarrow z \Longleftrightarrow x \cdot y \leq z,$
(pGa$^{\geq}$) $\quad x \geq y \rightarrow z \Longleftrightarrow y \geq x \rightsquigarrow z \Longleftrightarrow x \cdot y \geq z.$

Denote by **X-po-i-group** the class of all X-po-i-groups, namely:
- by **X-po-i-group** the class of all left-X-po-i-groups and
- by **X-po-i-group**$^R$ the class of all right-X-po-i-groups.

First, we give some general results as a general lemma:

**Lemma 1.5.58** *([116], Lemma 7.2.17) (see Lemma 1.5.38 and Lemma 1.4.17)*
*Let $\mathcal{A} = (A, \leq, \rightarrow, \rightsquigarrow, \cdot)$ (or $\mathcal{A} = (A, \leq, \cdot, \rightarrow, \rightsquigarrow)$) be a structure such that:*
*(a) $(A, \leq)$ is a poset;*
*(b) the properties (pGa$^{\leq}$) and (pGa$^{\geq}$) hold.*
*Then, we have:*
*(i) The following properties hold: for all $x, y, z \in A$,*
*(pGa$^{=}$)* $\quad x \cdot y = z \Longleftrightarrow x = y \rightarrow z \Longleftrightarrow y = x \rightsquigarrow z,$
*($\alpha$)* $\quad (x \rightarrow y) \cdot x = y = x \cdot (x \rightsquigarrow y),$
*($\beta$)* $\quad x \rightarrow (y \cdot x) = y = x \rightsquigarrow (x \cdot y).$
*(pEqCp')* $\quad x \leq y \Longleftrightarrow z \cdot x \leq z \cdot y \Longleftrightarrow x \cdot z \leq y \cdot z,$
*(p*')* $\quad x \leq y \Longrightarrow (z \rightarrow x \leq z \rightarrow y,\ z \rightsquigarrow x \leq z \rightsquigarrow y),$
*(p**')* $\quad x \leq y \Longrightarrow (y \rightarrow z \leq x \rightarrow z,\ y \rightsquigarrow z \leq x \rightsquigarrow z).$
*(ii) The following equivalence holds:*

$$(pGa^{\leq}) + (pGa^{\geq}) \Longleftrightarrow (pGa^{=}) + (pCp') + (p*').$$

**Proof.** (i): (pGa$^{=}$): The properties (pGa$^{\leq}$) and (pGa$^{\geq}$) imply the property (pGa$^{=}$) since: $\quad$ ($p \leftrightarrow q$ and $r \leftrightarrow s$) imply ($p$ and $r$)$\leftrightarrow$($q$ and $s$)
and since the order relation $\leq$ is antisymmetrique.

($\alpha$) and ($\beta$): By Lemma 1.5.38, since (pGa$^{=}$) holds, it follows that ($\alpha$) and ($\beta$) hold too.

(pEqCp'): Suppose $x \leq y$; by ($\beta$), $y = z \rightsquigarrow (z \cdot y)$, hence $x \leq z \rightsquigarrow (z \cdot y)$ and $x \leq z \rightsquigarrow (z \cdot y) \overset{(pGa^{\leq})}{\Longleftrightarrow} z \cdot x \leq z \cdot y$, i.e. the first part of (pCp') holds. Conversely, $z \cdot x \leq z \cdot y \overset{(pGa^{\leq})}{\Longleftrightarrow} x \leq z \rightsquigarrow (z \cdot y) \overset{(\beta)}{=} y$.
Suppose again $x \leq y$; by ($\beta$), $y = z \rightarrow (y \cdot z)$, hence $x \leq z \rightarrow (y \cdot z)$ and $x \leq z \rightarrow (y \cdot z) \overset{(pGa^{\leq})}{\Longleftrightarrow} x \cdot z \leq y \cdot z$; thus (pCp') holds. Conversely, $x \cdot z \leq y \cdot z \overset{(pGa^{\leq})}{\Longleftrightarrow}$ $x \leq z \rightarrow (y \cdot z) \overset{(\beta)}{=} y$.

(p*'): By Lemma 1.4.17, since (pGa$^{\leq}$) holds.

(p**'): By Lemma 1.4.17, since (pGa$^{\leq}$) holds.

(ii): By (i), it remains to prove that (pCp') + (p*') + (pGa$^{=}$) imply (pGa$^{\leq}$) + (pGa$^{\geq}$). Indeed, (pCp') + (p*') + (pGa$^{=}$) imply (pGa$^{\leq}$):
- Suppose $x \cdot y \leq z$; then by (p*'), $y \rightarrow (x \cdot y) \leq y \rightarrow z$ and $x \rightsquigarrow (x \cdot y) \leq x \rightsquigarrow z$; hence, by ($\beta$), we obtain that $x \leq y \rightarrow z$ and $y \leq x \rightsquigarrow z$.
- Conversely, suppose $x \leq y \rightarrow z$; then by (pCp'), $x \cdot y \leq (y \rightarrow z) \cdot y \overset{(\alpha)}{=} z$; suppose

$y \leq x \rightsquigarrow z$; then by (pCp'), $x \cdot y \leq x \cdot (x \rightsquigarrow z) \overset{(\alpha)}{=} z$.
Similarly, (pCp') + (p*') + (pGa$^=$) imply (pGa$^\geq$). □

**Corollary 1.5.59** *Let $(G, \leq, \rightarrow, \rightsquigarrow, \cdot, 1)$ be an X-po-i-group. Then, $(G, \rightarrow, \rightsquigarrow, \cdot, 1)$ is an X-i-group.*

**Proof.** Obviously, by Lemma 1.5.58. □

**Proposition 1.5.60** *Let $(G, \leq, \rightarrow, \rightsquigarrow, \cdot, 1)$ be an X-po-i-group. Then, the properties (pGa$^=$), ($\alpha$), ($\beta$), (pEqCp'), (p*'), (p**') hold.*

**Proof.** By Lemma 1.5.58. □

**Proposition 1.5.61** *Let $(G, \leq, \rightarrow, \rightsquigarrow, \cdot, 1)$ be an X-po-i-group. Then, the properties (Pass), (PU), (pD=), (pBB=), (Id), (IdEqR), (pEqrelR=) hold.*

**Proof.** By Corollary 1.5.59, $(G, \rightarrow, \rightsquigarrow, \cdot, 1)$ is an X-i-group, hence, by Proposition 1.5.41, (Pass), (PU), (pD=), (pBB=), (Id), (IdEqR), (pEqrelR=) hold. □

### 1.5.2.4 The equivalences (EI2)

Here we present the equivalences (EI2) between the po-i-groups and the X-po-i-groups. We prove that the po-i-groups are termwise equivalent to the X-po-i-groups.

**Theorem 1.5.62** *(See Theorem 1.5.47)*

*(1) Let $\mathcal{G} = (G, \leq, \rightarrow, \rightsquigarrow, 1)$ be a po-i-group.*
*Define $\pi'(\mathcal{G}) \overset{def.}{=} (G, \leq, \rightarrow, \rightsquigarrow, \cdot, 1)$, where $(G, \rightarrow, \rightsquigarrow, \cdot, 1) = \pi(G, \rightarrow, \rightsquigarrow, 1)$ from Theorem 1.5.47 (1).*

*Then, $\pi'(\mathcal{G})$ is an X-po-i-group.*

*(1') Conversely, let $\mathcal{G} = (G, \leq, \rightarrow, \rightsquigarrow, \cdot, 1)$ be an X-po-i-group.*
*Define $\pi^{*'}(\mathcal{G}) \overset{def.}{=} (G, \leq, \rightarrow, \rightsquigarrow, 1)$.*

*Then, $\pi^{*'}(\mathcal{G})$ is a po-i-group.*

*(2) The above defined mappings $\pi'$ and $\pi^{*'}$ are mutually inverse.*

**Proof.** (1): By Theorems 1.5.47 (1) and 1.5.53.

(1'): By Corollary 1.5.59, $(G, \rightarrow, \rightsquigarrow, \cdot, 1)$ is an X-i-group and then, by Theorem 1.5.47 (1'), $(G, \rightarrow, \rightsquigarrow, 1)$ is an i-group. It remains to prove that (p*) holds, which follows by above Proposition 4.2.14.

(2) follows by Theorem 1.5.47 (2). □

Hence, we have the equivalences (EI2):

(EI2) **po-i-group** $\Longleftrightarrow$ **X-po-i-group** and, dually,
**po-i-group**$^R$ $\Longleftrightarrow$ **X-po-i-group**$^R$.

### 1.5.3 $l$-implicative-groups

#### 1.5.3.1 The $l$-implicative-group

**Definition 1.5.63** Let $\mathcal{G} = (G, \leq, \rightarrow, \rightsquigarrow, 1)$ be a po-i-group. If the partial order relation $\leq$ is a lattice order relation, with the lattice operations $\wedge$ and $\vee$ ($x \leq y \Leftrightarrow x \wedge y = x \Leftrightarrow x \vee y = y$), then $\mathcal{G}$ is a *lattice-ordered implicative-group*, or an *l-implicative-group* or even an *l-i-group* for short, denoted by:

$$\mathcal{G} = (G, \vee, \wedge, \rightarrow, \rightsquigarrow, 1).$$

Denote by **$l$-i-group** the class of all $l$-i-groups, namely:
- by **$l$-i-group**, the class of all left-$l$-i-groups and
- by **$l$-i-group$^R$**, the class of all right-$l$-i-groups.

Note that an $l$-i-group may be linearly-ordered or not, while a linearly-ordered i-group is an $l$-i-group.

**Remark 1.5.64** *Note that an l-i-group may be defined equivalently as an algebra*

$$\mathcal{G} = (G, \wedge, \vee, \rightarrow, \rightsquigarrow, \cdot, ^{-1}, 1),$$

*by Remark 1.5.49.*

#### 1.5.3.2 The negative and the positive cones of $l$-i-groups

**Lemma 1.5.65** *Let $\mathcal{G} = (G, \wedge, \vee, \rightarrow, \rightsquigarrow, 1)$ be an l-i-group (left-l-i-group or right-l-i-group). Then, $G^-$ and $G^+$ are closed under $\wedge$ and $\vee$.*

**Proof.** See the proof of Lema 5.4.51. □

**Lemma 1.5.66** *(See Lemma 1.5.56)*

*Let $\mathcal{G} = (G, \wedge, \vee, \rightarrow, \rightsquigarrow, 1)$ be an l-i-group (left-l-i-group or right-l-i-group).*

*(1) Define, for all $x, y \in G^-$:*

$$x \rightarrow^L y \overset{def.}{=} (x \rightarrow y) \wedge 1, \quad x \rightsquigarrow^L y \overset{def.}{=} (x \rightsquigarrow y) \wedge 1. \tag{1.43}$$

*Then, we have:*
*(i) If $x \in G^-$, then $x \rightarrow^L 1 = 1$, $x \rightsquigarrow^L 1 = 1 \in G^-$;*
*(ii) $G^-$ is closed under $\rightarrow^L$, $\rightsquigarrow^L$ and we have, for all $x, y, z \in G^-$:*

$$x \leq y \rightarrow z \iff x \leq y \rightarrow^L z, \quad y \leq x \rightsquigarrow z \iff y \leq x \rightsquigarrow^L z, \tag{1.44}$$

$$x \leq y \rightarrow^L z \iff y \leq x \rightsquigarrow^L z. \tag{1.45}$$

*(1') Dually, define, for all $x, y \in G^+$:*

$$x \rightarrow^R y \overset{def.}{=} (x \rightarrow y) \vee 1, \quad x \rightsquigarrow^R y \overset{def.}{=} (x \rightsquigarrow y) \vee 1. \tag{1.46}$$

*Then, we have:*
*(i') If* $x \in G^+$*, then* $x \to^R 1 = 1$, $x \rightsquigarrow^R 1 = 1 \in G^+$*;*
*(ii')* $G^+$ *is closed under* $\to^R$, $\rightsquigarrow^R$ *and we have, for all* $x, y, z \in G^+$*:*

$$x \geq y \to z \iff x \geq y \to^R z, \quad y \geq x \rightsquigarrow z \iff y \geq x \rightsquigarrow^R z, \tag{1.47}$$

$$x \geq y \to^R z \iff y \geq x \rightsquigarrow^R z. \tag{1.48}$$

**Proof.** (1) (i): Let $x \in G^-$, i.e. $x \leq 1$; then, $1 = x \to x \leq x \to 1$ and $1 = x \rightsquigarrow x \leq x \rightsquigarrow 1$, by (p*), (pRe); hence, $x \to^L 1 \stackrel{(1.43)}{=} (x \to 1) \wedge 1 = 1$ and $x \rightsquigarrow^L 1 = (x \rightsquigarrow 1) \wedge 1 = 1$.

(ii): Let $x, y \in G^-$, i.e. $x \leq 1$ and $y \leq 1$. Then, $x \to y \leq x \to 1$ and $x \rightsquigarrow y \leq x \rightsquigarrow 1$, by (p*). Then, $x \to^L y \stackrel{(1.43)}{=} (x \to y) \wedge 1 \leq (x \to 1) \wedge 1 = 1$ and $x \rightsquigarrow^L y = (x \rightsquigarrow y) \wedge 1 \leq (x \rightsquigarrow 1) \wedge 1 = 1$; hence $x \to^L y \in G^-$ and $x \rightsquigarrow^L y \in G^-$.

(1.44): If $x \leq y \to z$ then, since $x \leq 1$, we obtain $x \leq (y \to z) \wedge 1 = y \to^L z$, i.e. $x \leq y \to^L z$; conversely, if $x \leq y \to^L z$ then, since $y \to^L z = (y \to z) \wedge 1 \leq y \to z$, we obtain $x \leq y \to z$; thus, $x \leq y \to z \iff x \leq y \to^L z$. The second part has a similar proof. Thus, (1.44) holds.

(1.45): By Theorem 1.5.53, the property (pGa$^{\leq}$) holds; then apply (1.44).

(1'): dually. □

### 1.5.3.3 The X-$l$-implicative-group - an intermediary notion

**Definition 1.5.67** Let $\mathcal{G} = (G, \leq, \to, \rightsquigarrow, \cdot, 1)$ be an X-po-i-group. If the partial order $\leq$ is a lattice order, then the X-po-i-group $\mathcal{G}$ is called an *X-l-i-group* and is denoted by:

$$\mathcal{G} = (G, \wedge, \vee, \to, \rightsquigarrow, \cdot, 1).$$

Denote by **X-$l$-i-group** the class of all X-$l$-i-groups, namely:
- by **X-$l$-i-group**, the class of all left-X-$l$-i-groups and
- by **X-$l$-i-group$^R$**, the class of all right-X-$l$-i-groups.

- **The equivalences (EI3)**

Here we present the equivalences (EI3) between the $l$-i-groups and the X-$l$-i-groups. We obtain obviously, by (EI2), the equivalences (EI3):

(EI3) **$l$-i-group** $\iff$ **X-$l$-i-group** and, dually,
**$l$-i-group$^R$** $\iff$ **X-$l$-i-group$^R$**,

i.e. the $l$-i-groups are term-equivalent to the X-$l$-i-groups.

## 1.5.4 Examples

### 1.5.4.1 Commutative finite examples

• ([116], Chapter 18) There exists one commutative implicative-group with one element, called the *trivial* implicative-group: $\mathcal{G}_{1igr} = (A_1 = \{1\}, \rightarrow, 1)$, where the table (matrix) of $\rightarrow$ has the following form:

$\mathcal{G}_{1igr}$

| $\rightarrow$ | 1 |
|---|---|
| 1 | 1 |

, hence

| $x$ | $x^{-1} = x \rightarrow 1$ |
|---|---|
| **1** | **1** |

.

Note that $\mathcal{G}_{1igr}$ is term equivalent, by (Eq1), to the commutative group with one elements $\mathcal{G}_{1gr}$ presented in Section 5.4.

• ([116], Chapter 18) There exists one commutative implicative-group with two elements: $\mathcal{G}_{2igr} = (A_2 = \{a, 1\}, \rightarrow, 1)$, where the table (matrix) of $\rightarrow$ has the following form:

$\mathcal{G}_{2igr}$

| $\rightarrow$ | a | 1 |
|---|---|---|
| a | 1 | a |
| 1 | a | 1 |

, hence

| $x$ | $x^{-1} = x \rightarrow 1$ |
|---|---|
| a | a |
| **1** | **1** |

.

Note that $\mathcal{G}_{2igr}$ is term equivalent, by (Eq1), to the commutative group with two elements $\mathcal{G}_{2gr}$ presented in Section 5.4.

• ([116], Chapter 18) There is **only one commutative implicative-group** with 3 elements $\mathcal{G}_{3igr} = (A_3, \rightarrow, ^{-1}, 1)$:

$\mathcal{G}_{3igr}$

| $\rightarrow$ | a | b | **1** |
|---|---|---|---|
| a | **1** | *a* | *b* |
| b | *b* | **1** | *a* |
| **1** | a | b | **1** |

| $x$ | $x^{-1} = x \rightarrow 1$ |
|---|---|
| a | b |
| b | a |
| **1** | **1** |

,

that is term equivalent (by (Eq1)) to the commutative group with 3 elements $\mathcal{G}_{3gr}$ presented in Section 5.4.

• ([116], Chapter 18) There are four commutative implicative-groups with four elements:
$\mathcal{G}^1_{4igr} = (A_4, \rightarrow_1, ^{-1}, 1)$, $\mathcal{G}^2_{4igr} = (A_4, \rightarrow_2, ^{-1}, 1)$, $\mathcal{G}^3_{4igr} = (A_4, \rightarrow_3, ^{-1}, 1)$, $\mathcal{G}^4_{4igr} = (A_4, \rightarrow_4, ^{-1}, 1)$, where $x^{-1} = x \rightarrow_i 1$, $i = 1, 4$:

$\mathcal{G}^1_{4igr}$

| $\rightarrow_1$ | a | b | c | **1** |
|---|---|---|---|---|
| a | **1** | *a* | *b* | *c* |
| b | *c* | **1** | *a* | b |
| c | *b* | *c* | **1** | *a* |
| **1** | a | b | c | **1** |

,

| $x$ | $x^{-1}$ |
|---|---|
| a | c |
| b | b |
| c | a |
| **1** | **1** |

$\mathcal{G}^2_{4igr}$

| $\rightarrow_2$ | a | b | c | **1** |
|---|---|---|---|---|
| a | **1** | *c* | *a* | *b* |
| b | *c* | **1** | *b* | *a* |
| c | *b* | *a* | **1** | c |
| **1** | a | b | c | **1** |

,

| $x$ | $x^{-1}$ |
|---|---|
| a | b |
| b | a |
| c | c |
| **1** | **1** |

$\mathcal{G}^3_{4igr}$

| $\rightarrow_3$ | a | b | c | **1** |
|---|---|---|---|---|
| a | **1** | *c* | *b* | *a* |
| b | *b* | **1** | *a* | *c* |
| c | *c* | *a* | **1** | *b* |
| **1** | a | b | c | **1** |

,

| $x$ | $x^{-1}$ |
|---|---|
| a | a |
| b | c |
| c | b |
| **1** | **1** |

$\mathcal{G}^4_{4igr}$

| $\rightarrow_4$ | a | b | c | **1** |
|---|---|---|---|---|
| a | **1** | *c* | *b* | *a* |
| b | *c* | **1** | *a* | b |
| c | *b* | *a* | **1** | c |
| **1** | a | b | c | **1** |

,

| $x$ | $x^{-1}$ |
|---|---|
| a | a |
| b | b |
| c | c |
| **1** | **1** |

.

Note that the commutative implicative-groups with four elements $\mathcal{G}^1_{4igr}$, $\mathcal{G}^2_{4igr}$,

$\mathcal{G}^3_{4igr}$, $\mathcal{G}^4_{4igr}$ are term equivalent (by (Eq1)) to the commutative groups with four elements $\mathcal{G}^1_{4gr}$, $\mathcal{G}^2_{4gr}$, $\mathcal{G}^3_{4gr}$, $\mathcal{G}^4_{4gr}$, respectively, from Section 5.4.

• (By *Mace4 program*) There exists one commutative implicative-group with five elements: $\mathcal{G}_{5igr} = (A_5 = \{a, b, c, d, 1\}, \rightarrow, 1)$, where the table (matrix) of $\rightarrow$ has the following form:

$\mathcal{G}_{5igr}$

| $\rightarrow$ | a | b | c | d | 1 |
|---|---|---|---|---|---|
| a | 1 | c | d | a | b |
| b | d | 1 | b | c | a |
| c | c | a | 1 | b | d |
| d | b | d | a | 1 | c |
| 1 | a | b | c | d | 1 |

,

| $x$ | $x^{-1}$ |
|---|---|
| a | b |
| b | a |
| c | d |
| d | c |
| 1 | 1 |

.

Note that $\mathcal{G}_{5igr}$ is term equivalent, by (Eq1), to the commutative group with five elements $\mathcal{G}_{5gr}$, presented in Section 5.4.

### 1.5.4.2 Non-commutative finite examples

By *Mace4 program* we can obtain the proper i-groups (i.e. $\rightarrow \neq \rightsquigarrow$) with 6, 8, 10, 12, 14 elements etc. Here is the example of proper i-group with 6 elements $\mathcal{G}_{6gr} = (G_6 = \{a, b, c, d, f, 1\}, \rightarrow, \rightsquigarrow, 1)$ given in [48]:

$\mathcal{G}_{6igr}$

| $\rightarrow$ | a | b | c | d | f | **1** |
|---|---|---|---|---|---|---|
| a | **1** | d | f | b | c | a |
| b | c | **1** | a | f | d | b |
| c | f | a | **1** | c | b | d |
| d | b | f | d | **1** | a | c |
| f | d | c | b | a | **1** | f |
| **1** | a | b | c | d | f | **1** |

| $\rightsquigarrow$ | a | b | c | d | f | **1** |
|---|---|---|---|---|---|---|
| a | **1** | c | b | f | d | a |
| b | d | **1** | f | a | c | b |
| c | b | f | **1** | c | a | d |
| d | f | a | d | **1** | b | c |
| f | c | d | a | b | **1** | f |
| **1** | a | b | c | d | f | **1** |

,

| $x$ | $x^{-1}$ |
|---|---|
| a | a |
| b | b |
| c | d |
| d | c |
| f | f |
| **1** | **1** |

.

Note that $\mathcal{G}_{6igr}$ is term equivalent, by (Eq1), to the non-commutative group with six elements $\mathcal{G}_{6gr}$, presented in Section 5.4.

## 1.6 The p-semisimple (or discrete) property

In Subsection 1.6.1, we present p-semisimple pseudo-structures: p-semisimple pBCI algebras, p-semisimple pBCI(pRP') algebras and p-semisimple pME algebras (which are implicative-groups). In Subsection 1.6.2, we present p-semisimple partially ordered pi-magmas: p-semisimple po-pes, po-prs, po-mprs and p-semisimple po-pes(pRP'), po-prs(pRP'), po-mprs(pRP') (which are all implicative-groups too).

Note that W. Rump uses in [193] the name 'discrete' for 'p-semisimple'.

### 1.6.1 p-semisimple pseudo-structures

G. Dymek [47] (see [146] for the commutative case) made the connection between the pseudo-BCI algebras (defined as reversed right-structures $(A, \leq, \star, \circ, 0)$) and the groups (defined additively), by introducing, as in the commutative case [146], the subclass of p-semisimple pseudo-BCI algebras and by proving that these are equivalent with the groups. Here, we make the direct connection with the implicative-groups.

Starting from the equivalent notion of *p-semisimple pseudo-BCI algebra* used in [47], we have introduced the following general definition.

**Definitions 1.6.1** ([116], Definition 2.1.3)

(i) We say that a left-pseudo-structure is *p-semisimple*, if the following property holds: for all $x, y$,

(p-s) $\qquad x \leq y \Longrightarrow x = y.$

(i') We say that a right-pseudo-structure is *p-semisimple*, if the following property holds: for all $x, y$,

(p-s$^R$) $\qquad x \geq y \Longrightarrow x = y.$

**Remark 1.6.2** *(The dual one is omitted)*

*If a left-pseudo-structure $\mathcal{A}^L$ verifies (La'), then (p-s) implies $A^L = \{1\}$. Indeed, any $x \in A^L$ verifies $x \leq 1$, by (La'), hence $x = 1$, by (p-s). Consequently, p-semisimple left-pBCK algebras are trivial.*

- **p-semisimple pseudo-BCI algebras**

**Remarks 1.6.3** *([116], Remarks 13.1.7)*

*Recall the following two equivalent definitions of left-pBCI algebras as structures.*

*- Definition 1' of pBCI algebras (Definition 1.2.2):*

*A left-pBCI algebra is a structure $\mathcal{A}^L = (A^L, \leq, \rightarrow, \rightsquigarrow, 1)$, where $\leq$ is a binary relation on $A^L$, $\rightarrow$ and $\rightsquigarrow$ are binary operations on $A^L$ and 1 is an element of $A^L$ verifying: for all $x, y, z \in A^L$,*

*(pBB')* $\quad y \rightarrow z \leq (z \rightarrow x) \rightsquigarrow (y \rightarrow x), \; y \rightsquigarrow z \leq (z \rightsquigarrow x) \rightarrow (y \rightsquigarrow x),$

*(pD')* $\quad y \leq (y \rightarrow x) \rightsquigarrow x, \; y \leq (y \rightsquigarrow x) \rightarrow x,$

*(Re')* $\quad x \leq x$ *(Reflexivity),*

*(An')* $\quad x \leq y$ *and* $y \leq x \Longrightarrow x = y$ *(Antisymmetry),*

*(IdEqR)* $\quad x \rightarrow y = 1 \Longleftrightarrow x \rightsquigarrow y = 1,$

*(pEqrelR)* $\quad x \leq y \Longleftrightarrow x \rightarrow y = 1 \; (\stackrel{(IdEqR)}{\Longleftrightarrow} x \rightsquigarrow y = 1).$

*- Definition 3' of pBCI algebras (Definition 1.2.12):*

*A left-pBCI algebra is a structure $\mathcal{A}^L = (A^L, \leq, \rightarrow, \rightsquigarrow, 1)$ such that $(A^L, \leq)$ is a poset and the properties (pBB'), (pM), (IdEqR), (pEqrelR) hold.*

*(1) Following Definition 1' of pBCI algebras (in left-case and in the right-case), a p-semisimple pBCI algebra is then an algebra $(A, \rightarrow, \rightsquigarrow, 1)$ verifying: for all $x, y, z \in A$,*

*(pBB=)* $\quad y \rightarrow z = (z \rightarrow x) \rightsquigarrow (y \rightarrow x), \; y \rightsquigarrow z = (z \rightsquigarrow x) \rightarrow (y \rightsquigarrow x),$

*(pD=)* $\quad y = (y \rightarrow x) \rightsquigarrow x, \; y = (y \rightsquigarrow x) \rightarrow x,$

*(IdEqR)* $\quad x \rightarrow y = 1 \Longleftrightarrow x \rightsquigarrow y = 1,$

*(pEqrelR=)* $\quad x = y \Longleftrightarrow x \rightarrow y = 1 \; (\stackrel{(IdEqR)}{\Longleftrightarrow} x \rightsquigarrow y = 1).$

*Note that such an algebra is just the implicative-group (Definition 2).*

*(2) Following Definition 3' of pBCI algebras (in left-case and in the right-case), a p-semisimple pBCI algebra is then an algebra* $(A, \to, \rightsquigarrow, 1)$ *verifying (pBB=), (pM), (IdEqR), (pEqrelR=).*

*Note that such an algebra is just the implicative-group (Definition 1').*

*(3) Following another equivalent definition of pBCI algebras, we obtain the corresponding equivalent definition of p-semisimple pBCI algebras, i.e. of implicative-groups.*

We then obtain immediately the following result (which follows from the Dymek's result and from the term-equivalence (recalled in Part III)) between groups and i-groups).

**Theorem 1.6.4** *(See ([116], Theorem 13.1.6))*
*The p-semisimple left-pBCI algebras coincide with the implicative-groups and, dually, the p-semisimple right-pBCI algebras coincide with the implicative-groups too.*

Hence, we have:

| | | | | |
|---|---|---|---|---|
| **pBCI** | + | (p-s) | = | **i-group** and, dually, |
| **pBCI**$^R$ | + | (p-s$^R$) | = | **i-group**. |

- **p-semisimple pBCI(pRP') algebras and their duals**

**Remarks 1.6.5**
*(1) Consider (Definition 4' (Definition 1.2.14)) that a pBCI algebra is a structure* $\mathcal{A}^L = (A^L, \leq, \to, \rightsquigarrow, 1)$ *such that* $(A^L, \leq)$ *is a poset and the properties (pM), (pEx), (p*'), (IdEqR), (pEqrelR) hold.*

*Hence, a pBCI(pRP') algebra is a structure* $\mathcal{A}^L = (A^L, \leq, \to, \rightsquigarrow, \odot, 1)$ *such that the reduct* $(A^L, \leq)$ *is a poset, the reduct* $(A^L, \leq, \to, \rightsquigarrow, 1)$ *verifies the properties (pM), (pEx), (p*'), (IdEqR), (pEqrelR) and the following property holds:*
*(pRP')* $\quad x \leq y \to z \Longleftrightarrow y \leq x \rightsquigarrow z \Longleftrightarrow x \odot y \leq z.$

*Consequently, a p-semisimple pBCI(pRP') algebra is an algebra* $(A, \to, \rightsquigarrow, \odot, 1)$ *verifying (pM), (pEx), (IdEqR), (pEqrelR=) and:*
*(pGa$^=$)* $\quad x = y \to z \Longleftrightarrow y = x \rightsquigarrow z \Longleftrightarrow x \odot y = z.$

*Note that such an algebra is just the X-implicative-group (Definition 2).*

*(2) Following another equivalent definition of pBCI algebras, we obtain the corresponding equivalent definition of p-semisimple pBCI(pRP') algebras, i.e. of X-implicative-groups.*

We then obtain immediately the following result.

**Theorem 1.6.6** *([116], Theorem 13.1.14)*
*The p-semisimple pBCI(pRP') algebras coincide with the X-implicative-groups and, dually, the p-semisimple pBCI(pcoRS') algebras coincide with the X-implicative-groups too.*

Hence, we have:

| | | | | |
|---|---|---|---|---|
| **pBCI(pRP')** | + | (p-s) | = | **X-i-group** and, dually, |
| **pBCI(pcoRS')** | + | (p-s$^R$) | = | **X-i-group**. |

• **Resuming connections for the family of pBCI algebras**

We have the connections from Figure 1.3. Note that pBCI(pRP') algebras are equivalent to pBCI(pP') algebras (by Theorem 1.4.33) and that i-groups are equivalent to X-i-groups, by (EI1).

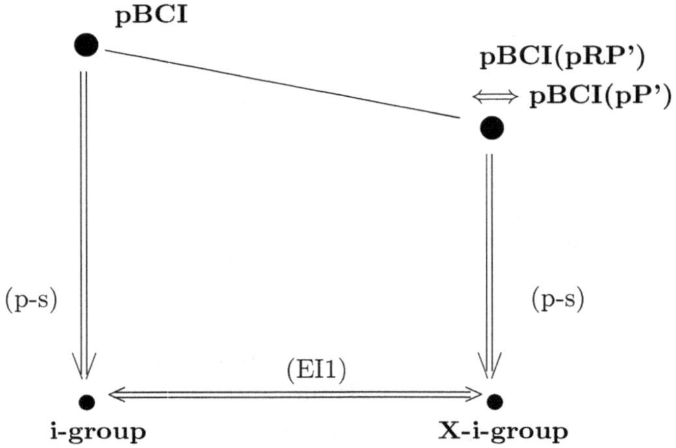


Figure 1.3: p-semisimple algebras from the family of pBCI algebras

• **p-semisimple pME algebras**

Let $(A^L, \leq, \to, \rightsquigarrow, 1)$ be a left-pME algebra (Definition 1.1.9), i.e. (pM), (pEx), (IdEqR), (pEqrelR) hold.

It follows that a p-semisimple left-pME algebra is then an algebra $(A, \to, \rightsquigarrow, 1)$ verifying (pM), (pEx), (IdEqR), (pEqrelR=).

Note that such an algebra is just the implicative-group (Definition 4' - Definition 1.5.17).

We then obtain immediately the following result.

**Theorem 1.6.7**
*The p-semisimple left-pME algebras coincide with the implicative-groups and, dually, the p-semisimple right-pME algebras coincide with the implicative-groups too.*

Hence, we have:

**pME** + (p-s) = **i-group** and, dually,
**pME**$^R$ + (p-s$^R$) = **i-group**.

**Corollary 1.6.8** *[116] The p-semisimple pRME = pCI, pre-pBCI, pBCH algebras also coincide with the implicative-groups.*

## 1.6.2 p-semisimple partially ordered pi-magmas

We introduce the following general definition.

**Definition 1.6.9** (Definition of po-pi-magmas)

(i) We say that a left-po-pi-magma $(A^L, \leq, \rightarrow, \rightsquigarrow)$ is *p-semisimple* (or *discrete*), if the following property holds: for all $x, y \in A^L$,

(p-s) $\qquad x \leq y \implies x = y.$

(i') We say that a right-po-pi-magma $(A^R, \geq, \rightarrow^R, \rightsquigarrow^R)$ is *p-semisimple* (or *discrete*), if the following property holds: for all $x, y \in A^R$,

(p-s$^R$) $\qquad x \geq y \implies x = y.$

**Remark 1.6.10** *(The dual one is omitted)*

*If a left-po-pi-magma $\mathcal{A}^L$ verifies (La'), then (p-s) implies $A^L = \{1\}$. Indeed, any $x \in A^L$ verifies $x \leq 1$, by (La'), hence $x = 1$, by (p-s). Consequently, p-semisimple left-po-iprs are trivial.*

• **p-semisimple po-pes (= discrete quantum-B algebras)**

Let $(A^L, \leq, \rightarrow, \rightsquigarrow)$ be a left-po-pe (= quantum-B algebra), i.e. $(A^L, \leq)$ is a poset and the properties (pEx), (p$*_{\rightarrow}$'), (pEq#') hold.

It follows that a p-semisimple left-po-pe (= discrete quantum-B algebra) is then an algebra $(A^L, \rightarrow, \rightsquigarrow)$ verifying (pEx), (pEq#=).

Note that such an algebra is just the implicative-group (Definition 5).

We then obtain immediately the following result (which follows from Rump's result ([193], Theorem 3)).

**Theorem 1.6.11**

*The p-semisimple left-po-pes (= discrete QB algebras) coincide with the implicative-groups and, dually, the p-semisimple right-po-pes (= discrete right-QB algebras) coincide with the implicative-groups too.*

Hence, we have:

| | | | | |
|---|---|---|---|---|
| **po-pe** (= **QB**) | + | (p-s) | = | **i-group** and, dually, |
| **po-pe**$^R$ (= **QB**$^R$) | + | (p-s$^R$) | = | **i-group**. |

• **p-semisimple po-prs (= discrete unital quantum-B algebras)**

Let $(A^L, \leq, \rightarrow, \rightsquigarrow, 1)$ be a left-po-pr (= unital quantum-B algebra), i.e. $(A^L, \leq)$ is a poset and the properties (pEx), (pM), (p$*_{\rightarrow}$'), (pEq#') hold.

It follows that a p-semisimple left-po-pr (= discrete unital quantum-B algebra) is then an algebra $(A^L, \rightarrow, \rightsquigarrow, 1)$ verifying (pEx), (pM), (pEq#=).

Note that such an algebra is just the implicative-group (Definition 5).

We then obtain immediately the following result.

**Theorem 1.6.12**

*The p-semisimple left-po-prs (= discrete unital QB algebras) coincide with the implicative-groups and, dually, the p-semisimple right-po-pes (= discrete unital right-QB algebras) coincide with the implicative-groups too.*

**• p-semisimple po-mprs (= discrete maximal quantum-B algebras) ($\Longleftrightarrow$ p-semisimple pBCI algebras)**

Let $(A^L, \leq, \rightarrow, \rightsquigarrow, 1)$ be a left-po-mpr (= maximal quantum-B algebra), i.e. $(A^L, \leq, 1)$ is a poset with 1 as maximal element (i.e. (N') holds) and the properties (pEx), (pM), ($p^*_{\rightarrow}$'), (pEq#') hold.

It follows that a p-semisimple left-po-mpr (= discrete maximal quantum-B algebra) is then an algebra $(A^L, \rightarrow, \rightsquigarrow, 1)$ verifying (pEx), (pM), (pEq#=).

Note that such an algebra is just the implicative-group (Definition 5).

We then obtain immediately the following result.

**Theorem 1.6.13**

*The p-semisimple left-po-mprs (= discrete maximal QB algebras) coincide with the implicative-groups and, dually, the p-semisimple right-po-pes (= discrete maximal right-QB algebras) coincide with the implicative-groups too.*

**• p-semisimple po-pe(pRP') (= discrete quantum-B algebras with (pRP')) and their duals**

Let $(A^L, \leq, \rightarrow, \rightsquigarrow, \odot)$ be a po-pe(pRP'), i.e. the reduct $(A^L, \leq)$ is a poset, (pEx) holds and the following property holds:

(pRP') $\quad x \leq y \rightarrow z \Longleftrightarrow y \leq x \rightsquigarrow z \Longleftrightarrow x \odot y \leq z.$

It follows that a p-semisimple po-pe(pRP') (= discrete QB algebra with (pRP')) is then an algebra $(A, \rightarrow, \rightsquigarrow, \odot)$ verifying (pEx) and

($pGa^=$) $\quad x = y \rightarrow z \Longleftrightarrow y = x \rightsquigarrow z \Longleftrightarrow x \odot y = z.$

Note that such an algebra is just the X-implicative-group (Definition 3).

We then obtain immediately the following result.

**Theorem 1.6.14**

*The p-semisimple po-pes(pRP') (= discrete quantum-B algebras with (pRP')) coincide with the X-implicative-groups and, dually, the p-semisimple po-pes(pcoRS') (= discrete dual quantum-B algebras with (pcoRS')) coincide with the X-implicative-groups too.*

Hence, we have:

| | | | |
|---|---|---|---|
| **po-pe(pRP')** (= **QB** + (pRP')) | + | (p-s) | = **X-i-group** and, dually, |
| **po-pe(pcoRS')** (= **QB** + (pcoRS')) | + | (p-$s^R$) | = **X-i-group**. |

**• Resuming connections for the family of po-pes (=QB algebras)**

We have the connections from Figure 1.4. Note that po-pes(pRP') are equivalent to po-pes(pP') (by Theorem 1.4.33) and that i-groups are equivalent to X-i-groups, by (EI1).

**• p-semisimple po-pr(pRP') (= discrete unital quantum-B algebras with (pRP')) and their duals**

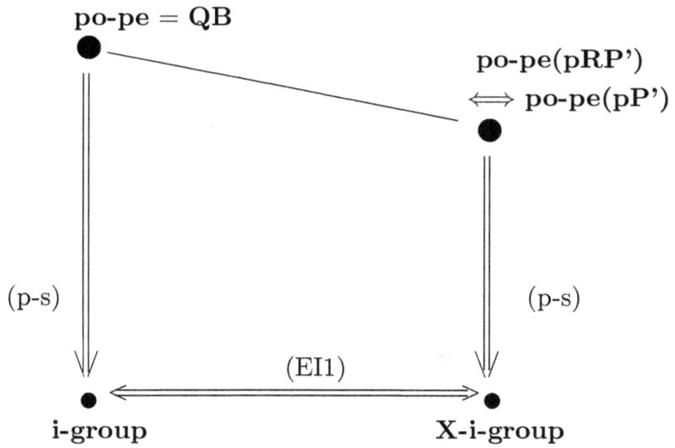

Figure 1.4: p-semisimple algebras from the family of QB algebras

Let $(A^L, \leq, \rightarrow, \rightsquigarrow, \odot, 1)$ be a po-pr(pRP'), i.e. the reduct $(A^L, \leq)$ is a poset, (pEx), (pM) hold and (pRP') holds.

It follows that a p-semisimple po-pr(pRP') (= discrete unital QB algebra with (pRP')) is then an algebra $(A, \rightarrow, \rightsquigarrow, \odot, 1)$ verifying (pEx), (pM) and ($pGa^=$).

Note that such an algebra is just the X-implicative-group (Definition 1).

We then obtain immediately the following result.

**Theorem 1.6.15**

*The p-semisimple po-prs(pRP') (= discrete unital quantum-B algebras with (pRP')) coincide with the X-implicative-groups and, dually, the p-semisimple po-pes(pcoRS') (= discrete dual unital quantum-B algebras with (pcoRS')) coincide with the X-implicative-groups too.*

**• p-semisimple po-mpr(pRP') (= discrete maximal quantum-B algebras with (pRP')) (⟺ p-semisimple pBCI(pRP') algebras) and their duals**

Let $(A^L, \leq, \rightarrow, \rightsquigarrow, \odot, 1)$ be a po-mpr(pRP'), i.e. the reduct $(A^L, \leq)$ is a poset, (pEx), (pM), (N') hold and (pRP') holds.

It follows that a p-semisimple po-mpr(pRP') (= discrete maximal QB algebra with (pRP')) is then an algebra $(A, \rightarrow, \rightsquigarrow, \odot, 1)$ verifying (pEx), (pM) and ($pGa^=$).

Note that such an algebra is just the X-implicative-group (Definition 1).

We then obtain immediately the following result.

**Theorem 1.6.16**

*The p-semisimple po-mprs(pRP') (= discrete maximal quantum-B algebras with (pRP')) coincide with the X-implicative-groups and, dually, the p-semisimple po-mprs(pcoRS') (= discrete dual maximal quantum-B algebras with (pcoRS')) coincide with the X-implicative-groups too.*

## 1.7 Logics

Consider the following set of axioms:

| | | |
|---|---|---|
| (pB) | = | (pB$\rightarrow$) + (pB$\rightsquigarrow$) (prefixing), where: |
| | (pB$\rightarrow$) | $(\psi \rightarrow \chi) \rightarrow ((\varphi \rightarrow \psi) \rightarrow (\varphi \rightarrow \chi))$, |
| | (pB$\rightsquigarrow$) | $(\psi \rightsquigarrow \chi) \rightarrow ((\varphi \rightsquigarrow \psi) \rightsquigarrow (\varphi \rightsquigarrow \chi))$; |
| (pBB) | = | (pBB1) + (pBB2) (suffixing), where: |
| | (pBB1) | $(\varphi \rightarrow \psi) \rightarrow ((\psi \rightarrow \chi) \rightsquigarrow (\varphi \rightarrow \chi))$, |
| | (pBB2) | $(\varphi \rightsquigarrow \psi) \rightarrow ((\psi \rightsquigarrow \chi) \rightarrow (\varphi \rightsquigarrow \chi))$, |
| (pC) | = | (pC1) + (pC2) (exchange), where: |
| | (pC1) | $(\varphi \rightarrow (\psi \rightsquigarrow \chi)) \rightarrow (\psi \rightsquigarrow (\varphi \rightarrow \chi))$, |
| | (pC2) | $(\varphi \rightsquigarrow (\psi \rightarrow \chi)) \rightarrow (\psi \rightarrow (\varphi \rightsquigarrow \chi))$; |
| (pK) | = | (pK$\rightarrow$) + (pK$\rightsquigarrow$), where: |
| | (pK$\rightarrow$) | $\varphi \rightarrow (\psi \rightarrow \varphi)$, |
| | (pK$\rightsquigarrow$) | $\varphi \rightarrow (\psi \rightsquigarrow \varphi)$; |
| (I$\rightarrow$) | | $\varphi \rightarrow \varphi$ (identity principle). |

Consider the following set of inference (deduction) rules:

$$(IR) = (IR1) + (IR2)\,, where: \quad (IR1)\ \frac{\varphi \rightarrow \psi}{\varphi \rightsquigarrow \psi} \quad and \quad (IR2)\ \frac{\varphi \rightsquigarrow \psi}{\varphi \rightarrow \psi},$$

$$(MP)\ \frac{\varphi, \varphi \rightarrow \psi}{\psi}\ (modus\ ponens);$$

$$(MK)\ \frac{\psi}{\varphi \rightarrow \psi};$$

$$(MKK)\ \frac{\varphi, \psi}{\varphi \rightarrow \psi}.$$

Consider the following related logics.

• The *pseudo-BCK logic* was introduced and studied by J. Kühr [145], as non-commutative generalization of the *BCK logic* which has two different implications, $\rightarrow$ and $\rightsquigarrow$. The *pseudo-BCK logic* has the axioms: (pB), (pC), (pK) and the inference rules: (IR), (MP). Kühr noted that the *pseudo-BCK logic* is an algebraizable logic in the sense of Block-Pigozzi [6], namely the *pseudo-BCK logic* is algebraizable with the set of equivalence formulas $\Delta = \{x \rightarrow y, y \rightarrow x\}$ and defining equation $x = x \rightarrow x$. Moreover, we have the following analogue of ([6], Theorem 5.11): "The quasivariety of pseudo-BCK algebras is (termwise equivalent to) the equivalent quasivariety semantics for the *pseudo-BCK logic*". In other words, the class of pseudo-BCK algebras forms an algebraic semantics for the *pseudo-BCK logic* (are the models of the *pseudo-BCK logic*).

• The *pseudo-BCI logic* was introduced and studied by Grzegorz Dymek and Anna Kozanecka-Dymek [49], as non-commutative generalization of the *BCI logic*. In [49], the authors present a non-commutative version of the *BCI logic*, called *pseudo-BCI logic*, which has two different implications, $\rightarrow$ and $\rightsquigarrow$. The *pseudo-BCI logic* has the axioms: (pB), (pC), (I$\rightarrow$) and the inference rules: (IR), (MP). Using advanced methods and techniques of [6], it can be proved that the *pBCI logic* is not algebraizable (in the sense of [6]), but it is order algebraizable (in the sense of [178]). Since it is not algebraizable, they [49] extended it (by adding the inference rule (MKK)) to *pBCI' logic*, which is algebraizable.

**Theorem 1.7.1** *([49], Theorem 3.1). The logic pBCI' is algebraizable with the set of equivalence formulas $\Delta = \{x \to y, y \to x\}$ and defining equation $x = x \to x$.*

Moreover, they [49] show that pseudo-BCI algebras are the models of *pBCI' logic* - as it is done in [145] for *pseudo-BCK logic*. The class of pseudo-BCI algebras forms an algebraic semantics for the logic pBCI'.

**Theorem 1.7.2** *([49], Theorem 3.2) The quasivariety of pseudo-BCI algebras is definitionally equivalent to the equivalent quasivariety semantics for the logic pBCI'.*

So, we have the following definitions.

**Definitions 1.7.3**

(1) The *pseudo-BCK logic*, or *pBCK logic* for short (denoted by $p\mathcal{BCK}$), has the axioms: (pB), (pC), (pK) and the inference rules: (IR), (MP).

(2) The *pseudo-BCI logic*, or *pBCI logic* for short (denoted by $p\mathcal{BCI}$), has the axioms: (pB), (pC), (I$\to$) and the inference rules: (IR), (MP).

Note that we could consider above, equivalently, the axioms (pBB) instead of axioms (pB).

Consider the following reducts.

**Definitions 1.7.4**

(3) The reduct of *pBCI logic*, called *pBCI$\to$ logic*, or *quantum-B logic*, or *QB logic* in [193] (denoted by $p\mathcal{BCI}\to$ (or $\mathcal{QB}$)), has the axioms: (pB$\to$), (pC), (I$\to$) and the inference rules: (IR), (MP), i.e.

$$p\mathcal{BCI} \;=\; p\mathcal{BCI}{\to} \;+\; (pB{\rightsquigarrow}) \;=\; \mathcal{QB} \;+\; (pB{\rightsquigarrow}).$$

(4) The reduct of *pBCK logic*, called *pBCK$\to$ logic* (denoted by $p\mathcal{BCK}\to$), has the axioms: (pB$\to$), (pC), (pK) and the inference rules: (IR), (MP), i.e.

$$p\mathcal{BCK} \;=\; p\mathcal{BCK}{\to} \;+\; (pB{\rightsquigarrow}).$$

Note that we could consider above, equivalently, the axiom (pB$\rightsquigarrow$) instead of the axiom (pB$\to$).

Consider the following extensions.

**Definitions 1.7.5**

(5) The extension of *pBCI logic* called *pBCI' logic* (denoted by $p\mathcal{BCI}$') has the axioms: (pB), (pC), (I$\to$) and the inference rules: (IR), (MP) and (MKK), i.e.

$$p\mathcal{BCI}' \;=\; p\mathcal{BCI} \;+\; (MKK).$$

(6) The extension of *pBCI logic* called *pBCI" logic* (denoted by $p\mathcal{BCI}$") has the axioms: (pB), (pC), (I→) and the inference rules: (IR), (MP) and (MK), i.e.

$$p\mathcal{BCI}'' = p\mathcal{BCI} + (MK).$$

(7) The extension of *pBCI→ logic* (or *QB logic*) called *pBCI→' logic* (or *QB' logic*) (denoted by $p\mathcal{BCI} \to$' (or $\mathcal{QB}$')) has the axioms: (pB→), (pC), (I→) and the inference rules: (IR), (MP) and (MKK), i.e.

$$p\mathcal{BCI} \to' (or\ \mathcal{QB}') = p\mathcal{BCI} \to (or\ \mathcal{QB}) + (MKK).$$

(8) The extension of *pBCI→ logic* (or *QB logic*) called *pBCI→" logic* (or *QB" logic*) (denoted by $p\mathcal{BCI} \to$" (or $\mathcal{QB}$")) has the axioms: (pB→), (pC), (I→) and the inference rules: (IR), (MP) and (MK), i.e.

$$p\mathcal{BCI} \to'' (or\ \mathcal{QB}'') = p\mathcal{BCI} \to (or\ \mathcal{QB}) + (MK).$$

**Proposition 1.7.6** *We have:*
*(a) (pC) + (pK) + (IR) + (MP) ⟹ (I→),*
*(b) (pK) + (MP) ⟹ (MK),*
*(c) (pC) + (I→) + (IR) + (MP) + (MK) ⟹ (pK),*
*(d) (MK) ⟹ (MKK).*

**Proof.** (a): We have:

$\vdash \varphi \to ((\varphi \to (\varphi \to \varphi) \rightsquigarrow \varphi)$ (pK⇝)
$\vdash (\varphi \to ((\varphi \to (\varphi \to \varphi) \rightsquigarrow \varphi)) \to ((\varphi \to (\varphi \to \varphi)) \rightsquigarrow (\varphi \to \varphi))$ (pC1)
$\vdash (\varphi \to (\varphi \to \varphi)) \rightsquigarrow (\varphi \to \varphi)$ (MP)
$\vdash (\varphi \to (\varphi \to \varphi)) \to (\varphi \to \varphi)$ (IR2)
$\vdash \varphi \to (\varphi \to \varphi)$ (pK→)
$\vdash \varphi \to \varphi$ (MP).

Thus, (I→) holds.

(b): We have:

$\vdash \psi$ hypothesis
$\vdash \psi \to (\varphi \to \psi)$ (pK→)
$\vdash \varphi \to \psi$ (MP).

Thus, (MK) holds.

(c): We have:

$\vdash \varphi \to \varphi$ (I→)
$\vdash \varphi \rightsquigarrow \varphi$ (IR1)
$\vdash \psi \to (\varphi \rightsquigarrow \varphi)$ (MK)
$\vdash (\psi \to (\varphi \rightsquigarrow \varphi)) \to (\varphi \rightsquigarrow (\psi \to \varphi))$ (pC1)
$\vdash \varphi \rightsquigarrow (\psi \to \varphi)$ (MP)
$\vdash \varphi \to (\psi \to \varphi)$ (IR2)

thus, (pK→) holds. Similarly,

$\vdash \varphi \to \varphi$ (I→)
$\vdash \psi \to (\varphi \to \varphi)$ (MK)
$\vdash \psi \rightsquigarrow (\varphi \to \varphi)$ (IR1)
$\vdash (\psi \rightsquigarrow (\varphi \to \varphi)) \to (\varphi \to (\psi \rightsquigarrow \varphi))$ (pC2)
$\vdash \varphi \to (\psi \rightsquigarrow \varphi)$ (MP)

thus, (pK⇝) holds too. Thus, (pK) holds.

(d): We have:

| | |
|---|---|
| $\vdash \varphi$ | hypothesis |
| $\vdash \chi \to \varphi$ | (MK) |
| $\vdash \psi \to \varphi$ | substitution |
| $\vdash \psi$ | hypothesis |
| $\vdash \chi \to \psi$ | (MK) |
| $\vdash \varphi \to \psi$ | substitution. |

Thus, $\frac{\varphi,\psi}{\psi\to\varphi,\varphi\to\psi}$, or $\frac{\varphi,\psi}{\varphi\to\psi}$, i.e. (MKK) holds. □

**Theorem 1.7.7**
*(i) Consider the pBCK→ logic. Then, (I→) and (MK) hold.*
*(i') Consider the pBCI→" logic (or QB" logic). Then, (pK) holds.*

**Proof.** By Proposition 1.7.6 (a), (b), (c). □

Note that Theorem 1.7.7 says that:

$$p\mathcal{BCK} \to \iff p\mathcal{BCI} \to'' \text{ (or } \mathcal{QB}''). \tag{1.49}$$

Consequently,

$$p\mathcal{BCK} \iff p\mathcal{BCI}''. \tag{1.50}$$

**Theorem 1.7.8** *Consider the pBCI→ logic (or QB logic). Then*

$$(MK) \implies (MKK).$$

**Proof.** By Proposition 1.7.6 (d). □

Note that Theorem 1.7.8 says that: $p\mathcal{BCI}$ →" (or $\mathcal{QB}$") is an extension of $p\mathcal{BCI}$ →' (or $\mathcal{QB}$').

Consequently, $p\mathcal{BCI}$" is an extension of $p\mathcal{BCI}$'.

By above definitions and results, we obtain the connections between the above logics reprezented in 'algebraic style' by the Hasse-type diagram from Figure 1.5.

In the following subsections, we present the sintax and the Lindenbaum-Tarski algebras for the above logics.

### 1.7.1 Pseudo-BCK→ logic

**Definition 1.7.9** The *alphabet* of the formal system of propositional pseudo-BCK→ logic is made of:
1) propositional variables, denoted: $x, y, z$...; let $V$ be the infinite set of propositional variables;
2) logical symbols (connectors): $\to$ and $\rightsquigarrow$, two implications;
3) parantheses (,).

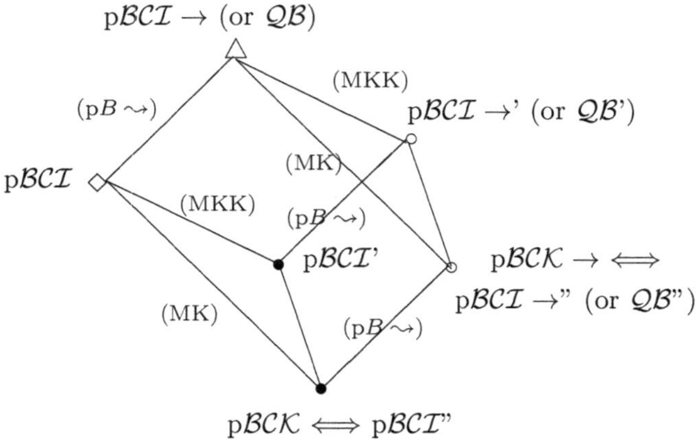


Figure 1.5: Connections in 'algebraic style' between the logics

**Definition 1.7.10** The set $L$ of *propositions* is defined by:
1) each variable belongs to $L$, i.e. $V \subset L$;
2) if $\varphi, \psi \in L$, then $\varphi \to \psi \in L$ and $\varphi \rightsquigarrow \psi \in L$.

**Definition 1.7.11** An *axiom* of the formal system of propositional pseudo-BCK→ logic is a proposition which has one of the following forms:

(pB→) $(\psi \to \chi) \to ((\varphi \to \psi) \to (\varphi \to \chi))$,
(pC1) $(\varphi \to (\psi \rightsquigarrow \chi)) \to (\psi \rightsquigarrow (\varphi \to \chi))$,
(pC2) $(\varphi \rightsquigarrow (\psi \to \chi)) \to (\psi \to (\varphi \rightsquigarrow \chi))$,
(pK→) $\varphi \to (\psi \to \varphi)$,
(pK⇝) $\varphi \to (\psi \rightsquigarrow \varphi)$.

**Definition 1.7.12** The *inference rules* of the formal system of propositional pBCK→ logic are the following:

$$(IR1)\ \frac{\varphi \to \psi}{\varphi \rightsquigarrow \psi} \quad \textit{and} \quad (IR2)\ \frac{\varphi \rightsquigarrow \psi}{\varphi \to \psi},$$

$$(MP)\ (\textit{modus ponens})\ \frac{\varphi, \varphi \to \psi}{\psi}.$$

**Definition 1.7.13** A proposition $\varphi \in L$ is *derivable* from a subset $\Gamma \subseteq L$, fact denoted by

$$\Gamma \vdash \varphi,$$

if $\varphi$ is obtained in finitely many steps as follows:
(1) $\varphi$ is an axiom,
(2) $\varphi \in \Gamma$,
(3) if $\psi$ is derivable, then $\varphi$ is obtained by substitution of the variables in $\psi$ by arbitrary propositions,
(4) $\psi$ and $\psi \to \varphi$ are derivable,
(5) $\psi \to \varphi$ is derivable iff $\psi \rightsquigarrow \varphi$ is derivable.

The set of all propositions $\varphi \in L$ derivable from $\Gamma$ will be denoted by $L_\Gamma$:

$$L_\Gamma \stackrel{def.}{=} \{\varphi \in L \mid \Gamma \vdash \varphi\}$$

(1.51) $$\varphi \in L_\Gamma \Longleftrightarrow \Gamma \vdash \varphi.$$

If $\Gamma = \emptyset \subset L$, we write $\vdash \varphi$ instead of $\emptyset \vdash \varphi$ and

$$L_\emptyset \stackrel{def.}{=} \{\varphi \in L \mid \emptyset \vdash \varphi\}$$

(1.52) $$\varphi \in L_\emptyset \Longleftrightarrow \vdash \varphi.$$

- **Sintactical properties of pBCK$\rightarrow$ logic**

**Proposition 1.7.14** *For any $\varphi, \psi, \chi \in L$, we have the following properties:*

*(pB$\rightarrow_\Gamma$)* $\Gamma \vdash (\psi \rightarrow \chi) \rightarrow ((\varphi \rightarrow \psi) \rightarrow (\varphi \rightarrow \chi))$,
*(pC1$_\Gamma$)* $\Gamma \vdash (\varphi \rightarrow (\psi \rightsquigarrow \chi)) \rightarrow (\psi \rightsquigarrow (\varphi \rightarrow \chi))$,
*(pC2$_\Gamma$)* $\Gamma \vdash (\varphi \rightsquigarrow (\psi \rightarrow \chi)) \rightarrow (\psi \rightarrow (\varphi \rightsquigarrow \chi))$,
*(pK$\rightarrow_\Gamma$)* $\Gamma \vdash \varphi \rightarrow (\psi \rightarrow \varphi)$,
*(pK$\rightsquigarrow_\Gamma$)* $\Gamma \vdash \varphi \rightarrow (\psi \rightsquigarrow \varphi)$;

*(IdEqR$_\Gamma$)* $\Gamma \vdash \varphi \rightarrow \psi \Longleftrightarrow \Gamma \vdash \varphi \rightsquigarrow \psi$;

*(MK$_\Gamma$)* $\Gamma \vdash \psi \Longrightarrow \Gamma \vdash \varphi \rightarrow \psi$;
*(MKK$_\Gamma$)* $\Gamma \vdash \varphi, \psi \Longrightarrow \Gamma \vdash \varphi \rightarrow \psi$;
*(I$\rightarrow_\Gamma$)* $\Gamma \vdash \varphi \rightarrow \varphi$;
*(Tr$_\Gamma$)* $\Gamma \vdash \varphi \rightarrow \psi$ *and* $\Gamma \vdash \psi \rightarrow \chi \Longrightarrow \Gamma \vdash \varphi \rightarrow \chi$,
*(pEq#$_\Gamma$)* $\Gamma \vdash \varphi \rightsquigarrow (\psi \rightarrow \chi) \Longleftrightarrow \Gamma \vdash \psi \rightarrow (\varphi \rightsquigarrow \chi)$,
*(pD$_\Gamma$)* $\Gamma \vdash \varphi \rightarrow ((\varphi \rightarrow \psi) \rightsquigarrow \psi)$ *and* $\Gamma \vdash \varphi \rightarrow ((\varphi \rightsquigarrow \psi) \rightarrow \psi)$,
*(p*$\rightarrow_\Gamma$)* $\Gamma \vdash \psi \rightarrow \chi \Longrightarrow \Gamma \vdash (\varphi \rightarrow \psi) \rightarrow (\varphi \rightarrow \chi)$,
*(p*$\rightsquigarrow_\Gamma$)* $\Gamma \vdash \psi \rightarrow \chi \Longrightarrow \Gamma \vdash (\varphi \rightsquigarrow \psi) \rightarrow (\varphi \rightsquigarrow \chi)$,
*(p**$\rightarrow_\Gamma$)* $\Gamma \vdash \psi \rightarrow \chi \Longrightarrow \Gamma \vdash (\chi \rightarrow \varphi) \rightarrow (\psi \rightarrow \varphi)$,
*(p**$\rightsquigarrow_\Gamma$)* $\Gamma \vdash \psi \rightarrow \chi \Longrightarrow \Gamma \vdash (\chi \rightsquigarrow \varphi) \rightarrow (\psi \rightsquigarrow \varphi)$,
*(pBB1$_\Gamma$)* $\Gamma \vdash (\varphi \rightarrow \psi) \rightarrow ((\psi \rightarrow \chi) \rightsquigarrow (\varphi \rightarrow \chi))$.

**Proof.** (pB$\rightarrow_\Gamma$): By Definition 1.7.13 and (1.51), since (pB$\rightarrow$) is an axiom.
(pC1$_\Gamma$): By Definition 1.7.13 and (1.51), since (pC1) is an axiom.
(pC2$_\Gamma$): By Definition 1.7.13 and (1.51), since (pC2) is an axiom.
(pK$\rightarrow_\Gamma$): By Definition 1.7.13 and (1.51), since (pK$\rightarrow_\Gamma$) is an axiom.
(pK$\rightsquigarrow_\Gamma$): By Definition 1.7.13 and (1.51), since (pK$\rightsquigarrow_\Gamma$) is an axiom.
(IdEqR$_\Gamma$): By Definition 1.7.13 and (1.51), and by (IR1) and (IR2).
(MK$_\Gamma$): We have:

| | |
|---|---|
| $\Gamma \vdash \psi$ | hypothesis |
| $\Gamma \vdash \psi \rightarrow (\varphi \rightarrow \psi)$ | (pK$\rightarrow_\Gamma$) |
| $\Gamma \vdash \varphi \rightarrow \psi$ | (MP). |

$(MKK_\Gamma)$: We have:

$$\begin{array}{ll}
\Gamma \vdash \varphi & \text{hypothesis} \\
\Gamma \vdash \chi \to \varphi & (MK_\Gamma) \\
\Gamma \vdash \psi \to \varphi & \text{substitution} \\
\Gamma \vdash \psi & \text{hypothesis} \\
\Gamma \vdash \chi \to \psi & (MK_\Gamma) \\
\Gamma \vdash \varphi \to \psi & \text{substitution.}
\end{array}$$

Thus, $\frac{\varphi,\psi}{\psi\to\varphi,\varphi\to\psi}$, or $\frac{\varphi,\psi}{\varphi\to\psi}$, i.e. $(MKK_\Gamma)$ holds.

$(I{\to}_\Gamma)$: We have:

$$\begin{array}{ll}
\Gamma \vdash \varphi \to ((\varphi \to (\varphi \to \varphi) \rightsquigarrow \varphi) & (pK{\rightsquigarrow}_\Gamma) \\
\Gamma \vdash (\varphi \to ((\varphi \to (\varphi \to \varphi) \rightsquigarrow \varphi)) \to ((\varphi \to (\varphi \to \varphi)) \rightsquigarrow (\varphi \to \varphi)) & (pC1_\Gamma) \\
\Gamma \vdash (\varphi \to (\varphi \to \varphi)) \rightsquigarrow (\varphi \to \varphi) & (MP) \\
\Gamma \vdash (\varphi \to (\varphi \to \varphi)) \to (\varphi \to \varphi) & (IR2) \\
\Gamma \vdash \varphi \to (\varphi \to \varphi) & (pK{\to}_\Gamma) \\
\Gamma \vdash \varphi \to \varphi & (MP).
\end{array}$$

$(Tr_\Gamma)$: We have:

$$\begin{array}{ll}
\Gamma \vdash \psi \to \chi & \text{hypothesis} \\
\Gamma \vdash (\psi \to \chi) \to ((\varphi \to \psi) \to (\varphi \to \chi)) & (pB{\to}_\Gamma) \\
\Gamma \vdash (\varphi \to \psi) \to (\varphi \to \chi) & (MP) \\
\Gamma \vdash \varphi \to \psi & \text{hypothesis} \\
\Gamma \vdash \varphi \to \chi & (MP).
\end{array}$$

$(pEq\#_\Gamma)$: We have:

$$\begin{array}{ll}
\Gamma \vdash \varphi \rightsquigarrow (\psi \to \chi) & \text{hypothesis} \\
\Gamma \vdash (\varphi \rightsquigarrow (\psi \to \chi)) \to (\psi \to (\varphi \rightsquigarrow \chi)) & (pC2_\Gamma) \\
\Gamma \vdash \psi \to (\varphi \rightsquigarrow \chi) & (MP).
\end{array}$$

Conversely, we have:

$$\begin{array}{ll}
\Gamma \vdash \psi \to (\varphi \rightsquigarrow \chi) & \text{hypothesis} \\
\Gamma \vdash (\psi \to (\varphi \rightsquigarrow \chi)) \to (\varphi \rightsquigarrow (\psi \to \chi)) & (pC1_\Gamma) \\
\Gamma \vdash \varphi \rightsquigarrow (\psi \to \chi) & (MP).
\end{array}$$

$(pD_\Gamma)$: We have:

$$\begin{array}{ll}
\Gamma \vdash (\varphi \to \psi) \to (\varphi \to \psi) & (I{\to}_\Gamma) \\
\Gamma \vdash (\varphi \to \psi) \rightsquigarrow (\varphi \to \psi) & (IR1) \\
\Gamma \vdash \varphi \to ((\varphi \to \psi) \rightsquigarrow \psi) & (pEq\#_\Gamma).
\end{array}$$

Similarly, we have:

$$\begin{array}{ll}
\Gamma \vdash (\varphi \rightsquigarrow \psi) \to (\varphi \rightsquigarrow \psi) & (I{\to}_\Gamma) \\
\Gamma \vdash \varphi \rightsquigarrow ((\varphi \rightsquigarrow \psi) \to \psi) & (pEq\#_\Gamma) \\
\Gamma \vdash \varphi \to ((\varphi \rightsquigarrow \psi) \to \psi) & (IR2).
\end{array}$$

$(p^*{\to}_\Gamma)$: We have:

$$\begin{array}{ll}
\Gamma \vdash \psi \to \chi & \text{hypothesis} \\
\Gamma \vdash (\psi \to \chi) \to ((\varphi \to \psi) \to (\varphi \to \chi)) & (pB_{\to\Gamma}) \\
\Gamma \vdash (\varphi \to \psi) \to (\varphi \to \chi) & (MP).
\end{array}$$

$(p^*{\rightsquigarrow}_\Gamma)$: We have:

(1) $\Gamma \vdash \psi \to \chi$ hypothesis
(2) $\Gamma \vdash ((\varphi \rightsquigarrow \psi) \to \psi) \to ((\varphi \rightsquigarrow \psi) \to \chi)$ (p*$\to_\Gamma$)
(3) $\Gamma \vdash \varphi \to ((\varphi \rightsquigarrow \psi) \to \psi)$ (pD$_\Gamma$)
(4) $\Gamma \vdash \varphi \to ((\varphi \rightsquigarrow \psi) \to \chi)$ (3), (2), (Tr$_\Gamma$)
(5) $\Gamma \vdash \varphi \rightsquigarrow ((\varphi \rightsquigarrow \psi) \to \chi)$ (IR1)
(6) $\Gamma \vdash (\varphi \rightsquigarrow \psi) \to (\varphi \rightsquigarrow \chi)$ (pEq#$_\Gamma$).

(p**$\to_\Gamma$): We have:
$\Gamma \vdash \psi \to \chi$ hypothesis
$\Gamma \vdash \chi \to ((\chi \to \varphi) \rightsquigarrow \varphi)$ (pD$_\Gamma$)
$\Gamma \vdash \psi \to ((\chi \to \varphi) \rightsquigarrow \varphi)$ (Tr$_\Gamma$)
$\Gamma \vdash (\chi \to \varphi) \rightsquigarrow (\psi \to \varphi)$ (pEq#$_\Gamma$)
$\Gamma \vdash (\chi \to \varphi) \to (\psi \to \varphi)$ (IR2).

(p**$\rightsquigarrow_\Gamma$): We have:
$\Gamma \vdash \psi \to \chi$ hypothesis
$\Gamma \vdash \chi \to ((\chi \rightsquigarrow \varphi) \to \varphi)$ (pD$_\Gamma$)
$\Gamma \vdash \psi \to ((\chi \rightsquigarrow \varphi) \to \varphi)$ (Tr$_\Gamma$)
$\Gamma \vdash \psi \rightsquigarrow ((\chi \rightsquigarrow \varphi) \to \varphi)$ (IR1)
$\Gamma \vdash (\chi \rightsquigarrow \varphi) \to (\psi \rightsquigarrow \varphi)$ (pEq#$_\Gamma$).

(pBB$_{1\Gamma}$): We have:
$\Gamma \vdash (\psi \to \chi) \to ((\varphi \to \psi) \to (\varphi \to \chi))$ (pB$\to_\Gamma$)
$\Gamma \vdash (\psi \to \chi) \rightsquigarrow ((\varphi \to \psi) \to (\varphi \to \chi))$ (IR1)
$\Gamma \vdash ((\psi \to \chi) \rightsquigarrow ((\varphi \to \psi) \to (\varphi \to \chi))) \to$
$((\varphi \to \psi) \to ((\psi \to \chi) \rightsquigarrow (\varphi \to \chi)))$ (pC2$_\Gamma$)
$\Gamma \vdash (\varphi \to \psi) \to ((\psi \to \chi) \rightsquigarrow (\varphi \to \chi))$ (MP).

□

**Remark 1.7.15** *The axiom*
*(pB$\to$)* $(\psi \to \chi) \to ((\varphi \to \psi) \to (\varphi \to \chi))$
*of the pBCK$\to$ logic does not imply the 'twin' axiom,*
*(pB$\rightsquigarrow$)* $(\psi \rightsquigarrow \chi) \to ((\varphi \rightsquigarrow \psi) \rightsquigarrow (\varphi \rightsquigarrow \chi))$,
*of the pBCK logic. We only have the result from the following proposition, obtained by Prover9.*

**Proposition 1.7.16** *For any $\varphi, \psi, \chi \in L$, we have:*

$$\Gamma \vdash (\varphi \rightsquigarrow \psi) \to (\varphi \rightsquigarrow ((\psi \rightsquigarrow \chi) \to \chi)). \tag{1.53}$$

**Proof.** (By **Prover9**) First, we prove:

$$\Gamma \vdash (\varphi \to \psi) \to (\varphi \to ((\psi \rightsquigarrow \chi) \to \chi)). \tag{1.54}$$

Indeed, we have:
$\Gamma \vdash \psi \to ((\psi \rightsquigarrow \chi) \to \chi)$ (pD$_\Gamma$)
$\Gamma \vdash (\psi \to ((\psi \rightsquigarrow \chi) \to \chi)) \to$
$((\varphi \to \psi) \to (\varphi \to ((\psi \rightsquigarrow \chi) \to \chi)))$ (pB$\to_\Gamma$)
$\Gamma \vdash (\varphi \to \psi) \to (\varphi \to ((\psi \rightsquigarrow \chi) \to \chi))$ (MP).

Then, we prove:

$$\Gamma \vdash (((\varphi \rightsquigarrow \psi) \to \psi) \to \chi) \to (\varphi \to \chi). \tag{1.55}$$

Indeed, we have:

$\Gamma \vdash \varphi \to ((\varphi \rightsquigarrow \psi) \to \psi)$ (pD$_\Gamma$)
$\Gamma \vdash (\varphi \to ((\varphi \rightsquigarrow \psi) \to \psi)) \to$
$((((\varphi \rightsquigarrow \psi) \to \psi) \to \chi) \rightsquigarrow (\varphi \to \chi))$ (pBB$_{1\Gamma}$)
$\Gamma \vdash (((\varphi \rightsquigarrow \psi) \to \psi) \to \chi) \rightsquigarrow (\varphi \to \chi)$ (MP)
$\Gamma \vdash (((\varphi \rightsquigarrow \psi) \to \psi) \to \chi) \to (\varphi \to \chi)$ (IR2).

Now, we prove (1.53):

$\Gamma \vdash ((\varphi \rightsquigarrow \psi) \to \psi) \to ((\varphi \rightsquigarrow \psi) \to ((\psi \rightsquigarrow \chi) \to \chi))$ (1.54)
$\Gamma \vdash (((\varphi \rightsquigarrow \psi) \to \psi) \to ((\varphi \rightsquigarrow \psi) \to ((\psi \rightsquigarrow \chi) \to \chi))) \to$
$(\varphi \to ((\varphi \rightsquigarrow \psi) \to ((\psi \rightsquigarrow \chi) \to \chi)))$ (1.55)
$\Gamma \vdash \varphi \to ((\varphi \rightsquigarrow \psi) \to ((\psi \rightsquigarrow \chi) \to \chi))$ (MP)
$\Gamma \vdash \varphi \rightsquigarrow ((\varphi \rightsquigarrow \psi) \to ((\psi \rightsquigarrow \chi) \to \chi))$ (IR1)
$\Gamma \vdash (\varphi \rightsquigarrow ((\varphi \rightsquigarrow \psi) \to ((\psi \rightsquigarrow \chi) \to \chi))) \to$
$((\varphi \rightsquigarrow \psi) \to (\varphi \rightsquigarrow ((\psi \rightsquigarrow \chi) \to \chi)))$ (pC2$_\Gamma$)
$\Gamma \vdash (\varphi \rightsquigarrow \psi) \to (\varphi \rightsquigarrow ((\psi \rightsquigarrow \chi) \to \chi))$ (MP).

□

## • The algebra of pBCK→ logic: the pBCK algebra

Let $\Gamma \subseteq L$. We define a binary relation $\leq_\Gamma$ on $L$ by: for all $\varphi, \psi \in L$,

$$\varphi \leq_\Gamma \psi \overset{def.}{\iff} \Gamma \vdash \varphi \to \psi \quad (\overset{(IdEqR_\Gamma)}{\iff} \Gamma \vdash \varphi \rightsquigarrow \psi). \tag{1.56}$$

By Proposition 1.7.14 and by (1.56), we obtain immediately the following

**Proposition 1.7.17** *For any $\varphi, \psi, \chi \in L$,*

*(pB→$_\Gamma$')* $\psi \to \chi \leq_\Gamma (\varphi \to \psi) \to (\varphi \to \chi)$,
*(pC1$_\Gamma$')* $\varphi \to (\psi \rightsquigarrow \chi) \leq_\Gamma \psi \rightsquigarrow (\varphi \to \chi)$,
*(pC2$_\Gamma$')* $\varphi \rightsquigarrow (\psi \to \chi) \leq_\Gamma \psi \to (\varphi \rightsquigarrow \chi)$,
*(pK→$_\Gamma$')* $\varphi \leq_\Gamma \psi \to \varphi$,
*(pK⇝$_\Gamma$')* $\varphi \leq_\Gamma \psi \rightsquigarrow \varphi$;

*(MK$_\Gamma$')* $\Gamma \vdash \psi \implies \varphi \leq_\Gamma a\psi$;
*(MKK$_\Gamma$')* $\Gamma \vdash \varphi, \psi \implies \varphi \leq_\Gamma \psi$;
*(I→$_\Gamma$')* $\varphi \leq_\Gamma \varphi$;
*(Tr$_\Gamma$')* $\varphi \leq_\Gamma \psi$ *and* $\psi \leq_\Gamma \chi \implies \varphi \leq_\Gamma \chi$,
*(pEq#$_\Gamma$')* $\varphi \leq_\Gamma \psi \to \chi \iff \psi \leq_\Gamma \varphi \rightsquigarrow \chi$,
*(pD$_\Gamma$')* $\varphi \leq_\Gamma (\varphi \to \psi) \rightsquigarrow \psi$ *and* $\varphi \leq_\Gamma (\varphi \rightsquigarrow \psi) \to \psi$,
*(p*→$_\Gamma$')* $\psi \leq_\Gamma \chi \implies \varphi \to \psi \leq_\Gamma \varphi \to \chi$,
*(p*⇝$_\Gamma$')* $\psi \leq_\Gamma \chi \implies \varphi \rightsquigarrow \psi \leq_\Gamma \varphi \rightsquigarrow \chi$,
*(p**→$_\Gamma$')* $\psi \leq_\Gamma \chi \implies \chi \to \varphi \leq_\Gamma \psi \to \varphi$,
*(p**⇝$_\Gamma$')* $\psi \leq_\Gamma \chi \implies \chi \rightsquigarrow \varphi \leq_\Gamma \psi \rightsquigarrow \varphi$,
*(pBB1$_\Gamma$')* $\varphi \to \psi \leq_\Gamma (\psi \to \chi) \rightsquigarrow (\varphi \to \chi)$.

**Proposition 1.7.18** *The binary relation $\leq_\Gamma$ is a preorder.*

**Proof.** - $\leq_\Gamma$ is *reflexive*, i.e. for all $\varphi \in L$, $\varphi \leq_\Gamma \varphi$, which is true by (I→$_\Gamma$').

- $\leq_\Gamma$ is *transitive*, i.e. for all $\varphi, \psi, \chi \in L$, $\varphi \leq_\Gamma \psi$ and $\psi \leq_\Gamma \chi$ imply $\varphi \leq_\Gamma \chi$, which is true by ($Tr_\Gamma$'). □

Let $\Gamma \subseteq L$. We define a binary relation $\sim_\Gamma$ on $L$ by: for all $\varphi, \psi \in L$,

$$\varphi \sim_\Gamma \psi \overset{def.}{\Longleftrightarrow} (\varphi \leq_\Gamma \psi \quad and \quad \psi \leq_\Gamma \varphi). \tag{1.57}$$

**Proposition 1.7.19** *The binary relation $\sim_\Gamma$ is an equivalence relation.*

**Proof.** - $\sim_\Gamma$ is *reflexive*, i.e. for all $\varphi \in L$, $\varphi \sim_\Gamma \varphi$; indeed, $\varphi \sim_\Gamma \varphi$ means, by definition, $\varphi \leq_\Gamma \varphi$, which is true since $\leq_\Gamma$ is reflexive, by ($I{\to}_\Gamma$').

- $\sim_\Gamma$ is *transitive*, i.e. for all $\varphi, \psi, \chi \in L$, $\varphi \sim_\Gamma \psi$ and $\psi \sim_\Gamma \chi$ imply $\varphi \sim_\Gamma \chi$; indeed, $\varphi \sim_\Gamma \psi$ and $\psi \sim_\Gamma \chi$ mean, by definition,
($\varphi \leq_\Gamma \psi$ and $\psi \leq_\Gamma \varphi$) and ($\psi \leq_\Gamma \chi$ and $\chi \leq_\Gamma \psi$);
since $\varphi \leq_\Gamma \psi$ and $\psi \leq_\Gamma \chi$, it follows that $\varphi \leq_\Gamma \chi$, and since $\chi \leq_\Gamma \psi$ and $\psi \leq_\Gamma \varphi$, it follows that $\chi \leq_\Gamma \varphi$, by ($Tr_\Gamma$'); thus, $\varphi \sim_\Gamma \chi$.

- $\sim_\Gamma$ is *symmetric*, i.e. for all $\varphi, \psi \in L$, $\varphi \sim_\Gamma \psi$ implies $\psi \sim_\Gamma \varphi$: obviously. □

**Proposition 1.7.20** *The binary relation $\sim_\Gamma$ is a congruence relation.*

**Proof.** Since $\sim_\Gamma$ is an equivalence relation, it remains to prove that $\sim_\Gamma$ is compatible with $\to$ and $\rightsquigarrow$, i.e.
if $\varphi \sim_\Gamma \varphi'$ and $\psi \sim_\Gamma \psi'$, then $(\varphi \to \psi) \sim_\Gamma (\varphi' \to \psi')$ and $(\varphi \rightsquigarrow \psi) \sim_\Gamma (\varphi' \rightsquigarrow \psi')$.

Indeed, $\varphi \sim_\Gamma \varphi'$ and $\psi \sim_\Gamma \psi'$ mean, by definition, ($\varphi \leq_\Gamma \varphi'$ and $\varphi' \leq_\Gamma \varphi$) and ($\psi \leq_\Gamma \psi'$ and $\psi' \leq_\Gamma \psi$).

Since $\psi \leq_\Gamma \psi'$, it follows, by ($p^*{\to}_\Gamma$'), that $\varphi \to \psi \leq_\Gamma \varphi \to \psi'$, and since $\varphi' \leq_\Gamma \varphi$, it follows, by ($p^{**}{\to}_\Gamma$'), that $\varphi \to \psi' \leq_\Gamma \varphi' \to \psi'$; then, by ($Tr_\Gamma$'), it follows that $\varphi \to \psi \leq_\Gamma \varphi' \to \psi'$.
Since $\psi' \leq_\Gamma \psi$, it follows, by ($p^*{\to}_\Gamma$'), that $\varphi \to \psi' \leq_\Gamma \varphi \to \psi$, and since $\varphi \leq_\Gamma \varphi'$, it follows, by ($p^{**}{\to}_\Gamma$'), that $\varphi' \to \psi' \leq_\Gamma \varphi \to \psi'$; then, by ($Tr_\Gamma$'), it follows that $\varphi' \to \psi' \leq_\Gamma \varphi \to \psi$.
Thus, $(\varphi \to \psi) \sim_\Gamma (\varphi' \to \psi')$.

Similarly, since $\psi \leq_\Gamma \psi'$, it follows, by ($p^*{\rightsquigarrow}_\Gamma$'), that $\varphi \rightsquigarrow \psi \leq_\Gamma \varphi \rightsquigarrow \psi'$, and since $\varphi' \leq_\Gamma \varphi$, it follows, by ($p^{**}{\rightsquigarrow}_\Gamma$'), that $\varphi \rightsquigarrow \psi' \leq_\Gamma \varphi' \rightsquigarrow \psi'$; then, by ($Tr_\Gamma$'), it follows that $\varphi \rightsquigarrow \psi \leq_\Gamma \varphi' \rightsquigarrow \psi'$.
Since $\psi' \leq_\Gamma \psi$, it follows, by ($p^*{\rightsquigarrow}_\Gamma$'), that $\varphi \rightsquigarrow \psi' \leq_\Gamma \varphi \rightsquigarrow \psi$, and since $\varphi \leq_\Gamma \varphi'$, it follows, by ($p^{**}{\rightsquigarrow}_\Gamma$'), that $\varphi' \rightsquigarrow \psi' \leq_\Gamma \varphi \rightsquigarrow \psi'$; then, by ($Tr_\Gamma$'), it follows that $\varphi' \rightsquigarrow \psi' \leq_\Gamma \varphi \rightsquigarrow \psi$.
Thus, $(\varphi \rightsquigarrow \psi) \sim_\Gamma (\varphi' \rightsquigarrow \psi')$. Hence, $\sim_\Gamma$ is a congruence relation. □

Since $\sim_\Gamma$ is an equivalence relation on $L$, let $|\varphi| \overset{notation}{=} |\varphi|_\Gamma$ be the equivalence class of $\varphi \in L$:

$$|\varphi| \overset{notation}{=} |\varphi|_\Gamma \overset{def.}{=} \{\psi \in L \mid \psi \sim_\Gamma \varphi\};$$

let $L/\Gamma \overset{notation}{=} L/{\sim_\Gamma}$ be the quotient set:

$$L/\Gamma \overset{notation}{=} L/{\sim_\Gamma} \overset{def.}{=} \{|\varphi| \mid \varphi \in L\}$$

and let

$$L_\Gamma/\Gamma \overset{notation}{=} L_\Gamma/{\sim_\Gamma} \overset{def.}{=} \{\mid \varphi \mid \mid \varphi \in L_\Gamma\}.$$

Note that

$$(1.58) \qquad \mid \varphi \mid \in L_\Gamma/\Gamma \iff \varphi \in L_\Gamma.$$

**Lemma 1.7.21** *The equivalence classes do not depend on their representants, i.e. for all* $\varphi, \psi \in L$,

$$\mid \varphi \mid = \mid \psi \mid \iff \varphi \sim_\Gamma \psi.$$

**Proof.** $\Longrightarrow$: Suppose $\mid \varphi \mid = \mid \psi \mid$; since $\varphi \in \mid \varphi \mid$, it follows that $\varphi \in \mid \psi \mid$, i.e. $\varphi \sim_\Gamma \psi$.

$\Longleftarrow$: Suppose $\varphi \sim_\Gamma \psi$ and let $\chi \in \mid \varphi \mid$, i.e. $\chi \sim_\Gamma \varphi$; it follows that $\chi \sim_\Gamma \psi$, i.e. $\chi \in \mid \psi \mid$; hence, $\mid \varphi \mid \subseteq \mid \psi \mid$. Similarly, $\mid \psi \mid \subseteq \mid \varphi \mid$. Thus, $\mid \varphi \mid = \mid \psi \mid$. □

Let us define on the quotient set $L/\Gamma$ two binary operations $\Rightarrow$ and $\approx>$ by: for all $\mid \varphi \mid, \mid \psi \mid \in L/\Gamma$,

$$\mid \varphi \mid \Rightarrow \mid \psi \mid \overset{def.}{=} \mid \varphi \to \psi \mid,$$

$$\mid \varphi \mid \approx> \mid \psi \mid \overset{def.}{=} \mid \varphi \rightsquigarrow \psi \mid.$$

By Proposition 1.7.20, the operations are well defined.

Let us define also on $L/\Gamma$ a binary relation $\leq$ by: for all $\mid \varphi \mid, \mid \psi \mid \in L/\Gamma$,

$$(1.59) \qquad \mid \varphi \mid \leq \mid \psi \mid \overset{def.}{\iff} \varphi \leq_\Gamma \psi.$$

Note that we have:

$$\begin{aligned}
\mid \varphi \mid \leq \mid \psi \mid &\overset{def.}{\iff} \varphi \leq_\Gamma \psi \\
&\overset{def.}{\iff} \Gamma \vdash \varphi \to \psi \quad (\overset{(IdEqR_\Gamma)}{\iff} \Gamma \vdash \varphi \rightsquigarrow \psi) \\
&\overset{(1.51)}{\iff} \varphi \to \psi \in L_\Gamma \quad (\iff \varphi \rightsquigarrow \psi \in L_\Gamma) \\
&\overset{(1.58)}{\iff} \mid \varphi \to \psi \mid \in L_\Gamma/\Gamma \quad (\overset{(IdEqR_{L_\Gamma/\Gamma})}{\iff} \mid \varphi \rightsquigarrow \psi \mid \in L_\Gamma/\Gamma) \\
&\overset{def.}{\iff} \mid \varphi \mid \Rightarrow \mid \psi \mid \in L_\Gamma/\Gamma \quad (\overset{(IdEqR_{L_\Gamma/\Gamma})}{\iff} \mid \varphi \mid \approx> \mid \psi \mid \in L_\Gamma/\Gamma),
\end{aligned}$$

hence, we have:

(1.60) $(IdEqR_{L_\Gamma/\Gamma})$ $\quad \mid \varphi \mid \Rightarrow \mid \psi \mid \in L_\Gamma/\Gamma \iff \mid \varphi \mid \approx> \mid \psi \mid \in L_\Gamma/\Gamma$ *and*

(1.61)

$(pEqrelR_{L_\Gamma/\Gamma})$ $\mid \varphi \mid \leq \mid \psi \mid \iff \mid \varphi \mid \Rightarrow \mid \psi \mid \in L_\Gamma/\Gamma$ $(\overset{(IdEqR_{L_\Gamma/\Gamma})}{\iff} \mid \varphi \mid \approx> \mid \psi \mid \in L_\Gamma/\Gamma)$.

By Proposition 1.7.17 and by (1.59), we obtain immediately the following.

**Proposition 1.7.22** *For any* $|\varphi|, |\psi|, |\chi| \in L/\Gamma$,

| | |
|---|---|
| *(pB$_{\Rightarrow}$')* | $|\psi| \Rightarrow |\chi| \leq (|\varphi| \Rightarrow |\psi|) \Rightarrow (|\varphi| \Rightarrow |\chi|)$, |
| *(pC$_1$')* | $|\varphi| \Rightarrow (|\psi| \approx> |\chi|) \leq |\psi| \approx> (|\varphi| \Rightarrow |\chi|)$, |
| *(pC$_2$')* | $|\varphi| \approx> (|\psi| \Rightarrow |\chi|) \leq |\psi| \Rightarrow (|\varphi| \approx> |\chi|)$, |
| *(pK$_{\Rightarrow}$')* | $|\varphi| \leq |\psi| \Rightarrow |\varphi|$, |
| *(pK$_{\approx>}$')* | $|\varphi| \leq |\psi| \approx> |\varphi|$; |

| | |
|---|---|
| *(MK')* | $\Gamma \vdash \psi \Longrightarrow |\varphi| \leq |\psi|$; |
| *(MKK')* | $\Gamma \vdash \varphi, \psi \Longrightarrow |\varphi| \leq |\psi|$; |
| *(I$_{\Rightarrow}$') = (Re')* | $|\varphi| \leq |\varphi|$; |
| *(Tr')* | $|\varphi| \leq |\psi|$ *and* $|\psi| \leq |\chi| \Longrightarrow |\varphi| \leq |\chi|$, |
| *(pEq#')* | $|\varphi| \leq |\psi| \Rightarrow |\chi| \Longleftrightarrow |\psi| \leq |\varphi| \approx> |\chi|$, |
| *(pD')* | $|\varphi| \leq (|\varphi| \Rightarrow |\psi|) \approx> |\psi|$ *and* $|\varphi| \leq (|\varphi| \approx> |\psi|) \Rightarrow |\psi|$, |
| *(p*$_{\Rightarrow}$')* | $|\psi| \leq |\chi| \Longrightarrow |\varphi| \Rightarrow |\psi| \leq |\varphi| \Rightarrow |\chi|$, |
| *(p*$_{\approx>}$')* | $|\psi| \leq |\chi| \Longrightarrow |\varphi| \approx> |\psi| \leq |\varphi| \approx> |\chi|$, |
| *(p**$_{\Rightarrow}$')* | $|\psi| \leq |\chi| \Longrightarrow |\chi| \Rightarrow |\varphi| \leq |\psi| \Rightarrow |\varphi|$, |
| *(p**$_{\approx>}$')* | $|\psi| \leq |\chi| \Longrightarrow |\chi| \approx> |\varphi| \leq |\psi| \approx> |\varphi|$, |
| *(pBB$_1$')* | $|\varphi| \Rightarrow |\psi| \leq (|\psi| \Rightarrow |\chi|) \approx> (|\varphi| \Rightarrow |\chi|)$. |

**Proposition 1.7.23** *The binary relation* $\leq$ *on* $L/\Gamma$ *is an order.*

**Proof.** - *Reflexivity*: for any $|\varphi| \in L/\Gamma$, $|\varphi| \leq |\varphi|$; that is true by (Re').

- *Transitivity*: for any $|\varphi|, |\psi|, |\chi| \in L/\Gamma$, $|\varphi| \leq |\psi|$ and $|\psi| \leq |\chi|$ imply $|\varphi| \leq |\chi|$; that is true by (Tr').

- *Antisymmetry*: $|\varphi| \leq |\psi|$ and $|\psi| \leq |\varphi|$ mean, by definition, $\varphi \leq_\Gamma \psi$ and $\psi \leq_\Gamma \varphi$, i.e. $\varphi \sim_\Gamma \psi$, hence, $|\varphi| = |\psi|$, by Lemma 1.7.21. □

**Proposition 1.7.24** $L_\Gamma$ *is an equivalence class, denoted by* $\mathbf{1}$, *and*

$$|\varphi| = \mathbf{1} \quad \stackrel{def.}{\Longleftrightarrow} \quad |\varphi| = L_\Gamma \quad \Longleftrightarrow \quad \varphi \in L_\Gamma \quad (\Longleftrightarrow \Gamma \vdash \varphi).$$

**Proof.**

$\Longrightarrow$: Suppose $|\varphi| = L_\Gamma$; then, for any $\psi \in L_\Gamma$, we have $\psi \in |\varphi|$, i.e. $\psi \sim_\Gamma \varphi$, i.e. $\psi \to \varphi \in L_\Gamma$ and $\varphi \to \psi \in L_\Gamma$; it follows, by (MP), that $\varphi \in L_\Gamma$.

$\Longleftarrow$: Suppose $\varphi \in L_\Gamma$; we must prove that $|\varphi| = L_\Gamma$. Indeed,

· $|\varphi| \subseteq L_\Gamma$: let $\psi \in |\varphi|$, i.e. $\psi \sim_\Gamma \varphi$, i.e. $\psi \to \varphi \in L_\Gamma$ and $\varphi \to \psi \in L_\Gamma$; since $\varphi \in L_\Gamma$, it follows, by (MP), that $\psi \in L_\Gamma$; thus, $|\varphi| \subseteq L_\Gamma$.

· $L_\Gamma \subseteq |\varphi|$: let $\psi \in L_\Gamma$, i.e. $\Gamma \vdash \psi$; then, by (MK$_\Gamma$), $\Gamma \vdash \chi \to \psi$, for all $\chi \in L$, hence for $\varphi \in L_\Gamma$; since $\varphi \in L_\Gamma$, i.e. $\Gamma \vdash \varphi$, then, by (MK$_\Gamma$) again, $\Gamma \vdash \chi \to \varphi$, for all $\chi \in L$, hence for $\psi \in L_\Gamma$; hence, $\Gamma \vdash \varphi \to \psi$ and $\Gamma \vdash \psi \to \varphi$, i.e. $\varphi \sim_\Gamma \psi$, hence $\psi \in |\varphi|$; thus, $L_\Gamma \subseteq |\varphi|$. □

**Remark 1.7.25** *Note that in this case* ($L_\Gamma$ *is an equivalence class, denoted by* $\mathbf{1}$), *the equivalences (1.60) and (1.61) become*

(1.62) $\qquad (IdEqR) \quad |\varphi| \Rightarrow |\psi| = \mathbf{1} \Longleftrightarrow |\varphi| \approx> |\psi| = \mathbf{1}$ *and*

(1.63) $\quad (pEqrelR) \quad |\varphi| \leq |\psi| \Longleftrightarrow |\varphi| \Rightarrow |\psi| = \mathbf{1} \quad (\stackrel{(IdEqR)}{\Longleftrightarrow} |\varphi| \approx> |\psi| = \mathbf{1})$.

Now, we can prove the following result.

**Theorem 1.7.26** *The structure $\mathcal{L}/\Gamma = (L/\Gamma, \leq, \Rightarrow, \approx>, \mathbf{1})$ is a pseudo-BCK algebra, called the Lindenbaum-Tarski algebra of the propositional pBCK→ logic.*

**Proof.** By Proposition 1.7.22, (pB$_{\Rightarrow}$'), (pC')= (pC$_1$') + (pC$_2$'), (pK')= (pK$_{\Rightarrow}$') + (pK$_{\approx>}$') hold; by Proposition 1.7.23, (An') holds; (1.62) and (1.63) hold; thus, $\mathcal{L}/\Gamma$ is a pBCK algebra (Definition 7'). □

**Remark 1.7.27** *If we consider the pBCK logic (extension of pBCK→ logic), the Lindenbaum-Tarski construction is the same and the Lindenbaum-Tarski algebra is again a pBCK algebra.*

### 1.7.2 Pseudo-BCI→ (or quantum-B) logic

**Definition 1.7.28** An *axiom* of the formal system of propositional pseudo-BCI→ logic (or QB logic) is a proposition which has one of the following forms:
(pB→) $(\psi \to \chi) \to ((\varphi \to \psi) \to (\varphi \to \chi))$,
(pC1) $(\varphi \to (\psi \rightsquigarrow \chi)) \to (\psi \rightsquigarrow (\varphi \to \chi))$,
(pC2) $(\varphi \rightsquigarrow (\psi \to \chi)) \to (\psi \to (\varphi \rightsquigarrow \chi))$,
(I→) $\varphi \to \varphi$ (identity principle).

**Definition 1.7.29** The *inference rules* of the formal system of propositional pseudo-BCI→ logic (or quantum-B logic) are the following:

$$(MP)\ (modus\ ponens)\quad \frac{\varphi, \varphi \to \psi}{\psi},$$

$$(IR1)\quad \frac{\varphi \to \psi}{\varphi \rightsquigarrow \psi} \quad and \quad (IR2)\quad \frac{\varphi \rightsquigarrow \psi}{\varphi \to \psi}.$$

**Definition 1.7.30** A proposition $\varphi \in L$ is *derivable* from a subset $\Gamma \subseteq L$, fact denoted by

$$\Gamma \vdash \varphi,$$

if $\varphi$ is obtained in finitely many steps as follows:
(1) $\varphi$ is an axiom,
(2) $\varphi \in \Gamma$,
(3) if $\psi$ is derivable, then $\varphi$ is obtained by substitution of the variables in $\psi$ by arbitrary propositions,
(4) $\varphi$ is obtained by an inference rule.

The set of all propositions $\varphi \in L$ derivable from $\Gamma$ will be denoted by $L_\Gamma$:

$$L_\Gamma \overset{def.}{=} \{\varphi \in L \mid \Gamma \vdash \varphi\}$$

(1.64) $$\varphi \in L_\Gamma \Longleftrightarrow \Gamma \vdash \varphi.$$

If $\Gamma = \emptyset \subset L$, we write $\vdash \varphi$ instead of $\emptyset \vdash \varphi$ and

$$L_\emptyset \stackrel{def.}{=} \{\varphi \in L \mid \emptyset \vdash \varphi\}$$

$$\varphi \in L_\emptyset \Longleftrightarrow \vdash \varphi. \tag{1.65}$$

- **Sintactical properties of pBCI→ logic (or QB logic)**

**Proposition 1.7.31** *For any $\varphi, \psi, \chi \in L$, we have the following properties:*

(pB→$_\Gamma$) $\Gamma \vdash (\psi \to \chi) \to ((\varphi \to \psi) \to (\varphi \to \chi))$,
(pC1$_\Gamma$) $\Gamma \vdash (\varphi \to (\psi \rightsquigarrow \chi)) \to (\psi \rightsquigarrow (\varphi \to \chi))$,
(pC2$_\Gamma$) $\Gamma \vdash (\varphi \rightsquigarrow (\psi \to \chi)) \to (\psi \to (\varphi \rightsquigarrow \chi))$,
(I→$_\Gamma$) $\Gamma \vdash \varphi \to \varphi$;

(IdEqR$_\Gamma$) $\Gamma \vdash \varphi \to \psi \iff \Gamma \vdash \varphi \rightsquigarrow \psi$;

(Tr$_\Gamma$) $\Gamma \vdash \varphi \to \psi$ *and* $\Gamma \vdash \psi \to \chi \implies \Gamma \vdash \varphi \to \chi$,
(pEq#$_\Gamma$) $\Gamma \vdash \varphi \rightsquigarrow (\psi \to \chi) \iff \Gamma \vdash \psi \to (\varphi \rightsquigarrow \chi)$,
(pD$_\Gamma$) $\Gamma \vdash \varphi \to ((\varphi \to \psi) \rightsquigarrow \psi)$ *and* $\Gamma \vdash \varphi \to ((\varphi \rightsquigarrow \psi) \to \psi)$,
(p*→$_\Gamma$) $\Gamma \vdash \psi \to \chi \implies \Gamma \vdash (\varphi \to \psi) \to (\varphi \to \chi)$,
(p*⇝$_\Gamma$) $\Gamma \vdash \psi \to \chi \implies \Gamma \vdash (\varphi \rightsquigarrow \psi) \to (\varphi \rightsquigarrow \chi)$,
(p**→$_\Gamma$) $\Gamma \vdash \psi \to \chi \implies \Gamma \vdash (\chi \to \varphi) \to (\psi \to \varphi)$,
(p**⇝$_\Gamma$) $\Gamma \vdash \psi \to \chi \implies \Gamma \vdash (\chi \rightsquigarrow \varphi) \to (\psi \rightsquigarrow \varphi)$,
(pBB1$_\Gamma$) $\Gamma \vdash (\varphi \to \psi) \to ((\psi \to \chi) \rightsquigarrow (\varphi \to \chi))$.

**Proof.** (I→$_\Gamma$): By Definition 1.7.30 and (1.64), since (I→) is an axiom.
For the rest of proofs, see Proposition 1.7.14. □

**Lemma 1.7.32** *For any $\varphi \in L$,*

$$\Gamma \vdash ((\varphi \to \varphi) \rightsquigarrow \varphi) \to \varphi \quad and \quad \Gamma \vdash ((\varphi \rightsquigarrow \varphi) \to \varphi) \rightsquigarrow \varphi.$$

**Proof.** We have:
$\Gamma \vdash \varphi \to \varphi$ (I→$_\Gamma$)
$\Gamma \vdash (\varphi \to \varphi) \to (((\varphi \to \varphi) \rightsquigarrow \varphi) \to \varphi)$ (pD$_\Gamma$)
$\Gamma \vdash ((\varphi \to \varphi) \rightsquigarrow \varphi) \to \varphi$ (MP)
and, similarly,
$\Gamma \vdash \varphi \to \varphi$ (I→$_\Gamma$)
$\Gamma \vdash \varphi \rightsquigarrow \varphi$ (IR1)
$\Gamma \vdash (\varphi \rightsquigarrow \varphi) \to (((\varphi \rightsquigarrow \varphi) \to \varphi) \rightsquigarrow \varphi)$ (pD$_\Gamma$)
$\Gamma \vdash ((\varphi \rightsquigarrow \varphi) \to \varphi) \rightsquigarrow \varphi$ (MP). □

**Lemma 1.7.33** *For any $\varphi, \psi \in L$,*

$$\Gamma \vdash (\varphi \to \varphi) \rightsquigarrow \psi \implies \Gamma \vdash (\psi \to \psi) \rightsquigarrow \psi.$$

**Proof.** We have:

| | |
|---|---|
| (1) $\Gamma \vdash (\varphi \to \varphi) \rightsquigarrow \psi$ | hypothesis |
| (2) $\Gamma \vdash (\varphi \to \varphi) \to \psi$ | (IR2) |
| (3) $\Gamma \vdash \varphi \to \varphi$ | ($I{\to}_\Gamma$) |
| (4) $\Gamma \vdash \psi$ | (3), (2), (MP) |
| (5) $\Gamma \vdash \psi \to ((\psi \to \psi) \rightsquigarrow \psi)$ | ($pD_\Gamma$) |
| (6) $\Gamma \vdash (\psi \to \psi) \rightsquigarrow \psi$ | (4), (5), (MP). |

□

**Remark 1.7.34** *The axiom*
*(pB→)* $(\psi \to \chi) \to ((\varphi \to \psi) \to (\varphi \to \chi))$
*of the QB logic does not imply the 'twin' axiom,*
*(pB⇝)* $(\psi \rightsquigarrow \chi) \to ((\varphi \rightsquigarrow \psi) \rightsquigarrow (\varphi \rightsquigarrow \chi))$,
*of the pBCI logc, as happens for pBCI algebras (see Theorem 1.2.18). We only have the result from the following proposition, obtained by Prover9.*

**Proposition 1.7.35** *For any* $\varphi, \psi, \chi \in L$, *we have:*

$$\Gamma \vdash (\varphi \rightsquigarrow \psi) \to (\varphi \rightsquigarrow ((\psi \rightsquigarrow \chi) \to \chi)). \tag{1.66}$$

**Proof.** See Proposition 1.7.16. □

**• The 'algebra' of pBCI→ logic (or QB logic): the locally unital QB algebra**

Let $\Gamma \subseteq L$. We define a binary relation $\leq_\Gamma$ on $L$ by: for all $\varphi, \psi \in L$,

$$\varphi \leq_\Gamma \psi \overset{def.}{\Longleftrightarrow} \Gamma \vdash \varphi \to \psi \quad (\overset{(IdEqR_\Gamma)}{\Longleftrightarrow}) \quad \Gamma \vdash \varphi \rightsquigarrow \psi). \tag{1.67}$$

By Proposition 1.7.31 and by (1.67), we obtain immediately the following.

**Proposition 1.7.36** *For any* $\varphi, \psi, \chi \in L$,
*($pB{\to}_\Gamma$')* $\psi \to \chi \leq_\Gamma (\varphi \to \psi) \to (\varphi \to \chi)$,
*($pC1_\Gamma$')* $\varphi \to (\psi \rightsquigarrow \chi) \leq_\Gamma \psi \rightsquigarrow (\varphi \to \chi)$,
*($pC2_\Gamma$')* $\varphi \rightsquigarrow (\psi \to \chi) \leq_\Gamma \psi \to (\varphi \rightsquigarrow \chi)$,
*($I{\to}_\Gamma$')* $\varphi \leq_\Gamma \varphi$;

*($Tr_\Gamma$')* $\varphi \leq_\Gamma \psi$ *and* $\psi \leq_\Gamma \chi \implies \varphi \leq_\Gamma \chi$,
*($pEq\#_\Gamma$')* $\varphi \leq_\Gamma \psi \to \chi \iff \psi \leq_\Gamma \varphi \rightsquigarrow \chi$,
*($pD_\Gamma$')* $\varphi \leq_\Gamma (\varphi \to \psi) \rightsquigarrow \psi$ *and* $\varphi \leq_\Gamma (\varphi \rightsquigarrow \psi) \to \psi$,
*($p^*{\to}_\Gamma$')* $\psi \leq_\Gamma \chi \implies \varphi \to \psi \leq_\Gamma \varphi \to \chi$,
*($p^*{\rightsquigarrow}_\Gamma$')* $\psi \leq_\Gamma \chi \implies \varphi \rightsquigarrow \psi \leq_\Gamma \varphi \rightsquigarrow \chi$,
*($p^{**}{\to}_\Gamma$')* $\psi \leq_\Gamma \chi \implies \chi \to \varphi \leq_\Gamma \psi \to \varphi$,
*($p^{**}{\rightsquigarrow}_\Gamma$')* $\psi \leq_\Gamma \chi \implies \chi \rightsquigarrow \varphi \leq_\Gamma \psi \rightsquigarrow \varphi$,
*($pBB1_\Gamma$')* $\varphi \to \psi \leq_\Gamma (\psi \to \chi) \rightsquigarrow (\varphi \to \chi)$.

By Lemma 1.7.32 and by (1.67), we obtain immediately the following.

**Lemma 1.7.37** *For any* $\varphi \in L$,

$$(\varphi \to \varphi) \rightsquigarrow \varphi \ \leq_\Gamma \ \varphi \quad \textit{and} \quad (\varphi \rightsquigarrow \varphi) \to \varphi \ \leq_\Gamma \ \varphi.$$

By Lemma 1.7.33 and by (1.67), we obtain immediately the following.

**Lemma 1.7.38** *For any* $\varphi, \psi \in L$,

$$\varphi \to \varphi \ \leq_\Gamma \ \psi \quad \Longrightarrow \quad \psi \to \psi \ \leq_\Gamma \ \psi.$$

By Proposition 1.7.35 and by (1.67), we obtain immediately the following.

**Proposition 1.7.39** *For any* $\varphi, \psi, \chi \in L$, *we have:*

(1.68) $$\varphi \rightsquigarrow \psi \ \leq_\Gamma \ \varphi \rightsquigarrow ((\psi \rightsquigarrow \chi) \to \chi).$$

**Proposition 1.7.40** *The binary relation* $\leq_\Gamma$ *is a preorder.*

**Proof.** See Proposition 1.7.18. □

Let $\Gamma \subseteq L$. We define a binary relation $\sim_\Gamma$ on $L$ by: for all $\varphi, \psi \in L$,

(1.69) $$\varphi \sim_\Gamma \psi \overset{def.}{\Longleftrightarrow} (\varphi \leq_\Gamma \psi \quad \textit{and} \quad \psi \leq_\Gamma \varphi).$$

**Proposition 1.7.41** *The binary relation* $\sim_\Gamma$ *is an equivalence relation.*

**Proof.** See Proposition 1.7.19. □

**Proposition 1.7.42** *The binary relation* $\sim_\Gamma$ *is a congruence relation.*

**Proof.** See Proposition 1.7.20. □

Since $\sim_\Gamma$ is an equivalence relation on $L$, let $| \varphi | \overset{notation}{=} | \varphi |_\Gamma$ be the equivalence class of $\varphi \in L$:

$$| \varphi | \overset{notation}{=} | \varphi |_\Gamma \overset{def.}{=} \{\psi \in L \mid \psi \sim_\Gamma \varphi\};$$

let $L/\Gamma \overset{notation}{=} L/\sim_\Gamma$ be the quotient set:

$$L/\Gamma \overset{notation}{=} L/\sim_\Gamma \overset{def.}{=} \{| \varphi | \mid \varphi \in L\}$$

and let

$$L_\Gamma/\Gamma \overset{notation}{=} L_\Gamma/\sim_\Gamma \overset{def.}{=} \{| \varphi | \mid \varphi \in L_\Gamma\}.$$

Note that

(1.70) $$| \varphi | \in L_\Gamma/\Gamma \quad \Longleftrightarrow \quad \varphi \in L_\Gamma.$$

**Lemma 1.7.43** *The equivalence classes do not depend on their representants, i.e. for all* $\varphi, \psi \in L$,

$$| \varphi | = | \psi | \Longleftrightarrow \varphi \sim_\Gamma \psi.$$

**Proof.** See Lemma 1.7.21. □

Let us define on the quotient set $L/\Gamma$ two binary operations $\Rightarrow$ and $\approx>$ by: for all $|\varphi|, |\psi| \in L/\Gamma$,

$$|\varphi| \Rightarrow |\psi| \stackrel{def.}{=} |\varphi \to \psi|,$$

$$|\varphi| \approx> |\psi| \stackrel{def.}{=} |\varphi \rightsquigarrow \psi|.$$

By Proposition 1.7.42, the operations are well defined.

Let us define also on $L/\Gamma$ a binary relation $\leq$ by: for all $|\varphi|, |\psi| \in L/\Gamma$,

$$(1.71) \qquad |\varphi| \leq |\psi| \stackrel{def.}{\iff} \varphi \leq_\Gamma \psi.$$

Note that we have:

$$\begin{aligned}
|\varphi| \leq |\psi| &\stackrel{def.}{\iff} \varphi \leq_\Gamma \psi \\
&\stackrel{def.}{\iff} \Gamma \vdash \varphi \to \psi \quad (\stackrel{(IdEqR_\Gamma)}{\iff} \Gamma \vdash \varphi \rightsquigarrow \psi) \\
&\stackrel{(1.64)}{\iff} \varphi \to \psi \in L_\Gamma \quad (\iff \varphi \rightsquigarrow \psi \in L_\Gamma) \\
&\stackrel{(1.70)}{\iff} |\varphi \to \psi| \in L_\Gamma/\Gamma \quad (\stackrel{(IdEqR_{L_\Gamma/\Gamma})}{\iff} |\varphi \rightsquigarrow \psi| \in L_\Gamma/\Gamma) \\
&\stackrel{def.}{\iff} |\varphi| \Rightarrow |\psi| \in L_\Gamma/\Gamma \quad (\stackrel{(IdEqR_{L_\Gamma/\Gamma})}{\iff} |\varphi| \approx> |\psi| \in L_\Gamma/\Gamma),
\end{aligned}$$

hence, we have:

$$(1.72) \ (IdEqR_{L_\Gamma/\Gamma}) \quad |\varphi| \Rightarrow |\psi| \in L_\Gamma/\Gamma \iff |\varphi| \approx> |\psi| \in L_\Gamma/\Gamma \quad \textit{and}$$

(1.73)

$$(pEqrelR_{L_\Gamma/\Gamma}) \ |\varphi| \leq |\psi| \iff |\varphi| \Rightarrow |\psi| \in L_\Gamma/\Gamma \ (\stackrel{(IdEqR_{L_\Gamma/\Gamma})}{\iff} |\varphi| \approx> |\psi| \in L_\Gamma/\Gamma).$$

By Proposition 1.7.36 and by (1.71), we obtain immediately the following

**Proposition 1.7.44** *For any* $|\varphi|, |\psi|, |\chi| \in L/\Gamma$,

$(pB_\Rightarrow')$ $\quad |\psi| \Rightarrow |\chi| \leq (|\varphi| \Rightarrow |\psi|) \Rightarrow (|\varphi| \Rightarrow |\chi|)$,
$(pC_1')$ $\quad |\varphi| \Rightarrow (|\psi| \approx> |\chi|) \leq |\psi| \approx> (|\varphi| \Rightarrow |\chi|)$,
$(pC_2')$ $\quad |\varphi| \approx> (|\psi| \Rightarrow |\chi|) \leq |\psi| \Rightarrow (|\varphi| \approx> |\chi|)$,
$(I_\Rightarrow') = (Re')$ $\quad |\varphi| \leq |\varphi|$;
$(Tr')$ $\quad |\varphi| \leq |\psi|$ *and* $|\psi| \leq |\chi| \implies |\varphi| \leq |\chi|$,
$(pEq\#')$ $\quad |\varphi| \leq |\psi| \Rightarrow |\chi| \iff |\psi| \leq |\varphi| \approx> |\chi|$,
$(pD')$ $\quad |\varphi| \leq (|\varphi| \Rightarrow |\psi|) \approx> |\psi|$ *and* $|\varphi| \leq (|\varphi| \approx> |\psi|) \Rightarrow |\psi|$,
$(p*_\Rightarrow')$ $\quad |\psi| \leq |\chi| \implies |\varphi| \Rightarrow |\psi| \leq |\varphi| \Rightarrow |\chi|$,
$(p*_{\approx>}')$ $\quad |\psi| \leq |\chi| \implies |\varphi| \approx> |\psi| \leq |\varphi| \approx> |\chi|$,
$(p**_\Rightarrow')$ $\quad |\psi| \leq |\chi| \implies |\chi| \Rightarrow |\varphi| \leq |\psi| \Rightarrow |\varphi|$,
$(p**_{\approx>}')$ $\quad |\psi| \leq |\chi| \implies |\chi| \approx> |\varphi| \leq |\psi| \approx> |\varphi|$,
$(pBB_1')$ $\quad |\varphi| \Rightarrow |\psi| \leq (|\psi| \Rightarrow |\chi|) \approx> (|\varphi| \Rightarrow |\chi|)$.

By Lemma 1.7.37 and by (1.71), we obtain immediately the following

**Lemma 1.7.45** *For any* $|\varphi| \in L/\Gamma$,

$$(|\varphi| \Rightarrow |\varphi|) \approx> |\varphi| \ \leq \ |\varphi| \quad \textit{and} \quad (|\varphi| \approx> |\varphi|) \Rightarrow |\varphi| \ \leq \ |\varphi|\,.$$

By Lemma 1.7.38 and by (1.71), we obtain immediately the following

**Lemma 1.7.46** *For any* $|\varphi|, |\psi| \in L/\Gamma$,

$$|\varphi| \Rightarrow |\varphi| \ \leq \ |\psi| \quad \Longrightarrow \quad |\psi| \Rightarrow |\psi| \ \leq \ |\psi|\,.$$

By Proposition 1.7.39 and by (1.71), we obtain immediately the following

**Proposition 1.7.47** *For any* $|\varphi|, |\psi|, |\chi| \in L/\Gamma$, *we have:*

$$|\varphi| \approx> |\psi| \ \leq \ |\varphi| \approx> ((|\psi| \approx> |\chi|) \Rightarrow |\chi|). \tag{1.74}$$

**Proposition 1.7.48** *The binary relation* $\leq$ *on* $L/\Gamma$ *is an order.*

**Proof.** See Proposition 1.7.23. □

**Corollary 1.7.49** *For any* $|\varphi|, |\psi|, |\chi| \in L/\Gamma$, *we have:*
*(pEx)* $\quad |\varphi| \Rightarrow (|\psi| \approx> |\chi|) = |\psi| \approx> (|\varphi| \Rightarrow |\chi|)$,
*(pB$_{\approx>}$')* $\quad |\psi| \approx> |\chi| \ \leq \ (|\varphi| \approx> |\psi|) \approx> (|\varphi| \approx> |\chi|)$,
*(pBB$_2$')* $\quad |\varphi| \approx> |\psi| \ \leq \ (|\psi| \approx> |\chi|) \Rightarrow (|\varphi| \approx> |\chi|)$.

**Proof.** (pEx): By (pC$_1$') and (pC$_2$'), we have
$|\varphi| \Rightarrow (|\psi| \approx> |\chi|) \ \leq \ |\psi| \approx> (|\varphi| \Rightarrow |\chi|)$ and
$|\psi| \approx> (|\varphi| \Rightarrow |\chi|) \ \leq \ |\varphi| \Rightarrow (|\psi| \approx> |\chi|)$,
hence, by antisymmetry,
$|\varphi| \Rightarrow (|\psi| \approx> |\chi|) = |\psi| \approx> (|\varphi| \Rightarrow |\chi|)$, i.e. (pEx) holds.

(pB$_{\approx>}$'): By (1.74) of Proposition 1.7.47, we have:
$|\varphi| \approx> |\psi| \ \leq \ |\varphi| \approx> ((|\psi| \approx> |\chi|) \Rightarrow |\chi|)$,
hence, by (pEx), we obtain:
$|\varphi| \approx> |\psi| \ \leq \ (|\psi| \approx> |\chi|) \Rightarrow (|\varphi| \approx> |\chi|)$,
which by (pEq#') gives:
$|\psi| \approx> |\chi| \ \leq \ (|\varphi| \approx> |\psi|) \approx> (|\varphi| \approx> |\chi|)$, i.e. (pB$_{\approx>}$') holds.

(pBB$_2$'): By (pB$_{\approx>}$'), $|\psi| \approx> |\chi| \ \leq \ (|\varphi| \approx> |\psi|) \approx> (|\varphi| \approx> |\chi|)$;
then, by (pEq#'), $|\varphi| \approx> |\psi| \ \leq \ (|\psi| \approx> |\chi|) \Rightarrow (|\varphi| \approx> |\chi|)$, i.e. (pBB$_2$') holds. □

**Corollary 1.7.50** *For any* $|\varphi| \in L/\Gamma$,

$$(|\varphi| \Rightarrow |\varphi|) \approx> |\varphi| \ = \ |\varphi| \quad \textit{and} \quad (|\varphi| \approx> |\varphi|) \Rightarrow |\varphi| \ = \ |\varphi|\,.$$

**Proof.** By (pD'), $|\varphi| \leq (|\varphi| \Rightarrow |\varphi|) \approx> |\varphi|$.
By Lemma 1.7.45, $(|\varphi| \Rightarrow |\varphi|) \approx> |\varphi| \leq |\varphi|$.
Then, by antisymmetry, it follows that $(|\varphi| \Rightarrow |\varphi|) \approx> |\varphi| = |\varphi|$.

Similarly, $(|\varphi| \approx> |\varphi|) \Rightarrow |\varphi| = |\varphi|$. □

Now, we can prove the following result.

**Theorem 1.7.51** *The structure $\mathcal{L}/\Gamma = (L/\Gamma, \leq, \Rightarrow, \approx>, L/\Gamma^+)$ is a locally unital quantum-B algebra, called the Lindenbaum-Tarski 'algebra' of the propositional pBCI→ logic (or QB logic), and $L/\Gamma^+ = L_\Gamma/\Gamma$.*

**Proof.** $(L/\Gamma, \leq)$ is a poset, by Proposition 1.7.48. The properties (pEx), ($p^*_{\Rightarrow}$'), (pEq#') hold; indeed: (pEx) holds by above Corollary; ($p^*_{\Rightarrow}$') holds by Proposition 1.7.44; (pEq#') holds by Proposition 1.7.44.

Thus, $(L/\Gamma, \leq, \Rightarrow, \approx>)$ is a quantum-B algebra (Definition 2).

We prove now that it is locally unital, i.e. that (1.13) and (1.14) hold.

(1.13): We must prove that, for any $|\varphi| \in L/\Gamma$, we have:
$(|\varphi| \Rightarrow |\varphi|) \approx> |\varphi| = |\varphi|$ *and* $(|\varphi| \approx> |\varphi|) \Rightarrow |\varphi| = |\varphi|$,
and this is true by Corollary 1.7.50.

(1.14): We must prove that, for any $|\varphi|, |\psi| \in L/\Gamma$, we have:
$|\varphi| \Rightarrow |\varphi| \leq |\psi| \Longrightarrow |\psi| \Rightarrow |\psi| \leq |\psi|$,
and this is true by Lemma 1.7.46.

Define $L/\Gamma^+ \stackrel{def.}{=} \{|\varphi| \in L/\Gamma \mid |\varphi| \Rightarrow |\varphi| \leq |\varphi|\} \subset L/\Gamma$. We prove that

$$L/\Gamma^+ = L_\Gamma/\Gamma. \tag{1.75}$$

Indeed,

$$\begin{array}{lll}
|\varphi| \in L/\Gamma^+ & \stackrel{def.}{\Longleftrightarrow} & |\varphi| \Rightarrow |\varphi| \leq |\varphi| \\
& \stackrel{def.}{\Longleftrightarrow} & |\varphi \to \varphi| \leq |\varphi| \\
& \stackrel{(1.71)}{\Longleftrightarrow} & \varphi \to \varphi \leq_\Gamma \varphi \\
& \stackrel{(1.67),(1.64)}{\Longleftrightarrow} & (\varphi \to \varphi) \rightsquigarrow \varphi \in L_\Gamma \\
& \stackrel{(1.70)}{\Longleftrightarrow} & |(\varphi \to \varphi) \rightsquigarrow \varphi| \in L_\Gamma/\Gamma \\
& \stackrel{def.}{\Longleftrightarrow} & (|\varphi| \Rightarrow |\varphi|) \approx> |\varphi| \in L_\Gamma/\Gamma \\
& \stackrel{Prop. 1.7.50}{\Longleftrightarrow} & |\varphi| \in L_\Gamma/\Gamma.
\end{array}$$

□

Note that, by (1.72) and (1.73), we have (see Proposition 1.3.9):

(1.76)
$$(IdEqR_{L/\Gamma^+}) \quad |\varphi| \Rightarrow |\psi| \in L/\Gamma^+ \iff |\varphi| \approx> |\psi| \in L/\Gamma^+ \quad \textit{and}$$

(1.77)
$$(pEqrelR_{L/\Gamma^+}) \ |\varphi| \leq |\psi| \Longleftrightarrow |\varphi| \Rightarrow |\psi| \in L/\Gamma^+ \ (\stackrel{(IdEqR_{L/\Gamma^+})}{\Longleftrightarrow} |\varphi| \approx> |\psi| \in L/\Gamma^+).$$

**Remark 1.7.52** *If we consider the pBCI logic (extension of pBCI→ logic (or QB logic)), the Lindenbaum-tarski construction is the same and the Lindenbaum-Tarski 'algebra' is again a locally unital QB algebra.*

### 1.7.3 Pseudo-BCI→' (or QB') logic

The pseudo-BCI→' logic (or quantum-B' logic) is the logic pBCI→ (or QB logic) with the additional inference rule (MKK): $\frac{\varphi, \psi}{\varphi \to \psi}$. Hence, we have additional results for the 'algebra' of pBCI→' logic (or QB logic).

**Proposition 1.7.53** *$L_\Gamma$ is an equivalence class, denoted by* **1**, *and*

$$|\varphi| = \mathbf{1} \quad \overset{def.}{\Longleftrightarrow} \quad |\varphi| = L_\Gamma \quad \Longleftrightarrow \quad \varphi \in L_\Gamma \quad (\Longleftrightarrow \Gamma \vdash \varphi).$$

**Proof.** $\Longrightarrow$: Suppose $|\varphi| = L_\Gamma$; then, for any $\psi \in L_\Gamma$, we have $\psi \in |\varphi|$, i.e. $\psi \sim_\Gamma \varphi$, i.e. $\psi \to \varphi \in L_\Gamma$ and $\varphi \to \psi \in L_\Gamma$; it follows, by (MP), that $\varphi \in L_\Gamma$.

$\Longleftarrow$: Suppose $\varphi \in L_\Gamma$; we must prove that $|\varphi| = L_\Gamma$. Indeed,
$\cdot\ |\varphi| \subseteq L_\Gamma$: le t $\psi \in |\varphi|$, i.e. $\psi \sim_\Gamma \varphi$, i.e. $\psi \to \varphi \in L_\Gamma$ and $\varphi \to \psi \in L_\Gamma$; since $\varphi \in L_\Gamma$, it follows, by (MP), that $\psi \in L_\Gamma$; thus, $|\varphi| \subseteq L_\Gamma$.
$\cdot\ L_\Gamma \subseteq |\varphi|$: let $\psi \in L_\Gamma$; snce $\varphi \in L_\Gamma$, it follows by (MKK), that $\varphi \to \psi \in L_\Gamma$ and $\psi \to \varphi \in L_\Gamma$, i.e. $\varphi \sim_\Gamma \psi$, hence $\psi \in |\varphi|$; thus, $L_\Gamma \subseteq |\varphi|$. □

**Remark 1.7.54** *Note that, in this case ($L_\Gamma$ is an equivalence class, denoted by* **1**), *we have*

$$L/\Gamma^+ \;=\; L_\Gamma/\Gamma \;=\; \{\mathbf{1}\}.$$

*Consequently, the equivalences (1.76) and (1.77) become*

(1.78) $\quad (IdEqR) \quad |\varphi| \Rightarrow |\psi| = \mathbf{1} \Longleftrightarrow |\varphi| \approx> |\psi| = \mathbf{1}$ *and*

(1.79) $\quad (pEqrelR) \quad |\varphi| \leq |\psi| \Longleftrightarrow |\varphi| \Rightarrow |\psi| = \mathbf{1} \quad (\overset{(IdEqR)}{\Longleftrightarrow} |\varphi| \approx> |\psi| = \mathbf{1}).$

**Proposition 1.7.55** *We have: for all $|\psi| \in L/\Gamma$,*
*(pM)* $\mathbf{1} \Rightarrow |\psi| = |\psi|$,
*(N')* $\mathbf{1} \leq |\psi| \Longrightarrow |\psi| = \mathbf{1}$.

**Proof.** (pM): By Proposition 1.7.53, $|\varphi| = \mathbf{1} \overset{def.}{\Longleftrightarrow} |\varphi| = L_\Gamma \Longleftrightarrow \varphi \in L_\Gamma$.
Hence, for any $\psi \in L$ and for any $\varphi \in L_\Gamma$, we have:

$$\begin{aligned} \mathbf{1} \Rightarrow |\psi| = |\psi| &\Longleftrightarrow |\varphi| \Rightarrow |\psi| = |\psi| \\ &\Longleftrightarrow |\varphi \to \psi| = |\psi| \\ &\Longleftrightarrow (\varphi \to \psi) \sim_\Gamma \psi \\ &\Longleftrightarrow \varphi \to \psi \leq_\Gamma \psi \text{ and } \psi \leq_\Gamma \varphi \to \psi \\ &\Longleftrightarrow \Gamma \vdash (\varphi \to \psi) \to \psi \text{ and } \Gamma \vdash \psi \to (\varphi \to \psi). \end{aligned}$$

So, we must prove that $\Gamma \vdash (\varphi \to \psi) \to \psi$ and $\Gamma \vdash \psi \to (\varphi \to \psi)$. Indeed,

| | |
|---|---|
| $\Gamma \vdash \varphi$ | by hypothesis |
| $\Gamma \vdash \varphi \to ((\varphi \to \psi) \rightsquigarrow \psi)$ | by (p$D_\Gamma$) |
| $\Gamma \vdash (\varphi \to \psi) \rightsquigarrow \psi$ | by (MP) |
| $\Gamma \vdash (\varphi \to \psi) \to \psi$ | by (IR2) |

and

| | |
|---|---|
| $\Gamma \vdash \varphi$ | by hypothesis |
| $\Gamma \vdash \psi \to \psi$ | by ($I{\to}_\Gamma$) |
| $\Gamma \vdash \psi \rightsquigarrow \psi$ | by (IR1), |
| $\Gamma \vdash \varphi \to (\psi \rightsquigarrow \psi)$ | by (MKK) |
| $\Gamma \vdash \psi \rightsquigarrow (\varphi \to \psi)$ | by (pEq$\#_\Gamma$) |
| $\Gamma \vdash \psi \to (\varphi \to \psi)$ | by (IR2). |

(N'): We must prove that: $\mathbf{1} \leq | \psi | \Longrightarrow | \psi | = \mathbf{1}$, i.e.
for all $\varphi \in L_\Gamma$, $\psi \in L$,

$$
\begin{array}{lll}
 & | \varphi | \leq | \psi | & \Longrightarrow \quad | \psi | = | \varphi |, \\
\Longleftrightarrow & (\varphi \leq_\Gamma \psi & \Longrightarrow \quad \psi \sim_\Gamma \varphi), \\
\Longleftrightarrow & (\Gamma \vdash \varphi \rightarrow \psi & \Longrightarrow \quad \psi \leq_\Gamma \varphi \text{ and } \varphi \leq_\Gamma \psi), \\
\Longleftrightarrow & (\Gamma \vdash \varphi \rightarrow \psi & \Longrightarrow \quad \Gamma \vdash \psi \rightarrow \varphi \text{ and } \Gamma \vdash \varphi \rightarrow \psi).
\end{array}
$$

So, it remains to prove that $\Gamma \vdash \psi \rightarrow \varphi$. Indeed,

(1) $\Gamma \vdash \varphi$ by hypothesis
(2) $\Gamma \vdash \varphi \rightarrow \psi$ by hypothesis
(3) $\Gamma \vdash \psi$ by (MP,
(4) $\Gamma \vdash \psi \rightarrow \varphi$ by (3), (1), (MKK).

□

Now, we can prove the following result.

**Theorem 1.7.56** *The structure $\mathcal{L}/\Gamma = (L/\Gamma, \leq, \Rightarrow, \approx>, \mathbf{1})$ is a pBCI algebra, called the Lindenbaum-Tarski algebra of the propositional pBCI→' logic (or QB' logic).*

**Proof.** We present two proofs.

**Proof 1**: By Proposition 1.7.44, (pB$_\Rightarrow$'), (pC')= (pC$_1$') + (pC$_2$'), (Re') hold; by Proposition 1.7.48, (An') holds; (1.78) and (1.79) hold; hence, $\mathcal{L}/\Gamma$ is a pBCI algebra (Definition 7').

**Proof 2**: By Theorem 1.7.51, the structure $\mathcal{L}/\Gamma$ is a locally unital quantum-B algebra; by Proposition 1.7.55, (pM) and (N') hold, hence, $\mathcal{L}/\Gamma$ is a maximal quantum-B algebra, hence is a pBCI algebra, by Proposition 1.3.18. □

**Remark 1.7.57** *If we consider the pBCI' logic (extension of pBCI→' logic), the Lindenbaum-Tarski construction is the same as above and the Lindenbaum-Tarski algebra is again a pBCI algebra.*

### 1.7.4 Pseudo-BCI→" (or QB") logic (⟺ pBCK→ logic)

The pseudo-BCI→" logic (or quantum-B" logic) is the pBCI→ logic (or QB logic) with the additional inference rule (MK): $\frac{\psi}{\varphi \rightarrow \psi}$. Hence, we have additional results for the algebra of pBCI→" logic (or QB" logic).

**Proposition 1.7.58** *We have: for all $| \psi | \in L/\Gamma$,*
*(La')* $| \psi | \leq \mathbf{1}$.

**Proof.** We have:
$| \psi | \leq \mathbf{1} \Longleftrightarrow | \psi | \leq | \varphi | \Longleftrightarrow \psi \leq_\Gamma \varphi \Longleftrightarrow \Gamma \vdash \psi \rightarrow \varphi$.

So, we must prove that, for all $\varphi \in L_\Gamma$, $\psi \in L$, we have $\Gamma \vdash \psi \rightarrow \varphi$.
Indeed,

$\Gamma \vdash \varphi$ by hypothesis
$\Gamma \vdash \psi \rightarrow \varphi$ by (MK).

□

Now, we can prove the following result.

**Theorem 1.7.59** *The structure* $\mathcal{L}/\Gamma = (L/\Gamma, \leq, \Rightarrow, \approx>, \mathbf{1})$ *is a pBCK algebra, called the Lindenbaum-Tarski algebra of the pBCI→" logic (or QB" logic).*

**Proof.** We present two proofs.

**Proof 1**: By (1.49), $p\mathcal{BCI} \rightarrow''$ $(or \mathcal{QB}'')$ $\Longleftrightarrow$ $p\mathcal{BCK} \rightarrow$. Then, apply Theorem 1.7.26.

**Proof 2**: By Theorem 1.7.8, since (MK) holds, then the inference rule (MKK) holds. It follows that pseudo-BCI→" logic is an extension of pseudo-BCI→' logic. Then, its Lindenbaum-Tarski algebra is a pBCI algebra (i.e., by Theorem 1.7.56, the structure $\mathcal{L}/\Gamma$ is a pBCI algebra); by Proposition 1.7.58, the property (La') holds, hence, $\mathcal{L}/\Gamma$ is a pBCK algebra. □

**Remark 1.7.60** *If we consider the pBCI" logic (extension of pBCI→" logic (or QB" logic), the Lindenbaum-Tarski construction is the same as above and the Lindenbaum-Tarski algebra is again a pBCK algebra.*

### 1.7.5 Connections between logics and their algebras

By the results from this section, we obtain the connections between the logics and their Lindenbaum-Tarski algebras (i.e. algebras obtained by Lindenbaum-Tarski constructions) presented, in 'algebraic style', in Figure 1.6.

### 1.7.6 The commutative case

Note that the pBCK→ logic and the pBCK logic become, in the commutative case, the BCK logic.

Note also that the pBCI→ logic (or QB logic) and the pBCI logic become, in the commutative case, the BCI logic. This is the argument which motivates the alternative name 'pBCI→ logic' for the name 'quantum-B logic' (or QB logic, for short) (name given in [193]).

The BCK/BCI logics were mentioned by A. N. Prior in [176] the second edition of his Formal Logic of 1962, where it was credited to C. A. Meredith and dated in 1956. Cf. [49], its significance is due to a certain correspondence between combinators and implicational formulas (see [37] and [154]).

The BCI logic is the propositional logic with the axioms:
(B) $(\varphi \to \psi) \to ((\psi \to \chi) \to (\varphi \to \chi))$,
(C) $(\varphi \to (\psi \to \chi)) \to ((\psi \to (\varphi \to \chi))$,
(I) $\varphi \to \varphi$
and the only inference rule:
(MP) $\frac{\varphi, \varphi \to \psi}{\psi}$,
while the BCK logic is the propositional logic with the axioms:
(B) $(\varphi \to \psi) \to ((\psi \to \chi) \to (\varphi \to \chi))$,
(C) $(\varphi \to (\psi \to \chi)) \to ((\psi \to (\varphi \to \chi))$,
(K) $\varphi \to (\psi \to \varphi)$
and the only inference rule (MP).

In 1966, K. Iseki introduced the concepts of BCK/BCI algebra as an algebraic counterparts of the BCK/BCI logic (see [126]). Unfortunately, BCI algebras fails

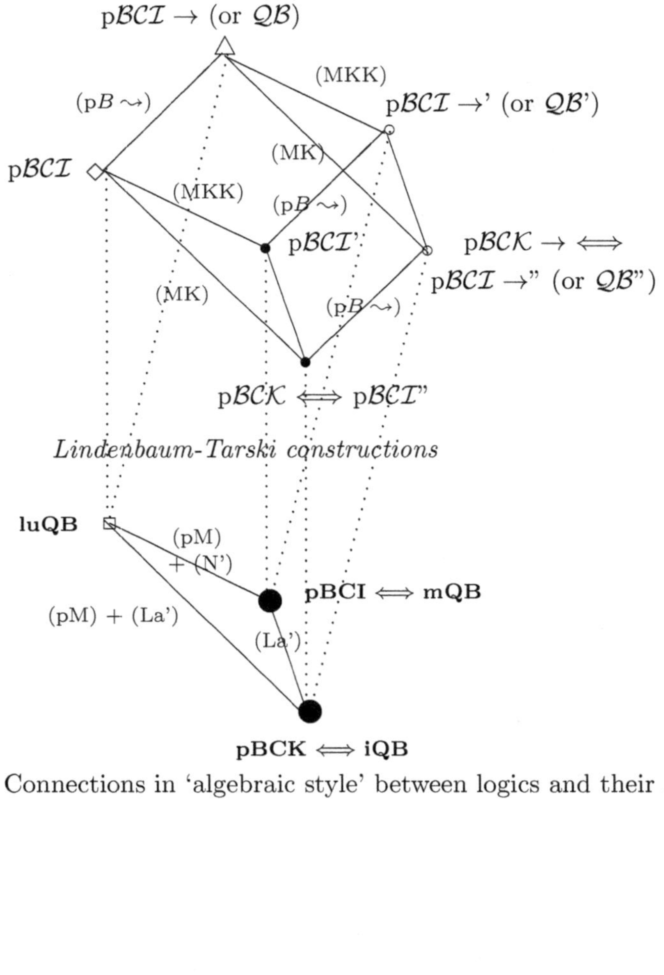


Figure 1.6: Connections in 'algebraic style' between logics and their algebras

to be the models of the BCI logic. W. J. Blok and D. Pigozzi proved that the BCI logic is not algebraizable (see Theorem 5.9 of [6]). The 'algebra' of the BCI logic is the locally unital commutative quantum-B algebra, a structure, introduced, in the non-commutative case, by W. Rump [193].

The BCI algebra is model of the *BCI' logic*, which is the BCI logic extended on one additional inference rule (see[132]):
(MKK) $\frac{\varphi, \psi}{\varphi \to \psi}$.

In the commutative case, we then obtain the connections between the logics and their algebras (algebras obtained by Lindenbaum-Tarski constructions) presented, in an 'algebraic style', in Figure 1.7.

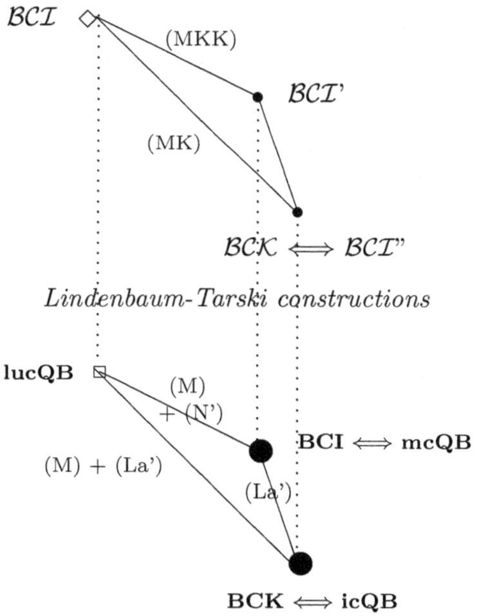


Figure 1.7: Connections in 'algebraic style' between logics and their algebras, in the commutative case

# 1.8 Prealgebras/prestructures

The study of *preboolean sets* in [173], [175], of *Nelson and Łukasiewicz prealgebras* in [158], on the one hand, and the study of $\mathcal{S}$-algebras in [179], on the other hand, led to $\mathcal{S}$*-prealgebras*. In [94], the notion of $\mathcal{S}$*-prealgebra* was introduced and proved a theorem from which Theorems 6.5, 9.2 and a generalization of 13.10 from ([179], Chapter VIII) follow as corollaries.

Following closely the study of *Hilbert prealgebras* in [42], the notion of *Boolean prealgebra* was introduced and studied in [105] - see [63].

Here, following closely the study of Boolean prealgebras from [63], we introduce and study the *pBCK, pBCI' prealgebras* and the *pBCI, QB prestructures.*

The *Lindenbaum-Tarski constructions* from the previous Section 1.7 are followed closely in the so-called *Lindenbaum-Tarski type constructions* from this Section.

Consider a structure $\mathcal{X} = (X, \to, \rightsquigarrow, D)$, where $\emptyset \neq D \subseteq X$ and $\to$ and $\rightsquigarrow$ are binary operations on $X$.

Consider the following basic properties:

| | | |
|---|---|---|
| (pB$_D$) | = | (pB$\to_D$) + (pB$\rightsquigarrow_D$), where: |
| | (pB$\to_D$) | $(y \to z) \to ((x \to y) \to (x \to z)) \in D$, |
| | (pB$\rightsquigarrow_D$) | $(y \rightsquigarrow z) \to ((x \rightsquigarrow y) \rightsquigarrow (x \rightsquigarrow z)) \in D$; |
| (pBB$_D$) | = | (pBB1$_D$) + (pBB2$_D$), where: |
| | (pBB1$_D$) | $(x \to y) \to ((y \to z) \rightsquigarrow (x \to z)) \in D$, |
| | (pBB2$_D$) | $(x \rightsquigarrow y) \to ((y \rightsquigarrow z) \to (x \rightsquigarrow z)) \in D$; |
| (pC$_D$) | = | (pC1$_D$) + (pC2$_D$), where: |
| | (pC1$_D$) | $(x \to (y \rightsquigarrow z)) \to (y \rightsquigarrow (x \to z)) \in D$, |
| | (pC2$_D$) | $(x \rightsquigarrow (y \to z)) \to (y \to (x \rightsquigarrow z)) \in D$; |
| (pK$_D$) | = | (pK$\to_D$) + (pK$\rightsquigarrow_D$), where: |
| | (pK$\to_D$) | $x \to (y \to x) \in D$, |
| | (pK$\rightsquigarrow_D$) | $x \to (y \rightsquigarrow x) \in D$; |
| (I$\to_D$) | $x \to x \in D$; | |

(IdEqR$_D$) $x \to y \in D \Longleftrightarrow x \rightsquigarrow y \in D$,
(MP$_D$) if $x \in D$ and $x \to y \in D \Longrightarrow y \in D$;
(MK$_D$) $y \in D \Longrightarrow x \to y \in D$;
(MKK$_D$) $x, y \in D \Longrightarrow x \to y \in D$.

**Remark 1.8.1** *The properties (pB$_D$), (pBB$_D$), (pC$_D$), (pK$_D$), (I$\to_D$) are the algebraic counterpart of the axioms (pB), (pBB), (pC), (pK), (I$\to$), respectively, while the properties (IdEqR$_D$), (MP$_D$), (MK$_D$), (MKK$_D$) are the algebraic counterpart of the inference rules (IR), (MP), (MK), (MKK), respectively, of the logics presented in previous Section 1.7.*

We introduce then the following definitions (see the analogous definitions of logics in Section 1.7). We use the name 'prealgebra' for the analogous 'algebraizable logics' and the name 'prestructure' for the analogous 'non-algebraizable logics'.

**Definitions 1.8.2** The structure $\mathcal{X} = (X, \to, \rightsquigarrow, D)$ is:

(1) a *pseudo-BCK prealgebra*, or *pBCK prealgebra* for short, if (pB$_D$), (pC$_D$), (pK$_D$), (IdEqR$_D$), (MP$_D$) hold.

(2) a *pseudo-BCI prestructure*, or *pBCI prestructure* for short, if (pB$_D$), (pC$_D$), (I$\to_D$), (IdEqR$_D$), (MP$_D$) hold.

(3) a *pBCI$\to$ prestructure*, or *quantum-B prestructure*, or *QB prestructure* for short, if (pB$\to_D$), (pC$_D$), (I$\to_D$), (IdEqR$_D$), (MP$_D$) hold,

(4) a *pBCK$\to$ prealgebra*, if (pB$\to_D$), (pC$_D$), (pK$_D$), (IdEqR$_D$), (MP$_D$) hold,

(5) a *pBCI' prealgebra*, if (pB$_D$), (pC$_D$), (I$\to_D$), (IdEqR$_D$), (MP$_D$) and (MKK$_D$) hold,

(6) a *pBCI" prealgebra*, if (pB$_D$), (pC$_D$), (I$\rightarrow_D$), (IdEqR$_D$), (MP$_D$) and (MK$_D$) hold,

(7) a *pBCI$\rightarrow$' prealgebra* (or *QB' prealgebra*), if (pB$\rightarrow_D$), (pC$_D$), (I$\rightarrow_D$), (IdEqR$_D$), (MP$_D$) and (MKK$_D$) hold,

(8) a *pBCI$\rightarrow$" prealgebra* (or *QB" prealgebra*), if (pB$\rightarrow_D$), (pC$_D$), (I$\rightarrow_D$), (IdEqR$_D$), (MP$_D$) and (MK$_D$) hold.

Denote by bold letters the classes of these structures. So, we have:

**pBCI prestructure** = **pBCI$\rightarrow$ (or QB) prestructure** + $(pB{\rightsquigarrow}_D)$,

**pBCK prealgebra** = **pBCK$\rightarrow$ prealgebra** + $(pB{\rightsquigarrow}_D)$,

**pBCI$'$ prealgebra** = **pBCI prestructure** + $(MKK_D)$,

**pBCI$''$ prealgebra** = **pBCI prestructure** + $(MK_D)$,

**pBCI $\rightarrow'$ (or QB$'$) prealgebra** = **pBCI $\rightarrow$ (or QB) prestructure** + $(MKK_D)$,

**pBCI $\rightarrow''$ (or QB$''$) prealgebra** = **pBCI $\rightarrow$ (orQB) prestructure** + $(MK_D)$.

**Proposition 1.8.3** *We have:*
*(a) (pC$_D$) + (pK$_D$) + (IdEqR$_D$) + (MP$_D$) $\Longrightarrow$ (I$\rightarrow_D$),*
*(b) (pK$_D$) + (MP$_D$) $\Longrightarrow$ (MK$_D$),*
*(c) (pC$_D$) + (I$\rightarrow_D$) + (IdEqR$_D$) + (MP$_D$) + (MK$_D$) $\Longrightarrow$ (pK$_D$),*
*(d) (MK$_D$) $\Longrightarrow$ (MKK$_D$).*

**Proof.** (a): We have:

$x \rightarrow ((x \rightarrow (x \rightarrow x) \rightsquigarrow x) \in D$, by (pK$\rightsquigarrow_D$),
$(x \rightarrow ((x \rightarrow (x \rightarrow x) \rightsquigarrow x)) \rightarrow ((x \rightarrow (x \rightarrow x)) \rightsquigarrow (x \rightarrow x)) \in D$, by (pC1$_D$),
$(x \rightarrow (x \rightarrow x)) \rightsquigarrow (x \rightarrow x) \in D$, by (MP$_D$),
$(x \rightarrow (x \rightarrow x)) \rightarrow (x \rightarrow x) \in D$, by (IdEqR$_D$),
$x \rightarrow (x \rightarrow x) \in D$, by (pK$\rightarrow_D$),
$x \rightarrow x \in D$, by (MP$_D$).

Thus, (I$\rightarrow_D$) holds.

(b): We have:

$y \in D$, by hypothesis,
$y \rightarrow (x \rightarrow y) \in D$, by (pK$\rightarrow_D$),
$x \rightarrow y \in D$, by (MP$_D$).

Thus, (MK$_D$) holds.

(c): We have:

$x \rightarrow x \in D$, by (I$\rightarrow_D$),
$x \rightsquigarrow x \in D$, by (IdEqR$_D$),
$y \rightarrow (x \rightsquigarrow x) \in D$, by (MK$_D$)
$(y \rightarrow (x \rightsquigarrow x)) \rightarrow (x \rightsquigarrow (y \rightarrow x)) \in D$, by (pC1$_D$),
$x \rightsquigarrow (y \rightarrow x) \in D$, by (MP$_D$),
$x \rightarrow (y \rightarrow x) \in D$, by (IdEqR$_D$);

thus, (pK$\rightarrow_D$) holds. Similarly,

$xi \to x \in D$, by (I$\to_D$),
$y \to (x \to x) \in D$, by (MK$_D$),
$y \rightsquigarrow (x \to x) \in D$, by (IdEqR$_D$),
$(y \rightsquigarrow (x \to x)) \to (x \to (y \rightsquigarrow x)) \in D$, by (pC2$_D$),
$x \to (y \rightsquigarrow x) \in D$, by (MP$_D$);
thus, (pK$\rightsquigarrow_D$) holds too. Thus, (pK$_D$) holds.

(d): We have:
$x \in D$, by hypothesis,
$z \to x \in D$, by (MK$_D$),
$y \to x \in D$, by substitution,
$y \in D$, by hypothesis,
$z \to y \in D$, by (MK$_D$),
$x \to y \in D$, by substitution.
Thus, $x, y \in D$ imply $x \to y \in D$, i.e. (MKK$_D$) holds. □

**Theorem 1.8.4**

*(i) Consider the pBCK$\to$ prealgebra. Then, (I$\to_D$) and (MK$_D$) hold.*

*(i') Consider the pBCI$\to$" (or QB") prealgebra. Then, (pK$_D$) holds.*

**Proof.** By Proposition 1.8.3 (a), (b), (c). □

Note that Theorem 1.8.4 says that:

(1.80) $\quad$ **pBCK $\to$ prealgebra $\Longleftrightarrow$ pBCI $\to''$ (or QB$''$) prealgebra.**

Consequently,

(1.81) $\quad$ **pBCK prealgebra $\Longleftrightarrow$ pBCI$''$ prealgebra.**

**Theorem 1.8.5** *Consider the pBCI$\to$ (or QB) prestructure. Then*

$$(MK_D) \Longrightarrow (MKK_D).$$

**Proof.** By Proposition 1.8.3 (d). □

Note that Theorem 1.8.5 says that:
**pBCI$\to$" (or QB") prealgebra $\subset$ pBCI$\to$' (or QB') prealgebra.**
Consequently,
**pBCI" prealgebra $\subset$ pBCI' prealgebra.**

By above definitions and results, we obtain the connections between the above prealgebras/prestructures reprezented by the Hasse-type diagram from Figure 1.8.

Consider also the following properties:

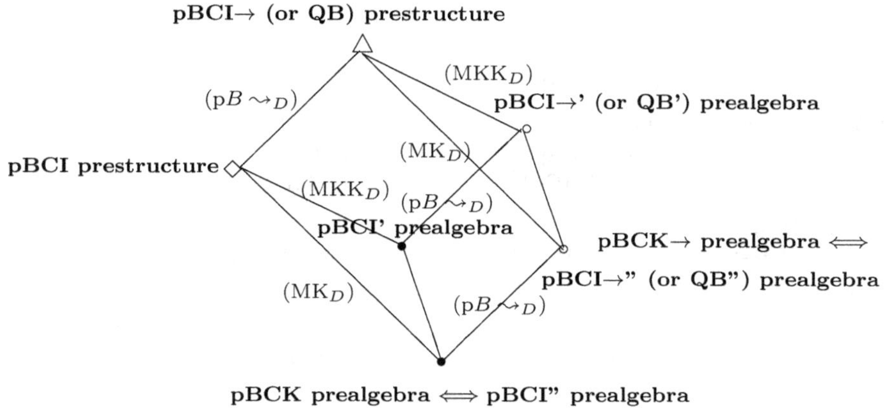


Figure 1.8: Connections between prealgebras/prestructures

| | |
|---|---|
| (Tr$_D$) | $x \to y \in D$ *and* $y \to z \in D \implies x \to z \in D$, |
| (pEq#$_D$) | $x \rightsquigarrow (y \to z) \in D \iff y \to (x \rightsquigarrow z) \in D$, |
| (pD$_D$) | $x \to ((x \to y) \rightsquigarrow y) \in D$ *and* $x \to ((x \rightsquigarrow y) \to y) \in D$, |
| (p*→$_D$) | $y \to z \in D \implies (x \to y) \to (x \to z) \in D$, |
| (p*⇝$_D$) | $y \rightsquigarrow z \in D \implies (x \rightsquigarrow y) \to (x \rightsquigarrow z) \in D$, |
| (p**→$_D$) | $y \to z \in D \implies (z \to x) \to (y \to x) \in D$, |
| (p**⇝$_D$) | $y \to z \in D \implies (z \rightsquigarrow x) \to (y \rightsquigarrow x) \in D$, |
| (pBB1$_D$) | $(x \to y) \to ((y \to z) \rightsquigarrow (x \to z)) \in D$. |

**Proposition 1.8.6** *We have:*
*(pre1) (pB→$_D$) + (MP$_D$) $\implies$ (Tr$_D$),*
*(pre2) (pC$_D$) + (MP$_D$) $\implies$ (pEq#$_D$),*
*(pre3) (I→$_D$) + (pEq#$_D$) + (IdEqR$_D$) $\implies$ (pD$_D$),*
*(pre4) (pB→$_D$) + (MP$_D$) $\implies$ (p*→$_D$),*
*(pre5) (p*→$_D$) + (pD$_D$ ) + (Tr$_D$) + (pEq#$_D$) + (IdEqR$_D$) $\implies$ (p*⇝$_D$),*
*(pre6) (pD$_D$ ) + (Tr$_D$) + (pEq#$_D$) + (IdEqR$_D$) $\implies$ (p**→$_D$),*
*(pre7) (pD$_D$ ) + (Tr$_D$) + (pEq#$_D$) + (IdEqR$_D$) $\implies$ (p**⇝$_D$),*
*(pre8) (pB→$_D$) + (pC2$_D$) + (IdEqR$_D$) + (MP$_D$) $\implies$ (pBB1$_D$).*

**Proof.** (pre1): Since $y \to z \in D$, by hypothesis, and
$(y \to z) \to ((x \to y) \to (x \to z)) \in D$, by (pB→$_D$), it follows that
$(x \to y) \to (x \to z) \in D$, by (MP$_D$);
since also $x \to y \in D$, by hypothesis, it follows
$x \to z \in D$, by (MP$_D$) again. Thus, (Tr$_D$) holds.

(pre2): Suppose that $x \rightsquigarrow (y \to z) \in D$; since
$(x \rightsquigarrow (y \to z)) \to (y \to (x \rightsquigarrow z)) \in D$, by (pC2$_D$), it follows that
$y \to (x \rightsquigarrow z) \in D$, by (MP$_D$).

Conversely, suppose that $y \to (x \rightsquigarrow z) \in D$; since
$(y \to (x \rightsquigarrow z)) \to (x \rightsquigarrow (y \to z)) \in D$, by (pC1$_D$), it follows that
$x \rightsquigarrow (y \to z) \in D$, by (MP$_D$). Thus, (pEq#$_D$) holds.

(pre3): We have: $(x \to y) \to (x \to y) \in D$, by (I$\to_D$); hence, $(x \to y) \rightsquigarrow (x \to y) \in D$, by (IdEqR$_D$); it follows that $x \to ((x \to y) \rightsquigarrow y) \in D$, by (pEq#$_D$).

Similarly, we have: $(x \rightsquigarrow y) \to (x \rightsquigarrow y) \in D$, by (I$\to_D$); hence, $x \rightsquigarrow ((x \rightsquigarrow y) \to y) \in D$, by (pEq#$_D$); it follows that $x \to ((x \rightsquigarrow y) \to y) \in D$, by (IdEqR$_D$). Thus, (pD$_D$ holds.

(pre4): If $y \to z \in D$, then, since $(y \to z) \to ((x \to y) \to (x \to z)) \in D$, by (pB$_{\to D}$), it follows that $(x \to y) \to (x \to z) \in D$, by (MP$_D$). Thus, (p*$\to_D$) holds.

(pre5): If $y \to z \in D$, then $(((x \rightsquigarrow y) \to y) \to ((x \rightsquigarrow y) \to z) \in D$, by (p*$\to_D$); but, $x \to ((x \rightsquigarrow y) \to y) \in D$, by (pD$_D$); it follows that $x \to ((x \rightsquigarrow y) \to z) \in D$, by (Tr$_D$); hence, $x \rightsquigarrow ((x \rightsquigarrow y) \to z) \in D$, by (IdEqR$_D$), hence, $(x \rightsquigarrow y) \to (x \rightsquigarrow z) \in D$, by (pEq#$_D$). Thus, (p*$\rightsquigarrow_D$) holds.

(pre6): If $y \to z \in D$, then, since $z \to ((z \to x) \rightsquigarrow x) \in D$, by (pD$_D$), it follows that $y \to ((z \to x) \rightsquigarrow x) \in D$, by (Tr$_D$); then, $(z \to x) \rightsquigarrow (y \to x) \in D$, by (pEq#$_D$), hence, $(z \to x) \to (y \to x) \in D$, by (IdEqR$_D$). Thus, (p**$\to_D$) holds.

(pre7): If $y \to z \in D$, then, since $z \to ((z \rightsquigarrow x) \to x) \in D$, by (pD$_D$), it follows that $y \to ((z \rightsquigarrow x) \to x) \in D$, by (Tr$_D$), hence, $y \rightsquigarrow ((z \rightsquigarrow x) \to x) \in D$, by (IdEqR$_D$), hence, $(z \rightsquigarrow x) \to (y \rightsquigarrow x) \in D$, by (pEq#$_D$). Thus, (p**$\rightsquigarrow_D$) holds.

(pre8): $(y \to z) \to ((x \to y) \to (x \to z)) \in D$, by (pB$\to_D$); hence, $(y \to z) \rightsquigarrow ((x \to y) \to (x \to z)) \in D$, by (IdEqR$_D$); then, since $((y \to z) \rightsquigarrow ((x \to y) \to (x \to z))) \to ((x \to y) \to ((y \to z) \rightsquigarrow (x \to z))) \in D$, by (pC2$_D$), it follows that $(x \to y) \to ((y \to z) \rightsquigarrow (x \to z)) \in D$, by (MP$_D$). Thus, (pBB1$_D$) holds. □

### 1.8.1 Pseudo-BCK→ prealgebras

Starting from the definition of a pBCK algebra as an algebra $(A, \to, \rightsquigarrow, 1)$ with the axioms (pB$_\to$), (pC), (pK), (An), (IdEqR) (Definition 7), we just introduced the following definition.

**Definition 1.8.7** The structure $\mathcal{X} = (X, \to, \rightsquigarrow, D)$, where $\emptyset \neq D \subseteq X$ and $\to$ and $\rightsquigarrow$ are binary operations on $X$, is a *pBCK→ prealgebra*, if the following axioms/properties hold: for all $x, y, z \in X$,

(pB$\to_D$) $(y \to z) \to ((x \to y) \to (x \to z)) \in D$,
(pC1$_D$) $(x \to (y \rightsquigarrow z)) \to (y \rightsquigarrow (x \to z)) \in D$,
(pC2$_D$) $(x \rightsquigarrow (y \to z)) \to (y \to (x \rightsquigarrow z)) \in D$,
(pK$\to_D$) $x \to (y \to x) \in D$,
(pK$\rightsquigarrow_D$) $x \to (y \rightsquigarrow x) \in D$,
(IdEqR$_D$) $x \to y \in D \Longleftrightarrow x \rightsquigarrow y \in D$,
(MP$_D$) if $x \in D$ and $x \to y \in D$, then $y \in D$.

We have denoted by **pBCK→ prealgebra** the class of all pBCK→ prealgebras.

**Remarks 1.8.8**
*(i) The pBCK→ propositional logic* $(L, \to, \rightsquigarrow, L_\Gamma)$ *(presented in Subsection 1.7.1) is an example of pBCK→ prealgebra, where L is the set of all propositions and* $L_\Gamma$ *is the set of formal theorems (propositions derivable from* $\Gamma$*).*
*(ii) The pBCK algebra, defined as algebra* $(A, \to, \rightsquigarrow, 1)$ *satisfying the axioms (pB$_\to$), (pC), (pK), (An), (IdEqR) (Definition 7 - see Definition 1.2.24) defines the pBCK→ prealgebra* $(A, \to, \rightsquigarrow, D)$*, with* $D = \{1\}$*. Note that the binary relation* $\leq$ *defined by:* $x \leq y \Longleftrightarrow x \to y \in \{1\} \Longleftrightarrow x \to y = 1$ *is an order, not a preorder.*
*(iii) Given a pBCK algebra* $\mathcal{A} = (A, \to, \rightsquigarrow, 1)$ *and a compatible deductive system* $D \neq \{1\}$ *of* $\mathcal{A}$*, then* $(A, \to, \rightsquigarrow, D)$ *is a pBCK→ prealgebra, since* $1 \in D$ *(see [94]).*

**Proposition 1.8.9** *For any* $x, y, z \in X$*, we have the following properties:*
*(MK$_D$)* $y \in D \implies x \to y \in D$;
*(MKK$_D$)* $x, y \in D \implies x \to y \in D$;
*(I$\to_D$)* $x \to x \in D$;

*(Tr$_D$)* $x \to y \in D$ *and* $y \to z \in D \implies x \to z \in D$,
*(pEq#$_D$)* $x \rightsquigarrow (y \to z) \in D \Longleftrightarrow y \to (x \rightsquigarrow z) \in D$,
*(pD$_D$)* $x \to ((x \to y) \rightsquigarrow y) \in D$ *and* $x \to ((x \rightsquigarrow y) \to y) \in D$,
*(p*$\to_D$)* $y \to z \in D \implies (x \to y) \to (x \to z) \in D$,
*(p*$\rightsquigarrow_D$)* $y \to z \in D \implies (x \rightsquigarrow y) \to (x \rightsquigarrow z) \in D$,
*(p**$\to_D$)* $y \to z \in D \implies (z \to x) \to (y \to x) \in D$,
*(p**$\rightsquigarrow_D$)* $y \to z \in D \implies (z \rightsquigarrow x) \to (y \rightsquigarrow x) \in D$,
*(pBB1$_D$)* $(x \to y) \to ((y \to z) \rightsquigarrow (x \to z)) \in D$.

**Proof.** By Propositions 1.8.3, 1.8.6. □

**Remark 1.8.10** *The property (pB$\to_D$)* $(y \to z) \to ((x \to y) \to (x \to z)) \in D$ *of the pBCK→ prealgebra does not imply the 'twin' property,*
*(pB$\rightsquigarrow_D$)* $(y \rightsquigarrow z) \rightsquigarrow ((x \rightsquigarrow y) \rightsquigarrow (x \rightsquigarrow z)) \in D$,
*as in the case of pBCK algebras (see Theorem 1.2.23). We only have the result from the following proposition, obtained by Prover9 (see Theorem 1.2.18).*

**Proposition 1.8.11** *For any* $x, y, z \in X$*, we have:*

$$(x \rightsquigarrow y) \to (x \rightsquigarrow ((y \rightsquigarrow z) \to z)) \in D. \quad (1.82)$$

**Proof.** (By **Prover9**) First, we prove:

$$(1.83) \qquad (x \to y) \to (x \to ((y \rightsquigarrow z) \to z)) \in D.$$

Indeed, we have:
$y \to ((y \rightsquigarrow z) \to z) \in D$, by ($pD_D$),
$(y \to ((y \rightsquigarrow z) \to z)) \to ((x \to y) \to (x \to ((y \rightsquigarrow z) \to z))) \in D$, by ($pB{\to}_D$),
$(x \to y) \to (x \to ((y \rightsquigarrow z) \to z)) \in D$, by ($MP_D$).

Then, we prove:

$$(1.84) \qquad (((x \rightsquigarrow y) \to y) \to z) \to (x \to z) \in D.$$

Indeed, we have:
$x \to ((x \rightsquigarrow y) \to y) \in D$, by ($pD_D$),
$(x \to ((x \rightsquigarrow y) \to y)) \to ((((x \rightsquigarrow y) \to y) \to z) \rightsquigarrow (x \to z)) \in D$, by ($pBB_{1D}$),
$(((x \rightsquigarrow y) \to y) \to z) \rightsquigarrow (x \to z) \in D$, by ($MP_D$),
$(((x \rightsquigarrow y) \to y) \to z) \to (x \to z) \in D$, by ($IdEqR_D$).

Now, we prove (1.82):
$((x \rightsquigarrow y) \to y) \to ((x \rightsquigarrow y) \to ((y \rightsquigarrow z) \to z)) \in D$, by (1.83),
$(((x \rightsquigarrow y) \to y) \to ((x \rightsquigarrow y) \to ((y \rightsquigarrow z) \to z))) \to$
$(x \to ((x \rightsquigarrow y) \to ((y \rightsquigarrow z) \to z))) \in D$, by (1.84),
$x \to ((x \rightsquigarrow y) \to ((y \rightsquigarrow z) \to z)) \in D$, by ($MP_D$),
$x \rightsquigarrow ((x \rightsquigarrow y) \to ((y \rightsquigarrow z) \to z)) \in D$, by ($IdEqR_D$),
$(x \rightsquigarrow ((x \rightsquigarrow y) \to ((y \rightsquigarrow z) \to z))) \to ((x \rightsquigarrow y) \to$
$(x \rightsquigarrow ((y \rightsquigarrow z) \to z))) \in D$, by ($pC2_D$);
$(x \rightsquigarrow y) \to (x \rightsquigarrow ((y \rightsquigarrow z) \to z)) \in D$, by ($MP_D$).

□

### • The Lindenbaum-Tarski type algebra: the pBCK algebra

Let $(X, \to, \rightsquigarrow, D)$ be a pBCK→ prealgebra throughout this subsection.

We define on $X$ a binary relation $\leq_D$ by: for all $x, y \in X$,

$$(1.85) \qquad x \leq_D y \overset{def.}{\Longleftrightarrow} x \to y \in D \quad (\overset{(IdEqR_D)}{\Longleftrightarrow} x \rightsquigarrow y \in D).$$

By Definition 1.8.7, Proposition 1.8.9 and by (1.85), we obtain immediately the following

**Proposition 1.8.12** *For any $x, y, z \in X$,*

(pB→$_D$') $y \to z \ \leq_D \ (x \to y) \to (x \to z)$,
(pC1$_D$') $x \to (y \rightsquigarrow z) \ \leq_D \ y \rightsquigarrow (x \to z)$,
(pC2$_D$') $x \rightsquigarrow (y \to z) \ \leq_D \ y \to (x \rightsquigarrow z)$,
(pK→$_D$') $x \ \leq_D \ y \to x$,
(pK⇝$_D$') $x \ \leq_D \ y \rightsquigarrow x$,
(MP$_D$') *if* $x \in D$ *and* $x \ \leq_D \ y$, *then* $y \in D$;

(MK$_D$') $y \in D \implies x \leq_D y$;
(MKK$_D$') $x, y \in D \implies x \leq_D y$;
(I→$_D$') $x \ \leq_D \ x$;
(Tr$_D$') $x \ \leq_D \ y$ *and* $y \ \leq_D \ z \implies x \ \leq_D \ z$,
(pEq#$_D$') $x \ \leq_D \ y \to z \iff y \ \leq_D \ x \rightsquigarrow z$,
(pD$_D$') $x \ \leq_D \ (x \to y) \rightsquigarrow y$ *and* $x \ \leq_D \ (x \rightsquigarrow y) \to y$,
(p*→$_D$') $y \ \leq_D \ z \implies x \to y \ \leq_D \ x \to z$,
(p*⇝$_D$') $y \ \leq_D \ z \implies x \rightsquigarrow y \ \leq_D \ x \rightsquigarrow z$,
(p**→$_D$') $y \ \leq_D \ z \implies z \to x \ \leq_D \ y \to x$,
(p**⇝$_D$') $y \ \leq_D \ z \implies z \rightsquigarrow x \ \leq_D \ y \rightsquigarrow x$,
(pBB1$_D$') $x \to y \ \leq_D \ (y \to z) \rightsquigarrow (x \to z)$.

**Proposition 1.8.13** *The binary relation $\leq_D$ is a preorder.*

**Proof.** - $\leq_D$ is *reflexive*, i.e. for all $x \in X$, $x \leq_D x$, which is true by (I→$_D$').
- $\leq_D$ is *transitive*, i.e. for all $x, y, z \in X$, $x \leq_D y$ and $y \leq_D z$ imply $x \leq_D z$, which is true by (Tr$_D$'). □

We define on $X$ a binary relation $\sim_D$ by: for all $x, y \in X$,

$$x \sim_D y \overset{def.}{\iff} (x \leq_D y \quad and \quad y \leq_D x). \tag{1.86}$$

**Proposition 1.8.14** *The binary relation $\sim_D$ is an equivalence relation.*

**Proof.** - $\sim_D$ is *reflexive*, i.e. for all $x \in X$, $x \sim_D x$; indeed, $x \sim_D x$ means, by definition, $x \leq_D x$, which is true since $\leq_D$ is reflexive, by (I→$_D$').
- $\sim_D$ is *transitive*, i.e. for all $x, y, z \in X$, $x \sim_D y$ and $y \sim_D z$ imply $x \sim_D z$; indeed, $x \sim_D y$ and $y \sim_D z$ mean, by definition,
($x \leq_D y$ and $y \leq_D x$) and ($y \leq_D z$ and $z \leq_D y$);
since $x \leq_D y$ and $y \leq_D z$, it follows that $x \leq_D z$, and since $z \leq_D y$ and $y \leq_D x$, it follows that $z \leq_D x$, by (Tr$_D$'); thus, $x \sim_D z$.
- $\sim_D$ is *symmetric*, i.e. for all $x, y \in X$, $x \sim_D y$ implies $y \sim_D x$: obviously. □

**Proposition 1.8.15** *The binary relation $\sim_D$ is a congruence relation.*

**Proof.** Since $\sim_D$ is an equivalence relation, it remains to prove that $\sim_D$ is compatible with $\to$ and $\rightsquigarrow$, i.e.
if $x \sim_D x'$ and $y \sim_D y'$, then $(x \to y) \sim_D (x' \to y')$ and $(x \rightsquigarrow y) \sim_D (x' \rightsquigarrow y')$.

Indeed, $x \sim_D x'$ and $y \sim_D y'$ mean, by definition, ($x \leq_D x'$ and $x' \leq_D x$) and ($y \leq_D y'$ and $y' \leq_D y$).

Since $y \leq_D y'$, it follows, by (p*$\to_D$'), that $x \to y \leq_D x \to y'$, and since $x' \leq_D x$, it follows, by (p**$\to_D$'), that $x \to y' \leq_D x' \to y'$; then, by (Tr$_D$'), it follows that $x \to y \leq_D x' \to y'$.
Since $y' \leq_D y$, it follows, by (p*$\to_D$'), that $x \to y' \leq_D x \to y$, and since $x \leq_D x'$, it follows, by (p**$\to_D$'), that $x' \to y' \leq_D x \to y'$; then, by (Tr$_D$'), it follows that $x' \to y' \leq_D x \to y$.
Thus, $(x \to y) \sim_D (x' \to y')$.

Similarly, since $y \leq_D y'$, it follows, by (p*$\rightsquigarrow_D$'), that $x \rightsquigarrow y \leq_D x \rightsquigarrow y'$, and since $x' \leq_D x$, it follows, by (p**$\rightsquigarrow_D$'), that $x \rightsquigarrow y' \leq_D x' \rightsquigarrow y'$; then, by (Tr$_D$'), it follows that $x \rightsquigarrow y \leq_D x' \rightsquigarrow y'$.
Since $y' \leq_D y$, it follows, by (p*$\rightsquigarrow_D$'), that $x \rightsquigarrow y' \leq_D x \rightsquigarrow y$, and since $x \leq_D x'$, it follows, by (p**$\rightsquigarrow_D$'), that $x' \rightsquigarrow y' \leq_D x \rightsquigarrow y'$; then, by (Tr$_D$'), it follows that $x' \rightsquigarrow y' \leq_D x \rightsquigarrow y$.
Thus, $(x \rightsquigarrow y) \sim_D (x' \rightsquigarrow y')$. Hence, $\sim_D$ is a congruence relation. □

Since $\sim_D$ is an equivalence relation on $X$, let $| x | \overset{notation}{=} | x |_D$ be the equivalence class of $x \in X$:

$$| x | \overset{notation}{=} | x |_D \overset{def.}{=} \{y \in X \mid y \sim_D x\};$$

let $X/D \overset{notation}{=} X/\sim_D$ be the quotient set:

$$X/D \overset{notation}{=} X/\sim_D \overset{def.}{=} \{| x | \mid x \in X\}$$

and let

$$D/D \overset{notation}{=} D/\sim_D \overset{def.}{=} \{| x | \mid x \in D\}.$$

Note that

$$| x | \in D/D \iff x \in D. \tag{1.87}$$

**Lemma 1.8.16** *The equivalence classes do not depend on their representants, i.e. for all $x, y \in X$,*

$$| x | = | y | \iff x \sim_D y.$$

**Proof.** $\Longrightarrow$: Suppose $| x | = | y |$; since $x \in | x |$, it follows that $x \in | y |$, i.e. $x \sim_D y$.
$\Longleftarrow$: Suppose $x \sim_D y$ and let $z \in | x |$, i.e. $z \sim_D x$; it follows that $z \sim_D y$, i.e. $z \in | y |$; hence, $| x | \subseteq | y |$. Similarly, $| y | \subseteq | x |$. Thus, $| x | = | y |$. □

Let us define on the quotient set $X/D$ two binary operations $\Rightarrow$ and $\approx>$ by: for all $| x |, | y | \in X/D$,

$$| x | \Rightarrow | y | \overset{def.}{=} | x \to y |,$$

$$| x | \approx> | y | \overset{def.}{=} | x \rightsquigarrow y | .$$

By Proposition 1.8.15, the operations are well defined.

Let us define also on $X/D$ a binary relation $\leq$ by: for all $|x|, |y| \in X/D$,

$$(1.88) \qquad |x| \leq |y| \stackrel{def.}{\Longleftrightarrow} x \leq_D y.$$

Note that we have:

$$\begin{array}{lll} |x| \leq |y| & \stackrel{def.}{\Longleftrightarrow} & x \leq_D y \\ & \stackrel{def.}{\Longleftrightarrow} & x \to y \in D \quad (\stackrel{(IdEqR_D)}{\Longleftrightarrow} \quad x \rightsquigarrow y \in D) \\ & \stackrel{(1.87)}{\Longleftrightarrow} & |x \to y| \in D/D \quad (\stackrel{(IdEqR_{D/D})}{\Longleftrightarrow} |x \rightsquigarrow y| \in D/D) \\ & \stackrel{def.}{\Longleftrightarrow} & |x| \Rightarrow |y| \in D/D \quad (\stackrel{(IdEqR_{D/D})}{\Longleftrightarrow} |x| \approx> |y| \in D/D), \end{array}$$

hence, we have:

$$(1.89) \quad (IdEqR_{D/D}) \quad |x| \Rightarrow |y| \in D/D \iff |x| \approx> |y| \in D/D \quad and$$

(1.90)

$$(pEqrelR_{D/D}) \; |x| \leq |y| \Longleftrightarrow |x| \Rightarrow |y| \in D/D \; (\stackrel{(IdEqR_{D/D})}{\Longleftrightarrow} |x| \approx> |y| \in D/D).$$

By Proposition 1.8.12 and by (1.88), we obtain immediately the following.

**Proposition 1.8.17** *For any* $|x|, |y|, |z| \in X/D$,

$(pB_{\Rightarrow}')$ $\quad |y| \Rightarrow |z| \leq (|x| \Rightarrow |y|) \Rightarrow (|x| \Rightarrow |z|)$,
$(pC_1')$ $\quad |x| \Rightarrow (|y| \approx> |z|) \leq |y| \approx> (|x| \Rightarrow |z|)$,
$(pC_2')$ $\quad |x| \approx> (|y| \Rightarrow |z|) \leq |y| \Rightarrow (|x| \approx> |z|)$,
$(pK_{\Rightarrow}')$ $\quad |x| \leq |y| \Rightarrow |x|$,
$(pK_{\approx>}')$ $\quad |x| \leq |y| \approx> |x|$,
*(MP')* $\quad$ *if* $|x| \in D/D$ *and* $|x| \leq |y|$, *then* $|y| \in D/D$;

*(MK')* $\quad |y| \in D/D \Longrightarrow |x| \leq |y|$;
*(MKK')* $\quad |x|, |y| \in D/D \Longrightarrow |x| \leq |y|$;
$(I_{\Rightarrow}') = (Re')$ $\quad |x| \leq |x|$;
*(Tr')* $\quad |x| \leq |y|$ *and* $|y| \leq |z| \Longrightarrow |x| \leq |z|$,
*(pEq#')* $\quad |x| \leq |y| \Rightarrow |z| \iff |y| \leq |x| \approx> |z|$,
*(pD')* $\quad |x| \leq (|x| \Rightarrow |y|) \approx> |y|$ *and* $|x| \leq (|x| \approx> |y|) \Rightarrow |y|$,
$(p^*_{\Rightarrow}')$ $\quad |y| \leq |z| \Longrightarrow |x| \Rightarrow |y| \leq |x| \Rightarrow |z|$,
$(p^*_{\approx>}')$ $\quad |y| \leq |z| \Longrightarrow |x| \approx> |y| \leq |x| \approx> |z|$,
$(p^{**}_{\Rightarrow}')$ $\quad |y| \leq |z| \Longrightarrow |z| \Rightarrow |x| \leq |y| \Rightarrow |x|$,
$(p^{**}_{\approx>}')$ $\quad |y| \leq |z| \Longrightarrow |z| \approx> |x| \leq |y| \approx> |x|$,
$(pBB_1')$ $\quad |x| \Rightarrow |y| \leq (|y| \Rightarrow |z|) \approx> (|x| \Rightarrow |z|)$.

**Proposition 1.8.18** *The binary relation* $\leq$ *on* $X/D$ *is an order.*

**Proof.** - *Reflexivity*: for any $|x| \in X/D$, $|x| \leq |x|$; that is true by (Re').

- *Transitivity*: for any $|x|, |y|, |z| \in X/D$, $|x| \leq |y|$ and $|y| \leq |z|$ imply $|x| \leq |z|$; that is true by (Tr').

- *Antisymmetry*: $|x| \leq |y|$ and $|y| \leq |x|$ mean, by definition, $x \leq_D y$ and $y \leq_D x$, i.e. $x \sim_D y$, hence, $|x| = |y|$, by Lemma 1.8.16. □

**Proposition 1.8.19** *D is an equivalence class, denoted by* **1**, *and*

$$| x |= \mathbf{1} \quad \overset{def.}{\Longleftrightarrow} \quad | x |= D \quad \Longleftrightarrow \quad x \in D.$$

**Proof.** $\Longrightarrow$: Suppose $| x |= D$; then, for any $y \in D$, we have $y \in| x |$, i.e. $y \sim_D x$, i.e. $y \to x \in D$ and $x \to y \in D$; it follows, by (MP$_D$), that $x \in D$.

$\Longleftarrow$: Suppose $x \in D$; we must prove that $| x | = D$. Indeed,
$\cdot$ $| x | \subseteq D$: let $y \in| x |$, i.e. $y \sim_D x$, i.e. $y \to x \in D$ and $x \to y \in D$; since $x \in D$, it follows, by (MP$_D$), that $y \in D$; thus, $| x | \subseteq D$.
$\cdot$ $D \subseteq | x |$: let $y \in D$, i.e. $y$; then, by (MK$_D$), $z \to y$, for all $z \in X$, hence for $x \in D$; since $x \in D$, i.e. $x$, then, by (MK$_D$) again, $z \to x$, for all $z \in X$, hence for $y \in D$; hence, $x \to y$ and $y \to x$, i.e. $x \sim_D y$, hence $y \in | x |$; thus, $D \subseteq | x |$. □

**Remark 1.8.20** *If $\mathcal{A} = (A, \to, \rightsquigarrow, 1)$ is a pBCK algebra and $D \neq \{1\}$ is a deductive system of $\mathcal{A}$, then, for the pBCK→ prealgebra $(A, \to, \rightsquigarrow, D)$, we have*

$$\mathbf{1} = D =| 1 |,$$

*since $1 \in D$.*

**Remarks 1.8.21** *Note that in this case (D is an equivalence class, denoted by* **1**), *we have $D/D = \mathbf{1}$. Hence,*

*(1) (MP') becomes:*
*(MP') if $| x | = \mathbf{1}$ and $| x | \leq | y |$, then $| y | = \mathbf{1}$,*
*i.e. (MP') is just (N');*

*(1') (MK') becomes:*
*(MK') $| y |= \mathbf{1} \Longrightarrow | x | \leq \mathbf{1}$,*
*i.e. (Mk') is just (La');*

*(1") (MKK') becomes:*
*(MKK') $| x |, | y |= \mathbf{1} \Longrightarrow \mathbf{1} \leq \mathbf{1}$;*

*(2) the equivalences (1.89) and (1.90) become:*

(1.91) $\quad (IdEqR) \quad | x |\Rightarrow| y |= \mathbf{1} \Longleftrightarrow | x |\approx>| y |= \mathbf{1}$ *and*

(1.92) $\quad (pEqrelR) \quad | x | \leq | y | \Longleftrightarrow | x |\Rightarrow| y |= \mathbf{1} \quad (\overset{(IdEqR)}{\Longleftrightarrow} | x |\approx>| y |= \mathbf{1}).$

We can now prove the following result.

**Theorem 1.8.22** *The structure $\mathcal{X}/D = (X/D, \leq, \Rightarrow, \approx>, \mathbf{1})$ is a pseudo-BCK algebra, called the Lindenbaum-Tarski type algebra of the pBCK→ prealgebra $\mathcal{X}$.*

**Proof.** By Proposition 1.8.17, (pB$_\Rightarrow$'), (pC')= (pC$_1$') + (pC$_2$'), (pK')= (pK$_\Rightarrow$') + (pK$_{\approx>}$') hold; by Proposition 1.8.18, (An') holds; (1.91) and (1.92) hold; thus, $\mathcal{X}/D$ is a pBCK algebra (Definition 7'). □

**Remark 1.8.23** *Equivalently, the algebra $\mathcal{X}/D = (X/D, \Rightarrow, \approx>, \mathbf{1})$ is a pBCK algebra (verifying (pB$_\Rightarrow$), (pC), (pK), (An), (IdEqR) (Definition 7)), called the Lindenbaum-Tarski type algebra of $\mathcal{X} = (X, \to, \rightsquigarrow, D)$.*

**Remark 1.8.24** *If we start from the definition of a pBCK algebra as an algebra with the axioms (pB), (pC), (pK), (An), (IdEqR) (Definition 5), then we obtain the definition of the pBCK prealgebra. We have denoted by* **pBCK prealgebra** *the class of all pBCK prealgebras. We then obviously have:*

**pBCK prealgebra** = **pBCK$\rightarrow$ prealgebra** + *($pB\rightsquigarrow_D$).*

*Then, the Lindenbaum-Tarski type construction is the same as for the above pBCK$\rightarrow$ prealgebras and the Lindenbaum-Tarski type algebra is again a pBCK algebra.*

### 1.8.2 Pseudo-BCI$\rightarrow$ (or QB) prestructures

Starting from the definition of a pBCI algebra as an algebra $(A, \rightarrow, \rightsquigarrow, 1)$ with the axioms ($pB_\rightarrow$), (pC), (Re), (An), (IdEqR) (Definition 7), we introduced the following definition.

**Definition 1.8.25** The structure $\mathcal{X} = (X, \rightarrow, \rightsquigarrow, D)$, with $\emptyset \neq D \subseteq X$ and $\rightarrow$, $\rightsquigarrow$ binary operations on $X$, is a *pBCI$\rightarrow$ prestructure*, or a *QB prestructure*, if the following axioms/properties hold: for all $x, y, z \in X$,

($pB\rightarrow_D$) $(y \rightarrow z) \rightarrow ((x \rightarrow y) \rightarrow (x \rightarrow z)) \in D$,
($pC1_D$) $(x \rightarrow (y \rightsquigarrow z)) \rightarrow (y \rightsquigarrow (x \rightarrow z)) \in D$,
($pC2_D$) $(x \rightsquigarrow (y \rightarrow z)) \rightarrow (y \rightarrow (x \rightsquigarrow z)) \in D$,
($I\rightarrow_D$) $x \rightarrow x \in R$,
($IdEqR_D$) $x \rightarrow y \in D \Longleftrightarrow x \rightsquigarrow y \in D$,
($MP_D$) if $x \in D$ and $x \rightarrow y \in D$, then $y \in D$.

Denote by **pBCI$\rightarrow$ (or QB) prestructure** the class of all pBCI$\rightarrow$ prestructures (or QB prestructures).

**Remarks 1.8.26**

*(i) The pBCI$\rightarrow$ propositional logic (or QB propositional logic) $(L, \rightarrow, \rightsquigarrow, L_\Gamma)$ (presented in Subsection 1.7.2) is an example of pBCI$\rightarrow$ (or QB) prestructure, where $L$ is the set of all propositions and $L_\Gamma$ is the set of formal theorems (propositions derived from $\Gamma$).*

*(ii) The pBCI algebra, defined as algebra $(A, \rightarrow, \rightsquigarrow, 1)$ satisfying the axioms ($pB_\rightarrow$), (pC), (Re), (An), (IdEqR) (Definition 7 - see Definition 1.2.19) defines the pBCI$\rightarrow$ (or QB) prestructure $(A, \rightarrow, \rightsquigarrow, D)$, with $D = \{1\}$. Note that the binary relation $\leq$ defined by: $x \leq y \Longleftrightarrow x \rightarrow y \in \{1\} \Longleftrightarrow x \rightarrow y = 1$ is an order, not a preorder.*

*(iii) Given a pBCI algebra $\mathcal{A} = (A, \rightarrow, \rightsquigarrow, 1)$ and a compatible deductive system $D \neq \{1\}$ of $\mathcal{A}$, then $(A, \rightarrow, \rightsquigarrow, D)$ is a pBCI$\rightarrow$ (or QB) prestructure, since $1 \in D$.*

**Remark 1.8.27** *Any pBCK$\rightarrow$ prealgebra (Definition 2) is a pBCI$\rightarrow$ (or QB) prestructure verifying additionaly ($MK_D$). We write:*

**pBCK$\rightarrow$ prealgebra** $\Longleftrightarrow$ **pBCI$\rightarrow$ (or QB) prestructure** + *($MK_D$)* = **pBCI$\rightarrow$" (or QB") prealgebra**.

**Proposition 1.8.28** *For any $x, y, z \in X$, we have the following properties:*

(*Tr*$_D$) $x \to y \in D$ *and* $y \to z \in D \implies x \to z \in D$,
(*pEq#*$_D$) $x \rightsquigarrow (y \to z) \in D \iff y \to (x \rightsquigarrow z) \in D$,
(*pD*$_D$) $x \to ((x \to y) \rightsquigarrow y) \in D$ *and* $x \to ((x \rightsquigarrow y) \to y) \in D$,
(*p*$^*\!\to_D$) $y \to z \in D \implies (x \to y) \to (x \to z) \in D$,
(*p*$^*\!\rightsquigarrow_D$) $y \to z \in D \implies (x \rightsquigarrow y) \to (x \rightsquigarrow z) \in D$,
(*p*$^{**}\!\to_D$) $y \to z \in D \implies (z \to x) \to (y \to x) \in D$,
(*p*$^{**}\!\rightsquigarrow_D$) $y \to z \in D \implies (z \rightsquigarrow x) \to (y \rightsquigarrow x) \in D$,
(*pBB1*$_D$) $(x \to y) \to ((y \to z) \rightsquigarrow (x \to z)) \in D$.

**Proof.** By Proposition 1.8.6. □

**Lemma 1.8.29** *For any $x \in X$,*

$$((x \to x) \rightsquigarrow x) \to x \in D \quad \textit{and} \quad ((x \rightsquigarrow x) \to x) \rightsquigarrow x \in D.$$

**Proof.** We have:
$x \to x \in D$, by (I$\to_D$);
$(x \to x) \to (((x \to x) \rightsquigarrow x) \to x) \in D$, by (pD$_D$);
$((x \to x) \rightsquigarrow x) \to x \in D$, by (MP$_D$)
and, similarly,
$x \to x \in D$, by (I$\to_D$),
$x \rightsquigarrow x \in D$, by (IdEqR$_D$);
$(x \rightsquigarrow x) \to (((x \rightsquigarrow x) \to x) \rightsquigarrow x) \in D$, by (pD$_D$);
$((x \rightsquigarrow x) \to x) \rightsquigarrow x \in D$, by (MP$_D$).

□

**Lemma 1.8.30** *For any $x, y \in X$,*

$$(x \to x) \rightsquigarrow y \in D \implies (y \to y) \rightsquigarrow y \in D.$$

**Proof.** Suppose that $(x \to x) \rightsquigarrow y \in D$; then,
$(x \to x) \to y \in D$, by (IdEqR$_D$);
$x \to x \in D$, by (I$\to_D$); hence
$y \in D$, by (MP$_D$);
$y \to ((y \to y) \rightsquigarrow y) \in D$, by (pD$_D$), hence,
$(y \to y) \rightsquigarrow y \in D$, by (MP$_D$).

□

**Remark 1.8.31** *The property*
(*pB*$\to_D$) $(y \to z) \to ((x \to y) \to (x \to z)) \in D$
*of the pBCI→ (or QB) prestructure does not imply the 'twin' property,*
(*pB*$\rightsquigarrow_D$) $(y \rightsquigarrow z) \rightsquigarrow ((x \rightsquigarrow y) \rightsquigarrow (x \rightsquigarrow z)) \in D$,
*as in the case of pBCI algebras (see Theorem 1.2.18). We only have the result from the following proposition, obtained by Prover9.*

**Proposition 1.8.32** *For any $x, y, z \in X$, we have:*

(1.93) $$(x \rightsquigarrow y) \to (x \rightsquigarrow ((y \rightsquigarrow z) \to z)) \in D.$$

**Proof.** See Proposition 1.8.11. □

• **The Lindenbaum-Taski type 'algebra': the locally unital QB algebra**

Let $(X, \rightarrow, \rightsquigarrow, D)$ be a pBCI→ (or QB) prestructure throughout this subsection. We define on $X$ a binary relation $\leq_D$ by: for all $x, y \in X$,

$$x \leq_D y \stackrel{def.}{\Longleftrightarrow} x \rightarrow y \in D \quad (\stackrel{(IdEqR_D)}{\Longleftrightarrow} \quad x \rightsquigarrow y \in D). \tag{1.94}$$

By Definition 1.8.25, Proposition 1.8.28 and by (1.94), we obtain immediately the following.

**Proposition 1.8.33** *For any $x, y, z \in X$,*

(pB→$_D$') $y \rightarrow z \leq_D (x \rightarrow y) \rightarrow (x \rightarrow z)$,
(pC1$_D$') $x \rightarrow (y \rightsquigarrow z) \leq_D y \rightsquigarrow (x \rightarrow z)$,
(pC2$_D$') $x \rightsquigarrow (y \rightarrow z) \leq_D y \rightarrow (x \rightsquigarrow z)$,
(I→$_D$') $x \leq_D x$,
(MP$_D$') *if* $x \in D$ *and* $x \leq_D y$, *then* $y \in D$;

(Tr$_D$') $x \leq_D y$ *and* $y \leq_D z \Longrightarrow x \leq_D z$,
(pEq#$_D$') $x \leq_D y \rightarrow z \Longleftrightarrow y \leq_D x \rightsquigarrow z$,
(pD$_D$') $x \leq_D (x \rightarrow y) \rightsquigarrow y$ *and* $x \leq_D (x \rightsquigarrow y) \rightarrow y$,
(p*→$_D$') $y \leq_D z \Longrightarrow x \rightarrow y \leq_D x \rightarrow z$,
(p*⇝$_D$') $y \leq_D z \Longrightarrow x \rightsquigarrow y \leq_D x \rightsquigarrow z$,
(p**→$_D$') $y \leq_D z \Longrightarrow z \rightarrow x \leq_D y \rightarrow x$,
(p**⇝$_D$') $y \leq_D z \Longrightarrow z \rightsquigarrow x \leq_D y \rightsquigarrow x$,
(pBB1$_D$') $x \rightarrow y \leq_D (y \rightarrow z) \rightsquigarrow (x \rightarrow z)$.

By Lemma 1.8.29 and by (1.94), we obtain immediately the following

**Lemma 1.8.34** *For any $x \in X$,*

$$(x \rightarrow x) \rightsquigarrow x \leq_D x \quad and \quad (x \rightsquigarrow x) \rightarrow x \leq_D x.$$

By Lemma 1.8.30 and by (1.94), we obtain immediately the following

**Lemma 1.8.35** *For any $x, y \in X$,*

$$x \rightarrow x \leq_D y \quad \Longrightarrow \quad y \rightarrow y \leq_D y.$$

By Proposition 1.8.32 and by (1.94), we obtain immediately the following

**Proposition 1.8.36** *For any $x, y, z \in X$, we have:*

$$x \rightsquigarrow y \leq_D x \rightsquigarrow ((y \rightsquigarrow z) \rightarrow z). \tag{1.95}$$

**Proposition 1.8.37** *The binary relation $\leq_D$ is a preorder.*

**Proof.** See Proposition 1.8.13. □

We define on $X$ a binary relation $\sim_D$ by: for all $x, y \in X$,

$$x \sim_D y \overset{def.}{\Longleftrightarrow} (x \leq_D y \quad and \quad y \leq_D x). \tag{1.96}$$

**Proposition 1.8.38** *The binary relation $\sim_D$ is an equivalence relation.*

**Proof.** See Proposition 1.8.14. □

**Proposition 1.8.39** *The binary relation $\sim_D$ is a congruence relation.*

**Proof.** See Proposition 1.8.15. □

Since $\sim_D$ is an equivalence relation on $X$, let $| x | \overset{notation}{=} | x |_D$ be the equivalence class of $x \in X$:

$$| x | \overset{notation}{=} | x |_D \overset{def.}{=} \{y \in X \mid y \sim_D x\};$$

let $X/D \overset{notation}{=} X/\sim_D$ be the quotient set:

$$X/D \overset{notation}{=} X/\sim_D \overset{def.}{=} \{| x | \mid x \in X\}$$

and let

$$D/D \overset{notation}{=} D/\sim_D \overset{def.}{=} \{| x | \mid x \in D\}.$$

Note that

$$| x | \in D/D \quad \Longleftrightarrow \quad x \in D. \tag{1.97}$$

**Lemma 1.8.40** *The equivalence classes do not depend on their representants, i.e. for all $x, y \in X$,*

$$| x | = | y | \Longleftrightarrow x \sim_D y.$$

**Proof.** See Lemma 1.8.16. □

Let us define on the quotient set $X/D$ two binary operations $\Rightarrow$ and $\approx>$ by: for all $| x |, | y | \in X/D$,

$$| x | \Rightarrow | y | \overset{def.}{=} | x \to y |,$$

$$| x | \approx> | y | \overset{def.}{=} | x \rightsquigarrow y |.$$

By Proposition 1.8.39, the operations are well defined.

Let us define also on $X/D$ a binary relation $\leq$ by: for all $| x |, | y | \in X/D$,

$$| x | \leq | y | \overset{def.}{\Longleftrightarrow} x \leq_D y. \tag{1.98}$$

Note that we have:

$$\begin{array}{llll} \mid x \mid \leq \mid y \mid & \stackrel{def.}{\Longleftrightarrow} & x \leq_D y & \\ & \stackrel{def.}{\Longleftrightarrow} & x \to y \in D & (\stackrel{(IdEqR_D)}{\Longleftrightarrow} x \rightsquigarrow y \in D) \\ & \stackrel{(1.97)}{\Longleftrightarrow} & \mid x \to y \mid \in D/D & (\stackrel{(IdEqR_{D/D})}{\Longleftrightarrow} \mid x \rightsquigarrow y \mid \in D/D) \\ & \stackrel{def.}{\Longleftrightarrow} & \mid x \mid \Rightarrow \mid y \mid \in D/D & (\stackrel{(IdEqR_{D/D})}{\Longleftrightarrow} \mid x \mid \approx> \mid y \mid \in D/D), \end{array}$$

hence, we have:

(1.99) $(IdEqR_{D/D})\quad \mid x \mid \Rightarrow \mid y \mid \in D/D \iff \mid x \mid \approx> \mid y \mid \in D/D$ *and*

(1.100)

$$(pEqrelR_{D/D})\ \mid x \mid \leq \mid y \mid \Longleftrightarrow \mid x \mid \Rightarrow \mid y \mid \in D/D\ (\stackrel{(IdEqR_{D/D})}{\Longleftrightarrow} \mid x \mid \approx> \mid y \mid \in D/D).$$

By Proposition 1.8.33 and by (1.98), we obtain immediately the following

**Proposition 1.8.41** *For any* $\mid x \mid, \mid y \mid, \mid z \mid \in X/D$,

(pB⇒') $\mid y \mid \Rightarrow \mid z \mid \ \leq\ (\mid x \mid \Rightarrow \mid y \mid) \Rightarrow (\mid x \mid \Rightarrow \mid z \mid)$,
(pC₁') $\mid x \mid \Rightarrow (\mid y \mid \approx> \mid z \mid)\ \leq\ \mid y \mid \approx> (\mid x \mid \Rightarrow \mid z \mid)$,
(pC₂') $\mid x \mid \approx> (\mid y \mid \Rightarrow \mid z \mid)\ \leq\ \mid y \mid \Rightarrow (\mid x \mid \approx> \mid z \mid)$,
(I⇒') = (Re') $\mid x \mid \leq \mid x \mid$,
(MP') *if* $\mid x \mid \in D/D$ *and* $\mid x \mid \leq \mid y \mid$, *then* $\mid y \mid \in D/D$;
(Tr') $\mid x \mid \leq \mid y \mid$ *and* $\mid y \mid \leq \mid z \mid \implies \mid x \mid \leq \mid z \mid$,
(pEq#') $\mid x \mid \leq \mid y \mid \Rightarrow \mid z \mid \iff \mid y \mid \leq \mid x \mid \approx> \mid z \mid$,
(pD') $\mid x \mid \leq (\mid x \mid \Rightarrow \mid y \mid) \approx> \mid y \mid$ *and* $\mid x \mid \leq (\mid x \mid \approx> \mid y \mid) \Rightarrow \mid y \mid$,
(p*⇒') $\mid y \mid \leq \mid z \mid \implies \mid x \mid \Rightarrow \mid y \mid \leq \mid x \mid \Rightarrow \mid z \mid$,
(p*≈>') $\mid y \mid \leq \mid z \mid \implies \mid x \mid \approx> \mid y \mid \leq \mid x \mid \approx> \mid z \mid$,
(p**⇒') $\mid y \mid \leq \mid z \mid \implies \mid z \mid \Rightarrow \mid x \mid \leq \mid y \mid \Rightarrow \mid x \mid$,
(p**≈>') $\mid y \mid \leq \mid z \mid \implies \mid z \mid \approx> \mid x \mid \leq \mid y \mid \approx> \mid x \mid$,
(pBB₁') $\mid x \mid \Rightarrow \mid y \mid \ \leq\ (\mid y \mid \Rightarrow \mid z \mid) \approx> (\mid x \mid \Rightarrow \mid z \mid)$.

By Lemma 1.8.34 and by (1.98), we obtain immediately the following

**Lemma 1.8.42** *For any* $\mid x \mid \in X/D$,

$$(\mid x \mid \Rightarrow \mid x \mid) \approx> \mid x \mid\ \leq\ \mid x \mid \quad \textit{and} \quad (\mid x \mid \approx> \mid x \mid) \Rightarrow \mid x \mid\ \leq\ \mid x \mid .$$

By Lemma 1.8.35 and by (1.98), we obtain immediately the following

**Lemma 1.8.43** *For any* $\mid x \mid, \mid y \mid \in X/D$,

$$\mid x \mid \Rightarrow \mid x \mid\ \leq\ \mid y \mid \implies \mid y \mid \Rightarrow \mid y \mid\ \leq\ \mid y \mid .$$

By Proposition 1.8.36 and by (1.98), we obtain immediately the following

**Proposition 1.8.44** *For any* $\mid x \mid, \mid y \mid, \mid z \mid \in X/D$, *we have:*

(1.101) $$\mid x \mid \approx> \mid y \mid\ \leq\ \mid x \mid \approx> ((\mid y \mid \approx> \mid z \mid) \Rightarrow \mid z \mid).$$

**Proposition 1.8.45** *The binary relation* $\leq$ *on* $X/D$ *is an order.*

**Proof.** See Proposition 1.8.18. □

**Remark 1.8.46** *D is not always an equivalence class anymore; it is a set of equivalence classes.*

**Corollary 1.8.47** *For any* $\mid x \mid, \mid y \mid, \mid z \mid \in X/D$, *we have:*

(pEx) $\mid x \mid \Rightarrow (\mid y \mid \approx> \mid z \mid) \; = \; \mid y \mid \approx> (\mid x \mid \Rightarrow \mid z \mid)$,

($pB_{\approx>}$') $\mid y \mid \approx> \mid z \mid \; \leq \; (\mid x \mid \approx> \mid y \mid) \approx> (\mid x \mid \approx> \mid z \mid)$,

($pBB_2$') $\mid x \mid \approx> \mid y \mid \; \leq \; (\mid y \mid \approx> \mid z \mid) \Rightarrow (\mid x \mid \approx> \mid z \mid)$.

**Proof.** (pEx): By ($pC_1$') and ($pC_2$'), we have
$\mid x \mid \Rightarrow (\mid y \mid \approx> \mid z \mid) \; \leq \; \mid y \mid \approx> (\mid x \mid \Rightarrow \mid z \mid)$ and
$\mid y \mid \approx> (\mid x \mid \Rightarrow \mid z \mid) \; \leq \; \mid x \mid \Rightarrow (\mid y \mid \approx> \mid z \mid)$,
hence, by antisymmetry,
$\mid x \mid \Rightarrow (\mid y \mid \approx> \mid z \mid) = \mid y \mid \approx> (\mid x \mid \Rightarrow \mid z \mid)$, i.e. (pEx) holds.

($pB_{\approx>}$'): By (1.101) of Proposition 1.8.44, we have:
$\mid x \mid \approx> \mid y \mid \; \leq \; \mid x \mid \approx> ((\mid y \mid \approx> \mid z \mid) \Rightarrow \mid z \mid)$,
hence, by (pEx), we obtain:
$\mid x \mid \approx> \mid y \mid \; \leq \; (\mid y \mid \approx> \mid z \mid) \Rightarrow (\mid x \mid \approx> \mid z \mid)$,
which by (pEq#') gives:
$\mid y \mid \approx> \mid z \mid \; \leq \; (\mid x \mid \approx> \mid y \mid) \approx> (\mid x \mid \approx> \mid z \mid)$, i.e. ($pB_{\approx>}$') holds.

($pBB_2$'): By ($pB_{\approx>}$'), $\mid y \mid \approx> \mid z \mid \; \leq \; (\mid x \mid \approx> \mid y \mid) \approx> (\mid x \mid \approx> \mid z \mid)$;
then, by (pEq#'), $\mid x \mid \approx> \mid y \mid \; \leq \; (\mid y \mid \approx> \mid z \mid) \Rightarrow (\mid x \mid \approx> \mid z \mid)$, i.e. ($pBB_2$') holds. □

**Corollary 1.8.48** *For any* $\mid x \mid \in X/D$,

$$(\mid x \mid \Rightarrow \mid x \mid) \approx> \mid x \mid \; = \; \mid x \mid \quad and \quad (\mid x \mid \approx> \mid x \mid) \Rightarrow \mid x \mid \; = \; \mid x \mid .$$

**Proof.** By (pD'), $\mid x \mid \leq (\mid x \mid \Rightarrow \mid x \mid) \approx> \mid x \mid$.
By Lemma 1.8.42, $(\mid x \mid \Rightarrow \mid x \mid) \approx> \mid x \mid \leq \mid x \mid$.
Then, by antisymmetry, it follows that $(\mid x \mid \Rightarrow \mid x \mid) \approx> \mid x \mid = \mid x \mid$.
Similarly, $(\mid x \mid \approx> \mid x \mid) \Rightarrow \mid x \mid = \mid x \mid$. □

We can now prove the following result.

**Theorem 1.8.49** *The structure* $\mathcal{X}/D = (X/D, \leq, \Rightarrow, \approx>, X/D^+)$ *is a locally unital quantum-B algebra, called the Lindenbaum-Tarsli type 'algebra' of* $\mathcal{X}$, *the pBCI*→ *(or QB) prestructure, and* $X/D^+ = D/D$.

**Proof.** $(X/D, \leq)$ is a poset, by Proposition 1.8.45. The properties (pEx), ($p^*_{\Rightarrow}$'), (pEq#') hold; indeed: (pEx) holds by above Corollary; ($p^*_{\Rightarrow}$') holds by Proposition 1.8.41; (pEq#') holds by Proposition 1.8.41.

Thus, $(X/D, \leq, \Rightarrow, \approx>)$ is a quantum-B algebra (Definition 2).

We prove now that it is locally unital, i.e. that (1.13) and (1.14) hold.

(1.13): We must prove that, for any $\mid x \mid \in X/D$, we have:
$(\mid x \mid \Rightarrow \mid x \mid) \approx> \mid x \mid = \mid x \mid$ *and* $(\mid x \mid \approx> \mid x \mid) \Rightarrow \mid x \mid = \mid x \mid$,
and this is true by Corollary 1.8.48.

(1.14): We must prove that, for any $\mid x \mid, \mid y \mid \in X/D$, we have:
$\mid x \mid \Rightarrow \mid x \mid \leq \mid y \mid \Longrightarrow \mid y \mid \Rightarrow \mid y \mid \leq \mid y \mid$,
and this is true by Lemma 1.8.43.

Define $X/D^+ \stackrel{def.}{=} \{\mid x \mid \in X/D \mid \mid x \mid \Rightarrow \mid x \mid \leq \mid x \mid\} \subset X/D$. We prove that

$$X/D^+ = D/D. \tag{1.102}$$

Indeed,

$$\begin{array}{lcl} \mid x \mid \in X/D^+ & \stackrel{def.}{\Longleftrightarrow} & \mid x \mid \Rightarrow \mid x \mid \leq \mid x \mid \\ & \stackrel{def.}{\Longleftrightarrow} & \mid x \to x \mid \leq \mid x \mid \\ & \stackrel{(1.98)}{\Longleftrightarrow} & x \to x \leq_D x \\ & \stackrel{(1.94))}{\Longleftrightarrow} & (x \to x) \rightsquigarrow x \in D \\ & \stackrel{(1.97)}{\Longleftrightarrow} & \mid (x \to x) \rightsquigarrow x \mid \in D/D \\ & \stackrel{def.}{\Longleftrightarrow} & (\mid x \mid \Rightarrow \mid x \mid) \approx> \mid x \mid \in D/D \\ & \stackrel{Prop.1.8.48}{\Longleftrightarrow} & \mid x \mid \in D/D. \end{array}$$

□

Note that, by (1.99) and (1.100), we have (see Proposition 1.3.9):
(1.103)

$$(IdEqR_{X/D^+}) \quad \mid x \mid \Rightarrow \mid y \mid \in X/D^+ \iff \mid x \mid \approx> \mid y \mid \in X/D^+ \quad \textit{and}$$

(1.104)

$$(pEqrelR_{X/D^+}) \ \mid x \mid \leq \mid y \mid \Longleftrightarrow \mid x \mid \Rightarrow \mid y \mid \in X/D^+ \ (\stackrel{(IdEqR_{X/D^+})}{\Longleftrightarrow} \mid x \mid \approx> \mid y \mid \in X/D^+).$$

**Remark 1.8.50** *If we start from the definition of a pBCI algebra as an algebra with the axioms (pB), (pC), (Re), (An), (IdEqR) (Definition 5), then we obtain the definition of the pBCI prestructure. We have denoted by* **pBCI prestructure** *the class of all pBCI prestructures. We then obviously have:*

**pBCI prestructure** = **pBCI→ (or QB) prestructure** + $(pB{\rightsquigarrow}_D)$.

*Then, the Lindenbaum-Tarski type construction is the same as above and the Lindenbaum-Tarski type 'algebra' is again a locally unital QB algebra.*

### 1.8.3 Pseudo-BCI→' (or QB') prealgebras

Starting again from the definition of a pBCI algebra as an algebra $(A, \to, \rightsquigarrow, 1)$ with the axioms ($pB_{\to}$), (pC), (Re), (An), (IdEqR) (Definition 7), we introduced also the following definition.

**Definition 1.8.51** The structure $\mathcal{X} = (X, \to, \rightsquigarrow, D)$, with $\emptyset \neq D \subseteq X$ and $\to$, $\rightsquigarrow$ binary operations on $X$, is a *pBCI→' prealgebra*, or a *QB' prealgebra*, if the following properties hold: for all $x, y, z \in X$,

(pB$\rightarrow_D$) $(y \rightarrow z) \rightarrow ((x \rightarrow y) \rightarrow (x \rightarrow z)) \in D$,
(pC1$_D$) $(x \rightarrow (y \rightsquigarrow z)) \rightarrow (y \rightsquigarrow (x \rightarrow z)) \in D$,
(pC2$_D$) $(x \rightsquigarrow (y \rightarrow z)) \rightarrow (y \rightarrow (x \rightsquigarrow z)) \in D$,
(I$\rightarrow_D$) $x \rightarrow x \in R$,
(IdEqR$_D$) $x \rightarrow y \in D \Longleftrightarrow x \rightsquigarrow y \in D$,
(MP$_D$) if $x \in D$ and $x \rightarrow y \in D$, then $y \in D$ and
(MKK$_D$) if $x, y \in D$, then $x \rightarrow y \in D$.

We have denoted by **pBCI→' (or QB') prealgebra** the class of all pBCI→' (or QB') prealgebras.

**Remark 1.8.52** *A pBCI→' (or QB') prealgebra is a pBCI→ (or QB) prestructure verifying additionally (MKK$_D$). We write:*
**pBCI→' (or QB') prealgebra** = **pBCI→ (or QB) prestructure** + *(MKK$_D$).*

Hence, we have additional results to the Lindenbaum-Tarski type 'algebra' of pBCI→ (or QB) prestructure.

**Proposition 1.8.53** *Let $(X, \rightarrow, \rightsquigarrow, D)$ be a pBCI→' (or QB') prealgebra. Then, $D$ is an equivalence class, denoted by* **1**, *and*

$$\mid x \mid = \mathbf{1} \quad \stackrel{def.}{\Longleftrightarrow} \quad \mid x \mid = D \quad \Longleftrightarrow \quad x \in D.$$

**Proof.** $\Longrightarrow$: Suppose $\mid x \mid = D$; then, for any $y \in D$, we have $y \in \mid x \mid$, i.e. $y \sim_D x$, i.e. $y \rightarrow x \in D$ and $x \rightarrow y \in D$; it follows, by (MP$_D$), that $x \in D$.
$\Longleftarrow$: Suppose $x \in D$; we must prove that $\mid x \mid = D$. Indeed,
· $\mid x \mid \subseteq D$: le t $y \in \mid x \mid$, i.e. $y \sim_D x$, i.e. $y \rightarrow x \in D$ and $x \rightarrow y \in D$; since $x \in D$, it follows, by (MP$_D$), that $y \in D$; thus, $\mid x \mid \subseteq D$.
· $D \subseteq \mid x \mid$: let $y \in D$; snce $x \in D$, it follows by (MKK), that $x \rightarrow y \in D$ and $y \rightarrow x \in D$, i.e. $x \sim_D y$, hence $y \in \mid x \mid$; thus, $D \subseteq \mid x \mid$. □

**Remarks 1.8.54** *Note that, in this case (D is an equivalence class, denoted by* **1**), *we have*

$$X/D^+ = D/D = \{\mathbf{1}\}.$$

*Consequently,*
*(1) the property (MP') from Proposition 1.8.41 becomes:*

(1.105) $$(MP') \quad if \mid x \mid = \mathbf{1} \quad and \quad \mid x \mid \leq \mid y \mid, \quad then \mid y \mid = \mathbf{1},$$

*i.e. (MP') is just (N'); thus, (N') holds.*
*(2) the equivalences (1.103) and (1.104) become*

(1.106) $$(IdEqR) \quad \mid x \mid \Rightarrow \mid y \mid = \mathbf{1} \Longleftrightarrow \mid x \mid \approx> \mid y \mid = \mathbf{1} \quad and$$

(1.107) $$(pEqrelR) \quad \mid x \mid \leq \mid y \mid \Longleftrightarrow \mid x \mid \Rightarrow \mid y \mid = \mathbf{1} \quad (\stackrel{(IdEqR)}{\Longleftrightarrow} \mid x \mid \approx> \mid y \mid = \mathbf{1}).$$

**Proposition 1.8.55** *We have: for all* $\mid y \mid \in X/D$,
*(pM)* $\quad \mathbf{1} \Rightarrow \mid y \mid \; = \; \mid y \mid$.

**Proof.** By Proposition 1.8.53, $\mid x \mid = \mathbf{1} \stackrel{def.}{\Longleftrightarrow} \mid x \mid = D \Longleftrightarrow x \in D$.
Hence, for any $y \in X$ and for any $x \in D$, we have:

$$\begin{aligned} \mathbf{1} \Rightarrow \mid y \mid = \mid y \mid \quad &\Longleftrightarrow \quad \mid x \mid \Rightarrow \mid y \mid = \mid y \mid \\ &\Longleftrightarrow \quad \mid x \to y \mid = \mid y \mid \\ &\Longleftrightarrow \quad (x \to y) \sim_D y \\ &\Longleftrightarrow \quad x \to y \leq_D y \text{ and } y \leq_D x \to y \\ &\Longleftrightarrow \quad (x \to y) \to y \in D \text{ and } y \to (x \to y) \in D. \end{aligned}$$

So, we must prove that $(x \to y) \to y \in D$ and $y \to (x \to y) \in D$. Indeed,
$x \in D$, by hypothesis,
$x \to ((x \to y) \rightsquigarrow y) \in D$, by (pD$_D$),
$(x \to y) \rightsquigarrow y \in D$, by (MP$_D$),
$(x \to y) \to y \in D$, by (IdEqR$_D$)
and
$x \in D$, by hypothesis,
$y \to y \in D$, by (I$\to_D$),
$y \rightsquigarrow y \in D$, by (IdEqR$_D$),
$x \to (y \rightsquigarrow y) \in D$, by (MKK$_D$),
$y \rightsquigarrow (x \to y) \in D$, by (pEq#$_D$),
$y \to (x \to y) \in D$, by (IdEqR$_D$).

□

Now, we can prove the following result.

**Theorem 1.8.56** *The structure* $\mathcal{X}/D = (X/D, \leq, \Rightarrow, \approx>, \mathbf{1})$ *is a pBCI algebra, called the Lindenbaum-Tarski type algebra of the pBCI→' (or QB') prealgebra* $\mathcal{X}$.

**Proof.** We present two proofs.

**Proof 1**: By Proposition 1.8.41, (pB$_\Rightarrow$'), (pC')= (pC$_1$') + (pC$_2$'), (Re') hold; by Proposition 1.8.45, (An') holds; (1.106) and (1.107) hold; hence, $\mathcal{X}/D$ is a pBCI algebra (Definition 7'). Note that (pB$_{\approx>}$') holds too, by Corollary 1.8.47, hence $\mathcal{X}/D$ is a pBCI algebra by Definition 5' too.

**Proof 2**: By Theorem 1.8.49, the structure $\mathcal{X}/D$ is a locally unital quantum-B algebra; by Proposition 1.8.55, (pM) holds and, by (1.105), (N') holds; hence, $\mathcal{X}/D$ is a maximal quantum-B algebra, hence is a pBCI algebra, by Proposition 1.3.18.
□

**Remark 1.8.57** *If we start from the definition of a pBCI algebra as an algebra with the axioms (pB), (pC), (Re), (An), (IdEqR) (Definition 5), then we obtain the definition of pBCI' prealgebra. We have denoted by* **pBCI' prealgebra** *the class of all pBCI' prealgebras. We then obviously have:*

**pBCI' prealgebra** = **pBCI→' (or QB') prealgebra** + *(pB*$\rightsquigarrow_D$*).*

*Then, the Lindenbaum-Tarski type construction is the same as above and the Lindenbaum-Tarski type algebra is again a pBCI algebra.*

### 1.8.4 Pseudo-BCI→" (or QB") prealgebras ($\Longleftrightarrow$ pBCK→ prealgebras)

Consider in this subsection that $\mathcal{X} = (X, \rightarrow, \rightsquigarrow, D)$ is a pBCI→ (or QB) prestructure verifying additionally ($MK_D$), i.e. is a *pBCI→" (or QB") prealgebra.* Hence, we have additional results for the Lindenbaum-Tarski type 'algebra' of pBCI→ (or QB) prestructure.

**Proposition 1.8.58** *Let $\mathcal{X} = (X, \rightarrow, \rightsquigarrow, D)$ be a pBCI→" (or QB") prealgebra. We have: for all $\mid y \mid \in X/D$,*
*(La')* $\quad \mid y \mid \leq \mathbf{1}$.

**Proof.** $\mid y \mid \leq \mathbf{1} \Longleftrightarrow \mid y \mid \leq \mid x \mid \Longleftrightarrow y \leq_D x \Longleftrightarrow y \rightarrow x \in D$.
So, we must prove that, for all $x \in D$, $y \in X$, we have $y \rightarrow x \in D$.
Indeed $x \in D$, by hypothesis, hence $y \rightarrow x \in D$, by ($MK_D$). □

Now, we can prove the following result.

**Theorem 1.8.59** *The structure $\mathcal{X}/D = (X/D, \leq, \Rightarrow, \approx>, \mathbf{1})$ is a pBCK algebra, called the Lindenbaum-Tarski type algebra of the pBCI→" (or QB") prealgebra $\mathcal{X}$.*

**Proof.** We present two proofs.

**Proof 1**: By Remark 1.8.27, $\mathcal{X}$ is a pBCK→ prealgebra. Then, apply Theorem 1.8.22.

**Proof 2**: By Theorem 1.8.5, ($MKK_D$) holds too. It follows that $\mathcal{X}$ is a pBCI→' prealgebra. Then, its Lindenbaun-Tarski type algebra is a pBCI algebra (i.e., by Theorem 1.8.56, the structure $\mathcal{X}/D$ is a pBCI algebra); by Proposition 1.8.58, the property (La') holds, hence, $\mathcal{X}/D$ is a pBCK algebra. □

**Remark 1.8.60** *If we start from the definition of a pBCI algebra as an algebra with the axioms (pB), (pC), (Re), (An), (IdEqR) (Definition 5), then we obtain the definition of the pBCI" prealgebra as a pBCI→" (or QB") prestructure verifying additionally ($pB{\rightsquigarrow}_D$). We have denoted by* **pBCI" prealgebra** *the class of all pBCI" prealgebras. We then obviously have:*

**pBCI" prealgebra** = **pBCI→" (or QB") prealgebra** + *($pB{\rightsquigarrow}_D$).*

*Consequently, we have:*

**pBCK prealgebra** $\Longleftrightarrow$ **pBCI" prealgebra**.

### 1.8.5 Connections between prealgebras/prestructures and their algebras/structures

By the results from this section, we obtain the connections between the prealgebras/prestructures and their Lindenbaum-Tarski type algebras (i.e. algebras obtained by Lindenbaum-Tarski type constructions) presented in Figure 1.9.

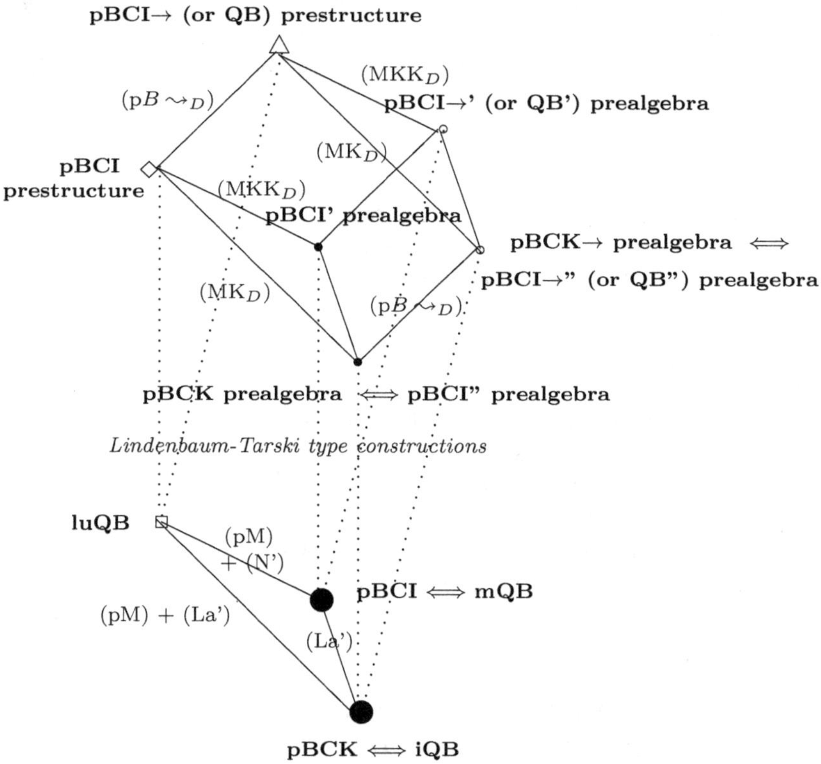


Figure 1.9: Connections between the prealgebras/prestructures and their algebras

### 1.8.6 The commutative case

Note that the pBCK→ prealgebra and the pBCK prealgebra become, in the commutative case, the BCK prealgebra.

Note also that the pBCI→ (or QB) prestructure and the pBCI prestructure become, in the commutative case, the BCI prestructure. This is the argument which motivates the alternative name 'pBCI→ prestructure' for the 'quantum-B prestructure' (or QB prestructure, for short) (name given here following the name of quantum-B logic in [193]).

The BCK prealgebra $(X, \rightarrow, D)$ has the axioms:
($B_D$) $(x \rightarrow y) \rightarrow ((y \rightarrow z) \rightarrow (x \rightarrow z)) \in D$,
($C_D$) $(x \rightarrow (y \rightarrow z)) \rightarrow (y \rightarrow (x \rightarrow z)) \in D$,
($K_D$) $x \rightarrow (y \rightarrow x) \in D$,
($MP_D$) if $x \in D$ and $x \rightarrow y \in D$, then $y \in D$.

The BCI prestructure $(X, \rightarrow, D)$ has the axioms:
($B_D$) $(x \rightarrow y) \rightarrow ((y \rightarrow z) \rightarrow (x \rightarrow z)) \in D$,
($C_D$) $(x \rightarrow (y \rightarrow z)) \rightarrow (y \rightarrow (x \rightarrow z)) \in D$,
($I_D$) $x \rightarrow x \in D$
($MP_D$) if $x \in D$ and $x \rightarrow y \in D$, then $y \in D$,

while the BCI' prealgebra $(X, \rightarrow, D)$ has the axioms:
($B_D$) $(x \rightarrow y) \rightarrow ((y \rightarrow z) \rightarrow (x \rightarrow z)) \in D$,
($C_D$) $(x \rightarrow (y \rightarrow z)) \rightarrow (y \rightarrow (x \rightarrow z)) \in D$,
($I_D$) $x \rightarrow x \in D$
($MP_D$) if $x \in D$ and $x \rightarrow y \in D$, then $y \in D$,
($MKK_D$) if $x, y \in D$, then $x \rightarrow y \in D$,

and the BCI" prealgebra $(X, \rightarrow, D)$ has the axioms:
($B_D$) $(x \rightarrow y) \rightarrow ((y \rightarrow z) \rightarrow (x \rightarrow z)) \in D$,
($C_D$) $(x \rightarrow (y \rightarrow z)) \rightarrow (y \rightarrow (x \rightarrow z)) \in D$,
($I_D$) $x \rightarrow x \in D$
($MP_D$) if $x \in D$ and $x \rightarrow y \in D$, then $y \in D$,
($MK_D$) if $y \in D$, then $x \rightarrow y \in D$.

In the commutative case, we then obtain the connections between the prealgebras/prestructures and their algebras (algebras obtained by Lindenbaum-Tarski type constructions) presented in Figure 1.10.

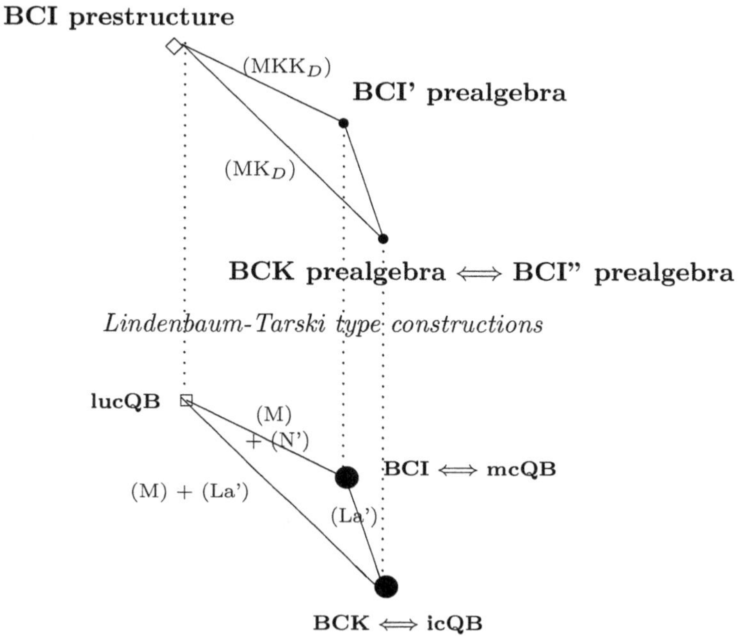


Figure 1.10: Connections between the prealgebras/prestructure and their algebras in the commutative case

### 1.8.7 Examples

**Example 1.8.61 Pseudo-BCK prealgebras**

Let us consider the pseudo-BCK(pP) algebra $\mathcal{A}_9 = (A_9 = \{0, a, c, d, m, s, a_1, a_2, 1\}, \to, \rightsquigarrow, 1)$, with the following tables of $\to$, $\rightsquigarrow$ and $\odot$ ([104], page 504).

| $\to$ | 0 | a | c | d | m | s | $a_1$ | $a_2$ | 1 |
|---|---|---|---|---|---|---|---|---|---|
| 0 | 1 | 1 | 1 | 1 | 1 | 1 | 1 | 1 | 1 |
| a | d | 1 | d | d | 1 | 1 | 1 | 1 | 1 |
| c | a | a | 1 | 1 | 1 | 1 | 1 | 1 | 1 |
| d | a | a | m | 1 | 1 | 1 | 1 | 1 | 1 |
| m | 0 | a | d | d | 1 | 1 | 1 | 1 | 1 |
| s | 0 | a | c | d | m | 1 | 1 | 1 | 1 |
| $a_1$ | 0 | a | c | d | m | $a_1$ | 1 | 1 | 1 |
| $a_2$ | 0 | a | c | d | m | $a_1$ | $a_1$ | 1 | 1 |
| 1 | 0 | a | c | d | m | s | $a_1$ | $a_2$ | 1 |

,

| $\rightsquigarrow$ | 0 | a | c | d | m | s | $a_1$ | $a_2$ | 1 |
|---|---|---|---|---|---|---|---|---|---|
| 0 | 1 | 1 | 1 | 1 | 1 | 1 | 1 | 1 | 1 |
| a | d | 1 | d | d | 1 | 1 | 1 | 1 | 1 |
| c | a | a | 1 | 1 | 1 | 1 | 1 | 1 | 1 |
| d | a | a | m | 1 | 1 | 1 | 1 | 1 | 1 |
| m | 0 | a | d | d | 1 | 1 | 1 | 1 | 1 |
| s | 0 | a | c | d | m | 1 | 1 | 1 | 1 |
| $a_1$ | 0 | a | c | d | m | $\mathbf{a_2}$ | 1 | 1 | 1 |
| $a_2$ | 0 | a | c | d | m | **s** | $a_1$ | 1 | 1 |
| 1 | 0 | a | c | d | m | s | $a_1$ | $a_2$ | 1 |

,

| $\odot$ | 0 | a | c | d | m | s | $a_1$ | $a_2$ | 1 |
|---|---|---|---|---|---|---|---|---|---|
| 0 | 0 | 0 | 0 | 0 | 0 | 0 | 0 | 0 | 0 |
| a | 0 | a | 0 | 0 | a | a | a | a | a |
| c | 0 | 0 | c | c | c | c | c | c | c |
| d | 0 | 0 | c | c | c | d | d | d | d |
| m | 0 | a | c | c | m | m | m | m | m |
| s | 0 | a | c | d | m | s | s | s | s |
| $a_1$ | 0 | a | c | d | m | s | s | **s** | $a_1$ |
| $a_2$ | 0 | a | c | d | m | s | $\mathbf{a_1}$ | $a_2$ | $a_2$ |
| 1 | 0 | a | c | d | m | s | $a_1$ | $a_2$ | 1 |

.

Since $x \leq y \Longleftrightarrow x \rightarrow y = 1 \Longleftrightarrow x \rightsquigarrow y = 1$, it follows that $0 \leq 0, a, c, d, m, s, a_1, a_2, 1$ $a \leq a, , m, s, a_1, a_2, 1$; ..., $1 \leq 1$; hence, the poset $(A_9, \leq)$, a bounded lattice, is represented by the Hasse diagram from Figure 1.11.

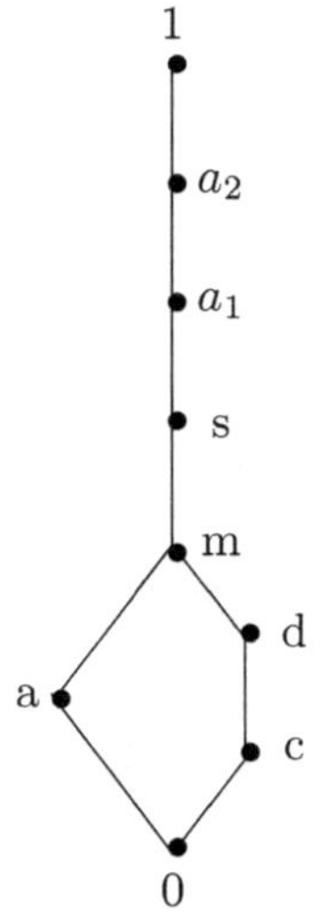


Figure 1.11: The bounded lattice $(A_9, \leq)$

$A_9$ has the following compatible deductive systems: $D_0 = \{1\}$, $D_1 = \{1, a_2, a_1, s\}$, $D_2 = \{1, a_2, a_1, s, m\}$, $D_3 = \{1, a_2, a_1, s, m, d, c\}$, $D_4 = \{1, a_2, a_1, s, m, a\}$, $D_5 =$

$A_9$. Note that $D = \{1, a_2\}$ is a deductive system, but not compatible, because $a_1 \to s = a_1 \notin D$, while $a_1 \rightsquigarrow s = a_2 \in D$.

• For $D_1 = \{1, a_2, a_1, s\}$, consider the pBCK prealgebra $(A_9, \to, \rightsquigarrow, D_1)$. We obtain:
$\mathbf{1} \stackrel{def.}{=} | 1 |= \{y \in A_9 \mid y \to 1 \in D_1, 1 \to y \in D_1\} = \{1, a_2, a_1, s\} =| a_2 |=| a_1 |=| s |$,
$\mathbf{m} \stackrel{def.}{=} | m |= \{y \in A_9 \mid y \to m \in D_1, m \to y \in D_1\} = \{m\}$,
$\mathbf{a} \stackrel{def.}{=} | a |= \{y \in A_9 \mid y \to a \in D_1, a \to y \in D_1\} = \{a\}$,
$\mathbf{d} \stackrel{def.}{=} | d |= \{y \in A_9 \mid y \to d \in D_1, d \to y \in D_1\} = \{d\}$,
$\mathbf{c} \stackrel{def.}{=} | c |= \{y \in A_9 \mid y \to c \in D_1, c \to y \in D_1\} = \{c\}$,
$\mathbf{0} \stackrel{def.}{=} | 0 |= \{y \in A_9 \mid y \to 0 \in D_1, 0 \to y \in D_1\} = \{0\}$.

Then, $\mathcal{A}_9/D_1 = (A_9/D_1, \Rightarrow, \approx>, \mathbf{1})$ is a pBCK(pP) algebra, where the tables of $\Rightarrow$, $\approx>$ and $\odot$ are the following:

| $\Rightarrow$ | 0 | a | c | d | m | 1 |
|---|---|---|---|---|---|---|
| 0 | 1 | 1 | 1 | 1 | 1 | 1 |
| a | d | 1 | d | d | 1 | 1 |
| c | a | a | 1 | 1 | 1 | 1 |
| d | a | a | m | 1 | 1 | 1 |
| m | 0 | a | d | d | 1 | 1 |
| 1 | 0 | a | c | d | m | 1 |

,

| $\approx>$ | 0 | a | c | d | m | 1 |
|---|---|---|---|---|---|---|
| 0 | 1 | 1 | 1 | 1 | 1 | 1 |
| a | d | 1 | d | d | 1 | 1 |
| c | a | a | 1 | 1 | 1 | 1 |
| d | a | a | m | 1 | 1 | 1 |
| m | 0 | a | d | d | 1 | 1 |
| 1 | 0 | a | c | d | m | 1 |

,

| $\odot$ | 0 | a | c | d | m | 1 |
|---|---|---|---|---|---|---|
| 0 | 0 | 0 | 0 | 0 | 0 | 0 |
| a | 0 | a | 0 | 0 | a | a |
| c | 0 | 0 | c | c | c | c |
| d | 0 | 0 | c | c | c | d |
| m | 0 | a | c | c | m | m |
| 1 | 0 | a | c | d | m | 1 |

.

Note that $\Rightarrow = \approx>$, hence $\mathcal{A}_9/D_1 = (A_9/D_1, \Rightarrow, \mathbf{1})$ is a BCK algebra.

Note that the poset $(A_9/D_1, \leq)$ is represented by the Hasse diagram from Figure 1.12.

• For $D_2 = \{1, a_2, a_1, s, m\}$, consider the pBCK prealgebra $(A_9, \to, \rightsquigarrow, D_2)$. We obtain:
$\mathbf{1} \stackrel{def.}{=} | 1 |= \{y \in A_9 \mid y \to 1 \in D_2, 1 \to y \in D_2\} = \{1, a_2, a_1, s, m\} =| a_2 |= ...$
$=| m |$,
$\mathbf{c} \stackrel{def.}{=} | d |= \{y \in A_9 \mid y \to d \in D_2, d \to y \in D_2\} = \{c, d\} =| c |$,
$\mathbf{a} \stackrel{def.}{=} | a |= \{y \in A_9 \mid y \to a \in D_2, a \to y \in D_2\} = \{a\}$,
$\mathbf{0} \stackrel{def.}{=} | 0 |= \{y \in A_9 \mid y \to 0 \in D_2, 0 \to y \in D_2\} = \{0\}$.

Then, $\mathcal{A}_9/D_2 = (A_9/D_2, \Rightarrow, \approx>, \mathbf{1})$ is a pBCK(pP) algebra, where the tables of $\Rightarrow$, $\approx>$ and $\odot$ are the following:

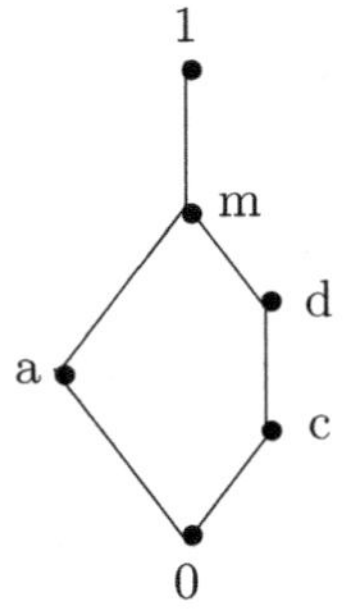


Figure 1.12: The bounded lattice $(A_9/D_1, \leq)$

| $\Rightarrow$ | **0** | **a** | **c** | **1** |
|---|---|---|---|---|
| **0** | **1** | **1** | **1** | **1** |
| **a** | **c** | **1** | **c** | **1** |
| **c** | **a** | **a** | **1** | **1** |
| **1** | **0** | **a** | **c** | **1** |

,

| $\approx>$ | **0** | **a** | **c** | **1** |
|---|---|---|---|---|
| **0** | **1** | **1** | **1** | **1** |
| **a** | **c** | **1** | **c** | **1** |
| **c** | **a** | **a** | **1** | **1** |
| **1** | **0** | **a** | **c** | **1** |

,

| $\bigodot$ | **0** | **a** | **c** | **1** |
|---|---|---|---|---|
| **0** | **0** | **0** | **0** | **0** |
| **a** | **0** | **a** | **0** | **a** |
| **c** | **0** | **0** | **c** | **c** |
| **1** | **0** | **a** | **c** | **1** |

.

Note that $\Rightarrow\ =\ \approx>$, hence $\mathcal{A}_9/D_2 = (A_9/D_2, \Rightarrow, \mathbf{1})$ is a BCK(P) algebra, namely is the non-linearly ordered Boolean algebra $\mathcal{L}_{2\times 2}$, represented by the Hasse diagram from Figure 1.13.

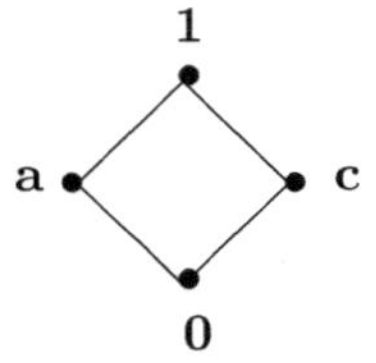


Figure 1.13: The bounded lattice $(A_9/D_2, \leq)$

• For $D_3 = \{1, a_2, a_1, s, m, d, c\}$, consider the pBCK prealgebra $(A_9, \rightarrow, \rightsquigarrow, D_3)$. We obtain:
$\mathbf{1} \stackrel{def.}{=} \mid 1 \mid = \{y \in A_9 \mid y \rightarrow 1 \in D_3, 1 \rightarrow y \in D_3\} = \{1, a_2, a_1, s, m, d, c\} = \mid a_2 \mid = \ldots = \mid c \mid$,
$\mathbf{0} \stackrel{def.}{=} \mid a \mid = \{y \in A_9 \mid y \rightarrow a \in D_3, a \rightarrow y \in D_3\} = \{0, a\} = \mid 0 \mid$.

Then, $\mathcal{A}_9/D_3 = (A_9/D_3, \Rightarrow, \approx>, \mathbf{1})$ is a pBCK(pP) algebra, where the tables of $\Rightarrow$, $\approx>$ and $\bigodot$ are the following:

| $\Rightarrow$ | **0** | **1** |
|---|---|---|
| **0** | **1** | **1** |
| **1** | **0** | **1** |

,

| $\approx>$ | **0** | **1** |
|---|---|---|
| **0** | **1** | **1** |
| **1** | **0** | **1** |

,

| $\bigodot$ | **0** | **1** |
|---|---|---|
| **0** | **0** | **0** |
| **1** | **0** | **1** |

.

Note that $\Rightarrow = \approx>$, hence $\mathcal{A}_9/D_3 = (A_9/D_3, \Rightarrow, \mathbf{1})$ is a BCK(P) algebra with 2 elements, namely is the standard Boolean algebra $\mathcal{L}_2$, represented by the Hasse diagram from Figure 1.14.

Figure 1.14: The bounded lattice $(\mathcal{A}_9/D_3, \leq)$

• For $D_4 = \{1, a_2, a_1, s, m, a\}$, consider the pBCK prealgebra $(A_9, \to, \rightsquigarrow, D_4)$. We obtain:
$\mathbf{1} \overset{def.}{=} | 1 | = \{y \in A_9 \mid y \to 1 \in D_4, 1 \to y \in D_4\} = \{1, a_2, a_1, s, m, a\} = | a_2 | = ...$
$= | a |$,
$\mathbf{0} \overset{def.}{=} | d | = \{y \in A_9 \mid y \to d \in D_4, d \to y \in D_4\} = \{0, c, d\} = | 0 | = | c |$.

Then, $\mathcal{A}_9/D_4$ is again the standard Boolean algebra $\mathcal{L}_2$.

• For $D_5 = A_9$, consider the pBCK prealgebra $(A_9, \to, \rightsquigarrow, D_5)$. We obtain:
$\mathbf{1} \overset{def.}{=} | 1 | = \{y \in A_9 \mid y \to 1 \in D_5, 1 \to y \in D_5\} = A_9 = | a_2 | = ... = | 0 |$.

Then, $\mathcal{A}_9/D_5 = (A_9/D_5, \Rightarrow, \approx>, \mathbf{1})$ is a pBCK(pP) algebra with only one element, $\mathbf{1}$, where the tables of $\Rightarrow$, $\approx>$ and $\bigodot$ coincide, i.e. is the BCK(P) algebra with one element, $\mathbf{1}$.

**Example 1.8.62 Pseudo-BCI prealgebras**

Consider the left-pBCI algebra $\mathcal{A}_8 = (A_8 = \{a, b, c, d, e, f, g, 1\}, \to, \rightsquigarrow, 1)$, with the following tables of $\to$, $\rightsquigarrow$ (found by *Mace4 program*).

| $\to$ | a | b | c | d | e | f | g | 1 |
|---|---|---|---|---|---|---|---|---|
| a | 1 | d | a | a | a | g | f | a |
| b | 1 | 1 | a | a | a | g | f | a |
| c | a | a | 1 | 1 | e | f | g | 1 |
| d | a | b | c | 1 | e | f | g | 1 |
| e | a | b | c | 1 | 1 | f | g | 1 |
| f | g | g | f | f | f | 1 | a | f |
| g | f | f | g | g | g | a | 1 | g |
| 1 | a | b | c | d | e | f | g | 1 |

and

| $\rightsquigarrow$ | a | b | c | d | e | f | g | 1 |
|---|---|---|---|---|---|---|---|---|
| a | 1 | c | a | a | a | g | f | a |
| b | 1 | 1 | a | a | a | g | f | a |
| c | a | a | 1 | 1 | e | f | g | 1 |
| d | a | a | c | 1 | e | f | g | 1 |
| e | a | a | c | 1 | 1 | f | g | 1 |
| f | g | g | f | f | f | 1 | a | f |
| g | f | f | g | g | g | a | 1 | g |
| 1 | a | b | c | d | e | f | g | 1 |

.

Since $x \leq y \iff x \to y = 1 \iff x \rightsquigarrow y = 1$, it follows that $a \leq a, 1$; $b \leq a, b$; $c \leq c, d, 1$; $d \leq d, 1$; $e \leq d, e, 1$; $f \leq f$; $g \leq g$; hence, the poset $(A_8, \leq)$ is represented by the Hasse diagram from Figure 1.15.

Note that, for example, $c \odot e = \min\{z \mid c \leq e \to z\} = \min\{c, d, e, 1\}$ does not exist, so the pBCI algebra $\mathcal{A}_8$ has no product $\odot$.

$\mathcal{A}_8$ has the following compatible deductive systems: $D_0 = \{1\}$, $D_1 = \{1, d\}$, $D_2 = \{1, c, d, e\}$, $D_3 = \{1, a, b, c, d, e\}$ and $D_4 = A_8$.

• For $D_1 = \{1, d\}$, consider the pBCI prealgebra $(A_8, \to, \rightsquigarrow, D_1)$. We obtain:
$\mathbf{1} \overset{def.}{=} | 1 | = \{y \in A_8 \mid y \to 1 \in D_1, 1 \to y \in D_1\} = \{1, d\} = | d |$,

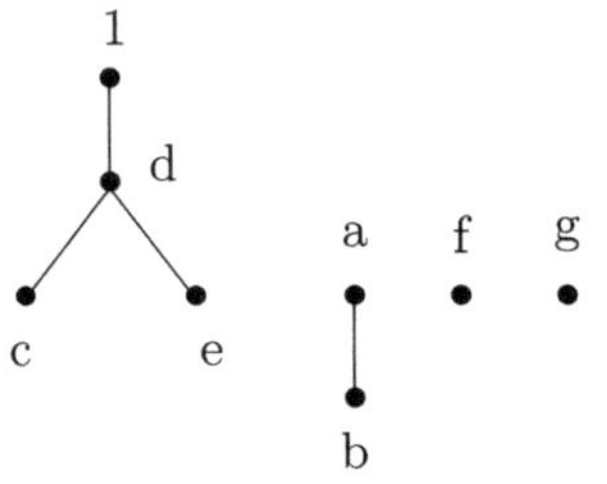

Figure 1.15: The poset $(A_8, \leq)$

$\mathbf{c} \overset{def.}{=} | c |= \{y \in A_8 \mid y \to c \in D_1, c \to y \in D_1\} = \{c\}$,
$\mathbf{e} \overset{def.}{=} | e |= \{y \in A_8 \mid y \to e \in D_1, e \to y \in D_1\} = \{e\}$,
$\mathbf{a} \overset{def.}{=} | a |= \{y \in A_8 \mid y \to a \in D_1, a \to y \in D_1\} = \{a, b\} =| b |$,
$\mathbf{f} \overset{def.}{=} | f |= \{y \in A_8 \mid y \to f \in D_1, f \to y \in D_1\} = \{f\}$,
$\mathbf{g} \overset{def.}{=} | g |= \{y \in A_8 \mid y \to g \in D_1, g \to y \in D_1\} = \{g\}$.

Then, $\mathcal{A}_8/D_1 = (A_8/D_1, \Rightarrow, \approx>, \mathbf{1})$ is a pBCI algebra, where the tables of $\Rightarrow$, $\approx>$ are the following:

| $\Rightarrow$ | a | c | e | f | g | 1 |
|---|---|---|---|---|---|---|
| a | 1 | a | a | g | f | a |
| c | a | 1 | e | f | g | 1 |
| e | a | c | 1 | f | g | 1 |
| f | g | f | f | 1 | a | f |
| g | f | g | g | a | 1 | g |
| 1 | a | c | e | f | g | 1 |

and

| $\approx>$ | a | c | e | f | g | 1 |
|---|---|---|---|---|---|---|
| a | 1 | a | a | g | f | a |
| c | a | 1 | e | f | g | 1 |
| e | a | c | 1 | f | g | 1 |
| f | g | f | f | 1 | a | f |
| g | f | g | g | a | 1 | g |
| 1 | a | c | e | f | g | 1 |

.

Note that $\Rightarrow = \approx>$, hence $\mathcal{A}_8/D_1 = (A_8/D_1, \Rightarrow, \mathbf{1})$ is a BCI algebra.

Note that the poset $(A_8/D_1, \leq)$ is represented by the Hasse diagram from Figure 1.16.

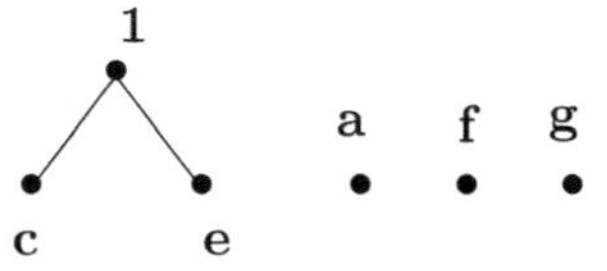

Figure 1.16: The poset $(A_8/D_1, \leq)$

• For $D_2 = \{1, c, d, e\}$, consider the pBCI prealgebra $(A_8, \to, \leadsto, D_2)$. We obtain:
$\mathbf{1} \overset{def.}{=} | 1 |= \{y \in A_8 \mid y \to 1 \in D_2, 1 \to y \in D_2\} = \{1, c, d, e\} =| c |=| d |=| e |$,
$\mathbf{a} \overset{def.}{=} | a |= \{y \in A_8 \mid y \to a \in D_2, a \to y \in D_2\} = \{a, b\} =| b |$,
$\mathbf{f} \overset{def.}{=} | f |= \{y \in A_8 \mid y \to f \in D_2, f \to y \in D_2\} = \{f\}$,
$\mathbf{g} \overset{def.}{=} | g |= \{y \in A_8 \mid y \to g \in D_2, g \to y \in D_2\} = \{g\}$.

Then, $\mathcal{A}_8/D_2 = (A_8/D_2, \Rightarrow, \approx>, \mathbf{1})$ is a pBCI algebra, where the tables of $\Rightarrow$, $\approx>$ are the following:

| $\Rightarrow$ | a | f | g | 1 |
|---|---|---|---|---|
| a | 1 | g | f | a |
| f | g | 1 | a | f |
| g | f | a | 1 | g |
| 1 | a | f | g | 1 |

and

| $\approx>$ | a | f | g | 1 |
|---|---|---|---|---|
| a | 1 | g | f | a |
| f | g | 1 | a | f |
| g | f | a | 1 | g |
| 1 | a | f | g | 1 |

.

Note that $\Rightarrow = \approx>$, hence $\mathcal{A}_8/D_2 = (A_8/D_2, \Rightarrow, \mathbf{1})$ is a BCI algebra, namely a p-semisimple BCI algebra, i.e. the implicative-group with 4 elements $\mathcal{G}^4_{4igr}$, which is term-equivalent, by (Eq1), to the commutative group with 4 elements $\mathcal{G}^4_{4gr}$:

$\mathcal{G}^4_{4igr}$

| $\Rightarrow_4$ | a | f | g | 1 |
|---|---|---|---|---|
| a | 1 | g | f | a |
| f | g | 1 | a | f |
| g | f | a | 1 | g |
| 1 | a | f | g | 1 |

,

| $x$ | $x^- = x$ |
|---|---|
| a | a |
| f | f |
| g | g |
| 1 | 1 |

$\mathcal{G}^4_{4gr}$

| $\odot_4$ | a | f | g | 1 |
|---|---|---|---|---|
| a | 1 | g | f | a |
| f | g | 1 | a | f |
| g | f | a | 1 | g |
| 1 | a | f | g | 1 |

,

| $x$ | $x^-$ |
|---|---|
| a | a |
| f | f |
| g | g |
| 1 | 1 |

.

Note that the poset $(A_8/D_2, \leq)$ is represented by the Hasse diagram from Figure 1.17.

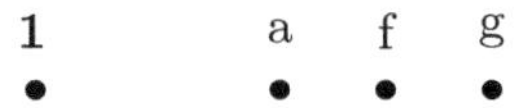


Figure 1.17: The poset $(A_8/D_2, \leq)$

• For $D_3 = \{1, a, b, c, d, e\}$, consider the pBCI prealgebra $(A_8, \rightarrow, \rightsquigarrow, D_3)$. We obtain:
$\mathbf{1} \stackrel{def.}{=} \mid 1 \mid = \{y \in A_8 \mid y \rightarrow 1 \in D_3, 1 \rightarrow y \in D_3\} = \{1, a, b, c, d, e\} = \mid a \mid = \ldots = \mid e \mid$,
$\mathbf{f} \stackrel{def.}{=} \mid f \mid = \{y \in A_8 \mid y \rightarrow f \in D_3, f \rightarrow y \in D_3\} = \{f, g\} = \mid g \mid$.

Then, $\mathcal{A}_8/D_3 = (A_8/D_3, \Rightarrow, \approx>, \mathbf{1})$ is a pBCI algebra, where the tables of $\Rightarrow$, $\approx>$ are the following:

| $\Rightarrow$ | f | 1 |
|---|---|---|
| f | 1 | f |
| 1 | f | 1 |

and

| $\approx>$ | f | 1 |
|---|---|---|
| f | 1 | f |
| 1 | f | 1 |

.

Note that $\Rightarrow = \approx>$, hence $\mathcal{A}_8/D_3 = (A_8/D_3, \Rightarrow, \mathbf{1})$ is a BCI algebra, namely a p-semisimple BCI algebra, i.e. the implicative-group with 2 elements $\mathcal{G}_{2igr}$, which is term-equivalent, by (Eq1), to the commutative group with 2 elements $\mathcal{G}_{2gr}$:

$\mathcal{G}_{2igr}$

| $\Rightarrow$ | f | 1 |
|---|---|---|
| f | 1 | f |
| 1 | f | 1 |

,

| $x$ | $x^- = x \Rightarrow 1$ |
|---|---|
| f | f |
| 1 | 1 |

$\mathcal{G}_{2gr}$

| $\odot$ | f | 1 |
|---|---|---|
| f | 1 | f |
| 1 | f | 1 |

,

| $x$ | $x^-$ |
|---|---|
| f | f |
| 1 | 1 |

.

Note that the poset $(A_8/D_3, \leq)$ is represented by the Hasse diagram from Figure

1.18.

Figure 1.18: The poset $(A_8/D_3, \leq)$

• For $D_4 = A_8$, consider the pBCI prealgebra $(A_8, \rightarrow, \rightsquigarrow, D_4)$. We obtain:
$\mathbf{1} \stackrel{def.}{=} \mid 1 \mid = \{y \in A_8 \mid y \rightarrow 1 \in D_4, 1 \rightarrow y \in D_4\} = \{1, a, b, c, d, e, f, g\} = A_8$.

Then, $\mathcal{A}_8/D_4 = (A_8/D_4, \Rightarrow, \approx>, \mathbf{1})$ is a pBCI algebra with only one element, $\mathbf{1}$, where the tables of $\Rightarrow$, $\approx>$ coincide, hence is a BCI algebra, namely a p-semisimple BCI algebra, i.e. the implicative-group with 1 element $\mathcal{G}_{1igr}$, which is term-equivalent, by (Eq1), to the commutative group with 1 element $\mathcal{G}_{1gr}$.

**Remark** All the compatible deductive systems $D_1$ - $D_4$ of the pBCI algebra $\mathcal{A}_8$ verify (MKK), therefore the Lindenbaum-Tarski algebras are pBCI algebras too.

**Open problem 1.8.63** *Find an example of pseudo-BCI algebra which has a compatible deductive system not verifying (MKK), hence which generates only a locally unital quantum-B algebra, and not a pseudo-BCI algebra.*

# Chapter 2

# M algebras

In this chapter, we make an overview in an unifying way of the most important 24 M algebras analysed in [114] and [116], with many clarifications and new results proved in the preprint [112].

In Section 2.1, we recall the list **A** of basic properties, the 24 M algebras and the "Big map" connecting them.

In Section 2.2, we recall the list **B** of particular properties and some other M algebras. We study in some detail the BCK algebras and the Hilbert algebras and we establish connections between classes of BCK algebras.

In Section 2.3, we discuss about (involutive) negations.

The content is taken mainly from [118].

## 2.1 List A of basic properties and M algebras

- **The basic natural internal binary relation: $\leq$**

Let $\mathcal{A}^L = (A^L, \rightarrow = \rightarrow^L, 1)$ ('L' comes from 'left') be an algebra of type $(2,0)$ throughout this chapter, where an *internal* binary relation $\leq$ can be defined by: for all $x, y \in A^L$,

$$(dfrelR) \qquad x \leq y \overset{def.}{\Longleftrightarrow} x \rightarrow y = 1.$$

Equivalently, let $\mathcal{A}^L = (A^L, \leq, \rightarrow, 1)$ be a structure, where $\leq$ is an *internal* binary relation on $A^L$, $\rightarrow$ is a binary operation on $A^L$ and $1 \in A^L$, all connected by the equivalence:

$$(EqrelR) \qquad x \leq y \Longleftrightarrow x \rightarrow y = 1.$$

Consider the following list **A** of basic properties that can be satisfied by $\mathcal{A}^L$ (in fact, the properties in the list are the most important properties satisfied by a BCK algebra), where almost each property is presented in two equivalent forms, determined by the corresponding two equivalent above definitions of $\mathcal{A}^L$:

(An) (Antisymmetry) $(x \to y = 1 \text{ and } y \to x = 1) \implies x = y$,
(An') (Antisymmetry) $(x \leq y \text{ and } y \leq x) \implies x = y$;

(B) $(y \to z) \to [(x \to y) \to (x \to z)] = 1$ (prefixing),
(B') $y \to z \leq (x \to y) \to (x \to z)$,

(BB) $(y \to z) \to [(z \to x) \to (y \to x)] = 1$ (suffixing),
(BB') $y \to z \leq (z \to x) \to (y \to x)$;

(*) $y \to z = 1 \implies (x \to y) \to (x \to z) = 1$,
(*') $y \leq z \implies x \to y \leq x \to z$;

(**) $y \to z = 1 \implies (z \to x) \to (y \to x) = 1$,
(**') $y \leq z \implies z \to x \leq y \to x$;

(C) $[x \to (y \to z)] \to [y \to (x \to z)] = 1$,
(C') $x \to (y \to z) \leq y \to (x \to z)$;

(D) $y \to [(y \to x) \to x] = 1$,
(D') $y \leq (y \to x) \to x$;

(Ex) (Exchange) $x \to (y \to z) = y \to (x \to z)$;

(Fi) (First element) $0 \to x = 1$,
(Fi') (First element) $0 \leq x$;

(H) $((y \to x) \to x) \to x = y \to x$;

(K) $x \to (y \to x) = 1$,
(K') $x \leq y \to x$;

(KLa) $x \to (y \to x) = y \to 1$;

(La) (Last element) $x \to 1 = 1$,
(La') (Last element) $x \leq 1$;

(M) $1 \to x = x$, i.e. 1 is unit;

(N) $1 \to x = 1 \implies x = 1$, i.e. 1 is maximal element,
(N') $1 \leq x \implies x = 1$, i.e. 1 is maximal element;

(Re) (Reflexivity) $x \to x = 1$ (we prefered the notation (Re) instead of the original (I)),
(Re') (Reflexivity) $x \leq x$;

(S) $x = y \implies x \to y = 1$,

(S') $x = y \implies x \leq y$;

(Tr) (Transitivity) $(x \to y = 1$ and $y \to z = 1) \implies x \to z = 1$,
(Tr') (Transitivity) $(x \leq y$ and $y \leq z) \implies x \leq z$;

(#) $x \to (y \to z) = 1 \Longrightarrow y \to (x \to z) = 1$,
(#') $x \leq y \to z \Longrightarrow y \leq x \to z$;
(Eq#) $x \to (y \to z) = 1 \Longleftrightarrow y \to (x \to z) = 1$,
(Eq#') $x \leq y \to z \Longleftrightarrow y \leq x \to z$;

($) $x \to (y \to z) = 1 \Longrightarrow (x \to y) \to (x \to z) = 1$,
($') $x \leq y \to z \Longrightarrow x \to y \leq x \to z$.

**Conventions.** In order to simplify the writting, if the algebra $(A^L, \to, 1)$ verifies the property (Re), for example, then we shall no more mention that the associated structure $(A^L, \leq, \to, 1)$ verifies the associated property (Re'). Also, we shall freely use (Re) or (Re') in a proof.

**Remarks 2.1.1**
*(i) The basic properties (M) and (Ex) are very special, because they have a unique form. The property (H) also has a unique form, but it is of secondary importance.*
*(ii) If $\leq$ is an* **internal** *relation, then (M) $\Longrightarrow$ (N).*

Dually, let $\mathcal{A}^R = (A^R, \to^R, 0)$ ('R' comes from 'right') be an algebra of type $(2, 0)$, where an *internal* binary relation $\geq$ can be defined by: for all $x, y \in A^R$,

$$(dfrelcoR) \qquad x \geq y \overset{def.}{\Longleftrightarrow} x \to^R y = 0.$$

Equivalently, let $\mathcal{A}^R = (A^R, \geq, \to^R, 0)$ be a structure, where $\geq$ is an *internal* binary relation on $A^R$, $\to^R$ is a binary operation on $A^R$ and $0 \in A^R$, all connected by the equivalence:

$$(EqrelcoR) \qquad x \geq y \Longleftrightarrow x \to^R y = 0.$$

The list of dual properties, (An$^R$), ... , ($$^R$), is omitted.

### 2.1.1 M algebras

Recall from [114], [116] the following definitions as algebras (the dual ones are omitted) (the equivalent definitions as structures are immediate):

• Algebras **without last element** (Hierarchies 1 and 1'):
**- without exchange property:**
An algebra $(A^L, \to, 1)$ is a:
- *left-M algebra*, if it verifies (M);
- *left-RM algebra*, if it verifies (M), (Re);
- *left-pre-BZ algebra*, if it verifies (M), (Re), (B);

- *left-aRM = left-BH algebra*, if it verifies (M), (Re), (An);
- *left-BZ algebra*, if it verifies (M), (Re), (An), (B).

**- with exchange property:**

An algebra $(A^L, \rightarrow, 1)$ is a:
- *left-ME algebra*, if it verifies (M), (Ex);
- *left-RME = left-CI algebra*, if it verifies (M), (Ex), (Re);
- *left-pre-BCI algebra*, if it verifies (M), (Ex), (Re), (B);
- *left-BCH algebra*, if it verifies (Ex), (Re), (An), hence (M);
- *left-BCI algebras*, if it verifies (BB), (D), (Re), (N), (An) (Definition 1) or, equivalently, (BB), (D), (Re), (An) (Definition 2), or, equivalently, (BB), (M), (An) (Definition 3), or, equivalently, (B), (C), (Re), (An) (Definition 4), or, equivalently, (M), (Re), (An), (Ex), (B) (Definition 5)

and, additionally,
- *left-RME** algebra*, if it is a left-RME=CI algebra verifying (**);
- *left-BCH** algebra*, if it is a left-BCH algebra verifying (**).

• Algebras **with last element** (Hierarchies 2 and 2'):

**- without exchange property:**

An algebra $(A^L, \rightarrow, 1)$ is a:
- *left-ML algebra*, if it verifies (M), (La);
- *left-RML algebra*, if it verifies (M), (La), (Re);
- *left-pre-BCC algebra*, if it verifies (M), (La), (Re), (B);
- *left-aRML algebra*, if it verifies (M), (La), (Re), (An);
- *left-BCC algebra*, if it verifies (M), (La), (Re), (B), (An).

**- with exchange property:**

An algebra $(A^L, \rightarrow, 1)$ is a:
- *left-MEL algebra*, if it verifies (M), (Ex), (La);
- *left-BE algebra*, if it verifies (M), (Ex), (La), (Re);
- *left-pre-BCK algebra*, if it verifies (M), (Ex), (La), (Re), (B);
- *left-aBE algebra*, if it verifies (M), (Ex), (La), (Re), (An);
- *left-BCK algebra*, if it verifies (BB), (D), (Re), (La), (An) (Definition 1 [126]) or, equivalently, (BB), (M), (La), (An) (Definition 2 [75]) or, equivalently, (B), (C), (K), (An) (Definition 3 [17], [114]) or, equivalently, (M), (Re), (La), (An), (Ex), (B) (Definition 4)

and, additionally,
- *left-BE** algebra*, if it is a left-BE algebra verifying (**),
- *left-aBE** algebra*, if it is a left-aBE algebra verifying (**).

Many other subclasses of M algebras (structures) of logic were defined in [114], with examples; their connections are presented there also.

**Remark 2.1.2** *The M algebras and the ME algebras are special algebras, because are defined only using the special basic properties (M) and (Ex). By their analogy with the unital commutative magmas and the monoids, respectively, it came the ideea to call ME algebras also as residoids (see Remarks 1.1.10).*

Denote by **M**, ..., **aBE**** the classes of all left-M algebras, ..., left-aBE** algebras, respectively.

**Proposition 2.1.3** *[114]*
*Let $(A^L, \rightarrow, 1)$ be an algebra of type $(2,0)$. Then the following are true:*
*(A0) (Re) $\Longrightarrow$ (S);*
*(A00) (M) $\Longrightarrow$ (N), if $\leq$ is internal;*
*(A1) (La) + (An) $\Longrightarrow$ (N);*
*(A2) (K) + (An) $\Longrightarrow$ (N);*
*(A3) (C) + (An) $\Longrightarrow$ (Ex);* *(A3') (Ex) + (Re) $\Longrightarrow$ (C);*
*(A4) (Re) + (Ex) $\Longrightarrow$ (D);* *(A4') (D) + (Re) + (An) $\Longrightarrow$ (N);*
*(A5) (Re) + (Ex) + (An) $\Longrightarrow$ (M);*
*(A6) (Re) + (K) $\Longrightarrow$ (La);*
*(A7) (N) + (K) $\Longrightarrow$ (La);* *(A7') (M) + (K) $\Longrightarrow$ (La);*
*(A8) (Re) + (La) + (Ex) $\Longrightarrow$ (K);*
*(A8') (C) + (Re) + (An) $\Longrightarrow$ (KLa),* *(A8") (KLa) $\Longrightarrow$ ((K) $\Leftrightarrow$ (La));*
*(A9) (M) + (La) + (B) $\Longrightarrow$ (K);* *(A9') (M) + (La) + (**) $\Longrightarrow$ (K);*
*(A10) (Ex) $\Longrightarrow$ ((B) $\Leftrightarrow$ (BB));*
*(A10') (Ex) + (B) $\Longrightarrow$ (BB);* *(A10") (Ex) + (BB) $\Longrightarrow$ (B);*
*(A11) (Re) + (Ex) + (*) $\Longrightarrow$ (BB),*
*(A11') (Ex) + (D') + (*') $\Longrightarrow$ (BB');*
*(A12) (N) + (B) $\Longrightarrow$ (*);* *(A12') (M) + (B) $\Longrightarrow$ (*);*
*(A13) (N) + (*) $\Longrightarrow$ (Tr);* *(A13') (M) + (*) $\Longrightarrow$ (Tr);*
*(A14) (N) + (B) $\Longrightarrow$ (Tr);* *(A14') (M) + (B) $\Longrightarrow$ (Tr);*
*(A15) (N) + (BB) $\Longrightarrow$ (**);* *(A15') (M) + (BB) $\Longrightarrow$ (**);*
*(A16) (N) + (**) $\Longrightarrow$ (Tr);* *(A16') (M) + (**) $\Longrightarrow$ (Tr);*
*(A17) (N) + (BB) $\Longrightarrow$ (Tr);* *(A17') (M) + (BB) $\Longrightarrow$ (Tr);*
*(A18) (M) + (BB) $\Longrightarrow$ (Re);* *(A18') (M) + (BB) $\Longrightarrow$ (D);*
*(A19) (M) + (B) $\Longrightarrow$ (Re);*
*(A20) (BB) + (D) + (N) $\Longrightarrow$ (C);* *(A20') (M) + (BB) $\Longrightarrow$ (C);*
*(A21) (BB) + (D) + (N) + (An) $\Longrightarrow$ (Ex);*
*(A21') (BB) + (D) + (La) + (An) $\Longrightarrow$ (Ex);*
*(A21") (M) + (BB) + (An) $\Longrightarrow$ (Ex);*
*(A22) (K) + (Ex) + (M) $\Longrightarrow$ (Re);*
*(A23) (C) + (K) + (An) $\Longrightarrow$ (Re);*
*(A24) (Re) + (Ex) + (Tr) $\Longrightarrow$ (**).*

**Proof.** (A0): Suppose $x = y$; then $x \rightarrow y = y \rightarrow y \overset{(Re)}{=} 1$; thus, (S) holds.

(A00): Suppose $1 \rightarrow x = 1$. Then, by (M), we get $x = 1$, i.e. (N) holds.

(A1): Suppose $1 \rightarrow x = 1$. By (La), we also have $x \rightarrow 1 = 1$. Hence, by (An), we get $x = 1$, i.e. (N) holds.

(A2): Suppose $1 \rightarrow x = 1$; by (K), we have $x \rightarrow (1 \rightarrow x) = 1$, then $x \rightarrow 1 = 1$. Hence, by (An), $x = 1$, i.e. (N) holds.

(A3): By (C), we have: $[x \rightarrow (y \rightarrow z)] \rightarrow [y \rightarrow (x \rightarrow z)] = 1$ and also $[y \rightarrow (x \rightarrow z)] \rightarrow [x \rightarrow (y \rightarrow z)] = 1$; hence, by (An), we obtain that: $x \rightarrow (y \rightarrow z) = y \rightarrow (x \rightarrow z)$, i.e. (Ex) holds.

(A3'): By (Ex), we have

$$x \to (y \to z) = y \to (x \to z); \tag{2.1}$$

by (A0), (Re) implies (S); hence, by (S), (2.1) implies $[x \to (y \to z)] \to [y \to (x \to z)] = 1$, i.e. (C) holds.

(A4): $y \to [(y \to x) \to x] \overset{(Ex)}{=} (y \to x) \to (y \to x) \overset{(Re)}{=} 1$, i.e. (D) holds.

(A4'): Suppose that:

$$1 \to x = 1; \tag{2.2}$$

we must prove that $x = 1$. By (D), $x \to [(x \to x) \to x] = 1$, hence by (Re), $x \to [1 \to x] = 1$; by (2.2), it follows that:

$$x \to 1 = 1. \tag{2.3}$$

From (2.2), (2.3) and (An), we obtain $x = 1$. Thus, (N) holds.

(A5): $x \to (1 \to x) \overset{(Ex)}{=} 1 \to (x \to x) \overset{(Re)}{=} 1 \to 1 \overset{(Re)}{=} 1$.
We shall prove now that $(1 \to x) \to x = 1$ also. Indeed, by (A4), (D) holds, consequently, $1 \to [(1 \to x) \to x] = 1$. By (A4'), (N) holds, hence we obtain that $(1 \to x) \to x = 1$.
Applying now (An), we obtain that $1 \to x = x$, i.e. (M) holds.

(A5'): By above (A5), (Re) + (Ex) + (An) imply (M); and by above (A0), (M) implies (N); hence, (Re) + (Ex) + (An) imply (N).

(A6): In (K) ($x \to (y \to x) = 1$), take $y = x$: we get $1 = x \to (x \to x) \overset{(Re)}{=} x \to 1$, i.e. (La) holds.

(A7): By (K), we have: $1 \to (x \to 1) = 1$; hence, by (N), we obtain that $x \to 1 = 1$, i.e. (La) holds.

(A7'): By above (A00), (M) implies (N); and by (A7), (N) + (K) imply (La); thus, (La) holds.

(A8): In (Ex) ($x \to (y \to z) = y \to (x \to z)$), take $z = x$: we get $x \to (y \to x) = y \to (x \to x) \overset{(Re)}{=} y \to 1 \overset{(La)}{=} 1$, i.e. (K) holds.

(A8'): By (C), $(x \to (y \to z)) \to (y \to (x \to z)) = 1$; then,
for $z := x$, we obtain $(x \to (y \to x)) \to (y \to 1) = 1$, by (Re), and,
for $z := y$, we obtain $(x \to 1) \to (y \to (x \to y)) = 1$, by (Re),
i.e., by interchanging $x$ with $y$, $(y \to 1) \to (x \to (y \to x)) = 1$;
then, by (An), we obtain $x \to (y \to x) = y \to 1$, which is (KLa).

(A8"): Obviously.

(A9): Take $y = 1$ in (B) ($(y \to z) \to [(x \to y) \to (x \to z)] = 1$); we obtain: $(1 \to z) \to [(x \to 1) \to (x \to z)] = 1$; then, by (M), we obtain: $z \to [(x \to 1) \to (x \to z)] = 1$; then by (La) and (M) again, we obtain $z \to (x \to z) = 1$, i.e. (K) holds.

(A9'): By (L'), $y \leq 1$ is true and hence, by (**), we get: $1 \to x \leq y \to x$, which by (M) means that $x \leq y \to x$, i.e. (K') holds.

(A10): $(y \to z) \to [(x \to y) \to (x \to z)] \overset{(Ex)}{=} (x \to y) \to [(y \to z) \to (x \to z)]$.
Hence, (B) $\Leftrightarrow$ (BB).

(A10'): Obviously, by (A10).

(A10"): Obviously, by (A10).

(A11): Since (Re) + (Ex) imply (D), i.e $y \to [(y \to z) \to z] = 1$, we apply (*) and we obtain:
$(x \to y) \to (x \to [(y \to z) \to z]) = 1$.
Then, by (Ex), we obtain:
$(x \to y) \to [(y \to z) \to (x \to z)] = 1$, i.e. (BB) holds.

(A11'): By (D'), $y \leq (y \to z) \to z$; then,
by (*'), $x \to y \leq x \to ((y \to z) \to z)$; hence,
by (Ex), $x \to y \leq (y \to z) \to (x \to z)$, which is (BB').

(A12): Suppose $y \to z = 1$; then, by (B) ($(y \to z) \to [(x \to y) \to (x \to z)] = 1$), it follows that $1 \to [(x \to y) \to (x \to z)] = 1$; hence, by (N), we obtain that $(x \to y) \to (x \to z) = 1$, i.e. (*) holds.

(A12'): By (A00), (M) implies (N); then apply above (A12).

(A13): Suppose $x \to y = 1 = y \to z$; then, by (*), we obtain: $(x \to y) \to (x \to z) = 1 = 1 \to (x \to z)$; hence by (N), we obtain: $x \to z = 1$, i.e. (Tr) holds.

(A13'): By (A00), (M) implies (N); then apply (A13).

(A14): Suppose $y \to z = 1$ and $x \to y = 1$; then, by (B), we obtain: $1 \to [1 \to (x \to z)] = 1$; then, by (N), we obtain: $1 \to (x \to z) = 1$; by (N) again, we obtain: $x \to z = 1$. Thus, (Tr) holds.

(A14'): By (A00), (M) implies (N); then, apply (A14).

(A15): Suppose $y \to z = 1$; then, by (BB) ($(y \to z) \to [(z \to x) \to (y \to x)] = 1$), it follows that $1 \to [(z \to x) \to (y \to x)] = 1$; hence, by (N), we obtain $(z \to x) \to (y \to x) = 1$, i.e. (**) holds.

(A15'): By (A00), (M) implies (N); then apply (A15).

(A16): Suppose $y \to z = 1 = z \to x$; then, by (**), we obtain: $1 \to (y \to x) = 1$; hence, by (N), we obtain: $y \to x = 1$, i.e. (Tr) holds.

(A16'): By (A00), (M) implies (N); then apply (A16).

(A17): Suppose $y \to z = 1$ and $z \to x = 1$; then, by (BB), we obtain: $1 \to [1 \to (y \to x)] = 1$; then, applying (N) twice, we obtain $y \to x = 1$. Thus, (Tr) holds.

(A17'): By (A00), (M) implies (N); then apply (A17).

(A18): In (BB) ($(y \to z) \to [(z \to x) \to (y \to x)] = 1$), take $y = z = 1$; we obtain: $(1 \to 1) \to [(1 \to x) \to (1 \to x)] = 1$, hence, by (M), $1 \to [x \to x] = 1$, hence by (M) again, $x \to x = 1$, i.e. (Re) holds.

(A18'): In (BB) ($(y \to z) \to [(z \to x) \to (y \to x)] = 1$), take $y = 1$; we obtain: $(1 \to z) \to [(z \to x) \to (1 \to x)] = 1$, i.e. $z \to [(z \to x) \to x] = 1$, by (M); thus, (D) holds.

(A19): In (B) ($(y \to z) \to [(x \to y) \to (x \to z)] = 1$), take $x = y = 1$; we obtain: $(1 \to z) \to [(1 \to 1) \to (1 \to z)] = 1$, hence, by (M), $z \to [1 \to z] = 1$, by (M) again, $z \to z = 1$; thus, (Re) holds.

(A20): (see [127], Theorem 1) We shall use $\leq$ for a better understanding.
By (BB'), we have: $y \to z \leq (z \to x) \to (y \to x)$. By (15.), (BB) + (N) imply (**). Then, by (**'), we obtain:

$$[(z \to x) \to (y \to x)] \to u \leq (y \to z) \to u. \tag{2.4}$$

We substitute in (2.4): $x$ by $u \to x$, $z$ by $z \to x$; $u$ by $(u \to z) \to (y \to (u \to x))$. Then, we obtain:

$$V \overset{notation}{=} [((z \to x) \to (u \to x)) \to (y \to (u \to x))] \to [(u \to z) \to (y \to (u \to x))] \leq$$

$$\leq (y \to (z \to x)) \to [(u \to z) \to (y \to (u \to x))] \overset{notation}{=} W.$$

Then, the left side $V = 1$, by (2.4) (with $Y = u$, $U = y \to (u \to x)$). Thus, $1 \leq W$; then, by (N'), $W = 1$, i.e.

$$y \to (z \to x) \leq (u \to z) \to (y \to (u \to x)). \tag{2.5}$$

Take now $u = z$, $z = y \to x$ in (2.5). Then, we obtain:

$$y \to ((y \to x) \to x) \leq [z \to (y \to x)] \to [y \to (z \to x)]. \tag{2.6}$$

By (D), from (2.6) we obtain:

$$1 \leq [z \to (y \to x)] \to [y \to (z \to x)]. \tag{2.7}$$

From (2.7), we obtain, by (N') again, that: $[z \to (y \to x)] \to [y \to (z \to x)] = 1$, i.e. (C) holds.

(A20'): (M) implies (N), by above (00.), and (M) + (BB) imply (D), by above (A18'). Hence, (M) + (BB) imply (BB) + (D) + (N), which imply (C), by above (A20)

(A21): By above (A20), (BB) + (D) + (N) imply (C); then, by (A3), (C) + (An) imply (Ex).

(A21'): By above (A1), (La) + (An) imply (N); then apply (A21).

(A21"): By above (A20'), (M) + (BB) imply (C); by above (A3), (C) + (An) imply (Ex); thus, (M) + (BB) + (An) imply (Ex).

(A22): $x \to x \overset{(M)}{=} 1 \to (x \to x) \overset{(Ex)}{=} x \to (1 \to x) \overset{(K)}{=} 1$, hence $x \to x = 1$, i.e. (Re) holds.

(23): By above (A2), (K) + (An) imply (N); by above (A7), (N) + (K) imply (La); by above (A3), (C) + (An) imply (Ex). Then, by (K), we have:

$$1 \to [(x \to x) \to 1] = 1. \tag{2.8}$$

From (2.8), by (N), we obtain:

$$(x \to x) \to 1 = 1. \tag{2.9}$$

On the other hand, $1 \to (x \to x) \overset{(Ex)}{=} x \to (1 \to x) \overset{(K)}{=} 1$, hence

$$1 \to (x \to x) = 1. \tag{2.10}$$

From (2.9) and (2.10), by applying (An), we obtain $x \to x = 1$, i.e. (Re) holds.

(A24): Suppose (Tr) holds, i.e. $X \to Y = 1$, $Y \to Z = 1$ imply $X \to Z = 1$. We must prove that (**) holds, i.e. $y \to z = 1$ implies $(z \to x) \to (y \to x) = 1$.

Suppose that $y \to z = 1$; we must prove that $H \overset{notation}{=} (z \to x) \to (y \to x) = 1$. Indeed, $H \overset{(Ex)}{=} y \to ((z \to x) \to x)$. Take $X = y$, $Z = (z \to x) \to x$ and $Y = z$; then we have $H = X \to Z$ and $X \to Y = y \to z = 1$, by hypothesis; $Y \to Z = z \to [(z \to x) \to x] \overset{(Ex)}{=} (z \to x) \to (z \to x) \overset{(Re)}{=} 1$. Hence, by (Tr), it follows that $X \to Z = 1$, i.e. $H = 1$. □

**Proposition 2.1.4** *[111]*

*Let $(A^L, \to, 1)$ be an algebra of type $(2, 0)$. Then, the following are true:*
*(A9") (M) + (La) + (BB) $\Longrightarrow$ (K);*
*(A18") (M) + (D) $\Longrightarrow$ (Re);*
*(A25) (D) + (K) + (N) + (An) $\Longrightarrow$ (M);*
*(A26) (#) $\Longleftrightarrow$ (Eq#);*
*(A27) (M) + (C) $\Longrightarrow$ (#);*
*(A28) (Ex) $\Longrightarrow$ (Eq#);*
*(A29) (BB) + (#) $\Longrightarrow$ (B); (A29') (B) + (#) $\Longrightarrow$ (BB);*
*(A30) (Re) + (B) + (Tr) + (#) $\Longrightarrow$ (C);*
*(A31) (Re) + (#) $\Longrightarrow$ (D) (see (A4));*
*(A32) (Re) + (#) + (An) $\Longrightarrow$ (M) (see (A5));*
*(A33) (D) + (BB) + (M) + (An) $\Longrightarrow$ (H).*

**Proof.** (A9"): $x \to (y \to x) \overset{(M)}{=} (1 \to x) \to (y \to x) \overset{(M)}{=} 1 \to [(1 \to x) \to (y \to x)] \overset{(La)}{=} (y \to 1) \to [(1 \to x) \to (y \to x)] \overset{(BB)}{=} 1$, i.e. (K) holds.

(A18"): $1 \to ((1 \to x) \to x) \overset{(D)}{=} 1$, hence, by (M), $x \to x = 1$.

(A25): $1 \to [(1 \to x) \to x] \overset{(D)}{=} 1$, hence, by (N), $(1 \to x) \to x = 1$. On the other hand, $x \to (1 \to x) \overset{(K)}{=} 1$. Then, by (An), $1 \to x = x$.

(A26): Obviously.

(A27): Suppose that $x \to (y \to z) = 1$; then, (C) ($[x \to (y \to z)] \to [y \to (x \to z)] = 1$) gives $1 \to [y \to (x \to z)] = 1$; hence, by (M), $y \to (x \to z) = 1$; thus (#) holds.

(A28): $1 = x \to (y \to z) \overset{(Ex)}{=} y \to (x \to z) = 1$.

(A29): (see the proof of (qW32) from [15])
By (BB), $(x \to y) \to [(y \to z) \to (x \to z)] = 1$; then, by (#), $(y \to z) \to [(x \to y) \to (x \to z)] = 1$, i.e. (B) holds.

(A29'): Similarly, by (B), $(y \to z) \to [(x \to y) \to (x \to z)] = 1$; then, by (#), $(x \to y) \to [(y \to z) \to (x \to z)] = 1$, i.e. (BB) holds.

(A30): (see the proof of (qW33) from [15])
Since $(y \to z) \to (y \to z) \overset{(Re)}{=} 1$, it follows by (#) that $y \to [(y \to z) \to z] = 1$. On the other hand, $[(y \to z) \to z] \to [(x \to (y \to z)) \to (x \to z)] \overset{(B)}{=} 1$.
Then, by (Tr), we obtain that $y \to [(x \to (y \to z)) \to (x \to z)] = 1$; hence, by (#), $[x \to (y \to z)] \to [y \to (x \to z)] = 1$, i.e. (C) holds.

(A31): $x \to y \overset{(Re)}{\leq} x \to y$ implies, by (#'), $x \leq (x \to y) \to y$.

(A32): First we prove: (a) $x \to (1 \to x) = 1$. Indeed, $1 \to (x \to x) \overset{(Re)}{=} 1 \to 1 \overset{(Re)}{=} 1$, hence by (#), we obtain (a).
Then, we prove: (b) $(1 \to x) \to x = 1$. Indeed, by (A31), (Re) + (#) $\Longrightarrow$ (D), and by (A4'), (D) + (Re) + (An) $\Longrightarrow$ (N); then $1 \to ((1 \to x) \to x) \overset{(D)}{=} 1$, hence by (N), we obtain (b).
Now, (a) + (b) + (An) imply (M); thus, (M) holds.

(A33): First, (a) $(x \to y) \to (((x \to y) \to y) \to y) \overset{(D)}{=} 1$.
Then, $(x \to ((x \to y) \to y)) \to ((((x \to y) \to y) \to y) \to (x \to y)) \overset{(BB)}{=} 1$,
hence, by (D), $1 \to ((((x \to y) \to y) \to y) \to (x \to y)) = 1$,
hence, by (M), (b) $(((x \to y) \to y) \to y) \to (x \to y) = 1$.
Finally, from (a), (b) and (An), we obtain (H). □

**Theorem 2.1.5** *([114], Theorem 1) (Generalization of ([19], Lemma 1.2 and Proposition 1.3))*
*If an algebra $(A^L, \to, 1)$ verifies the properties (Re), (M), (Ex), then:*

$$(B) \iff (BB) \iff (*).$$

**Proof.** By (A11), (Re) + (Ex) + (*) implies (BB). By (A10), (Ex) implies that (B) $\Leftrightarrow$ (BB). By (A12'), (M) + (B) implies (*). Hence, we have:
$(*) \Rightarrow (BB) \Leftrightarrow (B) \Rightarrow (*)$, thus $(BB) \Leftrightarrow (B) \Leftrightarrow (*)$. □

**Theorem 2.1.6** *([114], Theorem 2)*
*If an algebra $(A^L, \to, 1)$ verifies the properties (Re), (M), (Ex), then:*

$$(**) \iff (Tr).$$

**Proof.** By (A16'), (M) + (Re) + (**) imply (Tr). By (A24), (Re) + (Ex) + (Tr) imply (**). □

**Theorem 2.1.7** *([114], Theorem 3)*
*If an algebra $(A^L, \to, 1)$ verifies the properties (M), (B), (An), then:*

$$(Ex) \iff (BB).$$

**Proof.** By (A10'), (B) + (Ex) imply (BB). By (A21"), (M) + (An) + (BB) imply (Ex). □

It follows immediately that:

**Corollary 2.1.8** *In BZ, BCC algebras, we have: $(Ex) \Leftrightarrow (BB)$.*

**Proposition 2.1.9** [114] *An aBE** algebra verifying one of the equivalent properties (BB), (B), (*) is a BCK algebra.*

**Theorem 2.1.10** *(Michael Kinyon) In any algebra $(A^L, \to, 1)$ we have:*
*(i) (M) + (BB) imply (B),*
*(ii) (M) + (B) imply (**).*

**Proof.** (i): By (A18'), we have (M) + (BB) imply (D).
Next, if (BB) is $(x \to y) \to [(y \to z) \to (x \to z)] = 1$, set $x = u$ and $y = (u \to v) \to v$, to get:
$(u \to [(u \to v) \to v]) \to [(((u \to v) \to v) \to z) \to (u \to z)] \overset{(D)}{=}$
$1 \to [(((u \to v) \to v) \to z) \to (u \to z)] \overset{(M)}{=}$
$(((u \to v) \to v) \to z) \to (u \to z) = 1$.
After renaming variables, we get:
(a) $\quad (((x \to y) \to y) \to z) \to (x \to z) = 1$.
Next, in (BB) set $x = u \to v$ and $y = (v \to w) \to (u \to w)$, to get:
$((u \to v) \to [(v \to w) \to (u \to w)]) \to$
$[(((v \to w) \to (u \to w)) \to z) \to ((u \to v) \to z)] \overset{(BB)}{=}$
$1 \to [(((v \to w) \to (u \to w)) \to z) \to ((u \to v) \to z)] \overset{(M)}{=}$
$(((v \to w) \to (u \to w)) \to z) \to ((u \to v) \to z) = 1$.
After renaming variables, we get:
(b) $\quad (((x \to y) \to (u \to y)) \to z) \to ((u \to x) \to z) = 1$.
Taking $z = u \to y$ in (b), we get:
(c) $\quad (((x \to y) \to (u \to y)) \to (u \to y)) \to ((u \to x) \to (u \to y)) = 1$.
Now, in (a) set $x = v \to w$, $y = t \to w$, $z = (t \to v) \to (t \to w)$ to get:
$[(((v \to w) \to (t \to w)) \to (t \to w)) \to ((t \to v) \to (t \to w))] \to$
$((v \to w) \to ((t \to v) \to (t \to w))) \overset{(c)}{=}$
$1 \to ((v \to w) \to ((t \to v) \to (t \to w))) \overset{(M)}{=}$
$(v \to w) \to ((t \to v) \to (t \to w)) = 1$, i.e. (B) holds.

(ii): Suppose (B) is $(y \to z) \to [(x \to y) \to (x \to z)] = 1$.
If $x \to y = 1$, then we get, from (B):
$(y \to z) \to [1 \to (x \to z)] \overset{(M)}{=} (y \to z) \to (x \to z) = 1$, i.e. (**) holds. □

Remark that (A15') follows now by above Kinyon's Theorem 3.1.9.
By Kinyon's Theorem 3.1.9(i) and (A12'), we obtain immediately that:

**Corollary 2.1.11** (M) + (BB) imply (*).

Concluding, by above Kinyon's Theorem 3.1.9 and (A12'), (A13'), (A16'), we immediately obtain:

**Corollary 2.1.12** In any algebra $(A^L, \to, 1)$ verifying (M), we have:

$$(BB) \implies (B) \implies (*), (**) \implies (Tr).$$

### 2.1.2 The negative and positive cones of BCI algebras/lattices

Let $\mathcal{A}^L = (A^L, \leq, \to, 1)$ be a left-BCI algebra. Define the "negative (left) cone" and the "positive (right) cone" of $\mathcal{A}^L$ as follows, respectively:

$$A^{-L} \overset{def.}{=} \{x \in A^L \mid x \leq 1\} \quad and \quad A^{+L} \overset{def.}{=} \{x \in A^L \mid x \geq 1\}.$$

Then, $A^{-L}$ is closed under $\to$ and $A^{+L} = \{1\}$.

Dually, let $\mathcal{A}^R = (A^R, \geq, \to^R, 0)$ be a right-BCI algebra. Define the "positive (right) cone" and the "negative (left) cone" of $\mathcal{A}^R$ as follows, respectively:

$$A^{+R} \stackrel{def.}{=} \{x \in A^R \mid x \geq 0\} \quad and \quad A^{-R} \stackrel{def.}{=} \{x \in A^R \mid x \leq 0\}.$$

Then, $A^{+R}$ is closed under $\to^R$ and $A^{-R} = \{0\}$.

**Proposition 2.1.13** *[116]*

*(1) Let $\mathcal{A}^L = (A^L, \leq, \to, 1)$ be a left-BCI algebra. Then, $\mathcal{A}^{-L} = (A^{-L}, \leq, \to, 1)$ is a left-pBCK algebra.*

*(1') Dually, let $\mathcal{A}^R = (A^R, \geq, \to^R, 0)$ be a right-BCI algebra. Then, $\mathcal{A}^{+R} = (A^{+R}, \geq, \to^R, 0)$ is a right-BCK algebra.*

If the order relation $\leq$ is a lattice order, we denote by $\wedge$ and $\vee$ the Dedekind lattice operations: for all $x, y$, $x \leq y \Leftrightarrow x \wedge y = x \Leftrightarrow x \vee y = y$. (See Chapter 5) Note that a lattice is self-dual. We shall say that $(A, \wedge, \vee)$ is the *left-lattice*, while $(A, \vee, \wedge)$ is the *right-lattice*.

**Definition 2.1.14**

(i) A *left-BCI lattice* (*left-BCK lattice*) is a left-BCI algebra (left-BCK algebra) $\mathcal{A}^L = (A^L, \leq, \to, 1)$, where $\leq$ is a lattice order relation. It will be denoted by $(A^L, \wedge, \vee, \to, 1)$.

(i') Dually, a *right-BCI lattice* (*right-BCK lattice*) is a right-BCI algebra (right-BCK algebra) $\mathcal{A}^R = (A^R, \geq, \to^R, 0)$, where $\geq$ is a lattice order relation. It will be denoted by $(A^R, \vee, \wedge, \to^R, 0)$.

Denote by **BCI-L** (**BCK-L**) the class of all left-BCI lattices (left-BCK lattices, respectively) and, dually, denote by **BCI-L**$^R$ (**BCK-L**$^R$) the class of all right-BCI lattices (right-BCK lattices, respectively).

By Proposition 2.1.13, we obtain the following result.

**Proposition 2.1.15** *[116]*

*(1) Let $\mathcal{A}^L = (A^L, \wedge, \vee, \to, 1)$ be a left-BCI lattice. Then, $\mathcal{A}^{-L} = (A^{-L}, \wedge, \vee, \to, 1)$ is a left-BCK lattice.*

*(1') Dually, let $\mathcal{A}^R = (A^R, \vee, \wedge, \to^R, 0)$ be a right-BCI lattice. Then, $\mathcal{A}^{+R} = (A^{+R}, \vee, \wedge, \to^R, 0)$ is a right-BCK lattice.*

### 2.1.3 The "Big map": the global hierarchy of above M algebras

With analogous definitions as in Definitions 1.1.19 and with the same representations as in Remark 1.1.21, the above subclasses of M algebras are connected as in the following "Big map" (Figures 2.1 and 2.2 [114], [116]). Note that:

- the subclasses of M algebras *without last element*, namely the M, RM, pre-BZ, BH, BZ algebras and the ME, RME=CI, RME**, pre-BCI, BCH, BCH**, BCI algebras, determine the *Hierarchies 1* and *1'*,

- the subclasses of M algebras *with last element* 1, namely the ML, RML, pre-BCC, aRML, BCC algebras and the MEL, BE, BE**, pre-BCK, aBE, aBE**, BCK algebras, determine the *Hierarchies 2* and *2'*.

- **The p-semisimple property of algebras in Hierarchies 1 and 1'**

Recall also from [116], Definition 2.1.3, the following definition:

**Definition 2.1.16** The algebra $(A^L, \rightarrow, 1)$ verifying (Re) is said to be *p-semisimple*, if the following holds: for all $x, y \in A^L$,
(p-s) $\quad x \leq y \Longrightarrow x = y$
(or, equivalently, by (Re), $x \leq y \Longleftrightarrow x = y$).

Note that the p-semisimple algebras belonging to the Hierarchies 2 and 2' (i.e. verifying the property (La)) are **trivial**, because $A^L = \{1\}$ in these cases (see [116], Theorem 13.1.23).

## 2.2 List B of particular properties. Other M algebras

Consider the following list **B** of particular properties that can be satisfied by $\mathcal{A}^L$ [112]:

(impl) $\quad (x \rightarrow y) \rightarrow x = x$ (implicativity);

(pi) $\quad x \rightarrow (x \rightarrow y) = x \rightarrow y$ (general positive-implicativity),
(pimpl) $\quad x \rightarrow (y \rightarrow z) = (x \rightarrow y) \rightarrow (x \rightarrow z)$ (positive-implicativity),
(pimpl-1) $\quad [x \rightarrow (y \rightarrow z)] \rightarrow [(x \rightarrow y) \rightarrow (x \rightarrow z)] = 1$,
(pimpl-2) $\quad [(x \rightarrow y) \rightarrow (x \rightarrow z)] \rightarrow [x \rightarrow (y \rightarrow z)] = 1$;

($\vee$-comm) $\quad (x \rightarrow y) \rightarrow y = (y \rightarrow x) \rightarrow x$ ($\vee$-commutativity),
($\vee$-comm-1) $\quad [(x \rightarrow y) \rightarrow y] \rightarrow x = (x \rightarrow (y \rightarrow x)) \rightarrow (y \rightarrow x)$;

(L) $\quad (x \rightarrow y) \rightarrow (x \rightarrow z) = (y \rightarrow x) \rightarrow (y \rightarrow z)$.

**Definitions 2.2.1** (See [127] for BCK algebras)
Let $\mathcal{A}^L = (A^L, \rightarrow, 1)$ be an M algebra. We say that $\mathcal{A}^L$ is
- *positive-implicative*, if property (pimpl) is satisfied;
- *$\vee$-commutative*, if property ($\vee$-comm) is satisfied;
- *implicative*, if property (impl) is satisfied.

**Proposition 2.2.2** *(See [114] Remark 6.2, Propositions 6.4, 6.9, 6.8, Theorems 6.13, 6.16, Remarks 6.20 (ii), Proposition 6.21, Theorems 6.23, 6.25)*
*Let $(A^L, \rightarrow, 1)$ $((A^L, \leq, \rightarrow, 1))$ be an algebra (a structure). Then, we have:*
*(B0) (pimpl-1) + (pimpl-2) + (An) $\Longrightarrow$ (pimpl);*
*(B1) (pimpl) + (Re) $\Longrightarrow$ (La);*

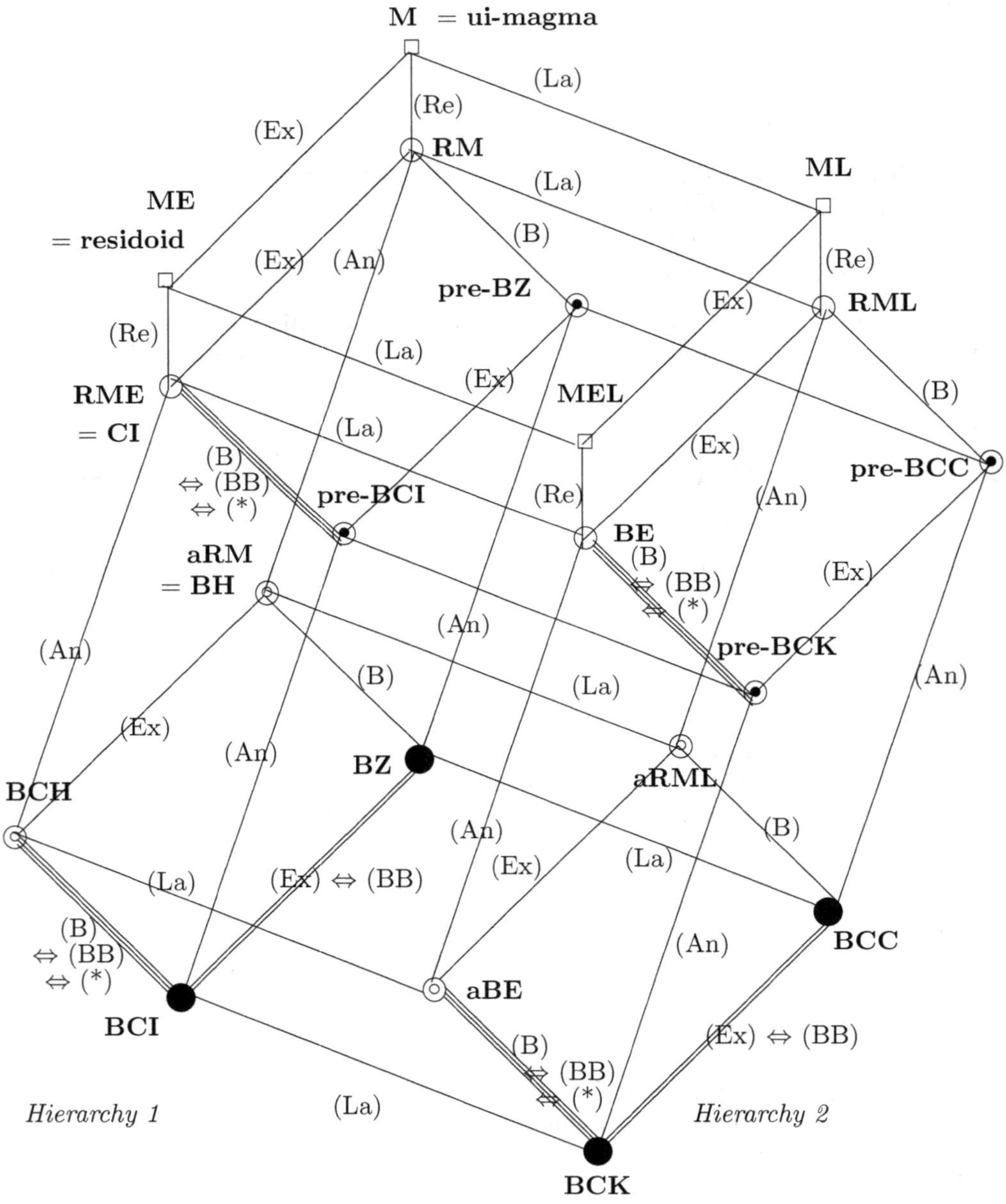


Figure 2.1: The “Big map”: the global hierarchy of M algebras

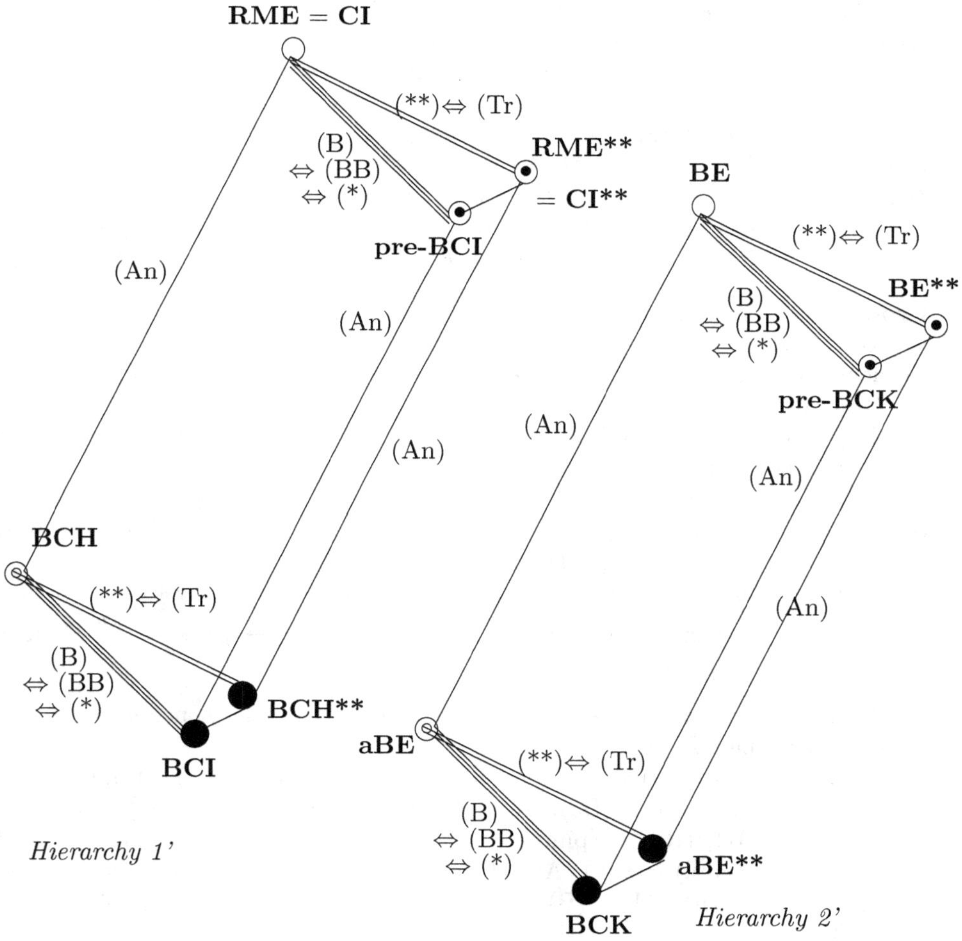


Figure 2.2: The "Big map": hierarchies determined by RME (= CI) and BE algebras

*(B2) (pi) + (Re) $\Longrightarrow$ (La);*
*(B3) (pimpl) + (Re) + (M) $\Longrightarrow$ (pi);*
*(B4) (pimpl) + (Re) + (La) $\Longrightarrow$ (K);*
*(B5) (pimpl) + (K) $\Longrightarrow$ (B);*
*(B6) (pimpl) + (Re) $\Longrightarrow$ (B);*
*(B7) (pimpl) + (Re) + (M) $\Longrightarrow$ (*), (**);*
*(B8) (pimpl) + (La) $\Longrightarrow$ (*) (Michael Kinyon);*
*(B9) (pimpl) + (Re) + (M) $\Longrightarrow$ (BB) (Michael Kinyon);*
*(B10) (pimpl) + (Re) + (M) $\Longrightarrow$ (C) (Michael Kinyon);*
*(B11) (pimpl) + (Re) + (M) + (An) $\Longrightarrow$ (Ex) (Michael Kinyon);*
*(B12) ($\vee$-comm) + (M) $\Longrightarrow$ (An);*
*(B13) (Re) + (La) + (Ex) + (**) $\Longrightarrow$ (pimpl-2);*
*(B14) (pi) + (Ex) + (B) + (*) $\Longrightarrow$ (pimpl-1);*
*(B15) (pi) + (Re) + (Ex) + (B) + (An) $\Longrightarrow$ (pimpl);*
*(B16) (pi) + (Re) + (M) + (B) + (An) $\Longrightarrow$ ((Ex) $\Leftrightarrow$ (BB) $\Leftrightarrow$ (pimpl));*
*(B17) (pi) + (Re) + (M) + (Ex) + (An) $\Longrightarrow$ ((BB) $\Leftrightarrow$ (B) $\Leftrightarrow$ (*) $\Leftrightarrow$ (pimpl)).*

**Proof.** (B0): Obviously.

(B1): Take $y = z = x$ in (pimpl); we obtain: $x \to (x \to x) = (x \to x) \to (x \to x)$, hence, by (Re), we obtain: $x \to 1 = 1 \to 1$, hence, by (Re) again, $x \to 1 = 1$, i.e. (La) holds.

(B2): Take $y = x$ in (pi); we obtain: $x \to (x \to x) = x \to x$, hence, by (Re), $x \to 1 = 1$, i.e. (La) holds.

(B3): Take $x = y$ in (pimpl); we obtain: $y \to (y \to z) = (y \to y) \to (y \to z)$; then, by (Re), we obtain: $y \to (y \to z) = 1 \to (y \to z)$; then, by (M), we obtain: $y \to (y \to z) = y \to z$, i.e. (pi) holds.

(B4): In (pimpl) $(x \to (y \to z) = (x \to y) \to (x \to z))$ take $z = x$; we obtain $x \to (y \to x) = (x \to y) \to (x \to x) \overset{(Re)}{=} (x \to y) \to 1 \overset{(La)}{=} 1$, hence $x \to (y \to x) = 1$, i.e. (K) holds.

(B5): $y \to z \overset{(K')}{\leq} x \to (y \to z) \overset{(pimpl)}{=} (x \to y) \to (x \to z)$; hence $y \to z \leq (x \to y) \to (x \to z)$, i.e. (B') holds.

(B6): By (B1), (Re) + (pimpl) $\Longrightarrow$ (La); by above (B4), (Re) + (La) + (pimpl) $\Longrightarrow$ (K); by above (B5), (K) + (pimpl) $\Longrightarrow$ (B); hence, (B6) holds.

(B7): By (B1), (Re) + (pimpl) $\Longrightarrow$ (La); by (B8), (La) + (pimpl) $\Longrightarrow$ (*); hence, (Re) + (pimpl) $\Longrightarrow$ (*). Also, by above (B6), (Re) + (pimpl) $\Longrightarrow$ (B); then, by Theorem 3.1.9 (ii), (M) + (B) $\Longrightarrow$ (**); hence, (Re) + (M) + (pimpl) $\Longrightarrow$ (**). Thus (B7) holds.

(B8): Suppose that $y \to z = 1$. We must prove that $(x \to y) \to (x \to z) = 1$. Take (pimpl) of the form: $x \to (y \to z) = (x \to y) \to (x \to z)$. We then obtain that $x \to (y \to z) = x \to 1 \overset{(La)}{=} 1$, hence $(x \to y) \to (x \to z) = 1$.

(B9): (Michael Kinyon) By (B1), (B3), (Re) + (pimpl) $\Longrightarrow$ (La) and (Re) + (M) + (pimpl) $\Longrightarrow$ (pi), and by (B4), (Re) + (La) + (pimpl) $\Longrightarrow$ (K).
Now, first prove that

$$((x \to y) \to z) \to (x \to (y \to u)) = (x \to y) \to (z \to (x \to u)). \tag{2.11}$$

Indeed, $(x \to y) \to (z \to (x \to u)) \overset{(pimpl)}{=} ((x \to y) \to z) \to ((x \to y) \to (x \to u)) \overset{(pimpl)}{=} ((x \to y) \to z) \to (x \to (y \to u))$; thus (2.11) holds.
We prove now that

$$(2.12) \qquad x \to ((y \to x) \to z) = x \to z.$$

Indeed, $x \to ((y \to x) \to z) \overset{(pimpl)}{=} (x \to (y \to x)) \to (x \to z) \overset{(K)}{=} 1 \to (x \to z) \overset{(M)}{=} x \to z$; thus (2.12) holds.
We prove now that

$$(2.13) \qquad x \to (y \to (z \to x)) = 1.$$

Indeed, $x \to (y \to (z \to x)) \overset{(pimpl)}{=} (x \to y) \to (x \to (z \to x)) \overset{(K)}{=} (x \to y) \to 1 \overset{(La)}{=} 1$; thus (2.13) holds.
We prove now that

$$(2.14) \qquad ((x \to y) \to z) \to (u \to (y \to z)) = 1.$$

Indeed, $((x \to y) \to z) \to (u \to (y \to z)) \overset{(2.12)}{=} ((x \to y) \to z) \to (u \to [y \to ((x \to y) \to z)]) \overset{(2.13)}{=} 1$, for $X = (x \to y) \to z$; thus, (2.14) holds.
We prove now that

$$(2.15) \qquad ((x \to y) \to (y \to z)) \to (u \to (y \to z)) = 1.$$

Indeed, $((x \to y) \to (y \to z)) \to (u \to (y \to z)) \overset{(pi)}{=} ((x \to y) \to (y \to z)) \to (u \to (y \to (y \to z))) \overset{(2.14)}{=} 1$, with $Z = y \to z$; thus (2.15) holds.
We are ready to prove now that (BB) holds, i.e. $(x \to y) \to ((y \to z) \to (x \to z)) = 1$. Indeed, for $Z = y \to z$ and $u = z$ in (2.11), we get $(x \to y) \to ((y \to z) \to (x \to z)) \overset{(2.11)}{=} ((x \to y) \to (y \to z)) \to (x \to (y \to z)) \overset{(2.15)}{=} 1$, for $u = x$; thus (BB) holds.

(B10): (Michael Kinyon) By (B1), (B3), properties (La) and (pi) hold, and by (B4), property (K) holds.

Now, first we prove that:

$$(2.16) \qquad x \to (y \to (x \to z)) = x \to (y \to z).$$

Indeed, $x \to (y \to z) \overset{(pimpl)}{=} (x \to y) \to (x \to z)$
$\overset{(pi)}{=} (x \to y) \to (x \to (x \to z)) \overset{(pimpl)}{=} x \to (y \to (x \to z))$.

Then, we prove that:

$$(2.17) \qquad y \to ((x \to y) \to z) = y \to z.$$

Indeed, $y \to ((x \to y) \to z) \overset{(pimpl)}{=} (y \to (x \to y)) \to (y \to z) \overset{(K)}{=} 1 \to (y \to z) \overset{(M)}{=} y \to z$.

Then, we prove that:

$$((x \to y) \to z) \to (y \to z) = 1. \tag{2.18}$$

Indeed, $((x \to y) \to z) \to (y \to z) \overset{(2.17)}{=} ((x \to y) \to z) \to (y \to ((x \to y) \to z)) \overset{(K)}{=} 1$.

Finally, we prove that property (C) holds, i.e.

$$(x \to (y \to z)) \to (y \to (x \to z)) = 1.$$

Indeed, $(x \to (y \to z)) \to (y \to (x \to z)) \overset{(pimpl)}{=} ((x \to y) \to (x \to z)) \to (y \to (x \to z)) \overset{(2.18)}{=} 1$; thus (C) holds.

(B11): By (A3), (An) + (C) $\Longrightarrow$ (Ex). Thus, property (B11) holds

(B12): Indeed, if $x \to y = 1 = y \to x$ in ($\vee$-comm), then we get $1 \to y = 1 \to x$, i.e. $y = x$, by (M).

(B13): By (A8), (Re) + (Ex) + (La) imply (K). By (K), we have: $y \to (x \to y) = 1$, hence, by (**), we obtain:
$[(x \to y) \to (x \to z)] \to [y \to (x \to z)] = 1$; hence, by (Ex), we obtain:
$[(x \to y) \to (x \to z)] \to [x \to (y \to z)] = 1$, i.e. (pimpl-2) holds.

(B14): We must prove that $[x \to (y \to z)] \to [(x \to y) \to (x \to z)] = 1$.
Denote $X = (x \to y) \to (x \to z)$. We obtain:
$X \overset{(pi)}{=} (x \to y) \to [x \to (x \to z)] \overset{(Ex)}{=} x \to [(x \to y) \to (x \to z)]$.
By (B), we have: $(y \to z) \to [(x \to y) \to (x \to z)] = 1$; then, by (*), we obtain:
$[x \to (y \to z)] \to [x \to [(x \to y) \to (x \to z)]] = 1 = [x \to (y \to z)] \to X$,
hence $[x \to (y \to z)] \to [(x \to y) \to (x \to z)] = 1$, i.e. (pimpl-1) holds.

(B15): By (B2), (pi) + (Re) $\Longrightarrow$ (La); by (A5), (Re) + (Ex) + (An) $\Longrightarrow$ (M); by (A12'), (M) + (B) $\Longrightarrow$ (*); by (A10'), (Ex) + (B) $\Longrightarrow$ (BB); by (A15'), (M) + (BB) $\Longrightarrow$ (**); by (B13), (Re) + (La) + (Ex) + (**) $\Longrightarrow$ (pimpl-2); by (B14), (pi) + (Ex) + (B) + (*) $\Longrightarrow$ (pimpl-1); by (B0), (pimpl-1) + (pimpl-2) + (An) $\Longrightarrow$ (pimpl).

(B16): First, by Theorem 2.1.7, if the algebra $(A, \to, 1)$ verifies (M), (B), (An), then: $(Ex) \Leftrightarrow (BB)$.

Then, by (B9), ((Re) + (M)) + (pimpl) $\Longrightarrow$ (BB) or, by (B11), ((Re) + (M) + (An)) + (pimpl) $\Longrightarrow$ (Ex). Note that (Re) + (pimpl) $\Longrightarrow$ (B), by (B6). Hence, if (Re), (M), (An) hold, then (pimpl) $\Rightarrow$ (Ex) ($\Leftrightarrow$ (BB)).

Conversely, by (B2), (Re) + (pi) $\Longrightarrow$ (La). By (A12'.), (M) + (B) $\Longrightarrow$ (*). By Kinyon's Theorem 3.1.9 (ii), (M) + (B) $\Longrightarrow$(**). By (B13), ((Re) + (La) + (**)) + (Ex) $\Longrightarrow$ (p-2). By (B14), ((B) + (*) + (pi)) + (Ex) $\Longrightarrow$ (p-1). Then, ((Re) + (M) + (B) + (pi) + (An)) + (Ex) $\Longrightarrow$ (pimpl-1) + (pimpl-2) + (An) $\Longrightarrow$ (pimpl). Consequently, if (Re), (M), (B), (An), (pi) hold, then ((BB) $\Leftrightarrow$ ) (Ex) $\Rightarrow$ (pimpl). Thus, if (Re), (M), (B), (An) and (pi) hold, we have $(BB) \Leftrightarrow (Ex) \Leftrightarrow (pimpl)$.

(B17): First, by Theorem 2.1.5, if the algebra $(A, \to, 1)$ verifies (Re), (M), (Ex), then we have:

$$(BB) \Leftrightarrow (B) \Leftrightarrow (*). \tag{2.19}$$

Then, by (B9), (Re) + (M) + (pimpl) $\Longrightarrow$ (BB). Hence, if (Re), (M) hold, then $(pimpl) \Longrightarrow (BB)(\Leftrightarrow (B) \Leftrightarrow (*))$.

Conversely, first, by (B2), (Re) + (pi) $\Longrightarrow$ (La). By (A15'), (M) + (BB) $\Longrightarrow$ (**). Hence, by (B13), (Re) + (La) + (Ex) + (**) $\Longrightarrow$ (pimpl-2). Hence, if (Re), (pi), (M), (Ex) hold, then $(BB) \Longrightarrow (pimpl-2)$.

On the other hand, (Ex) + (pi) + (BB) $\overset{(2.19)}{\Longrightarrow}$ (Ex) + (pi) + (B) ( $\Leftrightarrow$ (*)) $\Longrightarrow$ (p-1), by (B14). Hence, if (Re), (M), (Ex), (pi) hold, then $(BB) \Longrightarrow (pimpl-1)$.

Consequently, if (Re), (M), (Ex), (pi) and (An) hold, then, we obtain:
$(BB) \Longrightarrow ((pimpl-2) + (pimpl-1 + (An)) \Longrightarrow (pimpl)$, i.e. $(BB) \Longrightarrow (pimpl)$.

Thus, if (Re), (M), (Ex), (An) and (pi) hold, then we have:
$(BB) \Leftrightarrow (B) \Leftrightarrow (*) \Leftrightarrow (pimpl)$. □

**Proposition 2.2.3** *(See [112])*

*Let $(A^L, \rightarrow, 1)$ $((A^L, \leq, \rightarrow, 1))$ be an algebra (a structure). The following hold:*
*(B18) (K) + (**) + (C) + (Tr) $\Longrightarrow$ (pimpl-2);*
*(B19) (pimpl-1) + (K) + (N) $\Longrightarrow$ (Re);*
*(B19') (pimpl-1) + (K) + (M) $\Longrightarrow$ (Re);*
*(B20) (pimpl-1) + (La) + (N) $\Longrightarrow$ (*);*
*(B20') (pimpl-1) + (La) + (M) $\Longrightarrow$ (*);*
*(B21) (pimpl-1) + (La) + (N) $\Longrightarrow$ (Tr);*
*(B21') (pimpl-1) + (La) + (M) $\Longrightarrow$ (Tr);*
*(B22) (pimpl-1) + (K) + (Tr) $\Longrightarrow$ (B);*
*(B23) (pimpl-1) + (Re) + (*) + (K) + (M) $\Longrightarrow$ (D);*
*(B24) (pimpl-1) + (*) + (K) + (M) $\Longrightarrow$ (**);*
*(B25) (pimpl-1) + (K) + (**) + (Tr) $\Longrightarrow$ (C);*
*(B26) (pimpl-1) + (K) + (An) + (C) + (**) + (Tr) $\Longrightarrow$ (pimpl);*
*(B27) (pimpl-1) + (N) $\Longrightarrow$ ($);*
*(B27') (pimpl-1) + (M) $\Longrightarrow$ ($);*
*(B28) ($) + (K) + (Tr) $\Longrightarrow$ (#);*
*(B29) ($\vee$-comm) + (La) + (M) $\Longrightarrow$ (Re);*
*(B30) ($\vee$-comm) + (Re) + (M) + (Ex) $\Longrightarrow$ (La);*
*(B31) ($\vee$-comm) + (Re) + (M) + (Ex) $\Longrightarrow$ (*),*
*(B31') ($\vee$-comm) + (Re) + (Ex) + (La) $\Longrightarrow$ (BB);*
*(B32) ($\vee$-comm) + (M) + (BB) $\Longrightarrow$ (La);*
*(B33) ($\vee$-comm) + (K) + (BB) + (Tr) + (M) $\Longrightarrow$ (#);*
*(B34) ($\vee$-comm) $\Longrightarrow$ ($\vee$-comm-1);*
*(B35) (Re) + (Ex) + (B) + (M) + (An) $\Longrightarrow$ ((pimpl) $\Leftrightarrow$ (pi));*
*(B36) ($\vee$-comm) + (K) $\Longrightarrow$ (D).*

**Proof.** (B18): $y \overset{(K')}{\leq} x \rightarrow y$, then $(x \rightarrow y) \rightarrow (x \rightarrow z) \overset{(**')}{\leq} y \rightarrow (x \rightarrow z) \overset{(C)}{\leq} x \rightarrow (y \rightarrow z)$; hence, by (Tr), we obtain that $(x \rightarrow y) \rightarrow (x \rightarrow z) \leq x \rightarrow (y \rightarrow z)$, i.e. (pimpl-2) holds.

(B19): From (pimpl-1) ($[x \rightarrow (y \rightarrow z)] \rightarrow [(x \rightarrow y) \rightarrow (x \rightarrow z)] = 1$), for $y = x \rightarrow x$ and $z = x$ we obtain: $[x \rightarrow ((x \rightarrow x) \rightarrow x)] \rightarrow [(x \rightarrow (x \rightarrow x)) \rightarrow (x \rightarrow x)] = 1$; by (K), $x \rightarrow ((x \rightarrow x) \rightarrow x) = 1$ and $x \rightarrow (x \rightarrow x) = 1$, hence, $1 \rightarrow [1 \rightarrow (x \rightarrow x)] = 1$; then, applying (N) twice, we obtain that $x \rightarrow x = 1$, i.e. (Re) holds.

(B19'): By (A00), (M) $\Longrightarrow$ (N), then apply (B19).

(B20): Suppose that $y \to z = 1$; we must prove that $(x \to y) \to (x \to z) = 1$. Indeed, from (pimpl-1) ($[x \to (y \to z)] \to [(x \to y) \to (x \to z)] = 1$) we obtain: $(x \to 1) \to [(x \to y) \to (x \to z)] = 1$, hence, by (La) and (N), we obtain: $(x \to y) \to (x \to z) = 1$; thus (*) holds.

(B20'): By (A00), (M) $\Longrightarrow$ (N), then apply (B20).

(B21): By (B20), (pimpl-1) + (La) + (N) $\Longrightarrow$ (*) and by (A13), (N) + (*) $\Longrightarrow$ (Tr). Thus, (Tr) holds.

(B21'): By (A00), (M) $\Longrightarrow$ (N), then apply (B21).

(B22): $y \to z \overset{(K')}{\leq} x \to (y \to z) \overset{(pimpl-1)}{\leq} (x \to y) \to (x \to z)$; hence, by (Tr), we obtain $y \to z \leq (x \to y) \to (x \to z)$, i.e. (B') holds.

(B23): From (pimpl-1) ($[x \to (y \to z)] \to [(x \to y) \to (x \to z)] = 1$), for $x = y \to z$ we obtain:
$[(y \to z) \to (y \to z)] \to [((y \to z) \to y) \to ((y \to z) \to z)] = 1$; then, by (Re) and (M), we obtain:
$((y \to z) \to y) \to ((y \to z) \to z) = 1$ i.e. $(y \to z) \to y \leq (y \to z) \to z$; now, apply (*') and obtain:
$y \to [(y \to z) \to y] \leq y \to [(y \to z) \to z]$; then, by (K) and (M), we obtain:
$y \to [(y \to z) \to z] = 1$, i.e. (D) holds.

(B24): Suppose that $x \to y = 1$; we shall prove that $(y \to z) \to (x \to z) = 1$. Indeed,
(pimpl-1) ($[x \to (y \to z)] \to [(x \to y) \to (x \to z)] = 1$) gives:
$[x \to (y \to z)] \to [1 \to (x \to z)] = 1$; then, by (M), we obtain:
$[x \to (y \to z)] \to (x \to z) = 1$, i.e. $x \to (y \to z) \leq x \to z = 1$; now, by (*'), we obtain:
$(y \to z) \to [x \to (y \to z)] \leq (y \to z) \to (x \to z)$; now, by (K) and (M), we obtain:
$(y \to z) \to (x \to z) = 1$. Thus, (**) holds.

(B25): We have $x \to (y \to z) \overset{(pimpl-1')}{\leq} (x \to y) \to (x \to z)$. But, $y \overset{(K')}{\leq} x \to y$ implies, by (**'), $(x \to y) \to (x \to z) \leq y \to (x \to z)$.
Then, by (Tr), we obtain: $x \to (y \to z) \leq y \to (x \to z)$, i.e. (C') holds.

(B26): $y \overset{(K)}{\leq} x \to y$ implies, by (**'), $(x \to y) \to (x \to z) \leq y \to (x \to z)$. But $y \to (x \to z) \overset{(C')}{\leq} x \to (y \to z)$. Then, by (Tr), we obtain:
$(x \to y) \to (x \to z) \leq x \to (y \to z)$, on the one hand.
On the other hand, $x \to (y \to z) \overset{(pimpl-1')}{\leq} (x \to y) \to (x \to z)$.
Consequently, by (An), $x \to (y \to z) = (x \to y) \to (x \to z)$, i.e. (pimpl) holds.

(B27): By (pimpl-1), $[x \to (y \to z)] \to [(x \to y) \to (x \to z)] = 1$; if $x \to (y \to z) = 1$, then by (N), $(x \to y) \to (x \to z) = 1$, i.e. (\$) holds.

(B27'): By (A00), (M) $\Longrightarrow$ (N), then apply (B27).

(B28): If $x \leq y \to z$, then, by (\$'), $x \to y \leq x \to z$; but, by (K'), $y \leq x \to y$; hence, by (Tr), $y \leq x \to z$. Thus, (#) holds.

(B29): $x \to x \overset{(M)}{=} (1 \to x) \to x \overset{(\vee-comm)}{=} (x \to 1) \to 1 \overset{(La)}{=} 1$, i.e. (Re) holds.

(B30): $(x \to 1) \to 1 \stackrel{(\vee-comm)}{=} (1 \to x) \to x \stackrel{(M)}{=} x \to x \stackrel{(Re)}{=} 1$; then, $x \to 1 = x \to [(x \to 1) \to 1] \stackrel{(Ex)}{=} (x \to 1) \to (x \to 1) \stackrel{(Re)}{=} 1$, i.e. (La) holds.

(B31): By (B30), ($\vee$-comm) + (Re) + (M) + (Ex) $\Longrightarrow$ (La). Suppose now that $y \to z = 1$; then, $z \stackrel{(M)}{=} 1 \to z = (y \to z) \to z \stackrel{(\vee-comm)}{=} (z \to y) \to y$, hence $x \to z = x \to [(z \to y) \to y] \stackrel{(Ex)}{=} (z \to y) \to (x \to y)$. Then, $(x \to y) \to (x \to z) = (x \to y) \to [(z \to y) \to (x \to y)] \stackrel{(Ex)}{=} (z \to y) \to [(x \to y) \to (x \to y)] \stackrel{(Re)}{=} (z \to y) \to 1 \stackrel{(La)}{=} 1$. Thus, (*) holds.

(B31'): $(z \to x) \to (y \to x) \stackrel{(Ex)}{=} y \to [(z \to x) \to x]$
$\stackrel{(\vee-comm)}{=} y \to [(x \to z) \to z] \stackrel{(Ex)}{=} (x \to z) \to (y \to z)$.
Then, $(y \to z) \to [(z \to x) \to (y \to x)] = (y \to z) \to [(x \to z) \to (y \to z)]$
$\stackrel{(Ex)}{=} (x \to z) \to [(y \to z) \to (y \to z)] \stackrel{(Re)}{=} (x \to z) \to 1 \stackrel{(La)}{=} 1$, hence (BB) holds.

(B32): By (A18), (M) + (BB) $\Longrightarrow$ (Re); then $x \to 1 \stackrel{(Re)}{=} x \to (x \to x)$
$\stackrel{(M)}{=} x \to ((1 \to x) \to x) \stackrel{(\vee-comm)}{=} x \to ((x \to 1) \to 1)$
$\stackrel{(M)}{=} (1 \to x) \to [(x \to 1) \to (1 \to 1)] \stackrel{(BB)}{=} 1$, i.e. (La) holds.

(B33): Firstly, note that $y \stackrel{(K')}{\leq} (z \to y) \to y$.
Second, suppose that $x \to (y \to z) = 1$; then,
$((z \to y) \to y) \to (x \to z) \stackrel{(\vee-comm)}{=} ((y \to z) \to z) \to (x \to z)$
$\stackrel{(M)}{=} 1 \to [((y \to z) \to z) \to (x \to z)]$
$= [x \to (y \to z)] \to [((y \to z) \to z) \to (x \to z)] \stackrel{(BB)}{=} 1$, i.e. $(z \to y) \to y \leq x \to z$.
Now, by (Tr), we obtain that $y \leq x \to z$, i.e. (#') holds.

(B34): (See the proof of (qW15) from [15]) $[(x \to y) \to y] \to x$
$\stackrel{(\vee-comm)}{=} [(y \to x) \to x] \to x \stackrel{(\vee-comm)}{=} (x \to (y \to x)) \to (y \to x)$, i.e. ($\vee$-comm-1) holds.

(B35): By (B3), (pimpl) + (Re) + (M) $\Longrightarrow$ (pi), and by (B15), (pi) + (Re) + (Ex) + (B) + (An) $\Longrightarrow$ (pimpl).

(B36): In (K) $(x \to (y \to x) = 1)$, take $Y := y \to x$, to obtain:
$x \to ((y \to x) \to x) = 1$; then, by ($\vee$-comm), we obtain:
$x \to ((x \to y) \to y) = 1$, i.e. (D) holds. □

**Proposition 2.2.4** *[112] We have:*
*(BIM1) (impl) $\Longrightarrow$ (pi);*
*(BIM2) ($\vee$-comm) + (pi) + (Re) + (K) + (M) $\Longrightarrow$ (impl);*
*(BIM2') ($\vee$-comm) + (Re) + (K) + (M) $\Longrightarrow$ ((pi) $\Leftrightarrow$ (impl));*
*(BIM3) (impl) + (Ex) + (B) + (An) $\Longrightarrow$ ($\vee$-comm);*
*(BIM4) (impl) + (Re) $\Longrightarrow$ (M);*
*(BIM5) (impl) + (La) $\Longrightarrow$ (M);*
*(BIM5') (impl) + (M) $\Longrightarrow$ (La);*
*(BIM5") (impl) $\Longrightarrow$ ((La) $\Leftrightarrow$ (M));*
*(BIM6) (impl) + (K) $\Longrightarrow$ (Re);*

*(BIM7) (Re) + (K) + (M) + (Ex) + (B) + (An) $\Longrightarrow$ ((impl) $\Leftrightarrow$ (($\vee$-comm) + (pi))).*

**Proof.** (BIM1): $x \to (x \to y) \overset{(impl)}{=} [(x \to y) \to x] \to (x \to y) \overset{(impl)}{=} x \to y$, i.e. (pi) holds.

(BIM2): By (B12), ($\vee$-comm) + (M) $\Longrightarrow$ (An). Then, $((x \to y) \to x) \to x \overset{(\vee-comm)}{=} [x \to (x \to y)] \to (x \to y) \overset{(pi)}{=} (x \to y) \to (x \to y) \overset{(Re)}{=} 1$. Hence, $(x \to y) \to x \leq x$. But, we also have that $x \overset{(K)}{\leq} (x \to y) \to x$. Then, by (An), it follows that $(x \to y) \to x = x$, i.e. (impl) holds.

(BIM2'): By (BIM2) and (BIM1);

(BIM3): $[(x \to y) \to y] \to [(y \to x) \to x]$
$\overset{(impl)}{=} [(x \to y) \to y] \to [(y \to x) \to [(x \to y) \to x]]$
$\overset{(Ex)}{=} (y \to x) \to [[(x \to y) \to y] \to [(x \to y) \to x]] \overset{(B)}{=} 1.$
Thus, we obtained that

$$[(x \to y) \to y] \to [(y \to x) \to x] = 1. \tag{2.20}$$

Similarly,

$$[(y \to x) \to x] \to [(x \to y) \to y] = 1. \tag{2.21}$$

From (2.20) and (2.21), by (An), we obtain $(x \to y) \to y = (y \to x) \to x$, i.e. ($\vee$-comm) holds.

(BIM4): $1 \to x \overset{(Re)}{=} (x \to x) \to x \overset{(impl)}{=} x$, i.e. (M) holds.

(BIM5): $1 \to x \overset{(La)}{=} (x \to 1) \to x \overset{(impl)}{=} x$, i.e. (M) holds.

(BIM5'): $y \to 1 \overset{(M)}{=} (1 \to y) \to 1 \overset{(impl)}{=} 1$, i.e. (La) holds.

(BIM5"): By (BIM5) and (BIM5').

(BIM6): $x \to x \overset{(impl)}{=} x \to [(x \to y) \to x] \overset{(K)}{=} 1$, i.e. (Re) holds.

(BIM7): By (BIM3), (impl) + (Ex) + (B) + (An) $\Longrightarrow$ ($\vee$-comm), and by (BIM1), (impl) $\Longrightarrow$ (pi). Conversely, by (BIM2), ($\vee$-comm) + (pi) + (Re) + (K) + (M) $\Longrightarrow$ (impl). □

**Corollary 2.2.5** *Let $\mathcal{A}^L = (A^L, \to, 1)$ be a $\vee$-commutative left-RME=CI algebra. Then, $\leq$ is an order.*

**Proof.** By (B31), ($\vee$-comm) + (Re) + (M) + (Ex) $\Longrightarrow$ (*); by (A13'), (M) + (*) $\Longrightarrow$ (Tr); by (B12), ($\vee$-comm) + (M) $\Longrightarrow$ (An). Thus, $\leq$ is an order. □

**Corollary 2.2.6** *Let $\mathcal{A}^L = (A^L, \to, 1)$ be a positive-implicative left-aRM=BH algebra. Then, $\leq$ is an order.*

**Proof.** By (B6), (pimpl) + (Re) $\Longrightarrow$ (B); by (A12'), (M) + (B) $\Longrightarrow$ (*); by (A13'), (M) + (*) $\Longrightarrow$ (Tr). Thus, $\leq$ is an order. □

Then, we obtain the following result [112]:

**Theorem 2.2.7** *$\vee$-commutative left-RME = CI, left-pre-BCI, left-BE, left-pre-BCK, left-BCH, left-BCI, left-aBE, left-BCK algebras coincide.*

**Proof.** Let $\mathcal{A}^L = (A^L, \rightarrow, 1)$ be a $\vee$-commutative left-RME (=CI) algebra, i.e. the properties (Re), (M), (Ex) and ($\vee$-comm) hold. Then, by (B12), ($\vee$-comm) + (M) $\Longrightarrow$ (An); by (B30), ($\vee$-comm) + (Re) + (M) + (Ex) $\Longrightarrow$ (La) and, by (B31), ($\vee$-comm) + (Re) + (M) + (Ex) $\Longrightarrow$ (*); hence, (An), (La), (*) hold. Then, by Theorem 2.1.5, (*) $\Leftrightarrow$ (B) $\Leftrightarrow$ (BB), hence (B), (BB) hold too. Thus, $\mathcal{A}^L$ is a ($\vee$-commutative) left-pre-BCI, left-BE, left-pre-BCK, left-BCH, left-BCI, left-aBE, left-BCK algebra. $\Box$

Moreover, we have the following two theorems.

**Theorem 2.2.8** *$\vee$-commutative MEL algebras coincide with $\vee$-commutative BCK algebras too.*

**Proof.** Let $\mathcal{A}^L = (A^L, \rightarrow, 1)$ be a $\vee$-commutative left-MEL algebra, i.e. (M), (Ex), (La), ($\vee$-comm) hold. Then, by (B29), ($\vee$-comm) + (La) + (M) $\Longrightarrow$ (Re); by (B12), ($\vee$-comm) + (M) $\Longrightarrow$ (An); by (B31'), ($\vee$-comm) + (Re) + (La) + (Ex) $\Longrightarrow$ (BB). Hence, $\mathcal{A}^L$ is a bounded $\vee$-commutative BCK algebra. The converse is obvious. $\Box$

**Theorem 2.2.9** *$\vee$-commutative BCC algebras coincide with $\vee$-commutative BCK algebras too.*

**Proof. (Following the idea of proof by** *Prover9***)**

Let $\mathcal{A}^L = (A^L, \rightarrow, 1)$ be a $\vee$-commutative left-BCC algebra, i.e. the properties (Re), (M), (La), (B), (An), ($\vee$-comm) hold.

By (A9), (M) + (La) + (B) $\Longrightarrow$ (K), and, by (B36), ($\vee$-comm) + (K) $\Longrightarrow$ (D).

Now, we prove

$$(x \rightarrow (y \rightarrow z)) \rightarrow (x \rightarrow ((u \rightarrow y) \rightarrow (u \rightarrow z))) = 1. \tag{2.22}$$

Indeed, in (B) $((x \rightarrow y) \rightarrow ((z \rightarrow x) \rightarrow (z \rightarrow y)) = 1)$,
take $X := y \rightarrow z$ and $Y := (u \rightarrow y) \rightarrow (u \rightarrow z)$, to obtain:
(a) $((y \rightarrow z) \rightarrow ((u \rightarrow y) \rightarrow (u \rightarrow z))) \rightarrow$
$((x \rightarrow (y \rightarrow z)) \rightarrow (x \rightarrow ((u \rightarrow y) \rightarrow (u \rightarrow z)))) = 1;$
but, the part $(y \rightarrow z) \rightarrow ((u \rightarrow y) \rightarrow (u \rightarrow z))$ from (a) equals 1, by (B), hence, (a) becomes:
(a') $1 \rightarrow ((x \rightarrow (y \rightarrow z)) \rightarrow (x \rightarrow ((u \rightarrow y) \rightarrow (u \rightarrow z)))) = 1,$
hence, by (M), (2.22) holds.

Now, we prove

$$((x \rightarrow y) \rightarrow y) \rightarrow ((z \rightarrow (y \rightarrow x)) \rightarrow (z \rightarrow x)) = 1. \tag{2.23}$$

Indeed, in (B) $((x \rightarrow y) \rightarrow ((z \rightarrow x) \rightarrow (z \rightarrow y)) = 1)$,
take $X := y \rightarrow x$ and $Y := x$, to obtain:
(b) $((y \rightarrow x) \rightarrow x) \rightarrow ((z \rightarrow (y \rightarrow x)) \rightarrow (z \rightarrow x)) = 1;$

but, the part $(y \to x) \to x$ from (c) equals $(x \to y) \to y$, by ($\vee$-comm); hence, (b) becomes:

(b') $\qquad ((x \to y) \to y) \to ((z \to (y \to x)) \to (z \to x)) = 1,$

i.e. (2.23) holds.

Now, we prove

$$(2.24) \qquad (x \to y) \to (x \to (z \to y)) = 1.$$

Indeed, in (2.22), take $y := 1$ to obtain:

(c) $\qquad (x \to (1 \to z)) \to (x \to ((u \to 1) \to (u \to z)) = 1,$

hence, by (M) and (La), we obtain:

(c') $\qquad (x \to z) \to (x \to (u \to z)) = 1$, i.e (2.24) holds.

Now, we prove

$$(2.25) \qquad x \to ((y \to (x \to z)) \to (y \to z)) = 1.$$

Indeed, again in (2.22), take $Y := x \to z$ and $U := y$, to obtain:

(d) $\qquad (x \to ((x \to z) \to z)) \to (x \to ((y \to (x \to z)) \to (y \to z))) = 1;$

but, the part $x \to ((x \to z) \to z)$ from (d) equals 1, by (D); hence, (d) becomes:

(d') $\qquad 1 \to (x \to ((y \to (x \to z)) \to (y \to z))) = 1,$

hence, by (M), (2.25) holds.

Now, we prove

(2.26)

$$(((x \to y) \to (x \to (z \to y))) \to (x \to (z \to y))) \to (1 \to (z \to (x \to y))) = 1.$$

Indeed, in (2.23), take $X := x \to y$, $Y := x \to (z \to y)$ and $Z := z$, to obtain:

(e) $$(((x \to y) \to (x \to (z \to y))) \to (x \to (z \to y))) \\ \to ((z \to ((x \to (z \to y)) \to (x \to y))) \to (z \to (x \to y))) = 1;$$

now, in (2.25), take $X := z$, $Y := x$, $Z := y$, to obtain:

(f) $\qquad z \to ((x \to (z \to y)) \to (x \to y)) = 1;$

now, note that the part $z \to ((x \to (z \to y)) \to (x \to y))$ from (e) equals 1, by (f); hence, (e) becomes

(e') $(((x \to y) \to (x \to (z \to y))) \to (x \to (z \to y))) \to (1 \to (z \to (x \to y))) = 1,$

i.e. (2.26) holds.

Now, from (2.26), by (M), we obtain:

(g) $\qquad (((x \to y) \to (x \to (z \to y))) \to (x \to (z \to y))) \to (z \to (x \to y)) = 1;$

but, the part $(x \to y) \to (x \to (z \to y))$ from (g) equals 1, by (2.24); hence, (g) becomes:

(g') $\qquad (1 \to (x \to (z \to y))) \to (z \to (x \to y)) = 1,$

hence, by (M),

(g") $\qquad (x \to (z \to y)) \to (z \to (x \to y)) = 1$, i.e. (C) holds.

Finally, by (A3), (C) + (An) $\Longrightarrow$ (Ex). Hence, $\mathcal{A}^L$ is a $\vee$-commutative left-BCK algebra.

Conversely, any $\vee$-commutative left-BCK algebra is obviously a $\vee$-commutative left-BCC algebra. The proof is complete. □

By Theorem 2.2.9, we obtain:

**Corollary 2.2.10** *$\vee$-commutative pre-BCC algebras coincide with $\vee$-commutative BCK algebras.*

**Proof.** Let $\mathcal{A}^L = (A^L, \rightarrow, 1)$ be a $\vee$-commutative left-pre-BCC algebra, i.e. the properties (Re), (M), (La), (B), ($\vee$-comm) hold. By (B12), ($\vee$-comm) + (M) $\Longrightarrow$ (An); hence, $\mathcal{A}^L$ is a $\vee$-commutative left-BCC algebra. Then, apply Theorem 2.2.9. $\Box$

**Proposition 2.2.11** [211]
*Let $(A^L, \rightarrow, 1)$ be an algebra verifying (Re), (D), (**) and (impl). Then, for all $x, y \in A^L$,*

$$y \leq x \Longrightarrow (x \rightarrow y) \rightarrow y \leq x.$$

**Definitions 2.2.12** [114]
(i) A *left-RML** algebra* is a left-RML algebra verifying (**).
(ii) An *left-aRML** algebra* is a left-aRML algebra verifying (**).

**Theorem 2.2.13** [211]
*Let $(A^L, \rightarrow, 1)$ be an implicative left-RML** algebra satisfying (D). Then, we have:*
*(wComm)* $\quad (x \rightarrow y) \rightarrow y \leq (y \rightarrow x) \rightarrow x.$

**Corollary 2.2.14** *Every implicative aRML** algebra satisfying (D) (that is, implicative weak BCK-algebra) is $\vee$-commutative.*

Since any aBE** algebra is an aRML** algebra verifying (Ex), hence (D), by (A4), it follows that we have:

**Corollary 2.2.15** *Every implicative aBE** algebra is $\vee$-commutative.*

### 2.2.1 Two new binary relations: $\leq^W$ and $\leq^B$

**Remark 2.2.16** *Starting from the equality from the property*

$$(\vee - comm) \qquad (y \rightarrow x) \rightarrow x = (x \rightarrow y) \rightarrow y,$$

*verified by a left-Wajsberg algebra, we could introduce two different 'twin' operations, $\vee^W$ ('W' comes from 'Wajsberg algebra') and $\vee^B$ ('B' comes from 'Boolean algebra'), by: for all $x, y$:*
$x \vee^W y = (y \rightarrow x) \rightarrow x$ *and* $x \vee^B y = (x \rightarrow y) \rightarrow y$.
*Then, ($\vee$-comm) would mean: either*
($\vee^W$*-comm)* $\quad x \vee^W y = y \vee^W x$ *or*
($\vee^B$*-comm)* $\quad x \vee^B y = y \vee^B x$.

*In left-Wajsberg algebras, the two operations $\vee^W$ and $\vee^B$ are equal, but in general, they are different; but note that: $x \vee^W y = y \vee^B x$, which means, in the finite case, that the matrix of $\vee^B$ is the transposed matrix of that of $\vee^W$ and vice-versa.*

Following the above Remark 2.2.16, we shall introduce in an algebra $\mathcal{A}^L = (A^L, \rightarrow, 1)$ the following new operations:

(2.27) $\quad x \vee^W y \overset{def.}{=} (y \rightarrow x) \rightarrow x$ *and, dually,* $x \wedge^W y \overset{def.}{=} (y \rightarrow^R x) \rightarrow^R x$

and

(2.28) $\quad x \vee^B y \overset{def.}{=} (x \rightarrow y) \rightarrow y$ *and, dually,* $x \wedge^B y \overset{def.}{=} (x \rightarrow^R y) \rightarrow^R y.$

Beside the old, natural binary relation $\leq$ ($x \leq y \overset{def.}{\Longleftrightarrow} x \rightarrow y = 1$) and its dual $\geq$ ($x \geq y \overset{def.}{\Longleftrightarrow} x \rightarrow^R y = 0$), we introduce two new binary relations: for all $x, y \in A^L$,

$x \leq^W y \overset{def.}{\Longleftrightarrow} x \vee^W y = y$ and, dually, $x \geq^W y \overset{def.}{\Longleftrightarrow} x \wedge^W y = y$
and
$x \leq^B y \overset{def.}{\Longleftrightarrow} x \vee^B y = y$ and, dually, $x \geq^B y \overset{def.}{\Longleftrightarrow} x \wedge^B y = y.$

**Proposition 2.2.17** *(The duall one is omitted)*
*Let $\mathcal{A}^L = (A^L, \rightarrow, 1)$ be a left-BE algebra. We have:*
*(1) $x \leq^B y \Longleftrightarrow x \leq y$, for all $x, y \in A^L$;*
*(2) the binary relation $\leq^W$ is reflexive and antisymmetric and $x \leq^W 1$.*

**Proof.** Let $\mathcal{A}^L = (A^L, \rightarrow, 1)$ be a left-BE algebra, i.e. (Re), (M), (Ex), (La) hold. Then, by (A4), (Re) + (Ex) $\Longrightarrow$ (D) and, by (A8), (Re) + (La) + (Ex) $\Longrightarrow$ (K), hence (D) and (K) hold too.

(1): Suppose $x \leq^B y$, i.e. $(x \rightarrow y) \rightarrow y = y$. Then, $x \rightarrow y = x \rightarrow ((x \rightarrow y) \rightarrow y) \overset{(D)}{=} 1$, i.e. $x \leq y$. Conversely, suppose $x \leq y$, i.e. $x \rightarrow y = 1$. Then, $x \vee^B y = (x \rightarrow y) \rightarrow y = 1 \rightarrow y \overset{(M)}{=} y$, i.e. $x \leq^B y$.

(2): *Reflexivity*: $x \leq^W x \overset{def.}{\Longleftrightarrow} (x \rightarrow x) \rightarrow x = x$ and, indeed, $(x \rightarrow x) \rightarrow x \overset{(Re)}{=} 1 \rightarrow x \overset{(M)}{=} x$.

*Antisymmetry*: $x \leq^W y$ and $y \leq^W x$ mean, by definition, $(y \rightarrow x) \rightarrow x = y$ and $(x \rightarrow y) \rightarrow y = x$. Then, $x \rightarrow y = x \rightarrow ((y \rightarrow x) \rightarrow x) \overset{(K)}{=} 1$, hence $x = (x \rightarrow y) \rightarrow y = 1 \rightarrow y \overset{(M)}{=} y$.

$x \leq^W 1 \overset{def.}{\Longleftrightarrow} (1 \rightarrow x) \rightarrow x = 1$ and, indeed, $(1 \rightarrow x) \rightarrow x \overset{(M)}{=} x \rightarrow x \overset{(Re)}{=} 1$. □

**Remark 2.2.18** *The equivalence $\leq^B \Longleftrightarrow \leq$ implies that $\leq^B$ is an order relation if and only if $\leq$ is an order relation. But, it does not imply that, if $\leq$ is a lattice order, then $\leq^B$ is a lattice order too with respect to $\vee^B, \wedge^B$.*

**Proposition 2.2.19** *(The duall one is omitted) (See Proposition 2.2.17)*
*Let $\mathcal{A}^L = (A^L, \rightarrow, 1)$ be a $\vee$-commutative left-BE algebra. We have, for all $x, y \in A^L$,*

$$x \leq y \Longleftrightarrow x \leq^W y.$$

**Proof.** $x \leq^W y \overset{def.}{\Longleftrightarrow} x \vee^W y = y \overset{def.}{\Longleftrightarrow} (y \rightarrow x) \rightarrow x = y$
$\overset{(\vee-comm)}{\Longleftrightarrow} (x \rightarrow y) \rightarrow y = y \overset{def.}{\Longleftrightarrow} x \vee^B y = y \overset{def.}{\Longleftrightarrow} x \leq^B y$ and
$x \leq^B y \Longleftrightarrow x \leq y$, by Proposition 2.2.17. □

### 2.2.2 BCK algebras

Note that BCK algebras satisfy all the properties from **List A**.

**Theorem 2.2.20** *([127], Theorem 8)*
*A BCK algebra is positive-implicative if and only if the property (pi) holds (or, in a BCK algebra the properties (pimpl) and (pi) are equivalent).*

**Proof.** By (B35), (Re) + (M) + (Ex) + (B) + (An) $\Longrightarrow$ ((pimpl) $\Leftrightarrow$ (pi)). □

Following ([127], Remark 1), $\vee$-commutative BCK algebras were introduced by S. Tanaka [203]; moreover, H. Yutani proved that [205]: *the class of $\vee$-commutative BCK algebras is a variety* and found the following short equivalent system of axioms [206]: ($\vee$-comm), (M), (Re), (Ex). Thus, the $\vee$-commutative BCK algebras coincide with the $\vee$-commutative RME=CI algebras.

In a BCK algebra $\mathcal{A}^L = (A^L, \leq, \rightarrow, 1)$ we define, for all $x, y \in A^L$ (see [126]):

$$x \vee y \stackrel{def}{=} (x \rightarrow y) \rightarrow y. \tag{2.29}$$

**Theorem 2.2.21** *[126] A BCK algebra is $\vee$-commutative iff it is a semilattice (with greatest element 1) with respect to $\vee$ (under $\geq$).*

**Theorem 2.2.22** *([127], Theorem 9)*
*In $\vee$-commutative BCK algebras, the properties (pi) and (impl) are equivalent.*

**Proof.** By (BIM2'), ($\vee$-comm) + (L) + (K) + (M) $\Longrightarrow$ ((pi) $\Leftrightarrow$ (impl)). □

**Theorem 2.2.23** *([127], Theorem 10)*
*Any implicative BCK algebra is $\vee$-commutative and positive-implicative.*

**Proof.** By (BIM3), (impl) + (Ex) + (B) + (An) $\Longrightarrow$ ($\vee$-comm). By (BIM1), (impl) $\Longrightarrow$ (pi), and by Theorem 2.2.20, (pi) and (pimpl) are equivalent. □

**Theorem 2.2.24** *[112]*
*Any $\vee$-commutative and positive-implicative BCK algebra is implicative.*

**Proof.** By Theorem 2.2.20, a BCK algebra is positive implicative if and only if the property (pi) holds. Then, by (BIM2), ($\vee$-comm) + (pi) + (L) + (K) + (M) $\Longrightarrow$ (impl). □

**Corollary 2.2.25** *[112] In a BCK algebra, we have:*

$$(impl) \iff ((\vee - comm) + (pi)\ (\Leftrightarrow (pimpl)))$$

**Proof.** By Theorems 2.2.23 and 2.2.24. □

• **Two orders in BCK algebras: $\leq$ and $\leq^W$**

Let $\mathcal{A}^L = (A^L, \rightarrow, 1)$ be a left-BCK algebra. Beside the old, natural binary relation $\leq$ defined on $A^L$ by: $x \leq y \stackrel{def.}{\Longleftrightarrow} x \rightarrow y = 1$, which is an order on BCK algebras, let us consider the binary relation: for all $x, y \in A^L$,

$$x \leq^W y \stackrel{def.}{\Longleftrightarrow} (x \vee^W y \stackrel{def.}{=}) \, (y \rightarrow x) \rightarrow x = y,$$

where 'W' comes from 'Wajsberg'.

**Theorem 2.2.26** *(See analogous Theorem 12.2.3)*
*In left-BCK algebras, the binary relation $\leq^W$ is transitive.*

**Proof. (By** *Prover9* **program)**
Let $\mathcal{A}^L = (A^L, \rightarrow, 0, 1)$ be a left-BCK algebra, i.e. (B), (C), (K), (An) hold. By (A3), (C) + (An) $\Longrightarrow$ (Ex); by (A23), (C) + (K) + (An) $\Longrightarrow$ (Re); by (A4), (Re) + (Ex) $\Longrightarrow$ (D); by (A10'), (Ex) + (B) $\Longrightarrow$ (BB), by (A5), (Re) + (Ex) + (An) $\Longrightarrow$ (M) and by (A33), (D) + (BB) + (M) + (An) $\Longrightarrow$ (H); thus, (Ex), (Re), (D), (H) hold too.

Suppose that $x \leq^W y$ and $y \leq^W z$, i.e.
(a) $(y \rightarrow x) \rightarrow x = y$ and (b) $(z \rightarrow y) \rightarrow y = z$;
we have to prove that $x \leq^W z$, i.e. (c) $(z \rightarrow x) \rightarrow x = z$.

First, note that we have (c1) $z \rightarrow ((z \rightarrow x) \rightarrow x) \stackrel{(D)}{=} 1$.
We shall prove (c2) $((z \rightarrow x) \rightarrow x) \rightarrow z = 1$, by absurdum hypothesis.

Suppose, by absurdum, that

$$((z \rightarrow x) \rightarrow x) \rightarrow z \neq 1. \tag{2.30}$$

From (a), we obtain $A \rightarrow y = A \rightarrow ((y \rightarrow x) \rightarrow x) \stackrel{(Ex)}{=} (y \rightarrow x) \rightarrow (A \rightarrow x)$, hence

$$(y \rightarrow x) \rightarrow (A \rightarrow x) = A \rightarrow y. \tag{2.31}$$

In (2.31), take $A := (B \rightarrow x) \rightarrow x$, to obtain:
$(y \rightarrow x) \rightarrow (((B \rightarrow x) \rightarrow x) \rightarrow x) = ((B \rightarrow x) \rightarrow x) \rightarrow y$,
hence, by (H),

$$(y \rightarrow x) \rightarrow (B \rightarrow x) = ((B \rightarrow x) \rightarrow x) \rightarrow y. \tag{2.32}$$

From (2.32), by (2.31), we obtain:

$$((B \rightarrow x) \rightarrow x) \rightarrow y = B \rightarrow y. \tag{2.33}$$

Now, in (2.33), take $B := z$ to obtain

$$((z \rightarrow x) \rightarrow x) \rightarrow y = z \rightarrow y. \tag{2.34}$$

From (b), we obtain $A \rightarrow z = A \rightarrow ((z \rightarrow y) \rightarrow y) \stackrel{(Ex)}{=} (z \rightarrow y) \rightarrow (A \rightarrow y)$, hence

$$(z \rightarrow y) \rightarrow (A \rightarrow y) = A \rightarrow z. \tag{2.35}$$

In (2.35), take $A := (z \to x) \to x$ to obtain:
$(z \to y) \to (((z \to x) \to x) \to y) = ((z \to x) \to x) \to z$,
hence, by (2.30), $(z \to y) \to (((z \to x) \to x) \to y) \neq 1$,
hence, by (2.34), $(z \to y) \to (z \to y) \neq 1$: contradiction, by (Re).

Thus, (c2) holds. From (c1), (c2) and (An), we obtain (c). □

**Corollary 2.2.27** *In left-BCK algebras, the binary relation $\leq^W$ is an order and $x \leq^W 1$, for any $x \in A^L$.*

**Proof.** Let $\mathcal{A}^L = (A^L, \to, 0, 1)$ be a left-BCK algebra, i.e. (B), (C), (K), (An) hold. Then, $\mathcal{A}^L$ is a left-BE algebra and, by Proposition 2.2.17, the binary relation $\leq^W$ is reflexive and antisymmetric and $x \leq^W 1$. By Theorem 2.2.26, $\leq^W$ is transitive, hence it is an order. □

### • BCK(P') algebras and BCK(RP') algebras

**Definitions 2.2.28** (Definition of BCK(P') algebras)

(i) A *left-BCK algebra with product*, or a *BCK(P') algebra* for short, is left-BCK algebra $\mathcal{A}^L = (A^L, \leq, \to, 1)$ verifying additionally:

(P') for all $x, y \in A^L$, $\exists\, x \odot y \overset{def.}{=} \min\{z \mid x \leq y \to z\}$.

(ii) If the partial order $\leq$ is a lattice order, then we say that we have a *BCK(P') lattice.*

Denote by **BCK(P')** the class of all BCK(P') algebras and by **BCK(P')-L** the class of all BCK(P') lattices.

**Definitions 2.2.29** (Definition of BCK(RP') algebras)

(i) A *left-BCK algebra with residuum and product*, or a *BCK(RP') algbera* for short, is a structure $\mathcal{A}^L = (A^L, \leq, \to, \odot, 1)$ such that the reduct $(A^L, \leq, \to, \rightsquigarrow, 1)$ is a left-BCK algebra verifying additionally :

(RP') for all $x, y, z \in A^L$, $\quad x \leq y \to z \Longleftrightarrow x \odot y \leq z$.

(ii) If the partial order $\leq$ is a lattice order, then we say that we have a *BCK(RP') lattice.*

Denote by **BCK(RP')** the class of all BCK(RP') algebras and by **BCK(RP')-L** the class of all BCK(RP') lattices.

By Corollaries 1.4.34 and 1.4.35, in the commutative case, we have:

$$\mathbf{BCK(P')} \Longleftrightarrow \mathbf{BCK(RP')} \quad \textit{and} \quad \mathbf{BCK(P')-L} \Longleftrightarrow \mathbf{BCK(RP')-L}.$$

**Theorem 2.2.30** *Let $\mathcal{A}^L = (A^L, \leq, \to, 1)$ be a $\vee$-commutative BCK(P') algebra. Then,*

*(1) $(A^L, \leq)$ is a lattice, where for any $x, y \in A^L$,*
$x \vee y = (y \to x) \to x$,
$x \wedge y = ([x \to (x \odot y)] \vee [y \to (x \odot y)]) \to (x \odot y)$.

*(2) for any $x, y \in A$, $(x \to y) \vee (y \to x) = 1$, i.e. (prel) is satisfied.*

Note that the part (1) of the above Theorem is just ([148], Corollary 1), while the part (2) follows by above (1) and by (2) of ([104], Theorem 2.1.22).

**Example 2.2.31** Consider the example of $\vee$-commutative BCK algebra $\mathcal{A}_K = (A^L, \leq, \rightarrow, 1)$ from [136], with $A^L = (a, b, c, 1)$ and $b, c < a < 1$, $b$ and $c$ being incomparable, and the following table of $\rightarrow$:

| $\rightarrow$ | a | b | c | 1 |
|---|---|---|---|---|
| a | 1 | a | a | 1 |
| b | 1 | 1 | a | 1 |
| c | 1 | a | 1 | 1 |
| 1 | a | b | c | 1 |

Note that $\mathcal{A}_K$ is a $\vee$-semilattice with 1, but it is not a lattice and thefore $\mathcal{A}_K$ is not with condition (P'), by above Theorem; indeed, for example $a \odot a = \min\{z \mid a \leq a \rightarrow z\} = \min\{a, b, c, 1\}$ does not exist. Thus, $\mathcal{A}_K$ is not a $\vee$-commutative BCK(P') algebra, and therefore (prel) is not satisfied.

### 2.2.3 Hilbert algebras

**Definition 2.2.32** (see [42]) (The dual one is omitted)

A *left-Hilbert algebra* is an algebra $\mathcal{A}^L = (A^L, \rightarrow, 1)$ of type $(2, 0)$, satisfying, for all $x, y, z \in A^L$:

(K) $\quad x \rightarrow (y \rightarrow x) = 1$,

(pimpl-1) $\quad (x \rightarrow (y \rightarrow z)) \rightarrow ((x \rightarrow y) \rightarrow (x \rightarrow z)) = 1$,

(An) $\quad (x \rightarrow y = 1$ and $y \rightarrow x = 1)$ imply $x = y$.

Denote by **Hilbert** the class of all left-Hilbert algebras.

**Theorem 2.2.33** ([104], *Remarks 2.1.32 (1)) (see [112] for a direct proof)*

*Hilbert algebras are definitionally equivalent to positive-implicative BCK algebras.*

**Proof.** (1): Let $\mathcal{A}^L = (A^L, \rightarrow, 1)$ be a left-Hilbert algebra, i.e. properties (K), (pimpl-1), (An) hold. We must prove that $\mathcal{A}^L$ is a positive-implicative left-BCK algebra, i.e. (B), (C), (K), (An), (pimpl) hold. It remains to prove that (B), (C), (pimpl) hold.

Indeed, by (A2), (K) + (An) $\Longrightarrow$ (N); by (A7), (N) + (K) $\Longrightarrow$ (L); by (B19), (pimpl-1) + (K) + (N) $\Longrightarrow$ (Re); by (B27), (pimpl-1) + (N) $\Longrightarrow$ ($); by (B21), (pimpl-1) + (L) + (N) $\Longrightarrow$ (Tr); by (B22), (pimpl-1) + (K) + (Tr) $\Longrightarrow$ (B); thus (B) holds. Then, by (B28), ($) + (K) + (Tr) $\Longrightarrow$ (#); by (A31), (Re) + (#) $\Longrightarrow$ (D); by (A25), (D) + (K) + (N) + (An) $\Longrightarrow$ (M), hence $\mathcal{A}$ is a regular algebra; by (A29'), (B) + (#) $\Longrightarrow$ (BB); by (A21), (BB) + (D) + (N) + (An) $\Longrightarrow$ (Ex); by (A3'), (Re) + (Ex) $\Longrightarrow$ (C); thus (C) holds.

Finally, by (A15), (N) + (BB) $\Longrightarrow$ (**); by (B13), (Re) + (L) + (Ex) + (**) $\Longrightarrow$ (pimpl-2); by (B0), (pimpl-1) + (pimpl-2) + (An) $\Longrightarrow$ (pimpl); thus (pimpl) holds too.

(1'): Let $\mathcal{A}^L = (A^L, \to, 1)$ be a positive-implicative left-BCK algebra, i.e. (B), (C), (K), (An), (pimpl) hold. We must prove that $\mathcal{A}^L$ is a left-Hilbert algebra, i.e. (K), (An), (pimpl-1) hold. It remains to prove (pimpl-1).

Indeed, by (A23), (C) + (K) + (An) $\Longrightarrow$ (Re) and by (A0), (Re) implies (S); then (pimpl) + (S) $\Longrightarrow$ (pimpl-1). □

**Corollary 2.2.34** *[112] Any $\vee$-commutative Hilbert algebra is implicative.*

**Proof.** By above Theorem 2.2.33 and by Theorem 2.2.24. □

### 2.2.4 Connections between classes of BCK algebras

The BCK algebras and their particular cases: the positive-implicative ones (i.e. the Hilbert algebras), the $\vee$-commutative ones, the implicative ones, on the one hand, and the BCK(P') algebras, the $\vee$-commutative BCK(P') algebras, the positive-implicative BCK(P') algebras (i.e. the Hilbert(P') algebras) and the implicative BCK(P') algebras, on the other hand, are connected by the hierarchies presented in next Figure 2.3.

## 2.3 The (involutive) negations

### 2.3.1 The case without last element

In the algebras $(A^L, \to, 1)$ where 1 is not the last element (*without last element*) (from Hierarchies 1 and 1'), we can define a *negation* (an *inverse*) as follows:

$$(dfneg1) \qquad x^{-1} \stackrel{def.}{=} x \to 1$$

and we have (N1) ($1^{-1} = 1$), by (M).

Alternatively, in an algebra $(A^L, \to, ^{-1}, 1)$ *without last element* such that $1^{-1} = 1$, the *negation* (*inverse*) $^{-1}$ is connected with $\to$ and 1 by the equality:

$$(Neg_1) \qquad x^{-1} = x \to 1.$$

**Remark 2.3.1** *(see [116], Remark 5.1.6)*

*Just as the equivalence $x \leq y \Leftrightarrow x \to y = 1$ was used:*
*- either as the definition (dfrelR) of the binary relation $\leq$ in the algebra $(A^L, \to, 1)$,*
*- or as the connection (EqrelR) between the binary relation $\leq$ and $\to, 1$ in the structure $(A^L, \leq, \to, 1)$,*

*the same, the equality $x^{-1} = x \to 1$ was used:*
*- either as the definition (dfneg1) of the negation $^{-1}$ in the algebra $(A^L, \to, 1)$ and in this case $1^{-1} = 1$,*
*- or as the connection ($Neg_1$) between the negation $^{-1}$ and $\to, 1$ in the algebra $(A^L, \to, ^{-1}, 1)$ such that $1^{-1} = 1$.*

If the negation $^{-1}$ verifies (DN) (Double Negation) ($(x^{-1})^{-1} = x$), we say that it is *involutive* and that the algebra is *involutive*.

The subclass of involutive algebras of the class **X** will be denoted by $\mathbf{X}_{(DN)}$.

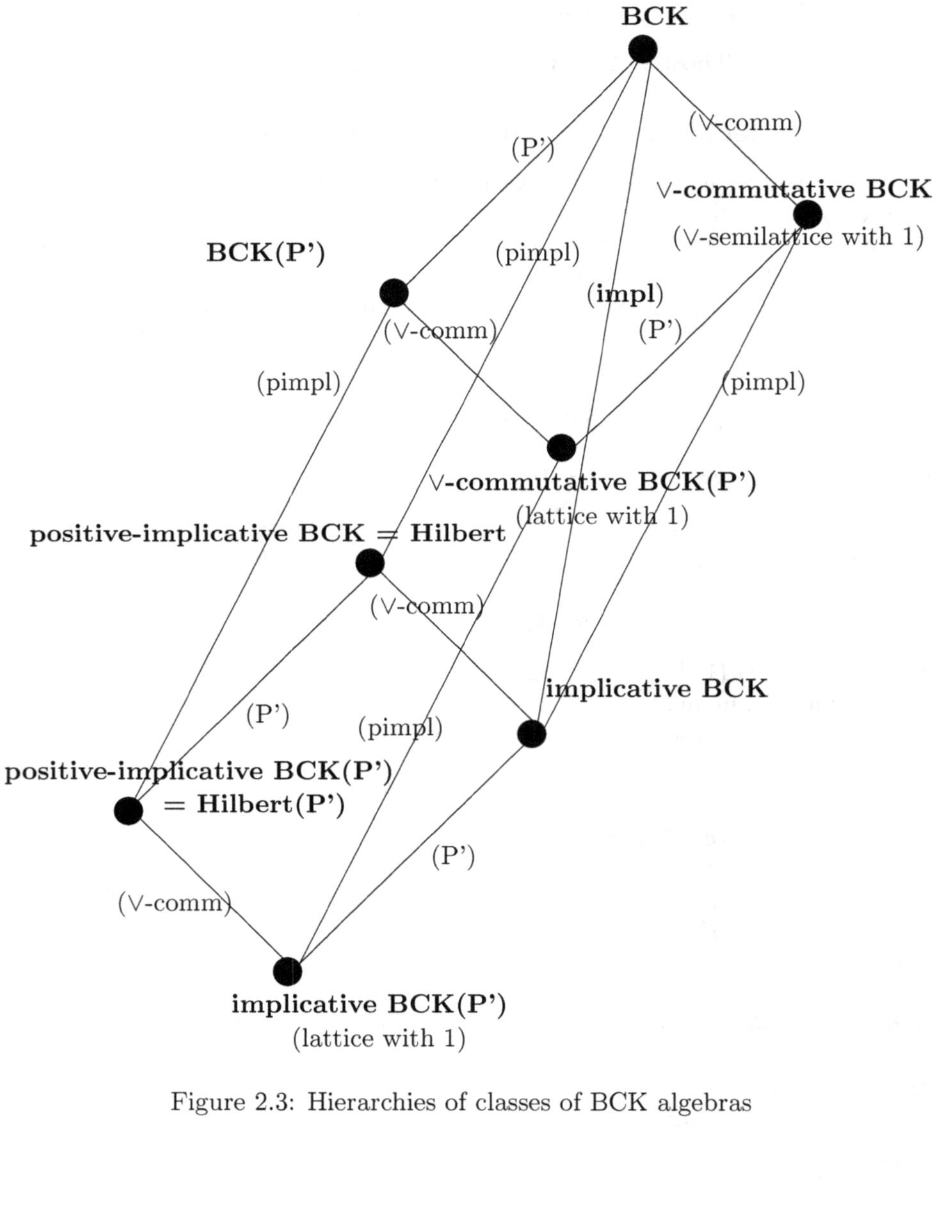


Figure 2.3: Hierarchies of classes of BCK algebras

### 2.3.2 The case with last element

In the algebras $(A^L, \rightarrow, 1)$ *with last element* 1 (from Hierarchies 2 and 2'), the above definition of the negation $^{-1}$ does not work, because it produces $x^{-1} = 1$, for all $x$, by (La). In this case, we add a first element, 0, verifying: (Fi) (First element) $0 \rightarrow x = 1$.

The algebra $(A^L, \rightarrow, 1)$ is said to be *bounded*, if beside the last element 1 (verifying (La)), there exists a first element also, 0 (verifying (Fi)); it is denoted by $(A^L, \rightarrow, 0, 1)$.

Denote by $\mathbf{M}^b$, $\mathbf{ME}^b$, $\mathbf{ML}^b$, ..., $\mathbf{BCK}^b$ the corresponding classes of bounded algebras (where 'b' means 'bounded').

In such algebra $(A^L, \rightarrow, 0, 1)$, a *negation* $^-$ can be defined as follows:

$$(dfneg) \qquad x^- \stackrel{def.}{=} x \rightarrow 0$$

and we have (Neg1-0) $1^- = 0$, by (M), and (Neg0-1) $0^- = 1$, by (Re).

Alternatively, in an algebra $(A^L, \rightarrow, ^-, 1)$ *with last element* 1 such that $0 \stackrel{def.}{=} 1^-$, the *negation* $^-$ is connected with $\rightarrow$ and 0 by the equality:

$$(Neg) \qquad x^- = x \rightarrow 0.$$

**Remark 2.3.2** *(see [116], Remark 5.1.6)*

*Just as the equivalence $x \leq y \Leftrightarrow x \rightarrow y = 1$ was used:*
*- either as the definition (dfrelR) of the binary relation $\leq$ in the algebra $(A^L, \rightarrow, 1)$,*
*- or as the connection (EqrelR) between the binary relation $\leq$ and $\rightarrow, 1$ in the structure $(A^L, \leq, \rightarrow, 1)$,*
*the same, the equality $x^- = x \rightarrow 0$ was used:*
*- either as the definition (dfneg) of the negation $^-$ in the algebra $(A^L, \rightarrow, 0, 1)$ and in this case $1^- = 0$ and $0^- = 1$,*
*- or as the connection (Neg) between the negation $^-$ and $\rightarrow, 0$ in the algebra $(A^L, \rightarrow, ^-, 1)$ such that $0 = 1^-$.*

If the negation $^-$ verifies (DN) ($(x^-)^- = x$), we say that it is *involutive* and that the algebra is *involutive*.

The subclass of involutive algebras of the class $\mathbf{X}$ will be denoted by $\mathbf{X}_{(DN)}$.

The subclasses of M algebras *with last element* 1 *and first element* 0, i.e. the bounded M, ML, RML, pre-BCC, aRML, BCC algebras and the bounded ME, MEL, BE, BE**, pre-BCK, aBE, aBE**, BCK algebras determine the *Hierarchies* $2^b$ *and* $2'^b$.

**Remark 2.3.3** *The special algebras, the M algebras and the ME algebras = residoids, can be considered belonging both to Hierarchy 1 and to Hierarchy $2^b$.*

The algebras without last element (Hierarchy 1) and the bounded algebras (Hierarchy $2^b$) are presented in the following Figure 2.4. The Hierarchy $2'^b$ is obvious and is omitted.

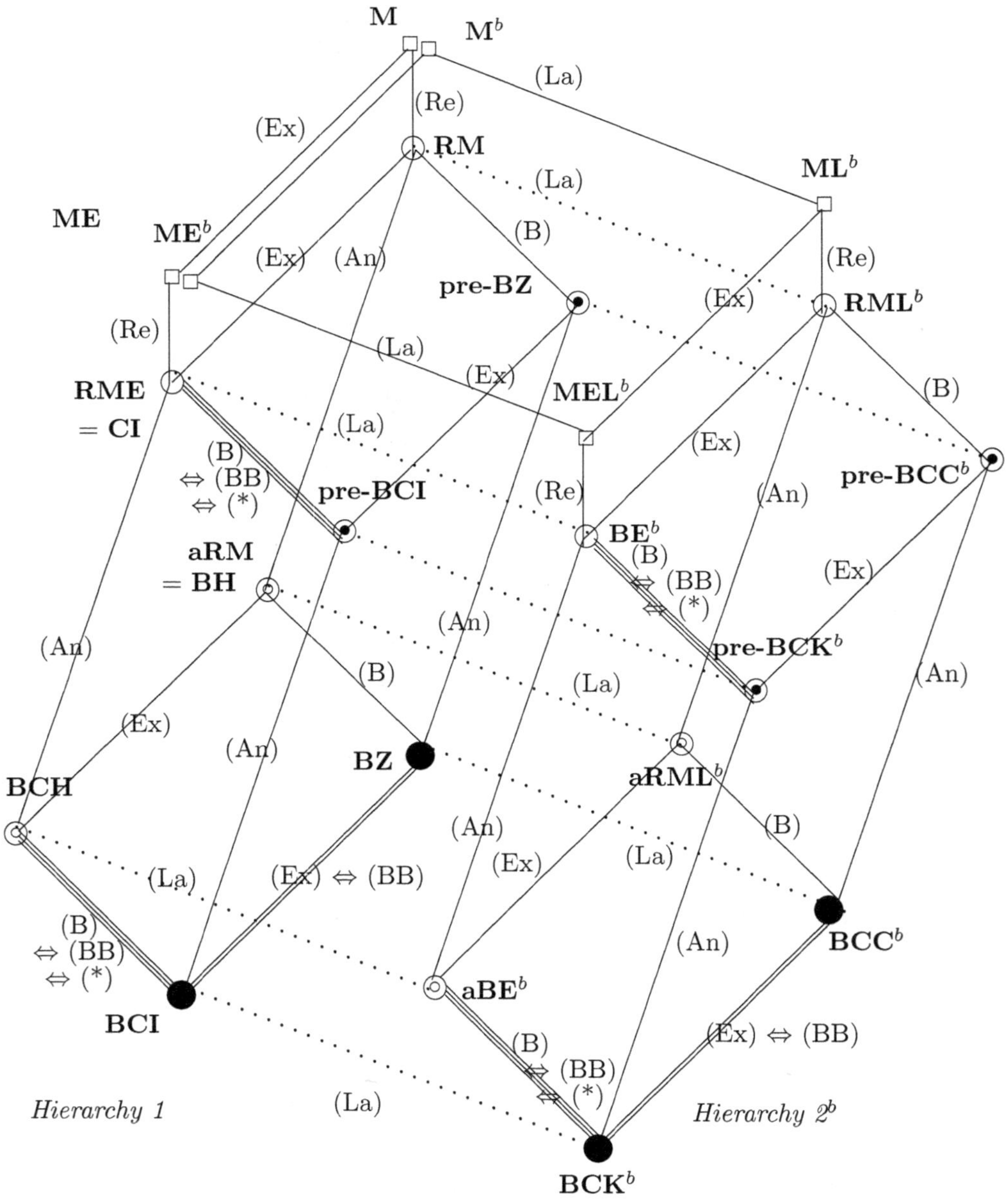


Figure 2.4: The Hierarchies 1 (without last element) and $2^b$ (with last element, and first element) of M algebras and $M^b$ algebras, respectively

# Chapter 3

# M algebras $(A, \rightarrow, 0, 1)$, or $(A, \rightarrow, ^-, 1)$, with $1^- = 0$ (Hierarchies $2^b$ and 2'$^b$)

In this chapter, we analyse in some detail the M algebras *with last element,* 1, with some additional operations, of the form $(A, \rightarrow, 0, 1)$ or $(A, \rightarrow, ^-, 1)$ with $1^- = 0$, which determine Hierarchies $2^b$, $2'^b$. We present new results, some of them proved in the preprint [113]. We introduce the new notion of *implicative-ortholattice* as a term-equivalent definition of the ortholattice recalled and studied in Chapter 6. We establish connections mainly between BE algebras, BCK algebras, Wajsberg algebras, implicative-ortholattices and implicative-Boolean algebras.

In Section 3.1, we present the list **C** of some properties of negation.

In Section 3.2, we discuss about involutive MEL algebras.

In Section 3.3, we discuss about involutive BCK algebras, namely about Wajsberg algebras, weak-$R_0$ algebras and $R_0$ algebras, implicative-Boolean algebras and about the connections existing between them.

In Section 3.4, we discuss about the implicative-ortholattice.

In Section 3.5, we present new results on L algebras introduced by W. Rump and some examples.

The content of this chapter is taken mainly from [118].

Let $\mathcal{A}^L = (A^L, \rightarrow, 0, 1)$ be a bounded M algebra *not verifying (Ex).* Let us introduce in this case the weaker condition:
(Ex0) $x \rightarrow (y \rightarrow 0) = y \rightarrow (x \rightarrow 0)$.

We shall say that $\mathcal{A}^L$ is *strong-0,* if it verifies (Ex0).

Note that (Ex) implies (Ex0).

## 3.1 List C of properties of the negation $^-$ and some results

Consider the following list **C** of properties that can be satisfied by an (involutive) negation $^-$ defined on $A^L$:

| | |
|---|---|
| (dfneg) | $x^- \overset{def.}{=} x \rightarrow 0$, |
| (Neg) | $x^- = x \rightarrow 0$; |
| | |
| (DN) | $(x^-)^- = x$ or $x^= = x$ (Double negation), |
| (DN0) | $(x \rightarrow 0) \rightarrow 0 = x$ (Double negation); |
| (DN01) | $(x \rightarrow 0)^- \rightarrow x = 1$, |
| (DN02) | $(x^- \rightarrow 0) \rightarrow x = 1$, |
| (DN03) | $x \rightarrow (x^- \rightarrow 0) = 1$, |
| (DN04) | $x^- \rightarrow 0 = x$; |
| | |
| (Neg1-0) | $1^- = 0$, |
| (Neg0-1) | $0^- = 1$; |
| | |
| (Neg1) | $(x \rightarrow y) \rightarrow (y^- \rightarrow x^-) = 1$, |
| (Neg1') | $x \rightarrow y \leq y^- \rightarrow x^-$; |
| (Neg2) | $x \rightarrow y = 1 \Longrightarrow y^- \rightarrow x^- = 1$ (contraposition), |
| (Neg2') | $x \leq y \Longrightarrow y^- \leq x^-$ (contraposition); |
| (Neg3) | $y \rightarrow x^- = x \rightarrow y^-$; |
| (Neg4) | $x \rightarrow (x^-)^- = 1$, |
| (Neg4') | $x \leq (x^-)^-$; |
| (Neg5) | $(x \rightarrow y)^- \rightarrow x = 1$, |
| (Neg5') | $(x \rightarrow y)^- \leq x$; |
| (Neg6) | $x \rightarrow x^- = x^-$; |
| (Neg7) = (TN) | $((x^-)^-)^- = x^-$ (Triple Negation); |
| | |
| (DN1) | $(y^- \rightarrow x^-) \rightarrow (x \rightarrow y) = 1$, |
| (DN1') | $y^- \rightarrow x^- \leq x \rightarrow y$; |
| (DN2) | $x \rightarrow y = y^- \rightarrow x^-$; |
| (DN3) | $y^- \rightarrow x = x^- \rightarrow y$; |
| (DN4) | $x \rightarrow y = 1 \Longleftrightarrow y^- \rightarrow x^- = 1$, |
| (DN4') | $x \leq y \Longleftrightarrow y^- \leq x^-$; |
| (DN5) | $x^- \rightarrow (x \rightarrow y) = 1$, |
| (DN5') | $x^- \leq x \rightarrow y$; |
| (i-G)=(DN6) | $x^- \rightarrow x = x$; |
| (@) | $(y^- \rightarrow x) \rightarrow y = x \rightarrow y$; |
| | |
| (WRe) | $x \wedge x^- = 0$, |
| (VRe) | $x \vee x^- = 1$. |

**Proposition 3.1.1** *We have:*

*(CE1) (dfneg) + (Ex0) $\Longrightarrow$ (Neg3),*
*(CE1') (dfneg) + (Neg3) $\Longrightarrow$ (Ex0),*
*(CE1") (dfneg) $\Longrightarrow$ ((Ex0) $\Longleftrightarrow$ (Neg3));*
*(CE2) (Neg) + (Ex0) $\Longrightarrow$ (Neg3),*
*(CE2') (Neg) + (Neg3) $\Longrightarrow$ (Ex0),*
*(CE2") (Neg) $\Longrightarrow$ ((Ex0) $\Longleftrightarrow$ (Neg3)).*

**Proof.** (CE1): $x \to y^- \stackrel{(dfneg)}{=} x \to (y \to 0) \stackrel{(Ex0)}{=} y \to (x \to 0) \stackrel{(dfneg)}{=} y \to x^-$; thus, (Neg3) holds.

(CE1'): $x \to (y \to 0) \stackrel{(dfneg)}{=} x \to y^- \stackrel{(Neg3)}{=} y \to x^- \stackrel{(dfneg)}{=} y \to (x \to 0)$; thus, (Ex0) holds.

(CE1"): By (CE1) and (CE1').

(CE2), (CE2'), (CE2"): similarly. □

**Proposition 3.1.2** *We have [113]:*
*(C0) (DN) + (Neg1-0) $\Longrightarrow$ (Neg0-1),   (C0') (DN) + (Neg0-1) $\Longrightarrow$ (Neg1-0),*
*(C0") (DN) $\Longrightarrow$ ((Neg1-0) $\Leftrightarrow$ (Neg0-1));*

*(C00) (Neg) + (DN) $\Longrightarrow$ (DN0),   (C00') (Neg) + (DN0) $\Longrightarrow$(DN),*
*(C00") (Neg) $\Longrightarrow$ ((DN) $\Leftrightarrow$ (DN0));*

*(C1) (DN) + (Neg1) $\Longrightarrow$ (DN1),   (C1') (DN) + (DN1) $\Longrightarrow$ (Neg1),*
*(C1") (DN) $\Longrightarrow$ ((Neg1) $\Leftrightarrow$ (DN1));*

*(C2) (DN) + (DN2) $\Longrightarrow$ (DN3),   (C2') (DN) + (DN3) $\Longrightarrow$ (DN2) ;*
*(C2") (DN) $\Longrightarrow$ ((DN2) $\Leftrightarrow$ (DN3));*

*(C3) (DN) + (Neg3) $\Longrightarrow$ (DN3),   (C3') (DN) + (DN3) $\Longrightarrow$ (Neg3),*
*(C3") (DN) $\Longrightarrow$ ((Neg3) $\Leftrightarrow$ (DN3));*

*(C4) (DN2) $\Longrightarrow$ (DN4);   (C4') (DN2) + (Re) $\Longrightarrow$ (DN1);*
*(C4") (DN) + (Neg2') $\Longrightarrow$ (DN4');*

*(C5) (DN) + (DN2) + (Neg5) $\Longrightarrow$ (DN5),*
*(C5') (DN) + (DN2) + (DN5) $\Longrightarrow$ (Neg5),*
*(C5") (DN) + (DN2) $\Longrightarrow$ ((Neg5) $\Leftrightarrow$ (DN5));*

*(C6) (DN) + (Neg6) $\Longrightarrow$ (DN6),   (C6') (DN) + (DN6) $\Longrightarrow$ (Neg6),*
*(C6") (DN) $\Longrightarrow$ ((Neg6) $\Leftrightarrow$ (DN6));*

*(C7) (DN) + (Neg) + (Ex) $\Longrightarrow$ (DN2);*
*(C8) (Neg0-1) + (DN1) + (La) + (M) $\Longrightarrow$ (Fi);*
*(C9) (DN1) + (DN0) + (Neg0-1) + (M) $\Longrightarrow$ (DN01);*
*(C10) (DN1) + (Neg0-1) + (M) $\Longrightarrow$ (DN02);*
*(C11) (DN1) +(DN01) + (M) $\Longrightarrow$ (DN03);*
*(C12) (DN02) + (DN03) + (An) $\Longrightarrow$ (DN04).*

**Proof.** (C0), (C0'), (C)"): immediately.

(C00): $(x \rightarrow 0) \rightarrow 0 \overset{(Neg)}{=} (x^-)^- \overset{(DN)}{=} x$.

(C00'): $(x^-)^- \overset{(Neg)}{=} (x \rightarrow 0) \rightarrow 0 \overset{(DN0)}{=} x$.

(C00"): By (C00) and (C00').

(C1): $y^- \rightarrow x^- \overset{(Neg1)}{\leq} (x^-)^- \rightarrow (y^-)^- \overset{(DN)}{=} x \rightarrow y$, i.e. (DN1) holds.

(C1'): $x \rightarrow y \overset{(DN)}{=} (x^-)^- \rightarrow (y^-)^- \overset{(DN1)}{\leq} y^- \rightarrow x^-$, i.e. (Neg1) holds.

(C1"): By (C1) and (C1').

(C2): $x^- \rightarrow y \overset{(DN)}{=} x^- \rightarrow (y^-)^- \overset{(DN2)}{=} y^- \rightarrow x$, i.e. (DN3) holds.

(C3): $y^- \rightarrow x \overset{(DN)}{=} y^- \rightarrow (x^-)^- \overset{(Neg3)}{=} x^- \rightarrow (y^-)^- \overset{(DN)}{=} x^- \rightarrow y$, i.e. (DN3) holds.

(C3'): $y \rightarrow x^- \overset{(DN)}{=} (y^-)^- \rightarrow x^- \overset{(DN3)}{=} (x^-)^- \rightarrow y^- \overset{(DN)}{=} x \rightarrow y^-$, i.e. (Neg3) holds.

(C3"): By (C3) and (C3').

(C4): By (DN2), $x \rightarrow y = y^- \rightarrow x^-$, hence $x \rightarrow y = 1 \Longleftrightarrow y^- \rightarrow x^- = 1$, i.e. (DN4) holds.

(C4'): $(y^- \rightarrow x^-) \rightarrow (x \rightarrow y) \overset{(DN2)}{=} (x \rightarrow y) \rightarrow (x \rightarrow y) \overset{(Re)}{=} 1$.

(C4"): $x \leq y \Longrightarrow y^- \leq x^- \Longrightarrow x^= \leq y^= \Longleftrightarrow x \leq y$.

(C5): $x^- \rightarrow (x \rightarrow y) \overset{(DN)}{=} x^- \rightarrow ((x \rightarrow y)^-)^- \overset{(DN2)}{=} (x \rightarrow y)^- \rightarrow x \overset{(Neg5)}{=} 1$, i.e. (DN5) holds.

(C5'): $(x \rightarrow y)^- \rightarrow x \overset{(DN)}{=} (x \rightarrow y)^- \rightarrow (x^-)^- \overset{(DN2)}{=} x^- \rightarrow (x \rightarrow y) \overset{(DN5)}{=} 1$, i.e. (Neg5) holds.

(C5"): By (C5) and (C5').

(C6): $x^- \rightarrow x \overset{(DN)}{=} x^- \rightarrow (x^-)^- \overset{(Neg6)}{=} (x^-)^- \overset{(DN)}{=} x$, i.e. (DN6) holds.

(C6'): $x \rightarrow x^- \overset{(DN)}{=} (x^-)^- \rightarrow x^- \overset{(DN6)}{=} x^-$, i.e. (Neg6) holds.

(C6"): By (C6) and (C6').

(C7): By (Neg), $x^- = x \rightarrow 0$, and by (DN), $x^= = x$, i.e. $(x \rightarrow 0) \rightarrow 0 = x$. Then, $y \rightarrow x = y \rightarrow ((x \rightarrow 0) \rightarrow 0) \overset{(Ex)}{=} (x \rightarrow 0) \rightarrow (y \rightarrow 0) = x^- \rightarrow y^-$, i.e. (DN2) holds.

(C8): In (DN1) $((x^- \rightarrow y^-) \rightarrow (y \rightarrow x) = 1)$, take $y := 0$, to obtain:
$(x^- \rightarrow 0^-) \rightarrow (0 \rightarrow x) = 1$, hence, by (Neg0-1),
$(x^- \rightarrow 1-) \rightarrow (0 \rightarrow x) = 1$, hence, by (La),
$1 \rightarrow (0 \rightarrow x) = 1$, hence, by (M),
$0 \rightarrow x = 1$, i.e. (Fi) holds.

(C9): In (DN1) $((x^- \rightarrow y^-) \rightarrow (y \rightarrow x) = 1)$, take $Y := x \rightarrow 0$, $X := 0$, to obtain:
$(0^- \rightarrow (x \rightarrow 0)^-) \rightarrow ((x \rightarrow 0) \rightarrow 0) = 1$, hence, by (Neg0-1), (DN0),
$(1 \rightarrow (x \rightarrow 0)^- \rightarrow x = 1$, hence, by (M),
$(x \rightarrow 0)^- \rightarrow x = 1$, i.e. (DN01) holds.

(C10): In (DN1) $((x^- \rightarrow y^-) \rightarrow (y \rightarrow x) = 1)$, take $y = 1$, to obtain:
$(x^- \rightarrow 1^-) \rightarrow (1 \rightarrow x) = 1$, hence, by (Neg1-0) and (M),
$(x^- \rightarrow 0) \rightarrow x = 1$, i.e. (DN02) holds.

(C11): In (DN1) $((x^- \to y^-) \to (y \to x) = 1)$, take $X := y^- \to 0$, to obtain:
$((y^- \to 0)^- \to y^-) \to (y \to (y^- \to 0)) = 1$, hence, by (DN01),
$1 \to (y \to (y^- \to 0)) = 1$, hence, by (M),
$y \to (y^- \to 0) = 1$, i.e. (DN03) holds.

(C12): Obviously. □

**Lemma 3.1.3** *Let $\mathcal{A}^L = (A^L, \to, ^-, 1)$ be an algebra of type $(2,1,0)$ verifying (M) and (Neg3). Define $0 \stackrel{def.}{=} 1^-$. Then, (Neg) holds.*

**Proof.** $x \to 0 \stackrel{def.}{=} x \to 1^- \stackrel{(Neg3)}{=} 1 \to x^- \stackrel{(M)}{=} x^-$; thus, (Neg) holds. □

The following two theorems allows us to say that, roughly speaking, the M algebras $(A^L, \to, 0, 1)$ and $(A^L, \to, ^-, 1)$ are definitionally equivalent.

**Theorem 3.1.4**

*(1) Let $\mathcal{A}^L = (A^L, \to, 0, 1)$ be an algebra of type $(2,0,0)$ verifying (M) and (Ex0). Define a negation $^-$ by (dfneg).*
*Then, $\alpha(\mathcal{A}^L) = (A^L, \to, ^-, 1)$ is an algebra of type $(2,1,0)$ verifying (M) and (Neg3).*

*(1') Let $\mathcal{A}^L = (A^L, \to, ^-, 1)$ be an algebra of type $(2,1,0)$ verifying (M) and (Neg3). Define $0 \stackrel{def.}{=} 1^-$.*
*Then, $\beta(\mathcal{A}^L) = (A^L, \to, 0, 1)$ is an algebra of type $(2,0,0)$ verifying (M) and (Ex0).*

*(2) The mapppings $\alpha$ and $\beta$ are inverse to each other.*

**Proof.** (1): By (CE1), (dfneg) + (Ex0) imply (Neg3); thus, (Neg3) holds.

(1'): By Lemma 3.1.3, (Neg) holds. Then, by (CE2'), (Neg) + (Neg3) imply (Ex0); thus, (Ex0) holds.

(2): Let

$$(A^L, \to, 0, 1) \quad \stackrel{\alpha}{\longrightarrow} \quad (A^L, \to, ^-, 1) \quad \stackrel{\beta}{\longrightarrow} \quad (A^L, \to, \mathbf{0}, 1).$$

Then, we have: $\mathbf{0} \stackrel{def.}{=} 1^- \stackrel{(dfneg)}{=} 1 \to 0 \stackrel{(M)}{=} 0$.

Conversely, let

$$(A^L, \to, ^-, 1) \quad \stackrel{\beta}{\longrightarrow} \quad (A^L, \to, 0, 1) \quad \stackrel{\alpha}{\longrightarrow} \quad (A^L, \to, ', 1).$$

Then, for all $x \in A^L$, we have: $x' \stackrel{(dfneg)}{=} x \to 0 \stackrel{def.}{=} x \to 1^- \stackrel{(Neg3)}{=} 1 \to x^- \stackrel{(M)}{=} x^-$.
□

**Theorem 3.1.5**

*(1) Let $\mathcal{A}^L = (A^L, \to, 0, 1)$ be an algebra of type $(2,0,0)$ verifying (M), (La), (Fi) and (Ex0). Define a negation $^-$ by (dfneg) such that (DN) holds.*
*Then, $\alpha(\mathcal{A}^L) = (A^L, \to, ^-, 1)$ is an algebra of type $(2,1,0)$ verifying (M), (La), (DN) and (Neg3).*

*(1') Let $\mathcal{A}^L = (A^L, \to, ^-, 1)$ be an algebra of type $(2,1,0)$ verifying (M), (La), (DN) and (Neg3). Define $0 \stackrel{def.}{=} 1^-$.*

*Then, $\beta(\mathcal{A}^L) = (A^L, \rightarrow, 0, 1)$ is an algebra of type $(2, 0, 0)$ verifying (M), (La), (Fi) (and (DN)) and (Ex0).*

*(2) The mappings $\alpha$ and $\beta$ are inverse to each other.*

**Proof.** By Theorem 3.1.4, it remains to prove that (Fi) holds, in the case (1'). Indeed, by (C3), (DN) + (Neg3) imply (DN3); then, $0 \rightarrow x = 1^- \rightarrow x \overset{(DN3)}{=} x^- \rightarrow 1 \overset{(La)}{=} 1$; thus, (Fi) holds. □

**Proposition 3.1.6** *We have (see [113]):*
*(CB0) (pimpl) + (Ex) + (DN6) $\Longrightarrow$ (@),*
*(CB1) ($\vee$-comm) + (Neg) + (Fi) + (M) $\Longrightarrow$ (DN);*
*(CB1') ($\vee$-comm) + (Fi) + (M) $\Longrightarrow$ (DN0);*
*(CB2) (impl) + (Neg) $\Longrightarrow$ (DN6);*
*(CB3) (pimpl) + (Neg) + (Re) + (M) $\Longrightarrow$ (Neg6);*
*(CB4) (pimpl-1) + (Neg) + (Re) + (M) + (K) + (An) $\Longrightarrow$ (Neg6);*
*(CB5) (DN) + (DN2) + (DN6) + (DN5) + (K) + (*) + (An) $\Longrightarrow$ (impl);*
*(CB6) ($\vee$-comm) + (DN04) + (Fi) + (M) $\Longrightarrow$ (Neg).*

**Proof.** (CB0): In (pimpl) $(x \rightarrow (y \rightarrow z) = (x \rightarrow y) \rightarrow (x \rightarrow z))$ take $x = x^-$ and $z = x$; we obtain:
$y \rightarrow x \overset{(DN6)}{=} y \rightarrow (x^- \rightarrow x) \overset{(Ex)}{=} x^- \rightarrow (y \rightarrow x)$
$\overset{(pimpl)}{=} (x^- \rightarrow y) \rightarrow (x^- \rightarrow x) \overset{(DN6)}{=} (x^- \rightarrow y) \rightarrow x$; thus, (@) holds.

(CB1): $(x^-)^- \overset{(Neg)}{=} (x \rightarrow 0) \rightarrow 0 \overset{(\vee-comm)}{=} (0 \rightarrow x) \rightarrow x \overset{(Fi)}{=} 1 \rightarrow x \overset{(M)}{=} x$, i.e. (DN) holds.

(CB1'): In ($\vee$-comm) $((x \rightarrow y) \rightarrow y = (y \rightarrow x) \rightarrow x)$, take $x := 0$, to obtain:
$(0 \rightarrow y) \rightarrow y = (y \rightarrow 0) \rightarrow 0$, hence, by (Fi),
$1 \rightarrow y = (y \rightarrow 0) \rightarrow 0$, hence, by (M),
$y = (y \rightarrow 0) \rightarrow 0$, i.e. (DN0) holds.

(CB2): Take $y = 0$ in (impl) $((x \rightarrow y) \rightarrow x = x)$; we obtain: $(x \rightarrow 0) \rightarrow x = x$, hence $x \overset{(Neg)}{=} x^- \rightarrow x$, i.e. (DN6) holds.

(CB3): In (pimpl) $(x \rightarrow (y \rightarrow z) = (x \rightarrow y) \rightarrow (x \rightarrow z))$, take $y = x$ and $z = 0$; we then obtain: $x \rightarrow (x \rightarrow 0) = (x \rightarrow x) \rightarrow (x \rightarrow 0)$, hence $x \rightarrow x^- = 1 \rightarrow x^- = x^-$, by (Neg), (Re), (M), i.e. (Neg6) holds.

(CB4): On the one hand, $x^- \leq x \rightarrow x^-$, by (K'). On the other hand,
$x \rightarrow x^- \overset{(Neg)}{=} x \rightarrow (x \rightarrow 0) \overset{(pimpl-1')}{\leq} (x \rightarrow x) \rightarrow (x \rightarrow 0) \overset{(Re)}{=} 1 \rightarrow (x \rightarrow 0) \overset{(M)}{=} x \rightarrow 0 \overset{(Neg)}{=} x^-$.
Now, by (An), we obtain that $x \rightarrow x^- = x^-$, i.e. (Neg6) holds.

(CB5): First, note that we have: (a) $x \leq (x \rightarrow y) \rightarrow x$, by (K').
Then, we prove: (b) $(x \rightarrow y) \rightarrow x \leq x$. Indeed, $(x \rightarrow y) \rightarrow x \overset{(DN2)}{=} x^- \rightarrow (x \rightarrow y)^-$. But (DN) + (DN2) + (DN5) imply (Neg5), by (C5'), i.e. $(x \rightarrow y)^- \leq x$; then, by (*'), $x^- \rightarrow (x \rightarrow y)^- \leq x^- \rightarrow x \overset{(DN6)}{=} x$, i.e. $x^- \rightarrow (x \rightarrow y)^- \leq x$. Hence, (b) holds.
Now, (a) + (b) + (An) imply $(x \rightarrow y) \rightarrow x = x$, i.e. (impl) holds.

(CB6): In ($\vee$-comm) $((x \to y) \to y = (y \to x) \to x)$, take $X := x^-$, $Y := 0$, to obtain:
$(x^- \to 0) \to 0 = (0 \to x^-) \to x^-$, hence, by (DN04) and (Fi),
$x \to 0 = 1 \to x^-$, hence, by (M),
$x \to 0 = x^-$, i.e. (Neg) holds. □

**Proposition 3.1.7** *We have:*
*(CC1) (@) + (DN2) + (DN3) $\Longrightarrow$ (H),*
*(CC2) (@) + (H) + (DN3) $\Longrightarrow$ ($\vee$-comm).*

**Proof.** (CC1): $((x \to y) \to y) \to y \stackrel{(DN2)}{=} ((y^- \to x^-) \to y) \to y$
$\stackrel{(@)}{=} (x^- \to y) \to y \stackrel{(DN3)}{=} (y^- \to x) \to y \stackrel{(@)}{=} x \to y$; thus, (H) holds.

(CC2): We must prove that $(x \to y) \to y = (y \to x) \to x$.
Indeed, $(x \to y) \to y \stackrel{(@)}{=} ((y^- \to x) \to y) \to y \stackrel{(DN3)}{=} ((x^- \to y) \to y) \to y$
$\stackrel{(H)}{=} x^- \to y$ and
$(y \to x) \to x \stackrel{(@)}{=} ((x^- \to y) \to x) \to x \stackrel{(DN3)}{=} ((y^- \to x) \to x) \to x$
$\stackrel{(H)}{=} y^- \to x \stackrel{(DN3)}{=} x^- \to y$; thus, ($\vee$-comm) holds. □

We add now the following results:

**Proposition 3.1.8** *(The ideea came from the analogous result (mCDN3) from Section 6) We have:*
*(i) (DN) + (Ex0) + (dfneg) $\Longrightarrow$ ((*) $\Leftrightarrow$ (**)),*
*(ii) (DN) + (Neg3) + (Neg) $\Longrightarrow$ ((*) $\Leftrightarrow$ (**)).*

**Proof.** (i): First note that (Ex0) + (dfneg) imply (Neg3), by (CE1); then, (DN) + (Neg3) imply (DN2), by (C3), (C2"); then, (DN2) implies (DN4), by (C4).
If $x \le y$, then by (DN4'), $y^- \le x^-$; then, by (*'), $z^- \to y^- \le z^- \to x^-$, hence $y \to z \le x \to z$, by (DN2); thus, (**') holds. Conversely, if $x \le y$, then $y^- \le x^-$, by (DN4'); then, by (**'), $x^- \to z^- \le y^- \to z^-$, hence $z \to x \le z \to y$, by (DN2); thus, (*') holds.

(ii): Similarly. □

**Theorem 3.1.9** *(This Theorem is inspired by Theorem 6.3.5)*

*If an algebra $(A, \to, 0, 1)$ verifies the properties (M), (Re), (Ex), (dfneg) and (DN), then:*

$$(B) \Leftrightarrow (BB) \Leftrightarrow (*) \Leftrightarrow (**) \Leftrightarrow (Tr).$$

**Proof.** By Theorems 2.1.5, 2.1.6 and Proposition 3.1.8 (i), since (Ex) implies (Ex0). □

**Open problem 3.1.10** *Does Theorem 3.1.9 remain valid for (Ex0) instead of (Ex)?*

By Propositions 3.1.8 and 2.1.9, we obtain the following result (inspired from Theorem 6.3.6):

**Proposition 3.1.11** *The involutive aBE** algebras coincide with the involutive BCK algebras.*

By Corollary 2.2.15 and (CB1), we obtain:

**Corollary 3.1.12** *Every bounded implicative aBE** algebra satisfies (DN) (i.e. it is involutive).*

By above Corollary 3.1.12 and Proposition 3.1.11, we obtain immediately that:

**Corollary 3.1.13** *The bounded implicative aBE** algebras coincide with the bounded implicative BCK algebras.*

**Proposition 3.1.14** *[126] In a bounded left-BCK algebra, the following properties hold : (Neg1-0), (Neg0-1), (Neg4'), (Neg1'), (Neg2'), (Neg3), (Neg7).*

**Definition 3.1.15** [97] Let $\mathcal{A}^L = (A^L, \le, \to, 0, 1)$ be a bounded left-BCK algebra. Define two new operations on $A$, $\cdot$ and $\oplus$, by: for all $x, y \in A^L$,

$$x \cdot y \overset{def}{=} (x \to y^-)^- \left(\overset{(Neg3)}{=} (y \to x^-)^-\right). \tag{3.1}$$

$$y \oplus x \overset{def}{=} (x^- \cdot y^-)^-. \tag{3.2}$$

**Lemma 3.1.16** *[95] Let $\mathcal{A}^L$ be a bounded left-BCK algebra. For every $x, y \in A^L$:*

$$x \le y \Rightarrow x \cdot z \le y \cdot z.$$

**Proof.** By (3.1), $x \cdot z = (x \to z^-)^-$ and $y \cdot z = (y \to z^-)^-$. If $x \le y$, then, by (**), $y \to z^- \le x \to z^-$. Then, by (Neg2'), $(x \to z^-)^- \le (y \to z^-)^-$, i.e. $x \cdot z \le y \cdot z$. □

**Proposition 3.1.17** *[95] Let $\mathcal{A}^L = (A^L, \le, \to, 0, 1)$ be a bounded left-BCK algebra. The following conditions are equivalent:*

$$x \cdot y \le z \Leftrightarrow x \le y \to z. \tag{3.3}$$

$$x \cdot y = \min\{z \mid x \le y \to z\}, \tag{3.4}$$

$$y \to z = \max\{x \mid x \cdot y \le z\}. \tag{3.5}$$

**Proof.** (3.3) $\Longleftrightarrow$ (3.4):
$\Longrightarrow$: Since $x \cdot y \le x \cdot y$, by (3.3) we get that $x \le y \to (x \cdot y)$. If $z$ verifies $x \le y \to z$, then by (3.3), $x \cdot y \le z$. Thus, $x \cdot y = \min\{z \mid x \le y \to z\}$.
$\Longleftarrow$: If $x \le y \to z$, then by (3.4), $x \cdot y \le z$. If $x \cdot y \le z$, then, it follows by (*), that $y \to (x \cdot y) \le y \to z$ and since we also have that $x \le y \to (x \cdot y)$, by (3.4), we get $x \le y \to z$.

(3.3) $\Longleftrightarrow$ (3.5):
$\Longrightarrow$: Since $y \to z \le y \to z$, by (3.3), we get $(y \to z) \cdot y \le z$. If $x$ verifies $x \cdot y \le z$, then, by (3.3), $x \le y \to z$. Thus, $y \to z = \max\{x \mid x \cdot y \le z\}$.
$\Longleftarrow$: If $x \cdot y \le z$, then by (3.5), $x \le y \to z$. If $x \le y \to z$, it follows, by Lemma 3.1.16, that $x \cdot y \le (y \to z) \cdot y$, and since we also have by (3.5) that $(y \to z) \cdot y \le z$, we get $x \cdot y \le z$. □

## 3.2 Involutive MEL algebras

Let $\mathcal{A}^L = (A^L, \rightarrow, 0, 1)$ be an involutive left-MEL algebra, i.e. (M), (Ex), (La), (Fi), (DN) hold, where $x^- \stackrel{def.}{=} x \rightarrow 0$, hence (Neg) holds too. We have $1^- = 1 \rightarrow 0 \stackrel{(M)}{=} 0$, hence (Neg1-0) holds too. By (C0), (DN) + (Neg1-0) imply (Neg0-1), hence (Neg0-1) holds too. By (C7), (DN) + (Neg) + (Ex) imply (DN2), hence (DN2) holds too.

Because of the axiom (DN), we can introduce the new operation $\rightarrow^R$, the dual of $\rightarrow$, by: for all $x, y \in A^L$,

$$x \rightarrow^R y \stackrel{def.}{=} (x^- \rightarrow y^-)^-.$$

**Proposition 3.2.1** *We have:*

*($M^R$)* $\quad 0 \rightarrow^R x = x$,
*($Ex^R$)* $\quad x \rightarrow^R (y \rightarrow^R z) = y \rightarrow^R (x \rightarrow^R z)$,
*($La^R$)* $\quad x \rightarrow^R 0 = 0$,
*(DeM1)* $\quad (x \rightarrow^R y)^- = x^- \rightarrow y^-$ *(De Morgan law 1),*
*(DeM2)* $\quad (x \rightarrow y)^- = x^- \rightarrow^R y^-$ *(De Morgan law 2) and, hence,*
$\quad x \rightarrow y = (x^- \rightarrow^R y^-)^-$.

**Proof.** Routine. □

**Corollary 3.2.2** *If $\mathcal{A}^L = (A^L, \rightarrow, 0, 1)$ is an involutive left-MEL algebra, then $(A^L, \rightarrow^R, 1, 0)$ is an involutive right-MEL algebra.*

**Proposition 3.2.3** *Let $\mathcal{A}^L = (A^L, \rightarrow, 0, 1)$ be an involutive left-MEL algebra. We have: for all $x, y \in A^L$,*

$$x \leq y \Longleftrightarrow y \geq x.$$

**Proof.** Suppose $x \leq y$, i.e. $x \rightarrow y = 1$; then $y^- \rightarrow x^- = 1$, by (DN2), hence $y \rightarrow^R x \stackrel{def.}{=} (y^- \rightarrow x^-)^- = 1^- \stackrel{(Neg1-0)}{=} 0$, i.e. $y \geq x$. Conversely, suppose $y \geq x$, i.e. $y \rightarrow^R x = 0$; then, by definition, $(y^- \rightarrow x^-)^- = 0$, hence $(x \rightarrow y)^- = 0$, by (DN2); then, $(x \rightarrow y)^= = 0^- \stackrel{(Neg0-1)}{=} 1$, hence $x \rightarrow y = 1$, by (DN), i.e. $x \leq y$. □

- **A new binary relation: $\leq^P$**

We shall define first on $\mathcal{A}^L$ the following two dual operations: for all $x, y \in A^L$,
$x \wedge^P y \stackrel{def.}{=} (x \rightarrow y^-)^-$ and, dually,
$x \vee^P y \stackrel{def.}{=} (x^- \wedge^P y^-)^- = (x^- \rightarrow y^=)^= \stackrel{(DN)}{=} x^- \rightarrow y$.

Beside the binary relations $\leq$ and $\leq^W$, $\leq^B$, we shall introduce another binary relation: for all $x, y \in A^L$,

$$(dfP) \quad x \leq^P y \stackrel{def.}{\Longleftrightarrow} x \wedge^P y = x \Longleftrightarrow (x \rightarrow y^-)^- = x \quad \textit{and, dually,}$$

$$(dfS) \quad x \geq^S y \stackrel{def.}{\Longleftrightarrow} (x \rightarrow^R y^-)^- = x.$$

**Proposition 3.2.4** *Let $\mathcal{A}^L$ be an involutive left-MEL algebra. The binary relation $\leq^P$ is antisymmetric and transitive and $0 \leq^P x \leq^P 1$, for all $x \in A^L$.*

**Proof.** (P-An) (*Antisymmetry*): for all $x, y \in A^L$, $x \leq^P y$ and $y \leq^P x$ imply $x = y$: indeed, by definition, $(x \to y^-)^- = x$ and $(y \to x^-)^- = y$, hence $x^- \stackrel{(DN)}{=} x \to y^- \stackrel{(Neg3)}{=} y \to x^- \stackrel{(DN)}{=} y^-$; hence, $x = y$, by (DN) again.

(P-Tr) (*Transitivity*): for all $x, y, z \in A^L$, $x \leq^P y$ and $y \leq^P z$ imply $x \leq^P z$: indeed, by definition, we have $(x \to y^-)^- = x$ and $(y \to z^-)^- = y$, hence $x \to y^- = x^-$ and $y \to z^- = y^-$, by (DN); then, $(x \to z^-)^- \stackrel{(Neg3)}{=} (z \to x^-)^-$ $= (z \to (x \to y^-))^- \stackrel{(Ex)}{=} (x \to (z \to y^-))^- \stackrel{(Neg3)}{=} (x \to (y \to z^-))^-$ $= (x \to y^-)^- = x$, i.e. $x \leq^P z$.

$0 \leq^P x \stackrel{def.}{\Longleftrightarrow} (0 \to x^-)^- = 0$, that is true by (Fi) and (Neg0-1).

$x \leq^P 1 \stackrel{def.}{\Longleftrightarrow} (x \to 1^-)^- = x$, that is true by (Neg1-0), (Neg), (DN). □

**Proposition 3.2.5** *Let $\mathcal{A}^L$ be an involutive left-MEL algebra. If (i-G)=(DN6) holds, then $\leq^P$ is an order.*

**Proof.** By above Proposition 3.2.4, it remains to prove the reflexivity of $\leq^P$.

(P-Re): for all $x \in A$, $x \leq^P x \stackrel{def.}{\Longleftrightarrow} (x \to x^-)^- = x$, that is true; indeed, in (DN6) $(x^- \to x = x)$, take $X := x^-$, to obtain $x^= \to x^- = x^-$, hence, by (DN), $x \to x^- = x^-$; then, $(x \to x^-)^- = x^= \stackrel{(DN)}{=} x$. □

## 3.3 Involutive BCK algebras and lattices

Denote by $\mathbf{BCK}_{(DN)}$ the class of all involutive left-BCK algebras (i.e. of all bounded left-BCK algebras verifying condition **(DN)**) and by $\mathbf{BCK\text{-}L}_{(DN)}$ the class of all involutive left-BCK lattices.

**Lemma 3.3.1** *(see [126]) Let $\mathcal{A}^L = (A^L, \to, 0, 1)$ be an involutive left-BCK algebra. Then, (DN4'), (DN2), (DN3) hold.*

Note that, by (C00), (Neg) + (DN) $\Longrightarrow$ (DN0), hence (DN0) holds too.

**Theorem 3.3.2** *[95] Let $\mathcal{A}^L = (A^L, \leq, \to, 0, 1)$ be an involutive left-BCK algebra. Then, for all $x, y \in A^L$,*

(3.6) $$x \to y = (x \cdot y^-)^-,$$

*where $\cdot$ is defined by (3.1).*

**Proof.** Since $x \cdot y = (x \to y^-)^- = (y \to x^-)^-$, by (3.1), and $(x^-)^- = x$, we get: $(x \cdot y^-)^- = ((x \to (y^-)^-)^-)^- = x \to y$. □

**Corollary 3.3.3** *[95] Let $\mathcal{A}^L$ be an involutive left-BCK algebra. Then, for all $x, y, z \in A^L$:*

$$x \cdot y \leq z \Leftrightarrow x \leq y \to z.$$

**Proof.** $x \cdot y \leq z \overset{(3.1)}{\Leftrightarrow} (x \to y^-)^- \leq z \overset{(DN4')}{\Leftrightarrow} z^- \leq x \to y^- \overset{(Eq\#)}{\Leftrightarrow} x \leq z^- \to y^- \overset{(DN2)}{=} (y^-)^- \to (z^-)^- = y \to z$. □

Consequently, we get the following important result:

**Theorem 3.3.4** *[95] Let $\mathcal{A}^L = (A^L, \leq, \to, 0, 1)$ be an involutive left-BCK algebra. Then $\mathcal{A}^L$ is with condition (P') and $x \odot y = x \cdot y$, for all $x, y \in A^L$; we have*

$$x \odot y \overset{notation}{=} \min\{z \mid x \leq y \to z\} = x \cdot y = (x \to y^-)^-.$$

**Proof.** By Corollary 3.3.3 and Proposition 3.1.17, $\mathcal{A}$ is with condition (P') and $x \odot y \overset{notation}{=} \min\{z \mid x \leq y \to z\} = x \cdot y$; hence, $x \odot y = x \cdot y = (x \to y^-)^-$, by (3.1). □

**Theorem 3.3.5** *(See Corollary 2.2.27)*
*Let $\mathcal{A}^L = (A^L, \leq, \to, 0, 1)$ be an involutive left-BCK algebra. Then, the binary relation $\leq^W$ is an order and $0 \leq^W x \leq^W 1$, for any $x \in A^L$.*

**Proof.** By Corollary 2.2.27, it remains to prove that $0 \leq^W x$. Indeed, $0 \leq^W x \Longleftrightarrow (x \to 0) \to 0 = x$, which is true by (DN0). □

**Proposition 3.3.6** *(see [126]) Let $\mathcal{A}^L$ be a bounded, $\vee$-commutative left-BCK algebra. Then, $\mathcal{A}^L$ verifies condition* **(DN)**.

**Proposition 3.3.7** *[126] Let $\mathcal{A}^L = (A^L, \to, 0, 1)$ be a bounded, $\vee$-commutative left-BCK algebra. Consider, for all $x, y \in A^L$:*
$x \vee y = x \vee^W y \overset{def.}{=} (y \to x) \to x$ *and* $x \wedge y = x \wedge^W y \overset{def.}{=} (x^- \vee y^-)^-$.
*Then, $\mathcal{A}^L$ is a bounded lattice with respect to $\vee$, $\wedge$ (under $x \leq y \Longleftrightarrow x \to y = 1$).*

**Remark 3.3.8** *In an involutive left-BCK algebra $\mathcal{A}^L = (A^L, \to, 0, 1)$:*
*- the initial binary relation, $\leq$ ($x \leq y \overset{def.}{\Longleftrightarrow} x \to y = 1$) ($x \leq y \Longleftrightarrow x \leq^B y$), is an* **order** *(since (Re), (An), (Tr) hold); it can be a lattice order, but we do not know when, in general, excepting the case of Wajsberg algebras;*
*- the binary relation $\leq^W$ ($x \leq^W y \overset{def.}{\Longleftrightarrow} x \vee^W y = y \overset{def.}{\Longleftrightarrow} (y \to x) \to x = y$) is an* **order**, *by Theorem 3.3.5, but not a lattice order in general, with respect to $\vee^W$, $\wedge^W$;*
*- the binary relation $\leq^P$ ($x \leq^P y \overset{def.}{\Longleftrightarrow} (x \to y^-)^- = x$) is only* **antisymmetric and transitive**, *by Proposition 3.2.4.*

### 3.3.1 Wajsberg algebras

Recall from [56] the definition of Wajsberg algebras.

**Definition 3.3.9** (Definition 1) (the dual case is omitted)
An algebra $(A, \to, ^-, 1)$ of type $(2, 1, 0)$ is a *left-Wajsberg algebra*, if the properties (M), (BB), (DN1), ($\vee$-comm) hold.

Denote by **W** the class of all left-Wajsberg algebras.

Consider the following properties:

(W1) $((x \rightarrow y) \rightarrow y) \rightarrow x = y \rightarrow x$,
(W2) $((x \rightarrow y) \rightarrow z) \rightarrow (y \rightarrow z) = 1$,
(W3) $(x^{-} \rightarrow 1^{-}) \rightarrow x = 1$,
(W4) $((x \rightarrow y) \rightarrow z) \rightarrow ((y^{-} \rightarrow x^{-}) \rightarrow z) = 1$,
(W5) $1^{-} \rightarrow 1^{=} = 1^{=}$,
(W6) $1^{=} \rightarrow 1^{-} = 1^{-}$,
(W7) $1^{-} \rightarrow x = 1$,
(W8) $x \rightarrow 1^{=} = 1$,
(W9) $1^{=} = 1$;
(W10) $x^{=} \rightarrow x = 1$.

**Proposition 3.3.10** *We have (by Prover9):*
*(BW1) ($\vee$-comm) + (K) + (M) $\Longrightarrow$ (W1),*
*(AW2) (BB) + (K) + (M) $\Longrightarrow$ (W2),*
*(CW3) (DN1) + (M) $\Longrightarrow$ (W3),*
*(CW4) (DN1) + (BB) + (M) $\Longrightarrow$ (W4);*
*(CW5) (W3) + (W1) + (M) $\Longrightarrow$ (W5),*
*(CW6) (W5) + (W1) + (Re) + (M) $\Longrightarrow$ (W6),*
*(CW7) (W3) + (W2) + (M) $\Longrightarrow$ (W7),*
*(CW8) (W6) + (W2) + ($\vee$-comm) + (W7) + (M) $\Longrightarrow$ (W8),*
*(CW9) (W8) + (W5) $\Longrightarrow$ (W9),*
*(CW10) (W3) + (W4) + (W9) + (M) $\Longrightarrow$ (W10),*
*(CW11) (W10) + (DN1) + (M) $\Longrightarrow$ (Neg4),*
*(CW12) (W10) + (Neg4) + (An) $\Longrightarrow$ (DN).*

**Proof.** (BW1): By ($\vee$-comm), $(x \rightarrow y) \rightarrow y = (y \rightarrow x) \rightarrow x$, hence:
$((x \rightarrow y) \rightarrow y) \rightarrow x = ((y \rightarrow x) \rightarrow x) \rightarrow x \overset{(\vee-comm)}{=} (x \rightarrow (y \rightarrow x)) \rightarrow (y \rightarrow x)$
$\overset{(K)}{=} 1 \rightarrow (y \rightarrow x) \overset{(M)}{=} y \rightarrow x$. Thus, (W1) holds.

(AW2): In (BB) ($(x \rightarrow y) \rightarrow ((y \rightarrow z) \rightarrow (x \rightarrow z)) = 1$), take $Y := y \rightarrow x$, to obtain:
$(x \rightarrow (y \rightarrow x)) \rightarrow (((y \rightarrow x) \rightarrow z) \rightarrow (x \rightarrow z)) = 1$, hence, by (K),
$1 \rightarrow (((y \rightarrow x) \rightarrow z) \rightarrow (x \rightarrow z)) = 1$, hence, by (M),
$((y \rightarrow x) \rightarrow z) \rightarrow (x \rightarrow z) = 1$, i.e. (W2) holds.

(CW3): In (DN1) ($(x^{-} \rightarrow y^{-}) \rightarrow (y \rightarrow x) = 1$), take $y := 1$, to obtain:
$(x^{-} \rightarrow 1^{-}) \rightarrow (1 \rightarrow x) \overset{(M)}{=} (x^{-} \rightarrow 1^{-}) \rightarrow x = 1$. Thus, (W3) holds.

(CW4): In (BB) ($(x \rightarrow y) \rightarrow ((y \rightarrow z) \rightarrow (x \rightarrow z)) = 1$), take $X := y^{-} \rightarrow x^{-}$, $Y := x \rightarrow y$, to obtain:
$((y^{-} \rightarrow x^{-}) \rightarrow (x \rightarrow y)) \rightarrow (((x \rightarrow y) \rightarrow z) \rightarrow ((y^{-} \rightarrow x^{-}) \rightarrow z)) = 1$, hence, by (DN1),
$1 \rightarrow (((x \rightarrow y) \rightarrow z) \rightarrow ((y^{-} \rightarrow x^{-}) \rightarrow z)) = 1$, hence, by (M),
$((x \rightarrow y) \rightarrow z) \rightarrow ((y^{-} \rightarrow x^{-}) \rightarrow z) = 1$, i.e. (W4) holds.

(CW5): In (W1) ($((x \rightarrow y) \rightarrow y) \rightarrow x = y \rightarrow x$), take $X := 1^{=}$, $Y := 1^{-}$, to obtain:

$((1^= \to 1^-) \to 1^-) \to 1^= = 1^- \to 1^=$, hence, by (W3),
$1 \to 1^= = 1^- \to 1^=$, hence, by (M),
$1^= = 1^- \to 1^=$, i.e. (W5) holds.

(CW6): In (W1) ($((x \to y) \to y) \to x = y \to x$), take $X := 1^-$, $Y := 1^=$, to obtain:
$((1^- \to 1^=) \to 1^=) \to 1^- = 1^= \to 1^-$, hence, by (W5),
$(1^= \to 1^=) \to 1^- = 1^= \to 1^-$, hence, by (Re),
$1 \to 1^- = 1^= \to 1^-$, hence, by (M),
$1^- = 1^= \to 1^-$, i.e. (W6) holds.

(CW7): In (W2) ($((x \to y) \to z) \to (y \to z) = 1$), take $X := z^-$, $Y := 1^-$, to obtain:
$((z^- \to 1^-) \to z) \to (1^- \to z) = 1$, hence, by (W3),
$1 \to (1^- \to z) = 1$, hence, by (M),
$1^- \to z = 1$, i.e. (W7) holds.

(CW8): In (W2) ($((x \to y) \to z) \to (y \to z) = 1$), take $Y := 1^=$, $Z := 1^-$, to obtain:
$((x \to 1^=) \to 1^-) \to (1^= \to 1^-) = 1$, hence, by (W6),
$((x \to 1^=) \to 1^-) \to 1^- = 1$, hence, by ($\vee$-comm),
$(1^- \to (x \to 1^=)) \to (x \to 1^=) = 1$, hence, by (W7),
$1 \to (x \to 1^=) = 1$, hence, by (M),
$x \to 1^= = 1$, i.e. (W8) holds.

(CW9): By (W5), $1^- \to 1^= = 1^=$, hence, by (W8), $1 = 1^=$, i.e. (W9) holds.

(CW10): In (W4) ($((x \to y) \to z) \to ((y^- \to x^-) \to z) = 1$), take $X := z^-$, $Y := 1^-$, to obtain:
$((z^- \to 1^-) \to z) \to ((1^= \to z^=) \to z) = 1$, hence, by (W3),
$1 \to ((1^= \to z^=) \to z) = 1$, hence, by (W9) and (M),
$z^= \to z = 1$, i.e. (W10) holds.

(CW11): In (DN1) ($(x^- \to y^-) \to (y \to x) = 1$), take $X := y^=$, to obtain:
$((y^-)^= \to y^-) \to (y \to y^=) = 1$, hence, by (W10),
$1 \to (y \to y^=) = 1$, hence, by (M),
$y \to y^= = 1$, i.e. (Neg4) holds.

(CW12): $x^= \to x = 1$ and $x \to x^= = 1$ give, by (An), $x^= = x$, i.e. (DN) holds.
□

**Theorem 3.3.11** *Wajsberg algebras are involutive.*

**Proof. (By *Prover9* program)**

We must prove that (DN) holds. Indeed,
by (B32), ($\vee$-comm) + (M) + (BB) $\Longrightarrow$ (La),
by (B29), ($\vee$-comm) + (La) + (M) $\Longrightarrow$ (Re),
by (B12), ($\vee$-comm) + (M) $\Longrightarrow$ (An,
by (A21"), (M) + (BB) + (An) $\Longrightarrow$ (Ex),
by (A8), (Re) + (La) + (Ex) $\Longrightarrow$ (K),
by (BW1), ($\vee$-comm) + (K) + (M) $\Longrightarrow$ (W1),
by (AW2), (BB) + (K) + (M) $\Longrightarrow$ (W2),
by (CW3), (DN1) + (M) $\Longrightarrow$ (W3), ...,

by (CW12), (W10) + (Neg4) + (An) $\Longrightarrow$ (DN). □

Recall that bounded $\vee$-commutative BCK algebras are definitionally equivalent to MV algebras [159] and that MV algebras are definitionally equivalent to Wajsberg algebras [56]. Hence we have:

**Theorem 3.3.12** *Bounded $\vee$-commutative BCK algebras are definitionally equivalent to Wajsberg algebras.*

**A direct proof** of above theorem is the following.

(1): Let $\mathcal{A}^L = (A^L, \to, 0, 1)$ be a bounded $\vee$-commutative left-BCK algebra, i.e. (BB), (D), (Re), (La), (An), (Fi), ($\vee$-comm) hold (Definition 1). We have to prove that (M) and (DN1) hold. Indeed,
by (A21'), (BB) + (D) + (La) + (An) $\Longrightarrow$ (Ex) and
by (A5), (RE) + (Ex) + (An) $\Longrightarrow$ (M), hence (M) holds.
Define $x^- = x \to 0$; hence, (Neg) holds. Then,
by (CB1), ($\vee$-comm) + (Neg) + (Fi) + (M) $\Longrightarrow$ (DN),
by (C7), (DN) + (Neg) + (Ex) $\Longrightarrow$ (DN2) and
by (C4'), (DN2) + (Re) $\Longrightarrow$ (DN1),
hence (DN1) holds. Thus, $(A^L, \to, ^-, 1)$ is a left-Wajsberg algebra.

(1'): Let $\mathcal{A}^L = (A^L, \to, ^-, 1)$ be a left-Wajsberg algebra, i.e. (M), (BB), (DN1), ($\vee$-comm) hold. We have to prove that (D), (La), (Re), (An), (Fi) hold. Indeed,
by (A18'), (M) + (BB) $\Longrightarrow$ (D),
by (B32), ($\vee$-comm) + (M) + (BB) $\Longrightarrow$ (La),
by (B29), ($\vee$-comm) + (La) + (M) $\Longrightarrow$ (Re),
by (B12), ($\vee$-comm) + (M) $\Longrightarrow$ (An), hence (D), (La), (Re), (An) hold.

It remains to prove that (Fi) ($0 \to x = 1$, for all $x \in A^L$) holds, where $0 \overset{def.}{=} 1^-$. First, we prove that $0^- = 1$, i.e. that $1^= = 1$, i.e. that (W9) holds. Indeed,
by (B29), ($\vee$-comm) + (La) + (M) $\Longrightarrow$ (Re),
by (A8), (Re) + (La) + (Ex) $\Longrightarrow$ (K),
by (BW1), ($\vee$-comm) + (K) + (M) $\Longrightarrow$ (W1),
by (AW2), (BB) + (K) + (M) $\Longrightarrow$ (W2),
by (CW3), (DN1) + (M) $\Longrightarrow$ (W3),
by (CW4), (DN1) + (BB) + (M) $\Longrightarrow$ (W4);
by (CW5), (W3) + (W1) + (M) $\Longrightarrow$ (W5),
by (CW6), (W5) + (W1) + (Re) + (M) $\Longrightarrow$ (W6),
by (CW7), (W3) + (W2) + (M) $\Longrightarrow$ (W7),
by (CW8), (W6) + (W2) + ($\vee$-comm) + (W7) + (M) $\Longrightarrow$ (W8),
by (CW9), (W8) + (W5) $\Longrightarrow$ (W9),
thus (W9) holds, hence $0^- = 1$, i.e. (Neg0-1) holds. Finally,
by (C8), (Neg0-1) + (DN1) + (La) + (M) $\Longrightarrow$ (Fi),
thua (Fi) holds. Thus, $(A^L, \to, 0, 1)$ is a bounded $\vee$-commutative left-BCK algebra.

(2): The two transformations are mutually inverse, obviously. □

We have seen in Proposition 2.1.15 the connection between BCI lattices and BCK lattices. The following theorem shows the deeper connection existing between commutative $l$-implicative-groups and BCK lattices:

**Theorem 3.3.13** *(see ([116], Theorem 14.1.1), in the non-commutative case)*
*Let $\mathcal{G} = (G, \vee, \wedge, \to, 1)$ be a commutative l-i-group (left-one or right-one). We have, for all $x, y \in G$: $x^{-1} \stackrel{def.}{=} x \to 1$ and $x \cdot y \stackrel{def.}{=} (x \to y^{-1})^{-1}$.*

*(1) Define, for all $x, y \in G^- \stackrel{def.}{=} \{z \in G \mid z \leq 1\}$:*

$$x \to^L y \stackrel{def.}{=} (x \to y) \wedge 1, \quad x \odot y \stackrel{def.}{=} x \cdot y.$$

*Then, $\mathcal{G}^- = (G^-, \wedge, \vee, \to^L, \mathbf{1} = 1)$ is a distributive BCK(P') lattice (with the product $\odot$) verifying: for all $x, y, z \in G^-$,*

*(CC)* $\quad x \vee y = (x \to^L y) \to^L y,$

*(HH)* $\quad x \to^L y = (x \odot z) \to^L (y \odot z).$

*(1') Dually, define for all $x, y \in G^+ \stackrel{def.}{=} \{z \in G \mid z \geq 1\}$:*

$$x \to^R y \stackrel{def.}{=} (x \to y) \vee 1, \quad x \oplus y \stackrel{def.}{=} x \cdot y.$$

*Then, $\mathcal{G}^+ = (G^+, \vee, \wedge, \to^R, \mathbf{0} = 1)$ is a distributive BCK(S') lattice (with the sum $\oplus$) verifying: for all $x, y, z \in G^+$,*

*($CC^R$)* $\quad x \wedge y = (x \to^R y) \to^R y,$

*($HH^R$)* $\quad x \to^R y = (x \oplus z) \to^R (y \oplus z).$

**Remarks 3.3.14** *(The dual case is omitted)*
*(i) Since $\mathcal{G}^-$ is a lattice, then $x \vee y = y \vee x$, for all $x, y \in G^-$, hence the property ($\vee$-comm) holds.*

*(ii) By ([116], Remarks 14.1.3) and (i), $(G^-, \to^L, \mathbf{1} = 1)$ is a left-L algebra (Section 3.5), hence, $\mathcal{G}^-$ is a distributive BCKL(P') lattice verifying ($\vee$-comm).*

*(iii) By ([116], Remark 14.1.2), $\mathcal{G}^-$ verifies (prel) and (div).*

Let us "bound" the BCK(P') lattice $\mathcal{G}^-$ and the BCK(S') lattice $\mathcal{G}^+$ from Theorem 3.3.13 with an "internal" element. We obtain the equivalent of known results:

**Corollary 3.3.15** *(see [116], Corollary 14.1.7), in the non-commutative case)*
*(1) Let $\mathcal{G}^- = (G^-, \wedge, \vee, \to^L, \ \mathbf{1} = 1)$ be the BCK(P') lattice (with the product $\odot = \cdot$) from Theorem 3.3.13 (1). Let us "bound" this algebra in the following way:*

*Let us take $u' < 1$ from $G^-$ and consider the interval $[u', 1] = \{x \in G^- \mid u' \leq x \leq 1\}$. Then, the algebra $\mathcal{G}_1^- = ([u', 1], \wedge, \vee, \to^L, \mathbf{0} = u', \mathbf{1} = 1)$ is a bounded distributive BCK(P') lattice (with the product $x \odot^L y \stackrel{def.}{=} (x \odot y) \vee u' = (x \cdot y) \vee u'$) with property (CC), i.e. is an equivalent definition of the* **left-Wajsberg algebra** *$\mathcal{G}_{1'}^- = ([u', 1], \to^L, {}^{-^L}, \mathbf{1} = 1)$, where: $x^{-^L} \stackrel{def.}{=} x \to^L u' = u' \cdot x^{-1}$.*

*(1') Dually, let $\mathcal{G}^+ = (G^+, \vee, \wedge, \to^R, \mathbf{0} = 1)$ be the BCK(S') lattice from Theorem 3.3.13 (1'). Let us "bound" this algebra in the following way:*

*Let us take $u > 1$ from $G^+$ and consider the interval $[1, u] = \{x \in G^+ \mid 1 \leq x \leq u\}$. Then, the algebra $\mathcal{G}_1^+ = ([1, u], \vee, \wedge, \to^R, \mathbf{0} = 1, \mathbf{1} = u)$ is a bounded BCK(S') lattice (with the sum $x \oplus^R y \stackrel{def.}{=} (x \oplus y) \wedge u = (x \cdot y) \wedge u$) with property ($CC^R$), i.e. is an equivalent definition of the* **right-Wajsberg algebra** *$\mathcal{G}_{1'}^+ = ([1, u], \to^R, {}^{-^R}, \mathbf{0} = 1)$, where: $x^{-^R} \stackrel{def.}{=} x \to^R u = u \cdot x^{-1}$.*

The ideea of the following two results came from Remark 6.5.2 concerning the definition of MV algebras.

**Corollary 3.3.16** *Bounded* $\vee$*-commutative MEL algebras coincide with bounded* $\vee$*-commutative BCK algebras.*

**Proof.** By Theorem 2.2.8. □

**Theorem 3.3.17** *Bounded* $\vee$*-commutative MEL algebras are definitionally equivalent to Wajsberg algebras.*

**Proof 1.** By Theorem 3.3.12 and Corollary 3.3.16. □

By above Theorem 3.3.17, we obtain a second equivalent definition of Wajsberg algebras.

**Definition 3.3.18** (Definition 2) (The dual case is omitted)
A *left-Wajsberg algebra* is a bounded $\vee$-commutative left-MEL algebra, i.e. an algebra $(A^L, \rightarrow, 0, 1)$ of type $(2, 0, 0)$ verifying: (M), (Ex), (La), (Fi), ($\vee$-comm).

**Proposition 3.3.19** *(The duall one is omitted)*
*Let* $\mathcal{A}^L = (A^L, \rightarrow, ^-, 1)$ *be a left-Wajsberg algebra. We have, for all* $x, y \in A^L$*:*
*(1)* $x \leq y \Longleftrightarrow x \leq^W y$,
*(2)* $\leq$ *and* $\leq^W$ *are both lattice orders with respect to* $\vee = \vee^W = \vee^B$, $x \wedge y = (x^- \vee y^-)^-$.

**Proof.** (1): follows by Proposition 2.2.19.
(2): follows by Proposition 3.3.7 and Theorem 3.3.12. □

**Remark 3.3.20** *In a left-Wajsberg algebra* $(A^L, \rightarrow, ^-, 1)$*:*
*- the initial binary relation* $\leq$ *(* $\Longleftrightarrow \leq^B$*) and the binary relation* $\leq^W$ *are both* **lattice orders** *w.r. to* $\vee = \vee^W = \vee^B$ *and are equivalent:* $x \leq y \Longleftrightarrow x \leq^W y$*, by Proposition 3.3.19;*
*- the binary relation* $\leq^P$ *is only* **antisymmetric and transitive***, by Proposition 3.2.4.*

### 3.3.2 Weak-$R_0$ algebras and $R_0$ algebras

Weak-$R_0$ algebras and $R_0$ algebras were introduced in 1997 by G.J. Wang [212] as follows:

**Definitions 3.3.21** [168]
(1) A *(left-) weak-*$R_0$ *algebra* is an algebra $\mathcal{M}^L = (M, \wedge, \vee, \rightarrow, ^-, 0, 1)$ of type $(2, 2, 2, 1, 0, 0)$, such that:
(i) $(M, \wedge, \vee, 0, 1)$ is a bounded lattice, $\leq^O$ being the lattice order relation, i.e. $x \leq^O y \Longleftrightarrow x \wedge y = x \Longleftrightarrow x \vee y = y$,
(ii) $^-$ is an order reversing involution with respect to $\leq^O$, i.e. (O-Neg2'), (DN) hold,

(iii) the following conditions hold: for all $x, y, z \in M$,
$(R_1) = (\text{DN2})$ $\quad x^- \to y^- = y \to x$,
$(R_2) = (\text{M})$ $\quad 1 \to x = x$,
$(R_3) = (\text{O-B'})$ $\quad y \to z \leq^O (x \to y) \to (x \to z)$,
$(R_4) = (\text{Ex})$ $\quad x \to (y \to z) = y \to (x \to z)$,
$(R_5) = (\text{prel}_{\to\vee})$ $\quad x \to (y \vee z) = (x \to y) \vee (x \to z)$.

(2) An *(left-) $R_0$ algebra* is a weak-$R_0$ algebra verifying the additional condition:
$(R_6)$ $(x \to y) \vee ((x \to y) \to (x^- \vee y)) = 1$.

$\mathrm{R}_0$ algebras play versus NM algebras the same role as Wajsberg algebras play versus MV algebras. Recall [168] that the IMTL algebras, introduced in 2001 by Esteva and Godo [50], are termwise equivalent to weak-$R_0$ algebras and that NM algebras, introduced also in 2001 by Esteva and Godo [50], are termwise equivalent to $R_0$ algebras.

The property $(R_5) = (\text{prel}_{\to\vee})$ is equivalent to:
(prel) (prelinearity) $((x \to y) \vee (y \to x) = 1)$ (see [104], Theorem 3.1.3).

The property $(R_6)$ is equivalent to:
(WNM) (Weak Nilpotent Minimum) $((x \odot y)^- \vee [(x \wedge y) \to (x \odot y)] = 1)$.

Denote by ${}_{(R_6)}\mathbf{W}$ the class of Wajsberg algebras verifying $(R_6)$; then, ${}_{(R_6)}\mathbf{W}$ is termwise equivalent to ${}_{(WNM)}\mathbf{MV}$, the class of MV algebras verifying (WNM) (See Part II).

We present only the following result.

**Proposition 3.3.22** *Let $\mathcal{M}^L = (M, \wedge, \vee, \to, ^-, 0, 1)$ be a (left-) weak-$R_0$ algebra. Then, it is an involutive left-BCK lattice.*

**Proof. (By *Prover9* program)**

(Neg1-0): First, $0 \leq^O x \Longleftrightarrow 0 \wedge x = 0$ (a).
Then, $x \leq^O 1 \overset{(O-Neg2')}{\Longrightarrow} 1^- \leq^O x^- \Longleftrightarrow 1^- \wedge x^- = 1^-$; for $X := x^-$, $1^- \wedge x^= = 1^- \overset{(DN)}{\Longleftrightarrow} 1^- \wedge x = 1^-$ (b).
From (a), for $x = 1^-$, we obtain $0 \wedge 1^- = 0$ and from (b), for $x := 0$, we obtain $1^- \wedge 0 = 1^-$, hence, $0 = 0 \wedge 1^- = 1^- \wedge 0 = 1^-$. Thus, (Neg1-0) holds.

(Neg): In (DN2), $x^- \to y^- = y \to x$, take $y := 1$ to obtain:
$x^- \to 1^- = 1 \to x \overset{(M)}{=} x$.
Then, for $X := x^-$, $x^= \to 1^- = x^-$, hence, by (DN),
$x^- = x \to 1^- \overset{(Neg1-0)}{=} x \to 0$. Thus, (Neg) holds.

(DN0): By (C00), (Neg) + (DN) $\Longrightarrow$ (DN0).

Now, we prove

$$0 \to 0 = 1. \tag{3.7}$$

Indeed, in (DN0) $(x \to 0) \to 0 = x$, take $x = 1$ to obtain $(1 \to 0) \to 0 = 1$, hence, by (M), $0 \to 0 = 1$; thus (3.7) holds.

(Re): From (O-B') $(x \to y) \wedge ((z \to x) \to (z \to y)) = x \to y$, by (Ex), we obtain:

$$(x \to y) \wedge (z \to ((z \to x) \to y)) = x \to y. \tag{3.8}$$

Now, in (3.8), take $X := 0$, $Y := 0$, $Z := x$, to obtain:
$(0 \rightarrow 0) \wedge (x \rightarrow ((x \rightarrow 0) \rightarrow 0)) = 0 \rightarrow 0$,
hence, by (3.7) and (DN0), we obtain $1 \wedge (x \rightarrow x) = 1$, hence, by (i), $x \rightarrow x = 1$; thus, (Re) holds.

Now, we prove

$$x \rightarrow (y \vee x) = 1. \tag{3.9}$$

Indeed, in $(R_5)$=(prel$_{\rightarrow\vee}$) $(x \rightarrow y) \vee (x \rightarrow z) = x \rightarrow (y \vee z)$, take $Z := x$ to obtain:
$(x \rightarrow y) \vee (x \rightarrow x) = x \rightarrow (y \vee x)$, hence, by (Re),
$(x \rightarrow y) \vee 1 = x \rightarrow (y \vee x)$, hence, by (i), $1 = x \rightarrow (y \vee x)$, i.e. (3.9) holds.

Now we prove that

$$x \leq^O y \Longleftrightarrow x \rightarrow y = 1 \ (\Longleftrightarrow x \leq y). \tag{3.10}$$

Indeed, If $x \leq^O y$, i.e. $x \vee y = y$, then $x \rightarrow y = x \rightarrow (x \vee y) \overset{(3.9)}{=} 1$.
Conversely, suppose (c) $C1 \rightarrow C2 = 1$; we must prove that $C1 \leq^O C2$.
In (3.8), take $X := C2$, $Y := x$, $Z := C1$, to obtain:
$(C2 \rightarrow x) \wedge (C1 \rightarrow ((C1 \rightarrow C2) \rightarrow x)) = C2 \rightarrow x$;
then, by hypothesis (c) and (M), we obtain:
$(C2 \rightarrow x) \wedge (C1 \rightarrow x) = C2 \rightarrow x$, i.e. $C2 \rightarrow x \leq^O C1 \rightarrow x$;
hence, for $x := 0$, we obtain, by (Neg), $C2^- \leq^O C1^-$;
then, by (O-Neg2'), we obtain $C1^= \leq^O C2^=$, hence, by (DN), $C1 \leq^O C2$.

(La): By (3.10), $x \rightarrow 1 = 1 \Longleftrightarrow x \leq^O 1$, that is true, by (i); thus, $x \rightarrow 1 = 1$, i.e. (La) holds.

Finally, by (3.10), (O-B') is (B') and $\leq$ is an order, hence (An) holds;
by (A3'), (Ex) + (Re) $\Longrightarrow$ (C);
by (A9), (M) + (La) + (B) $\Longrightarrow$ (K);
thus, since (B), (C), (K), (An) hold, $\mathcal{M}^L$ is an involutive left-BCK lattice. □

### 3.3.3 Implicative-Boolean algebras

There are many term-equivalent definitions of Boolean algebras. We shall use here the term-equivalent definition introduced in 2009 [105] and presented also in [63], motivated by the axioms system of the classical propositional logic (of truth).

**Definition 3.3.23** (Definition 1) (The dual case is omitted)
A *left-implicative-Boolean algebra*, or a *left-i-Boolean algebra* for short, is an algebra $\mathcal{B}^L = (B^L, \rightarrow, ^-, 1)$ of type $(2, 1, 0)$ verifying: (K), (pimpl-1), (DN1), (An).

Denote by **i-Boole** the class of all left-i-Boolean algebras.
We recall their term-equivalence (d.e.) with the Boolean algebras from [63]:

**Theorem 3.3.24** [63]
*(1) Let $\mathcal{B}^L = (B^L, \rightarrow, ^-, 1)$ be a left-i-Boolean algebra.*
*Define* $\Phi(\mathcal{B}^L) \overset{def.}{=} (B^L, \wedge, \vee, ^-, 0, 1)$ *as follows: for every* $x, y \in B^L$,
$x \wedge y \overset{def.}{=} (x \rightarrow y^-)^-$, $x \vee y \overset{def.}{=} (x^- \wedge y^-)^- = x^- \rightarrow y$, $0 \overset{def.}{=} 1^-$.

*Then $\Phi(\mathcal{B}^L)$ is a left-Boolean algebra.*

*(1') Conversely, let $\mathcal{B}^L = (B^L, \wedge, \vee, ^-, 0, 1)$ be a left-Boolean algebra. Define $\Psi(\mathcal{B}^L) \stackrel{def.}{=} (B^L, \rightarrow, ^-, 1)$ as follows: for every $x, y \in B^L$, $x \rightarrow y \stackrel{def.}{=} (x \wedge y^-)^- = x^- \vee y$.*

*Then $\Psi(\mathcal{B}^L)$ is a left-i-Boolean algebra.*

*(2) The mappings $\Phi$ and $\Psi$ are mutually inverse.*

**Theorem 3.3.25** *(See [127], Theorem 12)*

*Bounded implicative BCK algebras are definitionally equivalent to Boolean algebras.*

By this Theorem and above Theorem 3.3.24, it follows that:

**Corollary 3.3.26** *Bounded implicative BCK algebras are definitionally equivalent to i-Boolean algebras.*

By this Corollary 3.3.26 and the Corollaries 2.2.15, 3.1.12, we obtain:

**Theorem 3.3.27** *Bounded implicative aBE** algebras are definitionally equivalent to i-Boolean algebras.*

**Theorem 3.3.28** *Bounded positive-implicative BCK algebras with involutive negation are definitionally equivalent to i-Boolean algebras.*

It is known that any bounded Hilbert algebra with involutive negation is a Boolean algebra [18]. Moreover, we have the following result.

**Corollary 3.3.29** *Bounded Hilbert algebras with involutive negation are definitionally equivalent to i-Boolean algebras.*

### 3.3.4 The connection between Wajsberg algebras and i-Boolean algebras

We now make the connection between Wajsberg algebras and i-Boolean algebras:

**Theorem 3.3.30** *Wajsberg algebras verifying the property (DN6) = (i-G) are d.e. to i-Boolean algebras.*

**Proof.** (1): Let $\mathcal{A}^L = (A^L, \rightarrow, ^-, 1)$ be a left-Wajsberg algebra verifying (DN6). By Theorem 3.3.12, $\mathcal{A}^L$ is a bounded $\vee$-commutative BCK algebra. Then, by (CB5), (DN) + (DN2) + (DN5) + (K) + (*) + (An) + (DN6) $\Longrightarrow$ (impl), hence $\mathcal{A}^L$ is a bounded implicative left-BCK algebra. Then, by Corollary 3.3.26, $\mathcal{A}^L$ is a left-i-Boolean algebra.

(1'): Let $\mathcal{A}^L = (A^L, \rightarrow, ^-, 1)$ be a left-i-Boolean algebra, i.e. (K), (pimpl-1), (DN1), (An) hold. By Corollary 3.3.26, $(A^L, \rightarrow, 0, 1)$ is a bounded implicative left-BCK algebra, i.e. (BB), (M), (Neg), (impl) hold, where $0 = 1^-$. By Theorem 2.2.23, ($\vee$-comm) holds. By (CB2), (impl) + (Neg) imply (DN6). Thus, $\mathcal{A}^L$ satisfy the properties (M), (BB), ($\vee$-comm), (DN1) and (DN6), i.e. it is a left-Wajsberg algebra (Definition 1) verifying (DN6).

(2): The two transformations are mutually inverse, obviously. □

### 3.3.5 Resuming connections

Hence, we have the connections from Figure 3.1 (see Figure 12.4), where:
(div) (divisibility) $x \wedge y = x \odot (x \to y)$.

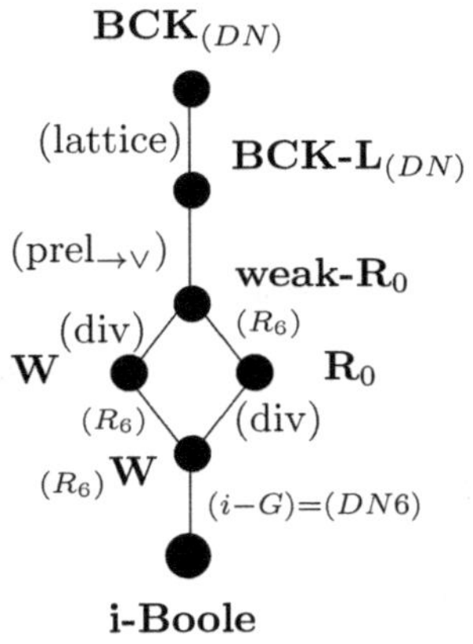


Figure 3.1: The connections between $\mathbf{BCK}_{(DN)}$, $\mathbf{BCK\text{-}L}_{(DN)}$, **weak-**$R_0$, **W**, $R_0$ and **i-Boole**

## 3.4 Implicative-ortholattices: i-OL

We introduce here the notion of implicative-ortholattice as the term-equivalent definition of the ortholattice recalled and studied further in this book.

**Definitions 3.4.1** (Definition 1)

(i) A *left-implicative-ortholattice*, or a *left-i-ortholattice* for short, is an algebra $\mathcal{A}^L = (A^L, \to = \to^L, ^- = ^{-^L}, 1)$ of type $(2, 1, 0)$ such that $x^- = x \to 0$ (i.e. (Neg) holds), where $0 \overset{def.}{=} 1^-$ (i.e. (Neg1-0) holds), and (M), (Ex), (Re), (La), (impl), (DN) hold.

(i') Dually, a *right-implicative-ortholattice*, or a *right-i-ortholattice* for short, is an algebra $\mathcal{A}^R = (A^R, \to^R, ^- = ^{-^R}, 0)$ of type $(2, 1, 0)$ such that $x^- = x \to^R 1$ (i.e. (Neg$^R$) holds), where $1 \overset{def.}{=} 0^-$ (i.e. (Neg0-1) holds), and (M$^R$), (Ex$^R$), (Re$^R$), (F$^R$), (impl$^R$), (DN) hold.

**Remark 3.4.2** *The name 'implicative-ortholattice' means an ortholattice defined equivalently with implication. Note that an implicative-ortholattice is implicative indeed.*

Note that we have immediately the following equivalent definition:

**Definition 3.4.3** (Definition 2) (the dual case is omitted)

A *left-implicative-ortholattice*, or a *left-i-ortholattice* for short, is an algebra $\mathcal{A}^L = (A^L, \to, 0, 1)$ of type $(2, 0, 0)$ such that (M), (Ex), (Re), (Fi), (La), (impl) and (DN) hold, where $x^- \stackrel{def.}{=} x \to 0$ by (dfneg), i.e. it is an *implicative involutive left-BE algebra*.

Denote by **i-OL** the class of all left-i-ortholattices.

**Proposition 3.4.4** *Let $\mathcal{A}^L = (A^L, \to, ^-, 1)$ be a left-i-ortholattice. Then, the following properties also hold: (pi), (Neg0-1), (Neg3), (DN3), (DN2), (i-G)=(DN6), (Neg6).*

**Proof.** By (BIM1), (pi) holds. By (C0), (Neg0-1) holds. (Ex) implies (Ex0), then by (CE2), (Neg) + (Ex0) imply (Neg3), hence (Neg3) holds. By (C3), (DN3) holds. By (C2'), (DN2) holds. By (CB2), (DN6) holds. By (C6'), (Neg6) holds. □

**Proposition 3.4.5** *Let $\mathcal{A}^L = (A^L, \to, ^-, 1)$ be a left-i-ortholattice. Consider, for all $x, y \in A^L$, $x \leq^P y \stackrel{def.}{\Longleftrightarrow} x \wedge^P y = x \Longleftrightarrow (x \to y^-)^- = x$. Then, we have:*
*(i) $(A, \leq^P, 0, 1)$ is a bounded poset.*
*(ii) $x \leq^P y \Longrightarrow x \leq y$ (i.e. $x \to y = 1$, by (dfrelR)).*
*(iii) The property (P-**') (if $x \leq^P y$ then $y \to z \leq^P x \to z$) holds.*
*(iv) The property (P-K') ($x \leq^P y \to x$) holds.*

**Proof.** (i): By Propositions 3.2.4, 3.2.5.
(ii): Suppose $x \leq^P y$, i.e. $(x \to y^-)^- = x$. Then, $x \to y = (x \to y^-)^- \to y \stackrel{(DN3)}{=} y^- \to (x \to y^-) \stackrel{(Ex)}{=} x \to (y^- \to y^-) \stackrel{(Re)}{=} x \to 1 \stackrel{(La)}{=} 1$.
(iii): Suppose $x \leq^P y$, i.e. $(x \to y^-)^- = x$; we must prove that $y \to a \leq^P x \to a$, i.e. $(y \to a) \to (x \to a)^- = (y \to a)^-$, by (DN).
Indeed, $x \to a = (x \to y^-)^- \to a \stackrel{(DN3)}{=} a^- \to (x \to y^-)$
$\stackrel{(Ex)}{=} x \to (a^- \to y^-) \stackrel{(DN2)}{=} x \to (y \to a)$.
Then, $(y \to a) \to (x \to a)^- \stackrel{(Neg3)}{=} (x \to a) \to (y \to a)^-$
$= [x \to (y \to a)] \to (y \to a)^- \stackrel{(DN6)}{=} [x \to ((y \to a)^- \to (y \to a))] \to (y \to a)^-$
$\stackrel{(Ex)}{=} [(y \to a)^- \to (x \to (y \to a))] \to (y \to a)^- \stackrel{(impl)}{=} (y \to a)^-$.
(iv): We must prove that $x \leq^P y \to x$, i.e. $x \to (y \to x)^- = x^-$, by (DN).
Indeed, $x \to (y \to x)^- \stackrel{(Neg3)}{=} (y \to x) \to x^- \stackrel{(DN2)}{=} (x^- \to y^-) \to x^- \stackrel{(impl)}{=} x^-$. □

**Theorem 3.4.6** *Let $\mathcal{A}^L = (A^L, \to, ^-, 1)$ be a left-i-ortholattice. Consider, for all $x, y \in A^L$, $x \wedge^P y \stackrel{def.}{=} (x \to y^-)^-$ and $x \vee^P y \stackrel{def.}{=} (x^- \wedge^P y^-)^- = x^- \to y$. Then, $(A^L, \wedge^P, \vee^P, 0, 1)$ is a bounded lattice.*

**Proof.** For all $x, y, z \in A^L$, we have:
(m-Wid): $x \wedge^P x = x \Longleftrightarrow (x \to x^-)^- = x$, that is true by reflexivity of $\leq^P$.
(m-Wcomm): $x \wedge^P y = (x \to y^-)^- \stackrel{(Neg3)}{=} (y \to x^-)^- = y \wedge^P x$.
(m-Wass): $x \wedge^P (y \wedge^P z) = x \wedge^P (y \to z^-)^- = (x \to (y \to z^-)^=)^-$
$\stackrel{(DN)}{=} (x \to (y \to z^-))^- \stackrel{(Ex)}{=} (y \to (x \to z^-))^-$ and also

$(x \wedge^P y) \wedge^P z = z \wedge^P (x \to y^-)^- = (z \to (x \to y^-)^=)^- = (z \to (x \to y^-))^- \stackrel{(Neg3)}{=}$
$(z \to (y \to x^-))^- \stackrel{(Ex)}{=} (y \to (z \to x^-))^- \stackrel{(Neg3)}{=} (y \to (x \to z^-))^-$.

(m-Vid): $x \vee^P x = x^- \to x \stackrel{(DN6)}{=} x$.

(m-Vcomm): $x \vee^P y = x^- \to y \stackrel{(DN3)}{=} y^- \to x = y \vee^P x$.

(m-Vass): $x \vee^P (y \vee^P z) = x \vee^P (y^- \to z) = x^- \to (y^- \to z)$ and also
$(x \vee^P y) \vee^P = z \vee^P (x^- \to y) = z^- \to (x^- \to y) \stackrel{(Ex)}{=} x^- \to (z^- \to y)$
$\stackrel{(DN3)}{=} x^- \to (y^- \to z)$.

(m-Wabs): $x \wedge^P (x \vee^P y) = x \wedge^P (x^- \to y) = (x \to (x^- \to y)^-)^-$
$\stackrel{(Neg3)}{=} ((x^- \to y) \to x^-)^- \stackrel{(impl)}{=} (x^-)^- = x$.

(m-Vabs): $x \vee^P (x \wedge^P y) = x \vee^P (x \to y^-)^- = x^- \to (x \to y^-)^-$
$\stackrel{(DN2)}{=} (x \to y^-) \to x \stackrel{(impl)}{=} x$. □

Note that $\leq$ is reflexive, by (Re), but it is not antisymmetric or transitive, as the following examples show.

**Examples 3.4.7** (See Examples 9.2.9)

Examples of left-i-ortholattices with six elements:

• **Example 1.**

The algebra $\mathcal{IO}_1 = (A_6 = \{0, a, b, c, d, 1\}, \to_1, ^{-1}, 1)$ with

| $\to_1$ | 0 | a | b | c | d | 1 |
|---|---|---|---|---|---|---|
| 0 | 1 | 1 | 1 | 1 | 1 | 1 |
| a | d | 1 | 1 | 1 | d | 1 |
| b | c | 1 | 1 | c | 1 | 1 |
| c | b | 1 | b | 1 | 1 | 1 |
| d | a | a | 1 | 1 | 1 | 1 |
| 1 | 0 | a | b | c | d | 1 |

and

| $x$ | $x^{-1}$ |
|---|---|
| 0 | 1 |
| a | d |
| b | c |
| c | b |
| d | a |
| 1 | 0 |

is an i-ortholattice with the bounded poset $(A_6, \leq^P, 0, 1)$ represented by the Hasse diagram from Figure 3.2. It does not verify (BB) for $(a, d, b)$, (B) for $(a, b, d)$, (**) for $(a, d, b)$, (*) for $(a, b, d)$, (An) for $(a, b)$, (Tr) for $(a, b, d)$, ($\vee$-comm) for $(a, b)$.

• **Example 2.**

The algebra $\mathcal{IO}_2 = (A_6 = \{0, a, b, c, d, 1\}, \to_2, ^{-2}, 1)$ with

| $\to_2$ | 0 | a | b | c | d | 1 |
|---|---|---|---|---|---|---|
| 0 | 1 | 1 | 1 | 1 | 1 | 1 |
| a | b | 1 | b | 1 | 1 | 1 |
| b | a | a | 1 | 1 | 1 | 1 |
| c | d | 1 | 1 | 1 | d | 1 |
| d | c | 1 | 1 | c | 1 | 1 |
| 1 | 0 | a | b | c | d | 1 |

and

| $x$ | $x^{-2}$ |
|---|---|
| 0 | 1 |
| a | b |
| b | a |
| c | d |
| d | c |
| 1 | 0 |

is an i-ortholattice with the bounded poset $(A_6, \leq^P, 0, 1)$ represented by the same Hasse diagram from Figure 3.2. It does not verify (BB) for $(a, b, c)$, (B) for $(a, c, b)$, (**) for $(a, b, c)$, (*) for $(a, c, b)$, (An) for $(b, c)$, (Tr) for $(a, c, b)$, ($\vee$-comm) for $(b, c)$.

Note that the i-ortholattices $\mathcal{IO}_1$ and $\mathcal{IO}_2$ correspond to the ortholattices $\mathcal{O}_1$ and $\mathcal{O}_2$, respectively, from Examples 9.2.9, by Corollary 17.1.6.

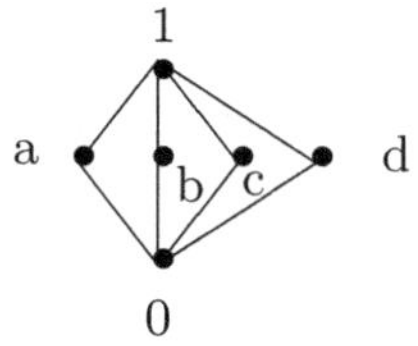


Figure 3.2: An i-ortholattice with 6 elements

**Remark 3.4.8** *The converse of Proposition 3.4.5 (ii) does not hold. For example, in the i-ortholattice $\mathcal{IO}_1$ from Examples 3.4.7, we have $a \leq b$ (i.e. $a \to b = 1$), but $a \not\leq^P b$, since $(a \to b^-)^- = (a \to c)^- = 1^- = 0 \neq a$.*

**Proposition 3.4.9** *Let $\mathcal{A}^L = (A^L, \to, ^-, 1)$ be a left-i-ortholattice.*
*If (@) $((y^- \to x) \to y = x \to y$, i.e. $(x \vee y) \wedge x^- = y \wedge x^-$, by $\Phi$ from Theorem 17.1.1) holds, then we have:*

| | |
|---|---|
| *(H)* | $((x \to y) \to y) \to y = x \to y$, |
| *($\vee$-comm)* | $(x \to y) \to y = (y \to x) \to x$, |
| *(EqPR)* | $x \leq^P y \Longleftrightarrow x \leq y$. |

**Proof.** By Proposition 3.4.4, (DN2) and (DN3) hold. If (@) holds, then, by (CC1), (@) + (DN2) + (DN3) imply (H), thus (H) holds.

By (CC2), (@) + (H) + (DN3) imply ($\vee$-comm), thus, ($\vee$-comm) holds.

By Proposition 3.4.5 (ii), $x \leq^P y$ implies $x \leq y$. Conversely, if $x \leq y$, i.e. $x \to y = 1$, then: $x \to y^- \overset{(@),(DN)}{=} (y \to x) \to y^- \overset{(DN2)}{=} (x^- \to y^-) \to y^- \overset{(\vee-comm)}{=} (y^- \to x^-) \to x^- \overset{(DN2)}{=} (x \to y) \to x^- = 1 \to x^- \overset{(M)}{=} x^-$, hence $(x \to y^-)^- = x$, by (DN), i.e. $x \leq^P y$. Thus, (EqPR) holds. □

**Remark 3.4.10** *In a left-i-ortholattice $(A^L, \to, ^-, 1)$:*
*- the initial binary relation $\leq$ ( $\leq \Longleftrightarrow \leq^B$, by Proposition 2.2.17) is only* **reflexive** *((Re) holds by definition of a left-BE algebra);*
*- the binary relation $\leq^W$ is only* **reflexive and antisymmetric**, *by Proposition 2.2.17;*
*- the binary relation $\leq^P$ is a* **lattice order**, *by Theorem 3.4.6.*

### 3.4.1 The connection between i-ortholattices and i-Boolean algebras

**Theorem 3.4.11** *Implicative-ortholattices satisfying (@) are d.e. to i-Boolean algebras.*

**Proof.** (1): Let $\mathcal{A}^L = (A^L, \rightarrow, ^-, 1)$ be a left-i-ortholattice satisfying the property (@). Then, by Proposition 3.4.9, ($\vee$-comm) holds. By (B31), ($\vee$-comm) + (Re) + (M) + (Ex) $\Longrightarrow$ (*). By Theorem 2.1.5, (BB) holds. By (B12), ($\vee$-comm) + (M) imply (An). Thus, (BB), (M), (La), (An) and (impl) hold, hence $(A^L, \rightarrow, 1)$ is an implicative left-BCK algebra (Definition 2), hence $(A^L, \rightarrow, 0, 1)$ is a bounded implicative left-BCK algebra; hence, $\mathcal{A}^L$ is a left-i-Boolean algebra (by Corollary 3.3.26).

(1'): Conversely, let $\mathcal{A}^L = (A^L, \rightarrow, ^-, 1)$ be a left-i-Boolean algebra. Then, $(A^L, \rightarrow, 0, 1)$ is a bounded, implicative left-BCK algebra, where $0 = 1^-$; hence, the properties (M), (Ex), (Re), (Fi), (La), (impl), (DN) hold; hence, (pimpl), (DN6) hold too. We prove that (@) holds. Indeed, by (CB0), (pimpl) + (Ex) + (DN6) imply (@). Thus, $\mathcal{A}^L$ is a left-i-ortholattice verifying (@). $\Box$

Theorem 3.4.11 gives us an equivalent definition of i-Boolean algebras:

**Definition 3.4.12** (Definition 2) (The dual case is omitted)

A *left-implicative-Boolean algebra*, or a *left-i-Boolean algebra* for short, is an involutive left-BE algebra verifying (impl) and (@), i.e. is an algebra $\mathcal{A}^L = (A^L, \rightarrow, ^-, 1)$ of type $(2, 1, 0)$ such that $0 \overset{def.}{=} 1^-$ (i.e. (Neg1-0) holds), $x^- = x \rightarrow 0$ (i.e. (Neg) holds) and (M), (Ex), (Re), (La), (impl), (DN) and (@) hold.

**Remark 3.4.13** *In a left-i-Boolean algebra* $(A^L, \rightarrow, ^-, 1)$
*(i.e. in a left-Wajsberg algebra verifying (DN6)=(i-G), by Theorem 3.3.30 or in a left-i-ortholattice verifying (@), by Theorem 3.4.11):*
*- the initial binary relation* $\leq$ $(\Longleftrightarrow \leq^B)$ *and the binary relation* $\leq^P$ *are equivalent:* $x \leq y \Longleftrightarrow x \leq^P y$*, by Proposition 3.4.9, and are* **lattice orders***, by Theorem 3.4.6;*
*- the initial binary relation* $\leq$ *and the binary relation* $\leq^W$ *are equivalent:* $x \leq y \Longleftrightarrow x \leq^W y$ *and a* **lattice orders***, by Theorem 3.3.30 and Proposition 3.3.19, hence we have:*

$$x \leq y \; (\Longleftrightarrow x \leq^B y) \iff x \leq^W y \iff x \leq^P y.$$

### 3.4.2 Resuming connections

Resuming, we have the connections from Figure 3.3.

**Remarks 3.4.14**

*(1) The Wajsberg algebras and the i-ortholattices are incomparable, since any Wajsberg algebra does not verify (impl) and any i-ortholattice does not verify ($\vee$-comm).*

*(2) The i-ortholattices are d.e. with the ortholattices, as we shall prove by Corollary 17.1.6.*

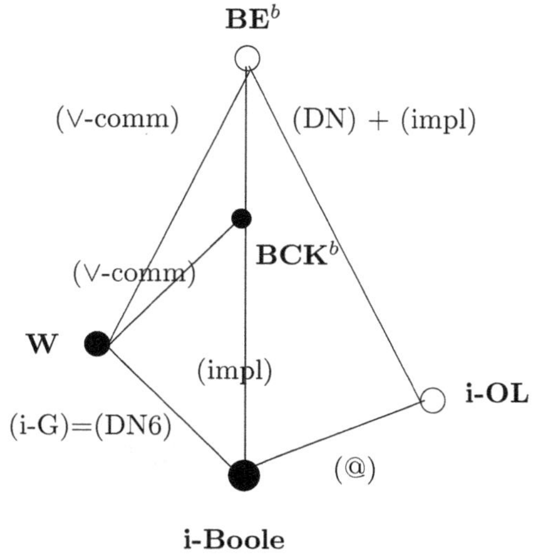


Figure 3.3: Connections between bounded BE algebras, bounded BCK algebras, Wajsberg algebras, i-ortholattices and i-Boolean algebras

## 3.5 L algebras

Every set X, with a binary operation $\cdot$ satisfying (L): $(x \cdot y) \cdot (x \cdot z) = (y \cdot x) \cdot (y \cdot z)$, corresponds to a solution of the quantum Yang-Baxter equation, if the left multiplication is bijective [52], [149], [181]. The same equation (L) becomes a true statement of propositional logic, if the binary operation is interpreted as implication. Equation (L) arose in algebraic logic in [12], [84], [13], [204], but it was never seriously pursued, and remained almost completely neglected. BCK algebras verifying (L) were investigated by Wroński [216] and Traczyk [204].

It was Wolfgang Rump who introduced and studied the L algebras [182], essentially defined by the mentioned equation (L). More informations on L algebras can be found in [183], [184], [185], [186], [187], [188], [196], [189], [190], [191] and [192].

**Definition 3.5.1** (The dual case is omitted)
A *left-L algebra* (or *L-algebra*, in [182]) is an algebra $\mathcal{A}^L = (A^L, \to, 1)$ of type $(2, 0)$ verifying: for all $x, y, z \in A^L$,
(Re) $x \to x = 1$,
(M) $1 \to x = x$,
(La) $x \to 1 = 1$,
(An) $x \to y = 1$ and $y \to x = 1$ imply $x = y$,
(L) $(x \to y) \to (x \to z) = (y \to x) \to (y \to z)$.

Denote by **L** the class of all left-L algebras.

Following the definition, any left-L algebra is a left-aRML algebra (see Figure 2.1). Define then the binary relation $\leq$ by: for all $x, y \in A^L$,

$$(dfrelR) \qquad x\le y \overset{def.}{\Longleftrightarrow} x\to y=1.$$

Recall some basic properties of an L algebra.

**Proposition 3.5.2** *[182] Let $\mathcal{A}^L=(A^L,\to,1)$ be a left-L algebra. Then,*
*(i) the following properties hold: for all $x,y,z\in A^L$,*
*(N')* $\quad 1\le x \Longrightarrow x=1,$
*(*')* $\quad y\le z \Longrightarrow x\to y\le x\to z,$
*(Tr')* $\quad x\le y$ *and* $y\le z \Longrightarrow x\le z;$
*(ii) $\le$ is an order.*

**Proof.** (i): (N'): By (A00), (M) $\Longrightarrow$ (N).
(*'): If $y\le z$, i.e. $y\to z=1$, then, by (L), $(x\to y)\to(x\to z)=(y\to x)\to(y\to z)$, we obtain $(x\to y)\to(x\to z)=(y\to x)\to 1 \overset{(La)}{=} 1$, hence $x\to y\le x\to z$.
(Tr'): By (A13'), (M) + (*) $\Longrightarrow$ (Tr).
(ii): Since (Re'), (Tr'), (An') hold, then $\le$ is an order relation. □

Note that, since (*) holds, then any left-L algebra is a left-*aRML algebra.
A left-L algebra $\mathcal{A}^L=(A^L,\to,1)$ is *bounded*, if there exists an element $0\in A^L$ verifying (Fi), which is unique; then, $\mathcal{A}^L$ will be denoted $(A^L,\to,0,1)$ and, hence, $(A^L,\le,0,1)$ is a bounded poset.
If $\mathcal{A}=(A^L,\to,0,1)$ is a bounded left-L algebra, then we can define a negation $^-$ by: for all $x\in A^L$,

$$x^- \overset{def.}{=} x\to 0.$$

We say that the bounded left-L algebra $\mathcal{A}^L$ is *involutive*, if the property (DN) ($(x^-)^-=x$) holds, i.e. $(x\to 0)\to 0=x$, for all $x\in A^L$.
We shall denote by $\mathbf{L}_{(DN)}$ the class of all involutive left-L algebras.

### 3.5.1 New results

Consider the properties:
(pimpl) $\quad x\to(y\to z)=(x\to y)\to(x\to z),$
(pi) $\quad y\to(y\to x)=y\to x,$
(L) $\quad (x\to y)\to(x\to z)=(y\to x)\to(y\to z).$

**Proposition 3.5.3** *We have*

$$(pimpl) \quad\Longrightarrow\quad ((Ex) \Leftrightarrow (L)).$$

**Proof.** Indeed, (pimpl) + (Ex) $\Longrightarrow$ (L):
$(x\to y)\to(x\to z) \overset{(pimpl)}{=} x\to(y\to z) \overset{(Ex)}{=} y\to(x\to z) \overset{(pimpl)}{=} (y\to x)\to(y\to z)$
and, also, (pimpl) + (L) $\Longrightarrow$ (Ex):
$x\to(y\to z) \overset{(pimpl)}{=} (x\to y)\to(x\to z) \overset{(L)}{=} (y\to x)\to(y\to z) \overset{(pimpl)}{=} y\to(x\to z)$. □

**Remark 3.5.4** *Suppose that the non-commutative generalizations of the properties (pimpl), (pi), (L) are the following properties, respectively:*

(ppimpl) $x \rightsquigarrow (y \to z) = (x \to y) \rightsquigarrow (x \to z)$ *and*
$x \to (y \rightsquigarrow z) = (x \rightsquigarrow y) \to (x \rightsquigarrow z)$,

(ppi) $y \rightsquigarrow (y \to x) = y \to x$ *and*
$y \to (y \rightsquigarrow x) = y \rightsquigarrow x$,

(pL) $(x \to y) \rightsquigarrow (x \to z) = (y \rightsquigarrow x) \to (y \rightsquigarrow z)$.

*Then, we have (see (B3))*
*(pB3) (ppimpl) + (Re) + (pM)* $\Longrightarrow$ *(ppi).*
*Indeed, take $x = y$ in (ppimpl) to obtain*
$y \rightsquigarrow (y \to z) = (y \to y) \rightsquigarrow (y \to z) \overset{(Re),(pM)}{=} y \to z$ *and*
$y \to (y \rightsquigarrow z) = (y \rightsquigarrow y) \to (y \rightsquigarrow z) \overset{(Re),(pM)}{=} y \rightsquigarrow z$.

*So, apparently, this (ppi) is the correct non-commutative generalization of (pi). But, note that $y \rightsquigarrow (y \to z) = y \to (y \rightsquigarrow z)$, by (pEx), so $y \to z = y \rightsquigarrow z$, hence $\to = \rightsquigarrow$, i.e. we have the commutative case.*

*Similarly, we have (see Proposition 3.5.3)*

$$(ppimpl) \quad \Longrightarrow ((pEx) \Leftrightarrow (pL)).$$

*Indeed, (ppimpl) + (pEx)* $\Longrightarrow$ *(pL):*
$(x \to y) \rightsquigarrow (x \to z) \overset{(pimpl)}{=} x \rightsquigarrow (y \to z) \overset{(pEx)}{=} y \to (x \rightsquigarrow z) \overset{(ppimpl)}{=} (y \rightsquigarrow x) \to (y \rightsquigarrow z)$
*and, also, (ppimpl) + (pL)* $\Longrightarrow$ *(pEx):*
$x \to (y \rightsquigarrow z) \overset{(ppimpl)}{=} (x \rightsquigarrow y) \to (x \rightsquigarrow z) \overset{(pL)}{=} (y \to x) \rightsquigarrow (y \to z) \overset{(ppimpl)}{=} y \rightsquigarrow (x \to z)$.

*So, apparently, (pL) is the correct non-commutative generalization of (L). But, note that, if we take $x = y$ in (pL), we obtain: $(y \to y) \rightsquigarrow (y \to z) = (y \rightsquigarrow y) \to (y \rightsquigarrow z)$, which, by (Re) and (pM), becomes: $y \to z = y \rightsquigarrow z$, hence $\to = \rightsquigarrow$, i.e. we have the commutative case.*

*It follows that:*
*- the correct non-commutative generalization of (pi) is:*
*(ppi) = ($pi_{\to}$) + ($pi_{\rightsquigarrow}$), where*
*($pi_{\to}$)* $y \to (y \to x) = y \to x$,
*($pi_{\rightsquigarrow}$)* $y \rightsquigarrow (y \rightsquigarrow x) = y \rightsquigarrow x$;
*- the correct non-commutative generalization of (L) is:*
*(pL) = ($L_{\to}$) + ($L_{\rightsquigarrow}$), where*
*($L_{\to}$)* $(x \to y) \to (x \to z) = (y \to x) \to (y \to z)$ *and*
*($L_{\rightsquigarrow}$)* $(x \rightsquigarrow y) \rightsquigarrow (x \rightsquigarrow z) = (y \rightsquigarrow x) \rightsquigarrow (y \rightsquigarrow z)$;
*- the correct non-commutative generalization of (pimpl) is:*
*(ppimpl) = ($pimpl_{\to}$) + ($pimpl_{\rightsquigarrow}$), where*
*($pimpl_{\to}$)* $x \to (y \to z) = (x \to y) \to (x \to z)$ *and*
*($pimpl_{\rightsquigarrow}$)* $x \rightsquigarrow (y \rightsquigarrow z) = (x \rightsquigarrow y) \rightsquigarrow (x \rightsquigarrow z)$,
*just as the non-commutative generalization of (B) is (pB) = ($B_{\to}$) + ($B_{\rightsquigarrow}$) and of (K) is ($K_{\to}$) + ($K_{\rightsquigarrow}$).*

*Note that, with these definitions, (pB3), i.e. (ppimpl) + (Re) + (pM) $\Longrightarrow$ (ppi) holds, while (ppimpl) $\Longrightarrow$ ((pEx) $\Leftrightarrow$ (pL)) does not hold.*

**Theorem 3.5.5** *Let $\mathcal{A}^L = (A^L, \rightarrow, 1)$ be a left-L algebra. Then,*

$$(Ex) \Longrightarrow (BB).$$

**Proof. (By *Prover9*, Length of proof 12)**
Let $\mathcal{A}^L = (A^L, \rightarrow, 1)$ be a left-L algebra, i.e. (An), (Re), (M), (La) and (L) hold.
First, by (A8), (Re) + (La) + (Ex) $\Longrightarrow$ (K). Then,

$$\begin{aligned}(y \rightarrow z) \rightarrow ((z \rightarrow x) \rightarrow (y \rightarrow x)) &\overset{(Ex)}{=} (z \rightarrow x) \rightarrow ((y \rightarrow z) \rightarrow (y \rightarrow x)) \\ &\overset{(L)}{=} (z \rightarrow x) \rightarrow ((z \rightarrow y) \rightarrow (z \rightarrow x)) \\ &\overset{(K)}{=} 1,\end{aligned}$$

i.e. (BB) holds. □

Note that, by above Theorem 3.5.5, a left-L algebra verifying (Ex) (called *CKL algebra* in [191]) verifies (BB), hence, it is a left-BCK algebra verifying (L), called a *left-BCKL algebra*. Conversely, a left-BCKL algebra is a left-L algebra verifying (Ex).

Denote by **BCKL** the class of all left-BCKL algebras.

Hence, we have proved the following result.

**Corollary 3.5.6**

$$\mathbf{BCKL} \iff \mathbf{L} + (Ex).$$

Recall that a left-Hilbert algebra is a positive-implicative left-BCK algebra, i.e. a left-BCK algebra verifying (pimpl) ($\Longleftrightarrow$ (pi)); hence, it verifies (Ex), hence, by Proposition 3.5.3, it verifies (L); hence, it is a left-BCKL algebra verifying (pimpl); hence, it is a left-L algebra verifying (pimpl). Conversely, a left-L algebra verifying (pimpl), by Proposition 3.5.3, verifies (Ex), hence, by Theorem 3.5.5, verifies (BB), hence it is a left-BCKL algebra verifying (pimpl), hence it is a left-Hilbert algebra.

Hence, we have proved the following result.

**Corollary 3.5.7**

$$\mathbf{Hilbert} \iff \mathbf{BCK} + (pimpl) \iff \mathbf{BCKL} + (pimpl) \iff \mathbf{L} + (pimpl).$$

Note that Proposition 3 from [191], saying: "Every Hilbert algebra is an L algebra. An L algebra $X$ is a Hilbert algebra if and only if verifies (pimpl).", follows by above Corollary 3.5.7.

**Theorem 3.5.8** *Let $\mathcal{A}^L = (A^L, \rightarrow, 1)$ be a left-L algebra. Then,*

$$(K) \Longrightarrow (B).$$

**Proof.** $(x \to y) \to ((z \to x) \to (z \to y)) \stackrel{(L)}{=} (x \to y) \to ((x \to z) \to (x \to y)) \stackrel{(K)}{=} 1$, i.e. (B) holds. □

Note that, by above Theorem 3.5.8, a left-L algebra verifying (K) (called *KL-algebra* in [182]) verifies (B), hence, is a left-BCC algebra verifying (L), called a *left-BCCL algebra*. Conversely, a left-BCCL algebra, i.e. an algebra $(A^L, \to, 1)$ verifying (An), (Re), (M), (La), (B), (L), is a left-L algebra and, by (A9), (M) + (La) + (B) $\Longrightarrow$ (K), hence it verifies (K).

Denote by **BCCL** the class of all left-BCCL algebras.

Hence, we have proved the following result.

**Corollary 3.5.9**

$$\mathbf{BCCL} \iff \mathbf{L} + (K).$$

### 3.5.2 Putting L algebras on the "map"

By above results, we have the connections from Figure 3.4.

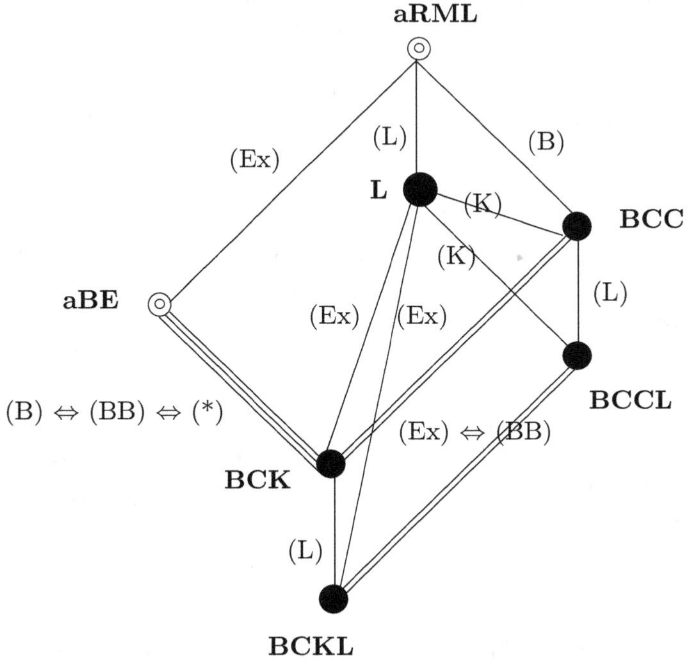


Figure 3.4: Hierarchies around L algebras

### 3.5.3 The involutive case

Recall the property (Ex0) $(x \to (y \to 0) = y \to (x \to 0))$, which appears in general 'bridge ' Theorem 17.1.1. We shall prove first that there are no proper involutive

L algebras (i.e. not being involutive BCK algebras) verifying (Ex0).

**Theorem 3.5.10** *Let $\mathcal{A}^L = (A^L, \rightarrow, 0, 1)$ be an involutive left-L algebra. Then,*

$$(Ex0) \implies (Ex).$$

**Proof. (By *Prover9*, Length of proof 28)**
First, we prove

$$x \rightarrow ((x \rightarrow 0) \rightarrow y) = 1. \tag{3.11}$$

Indeed, in (L) $((x \rightarrow y) \rightarrow (x \rightarrow z) = (y \rightarrow x) \rightarrow (y \rightarrow z))$, take $X := x \rightarrow 0$, $Y := 0$, $Z := y$, to obtain:
$((x \rightarrow 0) \rightarrow 0) \rightarrow ((x \rightarrow 0) \rightarrow y) = (0 \rightarrow (x \rightarrow 0)) \rightarrow (0 \rightarrow y)$,
hence, by (DN), (Fi), (Re), (3.11) holds.
Now, we prove

$$((x \rightarrow 0) \rightarrow y) \rightarrow x = y \rightarrow ((y \rightarrow (x \rightarrow 0)) \rightarrow 0). \tag{3.12}$$

Indeed, in (L), take $X := x \rightarrow 0$ and $Z := 0$, to obtain
$((x \rightarrow 0) \rightarrow y) \rightarrow ((x \rightarrow 0) \rightarrow 0) = (y \rightarrow (x \rightarrow 0)) \rightarrow (y \rightarrow 0)$,
hence, by (DN),
$((x \rightarrow 0) \rightarrow y) \rightarrow x = (y \rightarrow (x \rightarrow 0)) \rightarrow (y \rightarrow 0)$;
then, by (Ex0), we obtain immediately (3.12).
Now, we prove

$$(y \rightarrow 0) \rightarrow (y \rightarrow x) = 1. \tag{3.13}$$

Indeed, in (L), take $X := 0$ and $Z := x$, to obtain $(0 \rightarrow y) \rightarrow (0 \rightarrow x) = (y \rightarrow 0) \rightarrow (y \rightarrow x)$, hence, by (Fi) and (Re), (3.13) holds.
Now, we prove (K), i.e.

$$x \rightarrow (y \rightarrow x) = 1. \tag{3.14}$$

Indeed, by (3.11), for $Y := y \rightarrow 0$, we obtain
$x \rightarrow ((x \rightarrow 0) \rightarrow (y \rightarrow 0)) = 1$, hence, by (Ex0), $x \rightarrow (y \rightarrow ((x \rightarrow 0) \rightarrow 0)) = 1$, which, by (DN), gives (K).
Now, we prove

$$(y \rightarrow x) \rightarrow (y \rightarrow (z \rightarrow x)) = 1. \tag{3.15}$$

Indeed, in (L), take $Z := z \rightarrow x$, to obtain
$(x \rightarrow y) \rightarrow (x \rightarrow (z \rightarrow x)) = (y \rightarrow x) \rightarrow (y \rightarrow (z \rightarrow x))$; then, by (K) and (La), (3.15) holds.
Now, we prove

$$(x \rightarrow 0) \rightarrow (((y \rightarrow 0) \rightarrow x) \rightarrow y) = 1. \tag{3.16}$$

Indeed, by (3.12), $((y \rightarrow 0) \rightarrow x) \rightarrow y = x \rightarrow ((x \rightarrow (y \rightarrow 0)) \rightarrow 0)$;
in (3.13) $((x \rightarrow 0) \rightarrow (x \rightarrow y) = 1)$, take $Y := (x \rightarrow (y \rightarrow 0)) \rightarrow 0$ to obtain
$(x \rightarrow 0) \rightarrow (x \rightarrow (((x \rightarrow (y \rightarrow 0)) \rightarrow 0)) = 1$, hence, by above (3.12), (3.16) holds.

Now, we prove (D), i.e.

$$x \to ((x \to y) \to y) = 1. \tag{3.17}$$

Indeed, from (3.16), for $X := x \to 0$, we obtain
$((x \to 0) \to 0) \to (((y \to 0) \to (x \to 0)) \to y) = 1$, hence, by (DN) and (Ex0), $x \to ((x \to ((y \to 0) \to 0)) \to y) = 1$, hence, by (DN) again, (D) holds.

Now, we prove

$$x \to (y \to ((x \to z) \to z)) = 1. \tag{3.18}$$

Indeed, in (3.15) ($(x \to y) \to (x \to (z \to y)) = 1$), take $Y := (x \to z) \to z$ and $Z := y$, to obtain
$(x \to ((x \to z) \to z)) \to (x \to (y \to ((x \to z) \to z))) = 1$; hence, by (D), $1 \to (x \to (y \to ((x \to z) \to z))) = 1$; hence, by (M), (3.18) holds.

Now, we prove

$$x \to ((y \to (x \to z)) \to (y \to z)) = 1. \tag{3.19}$$

Indeed, by (L), for $X := x \to z$, we obtain
$(y \to (x \to z)) \to (y \to z) = ((x \to z) \to y) \to ((x \to z) \to z)$; hence, $x \to ((y \to (x \to z)) \to (y \to z)) = x \to (((x \to z) \to y) \to ((x \to z) \to z))$, which, by (3.18), gives
$x \to ((y \to (x \to z)) \to (y \to z)) = 1$, i.e. (3.19) holds.

Now, we prove (C), i.e.

$$(x \to (y \to z)) \to (y \to (x \to z)) = 1. \tag{3.20}$$

Indeed, in (3.19), take $X := x \to (y \to z)$ and $Z := x \to z$, to obtain
$(x \to (y \to z)) \to ((y \to ((x \to (y \to z)) \to (x \to z)) \to (y \to (x \to z))) = 1$, hence, by (3.19) again,
$(x \to (y \to z)) \to (1 \to (y \to (x \to z))) = 1$, hence, by (M), (C) holds.

Finally, by (A3), (C) + (An) $\Longrightarrow$ (Ex). □

Note that, by above Theorem 3.5.10, an involutive left-L algebra verifying (Ex0) verifies (Ex), hence, by Theorem 3.5.5, verifies (BB), hence, is an involutive left-BCKL algebra. Conversely, any involutive left-BCKL algebra is an involutive left-L algebra verifying (Ex0).

Hence, we have proved the following result.

**Corollary 3.5.11**

$$\mathbf{L_{(DN)}} + (Ex0) \iff \mathbf{BCKL_{(DN)}}.$$

Moreover, we shall prove that there are no proper involutive BCC algebras (i.e. not being involutive BCK algebras) verifying (Ex0).

**Theorem 3.5.12** *Let $\mathcal{A}^L = (A^L, \to, 0, 1)$ be an involutive left-BCC algebra. Then,*

$$(Ex0) \Longrightarrow (Ex).$$

**Proof. (By *Prover9*, Length of proof 15)**

Let $\mathcal{A}^L$ be an involutive left-BCC algebra, i.e. (An), (Re), (M), (La), (B), (Fi), (DN) ($(x \to 0) \to 0 = x$) hold.

First, we prove

$$(x \to (y \to z)) \to (x \to ((u \to y) \to (u \to z))) = 1. \tag{3.21}$$

Indeed, in (B) ($(x \to y) \to ((z \to x) \to (z \to y)) = 1$), take $X := y \to z$, $Y := (u \to y) \to (u \to z)$, $Z := x$, to obtain
(a) $((y \to z) \to ((u \to y) \to (u \to z))) \to ((x \to (y \to z)) \to (x \to ((u \to y) \to (u \to z)))) = 1$;
but, in (a), the part $(y \to z) \to ((u \to y) \to (u \to z))$ equals 1, by (B), hence (a) becomes
(a') $1 \to ((x \to (y \to z)) \to (x \to ((u \to y) \to (u \to z)))) = 1$, hence, by (M), (3.21) holds.

Now, we prove

$$(x \to 0) \to (((y \to 0) \to x) \to y) = 1. \tag{3.22}$$

Indeed, in (B) ($(x \to y) \to ((z \to x) \to (z \to y)) = 1$), take $X := x$, $Y := 0$, $Z := y \to 0$, to obtain
$(x \to 0) \to (((y \to 0) \to x) \to ((y \to 0) \to 0)) = 1$, hence, by (DN), (3.22) holds.

Now, we prove that (D) holds, i.e.

$$x \to ((x \to y) \to y) = 1. \tag{3.23}$$

Indeed, in (3.22), take $X := x \to 0$, to obtain
(b) $((x \to 0) \to 0) \to (((y \to 0) \to (x \to 0)) \to y) = 1$; but, in (b), the part $(x \to 0) \to 0$ equals $x$, by (DN), while the part $(y \to 0) \to (x \to 0)$ equals $x \to ((y \to 0) \to 0)$ by (Ex0), hence equals $x \to y$, by (DN); thus, (b) becomes $x \to ((x \to y) \to y) = 1$, i.e. (D) holds.

Now, we prove

$$x \to ((y \to (x \to z)) \to (y \to z)) = 1. \tag{3.24}$$

Indeed, in (3.21), take $X := x$, $Y := x \to z$, $Z := z$, $U := y$, to obtain
(c) $(x \to ((x \to z) \to z)) \to (x \to ((y \to (x \to z)) \to (y \to z))) = 1$;
but, in (c), the part $x \to ((x \to z) \to z)$ equals 1, by (D), hence (c) becomes
(c') $1 \to (x \to ((y \to (x \to z)) \to (y \to z))) = 1$;
then, by (M), (c') becomes (3.24).

Now, we prove (C), i.e.

$$(x \to (y \to z)) \to (y \to (x \to z)) = 1. \tag{3.25}$$

Indeed, in (3.24), take $X := x \to (y \to z)$ and $Z := x \to z$, to obtain
(d) $(x \to (y \to z)) \to ((y \to ((x \to (y \to z)) \to (x \to z))) \to (y \to (x \to z))) = 1$;
but, in (d), the part $y \to ((x \to (y \to z)) \to (x \to z))$ equals 1, by (3.24), hence (d) becomes

(d') $(x \to (y \to z)) \to (1 \to (y \to (x \to z))) = 1$, which, by (M), becomes (3.25), i.e. (C) holds.

Finally, by (A3), (C) + (An) $\Longrightarrow$ (Ex). □

Note that, by above Theorem 3.5.12, an involutive left-BCC algebra verifying (Ex0) verifies (Ex), hence, by Theorem 3.5.5, verifies (BB), hence, is an involutive left-BCK algebra. Conversely, any involutive left-BCK algebra is an involutive left-BCC algebra verifying (Ex0).

Hence, we have proved the following result.

**Corollary 3.5.13**

$$\mathbf{BCC_{(DN)}} + (Ex0) \iff \mathbf{BCK_{(DN)}}.$$

Consequently, we obtain:

$$\mathbf{BCCL_{(DN)}} + (Ex0) \iff \mathbf{BCKL_{(DN)}}.$$

We shall prove now, in two steps, that the involutive BCKL algebras are just the Wajsberg algebras. Recall that Wajsberg algebras are definitionally equivalent to bounded $\vee$-commutative BCK algebras and, also, that Wajsberg algebras are definitionally equivalent, by the 'bridge' Theorem 17.1.1, to the MV algebras.

**Theorem 3.5.14** *Let $\mathcal{A}^L = (A^L, \to, 0, 1)$ be a bounded $\vee$-commutative left-BCK algebra (i.e. an involutive left-BCK algebra verifying ($\vee$-comm) or a left-Wajsberg algebra). Then, (L) holds.*

**Proof. (By *Prover9*, in 11.30 seconds, Lengh of proof 16)**

Following Yutani's equivalent system of axioms for $\vee$-commutative BCK algebras, $\mathcal{A}^L$ verifies ($\vee$-comm), (M), (Re), (Ex) and (Fi). Let us define, for all $x \in A^L$, $x^- \stackrel{def.}{=} x \to 0$, i.e. (dfneg) holds and (Neg) holds too.

By (CB1), ($\vee$-comm) + (Neg) + (Fi) + (M) $\Longrightarrow$ (DN), i.e. we have

$$(x \to 0) \to 0 = x. \tag{3.26}$$

By (C7), (DN) + (Neg) + (Ex) $\Longrightarrow$ (DN2), i.e. we have

$$x \to y = (y \to 0) \to (x \to 0). \tag{3.27}$$

By (C2), (DN) + (DN2) $\Longrightarrow$ (DN3), i.e. we have

$$(x \to 0) \to y = (y \to 0) \to x. \tag{3.28}$$

Now, we prove

$$((x \to y) \to z) \to z = x \to ((x \to (z \to y)) \to y. \tag{3.29}$$

Indeed, in ($\vee$-comm) $((x \to y) \to y = (y \to x) \to x)$, take $X := x \to z$, to obtain $((x \to z) \to y) \to y = (y \to (x \to z)) \to (x \to z) \stackrel{(Ex)}{=} (x \to (y \to z)) \to (x \to z)$,

hence, by (Ex) again,
$((x \rightarrow z) \rightarrow y) \rightarrow y = x \rightarrow ((y \rightarrow (x \rightarrow z)) \rightarrow z)$, i.e. (3.29) holds.

Now, we prove

$$(3.30) \qquad x \rightarrow ((y \rightarrow 0) \rightarrow (z \rightarrow 0)) = z \rightarrow (x \rightarrow y).$$

Indeed, by (Ex), $x \rightarrow (z \rightarrow y) = z \rightarrow (x \rightarrow y)$, hence, by (3.27), (3.30) holds.

Now, we prove

$$(3.31) \qquad x \rightarrow ((x \rightarrow y) \rightarrow 0) = y \rightarrow ((y \rightarrow x) \rightarrow 0)$$

Indeed, in (3.29), take $X := y$, $Y := 0$, $Z := x \rightarrow 0$, to obtain
$((y \rightarrow 0) \rightarrow (x \rightarrow 0)) \rightarrow (x \rightarrow 0) = y \rightarrow ((y \rightarrow ((x \rightarrow 0) \rightarrow 0)) \rightarrow 0)$; hence, by 3.27),
$(x \rightarrow y) \rightarrow (x \rightarrow 0) = y \rightarrow ((y \rightarrow ((x \rightarrow 0) \rightarrow 0)) \rightarrow 0)$; then, by (Ex) and (3.26) = (DN), we obtain
$x \rightarrow ((x \rightarrow y) \rightarrow 0) = y \rightarrow ((y \rightarrow x) \rightarrow 0)$, (3.31) holds.

Now, we prove

$$(3.32) \qquad x \rightarrow ((y \rightarrow 0) \rightarrow z) = ((x \rightarrow y) \rightarrow 0) \rightarrow z.$$

Indeed, in (3.30), take $Z := z \rightarrow 0$ to obtain
$x \rightarrow ((y \rightarrow 0) \rightarrow ((z \rightarrow 0) \rightarrow 0)) = (z \rightarrow 0) \rightarrow (x \rightarrow y) \overset{(3.28)}{=} ((x \rightarrow y) \rightarrow 0) \rightarrow z$, hence, by (DN), (3.32) holds.

Now, we prove

$$(3.33) \qquad x \rightarrow ((((x \rightarrow y) \rightarrow 0) \rightarrow 0) \rightarrow z) = y \rightarrow ((((y \rightarrow x) \rightarrow 0) \rightarrow 0) \rightarrow z).$$

Indeed, in (3.32), take $Y := (x \rightarrow y) \rightarrow 0$, to obtain
(a) $((x \rightarrow ((x \rightarrow y) \rightarrow 0)) \rightarrow 0) \rightarrow z = x \rightarrow ((((x \rightarrow y) \rightarrow 0) \rightarrow 0) \rightarrow z)$;
but, in (a), $x \rightarrow ((x \rightarrow y) \rightarrow 0) = y \rightarrow ((y \rightarrow x) \rightarrow 0)$, by (3.31), hence,
(b) $((x \rightarrow ((x \rightarrow y) \rightarrow 0)) \rightarrow 0) \rightarrow z = ((y \rightarrow ((y \rightarrow x) \rightarrow 0)) \rightarrow 0) \rightarrow z$;
but, by (a), we obtain
(c) $((y \rightarrow ((y \rightarrow x) \rightarrow 0)) \rightarrow 0) \rightarrow z = y \rightarrow ((((y \rightarrow x) \rightarrow 0) \rightarrow 0) \rightarrow z)$;
by (a), (b), (c), (3.33) holds.

Finally, we prove that (L) holds. Indeed, from (3.33) and ($\vee$-comm), we obtain
$x \rightarrow (((0 \rightarrow (x \rightarrow y)) \rightarrow (x \rightarrow y)) \rightarrow z) = y \rightarrow (((0 \rightarrow (y \rightarrow x)) \rightarrow (y \rightarrow x)) \rightarrow z$,
hence, by (Fi) and (M), we obtain
$x \rightarrow ((x \rightarrow y) \rightarrow z) = y \rightarrow ((y \rightarrow x) \rightarrow z)$, hence, by (Ex),
$(x \rightarrow y) \rightarrow (x \rightarrow z) = (y \rightarrow x) \rightarrow (y \rightarrow z)$, i.e. (L) holds. □

**Lemma 3.5.15** *We have*

$$(L) + (Ex) \Longrightarrow (3.34),$$

*where*

$$(3.34) \qquad x \rightarrow ((x \rightarrow y) \rightarrow z) = y \rightarrow ((y \rightarrow x) \rightarrow z).$$

**Proof.**

$$x \to ((x \to y) \to z) \overset{(Ex)}{=} (x \to y) \to (x \to z) \overset{(L)}{=} (y \to x) \to (y \to z) \overset{(Ex)}{=} y \to ((y \to x) \to z).$$

□

**Theorem 3.5.16** *Let $\mathcal{A}^L = (A^L, \to, 0, 1)$ be an involutive left-BCK algebra verifying (L) (i.e. an involutive left-BCKL algebra). Then, ($\vee$-comm) holds.*

**Proof. (Following the ideea of proof by *Prover9*).**
Since $\mathcal{A}^L$ is an involutive left-BCK algebra, then (Ex), (DN), (DN2) hold, where $x^- \overset{def.}{=} x \to 0$. Also, (3.34) holds.
In (3.34) $(x \to ((x \to y) \to z) = y \to ((y \to x) \to z))$,
take $X := x \to 0$, $Y := y \to 0$, $Z := 0$, to obtain
$(x \to 0) \to (((x \to 0) \to (y \to 0)) \to 0) = (y \to 0) \to (((y \to 0) \to (x \to 0)) \to 0)$,
which by (DN2), becomes
$(x \to 0) \to ((y \to x) \to 0) = (y \to 0) \to ((x \to y) \to 0)$,
which by (Ex), becomes
$(y \to x) \to ((x \to 0) \to 0) = (x \to y) \to ((y \to 0) \to 0)$,
which by (DN) becomes
$(y \to x) \to x = (x \to y) \to y$, i.e. ($\vee$-comm) holds. □

By Theorems 3.5.14, 3.5.16, we obtain immediately the following result.

**Corollary 3.5.17**

$$\mathbf{BCK_{(DN)}} \implies ((\vee - comm) \Leftrightarrow (L)),$$

*or, equivalently,*

$$\mathbf{W} \iff \mathbf{BCK_{(DN)}} + (\vee - comm) \iff \mathbf{BCKL_{(DN)}}.$$

Consequenly, by Theorem 2.2.9, we have the following result.

**Corollary 3.5.18** *Bounded $\vee$-commutative BCC algebras coincide with bounded $\vee$-commutative BCK algebras, i.e. with Wajsberg algebras. For short,*

$$\mathbf{W} \iff \mathbf{BCK_{(DN)}} + (\vee - comm) \iff \mathbf{BCC_{(DN)}} + (\vee - comm).$$

By Corollaries 3.5.17 and 3.5.18, we then obtain:

**Corollary 3.5.19**

$$\mathbf{W} \iff \mathbf{BCK_{(DN)}} + (\vee - comm) \iff \mathbf{BCC_{(DN)}} + (\vee - comm) \iff \mathbf{BCKL_{(DN)}}.$$

Note that Proposition 5 from [191], saying: "... A bounded L algebra is an MV algebra with respect to $x \to y = (x \odot y^-)^-$ if and only if verifies (K) and ($\vee$-comm).", follows by above Corollary 3.5.19 and Corollary 3.5.9.

### 3.5.4 Examples

All the examples presented were found by *Mace4 program.*

**Example 3.5.20 Proper L algebra**, i.e. not verifying (Ex), (K)

The algebra $\mathcal{A}_4 = (A_4 = \{0, a, b, 1\}, \rightarrow, 1)$, with the following table of $\rightarrow$, verifies (An), (Re), (M), (La), (*), (Tr), (D) and (L), and does not verify (Ex) for $(a, b, a)$, (pimpl) for $(a, b, a)$, (pi) for $(a, b)$, (BB) for $(a, b, 0)$, (**) for $(a, b, 1)$, (B) for $(b, 0, a)$, (K) for $(a, b)$, ($\vee$-comm) for $(a, b)$, (DN) for $a$ (the table of negation $x^- = x \rightarrow 0$ is the column of 0 in the table of $\rightarrow$). Hence, $\mathcal{A}_4$ is a left-*aRML algebra verifying (L), hence is a **proper left-L algebra**. Since $x \leq y \Longleftrightarrow x \rightarrow y = 1$, it follows that we have $0 \leq 0, 1$; $a \leq a, 1$; $b \leq b, 1$; $1 \leq 1$, hence the poset $(A_4, \leq)$ is represented by the below Hasse diagram.

| $\rightarrow$ | 0 | a | b | 1 |
|---|---|---|---|---|
| 0 | 1 | a | b | 1 |
| a | 0 | 1 | b | 1 |
| b | 0 | b | 1 | 1 |
| 1 | 0 | a | b | 1 |

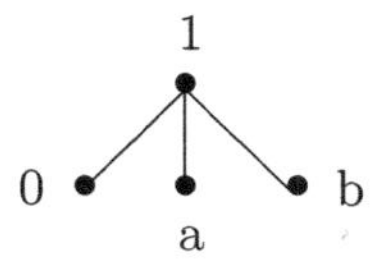


**Example 3.5.21 Proper BCC algebra**, i.e. not verifying (Ex), (L)

The algebra $\mathcal{A}_5 = (A_5 = \{0, a, b, c, 1\}, \rightarrow, 1)$, with the following table of $\rightarrow$, verifies (An), (Re), (M), (La), (B), (*), (**), (Tr) and does not verify (Ex) for $(0, a, c)$, (pimpl) for $(0, 0, b)$, (pi) for $(b, 0)$, (BB) for $(c, 1, 0)$, (D) for $(c, 0)$, ($\vee$-comm) for $(0, a)$, and (L) for $(0, a, b)$. Hence, $\mathcal{A}_5$ is a left-*aRML** algebra verifying (B), hence is a left-BCC algebra. Since (Ex) and (L) are not verified, then $\mathcal{A}_5$ is a **proper left-BCC algebra**.

| $\rightarrow$ | 0 | a | b | c | 1 |
|---|---|---|---|---|---|
| 0 | 1 | 1 | a | a | 1 |
| a | 0 | 1 | a | b | 1 |
| b | 0 | 1 | 1 | a | 1 |
| c | 0 | 1 | 1 | 1 | 1 |
| 1 | 0 | a | b | c | 1 |

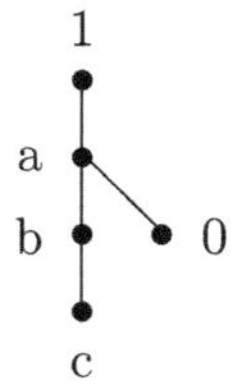


**Example 3.5.22 Proper BCK algebra**, i.e. not verifying (pimpl), (L)

The algebra $\mathcal{A}_5 = (A_5 = \{0, a, b, c, 1\}, \rightarrow, 1)$, with the following table of $\rightarrow$, verifies (An), (Re), (M), (La), (Ex), (BB), (B), (*), (**), (Tr), (D) and does not verify (L) for $(0, a, b)$, (pimpl) for $(0, 0, a)$, (pi) for $(a, 0)$, ($\vee$-comm) for $(0, c)$. Hence, $\mathcal{A}_5$ is a **proper left-BCK algebra**.

| $\rightarrow$ | 0 | a | b | c | 1 |
|---|---|---|---|---|---|
| 0 | 1 | 0 | c | 1 | 1 |
| a | 1 | 1 | c | 1 | 1 |
| b | 1 | 0 | 1 | 1 | 1 |
| c | 0 | a | 0 | 1 | 1 |
| 1 | 0 | a | b | c | 1 |

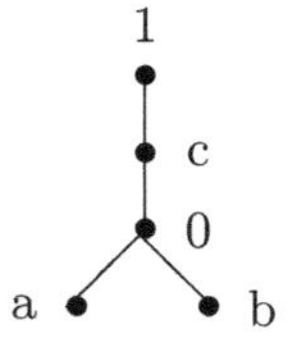

**Example 3.5.23 Proper BCCL algebra,** i.e. not verifying (Ex)

The algebra $\mathcal{A}_6 = (A_6 = \{0, a, b, c, d, 1\}, \rightarrow, 1)$, with the following table of $\rightarrow$, verifies (An), (Re), (M), (La), (B), (*), (**), (Tr) and (L), and does not verify (Ex) for $(d, a, b)$, (BB) for $(b, a.d)$, (D) for $(b, d)$, ($\vee$-comm) for $(0, a)$, (pimpl) for $(a, d, b)$, (pi) for $(b, d)$, (DN) for $a$ (the table of negation $x^- = x \rightarrow 0$ is the column of 0 in the table of $\rightarrow$). Hence, $\mathcal{A}_6$ is a left-*aRML** algebra verifying (B) and (L), hence is a left-BCC algebra verifying (L), hence is a left-BCCL algebra.

| $\rightarrow$ | 0 | a | b | c | d | 1 |
|---|---|---|---|---|---|---|
| 0 | 1 | 1 | 1 | 1 | 1 | 1 |
| a | 0 | 1 | b | c | d | 1 |
| b | 0 | 1 | 1 | 1 | c | 1 |
| c | 0 | a | b | 1 | d | 1 |
| d | 0 | 1 | a | 1 | 1 | 1 |
| 1 | 0 | a | b | c | d | 1 |

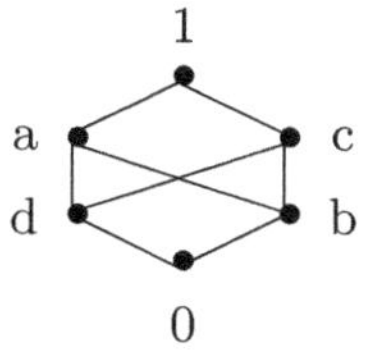


Since the property (Ex) is not verified for $(d, a, b)$: $a = d \rightarrow b = d \rightarrow (a \rightarrow b) \neq a \rightarrow (d \rightarrow b) = a \rightarrow a = 1$, it follows that $\mathcal{A}_6$ is not a left-BCKL algebra, hence it is a **proper left-BCCL algebra**.

**Example 3.5.24 Proper BCKL algebra,** i.e. not verifying (pimpl)

The algebra $\mathcal{A}_4 = (A_4 = \{0, a, b, 1\}, \rightarrow, 1)$, with the following table of $\rightarrow$, verifies (An), (Re), (M), (La), (Ex), (BB), (B), (*), (**). (Tr), (D) and (L), and does not verify (DN) for $a$, (pimpl) for $(0, 0, a)$, (pi) for $(a, 0)$, ($\vee$-comm) for $(0, b)$. Hence, $\mathcal{A}_4$ is a **proper left-BCKL algebra**.

| $\rightarrow$ | 0 | a | b | 1 |
|---|---|---|---|---|
| 0 | 1 | 0 | b | 1 |
| a | 1 | 1 | b | 1 |
| b | 1 | 1 | 1 | 1 |
| 1 | 0 | a | b | 1 |

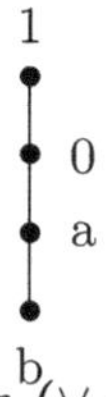


**Example 3.5.25 Proper BCK algebra verifying ($\vee$-comm) and not verifying (L)**

The algebra $\mathcal{A}_4 = (A_4 = \{0, a, b, 1\}, \rightarrow, 1)$, with the following table of $\rightarrow$, verifies (An), (Re), (M), (La), (Ex), (BB), (B), (*), (**). (Tr), (D) and ($\vee$-comm) and does not verify (DN) for $a$, (pimpl) for $(0, 0, a)$, (pi) for $(a, 0)$ and (L) for $(0, a, b)$.

| $\rightarrow$ | 0 | a | b | 1 |
|---|---|---|---|---|
| 0 | 1 | 0 | 0 | 1 |
| a | 1 | 1 | 0 | 1 |
| b | 1 | 0 | 1 | 1 |
| 1 | 0 | a | b | 1 |

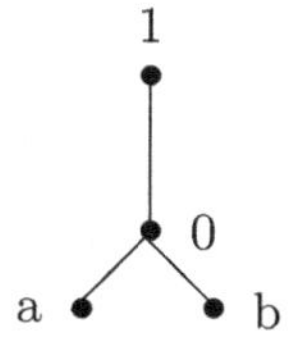


**Example 3.5.26 Proper BCKL algebra verifying ($\vee$-comm),** i.e. not verifying (pimpl)

The algebra $\mathcal{A}_4 = (A_4 = \{0, 2, 3, 1\}, \rightarrow, 1)$, with the following table of $\rightarrow$, verifies (An), (Re), (M), (La), (Ex), (BB), (B), (*), (**). (Tr), (D) and (L) and ($\vee$-comm) and does not verify (DN) for 2, (pimpl) for $(0, 0, 3)$, (pi) for $(3, 0)$.

| $\to$ | 0 | 2 | 3 | 1 |
|---|---|---|---|---|
| 0 | 1 | 2 | 0 | 1 |
| 2 | 0 | 1 | 3 | 1 |
| 3 | 1 | 2 | 1 | 1 |
| 1 | 0 | 2 | 3 | 1 |

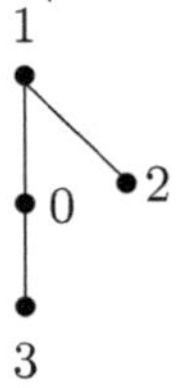


- **Involutive case**

**Example 3.5.27 The smallest proper involutive L algebra,** i.e. not verifying (Ex), (K)

The algebra $\mathcal{A}_4 = (A_4 = \{0, a, b, 1\}, \to, 1)$, with the following table of $\to$, verifies (An), (Re), (M), (La), (*), (Tr), (D), (DN) (the table of negation $x^- = x \to 0$ is the column of 0 in the table of $\to$) and (L), and does not verify (Ex) for $(a, b, 0)$, (pimpl) for $(a, a, 0)$, (pi) for $(0, a)$, (BB) for $(0, a, b)$, (**) for $(a, b, 1)$, (B) for $(a, 1, b)$, (K) for $(a, b)$, ($\vee$-comm) for $(a, b)$, Hence, $\mathcal{A}_4$ is the smallest **proper involutive left-L algebra**.

| $\to$ | 0 | a | b | 1 |
|---|---|---|---|---|
| 0 | 1 | 1 | 1 | 1 |
| a | a | 1 | a | 1 |
| b | b | b | 1 | 1 |
| 1 | 0 | a | b | 1 |

and

| $x$ | $x^-$ |
|---|---|
| 0 | 1 |
| a | a |
| b | b |
| 1 | 0 |

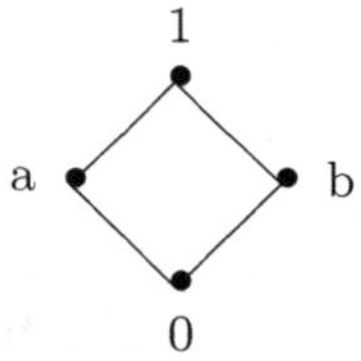


**Example 3.5.28 The smallest proper involutive BCC algebra,** i.e. not verifying (Ex)

The algebra $\mathcal{A}_8 = (A_8 = \{0, a, b, c, d, e, f, 1\}, \to, 1)$, with the following table of $\to$, verifies (An), (Re), (M), (La), (B), (*), (**), (Tr) and (DN), with $(0, a, b, c, d, e, f, 1)^- = (1, a, b, d, c, f, e, 0)$ - see the column of 0 in the table of $\to$) and does not verify (Ex) for $(a, b, 0)$, (BB) for $(0, a, b)$, (D) for $(a, b)$, ($\vee$-comm) for $(a, b)$, (pimpl) for $(a, a, 0)$, (pi) for $(0, a)$ and (L) for $(a, b, d)$. Hence, $\mathcal{A}_8$ is a left-*aRML** algebra verifying (B), hence is an involutive left-BCC algebra.

| $\to$ | 0 | a | b | c | d | e | f | 1 |
|---|---|---|---|---|---|---|---|---|
| 0 | 1 | 1 | 1 | 1 | 1 | 1 | 1 | 1 |
| a | a | 1 | c | 1 | a | 1 | c | 1 |
| b | b | e | 1 | 1 | e | 1 | b | 1 |
| c | d | e | b | 1 | d | e | b | 1 |
| d | c | 1 | 1 | 1 | 1 | 1 | c | 1 |
| e | f | a | c | c | a | 1 | f | 1 |
| f | e | 1 | 1 | 1 | e | 1 | 1 | 1 |
| 1 | 0 | a | b | c | d | e | f | 1 |

and

| $x$ | $x^-$ |
|---|---|
| 0 | 1 |
| a | a |
| b | b |
| c | d |
| d | c |
| e | f |
| f | e |
| 1 | 0 |

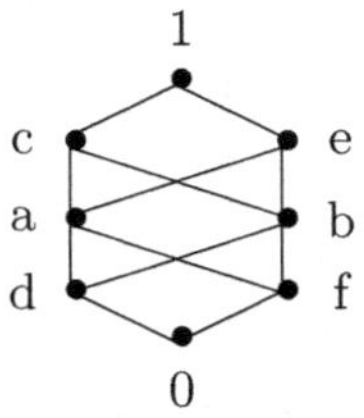


Since the property (Ex) is not verified for $(a, b, 0)$: $c = a \to b = a \to (b \to 0) \neq b \to (a \to 0) = b \to a = e$, it follows that $\mathcal{A}_8$ is not an involutive left-BCK algebra, hence it is a **proper involutive left-BCC algebra**.

**Example 3.5.29 Proper involutive BCK algebra,** i.e. not verifying (L), (pimpl)

The algebra $\mathcal{A}_4 = (A_4 = \{0, a, b, 1\}, \rightarrow, 1)$, with the following table of $\rightarrow$, verifies (An), (Re), (M), (La), (Ex), (BB), (B), (*), (**), (Tr), (D), (DN) (the table of negation $x^- = x \rightarrow 0$ is the column of 0 in the table of $\rightarrow$) and does not verify (L) for $(a, b, 0)$, (pimpl) for $(a, a, 0)$, (pi) for $(0, a)$, ($\vee$-comm) for $(a, b)$. Hence, $\mathcal{A}_4$ is a **proper involutive left-BCK algebra.**

| $\rightarrow$ | 0 | a | b | 1 |
|---|---|---|---|---|
| 0 | 1 | 1 | 1 | 1 |
| a | b | 1 | 1 | 1 |
| b | a | a | 1 | 1 |
| 1 | 0 | a | b | 1 |

and

| $x$ | $x^-$ |
|---|---|
| 0 | 1 |
| a | b |
| b | a |
| 1 | 0 |

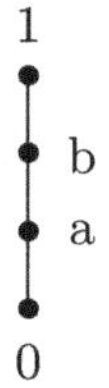


**Open problem 3.5.30** *Find an example of proper involutive BCCL algebra (i.e. not verifying (Ex)) or prove that there are no such examples.*

**Example 3.5.31 Proper involutive BCKL algebra**, i.e. not verifying (pimpl)

The algebra $\mathcal{A}_5 = (A_5 = \{0, a, b, c, 1\}, \rightarrow, 1)$, with the following table of $\rightarrow$, verifies (An), (Re), (M), (La), (Ex), (BB), (B), (*), (**), (Tr), (D), ($\vee$-comm), (L), (DN) (the table of negation - $x^- = x \rightarrow 0$ - is the column of 0 in the table of $\rightarrow$) and does not verify (pimpl) for $(a, a, 0)$, (pi) for $(0, a)$. Hence, $\mathcal{A}_5$ is a **proper involutive left-BCKL algebra.**

| $\rightarrow$ | 0 | a | b | c | 1 |
|---|---|---|---|---|---|
| 0 | 1 | 1 | 1 | 1 | 1 |
| a | a | 1 | 1 | b | 1 |
| b | c | b | 1 | a | 1 |
| c | b | 1 | 1 | 1 | 1 |
| 1 | 0 | a | b | c | 1 |

and

| $x$ | $x^-$ |
|---|---|
| 0 | 1 |
| a | a |
| b | c |
| c | b |
| 1 | 0 |

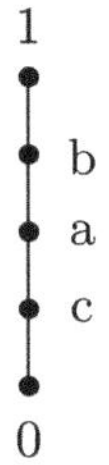


**Example 3.5.32 The smallest proper involutive aRML algebra verifying (Ex0)**

The algebra $\mathcal{A}_4 = (A_4 = \{0, a, b, 1\}, \rightarrow, 1)$, with the following table of $\rightarrow$, verifies (An), (Re), (M), (La), (Tr), (D), ($\vee$-comm), (DN), (Ex0) and does not verify (Ex) for $(a, b, a)$, (pimpl) for $(a, 0, b)$, (pi) for $(0, a)$, (BB) for $(a, b, 0)$, (**) for $(a, b, 1)$, (B) for $(a, 0, b)$, (*) for $(a, 0, b)$ and (L) for $(0, a, b)$. Since it does not verify (L), (B), (Ex), it follows that $\mathcal{A}_4$ is a **proper involutive left-aRML algebra verifyng (Ex0)**. Since $x \leq y \Longleftrightarrow x \rightarrow y = 1$, it follows that we have $0 \leq 0, a, b, 1$; $a \leq a, 1$; $b \leq b, 1$; $1 \leq 1$, hence we obtain the below Hasse-type diagram.

| $\rightarrow$ | 0 | a | b | 1 |
|---|---|---|---|---|
| 0 | 1 | 1 | 1 | 1 |
| a | a | 1 | 0 | 1 |
| b | b | 0 | 1 | 1 |
| 1 | 0 | a | b | 1 |

and

| $x$ | $x^-$ |
|---|---|
| 0 | 1 |
| a | a |
| b | b |
| 1 | 0 |

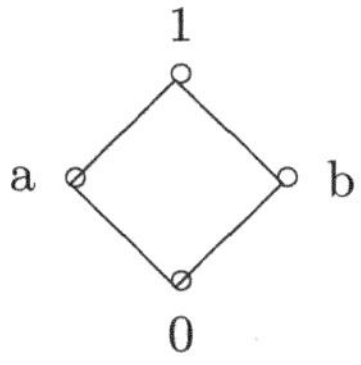

# Chapter 4

# M algebras $(A, \rightarrow, 1)$, or $(A, \rightarrow, ^{-1}, 1)$, with $1^{-1} = 1$ (Hierarchies 1 and 1')

In this chapter, we analyse in some detail the M algebras *without last element*, with some additional operations, of the form $(A, \rightarrow, 1)$ or $(A, \rightarrow, ^{-1}, 1)$ with $1^{-1} = 1$, which determine Hierarchies 1, 1'. We present new results and the connection with the implicative-group and the implicative-goop introduced in [116].

In Section 4.1, we present the list $\mathbf{C}_1$ of properties of the inverse $^{-1}$ and some results.

In Section 4.2, we discuss about the p-semisimple property (p-s) and the p-semisimple algebras.

The content of this chapter is taken mainly from [118].

Let $\mathcal{A}^L = (A, \rightarrow, 1)$ be an M algebra *without last element and not verifying (Ex)*.

Let us introduce the weaker condition [116]:

(Ex1) $\qquad x \rightarrow (y \rightarrow 1) = y \rightarrow (x \rightarrow 1)$.

We shall say that $\mathcal{A}^L$ is *strong*, if it verifies (Ex1) [116].

Note that (Ex) implies (Ex1).

## 4.1 List $\mathbf{C}_1$ of properties of the inverse $^{-1}$ and some results

Consider the following list $\mathbf{C}_1$ of properties that can be satisfied by the negation (inverse) $^{-1}$ defined on $A^L$:

(dfneg1) $\quad x^{-1} \overset{def.}{=} x \rightarrow 1$,

($\mathrm{Neg}_1$) $\quad x^{-1} = x \rightarrow 1$;

(DN) $(x^{-1})^{-1} = x$ (Double Negation);
(N1) $1^{-1} = 1$,

(Neg1) $(x \rightarrow y) \rightarrow (y^{-1} \rightarrow x^{-1}) = 1$,
(Neg1') $x \rightarrow y \leq y^{-1} \rightarrow x^{-1}$;
(Neg2) $x \rightarrow y = 1 \Longrightarrow y^{-1} \rightarrow x^{-1} = 1$ (contraposition),
(Neg2') $x \leq y \Longrightarrow y^{-1} \leq x^{-1}$ (contraposition);
(Neg3) $x \rightarrow y^{-1} = y \rightarrow x^{-1}$;
(Neg4) $x \rightarrow (x^{-1})^{-1} = 1$,
(Neg4') $x \leq (x^{-1})^{-1}$;
(Neg5) $(x \rightarrow y)^{-1} \rightarrow x = 1$,
(Neg5') $(x \rightarrow y)^{-1} \leq x$;
(Neg6) $x \rightarrow x^{-1} = x^{-1}$;

(DN1) $(y^{-1} \rightarrow x^{-1}) \rightarrow (x \rightarrow y) = 1$,
(DN1') $y^{-1} \rightarrow x^{-1} \leq x \rightarrow y$;
(DN2) $x \rightarrow y = y^{-1} \rightarrow x^{-1}$;
(DN3) $x^{-1} \rightarrow y = y^{-1} \rightarrow x$;
(DN4) $x \rightarrow y = 1 \Longleftrightarrow y^{-1} \rightarrow x^{-1} = 1$,
(DN4') $x \leq y \Longleftrightarrow y^{-1} \leq x^{-1}$;
(DN5) $x^{-1} \rightarrow (x \rightarrow y) = 1$,
(DN5') $x^{-1} \leq x \rightarrow y$;
(DN6) $x^{-1} \rightarrow x = x$.

**Proposition 4.1.1** *We have:*
*(1CE1) (dfneg1) + (Ex1) $\Longrightarrow$ (Neg3),*
*(1CE1') (dfneg1) + (Neg3) $\Longrightarrow$ (Ex1),*
*(1CE1") (dfneg1) $\Longrightarrow$ ((Ex1) $\Leftrightarrow$ (Neg3));*
*(1CE2) ($Neg_1$) + (Ex1) $\Longrightarrow$ (Neg3),*
*(1CE2') ($Neg_1$) + (Neg3) $\Longrightarrow$ (Ex1),*
*(1CE2") ($Neg_1$) $\Longrightarrow$ ((Ex1) $\Leftrightarrow$ (Neg3)).*

**Proof.** (1CE1): $x \rightarrow y^{-1} \overset{(dfneg1)}{=} x \rightarrow (y \rightarrow 1) \overset{(Ex1)}{=} y \rightarrow (x \rightarrow 1) = y \rightarrow x^{-1}$; thus, (Neg3) holds.

(1CE1'): $x \rightarrow (y \rightarrow 1) \overset{(dfneg1)}{=} x \rightarrow y^{-1} \overset{(Neg3)}{=} y \rightarrow x^{-1} \overset{(dfneg1)}{=} y \rightarrow (x \rightarrow 1)$; thus, (Ex1) holds.

(1CE1"): By (1CE1) and (1CE1').

(1CE2), (1CE2'), (1CE2"): similarly. □

**Proposition 4.1.2** *(See Proposition 3.1.17)*

*We have:*

*(1C1) (DN) + (Neg1) $\Longrightarrow$ (DN1),* *(1C1') (DN) + (DN1) $\Longrightarrow$ (Neg1),*
*(1C1") (DN) $\Longrightarrow$ ((Neg1) $\Leftrightarrow$ (DN1));*

*(1C2) (DN) + (DN2) $\Longrightarrow$ (DN3),* *(1C2') (DN) + (DN3) $\Longrightarrow$ (DN2) ;*
*(1C2") (DN) $\Longrightarrow$ ((DN2) $\Leftrightarrow$ (DN3));*

*(1C3) (DN) + (Neg3)* $\Longrightarrow$ *(DN3),* *(1C3') (DN) + (DN3)* $\Longrightarrow$ *(Neg3),*
*(1C3") (DN)* $\Longrightarrow$ *((Neg3)* $\Leftrightarrow$ *(DN3));*

*(1C4) (DN2)* $\Longrightarrow$ *(DN4);*

*(1C5) (DN) + (DN2) + (Neg5)* $\Longrightarrow$ *(DN5),*
*(1C5') (DN) + (DN2) + (DN5)* $\Longrightarrow$ *(Neg5),*
*(1C5") (DN) + (DN2)* $\Longrightarrow$ *((Neg5)* $\Leftrightarrow$ *(DN5));*

*(1C6) (DN) + (Neg6)* $\Longrightarrow$ *(DN6),* *(1C6') (DN) + (DN6)* $\Longrightarrow$ *(Neg6),*
*(1C6") (DN)* $\Longrightarrow$ *((Neg6)* $\Leftrightarrow$ *(DN6)).*

**Proof.** The same proof as for Proposition 3.1.17. □

**Lemma 4.1.3** *Let* $\mathcal{A}^L = (A^L, \to, ^{-1}, 1)$ *be an algebra of type* $(2,1,0)$ *verifying (M) and (Neg3) and such that* $1^{-1} = 1$*. Then, (*$Neg_1$*) holds.*

**Proof.** $x \to 1 = x \to 1^{-1} \overset{(Neg3)}{=} 1 \to x^{-1} \overset{(M)}{=} x^{-1}$; thus, ($\text{Neg}_1$) holds. □

The following theorem allows us to say that, roughly speaking, the M algebras $(A^L, \to, 1)$ and $(A^L, \to, ^{-1}, 1)$ are definitionally equivalent.

**Theorem 4.1.4**

*(1) Let* $\mathcal{A}^L = (A^L, \to, 1)$ *be an algebra of type* $(2,0)$ *verifying (M) and (Ex1). Define a negation* $^{-1}$ *by (dfneg*$_1$*).*
*Then,* $\alpha_1(\mathcal{A}^L) = (A^L, \to, ^{-1}, 1)$ *is an algebra of type* $(2,1,0)$ *verifying (M) and (Neg3).*

*(1') Let* $\mathcal{A}^L = (A^L, \to, ^{-1}, 1)$ *be an algebra of type* $(2,1,0)$ *verifying (M) and (Neg3) and such that* $1^{-1} = 1$*.*
*Then,* $\beta_1(\mathcal{A}^L) = (A^L, \to, 1)$ *is an algebra of type* $(2,0)$ *verifying (M) and (Ex1).*

*(2) The mappingss* $\alpha_1$ *and* $\beta_1$ *are inverse to each other.*

**Proof.** (1): By (1CE1), (dfneg1) + (Ex1) imply (Neg3); thus, (Neg3) holds.

(1'): By Lemma 4.1.3, ($\text{Neg}_1$) holds. Then, by (1CE2'), ($\text{Neg}_1$) + (Neg3) imply (Ex1); thus, (Ex1) holds.

(2): Let $(A^L, \to, 1) \overset{\alpha_1}{\longrightarrow} (A^L, \to, ^{-1}, 1) \overset{\beta_1}{\longrightarrow} (A^L, \to, 1)$.
Nothing to prove.

Conversely, let $(A^L, \to, ^{-1}, 1) \overset{\beta_1}{\longrightarrow} (A^L, \to, 1) \overset{\alpha_1}{\longrightarrow} (A^L, \to, ', 1)$.
Then, we have: $x' \overset{(dfneg_1)}{=} x \to 1 = x \to 1^{-1} \overset{(Neg3)}{=} 1 \to x^{-1} \overset{(M)}{=} x^{-1}$. □

**Proposition 4.1.5** *(See Proposition 3.1.8) We have:*
*(i) (DN) + (Ex1) + (dfneg1)* $\Longrightarrow$ *((*)* $\Leftrightarrow$ *(**)),*
*(ii) (DN) + (Ex1) + (*$Neg_1$*)* $\Longrightarrow$ *((*)* $\Leftrightarrow$ *(**)).*

**Proof.** (i): First note that (Ex1) + (dfneg1) imply (Neg3), by (1CE1); then, (DN) + (Neg3) imply (DN2), by (1C3) and (1C2"); then, (DN2) implies (DN4), by (1C4). If $x \leq y$, then by (DN4'), $y^{-1} \leq x^{-1}$; then, by (*'), $z^{-1} \to y^{-1} \leq z^{-1} \to x^{-1}$, hence $y \to z \leq x \to z$, by (DN2); thus, (**') holds. Conversely, if $x \leq y$, then $y^{-1} \leq x^{-1}$, by (DN4'); then, by (**'), $x^{-1} \to z^{-1} \leq y^{-1} \to z^{-1}$, hence $z \to x \leq z \to y$, by (DN2); thus, (*') holds.

(ii): Similarly. □

## 4.2 The p-semisimple algebras

We have seen in Chapter 1 that:

**Definition 4.2.1** The algebra $(A^L, \to, 1)$ verifying (Re) is said to be *p-semisimple*, if for all $x, y \in A^L$,
(p-s) $\quad x \leq y \Longrightarrow x = y$
(or, equivalently, by (Re), $x \leq y \Longleftrightarrow x = y$).

The subclass of p-semisimple algebras of the class **X** will be denoted by $\mathbf{X}_{(p-s)}$.
Consider the following additional properties of the list **A**:
(EqrelR=) $\quad x = y \Longleftrightarrow x \to y = 1$;
(BB=) $\quad y \to z = (z \to x) \to (y \to x)$,
(B=) $\quad y \to z = (x \to y) \to (x \to z)$,
(D=) $\quad x = (x \to y) \to y$,
(D=1) $\quad x = (x \to 1) \to 1$,
(DB=) $\quad x = (y \to 1) \to (y \to x)$.

**Remark 4.2.2** *The equivalence (EqrelR=) was introduced in* [116], *Chapter 8 (in the non-commutative case).*

**Proposition 4.2.3** *(See* [116], *Proposition 8.1.2)*
*Let $\mathcal{A}^L = (A^L, \to, 1)$ be an algebra verifying (Re). We have:*
*(0) (dfrelR) + (p-s) $\Longrightarrow$ (EqrelR=),*
*(0') (dfrelR) + (EqrelR=) $\Longrightarrow$ (p-s),*
*(0") (dfrelR) $\Longrightarrow$ ((p-s) $\Leftrightarrow$ (EqrelR=));*
*(1) (p-s) $\Longrightarrow$ (An'),*
*(1') (p-s) $\Longrightarrow$ (*'),*
*(1") (p-s) $\Longrightarrow$ (**'),*
*(1"') (p-s) $\Longrightarrow$ (Tr');*
*(2) (p-s) + (BB') $\Longrightarrow$ (BB=),*
*(2') (p-s) + (B') $\Longrightarrow$ (B=),*
*(2") (p-s) + (D') $\Longrightarrow$ (D=);*
*(3) (Ex1) + (Re) + (EqrelR=) + (dfneg1) $\Longrightarrow$ (DN);*
*(4) (EqrelR=) $\Longrightarrow$ (Re).*

**Proof.** (0), (0'), (0"): Obviously. (1), (1'), (1"), (1"'): Obviously.
(2), (2'), (2"): Immediately.

(3): $x \to (x^{-1})^{-1} \overset{(dfneg1)}{=} x \to (x^{-1} \to 1) \overset{(Ex1)}{=} x^{-1} \to (x \to 1)$
$= x^{-1} \to x^{-1} \overset{(Re)}{=} 1$; then, by (EqrelR=), we have: $(x^{-1})^{-1} = x$, i.e. (DN) holds.

(4): (Re) $\overset{def.}{\Longleftrightarrow} x \to x = 1 \overset{(EqrelR=)}{\Longleftrightarrow} x = x$, that is true. □

**Proposition 4.2.4** *(See* [116], *Proposition 8.1.2)*

*We have:*

*(1A1) (Ex) + (Re) + (EqrelR=) $\Longrightarrow$ (D=),*
*(1A1') (Ex) + (Re) + (EqrelR=) $\Longrightarrow$ (BB=); (Ex) + (D=) $\Longrightarrow$ (BB=);*
*(1A2) (Ex) + (EqrelR=) + (BB=) $\Longrightarrow$ (B=),*
*(1A2') (Ex) + (EqrelR=) + (B=) $\Longrightarrow$ (BB=),*
*(1A2") (Ex) + (EqrelR=) $\Longrightarrow$ ((B=) $\Leftrightarrow$ (BB=));*
*(1A3) (D=) + (EqrelR=) $\Longrightarrow$ (M);*
*(1A4) (M) + (BB=) $\Longrightarrow$ (Re),*
*(1A5) (M) + (BB=) $\Longrightarrow$ (D=),*
*(1A6) (Re) + (D=) $\Longrightarrow$ (EqrelR=),*
*(1A7) (M) + (BB=) $\Longrightarrow$ (Ex);*
*(1A8) (M) + (B) + (An) + (D=1) $\Longrightarrow$ (p-s).*

**Proof.** (1A1): $y \to [(y \to x) \to x] \overset{(Ex)}{=} (y \to x) \to (y \to x) \overset{(Re)}{=} 1$; then, by (EqrelR=), it follows that $y = (y \to x) \to x$, i.e. (D=) holds.

(1A1'): By (1A1), (Ex) + (Re) + (EqrelR=) $\Longrightarrow$ (D=), i.e. $z = (z \to x) \to x$; then, $y \to z = y \to [(z \to x) \to x] \overset{(Ex)}{=} (z \to x) \to (y \to x)$; thus, (BB=) holds.

(1A2): By (BB=), $y \to z = (z \to x) \to (y \to x)$; then, by (EqrelR=), $(y \to z) \to [(z \to x) \to (y \to x)] = 1$; then, by (Ex), $(z \to x) \to [(y \to z) \to (y \to x)] = 1$; then, by (EqrelR=) again, $z \to x = (y \to z) \to (y \to x)$, i.e. (B=) holds.

(1A2'): By (B=), $z \to x = (y \to z) \to (y \to x)$; then, by (EqrelR=), $(z \to x) \to [(y \to z) \to (y \to x)] = 1$; then, by (Ex), $(y \to z) \to [(z \to x) \to (y \to x)] = 1$; then, by (EqrelR=) again, $y \to z = (z \to x) \to (y \to x)$, i.e. (BB=) holds.

(1A2"): By (1A1) and (1A2').

(1A3): $1 \to x = x \overset{(EqrelR=)}{\Longleftrightarrow} (1 \to x) \to x = 1 \overset{(D=)}{\Longleftrightarrow} 1 = 1$ that is true.

(1A4): $x \to x \overset{(M)}{=} (1 \to x) \to (1 \to x) \overset{(BB=}{=} 1 \to 1 \overset{(M}{=} 1$, i.e. (Re) holds.

(1A5): $(x \to y) \to y \overset{(M)}{=} (x \to y) \to (1 \to y) \overset{(BB=)}{=} 1 \to x \overset{(M)}{=} x$, i.e. (D=) holds.

(1A6): If $x = y$, then $x \to y = y \to y \overset{(Re)}{=} 1$. Conversely, if $x \to y = 1$, then $x \overset{(D=)}{=} (x \to y) \to y = 1 \to y \overset{(Re)}{=} (y \to y) \to y \overset{(D=)}{=} y$. Thus, (EqrelR=) holds.

(1A7): By (1A4), (M) + (BB=) $\Longrightarrow$ (Re), thus (Re) holds.
By (1A5), (M) + (BB=) $\Longrightarrow$ (D=), thus (D=) holds.
By (1A6), (Re) + (D=) $\Longrightarrow$ (EqrelR=), thus (EqrelR=) holds.
By (BB=), $Y \to Z = (Z \to X) \to (Y \to X)$;
take $X = x \to z$, $Y = y$ and $Z = (y \to z) \to z \overset{(D=)}{=} y$; we obtain:
$1 \overset{(Re)}{=} y \to y \overset{(D=)}{=} y \to [(y \to z) \to z]$
$\overset{(BB=)}{=} [((y \to z) \to z) \to (x \to z)] \to [y \to (x \to z)]$

$\overset{(BB=)}{=} [x \rightarrow (y \rightarrow z)] \rightarrow [y \rightarrow (x \rightarrow z)]$;
then, by (EqrelR=), $x \rightarrow (y \rightarrow z) = y \rightarrow (x \rightarrow z)$, i.e. (Ex) holds.

(1A8): (by *Prover9*)
In (B) $((z \rightarrow y) \rightarrow ((x \rightarrow z) \rightarrow (x \rightarrow y)) = 1)$, take $z := 1$ to obtain:
$(1 \rightarrow y) \rightarrow ((x \rightarrow 1) \rightarrow (x \rightarrow y)) = 1$, hence, by (M):
(a) $y \rightarrow ((x \rightarrow 1) \rightarrow (x \rightarrow y)) = 1$.
Suppose now that $x \leq y$, i.e. $x \rightarrow y = 1$; we must prove that $x = y$. Indeed, suppose by absurdum hypothesis that $x \neq y$; then, by (An), we obtain:
(b) $y \rightarrow x \neq 1$.
By (a), since $x \rightarrow y = 1$, we obtain:
$y \rightarrow ((x \rightarrow 1) \rightarrow 1) = 1$, hence $y \rightarrow x = 1$, by (D=1): contradiction with (b). Thus, $x = y$ and (p-s) holds. □

**Corollary 4.2.5** *Any involutive BZ algebra is p-semisimple.*

**Proof.** Let $\mathcal{A}^L$ be an involutive left-BZ algebra, i.e. (M), (Re), (B), (An) and (DN), hence (D=1), hold. Then, by (1A8), (M) + (B) + (An) + (D=1) $\Longrightarrow$ (p-s). □

**Proposition 4.2.6** *(Following the numbering from Proposition 4.2.4)*
*We have:*
*(1AB0) (dfneg1) + (D=1) $\Longrightarrow$ (DN),*
*(1AB0') ($Neg_1$) + (D=1) $\Longrightarrow$ (DN),*
*(1AB0") ($Neg_1$) + (DN) $\Longrightarrow$ (D=1),*
*(1AB0"') ($Neg_1$) $\Longrightarrow$ ((DN) $\Leftrightarrow$ (D=1));*
*(1AB00) (D=) $\Longrightarrow$ (D=1);*
*(1AB1) (Ex) + (Re) + (EqrelR=) $\Longrightarrow$ (DB=),*
*(1AB1') (Ex) + (Re) + (EqrelR=) $\Longrightarrow$ (B=);*
*(1AB5) (B=) + (DB=) $\Longrightarrow$ (M),*
*(1AB5') (B=) + (M) $\Longrightarrow$ (DB=),*
*(1AB5") (B=) $\Longrightarrow$ ((DB=) $\Leftrightarrow$ (M));*
*(1AB6) (Re) + (DB=) $\Longrightarrow$ (EqrelR=),*
*(1AB6') (Re) + (DB=) $\Longrightarrow$ (D=1).*

**Proof.** (1AB0): Immediately.

(1AB0'): $(x^{-1})^{-1} = (x \rightarrow 1) \rightarrow 1 \overset{(D=1)}{=} x$.

(1AB0"): $(x \rightarrow 1) \rightarrow 1 = (x^{-1})^{-1} \overset{(DN)}{=} x$.

(1AB0"'): By (1AB0') and (1AB0").

(1AB00): Take $y = 1$ in (D=).

(1AB1): $y \rightarrow [(x \rightarrow 1) \rightarrow (x \rightarrow y)] \overset{(Ex)}{=} (x \rightarrow 1) \rightarrow (y \rightarrow (x \rightarrow y))$
$\overset{(Ex)}{=} (x \rightarrow 1) \rightarrow [x \rightarrow (y \rightarrow y)] \overset{(Re)}{=} (x \rightarrow 1) \rightarrow (x \rightarrow 1) \overset{(Re)}{=} 1$;
hence, by (EqrelR=), $y = (x \rightarrow 1) \rightarrow (x \rightarrow y)$, i.e. (DB=) holds.

(1AB1'): By (1A1), (Ex) + (Re) + (EqrelR=) imply (D=);
by (1AB1), (Ex) + (Re) + (EqrelR=) imply (DB=).
By (DB=), $z = (x \rightarrow 1) \rightarrow (x \rightarrow z)$; then,

$y \to z = y \to [(x \to 1) \to (x \to z)] \overset{(Ex)}{=} (x \to 1) \to [y \to (x \to z)]$
$\overset{(Re)}{=} (x \to (y \to y)) \to [y \to (x \to z)] \overset{(Ex)}{=} [y \to (x \to y)] \to [y \to (x \to z)]$.
It remains to prove that

$$[y \to (x \to y)] \to [y \to (x \to z)] = (x \to y) \to (x \to z),$$

i.e., by (EqrelR=), that

$$A \overset{notation}{=} [(x \to y) \to (x \to z)] \to [(y \to (x \to y)) \to (y \to (x \to z))] = 1 \tag{4.1}$$

Indeed, $A \overset{(Ex)}{=} [y \to (x \to y)] \to [((x \to y) \to (x \to z)) \to (y \to (x \to z))]$
$\overset{(Ex)}{=} [y \to (x \to y)] \to [y \to (((x \to y) \to (x \to z)) \to (x \to z))]$
$\overset{(D=)}{=} [y \to (x \to y)] \to [y \to (x \to y)] \overset{(Re)}{=} 1$, hence (6.1) holds.
Thus, $y \to z = (x \to y) \to (x \to z)$, i.e. (B=) holds.

(1AB5): $1 \to x \overset{(B=)}{=} (y \to 1) \to (y \to x) \overset{(DB=)}{=} x$, i.e. (M) holds.

(1AB5'): $(x \to 1) \to (x \to y) \overset{(B=)}{=} 1 \to y \overset{(M)}{=} y$, i.e. (DB=) holds.

(1AB5"): By (1AB5) and (1AB5').

(1AB6): If $x = y$, then $x \to y = y \to y \overset{(Re)}{=} 1$. Conversely, if $x \to y = 1$, then $y \overset{(DB=)}{=} (x \to 1) \to (x \to y) = (x \to 1) \to 1 \overset{(Re)}{=} (x \to 1) \to (x \to x) \overset{(DB=)}{=} x$. Thus, (EqrelR=) holds.

(1AB6'): $x \overset{(DB=)}{=} (y \to 1) \to (y \to x)$ and for $y = x$, we obtain:
$x = (x \to 1) \to (x \to x) \overset{(Re)}{=} (x \to 1) \to 1$; thus, (D=1) holds. □

**Theorem 4.2.7** *(See Proposition 4.2.9) If (Ex), (EqrelR=) hold, then:*

$$(B =) \Longleftrightarrow (BB =).$$

**Proof.** By (1A2"). □

**Proposition 4.2.8** *We have:*
*(1CN1) (Ex) + (Re) + (EqrelR=) + (dfneg1) $\Longrightarrow$ (DN),*
*(1CN2) (DN) + (Ex) + (DN3) + (Re) + (dfneg1) $\Longrightarrow$ (EqrelR=).*

**Proof.** (1CN1): By (1A1), (Ex) + (Re) + (EqrelR=) imply (D=); then, (D=) implies (D=1), by (1AB00), and (D=1) + (dfneg1) imply (DN), by (1AB0); thus, (DN) holds.

(1CN2): If $x = y$, then $x \to y = x \to x \overset{(Re)}{=} 1$. Conversely, if $x \to y = 1$, then: $x \overset{(DN)}{=} (x^{-1})^{-1} \overset{(dfneg1)}{=} x^{-1} \to 1 = x^{-1} \to (x \to y) \overset{(Ex)}{=} x \to [x^{-1} \to y] \overset{(DN3)}{=} x \to [y^{-1} \to x] \overset{(Ex)}{=} y^{-1} \to (x \to x) \overset{(Re)}{=} y^{-1} \to 1 = (y^{-1})^{-1} \overset{(DN)}{=} y$. Thus, (EqrelR=) holds. □

In [116], the notion of *implicative-group* was introduced, in the non-commutative case, as a term equivalent definition of the group; there are eight equivalent definitions of implicative-groups (see Section 1.5):

A *commutative implicative-group*, or a *commutative i-group* for short, is an algebra $\mathcal{G} = (G, \to, 1)$ of type $(2, 0)$ verifying one of the following:
- (M), (BB=) (Definition 1),
- (M), (BB=), (EqrelR=) (Definition 1'),
- (BB=), (D=), (EqrelR=) (Definition 2),
- (Ex), (EqrelR=) (Definition 3),
- (Ex), (M), (Re), (EqrelR=) (Definition 4),
- (Ex), (Eq$\#^=$) (Definition 5),
- (B=), (Eq$\#^=$) (Definition 6),
- (EqrelR=), (M) (not necessary), (Ex), (DN) (not necessary), (Neg) - for an algebra $(G, \to, ^-, 1)$ (Definition 7).

We prove the following results, their ideea coming from Remarks 7.1.1 (4).

**Proposition 4.2.9** *The p-semisimple left-RME = left-CI algebra coincides with the commutative i-group.*

**Proof.** (1): Let $\mathcal{G} = (G, \to, 1)$ be a commutative i-group, i.e., by Definition 3, (Ex), (EqrelR=) hold. We must prove that (M), (Re) and (p-s) hold. Indeed, by Proposition 4.2.3 (4), (EqrelR=) implies (Re); thus, (Re) holds; also, by Proposition 4.2.3 (0'), (dfrelR) + (EqrelR=) imply (p-s), thus (p-s) holds; then, by (1A1), (Ex) + (Re) + (EqrelR=) imply (D=) and by (1A3), (D=) + (EqrelR=) imply (M), thus (M) holds. Thus, $\mathcal{G}$ is a p-semisimple left-CI algebra.

(2): Let $\mathcal{A}^L = (A^L, \to, 1)$ be a p-semisimple left-CI algebra, i.e. (M), (Ex), (Re), (p-s) hold. We prove that (EqrelR=) holds. Indeed, by Proposition 4.2.3, (0), (dfrelR) + (p-s) imply (EqrelR=). Thus, $\mathcal{A}^L$ is a commutative i-group, by Definition 3. □

**Theorem 4.2.10** *We have:*
*(i) Any p-semisimple left-CI algebra is involutive.*
*(ii) Any involutive left-CI algebra is p-semisimple.*
*(iii) The p-semisimple left-CI algebras coincide with the involutive left-CI algebras.*

**Proof.** (i): Let $\mathcal{A}^L = (A^L, \to, 1)$ be a p-semisimple left-CI algebra, i.e. (M), (Ex), (Re), (p-s) hold. Then, by Proposition 4.2.3 (0), (dfrelR) + (p-s) imply (EqrelR=); by (1CN1), (Ex) + (Re) + (EqrelR=) + (dfneg1) $\Longrightarrow$ (DN), thus (DN) holds, where $x^{-1} \overset{def.}{=} x \to 1$.

(ii): Let $\mathcal{A}^L = (A^L, \to, 1)$ be an involutive left-CI algebra, i.e. (M), (Ex), (Re), ($\text{Neg}_1$), (DN) hold, where $x^{-1} \overset{def.}{=} x \to 1$, by (dfneg1). Then, ($\text{Neg}_1$) + (Ex) imply (Neg3) and (DN) + (Neg3) imply (DN3), by (1C3). Now, by (1CN2), (DN) + (Ex) + (DN3) + (Re) + (dfneg1) imply (EqrelR=) and by Proposition 4.2.3 (0'), (dfrelR) + (EqrelR=) imply (p-s); thus, (p-s) holds.

(iii): By above (i), (ii). □

We write: $\mathbf{CI}_{(p-s)} = \mathbf{CI}_{(DN)} =$ **commutative i-group**.

**Corollary 4.2.11** *The p-semisimple CI**, pre-BCI, BCH, BCH**, BCI algebras coincide with the involutive corresponding ones.*

**Proof.** All these algebras are p-semisimple CI algebras, by definition; then, apply Theorem 4.2.10 (iii). □

**Corollary 4.2.12**
*(i) The involutive CI algebras coincide with the involutive CI** algebras.*
*(ii) The involutive BCH algebras coincide with the involutive BCH** algebras.*

**Proof.** (i): Let $\mathcal{A}^L = (A^L, \rightarrow, 1)$ be an involutive left-CI algebra, i.e. (M), (Ex), (Re), (Neg$_1$), (DN) hold, where $x^{-1} = x \rightarrow 1$, by (dfneg1). By Theorem 4.2.10 (ii), (p-s) holds and by Proposition 4.2.3 (1"), (**) holds; thus, $\mathcal{A}$ is a left-CI** algebra. Since, conversely, any involutive left-CI** algebra is an involutive left-CI algebra, they coincide.

(ii): Let $\mathcal{A}^L = (A^L, \rightarrow, 1)$ be an involutive left-BCH algebra; then, $\mathcal{A}^L$ is an involutive left-CI algebra, by definition; then, by above (i), (p-s) and (**) hold; thus, $\mathcal{A}$ is a left-BCH** algebra. Since, conversely, any involutive left-BCH** algebra is an involutive left-BCH algebra, they coincide. □

Moreover, we have:

**Proposition 4.2.13** *Let $\mathcal{A}^L = (A^L, \rightarrow, 1)$ be a p-semisimple left-RME = left-CI algebra. Then, $\mathcal{A}^L$ is a (p-semisimple) left-BCI algebra.*

**Proof.** Let $\mathcal{A}^L = (A^L, \rightarrow, 1)$ be a p-semisimple left-RME= left-CI algebra, i.e. (M), (Ex), (Re) and (p-s) hold. We prove that (An) and (BB=) hold. Indeed, by Proposition 4.2.3 (1), (p-s) implies (An'), thus (An) holds. By Proposition 4.2.3 (0), (dfrelR) + (p-s) imply (EqrelR=) and by (1A1'), (Ex) + (Re) + (EqrelR=) imply (BB=), thus (BB=) holds. Thus, $\mathcal{A}^L$ is a p-semisimple left-BCI algebra, by Definition 3. □

Since conversely, any p-semisimple left-BCI algebra is a p-semisimple left-CI algebra, then, by Propositions 4.2.9, 4.2.13 and Theorem 4.2.10, we obtain:

**Corollary 4.2.14** *(See* [116], *Remark 13.1.16) (See the inspireing Corollary 7.3.13)*
*The p-semisimple (and involutive) left-CI, left-CI**, left-pre-BCI, left-BCH, left-BCH** and left-BCI algebras coincide (with the commutative i-groups).*

We write:
$\mathbf{CI}_{(p-s)} = \mathbf{CI}_{(DN)} = \mathbf{CI^{**}}_{(p-s)} = \mathbf{CI^{**}}_{(DN)} = \textbf{pre-BCI}_{(p-s)} = \textbf{pre-BCI}_{(DN)} = \mathbf{BCH}_{(p-s)} = \mathbf{BCH}_{(DN)} = \mathbf{BCH^{**}}_{(p-s)} = \mathbf{BCH^{**}}_{(DN)} = \mathbf{BCI}_{(p-s)} = \mathbf{BCI}_{(DN)}$ = **commutative i-group**.

By Corollary 4.2.14, it follows that the left-CI, left-CI**, left-pre-BCI, left-BCH, left-BCH**, left-BCI algebras are all generalizations *with the exchange property (Ex)* of the commutative i-group, while the left-RM, left-pre-BZ, left-BH, left-BZ algebras are all generalizations *without the exchange property (Ex)* of the commutative i-group.

**Proposition 4.2.15**
*(i) The p-semisimple RM and aRM = BH algebras coincide.*
*(ii) The p-semisimple pre-BZ and BZ algebras coincide.*

**Proof.** By Proposition 4.2.3 (1), (p-s) implies (An), then use the definitions. □

We write: $\mathbf{RM}_{(p-s)} = \mathbf{BH}_{(p-s)}$ and $\mathbf{pre\text{-}BZ}_{(p-s)} = \mathbf{BZ}_{(p-s)}$.

**Remark 4.2.16** *The notion of* **B-algebra** *was introduced by J. Neggers and H.S. Kim in 2002 [161] (in a dual reversed form), as an algebra $\mathcal{A}^L = (A^L, \to, 1)$ of type $(2,0)$ verifying the properties: (Re), (M) and $z \to (y \to x) = ((y \to 1) \to z) \to x$. A. Walendziak found [208] the equivalent properties: (Re), $(x \to 1) \to 1 = x$ and (B=); thus, by (dfneg1) ($x^{-1} \overset{def.}{=} x \to 1$), $\mathcal{A}^L$ verifies (DN); resuming, it verifies (M), (Re), (B=) and (DN), hence a B-algebra is an involutive p-semisimple pre-BZ algebra, hence an* **involutive p-semisimple BZ algebra**, *by above Proposition. Note that a B-algebra does not verify (Ex1), hence (Ex) - see the following example, found by the Mace4 program. Consider the B-algebra $\mathcal{A}_6^L = (A_6 = \{0,2,3,4,5,1\}, \to, 1)$ with the following table of $\to$:*

| $\to$ | 0 | 2 | 3 | 4 | 5 | 1 |
|---|---|---|---|---|---|---|
| 0 | 1 | 3 | 2 | 5 | 4 | 0 |
| 2 | 3 | 1 | 5 | 2 | 0 | 4 |
| 3 | 4 | 5 | 1 | 0 | 2 | 3 |
| 4 | 5 | 4 | 0 | 1 | 3 | 2 |
| 5 | 2 | 0 | 4 | 3 | 1 | 5 |
| 1 | 0 | 2 | 3 | 4 | 5 | 1 |

, *with*

| $x$ | $x^{-1} \overset{def.}{=} x \to 1$ |
|---|---|
| 0 | 0 |
| 2 | 4 |
| 3 | 3 |
| 4 | 2 |
| 5 | 5 |
| 1 | 1 |

.

*It does not verify (Ex1) (hence (Ex)) for $(x,y) = (0,2)$:*
$5 = 0 \to 4 = 0 \to (2 \to 1) \neq 2 \to (0 \to 1) = 2 \to 0 = 3.$
*Thus, $\mathcal{A}_6^L$ is not a commutative implicative-group (and Theorem 17.2.1 can not be applied). Consequently,* **it is not true** *that the class of B-algebras coincides with the class of (commutative) groups, as paper [141] states.*

In [116] also, the notions of *implicative-hub* ([116] Definition 11.1.1), *implicative-moon* ([116] Definition 11.1.10) and *implicative-goop* ([116] Definition 11.1.35) were introduced and studied, in the non-commutative case, as generalizations *without the exchange property* of the i-group. In the commutative case, we have the following definitions:

(i) A *commutative implicative-hub*, or a *commutative i-hub* for short, is an algebra $\mathcal{G} = (G, \to, 1)$ of type $(2,0)$ verifying (M), (EqrelR=).

(ii) A *commutative implicative-moon*, or a *commutative i-moon* for short, is an algebra $\mathcal{G} = (G, \to, ^{-1}, 1)$ of type $(2,1,0)$ verifying (M), (EqrelR=) and ($\mathrm{Neg}_1$).

(iii) A *commutative implicative-goop*, or a *commutative i-goop* for short, is a strong commutative i-moon (i.e. verifying (Ex1)).

**Proposition 4.2.17** *The commutative i-hubs are definitionally equivalent (d.e.) to the commutative i-moons.*

**Proof.** Immediately. □

**Proposition 4.2.18** *The p-semisimple left-BH algebras are d.e. (coincide) with the commutative i-hubs.*

**Proof.** (1): Let $\mathcal{A}^L = (A^L, \rightarrow, 1)$ be a p-semisimple left-BH algebra, i.e. (M), (Re), (An) and (p-s) hold. By Proposition 4.2.3 (0), (dfrelR) + (p-s) imply (EqrelR=); thus, $\mathcal{A}^L$ is a commutative i-hub.

(1'): Conversely, let $\mathcal{A}^L = (A^L, \rightarrow, 1)$ be a commutative i-hub, i.e. (M) and (EqrelR=) hold. By Proposition 4.2.3 (4), (EqrelR=) implies (Re), thus (Re) holds; by Proposition 4.2.3 (0'), (dfrelR) + (EqrelR=) imply (p-s) and by (1), (p-s) implies (An'), thus (An) holds. Thus, $\mathcal{A}^L$ is a p-semisimple left-BH algebra. □

**Corollary 4.2.19·** *The strong p-semisimple left-BH algebras are d.e. (coincide) with the commutative i-goops.*

**Proof.** Since the p-semisimple left-BH algebras are d.e. to the commutative i-moons, by Propositions 4.2.18 and 4.2.17. □

**Proposition 4.2.20** *Any strong p-semisimple left-BH algebra (i.e. any commutative i-goop) is involutive.*

**Proof.** Let $\mathcal{A}^L = (A^L, \rightarrow, 1)$ be a strong p-semisimple left-BH algebra, i.e. (M), (Re), (An), (p-s), (Ex1) hold. By Proposition 4.2.3 (0), (3), (dfrelR) + (p-s) imply (EqrelR=) and (Ex1) + (Re) + (EqrelR=) imply (DN), thus (DN) holds, where $x^{-1} \overset{def.}{=} x \rightarrow 1$. □

The converse of Proposition 4.2.20 does not hold: there are involutive p-semisimple left-BH algebras (i.e. involutive commutative i-moons) that are not strong; see the following Example 2.

**Examples 4.2.21** (See Examples 7.3.17)

• **Example 1.**

The algebra $\mathcal{A}_4^1 = (A_4 = \{a, b, c, 1\}, \rightarrow_1, ^{-1}, 1)$, with the following tables of $\rightarrow_1$, $^{-1}$, is a **commutative i-moon that is not involutive**, because (DN) is not verified for $x = b$.

| $\rightarrow_1$ | a | b | c | 1 |
|---|---|---|---|---|
| a | 1 | a | a | a |
| b | a | 1 | a | a |
| c | b | c | 1 | a |
| 1 | a | b | c | 1 |

and

| $x$ | $x^{-1}$ |
|---|---|
| a | a |
| b | a |
| c | a |
| 1 | 1 |

Equivalently, $(A_4 = \{a, b, c, 1\}, \rightarrow_1, 1)$, with the above table of $\rightarrow_1$, is a **p-semisimple left-BH algebra that is not involutive**, since (M), (Re), (An), (p-s) hold, while (Ex) does not hold for $(a, b, a)$, (Ex1) for $(a, b)$, (B) for $(a, b, c)$, (BB) for $(a, b, 1)$, (DN) for $x = b$, where $x^{-1} \overset{def.}{=} x \rightarrow 1$.

• **Example 2.**

The algebra $\mathcal{A}_4^2 = (A_4 = \{a, b, c, 1\}, \rightarrow_2, ^{-1}, 1)$, with the following tables of $\rightarrow_2$, $^{-1}$, is an **involutive commutative i-moon that is not strong** (see [116], Examples

11.1.40), because (Ex1) is not verified for $(x, y) = (a, b)$: $a = a \to b = a \to (b \to 1) \neq b \to (a \to 1) = b \to c = b$.

| $\to_2$ | a | b | c | 1 |
|---|---|---|---|---|
| a | 1 | a | b | c |
| b | a | 1 | b | b |
| c | a | a | 1 | a |
| 1 | a | b | c | 1 |

and

| $x$ | $x^{-1}$ |
|---|---|
| a | c |
| b | b |
| c | a |
| 1 | 1 |

Equivalently, $(A_4 = \{a, b, c, 1\}, \to_2, 1)$, with the above table of $\to_2$, is an **involutive p-semisimple left-BH algebra that is not strong**, since (M), (Re), (An), (p-s), (DN) hold, where $x^{-1} \overset{def.}{=} x \to 1$, while (Ex) does not hold for $(a, b, a)$, (Ex1) for $(a, b)$, (B) for $(a, b, a)$, (BB) for $(a, a, b)$. Note that Theorem 17.2.1 cannot be applied, because (Ex1) is not verified.

• **Example 3.**
The algebra $\mathcal{A}_4^3 = (A_4 = \{a, b, c, 1\}, \to_3, ^{-1}, 1)$, with the following tables of $\to_3$, $^{-1}$, is a **commutative i-goop**.

| $\to_3$ | a | b | c | 1 |
|---|---|---|---|---|
| a | 1 | b | b | b |
| b | b | 1 | a | a |
| c | a | b | 1 | c |
| 1 | a | b | c | 1 |

and

| $x$ | $x^{-1}$ |
|---|---|
| a | b |
| b | a |
| c | c |
| 1 | 1 |

Equivalently, $(A_4 = \{a, b, c, 1\}, \to_3, 1)$, with the above table of $\to_3$, is a **strong p-semisimple left-BH algebra**, since (M), (Re), (An), (p-s), (Ex1) (hence (DN)) hold (where $x^{-1} \overset{def.}{=} x \to 1$), while (Ex) does not hold for $(a, b, a)$, (B) for $(a, b, a)$, (BB) for $(a, a, b)$. Note that Theorem 17.2.1 can now be applied, because (Ex1) is verified, and we obtain the algebra from Examples 7.3.17 (3m).

Thus, the connections between the p-semisimple left-BH algebras (commutative i-moons), the involutive p-semisimple left-BH algebras (involutivecommutative i-moons) and the strong p-semisimple left-BH algebras (commutative i-goops) are presented in the following Figure 4.1 (see [116], Figure 11.4, in the non-commutative case).

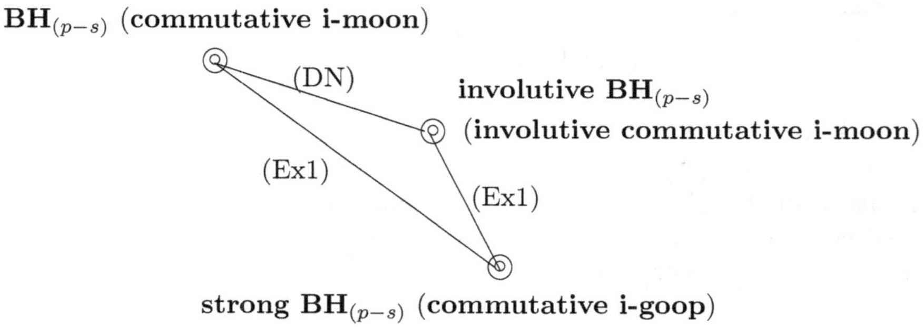


Figure 4.1: The hierarchy between commutative i-moons, involutive commutative i-moons and commutative i-goops

# Part II

# Magmas. Commutative magmas

# Chapter 5

# Preliminaries/Miscellany

In this chapter, we recall some notions needed in Part II (in commutative case) or related to notions from Chapter 1 (in non-commutative case).

In Section 5.1, we recall the notions of magma and commutative magma and discuss about semigroups and monoids.

In Section 5.2, we recall definitions and facts about lattices, namely Ore and Dedekind equivalent definitions of lattices.

In Section 5.3, we discuss about partially ordered magmas (po-magmas) and about lattice ordered magmas ($l$-magmas), namely about po-s ($l$-s), po-m ($l$-m), po-mm ($l$-mm), po-im ($l$-im). We also discuss about po-magmas with (pR') and with (pPR') and about the connections between them.

In Section 5.4, we recall things about groups, partially ordered groups, lattice ordered groups from [116] and we present some examples.

In Section 5.5, we discuss about the m-p-semisimple property (m-p-s) and about m-p-semisimple po-magmas.

## 5.1 Magmas. Commutative magmas

Recall the following well-known definitions.

**Definitions 5.1.1**

- A *magma* is an algebra $(A, \otimes)$ of type (2).

- A *unital magma* is an algebra $(A, \otimes, u)$ of type $(2, 0)$ verifying the axiom:
(PU) $u \otimes x = x = x \otimes u$, for all $x \in A$.

We have a *left-notation* and a *right-notation* for magmas. A magma in left-notation (right-notation) will be called a *left-magma* (*right-magma*, respectively). Thus, we have, for example:

**Definitions 5.1.2**

- *A unital left-magma is an algebra* $(A^L, \odot, 1)$ *of type* $(2, 0)$ *verifying the axiom: for all* $x, y \in A^L$,
*(PU)* $1 \odot x = x = 1 \odot x$, *i.e.* 1 *is the unit element of* $A^L$.
- *A unital right-magma is an algebra* $(A^R, \oplus, 0)$ *of type* $(2, 0)$ *verifying the axiom: for all* $x, y \in A^R$,
*(*$PU^R$*)* $0 \oplus x = x = 0 \oplus x$, *i.e.* 0 *is the unit element of* $A^R$.

Denote by **magma** the class of all left-magmas and by **uagma** the class of all unital left-magmas. We have: **umagma** $\subset$ **magma**.

**Definitions 5.1.3**
- A *commutative left-magma* is a left-magma $(A^L, \odot)$ verifying: for all $x, y \in A^L$,
(Pcomm) $x \odot y = y \odot x$.
- A *unital commutative left-magma* is an algebra $(A^L, \odot, 1)$ of type $(2, 0)$ verifying the axioms: for all $x, y \in A^L$,
(Pcomm) $x \odot y = y \odot x$,
(U) $1 \odot x = x$.

Denote by **c. magma** the class of all commutative left-magmas and by **uc. magma** the class of all unital commutative left-magmas. We have: **uc. magma** $\subset$ **c. magma**.

### 5.1.1 Semigroups. Monoids

Recall the following definitions.

**Definitions 5.1.4** (In left-notation)
(1) A *left-semigroup* is a left-magma $\mathcal{A}^L = (A^L, \odot)$ verifying the axiom: for all $x, y, z \in A^L$,
(Pass) (associativity of product) $x \odot (y \odot z) = (x \odot y) \odot z$.
(2) A *left-monoid* is a left-magma $\mathcal{A}^L = (A^L, \odot, 1)$ verifying the axioms (Pass) and (PU), i.e. is a unital left-semigroup.

Denote by **semigroup** the class of all left-semigroups and by **monoid** the class of all left-monoids. Hence we have:

**semigroup** = **magma** + (Pass) and **monoid** = **umagma** + (Pass),

(5.1) **monoid** $\subset$ **semigroup** $\subset$ **magma**.

**Definitions 5.1.5** (In left-notation)
(1) A *commutative left-semigroup* is a commutative left-magma $\mathcal{A}^L = (A^L, \odot)$ verifying the axiom: for all $x, y, z \in A^L$,
(Pass) (associativity) $x \odot (y \odot z) = (x \odot y) \odot z$.
(2) A *commutative left-monoid* is a commutative left-magma $\mathcal{A}^L = (A^L, \odot, 1)$ verifying the axioms (Pass) and (PU), i.e. is a unital commutative left-semigroup.

Denote by **c. semigroup** the class of all commutative left-semigroups and by **c. monoid** the class of all commutative left-monoids. Hence we have:

**c. semigroup** = **c. magma** + (Pass) and **c. monoid** = **uc. magma** + (Pass),

(5.2) $$\textbf{c.monoid} \subset \textbf{c.semigroup} \subset \textbf{c.magma}.$$

# 5.2 Lattices

## 5.2.1 Ore and Dedekind lattices. Their equivalence

**Definition 5.2.1** A poset (partially ordered set) $\mathcal{A} = (A, \leq^O)$ will be said to be an *Ore lattice*, if for each two elements $x, y \in A$, there exist $\inf(x, y)$ and $\sup(x, y)$.

Moreover, if there exist $0, 1 \in A$ such that $0 \leq^O x \leq^O 1$ for all $x \in A$, then $\mathcal{A}$ is said to be a *bounded* Ore lattice (with last (top) element 1 and first (bottom) element 0) and is denoted by $\mathcal{A} = (A, \leq^O, 0, 1)$.

In an Ore lattice, the following are equivalent: for all $x, y \in A$,
(i) $x \leq^O y$, (ii) $\inf(x, y) = x$, (iii) $\sup(x, y) = y$.

**Definition 5.2.2** An algebra $\mathcal{A} = (A, \wedge, \vee)$ or, dually, $\mathcal{A} = (A, \vee, \wedge)$, of type $(2, 2)$, will be said to be a *Dedekind lattice*, if the following properties hold: for all $x, y, z \in A$,

| | | |
|---|---|---|
| (m-Wid) | (idempotency of $\wedge$) | $x \wedge x = x$, |
| (m-Wcomm) | (commutativity of $\wedge$) | $x \wedge y = y \wedge x$, |
| (m-Wass) | (associativity of $\wedge$) | $x \wedge (y \wedge z) = (x \wedge y) \wedge z$, |
| (m-Wabs) | (absorbtion of $\wedge$ over $\vee$) | $x \wedge (x \vee y) = x$, and also |
| (m-Vid) | (idempotency of $\vee$) | $x \vee x = x$, |
| (m-Vcomm) | (commutativity of $\vee$) | $x \vee y = y \vee x$, |
| (m-Vass) | (associativity of $\vee$) | $(x \vee y) \vee z = x \vee (y \vee z)$, |
| (m-Vabs) | (absorbtion of $\vee$ over $\wedge$) | $x \vee (x \wedge y) = x$, |

where 'W' comes from 'wedge' (the LaTeX command for the meet, $\wedge$), 'V' comes from 'vee' (the LaTeX command for the join, $\vee$) and 'm' comes from 'magma'.

Moreover, if there exist $0, 1 \in A$ such that: for all $x \in A$,
(m-WU) $1 \wedge x = x$ and, dually,
(m-VU) $0 \vee x = x$,
then $\mathcal{A}$ is said to be a *bounded* Dedekind lattice (with last element 1 and first element 0) and is denoted by $\mathcal{A} = (A, \wedge, \vee, 0, 1)$ or, dually, by $\mathcal{A} = (A, \vee, \wedge, 0, 1)$.

Note that a *bounded Dedekind lattice* is (a pair of) both a *commutative left-monoid* (for $\odot = \wedge$, since (m-WU), (m-Wcomm), (m-Wass) hold) and a *commutative right-monoid* (for $\oplus = \vee$, since (m-VU), (m-Vcomm), (m-Vass) hold), plus additional properties.

In a Dedekind lattice, we have the equivalence: for all $x, y \in A$,

(EqWV) $$x \wedge y = x \Longleftrightarrow x \vee y = y.$$

**Naming convention for the dual lattices** We have given names to the dual lattices: $(A, \wedge, \vee)$ is the *left-lattice* and $(A, \vee, \wedge)$ is the *right-lattice* (names coming from the *left-continuity* of a t-norm and the *right-continuity* of a t-conorm; see more on *left- and right- algebras* in [104]).

Recall finally that the two kinds of lattices are *definitionally equivalent* (d.e.):

**Theorem 5.2.3**
*(1) Let $\mathcal{A} = (A, \leq^O)$ be an Ore lattice. Define $\delta(\mathcal{A}) \stackrel{def.}{=} (A, \wedge, \vee)$, where for all $x, y \in A$:*

$$x \wedge y \stackrel{def.}{=} \inf(x, y) \quad \text{and} \quad x \vee y \stackrel{def.}{=} \sup(x, y).$$

*Then, $\delta(\mathcal{A})$ is a Dedekind lattice.*

*(1') Let $\mathcal{A} = (A, \wedge, \vee)$ be a Dedekind lattice. Define $\omega(\mathcal{A}) \stackrel{def.}{=} (A, \leq^O)$, where for all $x, y \in A$:*

*(m-dfOW)* $x \leq^O y \stackrel{def.}{\Longleftrightarrow} x \wedge y = x$ *or, equivalently,*

*(m-dfOV)* $x \leq^O y \stackrel{def.}{\Longleftrightarrow} x \vee y = y.$

*Then, $\omega(\mathcal{A})$ is an Ore lattice.*

*(2) The two mappingss, $\delta$ and $\omega$, are mutually inverse.*

**Corollary 5.2.4** *Let $\mathcal{A} = (A, \wedge, \vee, 0, 1)$ be a bounded Dedekind lattice. Then, we have the equivalences:*

$$(m\text{-}WU) \Longleftrightarrow (m\text{-}VL) \text{ and } (m\text{-}VU) \Longleftrightarrow (m\text{-}WL),$$

*where:*

*(m-WL)* $0 \wedge x = 0$ *and, dually,*

*(m-VL)* $1 \vee x = 1.$

In this book, we shall work in general with Dedekind lattices, simply called lattices for short.

**Proposition 5.2.5** *In a lattice $\mathcal{A} = (A, \wedge, \vee)$, the following properties are equivalent:*

| | | |
|---|---|---|
| *(m-Wdis)* | *(distributivity of $\vee$ over $\wedge$)* | $z \wedge (x \vee y) = (z \wedge x) \vee (z \wedge y)$ |
| *(m-Vdis)* | *(distributivity of $\wedge$ over $\vee$)* | $z \vee (x \wedge y) = (z \vee x) \wedge (z \vee y),$ |
| *(m-Dis)* | | $(x \vee y) \wedge z \leq^O x \vee (y \wedge z).$ |

A lattice is *distributive*, if one of the equivalent properties from Proposition 5.2.5 holds.

## 5.3 Partially (lattice) ordered magmas

Most of the definitions and results from this section are taken from [116].

### 5.3.1 Partially (lattice) ordered magmas: po-s ($l$-s), po-m, po-mm, po-im ($l$-m, $l$-mm, $l$-im)

**Definitions 5.3.1**

(i) A *partially ordered magma*, or a po-magma for short, is a structure $(A, \leq, \otimes)$, such that the reduct $(A, \otimes)$ is a magma and $(A, \leq)$ is a poset (i.e. $\leq$ is a partial order on $A$). No compatibility between $\leq$ and $\otimes$ is assumed.

(ii) If the partial order $\leq$ is a lattice order, then we say that we have a *lattice ordered magma*, or an *l-magma* for short.

Since $\leq$ is an order relation, we shall denote by (Re') ($x \leq x$, for all $x$) the reflexivity, by (Tr') ($x \leq y$ and $y \leq z$ imply $x \leq z$, for all $x, y, z$) the transitivity and by (An') ($x \leq y$ and $y \leq x$ imply $x = y$, for all $x, y$) the antisymmetry.

Note that if $\leq$ is a lattice order, with lattice operations $\wedge$ and $\vee$, then the $l$-magma is an *algebra* (in the universal algebra sense).

We recall/introduce the following po-magmas: po-ss and po-ms, po-mms, po-ims, and their corresponding lattice ordered ones.

- **po-ss ($l$-ss)**

**Definitions 5.3.2** (Definition of po-s ($l$-s))

(i) A *left-partially ordered semigroup*, or a *left-po-s* for short, is a structure

$$\mathcal{A}^L = (A^L, \leq^e, \odot)$$

such that the reduct $(A^L, \odot)$ is a semigroup (i.e. (Pass) holds), $(A^L, \leq^e)$ is a poset ('L' comes from 'left' and 'e' means *external* partial order), and $\leq^e$ is compatible with $\odot$, i.e. the following property holds: for all $x, y, a \in A^L$,

(pCp') $\qquad x \leq^e y \Longrightarrow (a \odot x \leq^e a \odot y$ and $x \odot a \leq^e y \odot a)$.

(i') Dually, a *right-partially ordered semigroup*, or a *right-po-s* for short, is a structure

$$\mathcal{A}^R = (A^R, \geq^e, \oplus)$$

such that the reduct $(A^R, \oplus)$ is a semigroup (i.e. (Sass) holds), $(A^R, \leq^e)$ is a poset ('R' comes from 'right' and $x \leq^e y \Longleftrightarrow y \geq^e x$, for all $x, y \in A^R$), and $\geq^e$ is compatible with $\oplus$, i.e. the following property holds: for all $x, y, a \in A^R$,

(pCp') $\qquad x \geq^e y \Longrightarrow (a \oplus x \geq^e a \oplus y$ and $x \oplus a \geq^e y \oplus a)$.

(ii) If the partial *external* order $\leq^e$, or $\geq^e$, is a lattice order, then we say that we have a *left-lattice ordered semigroup* (*left-l-s*), denoted by $(A^L, \wedge, \vee, \odot)$, or a *right-lattice ordered semigroup* (*right-l-s*), denoted by $(A^R, \vee, \wedge, \oplus)$, respectively.

Denote by **po-s** the class of all left-po-ss and by **po-s**$^R$ the class of all right-po-ss. In lattice ordered case, denote by $l$-**s** the class of all left-$l$-ss and by $l$-**s**$^R$ the class of all right-$l$-ss.

- **po-ms, po-mms, po-ims ($l$-ms, $l$-mms, $l$-ims)**

**Definitions 5.3.3** (Definition of po-m ($l$-m)) (The dual ones are omitted)

(i) A *left-partially ordered monoid*, or a *left-po-m* for short, is a structure

$$\mathcal{A}^L = (A^L, \leq, \odot, 1)$$

such that the reduct $(A^L, \odot, 1)$ is a monoid (i.e. (Pass), (PU) hold), $(A^L, \leq)$ is a poset and the property (pCp') holds, i.e. $\mathcal{A}^L$ is a unital left-po-s.

(ii) If the partial order $\leq$ is a lattice order, then we say that we have a *left-lattice ordered monoid (left-l-m)*.

Denote by **po-m** (**$l$-m**) the class of all left-po-ms (left-$l$-ms, respectively).

We obviously have the inclusions (the dual ones are omitted):

$$\mathbf{po-m} \subset \mathbf{po-s} \quad \textit{and} \quad \mathbf{l-m} \subset \mathbf{l-s}. \tag{5.3}$$

**Definitions 5.3.4** (Definition of po-mm ($l$-mm)) (The dual ones are omitted)

(i) A *left-partially ordered maximal monoid*, or a *left-po-mm* for short, is a left-po-m

$$\mathcal{A}^L = (A^L, \leq, \odot, 1),$$

i.e. (Pass), (PU), (pCp') hold, such that $(A^L, \leq, 1)$ is a poset with 1 as maximal element, i.e. the additional property (N') holds: for all $x \in A^L$,

(N') $1 \leq x$ implies $x = 1$;

*maximal* means that the unit element of the monoid is the maximal element of the poset.

(ii) If the partial order $\leq$ is a lattice order, then we say that we have a *left-lattice ordered maximal monoid (left-l-mm)*

Denote by **po-mm** (**$l$-mm**) the class of all left-po-mms (left-$l$-mms, respectively).

Hence, a left-po-mm is a left-po-m verifying (N') and, similarly, in the lattice ordered case. We write:

$$\mathbf{po\text{-}mm} = \mathbf{po\text{-}m} + \text{(N')} \qquad \text{and} \qquad \boldsymbol{l}\mathbf{\text{-}mm} = \boldsymbol{l}\mathbf{\text{-}m} + \text{(N')}.$$

Hence, we obviously have the inclusions (the dual ones are omitted):

$$\mathbf{po-mm} \subset \mathbf{po-m} \quad \textit{and} \quad \mathbf{l-mm} \subset \mathbf{l-m}. \tag{5.4}$$

**Definitions 5.3.5** (Definition of po-im ($l$-im)) (The dual ones are omitted)

(i) A *left-partially ordered integral monoid*, or a *left-po-im* for short, is a left-po-m

$$\mathcal{A}^L = (A^L, \leq, \odot, 1),$$

i.e. (Pass), (PU), (pCp') hold, such that $(A^L, \leq, 1)$ is a poset with 1 as greatest element, i.e. the additional property (La') holds: for all $x \in A^L$,

(La') (Last element) $x \leq 1$;

*integral* means that the unit element of the monoid is the greatest element of the poset.

(ii) If the partial order $\leq$ is a lattice order, then we say that we have a *left-lattice ordered integral monoid (left-l-im)*.

Denote by **po-im** the class of all left-po-ims. In lattice ordered case, denote by **$l$-im** the class of all left-$l$-ims.

A left-$l$-m (left-$l$-mm, left-$l$-im) will be denoted by $(A^L, \wedge, \vee, \odot, 1)$. A right-$l$-m (right-$l$-mm, right-$l$-im) will be denoted by $(A^R, \vee, \wedge, \oplus, 0)$.

Hence, a left-po-im is a left-po-m verifying (La') and, similarly, in the lattice ordered case. We write:

$$\textbf{po-im} = \textbf{po-m} + \text{(La')} \qquad \text{and} \qquad l\textbf{-im} = l\textbf{-m} + \text{(La')}.$$

**Lemma 5.3.6** *Let $\mathcal{A}^L = (A^L, \leq, \odot, 1)$ be a left-po-im. Then, $\mathcal{A}^L$ is a left-po-mm.*

**Proof.** We have to prove that (La') implies (N'). Indeed, by (A1), (La') + (An') imply (N'). $\Box$

We then have, by above Lemma 5.3.6, the inclusions (the dual ones are omitted):

$$\textbf{po}-\textbf{im} \subset \textbf{po}-\textbf{mm} \quad \textit{and} \quad \textbf{l}-\textbf{im} \subset \textbf{l}-\textbf{mm}. \tag{5.5}$$

Finally, by inclusions 5.3, 5.4, 5.5, we have the inclusions (the dual case is omitted):

$$\textbf{po}-\textbf{im} \subset \textbf{po}-\textbf{mm} \subset \textbf{po}-\textbf{m} \subset \textbf{po}-\textbf{s} \quad \textit{and} \tag{5.6}$$

$$\textbf{l}-\textbf{im} \subset \textbf{l}-\textbf{mm} \subset \textbf{l}-\textbf{m} \subset \textbf{l}-\textbf{s}.$$

- **The commutative case**

Denote by **po-cs** (**$l$-cs**) the class of all commutative left-po-ss (left-$l$-ss, respectively). Denote by **po-cm** (**$l$-cm**) the class of all commutative left-po-ms (left-$l$-ms, respectively). Denote by **po-mcm** (**$l$-mcm**) the class of all commutative left-po-mms (left-$l$-mms, respectively). Denote by **po-cim** (**$l$-cim** ) the class of all commutative left-po-ims (left-$l$-ims, respectively).

### 5.3.2 Po($l$)-magmas with pseudo-residuum/pseudo-coresiduum

We shall study here particular left-po-magmas (left-$l$-magmas) which have also a pseudo-residuum $(\rightarrow, \rightsquigarrow)$ (residuum, $\rightarrow$, in the commutative case).

We introduce now the following two general definitions:

**Definitions 5.3.7** (Definition of structure with (pR')/(pcoR'))

(i) Let $\mathcal{A}^L = (A^L, \leq, \odot)$ be a structure such that the reduct $(A^L, \leq)$ is a poset (i.e. (Re'), (An'), (Tr') hold) and $\odot$ is a binary operation. We say that $\mathcal{A}^L$ is a *structure with pseudo-residuum*, or a *structure with (pR')*, if for any $y, z \in A^L$, there exist the greatest elements of the sets $\{x \in A^L \mid x \odot y \leq z\}$ and $\{x \in A^L \mid y \odot x \leq z\}$ and they define two new operations, $\rightarrow, \rightsquigarrow$ (the pair $(\rightarrow, \rightsquigarrow)$ is called *pseudo-residuum*), i.e. $\mathcal{A}^L$ satisfies the following property:

(pR') for all $y, z \in A^L$, $\exists\, y \to z \stackrel{def.}{=} \max\{x \in A^L \mid x \odot y \leq z\}$,
$\exists\, y \rightsquigarrow z \stackrel{def.}{=} \max\{x \in A^L \mid y \odot x \leq z\}$.

(i') Dually, let $\mathcal{A}^R = (A^R, \geq, \oplus)$ be a structure such that the reduct $(A^R, \geq)$ is a poset (i.e. (Re'), (An'), (Tr') hold) and $\oplus$ is a binary operation. We say that $\mathcal{A}^R$ is a *structure with pseudo-coresiduum*, or a *structure with (pcoR')*, if for any $y, z \in A^R$, there exist the smallest elements of the sets $\{x \in A^R \mid x \oplus y \geq z\}$ and $\{x \in A^R \mid y \oplus x \geq z\}$ and they define two new operations, $\to^R, \rightsquigarrow^R$ (the pair $(\to^R, \rightsquigarrow^R)$ is called *pseudo-coresiduum*), i.e. $\mathcal{A}^R$ satisfies the following property:

(pcoR') for all $y, z \in A^R$, $\exists\, y \to^R z \stackrel{def.}{=} \min\{x \in A^R \mid x \oplus y \geq z\}$,
$\exists\, y \rightsquigarrow^R z \stackrel{def.}{=} \min\{x \in A^R \mid y \oplus x \geq z\}$.

(ii) If the partial order $\leq$, or $\geq$, is a lattice order, then we say that we have a *lattice structure with (pR')/(pcoR')*.

**Remark 5.3.8** *(The dual case is omitted)*
*By Definition 5.3.7 (i), we have*
$$y \to z \in \{x \mid x \odot y \leq z\} \text{ and } y \rightsquigarrow z \in \{x \mid y \odot x \leq z\},$$
*hence we have the property:*
*(pRR') for all* $y, z$, $(y \to z) \odot y \leq z, \quad y \odot (y \rightsquigarrow z) \leq z.$

*Consequently, (pR')* $\Longrightarrow$ *(pRR').*

**Definitions 5.3.9** (Definition of structure with (pPR')/(pScoR'))
(i) Let $\mathcal{A}^L = (A^L, \leq, \odot, \to, \rightsquigarrow)$ be a structure such that the reduct $(A^L, \leq)$ is a partially preordered set (i.e. only (Re'), (Tr') hold) and $\odot$, $\to$, $\rightsquigarrow$ are binary operations. We say that $\mathcal{A}^L$ is a *structure with pseudo-product and pseudo-residuum*, or a *structure with (pPR')*, if $\leq$, $\odot$ and $\to, \rightsquigarrow$ are connected by the property:

(pPR') for all $x, y, z \in A^L$, $x \odot y \leq z \Longleftrightarrow x \leq y \to z \Longleftrightarrow y \leq x \rightsquigarrow z$.

(i') Dually, let $\mathcal{A}^R = (A^R, \geq, \oplus, \to^R, \rightsquigarrow^R)$ be a structure such that the reduct $(A^R, \geq)$ is a partially preordered set (i.e. only (Re'), (Tr') hold) and $\oplus$, $\to^R$, $\rightsquigarrow^R$ are binary operations. We say that $\mathcal{A}^R$ is a *structure with pseudo-sum and pseudo-coresiduum*, or a *structure with (pScoR')*, if $\geq$, $\oplus$ and $\to^R, \rightsquigarrow^R$ are connected by the property:

(pScoR') for all $x, y, z \in A^R$, $x \oplus y \geq z \Longleftrightarrow x \geq y \to^R z \Longleftrightarrow y \geq x \rightsquigarrow^R z$.

(ii) If the partial order $\leq$, or $\geq$, is a lattice order, then we say that we have a *lattice structure with (pPR')/(pScoR')*.

Note that condition (pPR') needs not $\leq$ be an order.

Note that the properties (pRP') (from Section 2.4) and (pPR') coincide (we write (pRP')=(pPR')) and that this property (often called the *residuation property*) reflects a non-commutative *Galois connection*.

**Remark 5.3.10**

$$(pPR') \Longrightarrow (pEq\#') \quad and \quad (pScoR') \Longrightarrow (pEq\#^{R'}),$$

*where:*
(pEq#') $x \leq y \to z \iff y \leq x \rightsquigarrow z$,
(pEq#$^R$') $x \geq y \to^R z \iff y \geq x \rightsquigarrow^R z$.

**Remark 5.3.11** *Just as the equivalences*

$$x \leq y \Longleftrightarrow x \to y = 1 \Longleftrightarrow x \rightsquigarrow y = 1$$

*were used either:*
*- to define the binary relation $\leq$ in the algebra $(A, \to, \rightsquigarrow, 1)$, by*
*(pdfrelR) ($x \leq y \overset{def.}{\Longleftrightarrow} x \to y = 1 \Longleftrightarrow x \rightsquigarrow y = 1$), or*
*- to connect $\leq$ and $\to, \rightsquigarrow$, 1 in the structure $(A, \leq, \to, \rightsquigarrow, 1)$, by*
*(pEqrelR) ($x \leq y \Longleftrightarrow x \to y = 1 \Longleftrightarrow x \rightsquigarrow y = 1$),*
*the same happen with the equalities*

$$y \to z = \max\{x \mid x \odot y \leq z\} \quad and \quad y \rightsquigarrow z = \max\{x \mid y \odot x \leq z\},$$

*that can be used either:*
*- to define two binary operations $\to, \rightsquigarrow$ in the structure $(A^L, \leq, \odot)$, by (pR'), or*
*- to connect $\to, \rightsquigarrow$ and $\leq, \odot$ in the structure $(A^L, \leq, \odot, \to, \rightsquigarrow)$, by (pPR').*

We prove now the following important general lemma.

**Lemma 5.3.12** *(See the Galois dual Lemma 1.4.16)*
*Let $\mathcal{A}^L = (A^L, \leq, \odot)$ be a structure with (pR') (i.e. (Re'), (Tr'), (An'), (pR') hold), where the pseudo-residuum is $(\to, \rightsquigarrow)$. If the property (pCp') also holds, then $\mathcal{A}^L = (A^L, \leq, \odot, \to, \rightsquigarrow)$ is a structure with (pPR') (i.e. (Re'), (Tr'), (pPR') hold).*

**Proof.** We must prove that (pPR') holds. Indeed, if $x \odot y \leq z$, then, by (pR'), $x \leq y \to z$; if $x \leq y \to z$, it follows, by (pCp'), that $x \odot y \leq (y \to z) \odot y$; and, since we also have, by (pR'), that (pRR') holds, i.e. $(y \to z) \odot y \leq z$, we obtain, by (Tr'), that $x \odot y \leq z$. Also, if $x \odot y \leq z$, then, by (pR') again, $y \leq x \rightsquigarrow z$; if $y \leq x \rightsquigarrow z$, it follows, by (pCp'), that $x \odot y \leq x \odot (x \rightsquigarrow z)$; and, since we also have, by (pR'), that (pRR') holds, i.e. $x \odot (x \rightsquigarrow z) \leq z$, we obtain, by (Tr'), that $x \odot y \leq z$. Thus, (pPR') holds. □

Now the following general theorem follows immediately:

**Theorem 5.3.13** *We have:*

$$(pR') + (pCp') \Longleftrightarrow (pPR').$$

**Proof.** $\Longrightarrow$: By Lemma 5.3.12.
$\Longleftarrow$: By Lemma 1.4.17, (pR') and (pCp') hold. □

### 5.3.3 Po-magmas ($l$-magmas) with (pR')/(pcoR')

We recall/introduce the po-ss(pR') and po-ms(pR'), po-mms(pR'), po-ims(pR'), and their corresponding lattice ordered ones.

- **Po-ss(pR') ($l$-ss(pR'))**

By general Definitions 5.3.7, we obtain the following definitions (the dual ones are omitted).

**Definitions 5.3.14** (Definition of po-s(pR') ($l$-s(pR')))

(i) A *left-po-s with pseudo-residuum*, or a *po-s(pR')* for short, is a left-po-s

$$\mathcal{A}^L = (A^L, \leq^e, \odot)$$

verifying additionally:

(pR') for all $y, z \in A^L$, $\quad \exists\, y \to z \overset{def.}{=} \max\{x \in A^L \mid x \odot y \leq^e z\}$,

$\exists\, y \rightsquigarrow z \overset{def.}{=} \max\{x \in A^L \mid y \odot x \leq^e z\}$.

(ii) If the partial order $\leq^e$ is a lattice order, then we say that we have a *left-lattice ordered semigroup with pseudo-residuum ($l$-s(pR'))*.

Denote by **po-s(pR')** (**$l$-s(pR')**) the class of all po-ss(pR') ($l$-ss(pR'), respectively).

- **Po-ms(pR'), po-mms(pR'), po-ims(pR')**
  **($l$-ms(pR'), $l$-mms(pR'), $l$-ims(pR'))**

**Definitions 5.3.15** (Definition of po-m(pR'), po-mm(pR'), po-im(pR') ($l$-m(pR'), $l$-mm(pR'), $l$-im(pR')))

(i) A *left-po-m (left-po-mm, left-po-im) with pseudo-residuum*, or a *po-m(pR') (po-mm(pR'), po-im(pR'))* for short, is a left-po-m (left-po-mm, left-po-im, respectively)

$$\mathcal{A}^L = (A^L, \leq, \odot, 1)$$

verifying additionally:

(pR') for all $y, z \in A^L$, $\quad \exists\, y \to z \overset{def.}{=} \max\{x \in A^L \mid x \odot y \leq z\}$,

$\exists\, y \rightsquigarrow z \overset{def.}{=} \max\{x \in A^L \mid y \odot x \leq z\}$.

(ii) If the partial order $\leq$ is a lattice order, then we say that we have a *$l$-m(pR') ($l$-mm(pR'), $l$-im(pR'))*.

Denote by **po-m(pR')** the class of all po-ms(pR').
Denote by **po-mm(pR')** the class of all po-mms(pR').
Denote by **po-im(pR')** the class of all po-ims(pR').
Denote analogously the lattice ordered case.
By inclusions 5.6, we have the inclusions:

$$(5.7) \quad \mathbf{po-im(pR')} \subset \mathbf{po-mm(pR')} \subset \mathbf{po-m(pR')} \subset \mathbf{po-s(pR')},$$

$$\mathbf{l-im(pR')} \subset \mathbf{l-mm(pR')} \subset \mathbf{l-m(pR')} \subset \mathbf{l-s(pR')}.$$

### 5.3.4 Po-magmas ($l$-magmas) with (pPR')/(pScoR')

We recall/introduce the po-ss(pPR') and po-ms(pPR'), po-mms(pPR'), po-ims(pPR'), and their corresponding lattice ordered ones.

By general Definitions 5.3.9, we obtain the following definitions (the dual ones are omitted).

- **po-ss(pPR') ($l$-ss(pPR'))**

**Definitions 5.3.16** (Definition of po-s(pPR') ($l$-s(pPR')))

(i) A *left-po-s with the property (pPR')*, or a *po-s(pPR')* for short, is a structure

$$\mathcal{A}^L = (A^L, \leq^e, \odot, \to, \leadsto)$$

such that $(A^L, \leq^e)$ is a poset, $(A^L, \odot)$ is a semigroup (i.e. (Pass) holds) and the pseudo-residuum $(\to, \leadsto)$ is connected to $\leq^e$, $\odot$ by the following property:
(pPR') for all $x, y, z \in A^L$, $\quad x \odot y \leq^e z \Longleftrightarrow x \leq^e y \to z \Longleftrightarrow y \leq^e x \leadsto z.$

(ii) If the partial order $\leq^e$ is a lattice order, then we say that we have a *left-lattice ordered semigroup with (pRP') (l-s(pRP'))*.

Denote by **po-s(pPR')** the class of all po-ss(pPR'). Analogously, in lattice order case.

- **po-ms(pPR'), po-mms(pPR'), po-ims(pPR')**
  **($l$-ms(pPR'), $l$-mms(pPR'), $l$-ims(pPR'))**

**Definitions 5.3.17** (Definition of po-m(pPR'), po-mm(pPR'), po-im(pPR') ($l$-m(pPR'), $l$-mm(pPR'), $l$-im(pPR')))

(i) A *left-po-m (left-po-mm, left-po-im) with the property (pPR')*, or a *po-m(pPR') (po-mm(pPR'), po-im(pPR'))* for short, is a structure

$$\mathcal{A}^L = (A^L, \leq, \odot, \to, \leadsto, 1)$$

such that $(A^L, \leq)$ is a poset ($(A^L, \leq, 1)$ is a poset with maximal element 1, with greatest element 1, respectively), $(A^L, \odot, 1)$ is a monoid (i,e, (Pass), (PU) hold) and the pseudo-residuum $(\to, \leadsto)$ is connected to $\leq$, $\odot$ by the following property:
(pPR') for all $x, y, z \in A^L$, $\quad x \odot y \leq z \Longleftrightarrow x \leq y \to z \Longleftrightarrow y \leq x \leadsto z.$

(ii) If the partial order $\leq$ is a lattice order, then we say that we have a *l-m(pPR') (l-mm(pPR'), l-im(pPR'))*.

Denote by **po-m(pPR')** the class of all po-ms(pPR').
Denote by **po-mm(pPR')** the class of all po-mms(pPR').
Denote by **po-im(pPR')** the class of all po-ims(pPR').
Denote analogously in lattice ordered case.
By inclusions (5.6), we have the inclusions:
(5.8)

$$\mathbf{po-im(pPR')} \subset \mathbf{po-mm(pPR')} \subset \mathbf{po-m(pPR')} \subset \mathbf{po-s(pPR')},$$

$$\mathbf{l-im(pPR')} \subset \mathbf{l-mm(pPR')} \subset \mathbf{l-m(pPR')} \subset \mathbf{l-s(pPR')}.$$

**Proposition 5.3.18** *(The dual case is omitted)*
*Any po-s(pPR'), po-m(pPR'), po-mm(pPR'), po-im(pPR') verify the properties (pEq#'), (pP'), (pR'), (pPP'), (pRR'), (pCp'), (p*'), (p**').*

**Proof.** Obviously, by Lemma 1.4.17. □

**Corollary 5.3.19** *We have:*
*(1) Any po-s(pPR') is a left-po-s.*
*(2) Any po-m(pPR') is a left-po-m.*
*(3) Any po-mm(pPR') is a left-po-mm.*
*(4) Any po-im(pPR') is a left-po-im.*

**Proof.** Obviously, since (pCp') holds, by Proposition 5.3.18. □

**Proposition 5.3.20** *(See Proposition 1.4.27)*
*Let* $\mathcal{A}^L = (A^L, \leq, \odot, \to, \rightsquigarrow, 1)$ *be a left-po-m(pPR'). Then, we have:*
*(IdEq)* $1 \leq x \to y \iff 1 \leq x \rightsquigarrow y$ *and*
*(pEqrel)* $x \leq y \iff 1 \leq x \to y \quad (\overset{(IdEq)}{\iff} 1 \leq x \rightsquigarrow y).$

**Proof.** $x \leq y \overset{(PU)}{\iff} 1 \odot x \leq y \overset{(pPR')}{\iff} 1 \leq x \to y$
and $x \leq y \overset{(PU)}{\iff} x \odot 1 \leq y \overset{(pPR')}{\iff} 1 \leq x \rightsquigarrow y$. Thus, (IdEq) and (pEqrel) hold. □

**Proposition 5.3.21** *(See Proposition 1.4.28)*
*Let* $\mathcal{A}^L = (A^L, \leq, \odot, \to, \rightsquigarrow 1)$ *be a left-po-mm(pPR'). Then, we have:*
*(IdEqR)* $\quad x \to y = 1 \iff x \rightsquigarrow y = 1$ *and*
*(pEqrelR)* $\quad x \leq y \iff x \to y = 1 \quad (\overset{(IdEqR)}{\iff} x \rightsquigarrow y = 1).$

**Proof.** Since $\mathcal{A}^L$ is also a left-po-m(pPR'), it follows that (IdEq) and (pEqrel) hold. Since $\mathcal{A}^L$ is maximal, then (N') holds, i.e. for all $x \in A^L$, $1 \leq x$ implies $x = 1$. It follows that (IdEqR) and (pEqrelR) hold. □

**Proposition 5.3.22** *(See Proposition 1.4.29)*
*Let* $\mathcal{A}^L = (A^L, \leq, \odot, \to, \rightsquigarrow 1)$ *be a left-po-im(pPR'). Then, we also have (IdEqR) and (pEqrelR).*

**Proof.** By (pA1), (La') + (An') imply (N'); it follows that $\mathcal{A}^L$ is also a left-po-mm(pPR'), hence (IdEqR) and (pEqrelR) hold. □

### • Pseudo-residuated lattices

Residuated lattices, the algebraic counterpart of logics without contraction rule, were introduced in 1924 by Krull; they have been investigated (cf. Kowalski-Ono [142] ) by Krull [144], Dilworth [43], Ward and Dilworth [214], Ward [213], Balbes and Dwinger [3], Pavelka [167], Idziak [92] and others. Residuated lattices have been known under many names; they have been called (cf. [142]) *BCK lattices* in [92], *full BCK-algebras* in [163], $FL_{ew}$*-algebras* in [164] and *integral, residuated, commutative l-monoids* in [88]; some of those definitions are free of 0.

Their non-commutative extensions, called "pseudo-residuated lattices" here, were investigated in many papers, as for examples in [129], [58]. Let us consider the following definition free of 0 (the smallest element), i.e. as posets with greatest element 1.

**Definition 5.3.23** [104] (The dual one is omitted)
A *non-commutative residuated left-lattice* or a *pseudo-residuated left-lattice* is an algebra

$$\mathcal{A}^L = (A^L, \wedge, \vee, \odot, \to, \rightsquigarrow, 1)$$

of type $(2, 2, 2, 2, 2, 0)$ such that the reduct $(A^L, \wedge, \vee, 1)$ is a left-lattice with last element 1 (under lattice order $\leq$), the reduct $(A^L, \odot, 1)$ is a monoid and the following property holds:
(pPR') for all $x, y, z \in A^L$, $\quad x \odot y \leq z \Longleftrightarrow x \leq y \to z \Longleftrightarrow y \leq x \rightsquigarrow z.$

Let **pR-L** denote the class of left-pseudo-residuated lattices.
Note that we have: ***l*-im(pPR')** $=$ **pR-L**.

### 5.3.5 Connections (pR') - (pPR')

By general Theorem 5.3.13, we obtain immediately the following theorem.

**Theorem 5.3.24** *(The dual case is omitted)*
*(1)* $(A^L, \leq^e, \odot)$ *is a po-s(pR') iff* $(A^L, \leq^e, \odot, \to, \rightsquigarrow)$*is a po-s(pPR').*
*(2)* $(A^L, \leq, \odot, 1)$ *is a po-m(pR') iff* $(A^L, \leq, \odot, \to, \rightsquigarrow, 1)$ *is a po-m(pPR').*
*(3)* $(A^L, \leq, \odot, 1)$ *is a po-mm(pR') iff* $(A^L, \leq, \odot, \to, \rightsquigarrow, 1)$ *is a po-mm(pPR').*
*(4)* $(A^L, \leq, \odot, 1)$ *is a po-im(pR') iff* $(A^L, \leq, \odot, \to, \rightsquigarrow, 1)$ *is a po-im(pPR').*

By Theorem 5.3.24, we have the equivalences:

**Corollary 5.3.25**

$$\mathbf{po-s(pR')} \Longleftrightarrow \mathbf{po-s(pPR')},$$

$$\mathbf{po-m(pR')} \Longleftrightarrow \mathbf{po-m(pPR')},$$

$$\mathbf{po-mm(pR')} \Longleftrightarrow \mathbf{po-mm(pPR')},$$

$$\mathbf{po-im(pR')} \Longleftrightarrow \mathbf{po-im(pPR')}.$$

and also the similar equivalences, in the lattice ordered case:

**Corollary 5.3.26**

$$\mathbf{l-s(pR')} \Longleftrightarrow \mathbf{l-s(pPR')},$$

$$\mathbf{l-m(pR')} \Longleftrightarrow \mathbf{l-m(pPR')},$$

$$\mathbf{l-mm(pR')} \Longleftrightarrow \mathbf{l-mm(pPR')},$$

$$\mathbf{l-im(pR')} \Longleftrightarrow \mathbf{l-im(pPR')} = \mathbf{pR-L}.$$

**Proposition 5.3.27** *(The dual case is omitted)*
*Let* $\mathcal{A}^L = (A^L, \leq = \leq^e, \odot, \to, \rightsquigarrow)$ *be a po-s(pPR'). Then,* $(A^L, \to, \rightsquigarrow)$ *is a pseudo-exchange (i.e. (pEx) holds).*

**Proof.**

$$\begin{array}{rcl} a \leq x \to (y \rightsquigarrow z) & \stackrel{(pPR')}{\Longleftrightarrow} & a \odot x \leq y \rightsquigarrow z \\ & \stackrel{(pPR')}{\Longleftrightarrow} & y \odot (a \odot x) \leq z \\ & \stackrel{(Pass)}{\Longleftrightarrow} & (y \odot a) \odot x \leq z \\ & \stackrel{(pPR')}{\Longleftrightarrow} & y \odot a \leq x \to z \\ & \stackrel{(pPR')}{\Longleftrightarrow} & a \leq y \rightsquigarrow (x \to z); \end{array}$$

hence, $x \to (y \rightsquigarrow z) = y \rightsquigarrow (x \to z)$, by the reflexivity and antisymmetry of $\leq$. □

**Proposition 5.3.28** *(The dual case is omitted)*
*Let* $\mathcal{A}^L = (A^L, \leq, \odot, \to, \rightsquigarrow, 1)$ *be a po-m(pPR'). Then,* $(A^L, \to, \rightsquigarrow, 1)$ *is a pseudo-residoid (i.e. (pEx), (pM) hold).*

**Proof.** The reduct $(A^L, \leq, \odot, \to, \rightsquigarrow)$ is a po-s(pPR'), hence, the property (pEx) holds, by Proposition 5.3.27. It remains to prove (pM). Indeed,
$a \leq 1 \to x \stackrel{(pPR')}{\Longleftrightarrow} a \odot 1 \leq x \stackrel{(PU)}{\Longleftrightarrow} a \leq x$ and $a \leq 1 \rightsquigarrow x \stackrel{(pPR')}{\Longleftrightarrow} 1 \odot a \leq x \stackrel{(PU)}{\Longleftrightarrow} a \leq x$,
hence, $1 \to x = x$ and $1 \rightsquigarrow x = x$, by reflexivity and antisymmetry of $\leq$ (which is an order). □

Connections between these notions and the corresponding one from the 'world' of pi-magmas (presented in Part I) are presented in Part III (Section 16.1).

## 5.4 Groups, po-groups, *l*-groups

In this section, we recall from [107], [110] the group, the po-group and the *l*-group, the intermediary notions of *X-po-group*, *X-l-group* and *X-group*, respectively, and prove the equivalences (EG1), (EG2), (EG3) between these notions, respectively.

The section has three subsections.

### 5.4.1 Groups

#### 5.4.1.1 The group

The most used definition of the group is the following:

**Definition 5.4.1** (Definition 1 of groups)
A *group* is an algebra (in multiplicative (left) notation) $\mathcal{G} = (G, \cdot, ^{-1}, 1)$ of type $(2, 1, 0)$, verifying: for all $x, y, z \in G$,
(PU) $\quad x \cdot 1 = x = 1 \cdot x$,
(Pass) $\quad x \cdot (y \cdot z) = (x \cdot y) \cdot z$,
(pIv) = ($m_1$-pRe) $\quad x \cdot x^{-1} = 1 = x^{-1} \cdot x$,
hence is a monoid verifying (pIv).

Denote by **group** the class of all groups.

The group is said to be *commutative*, or *abelian*, if (Pcomm) ($x \cdot y = y \cdot x$, for all $x, y \in G$) holds.

Recall now the following equivalent definition of the group.

**Definition 5.4.2** (Definition 1' of groups)

A *group* is an algebra $\mathcal{G} = (G, \cdot, 1)$ of type $(2, 0)$, verifying: for all $x, y, z \in G$,
(PU) $x \cdot 1 = x = 1 \cdot x$,
(Pass) $x \cdot (y \cdot z) = (x \cdot y) \cdot z$,
(pIv') for all $x \in G$, there exists $y \in G$, such that $x \cdot y = 1 = y \cdot x$, i.e. each element $x$ has an inverse element, $y$.

Note that the inverse element of $x$ is **unique**, because of (PU) and (Pass); indeed, if there exist $x^-$ and $x^\sim$ such that $x \cdot x^- = x^- \cdot x = 1$ and $x \cdot x^\sim = x^\sim \cdot x = 1$, then:
$x^- \overset{(PU)}{=} x^- \cdot 1 = x^- \cdot (x \cdot x^\sim) \overset{(Pass)}{=} (x^- \cdot x) \cdot x^\sim = 1 \cdot x^\sim \overset{(PU)}{=} x^\sim$.
The unique inverse of $x$ is denoted by $x^{-1}$, thus there exists an unary operation on $G$, denoted by $^{-1}$, introduced in the signature of $\mathcal{G}$.

**Proposition 5.4.3** *Let $\mathcal{G} = (G, \cdot, ^{-1}, 1)$ be a group.*
*Then, we have: for all $x, y \in G$,*

| | |
|---|---|
| *(pIvP)* | $x \cdot y = 1 (\Longleftrightarrow y \cdot x = 1) \Longleftrightarrow y = x^{-1}$, |
| *(DN)* | $(x^{-1})^{-1} = x$, |
| *(N1)* | $1^{-1} = 1$, |
| *(pIvG)* | $(x \cdot y^{-1})^{-1} = y \cdot x^{-1}, \quad (y^{-1} \cdot x)^{-1} = x^{-1} \cdot y$, |
| *(NegP)* | $(x \cdot y)^{-1} = y^{-1} \cdot x^{-1}$, |
| *(qm-NegP)* | $(1 \cdot x)^{-1} = 1 \cdot x^{-1} = x^{-1} \cdot 1 = (x \cdot 1)^{-1}$, |
| *(IdEqP)* | $y \cdot x^{-1} = 1 \Longleftrightarrow x^{-1} \cdot y = 1$, |
| *(pEqP=)* | $x = y \Longleftrightarrow y \cdot x^{-1} = 1 \overset{(IdEqP)}{\Longleftrightarrow} x^{-1} \cdot y = 1$. |

*We also have the equivalence:*

$$(pIvP) \Longleftrightarrow (pEqP=).$$

**Proof.** (pIvP): Suppose that $x \cdot y = 1$; then, $y \overset{(PU)}{=} 1 \cdot y \overset{(pIv)}{=} (x^{-1} \cdot x) \cdot y \overset{(Pass)}{=} x^{-1} \cdot (x \cdot y) = x^{-1} \cdot 1 \overset{(PU)}{=} x^{-1}$. Conversely, suppose that $y = x^{-1}$; then, $x \cdot y = x \cdot x^{-1} \overset{(pIv)}{=} 1$. Thus, $x \cdot y = 1 \Longleftrightarrow y = x^{-1}$. Similarly, $y \cdot x = 1 \Longleftrightarrow y = x^{-1}$. So, (pIvP) holds.

(DN): By (pIv), $x^{-1} \cdot x = 1$; then, by (pIvP), $x = (x^{-1})^{-1}$; thus, (DN) holds.

(N1): $1^{-1} \overset{(PU)}{=} 1 \cdot 1^{-1} \overset{(pIv)}{=} 1$.

(pIvG): $(x \cdot y^{-1}) \cdot (y \cdot x^{-1}) \overset{(Pass)}{=} x \cdot (y^{-1} \cdot y) \cdot x^{-1} \overset{(pIv)}{=} x \cdot 1 \cdot x^{-1} \overset{(Pass)}{=} (x \cdot 1) \cdot x^{-1} \overset{pU)}{=} x \cdot x^{-1} \overset{(pIv)}{=} 1$ and
$(y \cdot x^{-1}) \cdot (x \cdot y^{-1}) \overset{(Pass)}{=} y \cdot (x^{-1} \cdot x) \cdot y^{-1} \overset{(pIv)}{=} y \cdot 1 \cdot y^{-1} \overset{(Pass)}{=} (y \cdot 1) \cdot y^{-1} \overset{(PU)}{=} y \cdot y^{-1} \overset{(pIv)}{=} 1$; then, by (pIvP), (pIvG) holds.

(NegP): $(x \cdot y)^{-1} \overset{(DN)}{=} (x \cdot (y^{-1})^{-1})^{-1} \overset{(pIvG)}{=} y^{-1} \cdot x^{-1}$.

(qm-NegP): Obviously, by (PU).

(IdEqP): Suppose $y \cdot x^{-1} = 1$; then, $y^{-1} \overset{(PU)}{=} y^{-1} \cdot 1 = y^{-1} \cdot (y \cdot x^{-1}) \overset{(Pass)}{=} (y^{-1} \cdot y) \cdot x^{-1} \overset{(pIv)}{=} 1 \cdot x^{-1} \overset{(PU)}{=} x^{-1}$; then, $x^{-1} \cdot y = y^{-1} \cdot y \overset{(pIv)}{=} 1$. Conversely, suppose $x^{-1} \cdot y = 1$; then, $y^{-1} \overset{(PU)}{=} 1 \cdot y^{-1} = (x^{-1} \cdot y) \cdot y^{-1} \overset{(Pass)}{=} x^{-1} \cdot (y \cdot y^{-1}) \overset{(pIv)}{=} x^{-1} \cdot 1 \overset{(PU)}{=} x^{-1}$; then, $y \cdot x^{-1} = y \cdot y^{-1} \overset{(pIv)}{=} 1$. Thus, (IdEqP) holds.

(pEqP=): If $x = y$, then $y \cdot x^{-1} = x \cdot x^{-1} \overset{(pIv)}{=} 1$. Conversely, if $y \cdot x^{-1} = 1$, then
$x \overset{(PU)}{=} 1 \cdot x = (y \cdot x^{-1}) \cdot x \overset{(Pass)}{=} y \cdot (x^{-1} \cdot x) \overset{(pIv)}{=} y \cdot 1 = y$. Thus, (pEqP=) holds.

(pIvP) $\Longleftrightarrow$ (pEqP=), by (DN). □

Hence, the group is an involutive algebra.

- **New operations in groups: $\rightarrow$ and $\rightsquigarrow$**

The groups have the following very interesting property.

**Proposition 5.4.4** *Let $(G, \cdot, ^{-1}, 1)$ be a group. For all $x \in G$, define the mappings $f_x : G \longrightarrow G$ and ${}_xf : G \longrightarrow G$ by: for all $g \in G$,*

$$f_x(g) = g \cdot x, \quad {}_xf(g) = x \cdot g.$$

*Then, $f_x$ and ${}_xf$ are bijective.*

**Proof.** - *Injectivity*: Let $g_1, g_2 \in G$, $g_1 \neq g_2$. We must prove that $f_x(g_1) \neq f_x(g_2)$ and ${}_xf(g_1) \neq {}_xf(g_2)$.
Suppose, by absurdum hypothesis, that $f_x(g_1) = f_x(g_2)$, i.e. $g_1 \cdot x = g_2 \cdot x$; then, $(g_1 \cdot x) \cdot x^{-1} = (g_2 \cdot x) \cdot x^{-1}$ and hence, by (Pass), $g_1 = g_1 \cdot 1 = g_1 \cdot (x \cdot x^{-1}) = g_2 \cdot (x \cdot x^{-1}) = g_2 \cdot 1 = g_2$: contradiction.
Similarly, suppose, by absurdum hypothesis, that ${}_xf(g_1) = {}_xf(g_2)$, i.e. $x \cdot g_1 = x \cdot g_2$; then, $x^{-1} \cdot (x \cdot g_1) = x^{-1} \cdot (x \cdot g_2)$ and hence, by (Pass), $g_1 = 1 \cdot g_1 = (x^{-1} \cdot x) \cdot g_1 = (x^{-1} \cdot x) \cdot g_2 = 1 \cdot g_2 = g_2$: contradiction.

- *Surjectivity* of $f_x$: We must prove that for all $y \in G$, there exists $z \in G$, such that $y = f_x(z) = z \cdot x$. Indeed, let $y \in G$ and take $z = y \cdot x^{-1}$; then, $f_x(z) = z \cdot x = (y \cdot x^{-1}) \cdot x \overset{(Pass)}{=} y \cdot (x^{-1} \cdot x) \overset{(pIv)}{=} y \cdot 1 \overset{(PU)}{=} y$.

- *Surjectivity* of ${}_xf$: We must prove that for all $y \in G$, there exists $u \in G$, such that $y = {}_xf(u) = x \cdot u$. Indeed, let $y \in G$ and take $u = x^{-1} \cdot y$; then, ${}_xf(u) = x \cdot u = x \cdot (x^{-1} \cdot y) \overset{(Pass)}{=} (x \cdot x^{-1}) \cdot y \overset{(pIv)}{=} 1 \cdot y \overset{(PU)}{=} y$. □

We then have immediately the following corollary.

**Corollary 5.4.5** *Let $(G, \cdot, ^{-1}, 1)$ be a group. For all $x, y \in G$,*
*- there exists a unique $z \in G$, $z = y \cdot x^{-1}$, such that $y = f_x(z) = z \cdot x$ and*
*- there exists a unique $u \in G$, $u = x^{-1} \cdot y$, such that $y = {}_xf(u) = x \cdot u$.*

By above Corollary 5.4.5, we can introduce two binary operations on $(G, \cdot, ^{-1}, 1)$ as follows: for all $x, y \in G$,

$$x \rightarrow y \overset{def.}{=} z = y \cdot x^{-1}, \quad x \rightsquigarrow y \overset{def.}{=} u = x^{-1} \cdot y,$$

and they satisfy the following property: for all $x, y \in G$,

(pR=) $\exists\, x \to y = z$ such that $z \cdot x = y$ and $\exists\, x \rightsquigarrow y = u$ such that $x \cdot u = y$.

**Remark 5.4.6** *Recall the dual properties: for all $x, y$,*
*(pR') $\exists\, x \to^L y = \max\{z \mid z \odot x \leq y\}$ and $\exists\, x \rightsquigarrow^L y = \max\{u \mid x \odot u \leq y\}$,*
*(pcoR') $\exists\, x \to^R y = \min\{z \mid z \oplus x \geq y\}$ and $\exists\, x \rightsquigarrow^R y = \min\{u \mid x \oplus u \leq y\}$.*
*Recall also the property (p-s) ($x \leq y \Longrightarrow x = y$). Then, note that:*
*(pR') + (p-s) $\Longrightarrow$ (pR=) and, dually, (pcoR') + (p-s) $\Longrightarrow$ (pR=).*

Since the group is an involutive non-commutative algebra, then, we introduce the new operations $\to$ and $\rightsquigarrow$ on $G$ (the pair $(\to, \rightsquigarrow)$ is called "pseudo-residuum") defined by: for all $x, y \in G$,

$$(5.9)\quad x \to y \overset{def.}{=} (x \cdot y^{-1})^{-1} \overset{(pIvG)}{=} y \cdot x^{-1}, \quad x \rightsquigarrow y \overset{def.}{=} (y^{-1} \cdot x)^{-1} \overset{(pIvG)}{=} x^{-1} \cdot y.$$

Note that:

(i) in additive (right) notation, a group is an algebra $(G, +, -, 0)$ and, hence, (5.9) becomes:

$$x \to y \overset{def.}{=} -(x + (-y)), \quad x \rightsquigarrow y \overset{def.}{=} -(-y + x).$$

(ii) In ([58], pag. 161), the implication $\rightsquigarrow$ is denoted by $\backslash$ ($x \backslash y = x \rightsquigarrow y$) and the implication $\to$ is replaced by its inverse, denoted by $/$ (i.e. $x/y = y \to x$).

(iii) If the group is commutative, then the two implications coincide: $\to = \rightsquigarrow$.

(iv) The group may be defined equivalently as an algebra $(G, \cdot, \to, \rightsquigarrow, ^{-1}, 1)$, where $\to$ and $\rightsquigarrow$ are expressed in terms of $\cdot$ and $^{-1}$:

$$x \to y = (x \cdot y^{-1})^{-1} = y \cdot x^{-1}, \quad x \rightsquigarrow y = (y^{-1} \cdot x)^{-1} = x^{-1} \cdot y.$$

**Remarks 5.4.7** *Since the group is an involutive algebra (with an unique negation), then we can define the analogous additional operations: $\oplus$ and $\Rightarrow$, $\approx>$ in two alternative ways.*

*(1) The first way: for all $x, y \in G$,*
$x \oplus y \overset{def.}{=} (y^{-1} \cdot x^{-1})^{-1} = x \cdot y$ *and*
$x \Rightarrow y \overset{def.}{=} (x \oplus y^-)^- = (x \cdot y^{-1})^{-1} = y \cdot x^{-1} = x \to y$ *and*
$x \approx> y \overset{def.}{=} (y^- \oplus x)^- = (y^{-1} \cdot x)^{-1} = x^{-1} \cdot y = x \rightsquigarrow y$.

*Hence, the old operations $\to$, $\rightsquigarrow$ are expressed in term of the "new" operations $\to$, $\rightsquigarrow$:*
$x \to y = (x^- \approx> y^-)^- = (x^{-1} \rightsquigarrow y^{-1})^{-1}$ *and*
$x \rightsquigarrow y = (x^- \Rightarrow y^-)^- = (x^{-1} \to y^{-1})^{-1}$.

*(2) The second alternative way: for all $x, y \in G$,*
$x \oplus y \overset{def.}{=} (x^{-1} \cdot y^{-1})^{-1} = y \cdot x$, *and*
$x \Rightarrow y \overset{def.}{=} (x \oplus y^-)^- = (y^{-1} \cdot x)^{-1} = x^{-1} \cdot y = x \rightsquigarrow y$ *and*
$x \approx> y \overset{def.}{=} (y^- \oplus x)^- = (x \cdot y^{-1})^{-1} = y \cdot x^{-1} = x \to y$.

*Hence, the old operations $\to$, $\rightsquigarrow$ are expressed in term of the "new" operations $\to$, $\rightsquigarrow$:*

$x \to y = (x^- \Rightarrow y^-)^- = (x^{-1} \rightsquigarrow y^{-1})^{-1}$,
$x \rightsquigarrow y = (x^- \approx> y^-)^- = (x^{-1} \to y^{-1})^{-1}$.

*Note that, in both ways, the new additional operations coincide with the old operations. Hence, because of an unique negation, we have an unique addition and two unique implications. We can say that the addition* $\cdot$ *is selfdual and that the dual of* $\to$ *is* $\rightsquigarrow$ *and the dual of* $\rightsquigarrow$ *is* $\to$. *Moreover, the implication* $\to$ *can be expressed in terms of* $\rightsquigarrow$ *and viceversa:*

$$(5.10) \qquad x \to y = (x^{-1} \rightsquigarrow y^{-1})^{-1}, \quad x \rightsquigarrow y = (x^{-1} \to y^{-1})^{-1}.$$

*Consequently, now one may better understand the results from chapters [57] and [85] concerning algebras* $(G, \circ)$ *of type* (2), *with two (one respectively) equations, that are termwise equivalent to groups.*

**Remark 5.4.8** *Note that, in an involutive BCK(P') algebra, (5.9) does not hold, i.e.* $(x \odot y^-)^- \neq y \odot x^-$. *Indeed, take as example the following particular case (see [104], pag. 163): the Boolean algebra* $L_{2\times 2} = \{0, a, b, 1\}$, *with* $0 < a, b < 1$, *organized as a BCK(P') algebra with the operation* $\to$ *and* $\odot$ *as in the following tables* ($x^- = x \to 0$):

$\mathcal{L}_{2\times 2}$

| $\to$ | 0 | a | b | 1 |
|---|---|---|---|---|
| 0 | 1 | 1 | 1 | 1 |
| a | b | 1 | b | 1 |
| b | a | a | 1 | 1 |
| 1 | 0 | a | b | 1 |

| $\odot$ | 0 | a | b | 1 |
|---|---|---|---|---|
| 0 | 0 | 0 | 0 | 0 |
| a | 0 | a | 0 | a |
| b | 0 | 0 | b | b |
| 1 | 0 | a | b | 1 |

*Then, for* $x = a$ *and* $y = 1$, *we obtain* $1 = 0^- = (a \odot 0)^- = (a \odot 1^-)^- \neq 1 \odot a^- = 1 \odot b = b$.

**Remarks 5.4.9** *Let* $(G, \cdot, ^{-1}, 1)$ *be a group.*

*(1) We have the special property: for all* $x, y \in G$,

$$(5.11) \qquad x \cdot y = (x \to y^{-1})^{-1} = y^{-1} \to x, \quad x \cdot y = (y \rightsquigarrow x^{-1})^{-1} = x^{-1} \rightsquigarrow y.$$

*Indeed,* $(x \to y^{-1})^{-1} = (y^{-1} \cdot x^{-1})^{-1} \overset{(pIv)}{=} x \cdot y$ *and* $y^{-1} \to x = x \cdot (y^{-1})^{-1} \overset{(DN)}{=} x \cdot y$ *and* $(y \rightsquigarrow x^{-1})^{-1} = (y^{-1} \cdot x^{-1})^{-1} = x \cdot y$ *and* $x^{-1} \rightsquigarrow y = (x^{-1})^{-1} \cdot y = x \cdot y$.

*(2) It follows that we also have the property: for all* $x, y \in G$,
*(pDNeg2)* $y^{-1} \to x = x^{-1} \rightsquigarrow y (= x \cdot y)$.

**Theorem 5.4.10** *Let* $(G, \cdot, ^{-1}, 1)$ *be a group and consider the pseudo-residuum* $(\to, \rightsquigarrow)$ *defined by (5.9). Then, the property* ($pGa^=$) *holds, where for all* $x, y, z \in G$:
($pGa^=$) $x \cdot y = z \iff x = y \to z \iff y = x \rightsquigarrow z$ *(see [58], page 161).*

**Proof.** $x \cdot y = z$ implies $x = z \cdot y^{-1} = y \to z$ and $y = x^{-1} \cdot z = x \rightsquigarrow z$, by (pIv); conversely, $x = y \to z$, i.e. $x = z \cdot y^{-1}$, implies $x \cdot y = z$ and, similarly, $y = x \rightsquigarrow z$ implies $x \cdot y = z$ too, by (pIv). □

**Remark 5.4.11**

$$(pGa^=) \implies (pEq\#^=),$$

*where:*
*(pEq#=)* $\qquad x = y \to z \iff y = x \rightsquigarrow z.$

The following result, by its Corollary, will be useful in Part III to prove the definitionally equivalence between groups and i-groups.

**Proposition 5.4.12** *Let $(G, \cdot, ^{-1}, 1)$ be a group. Then, we have: for all $x, y, z \in G$,*
*(Id)* $\quad x \to 1 = x \rightsquigarrow 1,$
*(pBB=)* $\quad y \to z = (z \to x) \rightsquigarrow (y \to x), \quad y \rightsquigarrow z = (z \rightsquigarrow x) \to (y \rightsquigarrow x),$
*(pD=)* $\quad (y \to x) \rightsquigarrow x = y = (y \rightsquigarrow x) \to x,$
*(IdEqR)* $\quad x \to y = 1 \iff x \rightsquigarrow y = 1,$
*(pEqrelR=)* $\quad x = y \iff x \to y = 1 \; (\overset{(IdEq)}{\iff} x \rightsquigarrow y = 1),$
*(pM)* $\quad 1 \to x = x = 1 \rightsquigarrow x,$
*(pEx)* $\quad z \rightsquigarrow (y \to x) = y \to (z \rightsquigarrow x),$
*(pB=)* $\quad z \to x = (y \to z) \to (y \to x), \quad z \rightsquigarrow x = (y \rightsquigarrow z) \rightsquigarrow (y \rightsquigarrow x),$
*(pRe)* $\quad x \to x = 1 = x \rightsquigarrow x.$

**Proof.** (Id): $x \to 1 \overset{def.}{=} 1 \cdot x^{-1} \overset{(PU)}{=} x^{-1}$ and $x \rightsquigarrow 1 \overset{def.}{=} x^{-1} \cdot 1 \overset{(PU)}{=} x^{-1}$. Thus, (Id) holds.

(pBB=): $(z \to x) \rightsquigarrow (y \to x) = (x \cdot z^{-1}) \rightsquigarrow (x \cdot y^{-1}) = (x \cdot z^{-1})^{-1} \cdot (x \cdot y^{-1}) = (z \cdot x^{-1}) \cdot (x \cdot y^{-1}) \overset{(Pass),(pIv)}{=} z \cdot 1 \cdot y^{-1} \overset{(Pass),(PU)}{=} z \cdot y^{-1} = y \to z$ and
$(z \rightsquigarrow x) \to (y \rightsquigarrow x) = (z^{-1} \cdot x) \to (y^{-1} \cdot x) = (y^{-1} \cdot x) \cdot (z^{-1} \cdot x)^{-1} = (y^{-1} \cdot x) \cdot (x^{-1} \cdot z) \overset{(Pass),(pIv)}{=} y^{-1} \cdot 1 \cdot z \overset{(Pass),(PU)}{=} y^{-1} \cdot z = y \rightsquigarrow z.$

(pD=): $(y \to x) \rightsquigarrow x = (x \cdot y^{-1}) \rightsquigarrow x = (x \cdot y^{-1})^{-1} \cdot x = (y \cdot x^{-1}) \cdot x = y$ and $(y \rightsquigarrow x) \to x = (y^{-1} \cdot x) \to x = x \cdot (y^{-1} \cdot x)^{-1} = x \cdot (x^{-1} \cdot y) = y.$

(IdEq): If $x \to y = 1$, i.e. $y \cdot x^{-1} = 1$, then $y \overset{(PU)}{=} y \cdot 1 \overset{(pIv)}{=} y \cdot (x^{-1} \cdot x) \overset{(Pass)}{=} (y \cdot x^{-1}) \cdot x = 1 \cdot x \overset{(PU)}{=} x$, and hence $x \rightsquigarrow y = x^{-1} \cdot y = x^{-1} \cdot x = 1$. Similarly, if $x \rightsquigarrow y = 1$, then $x \to y = 1$.

(pEqrelR=): If $x = y$, then $1 \overset{(pIv)}{=} x \cdot x^{-1} = y \cdot x^{-1} = x \to y$; if $x \to y = 1$, i.e. $y \cdot x^{-1} = 1$, then $y \overset{(PU)}{=} y \cdot 1 \overset{(pIv)}{=} y \cdot (x^{-1} \cdot x) \overset{(Pass)}{=} (y \cdot x^{-1}) \cdot x = 1 \cdot x \overset{(PU)}{=} x$. Similarly, $x = y \iff x \rightsquigarrow y = 1$.

(pM): $1 \to x = x \cdot 1^{-1} \overset{(N1)}{=} x \cdot 1 \overset{(PU)}{=} x$ and $1 \rightsquigarrow x = 1^{-1} \cdot x \overset{(N1)}{=} 1 \cdot x \overset{(PU)}{=} x.$

(pEx): $z \rightsquigarrow (y \to x) = z \rightsquigarrow (x \cdot y^{-1}) = z^{-1} \cdot (x \cdot y^{-1})$; $y \to (z \rightsquigarrow x) = y \to (z^{-1} \cdot x) = (z^{-1} \cdot x) \cdot y^{-1} \overset{(Pass)}{=} z^{-1} \cdot (x \cdot y^{-1})$; thus (pEx) holds.

(pB=): $(y \to z) \to (y \to x) = (x \cdot y^{-1}) \cdot (z \cdot y^{-1})^{-1} = (x \cdot y^{-1}) \cdot (y \cdot z^{-1}) = x \cdot z^{-1} = z \to x$ and $(y \rightsquigarrow z) \rightsquigarrow (y \rightsquigarrow x) = (y^{-1} \cdot z)^{-1} \cdot (y^{-1} \cdot x) = (z^{-1} \cdot y) \cdot (y^{-1} \cdot x) = z^{-1} \cdot x = z \rightsquigarrow x$, by (Pass).

(pRe): $x \to x = x \cdot x^{-1} \overset{(pIv)}{=} 1$ and $x \rightsquigarrow x = x^{-1} \cdot x \overset{(pIv)}{=} 1.$ □

**Corollary 5.4.13** *Let $(G, \cdot, ^{-1}, 1)$ be a group. Then, $(G, \to, \rightsquigarrow, 1)$ is an i-group.*

**Proof.** By above Proposition 5.4.12, (pM) and (pBB=) hold, hence $(G, \to, \rightsquigarrow, 1)$ is an i-group (Definition 1). □

**Proposition 5.4.14** *Let* $(G, \cdot, ^{-1}, 1)$ *be a group. Then, we have: for all* $x, y, z \in G$,

*(pNg)* $(x \to y)^{-1} = y \to x$ *and* $(x \rightsquigarrow y)^{-1} = y \rightsquigarrow x$,
*(q-pNeg)* $x \to 1 = 1 \to x^- = (1 \rightsquigarrow x)^- = x \rightsquigarrow 1 = 1 \rightsquigarrow x^- = (1 \to x)^-$,
*(pNg1)* $x \to y^{-1} = y \rightsquigarrow x^{-1}$,
*(pDNeg1)* $x \to y = y^{-1} \rightsquigarrow x^{-1}$ *and* $x \rightsquigarrow y = y^{-1} \to x^{-1}$,
*(pDNeg2)* $y^{-1} \to x = x^{-1} \rightsquigarrow y (= x \cdot y)$.

**Proof.** (pNg): $(x \to y)^{-1} \stackrel{def.}{=} (y \cdot x^{-1})^{-1} \stackrel{(NegP)}{=} (x^{-1})^{-1} \cdot y^{-1} \stackrel{(DN)}{=} x \cdot y^{-1} \stackrel{def.}{=} y \to x$ and $(x \rightsquigarrow y)^{-1} \stackrel{def.}{=} (x^{-1} \cdot y)^{-1} \stackrel{(NegP)}{=} y^{-1} \cdot (x^{-1})^{-1} \stackrel{(DN)}{=} y^{-1} \cdot x \stackrel{def.}{=} y \rightsquigarrow x$.

(q-pNeg): Routine, by using (PU), (N1), (DN) and (qm-NegP).

(pNg1): $x \rightsquigarrow y^{-1} = x^{-1} \cdot y^{-1} = y \to x^{-1}$.

(pDNeg1): $y^{-1} \to x^{-1} = x^{-1} \cdot (y^{-1})^{-1} = x^{-1} \cdot y = x \rightsquigarrow y$ and $y^{-1} \rightsquigarrow x^{-1} = (y^{-1})^{-1} \cdot x^{-1} = y \cdot x^{-1} = x \to y$.

(pDNeg2): $(y^{-1}) \to x = x \cdot (y^{-1})^{-1} = x \cdot y$ and $(x^{-1}) \rightsquigarrow y = (x^{-1})^{-1} \cdot y = x \cdot y$, by (DN). □

**Proposition 5.4.15** *Let* $(G, \cdot, ^{-1}, 1)$ *be a group. Then, we have: for all* $x, y, z \in G$,

(5.12) $$x \to y = (x \cdot z) \to (y \cdot z), \quad x \rightsquigarrow y = (z \cdot x) \rightsquigarrow (z \cdot y);$$

(5.13) $$x \cdot y = (x \to 1) \rightsquigarrow y = (y \rightsquigarrow 1) \to x,$$

(5.14) $$(x \cdot y) \to z = x \to (y \to z), \quad (y \cdot x) \rightsquigarrow z = x \rightsquigarrow (y \rightsquigarrow z),$$

(5.15) $$(z \to x) \cdot y = x \cdot (z \rightsquigarrow y),$$

(5.16) $$z \to (x \cdot y) = x \cdot (z \to y), \quad z \rightsquigarrow (x \cdot y) = (z \rightsquigarrow x) \cdot y,$$

(5.17)
$[(y \to x) \rightsquigarrow x] \to x = y \to x$, $[(y \rightsquigarrow x) \to x] \rightsquigarrow x = y \rightsquigarrow x$, *i.e.* $(pH)$ *from List* **pA**,

(5.18) $$x^{-1} = x \to 1 = x \rightsquigarrow 1 \quad ,$$

(5.19) $$x = y \Longleftrightarrow y^{-1} = x^{-1},$$

(5.20) $$(x \to 1)^{-1} = x = (x \rightsquigarrow 1)^{-1},$$

(5.21) $$(y \cdot x) \to x = y^{-1} = (x \cdot y) \rightsquigarrow x,$$

**Proof.** (5.12): $(x \cdot z) \to (y \cdot z) = (y \cdot z) \cdot (x \cdot z)^{-1} = (y \cdot z) \cdot (z^{-1} \cdot x^{-1}) = y \cdot x^{-1} = x \to y$, $(z \cdot x) \rightsquigarrow (z \cdot y) = (z \cdot x)^{-1} \cdot (z \cdot y) = (x^{-1} \cdot z^{-1}) \cdot (z \cdot y) = x^{-1} \cdot y = x \rightsquigarrow y$.

(5.13): $x \cdot y = (x^{-1})^{-1} \cdot y = (1 \cdot x^{-1})^{-1} \cdot y = (x \to 1) \rightsquigarrow y$ and $x \cdot y = x \cdot (y^{-1})^{-1} = x \cdot (y^{-1} \cdot 1)^{-1} = (y \rightsquigarrow 1) \to x$.

(5.14): $x \to (y \to z) = x \to (z \cdot y^{-1}) = (z \cdot y^{-1}) \cdot x^{-1} \overset{(Pass)}{=} z \cdot (y^{-1} \cdot x^{-1}) = z \cdot (x \cdot y)^{-1} = (x \cdot y) \to z$ and
$x \rightsquigarrow (y \rightsquigarrow z) = x \rightsquigarrow (y^{-1} \cdot z) = x^{-1} \cdot (y^{-1} \cdot z) = [x^{-1} \cdot y^{-1}] \cdot z = (y \cdot x)^{-1} \cdot z = (y \cdot x) \rightsquigarrow z$.

(5.15): $(z \to x) \cdot y = (x \cdot z^{-1}) \cdot y \overset{(Pass)}{=} x \cdot (z^{-1} \cdot y) = x \cdot (z \rightsquigarrow y)$.

(5.16): $z \to (x \cdot y) = (x \cdot y) z^{-1} \overset{(Pass)}{=} x \cdot (y \cdot z^{-1}) = x \cdot (z \to y)$ and
$z \rightsquigarrow (x \cdot y) = (z^{-1}) \cdot (x \cdot y) \overset{(Pass)}{=} ((z^{-1}) \cdot x) \cdot y = (z \rightsquigarrow x) \cdot y$.

(5.17): $[(y \to x) \rightsquigarrow x] \to x = [(x \cdot y^{-1}) \rightsquigarrow x] \to x = [(x \cdot y^{-1})^{-1} \cdot x] \to x = [(y \cdot x^{-1}) \cdot x] \to x = [y \cdot (x^{-1} \cdot x)] \to x = [y \cdot 1] \to x = y \to x$ and
$[(y \rightsquigarrow x) \to x] \rightsquigarrow x = [(y^{-1} \cdot x) \to x] \rightsquigarrow x = [x \cdot (y^{-1} \cdot x)^{-1}] \rightsquigarrow x = [x \cdot (x^{-1} \cdot y)] \rightsquigarrow x = [(x \cdot (x^{-1})) \cdot y] \rightsquigarrow x = [1 \cdot y] \rightsquigarrow x = y \rightsquigarrow x$.

(5.18): $x \to 1 = 1 \cdot x^{-1} = x^{-1}$ and $x \rightsquigarrow 1 = x^{-1} \cdot 1 = x^{-1}$, by (PU).

(5.19): (i) If $x = y$, then, obviously, $y^{-1} = x^{-1}$; (ii) if $y^{-1} = x^{-1}$, then, by (i), $(x^{-1})^{-1} = (y^{-1})^{-1}$, i.e. $x = y$, by (DN).

(5.20): $(x \to 1)^{-1} = (1 \cdot x^{-1})^{-1} = x \cdot 1^{-1} \overset{(N1)}{=} x \cdot 1 \overset{(PU)}{=} x$ and $(x \rightsquigarrow 1)^{-1} = (x^{-1} \cdot 1)^{-1} \overset{(PU)}{=} (x^{-1})^{-1} \overset{(DN)}{=} x$.

(5.21): by (5.14). □

**Proposition 5.4.16** *Let $(G, \cdot, ^{-1}, 1)$ be a group. Then, we have: for all $x, y \in G$,*
*($\alpha$) $(x \to y) \cdot x = y = x \cdot (x \rightsquigarrow y)$,*
*($\beta$) $x \to (y \cdot x) = y = x \rightsquigarrow (x \cdot y)$,*

$$y \to (x \to (y \cdot x)) = 1 = y \rightsquigarrow (x \rightsquigarrow (x \cdot y)). \tag{5.22}$$

**Proof.** ($\alpha$): by (5.15).

($\beta$): by (5.16).

(5.22): $y \to (x \to (y \cdot x)) = y \to y = 1$ and $y \rightsquigarrow (x \rightsquigarrow (x \cdot y)) = y \rightsquigarrow y = 1$, by ($\beta$) and (pRe). □

**Remark 5.4.17** *Note that:*
*- ($pGa^{=}$) is a join of the analogous of the properties: (pPR') and (pScoR'),*
*- ($\alpha$) is the analogous of both the properties (pRR') and (pcoRR'),*
*- ($\beta$) is the analogous of both the properties (pPP') and (pSS').*

**Proposition 5.4.18** *In a group $(G, \cdot, ^{-1}, 1)$, the following properties hold: for all $x, y, z, x_1, x_2, \ldots, x_n \in G$,*
*(a) $(y \to z) \cdot (x \to y) = x \to z$, $(x \rightsquigarrow y) \cdot (y \rightsquigarrow z) = x \rightsquigarrow z$,*
*(b) $(x_{n-1} \to x_n) \cdot \ldots \cdot (x_2 \to x_3) \cdot (x_1 \to x_2) = x_1 \to x_n$,*
*$(x_1 \rightsquigarrow x_2) \cdot (x_2 \rightsquigarrow x_3) \cdot \ldots \cdot (x_{n-1} \rightsquigarrow x_n) = x_1 \rightsquigarrow x_n$.*

**Proof.** (a): $(x \rightsquigarrow y) \cdot (y \rightsquigarrow z) = (x^{-1} \cdot y) \cdot (y^{-1} \cdot z) = x^{-1} \cdot (y \cdot y^{-1}) \cdot z = x^{-1} \cdot 1 \cdot z = x^{-1} \cdot z = x \rightsquigarrow z$. The rest has similar proof. □

### 5.4.1.2 The X-group - an intermediary notion

**Definition 5.4.19** We shall name *X-group* an algebra $(G, \cdot, \to, \rightsquigarrow, 1)$ of type $(2, 2, 2, 0)$ such that ((IdEqR) follows) $(G, \cdot, 1)$ is a monoid (i.e. (Pass) and (PU) hold) and the operation $\cdot$ is connected to the pseudo-implication $(\to, \rightsquigarrow)$ by the following property: for all $x, y, z \in G$,
(pGa$^=$) $x \cdot y = z \Longleftrightarrow x = y \to z \Longleftrightarrow y = x \rightsquigarrow z$.

**Remark 5.4.20** *Note that (pGa$^=$) is a particular non-commutative Galois connection, with $=$ instead of $\leq$ or $\geq$. 'Ga' comes from 'Galois'.*

Denote by **X-group** the class of all X-groups.

**Proposition 5.4.21** *Let $(G, \cdot, \to, \rightsquigarrow, 1)$ be an X-group. Then, the properties $(\alpha)$ and $(\beta)$ hold and (pGa$^=$) $\Longleftrightarrow$ $(\alpha)$ + $(\beta)$.*

**Proof.** By Lemma 1.5.38. □

**Proposition 5.4.22** *Let $(G, \cdot, \to, \rightsquigarrow, 1)$ be an X-group. Then, the following properties hold: for all $x, y, z \in G$:*

| | |
|---|---|
| *(Id)* | $x \to 1 = x \rightsquigarrow 1$, |
| *(pEx)* | $x \to (y \rightsquigarrow z) = y \rightsquigarrow (x \to z)$, |
| *(pM)* | $1 \to x = x = 1 \rightsquigarrow x$, |
| *(IdEqR)* | $x \to y = 1 \Longleftrightarrow x \rightsquigarrow y = 1$, |
| *(pEqrelR=)* | $x = y \Longleftrightarrow x \to y = 1$ ($\stackrel{(IdEq)}{\Longleftrightarrow}$ $x \rightsquigarrow y = 1$). |

**Proof.** (Id): $x \to 1 \stackrel{(PU)}{=} (x \to 1) \cdot 1 \stackrel{(\alpha)}{=} (x \to 1) \cdot [x \cdot (x \rightsquigarrow 1)] \stackrel{(Pass)}{=} [(x \to 1) \cdot x] \cdot (x \rightsquigarrow 1) \stackrel{(\alpha)}{=} 1 \cdot (x \rightsquigarrow 1) \stackrel{(PU)}{=} x \rightsquigarrow 1$.

(pEx): $a = x \to (y \rightsquigarrow z) \stackrel{(pGa^=)}{\Longleftrightarrow} a \cdot x = y \rightsquigarrow z \stackrel{(pGa^=)}{\Longleftrightarrow} y \cdot (a \cdot x) = z \stackrel{(Pass)}{\Longleftrightarrow} (y \cdot a) \cdot x = z \stackrel{(pGa^=)}{\Longleftrightarrow} y \cdot a = x \to z \stackrel{(pGa^=)}{\Longleftrightarrow} a = y \rightsquigarrow (x \to z)$; thus, (pEx) holds.

(pM): $1 \to x \stackrel{(PU)}{=} (1 \to x) \cdot 1 = x \stackrel{(pGa^=)}{\Longleftrightarrow} 1 \to x = 1 \to x$, that is true.

Also, $1 \rightsquigarrow x \stackrel{(PU)}{=} 1 \cdot (1 \rightsquigarrow x) = x \stackrel{(pGa^=)}{\Longleftrightarrow} 1 \rightsquigarrow x = 1 \rightsquigarrow x$, that is true. Thus, (pM) holds.

(IdEq): $x = y \stackrel{(PU)}{\Longleftrightarrow} 1 \cdot x = y \stackrel{(pGa^=)}{\Longleftrightarrow} 1 = x \to y$ and
$x = y \stackrel{(PU)}{\Longleftrightarrow} x \cdot 1 = y \stackrel{(pGa^=)}{\Longleftrightarrow} 1 = x \rightsquigarrow y$. Thus, (IdEqR) holds.

(pEq=): By above proof. □

Note that the X-group may be defined equivalently as an algebra $(G, \cdot, \to, \rightsquigarrow, ^{-1}, 1)$, where $^{-1}$ can be expressed in terms of $\to$, $\rightsquigarrow$ and 1: $x^{-1} = x \to 1 \stackrel{(Id)}{=} x \rightsquigarrow 1$.

### 5.4.1.3 The basic equivalence (EG1)

Here we present the basic equivalence (EG1) between groups and X-groups.

We prove that the groups are termwise equivalent to the X-groups (result which can be found in ([58], page 161) in a different form; we prefer the following form):

**Theorem 5.4.23**
*(1) Let* $\mathcal{G} = (G, \cdot, ^{-1}, 1)$ *be a group.*
*Define* $\rho(\mathcal{G}) \stackrel{def.}{=} (G, \cdot, \to, \rightsquigarrow, 1)$, *where* $x \to y \stackrel{def.}{=} y \cdot x^{-1}$ *and* $x \rightsquigarrow y \stackrel{def.}{=} x^{-1} \cdot y$.
*Then,* $\rho(\mathcal{G})$ *is an X-group.*
*(1') Conversely, let* $\mathcal{G} = (G, \cdot, \to, \rightsquigarrow, 1)$ *be an X-group.*
*Define* $\rho^*(\mathcal{G}) \stackrel{def.}{=} (G, \cdot, ^{-1}, 1)$, *where* $x^{-1} \stackrel{def.}{=} x \to 1 \stackrel{(Id)}{=} x \rightsquigarrow 1$.
*Then,* $\rho^*(\mathcal{G})$ *is a group.*
*(2) The above defined mappings* $\rho$ *and* $\rho^*$ *are mutually inverse.*

**Proof.** (1): By Proposition 5.4.12.

(1'): Since $(G, \cdot, 1)$ is monoid, it follows that (Pass), (PU) hold. It remains to prove that (pIv) holds too. Indeed, $(x^{-1}) \cdot x = (x \to 1) \cdot x \stackrel{(\alpha)}{=} 1$ and $x \cdot (x^{-1}) = x \cdot (x \rightsquigarrow 1) \stackrel{(\alpha)}{=} 1$; thus, (pIv) holds.

(2): If $(G, \cdot, ^{-1}, 1) \stackrel{\rho}{\longrightarrow} (G, \cdot, \to, \rightsquigarrow, 1) \stackrel{\rho^*}{\longrightarrow} (G, \cdot, \sim, 1)$, then $\sim x = x \to 1 = 1 \cdot x^{-1} = x^{-1}$.

Let now $(G, \cdot, \to, \rightsquigarrow, 1) \stackrel{\rho^*}{\longrightarrow} (G, \cdot, ^{-1}, 1) \stackrel{\rho}{\longrightarrow} (G, \cdot, \Rightarrow, \approx>, 1)$.
First we prove that $x \Rightarrow y = x \to y$. Indeed,
$x \Rightarrow y = y \cdot x^{-1} = y \cdot (x^{-1}) = y \cdot (x \to 1) \stackrel{(Id)}{=} y \cdot (x \rightsquigarrow 1) \stackrel{(\alpha)}{=}$
$[(x \to y) \cdot x] \cdot (x \rightsquigarrow 1) \stackrel{(Pass)}{=} (x \to y) \cdot [x \cdot (x \rightsquigarrow 1)] \stackrel{(\alpha)}{=} (x \to y) \cdot 1 \stackrel{(PU)}{=} x \to y$.
Now we shall prove that $x \approx> y = x \rightsquigarrow y$. Indeed,
$x \approx> y = x^{-1} \cdot y = (x \to 1) \cdot y \stackrel{(\alpha)}{=} (x \to 1) \cdot [x \cdot (x \rightsquigarrow y)] \stackrel{(Pass)}{=}$
$[(x \to 1) \cdot x] \cdot (x \rightsquigarrow y) \stackrel{(\alpha)}{=} 1 \cdot (x \rightsquigarrow y) \stackrel{(PU)}{=} x \rightsquigarrow y$. □

Hence, we have the basic equivalence:

(EG1) **X-group** $\Longleftrightarrow$ **group.**

## 5.4.2 Po-groups

### 5.4.2.1 The po-group

**Definition 5.4.24** A *partially-ordered group*, or a *po-group* for short, is a structure $\mathcal{G} = (G, \leq, \cdot, ^{-1}, 1)$, such that $(G, \cdot, ^{-1}, 1)$ is a group, $(G, \leq)$ is a poset and $\leq$ is compatible with $\cdot$, i.e. the property (pCp) holds: for all $x, y \in G$,
(pCp) $x \leq y \Longrightarrow a \cdot x \leq a \cdot y$ and $x \cdot a \leq y \cdot a$.

**Remarks 5.4.25** *The presence of the order relation* $\leq$ *implies the presence of the duality principle. It follows that there are two dual po-groups,* $\mathcal{G}_1$ *and* $\mathcal{G}_2$.

*(i) If their support sets coincide ($G = G_1 = G_2$), we shall say that $G$ is self-dual, i.e. the dual of $(G, \leq, \cdot, ^{-1}, 1)$ is $(G, \geq, \cdot, ^{-1}, 1)$ and vice-versa. Namely, if $(G, \leq, \cdot, ^{-1}, 1)$ is a po-group, then note that $\cdot$ is a pseudo-t-norm on the poset $(G, \leq)$ containing 1, while in the dual po-group $(G, \geq, \cdot, ^{-1}, 1)$, $\cdot$ is a pseudo-t-conorm on the poset $(G, \geq)$ containing 1. It follows that we can say that $(G, \leq, \cdot, ^{-1}, 1)$ is the left-po-group, while $(G, \geq, \cdot, ^{-1}, 1)$ is the right-po-group.*

*(ii) If their support sets differ ($G_1 \neq G_2$), then their Unit elements $1_1$, $1_2$ differ and suppose that $1_1 < 1_2$ in the union set $G_1 \cup G_2$; we shall then call $\mathcal{G}_1$ as the left-po-group and $\mathcal{G}_2$ as the right-po-group.*

Denote by **po-group** the class of all po-groups, namely:
- by **po-group**, the class of all left-po-groups and
- by **po-group**$^R$, the class of all right-$l$-groups.

**Proposition 5.4.26** *Let $\mathcal{G}$ be a po-group. Then, we have: for all $x, y, a, b \in G$, if $x \leq y$ and $a \leq b$, then $x \cdot a \leq y \cdot b$ and $a \cdot x \leq b \cdot y$.*

**Proposition 5.4.27** *Let $\mathcal{G}$ be a po-group. Then (we recall the old results from literature): for all $x, y, a \in G$,*
*(Neg2')* $\quad x \leq y \Longrightarrow y^{-1} \leq x^{-1}$,
*(DN4')* $\quad x \leq y \Longleftrightarrow y^{-1} \leq x^{-1}$,
*(pEqCp)* $\quad x \leq y \Longleftrightarrow a \cdot x \leq a \cdot y \Longleftrightarrow x \cdot a \leq y \cdot a$.

**Proof.** We recall only the proof of (Neg2'): let $x \leq y$; then, by (pCp), (pIv), $1 = x \cdot (x^{-1}) \leq y \cdot (x^{-1})$; then, by (pCp) again and (PU), (Pass), (pIv), (PU), $y^{-1} = (y^{-1}) \cdot 1 \leq (y^{-1}) \cdot [y \cdot (x^{-1})] = [(y^{-1}) \cdot y] \cdot (x^{-1}) = 1 \cdot (x^{-1}) = x^{-1}$. $\square$

**Proposition 5.4.28** *Let $\mathcal{G}$ be a po-group. Then, we have: for all $x, y, a, z \in G$,*
*(pGa$^{\leq}$)* $\quad x \cdot y \leq z \Longleftrightarrow x \leq y \to z \Longleftrightarrow y \leq x \rightsquigarrow z$ *and, dually,*
*(pGa$^{\geq}$)* $\quad x \cdot y \geq z \Longleftrightarrow x \geq y \to z \Longleftrightarrow y \geq x \rightsquigarrow z$.

**Proof.**

(pGa$^{\leq}$): $x \leq y \to z \Longleftrightarrow x \leq z \cdot y^{-1} \Longrightarrow x \cdot y \leq (z \cdot y^{-1}) \cdot y = z \cdot (y^{-1} \cdot y) = z \cdot 1 = z$, by (pCp), (Pass), (pIv), (PU) and $x \cdot y \leq z \Longrightarrow x = x \cdot 1 = x \cdot (y \cdot y^{-1}) = (x \cdot y) \cdot (y^{-1}) \leq z \cdot (y^{-1}) = y \to z$, by (pCp). Similarly, $y \leq x \rightsquigarrow z \Longleftrightarrow y \leq x^{-1} \cdot z \Longleftrightarrow x \cdot y \leq z$, by (pCp).

(pGa$^{\geq}$): dually. $\square$

**Remark 5.4.29** *Note that:*
- *(pGa$^{\leq}$) is the analogous of the property (pPR')=(pRP'),*
- *(pGa$^{\geq}$) is the analogous of the property (pScoR')=(pcoRS').*

**Theorem 5.4.30**

*(1) Let $\mathcal{G} = (G, \leq, \cdot, ^{-1}, 1)$ be a po-group. Define $\phi(\mathcal{G}) \overset{def.}{=} (G, \leq, \cdot, \to, \rightsquigarrow, 1)$ by*

$$x \to y \overset{def.}{=} (x \cdot y^{-1})^{-1} = y \cdot x^{-1}, \quad x \rightsquigarrow y \overset{def.}{=} (y^{-1} \cdot x)^{-1} = x^{-1} \cdot y.$$

*Then, $\phi(\mathcal{G})$ is a po-m ($pGa^{\leq}$) (verifying (pIv), where $x^{-1} = x \to 1 = x \rightsquigarrow 1$, by (5.18)).*

*(1') Let $\mathcal{G} = (G, \leq, \cdot, \to^L, \rightsquigarrow^L, 1)$ be a po-m(pPR'), verifying (pIv), where $x^{-1} = x \to^L 1$. Define $\psi(\mathcal{G}) \overset{def.}{=} (G, \leq, \cdot, ^{-1}, 1)$.*
*Then, $\psi(\mathcal{G})$ is a po-group.*

*(2) The above defined mappings $\phi$ and $\psi$ are mutually inverse and (pPR')= ($pGa^{\leq}$).*

**Proof.** (1): By Proposition 5.4.28, the property ($\text{pGa}^{\leq}$) holds, hence $\phi(\mathcal{G})$ is a po-m(pPR'), with (pPR')=($\text{pGa}^{\leq}$) .

(1'): Let $\mathcal{G} = (G, \leq, \cdot, \to^L, \rightsquigarrow^L, 1)$ be a po-m(pPR'), verifying (pIv) ($x \cdot x^{-1} = 1 = x^{-1} \cdot x$), where $x^{-1} = x \to^L 1$. We must prove that $\psi(\mathcal{G}) \overset{def.}{=} (G, \leq, \cdot, ^{-1}, 1)$ is a po-group. It remains to prove that (pCp') holds. Indeed, by Lemma 1.4.17, (pCp') holds.

(2): First, we have:

| | | |
|---|---|---|
| $(G, \leq, \cdot, ^{-1}, 1) \overset{\phi}{\longrightarrow}$ | $(G, \leq, \cdot, \to, \rightsquigarrow, 1) \overset{\psi}{\longrightarrow}$ | $(G, \leq, \cdot, ', 1)$ |
| po-group | po-m(pPR')<br>where<br>$x \to y = y \cdot x^{-1}$<br>$x \rightsquigarrow y = x^{-1} \cdot y$ | po-group<br>where<br>$x' = x \to 1$ |

We must prove that $x' = x^{-1}$, for all $x \in G$. Indeed, $x' = x \to 1 = 1 \cdot (x^{-1}) \overset{(PU)}{=} x^{-1}$.

Then, we have:

| | | |
|---|---|---|
| $(G, \leq, \cdot, \to^L, \rightsquigarrow^L, 1) \overset{\psi}{\longrightarrow}$ | $(G, \leq, \cdot, ^{-1}, 1) \overset{\phi}{\longrightarrow}$ | $(G, \leq, \cdot, \to, \rightsquigarrow, 1)$ |
| po-m(pPR')<br>verifying (pIv), where<br>$x^{-1} = x \to^L 1$ | po-group<br>where<br>$x^{-1} = x \to^L 1$ | po-m($\text{pGa}^{\leq}$)<br>verifying (pIv), where<br>$x^{-1} = x \to 1 = x \rightsquigarrow 1$ |

where:
(pPR') $x \cdot y \leq z \Longleftrightarrow x \leq y \to^L z \Longleftrightarrow y \leq x \rightsquigarrow^L z$
($\text{pGa}^{\leq}$) $x \cdot y \leq z \Longleftrightarrow x \leq y \to z \Longleftrightarrow y \leq x \rightsquigarrow z$.
We must prove that:

$$x \to y = x \to^L y, \quad x \rightsquigarrow y = x \rightsquigarrow^L y.$$

We prove that $x \to y = x \to^L y$. Indeed,
$x \to y \leq x \to^L y \overset{(pPR')}{\Longleftrightarrow} y \overset{(\alpha)}{=} (x \to y) \cdot x \leq y$, that is true by the reflexivity of $\leq$;
$x \to^L y \leq x \to y \overset{(pGa^{\leq})}{\Longleftrightarrow} (x \to^L y) \cdot x \leq y$, that is true by (pRR'), by Lemma 1.4.17.
Hence, by the antisymmetry of $\leq$, we obtain that $x \to y = x \to^L y$.

We prove now that $x \rightsquigarrow y = x \rightsquigarrow^L y$. Indeed,
$x \rightsquigarrow y \leq x \rightsquigarrow^l y \overset{(pPR')}{\Longleftrightarrow} y \overset{(\alpha)}{=} x \cdot (x \rightsquigarrow y) \leq y$, that is true by the reflexivity of $\leq$;
$x \rightsquigarrow^L y \leq x \rightsquigarrow y \overset{(pGa^{\leq})}{\Longleftrightarrow} x \cdot (x \rightsquigarrow^L y) \leq y$, that is true by (pRR').
Hence, we obtain that $x \rightsquigarrow y = x \rightsquigarrow^L y$.

Consequently, (pGa$^{\leq}$) = (pPR'). □

**Remark 5.4.31** *By above Theorem 5.4.30 and its dual, we have the dual equivalences:*

*(EGM)* **po-group** $\Longleftrightarrow$ **po-m(pGa$^{\leq}$)** + *(pIv) and, dually,*
**po-group$^R$** $\Longleftrightarrow$ **po-m(pGa$^{\geq}$)** + *(pIv)* .

**Corollary 5.4.32** *A po-group is a po-m(pR').*

**Proof.** Let $(G, \leq, \cdot, ^{-1}, 1)$ be a po-group. By Theorem 5.4.30 (1), $(G, \leq, \cdot, \to, \rightsquigarrow, 1)$ is a po-m(pPR'), hence $(G, \leq, \cdot, 1)$ is a po-m(pR'), by Theorem 5.3.24 (2). □

**Corollary 5.4.33** *Let $\mathcal{G}$ be a po-group. Then, for all $x, y \in G$, we have:*
*(i)* $y \leq 1 \Longleftrightarrow x \leq y \to x \Longleftrightarrow x \leq y \rightsquigarrow x$ *and, dually,*
*(i')* $y \geq 1 \Longleftrightarrow x \geq y \to x \Longleftrightarrow x \geq y \rightsquigarrow x$;
*(ii)* $y \leq z \Longleftrightarrow 1 \leq y \to z \Longleftrightarrow 1 \leq y \rightsquigarrow z$ *and, dually,*
*(ii')* $y \geq z \Longleftrightarrow 1 \geq y \to z \Longleftrightarrow 1 \geq y \rightsquigarrow z$.

**Proof.** (i): take $z = x$ and then $z = y$ in (pGa$^{\leq}$). (i'): take $z = x$ and then $z = y$ in (pGa$^{\geq}$).

(ii): take $x = 1$ and then $y = 1$ in (pGa$^{\leq}$). (ii'): take $x = 1$ and then $y = 1$ in (pGa$^{\geq}$). □

Note that the properties (DN) and (DN4') make the operation $^{-1}$ be an *involution.*

**Proposition 5.4.34** *Let $\mathcal{G}$ be a po-group. Then, for all $x, y, z \in G$, we have:*
*(p*')* $x \leq y \Longleftrightarrow z \to x \leq z \to y$ *and* $z \rightsquigarrow x \leq z \rightsquigarrow y$,
*(p**')* $x \leq y \Longleftrightarrow y \to z \leq x \to z$ *and* $y \rightsquigarrow z \leq x \rightsquigarrow z$.

**Proof.** (p*'): Let $x \leq y$; then $z \to x = x \cdot z^{-1} \overset{(pCp')}{\leq} y \cdot z^{-1} = z \to y$ and $z \rightsquigarrow x = z^{-1} \cdot x \overset{(pCp)}{\leq} z^{-1} \cdot y = z \rightsquigarrow y$.

(p**'): Let $x \leq y$; then by (Neg2'), $y^{-1} \leq x^{-1}$; hence,
$y \to z = z \cdot y^{-1} \overset{(pCp')}{\leq} z \cdot x^{-1} = x \to z$ and $y \rightsquigarrow z = y^{-1} \cdot z \overset{(pCp')}{\leq} x^{-1} \cdot z = x \rightsquigarrow z$.
□

**Corollary 5.4.35** *Let $\mathcal{G}$ be a po-group. Then, we have: for all $x, y \in G$, if $x \leq y$ then $x \to y \geq 1$, $x \rightsquigarrow y \geq 1$ and $y \to x \leq 1$, $y \rightsquigarrow x \leq 1$.*

**Proof.** Let $x \leq y$; then,
- by (p$^{*L}$), $x \to x \leq x \to y$, i.e., by (pRe), $1 \leq x \to y$; similarly, $1 \leq x \rightsquigarrow y$;
- by (p$^{**L}$), $y \to x \leq x \to x = 1$, i.e. $y \to x \leq 1$; similarly, $y \rightsquigarrow x \leq 1$. □

### 5.4.2.2 The negative and the positive cones of po-groups

Let $\mathcal{G} = (G, \leq, \cdot, ^{-1}, 1)$ be a po-group (left-po-group or right-po-group).
Define the *negative cone* of $\mathcal{G}$ as follows: $G^- \stackrel{def.}{=} \{x \in G \mid x \leq 1\}$.
Define the *positive cone* of $\mathcal{G}$ as follows: $G^+ \stackrel{def.}{=} \{x \in G \mid x \geq 1\}$.
Then, we have the following result.

**Lemma 5.4.36** *$G^-$ and $G^+$ are closed under $\cdot$.*

**Proof.** Let $x, y \in G^-$, i.e. $x \leq 1$, $y \leq 1$; then, by (pCp), (PU), $x \cdot y \leq 1 \cdot y = y$; since $y \leq 1$, it follows, by transitivity of $\leq$, that $x \cdot y \leq 1$, i.e. $x \cdot y \in G^-$. Similarly, $G^+$ is closed under $\cdot$. □

Note that $G^-$ and $G^+$ are not closed under $^{-1}$.

Note that a po-group $(G, \leq, \cdot, ^{-1}, 1)$ is a po-m $(G, \leq, \cdot, 1)$ verifying (pIv). By Lemma 5.4.36, we immediately obtain:

**Proposition 5.4.37** *Let $\mathcal{G} = (G, \leq, \cdot, ^{-1}, 1)$ be a po-group. Then*
*(1) $\mathcal{G}^- = (G^-, \leq, \odot = \cdot, \mathbf{1} = 1)$ is a left-po-im;*
*(1') $\mathcal{G}^+ = (G^+, \leq, \oplus = \cdot, \mathbf{0} = 1)$ is a right-po-im.*

If the partial order $\leq$ is linear (total), then $\mathcal{G}$ is a *linearly-ordered group* or a *totally-ordered group*. Note that in a linearly-ordered group $\mathcal{G}$ we have either $x \leq 1$ or $x \geq 1$ (i.e. $1 \leq x$), for any $x \in G$, hence $G = G^- \cup G^+$.

### 5.4.2.3 The X-po-group - an intermediary notion

**Definition 5.4.38** We shall name *X-po-group* a structure $(G, \leq, \cdot, \rightarrow, \rightsquigarrow, 1)$ such that $(G, \cdot, 1)$ is a monoid (i.e. (Pass), (PU) hold), $(G, \leq)$ is a poset and the operation $\cdot$ and the pseudo-implication $(\rightarrow, \rightsquigarrow)$ verify the following two properties: for all $x, z \in G$,
(pGa$^{\leq}$) $\quad x \cdot y \leq z \Longleftrightarrow x \leq y \rightarrow z \Longleftrightarrow y \leq x \rightsquigarrow z,$
(pGa$^{\geq}$) $\quad x \cdot y \geq z \Longleftrightarrow x \geq y \rightarrow z \Longleftrightarrow y \geq x \rightsquigarrow z.$

Denote by **X-po-group** the class of all X-po-groups, namely:
- by **X-po-group**, the class of all left-X-po-groups,
- by **X-po-group**$^R$, the class of all right-X-po-groups.

**Remark 5.4.39** *Note that an X-po-group is in the same time a po-m(pGa$^{\leq}$) and a po-m(pGa$^{\geq}$). Hence, we can write:*

$$\textbf{X-po-group} = \textbf{po-m(pGa}^{\leq}\textbf{)} + \textbf{po-m(pGa}^{\geq}\textbf{)}.$$

Note that if $(G, \leq, \cdot, \rightarrow, \rightsquigarrow, 1)$ is an X-po-group, then $(G, \cdot, \rightarrow, \rightsquigarrow, 1)$ is an X-group, by Lemma 1.5.58.

**Proposition 5.4.40** *Let* $(G, \leq, \cdot, \to, \rightsquigarrow, 1)$ *be an X-po-group. Then the properties (pGa*$^=$*), (*$\alpha$*), (*$\beta$*), (pEqCp), (p*$^{*L}$*), (p*$^{**L}$*) and (Id), (pBB=), (pM), (pEq*$^=$*) hold.*

**Proof.** By Lemma 1.5.58, the properties (pGa$^=$), ($\alpha$), ($\beta$), (pEqCp'), (p*'), (p**') hold. Hence, $(G, \cdot, \to, \rightsquigarrow, 1)$ is an X-group and hence, by Proposition 5.4.22, the properties (Id), (pBB=), (pM), (pEq$^=$) also hold. □

### 5.4.2.4 The equivalences (EG2)

Here we present the equivalences (EG2) between the po-groups and the X-po-groups.

We prove that the po-groups are termwise equivalent to the X-po-groups.

**Theorem 5.4.41** *(See Theorem 5.4.23)*

*(1) Let* $\mathcal{G} = (G, \leq, \cdot, ^{-1}, 1)$ *be a po-group.*

*Define* $\rho'(\mathcal{G}) \overset{def.}{=} (G, \leq, \cdot, \to, \rightsquigarrow, 1)$*, with* $(G, \cdot, \to, \rightsquigarrow, 1) = \rho(G, \cdot, ^{-1}, 1)$ *from Theorem 5.4.23 (1).*

*Then,* $\rho'(\mathcal{G})$ *is an X-po-group.*

*(1') Conversely, let* $\mathcal{G} = (G, \leq, \cdot, \to, \rightsquigarrow, 1)$ *be an X-po-group.*

*Define* $\rho^{*'}(\mathcal{G}) \overset{def.}{=} (G, \leq, \cdot, ^{-1}, 1)$*, with* $(G, \cdot, ^{-1}, 1) = \rho^*(G, \cdot, \to, \rightsquigarrow, 1)$ *from Theorem 5.4.23 (1').*

*Then,* $\rho^{*'}(\mathcal{G})$ *is a po-group.*

*(2) The above defined mappings* $\rho'$ *and* $\rho^{*'}$ *are mutually inverse.*

**Proof.** (1): follows by Theorem 5.4.23 (1) and by Proposition 5.4.28.

(1'): by Theorem 5.4.23 (1'), $(G, \cdot, ^{-1}, 1)$ is a group; (pCp) holds by Lemma 1.5.58.

(2): follows by Theorem 5.4.23 (2). □

Hence, we have the dual equivalences (EG2) (the analogous of (iEM5)):

(EG2) **X-po-group** $\Longleftrightarrow$ **po-group** and, dually,
**X-po-group**$^R$ $\Longleftrightarrow$ **po-group**$^R$.

**Remarks 5.4.42**

*Note that (EG2) can be written as follows, by using Remark 5.4.39:*

**po-group** $\Longleftrightarrow$ **po-m(pGa**$^{\leq}$**)** + **po-m(pGa**$^{\geq}$**)**.

*In the same time, note that, by (EGM), we have:*

**po-group** $\Longleftrightarrow$ **po-m(pGa**$^{\leq}$**)** + *(pIv) and, dually,*
**po-group**$^R$ $\Longleftrightarrow$ **po-m(pGa**$^{\geq}$**)** + *(pIv).*

## 5.4.3 $l$-groups

### 5.4.3.1 The $l$-group

**Definition 5.4.43** Let $\mathcal{G} = (G, \leq, \cdot, ^{-1}, 1)$ be a po-group. If the partial order $\leq$ is a lattice order, then the po-group is called *lattice-ordered group*, or *l-group* for

short, and is denoted generally by: $\mathcal{G} = (G, \vee, \wedge, \cdot, ^{-1}, 1)$, namely:
- by $\mathcal{G}^L = (G^L, \wedge, \vee, \odot, ^{-1}, 1)$, if it is a left-$l$-group,
- by $\mathcal{G}^R = (G^R, \vee, \wedge, \oplus, -, 0)$, if it is a right-$l$-group.

Denote by $l$-**group** the class of all $l$-groups, namely:
- by $l$-**group** the class of all left-$l$-groups and
- by $l$-**group**$^R$ the class of all right-$l$-groups.

An introduction in l-groups is [1], see also [36].

Note that an $l$-group may be linearly-ordered or not, while a linearly-ordered group is an $l$-group.

**Proposition 5.4.44** *Let $\mathcal{G}$ be an $l$-group. Then the following properties hold (old results from the literature): for all $x, y, a, b \in G$,*
*(1)* $a \cdot (x \vee y) \cdot b = (a \cdot x \cdot b) \vee (a \cdot y \cdot b)$ *and, dually,*
*(1')* $a \cdot (x \wedge y) \cdot b = (a \cdot x \cdot b) \wedge (a \cdot y \cdot b)$;
*(2)* $(x \vee y)^{-1} = (x^{-1}) \wedge (y^{-1})$ *and, dually,*
*(2')* $(x \wedge y)^{-1} = (x^{-1}) \vee (y^{-1})$;
*(3)* $x \vee y = x \cdot (x \wedge y)^{-1} \cdot y = [(x \wedge y) \to x] \cdot y = x \cdot [(x \wedge y) \rightsquigarrow y]$ *and, dually,*
*(3')* $x \wedge y = x \cdot (x \vee y)^{-1} \cdot y = [(x \vee y) \to x] \cdot y = x \cdot [(x \vee y) \rightsquigarrow y]$;
*(4) The lattice* $(G, \vee, \wedge)$ *is distributive.*

**Theorem 5.4.45** *(See Theorem 5.4.30)*

*(1) Let* $\mathcal{G} = (G, \vee, \wedge, \cdot, ^{-1}, 1)$ *be an $l$-group.*
*Define* $\phi(\mathcal{G}) \overset{def.}{=} (G, \vee, \wedge, \cdot, \to, \rightsquigarrow, 1)$ *by:*

$$x \to y \overset{def.}{=} (x \cdot y^{-1})^{-1} = y \cdot x^{-1}, \quad x \rightsquigarrow y \overset{def.}{=} (y^{-1} \cdot x)^{-1} = x^{-1} \cdot y.$$

*Then, $\phi(\mathcal{G})$ is an $l$-m (pGa$^{\leq}$) (verifying (pIv), where $x^{-1} = x \to 1 = x \rightsquigarrow 1$, by (5.18)).*

*(1') Let* $\mathcal{G} = (G, \vee, \wedge, \cdot, \to^L, \rightsquigarrow^L, 1)$ *be a $l$-m(pPR'), verifying (pIv), where* $x^{-1} = x \to^L 1$.
*Define* $\psi(\mathcal{G}) \overset{def.}{=} (G, \leq, \cdot, ^{-1}, 1)$.
*Then, $\psi(\mathcal{G})$ is an $l$-group.*

*(2) The above defined mappings $\phi$ and $\psi$ are mutually inverse and (pPR')= (pGa$^{\leq}$).*

**Corollary 5.4.46** *(See [65])*

*The $l$-groups are term equivalent to residuated lattices verifying (pIv).*

**Proof.** By Theorem 5.4.45, the $l$-groups are term equivalent to l-ms(pPR') verifying (pIv), while **l-m(pPR')** = **pR-L**. □

**Remark 5.4.47** *(See Remark 5.4.31) By above Theorem 5.4.45, we have the dual equivalences:*

*(EGM-l)* $l$-**group** $\Longleftrightarrow$ $l$-**m(pGa$^{\leq}$)** + *(pIv) and, dually,*
$l$-**group**$^R$ $\Longleftrightarrow$ $l$-**m(pGa$^{\geq}$)** + *(pIv).*

By Corollary 5.4.32, we immediately obtain:

**Corollary 5.4.48** *An l-group is a l-m(pR').*

**Proposition 5.4.49** *In an l-group $(G, \vee, \wedge, \cdot, ^{-1}, 1)$, the following properties hold: for all $x, y, z \in G$,*

$$(5.23) \quad z \to (x \vee y) = (z \to x) \vee (z \to y), \quad z \rightsquigarrow (x \vee y) = (z \rightsquigarrow x) \vee (z \rightsquigarrow y)$$

*and, dually,*

$$(5.24) \quad z \to (x \wedge y) = (z \to x) \wedge (z \to y), \quad z \rightsquigarrow (x \wedge y) = (z \rightsquigarrow x) \wedge (z \rightsquigarrow y);$$

$$(5.25) \quad (x \vee y) \to z = (x \to z) \wedge (y \to z), \ (x \vee y) \rightsquigarrow z = (x \rightsquigarrow z) \wedge (y \rightsquigarrow z)$$

*and, dually,*

$$(5.26) \quad (x \wedge y) \to z = (x \to z) \vee (y \to z), \quad (x \wedge y) \rightsquigarrow z = (x \rightsquigarrow z) \vee (y \rightsquigarrow z).$$

**Proof.** (5.23):
$y \to (x \vee z) = (x \vee z) \cdot y^{-1} = (x \cdot y^{-1}) \vee (z \cdot y^{-1}) = (y \to x) \vee (y \to z)$,
$y \rightsquigarrow (x \vee z) = y^{-1} \cdot (x \vee z) = (y^{-1} \cdot x) \vee (y^{-1} \cdot z) = (y \rightsquigarrow x) \vee (y \rightsquigarrow z)$.
(5.24):
$y \to (x \wedge z) = (x \wedge z) \cdot y^{-1} = (x \cdot y^{-1}) \wedge (z \cdot y^{-1}) = (y \to x) \wedge (y \to z)$,
$y \rightsquigarrow (x \wedge z) = y^{-1} \cdot (x \wedge z) = (y^{-1} \cdot x) \wedge (y^{-1} \cdot z) = (y \rightsquigarrow x) \wedge (y \rightsquigarrow z)$.
(5.25):
$(x \vee z) \to y = y \cdot (x \vee z)^{-1} = y \cdot [(x^{-1}) \wedge (z^{-1})] = (y \cdot x^{-1}) \wedge (y \cdot z^{-1}) = (x \to y) \wedge (z \to y)$,
$(x \vee z) \rightsquigarrow y = (x \vee z)^{-1} \cdot y = [(x^{-1}) \wedge (z^{-1})] \cdot y = (x^{-1} \cdot y) \wedge (z^{-1} \cdot y) = (x \rightsquigarrow y) \wedge (z \rightsquigarrow y)$.
(5.26):
$(x \wedge z) \to y = y \cdot (x \wedge z)^{-1} = y \cdot [(x^{-1}) \vee (z^{-1})] = (y \cdot x^{-1}) \vee (y \cdot z^{-1}) = (x \to y) \vee (z \to y)$,
$(x \wedge z) \rightsquigarrow y = (x \wedge z)^{-1} \cdot y = [(x^{-1}) \vee (z^{-1})] \cdot y = (x^{-1} \cdot y) \vee (z^{-1} \cdot y) = (x \rightsquigarrow y) \vee (z \rightsquigarrow y)$. □

**Proposition 5.4.50** *In an l-group $(G, \vee, \wedge, \cdot, ^{-1}, 1)$, the following properties also hold: for all $x, y, z \in G$,*

$$(5.27) \quad [(x \wedge 1) \rightsquigarrow 1] \wedge 1 = 1, \quad [(x \wedge 1) \to 1] \wedge 1 = 1$$

*and, dually,*

$$(5.28) \quad [(x \vee 1) \rightsquigarrow 1] \vee 1 = 1, \quad [(x \vee 1) \to 1] \vee 1 = 1.$$

$$(5.29) \quad (x \vee y) \to (x \wedge y) = (x \to y) \wedge (y \to x) \wedge 1 \leq 1 \text{ and}$$

$$(5.30) \quad (x \vee y) \rightsquigarrow (x \wedge y) = (x \rightsquigarrow y) \wedge (y \rightsquigarrow x) \wedge 1 \leq 1$$

*and, dually,*

(5.31) $$(x \wedge y) \to (x \vee y) = (x \to y) \vee (y \to x) \vee 1 \geq 1 \text{ and}$$

(5.32) $$(x \wedge y) \rightsquigarrow (x \vee y) = (x \rightsquigarrow y) \vee (y \rightsquigarrow x) \vee 1 \geq 1;$$

(5.33) $$x \to (x \wedge y) = 1 \wedge (x \to y), \quad x \rightsquigarrow (x \wedge y) = 1 \wedge (x \rightsquigarrow y),$$

(5.34) $$(x \wedge y) \to x = 1 \vee (y \to x), \quad (x \wedge y) \rightsquigarrow x = 1 \vee (y \rightsquigarrow x).$$

**Proof.** (5.27):
$[(x \wedge 1) \rightsquigarrow 1] \wedge 1 = [(x \wedge 1)^{-1} \cdot 1] \wedge 1 = [(x^{-1} \vee -1) \cdot 1] \wedge 1 = [x^{-1} \vee 1] \wedge 1 \overset{absorbtion}{=} 1,$
$[(x \wedge 1) \to 1] \wedge 1 = [1(x \wedge 1)^{-1}] \wedge 1 = [1 \cdot (x^{-1} \vee -1)] \wedge 1 = [x^{-1} \vee 1] \wedge 1 \overset{absorbtion}{=} 1.$
(5.28):
$[(x \vee 1) \rightsquigarrow 1] \vee 1 = [(x \vee 1)^{-1} \cdot 1] \vee 1 = [(x^{-1} \wedge -1) \cdot 1] \vee 1 = [x^{-1} \wedge 1] \vee 1 \overset{absorbtion}{=} 1,$
$[(x \vee 1) \to 1] \vee 1 = [1 \cdot (x \vee 1)^{-1}] \vee 1 = [1 \cdot (x^{-1} \wedge -1)] \vee 1 = [x^{-1} \wedge 1] \vee 1 \overset{absorbtion}{=} 1.$

(9.2): $(x \vee y) \to (x \wedge y) = [x \to (x \wedge y)] \wedge [y \to (x \wedge y)] = [(x \to x) \wedge (x \to y)] \wedge [(y \to x) \wedge (y \to y)] = [1 \wedge (x \to y)] \wedge [(y \to x) \wedge 1] = (x \to y) \wedge (y \to x) \wedge 1 \leq 1$, by (5.24), (5.25).

(5.30): similarly.

(5.31): similarly, by (5.23), (5.26).

(5.32): similarly.

(5.33): by (5.24).

(5.34): by (5.26). □

### 5.4.3.2 The negative and the positive cones of $l$-groups

**Lemma 5.4.51** *Let $\mathcal{G} = (G, \vee, \wedge, \cdot, ^{-1}, 1)$ be an l-group (left-l-group or right-l-group). Then, $G^-$ and $G^+$ are closed under $\vee$ and $\wedge$.*

**Proof.** Let $x, y \in G^-$, i.e. $x \leq 1$ and $y \leq 1$. Then, 1 is an upper bound of $\{x, y\}$; hence, $1 \geq x \vee y$, i.e. $x \vee y \in G^-$; since $1 \geq x \vee y \geq x \wedge y$, it follows that $1 \geq x \wedge y$, i.e. $x \wedge y \in G^-$ too.

Let now $x, y \in G^+$, i.e. $x \geq 1$ and $y \geq 1$. Then, 1 is a lower bound of $\{x, y\}$; hence, $1 \leq x \wedge y$, i.e. $x \wedge y \in G^+$; since $1 \leq x \wedge y \leq x \vee y$, it follows that $1 \leq x \vee y$, i.e. $x \vee y \in G^+$ too. □

**Lemma 5.4.52** *(See Lemma 5.4.36)*
*Let $\mathcal{G} = (G, \vee, \wedge, \cdot, ^{-1}, 1)$ be an l-group (left-l-group or right-l-group). Then, $G^-$ and $G^+$ are closed under $\cdot$.*

**Corollary 5.4.53** *Let $\mathcal{G}$ be an l-group. Then we have: for all $x, y \in G$,*
*(i) $x, y \in G^-$ imply $x \cdot y \leq x \wedge y$,*
*(i') $x, y \in G^+$ imply $x \cdot y \geq x \vee y$.*

**Proof.** By Proposition 5.4.44 (3') and (3),
$x \wedge y = [(x \vee y) \to x] \cdot y$ and $x \vee y = [(x \wedge y) \to x] \cdot y$.
(i): if $x, y \leq 1$, then $x \vee y \leq 1$, hence by Corollary 5.4.33, $x \leq (x \vee y) \to x$; then, by (pCp), we obtain that $x \cdot y \leq [(x \vee y) \to x] \cdot y = x \wedge y$.
(i'): if $x, y \geq 1$, then $x \wedge y \geq 1$, hence by Corollary 5.4.33, $x \geq (x \wedge y) \to x$; then, by (pCp), we obtain that $x \cdot y \geq [(x \wedge y) \to x] \cdot y = x \vee y$. □

We have obviously the following result.

**Proposition 5.4.54** *(See Proposition 5.4.37)*
*Let $\mathcal{G} = (G, \vee, \wedge, \cdot, ^{-1}, 1)$ be an l-group. Then:*
*(1) $\mathcal{G}^- = (G^-, \wedge, \vee, \odot = \cdot, \mathbf{1} = 1)$ is a left-l-im;*
*(1') $\mathcal{G}^+ = (G^+, \vee, \wedge, \oplus = \cdot, \mathbf{0} = 1)$ is a right-l-im.*

### 5.4.3.3 The X-$l$-group - an intermediary notion

**Definition 5.4.55** An *X-l-group* is an X-po-group $\mathcal{G} = (G, \leq, \cdot, \to, \rightsquigarrow, 1)$, where the partial order $\leq$ is a lattice order. An *X-l-group* is denoted by

$$(G, \vee, \wedge, \cdot, \to, \rightsquigarrow, 1).$$

Denote by **X-$l$-group** the class of all X-$l$-groups, namely:
- by **X-$l$-group**, the class of all left-X-$l$-groups and
- by **X-$l$-group$^R$**, the class of all right-X-$l$-groups.

**Remark 5.4.56** *(See Remark 5.4.39*
*Note that an X-l-group is in the same time an l-m(pGa$^{\leq}$) and an l-m(pGa$^{\geq}$). Hence, we can write:*

$$\textbf{X-}l\textbf{-group} = l\textbf{-m(pGa}^{\leq}\textbf{)} + l\textbf{-m(pGa}^{\geq}\textbf{)}.$$

### 5.4.3.4 The equivalences (EG3)

Here we present the equivalences (EG3) between X-$l$-groups and $l$-groups.

We obviously obtain, by (EG2), the dual equivalences (EG3) (the analogous of (iEM6)):

(EG3) **X-$l$-group** $\Longleftrightarrow$ **$l$-group** and, dually,
**X-$l$-group$^R$** $\Longleftrightarrow$ **$l$-group$^R$**.

**Remarks 5.4.57** *(See Remarks 5.4.42)*
*Note that (EG3) can be written as follows, by using Remark 5.4.56:*

$$l\textbf{-group} \Longleftrightarrow l\textbf{-m(pGa}^{\leq}\textbf{)} + l\textbf{-m(pGa}^{\geq}\textbf{)}.$$

*In the same time, note that, by (EGM-l), we have:*

$$l\textbf{-group} \Longleftrightarrow l\textbf{-m(pGa}^{\leq}\textbf{)} + (pIv).$$

## 5.4.4 Examples

### 5.4.4.1 Commutative finite examples

• ([116], Chapter 18) There exists one commutative group with one element, called the *trivial* group: $\mathcal{G}_{1gr} = (A_1 = \{1\}, \cdot, ^{-1}, 1)$, where the tables of $\cdot$ and $^{-1}$ are:

$\mathcal{G}_{1gr}$

| $\cdot$ | 1 |
|---|---|
| 1 | 1 |

,

| $x$ | $x^{-1}$ |
|---|---|
| 1 | 1 |

.

• [116], Chapter 18) There exists one commutative group with two elements: $\mathcal{G}_{2gr} = (A_2 = \{a, 1\}, \cdot, ^{-1}1)$, where the tables of $\cdot$ and $^{-1}$ are:

$\mathcal{G}_{2gr}$

| $\cdot$ | a | 1 |
|---|---|---|
| a | 1 | a |
| 1 | a | 1 |

,

| $x$ | $x^{-1}$ |
|---|---|
| a | a |
| 1 | 1 |

.

• [116], Chapter 18) There exists only one commutative group with three elements, $\mathcal{G}_{3gr} = (A_3, \cdot, ^{-1}, 1)$, where the tables of $\cdot$ and $^{-1}$ are:

$\mathcal{G}_{3gr}$

| $\cdot$ | a | b | 1 |
|---|---|---|---|
| a | b | 1 | a |
| b | 1 | a | b |
| 1 | a | b | 1 |

,

| $x$ | $x^{-1}$ |
|---|---|
| a | b |
| b | a |
| 1 | 1 |

.

• [116], Chapter 18) There are four commutative groups with four elements: $\mathcal{G}^1_{4gr} = (A_4, \cdot_1, ^{-1}, 1)$, $\mathcal{G}^2_{4gr} = (A_4, \cdot_2, ^{-1}, 1)$, $\mathcal{G}^3_{4gr} = (A_4, \cdot_3, ^{-1}, 1)$, $\mathcal{G}^4_{4gr} = (A_4, \cdot_4, ^{-1}, 1)$, namely:

$\mathcal{G}^1_{4gr}$

| $\cdot_1$ | a | b | c | 1 |
|---|---|---|---|---|
| a | b | c | 1 | a |
| b | c | 1 | a | b |
| c | 1 | a | b | c |
| 1 | a | b | c | 1 |

,

| $x$ | $x^{-1}$ |
|---|---|
| a | c |
| b | b |
| c | a |
| 1 | 1 |

$\mathcal{G}^2_{4gr}$

| $\cdot_2$ | a | b | c | 1 |
|---|---|---|---|---|
| a | c | 1 | b | a |
| b | 1 | c | a | b |
| c | b | a | 1 | c |
| 1 | a | b | c | 1 |

,

| $x$ | $x^{-1}$ |
|---|---|
| a | b |
| b | a |
| c | c |
| 1 | 1 |

$\mathcal{G}^3_{4gr}$

| $\cdot_3$ | a | b | c | 1 |
|---|---|---|---|---|
| a | 1 | c | b | a |
| b | c | a | 1 | b |
| c | b | 1 | a | c |
| 1 | a | b | c | 1 |

,

| $x$ | $x^{-1}$ |
|---|---|
| a | a |
| b | c |
| c | b |
| 1 | 1 |

$\mathcal{G}^4_{4gr}$

| $\cdot_4$ | a | b | c | 1 |
|---|---|---|---|---|
| a | 1 | c | b | a |
| b | c | 1 | a | b |
| c | b | a | 1 | c |
| 1 | a | b | c | 1 |

,

| $x$ | $x^{-1}$ |
|---|---|
| a | a |
| b | b |
| c | c |
| 1 | 1 |

.

• (By *Mace4 program*) There exists one commutative group with five elements, $\mathcal{G}_{5gr}$, where the tables of $\cdot$ and $^{-1}$ are:

$\mathcal{G}_{5gr}$

| $\cdot$ | a | b | c | d | 1 |
|---|---|---|---|---|---|
| a | d | 1 | b | c | a |
| b | 1 | c | d | a | b |
| c | b | d | a | 1 | c |
| d | c | a | 1 | b | d |
| 1 | a | b | c | d | 1 |

,

| $x$ | $x^{-1}$ |
|---|---|
| a | b |
| b | a |
| c | d |
| d | c |
| 1 | 1 |

.

#### 5.4.4.2 Non-commutative finite examples

• Here is the commutative group with six elements, $\mathcal{G}_{6gr}$, from [48], where the tables of $\cdot$ and $^{-1}$ are:

$$\mathcal{G}_{6gr} \quad \begin{array}{c|cccccc} \cdot & a & b & c & d & f & \mathbf{1} \\ \hline a & \mathbf{1} & c & b & f & d & a \\ b & d & \mathbf{1} & f & a & c & b \\ c & f & a & d & \mathbf{1} & b & c \\ d & b & f & \mathbf{1} & c & a & d \\ f & c & d & a & b & \mathbf{1} & f \\ \mathbf{1} & a & b & c & d & f & \mathbf{1} \end{array}, \qquad \begin{array}{c|c} x & x^{-1} \\ \hline a & a \\ b & b \\ c & d \\ d & c \\ f & f \\ \mathbf{1} & \mathbf{1} \end{array}.$$

## 5.5 The m-p-semisimple (or discrete) property

### 5.5.1 m-p-semisimple partially ordered magmas

**Definitions 5.5.1** [116]

(i) We say that a left-po-magma is *m-p-semisimple*, if the following property holds: for all $x, y$,

(m-p-s) $\qquad x \leq y \implies x = y.$

(i') We say that a right-po-magma is *m-p-semisimple*, if the following property holds: for all $x, y$,

(m-p-$\text{s}^R$) $\qquad x \geq y \implies x = y.$

**Remark 5.5.2** *(The dual one is omitted)*

*If a left-po-magma $\mathcal{A}^L$ verifies (La'), then (m-p-s) implies $A^L = \{1\}$, i.e. $\mathcal{A}^L$ is trivial. Indeed, any $x \in A^L$ verifies $x \leq 1$, by (La'), hence $x = 1$, by (m-p-s). Consequently, m-p-semisimple left-po-ims are trivial.*

• **m-p-semisimple po-ss**

Let $\mathcal{A}^L = (A^L, \leq^e, \odot)$ be a left-po-s (left-partially ordered semigroup), i.e. the reduct $(A^L, \odot)$ is a semigroup (i.e. (Pass) holds), $(A^L, \leq^e)$ is a poset and the following property holds:

(pCp') $\qquad x \leq^e y \Longrightarrow (a \odot x \leq^e a \odot y$ and $x \odot a \leq^e y \odot a).$

It follows that the m-p-semisimple $\mathcal{A}^L$ is an algebra $(A^L, \odot)$ such that (Pass) holds, i.e. is a left-semigroup.

Thus, we have proved the following result.

**Theorem 5.5.3** *The m-p-semisimple po-ss coincide with the semigroups.*

• **m-p-semisimple po-ms**

Let $\mathcal{A}^L = (A^L, \leq, \odot, 1)$ be a left-po-m (left-partially ordered monoid), i.e. the reduct $(A^L, \odot, 1)$ is a monoid (i.e. (Pass), (PU) hold), $(A^L, \leq)$ is a poset and the property (pCp') holds.

It follows that the m-p-semisimple $\mathcal{A}^L$ is an algebra $(A^L, \odot, 1)$ such that (Pass), (PU) hold, i.e. is a left-monoid.

Thus, we have proved the following result.

**Theorem 5.5.4** *The m-p-semisimple po-ms coincide with the monoids.*

- **m-p-semisimple po-mms**

Let $\mathcal{A}^L = (A^L, \leq, \odot, 1)$ be a left-po-mm (left-partially ordered maximal monoid), i.e. the reduct $(A^L, \odot, 1)$ is a monoid (i.e. (Pass), (PU) hold), $(A^L, \leq, 1)$ is a poset with 1 as maximal element (i.e. (N') holds) and the property (pCp') holds.

It follows that the m-p-semisimple $\mathcal{A}^L$ is an algebra $(A^L, \odot, 1)$ such that (Pass), (PU) hold, i.e. is a left-monoid.

Thus, we have proved the following result.

**Theorem 5.5.5** *The m-p-semisimple po-mms coincide with the monoids.*

- **m-p-semisimple po-ss(pPR')**

Let $\mathcal{A}^L = (A^L, \leq^e, \odot, \rightarrow, \rightsquigarrow)$ be a po-s(pPR'), i.e. the reduct $(A^L, \odot)$ is a semigroup (i.e. (Pass) holds), $(A^L, \leq^e)$ is a poset and the following property holds:
(pPR') for all $x, y, z \in A^L$, $\quad x \odot y \leq^e z \Longleftrightarrow x \leq^e y \rightarrow z \Longleftrightarrow y \leq^e x \rightsquigarrow z$.

It follows that the m-p-semisimple $\mathcal{A}^L$ is an algebra $(A^L, \odot, \rightarrow, \rightsquigarrow)$ such that (Pass) holds and the following property holds:
(pGa$^=$) for all $x, y, z \in A^L$, $\quad x \odot y = z \Longleftrightarrow x = y \rightarrow z \Longleftrightarrow y = x \rightsquigarrow z$.
Then, (pEq#$^=$) holds, i.e. $x = y \rightarrow z \Longleftrightarrow y = x \rightsquigarrow z$.
Also, (pEx) holds; indeed,
$a = x \rightarrow (y \rightsquigarrow z) \overset{(pGa^=)}{\Longleftrightarrow} a \cdot x = y \rightsquigarrow z \overset{(pGa^=)}{\Longleftrightarrow} y \cdot (a \cdot x) = z$
$\overset{(Pass)}{\Longleftrightarrow} (y \cdot a) \cdot x = z \overset{(pGa^=)}{\Longleftrightarrow} y \cdot a = x \rightarrow z \overset{(pGa^=)}{\Longleftrightarrow} a = y \rightsquigarrow (x \rightarrow z)$; thus, (pEx) holds.
Then, by (pA#1=), (pEq#$^=$) $\Longrightarrow$ (pD=),
by Proposition 1.5.1 (11), (pEx) + (pD=) $\Longrightarrow$ (pBB=),
by (pA#3=), (pEq#$^=$) + (pD=) + (pBB=) $\Longrightarrow$ there exists $1 \in A^L$, such that (pRe) + (pM) hold;
we then prove that (PU) holds; indeed,
$x \odot 1 = a \overset{(pGa^=)}{\Longleftrightarrow} x = 1 \rightarrow a \overset{(pM)}{\Longleftrightarrow} x = a$ and
$1 \odot x = a \overset{(pGa^=)}{\Longleftrightarrow} x = 1 \rightsquigarrow a \overset{(pM)}{\Longleftrightarrow} x = a$; thus, (PU) holds.
It follows that $(A^L, \odot, \rightarrow, \rightsquigarrow, 1)$ is an X-group.

Thus, we have proved the following result.

**Theorem 5.5.6** *The m-p-semisimple po-ss(pPR') coincide with the X-groups.*

We have the connections from Figure 5.1.

- **m-p-semisimple po-ms(pPR')**

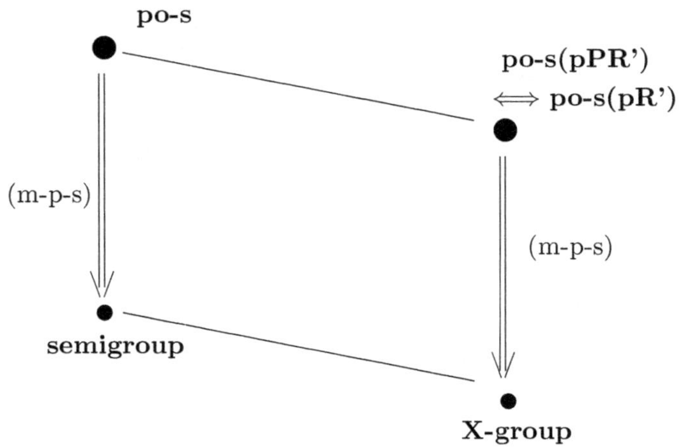


Figure 5.1: m-p-semisimple algebras from the family of po-ss

Let $\mathcal{A}^L = (A^L, \leq, \odot, \rightarrow, \rightsquigarrow, 1)$ be a po-m(pPR'), i.e. the reduct $(A^L, \odot, 1)$ is a monoid (i.e. (Pass), (PU) hold), $(A^L, \leq)$ is a poset and the property (pPR') holds.

It follows that the m-p-semisimple $\mathcal{A}^L$ is an algebra $(A^L, \odot, \rightarrow, \rightsquigarrow, 1)$ such that (Pass), (PU) hold and the property (pGa$^=$) holds: hence, it is an X-group.

Thus, we have proved the following result.

**Theorem 5.5.7** *The m-p-semisimple po-ms(pPR') coincide with the X-groups.*

- **m-p-semisimple po-mms(pPR')**

Let $\mathcal{A}^L = (A^L, \leq, \odot, \rightarrow, \rightsquigarrow, 1)$ be a po-mm(pPR'), i.e. the reduct $(A^L, \odot, 1)$ is a monoid (i.e. (Pass), (PU) hold), $(A^L, \leq, 1)$ is a poset with 1 as maximal element (i.e. (N') holds) and the property (pPR') holds.

It follows that the m-p-semisimple $\mathcal{A}^L$ is an algebra $(A^L, \odot, \rightarrow, \rightsquigarrow, 1)$ such that (Pass), (PU) hold and the property (pGa$^=$) holds; hence, it is an X-group.

Thus, we have proved the following result.

**Theorem 5.5.8** *The m-p-semisimple po-mms(pPR') coincide with the X-groups.*

And recall that the X-groups are d.e. with the groups.

# Chapter 6

# m-M algebras (Hierarchies m-2 and m-2')

In this chapter, in Section 6.1, we introduce 12 new algebras, called *m-M algebras*, the analogous of those 12 bounded M algebras from Chapter 2, as particular cases of unital commutative magmas with some additional operation of the form $(A, \odot, ^-, 1)$ with $1^- = 0$, which determine Hierarchies m-2, m-2'.

In Section 6.2, we present some particular properties of m-M algebras.

In Section 6.3, we present properties of the negation.

In Section 6.4, we study in some details the involutive m-MEL and m-BE algebras.

In Section 6.5, we study the connections between m-M algebras and MV algebras.

Most of the results from this chapter are taken from [118] and [124].

## 6.1 List m-A of basic properties and m-M algebras

• **The basic natural internal binary relation:** $\leq_m$

Let $\mathcal{A}^L = (A^L, \odot, ^- = ^{-^L}, 1)$ be an algebra of type $(2, 1, 0)$ and define $0 \stackrel{def.}{=} 1^-$. Define an *internal* binary relation $\leq_m$ on $A^L$ by: for all $x, y \in A^L$,

$$(m - dfrelP) \qquad x \leq_m y \stackrel{def.}{\Longleftrightarrow} x \odot y^- = 0.$$

Equivalently, consider a structure $\mathcal{A}^L = (A^L, \leq_m, \odot, ^-, 1)$ such that, if $0 \stackrel{def.}{=} 1^-$, then the *internal* binary relation $\leq_m$, the binary operation $\odot$, the unary operation $^-$ and the element 0 are connected by the equivalence ('m' comes from 'magma'):

$$(m - EqrelP) \qquad x \leq_m y \Longleftrightarrow x \odot y^- = 0.$$

Consider the following list **m-A** of basic properties that can be satisfied by $\mathcal{A}^L$ (in fact, the properties in the list are the most important properties satisfied by a left-m-BCK algebra, a new algebra that will be introduced in this section), where almost all properties can be presented in two equivalent forms, determined by the corresponding two equivalent above definitions of $\mathcal{A}^L$:

(PU) $1 \odot x = x = x \odot 1$ (unit element of product, the *identity*);
(Pcomm) $x \odot y = y \odot x$ (commutativity of product),
(Pass) $x \odot (y \odot z) = (x \odot y) \odot z$ (associativity of product);

(Neg1-0) $1^- = 0$;
(Neg0-1) $0^- = 1$;

(m-An) $(x \odot y^- = 0$ and $y \odot x^- = 0) \Longrightarrow x = y$ (antisymmetry),
(m-An') $(x \leq_m y$ and $y \leq_m x) \Longrightarrow x = y$ (antisymmetry);

(m-B) $[(x \odot y^-)^- \odot (x \odot z)] \odot (y \odot z)^- = 0$,
(m-B') $(x \odot y^-)^- \odot (x \odot z) \leq_m y \odot z$;

(m-BB) $[(z \odot x)^- \odot (y \odot x)] \odot (y \odot z^-)^- = 0$,
(m-BB') $(z \odot x)^- \odot (y \odot x) \leq_m y \odot z^-$;

(m-*) $x \odot y^- = 0 \Longrightarrow (z \odot y^-) \odot (z \odot x^-)^- = 0$,
(m-*') $x \leq_m y \Longrightarrow z \odot y^- \leq_m z \odot x^-$;

(m-**) $x \odot y^- = 0 \Longrightarrow (x \odot z) \odot (y \odot z)^- = 0$,
(m-**') $x \leq_m y \Longrightarrow x \odot z \leq_m y \odot z$;

(m-C) $[y \odot (x \odot z)] \odot [x \odot (y \odot z)]^- = 0$,
(m-C') $y \odot (x \odot z) \leq_m x \odot (y \odot z)$;

(m-D) $[(y^- \odot x)^- \odot x] \odot y^- = 0$,
(m-D') $(y^- \odot x)^- \odot x \leq_m y$;

(m-Fi) $0 \odot x^- = 0$ (first element),
(m-Fi') $0 \leq_m x$ (first element);

(m-K) $x \odot [(y \odot x^-)^-]^- = 0$,
(m-K') $x \leq_m (y \odot x^-)^-$;

(m-La) $x \odot 0 = 0$ (last element),
(m-La') $x \leq_m 1$ (last element);

(m-N) $1 \odot x^- = 0 \Longrightarrow x = 1$,
(m-N') $1 \leq_m x \Longrightarrow x = 1$;

(m-V) $x \odot 1 = 0 \Longrightarrow x = 0$,
(m-V') $x \leq_m 0 \Longrightarrow x = 0$;

(m-Re) $x \odot x^- = 0$ (reflexivity),
(m-Re') $x \leq_m x$ (reflexivity);

(m-Tr) $(x \odot y^- = 0$ and $y \odot z^- = 0) \Longrightarrow x \odot z^- = 0$ (transitivity),
(m-Tr') $(x \leq_m y$ and $y \leq_m z) \Longrightarrow x \leq_m z$ (transitivity);

(m-P- -) $x \odot y^- = 0$ and $a \odot b^- = 0 \Longrightarrow (x \odot a) \odot (y \odot b)^- = 0$,
(m-P- -') $x \leq_m y$ and $a \leq_m b \Longrightarrow x \odot a \leq_m y \odot b$;

(m-Pleq) $(x \odot y) \odot x^- = 0$ and $(x \odot y) \odot y^- = 0$,
(m-Pleq') $x \odot y \leq_m x, y$.

Dually,
let $\mathcal{A}^R = (A^R, \oplus, ^- = ^{-^R}, 0)$ be an algebra of type $(2,1,0)$ and define $1 \stackrel{def.}{=} 0^- \neq 0$. Define an *internal* binary relation $\geq_m$ on $A^R$ by: for all $x, y \in A^R$,

$$(m - dfrelS) \qquad x \geq_m y \stackrel{def.}{\Longleftrightarrow} x \oplus y^- = 1.$$

Equivalently, consider a structure $\mathcal{A}^R = (A^R, \geq_m, \oplus, ^-, 0)$ such that, if $1 \stackrel{def.}{=} 0^- \neq 0$, then the *internal* binary relation $\geq_m$, the binary operation $\oplus$, the unary operation $^-$ and the element 1 are connected by the equivalence:

$$(m - EqrelS) \qquad x \geq_m y \Longleftrightarrow x \oplus y^- = 1.$$

The list of dual properties (SU), (Scomm), (Sass) etc. is omitted.

**Remark 6.1.1** *If $\odot = \wedge$ (Wedge) and/or $\oplus = \vee$ (Vee), then the property (m-An) becomes (m-WAn) and/or the dual property (m-An$^R$) becomes (m-VAn) and so on.*

### 6.1.1 m-M algebras

Let $\mathcal{A}^L = (A^L, \odot, ^- = ^{-^L}, 1)$ be an algebra of type $(2,1,0)$ throughout this chapter. Define $0 \stackrel{def.}{=} 1^-$ (hence (Neg1-0) holds) and suppose that $0^- = 1$ (hence (Neg0-1) holds too).

We now introduce the following notions, the analogous of those *bounded M algebras* from Chapter 2 (Hierarchies $2^b$ and $2'^b$) (the dual ones are omitted).

**Algebras with last element, without associativity**:
$\mathcal{A}^L$ is a:
- *left-m-M algebra*, if it verifies (PU), (Pcomm);
- *left-m-ML algebra*, if it verifies (PU), (Pcomm), (m-La);
- *left-m-RML algebra*, if it verifies (PU), (Pcomm), (m-La), (m-Re);
- *left-pre-m-BCC algebra*, if it verifies (PU), (Pcomm), (m-La), (m-Re), (m-B);
- *left-m-aRML algebra*, if it verifies (PU), (Pcomm), (m-La), (m-Re), (m-An);

- *left-m-BCC algebra*, if it verifies (PU), (Pcomm), (m-La), (m-Re), (m-B), (m-An).

**Algebras with last element, with associativity**:
$\mathcal{A}^L$ is a:
- *left-m-ME algebra*, if it verifies (PU), (Pcomm), (Pass),
- *left-m-MEL algebra*, if it verifies (PU), (Pcomm), (Pass), (m-La),
- *left-m-BE algebra*, if it verifies (PU), (Pcomm), (Pass), (m-La), (m-Re), i.e. if it is a left-m-MEL algebra verifying the additional axiom (m-Re) of reflexivity;
- *left-pre-m-BCK algebra*, if it verifies (PU), (Pcomm), (Pass), (m-La), (m-Re) and (m-BB), i.e. if it is a left-m-BE algebra verifying the additional axiom (m-BB);
- *left-m-aBE algebra*, if it verifies (PU), (Pcomm), (Pass), (m-La), (m-Re), (m-An), i.e. if it is a left-m-BE algebra verifying the additional axiom (m-An);
- *left-m-BCK algebra*, if it verifies (PU), (Pcomm), (Pass), (m-La), (m-Re), (m-An) and (m-BB), i.e. it is a left-m-aBE algebra verifying the additional axiom (m-BB);
and, additionally,
- *left-m-BE* algebra*, if it verifies (PU), (Pcomm), (Pass), (m-La), (m-Re) and (m-*), i.e. if it is a left-m-BE algebra verifying the additional axiom (m-*),
- *left-m-aBE* algebra*, if it verifies (PU), (Pcomm), (Pass), (m-La), (m-Re), (m-An) and (m-*), i.e. if it is a left-m-aBE algebra verifying the additional axiom (m-*).

Denote by **m-M**, ... **m-aBE***, the classes of left-m-M, ..., left-m-aBE* algebras, respectively.

Note that sometimes we shall simply write *m-M algebra* instead of *left-m-M algebra* and so on.

### 6.1.2 Connections between the properties

**Proposition 6.1.2** *We have (the dual case is omitted):*
*(mA0) (m-La) + (Pcomm) $\Longrightarrow$ (m-Fi);*
*(mA1) (PU) + (Neg0-1) $\Longrightarrow$ (m-V'),*
*(mA2) (m-C) + (m-An) + (Pcomm) $\Longrightarrow$ (Pass),*
*(mA3) (PU) + (m-BB) + (Pcomm) + (Neg1-0) + (Neg0-1) $\Longrightarrow$ (m-Re),*
*(mA4) (PU) + (m-BB) + (Pcomm) + (Neg0-1) $\Longrightarrow$ (m-**),*
*(mA5) (PU) + (m-B) + (Pcomm) + (Neg0-1) $\Longrightarrow$ (m-*);*
*(mA6) (PU) + (m-**) + (Neg0-1) $\Longrightarrow$ (m-Tr),*
*(mA7) (PU) + (m-*) + (Neg0-1) $\Longrightarrow$ (m-Tr);*
*(mA8) (m-Re) + (Pass) + (Pcomm) $\Longrightarrow$ (m-D),*
*(mA9) (m-Tr) + (Pass) + (Pcomm) + (m-Re) $\Longrightarrow$ (m-*),*
*(mA10) (m-Re) + (Pass) + (Pcomm) + (m-**) $\Longrightarrow$ (m-B);*
*(mA11) (Pass) + (Pcomm) $\Longrightarrow$ ((m-B) $\Leftrightarrow$ (m-BB));*
*(mA12) (m-**) + (Pcomm) + (m-Tr) $\Longrightarrow$ (m-P- -),*
*(mA13) (m-**) + (PU) + (Pcomm) + (m-La) $\Longrightarrow$ (m-Pleq);*
*(mA14) (Pcomm) + (Pass) + (m-La) + (m-Re) $\Longrightarrow$ (m-Pleq).*

**Proof.** (mA0): $0 \odot x^- \overset{(Pcomm)}{=} x^- \odot 0 \overset{(m-La)}{=} 0$, i.e. $0 \leq_m x$; thus, (m-Fi) holds.

(mA1): Suppose $x \leq_m 0$, i.e. $x \odot 0^- = 0$; then, by (Neg0-1), $x \odot 1 = 0$, hence $x = 0$, by (PU); thus, (m-V') holds.

(mA2): By (m-C'), $x \odot (z \odot y) \leq_m z \odot (x \odot y)$ and also $z \odot (x \odot y) \leq_m x \odot (z \odot y)$; then, by (m-An), $x \odot (z \odot y) = z \odot (x \odot y)$, which gives $x \odot (y \odot z) = (x \odot y) \odot z$, by (Pcomm).

(mA3): In (m-BB) take $y = z = 1$; we obtain $0 = [(1 \odot x)^- \odot (1 \odot x)] \odot (1 \odot 1^-)^- \overset{(PU),(Neg1-0)}{=} [x^- \odot x] \odot 0^- \overset{(Neg0-1)}{=} [x^- \odot x] \odot 1 \overset{(PU)}{=} x^- \odot x \overset{(Pcomm)}{=} x \odot x^-$; thus, (m-Re) holds.

(mA4): Suppose $y \leq_m z$, i.e. $y \odot z^- = 0$; then, by (m-BB), $0 = [(z \odot x)^- \odot (y \odot x)] \odot 0^- \overset{(Neg0-1)}{=} [(z \odot x)^- \odot (y \odot x)] \odot 1 \overset{(Pcomm),(PU)}{=} (z \odot x)^- \odot (y \odot x) \overset{(Pcomm)}{=} (y \odot x) \odot (z \odot x)^-$, i.e. $y \odot x \leq_m z \odot x$; thus, (m-**) holds.

(mA5): Suppose $y \leq_m z$, i.e. $y \odot z^- = 0$; then, by (m-B), $0 = [(x \odot y^-)^- \odot (x \odot z^-)] \odot (y \odot z^-)^-$
$\overset{(Neg0-1)}{=} [(x \odot y^-)^- \odot (x \odot z^-)] \odot 1 \overset{(PU)}{=} (x \odot y^-)^- \odot (x \odot z^-) \overset{(Pcomm)}{=} (x \odot z^-) \odot (x \odot y^-)^-$, i.e. $x \odot z^- \leq_m x \odot y^-$; thus, (m-*) holds.

(mA6): Suppose $x \leq_m y$ and $y \leq_m z$, i.e. $x \odot y^- = 0$ and $y \odot z^- = 0$.
Then, by (m-**'), $x \odot z^- \leq_m y \odot z^-$, i.e. $(x \odot z^-) \odot (y \odot z^-)^- = 0$; hence, $(x \odot z^-) \odot 0^- = 0$.
Then, by (Neg0-1), (PU), we obtain $x \odot z^- = 0$, i.e. $x \leq_m z$. Thus, (m-Tr') holds.

(mA7): Suppose $x \leq_m y$ and $y \leq_m z$, i.e. $x \odot y^- = 0$ and $y \odot z^- = 0$.
Then, by (m-*'), $x \odot z^- \leq_m x \odot y^-$, i.e. $(x \odot z^-) \odot (x \odot y^-)^- = 0$; then,
$0 = (x \odot z^-) \odot 0^- \overset{(Neg0-1)}{=} (x \odot z^-) \odot 1 \overset{(PU)}{=} x \odot z^-$; it follows that $x \leq_m z$; thus, (m-Tr) holds.

(mA8): $[(y^- \odot x)^- \odot x] \odot y^- \overset{(Pass)}{=} (y^- \odot x)^- \odot (x \odot y^-) \overset{(Pcomm)}{=} (x \odot y^-) \odot (x \odot y^-)^- \overset{(m-Re)}{=} 0$; thus, (m-D) holds.

(mA9): Suppose $y \leq_m z$; by (mA8), (m-Re) + (Pass) + (Pcomm) $\Longrightarrow$ (m-D), hence $(y^- \odot x)^- \odot x \leq_m y$; then, by (m-Tr), $(y^- \odot x)^- \odot x \leq_m z$, i.e. $[(y^- \odot x)^- \odot x] \odot z^- = 0$; then, by (Pass), $0 = (y^- \odot x)^- \odot (x \odot z^-) \overset{(Pcomm)}{=} (x \odot z^-) \odot (x \odot y^-)^-$; it follows that $x \odot z^- \leq_m x \odot y^-$; thus, (m-*) holds.

(mA10): By (mA8), (m-Re) + (Pass) + (Pcomm) $\Longrightarrow$ (m-D), hence $(y^- \odot x)^- \odot x \leq_m y$;
then, by (m-**), $[(y^- \odot x)^- \odot x] \odot z \leq_m y \odot z$; hence, $(y^- \odot x)^- \odot (x \odot z) \leq_m y \odot z$, by (Pass);
it follows $(x \odot y^-)^- \odot (x \odot z) \leq_m y \odot z$, by (Pcomm); thus, (m-B') holds.

(mA11): If (m-BB) holds, i.e. $[(z \odot x)^- \odot (y \odot x)] \odot (y \odot z^-)^- = 0$, then, by (Pass), we obtain:
$(z \odot x)^- \odot [(y \odot x) \odot (y \odot z^-)^-] = 0$, hence, by (Pcomm), we obtain:
$[(y \odot z^-)^- \odot (y \odot x)] \odot (z \odot x)^- = 0$; thus, (m-B) holds.

Conversely, if (m-B) holds, i.e. $[(x \odot y^-)^- \odot (x \odot z)] \odot (y \odot z)^- = 0$, then, by (Pass), we obtain:
$(x \odot y^-)^- \odot [(x \odot z) \odot (y \odot z)^-] = 0$, hence, by (Pcomm), we obtain:
$[(y \odot z)^- \odot (x \odot z)] \odot (x \odot y^-)^- = 0$; thus, (m-BB) holds.

(mA12): Suppose $x \leq_m y$ and $a \leq_m b$. By (m-**'), $a \leq_m b$ implies $a \odot x \leq_m b \odot x$

and $x \leq_m y$ implies $x \odot b \leq_m y \odot b$; hence, $x \odot a \overset{(Pcomm)}{=} a \odot x \leq_m b \odot x \overset{(Pcomm)}{=} x \odot b \leq_m y \odot b$; then, $x \odot a \leq_m y \odot b$, by (m-Tr').

(mA13): Since $x \leq_m 1$, by (m-La'), it follows, by (m-**'), that $x \odot y \leq_m 1 \odot y \overset{(PU)}{=} y$. Similarly, we prove that $x \odot y \leq_m x$, by (Pcomm).

(mA14): $x \odot y \leq_m x \overset{def.}{\Longleftrightarrow} (x \odot y) \odot x^- = 0 \overset{(Pcomm)}{\Longleftrightarrow} (y \odot x) \odot x^- = 0 \overset{(Pass)}{\Longleftrightarrow} y \odot (x \odot x^-) = 0 \overset{(m-Re)}{\Longleftrightarrow} y \odot 0 = 0$, that is true by (m-La). Thus, $x \odot y \leq_m x$. Similarly, we prove that $x \odot y \leq_m y$. □

**Remark 6.1.3** *By (mA0), all above defined algebras verify both (m-La) and (m-Fi), i.e. are bounded: for all $x$, $0 \leq_m x \leq_m 1$.*

We can obtain now the following two important results (see the corresponding Theorems 2.1.5 and 2.1.6 from Chapter 2).

**Theorem 6.1.4** *Let $\mathcal{A}^L = (A^L, \odot, ^-, 1)$ be an algebra of type $(2,1,0)$ and let $0 \overset{def.}{=} 1^-$, hence (Neg1-0) holds. If (Neg0-1), (PU), (Pcomm), (Pass), (m-Re) hold, then*

$$(m-B) \Longleftrightarrow (m-BB) \Longleftrightarrow (m-**).$$

**Proof.** By (mA11), (Pass) + (Pcomm) $\Longrightarrow$ ((m-B) $\Leftrightarrow$ (m-BB));
by (mA4), (PU) + (m-BB) + (Pcomm) + (Neg0-1) $\Longrightarrow$ (m-**);
by (mA10), (m-Re) + (Pass) + (Pcomm) + (m-**) $\Longrightarrow$ (m-B). □

**Theorem 6.1.5** *Let $\mathcal{A}^L = (A^L, \odot, ^-, 1)$ be an algebra of type $(2,1,0)$ and let $0 \overset{def.}{=} 1^-$, hence (Neg1-0) holds. If (Neg0-1), (PU), (Pcomm), (Pass), (m-Re) hold, then*

$$(m-Tr) \Longleftrightarrow (m-*).$$

**Proof.** By (mA7) and (mA9). □

## 6.2 List m-B of particular properties. Other new algebras

Consider now the following list **m-B** of particular properties:

| | |
|---|---|
| ($\wedge_m$-comm) ($\wedge_m$-commutativity) | $(x^- \odot y)^- \odot y = (y^- \odot x)^- \odot x$, |
| (m-Pimpl) (m-Pimplicativity) | $[(x \odot y^-)^- \odot x^-]^- = x$, |
| (m-Wimpl) (m-Wimplicativity) | $[(x \wedge y^-)^- \wedge x^-]^- = x$, |
| (m-Pimpl-1) | $(x^- \odot y)^- \odot x = x$, |
| (m-DN6) | $(x^- \odot x^-)^- = x$, |
| (G) | $x \odot x = x$; |
| (m-Pdis) (distributivity of $\odot$ over $\oplus$) | $z \odot (x \oplus y) = (z \odot x) \oplus (z \odot y)$, |
| (m-Pabs) (absorbtion of $\odot$ over $\oplus$) | $x \odot (x \oplus y) = x$ |

and, dually,

(${\vee}_m$-comm) (${\vee}_m$-commutativity) $(x^- \oplus y)^- \oplus y = (y^- \oplus x)^- \oplus x$;
(m-Simpl) (m-Simplicativity) $[(x \oplus y^-)^- \oplus x^-]^- = x$,
(m-Vimpl) (m-Vimplicativity) $[(x \vee y^-)^- \vee x^-]^- = x$,
(m-Simpl-1) $(x^- \oplus y)^- \oplus x = x$,
(m-DN6$^R$) $(x^- \oplus x^-)^- = x$,
(G$^R$) $x \oplus x = x$;
(m-Sdis) (distributivity of $\oplus$ over $\odot$) $z \oplus (x \odot y) = (z \oplus x) \odot (z \oplus y)$,
(m-Sabs) (absorbtion of $\oplus$ over $\odot$) $x \oplus (x \odot y) = x$.

Note that (G) becomes (m-Wid), for $\odot = \wedge$, while (G$^R$) becomes (m-Vid), for $\oplus = \vee$, where (m-Wid) and (m-Vid) are axioms of Dedekind lattices.

**Definitions 6.2.1**

An algebra $(A^L, \odot, ^-, 1)$ is said to be:
(i) $\wedge_m$-*commutative*, if it verifies the property ($\wedge_m$-comm).
(ii) *m-Pimplicative*, if it verifies the property (m-Pimpl).

Dually, an algebra $(A^R, \oplus, ^-, 0)$ is said to be:
(i') $\vee_m$-*commutative*, if it verifies the property ($\vee_m$-comm).
(ii') *m-Simplicative*, if it verifies the property (m-Simpl).

**Proposition 6.2.2** *We have (the dual case is omitted):*
*(mB1) ($\wedge_m$-comm) + (Neg0-1) + (PU) + (Pcomm) + (m-La) $\Longrightarrow$ (m-Re),*
*(mB2) ($\wedge_m$-comm) + (Neg0-1) + (PU) + (Pcomm) $\Longrightarrow$ (m-An),*
*(mB3) (m-Pimpl) + (Neg0-1) + (PU) $\Longrightarrow$ (m-DN6).*

**Proof.**

(mB1): $x \odot x^- \overset{(Pcomm),(PU)}{=} (1 \odot x)^- \odot x \overset{(Neg0-1)}{=} (0^- \odot x)^- \odot x \overset{(\wedge_m-comm)}{=} (x^- \odot 0)^- \odot 0 \overset{(m-La)}{=} 0$; thus, (m-Re) holds.

(mB2): Suppose $x \leq_m y$ and $y \leq_m x$, i.e. $x \odot y^- = 0$ and $y \odot x^- = 0$; then, by (Pcomm), ($\wedge_m$-comm), (Neg0-1), we obtain: $1 \odot x = 1 \odot y$, hence $x = y$, by (PU); thus, (m-An') holds.

(mB3): Take $y = 0$ in (m-Pimpl); we obtain: $x = [(x \odot 0^-)^- \odot x^-]^- \overset{(Neg0-1)}{=} [(x \odot 1)^- \odot x^-]^- \overset{(PU)}{=} [x^- \odot x^-]^-$; thus, (m-DN6) holds. □

## 6.3 List m-C of properties of the negation

Consider now the following list **m-C** of properties of the negation $^-$:

(m-Neg7) $((x^-)^-)^- = x^-$;

(DN) (Double Negation) $(x^-)^- = x$;

(m-Neg2') $x \leq_m y \Longrightarrow y^- \leq_m x^-$,
(m-Neg4') $(x^-)^- \leq_m x$,
(m-DN4') $x \leq_m y \Longleftrightarrow y^- \leq_m x^-$,
(m-WRe) $x \wedge x^- = 0$,
(m-Pdiv) $x \odot (x \odot y^-)^- = x \odot y$,
(m-@) = (m-Wdiv) $x \wedge (x \wedge y^-)^- = x \wedge y$
and, dually,
(m-Neg2$^{R}$') $x \geq_m y \Longrightarrow y^- \geq_m x^-$,
(m-Neg4$^{R}$') $(x^-)^- \geq_m x$,
(m-DN4$^{R}$') $x \geq_m y \Longleftrightarrow y^- \geq_m x^-$,
(m-VRe) $x \vee x^- = 1$,
(m-Sdiv) $x \oplus (x \oplus y^-)^- = x \oplus y$,
(m-@$^R$) = (m-Vdiv) $x \vee (x \vee y^-)^- = x \vee y$.

**Proposition 6.3.1** *We have (the dual case is omitted):*
*(mC1) (m-*) + (PU) $\Longrightarrow$ (m-Neg2'),*
*(mC2) (m-Re) + (Pcomm) $\Longrightarrow$ (m-Neg4'),*
*(mC2') (m-D) + (PU) $\Longrightarrow$ (m-Neg4');*
*(mC3) (m-Neg4') + (m-*) + (PU) + (m-An) $\Longrightarrow$ (m-Neg7).*

**Proof.** (mC1): Suppose $x \leq_m y$; then, by (m-*), $z \odot y^- \leq_m z \odot x^-$; then, for $z = 1$, we obtain $1 \odot y^- \leq_m 1 \odot x^-$; hence, $y^- \leq_m x^-$, by (PU); thus, (m-Neg2') holds.

(mC2): $(x^-)^- \odot x^- \stackrel{(Pcomm)}{=} x^- \odot (x^-)^- \stackrel{(m-Re)}{=} 0$, i.e. $(x^-)^- \leq_m x$; thus, (m-Neg4') holds.

(mC2'): Take $x = 1$ in (m-D); we obtain $(y^- \odot 1)^- \odot 1 \leq_m y$; hence, $(y^-)^- \leq_m y$, by (PU); thus, (m-Neg4') holds.

(mC3): By (m-Neg4'), $((y^-)^-)^- \leq_m y^-$. On the other hand, by (m-Neg4'), $(y^-)^- \leq_m y$, hence $1 \odot y^- \leq_m 1 \odot ((y^-)^-)^-$, by (m-*); then, $y^- \leq_m ((y^-)^-)^-$, by (PU). Now apply (m-An) to obtain $((y^-)^-)^- = y^-$; thus, (m-Neg7) holds. □

**Definition 6.3.2** We shall say that an algebra $\mathcal{A}^L = (A^L, \odot, ^-, 1)$ is *involutive*, if it verifies (DN).

The subclass of involutive algebras of the class $\mathbf{X}$ will be denoted by $\mathbf{X}_{(DN)}$.

**Proposition 6.3.3** *We have (the dual case is omitted):*
*(mCBN1) ($\wedge_m$-comm)+(DN)+(Pass)+(Pcomm)+(m-Re)+(m-La) $\Longrightarrow$ (m-BB);*
*(mCIM1) (m-Pimpl) + (Neg1-0) + (Neg0-1) + (m-La) + (PU) $\Longrightarrow$ (DN),*
*(mCIM1') (m-Pimpl) + (m-Re) + (Neg0-1) + (PU) $\Longrightarrow$ (DN);*
*(mCIM2) (m-Pimpl) + (DN) $\Longrightarrow$ (m-Pimpl-1),*
*(mCIM3) (m-Pimpl-1) + (PU) + (DN) $\Longrightarrow$ (G),*
*(mCIM4) (m-Pimpl) + (DN) + (PU) $\Longrightarrow$ (G).*

**Proof.** (mCBN1): We must prove that $T \stackrel{notation}{=} [(z \odot x)^- \odot (y \odot x)] \odot (y \odot z^-)^- = 0$. Indeed, $(z \odot x)^- \odot (y \odot x) \stackrel{(Pcomm),(Pass)}{=} y \odot [(z \odot x)^- \odot x]$

$\overset{(DN),(\wedge_m-comm)}{=} y \odot [(x^- \odot z^-)^- \odot z^-]$. Then,
$T = (y \odot [(x^- \odot z^-)^- \odot z^-]) \odot (y \odot z^-)^-$
$\overset{(Pass),(Pcomm)}{=} (x^- \odot z^-)^- \odot [(y \odot z^-) \odot (y \odot z^-)^-]$
$\overset{(m-Re)}{=} (x^- \odot z^-)^- \odot 0 \overset{(m-La)}{=} 0$. Thus, (m-BB) holds.

(mCIM1): Take $y = 1$ in (m-Pimpl); we obtain: $x = [(x \odot 1^-)^- \odot x^-]^- \overset{(Neg1-0)}{=} [(x \odot 0)^- \odot x^-]^- \overset{(m-La)}{=} [0^- \odot x^-]^- \overset{(Neg0-1)}{=} [1 \odot x^-]^- \overset{(PU)}{=} (x^-)^-$; thus, (DN) holds.

(mCIM1'): Take $y = x$ in (m-Pimpl); we obtain: $x = [(x \odot x^-)^- \odot x^-]^- \overset{(m-Re)}{=} [0^- \odot x^-]^- \overset{(Neg0-1)}{=} [1 \odot x^-]^- \overset{(PU)}{=} (x^-)^-$.

(mCIM2): By (m-Pimpl), $[(x \odot y^-)^- \odot x^-]^- = x$; hence, by (DN), we obtain: $(x \odot y^-)^- \odot x^- = x^-$; then, by replacing $x$ by $x^-$ and $y$ by $y^-$, we obtain, by (DN) again: $(x^- \odot y)^- \odot x = x$; thus, (m-Pimpl-1) holds.

(mCIM3): Take $y = 1$ in (m-Pimpl-1); we obtain: $x = (x^- \odot 1)^- \odot x \overset{(PU)}{=} (x^-)^- \odot x \overset{(DN)}{=} x \odot x$; thus, (G) holds.

(mCIM4): By (mCIM2) and (mCIM3). □

**Proposition 6.3.4** *We have (the dual case is omitted):*
*(mCDN0) (DN) + (PU) + (Neg0-1) $\Longrightarrow$ (m-N'),*
*(mCDN1) (DN) $\Longrightarrow$ ((Neg1-0) $\Leftrightarrow$ (Neg0-1)),*
*(mCDN2) (DN) + (Pcomm) $\Longrightarrow$ (m-DN4),*
*(mCDN3) (DN) + (Pcomm) $\Longrightarrow$ ((m-*) $\Leftrightarrow$ (m-**)),*
*(mCDN4) (DN) + (Pass) + (m-Re) + (m-La) + (Pcomm) $\Longrightarrow$ (m-K);*
*(mCDN5) (PU) + (Pcomm) + (m-Re) + (m-*) + (m-An) $\Longrightarrow$ (DN);*
*(mCDN6) (Pcomm) + (m-Re) + (m-**) + (m-Fi) + (m-An) $\Longrightarrow$ (DN),*
*(mCDN6') (Pcomm) + (m-Re) + (m-**) + (m-V) + (m-An) $\Longrightarrow$ (DN).*

**Proof.** (mCDN0): Suppose $1 \leq_m x$, i.e. $1 \odot x^- = 0$; then, by (PU), $x^- = 0$, hence $x = 1$, by (DN) and (Neg0-1); thus, (m-N') holds.

(mCDN1): Suppose (Neg1-0) holds; then $0^- = (1^-)^- \overset{(DN)}{=} 1$, i.e. (Neg0-1) holds too. Similarly, (Neg0-1) implies (Neg1-0).

(mCDN2): $x \leq_m y \Longleftrightarrow x \odot y^- = 0 \overset{(DN)}{\Longleftrightarrow} (x^-)^- \odot y^- = 0 \overset{(Pcomm)}{\Longleftrightarrow} y^- \odot (x^-)^- = 0 \Longleftrightarrow y^- \leq_m x^-$; thus, (m-DN4') holds.

(mCDN3): By (mCDN2), (DN) + (Pcomm) $\Longrightarrow$ (m-DN4').
If $x \leq_m y$, then, by (m-DN4'), $y^- \leq_m x^-$; then, by (m-* '), $z \odot (x^-)^- \leq_m z \odot (y^-)^-$; then, by (DN), $z \odot x \leq_m z \odot y$; then, by (Pcomm), $x \odot z \leq_m y \odot z$; thus, (m-** ') holds.
Conversely, if $x \leq_m y$, then, by (m-DN4'), $y^- \leq_m x^-$; then, by (m-** '), $y^- \odot z \leq_m x^- \odot z$; then, by (Pcomm), $z \odot y^- \leq_m z \odot x^-$; thus, (m-* ') holds.

(mCDN4): $x \leq_m (y \odot x^-)^- \Longleftrightarrow x \odot ((y \odot x^-)^-)^- = 0$ and, indeed, $x \odot ((y \odot x^-)^-)^- \overset{(DN)}{=} x \odot (y \odot x^-) \overset{(Pcomm),(Pass)}{=} y \odot (x \odot x^-) \overset{(m-Re)}{=} y \odot 0 \overset{(m-La)}{=} 0$; thus, (m-K) holds.

(mCDN5): By (mC2), (m-Re) + (Pcomm) $\Longrightarrow$ (m-Neg4'), i.e. $(x^-)^- \leq_m x$. On the other hand, by (mC3), (m-Neg4') + (m-*) + (PU) + (m-An) $\Longrightarrow$ (m-Neg7), i.e. $((x^-)^-)^- = x^-$; then, $x \odot ((x^-)^-)^- = x \odot x^- \overset{(m-Re)}{=} 0$, i.e. $x \leq_m (x^-)^-$. Now, by (m-An), we obtain $(x^-)^- = x$, i.e. (DN) holds.

(mCDN6): $(x^-)^- \leq_m x \Longleftrightarrow (x^-)^- \odot x^- = 0$, that is true by (Pcomm) and (m-Re); thus, $(x^-)^- \leq_m x$. On the other hand, by above inequality, we have $((x^-)^-)^- \leq_m x^-$; then, by (m-**'),
$x \odot ((x^-)^-)^- \leq_m x \odot x^- \overset{(m-Re)}{=} 0$; by (m-Fi'), we also have $0 \leq_m x \odot ((x^-)^-)^-$; hence, by (m-An'), we obtain $x \odot ((x^-)^-)^- = 0$; thus, $x \leq_m (x^-)^-$.
From $(x^-)^- \leq_m x$ and $x \leq_m (x^-)^-$, we obtain $(x^-)^- = x$, by (m-An') again; thus, (DN) holds.

(mCDN6'): $(x^-)^- \leq_m x \Longleftrightarrow (x^-)^- \odot x^- = 0$, that is true by (Pcomm) and (m-Re); thus, $(x^-)^- \leq_m x$. On the other hand, by above inequality, we have $((x^-)^-)^- \leq_m x^-$, hence, by (m-**'),
$x \odot ((x^-)^-)^- \leq_m x \odot x^- \overset{(m-Re)}{=} 0$; by (m-V'), we obtain $x \odot ((x^-)^-)^- = 0$, hence, $x \leq_m (x^-)^-$.
From $(x^-)^- \leq_m x$ and $x \leq_m (x^-)^-$, we obtain $(x^-)^- = x$, by (m-An'); thus, (DN) holds. □

**Theorem 6.3.5** *Let $\mathcal{A}^L = (A^L, \odot, ^-, 1)$ be an algebra of type $(2,1,0)$ and let $0 \overset{def.}{=} 1^-$, hence (Neg1-0) holds. If (Neg0-1), (PU), (Pcomm), (Pass), (m-Re) and (DN) hold, then*

$$(m-BB) \Longleftrightarrow (m-B) \Longleftrightarrow (m-**) \Longleftrightarrow (m-*) \Longleftrightarrow (m-Tr).$$

**Proof.** By Theorems 6.1.4, 6.1.5 and (mCDN3). □

**Theorem 6.3.6** *We have:*
*(1) any left-m-aBE* algebra is involutive,*
*(2) any left-m-BCK algebra is involutive and*
*(3) the (involutive) left-m-aBE* algebras coincide with the (involutive) left-m-BCK algebras;*
*(4) any left-m-BCC algebra is involutive.*

**Proof.** (1): By (mCDN5), (PU) + (Pcomm) + (m-Re) + (m-*) + (m-An) $\Longrightarrow$ (DN).

(2): By (mA1), (m-V) holds; by Theorem 6.1.4, (m-**) holds; by (mCDN6'), (Pcomm) + (m-Re) + (m-**) + (m-V) + (m-An) $\Longrightarrow$ (DN).

(3): By (mCDN3), (DN) + (Pcomm) $\Longrightarrow$ ((m-*) $\Longleftrightarrow$ (m-**)).

(4): Let $\mathcal{A}^L = (A^L, \odot, ^-, 1)$ be a left-m-BCC algebra, i.e. (Neg1-0), (Neg0-1), (PU), (Pcomm), (m-La), (m-Re), (m-An), (m-B) hold. Then, by (mA5), (PU) + (m-B) + (Pcomm) + (Neg0-1) $\Longrightarrow$ (m-*), thus (m-*) holds; by (mCDN5), (PU) + (Pcomm) + (m-Re) + (m-*) + (m-An) $\Longrightarrow$ (DN), thus (DN) holds. □

By above Theorem 6.3.6, we write:
**m-aBE*=m-aBE***$_{(DN)}$= **m-BCK= m-BCK**$_{(DN)}$ , **m-BCC=m-BCC**$_{(DN)}$.

Note that an (involutive) m-BCK algebra satisfies all the properties in the first list **m-A** of properties and, additionally, (DN) and (m-DN4).

**Proposition 6.3.7** *We have:*

*(mCBB1)* (m-Pdiv) + (PU) $\Longrightarrow$ (DN),
*(mCBB2)* (m-Pdiv) + (Pcomm) $\Longrightarrow$ ($\wedge_m$-comm),
*(mCBB3)* (m-Pdiv) + (PU) + (m-La) + (Neg0-1) $\Longrightarrow$ (m-Re),
*(mCBB4)* (m-Pdiv) + (PU) + (Pcomm) + (m-Re) + (Neg0-1) $\Longrightarrow$ (G),
*(mCBB5)* (m-Pdiv) + (PU) + (Pcomm) + (Neg0-1) $\Longrightarrow$ (m-An),
*(mCBB6)* (m-Pdiv)+(DN)+ (Pcomm)+(Pass)+(m-La)+(m-Re) $\Longrightarrow$ (m-BB),
*(mCBB7)* (G) + (m-An) + (m-BB) + (DN) + (PU) + (Pcomm) + (Pass) + (m-La) + (Neg0-1) $\Longrightarrow$ (m-Pdiv),
*(mCBB8)* (DN)+(PU)+(Pcomm)+(Pass)+(m-La)+(m-Re)+(Neg0-1) $\Longrightarrow$ ((m-Pdiv) $\Leftrightarrow$ (G)+(m-An)+(m-BB)),
*(mCBB9)* (m-Pdiv) + (DN) + (PU) + (Pcomm) + (Pass) + (m-La) + (m-Re) + (Neg0-1) $\Longrightarrow$ (m-Pimpl),
*(mCBB10)* (m-Pimpl) + (PU) + (Pcomm) + (Pass) + (Neg0-1) + (m-Re) + (m-La) + (m-An) + (m-BB) $\Longrightarrow$ (m-Pdiv).

**Proof.** (mCBB1): In (m-Pdiv) ($x \odot (x \odot y^-)^- = x \odot y$) take $x = 1$ to obtain: $1 \odot (1 \odot y^-)^- = 1 \odot y$, i.e. $(y^-)^- = y$, by (PU); thus, (DN) holds.

(mCBB2): $(x^- \odot y)^- \odot y \overset{(Pcomm)}{=} y \odot (y \odot x^-)^- \overset{(m-Pdiv)}{=} y \odot x$ and, similarly, $(y^- \odot x)^- \odot x \overset{(Pcomm)}{=} x \odot (x \odot y^-)^- \overset{(m-Pdiv)}{=} x \odot y$;
then, $(x^- \odot y)^- \odot y = (y^- \odot x)^- \odot x$, by (Pcomm) again, hence ($\wedge_m$-comm) holds.

(mCBB3): Take $y = 0$ in (m-Pdiv) to obtain: $x \odot x^- = x \odot (x \odot 1)^- = x \odot 0 = 0$; thus, (m-Re) holds.

(mCBB4): $x \odot x \overset{(m-Pdiv)}{=} x \odot (x \odot x^-)^- \overset{(m-Re)}{=} x \odot 0^- \overset{(Neg0-1)}{=} x \odot 1 \overset{(Pcomm)}{=} 1 \odot x \overset{(PU)}{=} x$.

(mCBB5): Suppose $x \leq_m y$ and $y \leq_m x$, i.e. $x \odot y^- = 0$ and $y \odot x^- = 0$. Then, $x \odot y \overset{(m-Pdiv)}{=} x \odot (x \odot y^-)^- = x \odot 0^- \overset{(Neg0-1)}{=} x \odot 1 \overset{(Pcomm)}{=} 1 \odot x \overset{(PU)}{=} x$ and, hence, $x = x \odot y \overset{(Pcomm)}{=} y \odot x \overset{(m-Pdiv)}{=} y \odot (y \odot x^-)^- = y \odot 0^- \overset{(Neg0-1)}{=} y \odot 1 = y$; thus, (m-An) holds.

(mCBB6): $[(z \odot x)^- \odot (y \odot x)] \odot (y \odot z^-)^-$
$\overset{(Pass)}{=} (z \odot x)^- \odot y \odot x \odot (y \odot z^-)^-$
$\overset{(Pcomm),(Pass)}{=} (y \odot (y \odot z^-)^-) \odot (x \odot (x \odot z)^-)$
$\overset{(m-Pdiv),(DN)}{=} (y \odot z) \odot (x \odot z^-)$
$\overset{(Pass),(Pcomm)}{=} y \odot x \odot (z \odot z^-)$
$\overset{(m-Re)}{=} (y \odot x) \odot 0 \overset{(m-La)}{=} 0$; thus, (m-BB) holds.

(CBB7): We must prove that (m-Pdiv), i.e. $x \odot (x \odot y^-)^- = x \odot y$, holds; denote $X = x \odot (x \odot y^-)^-$ and $Y = x \odot y$. Then:
$X \odot Y^- = [x \odot (x \odot y^-)^-] \odot (x \odot y)^- = 0$; indeed,

in (m-BB) ($[(z \odot x)^- \odot (y \odot x)] \odot (y \odot z^-)^- = 0$) take $y = x$ to obtain:
$0 = [(z \odot x)^- \odot (x \odot x)] \odot (x \odot z^-)^-$
$\overset{(G)}{=} [(z \odot x)^- \odot x] \odot (x \odot z^-)^-$
$\overset{(Pass)}{=} (z \odot x)^- \odot [x \odot (x \odot z^-)^-]$
$\overset{(Pcomm)}{=} [x \odot (x \odot z^-)^-] \odot (x \odot z)^-$; hence, $X \odot Y^- = 0$.
$Y \odot X^- = (x \odot y) \odot [x \odot (x \odot y^-)^-]^- = 0$; indeed,
in (m-BB) ($[(z \odot x)^- \odot (y \odot x)] \odot (y \odot z^-)^- = 0$) take $z = (x \odot y^-)^-$ to obtain:
$0 \overset{(DN)}{=} [((x \odot y^-)^- \odot x)^- \odot (y \odot x)] \odot (y \odot (x \odot y^-))^-$
$\overset{(Pass),(Pcomm)}{=} [((x \odot y^-)^- \odot x)^- \odot (y \odot x)] \odot (x \odot (y \odot y^-))^-$
$\overset{(m-Re)}{=} [((x \odot y^-)^- \odot x)^- \odot (y \odot x)] \odot (x \odot 0)^-$
$\overset{(m-La)}{=} [((x \odot y^-)^- \odot x)^- \odot (y \odot x)] \odot 0^-$
$\overset{(Neg0-1)}{=} [((x \odot y^-)^- \odot x)^- \odot (y \odot x)] \odot 1$
$\overset{(Pcomm),(PU)}{=} ((x \odot y^-)^- \odot x)^- \odot (y \odot x)$
$\overset{(Pcomm)}{=} (x \odot y) \odot [x \odot (x \odot y^-)^-]^-$; hence, $Y \odot X^- = 0$.
Thus, $X \leq_m Y$ and $Y \leq_m X$, hence $X = Y$, by (m-An); so, (m-Pdiv) holds.

(mCBB8): By above (mCBB4) - (mCBB7).

(mCBB9): $[(x \odot y^-)^- \odot x^-]^- \overset{(Pcomm)}{=} [x^- \odot (x \odot y^-)^-]^-$
$\overset{(m-Pdiv),(DN)}{=} [x^- \odot (x^- \odot (x \odot y^-))^-]$ for $X = x^-$ and $Y = (x \odot y^-)^-$,
$\overset{(Pass)}{=} [x^- \odot ((x^- \odot x) \odot y^-)^-]^-$
$\overset{(Pcomm),(m-Re),(m-La)}{=} [x^- \odot 0^-]^- \overset{(Neg0-1)}{=} (x^- \odot 1)^- \overset{(PU)}{=} (x^-)^- \overset{(DN)}{=} x$; so, (m-Pimpl) holds.

(mCBB10): By (mA8), (m-Re) + (Pass) + (Pcomm) imply (m-D); by (m-D'),

$$x \odot (x \odot y^-)^- \leq_m y. \tag{6.1}$$

By (mA4), (PU) + (Pcomm) + (m-BB) + (Neg0-1) imply (m-**).
From (6.1), by (m-**'), we obtain $(x \odot (x \odot y^-)^-) \odot x \leq_m y \odot x$; hence, by (Pcomm), (Pass), we obtain

$$(x \odot x) \odot (x \odot y^-)^- \leq_m x \odot y. \tag{6.2}$$

By (mA0), (m-La) + (Pcomm) imply (m-Fi); by (mCDN6), (Pcomm) + (m-Re) + (m-**) + (m-Fi) + (m-An) imply (DN) (or by (mCIM1')). Then, by (mCIM2), (m-Pimpl) + (DN) imply (m-Pimpl-1);
by (mCIM3), (m-Pimpl-1) + (PU) + (DN) imply (G).
From (6.2) and (G), we obtain:

$$x \odot (x \odot y^-)^- \leq_m x \odot y. \tag{6.3}$$

On the other hand, by (mCDN4), (DN) + (Pass) + (m-Re) + (m-La) + (Pcomm) imply (m-K); by (m-K'), $y \leq_m (x \odot y^-)^-$; then, by (m-**'), we obtain:

$$x \odot y \leq_m x \odot (x \odot y^-)^-. \tag{6.4}$$

Now, from (6.3) and (6.4) and (m-An), we obtain $x \odot (x \odot y^-)^- = x \odot y$. Thus, (m-Pdiv) holds. □

With analogous definitions for the algebra $\mathcal{A} = (A, \odot, ^-, 1)$ and the associated binary relation $\leq_m$ as in Definitions 1.1.19 and with the same representations as in Remark 1.1.21, the left-m-M algebras defined in this chapter are connected (following their definitions and the above results) as in the *Hierarchy m-2* from Figure 8.1 and *Hierarchy m-2'* from Figure 8.2 - the "Big m-map", in Chapter 8.

# 6.4 Involutive m-MEL and m-BE algebras

## 6.4.1 Involutive m-MEL algebras

In the involutive case, we have naturally the following definition:

**Definitions 6.4.1**
(i) An *involutive left-m-MEL algebra* is a left-m-MEL algebra verifying (DN), i.e. is an algebra $\mathcal{A}^L = (A^L, \odot, ^- = ^{-^L}, 1)$ of type $(2, 1, 0)$ satisfying the following axioms:

(PU) $1 \odot x = x = x \odot 1$,
(Pcomm) $x \odot y = y \odot x$,
(Pass) $x \odot (y \odot z) = (x \odot y) \odot z$,
(m-La) $x \odot 0 = 0$, i.e. $x \leq_m 1$, where $0 \overset{def.}{=} 1^-$, i.e. (Neg1-0) holds,
(DN) $(x^-)^- = x$.

(i') Dually, an *involutive right-m-MEL algebra* is a right-m-MEL algebra verifying (DN), i.e. is an algebra $\mathcal{A}^R = (A^R, \oplus, ^- = ^{-^R}, 0)$ of type $(2, 1, 0)$ satisfying the following axioms:

(SU) $0 \oplus x = x = x \oplus 0$,
(Scomm) $x \oplus y = y \oplus x$,
(Sass) $x \oplus (y \oplus z) = (x \oplus y) \oplus z$,
(m-La$^R$) $x \oplus 1 = 1$, i.e. $x \geq_m 0$, where $1 \overset{def.}{=} 0^-$, i.e. (Neg0-1) holds,
(DN) $(x^-)^- = x$.

Note that, by (mCDN1), (DN) $\Longrightarrow$ ((Neg1-0) $\Longleftrightarrow$ (Neg0-1)), hence (Neg0-1) (dually, (Neg1-0)) holds too.

We shall denote by $\mathbf{m\text{-}MEL}_{(DN)}$ the class of all involutive left-m-MEL algebras and by $\mathbf{m\text{-}MEL}^R_{(DN)}$ the class of all involutive right-m-MEL algebras.

Let $\mathcal{A}^L = (A^L, \odot, ^-, 1)$ be an involutive left-m-MEL algebra.

Because of the axiom (DN), we can introduce the operation $\oplus$, the dual of $\odot$, by: for all $x, y \in A^L$,

(6.5) $$x \oplus y \overset{def.}{=} (x^- \odot y^-)^-.$$

**Proposition 6.4.2** *We have:*

(6.6) $$0 \oplus x = x = x \oplus 0, \quad \textit{i.e.} \quad (SU) \quad \textit{holds},$$

(6.7) $$x \oplus y = y \oplus x, \quad i.e. \quad (Scomm) \quad holds,$$

(6.8) $$x \oplus (y \oplus z) = (x \oplus y) \oplus z, \quad i.e. \quad (Sass) \quad holds,$$

(6.9) $$x \oplus 1 = 1, \quad i.e. \quad (m - La^R) \quad holds;$$

(6.10) $$(x \oplus y)^- = x^- \odot y^- \quad (De\ Morgan\ law\ 1),$$

(6.11) $$(x \odot y)^- = x^- \oplus y^- \quad (De\ Morgan\ law\ 2),\ and\ hence$$

(6.12) $$x \odot y = (x^- \oplus y^-)^-.$$

**Proof.** Routine. □

**Corollary 6.4.3** *If $\mathcal{A}^L = (A^L, \odot, ^-, 1)$ is an involutive left-m-MEL algebra, then $(A^L, \oplus, ^-, 0)$ is an involutive right-m-MEL algebra.*

**Proof.** By (6.6)–(6.9). □

**Proposition 6.4.4** *Let $\mathcal{A}^L = (A^L, \odot, ^-, 1)$ be an involutive left-m-MEL algebra. We have: (G) $\Longleftrightarrow$ ($G^R$), (m-Pabs) $\Longleftrightarrow$ (m-Sabs), (m-Pimpl) $\Longleftrightarrow$ (m-Simpl), (m-Pdis) $\Longleftrightarrow$ (m-Sdis), (m-Pdiv) $\Longleftrightarrow$ (m-Sdiv).*

**Proof.** Routine. □

**Proposition 6.4.5** *We have:*
*(mCIM5) (m-Pabs) + (SU) $\Longrightarrow$ (G),*
*(mCIM6) (m-Pimpl) + (DN) + (Pcomm) $\Longrightarrow$ (m-Pabs),*
*(mCIM7) (m-Pabs) + (DN) + (Pcomm) $\Longrightarrow$ (m-Pimpl),*
*(mCIM8) (DN) + (Pcomm) $\Longrightarrow$ ((m-Pabs) $\Leftrightarrow$ (m-Pimpl));*
*(mCIM9) (m-Pdis) + (PU) + (m-$La^R$) + (DN) $\Longrightarrow$ (G),*
*(mCIM10) (m-Pdis) + (PU) + (Scomm) + (DN) $\Longrightarrow$ (m-Pabs),*
*(mCIM11) (m-Pdis) + (PU) + (Pcomm) + (Scomm) + (DN) $\Longrightarrow$ (m-Pimpl);*
*(mCIM12) (m-Pdis) + (m-Re) $\Longrightarrow$ (m-Pdiv).*

**Proof.** Routine. □

**Proposition 6.4.6** *Let $\mathcal{A}^L = (A^L, \odot, ^-, 1)$ be an involutive left-m-MEL algebra. We have: (m-Pabs) $\Longleftrightarrow$ (m-Pimpl); (m-Pdis) $\Longrightarrow$ (m-Pimpl), (m-Pdis) $\Longrightarrow$ (DN).*

**Proof.** By Proposition 6.4.5 and (mCIM1). □

**Proposition 6.4.7** *Let $\mathcal{A}^L = (A^L, \odot, ^-, 1)$ be an involutive left-m-MEL algebra. Then:*

$$(\wedge_m - comm) \iff (\vee_m - comm).$$

**Proof.** Routine. □

**Proposition 6.4.8** *Let $\mathcal{A}^L = (A^L, \odot, ^-, 1)$ be an involutive left-m-MEL algebra. We have:*

$$x \leq_m y \iff y \geq_m x\,.$$

**Proof.** $x \leq_m y \overset{(m-dfrelP)}{\iff} x \odot y^- = 0 \overset{(DN),(Neg0-1)}{\iff} (x \odot y^-)^- = 1 \overset{(6.5),(DN)}{\iff}$ $x^- \oplus y = 1 \overset{(Scomm)}{\iff} y \oplus x^- = 1 \overset{(m-dfrelS)}{\iff} y \geq_m x$. □

- **A new binary relation: $\leq_m^P$**

Beside the old binary relation $\leq_m$ and its dual $\geq_m$, we shall introduce a new binary relation: for all $x, y \in A^L$,

(m-dfP) $\qquad x \leq_m^P y \overset{def.}{\iff} x \odot y = x \qquad$ and, dually,

(m-dfS) $\qquad x \geq_m^S y \overset{def.}{\iff} x \oplus y = x\,.$

**Proposition 6.4.9** *(See Proposition 3.2.4)*
*Let $\mathcal{A}^L = (A^L, \odot, ^-, 1)$ be an involutive left-m-MEL algebra. The binary relation $\leq_m^P$ is antisymmetric and transitive and $0 \leq_m^P x \leq_m^P 1$, for all $x \in A^L$, where $0 \overset{def.}{=} 1^-$.*

**Proof.** (m-P-An): If $x \leq_m^P y$ and $y \leq_m^P x$, i.e. $x \odot y = x$ and $y \odot x = y$, then $x = x \odot y \overset{(Pcomm)}{=} y \odot x = y$.

(m-P-Tr): If $x \leq_m^P y$ and $y \leq_m^P z$, i.e. $x \odot y = x$ and $y \odot z = y$, then $x \odot z = (x \odot y) \odot z \overset{(Pass)}{=} x \odot (y \odot z) = x \odot y = x$.

$0 \leq_m^P x$ means $0 \odot x = 0$, that is true by (Pcomm) and (m-La).

$x \leq_m^P 1$ means $x \odot 1 = x$, that is true by (Pcomm), (PU). □

**Proposition 6.4.10** *Let $\mathcal{A}^L = (A^L, \odot, ^-, 1)$ be an involutive left-m-MEL algebra. If (G) holds, then the binary relation $\leq_m^P$ is an order.*

**Proof.** (m-P-Re): $x \leq_m^P x$ means $x \odot x = x$, which is true by (G). □

### 6.4.2 Involutive m-BE algebras

In the involutive case, we have naturally the following definition:

**Definitions 6.4.11**

(i) An *involutive left-m-BE algebra* is a left-m-BE algebra verifying (DN), i.e. is an involutive left-m-MEL algebra verifying (m-Re).

(i') Dually, an *involutive right-m-BE algebra* is a right-m-BE algebra verifying (DN), i.e. is an involutive lright-m-MEL algebra verifying (m-Re$^R$).

We shall denote by $\mathbf{m\text{-}BE}_{(DN)}$ the class of all involutive left-m-BE algebras and by $\mathbf{m\text{-}BE}^R_{(DN)}$ the class of all involutive right-m-BE algebras.

**Proposition 6.4.12** *Let $\mathcal{A}^L = (A^L, \odot, ^-, 1)$ be an involutive left-m-BE algebra. We have:*

$$\text{(6.13)} \qquad if \quad x \odot y = 1, \quad then \quad x = y = 1,$$

$$\text{(6.14)} \qquad x \oplus x^- = 1, \quad i.e. \quad (m - Re^R) \quad holds,$$

$$\text{(6.15)} \qquad [y \odot (y^- \oplus x)] \odot x^- = 0\,.$$

**Proof.** (6.41): By (mA14), (Pcomm) + (Pass) + (m-La) + (m-Re) imply (m-Pleq) and, by (mCDN0), (DN) + (PU) + (Neg0-1) imply (m-N).
If $x \odot y = 1$ and since $x \odot y \leq_m x, y$, by (m-Pleq), it follows that $1 \leq_m x, y$, hence $x = y = 1$, by (m-N).

(6.48): $x \oplus x^- \overset{(6.5),(DN)}{=} (x^- \odot x)^- \overset{(Pcomm)}{=} (x \odot x^-)^- \overset{(m-Re)}{=} 0^- \overset{(Neg0-1)}{=} 1.$

(6.15): $[y \odot (y^- \oplus x)] \odot x^- \overset{(Pcomm),(Pass)}{=} (x^- \odot y) \odot (y^- \oplus x)$
$\overset{(Scomm)}{=} (x^- \odot y) \odot (x \oplus y^-) \overset{(6.5),(DN)}{=} (x^- \odot y) \odot (x^- \odot y)^- \overset{(m-Re)}{=} 0.$ □

**Corollary 6.4.13** *(See Corollary 6.4.3)*
*If $\mathcal{A}^L = (A^L, \odot, ^-, 1)$ is an involutive left-m-BE algebra, then $(A^L, \oplus, ^-, 0)$ is an involutive right-m-BE algebra.*

**Proof.** By (6.6)–(6.9) and (6.48). □

**Proposition 6.4.14** *Let $\mathcal{A}^L = (A^L, \odot, ^-, 1)$ be an involutive left-m-BE algebra. We have:*

*(m-Pdiv)* $\Longrightarrow$ *(m-Pimpl),*
*(m-Pdiv)* $\Longrightarrow$ *(DN),*
*(m-Pdiv)* $\Longleftrightarrow$ *(G) + (m-An) + (m-BB).*

**Proof.** By (mCBB9), (mCBB1) and (mCBB8). □

**Theorem 6.4.15** *Let $\mathcal{A}^L = (A^L, \odot, ^-, 1)$ be an involutive left-m-BE algebra. Then:*

$$(m - Pdis) \Longleftrightarrow (m - Pimpl) + (m - An)\,.$$

*Further, when these conditions hold, so does (m-Tr).*

**Proof.** (By PROVER9) Assume first that (m-Pdis) holds. Since (m-Re) also holds, so does (m-An) by Lemma 9.1.19. That (m-Pimpl) holds follows from Proposition 6.4.6.

Conversely, now assume (m-Pimpl) and (m-An) hold. We will prove (m-Sdis), which, in turn, is equivalent to (m-Pdis) by Proposition 6.4.4. We will also need the absorbtion laws (m-Pabs) and (m-Sabs), which hold by Propositions 6.4.6 and 6.4.4. The proof - made by PROVER9 - is rather long, so in the interest of readability, we will freely use the commutativity and associativity of $\odot$ and $\oplus$ (that is, (m-Pcomm), (m-Pass), (m-Scomm), (m-Sass)) without explicit reference, parenthesizing only for emphasis.

First, we will need

$$x \odot (y \oplus x)^- = 0 \,, \tag{6.16}$$

which we obtain as follows:
$x \odot (y \oplus x)^- \overset{(m-Pabs)}{=} x \odot (x \oplus y) \odot (y \oplus x)^- \overset{(m-Re)}{=} x \odot 0 \overset{(m-La)}{=} 0\,.$

Our next step is to verify

$$(x \odot y) \oplus (x \odot (x \odot y)^-) = x \,. \tag{6.17}$$

This will follow from applying (m-An) to the pair of identities

$$x \odot [(x \odot y) \oplus (x \odot (x \odot y)^-)]^- = 0 \,, \tag{6.18}$$

$$[(x \odot y) \oplus (x \odot (x \odot y)^-)] \odot x^- = 0 \,, \tag{6.19}$$

and so now we verify these. For (6.18), we have
$x \odot [(x \odot y) \oplus (x \odot (x \odot y)^-)]^- = x \odot (x \odot y)^- \odot (x \odot (x \odot y)^-)^- = 0,$
by (m-Re). For (6.19), set $z = (x \odot y)^-$ and compute
$[(x \odot y) \oplus (x \odot z)] \odot x^-$

$$\begin{aligned}
&\overset{(m-Sabs)}{=} [(x \odot y) \oplus (x \odot z)] \odot [x \oplus (x \odot z)]^- \\
&\overset{(m-Sabs)}{=} [(x \odot y) \oplus (x \odot z)] \odot [x \oplus (x \odot y) \oplus (x \odot z)]^- \\
&\overset{(6.16)}{=} 0\,.
\end{aligned}$$

Now, we verify

$$x \oplus (y \odot (y \odot x)^-) = x \oplus y \,. \tag{6.20}$$

This follows easily from (6.17):
$x \oplus (y \odot (y \odot x)^-) \overset{(m-Sabs)}{=} x \oplus (x \odot y) \oplus (y \odot (y \odot x)^-) \overset{(6.17)}{=} x \oplus y\,.$

Then, we verify

$$x \odot [(y \oplus x) \odot (z \oplus x)]^- = 0 \,. \tag{6.21}$$

Indeed,

$$\begin{aligned}
x \odot [(y \oplus x) \odot (z \oplus x)]^- &\overset{(m-Pabs)}{=} x \odot (z \oplus x) \odot [(y \oplus x) \odot (z \oplus x)]^- \\
&\overset{(m-Pabs)}{=} x \odot (y \oplus x) \odot (z \oplus x) \odot [(y \oplus x) \odot (z \oplus x)]^- \\
&\overset{(m-Re)}{=} x \odot 0 \\
&\overset{(m-La)}{=} 0.
\end{aligned}$$

Now, we verify

$$x \odot (y \oplus (x \oplus y)^-) = x \odot y\,, \tag{6.22}$$

$$x \oplus (y \odot (x \odot y)^-) = x \oplus y\,. \tag{6.23}$$

For (6.22), first, in (6.15), replace $y$ with $y \oplus x$:

$$[(y \oplus x) \odot ((y \oplus x)^- \oplus x)] \odot x^- = 0\,. \tag{6.24}$$

Second, in (6.21), replace $z$ with $(y \oplus x)^-$:

$$x \odot [(y \oplus x) \odot ((y \oplus x)^- \oplus x)]^- = 0\,. \tag{6.25}$$

From (6.24) and (6.25), we may apply (m-An) to get, by changing $x$, $y$ between them,

$$(x \oplus y) \odot (y \oplus (x \oplus y)^-) = y\,. \tag{6.26}$$

Finally, we have
$x \odot (y \oplus (x \oplus y)^-) \overset{(m-Pabs)}{=} x \odot (x \oplus y) \odot (y \oplus (x \oplus y)^-) \overset{(6.26)}{=} x \odot y\,,$
which is (6.22).

Note that (6.23) is just the dual of (6.22), hence it follows by duality (Corollary 6.4.13).

Next, we need the implication

$$x \odot y = 0 \quad \Longrightarrow \quad y^- \odot (x \oplus y) = x\,. \tag{6.27}$$

First, we have $(y \oplus (x \oplus y)^-)^- \odot x^- = (y \oplus (x \oplus y)^- \oplus x)^- \overset{(m-Re^R)}{=} 1^- = 0$. Thus, by (m-An) and (DN), it follows that

$$x \odot (y \oplus (x \oplus y)^-) = 0 \quad \Longrightarrow \quad (y \oplus (x \oplus y)^-)^- = x\,.$$

Applying (6.22) to the antecedent and rewriting the consequent, we obtain (6.27).

Next, we verify

$$(x^- \oplus (x \odot y)) \odot (x \oplus y) = y\,. \tag{6.28}$$

By (m-Re), we have $y \odot (x \odot (x \odot y)^-) = 0$. Applying (6.27) to this (with $X = y$, $Y = x \odot (x \odot y)^-$), we obtain
$(x \odot (x \odot y)^-)^- \odot (y \oplus (x \odot (x \odot y)^-)) = y$, i.e.
$(x^- \oplus (x \odot y)) \odot (y \oplus (x \odot (x \odot y)^-)) = y\,.$
Applying (6.23) in the left side, we get (6.28).

Next, we verify

$$x^- \oplus (x \odot y) = x^- \oplus y\,, \tag{6.29}$$

$$x^- \odot (x \oplus y) = x^- \odot y\,. \tag{6.30}$$

For (6.29), we have

$$\begin{aligned} x^- \oplus (x \odot y) &\overset{(m-Sabs)}{=} (x^- \oplus (x \odot y)) \oplus ((x^- \oplus (x \odot y)) \odot (x \oplus y)) \\ &\overset{(6.28)}{=} x^- \oplus (x \odot y) \oplus y \\ &\overset{(m-Sabs)}{=} x^- \oplus y\,. \end{aligned}$$

Note that (6.30) is just the dual of (6.29), hence it follows by duality (Corollary 6.4.13).

Next, we verify

$$x \oplus (y \odot (x \oplus z)) = x \oplus (y \odot z)\,. \tag{6.31}$$

We compute

$$\begin{aligned} x \oplus (y \odot (x \oplus z)) &\overset{(DN)}{=} (x^-)^- \oplus (y \odot (x \oplus z)) \\ &\overset{(6.29)}{=} (x^-)^- \oplus (x^- \odot y \odot (x \oplus z)) \\ &= (x^-)^- \oplus (x^- \odot (x \oplus z) \odot y) \\ &\overset{(6.30)}{=} (x^-)^- \oplus (x^- \odot z \odot y) \\ &\overset{(6.29)}{=} (x^-)^- \oplus (z \odot y) \\ &\overset{(DN)}{=} x \oplus (y \odot z)\,. \end{aligned}$$

Now, we can prove (m-Sdis). We have

$$\begin{aligned} x \oplus (y \odot z) &\overset{(6.31)}{=} x \oplus (y \odot (x \oplus z)) \\ &= x \oplus ((x \oplus z) \odot y) \\ &\overset{(6.31)}{=} x \oplus ((x \oplus z) \odot (x \oplus y)) \\ &\overset{(m-Pabs)}{=} (x \odot (x \oplus y)) \oplus ((x \oplus z) \odot (x \oplus y)) \\ &\overset{(m-Pabs)}{=} (x \odot (x \oplus z) \odot (x \oplus y)) \oplus ((x \oplus z) \odot (x \oplus y)) \\ &\overset{(m-Sabs)}{=} (x \oplus z) \odot (x \oplus y)\,. \end{aligned}$$

Finally, for the last claim of the theorem, suppose $a \odot b^- = 0 = b \odot c^-$ for some $a, b, c \in A^L$. Since (m-Re) holds, by Lemma 9.1.18, $a = a \odot b$ and $b = b \odot c$. Thus $a \odot c = (a \odot b) \odot c = a \odot (b \odot c) = a \odot b = a$. Applying Lemma 9.1.18 again, we have $a \odot c^- = 0$. Thus (m-Tr) holds. □

**Remark 6.4.16** *A finite involutive m-BE algebra $(A^L, \odot, ^-, 1)$ verifying (G) has an even number of elements. Indeed, suppose there is an odd number of elements, for example $n = 5$ and $A^L = \{0, a, b, c, 1\}$ with $0^- = 1$, $a^- = b$, $b^- = a$, $c^- = c$, $1^- = 0$. By (m-Re), $c \odot c^- = 0$, i.e. $c \odot c^- = c \odot c = 0$, but by (G), $c \odot c = c$: contradiction. Thus, $n$ cannot be an odd number.*

### 6.4.3 Two binary relations: $\leq_m^M$ and $\leq_m^B$

**Remark 6.4.17** *Starting from the equality from the property*

$$(\wedge_m - comm) \qquad (x^- \odot y)^- \odot y = (y^- \odot x)^- \odot x,$$

*verified by a left-MV algebra, we could introduce two different 'twin' operations, $\wedge_m^M$ ('M' comes from 'MV algebra') and $\wedge_m^B$ ('B' comes from 'Boolean algebra'), by: for all $x, y$:*
$x \wedge_m^M y = (x^- \odot y)^- \odot y$ *and* $x \wedge_m^B y = (y^- \odot x)^- \odot x$.
*Then, ($\wedge_m$-comm) would mean: either*
($\wedge_m^M$-*comm*) $x \wedge_m^M y = y \wedge_m^M x$ *or*
($\wedge_m^B$-*comm*) $x \wedge_m^B y = y \wedge_m^B x$.

*In left-MV algebras, the two operations $\wedge_m^M$ and $\wedge_m^B$ are equal, but in general, in an involutive left-m-MEL algebra, they are different; but note that: $x \wedge_m^M y = y \wedge_m^B x$, which means, in the finite case, that the matrix of $\wedge_m^B$ is the transposed matrix of that of $\wedge_m^M$ and vice-versa.*

Following the above Remark 6.4.17, we shall introduce in an involutive left-m-MEL algebra $\mathcal{A}^L = (A^L, \odot, ^-, 1)$ the following operations:

(6.32) $$x \wedge_m^M y \overset{def.}{=} (x^- \odot y)^- \odot y \overset{(Pcomm)}{=} y \odot (y \odot x^-)^- \quad \textit{and, dually,}$$

(6.33) $$x \vee_m^M y \overset{def.}{=} (x^- \wedge_m^M y^-)^-$$
$$= [(x \odot y^-)^- \odot y^-]^- (= (x \to y) \to y) = (x \odot y^-) \oplus y = (x^- \oplus y)^- \oplus y$$

and
(6.34)
$$x \wedge_m^B y \overset{def.}{=} (y^- \odot x)^- \odot x \overset{(Pcomm)}{=} x \odot (x \odot y^-)^- = x \odot (x \to y) = y \wedge_m^M x \quad \textit{and, dually,}$$

(6.35) $$x \vee_m^B y \overset{def.}{=} (x^- \wedge_m^B y^-)^-$$
$$= [(y \odot x^-)^- \odot x^-]^- (= (y \to x) \to x) = (y \odot x^-) \oplus x = (y^- \oplus x)^- \oplus x = y \vee_m^M x,$$

where $x \to y \overset{def.}{=} (x \odot y^-)^-$ (see Theorem 17.1.1).

**In what follows, we shall present only the properties of $\wedge_m^M$ and $\vee_m^M$.**

• **Two binary relations:** $\leq_m^M$ and $\leq_m^B$

Beside the old, natural binary relation $\leq_m$, and its dual $\geq_m$, and the binary relation $\leq_m^P$, and its dual $\geq_m^S$, we introduce two binary relations, the old $\leq_m^M$ (see [45]) and the new $\leq_m^B$: for all $x, y \in A^L$,

(m-dfWM) $\quad x \leq_m^M y \overset{def.}{\Longleftrightarrow} x \wedge_m^M y = x$ and, dually,
(m-dfVM) $\quad x \geq_m^M y \overset{def.}{\Longleftrightarrow} x \vee_m^M y = x$
and
(m-dfWB) $\quad x \leq_m^B y \overset{def.}{\Longleftrightarrow} x \wedge_m^B y = x \; (\Longleftrightarrow y \wedge_m^M x = x)$ and, dually,
(m-dfVB) $\quad x \geq_m^B y \overset{def.}{\Longleftrightarrow} x \vee_m^B y = x \; (\Longleftrightarrow y \vee_m^M x = x)$.

**Lemma 6.4.18** *Let $\mathcal{A}^L = (A^L, \odot, ^-, 1)$ be an involutive left-m-BE algebra. We have:*
*(1)* $x \odot y^- = 0 \Longleftrightarrow x \odot (x \odot y^-)^- = x$ *and, dually,*
*(1')* $x \oplus y^- = 1 \Longleftrightarrow x \oplus (x \oplus y^-)^- = x$.

**Proof.**
(1): Suppose that $x \odot y^- = 0$; then, $(x \odot y^-)^- = 1$, hence $x \odot (x \odot y^-)^- = x \odot 1 = x$.
Conversely, suppose that $x \odot (x \odot y^-)^- = x$; then,
$x \odot y^- = (x \odot (x \odot y^-)^-) \odot y^- \overset{(Pcomm),(Pass)}{=} (x \odot y^-) \odot (x \odot y^-)^- \overset{(m-Re)}{=} 0$.
(1'): Suppose that $x \oplus y^- = 1$; then, $(x \oplus y^-)^- = 0$, hence $x \oplus (x \oplus y^-)^- = x \oplus 0 = x$.
Conversely, suppose that $x \oplus (x \oplus y^-)^- = x$; then,
$x \oplus y^- = (x \oplus (x \oplus y^-)^-) \oplus y^- \overset{(Scomm),(Sass)}{=} (x \oplus y^-) \oplus (x \oplus y^-)^- \overset{(m-Re^R)}{=} 1$. □

**Proposition 6.4.19** *Let $\mathcal{A}^L = (A^L, \odot, ^-, 1)$ be an involutive left-m-BE algebra. We have:*
*(1)* $x \leq_m y \Longleftrightarrow x \leq_m^B y$ *and, dually*
*(1')* $x \geq_m y \Longleftrightarrow x \geq_m^B y$.
*(2)* *If ($\wedge_m$-comm) holds (i.e. $x \wedge_m^M y = y \wedge_m^M x$), then*
$$x \leq_m y \; (\Longleftrightarrow x \leq_m^B y) \;\Longleftrightarrow x \leq_m^M y.$$
*(2')* *If ($\wedge_m$-comm) holds, then ($\vee_m$-comm) holds (i.e. $x \vee_m^M y = y \vee_m^M x$) and*
$$x \geq_m y \; (\Longleftrightarrow x \geq_m^B y) \;\Longleftrightarrow x \geq_m^M y.$$

**Proof.** (1): By above Lemma 6.4.18 (1),
$x \leq_m y \overset{def.}{\Longleftrightarrow} x \odot y^- = 0 \Longleftrightarrow x \odot (x \odot y^-)^- = x \Longleftrightarrow x \wedge_m^B y = x \overset{def.}{\Longleftrightarrow} x \leq_m^B y$.
(1'): By above Lemma 6.4.18 (1'),
$x \geq_m y \overset{def.}{\Longleftrightarrow} x \oplus y^- = 1 \Longleftrightarrow x \oplus (x \oplus y^-)^- = x \Longleftrightarrow x \vee_m^B y = x \overset{def.}{\Longleftrightarrow} x \geq_m^B y$.
(2): By above (1),
$x \leq_m^M y \overset{def.}{\Longleftrightarrow} x \wedge_m^M y = x \overset{(\wedge_m-comm)}{\Longleftrightarrow} y \wedge_m^M x = x \Longleftrightarrow x \wedge_m^B y = x \overset{def.}{\Longleftrightarrow} x \leq_m^B y \Longleftrightarrow x \leq_m y$.
(2'): By Proposition 6.4.7 and above (1'). □

**Remark 6.4.20** *The equivalence $\leq_m \Longleftrightarrow \leq_m^B$ implies that $\leq_m$ is an order relation if and only if $\leq_m^B$ is an order relation. But, it does not imply that if $\leq_m$ is a lattice order w.r. to say $\wedge, \vee$, then $\leq_m^B$ is a lattice order too with respect to $\wedge_m^B, \vee_m^B$.*

**Proposition 6.4.21** *Let $\mathcal{A}^L = (A^L, \odot, ^-, 1)$ be an involutive left-m-BE algebra. Then,*
$$(x \leq_m^B y \Longleftrightarrow) \; x \leq_m y \Longleftrightarrow y \geq_m x \; (\Longleftrightarrow y \geq_m^B x).$$

**Proof.** By Propositions 6.4.19 and 6.4.8. □

**Proposition 6.4.22** *(See ([45], Proposition 2.1.2), in dual case)*
*Let $\mathcal{A}^L = (A^L, \odot, ^-, 1)$ be an involutive left-m-MEL algebra. We have:*

$$x \wedge_m^M 1 = x = 1 \wedge_m^M x, \quad x \wedge_m^M 0 = 0, \tag{6.36}$$

(6.37) $$x \vee_m^M 0 = x = 0 \vee_m^M x, \quad x \vee_m^M 1 = 1,$$

(6.38) $$(x \vee_m^M y)^- = x^- \wedge_m^M y^- \quad (\textit{De Morgan law } 1),$$

(6.39) $$(x \wedge_m^M y)^- = x^- \vee_m^M y^- \quad (\textit{De Morgan law } 2) \textit{ and, hence,}$$

(6.40) $$x \wedge_m^M y = (x^- \vee_m^M y^-)^-.$$

**Proposition 6.4.23** *(See ([45], Proposition 2.1.2), in dual case)*
*Let $\mathcal{A}^L = (A^L, \odot, ^-, 1)$ be an involutive left-m-BE algebra. We have:*

(6.41) $$\textit{if} \quad x \odot y = 1, \quad \textit{then} \quad x = y = 1,$$

(6.42) $$\textit{if} \quad x \wedge_m^M y = 1, \quad \textit{then} \quad x = y = 1,$$

(6.43) $$0 \wedge_m^M x = 0,$$

(6.44) $$1 \vee_m^M x = 1,$$

(6.45) $$x \wedge_m^M x = x, \quad x \vee_m^M x = x,$$

(6.46) $$\textit{if} \quad x \leq_m^M y, \quad \textit{then} \quad y \wedge_m^M x = x,$$

(6.47) $$\textit{if} \quad x \leq_m^M y, \quad \textit{then} \quad x \leq_m y.$$

**Proposition 6.4.24** *Let $\mathcal{A}^L = (A^L, \odot, ^-, 1)$ be an involutive left-m-BE algebra. We have:*

(6.48) $$x \oplus x^- = 1, \quad \textit{i.e.} \quad (m - Re^R) \quad \textit{holds;}$$

(6.49) $$x \odot (y \wedge_m^M x^-) = 0,$$

(6.50) $$x \odot (x^- \wedge_m^M y) = 0,$$

(6.51) $$(y \vee_m^M x) \wedge_m^M x = x,$$

(6.52) $$(y \wedge_m^M x) \vee_m^M x = x,$$

(6.53) $$\textit{if} \quad x \leq_m^M y, \quad \textit{then} \quad x \vee_m^M y = y,$$

(6.54) $$x \vee_m^M y = y \iff x \odot y^- = 0 \quad (\iff x \leq_m y),$$

(6.55) $$(x \odot y) \vee_m^M x = x,$$

(6.56) $$x \wedge_m^M (x \odot y) = x \odot y,$$

(6.57) $$x \wedge_m^M (y \wedge_m^M x) = y \wedge_m^M x.$$

**Proof.** (6.48): $x \oplus x^- \overset{(6.5)}{=} (x^- \odot x)^- \overset{(Pcomm)}{=} (x \odot x^-)^- \overset{(m-Re)}{=} 0^- \overset{(Neg0-1)}{=} 1.$

(6.49): $x \odot (y \wedge_m^M x^-) = x \odot [(y^- \odot x^-)^- \odot x^-]$
$\overset{(Pcomm)}{=} x \odot [x^- \odot (y^- \odot x^-)^-] \overset{(Pass)}{=} (x \odot x^-) \odot (y^- \odot x^-)^-$
$\overset{(m-Re)}{=} 0 \odot (y^- \odot x^-)^- \overset{(Pcomm)}{=} (y^- \odot x^-)^- \odot 0 \overset{(m-La)}{=} 0.$

(6.50): $x \odot (x^- \wedge_m^M y) = x \odot [(x \odot y)^- \odot y]$
$= x \odot [y \odot (x \odot y)^-] = (x \odot y) \odot (x \odot y)^- \overset{(m-Re)}{=} 0.$

(6.51): $(y \vee_m^M x) \wedge_m^M x \overset{(6.33)}{=} [(y \odot x^-)^- \odot x^-]^- \wedge_m^M x$
$\overset{(6.32),(DN)}{=} (((y \odot x^-)^- \odot x^-) \odot x)^- \odot x \overset{(Pass),(Pcomm)}{=} ((y \odot x^-)^- \odot (x \odot x^-))^- \odot x$
$\overset{(m-Re)}{=} ((y \odot x^-)^- \odot 0)^- \odot x \overset{(m-La)}{=} 0^- \odot x = 1 \odot x = x.$

(6.52): $(y \wedge_m^M x) \vee_m^M x \overset{(6.32)}{=} [(y^- \odot x)^- \odot x] \vee_m^M x \overset{(6.33)}{=} ([(y^- \odot x)^- \odot x] \odot x^-) \oplus x$
$\overset{(Pass)}{=} ((y^- \odot x)^- \odot (x \odot x^-)) \oplus x \overset{(m-Re)}{=} ((y^- \odot x)^- \odot 0) \oplus x \overset{(m-La)}{=} 0 \oplus x \overset{(SU)}{=} x.$

(6.53): Suppose $x \leq_m^M y$, i.e. $x \wedge_m^M y = x$; then, $x \vee_m^M y = (x \wedge_m^M y) \vee_m^M y \overset{(6.52)}{=} y.$

(6.54): If $y = x \vee_m^M y \overset{(6.33)}{=} (x \odot y^-) \oplus y$, then $y^- = ((x \odot y^-) \oplus y)^- = (x \odot y^-)^- \odot y^-$. Then,
$x \odot y^- = x \odot [(x \odot y^-)^- \odot y^-] \overset{(Pcomm),(Pass)}{=} (x \odot y^-) \odot (x \odot y^-)^- \overset{(m-Re)}{=} 0.$
Conversely, if $x \odot y^- = 0$, then $x \vee_m^M y \overset{(6.33)}{=} (x \odot y^-) \oplus y = 0 \oplus y \overset{(SU)}{=} y.$

(6.55: $(x \odot y) \vee_m^M x = [[(x \odot y) \odot x^-]^- n \odot x^-]^- \overset{(m-Re),(m-La)}{=} [0^- \odot x^-]^- = (x^-)^- = x.$

(6.56): $x \wedge_m^M (x \odot y) = [x^- \odot (x \odot y)]^- \odot (x \odot y) \overset{(m-Re),(m-La)}{=} 0^- \odot (x \odot y) = 1 \odot (x \odot y) = x \odot y.$

(6.57): $x \wedge_m^M (y \wedge_m^M x) = x \wedge_m^M (x \odot (x \odot y^-)^-)$
$= (x \odot (x \odot y^-)^-) \odot (x^- \odot (x \odot (x \odot y^-)^-))^-$
$\overset{(Pass)}{=} (x \odot (x \odot y^-)^-) \odot ((x^- \odot x) \odot (x \odot y^-)^-)^-$
$\overset{(Pcomm),(m-Re),(m-La)}{=} (x \odot (x \odot y^-)^-) \odot 0^- \overset{(Neg0-1),(PU)}{=} x \odot (x \odot y^-)^- = y \wedge_m^M x.$

□

**Corollary 6.4.25** *(See ([45], Corollary 2.1.3)) (See Proposition 2.2.17 (2))*

*Let $\mathcal{A}^L = (A^L, \odot, ^-, 1)$ be an involutive left-m-BE algebra. Then, the binary relation $\leq_m^M$ is reflexive and antisymmetric and $0 \leq_m^M x \leq_m^M 1$, for all $x \in A^L$, where $0 \overset{def.}{=} 1^-$.*

**Proof.** - *reflexivity*: $x \leq_M^M x \Longleftrightarrow x \wedge_m^M x = x$, that is true by (6.45);

- *antisymmetry*: $x \leq_m^M y$ and $y \leq_m^M x$ mean $y \wedge_m^M x = x$, by (6.46), and $y \wedge_m^M x = y$, by definition, so $x = y$.

- $0 \leq_m^M x \Longleftrightarrow 0 \wedge_m^M x = 0$, that is true by (6.43);

- $x \leq_m^M 1 \Longleftrightarrow x \wedge_m^M 1 = x$, that is true by (6.36). □

## 6.5 Connections of new algebras with MV algebras

### 6.5.1 MV algebras

The our days definition of MV algebras is the following:

**Definitions 6.5.1**
(i) A *left-MV algebra* is an algebra $\mathcal{A}^L = (A^L, \odot, ^- = ^{-^L}, 1)$ verifying: for all $x, y, z \in A^L$,
(PU) $1 \odot x = x (= x \odot 1)$,
(Pcomm) $x \odot y = y \odot x$,
(Pass) $x \odot (y \odot z) = (x \odot y) \odot z$,
(m-La) $x \odot 0 = 0$, where $0 \stackrel{def.}{=} 1^-$ (i.e. (Neg1-0) holds),
(DN) $(x^-)^- = x$,
($\wedge_m$-comm) $(x^- \odot y)^- \odot y = (y^- \odot x)^- \odot x$.
(i') A *right-MV algebra* is an algebra $\mathcal{A}^R = (A^R, \oplus, ^- = ^{-^R}, 0)$ verifying: for all $x, y, z \in A^R$,
(SU) $0 \oplus x = x (= x \oplus 0)$,
(Scomm) $x \oplus y = y \oplus x$,
(Sass) $x \oplus (y \oplus z) = (x \oplus y) \oplus z$,
(m-La$^R$) $x \oplus 1 = 1$, where $1 \stackrel{def.}{=} 0^-$ (i.e. (Neg0-1) holds),
(DN) $(x^-)^- = x$,
($\vee_m$-comm) $(x^- \oplus y)^- \oplus y = (y^- \oplus x)^- \oplus x$.

Denote by **MV** the class of all left-MV algebras and by $\mathbf{MV}^R$ the class of all right-MV algebras.

Note that, in a left-MV algebra, by (mCDN1), the property (Neg0-1) holds too, and in a right-MV algebra, by (mCDN1$^R$), the property (Neg1-0) holds too.

Note that the left-MV algebra verifies (DN) (it is *involutive*), hence it is self-dual, i.e. the dual of $(A^L, \odot, ^-, 1)$ is $(A^L, \oplus, ^-, 0)$, where $x \oplus y = (x^- \odot y^-)^-$, and vice-versa.

Dually, the right-MV algebra verifies (DN) (it is *involutive*), hence it is self-dual, i.e. the dual of $(A^R, \oplus, ^-, 0)$ is $(A^R, \odot, ^-, 1)$, where $x \odot y = (x^- \oplus y^-)^-$, and vice-versa.

We make the following important remark, which was the motivation of paper [118], and hence of this book (see the Preface):

**Remark 6.5.2**
*(i) Left-MV algebra is just the $\wedge_m$-commutative involutive left-m-MEL algebra.*
*(i') Right-MV algebra is just the $\vee_m$-commutative involutive right-m-MEL algebra.*

**Theorem 6.5.3** *Let $\mathcal{A}^L = (A^L, \odot, ^-, 1)$ be a left-MV algebra. Define*

$$x \wedge_m^M y \stackrel{def.}{=} (x^- \odot y)^- \odot y, \quad x \vee_m^M y \stackrel{def.}{=} (x^- \wedge_m^M y^-)^- = (x^- \oplus y)^- \oplus y,$$

*where* $x \oplus y \stackrel{def.}{=} (x^- \odot y^-)^-$, *and* $0 \stackrel{def.}{=} 1^-$.

*Then,* $(A^L, \wedge = \wedge_m^M, \vee = \vee_m^M, 0, 1)$ *is a bounded (Dedekind) lattice.*

**Proof.** (m-Wid): $x \wedge_m^M x = x$ follows by (6.45);
(m-Vid): $x \vee_m^M x = x$ follows by (6.45);
(m-Wcomm): $x \wedge_m^M y = y \wedge_m^M x$ follows by ($\wedge_m$-comm);
(m-Vcomm): $x \vee_m^M y = (x^- \wedge_m^M y^-)^- \stackrel{(\wedge_m-comm)}{=} (y^- \wedge_m^M x^-)^- = y \vee_m^M x$;
(m-Wass): $x \wedge_m^M (y \wedge_m^M z) = (x \wedge_m^M y) \wedge_m^M z \stackrel{(\wedge_m-comm)}{=} z \wedge_m^M (y \wedge_m^M x)$.
First, we prove:

$$x \wedge_m^M y = x \odot (y \oplus x^-). \tag{6.58}$$

Indeed, $x \wedge_m^M y = x \wedge_m^B y = x \odot (y^- \odot x)^- = x \odot (y \oplus x^-)$; thus, (6.58) holds.
Then, we prove:

$$x \wedge_m^M (y \oplus z) = x \odot (y \oplus (z \oplus x^-)). \tag{6.59}$$

Indeed, $x \wedge_m^M (y \oplus z) \stackrel{(6.58)}{=} x \odot ((y \oplus z) \oplus x^-) \stackrel{(Sass)}{=} x \odot (y \oplus (z \oplus x^-))$; thus, (6.59) holds.
Then, we prove:

$$x \wedge_m^M (y \odot z) = y \odot (z \wedge_m^M (x \oplus y^-)). \tag{6.60}$$

Indeed, $x \wedge_m^M (y \odot z) \stackrel{(\wedge_m-comm)}{=} (y \odot z) \wedge_m^M x \stackrel{(6.58)}{=} (y \odot z) \odot (x \oplus (y \odot z)^-) \stackrel{(Pass)}{=} y \odot (z \odot (x \oplus (y^- \oplus z^-))) \stackrel{(6.59)}{=} y \odot (z \wedge_m^M (x \oplus y^-))$; thus, (6.60) holds.
Finally,
$x \wedge_m^M (y \wedge_m^M z) \stackrel{(6.58)}{=} x \wedge_m^M (y \odot (z \oplus y^-)) \stackrel{(6.60)}{=} y \odot ((z \oplus y^-) \wedge_m^M (x \oplus y^-)) \stackrel{(\wedge_m-comm)}{=} y \odot ((x \oplus y^-) \wedge_m^M (z \oplus y^-)) \stackrel{(6.60)}{=} z \wedge_m^M (y \odot (x \oplus y^-)) \stackrel{(6.58)}{=} z \wedge_m^M (y \wedge_m^M x)$. Thus, (m-Wass) holds.
(m-Vass): $x \vee_m^M (y \vee_m^M z) = x \vee_m^M (y^- \wedge_m^M z^-)^- = (x^- \wedge_m^M (y^- \wedge_m^M z^-))^- \stackrel{(m-Wass)}{=} ((x^- \wedge_m^M y^-) \wedge_m^M z^-)^- = ((x \vee_m^M y)^- \wedge_m^M z^-)^- = (x \vee_m^M y) \vee_m^M z$;
(m-Wabs): $x \wedge_m^M (x \vee_m^M y) \stackrel{(m-Wcomm),(m-Vcomm)}{=} (y \vee_m^M x) \wedge_m^M x = x$ by (6.51);
(m-Vabs): $x \vee_m^M (x \wedge_m^M y) \stackrel{(m-Wcomm),(m-Vcomm)}{=} (y \wedge_m^M x) \vee_m^M x = x$ by (6.52);
(m-WU): $1 \wedge_m^M x = x$ follows by (6.36);
(m-VU): $0 \vee_m^M x = x$ follows by (6.37). □

**Proposition 6.5.4** *[61] Let* $\mathcal{A}^L = (A^L, \odot, ^-, 1)$ *be a left-MV algebra. The following properties hold:*

*(1)* $x \leq_m^M y \Longleftrightarrow y^- \leq_m^M x^-$,
*(2)* $x \leq_m^M y \Longrightarrow a \oplus x \leq_m^M a \oplus y$,
*(3)* $x \leq_m^M y \Longrightarrow a \odot x \leq_m^M a \odot y$,
*(4)* $x \odot y \leq_m^M z \Longleftrightarrow y \leq_m^M x^- \oplus z \Longleftrightarrow y \leq_m^M x \to z$,
*(5)* $x \odot y \leq_m^M x$, $y$ *and* $x, y \leq_m^M x \oplus y$,
*(6)* $x \odot y \leq_m^M x \wedge_m^M y \leq_m^M x \vee_m^M y \leq_m^M x \oplus y$.

Let $I$ be an arbitrary set throughout this chapter.

**Proposition 6.5.5** *[61] Let $\mathcal{A}^L = (A^L, \odot, ^-, 1)$ be a left-MV algebra. Denote $\wedge \stackrel{notation}{=} \wedge_m^M$ and $\vee \stackrel{notation}{=} \vee_m^M$. The following properties hold:*
*(m-DSW)* $a \oplus (\wedge_{i\in I} b_i) = \wedge_{i\in I}(a \oplus b_i)$,
*(m-DPV)* $a \odot (\vee_{i\in I} b_i) = \vee_{i\in I}(a \odot b_i)$,
*(m-Wdis)* $a \wedge (\vee_{i\in I} b_i) = \vee_{i\in I}(a \wedge b_i)$,
*(m-Vdis)* $a \vee (\wedge_{i\in I} b_i) = \wedge_{i\in I}(a \vee b_i)$,
*(m-DSV)* $a \oplus (\vee_{i\in I} b_i) = \vee_{i\in I}(a \oplus b_i)$,
*(m-DPW)* $a \odot (\wedge_{i\in I} b_i) = \wedge_{i\in I}(a \odot b_i)$.

Note that, by above (m-Wdis), the bounded lattice from Theorem 6.5.3 is distributive.

**Theorem 6.5.6** *Let $\mathcal{A}^L = (A^L, \odot, ^-, 1)$ be a left-MV algebra. If (G) holds, then $(G^R)$ holds too and*

$$x \wedge_m^M y = x \odot y, \quad x \vee_m^M y = x \oplus y.$$

**Proof.** $x \oplus x = (x^- \odot x^-)^- \stackrel{(G)}{=} (x^-)^- \stackrel{(DN)}{=} x$; thus, ($G^R$) holds too.

$x \wedge_m^M y = (y^- \odot x)^- \odot x \stackrel{(G)}{=} (y^- \odot x)^- \odot (x \odot x) \stackrel{(Pass)}{=} [(y^- \odot x)^- \odot x] \odot x = (x \wedge_m^M y) \odot x \stackrel{(m-DPW)}{=} (x \odot x) \wedge_m^M (y \odot x) \stackrel{(G)}{=} x \wedge_m^M (y \odot x) \stackrel{(PU)}{=} (1 \odot x) \wedge_m^M (y \odot x) \stackrel{(m-DPW)}{=} (1 \wedge_m^M y) \odot x \stackrel{(m-WU)}{=} y \odot x \stackrel{(Pcomm)}{=} x \odot y$. Thus, $x \wedge_m^M y = x \odot y$. By a dual proof, $x \vee_m^M y = x \oplus y$. □

**Proposition 6.5.7** *We have (the dual results are omitted):*
*(mCDN7) (G) + (DN) + (m-Re) + (Pcomm) $\Longrightarrow$ (m-WRe),*
*(mCDN8) (m-WRe) + (DN) + (m-Pleq) + (m-An) $\Longrightarrow$ (G).*

**Proof.** (mCDN7): $x \wedge x^- \stackrel{def.}{=} ((x^-)^- \odot x)^- \odot x \stackrel{(DN)}{=} (x \odot x)^- \odot x \stackrel{(G)}{=} x^- \odot x \stackrel{(Pcomm)}{=} x \odot x^- \stackrel{(m-Re)}{=} 0$.

(mCDN8): First, we have $x \odot x \leq_m x$, by (m-Pleq'). We prove that $x \leq_m x \odot x$ also. Indeed, $x \odot (x \odot x)^- \stackrel{(DN)}{=} x \odot (x \odot (x^-)^-)^- \stackrel{def.}{=} x \wedge x^- \stackrel{(m-WRe)}{=} 0$, hence $x \leq_m x \odot x$. Now, by (m-An'), we obtain (G). □

**Proposition 6.5.8** *Let $\mathcal{A}^L = (A^L, \odot, ^-, 1)$ be a left-MV algebra. Then, we have the equivalences:*

$$(G) \Longleftrightarrow (m - WRe), \qquad (G^R) \Longleftrightarrow (m - VRe).$$

**Proof.** By (mCDN1), (Neg0-1) holds and, by (mB1), (m-Re) holds. By (mA14), (m-Pleq) holds and, by (mB2), (m-An) holds.

By (mCDN7), (G) + (DN) + (m-Re) + (Pcomm) $\Longrightarrow$ (m-WRe), and by (mCDN8), (m-WRe) + (DN) + (m-Pleq) + (m-An) $\Longrightarrow$ (G). Thus, (G) $\Longleftrightarrow$ (m-WRe). Dually, ($G^R$) $\Longleftrightarrow$ (m-VRe). □

**Remark 6.5.9** *(See Remarks 11.2.13)*

*In a left-MV algebra* $\mathcal{A}^L = (A^L, \odot, ^-, 1)$*:*
*- the initial binary relation,* $\leq_m$ $(\Longleftrightarrow \leq_m^B)$*, is a* **lattice order relation** *w.r. to* $\wedge_m^B = \wedge_m^M, \vee_m^B = \vee_m^M$*, since (m-Re), (m-An), (m-Tr) hold and since* $\wedge_m^B$ *is commutative;*
*- the binary relation* $\leq_m^M$ *is a* **lattice order relation** *w.r. to* $\wedge_m^M, \vee_m^M$*, by Theorem 6.5.3, since* $\wedge_m^M$ *is commutative; the lattice is distributive;*
*-* $\leq_m$ *and* $\leq_m^M$ *are equivalent:* $\leq_m$ $(\Longleftrightarrow \leq_m^B) \Longleftrightarrow \leq_m^M$*, by Proposition 6.4.19;*
*- the binary relation* $\leq_m^P$ *is* **only antisymmetric and transitive**, *by Proposition 6.4.9.*

*In a left-MV algebra verifying (G)* $(x \odot x = x)$*, i.e. in a left-Boolean algebra,*

$$\leq_m \; (\Longleftrightarrow \leq_m^B) \Longleftrightarrow \leq_m^M \Longleftrightarrow \leq_m^P .$$

### 6.5.2 Connections with MV algebras

**Theorem 6.5.10** *The class of* $\wedge_m$*-commutative (involutive) left-m-BCK algebras is d.e. to the class of left-MV algebras.*

**Proof.** (1): Let $\mathcal{A}^L = (A^L, \odot, ^-, 1)$ be a $\wedge_m$-commutative (involutive) left-m-BCK algebra (see Theorem 6.3.6). By definition, $\mathcal{A}^L$ is a $\wedge_m$-commutative involutive left-m-MEL algebra, i.e. it is a left-MV algebra.

(1'): Conversely, let $\mathcal{A}^L = (A^L, \odot, ^-, 1)$ be a left-MV algebra, i.e. a $\wedge_m$-commutative involutive left-m-MEL algebra, i.e. (PU), (Pcomm), (Pass), (m-La), (Neg1-0), (DN), ($\wedge$-comm) hold. We must prove that $\mathcal{A}^L$ is a left-m-BCK algebra, i.e. (m-Re), (m-An), (m-BB) hold. Indeed,
by (mCDN1), (DN) + (Neg1-0) imply (Neg0-1);
by(mB1), ($\wedge_m$-comm) + (Neg0-1) + (PU) + (Pcomm) + (m-La) imply (m-Re);
by (mB2), ($\wedge_m$-comm) + (Neg0-1) + (PU) + (Pcomm) imply (m-An);
by (mCBN1), ($\wedge_m$-comm) + (DN) + (Pass) + (Pcomm) + (m-Re) + (m-La) imply (m-BB). □

We have seen in Proposition 5.4.54 (in the non-commutative case) the connection between commutative $l$-groups and $l$-cims. We shall see now the deeper connections existing between commutative $l$-groups and $l$-cims.

By the equivalences (Eq3) (from Chapter 16) between $l$-i-groups and $l$-groups and by the equivalences (from Subsection 16.1.2) between **BCK(P')-L** and **$l$-cim(R') = R-L**, we obtain the following analogous of Theorem 3.3.13 and Corollary 3.3.15 for unital commutative magmas.

**Theorem 6.5.11** *(See Theorem 3.3.13) (See ([116], Theorem 14.2.1), in the non-commutative case)*

*Let* $\mathcal{G} = (G, \vee, \wedge, \cdot, ^{-1}, 1)$ *be a commutative l-group (left-one or right-one). We have: for all* $x, y \in G$, $x \to y \stackrel{def.}{=} y \cdot x^{-1}$.

*(1) Define, for all* $x, y \in G^-$*:*

$$x \odot y \stackrel{def.}{=} x \cdot y, \quad x \to^L y \stackrel{def.}{=} (x \to y) \wedge 1.$$

*Then, $\mathcal{G}_m^- = (G^-, \wedge, \vee, \odot, \mathbf{1} = 1)$ is a distributive l-cim(R') (with the residuum $\to^L$) verifying the properties (CC) and (HH) (while the equivalent algebra (by Corollary 5.3.26) $\mathcal{G}_{m'}^- = (G^-, \wedge, \vee, \odot, \to^L, \mathbf{1} = 1)$ is a distributive residuated left-lattice verifying the properties (CC) and (HH)).*

*(1') Dually, define for all $x, y \in G^+$:*

$$x \oplus y \stackrel{def.}{=} x \cdot y, \quad x \to^R y \stackrel{def.}{=} (x \to y) \vee 1.$$

*Then, $\mathcal{G}_m^+ = (G^+, \vee, \wedge, \oplus, \mathbf{0} = 1)$ is a distributive l-cim(coR') (with the coresiduum $\to^R$) verifying the properties ($CC^R$) and ($HH^R$) (while the equivalent algebra $\mathcal{G}_{m'}^+ = (G^+, \vee, \wedge, \oplus, \to^R, \mathbf{0} = 1)$ is a distributive residuated right-lattice verifying the properties ($CC^R$) and ($HH^R$)).*

**Remark 6.5.12** $\mathcal{G}_m^-$ verifies the properties (prel) and (div) and $\mathcal{G}_m^+$ verifies the dual properties (prel$^R$) and (pdiv$^R$).

We shall "bound" the above algebras $\mathcal{G}_m^-$ and $\mathcal{G}_m^+$ and obtain known results [159].

**Corollary 6.5.13** *(See Corollary 3.3.15) (See ([116], Corollary 14.2.3), in the non-commutative case)*

*(1) Let $\mathcal{G}_m^- = (G^-, \wedge, \vee, \odot, \mathbf{1} = 1)$ be the l-cim(R') from above Theorem 6.5.11 (1). Let us "bound" this algebra in the following way:*

*Let us take $u' < 1$ from $G^-$ and consider the interval $[u', 1] = \{x \in G^- \mid u' \leq x \leq 1\}$. Define for all $x, y \in G^-$:*

$$x \odot^L y \stackrel{def.}{=} (x \odot y) \vee u' = (x \cdot y) \vee u', \quad x^{-^L} \stackrel{def.}{=} u' \cdot x^{-1}.$$

*Then, the algebra $\mathcal{G}_{m_1}^- = ([u', 1], \wedge, \vee, \odot^L, \mathbf{0} = u', \mathbf{1} = 1)$ is a bounded l-cim(R') verifying the property (CC), i.e. is an equivalent definition of the* **left-MV algebra** $\mathcal{G}_{m_1'}^- = ([u', 1], \odot^L, ^{-^L}, \mathbf{1} = 1)$.

*(1') Dually, let $\mathcal{G}_m^+ = (G^+, \vee, \wedge, \oplus, \mathbf{0} = 1)$ be the l-cim(coR') from above Theorem 6.5.11 (1'). Let us "bound" this algebra in the following way:*

*Let us take $u > 1$ from $G^+$ and consider the interval $[1, u] = \{x \in G^+ \mid 1 \leq x \leq u\}$. Define for all $x, y \in G^+$:*

$$x \oplus^R y \stackrel{def.}{=} (x \oplus y) \wedge u = (x \cdot y) \wedge u, \quad x^{-^R} \stackrel{def.}{=} u \cdot x^{-1}.$$

*Then, the algebra $\mathcal{G}_{m_1}^+ = ([1, u], \vee, \wedge, \oplus^R, \mathbf{0} = 1, \mathbf{1} = u)$ is a bounded l-cim(coR') verifying the property ($CC^R$), i.e. is an equivalent definition of the* **right-MV algebra** $\mathcal{G}_{m_1'}^+ = ([1, u], \oplus^R, ^{-^R}, \mathbf{0} = 1)$.

Note that, since any left-MV algebra $\mathcal{A}^L$ is a left-m-BCK algebra, it follows that all the properties in list **m-A** are satisfied by $\mathcal{A}^L$.

**Remarks 6.5.14** *(The dual ones are omitted)*

*(i) Since ($\wedge_m$-comm) implies (m-Re), by (mB1), it follows that* **any left-MV algebra is in fact an involutive left-m-BE algebra verifying ($\wedge_m$-comm).**

*(ii) Since ($\wedge_m$-comm) implies also (m-An) and (m-BB) ($\Leftrightarrow \ldots \Leftrightarrow$ (m-Tr)), by (mB2), (mCBN1), respectively, i.e. we have:*

$$(6.61) \qquad (\wedge_m - comm) \Longrightarrow (m - An) + (m - Tr),$$

*it follows that* **any left-MV algebra is in fact a left-m-BCK algebra**, *i.e. we have:*

$$\mathbf{MV} \subset \mathbf{m-BCK} = \mathbf{m-BCK_{(DN)}} (= \mathbf{m-taBE_{(DN)}}).$$

*(iii) Moreover, by Theorem 6.5.10, the class of left-MV algebras is d.e. with the class of $\wedge_m$-commutative (involutive) left-m-BCK algebras (i.e. left-m-BCK algebras verifying ($\wedge_m$-comm)).*

By Theorem 6.5.10 and Remark 6.5.2, we obtain immediately:

**Corollary 6.5.15** *$\wedge_m$-commutative involutive left-m-MEL, left-m-BE, left-m-BE*, left-pre-m-BCK, left-m-aBE, left-m-aBE* = left-m-BCK algebras coincide (with left-MV algebras).*

There are many examples of involutive m-M algebras in the sequel. Now, we present an example of non-involutive m-M algebra, found by *Mace4 program.*

**Example 6.5.16 Example of proper non-involutive m-BE algebra** (i.e. not verifying (m-An) and (m-BB))

The algebra $\mathcal{A}^L = (A_4 = \{0, 2, 3, 1\}, \odot, ^-, 1)$, with the following tables of $\odot$ and $^-$, verifies (Neg1-0), (Neg0-1), (PU), (Pcomm), (Pass), (m-La), (m-Re) and does not verify (DN) for 2, (m-B) for $(3, 2, 3)$, (m-BB) for $(2, 1, 2)$, (m-*) for $(2, 3, 1)$, (m-**) for $(3, 2, 3)$, (m-Tr) for $(1, 2, 3)$, (m-An) for $(2, 3)$, (m-Pimpl) for $(2, 0)$. Hence, it is a proper left-m-BE algebra.

| $\odot$ | 0 | 2 | 3 | 1 |
|---|---|---|---|---|
| 0 | 0 | 0 | 0 | 0 |
| 2 | 0 | 0 | 0 | 2 |
| 3 | 0 | 0 | 2 | 3 |
| 1 | 0 | 2 | 3 | 1 |

and

| $x$ | $x^-$ |
|---|---|
| 0 | 1 |
| 2 | 0 |
| 3 | 2 |
| 1 | 0 |

.

# Chapter 7

# $m_1$-M algebras (Hierarchies $m_1$-1 and $m_1$-1')

In this chapter, in Section 7.1, we introduce 10 new algebras, the analogous of those 10 M algebras *without last element* from Chapter 2, as particular cases of unital commutative magmas with some additional operation of the form $(A, \cdot, ^{-1}, 1)$ with $1^{-1} = 1$, called $m_1$-M algebras, which determine Hierarchies $m_1$-1, $m_1$-1'.

In Section 7.2, we present the properties of the negation $^{-1}$ and in Section 7.3, we study the $m_1$-p-semisimple algebras, in connection with results from Chapter 4. We make the connections with the commutative group, moon [116], goop [116].

The content of this chapter is taken mainly from [118].

Note that, in this chapter, the product operation is denoted by $\cdot$ instead of $\odot$ and the negation operation is denoted by $^{-1}$ (and is called the *inverse*) instead of $^{-}$.

## 7.1 List $m_1$-A of basic properties and $m_1$-M algebras

**• The basic natural internal binary relation: $\leq_{m_1}$**

Let $\mathcal{A}^L = (A^L, \cdot, ^{-1}, 1)$ be an algebra of type $(2, 1, 0)$ such that $1^{-1} = 1$. Define an *internal* binary relation $\leq_{m_1}$ on $A^L$ by: for all $x, y \in A^L$,

$$(m_1 - dfrelP) \qquad x \leq_{m_1} y \overset{def.}{\Longleftrightarrow} x \cdot y^{-1} = 1.$$

Equivalently, consider a structure $\mathcal{A}^L = (A^L, \leq_{m_1}, \cdot, ^{-1}, 1)$ such that $1^{-1} = 1$ and the *internal* binary relation $\leq_{m_1}$, the binary operation $\cdot$, the unary operation $^{-1}$ and the element 1 are connected by the equivalence:

$$(m_1 - EqrelP) \qquad x \leq_{m_1} y \Longleftrightarrow x \cdot y^{-1} = 1.$$

Consider the following list $\mathbf{m_1}$-**A** of basic properties that can be satisfied by $\mathcal{A}^L$, determined by the two above equivalent definitions of $\mathcal{A}^L$:

(PU) $1 \cdot x = x = x \cdot 1$ (unit element of product, the *identity*);
(Pcomm) $x \cdot y = y \cdot x$ (commutativity of product),
(Pass) $x \cdot (y \cdot z) = (x \cdot y) \cdot z$ (associativity of product);

(N1) $1^{-1} = 1$;

($m_1$-An) $(x \cdot y^{-1} = 1$ and $y \cdot x^{-1} = 1) \Longrightarrow x = y$ (antisymmetry),
($m_1$-An') $(x \leq_{m_1} y$ and $y \leq_{m_1} x) \Longrightarrow x = y$ (antisymmetry);

($m_1$-B) $[(x \cdot y^{-1})^{-1} \cdot (x \cdot z)] \cdot (y \cdot z)^{-1} = 1$,
($m_1$-B') $(x \cdot y^{-1})^{-1} \cdot (x \cdot z) \leq_{m_1} y \cdot z$;
($m_1$-B=) $(x \cdot y^{-1})^{-1} \cdot (x \cdot z) = y \cdot z$;

($m_1$-BB) $[(z \cdot x)^{-1} \cdot (y \cdot x)] \cdot (y \cdot z^{-1})^{-1} = 1$,
($m_1$-BB') $(z \cdot x)^{-1} \cdot (y \cdot x) \leq_{m_1} y \cdot z^{-1}$;
($m_1$-BB=) $(z \cdot x)^{-1} \cdot (y \cdot x) = y \cdot z^{-1}$;

($m_1$-*) $x \cdot y^{-1} = 1 \Longrightarrow (z \cdot y^{-1}) \cdot (z \cdot x^{-1})^{-1} = 1$,
($m_1$-*') $x \leq_{m_1} y \Longrightarrow z \cdot y^{-1} \leq_{m_1} z \cdot x^{-1}$;

($m_1$-**) $x \cdot y^{-1} = 1 \Longrightarrow (x \cdot z) \cdot (y \cdot z)^{-1} = 1$,
($m_1$-**') $x \leq_{m_1} y \Longrightarrow x \cdot z \leq_{m_1} y \cdot z$;

($m_1$-D) $[(y^{-1} \cdot x)^{-1} \cdot x] \cdot y^{-1} = 1$,
($m_1$-D') $(y^{-1} \cdot x)^{-1} \cdot x \leq_{m_1} y$;
($m_1$-D=) $(y^{-1} \cdot x)^{-1} \cdot x = y$;

($m_1$-Re) $x \cdot x^{-1} = 1$ (reflexivity),
($m_1$-Re') $x \leq_{m_1} x$ (reflexivity);

($m_1$-Tr) $(x \cdot y^{-1} = 1$ and $y \cdot z^{-1} = 1) \Longrightarrow x \cdot z^{-1} = 1$ (transitivity),
($m_1$-Tr') $(x \leq_{m_1} y$ and $y \leq_{m_1} z) \Longrightarrow x \leq_{m_1} z$ (transitivity);

($m_1$-p-s) $x \leq_{m_1} y \Longleftrightarrow x = y$ ($m_1$-p-semisimple);
(EqP=) $x = y \Longleftrightarrow x \cdot y^{-1} = 1$.

Dually,
let $\mathcal{A}^R = (A^R, +, -, 0)$ be an algebra of type $(2, 1, 0)$ such that $-0 = 0$. Define an *internal* binary relation $\geq_{m_0}$ on $A^R$ by: for all $x, y \in A^R$,

$$(m_0 - dfrelS) \qquad x \geq_{m_0} y \overset{def.}{\Longleftrightarrow} x + (-y) = 0.$$

Equivalently, consider a structure $\mathcal{A}^R = (A^R, \geq_{m_0}, +, -, 0)$ such that $-0 = 0$ and the *internal* binary relation $\geq_{m_0}$, the binary operation $+$, the unary operation

– and the element 0 are connected by the equivalence:

$$(m_0 - EqrelS) \qquad x \geq_{m_0} y \Longleftrightarrow x + (-y) = 0.$$

The list of basic dual properties of the dual algebra $\mathcal{A}^R = (A^R, +, -, 0)$ (structure $\mathcal{A}^R = (A^R, \geq_{m_0}, +, -, 0)$) is omitted.

### 7.1.1 $\mathrm{m_1}$-M algebras

Let $\mathcal{A}^L = (A^L, \cdot, ^{-1}, 1)$ be an algebra of type $(2, 1, 0)$ throughout this section such that (N1) holds ($1^{-1} = 1$).

We now introduce the following notions, the analogous of those *M algebras without last element* from Chapter 2 (Hierarchies 1 and 1') (the dual ones are omitted).

**- Algebras without last element, without associativity:**
$\mathcal{A}^L$ is a:
- *left-$m_1$-M algebra*, if it verifies (PU), (Pcomm);
- *left-$m_1$-RM algebra*, if it verifies (PU), (Pcomm), ($\mathrm{m_1}$-Re),
- *left-pre-$m_1$-BZ algebra*, if it verifies (PU), (Pcomm), ($\mathrm{m_1}$-Re), ($\mathrm{m_1}$-B),
- *left-$m_1$-aRM = left-$m_1$-BH algebra*, if it verifies (PU), (Pcomm), ($\mathrm{m_1}$-Re), ($\mathrm{m_1}$-An),
- *left-$m_1$-BZ algebra*, if it verifies (PU), (Pcomm), ($\mathrm{m_1}$-Re), ($\mathrm{m_1}$-An), ($\mathrm{m_1}$-B);

**- Algebras without last element, with associativity:**
$\mathcal{A}^L$ is a:
- *left-$m_1$-ME algebra*, if it verifies (PU), (Pcomm), (Pass);
- *left-$m_1$-RME =left-$m_1$-CI algebra*, if it verifies (PU), (Pcomm), (Pass), ($\mathrm{m_1}$-Re),
- *left-pre-$m_1$-BCI algebra*, if it verifies (PU), (Pcomm), (Pass), ($\mathrm{m_1}$-Re), ($\mathrm{m_1}$-B),
- *left-$m_1$-BCH algebra*, if it verifies (PU), (Pcomm), (Pass), ($\mathrm{m_1}$-Re), ($\mathrm{m_1}$-An),
- *left-$m_1$-BCI algebra*, if it verifies (PU), (Pcomm), (Pass), ($\mathrm{m_1}$-Re), ($\mathrm{m_1}$-An), ($\mathrm{m_1}$-B)

and, additionally:
- *left-$m_1$-RME* = left-$m_1$-CI* algebra*, if it is a left-$\mathrm{m_1}$-RME = left-$\mathrm{m_1}$-CI algebra verifying ($\mathrm{m_1}$-*),
- *left-$m_1$-BCH* algebra*, if it is a left-$\mathrm{m_1}$-BCH algebra verifying ($\mathrm{m_1}$-*).

Denote by $\mathbf{m_1}$**-M**, ... $\mathbf{m_1}$**-BCH***, the classes of left-$\mathrm{m_1}$-M, ..., left-$\mathrm{m_1}$-BCH* algebras, respectively.

Recall the following generalizations of groups introduced (in the non-commutative case) and studied in ([116], Definition 10.1.5 and 10.1.25, respectively):

An algebra $(A, \cdot, ^{-1}, 1)$ of type $(2, 1, 0)$ is:
- a *moon*, if it verifies
  (PU) $\qquad 1 \cdot x = x = x \cdot 1$ and
  (pIv)=($\mathrm{m_1}$-Re) $\qquad x \cdot x^{-1} = 1 = x^{-1} \cdot x$;

- a *goop*, if it verifies (PU) and

(IdEqP) $\quad y \cdot x^{-1} = 1 \Longleftrightarrow x^{-1} \cdot x^{-1} \cdot y = 1$ and

(pEqP=) $\quad x = y \Longleftrightarrow y \cdot x^{-1} = 1 (\overset{(IdEqP)}{\Longleftrightarrow} x^{-1} \cdot y = 1.$

**Remarks 7.1.1**

*(1) We do not know the relations existing between the classes:* $\mathbf{m_1}$**-M**, **m-M** *(from Chapter 6) and* **uc. magma**.

*(2) We do not know the relations existing between the classes:* $\mathbf{m_1}$**-ME**, **m-ME** *(from Chapter 6) and* **c. monoid**.

*(3) Note that the left-$m_1$-RM algebra is just the commutative moon, as defined in ([116], Definition 10.1.5). Hence, we have:* $\mathbf{m_1}$**-RM** = **c. moon**.

*(4) Note that the left-$m_1$-RME = left-$m_1$-CI algebra is just the commutative group (recalled and studied in ([116], Definition 7.1.1), see Chapter 5). Hence, we have:* $\mathbf{m_1}$**-RME** = $\mathbf{m_1}$**-CI** = **c. group**.

### 7.1.2 Connections between the properties

**Proposition 7.1.2** *(See Proposition 6.1.2)*

*We have (the dual case is omitted):*

*(1mA4) (PU) + (Pcomm) + (N1) + ($m_1$-BB) $\Longrightarrow$ ($m_1$-**),*
*(1mA5) (PU) + (Pcomm) + (N1) + ($m_1$-B) $\Longrightarrow$ ($m_1$-*);*
*(1mA6) (PU) + (N1) + ($m_1$-**) $\Longrightarrow$ ($m_1$-Tr),*
*(1mA7) (PU) + (N1) + ($m_1$-*) $\Longrightarrow$ ($m_1$-Tr).*

**Proof.** (1mA4): Suppose $y \leq_{m_1} z$, i.e. $y \cdot z^{-1} = 1$; then, by ($m_1$-BB),
$1 = [(z \cdot x)^{-1} \cdot (y \cdot x)] \cdot 1^{-1} \overset{(N1)}{=} [(z \cdot x)^{-1} \cdot (y \cdot x)] \cdot 1 \overset{(Pcomm),(PU)}{=} (z \cdot x)^{-1} \cdot (y \cdot x)$ $\overset{(Pcomm)}{=} (y \cdot x) \cdot (z \cdot x)^{-1}$, i.e. $y \cdot x \leq_{m_1} z \cdot x$; thus, ($m_1$-**') holds.

(1mA5): Suppose $y \leq_{m_1} z$, i.e. $y \cdot z^{-1} = 1$; then, by ($m_1$-B),
$1 = [(x \cdot y^{-1})^{-1} \cdot (x \cdot z^{-1})] \cdot (y \cdot z^{-1})^{-1} \overset{(N1)}{=} [(x \cdot y^{-1})^{-1} \cdot (x \cdot z^{-1})] \cdot 1$
$\overset{(Pcomm),(PU)}{=} (x \cdot y^{-1})^{-1} \cdot (x \cdot z^{-1}) \overset{(Pcomm)}{=} (x \cdot z^{-1}) \cdot (x \cdot y^{-1})^{-1}$,
i.e. $x \cdot z^{-1} \leq_{m_1} x \cdot y^{-1}$; thus, ($m_1$-*') holds.

(1mA6): Suppose $x \leq_{m_1} y$ and $y \leq_{m_1} z$, i.e. $x \cdot y^{-1} = 1$ and $y \cdot z^{-1} = 1$; then, by ($m_1$-**'), $x \cdot z^{-1} \leq_{m_1} y \cdot z^{-1}$, i.e. $(x \cdot z^{-1}) \cdot (y \cdot z^{-1})^{-1} = 1$; hence, $(x \cdot z^{-1}) \cdot 1^{-1} = 1$; then, by (N1), (PU), we obtain $x \cdot z^{-1} = 1$, i.e. $x \leq_{m_1} z$; thus, ($m_1$-Tr') holds.

(1mA7): Suppose $x \leq_{m_1} y$ and $y \leq_{m_1} z$, i.e. $x \cdot y^{-1} = 1$ and $y \cdot z^{-1} = 1$; then, by ($m_1$-*'), $x \cdot z^{-1} \leq_{m_1} x \cdot y^{-1}$, i.e. $(x \cdot z^{-1}) \cdot (x \cdot y^{-1})^{-1} = 1$; then,
$1 = (x \cdot z^{-1}) \cdot 1^{-1} \overset{(N1)}{=} (x \cdot z^{-1}) \cdot 1 \overset{(Pcomm),(PU)}{=} x \cdot z^{-1}$, i.e. $x \leq_{m_1} z$; thus, ($m_1$-Tr') holds. □

## 7.2 List $\mathbf{m_1}$-C of properties of the inverse

Consider now the following list $\mathbf{m_1}$**-C** of properties of the inverse $^{-1}$ (the dual one is omitted):

| | |
|---|---|
| (m$_1$-Neg4') | $(x^{-1})^{-1} \leq_{m_1} x$, |
| (m$_1$-Neg7) | $((x^{-1})^{-1})^{-1} = x^{-1}$; |
| (m$_1$-Neg2') | $x \leq_{m_1} y \Longrightarrow y^{-1} \leq_{m_1} x^{-1}$; |
| (DN) (Double Negation) | $(x^{-1})^{-1} = x$; |
| (m$_1$-DN4') | $x \leq_{m_1} y \Longleftrightarrow y^{-1} \leq_{m_1} x^{-1}$. |

**Proposition 7.2.1** *(See Proposition 6.3.1)*
*We have:*
*(m1C1) (m$_1$-*) + (PU) $\Longrightarrow$ (m$_1$-Neg2'),*
*(m1C2) (m$_1$-Re) + (Pcomm) $\Longrightarrow$ (m$_1$-Neg4'),*
*(m1C2') (m$_1$-D) + (PU) + (Pcomm) $\Longrightarrow$ (m$_1$-Neg4');*
*(m1C3) (m$_1$-Neg4') + (m$_1$-*) + (PU) + (m$_1$-An) $\Longrightarrow$ (m$_1$-Neg7).*

**Proof.** (m1C1): Suppose $x \leq_{m_1} y$; then, by (m$_1$-*), $z \cdot y^{-1} \leq_{m_1} z \cdot x^{-1}$; then, for $z = 1$, we obtain $1 \cdot y^{-1} \leq_{m_1} 1 \cdot x^{-1}$, hence, $y^{-1} \leq_{m_1} x^{-1}$, by (PU); thus, (m$_1$-Neg2') holds.

(m1C2): $(x^{-1})^{-1} \cdot x^{-1} \overset{(Pcomm)}{=} x^{-1} \cdot (x^{-1})^{-1} \overset{(m_1-Re)}{=} 1$, i.e. $(x^{-1})^{-1} \leq_{m_1} x$; thus, (m$_1$-Neg4') holds.

(m1C2'): Take $x = 1$ in (m$_1$-D); we obtain $(y^{-1} \cdot 1)^{-1} \cdot 1 \leq_{m_1} y$; hence, $(y^{-1})^{-1} \leq_{m_1} y$, by (Pcomm), (PU); thus, (m$_1$-Neg4') holds.

(m1C3): By (m$_1$-Neg4'), $((y^{-1})^{-1})^{-1} \leq_{m_1} y^{-1}$. On the other hand, by (m$_1$-Neg4'), $(y^{-1})^{-1} \leq_{m_1} y$, hence $1 \cdot y^{-1} \leq_{m_1} 1 \cdot ((y^{-1})^{-1})^{-1}$, by (m$_1$-*); then, $y^{-1} \leq_{m_1} ((y^{-1})^{-1})^{-1}$, by (PU). Now apply (m$_1$-An) to obtain $((y^{-1})^{-1})^{-1} = y^{-1}$; thus, (m$_1$-Neg7) holds. □

**Definition 7.2.2** We shall say that an algebra $\mathcal{A}^L = (A^L, \cdot, ^{-1}, 1)$ is *involutive*, if it verifies (DN).

The subclass of involutive algebras of the class **X** will be denoted by $\mathbf{X}_{(DN)}$.

**Proposition 7.2.3** *(See Proposition 6.3.4)*
*We have:*
*(m1CDN2) (DN) + (Pcomm) $\Longrightarrow$ (m$_1$-DN4),*
*(m1CDN3) (DN) + (Pcomm) $\Longrightarrow$ ((m$_1$-*) $\Leftrightarrow$ (m$_1$-**)),*
*(m1CDN5) (PU) + (Pcomm) + (m$_1$-Re) + (m$_1$-*) + (m$_1$-An) $\Longrightarrow$ (DN).*

**Proof.** (m1CDN2): $x \leq_{m_1} y \overset{(m_1-dfrelP)}{\Longleftrightarrow} x \cdot y^{-1} = 1 \overset{(DN)}{\Longleftrightarrow} (x^{-1})^{-1} \cdot y^{-1} = 1 \overset{(Pcomm)}{\Longleftrightarrow} y^{-1} \cdot (x^{-1})^{-1} = 1 \overset{(m_1-dfrelP)}{\Longleftrightarrow} y^{-1} \leq_{m_1} x^{-1}$; thus, (m$_1$-DN4') holds.

(m1CDN3): By (m1CDN2), (DN) + (Pcomm) $\Longrightarrow$ (m$_1$-DN4').
Suppose that (m$_1$-*') holds; then, if $x \leq_{m_1} y$, then, by (m$_1$-DN4'), $y^{-1} \leq_{m_1} x^{-1}$, hence, by (m$_1$-*'), $z \cdot (x^{-1})^{-1} \leq_{m_1} z \cdot (y^{-1})^{-1}$; then, by (DN), $z \cdot x \leq_{m_1} z \cdot y$; then, by (Pcomm), $x \cdot z \leq_{m_1} y \cdot z$; thus, (m$_1$-**') holds.
Conversely, suppose that (m$_1$-**') holds; then, if $x \leq_{m_1} y$, then, by (m$_1$-DN4'), $y^{-1} \leq_{m_1} x^{-1}$; then, by (m$_1$-** '), $y^{-1} \cdot z \leq_{m_1} x^{-1} \cdot z$; then, by (Pcomm), $z \cdot y^{-1} \leq_{m_1} z \cdot x^{-1}$; thus, (m$_1$-*') holds.

(m1CDN5): By (m1C2), ($m_1$-Re) + (Pcomm) $\Longrightarrow$ ($m_1$-Neg4'), i.e. $(x^{-1})^{-1} \leq_{m_1} x$.
On the other hand, by (m1C3), ($m_1$-Neg4') + ($m_1$-*) + (PU) + ($m_1$-An) $\Longrightarrow$ ($m_1$-Neg7), i.e. $((x^{-1})^{-1})^{-1} = x^{-1}$; then, $x \cdot ((x^{-1})^{-1})^{-1} = x \cdot x^{-1} \stackrel{(m_1-Re)}{=} 1$, i.e. $x \leq_{m_1} (x^{-1})^{-1}$.
Now, by ($m_1$-An), we obtain $(x^{-1})^{-1} = x$, i.e. (DN) holds. □

**Theorem 7.2.4** *Any left-$m_1$-BZ algebra is involutive.*

**Proof.** Let $\mathcal{A}^L = (A^L, \cdot, ^{-1}, 1)$ be a left-$m_1$-BZ algebra, i.e. (N1), (PU), (Pcomm), ($m_1$-Re), ($m_1$-An), ($m_1$-B) hold. Then, by (1mA5), (PU) + ($m_1$-B) + (Pcomm) + (N1) $\Longrightarrow$ ($m_1$-*), hence ($m_1$-*) holds too. By (m1CDN5), (PU) + (Pcomm) + ($m_1$-Re) + ($m_1$-*) + ($m_1$-An) $\Longrightarrow$ (DN), thus, (DN) holds. □

By above theorem, we shall write: $\mathbf{m_1\text{-}BZ} = \mathbf{m_1\text{-}BZ}_{(DN)}$.

## 7.3 The $m_1$-p-semisimple algebras

Let us introduce the following definition.

**Definition 7.3.1** (The dual one is omitted)
Let $\mathcal{A}^L = (A^L, \leq_{m_1}, \cdot, ^{-1}, 1)$ be a structure verifying ($m_1$-Re).
$\mathcal{A}^L$ is said to be *$m_1$-p-semisimple*, if for all $x, y \in A^L$,
($m_1$-p-s) $\qquad x \leq_{m_1} y \Longrightarrow x = y$
(or, equivalently, by ($m_1$-Re), $x \leq_{m_1} y \Longleftrightarrow x = y$).

The subclass of $m_1$-p-semisimple algebras of the class **X** will be denoted by $\mathbf{X}_{(m_1-p-s)}$.

**Remark 7.3.2** *(See Remark 5.5.2)*
*Note that the m-p-semisimple structures verifying the property (m-La') are trivial, because in this case $A^L = \{1\}$. Therefore, the m-p-semisimple property ($m_1$-p-semisimple) is studied only for structures without last element.*

The following equivalence was introduced in [116], Definition 10.1.24 (in the non-commutative case) to define a *sharp* moon:
(EqP=) $\qquad x = y \Longleftrightarrow x \cdot y^{-1} = 1.$

**Proposition 7.3.3** *(See Proposition 4.2.3)*
*Let $(A^L, \cdot, ^{-1}, 1)$ be an algebra verifying ($m_1$-Re). We have:*
*(0) ($m_1$-dfrelP) + ($m_1$-p-s) $\Longrightarrow$ (EqP=),*
*(0') ($m_1$-dfrelP) + (EqP=) $\Longrightarrow$ ($m_1$-p-s),*
*(0") ($m_1$-dfrelP) $\Longrightarrow$ (($m_1$-p-s) $\Leftrightarrow$ (EqP=));*
*(1) ($m_1$-p-s) $\Longrightarrow$ ($m_1$-An),*
*(1') ($m_1$-p-s) $\Longrightarrow$ ($m_1$-*'),*
*(1") ($m_1$-p-s) $\Longrightarrow$ ($m_1$-**'),*
*(1"') ($m_1$-p-s) $\Longrightarrow$ ($m_1$-Tr');*

*(2) ($m_1$-p-s) + ($m_1$-BB') $\Longrightarrow$ ($m_1$-BB=),*
*(2') ($m_1$-p-s) + ($m_1$-B') $\Longrightarrow$ ($m_1$-B=),*
*(2") ($m_1$-p-s) + ($m_1$-D') $\Longrightarrow$ ($m_1$-D=);*
*(3) (EqP=) + (Pcomm) + ($m_1$-Re) $\Longrightarrow$ (DN);*
*(4) (EqP=) $\Longrightarrow$ ($m_1$-Re);*
*(5) (Pcomm) + (DN) + ($m_1$-B=) $\Longrightarrow$ ($m_1$-BB=),*
*(5') (Pcomm) + (DN) + ($m_1$-BB=) $\Longrightarrow$ ($m_1$-B=),*
*(5") (Pcomm) + (DN) $\Longrightarrow$ (($m_1$-B=) $\Leftrightarrow$ ($m_1$-BB=)).*

**Proof.** (0), (0'), (0"): Obviously.
(1), (1'), (1"), (1"'): Obviously.
(2), (2'), (2"): Immediately.
(3): $(x^{-1})^{-1} = x \overset{(EqP=)}{\Longleftrightarrow} (x^{-1})^{-1} \cdot x^{-1} = 1$ that is true by (Pcomm) and ($m_1$-Re); thus, (DN) holds.
(4): ($m_1$-Re) $\overset{def.}{\Longleftrightarrow} x \cdot x^{-1} = 1 \overset{(EqP=)}{\Longleftrightarrow} x = x$, that is true.
(5): $(z \cdot x)^{-1} \cdot (y \cdot x) \overset{(Pcomm)}{=} (x \cdot z)^{-1} \cdot (x \cdot y) \overset{(DN)}{=} (x \cdot (z^{-1})^{-1})^{-1} \cdot (x \cdot y) \overset{(m_1-B=)}{=} z^{-1} \cdot y \overset{(Pcomm)}{=} y \cdot z^{-1}$; thus, ($m_1$-BB=) holds.
(5'): $(x \cdot y^{-1})^{-1} \cdot (x \cdot z) \overset{(Pcomm)}{=} (y^{-1} \cdot x)^{-1} \cdot (z \cdot x) \overset{(m_1-BB=)}{=} z \cdot (y^{-1})^{-1} \overset{(DN)}{=} z \cdot y \overset{(Pcomm)}{=} y \cdot z$; thus, ($m_1$-B=) holds.
(5"): By (5) and (5'). □

**Remark 7.3.4** *By above (0"), being* **$m_1$-p-semisimple** *is equivalent to being* **sharp**.

**Proposition 7.3.5** *(See Proposition 4.2.4)*
*Let $(A^L, \cdot, ^{-1}, 1)$ be an algebra. We have:*
*(m1A0) (Pass) + (Pcomm) + (PU) + ($m_1$-Re) $\Longrightarrow$ (EqP=),*
*(m1A1) (Pass) + (Pcomm) + ($m_1$-Re) + (EqP=) $\Longrightarrow$ ($m_1$-D=);*
*(m1A1') (Pass) + (Pcomm) + ($m_1$-Re) + (EqP=) $\Longrightarrow$ ($m_1$-BB=);*
*(m1A2) (Pass) + (Pcomm) + (EqP=) + ($m_1$-BB=) $\Longrightarrow$ ($m_1$-B=),*
*(m1A2') (Pass) + (Pcomm) + (EqP=) + ($m_1$-B=) $\Longrightarrow$ ($m_1$-BB=),*
*(m1A2") (Pass) + (Pcomm) + (EqP=) $\Longrightarrow$ (($m_1$-B=) $\Leftrightarrow$ ($m_1$-BB=));*
*(m1A3) ($m_1$-D=) + (EqP=) + (Pcomm) + (N1) $\Longrightarrow$ (PU),*
*(m1A4) (PU) + (Pcomm) + (N1) + ($m_1$-BB=) $\Longrightarrow$ ($m_1$-Re),*
*(m1A5) (PU) + (DN) + ($m_1$-BB=) $\Longrightarrow$ ($m_1$-D=),*
*(m1A6) (PU) + (Pcomm) + (N1) + ($m_1$-Re) + ($m_1$-D=) $\Longrightarrow$ (EqP=),*
*(m1A7) (PU) + (Pcomm) + (N1) + (DN) + ($m_1$-BB=) $\Longrightarrow$ (Pass),*
*(m1A8) (N1) + (PU) + ($m_1$-Re) + ($m_1$-An) + ($m_1$-B) + (DN) $\Longrightarrow$ ($m_1$-p-s).*

**Proof.** (m1A0): If $x = y$, then $x \cdot y^{-1} = y \cdot y^{-1} \overset{(m_1-Re)}{=} 1$. Conversely, if $x \cdot y^{-1} = 1$, then: $y \overset{(PU)}{=} 1 \cdot y = (x \cdot y^{-1}) \cdot y \overset{(Pass)}{=} x \cdot (y^{-1} \cdot y) \overset{(Pcomm)}{=} (y \cdot y^{-1}) \cdot x \overset{(m_1-Re)}{=} 1 \cdot x \overset{(PU)}{=} x$. Thus, (EqP=) holds.
(m1A1): $[(y^{-1} \cdot x)^{-1} \cdot x] \cdot y^{-1} \overset{(Pass)}{=} (y^{-1} \cdot x)^{-1} \cdot (x \cdot y^{-1}) \overset{(Pcomm)}{=} (x \cdot y^{-1}) \cdot (x \cdot y^{-1})^{-1} \overset{(m_1-Re)}{=} 1$, hence $(y^{-1} \cdot x)^{-1} \cdot x = y$, by (EqP=); thus, ($m_1$-D=) holds.

(m1A1'): By (m1A1), (Pass) + (Pcomm) + ($m_1$-Re) + (EqP=) imply ($m_1$-D=). Then,
$[(z \cdot x)^{-1} \cdot (y \cdot x)] \cdot (y \cdot z^{-1})^{-1}$
$\overset{(Pass)}{=} (z \cdot x)^{-1} \cdot [(y \cdot x) \cdot (y \cdot z^{-1})^{-1}]$
$\overset{(Pcomm)}{=} (z \cdot x)^{-1} \cdot [(z^{-1} \cdot y)^{-1} \cdot (y \cdot x)]$
$\overset{(Pass)}{=} (z \cdot x)^{-1} \cdot [((z^{-1} \cdot y)^{-1} \cdot y) \cdot x]$
$\overset{(m_1-D=)}{=} (z \cdot x)^{-1} \cdot (z \cdot x) \overset{(Pcomm)}{=} (z \cdot x) \cdot (z \cdot x)^{-1} \overset{(m_1-Re)}{=} 1.$
Then, by (EqP=), $(z \cdot x)^{-1} \cdot (y \cdot x) = y \cdot z^{-1}$, i.e. ($m_1$-BB=) holds.

(m1A2): $[(x \cdot y^{-1})^{-1} \cdot (x \cdot z)] \cdot (y \cdot z)^{-1} \overset{(Pass)}{=} (x \cdot y^{-1})^{-1} \cdot [(x \cdot z) \cdot (y \cdot z)^{-1}] \overset{(Pcomm)}{=} [(y \cdot z)^{-1} \cdot (x \cdot z)] \cdot (x \cdot y^{-1})^{-1} \overset{(m_1-BB=),(EqP=)}{=} 1$, hence $(x \cdot y^{-1})^{-1} \cdot (x \cdot z) = y \cdot z$, by (EqP=) again, i.e. ($m_1$-B=) holds.

(m1A2'): $[(z \cdot x)^{-1} \cdot (y \cdot x)] \cdot (y \cdot z^{-1})^{-1} \overset{(Pass)}{=} (z \cdot x)^{-1} \cdot [(y \cdot x) \cdot (y \cdot z^{-1})^{-1}] \overset{(Pcomm)}{=} [(y \cdot z^{-1})^{-1} \cdot (y \cdot x)] \cdot (z \cdot x)^{-1} \overset{(m_1-B=),(EqP=)}{=} 1$, hence $(z \cdot x)^{-1} \cdot (y \cdot x) = y \cdot z^{-1}$, by (EqP=) again, i.e. ($m_1$-BB=) holds.

(m1A2"): By (m1A2) and (m1A2').

(m1A3): $x = 1 \cdot x \overset{(EqP=)}{\Longleftrightarrow} x \cdot (1 \cdot x)^{-1} = 1 \overset{(Pcomm),(N1)}{\Longleftrightarrow} (1^{-1} \cdot x)^{-1} \cdot x = 1 \overset{(m_1-D=)}{\Longleftrightarrow} 1 = 1$, that is true; thus, (PU) holds.

(m1A4): $x \cdot x^{-1} \overset{(PU)}{=} (1 \cdot x) \cdot (1 \cdot x)^{-1} \overset{(Pcomm)}{=} (1 \cdot x)^{-1} \cdot (1 \cdot x) \overset{(m_1-BB=)}{=} 1 \cdot 1^{-1} \overset{(N1)}{=} 1 \cdot 1 \overset{(PU)}{=} 1$; thus, ($m_1$-Re) holds.

(m1A5): $(y^{-1} \cdot x)^{-1} \cdot x \overset{(PU)}{=} (y^{-1} \cdot x)^{-1} \cdot (1 \cdot x) \overset{(m_1-BB=)}{=} 1 \cdot (y^{-1})^{-1} \overset{(DN),(PU)}{=} y$; thus, ($m_1$-D=) holds.

(m1A6): If $x = y$, then $x \cdot y^{-1} = x \cdot x^{-1} \overset{(m_1-Re)}{=} 1$. Conversely, if $x \cdot y^{-1} = 1$, then $y \overset{(m_1-D=)}{=} (y^{-1} \cdot x)^{-1} \cdot x \overset{(Pcomm)}{=} (x \cdot y^{-1})^{-1} \cdot x = 1^{-1} \cdot x \overset{(N1)}{=} 1 \cdot x \overset{(PU)}{=} x$. Thus, (EqP=) holds.

(m1A7): By (m1A4), (PU) + (Pcomm) + (N1) + ($m_1$-BB=) imply ($m_1$-Re), by (m1A5), (PU) + (DN) + ($m_1$-BB=) imply ($m_1$-D=), and by (m1A6), (PU) + (Pcomm) + (N1) + ($m_1$-Re) + ($m_1$-D=) imply (EqP=); thus, ($m_1$-Re), ($m_1$-D=) and (EqP=) hold. Then,
$1 \overset{(m_1-Re)}{=} y \cdot y^{-1} \overset{(m_1-D=)}{=} [(y^{-1} \cdot x)^{-1} \cdot x] \cdot y^{-1}$
$\overset{(DN)}{=} ([(y^{-1} \cdot x)^{-1} \cdot x]^{-1})^{-1} \cdot y^{-1}$
$\overset{(Pcomm)}{=} y^{-1} \cdot ([(y^{-1} \cdot x)^{-1} \cdot x]^{-1})^{-1}$
$\overset{(m_1-BB=)}{=} ([(y^{-1} \cdot x)^{-1} \cdot x]^{-1} \cdot (z \cdot x))^{-1} \cdot (y^{-1} \cdot (z \cdot x))$
$\overset{(m_1-BB=)}{=} (z \cdot ((y^{-1} \cdot x)^{-1})^{-1})^{-1} \cdot (y^{-1} \cdot (z \cdot x))$
$\overset{(DN)}{=} (z \cdot (y^{-1} \cdot x))^{-1} \cdot (y^{-1} \cdot (z \cdot x))$
$\overset{(Pcomm)}{=} (y^{-1} \cdot (z \cdot x)) \cdot (z \cdot (y^{-1} \cdot x))^{-1}.$
Then, by (EqP=), $y^{-1} \cdot (z \cdot x) = z \cdot (y^{-1} \cdot x)$, hence, by (Pcomm), $y^{-1} \cdot (x \cdot z) = (y^{-1} \cdot x) \cdot z$, hence $y \cdot (x \cdot z) = (y \cdot x) \cdot z$, by (DN); thus, (Pass) holds.

(m1A8): Suppose $x \leq_{m_1} y$, i.e. $x \cdot y^{-1} = 1$; we must prove that $x = y$.

Indeed, by ($m_1$-B), $[(x \cdot y^{-1})^{-1} \cdot (x \cdot z)] \cdot (y \cdot z)^{-1} = 1$, hence, we obtain: $[1^{-1} \cdot (x \cdot z)] \cdot (y \cdot z)^{-1} = 1$; then, by (N1) and (PU), we obtain:
(a) $(x \cdot z) \cdot (y \cdot z)^{-1} = 1$; take now $z := x^{-1}$ in (a) to obtain:
$(x \cdot x^{-1}) \cdot (y \cdot x^{-1})^{-1} = 1$, hence, by ($m_1$-Re) and (PU), $(y \cdot x^{-1})^{-1} = 1$; then, by (DN) and (N1), we obtain $y \cdot x^{-1} = 1$, i.e. $y \leq_{m_1} x$; it follows, by ($m_1$-An), that $x = y$; thus, ($m_1$-p-s) holds. □

**Corollary 7.3.6** *Any (involutive) left-$m_1$-BZ algebra is $m_1$-p-semisimple.*

**Proof.** By (m1A8). □

**Proposition 7.3.7** *(See Proposition 4.2.6)*
*We have:*
*(m1AB0) ($m_1$-D=) + (PU) $\Longrightarrow$ (DN);*
*(m1AB1) (Pass) + (Pcomm) + ($m_1$-Re) + (EqP=) $\Longrightarrow$ ($m_1$-B=);*
*(m1AB7) (PU) + (Pcomm) + ($m_1$-B=) $\Longrightarrow$ ($m_1$-D=),*
*(m1AB7') (PU) + (Pcomm) + ($m_1$-B=) $\Longrightarrow$ (Pass),*
*(m1AB7") (PU) + (Pcomm) + ($m_1$-B=) $\Longrightarrow$ ($m_1$-BB=).*

**Proof.** (m1AB0): In ($m_1$-D=) ($(y^{-1} \cdot x)^{-1} \cdot x = y$) take $x = 1$; we obtain, by (PU), $(y^{-1})^{-1} = y$, i.e. (DN) holds.

(m1AB1): By (m1A1), (Pass) + (Pcomm) + ($m_1$-Re) + (EqP=) imply ($m_1$-D=), hence $(y^{-1} \cdot x)^{-1} \cdot x = y$; then, $[(y^{-1} \cdot x)^{-1} \cdot x] \cdot z = y \cdot z$, hence $(y^{-1} \cdot x)^{-1} \cdot (x \cdot z) = y \cdot z$, by (Pass); it follows, by (Pcomm), that $(x \cdot y^{-1})^{-1} \cdot (x \cdot z) = y \cdot z$, i.e. ($m_1$-B=) holds.

(m1AB7): By ($m_1$-B=), $(x \cdot y^{-1})^{-1} \cdot (x \cdot z) = y \cdot z$; then, for $z = 1$, we obtain $(x \cdot y^{-1})^{-1} \cdot (x \cdot 1) = y \cdot 1$, hence $(y^{-1} \cdot x)^{-1} \cdot x = y$, by (PU), (Pcomm); thus, ($m_1$-D=) holds.

(m1AB7'): By (m1AB7), (PU) + (Pcomm) + ($m_1$-B=) imply ($m_1$-D=), and by (m1AB0), ($m_1$-D=) + (PU) imply (DN). Then,
$(y \cdot z) \cdot x \overset{(Pcomm)}{=} x \cdot (y \cdot z) \overset{(m_1-B=)}{=} (X \cdot x^{-1})^{-1} \cdot (X \cdot (y \cdot z))$ and for $X = (y \cdot x^{-1})^{-1}$, we obtain:

$$(y \cdot z) \cdot x = ((y \cdot x^{-1})^{-1} \cdot x^{-1})^{-1} \cdot ((y \cdot x^{-1})^{-1} \cdot (y \cdot z))$$
$$\overset{(m_1-B=)}{=} ((y \cdot x^{-1})^{-1} \cdot x^{-1})^{-1} \cdot (x \cdot z)$$
$$\overset{(DN)}{=} (((y^{-1})^{-1} \cdot x^{-1})^{-1}) \cdot x^{-1})^{-1} \cdot (x \cdot z)$$
$$\overset{(m_1-D=)}{=} (y^{-1})^{-1} \cdot (x \cdot z) \overset{(DN)}{=} y \cdot (x \cdot z) \overset{(Pcomm)}{=} y \cdot (z \cdot x),$$ i.e. (Pass) holds.

(m1AB7"): By (m1AB7), (PU) + (Pcomm) + ($m_1$-B=) imply ($m_1$-D=), and by (m1AB0), ($m_1$-D=) + (PU) imply (DN); by Proposition 7.3.3 (5), (Pcomm) + (DN) + ($m_1$-B=) imply ($m_1$-BB=); thus, ($m_1$-BB=) holds. □

**Theorem 7.3.8** *(See Theorem 6.1.4) (See Proposition 4.2.4)*
*Let $\mathcal{A}^L = (A^L, \cdot, ^{-1}, 1)$ be an algebra of type $(2, 1, 0)$. If (PU), (Pcomm), (Pass), ($m_1$-Re) hold, then*

$$(m_1 - B =) \Longleftrightarrow (m_1 - BB =).$$

**Proof.** By (m1A0), (Pass) + (Pcomm) + (PU) + ($m_1$-Re) imply (EqP=), hence (EqP=) holds; then, by (m1A2"), (Pass) + (Pcomm) + (EqP=) $\Longrightarrow$ (($m_1$-B=) $\Leftrightarrow$ ($m_1$-BB=)). □

By above Propositions 7.3.3 and 7.3.5 we obtain:

**Theorem 7.3.9** *Any $m_1$-CI algebra (i.e. any commutative group) is $m_1$-p-semisimple (= sharp) and involutive.*

**Proof.** Let $\mathcal{A}^L = (A^L, \cdot, ^{-1}, 1)$ be a left-$m_1$-CI algebra, i.e. (N1), (PU), (Pcomm), (Pass), ($m_1$-Re) hold. Then, by (m1A0), (Pass) + (Pcomm) + (PU) + ($m_1$-Re) $\Longrightarrow$ (EqP=), hence (EqP=) holds, hence ($m_1$-p-s) holds, by Proposition 7.3.3 (0'); by Proposition 7.3.3 (3), (EqP=) + (Pcomm) + ($m_1$-Re) $\Longrightarrow$ (DN), hence (DN) holds too. □

We write this as follows:

$$\mathbf{m_1\text{-}CI} = \mathbf{m_1\text{-}CI}_{(m_1-p-s)} = \mathbf{m_1\text{-}CI}_{(DN)} = \mathbf{c.\ group}.$$

**Corollary 7.3.10** *The pre-$m_1$-BCI, $m_1$-BCH, $m_1$-BCI algebras and $m_1$-CI*, $m_1$-BCH* algebras are $m_1$-p-semisimple and involutive.*

**Proof.** All these algebras are $m_1$-CI algebras (=commutative groups), by definition; then apply Theorem 7.3.9. □

**Corollary 7.3.11** *(i) The $m_1$-CI algebras coincide with the $m_1$-CI* algebras.*
*(ii) The $m_1$-BCH algebras coincide with the $m_1$-BCH* algebras.*

**Proof.** (i): Let $\mathcal{A}^L$ be a left-$m_1$-CI algebra; by Theorem 7.3.9, ($m_1$-p-s) holds; then, by Proposition 7.3.3 (1'), ($m_1$-*) holds; hence, $\mathcal{A}^L$ is a left-$m_1$-CI* algebra. Since, conversely, any left-$m_1$-CI* algebra is a left-$m_1$-CI algebra, it follows that they coincide.

(ii): Let $\mathcal{A}^L$ be a left-$m_1$-BCH algebra; by Corollary 7.3.10, ($m_1$-p-s) holds; then, by Proposition 7.3.3 (1'), ($m_1$-*) holds; hence, $\mathcal{A}^L$ is a left-$m_1$-BCH* algebra. Since, conversely, any left-$m_1$-BCH* algebra is a left-$m_1$-BCH algebra, it follows that they coincide. □

Moreover, we have the following expected result.

**Proposition 7.3.12** *Any $m_1$-CI algebra (= commutative group) is an $m_1$-BCI algebra.*

**Proof.** By Corollary 7.3.10, any $m_1$-BCI algebra is $m_1$-p-semisimple; by Proposition 7.3.3 (2'), ($m_1$-p-s) + ($m_1$-B') imply ($m_1$-B=).

Let now $\mathcal{A}^L$ be an $m_1$-CI algebra, i.e. (N1), (PU), (Pcomm), (Pass), ($m_1$-Re) hold. We must prove that ($m_1$-An) and ($m_1$-B=) hold. Indeed, by Theorem 7.3.9, ($m_1$-p-s) holds; by Proposition 7.3.3 (1), ($m_1$-p-s) implies ($m_1$-An'), hence ($m_1$-An) holds. By Proposition 7.3.3 (0), ($m_1$-dfrelP) + ($m_1$-p-s) $\Longrightarrow$ (EqP=), hence (EqP=) holds; by (m1AB1), (Pass) + (Pcomm) + ($m_1$-Re) + (EqP=) $\Longrightarrow$ ($m_1$-B=); thus,

($m_1$-B=) holds too. Hence, $\mathcal{A}^L$ is an ($m_1$-p-semisimple) $m_1$-BCI algebra. □

Since, conversely, any $m_1$-BCI algebra is an $m_1$-CI algebra, by definition, we obtain:

**Corollary 7.3.13** *The ($m_1$-p-semisimple and involutive) left-$m_1$-CI, left-$m_1$-CI*, left-pre-$m_1$-BCI, left-$m_1$-BCH, left-$m_1$-BCH* and left-$m_1$-BCI algebras coincide (with the commutative groups).*

We write this as follows, for short:
**$m_1$-CI=$m_1$-CI*=pre-$m_1$-BCI=$m_1$-BCH=$m_1$-BCH*=$m_1$-BCI=c. group.**

By Corollary 7.3.13, it follows that the *algebras without associativity* (or *non-associative algebras*) defined in this section (the left-$m_1$-RM, left-pre-$m_1$-BZ, left-$m_1$-BH, left-$m_1$-BZ algebras) are all (non-associative) generalizations of the commutative group.

**Proposition 7.3.14** *(See Proposition 4.2.15)*
*(i) The $m_1$-p-semisimple $m_1$-RM algebras (= commutative moons) coincide with the $m_1$-p-semisimple $m_1$-BH algebras.*
*(ii) The $m_1$-p-semisimple pre-$m_1$-BZ algebras coincide with the $m_1$-p-semisimple $m_1$-BZ algebras.*

**Proof.** By Proposition 7.3.3 (1). □

**Proposition 7.3.15** *(See Proposition 4.2.20)*
*Any $m_1$-p-semisimple non-associative algebra is involutive.*

**Proof.** By Proposition 7.3.3 (0) and (3). □

The converse of Proposition 7.3.15 does not hold: there are involutive non-associative algebras that are not $m_1$-p-semisimple (see the Example (2m) below).

**Remark 7.3.16** *Note that the $m_1$-p-semisimple (hence involutive) $m_1$-BH algebra is just the commutative goop (defined in [116] Definition 10.1.25, see also Theorem 10.1.28).*

**Examples 7.3.17** (See Examples 4.2.21)
(1m) The algebra $\mathcal{A}_1 = (\{a, b, c, 1\}, \cdot_1, ^{-1_1}, 1)$, with

| $\cdot_1$ | a | b | c | 1 |
|---|---|---|---|---|
| a | 1 | 1 | 1 | a |
| b | 1 | a | a | b |
| c | 1 | a | a | c |
| 1 | a | b | c | 1 |

and

| $x$ | $x^{-1_1}$ |
|---|---|
| a | a |
| b | a |
| c | a |
| 1 | 1 |

,

is a **$m_1$-RM algebra (i.e. a commutative moon) that is not involutive.** (Pass) is not verified for $(a, a, b)$; ($m_1$-B) is not verified for $(a, b, b)$; ($m_1$-BB) is not verified for $(b, a, b)$; ($m_1$-An) for $(a, b)$.

(2m) The algebra $\mathcal{A}_2 = (\{a,b,c,1\}, \cdot_2, ^{-1_2}, 1)$, with

| $\cdot_2$ | a | b | c | 1 |
|---|---|---|---|---|
| a | a | 1 | 1 | a |
| b | 1 | a | 1 | b |
| c | 1 | 1 | 1 | c |
| 1 | a | b | c | 1 |

and

| $x$ | $x^{-1_2}$ |
|---|---|
| a | b |
| b | a |
| c | c |
| 1 | 1 |

,

is an **involutive m$_1$-RM algebra (i.e. an involutive commutative moon) that is not m$_1$-p-semisimple**. (Pass) is not verified for $(a,a,b)$; (m$_1$-B) is not verified for $(a,b,b)$; (m$_1$-BB) is not verified for $(a,a,c)$; (m$_1$-An) for $(a,b)$; (EqP=) $\Longleftrightarrow$ (m$_1$-p-s) for $(b,c)$: $b \neq c$ but $b \cdot {}_2c^{-1} = b \cdot {}_2c = 1$. Note that we can apply Theorem 17.2.1 and we obtain the involutive RM algebra $(\{a,b,c,1\}, \to, 1)$, with

| $\to$ | a | b | c | 1 |
|---|---|---|---|---|
| a | 1 | b | 1 | b |
| b | b | 1 | 1 | a |
| c | 1 | 1 | 1 | c |
| 1 | a | b | c | 1 |

and $x^{-1_2} \overset{def.}{=} x \to 1$, since (Re), (M), (DN), and (Ex1) hold; it does not verify (Ex) for $(a,b,a)$, (B) for $(a,b,a)$, (BB) for $(a,a,b)$, (*) for $(a,b,c)$, (**) for $(a,b,c)$, (Tr) for $(a,c,b)$, (An) for $(a,c)$.

(3m) The algebra $\mathcal{A}_3 = (\{a,b,c,1\}, \cdot_3, ^{-1_3}, 1)$, with

| $\cdot_3$ | a | b | c | 1 |
|---|---|---|---|---|
| a | a | 1 | a | a |
| b | 1 | a | b | b |
| c | a | b | 1 | c |
| 1 | a | b | c | 1 |

and

| $x$ | $x^{-1_3}$ |
|---|---|
| a | b |
| b | a |
| c | c |
| 1 | 1 |

,

is a **m$_1$-p-semisimple (involutive) m$_1$-BH algebra (i.e. a commutative goop)**. (Pass) is not verified for $(a,a,b)$; (m$_1$-B) is not verified for $(a,b,b)$; (m$_1$-BB) is not verified for $(a,a,c)$. Note that we can apply Theorem 17.2.1 and we obtain the algebra from Examples 4.2.21 (3).

The connections between the left-m$_1$-RM algebras (commutative moons), the involutive left-m$_1$-RM algebras (involutive moons) and the m$_1$-p-semisimple left-m$_1$-BH algebras (commutative goops) are presented in the following Figure 7.1 (see [116], Figure 10.4, in the non-commutative case).

**Theorem 7.3.18** *The m$_1$-p-semisimple (involutive) m$_1$-BZ algebras coincide with the m$_1$-BCI algebras (i.e. with the commutative groups).*

**Proof.** Let $\mathcal{A}^L = (A^L, \cdot, ^{-1}, 1)$ be a m$_1$-p-semisimple (involut ive) left-m$_1$-BZ algebra, i.e. (PU), (Pcomm), (m$_1$-Re), (m$_1$-An), (m$_1$-B),((DN)), (m$_1$-p-s) hold. We prove that (Pass) holds. Indeed, by Proposition 7.3.3 (2'), (m$_1$-p-s) + (m$_1$-B) imply (m$_1$-B=). Now, by (m1AB7'), (PU) + (Pcomm) + (m$_1$-B=) imply (Pass). Thus, $\mathcal{A}^L$ is a (m$_1$-p-semisimple) left-m$_1$-BCI algebra. Since the converse implication is obvious, the proof is complete. □

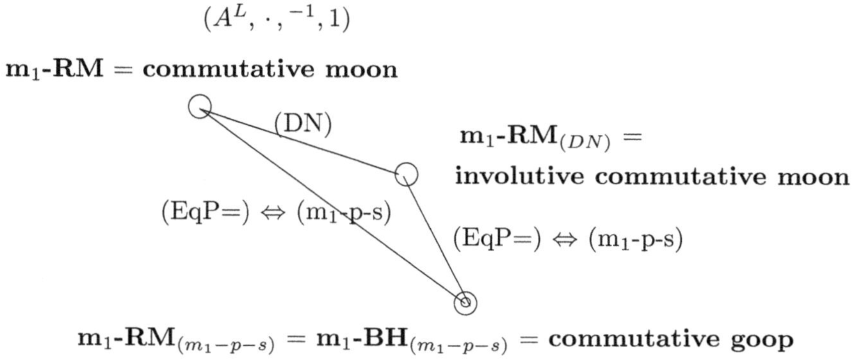


Figure 7.1: The hierarchy between commutative moons, involutive commutative moons and commutative goops

Hence, by Corollary 7.3.6, we have:

$$\mathbf{m_1 - BZ} \;=\; \mathbf{m_1 - BZ}_{(DN)} \;=\; \mathbf{m_1 - BZ}_{(m_1-p-s)} \;=\; \mathbf{c.\ group}.$$

All left-$m_1$-M algebras defined in this chapter are connected (following their definitions and the above results) as in the *Hierarchy $m_1$-1* from Figure 8.1 and *Hierarchy $m_1$-1'* from Figure 8.2 (the "Big m-map"), in Chapter 8.

# Chapter 8

# The "Big m-map" - final connections between the algebras defined in Chapters 6, 7

In this chapter, we establish the global hierarchies of the new algebras, containing Hierarchies $m_1$-1 and m-2, $m_1$-1' and m-2'.

The content of this chapter is taken mainly from [118].

With analogous definitions for $\leq_m$ and $\leq_{m_1}$ as in Definitions 1.1.19 and with the same representations as in Remark 1.1.21, the final connections between the subclasses of unital commutative magmas defined in Chapters 6, 7 are presented in the "Big m-map" from Figures 8.1 and 8.2.

Other m-M algebras can be defined following the other algebras defined in [114].

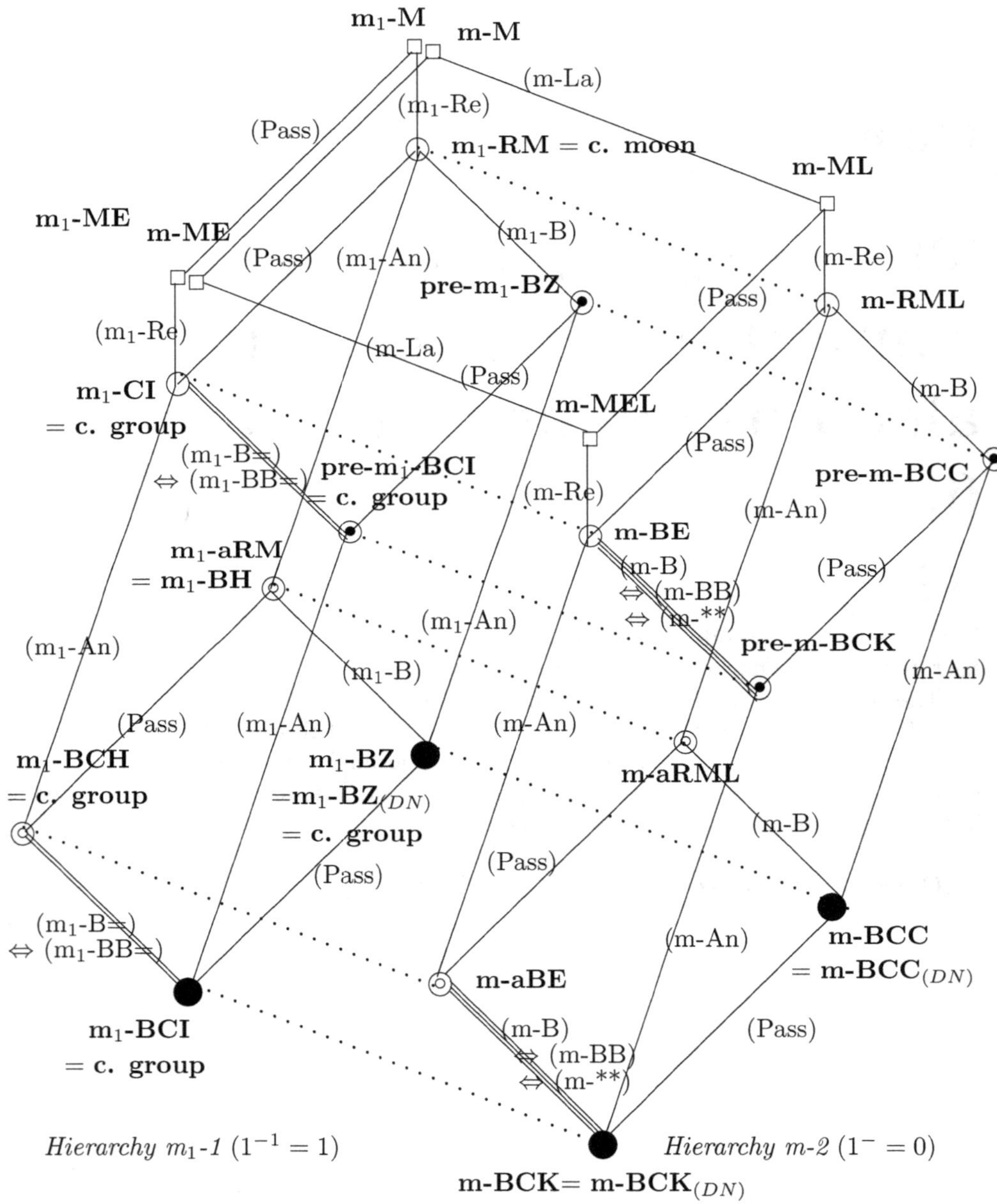


Figure 8.1: The "Big m-map": Hierarchies $m_1$-1 and m-2 of unital commutative magmas (**uc. magma**) with additional operation

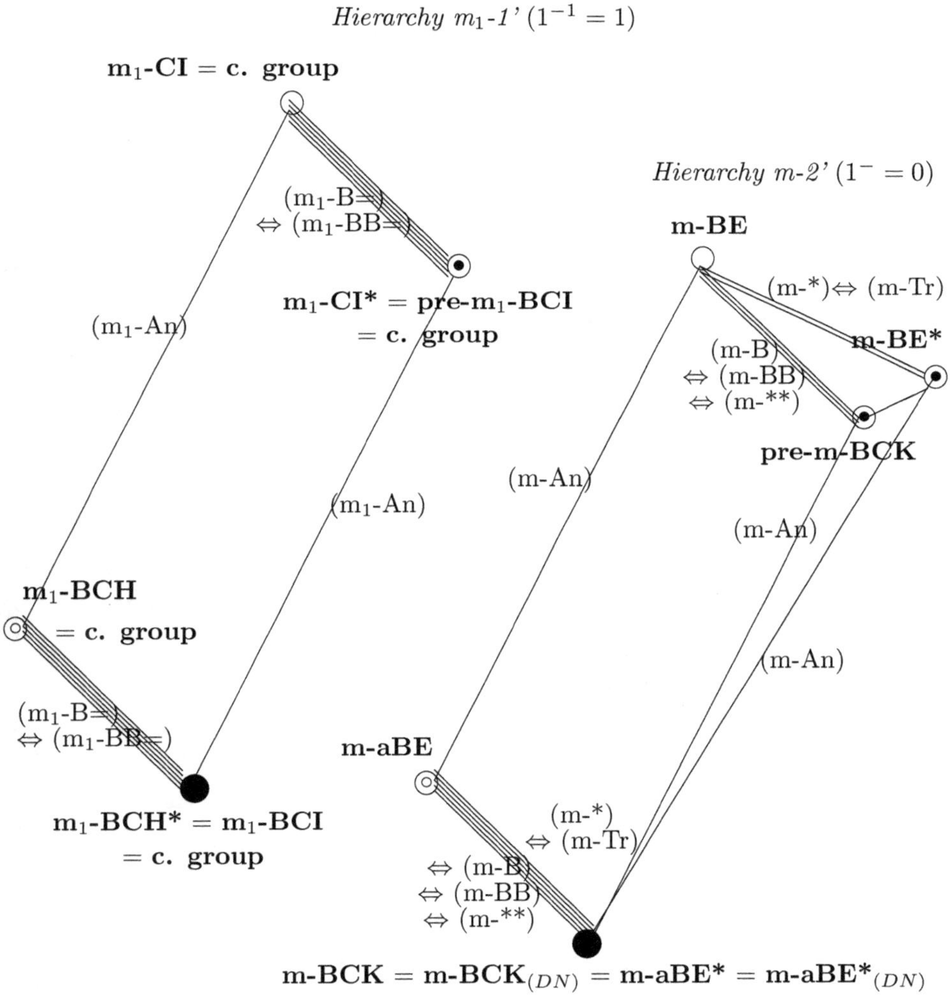


Figure 8.2: The "Big m-map": Hierarchies $m_1$-1' and m-2', determined by $m_1$-CI algebras (= commutative groups) and m-BE algebras, respectively

# Chapter 9

# Putting bounded involutive lattices, De Morgan algebras, ortholattices and Boolean algebras on the "map"

In this chapter, we continue the results from Chapter 6 in the "world" of *m-M algebras* verifying (Pass).

In Section 9.1, we introduce three new algebras (m-tMEL, m-aMEL and m-taMEL algebras), thus extending the initial "Big m-map" from Chapter 8 with a "Little m-map", and study the involutive case. We redefine the bounded involutive lattices (**BIL**) and the De Morgan algebras (**De Morgan**) as involutive m-MEL algebras. Finally, we put **BIL** and **De Morgan**, and their subclasses, on the "involutive Little m-map".

In Section 9.2, we redefine the ortholattices (**OL**) and the Boolean algebras (**Boole**) as involutive m-BE algebras. Finally, we put **OL**, and its subclass, and **Boole** on the "involutive Little m-map".

In Section 9.3, we present 18 examples of the involved algebras.

The content of this chapter is taken from [122] and [118].

## 9.1 Involutive m-MEL algebras

We need, for the purpose of this chapter, to extend the "Big m-map" (Figures 8.1 and 8.2) with subclasses of the class of m-MEL algebras not verifying (m-Re). Since, by (mA3), (m-BB)+(PU)+(Pcomm)+(Neg1-0)+(Neg0-1) imply (m-Re), it follows that the property (m-BB) cannot be verified in such a m-MEL algebra. By (m-A11), (Pcomm)+(Pass) imply ((m-B) $\Longleftrightarrow$ (m-BB)), so (m-B) also cannot be verified in such a m-MEL algebra. It remains to consider the properties (m-Tr) and (m-An). So, we introduce the following new algebras:

**Definitions 9.1.1** *(The dual ones are omitted)*

- *A* transitive left-m-MEL algebra, *or a* left-m-tMEL algebra *for short, is a left-m-MEL algebra verifying (m-Tr).*

- *An* antisymmetric left-m-MEL algebra, *or a* left-m-aMEL algebra *for short, is a left-m-MEL algebra verifying (m-An).*

- *A* transitive and antisymmetric left-m-MEL algebra, *or a* left-m-taMEL algebra *for short, is a left-m-MEL algebra verifying (m-An) and (m-Tr).*

Let **m-tMEL**, **m-aMEL** and **m-taMEL** denote their classes, respectively.

Then, these new classes are connected with those from the "Big m-map" by the following "Little m-map" from next Figure 9.1; the resulting "extended Big m-map" is not drawn anymore, being too complicated.

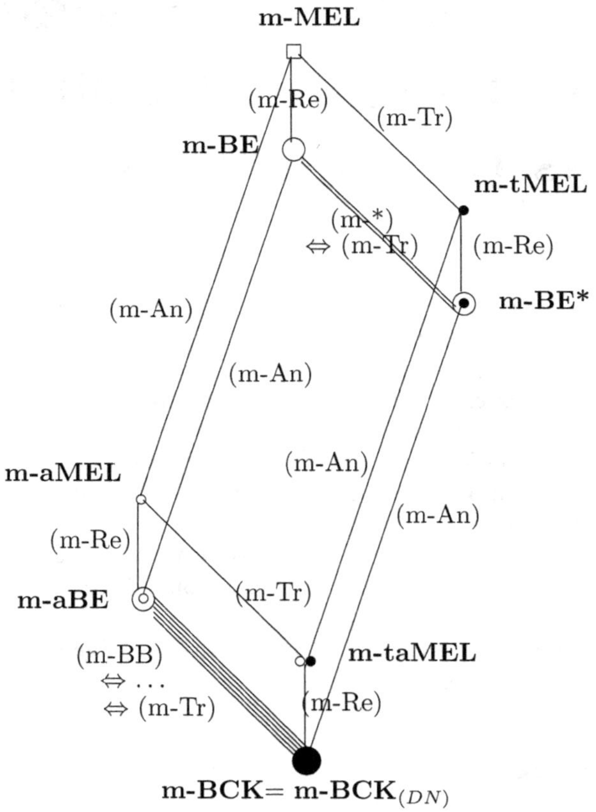


Figure 9.1: The "Little m-map": additional hierarchy to the "Big m-map"

**Definitions 9.1.2** *(The dual case is omitted)*

- *An* involutive left-m-tMEL algebra *is a left-m-tMEL algebra verifying (DN) or, equivalently, an involutive left-m-MEL algebra verifying (m-Tr).*

- *An* involutive left-m-aMEL algebra *is a left-m-aMEL algebra verifying (DN) or, equivalently, an involutive left-m-MEL algebra verifying (m-An).*

- *An* involutive left-m-taMEL algebra *is a left-m-taMEL algebra verifying (DN) or, equivalently, an involutive left-m-MEL algebra verifying (m-An) and (m-Tr).*

We shall denote by $\mathbf{m\text{-}tMEL}_{(DN)}$, $\mathbf{m\text{-}aMEL}_{(DN)}$, $\mathbf{m\text{-}taMEL}_{(DN)}$ their corresponding classes.

Since in an involutive m-BE algebra we have, by Theorem 6.3.5, the equivalences: (m-BB) $\Longleftrightarrow$ (m-B) $\Longleftrightarrow$ (m-**) $\Longleftrightarrow$ (m-*) $\Longleftrightarrow$ (m-Tr), it follows that $\mathbf{m\text{-}pre\text{-}BCK}_{(DN)} = \mathbf{m\text{-}BE^*}_{(DN)}$ and, consequently, we have the hierarchies from next Figure 9.2, the "involutive Little m-map".

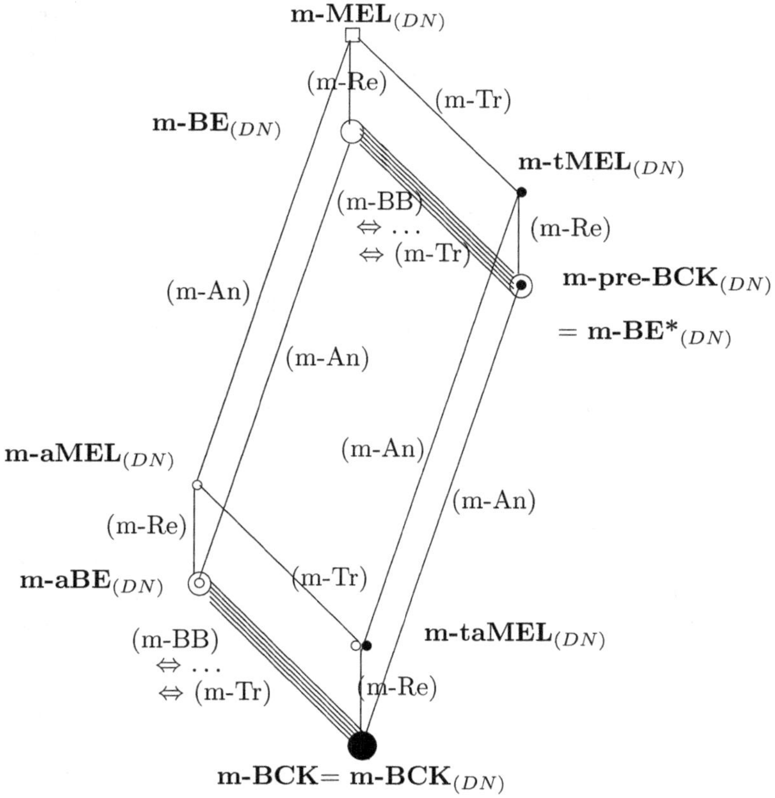


Figure 9.2: The "involutive Little m-map"

See Examples 9.3.1, 9.3.2, 9.3.3, 9.3.4 of proper involutive left-m-MEL, m-tMEL, m-aMEL, m-taMEL algebra, respectively.

**Proposition 9.1.3** *Let $\mathcal{A}^L = (A^L, \odot, ^-, 1)$ be an involutive left-m-MEL algebra. Define, for all $x, y \in A^L$,*

$$x \wedge y \stackrel{def.}{=} x \odot y, \quad x \vee y \stackrel{def.}{=} x \oplus y, \quad 0 \stackrel{def.}{=} 1^-,$$

*where $x \oplus y \stackrel{def.}{=} (x^- \odot y^-)^-$, by (6.5).*

*Then, the algebra $(A^L, \wedge, \vee, ^-, 0, 1)$ satisfies the following properties:*

| | | | |
|---|---|---|---|
| *(m-Wcomm)* | $x \wedge y = y \wedge x,$ | *(m-Vcomm)* | $x \vee y = y \vee x,$ |
| *(m-Wass)* | $x \wedge (y \wedge z) = (x \wedge y) \wedge z,$ | *(m-Vass)* | $x \vee (y \vee z) = (x \vee y) \vee z,$ |
| *(m-WU)* | $1 \wedge x = x,$ | *(m-VU)* | $0 \vee x = x,$ |
| *(m-WL)* | $0 \wedge x = 0,$ | *(m-VL)* | $1 \vee x = 1$ |

*and the unary operation $^-$ satisfies (DN), (DeM1), (DeM2).*

**Proof.** (m-Wcomm): $x \wedge y = x \odot y \stackrel{(Pcomm)}{=} y \odot x = y \wedge x$.
(m-Wass): $(x \wedge y) \wedge z = (x \odot y) \odot z \stackrel{(Pass)}{=} x \odot (y \odot z) = x \wedge (y \wedge z)$.
(m-WU): $1 \wedge x = 1 \odot x \stackrel{(PU)}{=} x$.
(m-WL): $0 \wedge x = 0 \odot x \stackrel{(Pcomm),(La)}{=} 0$.
(m-Vcomm): $x \vee y = x \oplus y = (x^- \odot y^-)^- \stackrel{(Pcomm)}{=} (y^- \odot x^-)^- = y \oplus x = y \vee x$.
( mVass): $(x \vee y) \vee z = (x \oplus y) \oplus z = (x^- \odot y^-)^- \oplus z \stackrel{(DN)}{=} [(x^- \odot y^-) \odot z^-]^- \stackrel{(Pass)}{=}$ $[x^- \odot (y^- \odot z^-)]^- \stackrel{(DN)}{=} [x^- \odot (y \oplus z)^-]^- = x \oplus (y \oplus z) = x \vee (y \vee z)$.
(m-VU): $0 \vee x = 0 \oplus x = (0^- \odot x^-)^- \stackrel{(Neg0-1)}{=} (1 \odot x^-)^- \stackrel{(PU)}{=} (x^-)^- \stackrel{(DN)}{=} x$.
(m-VL): $1 \oplus x = 1$.

The unary operation $^-$ satisfies (DN), by hypothesis.
(DeM1): $(x \vee y)^- = (x \oplus y)^- = x^- \odot y^- = x^- \wedge y^-$;
(DeM2): $(x \wedge y)^- = (x \odot y)^- = x^- \oplus y^- = x^- \vee y^-$. □

**Remark 9.1.4** *Let $\mathcal{A}^L = (A^L, \odot, ^-, 1)$ be an involutive left-m-MEL algebra. If (G) holds, then ($G^R$) holds too, by Proposition 6.4.4; then, by definitions from Proposition 9.1.3, the properties (m-Wid) and (m-Vid) hold, but (m-Wabs) and (m-Vabs) still do not hold, so the order relation $\leq^P_m$, by Proposition 6.4.10, is not a lattice order.*

**Proposition 9.1.5** *Let $\mathcal{A}^L = (A^L, \odot, ^-, 1)$ be an involutive left-m-MEL algebra. If (m-Pimpl) holds, then:*
*(1) the order relation $\leq^P_m$ is a lattice order (denoted by $\leq^O_m$, in the signature $\wedge$, $\vee$),*
*(2) $x \leq^P_m y \Longleftrightarrow y \geq^S_m x$,*
*(3) $x \leq^P_m y \Longrightarrow y^- \leq^P_m x^-$.*

**Proof.** (1): If (m-Pimpl) holds, then (G) holds, by (mCIM4), and (m-Pabs) holds, by (mCIM6); hence, ($G^R$ and (m-Sabs) holds, by Proposition 6.4.4. Then, by Proposition 9.1.3, the algebra $(A^L, \wedge, \vee)$ verifies also (m-Wid), (m-Vid) and (m-Wabs), (m-Vabs), hence is a lattice; consequently, the binary relation $\leq^P_m$ verifies:

$$x \leq_m^P y \overset{def.}{\Longleftrightarrow} x \odot y = x \Longleftrightarrow x \wedge y = x \overset{def.}{\Longleftrightarrow} x \leq_m^O y,$$

hence is a lattice order.

(2): Suppose that $x \leq_m^P y$, i.e. $x \odot y = x$; then, $y \oplus x = y \oplus (x \odot y) \overset{(Pcomm),(m-Sabs)}{=} y$, i.e. $y \geq_m^S x$. Conversely, suppose that $y \geq_m^S x$, i.e. $y \oplus x = y$; then, $x \odot y = x \odot (y \oplus x) \overset{(Scomm),(m-Pabs)}{=} x$, i.e. $x \leq_m^P y$.

(3): Suppose $x \leq_m^P y$, i.e. $x \odot y = x$; then, $y^- \odot x^- = y^- \odot (x \odot y)^- \overset{(6.5)}{=} y^- \odot (x^- \oplus y^-) \overset{(Scomm)}{=} y^- \odot (y^- \oplus x^-) \overset{(m-Pabs)}{=} y^-$, i.e. $y^- \leq_m^P x^-$. □

### 9.1.1 Connections between involutive m-MEL algebras and bounded involutive lattices (BIL)

Recall first the following definitions:

**Definitions 9.1.6** (Definition 1) (See [38] for an equivalent definition)

(i) A *bounded involutive (or involution) left-lattice*, or a *left-BIL* for short, is an algebra $\mathcal{A}^L = (A^L, \wedge, \vee, ^- = ^{-^L}, 0, 1)$ such that the reduct $(A^L, \wedge, \vee, 0, 1)$ is a bounded Dedekind left-lattice and the unary operation $^-$ (called *involution* or *generalized complement*) satisfies the following conditions:

(DN) $(x^-)^- = x$ or $x^= = x$ (Double Negation),

(m-O-Neg2') $x \leq_m^O y$ implies $y^- \leq_m^O x^-$ (contraposition),

where $x \leq_m^O y \overset{def.}{\Longleftrightarrow} x \wedge y = x$ ($\Longleftrightarrow x \vee y = y$) is the Ore left-lattice order relation.

(i') Dually, a *bounded involutive (or involution) right-lattice*, or a *right-BIL* for short, is an algebra $\mathcal{A}^R = (A^R, \vee, \wedge, ^- = ^{-^R}, 0, 1)$ such that the reduct $(A^R, \vee, \wedge, 0, 1)$ is a bounded Dedekind right-lattice and the unary operation $^-$ satisfies the following conditions:

(DN) $(x^-)^- = x$ (Double Negation),

(m-O-Neg2' $^R$) $x \geq_m^O y$ implies $y^- \geq_m^O x^-$ (contraposition),

where $x \geq_m^O y \overset{def.}{\Longleftrightarrow} x \vee y = x$ ($\Longleftrightarrow x \wedge y = y$) is the Ore right-lattice order relation.

We shall denote by **BIL** the class of all left-BILs and by **BIL**$^R$ the class of all right-BILs.

**Proposition 9.1.7** *Let $\mathcal{A}^L = (A^L, \wedge, \vee, ^-, 0, 1)$ be a left-BIL. Define, for all $x, y \in A^L$,*

*(m-dfOW)* $x \leq_m^O y \overset{def.}{\Longleftrightarrow} x \wedge y = x$ *or, equivalently,*

*(m-dfOV)* $x \leq_m^O y \overset{def.}{\Longleftrightarrow} x \vee y = y$.

*Then, $(A^L, \leq_m^O, 0, 1)$ is a bounded Ore left-lattice.*

**Proposition 9.1.8** *[38] Let $\mathcal{A}^L = (A^L, \wedge, \vee, ^-, 0, 1)$ be a left-BIL. The following are equivalent:*

*(m-O-Neg2')* $x \leq_m^O y \Longrightarrow y^- \leq_m^O x^-$ *(contraposition),*

*(DeM1)* $(x \vee y)^- = x^- \wedge y^-$ *(De Morgan law 1),*

*(DeM2)* $(x \wedge y)^- = x^- \vee y^-$ *(De Morgan law 2).*

By Proposition 9.1.8, we obtain the following equivalent definition:

**Definition 9.1.9** (Definition 1') (The dual one is omitted)

A *bounded involutive left-lattice*, or a *left-BIL* for short, is an algebra $\mathcal{A}^L = (A^L, \wedge, \vee, ^- = ^{-^L}, 0, 1)$ such that the reduct $(A^L, \wedge, \vee, 0, 1)$ is a bounded Dedekind left-lattice and the unary operation $^-$ satisfies (DN) and (DeM1) or, equivalently, (DeM2).

We have the following definitional equivalence:

**Theorem 9.1.10**

*(1) Let $\mathcal{A}^L = (A^L, \odot, ^-, 1)$ be an involutive left-m-MEL algebra verifying (m-Pimpl). Define, for all $x, y \in A^L$,*

$$x \wedge y \stackrel{def.}{=} x \odot y, \quad x \vee y \stackrel{def.}{=} x \oplus y, \quad 0 \stackrel{def.}{=} 1^-,$$

*where $x \oplus y \stackrel{def.}{=} (x^- \odot y^-)^-$, by (6.5).*
*Then, $f(\mathcal{A}^L) = (A^L, \wedge, \vee, ^-, 0, 1)$ is a left-BIL.*

*(1') Let $\mathcal{A}^L = (A^L, \wedge, \vee, ^-, 0, 1)$ be a left-BIL. Define, for all $x, y \in A^L$,*

$$x \odot y \stackrel{def.}{=} x \wedge y.$$

*Then, $g(\mathcal{A}^L) = (A^L, \odot, ^-, 1)$ is an involutive left-m-MEL algebra verifying (m-Pimpl).*

*(2) The mappings $f$ and $g$ are mutually inverse.*

**Proof.** (1): Let $\mathcal{A}^L = (A^L, \odot, ^-, 1)$ be an involutive left-m-MEL algebra verifying (m-Pimpl). Then, Proposition 9.1.3 holds; by (mCIM6), (mCIM4), (m-Pabs) and (G) hold; by Proposition 6.4.4, (m-Sabs) and ($G^R$) hold too; hence, (m-Wabs), (m-Vabs), (m-Wid), (m-Vid) hold, respectively; so, $(A^L, \wedge, \vee, 0, 1)$ is a bounded Dedekind left-lattice and $^-$ satisfy (DN), (DeM1) and (DeM2). By Proposition 9.1.5, the binary relation $\leq_m^P$ is a lattice order, denoted by $\leq_m^O$, and $x \leq_m^O y$ implies $y^- \leq_m^O x^-$, i.e. (m-O-Neg2') holds. Thus, $(A^L, \wedge, \vee, ^-, 0, 1)$ is a bounded involutive left-lattice.

(1'): Let $\mathcal{A}^L = (A^L, \wedge, \vee, ^-, 0, 1)$ be a bounded involutive left-lattice. Define $x \odot y \stackrel{def.}{=} x \wedge y$. Then, (PU) (=(m-WU)), (Pcomm) (=(m-Wcomm)), (Pass) (= (m-Wass)), (DN), (m-Pabs) (= (m-Wabs)) hold. By (mCIM7), (m-Pimpl) holds. It remains to prove that (m-La) holds; indeed, $0 \stackrel{(m-Vabs)}{=} 0 \vee (0 \wedge x) \stackrel{(m-VU)}{=} 0 \wedge x \stackrel{(m-Wcomm)}{=} x \wedge 0$, hence $x \odot 0 = 0$, i.e. (m-La) holds. Thus, $(A^L, \odot, ^-, 1)$ is an involutive left-m-MEL algebra verifying (m-Pimpl).

(2): Routine. □

Theorem 9.1.10 allows us to give a new, equivalent definition of bounded involutive lattices, as follows:

**Definition 9.1.11** *(Definition 2) (The dual one is omitted)*

*A* bounded involutive left-lattice, *or a* left-BIL *for short, is an (involutive) left-m-MEL algebra verifying (m-Pimpl).*

**Remark 9.1.12** *The bounded involutive left-lattices (Definition 2) may verify or not the property (m-An), i.e. they are either m-aMEL algebras or m-MEL algebras, respectively, as the examples given in Section 9.3 show.*

We shall denote by **tBIL** the class of all left-BILs verifying (m-Tr), by **aBIL** the class of all left-BILs verifying (m-An) and by **taBIL** the class of all left-BILs verifying (m-An) and (m-Tr).

See Examples 9.3.5, 9.3.6, 9.3.7, 9.3.8 of proper left-BIL, tBIL, aBIL, taBIL, respectively.

### 9.1.2 Connections between involutive m-MEL algebras and De Morgan algebras (De Morgan)

Recall first the following definitions:

**Definitions 9.1.13** [38] (Definition 1)

(i) A *left-De Morgan algebra* is an algebra $\mathcal{A}^L = (A^L, \wedge, \vee, ^- = ^{-^L}, 1)$ such that, if $0 \stackrel{def.}{=} 1^-$, then the reduct $(A^L, \wedge, \vee, 0, 1)$ is a bounded Dedekind left-lattice that is distributive, i.e. it verifies (m-Wdis) or, equivalently, (m-Vdis), and the unary operation $^-$ satisfies (DN), (DeM2).

(i') Dually, a *right-De Morgan algebra* is an algebra $\mathcal{A}^R = (A^R, \vee, \wedge, ^{-^R}, 0)$ such that, if $1 \stackrel{def.}{=} 0^{-^R}$, then the reduct $(A^R, \vee, \wedge, 0, 1)$ is a bounded Dedekind right-lattice that is distributive and the unary operation $^{-^R}$ satisfies (DN), (DeM1).

We shall denote by **De Morgan** the class of all left-De Morgan algebras.

We have the following definitional equivalence:

**Theorem 9.1.14** *(See Theorem 9.1.10)*

*(1) Let $\mathcal{A}^L = (A^L, \odot, ^-, 1)$ be an involutive left-m-MEL algebra verifying (m-Pdis). Define, for all $x, y \in A^L$,*

$$x \wedge y \stackrel{def.}{=} x \odot y, \quad x \vee y \stackrel{def.}{=} x \oplus y, \quad 0 \stackrel{def.}{=} 1^-,$$

*where $x \oplus y \stackrel{def.}{=} (x^- \odot y^-)^-$, by (6.5).*
*Then, $f(\mathcal{A}^L) = (A^L, \wedge, \vee, ^-, 0, 1)$ is a left-De Morgan algebra.*

*(1') Let $\mathcal{A}^L = (A^L, \wedge, \vee, ^-, 0, 1)$ be a left-De Morgan algebra. Define, for all $x, y \in A^L$,*

$$x \odot y \stackrel{def.}{=} x \wedge y.$$

*Then, $g(\mathcal{A}^L) = (A^L, \odot, ^-, 1)$ is an involutive left-m-MEL algebra verifying (m-Pdis).*

*(2) The mappings $f$ and $g$ are mutually inverse.*

**Proof.** (1): Let $\mathcal{A}^L = (A^L, \odot, ^-, 1)$ be an involutive left-m-MEL algebra verifying (m-Pdis). By Proposition 6.4.6, (m-Pdis) implies (m-Pimpl), hence $\mathcal{A}^L$ verifies (m-Pimpl); then, by Theorem 9.1.10 (1), $(A^L, \wedge, \vee, ^-, 0, 1)$ is a bounded involutive left-lattice which verifies (m-Wdis); hence, it is a left-De Morgan algebra.

(1'): Let $\mathcal{A}^L = (A^L, \wedge, \vee, ^-, 0, 1)$ be a left-De Morgan algebra; hence, $\mathcal{A}^L$ is a bounded involutive left-lattice verifying (m-Wdis). Define $x \odot y \stackrel{def.}{=} x \wedge y$. Then, $(A^L, \odot, ^-, 1)$ is an involutive left-m-MEL algebra verifying (m-Pimpl), by Theorem 9.1.10 (1'), and verifying (m-Pdis). Thus, $(A^L, \odot, ^-, 1)$ is an involutive left-m-MEL algebra verifying (m-Pdis).

(2): Routine. □

The above Theorem 9.1.14 allows us to give a new, equivalent definition of De Morgan algebras, as follows:

**Definition 9.1.15** *(Definition 2) (The dual one is omitted)*
*A* left-De Morgan algebra *is a (involutve) left-MEL algebra verifying (m-Pdis).*

**Remark 9.1.16** *The left-De Morgan algebras (Definition 2) may verify or not the property (m-An), i.e. they are either m-aMEL algebras or m-MEL algebras, respectively, as the examples given in Section 9.3 show.*

We shall denote by **a-De Morgan** the class of all left-De Morgan algebras verifying (m-An) and by **ta-De Morgan** the class of all left-De Morgan algebras verifying (m-An) and (m-Tr).

**Remark 9.1.17** *We wanted to denote by* **t-De Morgan** *the class of all left-De Morgan algebras verifying (m-Tr). But, proper* t-De Morgan algebras *do not exist, as the following important Theorem 9.1.20 says: a* t-De Morgan algebra *is in fact a* ta-De Morgan algebra.

**Lemma 9.1.18** *Let $\mathcal{A}^L = (A^L, \odot, ^-, 1)$ be an involutive left-m-MEL algebra satisfying (m-Pdis). Assume $b \odot b^- = 0$ for some $b \in A^L$. Then, for all $a \in A^L$,*

$$(a \leq_m b \overset{def.}{\Longleftrightarrow}) \quad a \odot b^- = 0 \quad \Longleftrightarrow \quad a \odot b = a \quad (\overset{def.}{\Longleftrightarrow} a \leq_m^P b).$$

**Proof.** (By PROVER9) First, suppose $a \odot b^- = 0$. Then,

$$a \overset{(PU)}{=} a \odot 1 \overset{(Neg0-1)}{=} a \odot 0^- = a \odot (b \odot b^-)^- = a \odot (b^- \oplus b)$$
$$\overset{(m-Pdis)}{=} (a \odot b^-) \oplus (a \odot b) = 0 \oplus (a \odot b) \overset{(SU)}{=} a \odot b\,.$$

Conversely, suppose $a \odot b = a$. Then $a \odot b^- = (a \odot b) \odot b^- = a \odot (b \odot b^-) = a \odot 0 = 0$, as claimed. □

**Lemma 9.1.19** *Let $\mathcal{A}^L = (A^L, \odot, ^-, 1)$ be an involutive left-m-MEL algebra satisfying (m-Pdis), and assume that (m-Re) or (m-Tr) holds. Then, (m-An) holds.*

**Proof.** Suppose $a \leq_m b$ and $b \leq_m a$, i.e. $a \odot b^- = 0 = b \odot a^-$, for some $a, b \in A$. Since (m-Re) or (m-Tr) hold, we obtain $a \odot a^- = b \odot b^- = 0$. By Lemma 9.1.18, $a \odot b = a$ and also $b \odot a = b$. By (Pcomm), $a = b$. This proves (m-An). □

**Theorem 9.1.20** *Let $\mathcal{A}^L = (A^L, \odot, ^-, 1)$ be an involutive left-m-MEL algebra. Then,*

$$(m - Pdis) \;+\; (m - Tr) \;\Longrightarrow\; (m - An)\,.$$

**Proof.** By Lemma 9.1.19. □

Note that Theorem 9.1.20 says that: **t-De Morgan** = **ta-De Morgan**.

See Examples 9.3.9, 9.3.10, 9.3.11 of proper left-De Morgan, a-De Morgan, ta-De Morgan algebra, respectively.

### 9.1.3 Putting BIL and De Morgan on the "involutive Little m-map"

The definitions (Definition 2) and the results from this section allow us to draw the hierarchies from the following Figure 9.3, thus putting the mentioned algebras on the "involutive Little m-map".

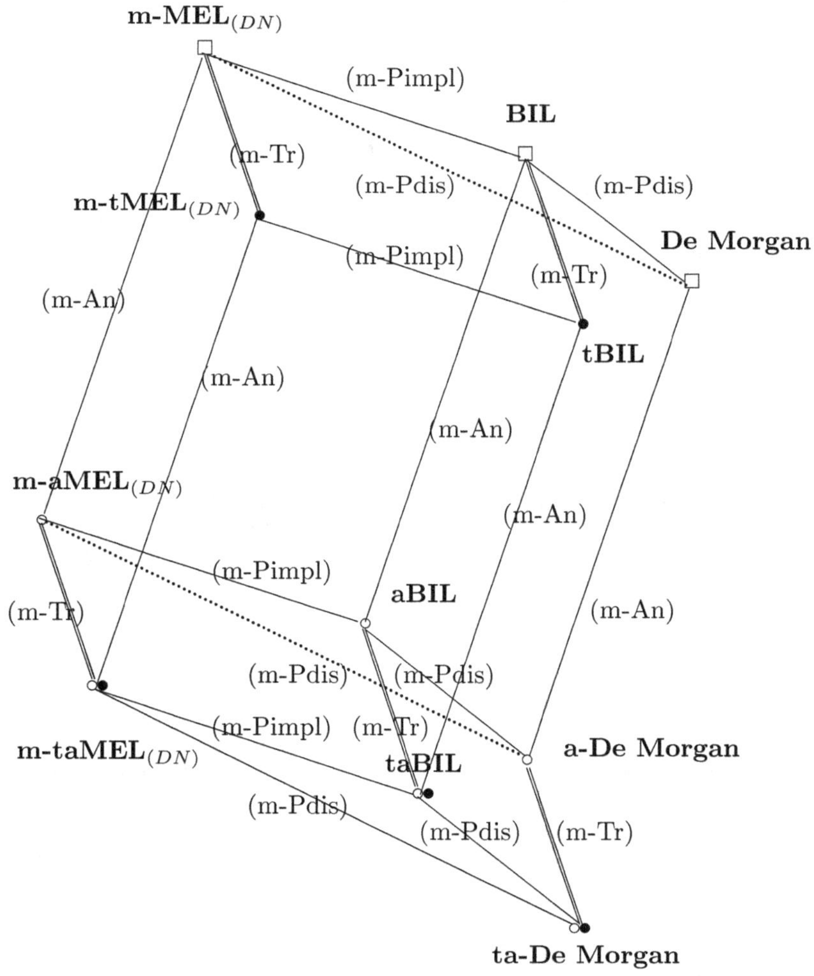


Figure 9.3: Hierarchies **BIL**, **De Morgan** vs. (m-Tr)

## 9.2 Involutive m-BE algebras

**Definitions 9.2.1**

(i) An *involutive left-m-aBE algebra* is an involutive left-m-aMEL algebra verifying additionally (m-Re) ($x \odot x^- = 0$, i.e. $x \leq_m x$).

(i') An *involutive right-m-aBE algebra* is an involutive right-m-aMEL algebra verifying additionally (m-Re$^R$) ($x \oplus x^- = 1$, i.e. $x \geq_m x$).

We shall denote by $\mathbf{m\text{-}aBE}_{(DN)}$ the class of all involutive left-m-aBE algebras and by $\mathbf{m\text{-}aBE}^R_{(DN)}$ the class of all involutive right-m-aBE algebras.

We shall denote by $\mathbf{m\text{-}pre\text{-}BCK}_{(DN)}$ = $\mathbf{pre\text{-}m\text{-}BCK}_{(DN)}$ (= $\mathbf{m\text{-}tBE}_{(DN)}$) the class of all involutive left-m-BE algebras verifying, by Theorem 6.3.5:

(m-BB) $\Longleftrightarrow$ (m-B) $\Longleftrightarrow$ (m-**) $\Longleftrightarrow$ (m-*) $\Longleftrightarrow$ (m-Tr).

Recall that $\mathbf{m\text{-}BCK}$ = $\mathbf{m\text{-}BCK}_{(DN)}$ (= $\mathbf{m\text{-}taBE}_{(DN)}$) is just the class of all involutive m-aBE algebras verifying, by Theorem 6.3.5:

(m-BB) $\Longleftrightarrow$ (m-B) $\Longleftrightarrow$ (m-**) $\Longleftrightarrow$ (m-*) $\Longleftrightarrow$ (m-Tr).

See Examples 9.3.12, 9.3.13, 9.3.14, 9.3.15 of proper involutive left- m-BE, m-pre-BCK, m-aBE, m-BCK algebra, respectively.

**Proposition 9.2.2** *(See Proposition 9.1.3)*

*Let $\mathcal{A}^L = (A^L, \odot, ^-, 1)$ be an involutive left-m-BE algebra. Define, for all $x, y \in A^L$,*

$$x \wedge y \stackrel{def.}{=} x \odot y, \quad x \vee y \stackrel{def.}{=} x \oplus y, \quad 0 \stackrel{def.}{=} 1^-,$$

*where $x \oplus y \stackrel{def.}{=} (x^- \odot y^-)^-$, by (6.5).*

*Then, the algebra $(A^L, \wedge, \vee, ^-, 0, 1)$ satisfies the following properties:*

| | | | |
|---|---|---|---|
| *(m-Wcomm)* | $x \wedge y = y \wedge x$, | *(m-Vcomm)* | $x \vee y = y \vee x$, |
| *(m-Wass)* | $x \wedge (y \wedge z) = (x \wedge y) \wedge z$, | *(m-Vass)* | $x \vee (y \vee z) = (x \vee y) \vee z$, |
| *(m-WU)* | $1 \wedge x = x$, | *(m-VU)* | $0 \vee x = x$, |
| *(m-WL)* | $0 \wedge x = 0$, | *(m-VL)* | $1 \vee x = 1$ |

*and the unary operation $^-$ satisfies (DN), (DeM1), (DeM2), (m-WRe) and (m-VRe).*

**Proof.** By Proposition 9.1.3, it remains to prove that (m-WRe) and (m-VRe) hold. Indeed,

(m-WRe): $x \wedge x^- = x \odot x^- \stackrel{(m-Re)}{=} 0$,
(m-VRe): $x \vee x^- = x \oplus x^- = 1$. □

### 9.2.1 Connections between involutive m-BE algebras and ortholattices (OL)

Recall first the following definition.

**Definitions 9.2.3** [38] A *(left-) bounded involutive poset*, or a *bounded involution poset*, is a structure $(A, \leq, ^-, 0, 1)$, where:

(i) $(A, \leq, 0, 1)$ is a bounded poset,
(ii) $^-$ is a unary operation (called *involution* or *generalizaed complement*) that satisfies the following conditions:
(DN) $(x^-)^- = x$ (Double Negation),
(Neg2') $x \leq y$ implies $y^- \leq x^-$ (contraposition).

A *bounded involutive lattice* is a bounded involutive poset that is also a lattice.

**Definitions 9.2.4** (Definition 1) (see [166]) (see also [38])

(i) A *left-ortholattice*, or a *left-OL* for short, is an algebra $\mathcal{A}^L = (A^L, \wedge, \vee, ^- = ^{-^L}, 0, 1)$ such that the reduct $(A^L, \wedge, \vee, 0, 1)$ is a bounded Dedekind lattice, i.e. the following properties hold:

| | | | |
|---|---|---|---|
| (m-Wid) | $x \wedge x = x,$ | (m-Vid) | $x \vee x = x,$ |
| (m-Wcomm) | $x \wedge y = y \wedge x,$ | (m-Vcomm) | $x \vee y = y \vee x,$ |
| (m-Wass) | $x \wedge (y \wedge z)$ $= (x \wedge y) \wedge z,$ | (m-Vass) | $x \vee (y \vee z)$ $= (x \vee y) \vee z,$ |
| (m-Wabs) | $x \wedge (x \vee y) = x,$ | (m-Vabs) | $x \vee (x \wedge y) = x;$ |
| (m-WU) | $1 \wedge x = x,$ | (m-VU) | $0 \vee x = x;$ |

and the unary operation $^-$ satisfies the law of Double Negation, the De Morgan laws and the complementation laws, where:
(DN) $(x^-)^- = x;$
(DeM1) $(x \vee y)^- = x^- \wedge y^-$ (De Morgan law 1),
(DeM2) $(x \wedge y)^- = x^- \vee y^-$ (De Morgan law 2);
(m-WRe) $x \wedge x^- = 0$ (noncontradiction principle),
(m-VRe) $x \vee x^- = 1$ (excluded middle principle).

(i') A *right-ortholattice*, or a *right-OL* for short, is an algebra $\mathcal{A}^R = (A^R, \vee, \wedge, ^- = ^{-^R}, 0, 1)$ such that the reduct $(A^R, \vee, \wedge, 0, 1)$ is a bounded Dedekind lattice and the unary operation $^-$ satisfies the law of Double Negation, the dual De Morgan laws and the dual complementation laws.

Denote by **OL** the class of all left-ortholattices and by $\mathbf{OL}^R$ the class of all right-ortholattices.

Note that a left-ortholattice verifies (DN) (it is *involutive*), hence it is self-dual, i.e. the dual of $(A^L, \wedge, \vee, ^{-^L}, 0, 1)$ is $(A^L, \vee, \wedge, ^{-^L}, 0, 1)$, and vice-versa.

Dually, a right-ortholattice verifies (DN) (it is *involutive*), hence it is self-dual, i.e. the dual of $(A^R, \vee, \wedge, ^{-^R}, 0, 1)$ is $(A^R, \wedge, \vee, ^{-^R}, 0, 1)$, and vice-versa.

Note [166] that it suffices to postulate only one of the De Morgan laws. For instance, if (DeM1) holds, then $(x \wedge y)^- \overset{(DN)}{=} ((x^-)^- \wedge (y^-)^-)^- \overset{(DeM1)}{=} ((x^- \vee y^-)^-)^- \overset{(DN)}{=} x^- \vee y^-$, i.e. (DeM2) holds too.

Roughly speaking, one could say that an ortholattice is a Boolean algebra without distributivity.

**Proposition 9.2.5** *Let* $\mathcal{A}^L = (A^L, \wedge, \vee, ^-, 0, 1)$ *be a left-ortholattice. Define, for all* $x, y \in A^L$, *(m-dfOW)* $x \leq_m^O y \stackrel{def.}{\Longleftrightarrow} x \wedge y = x$ *or, equivalently,*

*(m-dfOV)* $x \leq_m^O y \stackrel{def.}{\Longleftrightarrow} x \vee y = y$.

*Then,* $(A^L, \leq_m^O, 0, 1)$ *is a bounded Ore lattice.*

**Proof.** Obviously, by Theorem 5.2.3, since $(A^L, \wedge, \vee, 0, 1)$ is a bounded Dedekind lattice. □

**Proposition 9.2.6** *[38] Let* $\mathcal{A}^L = (A^L, \wedge, \vee, ^-, 0, 1)$ *be a left-ortholattice. The following are equivalent:*

*(m-O-Neg2')* $x \leq_m^O y \Longrightarrow y^- \leq_m^O x^-$,

*(DeM1)* $(x \vee y)^- = x^- \wedge y^-$,

*(DeM2)* $(x \wedge y)^- = x^- \vee y^-$.

**Remarks 9.2.7** *Let* $\mathcal{A}^L = (A^L, \wedge, \vee, ^-, 0, 1)$ *be a left-ortholattice. Note that:*
*(i) the Dedekind lattice operations* $\wedge$ *and* $\vee$ *are connected as follows: for al* $x, y \in A^L$,

$$x \vee y = (x^- \wedge y^-)^-, \quad x \wedge y = (x^- \vee y^-)^-;$$

*(ii)* $(A^L, \leq_m^O, ^-, 0, 1)$ *is a bounded involutive lattice.*

**Proposition 9.2.8** *Let* $\mathcal{A}^L = (A^L, \wedge, \vee, ^-, 0, 1)$ *be a left-ortholattice. Then, we have:*

*(Neg1-0)* $1^- = 0$,

*(Neg0-1)* $0^- = 1$,

*(m-Wimpl)* $[(x \wedge y^-)^- \wedge x^-]^- = x$,

*(m-WL)* $x \wedge 0 = 0$,

*(m-O-** ')* *If* $x \leq_m^O y$*, then* $x \wedge a \leq_m^O y \wedge a$.

**Proof.** (Neg1-0): $1^- \stackrel{(m-WU)}{=} 1 \wedge 1^- \stackrel{(m-WRe)}{=} 0$.

(Neg0-1): By (mCDN1).

(m-Wimpl): $[(x \wedge y^-)^- \wedge x^-]^- \stackrel{(DeM2)}{=} [(x^- \vee (y^-)^-) \wedge x^-]^- \stackrel{(m-Wcomm),(DN)}{=} [x^- \wedge (x^- \vee y)]^- \stackrel{(m-Wabs)}{=} [x^-]^- \stackrel{(DN)}{=} x$.

(m-WL): $x \wedge 0 = 0 \stackrel{(m-Wcomm)}{\Longleftrightarrow} 0 \wedge x = 0 \stackrel{(m-dfOW)}{\Longleftrightarrow} 0 \leq_m^O x$, that is true by Proposition 9.2.5; thus, (m-WL) holds.

(m-O-**'): Suppose $x \leq_m^O y$, i.e. $x \wedge y = x$; then, $(x \wedge a) \wedge (y \wedge a) \stackrel{(m-Wcomm),(m-Wass)}{=} (x \wedge y) \wedge (a \wedge a) \stackrel{(m-Wid)}{=} x \wedge a$, i.e. $x \wedge a \leq_m^O y \wedge a$; thus, (m-O-** ') holds. □

Consider the following binary relation:

(m-dfrelW) $x \leq_m y \stackrel{def.}{\Longleftrightarrow} x \wedge y^- = 0$.

Note that $\leq_m$ is reflexive, by (m-WRe), but it is not antisymmetric or transitive, as the following examples show.

**Examples 9.2.9** (See Examples 3.4.7)

Examples of ortholattices with six elements.

(m1) The algebra $\mathcal{O}_1 = (A_6 = \{0, a, b, c, d, 1\}, \wedge_1, \vee_1, ^{-1}, 0, 1)$ with

| $\wedge_1$ | 0 | a | b | c | d | 1 |
|---|---|---|---|---|---|---|
| 0 | 0 | 0 | 0 | 0 | 0 | 0 |
| a | 0 | a | 0 | 0 | 0 | a |
| b | 0 | 0 | b | 0 | 0 | b |
| c | 0 | 0 | 0 | c | 0 | c |
| d | 0 | 0 | 0 | 0 | d | d |
| 1 | 0 | a | b | c | d | 1 |

,

| $x$ | $x^{-1}$ |
|---|---|
| 0 | 1 |
| a | d |
| b | c |
| c | b |
| d | a |
| 1 | 0 |

and $x \vee_1 y = (x^{-1} \wedge_1 y^{-1})^{-1}$

is an ortholattice with the bounded poset $(A_6, \leq_m^O, 0, 1)$ represented by the Hasse diagram from Figure 9.4. It does not verify (m-WBB) for $(a, a, b)$, (m-WB) for $(a, b, a)$, (m-W**) for $(a, b, a)$, (m-W*) for $(a, b, c)$, (m-WAn) for $(a, b)$, (m-WTr) for $(a, b, d)$, ($\wedge_m$-comm) for $(a, b)$.

(m2) The algebra $\mathcal{O}_2 = (A_6 = \{0, a, b, c, d, 1\}, \wedge_2 = \wedge_1, \vee_2, ^{-2}, 0, 1)$, with $(0, a, b, c, d, 1)^{-2} = (1, b, a, d, c, 0)$ and $x \vee_2 y = (x^{-2} \wedge_2 y^{-2})^{-2}$, is an ortholattice with the bounded poset $(A_6, \leq_m^O, 0, 1)$ represented by the same Hasse diagram from Figure 9.4. It does not verify (m-WBB) for $(a, a, c)$, (m-WB) for $(a, c, a)$, (m-W**) for $(a, c, a)$, (m-W*) for $(a, c, d)$, (m-WAn) for $(a, c)$, (m-WTr) for $(a, c, b)$, ($\wedge_m$-comm) for $(a, c)$.

Note that the ortholattices $\mathcal{O}_1$ and $\mathcal{O}_2$ correspond to the i-ortholattices $\mathcal{IO}_1$ and $\mathcal{IO}_2$, respectively, from Examples 3.4.7, by Corollary 17.1.6.

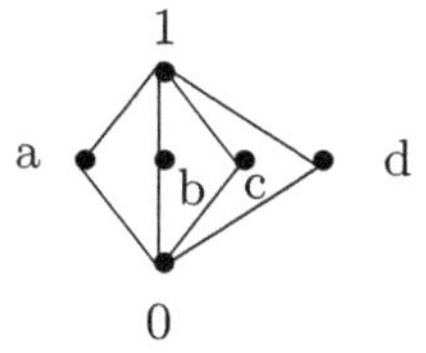


Figure 9.4: An ortholattice with 6 elements

**Proposition 9.2.10** *Let $\mathcal{A}^L = (A^L, \wedge, \vee, ^-, 0, 1)$ be a left-ortholattice. Then, we have:*

*(1)* $\qquad x \leq_m^O y \Longrightarrow x \leq_m y.$

*(2) If (m-@)=(m-Wdiv) $(x \wedge (x \wedge y^-)^- = x \wedge y)$ holds, then:*

*(m-EqOW)* $\qquad x \leq_m^O y \Longleftrightarrow x \leq_m y.$

**Proof.** (1): Suppose $x \leq_m^O y$, i.e. $x \wedge y = x$, by (m-dfOW); then, $x \wedge y^- = (x \wedge y) \wedge y^- \overset{(m-Wass)}{=} x \wedge (y \wedge y^-) \overset{(m-WRe)}{=} x \wedge 0 \overset{(m-WL)}{=} 0$; thus, $x \leq_m y$.

(2): If $x \leq_m y$, i.e. $x \wedge y^- = 0$, then $x \wedge y \overset{(m-Wdiv)}{=} x \wedge (x \wedge y^-)^- = x \wedge 0^- \overset{(Neg0-1)}{=} x \wedge 1 \overset{(m-Wcomm)}{=} 1 \wedge x \overset{(m-WU)}{=} x$, i.e. $x \leq_m^O y$, by (m-dfOW) again. Thus, $x \leq_m^O y \Longleftrightarrow x \leq_m y$, by above (1), i.e. (m-EqOW) holds. □

**Remark 9.2.11** *The converse of Proposition 9.2.10 (1) does not hold. For example, in the ortholattice $\mathcal{O}_1$ from Examples 9.2.9, we have $a \leq_m b$, i.e. $a \wedge_1 b^{-1} = a \wedge_1 c = 0$, but $a \not\leq_m^O b$.*

Consider the following equivalent (dual) properties of distributivity:
(m-Wdis) (distributivity of $\wedge$ over $\vee$) $\quad z \wedge (x \vee y) = (z \wedge x) \vee (z \wedge y)$
and, dually,
(m-Vdis) (distributivity of $\vee$ over $\wedge$) $\quad z \vee (x \wedge y) = (z \vee x) \wedge (z \vee y)$.

Then, we have the following important result (the dual one is omitted):

**Theorem 9.2.12** *Let $\mathcal{A}^L = (A^L, \wedge, \vee, ^-, 0, 1)$ be a left-ortholattice. The following are equivalent:*
*(m-@)=(m-Wdiv)* $\quad x \wedge (x \wedge y^-)^- = x \wedge y$;
*(m-Wdis)* $\quad z \wedge (x \vee y) = (z \wedge x) \vee (z \wedge y)$.

**Proof.** (m-Wdiv) $\Longrightarrow$ (m-Wdis): By Proposition 9.2.10 (2), (m-EqOW) holds.
We prove first that:

$$(z \wedge x) \vee (z \wedge y) \leq_m^O z \wedge (x \vee y). \tag{9.1}$$

Indeed, by (m-O-Vgeq'), $x,\ y \leq_m^O x \vee y$; then, by (m-Wcomm), (m-O-**), $z \wedge x,\ z \wedge y \leq_m^O z \wedge (x \vee y)$, i.e. $z \wedge (x \vee y)$ is a majorant of $\{z \wedge x, z \wedge y\}$; then, $(z \wedge x) \vee (z \wedge y) \leq_m^O z \wedge (x \vee y)$, i.e. (9.1) holds.
Now we prove that:

$$z \wedge (x \vee y) \leq_m^O (z \wedge x) \vee (z \wedge y). \tag{9.2}$$

Indeed, first: $x \vee y \overset{(m-O-Vgeq')}{\leq_m^O} z^- \vee (x \vee y) \overset{(DeM1),(DN)}{=} (z \wedge x^- \wedge y^-)^-$ $\overset{(m-Wid)}{=} (z \wedge z \wedge x^- \wedge y^-)^- \overset{(m-Wcomm),(m-Wass)}{=} ((z \wedge x^-) \wedge (z \wedge y^-))^-$ $\overset{(DN),(m-Wdiv)}{=} [(z \wedge (z \wedge x)^-) \wedge (z \wedge (z \wedge y)^-)]^-$ $\overset{(m-Wcomm),(m-Wass),(m-Wid)}{=} [z \wedge (z \wedge x)^- \wedge (z \wedge y)^-]^-$.
Then, by (m-EqOW), (DN), we obtain: $0 = (x \vee y) \wedge [z \wedge (z \wedge x)^- \wedge (z \wedge y)^-]$ $\overset{(m-Wass)}{=} [(x \vee y) \wedge z] \wedge [(z \wedge x)^- \wedge (z \wedge y)^-] \overset{(DeM1)}{=} [(x \vee y) \wedge z] \wedge [(z \wedge x) \vee (z \wedge y)]^-$.
Then, by (m-EqOW) again, $z \wedge (x \vee y) \overset{(m-Wcomm)}{=} (x \vee y) \wedge z \leq_m^O (z \wedge x) \vee (z \wedge y)$, i.e. (9.2) holds.
Finally, by (9.1), (9.2) and (m-O-An'), it follows that (m-Wdis) holds.

(m-Wdis) $\Longrightarrow$ (m-Wdiv): First, $x \wedge (x^- \vee y) \overset{(m-Wdis)}{=} (x \wedge x^-) \vee (x \wedge y) \overset{(m-WRe)}{=} 0 \vee (x \wedge y) \overset{(m-VU)}{=} x \wedge y$. Then, $x \wedge y = x \wedge (x^- \vee y) \overset{(DeM1)}{=} x \wedge ((x^-)^- \wedge y^-)^- \overset{(DN)}{=} x \wedge (x \wedge y^-)^-$, i.e. (m-Wdiv) holds. □

- **Connections with ortholattices**

**Proposition 9.2.13** *Let $\mathcal{A}^L = (A^L, \wedge, \vee, ^-, 0, 1)$ be a left-ortholattice. Define for all $x \in A^L$:*

$$x \odot y \stackrel{def.}{=} x \wedge y.$$

*Then, $(A^L, \odot, ^-, 1)$ is a m-Pimplicative involutive left-m-BE algebra.*

**Proof.** By Proposition 9.2.8, (Neg1-0), (Neg01), (m-WL) and (m-Wimpl) hold. Then:
(PU): $1 \odot x = 1 \wedge x \stackrel{(m-WU)}{=} x$.
(Pcomm): $x \odot y = x \wedge y \stackrel{(m-Wcomm)}{=} y \wedge x = y \odot x$.
(Pass): $x \odot (y \odot z) = x \wedge (y \wedge z) \stackrel{(m-Wass)}{=} (x \wedge y) \wedge z = (x \odot y) \odot z$.
(m-La): $x \odot 0 = x \wedge 0 \stackrel{(m-WL)}{=} 0$.
(m-Re): $x \odot x^- = x \wedge x^- \stackrel{(m-WRe)}{=} 0$.
(m-Pimpl): $[(x \odot y^-)^- \odot x^-]^- = [(x \wedge y^-)^- \wedge x^-]^- \stackrel{(m-Wimpl)}{=} x$.
(DN) holds by hypothesis. □

In order to establish the converse result, first note that we have:

**Proposition 9.2.14** *Let $\mathcal{A}^L = (A^L, \odot, ^-, 1)$ be an m-Pimplicative left-m-BE algebra. Then, (DN), (m-Pimpl-1) and (G) hold.*

**Proof.** By (mCIM1) or (mCIM1'), (DN) holds. Then, by (mCIM2), (m-Pimpl-1) holds and by (mCIM3), (G) holds. □

Now, we can establish the converse of Proposition 9.2.13.

**Proposition 9.2.15** *Let $\mathcal{A}^L = (A^L, \odot, ^-, 1)$ be an m-Pimplicative (involutive) left-m-BE algebra. Define, for all $x, y \in A^L$,*

$$x \wedge y \stackrel{def.}{=} x \odot y, \quad x \vee y \stackrel{def.}{=} x \oplus y, \quad 0 \stackrel{def.}{=} 1^-,$$

*where $x \oplus y \stackrel{def.}{=} (x^- \odot y^-)^-$.*

*Then, $(A^L, \wedge, \vee, ^-, 0, 1)$ is a left-ortholattice.*

**Proof.** By Proposition 9.2.14, (DN), (m-Pimpl-1) and (G) hold. By Proposition 9.2.2, it remains to prove (m-Wid), (m-Vid), (m-Wabs), (m-Vabs). Indeed:
(m-Wid): $x \wedge x = x \odot x \stackrel{(G)}{=} x$.
(m-Wabs): $x \wedge (x \vee y) = x \odot (x \oplus y) = x \odot (x^- \odot y^-)^- \stackrel{(m-Pimpl-1)}{=} x$.
(m-Vid): $x \vee x = x \oplus x = (x^- \odot x^-)^- \stackrel{(G)}{=} (x^-)^- \stackrel{(DN)}{=} x$.
(m-Vabs): $x \vee (x \wedge y) = x \oplus (x \odot y) = [x^- \odot (x \odot y)^-]^- \stackrel{(m-Pimpl)}{=} x$. □

The next result says that the class of m-Pimplicative (involutive) m-BE algebras is *definitionally equivalent* (d.e.) to the class of ortholattices (hence their categories are isomorphic).

**Theorem 9.2.16**
*(1) Let $\mathcal{A}^L = (A^L, \odot, ^-, 1)$ be an m-Pimplicative (involutive) left-m-BE algebra. Define*

$$x \wedge y \stackrel{def.}{=} x \odot y, \quad x \vee y \stackrel{def.}{=} x \oplus y, \quad 0 \stackrel{def.}{=} 1^-.$$

*Then,* $f(\mathcal{A}^L) = (A^L, \wedge, \vee, ^-, 0, 1)$ *is a left-ortholattice.*

*(1') Conversely, let* $\mathcal{A}^L = (A^L, \wedge, \vee, ^-, 0, 1)$ *be a left-ortholattice. Define*

$$x \odot y \stackrel{def.}{=} x \wedge y.$$

*Then,* $g(\mathcal{A}^L) = (A^L, \odot, ^-, 1)$ *is an m-Pimplicative (involutive) left-m-BE algebra.*

*(2) The mappings* $f$ *and* $g$ *are mutually inverse.*

**Proof.** By Proposition 9.2.13 and Proposition 9.2.15. □

The above Theorem 9.2.16 helps us to give an equivalent definition of ortholattices, as follows.

**Definitions 9.2.17** (Definition 2)

(i) A *left-ortholattice*, or a *left-OL* for short, is a (involutive) left-m-BE algebra verifying (m-Pimpl).

(i') A *right-ortholattice*, or a *right-OL* for short, is a (involutive) right-m-BE algebra verifying (m-Simpl).

We shall denote by **tOL** the class of left-ortholattices verifying (m-Tr) ($\Longleftrightarrow \ldots$ (m-BB)).

See Examples 9.3.16, 9.3.17 of proper left-OL, tOL, respectively.

**Remark 9.2.18** *We wanted to denote by* ***aOL*** *the class of all left-OLs verifying (m-An). But, proper* aOLs *do not exist, as the following important Theorem 9.2.19 says: an* aOL *is in fact a* Boolean algebra.

We have the following very important result.

**Theorem 9.2.19**

$$\boldsymbol{OL} \; + \; (m - An) \; = \; \boldsymbol{Boole}.$$

**Proof.** Let $(A^L, \wedge, \vee, ^-, 0, 1)$ be a left-ortholattice (Definition 1), hence $(A^L, \odot, ^-, 1)$ is an involutive m-BE algebra verifying (m-Pimpl), by Theorem 9.2.16. By adding (m-An), $(A^L, \odot, ^-, 1)$ verifies (m-Pdis), by Theorem 6.4.15. Then, by Proposition 6.4.4, it verifies also (m-Sdis). By Theorem 9.2.16, $(A^L, \wedge, \vee, ^-, 0, 1)$ is a left-ortholattice verifying (m-Wdis) and (m-Vdis), hence it is a Boolean algebra. For the converse, we follow the reverse way. □

Consequently, we have:

**tOL** + (m-An) = **OL** + (m-Tr) + (m-An)
= **OL** + (m-An) + (m-Tr) = **Boole** + (m-Tr) = **Boole**,
hence **aOL** = **taOL** = **Boole**.

**Remark 9.2.20** *In a left-OL* $\mathcal{A}^L = (A^L, \odot, ^-, 1)$ *(Definition 2):*
*- the initial binary relation,* $\leq_m$ *(*$x \leq_m y \Longleftrightarrow x \odot y^- = 0$*), is only* **reflexive** *((m-Re) holds, by definition of m-BE algebra);*
*- the binary relation* $\leq_m^M$ *(*$x \leq_m^M y \Longleftrightarrow x \wedge_m^M y = x$*) is only* **reflexive and antisymmetric***;*
*- the binary relation* $\leq_m^P$ *(*$x \leq_m^P y \Longleftrightarrow x \odot y = x$*) is a* **lattice order***, with respect to* $\wedge = \odot$, $\vee = \oplus$*, denoted* $\leq_m^O$*, by Proposition 9.1.5.*

### 9.2.2 Connections between involutive m-BE algebras and Boolean algebras (Boole)

Boolean algebras are only commutative. The most used definition is as a *complemented, distributive, bounded (Dedekind) lattice*, namely:

**Definitions 9.2.21** (Definition 1)

(i) A *left-Boolean algebra* is an algebra $\mathcal{A}^L = (A^L, \wedge = \wedge_m, \vee = \vee_m, ^- = {}^{-^L}, 0, 1)$ verifying: for all $x, y, z \in A^L$,

| | | | |
|---|---|---|---|
| (m-Wid) | $x \wedge x = x$, | (m-Vid) | $x \vee x = x$, |
| (m-Wcomm) | $x \wedge y = y \wedge x$, | (m-Vcomm) | $x \vee y = y \vee x$, |
| (m-Wass) | $x \wedge (y \wedge z) = (x \wedge y) \wedge z$, | (m-Vass) | $x \vee (y \vee z) = (x \vee y) \vee z$, |
| (m-Wabs) | $x \wedge (x \vee y) = x$, | (m-Vabs) | $x \vee (x \wedge y) = x$; |
| (m-WU) | $1 \wedge x = x$, | (m-VU) | $0 \vee x = x$, |
| (m-Wdis) | $x \wedge (y \vee z) = (x \wedge y) \vee (x \wedge z)$, | (m-Vdis) | $x \vee (y \wedge z) = (x \vee y) \wedge (x \vee z)$, |
| (m-WRe) | $x \wedge x^- = 0$, | (m-VRe) | $x \vee x^- = 1$. |

(i') Dually, a *right-Boolean algebra* is an algebra $\mathcal{A}^R = (A^R, \vee = \vee_m, \wedge = \wedge_m, ^- = {}^{-^R}, 0, 1)$ verifying: (m-Vid), (m-Vcomm), (m-Vass), (m-Vabs), (m-VU), (m-Vdis), (m-VRe) and, dually, (m-Wid), (m-Wcomm), (m-Wass), (m-Wabs), (m-WU), (m-Wdis), (m-WRe).

Denote by **Boole** the class of all left-Boolean algebras and by **Boole**$^R$ the class of all right-Boolean algebras.

**Proposition 9.2.22** *Let $\mathcal{A}^L = (A^L, \wedge, \vee, ^-, 0, 1)$ be a left-Boolean algebra. The following hold: (DN), (DeM1), (DeM2).*

By the above definitions and considerations, we obtain the connections:

**OL** = **BIL** + (m-WRe) + (m-VRe),
**De Morgan** = **BIL** + (m-Wdis) + (m-Vdis),
**Boole** = **OL** + (m-Wdis) + (m-Vdis) = **De Morgan** + (m-WRe) + (m-VRe).

Note that a left-Boolean algebra verifies (DN) (it is *involutive*), hence it is self-dual, i.e. the dual of $(A^L, \wedge, \vee, {}^{-^L}, 0, 1)$ is $(A^L, \vee, \wedge, {}^{-^L}, 0, 1)$, and vice-versa.

Dually, a right-Boolean algebra verifies (DN) (it is *involutive*), hence it is self-dual, i.e. the dual of $(A^R, \vee, \wedge, {}^{-^R}, 0, 1)$ is $(A^R, \wedge, \vee, {}^{-^R}, 0, 1)$, and vice-versa.

**• Connections between m-BCK, MV algebras, ortholattices and Boolean algebras**

The following result says that the class of m-Pimplicative (involutive) m-BCK algebras is *definitionally equivalent* (d.e.) with the class of Boolean algebras (hence their categories are isomorphic).

**Theorem 9.2.23**
*(1) Let $\mathcal{A}^L = (A^L, \odot, ^-, 1)$ be an m-Pimplicative (involutive) left-m-BCK algebra. Define, for all $x, y \in A^L$:*

$$x \wedge y \stackrel{def.}{=} x \odot y, \quad x \vee y \stackrel{def.}{=} x \oplus y, \quad 0 \stackrel{def.}{=} 1^-.$$

*Then, $f(\mathcal{A}^L) = (A^L, \wedge, \vee, ^-, 0, 1)$ is a left-Boolean algebra.*
*(1') Conversely, let $\mathcal{A}^L = (A^L, \wedge, \vee, ^-, 0, 1)$ be a left-Boolean algebra. Define*

$$x \odot y \stackrel{def.}{=} x \wedge y.$$

*Then, $g(\mathcal{A}^L) = (A^L, \odot, ^-, 1)$ is an m-Pimplicative (involutive) left-m-BCK algebra.*
*(2) The mappings $f$ and $g$ are mutually inverse.*

**Proof** (1): Let $\mathcal{A}^L = (A^L, \odot, ^-, 1)$ be an m-Pimplicative (involutive) left-m-BCK algebra. Then, $\mathcal{A}^L$ is an m-Pimplicative left-m-BE algebra, hence $(A^L, \wedge, \vee, ^-, 0, 1)$ is a left-ortholattice, by Proposition 9.2.15. By (mCBB10), (m-Pimpl) + (PU) + (Pcomm) + (Pass) + (Neg0-1) + (m-Re) + (m-La) + (m-An) + (m-BB) $\Longrightarrow$ (m-Pdiv), thus (m-@)=(m-Wdiv) holds. Then, (m-Wdis) holds, by Theorem 9.2.12. Thus, $f(\mathcal{A}^L)$ is a left-Boolean algebra.

(1'): Let $\mathcal{A}^L = (A^L, \wedge, \vee, ^-, 0, 1)$ be a left-Boolean algebra. We prove that $(A^L, \odot, ^-, 1)$ is an m-Pimplicative (involutive) left-m-BCK algebra. Indeed,
(PU): $1 \odot x = 1 \wedge x \stackrel{(m-WU)}{=} x$.
(Pcomm): $x \odot y = x \wedge y \stackrel{(m-Wcomm)}{=} y \wedge x = y \odot x$.
(Pass): $x \odot (y \odot z) = x \wedge (y \wedge z) \stackrel{(m-Wass)}{=} (x \wedge y) \wedge z = (x \odot y) \odot z$.
(m-La): $x \odot 0 = x \wedge 0 \stackrel{(m-WL)}{=} 0$.
(m-Re): $x \odot x^- = x \wedge x^- \stackrel{(m-WRe)}{=} 0$. Hence, $(A^L, \odot, ^-, 1)$ is until now a left-m-BE algebra.
(DN) holds and $\leq_m$ is an order relation, hence (m-An) holds.
(m-BB): by (m-Wdis), (m-@)=(m-Wdiv) holds, hence (m-Pdiv) holds; then, by (mCBB6), (DN) + (Pcomm) + (Pass) + (m-Re) + (m-La) + (m-Pdiv) $\Longrightarrow$ (m-BB). Hence, $(A^L, \odot, ^-, 1)$ is until now an (involutive) left-m-BCK algebra.
(m-Pimpl): $[(x \odot y^-)^- \odot x^-]^- = [(x \wedge y^-)^- \wedge x^-]^- = x \vee (x \wedge y^-) \stackrel{(m-Wabs)}{=} x$.
Hence, $g(\mathcal{A}^L)$ is an m-Pimplicative (involutive) left-m-BCK algebra.
(2): Routine. □

Recall the well-known result:

**Theorem 9.2.24** *The class of MV algebras of Gödel type (i.e. satisfying the property (G)) is d.e. to the class of Boolean algebras.*

**Theorem 9.2.25** *The class of ortholattices satisfying one of the equivalent conditions (m-@)=(m-Wdiv), (m-Wdis), (m-Vdis) is d.e. to the class of Boolean algebras.*

**Proof.** Obviously. □

**• Connections between involutive m-BE algebras and Boolean algebras**

We have the following definitional equivalence:

**Theorem 9.2.26** *(See Theorem 9.1.14) (See Theorem 9.2.16)*

*(1) Let $\mathcal{A}^L = (A^L, \odot, ^-, 1)$ be a (involutive) left-m-BE algebra verifying (m-Pdiv). Define, for all $x, y \in A^L$:*

$$x \wedge y \overset{def.}{=} x \odot y, \quad x \vee y \overset{def.}{=} x \oplus y, \quad 0 \overset{def.}{=} 1^- .$$

*Then, $f(\mathcal{A}^L) = (A^L, \wedge, \vee, ^-, 0, 1)$ is a left-Boolean algebra.*

*(1') Let $(A^L, \wedge, \vee, ^-, 0, 1)$ be a left-Boolean algebra. Define, for all $x, y \in A^L$:*

$$x \odot y \overset{def.}{=} x \wedge y .$$

*Then, $g(\mathcal{A}^L) = (A^L, \odot, ^-, 1)$ is a (involutive) left-m-BE algebra verifying (m-Pdiv).*

*(2) The mappings $f$ and $g$ are mutually inverse.*

**Proof.** (1): By Proposition 6.4.14, (m-Pimpl) holds, hence $\mathcal{A}^L$ is an involutive left-m-BE algebra verifying (m-Pimpl). Then, by Theorem 9.2.16 (1), $(A^L, \wedge, \vee, ^-, 0, 1)$ is a left-ortholattice and (m-Pdiv) becomes (m-Wdiv); by Theorem 9.2.12, (m-Wdiv) $\Leftrightarrow$ (m-Wdis); hence $(A^L, \wedge, \vee, ^-, 0, 1)$ is a left-ortholattice verifying (m-Wdis), i.e. is a left-Boolean algebra.

(1'): The fact that $\mathcal{A}^L$ is a left-Boolean algebra means that it is a left-ortholattice verifying (m-Wdis), hence verifying (m-Wdiv), by Theorem 9.2.12. Then, by Theorem 9.2.16 (1'), $(A^L, \odot, ^-, 1)$ is a (involutive) left-m-BE algebra verifying (m-Pimpl) and (m-Pdiv).

(2): Routine. □

This Theorem 9.2.26 allows us to give a new, equivalent definition of Boolean algebras, as follows:

**Definition 9.2.27** *(Definition 2) (The dual one is omitted)*

*A* left-Boolean algebra *is a (involutive) left-m-BE algebra verifying (m-Pdiv).*

We obtain also the following definitional equivalence:

**Theorem 9.2.28** *(See Theorem 9.2.26)*

*(1) Let $\mathcal{A}^L = (A^L, \odot, ^-, 1)$ be a (involutive) left-m-BE algebra verifying (m-Pdis). Define, for all $x, y \in A^L$:*

$$x \wedge y \overset{def.}{=} x \odot y, \quad x \vee y \overset{def.}{=} x \oplus y, \quad 0 \overset{def.}{=} 1^- .$$

*Then, $f(\mathcal{A}^L) = (A^L, \wedge, \vee, ^-, 0, 1)$ is a left-Boolean algebra.*

*(1') Let $(A^L, \wedge, \vee, ^-, 0, 1)$ be a left-Boolean algebra. Define, for all $x, y \in A^L$:*

$$x \odot y \overset{def.}{=} x \wedge y .$$

*Then, $g(\mathcal{A}^L) = (A^L, \odot, ^-, 1)$ is a (involutive) left-m-BE algebra verifying (m-Pdis).*

*(2) The mappings $f$ and $g$ are mutually inverse.*

**Proof.** By Theorems 6.4.15, 9.2.16 and 9.2.19. □

This Theorem 9.2.28 allows us to give a new, equivalent definition of Boolean algebras, as follows:

**Definition 9.2.29** *(Definition 3) (The dual one is omitted)*
*A* left-Boolean algebra *is a (involutive) left-m-BE algebra verifying (m-Pdis).*

**Proposition 9.2.30** *Let $\mathcal{A}^L = (A^L, \odot, ^-, 1)$ be a (involutive) left-m-BE algebra verifying (m-Pdis), i.e. a left-Boolean algebra (Definition 3). We have: for any $x, y \in A^L$,*

$$x \leq_m y \iff x \leq_m^P y.$$

**Proof.** By Lemma 9.1.18, we have $x \leq_m y \iff x \leq_m^P y$, since (m-Re) holds for any $x \in A^L$. By Proposition 6.4.6, (m-Pimpl) holds, and by Proposition 9.1.5, $\leq_m^P$ is a lattice order. □

**Remark 9.2.31** *In a left-Boolean algebra $(A^L, \odot, ^-, 1)$ (Definitions 2, 3), we can define three (four) order relations:*

$x \leq_m y \stackrel{def.}{\iff} x \odot y^- = 0, \quad x \leq_m^P y \stackrel{def}{\iff} x \odot y = x,$
$x \leq_m^M y \stackrel{def}{\iff} x \wedge_m^M y = x, \quad x \leq_m^B y \stackrel{def}{\iff} x \wedge_m^B y = x,$

*where $x \wedge_m^M y = (x^- \odot y)^- \odot y$ and $x \wedge_m^B y = (y^- \odot x)^- \odot x$.*
*We have $\wedge_m^M = \wedge_m^B = \odot$ and we have the equivalences:*

$$x \leq_m y \ (\iff x \leq_m^B y) \iff x \leq_m^M y \iff x \leq_m^P y.$$

See Example 9.3.18 of Boolean algebra.

### 9.2.3 Putting OL and Boole on the "involutive Little m-map"

The definitions (Definition 2) and the results from this section allow us to draw the hierarchies from the following Figure 9.5, thus putting the mentioned algebras on the "involutive Little m-map".

Based on the equivalent definitions (Definitions 2) of the algebras involved in this chapter and on the results and connections presented, we obtain the following final connections:

- if (m-An) does not hold:
**m-MEL**$_{(DN)}$ + (m-Re) = **m-BE**$_{(DN)}$,
**m-tMEL**$_{(DN)}$ + (m-Re) = **m-pre-BCK**$_{(DN)}$;
**BIL** + (m-Re) = **OL**,
**tBIL** + (m-Re) = **tOL**;
**De Morgan** + (m-Re) = **Boole** and,

- if (m-An) holds:
**m-aMEL**$_{(DN)}$ + (m-Re) = **m-aBE**$_{(DN)}$,
**m-taMEL**$_{(DN)}$ + (m-Re) = **m-BCK**$_{(DN)}$;
**aBIL** + (m-Re) = **Boole**,
**taBIL** + (m-Re) = **Boole**;
**a-De Morgan** + (m-Re) = **Boole**,
**ta-De Morgan** + (m-Re) = **Boole**.

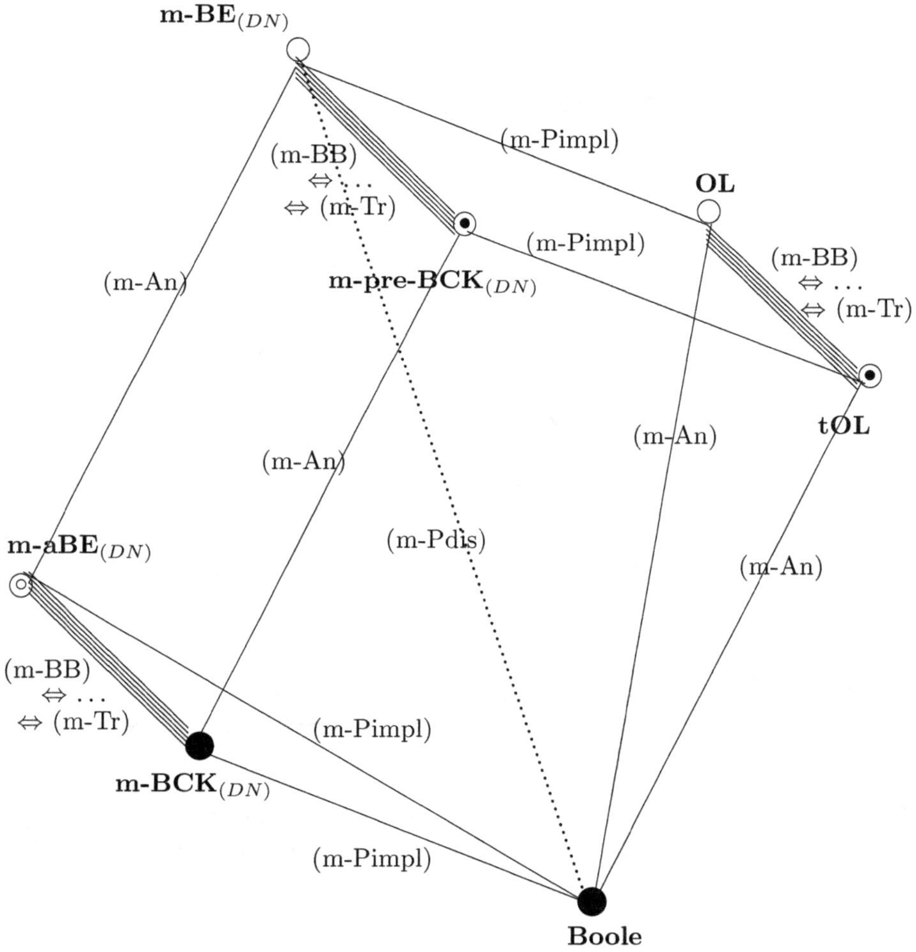


Figure 9.5: Hierarchies **OL**, **Boole** vs. (m-Tr) $\Longleftrightarrow \ldots$ (m-BB)

### 9.2.4 Connections between m-BE algebras and m-BCK, MV, OL, Boole

Resuming, between m-BE algebras, (involutive) m-BCK algebras, MV algebras, ortholattices and Boolean algebras there are the connections from the following Figure 9.6.

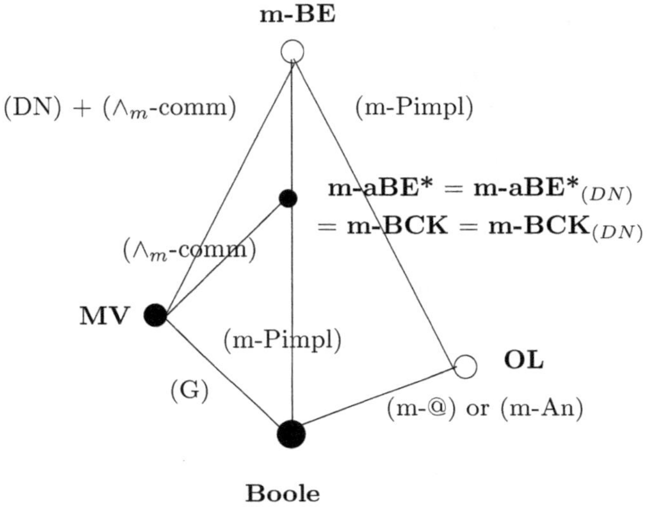


Figure 9.6: Connections between m-BE algebras, m-BCK algebras, MV algebras, ortholattices and Boolean algebras

**Remark 9.2.32** *MV algebras and ortholattices are incomparable: any MV algebra does not verify (m-Pimpl) and any ortholattice does not verify ($\wedge_m$-comm).*

**Open problem 9.2.33** *It is an open problem to define the non-associative MV algebras, the non-associative ortholattices and the non-associative Boolean algebras and to connect them with the non-associative algebras defined above (the m-ML, ..., m-BCC algebras).*

## 9.3 Examples

### 9.3.1 Examples of involutive m-MEL algebras

**Example 9.3.1 Proper involutive m-MEL algebra: m-MEL$_{(DN)}$**

By PASCAL program, we found that the algebra $\mathcal{A}^L = (A_6 = \{0, a, b, c, d, 1\}, \odot, ^-, 1)$, with the following tables of $\odot$ and $^-$ and of the additional operation $\oplus$, is a proper involutive left-m-MEL algebra (i.e. (PU),

(Pcomm), (Pass), (m-La), (DN) hold), since it does not verify (m-Re) for $a$, (G) for $a$, (m-Tr) for $(a, b, a)$, (m-An) for $(a, b)$, (m-Pimpl) for $(b, 0)$, (m-Pdis) for $(a, b, a)$.

| $\odot$ | 0 | a | b | c | d | 1 |
|---|---|---|---|---|---|---|
| 0 | 0 | 0 | 0 | 0 | 0 | 0 |
| a | 0 | 0 | 0 | 0 | a | a |
| b | 0 | 0 | c | 0 | 0 | b |
| c | 0 | 0 | 0 | 0 | 0 | c |
| d | 0 | 0 | 0 | 0 | d | d |
| 1 | 0 | a | b | c | d | 1 |

and

| $x$ | $x^-$ |
|---|---|
| 0 | 1 |
| a | d |
| b | c |
| c | b |
| d | a |
| 1 | 0 |

, with

| $\oplus$ | 0 | a | b | c | d | 1 |
|---|---|---|---|---|---|---|
| 0 | 0 | a | b | c | d | 1 |
| a | a | a | 1 | 1 | 1 | 1 |
| b | b | 1 | 1 | 1 | 1 | 1 |
| c | c | 1 | 1 | b | 1 | 1 |
| d | d | d | 1 | 1 | 1 | 1 |
| 1 | 1 | 1 | 1 | 1 | 1 | 1 |

.

**Example 9.3.2 Proper involutive transitive m-MEL algebra: m-tMEL$_{(DN)}$**

By PASCAL program, we found that the algebra $\mathcal{A}^L = (A_6 = \{0, a, b, c, d, 1\}, \odot, ^-, 1)$, with the following tables of $\odot$ and $^-$ and of the additional operation $\oplus$, is a proper involutive left-m-MEL algebra verifying (m-Tr), since it does not verify (m-Re) for $a$, (G) for $b$, (m-An) for $(b, c)$, (m-Pimpl) for $(a, 0)$, (m-Pdis) for $(a, a, b)$.

| $\odot$ | 0 | a | b | c | d | 1 |
|---|---|---|---|---|---|---|
| 0 | 0 | 0 | 0 | 0 | 0 | 0 |
| a | 0 | a | b | c | d | a |
| b | 0 | b | 0 | 0 | c | b |
| c | 0 | c | 0 | 0 | b | c |
| d | 0 | d | c | b | a | d |
| 1 | 0 | a | b | c | d | 1 |

and

| $x$ | $x^-$ |
|---|---|
| 0 | 1 |
| a | d |
| b | c |
| c | b |
| d | a |
| 1 | 0 |

, with

| $\oplus$ | 0 | a | b | c | d | 1 |
|---|---|---|---|---|---|---|
| 0 | 0 | a | b | c | d | 1 |
| a | a | d | c | b | a | 1 |
| b | b | c | 1 | 1 | b | 1 |
| c | c | b | 1 | 1 | c | 1 |
| d | d | a | b | c | d | 1 |
| 1 | 1 | 1 | 1 | 1 | 1 | 1 |

.

**Example 9.3.3 Proper involutive antisymmetric m-MEL algebra: m-aMEL$_{(DN)}$**

By PASCAL program, we found that the algebra $\mathcal{A}^L = (A_6 = \{0, a, b, c, d, 1\}, \odot, ^-, 1)$, with the following tables of $\odot$ and $^-$ and of the additional operation $\oplus$, is a proper involutive left-m-aMEL algebra, since it does not verify (m-Re) for $b$, (G) for $c$, (m-Tr) for $(a, b, d)$, (m-Pimpl) for $(a, 0)$, (m-Pdis) for $(a, a, a)$.

| $\odot$ | 0 | a | b | c | d | 1 |
|---|---|---|---|---|---|---|
| 0 | 0 | 0 | 0 | 0 | 0 | 0 |
| a | 0 | a | 0 | 0 | 0 | a |
| b | 0 | 0 | b | b | b | b |
| c | 0 | 0 | b | b | b | c |
| d | 0 | 0 | b | b | c | d |
| 1 | 0 | a | b | c | d | 1 |

and

| $x$ | $x^-$ |
|---|---|
| 0 | 1 |
| a | d |
| b | c |
| c | b |
| d | a |
| 1 | 0 |

, with

| $\oplus$ | 0 | a | b | c | d | 1 |
|---|---|---|---|---|---|---|
| 0 | 0 | a | b | c | d | 1 |
| a | a | b | c | c | 1 | 1 |
| b | b | c | c | c | 1 | 1 |
| c | c | c | c | c | 1 | 1 |
| d | d | 1 | 1 | 1 | d | 1 |
| 1 | 1 | 1 | 1 | 1 | 1 | 1 |

.

**Example 9.3.4 Proper involutive m-MEL algebra verifying (m-An), (m-Tr): m-taMEL$_{(DN)}$**

By MACE4, we found that the algebra $\mathcal{A}^L = (A_6 = \{0, a, b, c, d, 1\}, \odot, ^-, 1)$, with the following tables of $\odot$ and $^-$ and of the additional operation $\oplus$, is a proper involutive left-m-MEL algebra verifying (m-An), (m-Tr), since it does not verify (m-Re) for $a$, (G) for $b$, (m-Pimpl) for $(b, 0)$, (m-Pdis) for $(a, a, b)$.

| $\odot$ | 0 | a | b | c | d | 1 |
|---|---|---|---|---|---|---|
| 0 | 0 | 0 | 0 | 0 | 0 | 0 |
| a | 0 | a | c | c | a | a |
| b | 0 | c | 0 | 0 | b | b |
| c | 0 | c | 0 | 0 | c | c |
| d | 0 | a | b | c | 1 | d |
| 1 | 0 | a | b | c | d | 1 |

and

| $x$ | $x^-$ |
|---|---|
| 0 | 1 |
| a | a |
| b | b |
| c | d |
| d | c |
| 1 | 0 |

, with

| $\oplus$ | 0 | a | b | c | d | 1 |
|---|---|---|---|---|---|---|
| 0 | 0 | a | b | c | d | 1 |
| a | a | a | d | a | d | 1 |
| b | b | d | 1 | b | 1 | 1 |
| c | c | a | b | 0 | d | 1 |
| d | d | d | 1 | d | 1 | 1 |
| 1 | 1 | 1 | 1 | 1 | 1 | 1 |

.

**Example 9.3.5 Proper bounded involutive lattice: BIL**

By PASCAL program, we found that the algebra $\mathcal{A}^L = (A_6 = \{0, a, b, c, d, 1\}, \odot, ^-, 1)$, with the following tables of $\odot$ and $^-$ and of the additional operation $\oplus$, is an involutive left-m-MEL algebra verifying (m-Pimpl) and not verifying (m-Re) for $b$, (m-Tr) for $(a, b, d)$, (m-An) for $(a, b)$, (m-Pdis) for $(a, b, c)$, i.e. is a proper bounded involutive lattice (Definition 2).

| $\odot$ | 0 | a | b | c | d | 1 |
|---|---|---|---|---|---|---|
| 0 | 0 | 0 | 0 | 0 | 0 | 0 |
| a | 0 | a | 0 | 0 | 0 | a |
| b | 0 | 0 | b | b | 0 | b |
| c | 0 | 0 | b | c | 0 | c |
| d | 0 | 0 | 0 | 0 | d | d |
| 1 | 0 | a | b | c | d | 1 |

and

| $x$ | $x^-$ |
|---|---|
| 0 | 1 |
| a | d |
| b | c |
| c | b |
| d | a |
| 1 | 0 |

, with

| $\oplus$ | 0 | a | b | c | d | 1 |
|---|---|---|---|---|---|---|
| 0 | 0 | a | b | c | d | 1 |
| a | a | a | 1 | 1 | 1 | 1 |
| b | b | 1 | b | c | 1 | 1 |
| c | c | 1 | c | c | 1 | 1 |
| d | d | 1 | 1 | 1 | d | 1 |
| 1 | 1 | 1 | 1 | 1 | 1 | 1 |

.

**Example 9.3.6 Proper transitive bounded involutive lattice: tBIL**

By MACE4, we found that the algebra $\mathcal{A}^L = (A_8 = \{0, a, b, c, d, e, f, 1\}, \odot, ^-, 1)$, with the following tables of $\odot$ and $^-$ and of the additional operation $\oplus$, is an involutive m-MEL algebra verifying (m-Pimpl) and (m-Tr) and not verifying (m-Re) for $e$, (m-An) for $(a, c)$, (m-Pdis) for $(a, b, c)$, $(a, d, b)$. Hence, $\mathcal{A}^L$ is a bounded involutive lattice (Definition 2) verifying (m-Tr).

| $\odot$ | 0 | a | b | c | d | e | f | 1 |
|---|---|---|---|---|---|---|---|---|
| 0 | 0 | 0 | 0 | 0 | 0 | 0 | 0 | 0 |
| a | 0 | a | 0 | a | 0 | 0 | a | a |
| b | 0 | 0 | b | 0 | d | e | e | b |
| c | 0 | a | 0 | c | 0 | 0 | c | c |
| d | 0 | 0 | d | 0 | d | e | e | d |
| e | 0 | 0 | e | 0 | e | e | e | e |
| f | 0 | a | e | c | e | e | f | f |
| 1 | 0 | a | b | c | d | e | f | 1 |

,

| $\oplus$ | 0 | a | b | c | d | e | f | 1 |
|---|---|---|---|---|---|---|---|---|
| 0 | 0 | a | b | c | d | e | f | 1 |
| a | a | a | 1 | c | 1 | f | f | 1 |
| b | b | 1 | b | 1 | b | b | 1 | 1 |
| c | c | c | 1 | c | 1 | f | f | 1 |
| d | d | 1 | b | 1 | d | d | 1 | 1 |
| e | e | f | b | f | d | e | f | 1 |
| f | f | f | 1 | f | 1 | f | f | 1 |
| 1 | 1 | 1 | 1 | 1 | 1 | 1 | 1 | 1 |

and $(0, a, b, c, d, e, f, 1)^- = (1, b, a, d, c, f, e, 0)$.

**Example 9.3.7 Proper antisymmetric bounded involutive lattice: aBIL**

By MACE4, we found that the algebra $\mathcal{A}^L = (A_7 = \{0, a, b, c, d, e, 1\}, \odot, ^-, 1)$, with the following tables of $\odot$ and $^-$ and of the additional operation $\oplus$, is an involutive m-MEL algebra verifying (m-Pimpl) and (m-An) and not verifying (m-Re) for $a$, (m-Tr) for $(b, a, c)$, (m-Pdis) for $(a, c, b)$. Hence, $\mathcal{A}^L$ is a bounded involutive lattice (Definition 2) verifying (m-An).

| ⊙ | 0 | a | b | c | d | e | 1 |
|---|---|---|---|---|---|---|---|
| 0 | 0 | 0 | 0 | 0 | 0 | 0 | 0 |
| a | 0 | a | 0 | d | d | a | a |
| b | 0 | 0 | b | 0 | 0 | b | b |
| c | 0 | d | 0 | c | d | d | c |
| d | 0 | d | 0 | d | d | d | d |
| e | 0 | a | b | d | d | e | e |
| 1 | 0 | a | b | c | d | e | 1 |

, with

| ⊕ | 0 | a | b | c | d | e | 1 |
|---|---|---|---|---|---|---|---|
| 0 | 0 | a | b | c | d | e | 1 |
| a | a | a | e | 1 | a | e | 1 |
| b | b | e | b | 1 | e | e | 1 |
| c | c | 1 | 1 | c | c | 1 | 1 |
| d | d | a | e | c | d | e | 1 |
| e | e | e | e | 1 | e | e | 1 |
| 1 | 1 | 1 | 1 | 1 | 1 | 1 | 1 |

and $(0, a, b, c, d, e, 1)^- = (1, a, c, b, e, d, 0)$.

**Example 9.3.8 Proper transitive, antisymmetric bounded involutive lattice: taBIL**

By MACE4, we found that the algebra $\mathcal{A}^L = (A_7 = \{0, a, b, c, d, e, 1\}, \odot, ^-, 1)$, with the following tables of $\odot$ and $^-$ and of the additional operation $\oplus$, is an involutive left-m-MEL algebra verifying (m-Pimpl) and (m-An), (m-Tr), and not verifying (m-Re) for $a$, (m-Pdis) for$(a, b, c)$. Hence, $\mathcal{A}^L$ is a bounded involutive lattice (Definition 2) verifying (m-An) and (m-Tr).

| ⊙ | 0 | a | b | c | d | e | 1 |
|---|---|---|---|---|---|---|---|
| 0 | 0 | 0 | 0 | 0 | 0 | 0 | 0 |
| a | 0 | a | d | d | d | a | a |
| b | 0 | d | b | d | d | b | b |
| c | 0 | d | d | c | d | c | c |
| d | 0 | d | d | d | d | d | d |
| e | 0 | a | b | c | d | e | e |
| 1 | 0 | a | b | c | d | e | 1 |

, with

| ⊕ | 0 | a | b | c | d | e | 1 |
|---|---|---|---|---|---|---|---|
| 0 | 0 | a | b | c | d | e | 1 |
| a | a | a | e | e | a | e | 1 |
| b | b | e | b | e | b | e | 1 |
| c | c | e | e | c | c | e | 1 |
| d | d | a | b | c | d | e | 1 |
| e | e | e | e | e | e | e | 1 |
| 1 | 1 | 1 | 1 | 1 | 1 | 1 | 1 |

,

and $(0, a, b, c, d, e, 1)^- = (1, a, b, c, e, d, 0)$.

**Example 9.3.9 Proper De Morgan algebra: De Morgan**

By MACE4, we found that the algebra $\mathcal{A}^L = (A_4 = \{0, a, b, 1\}, \odot, ^-, 1)$, with the following tables of $\odot$ and $^-$ and of the additional operation $\oplus$, is an involutive left-m-MEL algebra verifying (m-Pdis) (hence (m-Pimpl) and (G)) and not verifying (m-Re) for $a$, (m-Tr) for $(a, b, a)$, (m-An) for $(a, b)$. Hence, it is a proper De Morgan algebra (Definition 2) (i.e. without (m-An), (m-Tr)).

| ⊙ | 0 | a | b | 1 |
|---|---|---|---|---|
| 0 | 0 | 0 | 0 | 0 |
| a | 0 | a | 0 | a |
| b | 0 | 0 | b | b |
| 1 | 0 | a | b | 1 |

and

| $x$ | $x^-$ |
|---|---|
| 0 | 1 |
| a | a |
| b | b |
| 1 | 0 |

, with

| ⊕ | 0 | a | b | 1 |
|---|---|---|---|---|
| 0 | 0 | a | b | 1 |
| a | a | a | 1 | 1 |
| b | b | 1 | b | 1 |
| 1 | 1 | 1 | 1 | 1 |

.

**Example 9.3.10 Proper antisymmetric De Morgan algebra: a-De Morgan**

By MACE4, we found that the algebra $\mathcal{A}^L = (A_8 = \{0, a, b, c, d, e, f, 1\}, \odot, ^-, 1)$, with the following tables of $\odot$ and $^-$ and of the additional operation $\oplus$, is an involutive left-m-MEL algebra verifying (m-Pdis) and (m-An) and not verifying (m-Re) for $a$, (m-B) for $(a, 0, a)$, (m-BB) for $(a, a, 0)$, (m-*) for $(0, 0, a)$, (m-**) for $(a, f, a)$, (m-Tr) for $(e, a, f)$, i.e. it is a proper antisymmetric De Morgan algebra (Definition 2) (i.e. without (m-Tr)).

| ⊙ | 0 | a | b | c | d | e | f | 1 |
|---|---|---|---|---|---|---|---|---|
| 0 | 0 | 0 | 0 | 0 | 0 | 0 | 0 | 0 |
| a | 0 | a | c | c | a | 0 | c | a |
| b | 0 | c | b | c | b | e | b | b |
| c | 0 | c | c | c | c | 0 | c | c |
| d | 0 | a | b | c | d | e | b | d |
| e | 0 | 0 | e | 0 | e | e | e | e |
| f | 0 | c | b | c | b | e | f | f |
| 1 | 0 | a | b | c | d | e | f | 1 |

,

| ⊕ | 0 | a | b | c | d | e | f | 1 |
|---|---|---|---|---|---|---|---|---|
| 0 | 0 | a | b | c | d | e | f | 1 |
| a | a | a | d | a | d | d | 1 | 1 |
| b | b | d | b | b | d | b | f | 1 |
| c | c | a | b | c | d | b | f | 1 |
| d | d | d | d | d | d | d | 1 | 1 |
| e | e | d | b | b | d | e | f | 1 |
| f | f | 1 | f | f | 1 | f | f | 1 |
| 1 | 1 | 1 | 1 | 1 | 1 | 1 | 1 | 1 |

and $(0, a, b, c, d, e, f, 1)^- = (1, a, b, d, c, f, e, 0)$.

**Example 9.3.11 Proper transitive, antisymmetric De Morgan algebra: ta-De Morgan**

By PASCAL program, we found that the algebra $\mathcal{A}^L = (A_6 = \{0, a, b, c, d, 1\}, \odot, ^-, 1)$, with the following tables of $\odot$ and $^-$ and of the additional operation $\oplus$, is an involutive left-m-MEL algebra verifying (m-Pdis) and (m-An), (m-Tr) and not verifying (m-Re) for $b$. Hence, it is a De Morgan algebra (Definition 2) verifying (m-An) and (m-Tr).

| ⊙ | 0 | a | b | c | d | 1 |
|---|---|---|---|---|---|---|
| 0 | 0 | 0 | 0 | 0 | 0 | 0 |
| a | 0 | a | 0 | a | 0 | a |
| b | 0 | 0 | b | b | b | b |
| c | 0 | a | b | c | b | c |
| d | 0 | 0 | b | b | d | d |
| 1 | 0 | a | b | c | d | 1 |

and

| $x$ | $x^-$ |
|---|---|
| 0 | 1 |
| a | d |
| b | c |
| c | b |
| d | a |
| 1 | 0 |

, with

| ⊕ | 0 | a | b | c | d | 1 |
|---|---|---|---|---|---|---|
| 0 | 0 | a | b | c | d | 1 |
| a | a | a | c | c | 1 | 1 |
| b | b | c | b | c | d | 1 |
| c | c | c | c | c | 1 | 1 |
| d | d | 1 | d | 1 | d | 1 |
| 1 | 1 | 1 | 1 | 1 | 1 | 1 |

.

## 9.3.2 Examples of involutive m-BE algebras

**Example 9.3.12 Proper involutive m-BE algebra: m-BE$_{(DN)}$**

By PASCAL program, we found that the algebra $\mathcal{A}^L = (A_6 = \{0, a, b, c, d, 1\}, \odot, ^-, 1)$, with the following tables of $\odot$ and $^-$ and of the additional operation $\oplus$, is a proper involutive left-m-BE algebra not verifying (m-B) for $(a, b, a)$, (m-BB) for $(a, a, b)$, (m-*) for $(b, a, c)$, (m-**) for $(a, b, a)$; (m-Tr) for $(a, b, d)$, (m-An) for $(a, b)$, (m-Pimpl) for $(a, 0)$, (m-Pdis) for $(a, a, a)$.

| ⊙ | 0 | a | b | c | d | 1 |
|---|---|---|---|---|---|---|
| 0 | 0 | 0 | 0 | 0 | 0 | 0 |
| a | 0 | a | 0 | 0 | 0 | a |
| b | 0 | 0 | 0 | 0 | 0 | b |
| c | 0 | 0 | 0 | 0 | b | c |
| d | 0 | 0 | 0 | b | b | d |
| 1 | 0 | a | b | c | d | 1 |

and

| $x$ | $x^-$ |
|---|---|
| 0 | 1 |
| a | d |
| b | c |
| c | b |
| d | a |
| 1 | 0 |

, with

| ⊕ | 0 | a | b | c | d | 1 |
|---|---|---|---|---|---|---|
| 0 | 0 | a | b | c | d | 1 |
| a | a | c | c | 1 | 1 | 1 |
| b | b | c | 1 | 1 | 1 | 1 |
| c | c | 1 | 1 | 1 | 1 | 1 |
| d | d | 1 | 1 | 1 | d | 1 |
| 1 | 1 | 1 | 1 | 1 | 1 | 1 |

.

**Example 9.3.13 Proper involutive m-pre-BCK algebra: m-pre-BCK$_{(DN)}$**

By PASCAL program, we found that the algebra $\mathcal{A}^L = (A_6 = \{0, a, b, c, d, 1\}, \odot, ^-, 1)$, with the following tables of $\odot$ and $^-$ and of the additional operation $\oplus$, is an involutive left-m-pre-BCK algebra not verifying (m-An) for $(a, b)$, (m-Pimpl) for $(a, 0)$, (m-Pdis) for $(a, c, a)$.

| $\odot$ | 0 | a | b | c | d | 1 |
|---|---|---|---|---|---|---|
| 0 | 0 | 0 | 0 | 0 | 0 | 0 |
| a | 0 | 0 | 0 | 0 | 0 | a |
| b | 0 | 0 | 0 | 0 | 0 | b |
| c | 0 | 0 | 0 | c | d | c |
| d | 0 | 0 | 0 | d | c | d |
| 1 | 0 | a | b | c | d | 1 |

and

| $x$ | $x^-$ |
|---|---|
| 0 | 1 |
| a | d |
| b | c |
| c | b |
| d | a |
| 1 | 0 |

, with

| $\oplus$ | 0 | a | b | c | d | 1 |
|---|---|---|---|---|---|---|
| 0 | 0 | a | b | c | d | 1 |
| a | a | b | a | 1 | 1 | 1 |
| b | b | a | b | 1 | 1 | 1 |
| c | c | 1 | 1 | 1 | 1 | 1 |
| d | d | 1 | 1 | 1 | 1 | 1 |
| 1 | 1 | 1 | 1 | 1 | 1 | 1 |

.

**Example 9.3.14 Proper involutive m-aBE algebra: m-aBE$_{(DN)}$**

By PASCAL program, we found that the algebra $\mathcal{A}^L = (A_6 = \{0, a, b, c, d, 1\}, \odot, ^-, 1)$, with the following tables of $\odot$ and $^-$ and of the additional operation $\oplus$, is an involutive left-m-aBE algebra not verifying (m-B) for $(d, b, d)$, (m-BB) for $(d, d, b)$, (m-*) for $(b, a, d)$, (m-**) for $(d, b, d)$; (m-Tr) for $(d, b, a)$, ($\wedge_m$-comm) for $(b, d)$, (m-Pimpl) for $(a, b)$, (m-Pdis) for $(a, a, b)$.

| $\odot$ | 0 | a | b | c | d | 1 |
|---|---|---|---|---|---|---|
| 0 | 0 | 0 | 0 | 0 | 0 | 0 |
| a | 0 | a | b | c | 0 | a |
| b | 0 | b | c | 0 | 0 | b |
| c | 0 | c | 0 | 0 | 0 | c |
| d | 0 | 0 | 0 | 0 | d | d |
| 1 | 0 | a | b | c | d | 1 |

and

| $x$ | $x^-$ |
|---|---|
| 0 | 1 |
| a | d |
| b | c |
| c | b |
| d | a |
| 1 | 0 |

, with

| $\oplus$ | 0 | a | b | c | d | 1 |
|---|---|---|---|---|---|---|
| 0 | 0 | a | b | c | d | 1 |
| a | a | a | 1 | 1 | 1 | 1 |
| b | b | 1 | 1 | 1 | b | 1 |
| c | c | 1 | 1 | b | c | 1 |
| d | d | 1 | b | c | d | 1 |
| 1 | 1 | 1 | 1 | 1 | 1 | 1 |

.

**Example 9.3.15 Proper (involutive) m-BCK algebra: m-BCK$_{(DN)}$**

By PASCAL program, we found that the algebra $\mathcal{A}^L = (A_6 = \{0, a, b, c, d, 1\}, \odot, ^-, 1)$, with the following tables of $\odot$ and $^-$ and of the additional operation $\oplus$, is a proper (involutive) left-m-BCK algebra, not verifying ($\wedge_m$-comm) for $(b, d)$, (m-Pimpl) for $(a, 0)$, (m-Pdis) for $(a, a, b)$. The relation $\leq_m$ is a lattice order, hence $\mathcal{A}^L$ is a proper (involutive) left-m-BCK lattice. For an example of proper (involutive) left-m-BCK algebra that is not a lattice, see [104], Example 7.2.1.

| $\odot$ | 0 | a | b | c | d | 1 |
|---|---|---|---|---|---|---|
| 0 | 0 | 0 | 0 | 0 | 0 | 0 |
| a | 0 | 0 | 0 | 0 | 0 | a |
| b | 0 | 0 | a | 0 | a | b |
| c | 0 | 0 | 0 | a | a | c |
| d | 0 | 0 | a | a | a | d |
| 1 | 0 | a | b | c | d | 1 |

and

| $x$ | $x^-$ |
|---|---|
| 0 | 1 |
| a | d |
| b | c |
| c | b |
| d | a |
| 1 | 0 |

, with

| $\oplus$ | 0 | a | b | c | d | 1 |
|---|---|---|---|---|---|---|
| 0 | 0 | a | b | c | d | 1 |
| a | a | d | d | d | 1 | 1 |
| b | b | d | d | 1 | 1 | 1 |
| c | c | d | 1 | d | 1 | 1 |
| d | d | 1 | 1 | 1 | 1 | 1 |
| 1 | 1 | 1 | 1 | 1 | 1 | 1 |

.

**Example 9.3.16 Proper ortholattice: OL**

By PASCAL program, we found that the algebra $\mathcal{A}^L = (A_8 = \{0, a, b, c, d, e, f, 1\}, \odot, ^-, 1)$, with the following tables of $\odot$ and $^-$ and of the additional operation $\oplus$, is an involutive left-m-BE algebra verifying (m-Pimpl) and not verifying (m-B) for $(a, b, a)$, (m-BB) for $(a, a, b)$, (m-*) for $(a, b, c)$, (m-**) for $(a, b, a)$; (m-Tr) for $(a, b, f)$, (m-An) for $(a, b)$, (m-Pdis) for $(a, b, c)$, i.e it is a proper left-ortholattice (Definition 2).

| $\odot$ | 0 | a | b | c | d | e | f | 1 |
|---|---|---|---|---|---|---|---|---|
| 0 | 0 | 0 | 0 | 0 | 0 | 0 | 0 | 0 |
| a | 0 | a | 0 | 0 | 0 | 0 | 0 | a |
| b | 0 | 0 | b | 0 | b | 0 | 0 | b |
| c | 0 | 0 | 0 | c | 0 | c | 0 | c |
| d | 0 | 0 | b | 0 | d | 0 | 0 | d |
| e | 0 | 0 | 0 | c | 0 | e | 0 | e |
| f | 0 | 0 | 0 | 0 | 0 | 0 | f | f |
| 1 | 0 | a | b | c | d | e | f | 1 |

,

| $\oplus$ | 0 | a | b | c | d | e | f | 1 |
|---|---|---|---|---|---|---|---|---|
| 0 | 0 | a | b | c | d | e | f | 1 |
| a | a | a | 1 | 1 | 1 | 1 | 1 | 1 |
| b | b | 1 | b | 1 | d | 1 | 1 | 1 |
| c | c | 1 | 1 | c | 1 | e | 1 | 1 |
| d | d | 1 | d | 1 | d | 1 | 1 | 1 |
| e | e | 1 | 1 | e | 1 | e | 1 | 1 |
| f | f | 1 | 1 | 1 | 1 | 1 | f | 1 |
| 1 | 1 | 1 | 1 | 1 | 1 | 1 | 1 | 1 |

and $(0, a, b, c, d, e, f, 1)^- = (1, f, e, d, c, b, a, 0)$.

**Example 9.3.17 Proper transitive ortholattice: tOL**

By MACE4, we found that the algebra $\mathcal{A}^L = (A_6 = \{0, a, b, c, d, 1\}, \odot, ^-, 1)$, with the following tables of $\odot$ and $^-$ and of the additional operation $\oplus$, is an involutive left-m-BE algebra verifying (m-Pimpl) and (m-Tr) and not verifying (m-An) for $(a, c)$, (m-Pdis) for $(a, b, c)$, $(b, a, d)$. Hence, $\mathcal{A}^L$ is a proper transitive left-ortholattice (Definition 2).

| $\odot$ | 0 | a | b | c | d | 1 |
|---|---|---|---|---|---|---|
| 0 | 0 | 0 | 0 | 0 | 0 | 0 |
| a | 0 | a | 0 | a | 0 | a |
| b | 0 | 0 | b | 0 | d | b |
| c | 0 | a | 0 | c | 0 | c |
| d | 0 | 0 | d | 0 | d | d |
| 1 | 0 | a | b | c | d | 1 |

and

| $x$ | $x^-$ |
|---|---|
| 0 | 1 |
| a | b |
| b | a |
| c | d |
| d | c |
| 1 | 0 |

, with

| $\oplus$ | 0 | a | b | c | d | 1 |
|---|---|---|---|---|---|---|
| 0 | 0 | a | b | c | d | 1 |
| a | a | a | 1 | c | 1 | 1 |
| b | b | 1 | b | 1 | b | 1 |
| c | c | c | 1 | c | 1 | 1 |
| d | d | 1 | b | 1 | d | 1 |
| 1 | 1 | 1 | 1 | 1 | 1 | 1 |

.

**Example 9.3.18 Boolean algebra: Boole**

The algebra $\mathcal{A}^L = (A^L = L_2 = \{0, 1\}, \odot, ^-, 1)$, with the following tables of $\odot$ and $^-$ and of the additional operation $\oplus$, is the linearly ordered left-Boolean algebra (Definition 2) with the smallest number of elements (i.e. (PU), (Pcomm), (Pass), (m-La), (m-Re), (DN), (m-Pdiv), (Pdis) hold).

| $\odot$ | 0 | 1 |
|---|---|---|
| 0 | 0 | 0 |
| 1 | 0 | 1 |

and

| $x$ | $x^-$ |
|---|---|
| 0 | 1 |
| 1 | 0 |

, with

| $\oplus$ | 0 | 1 |
|---|---|---|
| 0 | 0 | 1 |
| 1 | 1 | 1 |

.

# Chapter 10

# Two generalizations of bounded involutive lattices and of ortholattices

In this chapter, we continue the results from the previous Chapter 9.

In Section 10.1, we introduce and study two dual independent absorbtion laws: (m-Wabs-i) and (m-Vabs-i) and two generalizations of bounded lattices (**BL**): bounded softlattices (**BSL**) and bounded widelattices (**BWL**).

In Section 10.2, we introduce and study two corresponding generalizations of **BIL**: bounded involutive softlattices (**BISL**) and bounded involutive widelattices (**BIWL**). Finally, we put **BIL**, **BISL**, **BIWL**, and their subclasses, on the "involutive Little m-map".

In Section 10.3, we introduce and study two corresponding generalizations of **OL**: orthosoftlattices (**OSL**) and orthowidelattices (**OWL**). The core of the section is Theorem 10.3.14 (based on Theorem 10.3.13, proved by PROVER9), saying that transitive and antisymmetric orthowidelattices are MV algebras; thus, a proper subclass of MV algebras (**taOWL**) is obtained; a generalization of ($\wedge_m$-comm) property, called ($\Delta_m$), is introduced on this occasion. We put **OL**, **OSL**, **OWL**, and their subclasses, on the "involutive Little m-map".

In Section 10.4, we present 11 examples of the various algebras discussed herein.

The content of this chapter is taken from [123].

# 10.1 Two dual independent absorbtion laws and two generalizations of bounded lattices

## 10.1.1 Two dual independent absorbtion laws: (m-Wabs-i), (m-Vabs-i)

Since, in a lattice, the absorbtion laws (m-Wabs) and (m-Vabs) are not independent (they imply the idempotency laws (m-Wid) and (m-Vid)), we shall introduce the following two dual *independent absorbtion laws*:
(m-Wabs-i) $x \wedge (x \vee x \vee y) = x$ and, dually,
(m-Vabs-i) $x \vee (x \wedge x \wedge y) = x$ (the dual laws of independent absorbtion), and also
(m-Pabs-i) $x \odot (x \oplus x \oplus y) = x$ and, dually,
(m-Sabs-i) $x \oplus (x \odot x \odot y) = x$ (the dual laws of independent absorbtion).

We shall prove that the system of eight axioms: **L8-i**={(m-Wid), (m-Vid), (m-Wcomm), (m-Vcomm), (m-Wass), (M-Vass), (m-Wabs-i), (m-Vabs-i)} is equivalent with the "standard" system **L8** of axioms of the lattice. First, we prove the following Lemma.

**Lemma 10.1.1**
*(a) (m-Vid) + (m-Vass) $\Longrightarrow$ ((m-Wabs-i) $\Longleftrightarrow$ (m-Wabs)) and, dually,*
*(a') (m-Wid) + (m-Wass) $\Longrightarrow$ ((m-Vabs-i) $\Longleftrightarrow$ (m-Vabs)).*

**Proof.** First, (m-Vid) + (m-Vass) + (m-Wabs-i) imply (m-Wabs):
$x \wedge (x \vee y) = x \wedge ((x \vee x) \vee y) = x \wedge (x \vee x \vee y) = x$
and (m-Vid) + (m-Vass) + (m-Wabs) imply (m-Wabs-i):
$x \wedge (x \vee x \vee y) = x \wedge ((x \vee x) \vee y) = x \wedge (x \vee y) = x$; thus, (a) holds.

Then, dually, (m-Wid) + (m-Wass) + (m-Vabs-i) imply (m-Vabs):
$x \vee (x \wedge y) = x \vee ((x \wedge x) \wedge y) = x \vee (x \wedge x \wedge y) = x$
and (m-Wid) + (m-Wass) + (m-Vabs) imply (m-Vabs-i):
$x \vee (x \wedge x \wedge y) = x \vee ((x \wedge x) \wedge y) = x \vee (x \wedge y) = x$; thus, (a') holds. □

Now, the announced result follows by the above Lemma.

**Theorem 10.1.2** *We have:* **L8-i** $\Longleftrightarrow$ **L8**.

## 10.1.2 Two generalizations of bounded lattices (BL): BSL, BWL

### 10.1.2.1 Bounded softlattices: BSL

**Definitions 10.1.3** (The dual ones are omitted)

(1) A *left-softlattice* is an algebra $\mathcal{A}^L = (A^L, \wedge, \vee)$ of type $(2, 2)$ such that the axioms (m-Wid), (m-Vid), (m-Wcomm), (m-Vcomm), (m-Wass), (m-Vass) are satisfied.

(2) A *bounded left-softlattice* is an algebra $\mathcal{A}^L = (A^L, \wedge, \vee, 0, 1)$ of type $(2, 2, 0, 0)$ such that the reduct $(A^L, \wedge, \vee)$ is a left-softlattice and the elements 0 and 1 verify the axioms: for all $x \in A^L$,

(m-WU) $1 \wedge x = x$, (m-VU) $0 \vee x = x$,
(m-WL) $0 \wedge x = 0$, (m-VL) $1 \vee x = 1$
(i.e. such that the reduct $(A^L, \wedge, 0, 1)$ is a bounded meet-semilattice with top element 1 and the reduct $(A^L, \vee, 0, 1)$ is a bounded join-semilattice with bottom element 0).

We shall denote by **SL** the class of all left-softlattices and by **BSL** the class of all bounded left-softlattices.

Note that any left-lattice is a left-softlattice: **L** $\subset$ **SL**, the inclusion being strict, since there are left-softlattices that are not left-lattices - see the next example.

**Example 10.1.4 Proper softlattice**
The algebra $\mathcal{A}^L = (L_2 = \{a, b\}, \wedge, \vee)$, with the following tables of $\wedge$ and $\vee$, is a proper left-softlattice, because (m-Wabs-i) and (m-Vabs-i) are not verified for $(x, y) = (a, b)$.

| $\wedge$ | a | b |
|---|---|---|
| a | a | b |
| b | b | b |

and

| $\vee$ | a | b |
|---|---|---|
| a | a | b |
| b | b | b |

.

Hence, we have: **SL** + (m-Wabs-i) + (m-Vabs-i) = **L** = **SL** + (m-Wabs) + (m-Vabs).

The following example shows an algebra $\mathcal{A}^L = (A^L, \wedge, \vee, 0, 1)$ such that the reduct $(A^L, \wedge, \vee)$ is a left-softlattice, verifying (m-WU) and (m-VU), but not verifying (m-WL) and (m-VL), therefore it is not a bounded left-softlattice.

**Example 10.1.5** The algebra $\mathcal{A}^L = (L_3 = \{0, a, 1\}, \wedge, \vee, 0, 1)$, with the following tables of $\wedge$ and $\vee$, is a proper left-softlattice, because (m-Wabs-i) and (m-Vabs-i) are not verified for $(x, y) = (0, a)$. It verifies (m-WU) and (m-VU), but does not verify (m-WL) and (m-VL) for $x = a$.

| $\wedge$ | 0 | a | 1 |
|---|---|---|---|
| 0 | 0 | a | 0 |
| a | a | a | a |
| 1 | 0 | a | 1 |

and

| $\vee$ | 0 | a | 1 |
|---|---|---|---|
| 0 | 0 | a | 1 |
| a | a | a | a |
| 1 | 1 | a | 1 |

.

**Proposition 10.1.6** *Let $\mathcal{A}^L = (A^L, \wedge, \vee, 0, 1)$ be a bounded left-softlattice. Define, for all $x, y \in A^L$,*

*(m-dfO($\wedge$))* $x \leq_m^{O(\wedge)} y \stackrel{def.}{\Longleftrightarrow} x \wedge y = x$ *and, dually,*

*(m-dfO($\vee$))* $x \geq_m^{O(\vee)} y \stackrel{def.}{\Longleftrightarrow} x \vee y = x$.

*Then,*

*(1)* $(A^L, \leq_m^{O(\wedge)}, 1)$ *is an inf-semilattice with top element* 1,
*(1')* $(A^L, \geq_m^{O(\vee)}, 0)$ *is a sup-semilattice with bottom element* 0.
*(2)* $0 \leq_m^{O(\wedge)} x \leq_m^{O(\wedge)} 1$, *for all* $x \in A^L$, *i.e. the inf-semilattice is bounded,*
*(2')* $1 \geq_m^{O(\vee)} x \geq_m^{O(\vee)} 0$, *for all* $x \in A^L$, *i.e. the sup-semilattice is bounded.*

**Proof.** (1), (1'): Obviously.

(2): $0 \leq_m^{O(\wedge)} x \Longleftrightarrow 0 \wedge x = 0$, that is true by (m-WL).
$x \leq_m^{O(\wedge)} 1 \Longleftrightarrow x \wedge 1 = x$, that is true by (m-WU) and (m-Wcomm).
(2'): $1 \geq_m^{O(\vee)} x \Longleftrightarrow 1 \vee x = 1$, that is true by (m-VL).
$x \geq_m^{O(\vee)} 0 \Longleftrightarrow x \vee 0 = 0$, that is true by (m-VU) and (m-Vcomm). □

**Proposition 10.1.7** *Let $\mathcal{A}^L = (A^L, \wedge, \vee, 0, 1)$ be a bounded left-softlattice. Then,*

$$(m - Wabs) + (m - Vabs) \Longleftrightarrow (x \leq_m^{O(\wedge)} y \Leftrightarrow y \geq_m^{O(\vee)} x).$$

**Proof.** Suppose (m-Wabs) and (m-Vabs) hold. Then, if $x \leq_m^{O(\wedge)} y$, i.e. $x \wedge y = x$, then $y \vee x = y \vee (x \wedge y) \overset{(m-Wcomm)}{=} y \vee (y \wedge x) \overset{(m-Vabs)}{=} y$, i.e. $y \geq_m^{O(\vee)} x$. Conversely, if $y \geq_m^{O(\vee)} x$, i.e. $y \vee x = y$, then $x \wedge y = x \wedge (y \vee x) \overset{(m-Vcomm)}{=} x \wedge (x \vee y) \overset{(m-Wabs)}{=} x$, i.e. $x \leq_m^{O(\wedge)} y$.

Conversely, suppose now that $x \leq_m^{O(\wedge)} y \Leftrightarrow y \geq_m^{O(\vee)} x$, i.e. (a) $x \wedge y = x \Leftrightarrow y \vee x = y$ or (b) $y \wedge x = y \Leftrightarrow x \vee y = x$. Then, $x \wedge (x \vee y) = x \wedge (y \vee x) = x \wedge y = x$, by (a), and $x \vee (x \wedge y) = x \vee (y \wedge x) = x \vee y = x$, by (b). Thus, (m-Wabs) and (m-Vabs) hold. □

**Proposition 10.1.8** *Let $\mathcal{A}^L = (A^L, \wedge, \vee, 0, 1)$ be a bounded left-softlattice. We have: for all $x, y, z \in A^L$,*
*(1)* $x \wedge y \leq_m^{O(\wedge)} x, y$,
*(1')* $x \vee y \geq_m^{O(\vee)} x, y$;
*(2)* *if* $x \leq_m^{O(\wedge)} y$, *then* $x \wedge z \leq_m^{O(\wedge)} y \wedge z$,
*(2')* *if* $x \geq_m^{O(\vee)} y$, *then* $x \vee z \geq_m^{O(\vee)} y \vee z$.

**Proof.** Routine. □

### 10.1.2.2 Bounded widelattices: BWL

**Definitions 10.1.9** (The dual ones are omitted)

(1) A *left-widelattice* is an algebra $\mathcal{A}^L = (A^L, \wedge, \vee)$ of type $(2, 2)$ such that the axioms (m-Wcomm), (m-Vcomm), (m-Wass), (m-Vass), (m-Wabs-i), (m-Vabs-i) are satisfied.

(2) A *bounded left-widelattice* is an algebra $\mathcal{A}^L = (A^L, \wedge, \vee, 0, 1)$ of type $(2, 2, 0, 0)$ such that the reduct $(A^L, \wedge, \vee)$ is a left-widelattice and the elements 0 and 1 verify the axioms: for all $x \in A^L$,
(m-WU) $1 \wedge x = x$, (m-VU) $0 \vee x = x$.

We shall denote by **WL** the class of all left-widelattices and by **BWL** the class of all bounded left-widelattices.

Note that any left-lattice is a left-widelattice: $\mathbf{L} \subset \mathbf{WL}$, the inclusion being strict, since there are left-widelattices that are not left-lattices - see the next example.

**Example 10.1.10 Proper widelattice**

The algebra $\mathcal{A}^L = (L_3 = \{a, b, c\}, \wedge, \vee)$, with the following tables of $\wedge$ and $\vee$, is a proper left-widelattice, because (m-Wid) and (m-Vid) are not verified for $a$.

| $\wedge$ | a | b | c |
|---|---|---|---|
| a | c | a | c |
| b | a | b | c |
| c | c | c | c |

and

| $\vee$ | a | b | c |
|---|---|---|---|
| a | b | b | a |
| b | b | b | b |
| c | a | b | c |

.

Hence, we have: **WL** + (m-Wid) + (m-Vid) = **L**,

$$\mathbf{SL} \cap \mathbf{WL} = \mathbf{L} \quad \textit{and} \quad \mathbf{BSL} \cap \mathbf{BWL} = \mathbf{BL}.$$

**Corollary 10.1.11** *(See Corollary 5.2.4)*
*Let $\mathcal{A}^L = (A^L, \wedge, \vee, 0, 1)$ be a bounded left-widelattice. Then, we have the equivalences:*

$$\textit{(m-WU)} \Longleftrightarrow \textit{(m-VL)} \qquad \textit{and} \qquad \textit{(m-VU)} \Longleftrightarrow \textit{(m-WL)}.$$

**Proof.** First, (m-WU) + (m-Wass) + (m-Vabs-i) imply (m-VL):
indeed, in (m-Vabs-i) ($x \vee (x \wedge x \wedge y) = x$), take $x := 1$ to obtain:
$1 = 1 \vee (1 \wedge 1 \wedge y) = 1 \vee (1 \wedge (1 \wedge y)) = 1 \vee (1 \wedge y) = 1 \vee y$; thus, (m-VL) holds.
Conversely, (m-VL) + (m-Vass) + (m-Vcomm) + (m-Wcomm) + (m-Wabs-i) imply (m-WU):
indeed, in (m-Wabs-i) ($x \wedge (x \vee x \vee y) = x$) take $y := 1$ to obtain:
$x = x \wedge (x \vee x \vee 1) = x \wedge (x \vee (x \vee 1)) = x \wedge (x \vee (1 \vee x)) = x \wedge (x \vee 1) = x \wedge 1 = 1 \wedge x$; thus, (m-WU) holds.

Dually, first, (m-VU) + (m-Vass) + (m-Wabs-i) imply (m-WL):
indeed, in (m-Wabs-i), take $x := 0$ to obtain:
$0 = 0 \wedge (0 \vee 0 \vee y) = 0 \wedge (0 \vee (0 \vee y)) = 0 \wedge (0 \vee y) = 0 \wedge y$; thus, (m-WL) holds.
Conversely, (m-WL) + (m-Wass) + (m-Wcomm) + (m-Vcomm) + (m-Vabs-i) imply (m-VU):
indeed, in (m-Vabs-i), take $y := 0$ to obtain:
$x = x \vee (x \wedge x \wedge 0) = x \vee (x \wedge (x \wedge 0)) = x \vee (x \wedge (0 \wedge x)) = x \vee (x \wedge 0) = x \vee 0 = 0 \vee x$; thus, (m-VU) holds. □

### 10.1.3 More on involutive m-MEL algebras

Let $\mathcal{A}^L = (A^L, \odot, ^-, 1)$ be an involutive left-m-MEL algebra throughout this subsection.

**Proposition 10.1.12** *We have:*

$$(m - Pabs - i) \Longleftrightarrow (m - Sabs - i).$$

**Proof.** Routine. □

**Proposition 10.1.13** *We have:*
*(mCIM13) (m-Pabs) + (SU) $\Longrightarrow$ (m-Pabs-i),*
*(mCIM14) (G) + (m-Pabs-i) $\Longrightarrow$ (m-Pabs),*
*(mCIM15) (SU) $\Longrightarrow$ ((m-Pabs) $\Leftrightarrow$ (G) + (m-Pabs-i)).*

**Proof.** Routine.

**Proposition 10.1.14** *We have:*

$$(m - Pabs) \iff (G) + (m - Pabs - i).$$

**Proof.** By (mCIM15). □

**Proposition 10.1.15** *We have:*

$$(m - Pimpl) \iff (G) + (m - Pabs - i).$$

**Proof.** By Proposition 6.4.6, (m-Pimpl) $\iff$ (m-Pabs), then apply above Proposition 10.1.14. □

## 10.2 Two generalizations of bounded involutive lattices

We present, in the next two subsections, two generalizations of bounded involutive (involution) lattices: bounded involutive softlattices and bounded involutive widelattices.

### 10.2.1 Bounded involutive softlattices: BISL

**Definition 10.2.1** (Definition 1) (The dual one is omitted)
A *bounded involutive left-softlattice*, or a *left-BISL* for short, is an algebra $\mathcal{A}^L = (A^L, \wedge, \vee, ^- = ^{-^L}, 0, 1)$ such that the reduct $(A^L, \wedge, \vee, 0, 1)$ is a bounded left-softlattice (Definitions 10.1.3) and the unary operation $^-$ satisfies (DN), (DeM1) and (DeM2).

Recall that in a left-BIL we have: (m-WU) $\iff$ (m-VL) and (m-VU) $\iff$ (m-WL), by Corollary 5.2.4. We shall denote by **BISL** the class of all left-BISLs. We have:
**BISL** + (m-Wabs-i) + (m-Vabs-i) = **BIL** = **BISL** + (m-Wabs) + (m-Vabs).

**Proposition 10.2.2** *Let $\mathcal{A}^L = (A^L, \wedge, \vee, 0, 1)$ be a left-BISL. The following are equivalent:*
*(i)* $y \geq_m^{O(\vee)} x \iff y^- \leq_m^{O(\wedge)} x^-$;
*(ii)=(DeM1)* $(x \vee y)^- = x^- \wedge y^-$.

**Proof.** (i) $\Longrightarrow$ (ii): Suppose (i) holds. Since $x \vee y \geq_m^{O(\vee)} x$, by Proposition 10.1.8 (1'), then $(x \vee y)^- \leq_m^{O(\wedge)} x^-$, by (i), and, similarly, $(x \vee y)^- \leq_m^{O(\wedge)} y^-$; thus, $(x \vee y)^-$ is a lower bound of $\{x^-, y^-\}$. Suppose $z$ is a lower bound of $\{x^-, y^-\}$, i.e. $z \leq_m^{O(\wedge)} x^-, y^-$; then, by (i) and (DN), $z^- \geq_m^{O(\vee)} x, y$, hence $z^- = z^- \vee z^- \geq_m^{O(\vee)} x \vee z^- \geq_m^{O(\vee)} x \vee y$, by Proposition 10.1.8 (2'), hence $z \leq_m^{O(\wedge)} (x \vee y)^-$. Hence, $(x \vee y)^- = x^- \wedge y^-$, i.e. (ii)=(DeM1) holds.

(ii) $\Longrightarrow$ (i): Suppose $y \leq_M^{O(\vee)} x$, i.e. $y \vee x = y$; then, $y^- \wedge x^- \stackrel{(ii)}{=} (y \vee x)^- = y^-$, i.e. $y^- \leq_m^{O(\wedge)} x^-$.
Conversely, suppose $y^- \leq_m^{O(\wedge)} x^-$, i.e. $y^- \wedge x^- = y^-$; then, $y \vee x \stackrel{(DN)}{=} ((y \vee x)^-)^- \stackrel{(ii)}{=} (y^- \wedge x^-)^- = (y^-)^- \stackrel{(DN)}{=} y$, i.e. $y \geq_m^{O(\vee)} x$. □

We have the following definitionally equivalence:

**Theorem 10.2.3**
*(1) Let* $\mathcal{A}^L = (A^L, \odot, ^-, 1)$ *be an involutive left-m-MEL algebra verifying (G). Define, for all* $x, y \in A^L$,

$$x \wedge y \stackrel{def.}{=} x \odot y,\ x \vee y \stackrel{def.}{=} x \oplus y,\ 0 \stackrel{def.}{=} 1^-,$$

*where* $x \oplus y \stackrel{def.}{=} (x^- \odot y^-)^-$, *by (6.5).*
*Then,* $f(\mathcal{A}^L) = (A^L, \wedge, \vee, ^-, 0, 1)$ *is a left-BISL.*
*(1') Let* $\mathcal{A}^L = (A^L, \wedge, \vee, ^-, 0, 1)$ *be a left-BISL. Define, for all* $x, y \in A^L$,

$$x \odot y \stackrel{def.}{=} x \wedge y.$$

*Then,* $g(\mathcal{A}^L) = (A^L, \odot, ^-, 1)$ *is an involutive left-m-MEL algebra verifying (G).*
*(2) The mappings f and g are mutually inverse.*

**Proof.** (1): Let $\mathcal{A}^L = (A^L, \odot, ^-, 1)$ be an involutive left-m-MEL algebra verifying (G). Then, (m-Wcomm), (m-Wass), (m-WU), (m-WL) and (m-Vcomm), (m-Vass), (m-VU), (m-VL) and (DN), (DeM1), (DeM2) hold. By Proposition 6.4.4, since (G) holds, then ($G^R$) holds, hence (m-Wid) and (m-Vid) hold too. Thus, $(A^L, \wedge, \vee, ^-, 0, 1)$ is a left-BISL.

(1'): Let $\mathcal{A}^L = (A^L, \wedge, \vee, ^-, 0, 1)$ be a left-BISL. Define $x \odot y \stackrel{def.}{=} x \wedge y$. Then, (PU) (=(m-WU)), (Pcomm) (=(m-Wcomm)), (Pass) (= (m-Wass)), (m-La) (=(m-WL)), (DN), (G) (=(m-Wid)) hold. Thus, $(A^L, \odot, ^-, 1)$ is an involutive left-m-MEL algebra verifying (G).

(2): Routine. □

The above Theorem 10.2.3 allows us to give a new, equivalent definition of bounded involutive left-softlattices, as follows:

**Definition 10.2.4** (Definition 2) (The dual one is omitted)
A *bounded involutive left-softlattice*, or a *left-BISL* for short, is an involutive left-m-MEL algebra verifying (G).

Hence, **BISL** = **m-MEL**$_{(DN)}$ + (G).
We shall denote by **aBISL** the class of all left-BISLs verifying (m-An).
See Examples 10.4.1, 10.4.2 of proper BISL, aBISL respectively.

### 10.2.2 Bounded involutive widelattices: BIWL

**Definition 10.2.5** (Definition 1) (The dual one is omitted)
A *bounded involutive left-widelattice*, or a *left-BIWL* for short, is an algebra $\mathcal{A}^L = (A^L, \wedge, \vee, ^- = ^{-^L}, 0, 1)$ such that the reduct $(A^L, \wedge, \vee, 0, 1)$ is a bounded left-widelattice (Definitions 10.1.9) and the unary operation $^-$ satisfies (DN), (DeM1) and (DeM2).

Recall that in a left-BIL we have (m-WU) $\Longleftrightarrow$ (m-VL) and (m-VU) $\Longleftrightarrow$ (m-WL), by Corollary 5.2.4. We shall denote by **BIWL** the class of all left-BIWLs. We have:
**BIWL** + (m-Wid) + (m-Vid) = **BIL**.

We have the following definitionally equivalence:

**Theorem 10.2.6**

*(1) Let* $\mathcal{A}^L = (A^L, \odot, ^-, 1)$ *be an involutive left-m-MEL algebra verifying (m-Pabs-i). Define, for all* $x, y \in A^L$,

$$x \wedge y \stackrel{def.}{=} x \odot y,\ x \vee y \stackrel{def.}{=} x \oplus y,\ 0 \stackrel{def.}{=} 1^-,$$

*where* $x \oplus y \stackrel{def.}{=} (x^- \odot y^-)^-$, *by (6.5).*
*Then,* $f(\mathcal{A}^L) = (A^L, \wedge, \vee, ^-, 0, 1)$ *is a left-BIWL.*

*(1') Let* $\mathcal{A}^L = (A^L, \wedge, \vee, ^-, 0, 1)$ *be a left-BIWL. Define, for all* $x, y \in A^L$,

$$x \odot y \stackrel{def.}{=} x \wedge y.$$

*Then,* $g(\mathcal{A}^L) = (A^L, \odot, ^-, 1)$ *is an involutive left-m-MEL algebra verifying (m-Pabs-i).*

*(2) The mappings f and g are mutually inverse.*

**Proof.** (1): Let $\mathcal{A}^L = (A^L, \odot, ^-, 1)$ be an involutive left-m-MEL algebra verifying (m-Pabs-i). Then, (m-Wcomm), (m-Wass), (m-WU), (m-WL) and (m-Vcomm), (m-Vass), (m-VU), (m-VL) and (DN), (DeM1), (DeM2) hold. By Proposition 10.1.12, since (m-Pabs-i) holds, then (m-Sabs-i) holds, hence (m-Wabs-i) and (m-Vabs-i) hold too. Thus, $(A^L, \wedge, \vee, ^-, 0, 1)$ is a left-BIWL.

(1'): Let $\mathcal{A}^L = (A^L, \wedge, \vee, ^-, 0, 1)$ be a left-BIWL. Define $x \odot y \stackrel{def.}{=} x \wedge y$. Then, (PU) (=(m-WU)), (Pcomm) (=(m-Wcomm)), (Pass) (= (m-Wass)), (m-La) (=(m-WL)), (DN), (m-Pabs-i) (=(m-Wabs-i)) hold, by Corollary 10.1.11. Thus, $(A^L, \odot, ^-, 1)$ is an involutive left-m-MEL algebra verifying (m-Pabs-i).

(2): Routine. □

The above Theorem 10.2.6 allows us to give a new, equivalent definition of bounded involutive left-widelattices, as follows:

**Definition 10.2.7** (Definition 2) (The dual one is omitted)

A *bounded involutive left-widelattice*, or a *left-BIWL* for short, is an involutive left-m-MEL algebra verifying (m-Pabs-i).

Hence, **BIWL** = **m-MEL**$_{(DN)}$ + (m-Pabs-i).

We shall denote by **aBIWL** the class of left-BIWLs verifying (m-An).

See Examples 10.4.3, 10.4.4 of proper BIWL, aBIWL respectively.

### 10.2.3 Resuming connections between BIL, BISL, BIWL

We have obtained, based on Proposition 10.1.15, the following equivalent definitions (Definitions 2):

An involutive left-m-MEL algebra $\mathcal{A}^L = (A^L, \odot, ^-, 1)$ is a:
- *left-BIL*, if (m-Pimpl) holds (Definition 9.1.11),
- *left-BISL*, if (G) holds (Definition 10.2.4),
- *left-BIWL*, if (m-Pabs-i) holds (Definition 10.2.7),

i.e.
**BIL** = **m-MEL**$_{(DN)}$ + (m-Pimpl),
**BISL** = **m-MEL**$_{(DN)}$ + (G),
**BIWL** = **m-MEL**$_{(DN)}$ + (m-Pabs-i).

Hence, we have:

$$\mathbf{BIL} = \mathbf{BISL} \cap \mathbf{BIWL}, \tag{10.1}$$

i.e. we have the resuming basic connections from Figure 10.1, useful in the sequel.

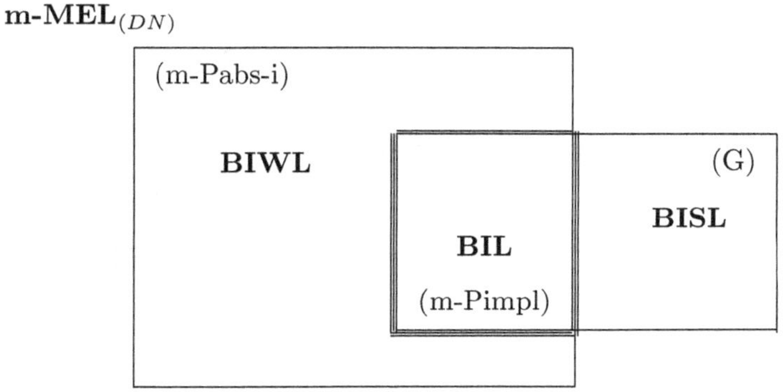


Figure 10.1: Basic connections between **BISL**, **BIWL** and **BILL**

### 10.2.4 Putting BIL and their two generalizations on the "involutive Little m-map"

The definitions (Definitions 2) and the results from this section allow us to draw the hierarchies from the following Figure 10.2, thus putting the mentioned algebras on the "involutive Little m-map".

## 10.3 Two generalizations of ortholattices

We present, in the next two subsections, two generalizations of ortholattices: orthosoftlattices and orthowidelattices.

### 10.3.1 Orthosoftlattices: OSL

**Definition 10.3.1** (Definition 1) (The dual one is omitted)

A *left-orthosoftlattice*, or a *left-OSL* for short, is an algebra $\mathcal{A}^L = (A^L, \wedge, \vee, ^- = {}^{-^L}, 0, 1)$ such that the reduct $(A^L, \wedge, \vee, 0, 1)$ is a bounded left-softlattice (Definitions 10.1.3) and the unary operation $^-$ satisfies (DN), (DeM1), (DeM2) and (m-WRe), (m-VRe).

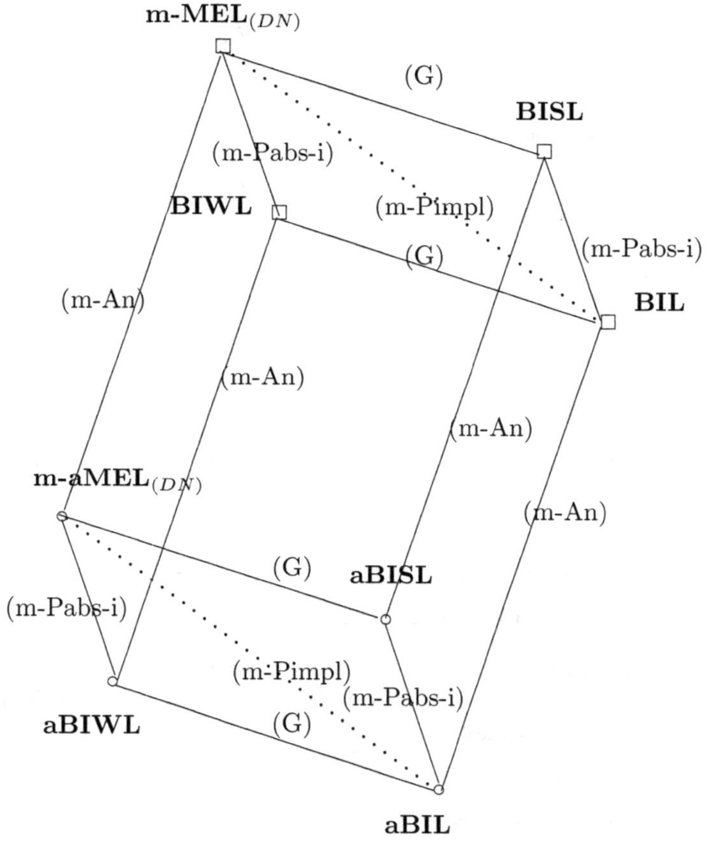


Figure 10.2: Hierarchies **BIWL** vs. **BISL**

We shall denote by **OSL** the class of all left-OSLs. We have:
**OSL** + (m-Wabs-i) + (m-Vabs-i) = **OL** = **OSL** + (m-Wabs) + (m-Vabs)
and, since **OL** = **BIL** + (m-WRe) + (m-VRe), then:
**OSL** = **BISL** + (m-WRe) + (m-VRe).

We have the following definitionally equivalence.

**Theorem 10.3.2** *(See Theorem 10.2.3)*

*(1) Let $\mathcal{A}^L = (A^L, \odot, ^-, 1)$ be an involutive left-m-BE algebra verifying (G). Define, for all $x, y \in A^L$:*

$$x \wedge y \stackrel{def.}{=} x \odot y, \quad x \vee y \stackrel{def.}{=} x \oplus y, \quad 0 \stackrel{def.}{=} 1^-.$$

*Then, $f(\mathcal{A}^L) = (A^L, \wedge, \vee, ^-, 0, 1)$ is a left-OSL.*

*(1') Conversely, let $\mathcal{A}^L = (A^L, \wedge, \vee, ^-, 0, 1)$ be a left-OSL. Define, for all $x, y \in A^L$:*

$$x \odot y \stackrel{def.}{=} x \wedge y.$$

*Then, $g(\mathcal{A}^L) = (A^L, \odot, ^-, 1)$ is an involutive left-m-BE algebra verifying (G).*

*(2) The mappings $f$ and $g$ are mutually inverse.*

**Proof.** (1): Let $\mathcal{A}^L = (A^L, \odot, ^-, 1)$ be an involutive left-m-BE algebra verifying (G). Hence, $\mathcal{A}^L$ is an involutive left-m-MEL algebra verifying (G) and, by Theorem 10.2.3 (1), $(A^L, \wedge, \vee, ^-, 0, 1)$ is a bounded involutive left-softlattice (left-BISL). Since (m-Re) and (m-Re$^R$) hold, then (m-WRe), (m-VRe) hold, hence $(A^L, \wedge, \vee, ^-, 0, 1)$ is a left-OSL.

(1'): Let $\mathcal{A}^L = (A^L, \wedge, \vee, ^-, 0, 1)$ be a left-OSL. Hence, $\mathcal{A}^L$ is a left-BISL verifying (m-WRe), (m-VRe). Define $x \odot y \stackrel{def.}{=} x \wedge y$. By Theorem 10.2.3 (1'), $(A^L, \odot, ^-, 1)$ is an involutive left-m-MEL algebra verifying (G). Since (m-WRe), (m-VRe) hold, it follows that (m-Re), (m-Re$^R$) hold, thus $(A^L, \odot, ^-, 1)$ is an involutive left-m-BE algebra verifying (G).

(2): Routine. □

The above Theorem 10.3.2 allows us to give a new, equivalent definition of left-orthosoftlattices, as follows.

**Definition 10.3.3** (Definition 2) (The dual one is omitted)

A *left-orthosoftlattice*, or a *left-OSL* for short, is an involutive left-m-BE algebra verifying (G).

Hence, **OSL** = **m-BE**$_{(DN)}$ + (G).

**Proposition 10.3.4** *Let $\mathcal{A}^L = (A^L, \wedge, \vee, ^-, 0, 1)$ be a left-OSL. Consider the following dual binary relations:*

*(m-dfrelW)* $\quad x \leq_m y \stackrel{def.}{\Longleftrightarrow} x \wedge y^- = 0$ *and*

*(m-dfrelV)* $\quad x \geq_m y \stackrel{def.}{\Longleftrightarrow} x \vee y^- = 1.$

*Then, we have:*

*(1)* $\quad x \leq_m y \Longleftrightarrow y \geq_m x;$

*(2)* $\quad x \leq_m^{O(\wedge)} y \Longrightarrow x \leq_m y$ *and*

*(2')* $\quad x \geq_m^{O(\vee)} y \Longrightarrow x \geq_m y.$

**Proof.** (1): By Theorem 10.3.2 and Proposition 6.4.8.

(2): Suppose $x \leq_m^{O(\wedge)} y$, i.e. $x \wedge y = x$ (see Proposition 10.1.6); then, $x \wedge y^- = (x \wedge y) \wedge y^- \overset{(m-Wass)}{=} x \wedge (y \wedge y^-) \overset{(m-WRe)}{=} x \wedge 0 \overset{(m-WL)}{=} 0$; thus, $x \leq_m y$.

(2'): Dually, suppose $x \geq_m^{O(\vee)} y$, i.e. $x \vee y = x$ (see Proposition 10.1.6); then, $x \vee y^- = (x \vee y) \vee y^- \overset{(m-Vass)}{=} x \vee (y \vee y^-) \overset{(m-VRe)}{=} x \vee 1 \overset{(m-VL)}{=} 1$; thus, $x \geq_m y$.
□

**Proposition 10.3.5** *Let $\mathcal{A}^L = (A^L, \wedge, \vee, 0, 1)$ be a left-OSL.*
*(1) If (m-Wdiv) $(x \wedge (x \wedge y^-)^- = x \wedge y)$ holds, then*

$$x \leq_m^{O(\wedge)} y \Longleftrightarrow x \leq_m y \ (\Longleftrightarrow y \geq_m x).$$

*(1') If (m-Vdiv) $(x \vee (x \vee y^-)^- = x \vee y)$ holds, then*

$$x \geq_m^{O(\vee)} y \Longleftrightarrow x \geq_m y \ (\Longleftrightarrow y \leq_m x).$$

**Proof.** (1): Suppose $x \leq_m y$, i.e. $x \wedge y^- = 0$; then,
$x \wedge y \overset{(m-Wdiv)}{=} x \wedge (x \wedge y^-)^- = x \wedge 0^- \overset{(Neg0-1)}{=} x \wedge 1 \overset{(m-Wcomm)}{=} 1 \wedge x \overset{(m-WU)}{=} x$,
i.e. $x \leq_m^{O(\wedge)} y$.
The converse follows by Proposition 10.3.4 (2).

(1'): Suppose $x \geq_m y$, i.e. $x \vee y^- = 1$; then,
$x \vee y \overset{(m-Vdiv)}{=} x \vee (x \vee y^-)^- = x \vee 1^- \overset{(Neg1-0)}{=} x \vee 0 \overset{(m-Vcomm)}{=} 0 \vee x \overset{(m-VU)}{=} x$,
i.e. $x \geq_m^{O(\vee)} y$.
The converse follows by Proposition 10.3.4 (2'). □

We shall denote by **tOSL** the class of all transitive left-OSLs (i.e. verifying (m-Tr)), by **aOSL** the class of all antisymmetric left-OSLs (i.e. verifying (m-An)) and by **taOSL** the class of all transitive and antisymmetric left-OSLs.

Note that, by Definition 10.3.3, Proposition 6.4.14 and Definition 9.2.27, we have:
**taOSL** = $\mathbf{m}$-$\mathbf{BE}_{(DN)}$ + (G) + (m-An) + (m-BB) = $\mathbf{m}$-$\mathbf{BE}_{(DN)}$ + (m-Pdiv) = **Boole**.

See Examples 10.4.5, 10.4.6, 10.4.7 of proper OSL, tOSL, aOSL respectively.

### 10.3.2 Orthowidelattices: OWL

**Definition 10.3.6** (Definition 1) (The dual one is omitted)
A *left-orthowidelattice*, or a *left-OWL* for short, is an algebra $\mathcal{A}^L = (A^L, \wedge, \vee, ^- = ^{-^L}, 0, 1)$ such that the reduct $(A^L, \wedge, \vee, 0, 1)$ is a bounded left-widelattice (Definition 10.1.9) and the unary operation $^-$ satisfies (DN), (DeM1), (DeM2) and (m-WRe), (m-VRe).

We shall denote by **OWL** the class of all left-OWLs. We have:

**OWL** + (m-Wid) + (m-Vid) = **OL**

and, since **OL** = **BIL** + (m-WRe) + (m-VRe), then:

**OWL** = **BIWL** + (m-WRe) + (m-VRe).

We have the following definitionally equivalence:

**Theorem 10.3.7** *(See Theorem 10.2.6)*

*(1) Let $\mathcal{A}^L = (A^L, \odot, ^-, 1)$ be an involutive left-m-BE algebra verifying (m-Pabs-i). Define, for all $x, y \in A^L$,*

$$x \wedge y \stackrel{def.}{=} x \odot y,\ x \vee y \stackrel{def.}{=} x \oplus y,\ 0 \stackrel{def.}{=} 1^-,$$

*where $x \oplus y \stackrel{def.}{=} (x^- \odot y^-)^-$, by (6.5).*
*Then, $f(\mathcal{A}^L) = (A^L, \wedge, \vee, ^-, 0, 1)$ is a left-OWL.*

*(1') Let $\mathcal{A}^L = (A^L, \wedge, \vee, ^-, 0, 1)$ be a left-OWL. Define, for all $x, y \in A^L$,*

$$x \odot y \stackrel{def.}{=} x \wedge y.$$

*Then, $g(\mathcal{A}^L) = (A^L, \odot, ^-, 1)$ is an involutive left-m-BE algebra verifying (m-Pabs-i).*

*(2) The mappings $f$ and $g$ are mutually inverse.*

**Proof.** (1): Let $\mathcal{A}^L = (A^L, \odot, ^-, 1)$ be an involutive left-m-BE algebra verifying (m-Pabs-i). By Proposition 9.2.2, (m-WRe) and (m-VRe) hold. Then, $\mathcal{A}^L$ is an involutive left-m-MEL algebra verifying (m-Pabs-i) and, by Theorem 10.2.6 (1), $(A^L, \wedge, \vee, ^-, 0, 1)$ is a bounded involutive left-widelattice (left-BIWL). Since (m-WRe), (m-VRe) hold, it follows that $(A^L, \wedge, \vee, ^-, 0, 1)$ is a left-orthowidelattice.

(1'): Let $\mathcal{A}^L = (A^L, \wedge, \vee, ^-, 0, 1)$ be a left-orthowidelattice. Hence, $\mathcal{A}^L$ is a bounded involutive left-widelattice verifying (m-WRe), (m-VRe). Define $x \odot y \stackrel{def.}{=} x \wedge y$. By Theorem 10.2.6 (1'), $(A^L, \odot, ^-, 1)$ is an involutive left-m-MEL algebra verifying (m-Pabs-i). Since (m-WRe), (m-VRe) hold, it follows that (m-Re), (m-Re$^R$) hold, thus $(A^L, \odot, ^-, 1)$ is an involutive left-m-BE algebra verifying (m-Pabs-i).

(2): Routine. □

The above Theorem 10.3.7 allows us to give a new, equivalent definition of left-orthowidelattices, as follows:

**Definition 10.3.8** (Definition 2) (The dual one is omitted)

A *left-orthowidelattice*, or a *left-OWL* for short, is an involutve left-m-BE algebra verifying (m-Pabs-i).

Hence, **OWL** = **m-BE**$_{(DN)}$ + (m-Pabs-i).

We shall denote by **tOWL** the class of all transitive left-OWLs (i.e. verifying (m-Tr)), by **aOWL** the class of all antisymmetric left-OWLs (i.e. verifying (m-An)) and by **taOWL** the class of all antisymmetric and transitive left-OWLs.

See Example 10.4.8 of proper OWL and Examples 10.4.9 of two proper tOWLs.

A problem we have not been able to resolve is the following.

**Open problem 10.3.9** *Find an example of antisymmetric orthowidelattice which does not verify (m-Tr) ($\Leftrightarrow$ ... (m-BB)), i.e. a proper element of* ***aOWL*** *(using* MACE4, *we have searched exhaustively for an example up through and including size 24), or prove that an involutive left-m-aBE algebra satisfying (m-Pabs-i) satisfies also (m-Tr) (we have also tried to find a proof using* PROVER9, *but despite letting it run for several days, it was unable to find one).*

- **Transitive, antisymmetric OWLs: taOWL**

Here we shall prove Theorem 10.3.14, saying that taOWLs are MV algebras, which is **the core of this chapter**. A direct proof of Theorem 10.3.14 by *Prover9* takes 58321 seconds (about 17 hours) and had 205 lines (length of the proof). The shortest proof that we were able to find has two steps:

**Step 1**: in an involutive m-BE algebra verifying (m-Pabs-i) and (m-BB) ($\Leftrightarrow \ldots$ (m-Tr)), but not necessarily (m-An), we have the property:

$$(\Delta_m) \qquad (x \wedge_m y) \odot (y \wedge_m x)^- = 0,$$

where $x \wedge_m y \overset{def.}{=} (x^- \odot y)^- \odot y$; this is Theorem 10.3.13, whose proof by *Prover9* had 'only' 178 lines (intermediary results = steps of the proof) which, after making the graph of their dependencies, were grouped into the below Lemmas 10.3.10, 10.3.11, 10.3.12 and Theorem 10.3.13, with their corresponding "humanized" proofs.

**Step 2**: in an involutive m-BE algebra verifying (m-Pabs-i), (m-BB) and (m-An), we have ($\wedge_m$-comm); this is Theorem 10.3.14, which follows easily by Theorem 10.3.13, since

$$(10.2) \qquad (\Delta_m) + (m - An) \Longleftrightarrow (\wedge_m - comm)$$

(indeed, ($\wedge_m$-comm) implies (m-An), by (mB2) from Section 6.2).

**Lemma 10.3.10** *Let $\mathcal{A}^L = (A^L, \odot, ^-, 1)$ be an involutive left-m-BE algebra. We have:*

$$(10.3) \qquad x \odot (y \odot z) = y \odot (x \odot z),$$

$$(10.4) \qquad x \odot (y \oplus x^-) = y \wedge_m x,$$

$$(10.5) \qquad (x \odot (y \odot z^-))^- = (x \odot y)^- \oplus z,$$

$$(10.6) \qquad x^- \odot (x \wedge_m y) = 0,$$

$$(10.7) \qquad (x \odot (y^- \odot z))^- = y \oplus (x \odot z)^-,$$

$$(10.8) \qquad x \odot ((y \oplus x^-_{\odot} z) = (y \wedge_m x) \odot z,$$

$$(10.9) \qquad x \odot (y \odot (z \oplus (x \odot y)^-)) = z \wedge_m (x \odot y),$$

$$(10.10) \qquad x \odot (y \odot (z \oplus x^-)) = y \odot (z \wedge_m x),$$

$$(10.11) \qquad x \oplus (y \oplus x)^- = (y \wedge_m x^-)^-,$$

(10.12) $$x \odot (x \oplus y)^- = 0,$$

(10.13) $$x \oplus (x \odot y)^- = 1,$$

(10.14) $$x^- \odot ((x \wedge_m y) \odot z) = 0,$$

(10.15) $$x \odot (y \wedge_m x^-) = 0,$$

(10.16) $$x \oplus (y \odot x)^- = 1,$$

(10.17) $$x \odot (y \odot x)^- = y^- \wedge_m x,$$

(10.18) $$x \oplus (x \wedge_m y)^- = 1,$$

(10.19) $$(x \oplus y^-) \odot (z \wedge_m y) = (x \wedge_m y) \odot (z \oplus y^-),$$

(10.20) $$x \odot (x \odot y)^- = y^- \wedge_m x,$$

(10.21) $$(x^- \wedge_m y^-)^- = y \oplus (x \odot y^-),$$

(10.22) $$x \odot (y \wedge_m x)^- = (y^- \odot x) \wedge_m x,$$

(10.23) $$x \oplus ((y \wedge_m x) \odot z)^- = 1,$$

(10.24) $$x \wedge_m (y \odot x) = y \odot x,$$

(10.25) $$(x \oplus (y \odot x^-))^- = y^- \wedge_m x^-,$$

(10.26) $$x \odot (y \oplus (z \odot x)^-) = (z^- \oplus y) \wedge_m x,$$

(10.27) $$x^- \wedge_m (x \oplus y)^- = (x \oplus y)^-,$$

(10.28) $$x \wedge_m ((x \wedge_m y) \odot z) = (x \wedge_m y) \odot z,$$

(10.29) $$x \odot ((x^- \oplus y) \wedge_m z) = y \wedge_m (x \odot z),$$

(10.30) $$(x^- \wedge_m y)^- \wedge_m x = x,$$

(10.31) $$x \odot (y \oplus (x^- \oplus z)) = (y \oplus z) \wedge_m x,$$

(10.32) $$x \odot ((x \odot y)^- \oplus z) = (y^- \oplus z) \wedge_m x,$$

(10.33) $$(x \wedge_m y)^- \wedge_m x^- = x^-,$$

(10.34) $$(x \oplus y) \wedge_m x = x,$$

(10.35) $$(x \wedge_m y^-)^- \wedge_m y = y,$$

(10.36) $$((x^- \odot y) \wedge_m y) \odot z = y \odot ((x \wedge_m y)^- \odot z),$$

(10.37) $$x \oplus (y \wedge_m (z \wedge_m x))^- = 1,$$

(10.38) $$x \wedge_m (y \wedge_m (x \wedge_m z)) = y \wedge_m (x \wedge_m z),$$

(10.39) $$x \wedge_m (y \odot (z \wedge_m (x \wedge_m u))) = y \odot (z \wedge_m (x \wedge_m u)),$$

(10.40) $$(x \oplus (y \odot x^-)) \wedge_m y = y,$$

(10.41) $$x \odot (y \oplus (z \wedge_m x)^-) = (y \oplus (z^- \odot x)) \wedge_m x,$$

(10.42) $$(x \odot y)^- \wedge_m x^- = x^-,$$

(10.43) $$x \odot (y \odot ((z \oplus x^-) \wedge_m u)) = y \odot (z \wedge_m (x \odot u)),$$

(10.44) $$x \wedge_m (y \wedge_m (z \wedge_m x)) = y \wedge_m (z \wedge_m x),$$

(10.45) $$x \wedge_m (y \wedge_m (z \wedge_m (x \wedge_m u))) = y \wedge_m (z \wedge_m (x \wedge_m u)).$$

**Proof.** We shall prove most of the above properties:

(10.14): $x^- \odot ((x \wedge_m y) \odot z) \overset{(10.8)}{=} x^- \odot [y \odot ((x \oplus y^-) \odot z)] = x^- \odot y \odot (x^- \odot y)^- \odot z \overset{(m-Re),(m-La)}{=} 0$.

(10.19): $(x \oplus y^-) \odot (z \wedge_m y) \overset{(10.10)}{=} y \odot ((x \oplus y^-) \odot (z \oplus y^-))$ and $(x \wedge_m y) \odot (z \oplus y^-) \overset{(10.8)}{=} y \odot ((x \oplus y^-) \odot (z \oplus y^-))$.

(10.22): $x \odot (y \wedge_m x)^- \overset{(10.11)}{=} x \odot (x^- \oplus (y \oplus x^-)^-) = x \odot (x^- \oplus (y^- \odot x))$ and $(y^- \odot x) \wedge_m x = x \odot (x^- \oplus (y^- \odot x))$.

(10.23): $x \oplus ((y \wedge_m x) \odot z)^- \overset{(10.8)}{=} x \oplus [x \odot ((y \oplus x^-) \odot z)]^- \overset{(10.13)}{=} 1$.

(10.24): $x \wedge_m (y \odot x) \overset{(10.9)}{=} y \odot (x \odot (x \oplus (y \odot x)^-)) \overset{(10.16)}{=} y \odot (x \odot 1) = y \odot x$.

(10.26): $(z^- \oplus y) \wedge_m x = x \odot ((z^- \oplus y)^- \odot x)^- = x \odot (z^- \oplus y \oplus x^-) = x \odot (y \oplus (z^- \oplus x^-)) = x \odot (y \oplus (z \odot x)^-)$.

(10.27): $x^- \wedge_m (x \oplus y)^- = x^- \wedge_m (x^- \odot y^-) = x^- \wedge_m (y^- \odot x^-) \overset{(10.24)}{=} y^- \odot x^- = (x \oplus y)^-$.

(10.28): Take $X := (x \wedge_m y) \odot z$ in (10.17); we obtain:
$x \wedge_m ((x \wedge_m y) \odot z) \overset{(10.17)}{=} X \odot (x^- \odot X)^- = X \odot (x^- \odot ((x \wedge_m y) \odot z))^- \overset{(10.14)}{=} X \odot 0^- = X \odot 1 = X = (x \wedge_m y) \odot z$.

(10.29): $y \wedge_m (x \odot z) \overset{(10.9)}{=} x \odot (z \odot (y \oplus (x \odot z)^-)) \overset{(10.26)}{=} x \odot ((x^- \oplus y) \wedge_m z)$.

(10.30): $(x^- \wedge_m y)^- \wedge_m x = x \odot (x^- \oplus (x^- \wedge_m y)^-) \overset{(10.18)}{=} x \odot 1 = x$.

(10.31): $x \odot (y \oplus (x^- \oplus z)) = x \odot (x^- \oplus (y \oplus x)) = (y \oplus z) \wedge_m x$.

(10.32): $x \odot ((x \odot y)^- \oplus z) \overset{(10.5)}{=} x \odot (x \odot (y \odot z^-))^- (y \odot z^-)^- \wedge_m x = (y^- \oplus z) \wedge_m x$.

(10.33): $(x \wedge_m y)^- \wedge_m x^- = x^- \odot (x^- \odot (x \wedge_m y))^- \overset{(10.6)}{=} x^- \odot 0^- = x^-$.

(10.34): $(x \oplus y) \wedge_m x = x \odot (x \odot (x \oplus y)^-)^- \overset{(10.12)}{=} x \odot 0^- = x \odot 1 = x$.

(10.35): $(x \wedge_m y^-)^- \wedge_m y = y \odot (y \odot (x \wedge_m y^-))^- \overset{(10.15)}{=} y \odot 0^- = y$.

(10.36): $((x^- \odot y) \wedge_m y) \odot z \overset{(10.22)}{=} (y \odot (x \wedge_m y)^-) \odot z$.

(10.37): $x \oplus (y \wedge_m (z \wedge_m x))^- \overset{(10.4)}{=} x \oplus (X \odot (y \oplus X^-))^- = x \oplus ((z \wedge_m x) \odot (y \oplus (z \wedge_m x)^-))^- \overset{(10.23)}{=} 1$.

(10.38): $x \wedge_m (y \wedge_m (x \wedge_m z)) \overset{(10.4)}{=} x \wedge_m (X \odot (y \oplus X^-)) = x \wedge_m ((x \wedge_m z) \odot (y \oplus (x \wedge_m z)^-)) \overset{(10.28)}{=} (x \wedge_m z) \odot (y \oplus (x \wedge_m z)^-) \overset{(10.4)}{=} y \wedge_m (x \wedge_m z)$.

(10.39): Put $X := x \wedge_m u$ in (10.10); we obtain:
$y \odot (z \wedge_m (x \wedge_m u)) \overset{(10.10)}{=} X \odot (y \odot (z \oplus X^-)) = (x \wedge_m u) \odot (y \odot (z \oplus (x \wedge_m u)^-))$
$\overset{(10.28)}{=} x \wedge_m ((x \wedge_m u) \odot Z) = x \wedge_m ((x \wedge_m u) \odot (y \odot (z \oplus (x \wedge_m u)^-)))$
$\overset{(10.10)}{=} x \wedge_m (y \odot (z \wedge_m X)) = x \wedge_m (y \odot (z \wedge_m (x \wedge_m u)))$, for $Z := y \odot (z \oplus (x \wedge_m u)^-)$ and $X := x \wedge_m u$ again.

(10.40): $(x \oplus (y \odot x^-)) \wedge_m y \overset{(10.21)}{=} (y^- \wedge_m x^-)^- \wedge_m y \overset{(10.30)}{=} y$.

(10.41): Put $Z := z^- \odot x$; then, we have:
$(y \oplus (z^- \odot x)) \wedge_m x \overset{(10.31)}{=} x \odot (y \oplus (x^- \oplus Z)) = x \odot (y \oplus (x^- \oplus (z^- \odot x)))$
$= x \odot (y \oplus x^- \oplus (z \oplus x^-)^-) \overset{(10.11)}{=} x \odot (y \oplus (z \wedge_m x)^-)$.

(10.42): $(x \odot y)^- \wedge_m x^- = (x^- \oplus y^-) \wedge_m x^- \overset{(10.34)}{=} x^-$.

(10.43): Put $Z := (z \oplus x^-) \wedge_m u$ to obtain:
$x \odot (y \odot ((z \oplus x^-) \wedge_m u)) \overset{(10.3)}{=} y \odot (x \odot Z)$
$= y \odot (x \odot ((z \oplus x^-) \wedge_m u)) = y \odot (x \odot ((x^- \oplus z) \wedge_m u)) \overset{(10.29)}{=} y \odot (z \wedge_m (x \odot u))$.

(10.44): $x \wedge_m (y \wedge_m (z \wedge_m x)) \overset{(10.4)}{=} (y \wedge_m (z \wedge_m x)) \odot (x \oplus (y \wedge_m (z \wedge_m x))^-)$
$\overset{(10.37)}{=} (y \wedge_m (z \wedge_m x)) \odot 1 = y \wedge_m (z \wedge_m x)$.
(10.45): $y \wedge_m (z \wedge_m (x \wedge_m u)) \overset{(10.38)}{=} y \wedge_m [x \wedge_m (z \wedge_m (x \wedge_m u))]$
$\overset{(10.38)}{=} x \wedge_m (y \wedge_m [x \wedge_m (z \wedge_m (x \wedge_m u))]) = x \wedge_m (y \wedge_m (z \wedge_m (x \wedge_m u)))$. □

**Lemma 10.3.11** *Let $\mathcal{A}^L = (A^L, \odot, ^-, 1)$ be an involutive left-m-BE algebra. If (m-BB) holds (i.e. if $\mathcal{A}^L$ is a left-m-pre-BCK$_{(DN)}$ algebra), then we have:*

(10.46) $$(x \odot y^-)^- \odot (z \odot ((y \odot z)^- \odot x)) = 0,$$

(10.47) $$(x^- \oplus y) \odot (z \odot ((y \odot z)^- \odot x)) = 0,$$

(10.48) $$(x^- \oplus y) \odot (z \odot (x \odot (y \odot z)^-)) = 0,$$

(10.49) $$x \odot ((x^- \oplus y) \odot (y^- \wedge_m z)) = 0,$$

(10.50) $$(x \wedge_m y) \odot (x^- \wedge_m z) = 0,$$

(10.51) $$(x^- \wedge_m y)^- \wedge_m (x \wedge_m z) = x \wedge_m z,$$

(10.52) $$x \odot ((x^- \wedge_m y) \wedge_m z) = 0,$$

(10.53) $$x^- \odot ((x \wedge_m y) \wedge_m z) = 0,$$

(10.54) $$(x \oplus (y \odot x^-)) \wedge_m (y \wedge_m z) = y \wedge_m z,$$

(10.55) $$x \wedge_m ((x \wedge_m y) \wedge_m z) = (x \wedge_m y) \wedge_m z,$$

(10.56) $$(x \wedge_m y)^- \odot (x \wedge_m (y \odot z)) = 0,$$

(10.57) $$x \oplus ((x \wedge_m y) \wedge_m z)^- = 1,$$

(10.58) $$x \wedge_m ((x \odot y) \wedge_m z) = (x \odot y) \wedge_m z,$$

(10.59) $$x \wedge_m ((y \wedge_m x) \wedge_m z) = (y \wedge_m x) \wedge_m z,$$

(10.60) $$(x \odot y) \wedge_m (y \odot (x \wedge_m z)) = y \odot (x \wedge_m z),$$

(10.61) $$x \wedge_m ((y \wedge_m (x \wedge_m z) \wedge_m u) = (y \wedge_m (x \wedge_m z)) \wedge_m u,$$

(10.62) $$(x \wedge_m y) \oplus (x \wedge_m (y \odot z))^- = 1,$$

(10.63) $$(x \wedge_m y)^- \odot (x \wedge_m (z \wedge_m y)) = 0,$$

(10.64) $$x \odot (y \odot (y \wedge_m x)^-) = 0,$$

(10.65) $$(x \oplus ((x \wedge_m y)^- \odot z)) \wedge_m z = z,$$

(10.66) $$x \wedge_m ((y \wedge_m (z \odot x)) \wedge_m u) = (y \wedge_m (z \odot x)) \wedge_m u,$$

(10.67) $$(x \wedge_m y) \wedge_m (x \wedge_m (z \wedge_m y)) = x \wedge_m (z \wedge_m y),$$

(10.68) $$(x \wedge_m y)^- \odot ((x \wedge_m z) \wedge_m y) = 0,$$

(10.69) $$(x \wedge_m y)^- \odot ((z \wedge_m x) \wedge_m y) = 0,$$

(10.70) $$x^- \wedge_m (y \odot (y \wedge_m x)^-) = y \odot (y \wedge_m x)^-,$$

(10.71) $$(x^- \oplus ((x \oplus y) \odot z)) \wedge_m z = z,$$

(10.72) $$((x \wedge_m y)^- \oplus (x \odot z)) \wedge_m z = z,$$

(10.73) $$(x \wedge_m y)^- \wedge_m (z \odot (z \wedge_m x)^-) = z \odot (z \wedge_m x)^-,$$

(10.74) $$((x \oplus y) \odot z) \wedge_m (x \odot z) = x \odot z,$$

(10.75) $$((x \oplus y) \odot z) \oplus ((x \odot z) \wedge_m u)^- = 1.$$

**Proof.** (10.49): $x \odot ((x^- \oplus y) \odot (y^- \wedge_m z)) \overset{(10.17)}{=} x \odot ((x^- \oplus y) \odot z \odot (y \odot z)^-) \overset{(10.48)}{=} 0.$

(10.50): $(x \wedge_m y) \odot (x^- \wedge_m z) \overset{(10.8)}{=} y \odot ((y^- \oplus x) \odot (x^- \wedge_m z)) \overset{(10.49)}{=} 0.$

(10.51): $(x^- \wedge_m y)^- \wedge_m (x \wedge_m z) \overset{(10.20)}{=} (x \wedge_m z) \odot ((x \wedge_m z) \odot (x^- \wedge_m y))^- \overset{(10.50)}{=} (x \wedge_m z) \odot 0^- = (x \wedge_m z) \odot 1 = x \wedge_m z.$

(10.52): $x \odot ((x^- \wedge_m y) \wedge_m z) \overset{(10.30)}{=} ((x^- \wedge_m y)^- \wedge_m x) \odot ((x^- \wedge_m y) \wedge_m z) \overset{(10.50)}{=} 0.$

(10.53): $x^- \odot ((x \wedge_m y) \wedge_m z) \overset{(10.33)}{=} ((x \wedge_m y)^- \wedge_m x^-) \odot ((x \wedge_m y) \wedge_m z) \overset{(10.50)}{=} 0.$

(10.54): $(x \oplus (y \odot x^-)) \wedge_m (y \wedge_m z) \overset{(10.21)}{=} (y^- \wedge_m x^-)^- \wedge_m (y \wedge_m z) \overset{(10.51)}{=} y \wedge_m z$.

(10.55): $x \wedge_m ((x \wedge_m y) \wedge_m z) \overset{(10.33)}{=} ((x \wedge_m y)^- \wedge_m x^-)^- \wedge_m ((x \wedge_m y) \wedge_m z) \overset{(10.51)}{=} (x \wedge_m y) \wedge_m z$.

(10.56): $(x \wedge_m y)^- \odot (x \wedge_m (y \odot z)) \overset{(10.43)}{=} y \odot ((x \wedge_m y)^- \odot ((x \oplus y^-) \wedge_m z))$
$y \odot ((x \wedge_m y)^- \odot ((x^- \odot y)^- \wedge_m z)) \overset{(10.36)}{=} ((x^- \odot y) \wedge_m y) \odot Z$
$= ((x^- \odot y) \wedge_m y) \odot ((x^- \odot y)^- \wedge_m z) \overset{(10.50)}{=} 0$, for $Z := (x^- \odot y)^- \wedge_m z$.

(10.57): $x \oplus ((x \wedge_m y) \wedge_m z)^- = x^- \odot ((x \wedge_m y) \wedge_m z) \overset{(10.52)}{=} 0$.

(10.58): $x \wedge_m ((x \odot y) \wedge_m z) \overset{(10.42)}{=} ((x \odot y)^- \wedge_m x^-)^- \wedge_m ((x \odot y) \wedge_m z) \overset{(10.51)}{=} (x \odot y) \wedge_m z$.

(10.59): $x \wedge_m ((y \wedge_m x) \wedge_m z) \overset{(10.35)}{=} ((y \wedge_m x)^- \wedge_m x^-)^- \wedge_m ((y \wedge_m x) \wedge_m z) \overset{(10.51)}{=} (y \wedge_m x) \wedge_m z$.

(10.60): $(x \odot y) \wedge_m (y \odot (x \wedge_m z)) \overset{(10.29)}{=} y \odot ((y^- \oplus (x \odot y)) \wedge_m (x \wedge_m z)) \overset{(10.54)}{=} y \odot (x \wedge_m z)$.

(10.61): $(y \wedge_m (x \wedge_m z)) \wedge_m u \overset{(10.38)}{=} (x \wedge_m (y \wedge_m (x \wedge_m z))) \wedge_m u$
$\overset{(10.55)}{=} x \wedge_m ((x \wedge_m (y \wedge_m (x \wedge_m z))) \wedge_m u) \overset{(10.38)}{=} x \wedge_m ((y \wedge_m (x \wedge_m z)) \wedge_m u)$.

(10.63): $(x \wedge_m y)^- \odot (x \wedge_m (z \wedge_m y)) \overset{(10.4)}{=} (x \wedge_m y)^- \odot (x \wedge_m (y \odot (z \oplus y^-))) \overset{(10.56}{=} 0$.

(10.64): $x \odot (y \odot (y \wedge_m x)^-) = (y \wedge_m x)^- \odot (x \odot y) \overset{(10.24)}{=} (y \wedge_m x)^- \odot (y \wedge_m (x \odot y)) \overset{(10.56)}{=} 0$.

(10.65): $(x \oplus ((x \wedge_m y)^- \odot z)) \wedge_m z \overset{(10.41)}{=} z \odot (x \oplus ((x \wedge_m y) \wedge_m z)^-) \overset{(10.57)}{=} z \odot 1 = z$.

(10.66): $(y \wedge_m (z \odot x)) \wedge_m u \overset{(10.59)}{=} (z \odot x) \wedge_m ((y \wedge_m (z \odot x)) \wedge_m u) = (x \odot z) \wedge_m ((y \wedge_m (z \odot x)) \wedge_m u)$
$\overset{(10.58)}{=} x \wedge_m ((z \odot x) \wedge_m ((y \wedge_m (z \odot x)) \wedge_m u) \overset{(10.59)}{=} x \wedge_m ((y \wedge_m (z \odot x)) \wedge_m u)$.

(10.67): $(x \wedge_m y) \wedge_m (x \wedge_m (z \wedge_m y)) \overset{(10.17)}{=} (x \wedge_m (z \wedge_m y)) \odot ((x \wedge_m y)^- \odot (x \wedge_m (z \wedge_m y)))^-$
$\overset{(10.63)}{=} (x \wedge_m (z \wedge_m y)) \odot 0^- = (x \wedge_m (z \wedge_m y)) \odot 1 = x \wedge_m (z \wedge_m y)$.

(10.68): $(x \wedge_m y)^- \odot ((x \wedge_m z) \wedge_m y) \overset{(10.55)}{=} (x \wedge_m y)^- \odot (x \wedge_m ((x \wedge_m z) \wedge_m y)) \overset{(10.63)}{=} 0$.

(10.69): $(x \wedge_m y)^- \odot ((z \wedge_m x) \wedge_m y) \overset{(10.59)}{=} (x \wedge_m y)^- \odot (x \wedge_m ((z \wedge_m x) \wedge_m y)) \overset{(10.63)}{=} 0$.

(10.70): $x^- \wedge_m (y \odot (y \wedge_m x)^-) \overset{(10.17)}{=} (y \odot (y \wedge_m x)^-) \odot (x \odot (y \odot (y \wedge_m x)^-))^-$
$\overset{(10.64)}{=} (y \odot (y \wedge_m x)^-) \odot 0^- = y \odot (y \wedge_m x)^-$.

(10.71): $(x^- \oplus ((x \oplus y) \odot z)) \wedge_m z \overset{(10.27)}{=} (x^- \oplus ((x^- \wedge_m (x \oplus y)^-)^- \odot z)) \wedge_m z \overset{(10.65)}{=} z$.

(10.72): $((x \wedge_m y)^- \oplus (x \odot z)) \wedge_m z \overset{(10.33)}{=} ((x \wedge_m y)^- \oplus (((x \wedge_m y)^- \wedge_m x^-)^- \odot z)) \wedge_m z \overset{(10.65)}{=} z$.

(10.73): $(x \wedge_m y)^- \wedge_m (z \odot (z \wedge_m x)^-) \overset{(10.70)}{=} (x \wedge_m y)^- \wedge_m [x^- \wedge_m (z \odot (z \wedge_m x)^-)]$
$\overset{(10.51)}{=} x^- \wedge_m (z \odot (z \wedge_m x)^-) \overset{(10.70)}{=} z \odot (z \wedge_m x)^-$.

(10.74): Put $X := x$, $Y := (x \oplus y) \odot z$ and $Z := z$ in (10.29) to obtain:
$((x \oplus y) \odot z) \wedge_m (x \odot z) \overset{(10.29)}{=} x \odot ((x^- \oplus ((x \oplus y) \odot z)) \wedge_m z) \overset{(10.71)}{=} x \odot z$.

(10.75): Put $X := (x \oplus y) \odot z$ and $Y := x \odot z$ in (10.74); hence, $X \wedge_m Y = x \odot z$. Then,
$((x \oplus y) \odot z) \oplus ((x \odot z) \wedge_m u)^- \overset{(10.74)}{=} ((x \oplus y) \odot z) \oplus ([((x \oplus y) \odot z) \wedge_m (x \odot z)] \wedge_m u)^- \overset{(10.57)}{=} 1$. □

**Lemma 10.3.12** *Let $\mathcal{A}^L = (A^L, \odot, ^-, 1)$ be an involutive left-m-BE algebra. If (m-Pabs-i) holds (i.e. if $\mathcal{A}^L$ is a left-OWL), then we have:*

$$x \odot (x \oplus (y \oplus x)) = x, \tag{10.76}$$

$$x \odot (x \oplus x) = x, \tag{10.77}$$

$$x \oplus (x \odot x) = x, \tag{10.78}$$

$$(x \oplus x) \odot (y \wedge_m x) = y \wedge_m x, \tag{10.79}$$

$$(x \odot y) \oplus (x \odot (y \odot (x \odot y))) = x \odot y, \tag{10.80}$$

$$x \odot (x \odot x) = x \odot x, \tag{10.81}$$

$$(x^- \oplus y) \wedge_m (x \odot x) = y \wedge_m (x \odot x), \tag{10.82}$$

$$x \odot (y \wedge_m (x \odot x)) = y \wedge_m (x \odot x), \tag{10.83}$$

$$(x \odot y) \wedge_m (y \odot x) = y \odot x. \tag{10.84}$$

**Proof.** (10.76): By (Scomm).

(10.77): From (m-Pabs-i), taking $y = 0$.

(10.78): It is the dual of (10.77).

(10.79): $(x \oplus x) \odot (y \wedge_m x) = (x \oplus x) \odot (x \odot (y^- \odot x^-)) \overset{(Pass)}{=} ((x \oplus x) \odot x) \odot (y^- \odot x)^- \overset{(10.77)}{=} x \odot (y^- \odot x)^- = y \wedge_m x$.

(10.80): By (10.78), for $X := x \odot y$.

(10.81): $x \odot (x \odot x) \overset{(10.78)}{=} (x \oplus (x \odot x)) \odot (x \odot x) \overset{(10.78)}{=} (x \oplus (x \odot x) \oplus (x \odot x)) \odot (x \odot x) \overset{(m-Pabs-i)}{=} x \odot x$.

(10.82): Put $X := x \odot x$ in (10.26); we obtain:
$(x^- \oplus y) \wedge_m (x \odot x) \overset{(10.26)}{=} X \odot (y \oplus (x \odot X)^-) = (x \odot x) \odot (y \oplus (x \odot (x \odot x))^-)$
$\overset{(10.81)}{=} (x \odot x) \odot (y \oplus (x \odot x)^-) \overset{(10.4)}{=} y \wedge_m (x \odot x)$.

(10.83): $y \wedge_m (x \odot x) \overset{(10.81)}{=} y \wedge_m (x \odot (x \odot x)) \overset{(10.29)}{=} x \odot ((x^- \oplus y) \wedge_m z)$
$= x \odot ((x^- \oplus y) \wedge_m (x \odot x)) \overset{(10.82)}{=} x \odot (y \wedge_m (x \odot x))$.

(10.84): $(x \odot y) \wedge_m (y \odot x) \overset{(10.80)}{=} [(y \odot x) \oplus (x \odot (y \odot (x \odot y)))] \wedge_m (y \odot x) \overset{(10.34)}{=} y \odot x$.
□

**Theorem 10.3.13** *Let $\mathcal{A}^L = (A^L, \odot, ^-, 1)$ be an involutive left-m-BE algebra. If (m-BB) and (m-Pabs-i) hold (i.e. if $\mathcal{A}^L$ is a left-tOWL), then $(\Delta_m)$ holds.*

**Proof.** The proof has 47 steps. Here they are:

(10.85) $$x^- \wedge_m ((x \wedge_m y) \odot (x \wedge_m y)) = 0.$$

(10.85): Put $X := x \wedge_m y$ and $Y := x^-$ in (10.83) to obtain:
$x^- \wedge_m ((x \wedge_m y) \odot (x \wedge_m y)) \overset{(10.83)}{=} (x \wedge_m y) \odot (x^- \wedge_m ((x \wedge_m y) \odot (x \wedge_m y))) \overset{(10.50)}{=} 0.$

(10.86) $$((x \oplus x)^- \wedge_m y) \wedge_m x = 0.$$

(10.86): Put $Y := (x \oplus x)^- \wedge_m y$ in (10.79) to obtain:
$((x \oplus x)^- \wedge_m y) \wedge_m x \overset{(10.79)}{=} (x \oplus x) \odot (Y \wedge_m x) = (x \oplus x) \odot (((x \oplus x)^- \wedge_m y) \wedge_m x) \overset{(10.52)}{=} 0.$

(10.87) $$((x \odot x) \wedge_m y) \wedge_m x^- = 0.$$

(10.87): $((x \odot x) \wedge_m y) \wedge_m x^- = ((x^- \oplus x^-)^- \wedge_m y) \wedge_m x^- \overset{(10.86)}{=} 0.$

(10.88) $$(x \odot y)^- \oplus ((y \odot x) \wedge_m x) = 1.$$

(10.88): $(x \odot y)^- \oplus ((y \odot x) \wedge_m x) \overset{(10.84)}{=} ((y \odot x) \wedge_m (x \odot y))^- \oplus ((y \odot x) \wedge_m x) \overset{(10.62)}{=} 1.$

(10.89) $$(x \oplus ((x \odot x) \wedge_m y))^- \wedge_m x^- = x^-.$$

(10.89): $(x \oplus ((x \odot x) \wedge_m y))^- \wedge_m x^- = (x^- \odot ((x \odot x) \wedge_m y)^-) \wedge_m x^-$
$= (((x \odot x) \wedge_m y)^- \odot x^-) \wedge_m x^- \overset{(10.22)}{=} x^- \odot [((x \odot x) \wedge_m y) \wedge_m x^-]^- \overset{(10.87)}{=} x^- \odot 0^- =$
$x^- \odot 1 = x^-.$

(10.90) $$(x \odot (x \wedge_m y)) \wedge_m x^- = 0.$$

(10.90): $(x \odot (x \wedge_m y)) \wedge_m x^- \overset{(10.60)}{=} [(x \odot x) \wedge_m (x \odot (x \wedge_m y))] \wedge_m x^- \overset{(10.87)}{=} 0.$

(10.91) $$(x^- \oplus ((x \odot y) \wedge_m y)) \wedge_m y = y.$$

(10.91): $(x^- \oplus ((x \odot y) \wedge_m y)) \wedge_m y \overset{(10.32)}{=} y \odot [(y \odot x)^- \oplus ((x \odot y) \wedge_m y)] \overset{(10.88)}{=} y \odot 1 = y.$

(10.92) $$((x \odot (x \wedge_m y)) \wedge_m z) \wedge_m x^- = 0.$$

(10.92): In (10.68), put $X := x \odot (x \wedge_m y)$, $Y := x^-$ and $Z := z$, to obtain:
$((x \odot (x \wedge_m y)) \wedge_m x^-)^- \odot [((x \odot (x \wedge_m y)) \wedge_m z) \wedge_m x^-] = 0$, hence, by (10.90), $0^- \odot [((x \odot (x \wedge_m y)) \wedge_m z) \wedge_m x^-] = 0$, i.e. $1 \odot [((x \odot (x \wedge_m y)) \wedge_m z) \wedge_m x^-] = 0$, thus $((x \odot (x \wedge_m y)) \wedge_m z) \wedge_m x^- = 0$.

(10.93) $$x \odot (((x \odot y) \wedge_m y)^- \odot (y \wedge_m z)) = 0.$$

(10.93): $A := x \odot (((x \odot y) \wedge_m y)^- \odot (y \wedge_m z)) \overset{(Pass)}{=} (x \odot ((x \odot y) \wedge_m y)^-) \odot (y \wedge_m z)$
$= (x^- \oplus ((x \odot y) \wedge_m y))^- \odot (y \wedge_m z)$; since $y \overset{(10.91)}{=} (x^- \oplus ((x \odot y)) \wedge_m y) \wedge_m y$, it

follows that
$A \overset{(10.91)}{=} (x^- \oplus ((x \odot y) \wedge_m y))^- \odot (((x^- \oplus ((x \odot y) \wedge_m y)) \wedge_m y) \wedge_m z) \overset{(10.53)}{=} 0.$

(10.94) $$((x \odot y)^- \odot (x \wedge_m z)) \wedge_m (y \odot y) = 0.$$

(10.94): Note that $y \overset{(10.72)}{=} ((x \odot y) \oplus (x \wedge_m z)^-) \wedge_m y = ((x \odot y)^- \odot (x \wedge_m z))^- \wedge_m y$; we put $X := ((x \odot y)^- \odot (x \wedge_m z))^-$, hence $y = X \wedge_m y$. Then,
$((x \odot y)^- \odot (x \wedge_m z)) \wedge_m (y \odot y) = X^- \wedge_m (y \odot y) = X^- \wedge_m ((X \wedge_m y) \odot (X \wedge_m y)) \overset{(10.85)}{=} 0$.

(10.95) $$(x \odot ((x \oplus y) \wedge_m z)) \wedge_m (x \oplus y)^- = 0.$$

(10.95): In (10.92), put $X := x \oplus y$, $Y := z$ and $Z := x \odot ((x \oplus y) \wedge_m z)$, to obtain:
$[((x \oplus y) \odot ((x \oplus y) \wedge_m z)) \wedge_m (x \odot ((x \oplus y) \wedge_m z))] \wedge_m (x \oplus y)^- = 0$, hence, by (10.74), we obtain:
$[x \odot ((x \oplus y) \wedge_m z)] \wedge_m (x \oplus y)^- = 0$.

(10.96) $$(x \wedge_m y)^- \odot (((z \wedge_m x) \wedge_m y) \wedge_m u) = 0.$$

(10.96): In (10.93), put $X := (x \wedge_m y)^-$, $Y := (z \wedge_m x) \wedge_m y$ and $Z := u$, to obtain:
$(x \wedge_m y)^- \odot [(((x \wedge_m y)^- \odot ((z \wedge_m x) \wedge_m y)) \wedge_m ((z \wedge_m x) \wedge_m y)]^- \odot (((z \wedge_m x) \wedge_m y) \wedge_m u) = 0$,
hence, by (10.69), $(x \wedge_m y)^- \odot [0 \wedge_m ((z \wedge_m x) \wedge_m y)]^- \odot (((z \wedge_m x) \wedge_m y) \wedge_m u) = 0$,
hence $(x \wedge_m y)^- \odot [0]^- \odot (((z \wedge_m x) \wedge_m y) \wedge_m u) = 0$, i.e. $(x \wedge_m y)^- \odot (((z \wedge_m x) \wedge_m y) \wedge_m u) = 0$.

(10.97) $$(x \odot y) \wedge_m (y^- \wedge_m x^-) = 0.$$

(10.97): In (10.95), put $X := x$, $Y := y \odot x^-$ and $Z := y$, to obtain:
$(x \odot ((x \oplus (y \odot x^-)) \wedge_m y) \wedge_m (x \oplus (y \odot x^-))^- = 0$, hence, by (10.40), to obtain:
$(x \odot y) \wedge_m (x \oplus (y \odot x^-)) = 0$, i.e. $(x \odot y) \wedge_m (y^- \wedge_m x^-) = 0$, by (10.25).

(10.98) $$(x \odot (y \wedge_m z)) \wedge_m (y^- \wedge_m x^-) = 0.$$

(10.98): In (10.95), put $X := x$, $Y := y \odot x^-$ and $Z := y \wedge_m z$, to obtain:
$(x \odot [(x \oplus (y \odot x^-)) \wedge_m (y \wedge_m z)]) \wedge_m (x \oplus (y \odot x^-))^- = 0$, hence, by (10.54), to obtain:
$(x \odot [y \wedge_m z]) \wedge_m (x \oplus (y \odot x^-))^- = 0$, hence, by (10.25), $(x \odot (y \wedge_m z)) \wedge_m (y^- \wedge_m x^-) = 0$.

(10.99) $$(x \oplus y)^- \wedge_m (y \wedge_m x) = 0.$$

(10.99): $(x \oplus y)^- \wedge_m (y \wedge_m x) = (x^- \odot y^-) \wedge_m (y \wedge_m x) \overset{(10.97)}{=} 0.$

(10.100) $$((x \oplus y)^- \wedge_m z) \wedge_m (y \wedge_m x) = 0.$$

(10.100): In (10.68), put $X := (x \oplus y)^-$, $Y := y \wedge_m x$ and $Z := z$, to obtain:
$[(x \oplus y)^- \wedge_m (y \wedge_m x)]^- \odot (((x \oplus y)^- \wedge_m z) \wedge_m (y \wedge_m x)) = 0$, hence, by (10.99), to

obtain:
$0^- \odot (((x \oplus y)^- \wedge_m z) \wedge_m (y \wedge_m x)) = 0$, i.e. $((x \oplus y)^- \wedge_m z) \wedge_m (y \wedge_m x) = 0$.

(10.101) $$(x \oplus y)^- \wedge_m ((y \wedge_m z) \wedge_m x) = 0.$$

(10.101): $(x \oplus y)^- \wedge_m ((y \wedge_m z) \wedge_m x) = (x^- \odot y^-) \wedge_m ((y \wedge_m z) \wedge_m x)$ $\overset{(10.33)}{=} (x^- \odot ((y \wedge_m z)^- \wedge_m y^-)) \wedge_m ((y \wedge_m z) \wedge_m x) \overset{(10.98)}{=} 0$.

(10.102) $$((x \odot y^-) \wedge_m z) \wedge_m (y \wedge_m x^-) = 0.$$

(10.102): $((x \odot y^-) \wedge_m z) \wedge_m (y \wedge_m x^-) = ((x^- \oplus y)^- \wedge_m z) \wedge_m (y \wedge_m x^-) \overset{(10.100)}{=} 0$.

(10.103) $$(x \wedge_m ((y \oplus x)^- \wedge_m z)) \wedge_m y = 0.$$

(10.103): $(x \wedge_m ((y \oplus x)^- \wedge_m z)) \wedge_m y \overset{(10.61)}{=} (y \oplus x)^- \wedge_m ((x \wedge_m ((y \oplus x)^- \wedge_m z)) \wedge_m y) \overset{(10.101)}{=} 0$.

(10.104) $$x^- \wedge_m ((x \odot x) \wedge_m y) = 0.$$

(10.104): $A := x^- \wedge_m ((x \odot x) \wedge_m y) \overset{(10.89)}{=} ((x \oplus ((x \odot x) \wedge_m y))^- \wedge_m x^-) \wedge_m ((x \odot x) \wedge_m y)$;
put $Y := (x \odot x) \wedge_m y$ and by (10.58), we obtain: $Y = x \wedge_m Y$; hence, we have:
$A \overset{(10.58)}{=} ((x \oplus Y)^- \wedge_m x^-) \wedge_m (x \wedge_m Y) \overset{(10.100)}{=} 0$.

(10.105) $$(x^- \odot (y \wedge_m z)) \wedge_m (x \wedge_m y^-) = 0.$$

(10.105): $(x^- \odot (y \wedge_m z)) \wedge_m (x \wedge_m y^-) \overset{(10.60)}{=} [(y \odot x^-) \wedge_m (x^- \odot (y \wedge_m z))] \wedge_m (x \wedge_m y^-) \overset{(10.102)}{=} 0$.

(10.106) $$(x \wedge_m ((y \odot x^-) \wedge_m z)) \wedge_m y^- = 0.$$

(10.106): $(x \wedge_m ((y \odot x^-) \wedge_m z)) \wedge_m y^- = (x \wedge_m ((y^- \oplus x)^- \wedge_m z)) \wedge_m y^- \overset{(10.103)}{=} 0$.

(10.107) $$(x \odot x) \wedge_m (x^- \wedge_m y) = 0.$$

(10.107): $(x \odot x) \wedge_m (x^- \wedge_m y) \overset{(10.38)}{=} x^- \wedge_m ((x \odot x) \wedge_m (x^- \wedge_m y)) \overset{(10.104)}{=} 0$.

(10.108) $$(x \oplus y)^- \wedge_m (x \wedge_m (y \wedge_m z)) = 0.$$

(10.108): In (10.105), put $X := x$, $Y := (y \wedge_m z)^-$ and $Z := y^-$, to obtain:
$(x^- \odot [(y \wedge_m z)^- \wedge_m y^-]) \wedge_m (x \wedge_m (y \wedge_m z)) = 0$, hence, by (10.33), $(x^- \odot y^-) \wedge_m (x \wedge_m (y \wedge_m z)) = 0$,
i.e. $(x \oplus y)^- \wedge_m (x \wedge_m (y \wedge_m z)) = 0$.

(10.109) $$(x \wedge_m (x^- \odot (y \wedge_m z))) \wedge_m y^- = 0.$$

(10.109): $(x \wedge_m (x^- \odot (y \wedge_m z))) \wedge_m y^- \overset{(10.60)}{=} (x \wedge_m [(y \odot x^-) \wedge_m (x^- \odot (y \wedge_m z))]) \wedge_m y^- \overset{(10.106)}{=} 0$.

(10.110) $$(x \odot x) \wedge_m (y \odot (y \wedge_m x)^-) = 0.$$

(10.110): $(x \odot x) \wedge_m (y \odot (y \wedge_m x)^-) \overset{(10.70)}{=} (x \odot x) \wedge_m (x^- \wedge_m (y \odot (y \wedge_m x)^-)) \overset{(10.107)}{=} 0$.

(10.111) $$(x \oplus y)^- \wedge_m (y \wedge_m (z \wedge_m x)) = 0.$$

(10.111): $(x \oplus y)^- \wedge_m (y \wedge_m (z \wedge_m x)) \overset{(10.44)}{=} (x \oplus y)^- \wedge_m (x \wedge_m (y \wedge_m (z \wedge_m x))) \overset{(10.108)}{=} 0$.

(10.112) $$(x \wedge_m (x \oplus y)^-) \wedge_m (y \wedge_m z) = 0.$$

(10.112): $(x \wedge_m (x \oplus y)^-) \wedge_m (y \wedge_m z) = (x \wedge_m (x^- \odot y^-)) \wedge_m (y \wedge_m z)$
$\overset{(10.33)}{=} (x \wedge_m (x^- \odot ((y \wedge_m z)^- \wedge_m y^-))) \wedge_m (y \wedge_m z) \overset{(10.109)}{=} 0$.

(10.113) $$(x \odot x) \wedge_m ((y \odot x)^- \odot (y \wedge_m z)) = 0.$$

(10.113): In (10.110), put $X := y \odot y$ and $Y := (x \odot y)^- \odot (x \wedge_m z)$, to obtain:
$((y \odot y) \odot (y \odot y)) \wedge_m (((x \odot y)^- \odot (x \wedge_m z)) \odot [((x \odot y)^- \odot (x \wedge_m z)) \wedge_m (y \odot y)]^-) = 0$,
hence, by (10.94), $((y \odot y) \odot (y \odot y)) \wedge_m (((x \odot y)^- \odot (x \wedge_m z)) \odot 0^-) = 0$,
hence $((y \odot y) \odot (y \odot y)) \wedge_m ((x \odot y)^- \odot (x \wedge_m z)) = 0$,
hence, by (Pass) and (10.81), $(y \odot y) \wedge_m ((x \odot y)^- \odot (x \wedge_m z)) = 0$.

(10.114) $$x \wedge_m ((y \oplus x)^- \wedge_m y) = 0.$$

(10.114): $x \wedge_m ((y \oplus x)^- \wedge_m y) \overset{(10.38)}{=} (y \oplus x)^- \wedge_m (x \wedge_m ((y \oplus x)^- \wedge_m y)) \overset{(10.111)}{=} 0$.

(10.115) $$(x \odot y)^- \odot (x \wedge_m ((y \odot y) \wedge_m z)) = 0.$$

(10.115): Put $X := y \odot y$, $Y := (x \odot y)^-$, $Z := x$ and $u := z$ to obtain:
$(x \odot y)^- \odot (x \wedge_m ((y \odot y) \wedge_m z)) \overset{(10.39)}{=} (y \odot y) \wedge_m [(x \odot y)^- \odot (x \wedge_m ((y \odot y) \wedge_m z))] \overset{(10.113)}{=} 0$.

(10.116) $$x \wedge_m (y \odot ((z \oplus x)^- \wedge_m z)) = 0.$$

(10.116): In (10.56), put $X := x$ and $Y := (y \oplus x)^- \wedge_m y$, to obtain:
$(x \wedge_m ((y \oplus x)^- \wedge_m y))^- \odot (x \wedge_m (z \odot ((y \oplus x)^- \wedge_m y))) = 0$,
hence, by (10.114), $0^- \odot (x \wedge_m (z \odot ((y \oplus x)^- \wedge_m y))) = 0$,
hence $x \wedge_m (z \odot ((y \oplus x)^- \wedge_m y)) = 0$, hence $x \wedge_m (y \odot ((z \oplus x)^- \wedge_m z)) = 0$.

(10.117) $$x \wedge_m (y \wedge_m ((z \wedge_m (x \oplus z)^-) \wedge_m u)) = 0.$$

(10.117): $x \wedge_m (y \wedge_m ((z \wedge_m (x \oplus z)^-) \wedge_m u)) \overset{(10.45)}{=} (z \wedge_m (x \oplus z)^-) \wedge_m (x \wedge_m (y \wedge_m ((z \wedge_m (x \oplus z)^-) \wedge_m u))) \overset{(10.112)}{=} 0$.

(10.118) $$x \wedge_m ((y \wedge_m z) \odot (z \odot (z \oplus x))^-) = 0.$$

(10.118): First, $A := (y \wedge_m z) \odot (z \odot (z \oplus x))^- = (y \wedge_m z) \odot (z^- \oplus (z \oplus x)^-) = (y \wedge_m z) \odot ((z \oplus x)^- \oplus z^-)$;
put $X := y$, $Y := z$, $Z := (z \oplus x)^-$ in (10.19) to obtain: $A \stackrel{(10.19)}{=} (y \oplus z^-) \odot ((z \oplus x)^- \wedge_m z)$;
then, $x \wedge_m ((y \wedge_m z) \odot (z \odot (z \oplus x))^-) = x \wedge_m A = x \wedge_m ((y \oplus z^-) \odot ((z \oplus x)^- \wedge_m z)) \stackrel{(10.116)}{=} 0$.

(10.119) $$x \wedge_m ((y \wedge_m ((x \wedge_m z) \oplus y)^- \wedge_m z) = 0.$$

(10.119): Put $X := x$, $Y := z$ and $Z := y \wedge_m ((x \wedge_m z) \oplus y)^-$ in the right side of (10.67); then,
$x \wedge_m ((y \wedge_m ((x \wedge_m z) \oplus y)^- \wedge_m z) \stackrel{(10.67)}{=} (x \wedge_m z) \wedge_m [x \wedge_m ((y \wedge_m ((x \wedge_m z) \oplus y)^-) \wedge_m z)] \stackrel{(10.117)}{=} 0$.

(10.120) $$(x \wedge_m y) \odot (y \odot (y \oplus x))^- = 0.$$

(10.120): Put $X := x$, $Y := y$ and $Z := (y \odot (y \oplus x))^-$ in the right side of (10.28); then,
$(x \wedge_m y) \odot (y \odot (y \oplus x))^- = (X \wedge_m Y) \odot Z \stackrel{(10.28)}{=} x \wedge_m [(x \wedge_m y) \odot (y \odot (y \oplus x))^-] \stackrel{(10.118)}{=} 0$.

(10.121) $$x \wedge_m ((y^- \wedge_m ((x \wedge_m z)^- \odot y)) \wedge_m z) = 0.$$

(10.121): $x \wedge_m ((y^- \wedge_m ((x \wedge_m z)^- \odot y)) \wedge_m z) = x \wedge_m ((y^- \wedge_m ((x \wedge_m z) \oplus y^-)^-) \wedge_m z) \stackrel{(10.119)}{=} 0$.

(10.122) $$(x \odot (x \oplus y)) \wedge_m (y \wedge_m x) = y \wedge_m x.$$

(10.122): Put $X := y \wedge_m x$ and $Y^- := x \odot (x \oplus y)$ in the right side of (10.20); then,
$(x \odot (x \oplus y)) \wedge_m (y \wedge_m x) = Y^- \wedge_m X \stackrel{(10.20)}{=} X \odot (X \odot Y)^-$
$= (y \wedge_m x) \odot [(y \wedge_m x) \odot (x \odot (x \oplus y))^-]^- \stackrel{(10.120)}{=} (y \wedge_m x) \odot 0^- = y \wedge_m x$.

(10.123) $$(x^- \wedge_m (x \odot (x \wedge_m y)^-)) \wedge_m y = 0.$$

(10.123): $A := (x^- \wedge_m (x \odot (x \wedge_m y)^-)) \wedge_m y \stackrel{(Pcomm)}{=} (x^- \wedge_m ((x \wedge_m y)^- \odot x)) \wedge_m y$;
put $X := x$, $Y := x^-$, $Z := (x \wedge_m y)^-$ and $U := y$ in the right side of (10.66); then,
$A = (Y \wedge_m (Z \odot X)) \wedge_m U \stackrel{(10.66)}{=} X \wedge_m ((Y \wedge_m (Z \odot X)) \wedge_m U) = x \wedge_m [(x^- \wedge_m ((x \wedge_m y)^- \odot x)) \wedge_m y] \stackrel{(10.121)}{=} 0$.

(10.124) $$(x \wedge_m y)^- \oplus ((y \oplus z) \odot (y \oplus x)) = 1.$$

(10.124): Since $x \wedge_m y \stackrel{(10.122)}{=} (y \odot (y \oplus x)) \wedge_m (x \wedge_m y)$, it follows that
$(x \wedge_m y)^- \oplus ((y \oplus z) \odot (y \oplus x)) \stackrel{(10.122)}{=} [(y \odot (y \oplus x)) \wedge_m (x \wedge_m y)]^- \oplus ((y \oplus z) \odot (y \oplus x))$
$\stackrel{(Scomm)}{=} ((y \oplus z) \odot (y \oplus x)) \oplus [(y \odot (y \oplus x)) \wedge_m (x \wedge_m y)]^- \stackrel{(10.75)}{=} 1$,
for $X := y$, $Y := z$, $Z := y \oplus x$ and $U := x \wedge_m y$.

(10.125) $$((x \wedge_m y) \odot ((x \wedge_m y) \wedge_m x)^-) \wedge_m x = 0.$$

(10.125): Put $X := x$, $Y := y$ and $Z := x \wedge_m y$ in (10.73); then,
$A := (x \wedge_m y) \odot ((x \wedge_m y) \wedge_m x)^- = Z \odot (Z \wedge_m X)^- \overset{(10.73)}{=} (X \wedge_m Y)^- \wedge_m (Z \odot (Z \wedge_m X)^-)$
$= (x \wedge_m y)^- \wedge_m ((x \wedge_m y) \odot ((x \wedge_m y) \wedge_m x)^-)$; then,
$((x \wedge_m y) \odot ((x \wedge_m y) \wedge_m x)^-) \wedge_m x = A \wedge_m x = [(x \wedge_m y)^- \wedge_m ((x \wedge_m y) \odot ((x \wedge_m y) \wedge_m x)^-)] \wedge_m x \overset{(10.123)}{=} 0$.

$$\text{(10.126)} \qquad ((x \oplus y) \odot (x \oplus z)) \wedge_m (z \wedge_m x) = z \wedge_m x.$$

(10.126): $((x \oplus y) \odot (x \oplus z)) \wedge_m (z \wedge_m x) \overset{def.}{=} (z \wedge_m x) \odot [(z \wedge_m x)^- \oplus ((x \oplus y) \odot (x \oplus z))] \overset{(10.124)}{=} (z \wedge_m x) \odot 1 = z \wedge_m x$.

$$\text{(10.127)} \qquad (x \wedge_m (y \wedge_m z)) \odot (x \odot (z \oplus y))^- = 0.$$

(10.127): Since $A := y \wedge_m z \overset{(10.126)}{=} ((z \oplus y) \odot (z \oplus y)) \wedge_m (y \wedge_m z)$, then,
$(x \wedge_m (y \wedge_m z)) \odot (x \odot (z \oplus y))^- \overset{(Pcomm)}{=} (x \odot (z \oplus y))^- \odot (x \wedge_m A)$
$= (x \odot (z \oplus y))^- \odot (x \wedge_m [((z \oplus y) \odot (z \oplus y)) \wedge_m (y \wedge_m z)]) \overset{(10.115)}{=} 0$, with $X := x$, $Y := z \oplus y$, $Z := y \wedge_m z$.

$$\text{(10.128)} \qquad (x \wedge_m y)^- \odot (y \wedge_m (y^- \wedge_m x)) = 0.$$

(10.128): $(x \wedge_m y)^- \odot (y \wedge_m (y^- \wedge_m x)) \overset{(10.4)}{=} (y \odot (x \oplus y^-))^- \odot (y \wedge_m (y^- \wedge_m x)) \overset{(Pcomm),(10.127)}{=} 0$ for $X := y$, $Y := y^-$, $Z := x$.

$$\text{(10.129)} \qquad (x \wedge_m y) \odot ((x \wedge_m y) \wedge_m x)^- = 0.$$

(10.129): Put $X := (x \wedge_m y) \odot ((x \wedge_m y) \wedge_m x)^-$ and $Y := x$ in (10.128) to obtain:
$[((x \wedge_m y) \odot ((x \wedge_m y) \wedge_m x)^-) \wedge_m x]^- \odot [x \wedge_m (x^- \wedge_m ((x \wedge_m y) \odot ((x \wedge_m y) \wedge_m x)^-))] = 0$;
then, by (10.125), $[0]^- \odot [x \wedge_m (x^- \wedge_m ((x \wedge_m y) \odot ((x \wedge_m y) \wedge_m x)^-))] = 0$,
hence $x \wedge_m (x^- \wedge_m ((x \wedge_m y) \odot ((x \wedge_m y) \wedge_m x)^-)) = 0$,
hence, by (10.70), $x \wedge_m ((x \wedge_m y) \odot ((x \wedge_m y) \wedge_m x)^-) = 0$, hence, by (10.28), $(x \wedge_m y) \odot ((x \wedge_m y) \wedge_m x)^- = 0$.

$$\text{(10.130)} \qquad ((x \wedge_m y) \wedge_m x) \wedge_m (x \wedge_m y) = x \wedge_m y.$$

(10.130): $((x \wedge_m y) \wedge_m x) \wedge_m (x \wedge_m y) \overset{(10.20)}{=} (x \wedge_m y) \odot ((x \wedge_m y) \odot ((x \wedge_m y) \wedge_m x)^-)^- \overset{(10.129)}{=} (x \wedge_m y) \odot 0^- = x \wedge_m y$.

$$\text{(10.131)} \qquad (\Delta_m) \quad (x \wedge_m y) \odot (y \wedge_m x)^- = 0.$$

(10.131): $(x \wedge_m y) \odot (y \wedge_m x)^- \overset{(Pcomm)}{=} (y \wedge_m x)^- \odot (x \wedge_m y) \overset{(10.130)}{=} (y \wedge_m x)^- \odot [((x \wedge_m y) \wedge_m x) \wedge_m (x \wedge_m y)] \overset{(10.96)}{=} 0$. □

Now, we are able to prove the following:

**Theorem 10.3.14** *Let $\mathcal{A}^L = (A^L, \odot, ^-, 1)$ be an involutive left-m-BE algebra. If (m-BB), (m-Pabs-i) and (m-An) hold (i.e. if $\mathcal{A}^L$ is a left-taOWL), then ($\wedge_m$-comm) ($x \wedge_m y = y \wedge_m x$) holds (i.e. $\mathcal{A}^L$ is a left-MV algebra).*

**Proof.** By Theorem 10.3.13, the property ($\Delta_m$) holds, hence, for any $x, y \in A^L$, $(x \wedge_m y) \odot (y \wedge_m x)^- = 0$ and $(y \wedge_m x) \odot (x \wedge_m y)^- = 0$, i.e. $x \wedge_m y \leq_m y \wedge_m x$ and $y \wedge_m x \leq_m x \wedge_m y$; since (m-An) holds, it follows that $x \wedge_m y = y \wedge_m x$, i.e. ($\wedge_m$-comm) holds. □

Thus, ($\Delta_m$) + (m-An) $\Longleftrightarrow$ ($\wedge_m$-comm) and, consequently, antisymmetric and transitive left-OWLs (= left-taOWLs) are particular left-MV algebras. See Examples 10.4.10 and 10.4.11 of taOWLs and proper MV algebra, respectively. Hence, we have: **taOWL** $\subset$ **MV**.

### 10.3.3 Resuming connections between OL, OSL, OWL

We have obtained, based on Proposition 10.1.15, the following equivalent definitions (Definitions 2):

An involutive left-m-BE algebra $\mathcal{A}^L = (A^L, \odot, ^-, 1)$ is a:

| | | |
|---|---|---|
| - *left-ortholattice* (left-OL), | if (m-Pimpl) holds | (Definition 9.2.17), |
| - *left-orthosoftlattice* (left-OSL), | if (G) holds | (Definition 10.3.3), |
| - *left-orthowidelattice* (left-OWL), | if (m-Pabs-i) holds | (Definition 10.3.8), |

i.e.
**OL** = **m-BE**$_{(DN)}$ + (m-Pimpl),
**OSL** = **m-BE**$_{(DN)}$ + (G),
**OWL** = **m-BE**$_{(DN)}$ + (m-Pabs-i).

Hence, we have:

$$\textbf{OL} = \textbf{OSL} \cap \textbf{OWL}, \tag{10.132}$$

i.e. we have the basic connections from Figure 10.3, useful in the sequel.

The resuming connections in m-BE$_{(DN)}$ algebras and in m-aBE$_{(DN)}$ algebras are presented in Figures 10.4 and 10.5, respectively.

### 10.3.4 Putting OL and their two generalizations on the "involutive Little m-map"

The definitions (Definitions 2) and the results from this section allow us to draw the hierarchies from the following three Figures, 10.6, 10.7 and 10.8, thus putting all the mentioned algebras on the "involutive Little m-map".

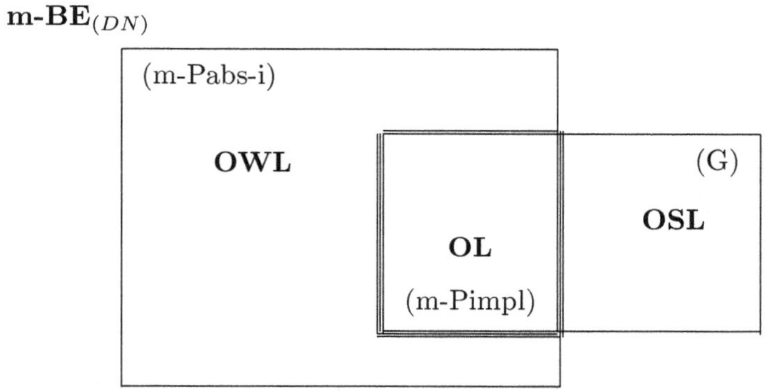


Figure 10.3: Basic connections between **OSL**, **OWL** and **OL**

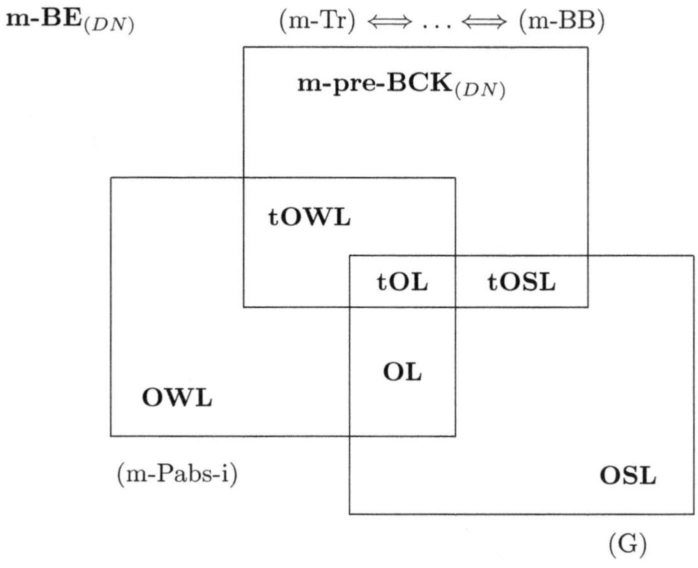


Figure 10.4: Resuming connections in **m-BE**$_{(DN)}$

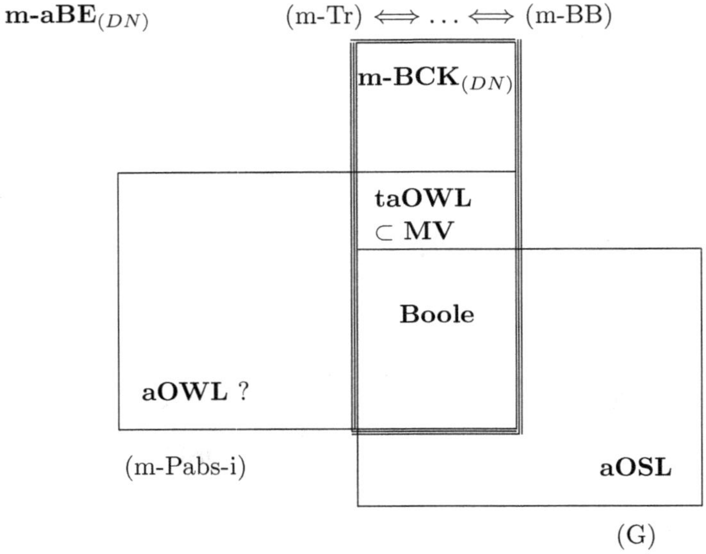


Figure 10.5: Resuming connections in $\mathbf{m\text{-}aBE}_{(DN)}$, where ? means that there is an open problem concerning **aOWL**

## 10.4 Examples

**Example 10.4.1 Proper BISL**

By a PASCAL program, we found that the algebra $\mathcal{A}^L = (A_6 = \{0, a, b, c, d, 1\}, \odot, ^-, 1)$, with the following tables of $\odot$ and $^-$ and of the additional operation $\oplus$, is an involutive left-m-MEL algebra verifying (G) and not verifying (m-Re) for $b$; (m-B), (m-BB), (m-*), (m-**); (m-Tr) for $(a, b, d)$, (m-An) for $(a, b)$, (m-Pimpl) for $(c, a)$, (m-Pabs-i) for $(b, a)$, (m-Pdis) for $(a, b, b)$. Hence, $\mathcal{A}^L$ is a proper bounded involutive left-softlattice (Definition 2) (i.e. without (m-An), (m-Tr)).

| $\odot$ | 0 | a | b | c | d | 1 |
|---|---|---|---|---|---|---|
| 0 | 0 | 0 | 0 | 0 | 0 | 0 |
| a | 0 | a | 0 | 0 | 0 | a |
| b | 0 | 0 | b | b | 0 | b |
| c | 0 | 0 | b | c | d | c |
| d | 0 | 0 | 0 | d | d | d |
| 1 | 0 | a | b | c | d | 1 |

and

| $x$ | $x^-$ |
|---|---|
| 0 | 1 |
| a | d |
| b | c |
| c | b |
| d | a |
| 1 | 0 |

, with

| $\oplus$ | 0 | a | b | c | d | 1 |
|---|---|---|---|---|---|---|
| 0 | 0 | a | b | c | d | 1 |
| a | a | a | a | 1 | 1 | 1 |
| b | b | a | b | c | 1 | 1 |
| c | c | 1 | c | c | 1 | 1 |
| d | d | 1 | 1 | 1 | d | 1 |
| 1 | 1 | 1 | 1 | 1 | 1 | 1 |

.

**Example 10.4.2 Proper aBISL**

By a PASCAL program, we found that the algebra $\mathcal{A}^L = (A_6 = \{0, a, b, c, d, 1\}, \odot, ^-, 1)$, with the following tables of $\odot$ and $^-$ and of the additional operation $\oplus$, is an involutive left-m-aMEL algebra verifying (G) and not verifying (m-Re) for $b$; (m-B), (m-BB), (m-*), (m-**); (m-Tr) for $(a, b, d)$, (m-

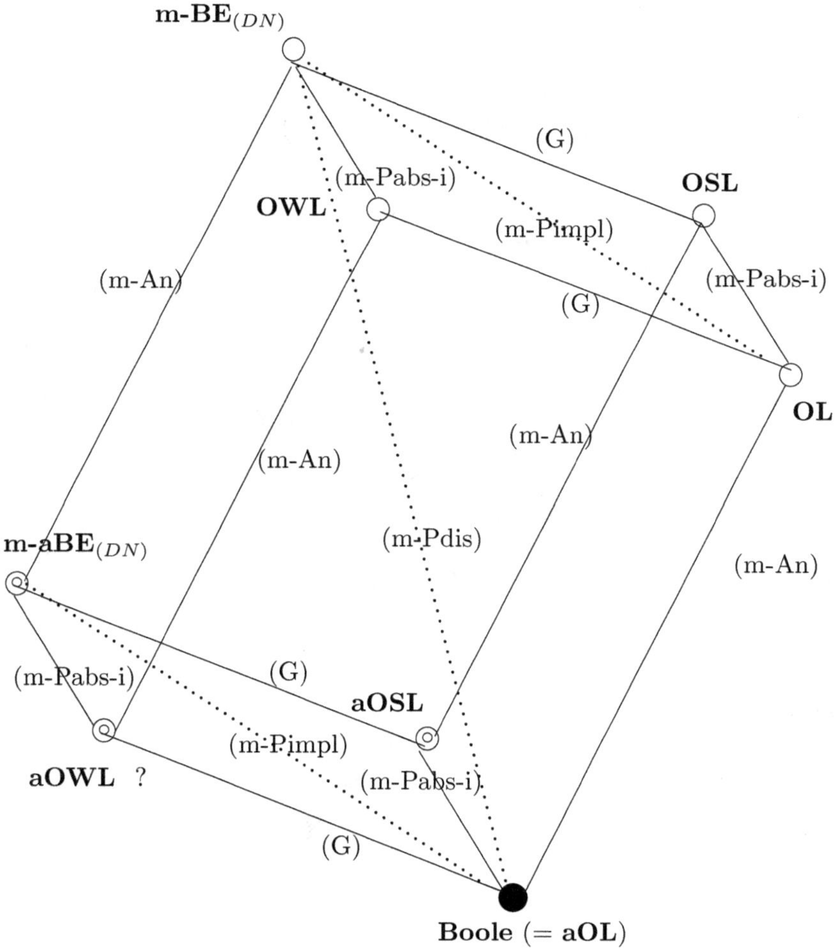


Figure 10.6: Hierarchies **OWL** vs. **OSL**, where ? means that there is an open problem concerning **aOWL**

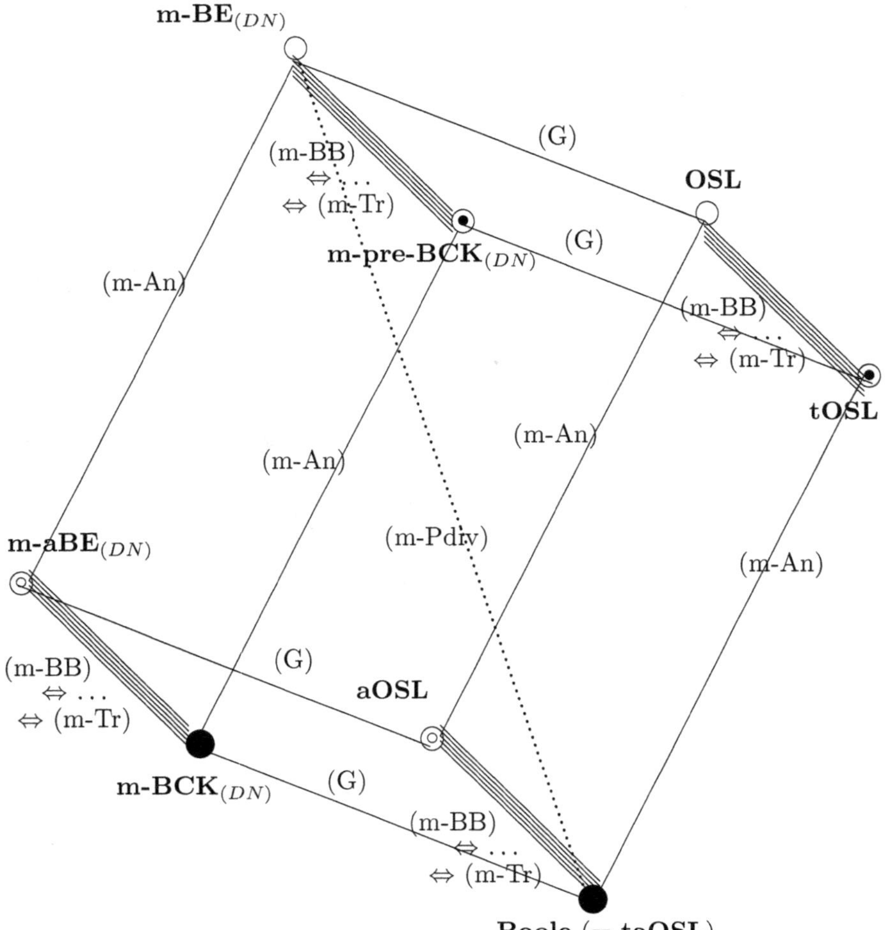


Figure 10.7: Hierarchies **OSL** vs. **m-pre-BCK**$_{(DN)}$

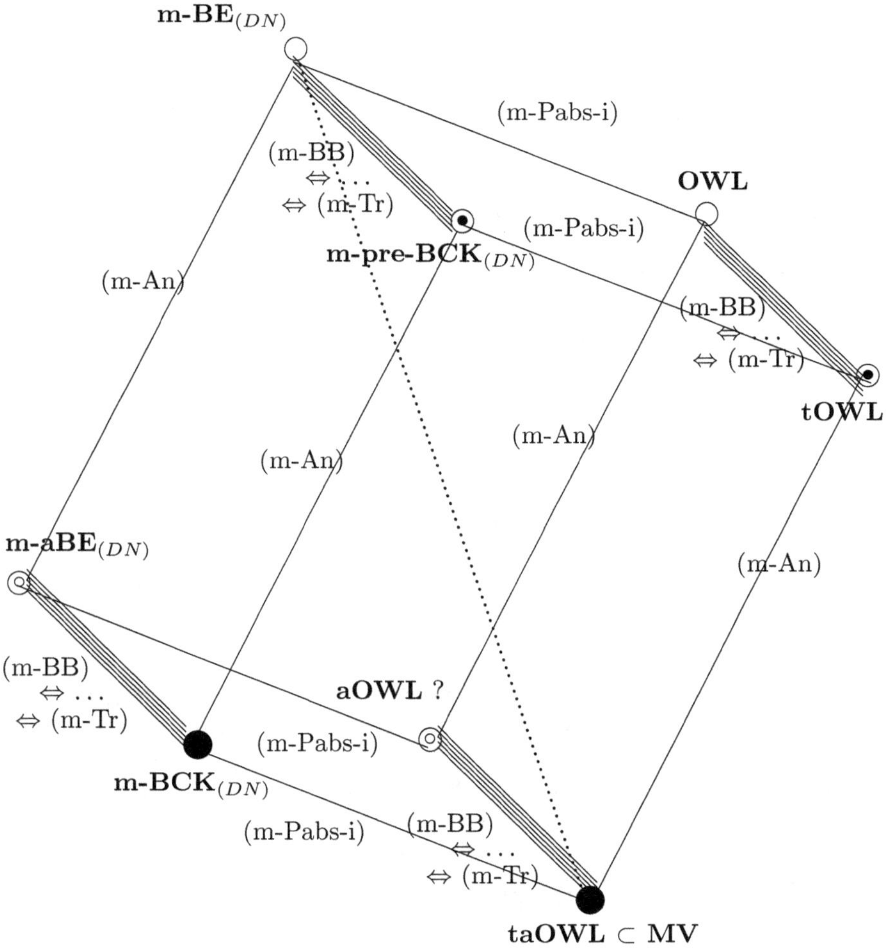


Figure 10.8: Hierarchies **OWL** vs. **m-pre-BCK**$_{(DN)}$ , where ? means that there is an open problem concerning **aOWL**

Pimpl) for $(b, a)$, (m-Pabs-i) for $(b, a)$, (m-Pdis) for $(a, b, b)$. Hence, it is a proper antisymmetric bounded involutive left-softlattice (Definition 2) (i.e. without (m-Tr)).

| $\odot$ | 0 | a | b | c | d | 1 |
|---|---|---|---|---|---|---|
| 0 | 0 | 0 | 0 | 0 | 0 | 0 |
| a | 0 | a | 0 | 0 | 0 | a |
| b | 0 | 0 | b | d | d | b |
| c | 0 | 0 | d | c | d | c |
| d | 0 | 0 | d | d | d | d |
| 1 | 0 | a | b | c | d | 1 |

and

| $x$ | $x^-$ |
|---|---|
| 0 | 1 |
| a | d |
| b | c |
| c | b |
| d | a |
| 1 | 0 |

, with

| $\oplus$ | 0 | a | b | c | d | 1 |
|---|---|---|---|---|---|---|
| 0 | 0 | a | b | c | d | 1 |
| a | a | a | a | a | 1 | 1 |
| b | b | a | b | a | 1 | 1 |
| c | c | a | a | c | 1 | 1 |
| d | d | 1 | 1 | 1 | d | 1 |
| 1 | 1 | 1 | 1 | 1 | 1 | 1 |

.

**Example 10.4.3 Proper BIWL**

By Mace4, we found that the algebra $\mathcal{A}^L = (A_4 = \{0, a, b, 1\}, \odot, ^-, 1)$, with the following tables of $\odot$ and $^-$ and of the additional operation $\oplus$, is an involutive left-m-MEL algebra verifying (m-Pabs-i) and not verifying (m-Re) for $a$, (m-B), (m-BB), (m-*), (m-**); (m-Tr) for $(a, b, a)$, (m-An) for $(a, b)$, (m-Pimpl) for $(b, 0)$, (G) for $b$, (m-Pdis) for$(a, b, b)$. Hence, it is a proper bounded involutive left-widelattice (Definition 2) (i.e. without (m-An), (m-Tr)).

| $\odot$ | 0 | a | b | 1 |
|---|---|---|---|---|
| 0 | 0 | 0 | 0 | 0 |
| a | 0 | a | 0 | a |
| b | 0 | 0 | 0 | b |
| 1 | 0 | a | b | 1 |

and

| $x$ | $x^-$ |
|---|---|
| 0 | 1 |
| a | a |
| b | b |
| 1 | 0 |

, with

| $\oplus$ | 0 | a | b | 1 |
|---|---|---|---|---|
| 0 | 0 | a | b | 1 |
| a | a | a | 1 | 1 |
| b | b | 1 | 1 | 1 |
| 1 | 1 | 1 | 1 | 1 |

.

**Example 10.4.4 Proper aBIWL**

By MACE4 program, we found that the algebra
$\mathcal{A}^L = (A_7 = \{0, a, b, c, d, e, 1\}, \odot, ^-, 1)$, with the following tables of $\odot$ and $^-$ and of the additional operation $\oplus$, is an involutive left-m-MEL algebra verifying (m-Pabs-i) and (m-An) and not verifying (m-Re) for $a$, (m-B) for $(a, 0, a)$, (m-BB) for $(a, a, 0)$, (m-*) for $(0, 0, a)$, (m-**) for $(a, e, a)$, (m-Tr) for $(d, a, e)$, (m-Pimpl) for $(a, 0)$, (G) for $a$, (m-Pdis) for $(a, a, a)$. Hence, it is a proper antisymmetric bounded involutive left-widelattice (Definition 2) (i.e. without (m-Tr)).

| $\odot$ | 0 | a | b | c | d | e | 1 |
|---|---|---|---|---|---|---|---|
| 0 | 0 | 0 | 0 | 0 | 0 | 0 | 0 |
| a | 0 | c | a | c | 0 | c | a |
| b | 0 | a | b | c | d | c | b |
| c | 0 | c | c | c | 0 | c | c |
| d | 0 | 0 | d | 0 | d | 0 | d |
| e | 0 | c | c | c | 0 | e | e |
| 1 | 0 | a | b | c | d | e | 1 |

with

| $\oplus$ | 0 | a | b | c | d | e | 1 |
|---|---|---|---|---|---|---|---|
| 0 | 0 | a | b | c | d | e | 1 |
| a | a | b | b | a | b | 1 | 1 |
| b | b | b | b | b | b | 1 | 1 |
| c | c | a | b | c | b | e | 1 |
| d | d | b | b | b | d | 1 | 1 |
| e | e | 1 | 1 | e | 1 | e | 1 |
| 1 | 1 | 1 | 1 | 1 | 1 | 1 | 1 |

and $(0, a, b, c, d, e, 1)^- = (1, a, c, b, e, d, 0)$.

**Example 10.4.5 Proper OSL**

By a PASCAL program, we found that the algebra $\mathcal{A}^L = (A_6 = \{0, a, b, c, d, 1\}, \odot, ^-, 1)$, with the following tables of $\odot$, $^-$ and of the additional operation $\oplus$, is a proper involutive left-m-BE algebra verifying (G) and not verifying (m-B) for $(a, b, a)$, (m-BB) for $(a, a, b)$, (m-*) for $(a, c, b)$, (m-**) for $(a, b, a)$, (m-Tr) for $(a, b, d)$, (m-An) for $(a, b)$, (m-Pimpl) for $(d, b)$, (m-Pabs-i) for $(a, b)$. Hence, it is a proper left-orthosoftlattice (Definition 2).

| $\odot$ | 0 | a | b | c | d | 1 |
|---|---|---|---|---|---|---|
| 0 | 0 | 0 | 0 | 0 | 0 | 0 |
| a | 0 | a | 0 | 0 | 0 | a |
| b | 0 | 0 | b | 0 | 0 | b |
| c | 0 | 0 | 0 | c | c | c |
| d | 0 | 0 | 0 | c | d | d |
| 1 | 0 | a | b | c | d | 1 |

and

| $x$ | $x^-$ |
|---|---|
| 0 | 1 |
| a | d |
| b | c |
| c | b |
| d | a |
| 1 | 0 |

, with

| $\oplus$ | 0 | a | b | c | d | 1 |
|---|---|---|---|---|---|---|
| 0 | 0 | a | b | c | d | 1 |
| a | a | a | b | 1 | 1 | 1 |
| b | b | b | b | 1 | 1 | 1 |
| c | c | 1 | 1 | c | 1 | 1 |
| d | d | 1 | 1 | 1 | d | 1 |
| 1 | 1 | 1 | 1 | 1 | 1 | 1 |

.

**Example 10.4.6 Proper tOSL**

By a PASCAL program, we found that the algebra $\mathcal{A}^L = (A_6 = \{0, a, b, c, d, 1\}, \odot, ^-, 1)$, with the following tables of $\odot$, $^-$ and of the additional operation $\oplus$, is an involutive left-m-BE algebra verifying (G) and (m-Tr), and not verifying (m-An) for $(a, c)$, (m-Pimpl) for $(b, a)$, (m-Pabs-i) for $(b, d)$. Hence, it is a proper transitive left-orthosoftlattice (Definition 2).

| $\odot$ | 0 | a | b | c | d | 1 |
|---|---|---|---|---|---|---|
| 0 | 0 | 0 | 0 | 0 | 0 | 0 |
| a | 0 | a | 0 | a | 0 | a |
| b | 0 | 0 | b | 0 | d | b |
| c | 0 | a | 0 | c | 0 | c |
| d | 0 | 0 | d | 0 | d | d |
| 1 | 0 | a | b | c | d | 1 |

and

| $x$ | $x^-$ |
|---|---|
| 0 | 1 |
| a | d |
| b | c |
| c | b |
| d | a |
| 1 | 0 |

, with

| $\oplus$ | 0 | a | b | c | d | 1 |
|---|---|---|---|---|---|---|
| 0 | 0 | a | b | c | d | 1 |
| a | a | a | 1 | a | 1 | 1 |
| b | b | 1 | b | 1 | d | 1 |
| c | c | a | 1 | c | 1 | 1 |
| d | d | 1 | d | 1 | d | 1 |
| 1 | 1 | 1 | 1 | 1 | 1 | 1 |

.

**Example 10.4.7 Proper aOSL**

By a PASCAL program, we found that the algebra $\mathcal{A}^L = (A_6 = \{0, a, b, c, d, 1\}, \odot, ^-, 1)$, with the following tables of $\odot$, $^-$ and of the additional operation $\oplus$, is a proper involutive left-m-aBE algebra verifying (G), and not verifying (m-B) for $(a, b, a)$, (m-BB) for $(a, a, b)$, (m-*) for $(b, d, a)$, (m-**) for $(a, b, a)$, (m-Tr) for $(a, b, d)$, (m-Pimpl) for $(d, b)$, (m-Pabs-i) for $(a, b)$. Hence, it is a proper antisymmetric left-orthosoftlattice (Definition 2).

| $\odot$ | 0 | a | b | c | d | 1 |
|---|---|---|---|---|---|---|
| 0 | 0 | 0 | 0 | 0 | 0 | 0 |
| a | 0 | a | 0 | 0 | 0 | a |
| b | 0 | 0 | b | 0 | b | b |
| c | 0 | 0 | 0 | c | c | c |
| d | 0 | 0 | b | c | d | d |
| 1 | 0 | a | b | c | d | 1 |

and

| $x$ | $x^-$ |
|---|---|
| 0 | 1 |
| a | d |
| b | c |
| c | b |
| d | a |
| 1 | 0 |

, with

| $\oplus$ | 0 | a | b | c | d | 1 |
|---|---|---|---|---|---|---|
| 0 | 0 | a | b | c | d | 1 |
| a | a | a | b | c | 1 | 1 |
| b | b | b | b | 1 | 1 | 1 |
| c | c | c | 1 | c | 1 | 1 |
| d | d | 1 | 1 | 1 | d | 1 |
| 1 | 1 | 1 | 1 | 1 | 1 | 1 |

.

**Example 10.4.8 Proper OWL**

Let $\mathcal{A}^L = (A_5 = \{0, a, b, c, 1\}, \odot, ^-, 1)$ be an algebra, where:

| $\odot$ | 0 | a | b | c | 1 |
|---|---|---|---|---|---|
| 0 | 0 | 0 | 0 | 0 | 0 |
| a | 0 | 0 | 0 | 0 | a |
| b | 0 | 0 | 0 | a | b |
| c | 0 | 0 | a | 0 | c |
| 1 | 0 | a | b | c | 1 |

and

| $x$ | $x^-$ |
|---|---|
| 0 | 1 |
| a | b |
| b | a |
| c | c |
| 1 | 0 |

, with

| $\oplus$ | 0 | a | b | c | 1 |
|---|---|---|---|---|---|
| 0 | 0 | a | b | c | 1 |
| a | a | 1 | 1 | b | 1 |
| b | b | 1 | 1 | 1 | 1 |
| c | c | b | 1 | 1 | 1 |
| 1 | 1 | 1 | 1 | 1 | 1 |

.

Then, $\mathcal{A}^L$ is an involutive left-m-BE algebra verifying (m-Pabs-i), while (m-B) does not hold for $(b, a, c)$, (m-BB) for $(b, c, b)$, (m-*) for $(a, c, b)$, (m-**) for $(b, a, c)$, (m-Tr) for $(b, a, c)$, (m-An) for $(a, b)$, (m-Pimpl) for $(a, 0)$, (G) for $a$. Hence, it is a proper left-orthowidelattice (Definition 2).

**Examples 10.4.9 Proper tOWLs**

**Example 1.**
By MACE4 program, we found the algebra $\mathcal{A}^L = (A_8 = \{0, a, b, c, d, e, f, 1\}, \odot, ^-, 1)$ with the following tables of $\odot$ and $^-$ and of the additional operation $\oplus$:

| $\odot$ | 0 | a | b | c | d | e | f | 1 |
|---|---|---|---|---|---|---|---|---|
| 0 | 0 | 0 | 0 | 0 | 0 | 0 | 0 | 0 |
| a | 0 | a | 0 | 0 | a | a | 0 | a |
| b | 0 | 0 | b | c | 0 | 0 | f | b |
| c | 0 | 0 | c | f | 0 | 0 | f | c |
| d | 0 | a | 0 | 0 | a | d | 0 | d |
| e | 0 | a | 0 | 0 | d | e | 0 | e |
| f | 0 | 0 | f | f | 0 | 0 | f | f |
| 1 | 0 | a | b | c | d | e | f | 1 |

with

| $\oplus$ | 0 | a | b | c | d | e | f | 1 |
|---|---|---|---|---|---|---|---|---|
| 0 | 0 | a | b | c | d | e | f | 1 |
| a | a | a | 1 | 1 | d | e | 1 | 1 |
| b | b | 1 | b | b | 1 | 1 | b | 1 |
| c | c | 1 | b | b | 1 | 1 | c | 1 |
| d | d | d | 1 | 1 | e | e | 1 | 1 |
| e | e | e | 1 | 1 | e | e | 1 | 1 |
| f | f | 1 | b | c | 1 | 1 | f | 1 |
| 1 | 1 | 1 | 1 | 1 | 1 | 1 | 1 | 1 |

and $(0, a, b, c, d, e, f, 1)^- = (1, b, a, d, c, f, e, 0)$.

$\mathcal{A}^L$ is an involutive left-m-BE algebra verifying (m-Pabs-i) and (m-BB) $\Longleftrightarrow \ldots$ (m-Tr) and not verifying (m-An) for $(a, d)$, (m-Pimpl) for $(c, 0)$, (G) for $c$. Hence, it is a proper left-tOWL (Definition 2).

**Example 2.**
By MACE4 program, we found the algebra
$\mathcal{A}^L = (A_{14} = \{0, a, b, c, d, e, f, g, h, i, j, k, m, 1\}, \odot, ^-, 1)$ with the following tables of $\odot$ and $^-$:

| $\odot$ | 0 | a | b | c | d | e | f | g | h | i | j | k | m | 1 |
|---|---|---|---|---|---|---|---|---|---|---|---|---|---|---|
| 0 | 0 | 0 | 0 | 0 | 0 | 0 | 0 | 0 | 0 | 0 | 0 | 0 | 0 | 0 |
| a | 0 | c | 0 | c | b | g | 0 | c | k | g | f | 0 | c | a |
| b | 0 | 0 | 0 | 0 | b | 0 | 0 | 0 | k | 0 | f | 0 | 0 | b |
| c | 0 | c | 0 | c | 0 | c | 0 | c | 0 | c | 0 | 0 | c | c |
| d | 0 | b | b | 0 | d | b | f | 0 | h | 0 | j | k | b | d |
| e | 0 | g | 0 | c | b | i | 0 | g | k | i | f | 0 | c | e |
| f | 0 | 0 | 0 | 0 | f | 0 | 0 | 0 | f | 0 | f | 0 | 0 | f |
| g | 0 | c | 0 | c | 0 | g | 0 | c | 0 | g | 0 | 0 | c | g |
| h | 0 | k | k | 0 | h | k | f | 0 | j | 0 | j | f | k | h |
| i | 0 | g | 0 | c | 0 | i | 0 | g | 0 | i | 0 | 0 | c | i |
| j | 0 | f | f | 0 | j | f | f | 0 | j | 0 | j | f | f | j |
| k | 0 | 0 | 0 | 0 | k | 0 | 0 | 0 | f | 0 | f | 0 | 0 | k |
| m | 0 | c | 0 | c | b | c | 0 | c | k | c | f | 0 | c | m |
| 1 | 0 | a | b | c | d | e | f | g | h | i | j | k | m | 1 |

and

| $x$ | $x^-$ |
|---|---|
| 0 | 1 |
| a | b |
| b | a |
| c | d |
| d | c |
| e | f |
| f | e |
| g | h |
| h | g |
| i | j |
| j | i |
| k | m |
| m | k |
| 1 | 0 |

.

$\mathcal{A}^L$ is an involutive left-m-BE algebra verifying (m-Pabs-i) and (m-BB) $\Longleftrightarrow \ldots$ (m-Tr) and not verifying (m-An) for $(a, e)$, (m-Pimpl) for $(a, 0)$, (G) for $a$. Hence, it is a proper left-tOWL (Definition 2).

**Examples 10.4.10 taOWLs (= MV algebras + (m-Pabs-i))**

**Example 1.**

The algebra $\mathcal{A}^L = (L_3 = \{0, a, 1\}, \odot, ^-, 1)$, with the following tables of $\odot$ and $^-$ and of the additional operation $\oplus$, is the linearly ordered left-MV algebra with the smallest number of elements verifying (m-Pabs-i) and not verifying (m-Pimpl) for $(a, 0)$, (G) for $a$, (m-Pdis) for $(a, a, a)$.

| $\odot$ | 0 | a | 1 |
|---|---|---|---|
| 0 | 0 | 0 | 0 |
| a | 0 | 0 | a |
| 1 | 0 | a | 1 |

and

| $x$ | $x^-$ |
|---|---|
| 0 | 1 |
| a | a |
| 1 | 0 |

, with

| $\oplus$ | 0 | a | 1 |
|---|---|---|---|
| 0 | 0 | a | 1 |
| a | a | 1 | 1 |
| 1 | 1 | 1 | 1 |

.

**Example 2.**

The algebra $\mathcal{A}^L = (A_6 = \{0, a, b, c, d, 1\}, \odot, ^-, 1)$, with the following tables of $\odot$ and $^-$ and of the additional operation $\oplus$, is a left-MV algebra verifying (m-Pabs-i) and not verifying (m-Pimpl) for $(b, 0)$, (G) for $b$, (m-Pdis) for $(b, b, b)$. It is isomorphic with $\mathcal{L}_{3\times 2}$ from ([104], page 165).

| $\odot$ | 0 | a | b | c | d | 1 |
|---|---|---|---|---|---|---|
| 0 | 0 | 0 | 0 | 0 | 0 | 0 |
| a | 0 | a | 0 | a | 0 | a |
| b | 0 | 0 | 0 | 0 | b | b |
| c | 0 | a | 0 | a | b | c |
| d | 0 | 0 | b | b | d | d |
| 1 | 0 | a | b | c | d | 1 |

and

| $x$ | $x^-$ |
|---|---|
| 0 | 1 |
| a | d |
| b | c |
| c | b |
| d | a |
| 1 | 0 |

, with

| $\oplus$ | 0 | a | b | c | d | 1 |
|---|---|---|---|---|---|---|
| 0 | 0 | a | b | c | d | 1 |
| a | a | a | c | c | 1 | 1 |
| b | b | c | d | 1 | d | 1 |
| c | c | c | 1 | 1 | 1 | 1 |
| d | d | 1 | d | 1 | d | 1 |
| 1 | 1 | 1 | 1 | 1 | 1 | 1 |

.

**Example 10.4.11 Proper MV algebra**

The algebra $\mathcal{A}^L = (L_4 = \{0, a, b, 1\}, \odot, ^-, 1)$, with the following tables of $\odot$ and $^-$ and of the additional operation $\oplus$, is a proper left-MV algebra, not verifying (m-

Pabs-i) for $(a, 0)$, (m-Pimpl) for $(a, 0)$, (G) for $b$, (m-Pdis) for $(a, a, b)$, (aWNM$_m$) for $(b, b)$. $\mathcal{A}^L$ is the chain $0 \leq_m a \leq_m b \leq_m 1$, i.e. $\mathcal{A}^L$ is $\mathcal{L}_4$ from [104].

| $\odot$ | 0 | a | b | 1 |
|---|---|---|---|---|
| 0 | 0 | 0 | 0 | 0 |
| a | 0 | 0 | 0 | a |
| b | 0 | 0 | a | b |
| 1 | 0 | a | b | 1 |

and

| $x$ | $x^-$ |
|---|---|
| 0 | 1 |
| a | b |
| b | a |
| 1 | 0 |

, with

| $\oplus$ | 0 | a | b | 1 |
|---|---|---|---|---|
| 0 | 0 | a | b | 1 |
| a | a | b | 1 | 1 |
| b | b | 1 | 1 | 1 |
| 1 | 1 | 1 | 1 | 1 |

.

# Chapter 11

# Putting quantum MV algebras on the "map"

In this chapter, we clarify, mainly, some aspects concerning the quantum-MV algebras as non-lattice generalizations of MV algebras.

In Section 11.1, we recall the initial definition of quantum-MV algebras given by Roberto Giuntini.

In Section 11.2, we redefine the QMV algebras as involutive m-BE algebras verifying the property (Pqmv), just as we have redefined the MV algebras as involutive m-BE algebras verifying the property ($\wedge_m$-comm). We prove that (Pqmv) is equivalent with only two properties, ($\Delta_m$) and (Pom), where ($\Delta_m$) is the largest non-antisymmetric generalization of ($\wedge_m$-comm), introduced in Chapter 10, and (Pom) is the property characterizing the orthomodular lattices among the ortholattices (Definitions 2).

In Section 11.3, we introduce three generalizations of QMV algebras (**QMV**): two new non-antisymmetric generalizations of MV algebras, the *pre-MV algebras* (**PreMV**) and the *metha-MV algebras* (**MMV**), and the *orthomodular algebras* (**OM**). The QMV algebra is then just an *orthomodular PreMV algebra* or an *orthomodular MMV algebra*; in other words, the QMV algebra is that non-antisymmetric generalization of MV algebra that is an orthomodular algebra. We shall also introduce and study the *transitive QMV algebras* (**tQMV**), the *transitive PreMV algebras* (**tPreMV**), the *transitive MMV algebras* (**tMMV**) and the *transitive OM algebras* (**tOM**). It was known that any MV algebra is a QMV algebra, but the exact connection between MV and QMV algebras it was not known. We clarify this problem, by proving that MV algebras coincide with the antisymmetric QMV algebras (**aQMV**) - but also with the antisymmetric preMV algebras (**aPreMV**) and with the antisymmetric MMV algebras (**aMMV**). Consequently, MV algebras and QMV algebras, and also tQMV algebras, will be put on the same "map" (involutive "Big m-map"). The *taOM algebra*, a proper generalization of MV algebra inside the class of m-BCK algebras, is put in evidence. By putting QMV (and tQMV) algebras on the "map", we prove again (see Chapter 9) the deep connections existing between the algebraic structures connected to the classical and non-classical logics and the algebraic structures connected to the quantum logics: they exist on the same "map", but at different levels (parallels), i.e. the QMV (and tQMV) algebras also belong to the "world" of left-m-M algebras.

In Section 11.4, we present nine examples of the involved algebras.

The content of this chapter is taken from [124].

## 11.1 The original definition of quantum-MV algebras. The property (Wom)

**Definition 11.1.1** ([45], Definition 2.1.1) [76]

A *supplement algebra*, or an *S algebra* for short, is an algebra $\mathcal{M} = (M, \oplus, ^-, 0, 1)$ consisting of a nonempty set $M$, two constant elements $0, 1$ in $M$, a unary operation $^-$ and a binary operation $\oplus$ on $M$ satisfying the following axioms: for all $x, y, z \in M$,
(S1) $x \oplus y = y \oplus x$, (S2) $x \oplus (y \oplus z) = (x \oplus y) \oplus z$, (S3) $x \oplus x^- = 1$, (S4) $x \oplus 0 = x$, (S5) $(x^-)^- = x$ (or $x^= = x$), (S6) $0^- = 1$, (S7) $x \oplus 1 = 1$.

On every S algebra, the following operations can be introduced:
$x \odot y := (x^- \oplus y^-)^-, \quad x \sqcap y := (x \oplus y^-) \odot y, \quad x \sqcup y := (x \odot y^-) \oplus y.$

**Remark 11.1.2** *An S algebra is just an involutive right-m-BE algebra.*

QMV algebras were introduced by Roberto Giuntini [66] as S algebras satisfying additionally five axioms. The equivalence of the five axioms with the next axiom (QMV) was proved in [71].

**Definition 11.1.3** ([45], Definition 2.3.1) A *quantum MV algebra*, or a *QMV algebra* for short, is an S algebra $\mathcal{M} = (M, \oplus, ^-, 0, 1)$ satisfying: for all $x, y, z \in M$,
(QMV) $x \oplus ((x^- \sqcap y) \sqcap (z \sqcap x^-)) = (x \oplus y) \sqcap (x \oplus z)$.

**Note that QMV algebras were originally defined as right-algebras** (see more on left- and right- algebras in [104], [116], [118]).

## 11.2 Redefining the QMV algebras: QMV

### 11.2.1 Redefining the QMV algebras as involutive m-BE algebras

Following the original definition (Definition 11.1.3) of QMV algebras, the definition of involutive m-BE algebras and Remark 11.1.2, we obtain the following redefinition of QMV algebras as involutive m-BE algebras, which helps us to put them on the "map" (the involutive "Big m-map"):

**Definitions 11.2.1**
(i) A *left-quantum-MV algebra*, or a *left-QMV algebra* for short, is an involutive left-m-BE algebra $\mathcal{A}^L = (A^L, \odot, ^- = ^{-_L}, 1)$ verifying the following axiom: for all $x, y, z \in A^L$,
(Pqmv) $\quad x \odot [(x^- \vee_m^M y) \vee_m^M (z \vee_m^M x^-)] = (x \odot y) \vee_m^M (x \odot z)$.
(i') A *right-quantum-MV algebra*, or a *right-QMV algebra* for short, is an involutive right-m-BE algebra (= S algebra) $\mathcal{A}^R = (A^R, \oplus, ^- = ^{-_R}, 0)$ verifying the following dual axiom: for all $x, y, z \in A^R$,
(Sqmv) = (QMV) $\quad x \oplus [(x^- \wedge_m^M y) \wedge_m^M (z \wedge_m^M x^-)] = (x \oplus y) \wedge_m^M (x \oplus z)$.

Note that the dual operations $\wedge_m^M$, $\vee_m^M$ (recalled in Chapter 6) are just $\sqcap$, $\sqcup$, respectively.

We shall denote by **QMV** the class of all left-QMV algebras and by $\mathbf{QMV}^R$ the class of all right-QMV algebras.

**Proposition 11.2.2** *Let $\mathcal{A}^L = (A^L, \odot, ^-, 1)$ be an involutive left-m-BE algebra. Then:*

$$(Pqmv) \Longleftrightarrow (Sqmv).$$

**Proof.** Suppose (Pqmv) holds; then, $x \oplus [(x^- \wedge_m^M y) \wedge_m^M (z \wedge_m^M x^-)]$
$\overset{(6.5)}{=} (x^- \odot [(x^- \wedge_m^M y) \wedge_m^M (z \wedge_m^M x^-)])^- \overset{(6.39)}{=} (x^- \odot [(x^- \wedge_m^M y)^- \vee_m^M (z \wedge_m^M x^-)^-])^-$
$\overset{(6.39)}{=} (x^- \odot [(x \vee_m^M y^-) \vee_m^M (z^- \vee_m^M x)])^- \overset{(Pqmv)}{=} ((x^- \odot y^-) \vee_m^M (x^- \odot z^-))^-$
$\overset{(6.38)}{=} (x^- \odot y^-)^- \wedge_m^M (x^- \odot z^-)^- \overset{(6.5)}{=} (x \oplus y) \wedge_m^M (x \oplus z)$, i.e. (Sqmv) holds.
Suppose (Sqmv) holds; then, $x \odot [(x^- \vee_m^M y) \vee_m^M (z \vee_m^M x^-)]$
$\overset{(6.12)}{=} (x^- \oplus [(x^- \vee_m^M y) \vee_m^M (z \vee_m^M x^-)])^- \overset{(6.38)}{=} (x^- \oplus [(x^- \vee_m^M y)^- \wedge_m^M (z \vee_m^M x^-)^-])^-$
$\overset{(6.38)}{=} (x^- \oplus [(x \wedge_m^M y^-) \wedge_m^M (z^- \wedge_m^M x)])^- \overset{(Sqmv)}{=} ((x^- \oplus y^-) \wedge_m^M (x^- \oplus z^-))^-$
$\overset{(6.39)}{=} (x^- \oplus y^-)^- \vee_m^M (x^- \oplus z^-)^- \overset{(6.12)}{=} (x \odot y) \vee_m^M (x \odot z)$, i.e. (Pqmv) holds. □

**Corollary 11.2.3** Let $\mathcal{A}^L = (A^L, \odot, ^-, 1)$ be a left-QMV algebra. Then, $(A^L, \oplus, ^-, 0)$ is a right-QMV algebra.

**Proof.** By Corollary 6.4.13 and Proposition 11.2.2. □

**Proposition 11.2.4** *(See ([45], Proposition 2.3.2), in dual case)*
*Let $\mathcal{A}^L = (A^L, \odot, ^-, 1)$ be a left-QMV algebra. We have:*

(11.1) $$x \odot (y \vee_m^M x^-) = x \odot y,$$

(*Pmv*) $$x \odot (x^- \vee_m^M y) = x \odot y,$$

(*Pq*) $$x \odot x \odot [y \vee_m^M (z \vee_m^M x^-)] = (x \odot y) \vee_m^M (x \odot z);$$

(11.2) $$x \odot y \leq_m^M x, \quad \text{i.e.} \quad (x \odot y) \wedge_m^M x = x \odot y,$$

(11.3) $$x \leq_m^M x \oplus y, \quad \text{i.e.} \quad x \wedge_m^M (x \oplus y) = x,$$

(11.4) $$x \wedge_m^M y \leq_m^M y, \quad i.e. \quad (x \wedge_m^M y) \wedge_m^M y = x \wedge_m^M y,$$

(11.5) $$y \leq_m^M x \vee_m^M y, \quad i.e. \quad y \wedge_m^M (x \vee_m^M y) = y,$$

(11.6) $$x \odot [y \vee_m^M (x \odot z)^-] = (x \odot y) \vee_m^M (x \odot (x \odot z)^-),$$

(11.7) $$x \vee_m^M (y \wedge_m^M x) = x,$$

(11.8) $$x \leq_m^M y \Longrightarrow y \vee_m^M x = y,$$

(11.9) $$x \leq_m^M y \Longrightarrow y^- \leq_m^M x^- \quad (order-reversibility\, of\, ^-),$$

(11.10) $$x \leq_m^M y \Longrightarrow x \oplus z \leq_m^M y \oplus z \quad (monotonicity\, of\, \oplus),$$

(11.11) $$x \leq_m^M y \Longrightarrow x \odot z \leq_m^M y \odot z \quad (monotonicity\, of\, \odot),$$

(11.12) $$(x \wedge_m^M y) \wedge_m^M z = (x \wedge_m^M y) \wedge_m^M (y \wedge_m^M z),$$

(11.13) $$(x \vee_m^M y) \vee_m^M z = (x \vee_m^M y) \vee_m^M (y \vee_m^M z),$$

(11.14) $$x \odot y = x \odot y \odot (x \oplus y),$$

(11.15) $$(x^- \odot y) \wedge_m^M (y^- \odot x) = 0.$$

**Remarks 11.2.5** *(i) Concerning (11.4), note that* $x \wedge_m^M y \not\leq_m^M x$. *For example, in the left-QMV algebra from Example 11.4.4,* $a \wedge_m^M c \not\leq_m^M a$. *Indeed,* $a \wedge_m^M c = c$, *while* $(a \wedge_m^M c) \wedge_m^M a = c \wedge_m^M a = a \neq c$.

*(ii) Concerning (11.5), note that* $x \not\leq_m^M x \vee_m^M y$. *For example, in the left-QMV algebra from Example 11.4.4,* $a \not\leq_m^M a \vee_m^M c$. *Indeed,* $a \vee_m^M c = c$, *while* $a \wedge_m^M (a \vee_m^M c) = a \wedge_m^M c = c \neq a$.

Recall now the following well known property (prel) (prelinearity) from a bounded residuated lattice $(A, \wedge, \vee, \odot, \rightarrow, 0, 1)$ [104]:

(prel) $$(x \rightarrow y) \vee (y \rightarrow x) = 1.$$

Here, we shall consider that $\vee \stackrel{def.}{=} \vee_m^B$, which is no more a lattice operation, therefore the property will be denoted by (prel$_m$):

(prel$_m$) $$(x \rightarrow y) \vee_m^B (y \rightarrow x) = 1,$$

where $x \rightarrow y \stackrel{def.}{=} (x \odot y^-)^-$ (see the map $\Psi$ from Chapter 17).

Note that, in MV algebras, (prel$_m$) and (prel) coincide.

Then, we have the following result:

**Corollary 11.2.6** *Any left-QMV algebra verifies the property (prel$_m$).*

**Proof.** $(x \to y) \vee_m^B (y \to x) = (x \odot y^-)^- \vee_m^B (y \odot x^-)^-$
$= (y \odot x^-)^- \vee_m^M (x \odot y^-)^- = ((y \odot x^-) \wedge_m^M (x \odot y^-))^- \stackrel{(Pcomm),(11.15)}{=} 0^- = 1.$ □

**Proposition 11.2.7** *Let $\mathcal{A}^L = (A^L, \odot, ^-, 1)$ be a left-QMV algebra. We have:*

$$x \vee_m^M y \leq_m^M x \oplus y, \tag{11.16}$$

$$x \odot y \leq_m^M x \wedge_m^M y. \tag{11.17}$$

**Proof.** (11.16): Since $x \odot y^- \leq_m^M x$, by (11.2), then $x \vee_m^M y = (x \odot y^-) \oplus y \leq_m^M x \oplus y$, by (11.10).

(11.17): Since $x^- \odot y \leq_m^M x^-$, by (11.2), then $x \leq_m^M (x^- \odot y)^-$, by (11.9) and (DN); hence, $x \odot y \leq_m^M (x^- \odot y)^- \odot y = x \wedge_m^M y$, by (11.11). □

**Proposition 11.2.8** *(See ([45], Proposition 2.3.5), in dual case)*
*Let $\mathcal{A}^L = (A^L, \odot, ^-, 1)$ be a left-QMV algebra. We have:*

$$x \wedge_m^M ((x \oplus y) \wedge_m^M z) = x \wedge_m^M z \quad (\textit{absorbtion law } 1), \tag{11.18}$$

$$x \vee_m^M ((x \odot y) \vee_m^M z) = x \vee_m^M z \quad (\textit{absorbtion law } 2), \tag{11.19}$$

$$x \leq_m^M z^-,\ y \leq_m^M z^-,\ x \oplus z = y \oplus z \Longrightarrow x = y \quad (\textit{cancellation law } 1), \tag{11.20}$$

$$z^- \leq_m^M x,\ z^- \leq_m^M y,\ x \odot z = y \odot z \Longrightarrow x = y \quad (\textit{cancellation law } 2), \tag{11.21}$$

$$x \leq_m^M y \Longrightarrow x \wedge_m^M z \leq_m^M y \wedge_m^M z \quad (\textit{monotonicity of } \wedge_m^M), \tag{11.22}$$

$$x \leq_m^M y \Longrightarrow x \vee_m^M z \leq_m^M y \vee_m^M z \quad (\textit{monotonicity of } \vee_m^M), \tag{11.23}$$

$$x \leq_m^M y,\ y \leq_m^M z \Longrightarrow x \leq_m^M z \quad (\textit{transitivity of } \leq_m^M). \tag{11.24}$$

**Corollary 11.2.9** *(See ([45], page 157))*
*Let $\mathcal{A}^L = (A^L, \odot, ^-, 1)$ be a left-QMV algebra. The binary relation $\leq_m^M$ is an order relation.*

We shall prove next **the first very important result of this chapter**, Theorem 11.2.10, saying that axiom (Pqmv) is equivalent to only two properties, the properties (Pmv) and (Pq) from Proposition 11.2.4:

(Pmv) $\quad x \odot (x^- \vee_m^M y) = x \odot y,$

(Pq) $\quad x \odot [y \vee_m^M (z \vee_m^M x^-)] = (x \odot y) \vee_m^M (x \odot z).$

Recall that, cf. ([45], Proposition 2.3.4), Giuntini proved that axiom (Pqmv) is equivalent to the properties (Pmv), (11.6), (11.7), (11.12) and (11.15).

**Theorem 11.2.10** *Let* $\mathcal{A}^L = (A^L, \odot, ^-, 1)$ *be an involutive left-m-BE algebra. Then,*

$$(Pqmv) \iff (Pmv) + (Pq).$$

**Proof.** By Proposition 11.2.4, the axioms of $\mathcal{A}^L$ and (Pqmv) imply (Pmv) and (Pq). To prove the converse, assume that (Pmv) and (Pq) are satisfied by $\mathcal{A}^L$. Then:

$x \odot [(x^- \vee_m^M y) \vee_m^M (z \vee_m^M x^-)]$
$\stackrel{(6.33)}{=} x \odot [((x^- \vee_m^M y) \odot (z \vee_m^M x^-)^-) \oplus (z \vee_m^M x^-)]$
$\stackrel{(6.38),(DN)}{=} x \odot [((x^- \vee_m^M y) \odot (z^- \wedge_m^M x)) \oplus (z \vee_m^M x^-)]$
$\stackrel{(6.32),(DN)}{=} x \odot [((x^- \vee_m^M y) \odot ((z \odot x)^- \odot x)) \oplus (z \vee_m^M x^-)]$
$\stackrel{(Pass),(Pcomm)}{=} x \odot [((x \odot (x^- \vee y)) \odot (z \odot x)^-) \oplus (z \vee_m^M x^-)]$
$\stackrel{(Pmv)}{=} x \odot [((x \odot y) \odot (z \odot x)^-) \oplus (z \vee_m^M x^-)]$
$\stackrel{(Pcomm),(Pass)}{=} x \odot [(y \odot ((z \odot x)^- \odot x)) \oplus (z \vee_m^M x^-)]$
$\stackrel{(6.32),(DN)}{=} x \odot [(y \odot (z^- \wedge_m^M x)) \oplus (z \vee_m^M x^-)]$
$\stackrel{(6.38),(DN)}{=} x \odot [(y \odot (z \vee_m^M x^-)^-) \oplus (z \vee_m^M x^-)]$
$\stackrel{(6.33)}{=} x \odot [y \vee_m^M (z \vee_m^M x^-)]$
$\stackrel{(Pq)}{=} (x \odot y) \vee_m^M (x \odot z)$; thus, (Pqmv) holds. □

**Proposition 11.2.11** *Let* $\mathcal{A}^L = (A^L, \odot, ^-, 1)$ *be a left-QMV algebra. Then, (Pom) holds.*

**Proof.** Take $y = 1$ in (Pqmv). □

**Proposition 11.2.12** *Let* $\mathcal{A}^L = (A^L, \odot, ^-, 1)$ *be a left-QMV algebra verifying (G)* $(x \odot x = x)$. *Then:*
*(1)* $\leq_m^P$ *is reflexive also, hence it is an* **order relation**.
*(2) We have the equivalence:*

$$(x \odot y = x \iff)\ x \leq_m^P y \iff x \leq_m^M y\ (\iff x \wedge_m^M y = x).$$

**Proof.** (1): $x \leq_m^P x \iff x \odot x = x$, that is true by (G).

(2): Suppose $x \leq_m^P y$, i.e. $x \odot y = x$. Then, by (11.2), $x = x \odot y \leq_m^M y$. Conversely, suppose $x \leq_m^M y$. Then, by (11.11), we have: $x \stackrel{(G)}{=} x \odot x \leq_m^M y \odot x \stackrel{(Pcomm)}{=} x \odot y$, and since we also have, by (11.2), that $x \odot y \leq_m^M x$, we obtain, by antisymmetry of $\leq_m^M$ (by Corollary 6.4.25), that $x \odot y = x$, i.e. $x \leq_m^P y$. □

**Remarks 11.2.13** *In a left-QMV algebra* $\mathcal{A}^L = (A^L, \odot, ^-, 1)$*:*
*- the initial binary relation,* $\leq_m$ $(x \leq_m y \iff x \odot y^- = 0)$ $(\leq_m \iff \leq_m^B)$*, is* **only reflexive** *((m-Re) holds, by definition of m-BE algebra);*
*- the binary relation* $\leq_m^M$ $(x \leq_m^M y \iff x \wedge_m^M y = x)$ *is an* **order**, *by Corollary 11.2.9, but not a lattice order with respect to* $\wedge_m^M$, $\vee_m^M$, *since* $x \wedge_m^M y \neq y \wedge_m^M x$*;*

*- the binary relation* $\leq_m^P$ $(x \leq_m^P y \Longleftrightarrow x \odot y = x)$ *is* **only antisymmetric and transitive**, *by Proposition 6.4.9.*

*In a left-QMV algebra verifying (G),* $\leq_m^M$ *and* $\leq_m^P$ *are order relations and* $\leq_m^M \Longleftrightarrow \leq_m^P$.

### 11.2.2 The equivalence between (Pq) and (Pom)

Consider now the properties:
(Pq) $\quad x \odot [y \vee_m^M (z \vee_m^M x^-)] = (x \odot y) \vee_m^M (x \odot z)$ and
(Pom) $\quad (x \odot y) \oplus ((x \odot y)^- \odot x) = x$ or, equivalently, $x \vee_m^M (x \odot y) = x$,
which characterizes the orthomodular lattices among ortholattices.

**Proposition 11.2.14** *Let* $\mathcal{A}^L = (A^L, \odot, ^-, 1)$ *be an involutive left-m-BE algebra. Then,*

$$(Pq) \implies (Pom).$$

**Proof.** In (Pq) take $y := 1$ to obtain: $x = x \vee_m^M (x \odot z)$, i.e. (Pom). □

**The converse result, the next Proposition 11.2.16 (saying that (Pom) implies (Pq)), was proved by** *Prover9* **in about an hour, only after changing the basic Prover9 options** *order* **from 'lpo' to 'kbo' and** *eq-defs* **from 'unfold' to 'fold' and after removing those axioms of the algebra containing 0, 1. The proof by** *Prover9* **had the length 54 (i.e. there were 54 steps); after proving the 54 steps from the chain of length 54, we have gouped the steps into the following Lemma 11.2.15 and Proposition 11.2.16.**

**Lemma 11.2.15** *Let* $\mathcal{A}^L = (A^L, \odot, ^-, 1)$ *be an involutive left-m-MEL algebra. We have:*

$$x^- \oplus (y \odot x) = y \vee_m^M x^-, \tag{11.25}$$

$$x \oplus (y \odot (z \odot x^-)) = (y \odot z) \vee_m^M x, \tag{11.26}$$

$$x^- \oplus ((y \odot x) \vee_m^M z) = y \vee_m^M (x^- \oplus z), \tag{11.27}$$

$$(x \odot y) \oplus (z \odot (x^- \wedge_m^M y)) = (z \odot y) \vee_m^M (x \odot y), \tag{11.28}$$

**Proof.** (11.25): $y \vee_m^M x^- = x^- \oplus (y \odot x)$, by definition and (DN).

(11.26): $x \oplus (y \odot (z \odot x^-)) \overset{(Pass)}{=} x \oplus ((y \odot z) \odot x^-) = (y \odot z) \vee_m^M x$.

(11.27): The left side: $x^- \oplus ((y \odot x) \vee_m^M z) \overset{(11.26)}{=} x^- \oplus (z \oplus (y \odot (x \odot z^-))) \overset{(Pass)}{=} x^- \oplus [z \oplus ((y \odot x) \odot z^-)]$.
The right side: $y \vee_m^M (x^- \oplus z) = (x^- \oplus z) \oplus (y \odot (x^- \oplus z)^-) = (x^- \oplus z) \oplus (y \odot (x \odot z^-)) \overset{(Pass),(Sass)}{=} x^- \oplus [z \oplus ((y \odot x) \odot z^-)]$. Hence, (11.27) holds.

(11.28): $(x \odot y) \oplus (z \odot (x^- \wedge_m^M y)) = (x \odot y) \oplus (z \odot (y \odot (x \odot y)^-)) \overset{(11.26)}{=} (z \odot y) \vee_m^M (x \odot y)$, for $X := x \odot y$, $Y := z$, $Z := y$ in (11.26). □

**Proposition 11.2.16** *Let $\mathcal{A}^L = (A^L, \odot, ^-, 1)$ be an involutive left-m-BE algebra. Then,*

$$(Pom) \implies (Pq).$$

**Proof.** The proof has 13 steps:

(11.29) $$x \wedge_m^M (y \oplus x) = x.$$

Indeed, $x^- \vee_m^M (y \oplus x)^- = x^- \vee_m^M (y^- \odot x^-) \stackrel{(Pom)}{=} x^-$; then, $x \wedge_m^M (y \oplus x) = (x^- \vee_m^M (y \oplus x)^-)^- = x^= \stackrel{(DN)}{=} x$.

(11.30) $$x \vee_m^M (y \wedge_m^M x) = x,$$

(11.31) $$x \wedge_m^M (y \vee_m^M x) = x.$$

Indeed, $x \vee_m^M (y \wedge_m^M x) = x \vee_m^M (x \odot (y \oplus x^-)) \stackrel{(Pom)}{=} x$; thus, (11.30) holds. (11.31) follows by duality.

(11.32) $$(x \odot y) \wedge_m^M y = x \odot y,$$

(11.33) $$(x \oplus y) \vee_m^M y = x \oplus y.$$

Indeed, $(x \odot y) \wedge_m^M y \stackrel{(Pom)}{=} (x \odot y) \wedge_m^M (y \vee_m^M (x \odot y)) \stackrel{(11.31)}{=} x \odot y$, with $X := x \odot y$; thus, (11.32) holds. (11.33) follows by duality.

Now, we prove

(11.34) $$(x \vee_m^M y) \odot (x^- \oplus y) = y,$$

(11.35) $$(x \wedge_m^M y) \oplus (x^- \odot y) = y.$$

Indeed, $(x \vee_m^M y) \odot (x^- \oplus y) = ((x^- \oplus y)^- \oplus y) \odot (x^- \oplus y) = y \wedge_m^M (x^- \oplus y) \stackrel{(11.29)}{=} y$; thus, (11.34) holds. (11.35) follows by duality.

(11.36) $$(x \odot y^-) \oplus (z \oplus (y \wedge_m^M x)) = z \oplus x.$$

Indeed, $(x \odot y^-) \oplus (z \oplus (y \wedge_m^M x)) \stackrel{(Scomm)}{=} (z \oplus (y \wedge_m^M x)) \oplus (x \odot y^-) \stackrel{(Sass)}{=} z \oplus ((y \wedge_m^M x) \oplus (x \odot y^-)) \stackrel{(Pcomm)}{=} z \oplus ((y \wedge_m^M x) \oplus (y^- \odot x)) \stackrel{(11.35)}{=} z \oplus x$.

(11.37) $$(x \oplus y) \vee_m^M (x \oplus (z \wedge_m^M y)) = x \oplus y.$$

Indeed, by (11.36), we have $(y \odot z^-) \oplus (x \oplus (z \wedge_m^M y)) = x \oplus y$; put $X := y \odot z^-$, $Y := x \oplus (z \wedge_m^M y)$; hence, we have $X \oplus Y = x \oplus y$; then, $x \oplus y = X \oplus Y \stackrel{(11.33)}{=} (X \oplus Y) \vee_m^M Y = (x \oplus y) \vee_m^M (x \oplus (z \wedge_m^M y))$.

(11.38) $$(x \oplus y) \vee_m^M (x \oplus (z \odot y)) = x \oplus y.$$

Indeed, $(x \oplus y) \vee_m^M (x \oplus (z \odot y)) \overset{(11.32)}{=} (x \oplus y) \vee_m^M (x \oplus ((z \odot y) \wedge_m^M y)) \overset{(11.37)}{=} x \oplus y$, where $Z := z \odot y$ in (11.37).

$$(11.39) \qquad x \vee_m^M ((y \odot x) \oplus (z \odot (y^- \wedge_m^M x))) = x.$$

Indeed, first, by (11.35), we have $(y \odot x) \oplus (y^- \wedge_m^M x) = x$; put $X := y \odot x$ and $Y := y^- \wedge_m^M x$, hence we have $X \oplus Y = x$; now, $x = X \oplus Y \overset{(11.38)}{=} (X \oplus Y) \vee_m^M (X \oplus (z \odot Y)) = x \vee_m^M ((y \odot x) \oplus (z \odot (y^- \wedge_m^M x)))$.

Now, we prove

$$(11.40) \qquad x \vee_m^M ((y \odot x) \vee_m^M (z \odot x)) = x.$$

Indeed, $x \vee_m^M ((y \odot x) \vee_m^M (z \odot x)) \overset{(11.28)}{=} x \vee_m^M [(z \odot x) \oplus (y \odot (z^- \wedge_m^M x))] \overset{(11.39)}{=} x$, with $Y := z$ and $Z := y$.

Finally, we prove (Pq), i.e. $x \odot [y \vee_m^M (z \vee_m^M x^-)] = (y \odot x) \vee_m^M (z \odot x)$.
Indeed, $x \odot [y \vee_m^M (z \vee_m^M x^-)]$
$\overset{(11.25)}{=} x \odot [y \vee_m^M (x^- \oplus (z \odot x))]$
$\overset{(11.27)}{=} x \odot [x^- \oplus ((y \odot x) \vee_m^M (z \odot x))]$
$\overset{(11.40)}{=} (x \vee_m^M [(y \odot x) \vee_m^M (z \odot x)]) \odot (x^- \oplus [(y \odot x) \vee_m^M (z \odot x)])$
$\overset{(11.34)}{=} (y \odot x) \vee_m^M (z \odot x) \overset{(Pcomm)}{=} (x \odot y) \vee_m^M (x \odot z)$. □

By Propositions 11.2.14 and 11.2.16, we obtain **the second very important result, the core of this chapter**, by its difficulty:

**Theorem 11.2.17** *Let $\mathcal{A}^L = (A^L, \odot, ^-, 1)$ be an involutive left-m-BE algebra. Then,*

$$(Pq) \iff (Pom).$$

Consequently, by Theorems 11.2.10 and 11.2.17, we obtain:

**Theorem 11.2.18** *Let $\mathcal{A}^L = (A^L, \odot, ^-, 1)$ be an involutive left-m-BE algebra. Then,*

$$(Pqmv) \iff (Pmv) + (Pom).$$

### 11.2.3 The property $(\Delta_m)$

Consider now the property introduced in Section 10.3 (the dual one is omitted):
$(\Delta_m)$ $\qquad (x \wedge_m^M y) \odot (y \wedge_m^M x)^- = 0.$

Note that $(\Delta_m)$ is the largest non-antisymmetric generalization of ($\wedge_m$-comm). It is equivalent to: $(y \odot (x^- \odot y)^-) \odot (x \odot (y^- \odot x)^-)^- = 0$.

**Proposition 11.2.19** *Let $\mathcal{A}^L = (A^L, \odot, ^-, 1)$ be an involutive left-m-BE algebra. Then,*

$$(Pmv) \implies (\Delta_m).$$

**Proof.** First, note that, by replacing $y$ with $y^-$, (Pmv) becomes, by (DN):
(a) $x \odot ((x^- \odot y)^- \odot y)^- = x \odot y^-$.

Now, consider $(\Delta_m)$, i.e. $(y \odot (x^- \odot y)^-) \odot (x \odot (y^- \odot x)^-)^- = 0$, and by interchanging $x$ with $y$, we obtain:
(b) $(x \odot (y^- \odot x)^-) \odot (y \odot (x^- \odot y)^-)^- = 0$. We shall prove (b).
Indeed, $(x \odot (y^- \odot x)^-) \odot (y \odot (x^- \odot y)^-)^-$
$\overset{(Pcomm),(Pass)}{=} (x \odot ((x^- \odot y)^- \odot y)^-) \odot (x \odot y^-)^-$
$\overset{(a)}{=} (x \odot y^-) \odot (x \odot y^-)^- \overset{(m-Re)}{=} 0$. Thus, $(\Delta_m)$ holds. □

The converse of Proposition 11.2.19 does not hold, in general; there are examples of involutive m-BE algebras verifying $(\Delta_m)$ and not verifying (Pmv), see Example 11.4.3.

**But, in particular, we have the following Proposition 11.2.22 (saying that if the involutive m-BE algebra verifies (Pom), then $(\Delta_m)$ implies (Pmv)), proved by *Prover9* in 2453 seconds, the length of the proof being 33; the proof by *Prover9* generated the proofs of the following Lemmas 11.2.20, 11.2.21 and Proposition 11.2.22.**

**Lemma 11.2.20** *Let $\mathcal{A}^L = (A^L, \odot, ^-, 1)$ be an involutive left-m-BE algebra verifying (Pom). Then,*

$$(x \odot y)^- \odot (x \odot (x \odot y)^-)^- = x^-, \tag{11.41}$$

$$(x \odot (y \odot z))^- \odot [x \odot (y \odot (x \odot (y \odot z))^-)]^- = (x \odot y)^-, \tag{11.42}$$

$$(x \odot y^-)^- \odot [x \odot ((y \odot z)^- \odot (x \odot y^-)^-)]^- = (x \odot (y \odot z)^-)^-. \tag{11.43}$$

**Proof.** (11.41): From (Pom), by (Pcomm).

(11.42: In (11.41), take $X := x \odot y$ and $Y := z$ to obtain:
$((x \odot y) \odot z)^- \odot ((x \odot y) \odot ((x \odot y) \odot z)^-)^- = (x \odot y)^-$; then, by (Pass), we obtain (11.42).

(11.43): In (11.42), take $X := x$, $Y := (y \odot z)^-$, $Z := (y \odot (y \odot z)^-)^-$ to obtain:
(a) $(X \odot (Y \odot Z))^- \odot [X \odot (Y \odot (X \odot (Y \odot Z))^-)]^- = (X \odot Y)^-$; but,
$X \odot (Y \odot Z) = x \odot ((y \odot z)^- \odot (y \odot (y \odot z)^-)^-) \overset{(11.41)}{=} x \odot y^-$; hence, (a) becomes:
$(x \odot y^-)^- \odot [x \odot ((y \odot z)^- \odot (x \odot y^-)^-)]^- = (x \odot (y \odot z)^-)^-$, that is (11.43). □

**Lemma 11.2.21** *Let $\mathcal{A}^L = (A^L, \odot, ^-, 1)$ be an involutive left-m-BE algebra verifying $(\Delta_m)$. Then,*

$$x \odot ((y^- \odot x)^- \odot (y \odot (x^- \odot y)^-)^-) = 0, \tag{11.44}$$

$$x \odot ((x \odot y^-)^- \odot (y \odot (y \odot x^-)^-)^-) = 0, \tag{11.45}$$

$$x^- \odot ((x^- \odot y^-)^- \odot (y \odot (y \odot x)^-)^-) = 0. \tag{11.46}$$

**Proof.** (11.44): Since $x \wedge_m y = y \odot (x^- \odot y)^-$, then $(\Delta_m)$ $((x \wedge_m y) \odot (y \wedge_m x)^- = 0)$ becomes:
(a) $(y \odot (x^- \odot y)^-) \odot (x \odot (y^- \odot x)^-)^- = 0$; then, interchanging $x$ with $y$ in (a), we obtain:
(b) $(x \odot (y^- \odot x)^-) \odot (y \odot (x^- \odot y)^-)^- = 0$; then, by (Pass), we obtain (11.44).
(11.45): From (11.44), by (Pcomm).
(11.46): From (11.45), by taking $X := x^-$ and by (DN). □

**Proposition 11.2.22** *Let $\mathcal{A}^L = (A^L, \odot, ^-, 1)$ be an involutive left-m-BE algebra. Then,*

$$(Pom) \ + \ (\Delta_m) \implies (Pmv).$$

**Proof.** First, we prove:

$$(x^- \odot (y \odot (y \odot x)^-)^-)^- = (x^- \odot y^-)^-. \tag{11.47}$$

Indeed, in (11.43), take $X := x^-$, $Y := y$ and $Z := (y \odot x)^-$ to obtain:
(a) $(x^- \odot y^-)^- \odot [x^- \odot ((y \odot (y \odot x)^-)^- \odot (x^- \odot y^-)^-)]^- = (x^- \odot (y \odot (y \odot x)^-)^-)^-$;
but, $x^- \odot ((y \odot (y \odot x)^-)^- \odot (x^- \odot y^-)^-) \overset{(11.46)}{=} 0$; hence, (a) becomes:
$(x^- \odot y^-)^- \odot [0]^- = (x^- \odot (y \odot (y \odot x)^-)^-)^-$, i.e. (11.47) holds, by (Neg0-1), (PU).
Next, from (11.47), it follows, by (DN) and (Pcomm):

$$x^- \odot ((x \odot y)^- \odot y)^- = x^- \odot y^-. \tag{11.48}$$

Finally, from (11.48), by taking $X := x^-$ and $Y := y^-$, we obtain, by (DN): $x \odot ((x^- \odot y^-)^- \odot y^-)^- = x \odot y$, that is (Pmv). □

Resuming, by Propositions 11.2.19, 11.2.22, we obtain **the third very important result of this chapter**:

**Theorem 11.2.23** *Let $\mathcal{A}^L = (A^L, \odot, ^-, 1)$ be an involutive left-m-BE algebra. Then,*

$$(Pom) \implies ((Pmv) \Leftrightarrow (\Delta_m)).$$

Consequently, by Theorems 11.2.18 and 11.2.23, we obtain:

**Theorem 11.2.24** *Let $\mathcal{A}^L = (A^L, \odot, ^-, 1)$ be an involutive left-m-BE algebra. Then,*

$$(Pqmv) \iff (\Delta_m) \ + \ (Pom).$$

## 11.3 Three generalizations of QMV algebras

Let $\mathcal{A}^L = (A^L, \odot, ^-, 1)$ be an involutive left-m-BE algebra throughout this section.

### 11.3.1 The three algebras

Consider the properties:
(Pom) $(x \odot y) \oplus ((x \odot y)^- \odot x) = x$ or, equivalently, $x \vee_m^M (x \odot y) = x$ and, dually,
(Som) $(x \oplus y) \odot ((x \oplus y)^- \oplus x) = x$ or, equivalently, $x \wedge_m^M (x \oplus y) = x$;
(Pmv) $x \odot (x^- \vee_m^M y) = x \odot y$ and, dually,
(Smv) $x \oplus (x^- \wedge_m^M y) = x \oplus y$;
($\Delta_m$) $(x \wedge_m^M y) \odot (y \wedge_m^M x)^- = 0$ and, dually,
($\nabla_m$) $(x \vee_m^M y) \oplus (y \vee_m^M x)^- = 1$.
We introduce the following notions:

**Definitions 11.3.1**
(i) An involutive left-m-BE algebra $\mathcal{A}^L = (A^L, \odot, ^-, 1)$ is:
- a *left-orthomodular algebra*, or a *left-OM algebra* for short, if it verifies (Pom),
- a *left-pre-MV algebra*, or a *left-PreMV algebra* for short, if it verifies (Pmv),
- a *left-metha-MV algebra*, or a *left-MMV algebra* for short, if it verifies ($\Delta_m$).
(i') Dually, an involutive right-m-BE algebra $\mathcal{A}^R = (A^R, \oplus, ^-, 0)$ is:
- a *right-orthomodular algebra*, or a *right-OM algebra* for short, if it verifies (Som),
- a *right-pre-MV algebra*, or a *right-PreMV algebra* for short, if it verifies (Smv),
- a *right-metha-MV algebra*, or a *right-MMV algebra* for short, if it verifies ($\nabla_m$).

We shall denote by **OM**, **PreMV**, **MMV** the classes of the corresponding left-algebras and by $\mathbf{OM}^R$, $\mathbf{PreMV}^R$, $\mathbf{MMV}^R$ the classes of the corresponding right-algebras.

See Examples 11.4.1, 11.4.2, 11.4.3 of left-OM, left-PreMV, left-MMV algebras, respectively, and Example 11.4.4 of left-QMV algebra.

By Propositions 11.2.11, 11.2.4, 11.2.19 and Theorems 11.2.18, 11.2.24, we obtain:

**Corollary 11.3.2** *We have:*

$\mathbf{QMV} \subset \mathbf{OM}$, $\mathbf{QMV} \subset \mathbf{PreMV} \subset \mathbf{MMV}$
*and*
$\mathbf{QMV} = \mathbf{PreMV} \cap \mathbf{OM} = \mathbf{MMV} \cap \mathbf{OM}$.

Note that we can say that QMV algebras are *orthomodular PreMV algebras*, or *orthomodular MMV algebras*.

Hence, we have the connections from Figure 11.1.

### 11.3.2 The transitive and/or antisymmetric algebras

We shall denote by **tOM**, **tPreMV**, **tMMV**, **tQMV** the classes of the corresponding transitive left-algebras. Note that these classes of algebras are contained in the class $\mathbf{m\text{-}pre\text{-}BCK}_{(DN)} = \mathbf{m\text{-}tBE}_{(DN)}$. See Examples 11.4.5, 11.4.6, 11.4.7, 11.4.8 of left-tOM, left–tPreMV, left-MMV, left-tQMV algebras, respectively.

By the previous Corollary 11.3.2, we obtain:

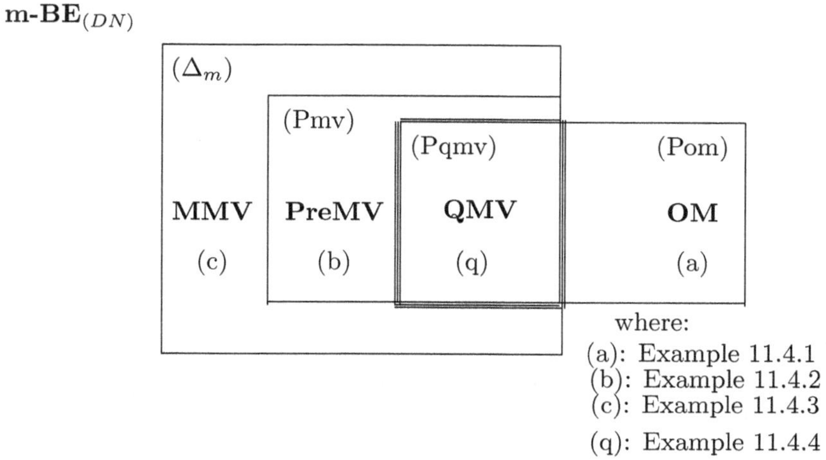


Figure 11.1: Resuming connections between **OM**, **PreMV**, **MMV** and **QMV**

**Corollary 11.3.3** *We have:*

$$\begin{array}{ccc} \mathbf{QMV} & \subset & \mathbf{OM} \\ \cup & & \cup \\ \mathbf{tQMV} & \subset & \mathbf{tOM}, \end{array} \qquad \begin{array}{ccccc} \mathbf{QMV} & \subset & \mathbf{PreMV} & \subset & \mathbf{MMV} \\ \cup & & \cup & & \cup \\ \mathbf{tQMV} & \subset & \mathbf{tPreMV} & \subset & \mathbf{tMMV} \end{array}$$

*and*

$$\mathbf{tQMV} = \mathbf{tPreMV} \cap \mathbf{tOM} = \mathbf{tMMV} \cap \mathbf{tOM}.$$

Hence, we have the connections from Figure 11.2.

We shall denote by **aOM**, **aPreMV**, **aMMV**, **aQMV** the classes of the corresponding antisymmetric left-algebras. Note that these classes of algebras are contained in the class $\mathbf{m\text{-}aBE}_{(DN)}$.

By Corollary 11.3.2 again, we obtain the analogous of Corollary 11.3.3, which by lack of space is omitted.

We shall denote by **taOM**, **taPreMV**, **taMMV**, **taQMV** the classes of the corresponding transitive and antisymmetric left-algebras. Note that these classes of algebras are contained in the class $\mathbf{m\text{-}BCK} = \mathbf{m\text{-}taBE}_{(DN)}$.

By Corollary 11.3.3 and its analogous, we then obtain:

**Corollary 11.3.4** *We have:*

$$\begin{array}{ccc} \mathbf{QMV} & \subset & \mathbf{OM} \\ \cup & & \cup \\ \mathbf{tQMV} & \subset & \mathbf{tOM} \\ \cup & & \cup \\ \mathbf{taQMV} & \subset & \mathbf{taOM}, \end{array} \qquad \begin{array}{ccccc} \mathbf{QMV} & \subset & \mathbf{PreMV} & \subset & \mathbf{MMV} \\ \cup & & \cup & & \cup \\ \mathbf{tQMV} & \subset & \mathbf{tPreMV} & \subset & \mathbf{tMMV} \\ \cup & & \cup & & \cup \\ \mathbf{taQMV} & \subseteq & \mathbf{taPreMV} & \subseteq & \mathbf{taMMV} \end{array}$$

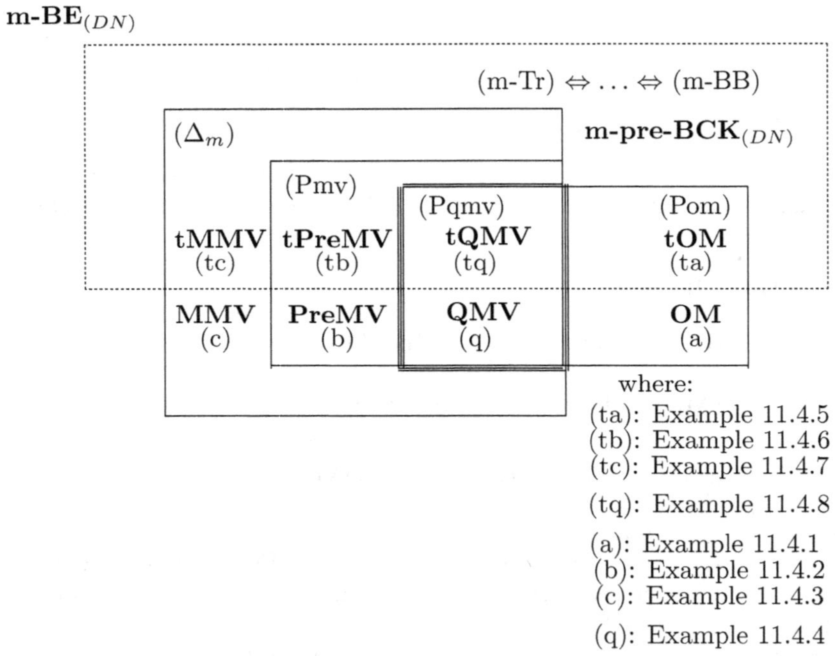


Figure 11.2: Resuming connections between **OM**, **PreMV**, **MMV**, **QMV** and (m-Tr)

*and*

$$\begin{array}{ccccccccc}
\mathbf{QMV} & \subset & \mathbf{OM} & \quad & \mathbf{QMV} & \subset & \mathbf{PreMV} & \subset & \mathbf{MMV} \\
\cup & & \cup & & \cup & & \cup & & \cup \\
\mathbf{aQMV} & \subset & \mathbf{aOM} & & \mathbf{aQMV} & \subseteq & \mathbf{aPreMV} & \subseteq & \mathbf{aMMV} \\
\cup & & \cup & & \cup & & \cup & & \cup \\
\mathbf{taQMV} & \subset & \mathbf{taOM}, & & \mathbf{taQMV} & \subseteq & \mathbf{taPreMV} & \subseteq & \mathbf{taMMV}
\end{array}$$

*and*

$$\mathbf{aQMV} = \mathbf{aPreMV} \cap \mathbf{aOM} = \mathbf{aMMV} \cap \mathbf{aOM},$$
$$\mathbf{taQMV} = \mathbf{taPreMV} \cap \mathbf{taOM} = \mathbf{taMMV} \cap \mathbf{taOM}.$$

### 11.3.3 Connections with MV algebras

We know (see ([45], Example 2.3.14)) that any MV algebra is a QMV algebra: $\mathbf{MV} \subset \mathbf{QMV}$, since:

$$(\wedge_m - comm) \Longrightarrow (Pqmv). \tag{11.49}$$

Consequently, we have:

$$(\wedge_m - comm) \Longrightarrow (Pom) + (Pmv) + (\Delta_m). \tag{11.50}$$

The next Theorems 11.3.6, 11.3.7 and 11.3.9 say that ($\wedge_m$-comm) is equivalent with some properties.

**Proposition 11.3.5** *Let* $\mathcal{A}^L = (A^L, \odot, ^-, 1)$ *be an involutive left-m-BE algebra. Then,*

$$(Pmv) + (m - An) \Longrightarrow (\wedge_m - comm).$$

**Proof.** Suppose (m-An) holds, i.e. $X \leq_m Y$ and $Y \leq_m X$ imply $X = Y$, which mean $X \odot Y^- = 0$ and $Y \odot X^- = 0$ imply $X = Y$.

Take $X \overset{notation}{=} x \wedge_m^M y \overset{(6.32)}{=} (x^- \odot y)^- \odot y$ and $Y \overset{notation}{=} y \wedge_m^M x \overset{(6.32)}{=} (y^- \odot x)^- \odot x$.

We have: $X \odot Y^- = [(x^- \odot y)^- \odot y] \odot [y \wedge_m^M x]^-$
$\overset{(6.39)}{=} [(x^- \odot y)^- \odot y] \odot [y^- \vee_m^M x^-] \overset{(Pass)}{=} (x^- \odot y)^- \odot (y \odot (y^- \vee_m^M x^-))$
$\overset{(Pmv)}{=} (x^- \odot y)^- \odot (y \odot x^-) \overset{(Pcomm)}{=} (y \odot x^-) \odot (y \odot x^-)^- \overset{(m-Re)}{=} 0.$

Similarly, we have: $Y \odot X^- = (y^- \odot x)^- \odot x \odot [x \wedge_m^M y]^-$
$(y^- \odot x)^- \odot x \odot (x^- \vee_m^M y^-) \overset{(Pmv)}{=} (y^- \odot x)^- \odot (x \odot y^-) = 0.$

By (m-An), we obtain $X = Y$, i.e. ($\wedge_m$-comm) holds. □

By Proposition 11.3.5 and (6.61), (11.50), we obtain:

**Theorem 11.3.6**

$$(Pmv) + (m - An) \Longleftrightarrow (\wedge_m - comm).$$

Recall again (10.2) saying:

**Theorem 11.3.7**

$$(\Delta_m) + (m - An) \iff (\wedge_m - comm).$$

**Proposition 11.3.8**

$$(Pqmv) + (m - An) \Longrightarrow (\wedge_m - comm).$$

**Proof.** By Propositions 11.2.4, 11.3.5, we obtain:
(Pqmv) + (m-An) $\Longrightarrow$ (Pmv) + (m-An) $\Longrightarrow$ ($\wedge_m$-comm). □

By Proposition 11.3.8 and by (6.61), (11.49), we obtain:

**Theorem 11.3.9**

$$(Pqmv) + (m - An) \iff (\wedge_m - comm).$$

By previous Theorems 11.3.6, 11.3.7, 11.3.9, we obtain **the fourth very important result of this chapter**:

**Corollary 11.3.10** *We have:*

| | | | | | | | |
|---|---|---|---|---|---|---|---|
| **PreMV** | + | *(m-An)* | = | **MV**, | *i.e.* | **aPreMV** | = **MV**, |
| **MMV** | + | *(m-An)* | = | **MV**, | *i.e.* | **aMMV** | = **MV**, |
| **QMV** | + | *(m-An)* | = | **MV**, | *i.e.* | **aQMV** | = **MV**. |

**Remark 11.3.11** *By (6.61), we have:*
**MV** = **aMV** = **tMV** = **taMV**, *hence*
**taPreMV** = **taMMV** = **taQMV** = **MV**.

By Corollaries 11.3.4, 11.3.10 and by Remark 11.3.11, we obtain:

**Corollary 11.3.12** *We have:*

$$\begin{array}{ccc} \mathbf{QMV} & \subset & \mathbf{OM} \\ \cup & & \cup \\ \mathbf{tQMV} & \subset & \mathbf{tOM} \\ \cup & & \cup \\ \mathbf{MV} & \subset & \mathbf{taOM}, \end{array} \qquad \begin{array}{ccccc} \mathbf{QMV} & \subset & \mathbf{PreMV} & \subset & \mathbf{MMV} \\ \cup & & \cup & & \cup \\ \mathbf{tQMV} & \subset & \mathbf{tPreMV} & \subset & \mathbf{tMMV} \\ \cup & & \cup & & \cup \\ \mathbf{MV} & \subseteq & \mathbf{MV} & \subseteq & \mathbf{MV} \end{array}$$

*and*

$$\begin{array}{ccc} \mathbf{QMV} & \subset & \mathbf{OM} \\ \cup & & \cup \\ \mathbf{MV} & \subset & \mathbf{aOM} \\ \cup & & \cup \\ \mathbf{MV} & \subset & \mathbf{taOM} \end{array} \qquad \begin{array}{ccccc} \mathbf{QMV} & \subset & \mathbf{PreMV} & \subset & \mathbf{MMV} \\ \cup & & \cup & & \cup \\ \mathbf{MV} & \subseteq & \mathbf{MV} & \subseteq & \mathbf{MV} \end{array}$$

*and*

**MV** = **MV** $\cap$ **aOM**,
**MV** = **MV** $\cap$ **taOM**.

Note that *taOM algebras* are proper generalizations of MV algebras inside the class of m-BCK algebras. See Example 11.4.9 of left-taOM algebra.

A problem we have not been able to resolve is the following.

**Open problem 11.3.13** *Find an example of antisymmetric orthomodular algebra (aOM) which does not verify (m-Tr) ($\Longleftrightarrow \ldots$ (m-BB)), i.e. a proper element of* ***aOM*** *(using* MACE4, *we have searched exhaustively for an example up through and including size* 20*), or prove that an involutive left-m-aBE algebra satisfying (Pom) satisfies also (m-Tr) (i.e.* **aOM** = **taOM***) (we have also tried to find a proof using* PROVER9, *but despite letting it run for several days, it was unable to find one).*

Hence, we have the connections from Figure 11.3.

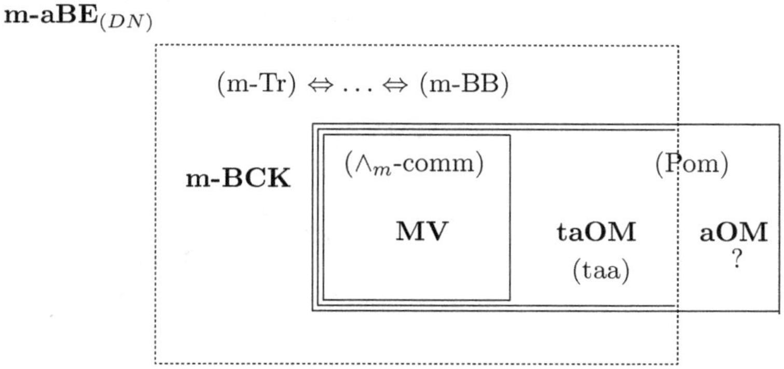


where:

(taa): Example 11.4.9

Figure 11.3: Resuming connections between **MV**, **taOM** and **aOM**, where ? means that there is an open problem concerning **aOM**

Note that, by Theorems 11.3.6, 11.3.7, 11.3.9 again, we obtain:

**Theorem 11.3.14** *Let* $\mathcal{A}^L = (A^L, \odot, ^-, 1)$ *be an involutive left-m-aBE algebra. Then,*

$$(\wedge_m - comm) \Longleftrightarrow (Pmv) \Longleftrightarrow (\Delta_m) \Longleftrightarrow (Pqmv).$$

### 11.3.4 Putting QMV and tQMV on the "map"

By the previous results, we are now able to put QMV algebras and tQMV algebras (and MV algebras) on the involutive "Big m-map" (and, hence, on the "map") - see the Figure 11.4.

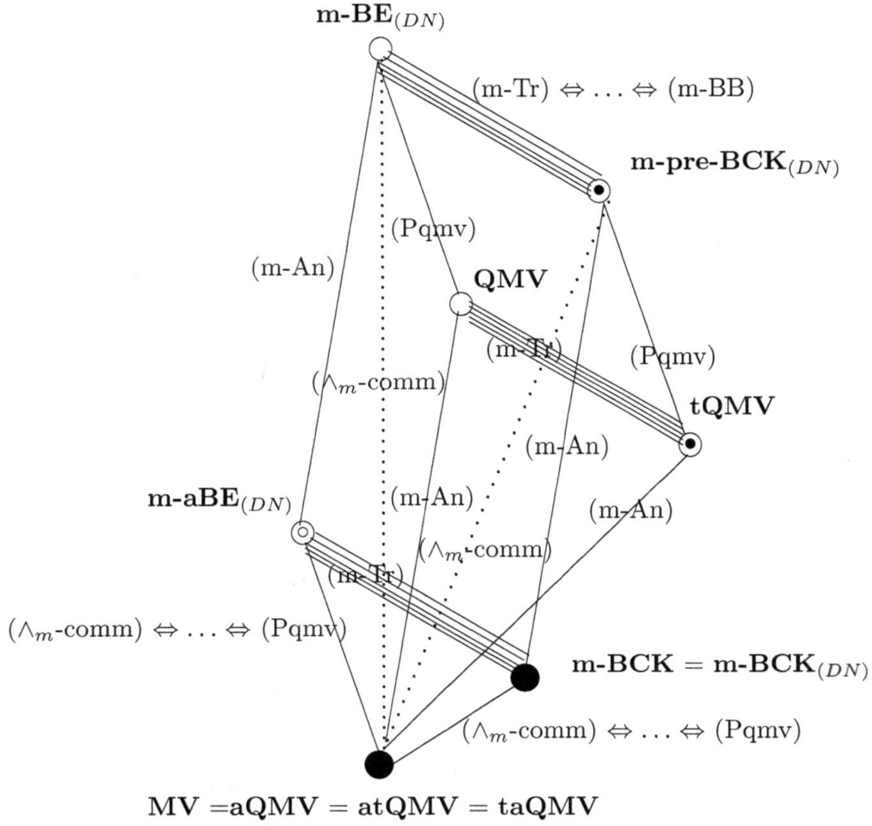


Figure 11.4: Putting QMV and tQMV algebras on the "map"

## 11.4 Examples

We introduce the following definition: an $X$ algebra is said to be *proper*, if it verifies the properties from its definition and does not verify the other properties from this chapter, except (prel$_m$).

**Example 11.4.1 Proper orthomodular algebra: OM**
By a PASCAL program, we found that the algebra
$\mathcal{A}^L = (A_6 = \{0, a, b, c, d, 1\}, \odot, ^-, 1)$, with the following tables of $\odot$ and $^-$ and of the additional operation $\oplus$, is an involutive left-m-BE algebra verifying (Pom) and (prel$_m$) and not verifying (m-B) for $(a, b, a)$, (m-BB) for $(a, a, b)$, (m-*) for $(b, d, a)$, (m-**) for $(a, b, a)$, (m-Tr) for $(a, b, d)$, (m-An) for $(a, b)$, (Pqmv) for $(d, d, 0)$, (Pmv) for $(d, d)$, $(\Delta_m)$ for $(a, d)$.

| $\odot$ | 0 | a | b | c | d | 1 |
|---|---|---|---|---|---|---|
| 0 | 0 | 0 | 0 | 0 | 0 | 0 |
| a | 0 | a | 0 | 0 | 0 | a |
| b | 0 | 0 | 0 | 0 | 0 | b |
| c | 0 | 0 | 0 | 0 | 0 | c |
| d | 0 | 0 | 0 | 0 | 0 | d |
| 1 | 0 | a | b | c | d | 1 |

and

| $x$ | $x^-$ |
|---|---|
| 0 | 1 |
| a | d |
| b | c |
| c | b |
| d | a |
| 1 | 0 |

, with

| $\oplus$ | 0 | a | b | c | d | 1 |
|---|---|---|---|---|---|---|
| 0 | 0 | a | b | c | d | 1 |
| a | a | 1 | 1 | 1 | 1 | 1 |
| b | b | 1 | 1 | 1 | 1 | 1 |
| c | c | 1 | 1 | 1 | 1 | 1 |
| d | d | 1 | 1 | 1 | d | 1 |
| 1 | 1 | 1 | 1 | 1 | 1 | 1 |

.

Note that $\leq_m^M$ is transitive, hence $\leq_m^M$ is an order relation, by Corollary 6.4.25, but not a lattice order w.r. to $\wedge_m^M$, $\vee_m^M$, since $\wedge_m^M$ is not commutative.

**Example 11.4.2 Proper PreMV algebra: PreMV**
By a PASCAL program, we found that the algebra
$\mathcal{A}^L = (A_6 = \{0, a, b, c, d, 1\}, \odot, ^-, 1)$, with the following tables of $\odot$ and $^-$ and of the additional operation $\oplus$, is an involutive left-m-BE algebra verifying (Pmv) (hence $(\Delta_m)$) and (prel$_m$) and not verifying (m-B) for $(a, d, c)$, (m-BB) for $(a, c, a)$, (m-*) for $(a, d, c)$, (m-**) for $(a, d, c)$, (m-Tr) for $(a, d, b)$, (m-An) for $(a, c)$, (Pqmv) for $(a, 1, c)$, (Pom) for $(a, c)$.

| $\odot$ | 0 | a | b | c | d | 1 |
|---|---|---|---|---|---|---|
| 0 | 0 | 0 | 0 | 0 | 0 | 0 |
| a | 0 | 0 | 0 | b | 0 | a |
| b | 0 | 0 | 0 | 0 | 0 | b |
| c | 0 | b | 0 | b | 0 | c |
| d | 0 | 0 | 0 | 0 | 0 | d |
| 1 | 0 | a | b | c | d | 1 |

and

| $x$ | $x^-$ |
|---|---|
| 0 | 1 |
| a | d |
| b | c |
| c | b |
| d | a |
| 1 | 0 |

, with

| $\oplus$ | 0 | a | b | c | d | 1 |
|---|---|---|---|---|---|---|
| 0 | 0 | a | b | c | d | 1 |
| a | a | 1 | 1 | 1 | 1 | 1 |
| b | b | 1 | c | 1 | c | 1 |
| c | c | 1 | 1 | 1 | 1 | 1 |
| d | d | 1 | c | 1 | 1 | 1 |
| 1 | 1 | 1 | 1 | 1 | 1 | 1 |

.

Note that $\leq_m^M$ is transitive, hence $\leq_m^M$ is an order relation, by Corollary 6.4.25, but not a lattice order w.r. to $\wedge_m^M$, $\vee_m^M$, since $\wedge_m^M$ is not commutative.

**Example 11.4.3 Proper MMV algebra: MMV**
By a PASCAL program, we found that the algebra
$\mathcal{A}^L = (A_6 = \{0, a, b, c, d, 1\}, \odot, ^-, 1)$, with the following tables of $\odot$ and $^-$ and of the additional operation $\oplus$, is an involutive left-m-BE algebra verifying $(\Delta_m)$ and not verifying (m-B) for $(a, b, a)$, (m-BB) for $(a, a, b)$, (m-*) for $(a, b, c)$, (m-**) for

$(a, b, a)$, (m-Tr) for $(a, b, d)$, (m-An) for $(a, b)$, (Pqmv) for $(d, 0, d)$, (Pom) for $(d, d)$, (Pmv) for $(d, d)$, ($\text{prel}_m$) for $(a, d)$.

| $\odot$ | 0 | a | b | c | d | 1 |
|---|---|---|---|---|---|---|
| 0 | 0 | 0 | 0 | 0 | 0 | 0 |
| a | 0 | a | 0 | 0 | 0 | a |
| b | 0 | 0 | 0 | 0 | 0 | b |
| c | 0 | 0 | 0 | b | 0 | c |
| d | 0 | 0 | 0 | 0 | b | d |
| 1 | 0 | a | b | c | d | 1 |

and

| $x$ | $x^-$ |
|---|---|
| 0 | 1 |
| a | d |
| b | c |
| c | b |
| d | a |
| 1 | 0 |

, with

| $\oplus$ | 0 | a | b | c | d | 1 |
|---|---|---|---|---|---|---|
| 0 | 0 | a | b | c | d | 1 |
| a | a | c | 1 | 1 | 1 | 1 |
| b | b | 1 | c | 1 | 1 | 1 |
| c | c | 1 | 1 | 1 | 1 | 1 |
| d | d | 1 | 1 | 1 | d | 1 |
| 1 | 1 | 1 | 1 | 1 | 1 | 1 |

.

Note that $\leq_m^M$ is transitive, hence $\leq_m^M$ is an order relation, by Corollary 6.4.25, but not a lattice order w.r.t. to $\wedge_m^M$, $\vee_m^M$, since $\wedge_m^M$ is not commutative.

**Example 11.4.4 Proper QMV algebra: QMV**

By a PASCAL program, we found that the algebra $\mathcal{A}^L = (A_6 = \{0, a, b, c, d, 1\}, \odot, ^-, 1)$, with the following tables of $\odot$ and $^-$ and of the additional operation $\oplus$, is a proper left-QMV algebra, i.e. (PU), (Pcomm), (Pass), (m-La), (m-Re), (Pqmv) (hence (Pom), (Pmv), ($\Delta_m$), ($\text{prel}_m$)), (DN) hold and it does not verify (m-B) for $(a, d, c)$, (m-BB) for $(a, c, a)$, (m-*) for $(a, d, c)$, (m-**) for $(a, d, c)$, (m-Tr) for $(a, d, b)$, (m-An) for $(a, c)$, ($\wedge_m$-comm) for $(a, c)$.

| $\odot$ | 0 | a | b | c | d | 1 |
|---|---|---|---|---|---|---|
| 0 | 0 | 0 | 0 | 0 | 0 | 0 |
| a | 0 | 0 | 0 | b | 0 | a |
| b | 0 | 0 | 0 | 0 | 0 | b |
| c | 0 | b | 0 | d | 0 | c |
| d | 0 | 0 | 0 | 0 | 0 | d |
| 1 | 0 | a | b | c | d | 1 |

and

| $x$ | $x^-$ |
|---|---|
| 0 | 1 |
| a | d |
| b | c |
| c | b |
| d | a |
| 1 | 0 |

, with

| $\oplus$ | 0 | a | b | c | d | 1 |
|---|---|---|---|---|---|---|
| 0 | 0 | a | b | c | d | 1 |
| a | a | 1 | 1 | 1 | 1 | 1 |
| b | b | 1 | a | 1 | c | 1 |
| c | c | 1 | 1 | 1 | 1 | 1 |
| d | d | 1 | c | 1 | 1 | 1 |
| 1 | 1 | 1 | 1 | 1 | 1 | 1 |

.

Note that $\leq_m^M$ is an order relation, by Corollary 11.2.9, but not a lattice order w.r.t. to $\wedge_m^M$, $\vee_m^M$, since $\wedge_m^M$ is not commutative.

**Example 11.4.5 Proper transitive OM algebra : tOM**

By MACE4 program, we found that the algebra $\mathcal{A}^L = (A_8 = \{0, a, b, c, d, e, f, 1\}, \odot, ^-, 1)$, with the following tables of $\odot$ and $^-$ and of the additional operation $\oplus$, is an involutive left-m-BE algebra verifying (Pom) and (m-Tr) $\Longleftrightarrow \ldots \Longleftrightarrow$ (m-BB), and also ($\text{prel}_m$), and not verifying (m-An) for $(c, d)$, (Pqmv) for $(b, b, 0)$, (Pmv) for $(b, b)$, ($\Delta_m$) for $(a, b)$.

| $\odot$ | 0 | a | b | c | d | e | f | 1 |
|---|---|---|---|---|---|---|---|---|
| 0 | 0 | 0 | 0 | 0 | 0 | 0 | 0 | 0 |
| a | 0 | a | 0 | c | d | c | d | a |
| b | 0 | 0 | 0 | 0 | 0 | 0 | 0 | b |
| c | 0 | c | 0 | 0 | 0 | 0 | 0 | c |
| d | 0 | d | 0 | 0 | 0 | 0 | 0 | d |
| e | 0 | c | 0 | 0 | 0 | b | b | e |
| f | 0 | d | 0 | 0 | 0 | b | b | f |
| 1 | 0 | a | b | c | d | e | f | 1 |

and

| $\oplus$ | 0 | a | b | c | d | e | f | 1 |
|---|---|---|---|---|---|---|---|---|
| 0 | 0 | a | b | c | d | e | f | 1 |
| a | a | 1 | 1 | 1 | 1 | 1 | 1 | 1 |
| b | b | 1 | b | e | f | e | f | 1 |
| c | c | 1 | e | a | a | 1 | 1 | 1 |
| d | d | 1 | f | a | a | 1 | 1 | 1 |
| e | e | 1 | e | 1 | 1 | 1 | 1 | 1 |
| f | f | 1 | f | 1 | 1 | 1 | 1 | 1 |
| 1 | 1 | 1 | 1 | 1 | 1 | 1 | 1 | 1 |

with $(0, a, b, c, d, e, f, 1)^- = (1, b, a, e, f, c, d, 0)$.

Note that $\leq_m^M$ is transitive, hence $\leq_m^M$ is an order relation, by Corollary 6.4.25, but not a lattice order w.r.t. to $\wedge_m^M$, $\vee_m^M$, since $\wedge_m^M$ is not commutative.

**Example 11.4.6 Proper transitive pre-MV algebra: tPreMV**

By a PASCAL program, we found that the algebra $\mathcal{A}^L = (A_6 = \{0, a, b, c, d, 1\}, \odot, ^-, 1)$, with the following tables of $\odot$ and $^-$ and of the additional operation $\oplus$, is an involutive left-m-BE algebra verifying (Pmv) (hence ($\Delta_m$)) and (m-Tr) $\Longleftrightarrow \ldots \Longleftrightarrow$ (m-BB), (prel$_m$) and not verifying (m-An) for $(b, c)$, (Pqmv) for $(b, 1, d)$, (Pom) for $(b, d)$.

| $\odot$ | 0 | a | b | c | d | 1 |
|---|---|---|---|---|---|---|
| 0 | 0 | 0 | 0 | 0 | 0 | 0 |
| a | 0 | 0 | 0 | 0 | 0 | a |
| b | 0 | 0 | 0 | 0 | a | b |
| c | 0 | 0 | 0 | 0 | a | c |
| d | 0 | 0 | a | a | b | d |
| 1 | 0 | a | b | c | d | 1 |

and

| $x$ | $x^-$ |
|---|---|
| 0 | 1 |
| a | d |
| b | c |
| c | b |
| d | a |
| 1 | 0 |

, with

| $\oplus$ | 0 | a | b | c | d | 1 |
|---|---|---|---|---|---|---|
| 0 | 0 | a | b | c | d | 1 |
| a | a | c | d | d | 1 | 1 |
| b | b | d | 1 | 1 | 1 | 1 |
| c | c | d | 1 | 1 | 1 | 1 |
| d | d | 1 | 1 | 1 | 1 | 1 |
| 1 | 1 | 1 | 1 | 1 | 1 | 1 |

.

Note that $\leq_m^M$ is transitive, hence $\leq_m^M$ is an order relation, by Corollary 6.4.25, but not a lattice order w.r.t. to $\wedge_m^M$, $\vee_m^M$, since $\wedge_m^M$ is not commutative.

**Example 11.4.7 Proper transitive MMV algebra : tMMV**

By a PASCAL program, we found that the algebra $\mathcal{A}^L = (A_6 = \{0, a, b, c, d, 1\}, \odot, ^-, 1)$, with the following tables of $\odot$ and $^-$ and of the additional operation $\oplus$, is an involutive left-m-BE algebra verifying ($\Delta_m$) and (m-Tr) $\Longleftrightarrow \ldots \Longleftrightarrow$ (m-BB), and also (prel$_m$), and not verifying (m-An) for $(a, b)$, (Pqmv) for $(b, 0, a)$, (Pom) for $(b, a)$, (Pmv) for $(b, a)$.

| $\odot$ | 0 | a | b | c | d | 1 |
|---|---|---|---|---|---|---|
| 0 | 0 | 0 | 0 | 0 | 0 | 0 |
| a | 0 | a | a | 0 | 0 | a |
| b | 0 | a | a | 0 | 0 | b |
| c | 0 | 0 | 0 | c | c | c |
| d | 0 | 0 | 0 | c | d | d |
| 1 | 0 | a | b | c | d | 1 |

and

| $x$ | $x^-$ |
|---|---|
| 0 | 1 |
| a | d |
| b | c |
| c | b |
| d | a |
| 1 | 0 |

, with

| $\oplus$ | 0 | a | b | c | d | 1 |
|---|---|---|---|---|---|---|
| 0 | 0 | a | b | c | d | 1 |
| a | a | a | b | 1 | 1 | 1 |
| b | b | b | b | 1 | 1 | 1 |
| c | c | 1 | 1 | d | d | 1 |
| d | d | 1 | 1 | d | d | 1 |
| 1 | 1 | 1 | 1 | 1 | 1 | 1 |

.

Note that $\leq_m^M$ is transitive, hence $\leq_m^M$ is an order relation, by Corollary 6.4.25, but not a lattice order w.r.t. to $\wedge_m^M$, $\vee_m^M$, since $\wedge_m^M$ is not commutative.

**Example 11.4.8 Proper transitive QMV algebra: tQMV**

By a PASCAL program, we found that the algebra $\mathcal{A}^L = (A_6 = \{0, a, b, c, d, 1\}, \odot, ^-, 1)$, with the following tables of $\odot$ and $^-$ and of the additional operation $\oplus$, is a proper left-tQMV algebra, i.e. (PU), (Pcomm), (Pass), (m-La), (m-Re), (Pqmv) (hence (Pom), (Pmv), ($\Delta_m$), (prel$_m$)), (DN), (m-Tr) $\Longleftrightarrow \ldots \Longleftrightarrow$ (m-BB) hold and it does not verify (m-An) for $(a, b)$.

| $\odot$ | 0 | a | b | c | d | 1 |
|---|---|---|---|---|---|---|
| 0 | 0 | 0 | 0 | 0 | 0 | 0 |
| a | 0 | 0 | 0 | 0 | 0 | a |
| b | 0 | 0 | 0 | 0 | 0 | b |
| c | 0 | 0 | 0 | a | b | c |
| d | 0 | 0 | 0 | b | a | d |
| 1 | 0 | a | b | c | d | 1 |

and

| $x$ | $x^-$ |
|---|---|
| 0 | 1 |
| a | d |
| b | c |
| c | b |
| d | a |
| 1 | 0 |

, with

| $\oplus$ | 0 | a | b | c | d | 1 |
|---|---|---|---|---|---|---|
| 0 | 0 | a | b | c | d | 1 |
| a | a | d | c | 1 | 1 | 1 |
| b | b | c | d | 1 | 1 | 1 |
| c | c | 1 | 1 | 1 | 1 | 1 |
| d | d | 1 | 1 | 1 | 1 | 1 |
| 1 | 1 | 1 | 1 | 1 | 1 | 1 |

.

Note that $\leq_m^M$ is an order relation, by Corollary 11.2.9, but not a lattice order w.r. to $\wedge_m^M$, $\vee_m^M$, since $\wedge_m^M$ is not commutative.

**Example 11.4.9 Transitive, antisymmetric OM algebra: taOM**

By a PASCAL program, we found that the algebra $\mathcal{A}^L = (A_4 = \{0, a, b, 1\}, \odot, ^-, 1)$, with the following tables of $\odot$ and $^-$ and of the additional operation $\oplus$, is a transitive, antisymmetric left-orthomodular algebra (= m-BCK algebra verifying (Pom)), i.e. (PU), (Pcomm), (Pass), (m-La), (m-Re), (m-An), (DN), (m-Tr) $\Longleftrightarrow \ldots \Longleftrightarrow$ (m-BB), (Pom), but also (prel$_m$) hold and it does not verify ($\wedge_m$-comm) for $(a, b)$, (Pqmv) for $(a, a, 0)$, (Pmv) for $(a, a)$, $(\Delta_m)$ for $(b, a)$.

| $\odot$ | 0 | a | b | 1 |
|---|---|---|---|---|
| 0 | 0 | 0 | 0 | 0 |
| a | 0 | 0 | 0 | a |
| b | 0 | 0 | b | b |
| 1 | 0 | a | b | 1 |

and

| $x$ | $x^-$ |
|---|---|
| 0 | 1 |
| a | b |
| b | a |
| 1 | 0 |

, with

| $\oplus$ | 0 | a | b | 1 |
|---|---|---|---|---|
| 0 | 0 | a | b | 1 |
| a | a | a | 1 | 1 |
| b | b | 1 | 1 | 1 |
| 1 | 1 | 1 | 1 | 1 |

.

Note that $\leq_m^M$ is transitive, hence $\leq_m^M$ is an order relation, by Corollary 6.4.25, but not a lattice order w.r. to $\wedge_m^M$, $\vee_m^M$, since $\wedge_m^M$ is not commutative.

Note that this algebra is the NM (Nilpotent Minimum) algebra $\mathcal{F}_4$ from [104] - see Section 15.3.

# Chapter 12

# Orthomodular algebras

In this chapter, we continue the results from the previous Chapter 11 in the "world" of involutive *m-M algebras*: we analyse in some details the orthomodular (OM) algebras defined in Chapter 11, with a special insight on taOM algebras.

In Section 12.1, we prove that almost all the properties of QMV algebras are also verified by orthomodular (OM) algebras; we put OM algebras on the "map".

In Section 12.2, we mainly prove that any m-BCK algebra verifies the property (trans) (the binary relation $\leq_m^M$ is transitive) and we introduce and analyse the so called *trans algebras*.

In Section 12.3, we prove the definitional equivalence between involutive residuated lattices and m-BCK lattices, thus putting IMTL, NM, MV and $_{(WNM)}$MV algebras on the same "map". Concerning the taOM algebras, all the finite examples we found are m-BCK lattices and the open problem is if any taOM algebra is an m-BCK lattice.

In Section 12.4, we present 15 examples of the involved algebras.

The content of this chapter is taken from [119].

## 12.1 Orthomodular algebras: OM

### 12.1.1 Properties of OM algebras

We shall see that almost all the properties verified by a QMV algebra are also verified by an OM algebra.

**Proposition 12.1.1** *(See Proposition 11.2.4 for QMV algebras)*
*Let $\mathcal{A}^L = (A^L, \odot, ^-, 1)$ be a left-OM algebra. We have:*

(12.1) $$x \odot (y \vee_m^M x^-) = x \odot y,$$

(12.2) $$x \odot y \leq_m^M x, \quad i.e. \quad (x \odot y) \wedge_m^M x = x \odot y,$$

(12.3) $$x \leq_m^M x \oplus y, \quad i.e. \quad x \wedge_m^M (x \oplus y) = x,$$

(12.4) $$x \wedge_m^M y \leq_m^M y, \quad i.e. \quad (x \wedge_m^M y) \wedge_m^M y = x \wedge_m^M y,$$

(12.5) $$y \leq_m^M x \vee_m^M y, \quad i.e. \quad y \wedge_m^M (x \vee_m^M y) = y,$$

(12.6) $$x \vee_m^M (y \wedge_m^M x) = x,$$

(12.7) $$x \leq_m^M y \Longrightarrow y \vee_m^M x = y,$$

(12.8) $$x \leq_m^M y \Longrightarrow y^- \leq_m^M x^- \quad (order-reversibility\ of\ ^-),$$

(12.9) $$x \leq_m^M y \Longrightarrow x \oplus z \leq_m^M y \oplus z \quad (monotonicity\ of\ \oplus),$$

(12.10) $$x \leq_m^M y \Longrightarrow x \odot z \leq_m^M y \odot z \quad (monotonicity\ of\ \odot),$$

(12.11) $$(x \wedge_m^M y) \wedge_m^M z = (x \wedge_m^M y) \wedge_m^M (y \wedge_m^M z),$$

(12.12) $$(x \vee_m^M y) \vee_m^M z = (x \vee_m^M y) \vee_m^M (y \vee_m^M z).$$

**Proof.** (12.1): $x \odot (y \vee_m^M x^-)$
$\stackrel{(DN),(6.33)}{=} x \odot [(y \odot x)^- \odot x]^-$
$\stackrel{(6.12)}{=} (x^- \oplus [(y \odot x)^- \odot x])^-$
$\stackrel{(Pom)}{=} ([(x \odot y) \oplus ((x \odot y)^- \odot x)]^- \oplus [(x \odot y)^- \odot x])^-$
$\stackrel{(6.10)}{=} ([(x \odot y)^- \odot ((x \odot y)^- \odot x)^-] \oplus ((x \odot y)^- \odot x))^-$ (put $X = (x \odot y)^-$ and $Y = x$)
$= ((X \odot Y) \oplus [(X \odot Y)^- \odot X])^-$
$\stackrel{(Pom)}{=} X^- = ((x \odot y)^-)^- \stackrel{(DN)}{=} x \odot y$.

(12.2): $x \odot y \stackrel{(12.1)}{=} x \odot (y \vee_m^M x^-) \stackrel{(6.33)}{=} x \odot [(y \odot x) \oplus x^-]$
$\stackrel{(Pcomm)}{=} x \odot [(x \odot y) \oplus x^-] \stackrel{(6.5)}{=} x \odot ((x \odot y)^- \odot x)^-$
$\stackrel{(Pcomm)}{=} ((x \odot y)^- \odot x)^- \odot x \stackrel{(6.32)}{=} (x \odot y) \wedge_m^M x$.

(12.3): $x \wedge_m^M (x \oplus y)$
$\stackrel{(6.5)}{=} x \wedge_m^M (x^- \odot y^-)^-$
$\stackrel{(6.32)}{=} (x^- \odot (x^- \odot y^-)^-)^- \odot (x^- \odot y^-)^-$
$\stackrel{(6.10)}{=} ((x^- \odot (x^- \odot y^-)^-) \oplus (x^- \odot y^-))^-$
$\stackrel{(Scomm),(Pcomm)}{=} ((x^- \odot y^-) \oplus ((x^- \odot y^-)^- \odot x^-))^-$
$\stackrel{(Pom)}{=} (x^-)^- \stackrel{(DN)}{=} x$, hence $x \leq_m^M (x \oplus y)$.

(12.4): $x \wedge_m^M y \stackrel{(6.32)}{=} (x^- \odot y)^- \odot y \stackrel{(Pcomm)}{=} y \odot (x^- \odot y) \leq_m^M y$, by (12.2).

(12.5): $x \vee_m^M y \stackrel{(6.33)}{=} (x \odot y^-) \oplus y \stackrel{(6.7)}{=} y \oplus (x \odot y^-) \geq^{\wedge_m^M} y$, by (12.3).

(12.6): $x \vee_m^M (y \wedge_m^M x) \overset{(6.33)}{=} (x \odot (y \wedge_m^M x)^-) \oplus (y \wedge_m^M x)$
$\overset{(6.39)}{=} (x \odot (y^- \vee_m^M x^-)) \oplus (y \wedge_m^M x)$
$\overset{(12.1)}{=} (x \odot y^-) \oplus (y \wedge_m^M x)$
$\overset{(6.5)}{=} ((x \odot y^-)^- \odot (y \wedge_m^M x)^-)^-$
$\overset{(6.32)}{=} [(x \odot y^-)^- \odot ((y^- \odot x)^- \odot x)^-]^-$
$\overset{(Pcomm)}{=} [(x \odot (y^- \odot x)^-)^- \odot (y^- \odot x)^-]^-$
$\overset{(6.32),(DN)}{=} [x^- \wedge_m^M (y^- \odot x)^-]^-$
$\overset{(6.5)}{=} [x^- \wedge_m^M (y \oplus x^-)]^-$
$\overset{(Scomm)}{=} [x^- \wedge_m^M (x^- \oplus y)]^- \overset{(12.3)}{=} (x^-)^- = x.$

(12.7): Since $x \leq_m^M y \Longleftrightarrow x \wedge_m^M y = x$, it follows that $y \vee_m^M x = y \vee_m^M (x \wedge_m^M y) \overset{(12.6)}{=} y$.

(12.8): If $x \leq_m^M y$, then $y \vee_m^M x = y$, by (12.7); then, $y^- = (y \vee_m^M x)^- \overset{(6.38)}{=} y^- \wedge_m^M x^-$, i.e. $y^- \leq_m^M x^-$.

(12.9): If $x \leq_m^M y$, then $y = y \vee_m^M x$, by (12.7). Then,
$(x \oplus z) \wedge_m^M (y \oplus z)$
$= (x \oplus z) \wedge_m^M ((y \vee_m^M x) \oplus z)$
$\overset{(6.33)}{=} (x \oplus z) \wedge_m^M (((y \odot x^-) \oplus x) \oplus z)$
$\overset{(Sass)}{=} (x \oplus z) \wedge_m^M ((y \odot x^-) \oplus (x \oplus z))$
$\overset{(Scomm)}{=} (x \oplus z) \wedge_m^M ((x \oplus z) \oplus (y \odot x^-))$
$\overset{(12.3)}{=} x \oplus z.$

(12.10): If $x \leq_m^M y$, then $y^- \leq_m^M x^-$, by (12.8); it follows, by (12.9), that: $(y \odot z)^- \overset{(6.5),(DN)}{=} y^- \oplus z^- \leq_m^M x^- \oplus z^- \overset{(6.5),(DN)}{=} (x \odot z)^-$;
hence, $x \odot z \leq_m^M y \odot z$, by (12.8) again.

(12.11): $(x \wedge_m^M y) \wedge_m^M (y \wedge_m^M z)$
$\overset{(6.32)}{=} [(x \wedge_m^M y)^- \odot (y \wedge_m^M z)]^- \odot (y \wedge_m^M z)$
$\overset{(6.5)}{=} [(x \wedge_m^M y) \oplus (y \wedge_m^M z)^-] \odot (y \wedge_m^M z)$
$\overset{(6.39),(6.32)}{=} [(x \wedge_m^M y) \oplus (y^- \vee_m^M z^-)] \odot ((y^- \odot z)^- \odot z)$
$\overset{(6.33)}{=} [(x \wedge_m^M y) \oplus ((y^- \odot z) \oplus z^-)] \odot (y^- \odot z)^- \odot z$
$\overset{(6.5)}{=} [(x \wedge_m^M y) \oplus ((y^- \odot z) \oplus z^-)] \odot (y \oplus z^-) \odot z$
$\overset{(Sass),(6.5),(Pass)}{=} [[(x \wedge_m^M y) \oplus z^- \oplus (y \oplus z^-)^-] \odot (y \oplus z^-)] \odot z$
$\overset{(6.5)}{=} ([((x \wedge_m^M y) \oplus z^-)^- \odot (y \oplus z^-)]^- \odot (y \oplus z^-)) \odot z$
$\overset{(6.32)}{=} (((x \wedge_m^M y) \oplus z^-) \wedge_m^M (y \oplus z^-)) \odot z$
$= ((x \wedge_m^M y) \oplus z^-) \odot z$
$\overset{(6.5)}{=} ((x \wedge_m^M y)^- \odot z)^- \odot z$
$\overset{(6.32)}{=} (x \wedge_m^M y) \wedge_m^M z,$
since $x \wedge_m^M y \leq_m^M y$, by (12.4), implies $(x \wedge_m^M y) \oplus z^- \leq_m^M y \oplus z^-$, by (12.9),

i.e. $((x \wedge_m^M y) \oplus z^-) \wedge_m^M (y \oplus z^-) = (x \wedge_m^M y) \oplus z^-$.

(12.12): $(x \vee_m^M y) \vee_m^M (y \vee_m^M z)$
$\overset{(6.33)}{=} [(x \vee_m^M y) \odot (y \vee_m^M z)^-] \oplus (y \vee_m^M z)$
$\overset{(6.38)}{=} [(x \vee_m^M y) \odot (y^- \wedge_m^M z^-)] \oplus (y \vee_m^M z)$
$\overset{(6.32),(DN)}{=} [(x \vee_m^M y) \odot ((y \odot z^-)^- \odot z^-)] \oplus (y \vee_m^M z)$
$\overset{(6.33),(Pcomm),(Pass)}{=} [((x \vee_m^M y) \odot z^-) \odot (y \odot z^-)^-] \oplus ((y \odot z^-) \oplus z)$
$\overset{(Sass)}{=} ([((x \vee_m^M y) \odot z^-) \odot (y \odot z^-)^-] \oplus (y \odot z^-)) \oplus z$
$\overset{(6.33)}{=} [((x \vee_m^M y) \odot z^-) \vee_m^M (y \odot z^-)] \oplus z$
$= [(x \vee_m^M y) \odot z^-] \oplus z$
$\overset{(6.33)}{=} (x \vee_m^M y) \vee_m^M z$,
since $y \leq_m^M x \vee_m^M y$, by (12.5), implies $y \odot z^- \leq_m^M (x \vee_m^M y) \odot z^-$, by (12.10), and hence
$((x \vee_m^M y) \odot z^-) \vee_m^M (y \odot z^-) = (x \vee_m^M y) \odot z^-$, by (12.7). □

Consider the following properties:

(Pq) $\quad x \odot [y \vee_m^M (z \vee_m^M x^-)] = (x \odot y) \vee_m^M (x \odot z)$ and

(Pqq) $\quad x \odot [y \vee_m^M (x \odot z)^-] = (x \odot y) \vee_m^M (x \odot (x \odot z)^-)$.

**Lemma 12.1.2** *Let $\mathcal{A}^L = (A^L, \odot, ^-, 1)$ be an involutive left-m-BE algebra. Then, (Pq) $\Longrightarrow$ (12.1) and (Pqq) $\Longrightarrow$ (12.1).*

**Proof.** Take $z = 0$ in (Pq) to obtain (12.1) and $z = 1$ in (Pqq) to obtain (12.1). □

**Theorem 12.1.3** *Let $\mathcal{A}^L = (A^L, \odot, ^-, 1)$ be an involutive left-m-BE algebra. Then,*

$$(Pq) \iff (Pqq).$$

**Proof.** First, we prove:

(a) $\quad z \vee_m^M x^- = (x \odot (x \odot z)^-)^-$.

Indeed, $z \vee_m^M x^- = (z \odot x^=) \oplus x^- \overset{(DN)}{=} (z \odot x) \oplus x^- = ((z \odot x)^- \odot x^=)^- = ((z \odot x)^- \odot x)^- \overset{(Pcomm)}{=} (x \odot (x \odot z)^-)^-$.

$(Pq) \Longrightarrow (Pqq)$: $(x \odot y) \vee_m^M (x \odot (x \odot z)^-)$
$\overset{(Pq)}{=} x \odot [y \vee_m^M (Z \vee_m^M x^-)]$, where $Z := (x \odot z)^-$,
$\overset{(a)}{=} x \odot [y \vee_m^M (x \odot (x \odot Z)^-)^-]$
$= x \odot [y \vee_m^M (x \odot (x \odot (x \odot z)^-)^-)^-]$
$\overset{(a)}{=} x \odot [y \vee_m^M (x \odot (z \vee_m^M x^-))^-]$
$\overset{(12.1)}{=} x \odot [y \vee_m^M (x \odot z)^-]$.

$(Pqq) \Longrightarrow (Pq)$: $x \odot [y \vee_m^M (z \vee_m^M x^-)]$
$\overset{(a)}{=} x \odot [y \vee_m^M (x \odot (x \odot z)^-)^-]$
$\overset{(Pqq)}{=} (x \odot y) \vee_m^M (x \odot (x \odot Z)^-)$, where $Z := (x \odot z)^-$,
$\overset{(a)}{=} (x \odot y) \vee_m^M (x \odot (x \odot (x \odot z)^-)^-)$

$\overset{(a)}{=} (x \odot y) \vee_m^M (x \odot (z \vee_m^M x^-))$
$\overset{(12.1)}{=} (x \odot y) \vee_m^M (x \odot z).$ □

By Theorem 12.1.3 and since (Pom) $\Longleftrightarrow$ (Pq), we obtain:

**Corollary 12.1.4** *Let* $\mathcal{A}^L = (A^L, \odot, ^-, 1)$ *be an involutive left-m-BE algebra. Then,*

$$(Pom) \iff (Pq) \iff (Pqq).$$

**Proposition 12.1.5** *(See Proposition 11.2.7 for QMV algebras)*
*Let* $\mathcal{A}^L = (A^L, \odot, ^-, 1)$ *be a left-OM algebra. We have:*

$$(12.13) \qquad x \vee_m^M y \ \leq_m^M x \oplus y,$$

$$(12.14) \qquad x \odot y \ \leq_m^M x \wedge_m^M y.$$

**Proof.**
(12.13): Since $x \odot y^- \ \leq_m^M \ x$, by (12.2), then $x \vee_m^M y = (x \odot y^-) \oplus y \ \leq_m^M \ x \oplus y$, by (12.9).
(12.14): Since $x^- \odot y \ \leq_m^M \ x^-$, by (12.2), then $x \ \leq_m^M \ (x^- \odot y)^-$, by (12.8) and (DN); hence, $x \odot y \ \leq_m^M \ (x^- \odot y)^- \odot y = x \wedge_m^M y$, by (12.10). □

**Proposition 12.1.6** *(See Proposition 11.2.8 for QMV algebras)*
*Let* $\mathcal{A}^L = (A^L, \odot, ^-, 1)$ *be a left-OM algebra. We have:*

$$(12.15) \qquad x \wedge_m^M ((x \oplus y) \wedge_m^M z) = x \wedge_m^M z \quad (\textit{absorbtion law } 1),$$

$$(12.16) \qquad x \vee_m^M ((x \odot y) \vee_m^M z) = x \vee_m^M z \quad (\textit{absorbtion law } 2),$$

$$(12.17) \quad x \ \leq_m^M \ z^-,\ y \ \leq_m^M \ z^-,\ x \oplus z = y \oplus z \implies x = y \quad (\textit{cancellation law } 1),$$

$$(12.18) \quad z^- \ \leq_m^M \ x,\ z^- \ \leq_m^M \ y,\ x \odot z = y \odot z \implies x = y \quad (\textit{cancellation law } 2),$$

$$(12.19) \qquad x \ \leq_m^M \ y \implies x \wedge_m^M z \ \leq_m^M \ y \wedge_m^M z \quad (\textit{monotonicity of } \wedge_m^M),$$

$$(12.20) \qquad x \ \leq_m^M \ y \implies x \vee_m^M z \ \leq_m^M \ y \vee_m^M z \quad (\textit{monotonicity of } \vee_m^M),$$

$$(12.21) \qquad x \ \leq_m^M \ y,\ y \ \leq_m^M \ z \implies x \ \leq_m^M \ z \quad (\textit{transitivity of } \leq_m^M).$$

**Proof.** The same as for QMV algebras, namely:
(12.15): $x \wedge_m^M ((x \oplus y) \wedge_m^M z)$
$\overset{(6.32)}{=} (x^- \odot ((x \oplus y) \wedge_m^M z))^- \odot ((x \oplus y) \wedge_m^M z)$
$\overset{(6.5),(6.32)}{=} (x \oplus ((x \oplus y) \wedge_m^M z)^-) \odot (((x \oplus y)^- \odot z)^- \odot z)$

$\overset{(6.39),(6.5)}{=} (x \oplus ((x \oplus y)^- \vee_m^M z^-)) \odot (((x \oplus y) \oplus z^-) \odot z)$
$\overset{(6.33)}{=} (x \oplus (((x \oplus y)^- \odot z) \oplus z^-)) \odot ((x \oplus y \oplus z^-) \odot z)$
$\overset{(Sass)}{=} ((x \oplus z^-) \oplus (x \oplus y \oplus z^-)^-) \odot ((x \oplus y \oplus z^-) \odot z)$
$\overset{(6.5)}{=} ((x \oplus z^-)^- \odot (x \oplus y \oplus z^-))^- \odot (x \oplus y \oplus z^-) \odot z$
$\overset{(Pass)}{=} [((x \oplus z^-)^- \odot (x \oplus y \oplus z^-))^- \odot (x \oplus y \oplus z^-)] \odot z$
$\overset{(6.32)}{=} ((x \oplus z^-) \wedge_m^M (x \oplus y \oplus z^-)) \odot z$
$= (x \oplus z^-) \odot z$
$\overset{(6.5)}{=} (x^- \odot z)^- \odot z$
$\overset{(6.32)}{=} x \wedge_m^M z,$
since $x \oplus z^- \leq_m^M x \oplus z^- \oplus y$, by (12.3), implies $(x \oplus z^-) \wedge_m^M (x \oplus z^- \oplus y) = x \oplus z^-$.

(12.16): $x \vee_m^M ((x \odot y) \vee_m^M z)$
$\overset{(6.33)}{=} (x \odot ((x \odot y) \vee_m^M z)^-) \oplus ((x \odot y) \vee_m^M z)$
$\overset{(6.38),(6.33)}{=} (x \odot ((x \odot y)^- \wedge_m^M z^-)) \oplus (((x \odot y) \odot z^-) \oplus z)$
$\overset{(6.32)}{=} (x \odot (((x \odot y) \odot z^-)^- \odot z^-)) \oplus ((x \odot y \odot z^-) \oplus z)$
$\overset{(Pass)}{=} ((x \odot z^-) \odot (x \odot y \odot z^-)^-) \oplus ((x \odot y \odot z^-) \oplus z)$
$\overset{(Sass)}{=} [((x \odot z^-) \odot (x \odot y \odot z^-)^-) \oplus (x \odot y \odot z^-)] \oplus z$
$\overset{(6.5)}{=} [(x \odot z^-) \vee_m^M (x \odot y \odot z^-)] \oplus z$
$\overset{(12.7)}{=} (x \odot z^-) \oplus z$
$\overset{(6.33)}{=} x \vee_m^M z,$
since $x \odot y \odot z^- \leq_m^M x \odot z^-$, by (12.2), implies $(x \odot z^-) \vee_m^M (x \odot y \odot z^-) = x \odot z^-$, by (12.7).

(12.17): $x \leq_m^M z^-$ and $y \leq_m^M z^-$ mean $x \wedge_m^M z^- = x$ and $y \wedge_m^M z^- = y$. Then, $x = x \wedge_m^M z^- \overset{(6.32)}{=} (x^- \odot z^-)^- \odot z^- \overset{(6.5)}{=} (x \oplus z) \odot z^- = (y \oplus z) \odot z^- \overset{(6.5)}{=} (y^- \odot z^-)^- \odot z^- \overset{(6.32)}{=} y \wedge_m^M z^- = y$.

(12.18): $z^- \leq_m^M x$ and $z^- \leq_m^M y$ imply $x \vee_m^M z^- = x$ and $y \vee_m^M z^- = y$, by (12.7). Then,
$x = x \vee_m^M z^- \overset{(6.33),(DN)}{=} (x \odot z) \oplus z^- = (y \odot z) \oplus z^- \overset{(6.33)}{=} y \vee_m^M z^- = y$.

(12.19): $x \leq_m^M y$ implies $x \oplus z^- \leq_m^M y \oplus z^-$, by (12.9), and hence $(x \oplus z^-) \odot z \leq_m^M (y \oplus z^-) \odot z$, by (12.10). Then,
$x \wedge_m^M z \overset{(6.32)}{=} (x^- \odot z)^- \odot z = (x \oplus z^-) \odot z \leq_m^M (y \oplus z^-) \odot z \overset{(6.5)}{=} (y^- \odot z)^- \odot z = y \wedge_m^M z$.

(12.20): $x \leq_m^M y$ implies $x \odot z^- \leq_m^M y \odot z^-$, by (12.10), hence $(x \odot z^-) \oplus z \leq_m^M (y \odot z^-) \oplus z$, by (12.9). Then,
$x \vee_m^M z \overset{(6.33)}{=} (x \odot z^-) \oplus z \leq_m^M (y \odot z^-) \oplus z \overset{(6.33)}{=} y \vee_m^M z$.

(12.21): $x \leq_m^M y$ and $y \leq_m^M z$ mean $x = x \wedge_m^M y$ and $y = y \wedge_m^M z$. Then,
$x = x \wedge_m^M y = (x \wedge_m^M y) \wedge_m^M (y \wedge_m^M z) \overset{(12.11)}{=} (x \wedge_m^M y) \wedge_m^M z = x \wedge_m^M z$; thus, $x \leq_m^M z$.
□

**Corollary 12.1.7** *Let $\mathcal{A}^L = (A^L, \odot, ^-, 1)$ be a left-OM algebra. The binary rela-*

*tion $\leq_m^M$ is an order relation.*

**Proof.** The reflexivity and the antisymmetry follow by Corollary 6.4.25, while the transitivity follows by (12.21). □

**Proposition 12.1.8** *Let $\mathcal{A}^L = (A^L, \odot, ^-, 1)$ be a left-QMV algebra. Then, $\mathcal{A}^L$ is a left-OM algebra.*

**Proof.** By Theorem 11.2.18 (1). □

**Proposition 12.1.9** *Let $\mathcal{A}^L = (A^L, \odot, ^-, 1)$ be a left-MV algebra. Then, $\mathcal{A}^L$ is a left-OM algebra, i.e.* **MV** $\subset$ **OM**.

**Proof.** Obviously. □

**Remark 12.1.10** *In a left-OM algebra $\mathcal{A}^L = (A^L, \odot, ^-, 1)$:*
*- the initial binary relation, $\leq_m$ ($x \leq_m y \Longleftrightarrow x \odot y^- = 0$), is only* **reflexive** *((m-Re) holds);*
*- the binary relation $\leq_m^M$ ($x \leq_m^M y \Longleftrightarrow x \wedge_m^M y = x$) is an* **order**, *by Corollary 12.1.7, but not a lattice order with respect to $\wedge_m^M$, $\vee_m^M$, since $x \wedge_m^M y \neq y \wedge_m^M x$;*
*- the binary relation $\leq_m^P$ ($x \leq_m^P y \Longleftrightarrow x \odot y = x$) is only* **antisymmetric and transitive**, *by Proposition 6.4.9.*

### 12.1.2 Putting orthomodular algebras on the "map"

We have the connections from Figure 12.1.

## 12.2 A new algebra: the trans algebra (TRANS)

Concerning the involutive m-BE algebras not verifying (m-An), note that the examples: Example 11.4.2 of PreMV algebra, Example 11.4.3 of MMV algebra, Example 11.4.6 of tPreMV algebra and Example 11.4.7 of tMMV algebra have the binary relation $\leq_m^M$ transitive, hence an order relation (that is not a lattice order since $x \wedge_m^M y \neq y \wedge_m^M x$ for some $x, y$). But this is not true in general. We present Example 12.4.3 of PreMV algebra, Example 12.4.4 of MMV algebra, Example 11.4.6 of tPreMV algebra and Example 12.4.8 of tMMV algebra whose binary relation $\leq_m^M$ is not transitive. We present also Example 12.4.2 of involutive m-BE algebra and Example 12.4.6 of involutive m-pre-BCK algebra having the binary relation $\leq_m^M$ transitive, hence an order relation (that is not a lattice order since $x \wedge_m^M y \neq y \wedge_m^M x$ for some $x, y$), but also Example 12.4.1 of involutive m-BE algebra and Example 12.4.5 of involutive m-pre-BCK algebra having $\leq_m^M$ not transitive.

Concerning the involutive m-aBE algebras, we have a surprising result, the next Theorem 12.2.3 (saying that any m-BCK algebra has the binary relation $\leq_m^M$ transitive, hence an order relation), **obtained by** ***Prover9*** **in about 4 seconds and in 29 steps (the length of proof is 29); we have grouped the steps of proof in the following Lemmas 12.2.1, 12.2.2 and Theorem 12.2.3.**

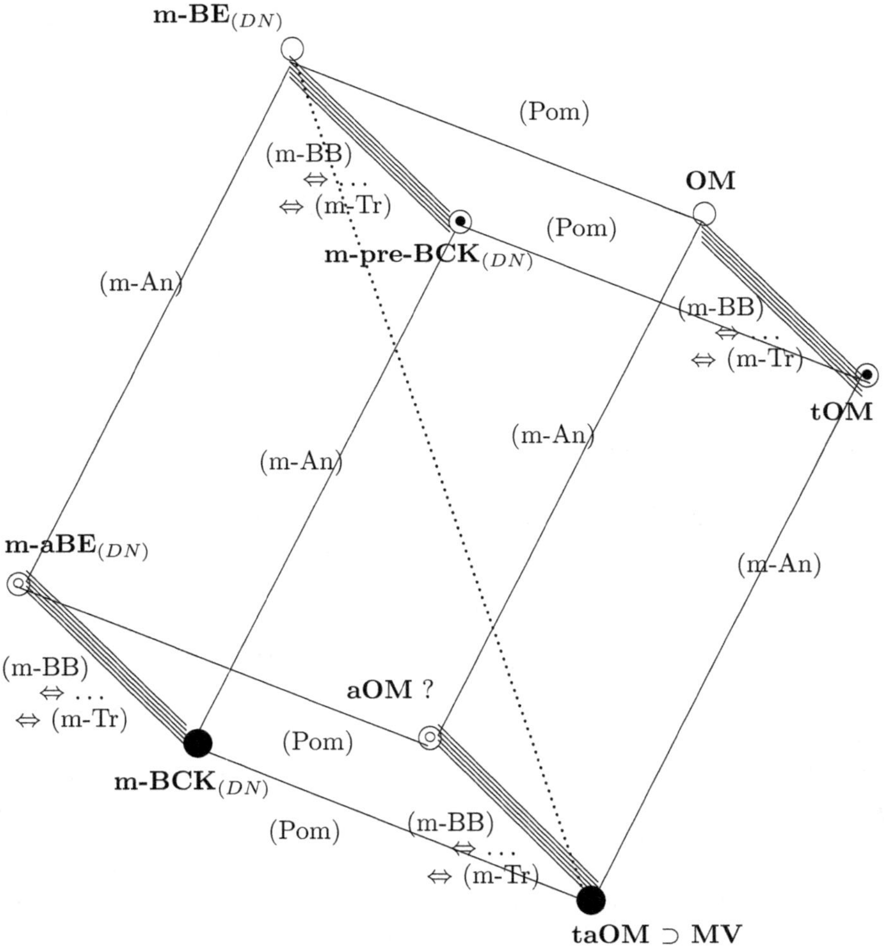


Figure 12.1: Putting **OM**, **tOM**, **aOM** and **taOM** on the "map", where ? means that there is an open problem concerning **aOM**

**Lemma 12.2.1** *Let $\mathcal{A}^L = (A^L, \odot, ^-, 1)$ be an involutive m-BE algebra. Then,*

(12.22) $$x \odot (y \odot (y \odot x)^-) = 0,$$

(12.23) $$(x \odot y)^- \odot (x \odot (x \odot (x \odot y)^-)^-) = 0.$$

**Proof.** (12.22): By (m-Re), $(x \odot y) \odot (x \odot y)^- = 0$, hence by (Pass) and (Pcomm), we obtain (12.22).

(12.23): In (12.22), take $X := (x \odot y)^-$ and $Y := z$ to obtain:
(a) $$(x \odot y)^- \odot (z \odot (z \odot (x \odot y)^-)^-) = 0;$$
take then $z := x$ in (a) to obtain (12.23). □

**Lemma 12.2.2** *Let $\mathcal{A}^L = (A^L, \odot, ^-, 1)$ be an involutive m-BE algebra verifying (m-BB) (i.e. an involutive m-tBE algebra (= m-pre-BCK algebra)). Then,*

(12.24) $$(x \odot y)^- \odot (z \odot (y \odot (z \odot x^-)^-)) = 0,$$

(12.25) $$(x \odot y)^- \odot (z \odot (x \odot (z \odot y^-)^-)) = 0,$$

(12.26) $$x \odot (y \odot (x \odot (z \odot (z \odot y)^-)^-)^-) = 0,$$

(12.27) $$x \odot (y \odot (x \odot (x \odot (x \odot y)^-)^-)^-) = 0.$$

**Proof.** (12.24): From (m-BB) ($((x \odot y)^- \odot (z \odot y)) \odot (z \odot x^-)^- = 0$), by (Pass) we obtain (12.24).

(12.25): In (12.24), interchange $x$ with $y$ and apply (Pcomm) to obtain (12.25).

(12.26): In (12.25), take $X := y$, $Y := z \odot (z \odot y)^-$ and $Z := x$ to obtain:
(a) $$(y \odot (z \odot (z \odot y)^-))^- \odot (x \odot (y \odot (x \odot (z \odot (z \odot y)^-)^-)^-)) = 0;$$
note that, in (a), $y \odot (z \odot (z \odot y)^-) \overset{(12.22)}{=} 0$, hence, by (Neg0-1), (PU), (a) becomes (12.26).

(12.27): In (12.26), take $z := x$ to obtain (12.27). □

**Theorem 12.2.3** *(See Theorem 2.2.26)*

*Let $\mathcal{A}^L = (A^L, \odot, ^-, 1)$ be a (involutive) m-BCK algebra. Then, the binary relation $\leq_m^M$ is transitive.*

**Proof.** Suppose $C1 \leq_m^M C2$ and $C2 \leq_m^M C3$, i.e. $C1 \wedge_m^M C2 = C1$ and $C2 \wedge_m^M C3 = C2$, i.e.:

(12.28) $$C2 \odot (C2 \odot C1^-)^- = C1 \quad \textit{and}$$

(12.29) $$C3 \odot (C3 \odot C2^-)^- = C2.$$

We must prove that $C1 \leq_m^M C3$, i.e. $C1 \wedge_m^M C3 = C1$, i.e.:

(12.30) $$C3 \odot (C3 \odot C1^-)^- = C1.$$

In (m-An) ($y \odot x^- = 0$ and $x \odot y^- = 0$ imply $x = y$), take $X := x \odot y$ and $Y := z$ to obtain:

$$(12.31) \qquad (x \odot y)^- \odot z = 0 \textit{ and } x \odot (y \odot z^-) = 0 \textit{ imply } x \odot y = z.$$

In (12.31), take $Z := x \odot (x \odot (x \odot y)^-)^-$ to obtain:

$$(12.32) \qquad (x \odot y)^- \odot [x \odot (x \odot (x \odot y)^-)^-] = 0 \textit{ and}$$

$$(12.33) \qquad x \odot (y \odot [x \odot (x \odot (x \odot y)^-)^-]^-) = 0 \textit{ imply}$$

$$(12.34) \qquad x \odot y = x \odot (x \odot (x \odot y)^-)^-.$$

Note that (12.32) is true by (12.23) and (12.33) is true by (12.27). It follows that (12.34) holds. It follows that:

$$(12.35) \qquad x \odot (x \odot (x \odot y)^-)^- = x \odot y.$$

Now, from (12.29), by multiplying by $x$ on the right side and by (Pcomm), we obtain:

$$(12.36) \qquad C3 \odot ((C3 \odot C2^-)^- \odot x) = C2 \odot x.$$

Next, in (12.35), take $X := C3$ and $Y := (C3 \odot C2^-)^- \odot x$ to obtain:

$$(12.37) \quad C3 \odot (C3 \odot (C3 \odot ((C3 \odot C2^-)^- \odot x))^-)^- = C3 \odot ((C3 \odot C2^-)^- \odot x).$$

Note that, in (12.37), we have twice that $C3 \odot ((C3 \odot C2^-)^- \odot x) \overset{(12.36)}{=} C2 \odot x$, hence (12.37) becomes:

$$(12.38) \qquad C3 \odot (C3 \odot (C2 \odot x)^-)^- = C2 \odot x.$$

Next, in (12.38), take $X := (C2 \odot C1^-)^-$ to obtain:

$$(12.39) \qquad C3 \odot (C3 \odot (C2 \odot (C2 \odot C1^-)^-)^-)^- = C2 \odot (C2 \odot C1^-)^-.$$

Note that, in (12.39), we have twice that $C2 \odot (C2 \odot C1^-)^- \overset{(12.28)}{=} C1$, hence (12.39) becomes (12.30). □

**Corollary 12.2.4** *Let $\mathcal{A}^L = (A^L, \odot, ^-, 1)$ be a (involutive) m-BCK algebra. Then, the binary relation $\leq_m^M$ is an order and $x \leq_m^M 1$, for all $x \in A^L$.*

**Proof.** By Corollary 6.4.25 and Theorem 12.2.3. □

We present, in Section 12.4, Example 12.4.11 of involutive m-aBE algebra having $\leq_m^M$ transitive (hence an order relation, by Corollary 12.2.4), but also Example 12.4.10 of involutive m-aBE algebra having $\leq_m^M$ not transitive.

Since the property of having $\leq_m^M$ transitive (hence an order relation) is so spread, we shall introduce the following new algebra:

**Definition 12.2.5** (The dual one is omitted)

A *left-trans algebra* is an involutive left-m-BE algebra $\mathcal{A}^L = (A^L, \odot, ^-, 1)$ having the binary relation $\leq_m^M$ transitive, i.e. verifying the property: for all $x, y, z \in A^L$,

(trans) $\quad x \leq_m^M y$ and $y \leq_m^M z$ imply $x \leq_m^M z$.

It follows that trans algebras are those involutive m-BE algebras having $\leq_m^M$ an order relation, by Corollary 6.4.25. We shall denote by **TRANS** the class of all left-trans algebras. Note that **taTRANS** = **m-BCK**, by above Theorem 12.2.3. Hence, we have the connections from Figures 12.2 and 12.3 (see Figures 11.2 and 11.3, respectively).

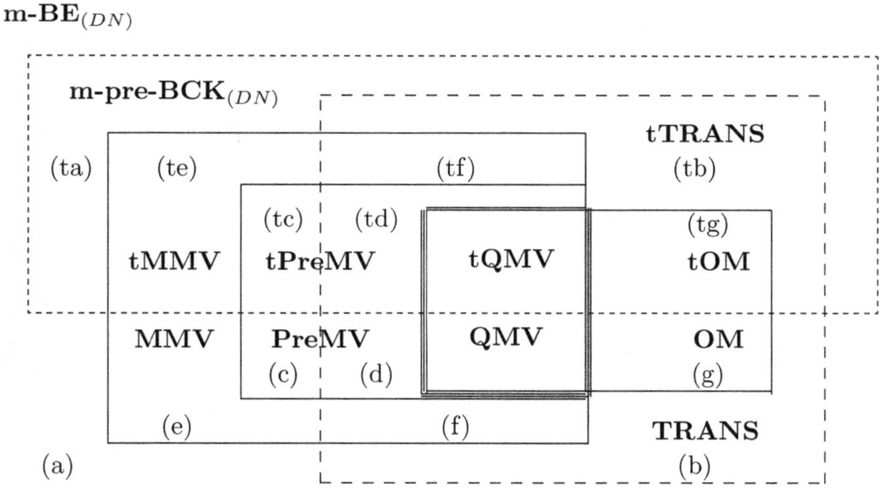


where:

(a): Example 12.4.1
(b): Example 12.4.2
(c): Example 12.4.3
(d): Example 11.4.2
(e): Example 12.4.4
(f): Example 11.4.3
(g): Example 12.4.9
(ta): Example 12.4.5
(tb): Example 12.4.6
(tc): Example 11.4.6
(td): Example 11.4.6
(te): Example 12.4.8
(tf): Example 11.4.7
(tg): Example 11.4.5

Figure 12.2: Resuming connections between **OM**, **PreMV**, **MMV**, **QMV** and **m-pre-BCK**$_{(DN)}$, **TRANS**

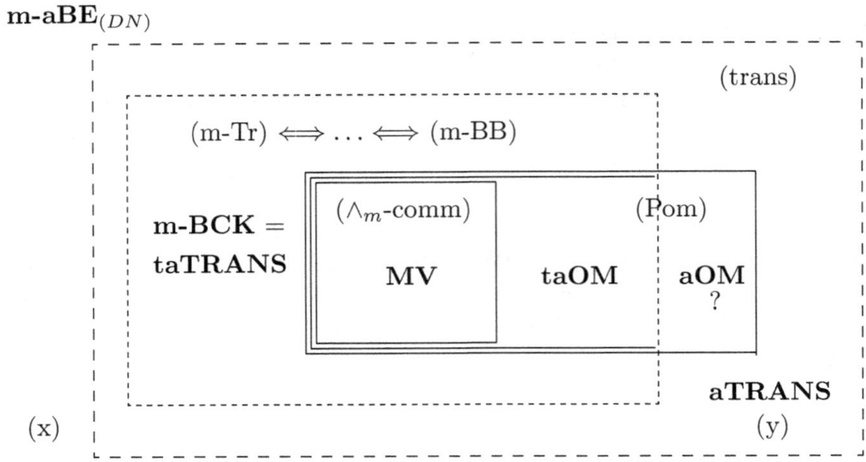


where:

(x): Example 12.4.10 (y): Example 12.4.11

Figure 12.3: Resuming connections between **MV**, **taOM**, **aOM**, **m-BCK** and **aTRANS**, where ? means that there is an open problem concerning **aOM**

**Remark 12.2.6** *In a left-trans algebra $\mathcal{A}^L = (A^L, \odot, ^-, 1)$:*
*- the initial binary relation, $\leq_m$ ($x \leq_m y \Longleftrightarrow x \odot y^- = 0$), is only* **reflexive** *((m-Re) holds);*
*- the binary relation $\leq_m^M$ ($x \leq_m^M y \Longleftrightarrow x \wedge_m^M y = x$) is an* **order**, *by definition and Corollary 6.4.25, but not a lattice order with respect to $\wedge_m^M$, $\vee_m^M$, since $x \wedge_m^M y \neq y \wedge_m^M x$;*
*- the binary relation $\leq_m^P$ ($x \leq_m^P y \Longleftrightarrow x \odot y = x$) is only* **antisymmetric and transitive**, *by Proposition 6.4.9.*

## 12.3 taOM algebras inside m-BCK algebras

Note that a *taOM algebra* is a transitive antisymmetric involutive m-BE algebra verifying (Pom), hence it is a (involutive) m-BCK algebra verifying (Pom), so we could say that it is an *orthomodular m-BCK algebra*.

First, we shall analyse more deeply the (involutive) m-BCK algebras.

### 12.3.1 (Involutive) m-BCK algebras and lattices

We have seen, in the previous section, that any m-BCK algebra has $\leq_m^M$ an order relation, by Theorem 12.2.3 and Corollary 6.4.25. This order relation $\leq_m^M$ is a lattice order if and only if the property ($\wedge_m$-comm) holds, i.e. if and only if the m-BCK algebra is an MV algebra, and in MV algebras, $x \leq_m^M y \Longleftrightarrow x \leq_m y$ and both $\leq_m^M$ and $\leq_m$ are distributive lattice orders.

But any m-BCK algebra has $\leq_m$ as an order relation too, since (m-Re), (m-An) and (m-Tr) hold. But **we do not know when (what property determines that) this order relation $\leq_m$ ($\Longleftrightarrow \leq_m^B$) is a lattice order**; we have examples of m-BCK algebras that are not lattices and examples that are lattices, distributive or not.

We shall denote by **m-BCK-L** the class of all left-m-BCK lattices.
We have: **m-BCK-L** $\subset$ **m-BCK**.

Recall the following definition ([104], Definition 1.2.9) (see Definition 5.3.23):

**Definition 12.3.1** (The dual one is omitted)
An *involutive residuated left-lattice*, or a *left-IRL* for short, is a bounded residuated left-lattice satisfying (DN), i.e. is an algebra $\mathcal{A}^L = (A^L, \wedge, \vee, \odot, \to, 0, 1)$ of type $(2, 2, 2, 2, 0, 0)$ such that:
(irl1) $(A^L, \wedge, \vee, 0, 1)$ is a bounded lattice w.r. to the lattice order $\leq$,
(irl2) $(A^L, \odot, 1)$ is an abelian monoid (i.e. (PU), (Pcomm), (Pass) hold),
(PR') for all $x, y, z \in A^L$, $\quad x \odot y \leq z \Longleftrightarrow x \leq y \to z$,
(DN) for all $x \in A^L$, $\quad (x^-)^- = x$ (or $x^= = x$), where $x^- \overset{def.}{=} x \to 0$.

Note that (irl1) means that:
- $\leq$ is an order (i.e. it is reflexive (Re), antisymmetric (An) and transitive (Tr)) and
- for all $x, y \in A^L$, $\exists\ x \wedge y = \inf(x, y)$ and $\exists\ x \vee y = \sup(x, y)$ and
- for all $x \in A^L$, $0 \leq x \leq 1$, i.e. we have:
(Fi') (first element) $0 \leq x$ and
(La') (last element) $x \leq 1$.

We shall denote by $\mathbf{R\text{-}L}_{(DN)}$ the class of all involutive residuated left-lattices. We shall prove that involutive residuated left-lattices ($\mathbf{R\text{-}L}_{(DN)}$) are definitionally equivalent (d. e.) to left-m-BCK-lattices (**m-BCK-L**). First, we prove some properties of left-IRLs.

**Proposition 12.3.2** *Let $\mathcal{A}^L = (A^L, \wedge, \vee, \odot, \to, 0, 1)$ be a left-IRL. We have: for all $x, y \in A^L$,*
*(V)* $x \leq 0 \Longleftrightarrow x = 0$,
*(N)* $1 \leq x \Longleftrightarrow x = 1$,
*(EqrelR)* $x \leq y \Longleftrightarrow x \to y = 1$,
*(irl3)* $x \leq y \Longleftrightarrow x \leq_m y$, *where* $x \leq_m y \overset{def.}{\Longleftrightarrow} x \odot y^- = 0$, *hence (Re)* $\Longleftrightarrow$ *(m-Re), (An)* $\Longleftrightarrow$ *(m-An), (Tr)* $\Longleftrightarrow$ *(m-Tr), (La)* $\Longleftrightarrow$ *(m-La) and* $x \wedge y = x \wedge_m y = \inf_m(x, y)$, $x \vee y = x \vee_m y = \sup_m(x, y)$,
*(Neg0-1)* $0^- = 1$,
*(Neg1-0)* $1^- = 0$,
*(m-La)* $x \odot 0 = 0$,
*(m-Re)* $x \odot x^- = 0$,
*(irl4)* $x \odot (x \to y) \leq y$,
*(irl5)* $x \to y = (x \odot y^-)^-$,
*(irl6)* $x \odot y = (x \to y^-)^-$,
*(irl7) the algebra* $(A^L, \odot, ^-, 1)$ *is a (involutive) left-m-BCK lattice, with the lattice order* $\leq_m \Longleftrightarrow \leq$ *and lattice operations* $\wedge_m = \wedge$, $\vee_m = \vee$.

**Proof.** (V): Since $0 \leq x$, by (Fi'), then, if $x \leq 0$, we obtain, by (An), $x = 0$. Conversely, if $x = 0$, then $x \leq 0$, by (Re').

(N): Since $x \leq 1$, by (La'), then, if $1 \leq x$, we obtain, by (An), $x = 1$. Conversely, if $x = 1$, then $1 \leq x$, by (Re').

(EqrelR) $x \to y = 1 \overset{(N)}{\Longleftrightarrow} 1 \leq x \to y \overset{(PR')}{\Longleftrightarrow} 1 \odot x \leq y \overset{(PU)}{\Longleftrightarrow} x \leq y$.

(irl3): $x \leq_m y \overset{def.}{\Longleftrightarrow} x \odot y^- = 0 \overset{(V)}{\Longleftrightarrow} x \odot y^- \leq 0 \overset{(PR')}{\Longleftrightarrow} x \leq y^- \to 0 = y^= \overset{(DN)}{=} y$; the rest follows by this equivalence.

(Neg0-1): $0^- \overset{def.}{=} 0 \to 0 = 1 \overset{(EqrelR)}{\Longleftrightarrow} 0 \leq 0$, which is true by (Re).

(Neg1-0): $1^- \overset{(Neg0-1)}{=} (0^-)^- \overset{(DN)}{=} 0$.

(m-La): (direct proof) $x \odot 0 = 0 \overset{(V)}{\Longleftrightarrow} x \odot 0 \leq 0 \overset{(PR')}{\Longleftrightarrow} x \leq 0 \to 0 = 0^- \overset{(Neg0-1)}{=} 1$, that is true by (La').

(m-Re): (direct proof) $x \odot x^- = 0 \overset{(V)}{\Longleftrightarrow} x \odot x^- \leq 0 \overset{(PR')}{\Longleftrightarrow} x \leq x^- \to 0 = x^= \overset{(DN)}{=} x$, that is true by (Re).

(irl4): $x \odot (x \to y) \leq y \overset{(Pcomm)}{\Longleftrightarrow} (x \to y) \odot x \leq y \overset{(PR')}{\Longleftrightarrow} x \to y \leq x \to y$, that is true by (Re').

(irl5): First, we prove (a): $x \to y \leq (x \odot y^-)^-$.
Indeed, $x \to y \leq (x \odot y^-)^- \overset{(irl3)}{\Longleftrightarrow} (x \to y) \odot (x \odot y^-)^= = 0 \overset{(DN)}{\Longleftrightarrow} (x \to y) \odot (x \odot y^-) = 0 \overset{(Pass),(Pcomm)}{\Longleftrightarrow} (x \odot (x \to y)) \odot y^- = 0 \overset{(irl3)}{\Longleftrightarrow} x \odot (x \to y) \leq y$, that is true by (irl4); thus, (a) holds.
Then, we prove (b): $(x \odot y^-)^- \leq x \to y$.
Indeed, $(x \odot y^-)^- \leq x \to y \overset{(PR')}{\Longleftrightarrow} (x \odot y^-)^- \odot x \leq y \overset{(irl3)}{\Longleftrightarrow} ((x \odot y^-)^- \odot x) \odot y^- = 0 \overset{(Pass)}{\Longleftrightarrow} (x \odot y^-)^- \odot (x \odot y^-) = 0$, that is true by (Pcomm) and (m-Re); thus, (b) holds.
By (a), (b) and (An), we obtain (irl5).

(irl6): $(x \to y^-)^- \overset{(irl5)}{=} (x \odot y^=)^= \overset{(DN)}{=} x \odot y$.

(irl7): By (Neg1-0), (irl2), (m-La), (DN), (m-Re), (m-An), (m-Tr) ($\Leftrightarrow \ldots \Leftrightarrow$ (m-BB)), $(A^L, \odot, ^-, 1)$ is a (involutive) left-m-BCK algebra; since $\leq$ is a lattice order, by (irl1), it follows that $\leq_m$ is a lattice order, by (irl3), hence, $(A^L, \odot, ^-, 1)$ is a (involutive) left-m-BCK lattice, with the lattice operations $\wedge_m = \wedge$ and $\vee_m = \vee$. □

Now, we prove the definitional equivalence (d. e.) between $\mathbf{R\text{-}L}_{(DN)}$ and **m-BCK-L**.

**Theorem 12.3.3**

*(1) Let $\mathcal{A}^L = (A^L, \wedge, \vee, \odot, \to, 0, 1)$ be an involutive residuated left-lattice. Define $f(\mathcal{A}^L) = (A^L, \odot, ^-, 1)$, where $x^- \overset{def.}{=} x \to 0$.*

*Then, $f(\mathcal{A}^L)$ is a (involutive) left-m-BCK lattice, with the lattice order $\leq_m \Longleftrightarrow \leq$ and the lattice operations $\wedge_m = \wedge$, $\vee_m = \vee$.*

*(1') Let $\mathcal{A}^L = (A^L, \odot, ^-, 1)$ be a (involutive) left-m-BCK lattice, with the lattice order $\leq_m$ ($x \leq_m y \overset{def.}{\Longleftrightarrow} x \odot y^- = 0$) and the lattice operations $x \wedge_m y = \inf_m(x, y)$*

*and* $x \vee_m y = \sup_m(x, y)$. *Define* $g(\mathcal{A}^L) = (A^L, \wedge_m, \vee_m, \odot, \rightarrow, 0, 1)$, *where* $x \rightarrow y \stackrel{def.}{=} (x \odot y^-)^-$ *and* $0 \stackrel{def.}{=} 1^-$.

*Then,* $g(\mathcal{A}^L)$ *is an involutive residuated left-lattice.*

*(2) The maps* $f$ *and* $g$ *are mutually inverse.*

**Proof.** (1): By Proposition 12.3.2, (irl7).

(1'): Let $\mathcal{A}^L = (A^L, \odot, ^-, 1)$ be an (involutive) left-m-BCK lattice w.r. to the lattice order $\leq_m$ and the lattice operations $\wedge_m$ and $\vee_m$. Then, (irl1), (irl2) and (DN) hold.

Note that $x \rightarrow 0 \stackrel{def.}{=} (x \odot 0^-)^- \stackrel{(Neg0-1)}{=} (x \odot 1)^- \stackrel{(PU)}{=} x^-$.

We prove that (PR') holds.

Indeed, first, by (mCDN2), (DN) + (Pcomm) $\Longrightarrow$ (DN4), where: (DN4) $x \leq_m y \Longleftrightarrow y^- \leq_m x^-$; then,

$x \leq_m y \rightarrow z \Longleftrightarrow x \leq_m (y \odot z^-)^- \stackrel{(DN4)}{\Longleftrightarrow} (y \odot z^-)^= \leq_m x^- \stackrel{(DN)}{\Longleftrightarrow} y \odot z^- \leq_m x^- \Longleftrightarrow (y \odot z^-) \odot x^= = 0 \stackrel{(DN),(Pcomm),(Pass)}{\Longleftrightarrow} (x \odot y) \odot z^- = 0 \Longleftrightarrow x \odot y \leq_m z$; thus, (PR') holds.

Hence, $(A^L, \wedge_m, \vee_m, \odot, \rightarrow, 0, 1)$ is an involutive residuated left-lattice.

(2): Routine, by Proposition 12.3.2. □

We write: $\mathbf{R\text{-}L}_{(DN)} \cong$ **m-BCK-L**.

**Remark 12.3.4** *In a left-m-BCK algebra* $\mathcal{A}^L = (A^L, \odot, ^-, 1)$*:*

*- the initial binary relation,* $\leq_m$ $(x \leq_m y \stackrel{def.}{\Longleftrightarrow} x \odot y^- = 0)$*, is an* **order** *(since (m-Re), (m-An), (m-Tr) hold); it can be a lattice order, but we do not know when, in general, excepting the case of MV algebras;*

*- the binary relation* $\leq_m^M$ $(x \leq_m^M y \stackrel{def.}{\Longleftrightarrow} x \wedge_m^M y = x)$ *is an* **order**, *by Corollary 6.4.25 and Theorem 12.2.3, but not a lattice order, in general, with respect to* $\wedge_m^M$, $\vee_m^M$, *since* $x \wedge_m^M y \neq y \wedge_m^M x$*; it is a distributive lattice order if and only if (*$\wedge_m$*-comm) holds (i.e.* $x \wedge_m^M y = y \wedge_m^M x$*), i.e. in the case of MV algebras, when* $\leq_m \Longleftrightarrow \leq_m^M$*;*

*- the binary relation* $\leq_m^P$ $(x \leq_m^P y \stackrel{def.}{\Longleftrightarrow} x \odot y = x)$ *is only* **antisymmetric and transitive**, *by Proposition 6.4.9; in Boolean algebras, it is a* **distributive lattice order** *and* $\leq_m^P \Longleftrightarrow \leq_m \Longleftrightarrow \leq_m^M$.

### 12.3.2 IMTL algebras and NM algebras

The IMTL (Involutive Monoidal t-norm based Logic) algebras and the NM (Nilpotent Minimum) algebras were introduced in [50], see also [51], [162].

**Definitions 12.3.5** (See ([104], Definition 1.2.32)) (Definitions 1) (The dual ones are omitted)

- A *left-IMTL algebra* is a left-IRL $\mathcal{A}^L = (A^L, \wedge, \vee, \odot, \rightarrow, 0, 1)$ verifying: for all $x, y \in A^L$,

(prel) (pre-linearity) $\qquad (x \rightarrow y) \vee (y \rightarrow x) = 1.$

- A *left-NM algebra* is a left-IMTL algebra $\mathcal{A}^L = (A^L, \wedge, \vee, \odot, \rightarrow, 0, 1)$ verifying: for all $x, y \in A^L$,
(WNM) (Weak Nilpotent Minimum) $\quad (x \odot y)^- \vee ((x \wedge y) \rightarrow (x \odot y)) = 1.$

We shall denote by **IMTL** the class of all left-IMTL algebras and by **NM** the class of all left-NM algebras. Hence, we have:
**IMTL** = $\mathbf{R\text{-}L}_{(DN)}$ + (prel) $\cong$ **m-BCK-L** + (prel),
**NM** = **IMTL** + (WNM).

Recall [104] that: **MV** = **IMTL** + (div), where:
(div) (divisibility) $x \wedge y = x \odot (x \rightarrow y) = x \wedge_m^B y$, for all $x, y$.

Recall also [104] that: $_{(WNM)}\mathbf{MV} \stackrel{def.}{=} \mathbf{MV}$ + (WNM) = **NM** + (div).

By the above d.e. $\mathbf{R\text{-}L}_{(DN)} \cong$ **m-BCK-L**, we obtain a second, equivalent definition of IMTL and NM algebras:

**Definitions 12.3.6** (Definitions 2) (The dual ones are omitted)

- A *left-IMTL algebra* is a (involutive) left-m-BCK lattice $\mathcal{A}^L = (A^L, \odot, ^-, 1)$, with the lattice order $\leq_m$ and the lattice operations $\wedge_m = \inf_m$, $\vee_m = \sup_m$, verifying: for all $x, y \in A^L$,
(prel) $\quad (x \rightarrow y) \vee_m (y \rightarrow x) = 1,$
where $x \rightarrow y \stackrel{def.}{=} (x \odot y^-)^-$.

- A *left-NM algebra* is a left-IMTL algebra $\mathcal{A}^L = (A^L, \odot, ^-, 1)$, with the lattice order $\leq_m$ and the lattice operations $\wedge_m$ and $\vee_m$, verifying: for all $x, y \in A^L$,
(WNM) $\quad (x \odot y)^- \vee_m ((x \wedge_m y) \rightarrow (x \odot y)) = 1,$
where $x \rightarrow y \stackrel{def.}{=} (x \odot y^-)^-$.

Hence, we have the connections from Figure 12.4.

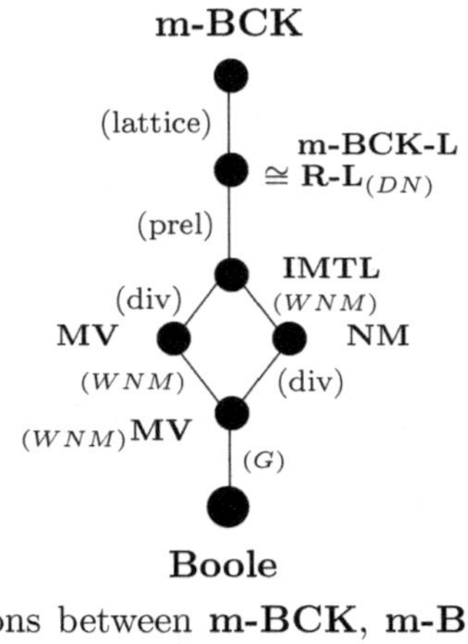


Figure 12.4: The connections between **m-BCK**, **m-BCK-L**, **IMTL**, **MV**, **NM** and **Boole**

**Remark 12.3.7** *There is a connection between above (prel) and (prel$_m$), where:*
*(prel$_m$)* $\quad (x \rightarrow y) \vee_m^B (y \rightarrow x) = 1$

*and between above (WNM) and (a$WNM_m$), where:*
(a$WNM_m$) $\quad (x \odot y) \odot [x \odot (x \odot y^-)^- \odot (x \odot y)^-]^- = x \odot y,$
*namely: in MV algebras, they coincide, (prel) = ($prel_m$) and (WNM) = (a$WNM_m$).*

### 12.3.3 taOM algebras

Note that all the examples of finite taOM algebras we found are lattices w.r. to $\leq_m$. The examples of proper taOM lattices (i.e. not being MV algebras or $_{(WNM)}$MV algebras) we found are either NM algebras, or proper IMTL algebras (i.e. not being NM algebras), or proper m-BCK lattices (i.e. not IMTL algebras) - distributive or not-distributive.

Hence, we have the following:

**Open problem 12.3.8** *Prove that any m-BCK algebra verifying (Pom) (i.e. taOM algebra) is a lattice w.r. to $\leq_m$ - or, equivalently, prove that any m-BCK algebra which is not a lattice does not verify (Pom) - or find an example of taOM algebra that is not a lattice. We have tried for several days to prove by Prover9 program that any m-BCK algebra which is not a lattice, i.e. which has the connections from Figure 12.5, does not verify (Pom).*

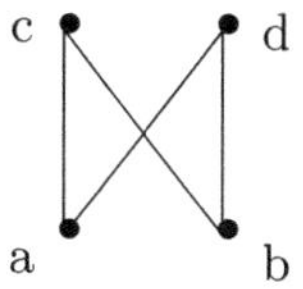


Figure 12.5: The Hasse diagram of the non-lattice connections of the order relation $\leq_m$

*By using the following Mace4 input file, we have tried in vain, by Mace4 program, to find an example of finite taOM algebra that is not a lattice: practically, there is no such a finite example.*

```
1 * x = x                      #label ("PU").
x * y = y * x                  #label ("Pcomm").
x * (y * z) = (x * y) * z      #label ("Pass").
x * -x = 0                     #label ("m-Re").
x * 0 = 0                      #label ("m-L").
- -x = x                       #label ("DN").
-0 = 1                         #label ("Neg0-1").
-1 = 0                         #label ("Neg1-0").
((x * -y = 0) & (y * -x=0)) - > (x=y) #label ("m-An").
(-(z * x) * (y * x)) * -(y * -z) = 0 #label ("m-BB").

-(-x * -y) = x + y #label ("sum").
-(-x + -y) = x * y #label ("product").
(x * y) + (-(x * y) * x) = x #label("Pom").
a != 0.
```

a != 1.
b != 0.
b != 1.
c != 0.
c != 1.
d != 0.
d !=1.
c != d.
c != a.
c != b.
a != d.
a != b.
b != d.
(c * d) * -a = 0.
(c * d) * -b = 0.
a * -b != 0.
b * -a != 0.
c * -d !=0.
d * -c !=0.
a * -c = 0.
a * -d = 0.
b * -c = 0.
b * -d=0.
a * -x = 0 & x * -c = 0 − > (x=a | x=c).
a * -x = 0 & x * -d = 0 − > (x=a | x=d).
b * -x = 0 & x * -c = 0 − > (x=b | x=c).
b * -x = 0 & x * -d = 0 − > (x=b | x=d).

where != means "≠" and | means "or".

**Open problem 12.3.9** *Prove that there is a connection between the Open problem 11.3.13 connected to the aOM algebras and the previous Open problem; most probably, there is an example of aOM algebra if and only if there is an example of taOM algebra that is not a lattice.*

*We believe that there are no such examples, we believe that any taOM algebra is a lattice w.r. to* $\leq_m$ *and that we have:* **MV** $\subset$ **taOM** $\subset$ **BCK-L**.

**Remark 12.3.10** *In a left-taOM algebra* $\mathcal{A}^L = (A^L, \odot, ^-, 1)$*:*
*- the initial binary relation,* $\leq_m$ $(x \leq_m y \Longleftrightarrow x \odot y^- = 0)$*, is an* **order**, *since (m-Re), (m-An), (m-Tr) hold, namely a (distributive or not distributive) lattice order in all the finite examples found;*
*- the binary relation* $\leq_m^M$ $(x \leq_m^M y \Longleftrightarrow x \wedge_m^M y = x)$ *is an* **order**, *by Corollary 12.1.7, but not a lattice order, in general, with respect to* $\wedge_m^M$, $\vee_m^M$, *since* $x \wedge_m^M y \neq y \wedge_m^M x$*; it is a distributive lattice order in MV algebras, where* $x \wedge_m^M y = y \wedge_m^M x$ *and* $\leq_m^M \Longleftrightarrow \leq_m$*;*
*- the binary relation* $\leq_m^P$ $(x \leq_m^P y \Longleftrightarrow x \odot y = x)$ *is only* **antisymmetric and**

**transitive**, *by Proposition 6.4.9; in Boolean algebras, it is a distributive lattice order and* $\leq_m^P \Longleftrightarrow \leq_m \Longleftrightarrow \leq_m^M$.

Resuming, the examples we found helped us to conclude that we have the connections from Figure 12.6.

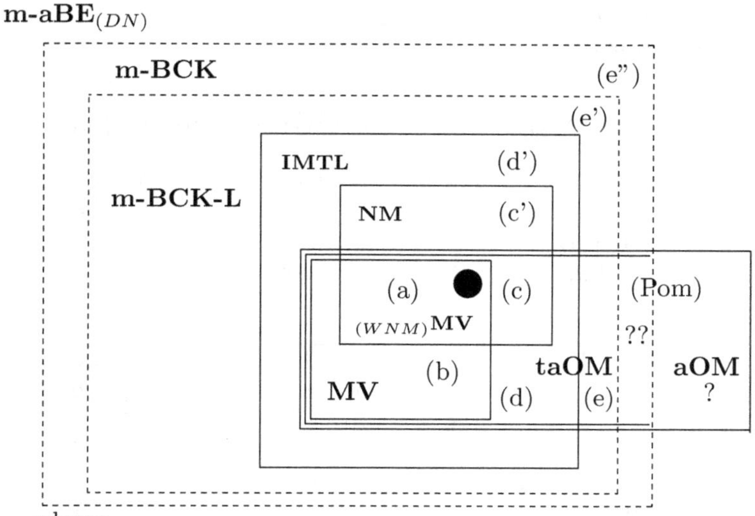


where:

● means **Boole**

(a): Examples 15.3.4, 15.3.6, n=1,2

(b): Examples 15.3.4, $n \geq 3$

(c): Examples 15.3.6, n=3 (c’): Examples 15.3.6, $n \geq 4$

(d): Examples 12.4.12, 1 (d’): Example 12.4.13

(e): Examples 12.4.12, 2, 3 (e’): Example 15.3.2

(e”): Example 15.3.1

Figure 12.6: Resuming connections between **MV**, **NM**, **IMTL**, **taOM** and **aOM**, where ‘? ’ means that there is an open problem concerning **aOM** and ‘??’ means that there is an open problem concerning taOM non-lattices

## 12.4 Examples

We introduce the following definition: an $X$ algebra is said to be *proper*, if it verifies the properties from its definition and does not verify the other properties (from this chapter and from the previous chapters, recalled below), except (prel$_m$) from Chapter 11 and (WNM$_m$), (aWNM$_m$) from Chapter 10.

(prel$_m$) $(x \to y) \vee_m^B (y \to x) = 1$,
where $x \to y \overset{def.}{=} (x \odot y^-)^-$;

(WNM$_m$) $(x \odot y)^- \vee [(x \wedge y) \to (x \odot y)] = 1$,
where $x \wedge y \overset{def.}{=} y \wedge_m^M x$, $x \vee y \overset{def.}{=} y \vee_m^M x$;
(aWNM$_m$) $(x \odot y) \odot [x \odot (x \odot y^-)^- \odot (x \odot y)^-]^- = x \odot y$;
(m-Pimpl) $[(x \odot y^-)^- \odot x^-]^- = x$ (Chapter 6),
(G) $x \odot x = x$ (Chapter 6),
(m-Pabs-i) $x \odot (x \oplus x \oplus y) = x$ (Chapter 10),
(m-Pdis) $z \odot (x \oplus y) = (z \odot x) \oplus (z \odot y)$ (Chapter 9).

### 12.4.1 At involutive m-BE algebras without (m-An) level

**Examples 12.4.1 Proper involutive m-BE algebras (not verifying (trans)): m-BE$_{(DN)}$**

**Example 1. (R. Giuntini)**

Let $\mathcal{A}^L = (A_7 = \{0, a, b, c, d, e, 1\}, \odot, ^-, 1)$ be an algebra, where:

| $\odot$ | 0 | a | b | c | d | e | 1 |
|---|---|---|---|---|---|---|---|
| 0 | 0 | 0 | 0 | 0 | 0 | 0 | 0 |
| a | 0 | 0 | 0 | 0 | c | 0 | a |
| b | 0 | 0 | a | 0 | e | c | b |
| c | 0 | 0 | 0 | 0 | 0 | 0 | c |
| d | 0 | c | e | 0 | 0 | 0 | d |
| e | 0 | 0 | c | 0 | 0 | 0 | e |
| 1 | 0 | a | b | c | d | e | 1 |

,

| $x$ | $x^-$ |
|---|---|
| 0 | 1 |
| a | a |
| b | c |
| c | b |
| d | e |
| e | d |
| 1 | 0 |

,

| $\oplus$ | 0 | a | b | c | d | e | 1 |
|---|---|---|---|---|---|---|---|
| 0 | 0 | a | b | c | d | e | 1 |
| a | a | 1 | 1 | 1 | 1 | b | 1 |
| b | b | 1 | 1 | 1 | 1 | 1 | 1 |
| c | c | 1 | 1 | a | b | d | 1 |
| d | d | 1 | 1 | b | 1 | 1 | 1 |
| e | e | b | 1 | d | 1 | 1 | 1 |
| 1 | 1 | 1 | 1 | 1 | 1 | 1 | 1 |

.

Then, $\mathcal{A}^L$ is a proper *involutive left-m-BE algebra*, since (PU), (Pcomm), (Pass), (m-La), (m-Re), (DN) hold, while (m-B) does not hold for $(a, c, d)$, (m-BB) for $(a, d, b)$, (m-*) for $(a, c, b)$, (m-**) for $(a, c, d)$, (m-Tr) for $(a, c, e)$, (m-An) for $(a, b)$, (m-Pimpl) for $(a, 0)$, (G) for $x = a$, (Pqmv) for $(a, 0, d)$, (Pom) for $(a, d)$, (Pmv) for $(a, e)$, $(\Delta_m)$ for $(a, e)$, (m-Pabs-i) for $(c, 0)$, (prel$_m$) for $(b, c)$, (WNM$_m$) and (aWNM$_m$) for $(c, c)$, (m-Pdis) for $(a, a, c)$.

Then, the tables of $\wedge_m^M$ and its dual, $\vee_m^M$, are the following:

| $\wedge_m^M$ | 0 | a | b | c | d | e | 1 |
|---|---|---|---|---|---|---|---|
| 0 | 0 | 0 | 0 | 0 | 0 | 0 | 0 |
| a | 0 | a | b | c | e | e | a |
| b | 0 | a | b | c | d | e | b |
| c | 0 | a | 0 | c | 0 | c | c |
| d | 0 | a | a | c | d | e | d |
| e | 0 | 0 | e | c | d | e | e |
| 1 | 0 | a | b | c | d | e | 1 |

and

| $\vee_m^M$ | 0 | a | b | c | d | e | 1 |
|---|---|---|---|---|---|---|---|
| 0 | 0 | a | b | c | d | e | 1 |
| a | a | a | b | c | d | d | 1 |
| b | b | a | b | 1 | b | 1 | 1 |
| c | c | a | b | c | d | e | 1 |
| d | d | 1 | b | d | d | e | 1 |
| e | e | a | b | a | d | e | 1 |
| 1 | 1 | 1 | 1 | 1 | 1 | 1 | 1 |

.

Note that the binary relation $\leq_m^M$ is not transitive (hence (trans) is not verified) for $(c, e, b)$.

Note that this algebra was found, in the dual form, by R. Giuntini, using the program MACE4, cf. [45], as an S algebra with a *non-transitive* binary relation $\leq_m^M$. Indeed, $c \leq_m^M e$ ($c \wedge_m^M e = c$) and $e \leq_m^M b$ ($e \wedge_m^M b = e$), but $c \not\leq_m^M b$ ($c \wedge_m^M b = 0 \neq c$).

**Example 2.**

By MACE4 program, we found that the algebra $\mathcal{A}^L = (A_8 = \{0, a, b, c, d, e, f, 1\}, \odot, ^-, 1)$, with the following tables of $\odot$ and $^-$ and of the additional operation $\oplus$, is an involutive left-m-BE algebra not verifying (m-B) for $(b, a, e)$, (m-BB) for $(b, e, b)$, (m-*) for $(a, e, b)$, (m-**) for $(b, a, e)$, (m-Tr) for $(b, a, e)$, (m-An) for $(a, b)$, (m-Pimpl) for $(a, 0)$, (m-Pabs-i) for $(d, 0)$, (G) for $a$, (Pqmv) for $(b, 0, c)$, (Pom) for $(b, c)$, (Pmv) for $(d, a)$, $(\Delta_m)$ for $(a, d)$, $(\text{prel}_m)$ for $(e, f)$, $(\text{WNM}_m)$ and $(\text{aWNM}_m)$ for $(e, b)$, (m-Pdis) for $(a, a, b)$.

| $\odot$ | 0 | a | b | c | d | e | f | 1 |
|---|---|---|---|---|---|---|---|---|
| 0 | 0 | 0 | 0 | 0 | 0 | 0 | 0 | 0 |
| a | 0 | 0 | 0 | a | 0 | 0 | 0 | a |
| b | 0 | 0 | 0 | f | 0 | a | 0 | b |
| c | 0 | a | f | c | 0 | e | f | c |
| d | 0 | 0 | 0 | 0 | 0 | 0 | 0 | d |
| e | 0 | 0 | a | e | 0 | 0 | a | e |
| f | 0 | 0 | 0 | f | 0 | a | 0 | f |
| 1 | 0 | a | b | c | d | e | f | 1 |

, with

| $\oplus$ | 0 | a | b | c | d | e | f | 1 |
|---|---|---|---|---|---|---|---|---|
| 0 | 0 | a | b | c | d | e | f | 1 |
| a | a | 1 | 1 | 1 | f | b | 1 | 1 |
| b | b | 1 | 1 | 1 | b | 1 | 1 | 1 |
| c | c | 1 | 1 | 1 | 1 | 1 | 1 | 1 |
| d | d | f | b | 1 | d | e | f | 1 |
| e | e | b | 1 | 1 | e | 1 | b | 1 |
| f | f | 1 | 1 | 1 | f | b | 1 | 1 |
| 1 | 1 | 1 | 1 | 1 | 1 | 1 | 1 | 1 |

and $(0, a, b, c, d, e, f, 1)^- = (1, b, a, d, c, e, f, 0)$.

The matrix of $\wedge_m^M$ is:

| $\wedge_m^M$ | 0 | a | b | c | d | e | f | 1 |
|---|---|---|---|---|---|---|---|---|
| 0 | 0 | 0 | 0 | 0 | 0 | 0 | 0 | 0 |
| a | 0 | a | b | f | d | a | f | a |
| b | 0 | a | b | f | d | e | f | b |
| c | 0 | a | b | c | d | e | f | c |
| d | 0 | 0 | 0 | 0 | d | 0 | 0 | d |
| e | 0 | a | 0 | e | d | e | 0 | e |
| f | 0 | a | b | f | d | a | f | f |
| 1 | 0 | a | b | c | d | e | f | 1 |

.

The binary relation $\leq_m^M$ is not transitive (hence (trans) is not verified) for $(a, e, c)$: $a \leq_m^M e$, $e \leq_m^M c$, but $a \not\leq_m^M c$, since $a \wedge_m^M c = f \neq a$.

**Example 12.4.2 Involutive m-BE algebra verifying (trans): TRANS**

By a PASCAL program, we found that the algebra $\mathcal{A}^L = (A_6 = \{0, a, b, c, d, 1\}, \odot, ^-, 1)$, with the following tables of $\odot$ and $^-$ and of the additional operation $\oplus$, is an *involutive left-m-BE algebra*, since (PU), (Pcomm), (Pass), (m-La), (m-Re), (DN) hold, while (m-B) does not hold for $(b, a, b)$, (m-BB) for $(b, b, a)$, (m-*) for $(a, c, b)$, (m-**) for $(b, a, b)$, (m-Tr) for $(b, a, c)$, (m-An) for $(a, b)$, (m-Pimpl) for $(a, 0)$, (Pqmv) for $(b, 0, b)$, (Pom) for $(b, b)$, (Pmv) for $(b, b)$, $(\Delta_m)$ for $(b, c)$, (m-Pabs-i) for $(a, 0)$, (G) for $a$, $(\text{prel}_m)$ for $(b, c)$, $(\text{WNM}_m)$ and $(\text{aWNM}_m)$ for $(c, c)$, (m-Pdis) for $(a, a, c)$.

| $\odot$ | 0 | a | b | c | d | 1 |
|---|---|---|---|---|---|---|
| 0 | 0 | 0 | 0 | 0 | 0 | 0 |
| a | 0 | 0 | 0 | 0 | 0 | a |
| b | 0 | 0 | a | 0 | 0 | b |
| c | 0 | 0 | 0 | a | a | c |
| d | 0 | 0 | 0 | a | a | d |
| 1 | 0 | a | b | c | d | 1 |

and

| $x$ | $x^-$ |
|---|---|
| 0 | 1 |
| a | d |
| b | c |
| c | b |
| d | a |
| 1 | 0 |

, with

| $\oplus$ | 0 | a | b | c | d | 1 |
|---|---|---|---|---|---|---|
| 0 | 0 | a | b | c | d | 1 |
| a | a | d | d | 1 | 1 | 1 |
| b | b | d | d | 1 | 1 | 1 |
| c | c | 1 | 1 | d | 1 | 1 |
| d | d | 1 | 1 | 1 | 1 | 1 |
| 1 | 1 | 1 | 1 | 1 | 1 | 1 |

.

Then, the tables of $\wedge_m^M$ and its dual, $\vee_m^M$, are the following:

| $\wedge_m^M$ | 0 | a | b | c | d | 1 |
|---|---|---|---|---|---|---|
| 0 | 0 | 0 | 0 | 0 | 0 | 0 |
| a | 0 | a | b | a | a | a |
| b | 0 | a | b | a | a | b |
| c | 0 | a | 0 | c | d | c |
| d | 0 | a | b | c | d | d |
| 1 | 0 | a | b | c | d | 1 |

and

| $\vee_m^M$ | 0 | a | b | c | d | 1 |
|---|---|---|---|---|---|---|
| 0 | 0 | a | b | c | d | 1 |
| a | a | a | b | c | d | 1 |
| b | b | a | b | 1 | d | 1 |
| c | c | d | d | c | d | 1 |
| d | d | d | d | c | d | 1 |
| 1 | 1 | 1 | 1 | 1 | 1 | 1 |

.

Note that $\leq_m^M$ is transitive (hence (trans) is verified), hence $\leq_m^M$ is an order relation, by Corollary 6.4.25, but not a lattice order w.r. to $\wedge_m^M$, $\vee_m^M$, since $\wedge_m^M$ is not commutative.

**Example 12.4.3 Proper PreMV algebra (not verifying (trans) and (m-Tr)): PreMV**

By MACE4 program, we found that the algebra $\mathcal{A}^L = (A_9 = \{0, a, b, c, d, e, f, g, 1\}, \odot, ^-, 1)$, with the following tables of $\odot$ and $^-$ and of the additional operation $\oplus$, is an involutive left-m-BE algebra verifying (Pmv) (hence ($\Delta_m$)) and also (prel$_m$), and not verifying (m-B) for $(b, g, b)$, (m-BB) for $(b, b, g)$, (m-*) for $(a, e, g)$, (m-**) for $(b, g, b)$, (m-Tr) for $(b, g, a)$, (m-An) for $(a, e)$, (m-Pimpl) for $(a, 0)$, (m-Pabs-i) for $(a, 0)$, (G) for $a$, (Pqmv) for $(b, 0, c)$, (Pom) for $(b, c)$, (WNM$_m$) and (aWNM$_m$) for $(b, b)$, (m-Pdis) for $(a, a, b)$.

| $\odot$ | 0 | a | b | c | d | e | f | g | 1 |
|---|---|---|---|---|---|---|---|---|---|
| 0 | 0 | 0 | 0 | 0 | 0 | 0 | 0 | 0 | 0 |
| a | 0 | 0 | 0 | 0 | 0 | 0 | 0 | 0 | a |
| b | 0 | 0 | c | a | e | 0 | c | 0 | b |
| c | 0 | 0 | a | 0 | 0 | 0 | a | 0 | c |
| d | 0 | 0 | e | 0 | 0 | 0 | a | 0 | d |
| e | 0 | 0 | 0 | 0 | 0 | 0 | 0 | 0 | e |
| f | 0 | 0 | c | a | a | 0 | c | e | f |
| g | 0 | 0 | 0 | 0 | 0 | 0 | e | 0 | g |
| 1 | 0 | a | b | c | d | e | f | g | 1 |

,

| $\oplus$ | 0 | a | b | c | d | e | f | g | 1 |
|---|---|---|---|---|---|---|---|---|---|
| 0 | 0 | a | b | c | d | e | f | g | 1 |
| a | a | d | 1 | f | b | d | 1 | 1 | 1 |
| b | b | 1 | 1 | 1 | 1 | 1 | 1 | 1 | 1 |
| c | c | f | 1 | 1 | 1 | b | 1 | 1 | 1 |
| d | d | b | 1 | 1 | 1 | b | 1 | 1 | 1 |
| e | e | d | 1 | b | b | d | 1 | f | 1 |
| f | f | 1 | 1 | 1 | 1 | 1 | 1 | 1 | 1 |
| g | g | 1 | 1 | 1 | 1 | f | 1 | 1 | 1 |
| 1 | 1 | 1 | 1 | 1 | 1 | 1 | 1 | 1 | 1 |

and $(0, a, b, c, d, e, f, g, 1)^- = (1, b, a, d, c, f, e, g, 0)$.

The matrix of $\wedge_m^M$ is:

| $\wedge_m^M$ | 0 | a | b | c | d | e | f | g | 1 |
|---|---|---|---|---|---|---|---|---|---|
| 0 | 0 | 0 | 0 | 0 | 0 | 0 | 0 | 0 | 0 |
| a | 0 | a | e | a | a | e | a | g | a |
| b | 0 | a | b | c | d | e | f | g | b |
| c | 0 | a | c | c | d | e | c | g | c |
| d | 0 | a | c | c | d | e | c | g | d |
| e | 0 | a | e | a | e | e | a | e | e |
| f | 0 | a | b | c | d | e | f | g | f |
| g | 0 | a | b | c | d | e | c | g | g |
| 1 | 0 | a | b | c | d | e | f | g | 1 |

.

The binary relation $\leq_m^M$ is not transitive (hence (trans) is not verified) for $(a, c, b)$: $a \leq_m^M c$, $c \leq_m^M b$, but $a \not\leq_m^M b$, since $a \wedge_m^M b = e \neq a$.

**Example 12.4.4 Proper MMV algebra (not verifying (Pmv), (m-Tr) and (trans)): MMV**

By MACE4 program, we found that the algebra $\mathcal{A}^L = (A_{10} = \{0, a, b, c, d, e, f, g, h, 1\}, \odot, ^-, 1)$, with the following tables of $\odot$ and $^-$ and of the additional operation $\oplus$, is an involutive left-m-BE algebra verifying $(\Delta_m)$ and not verifying (m-B) for $(b, g, b)$, (m-BB) for $(b, b, g)$, (m-*) for $(a, g, h)$, (m-**) for $(b, g, b)$, (m-Tr) for $(b, g, a)$, (m-An) for $(a, e)$, (m-Pimpl) for $(a, 0)$, (m-Pabs-i) for $(a, 0)$, (G) for $a$, (Pqmv) for $(b, 0, c)$, (Pom) for $(b, c)$, (Pmv) for $(g, h)$, (prel$_m$) for $(g, h)$, (WNM$_m$) and (aWNM$_m$) for $(b, b)$, (m-Pdis) for $(a, a, b)$.

| $\odot$ | 0 | a | b | c | d | e | f | g | h | 1 |
|---|---|---|---|---|---|---|---|---|---|---|
| 0 | 0 | 0 | 0 | 0 | 0 | 0 | 0 | 0 | 0 | 0 |
| a | 0 | 0 | 0 | 0 | 0 | 0 | 0 | 0 | 0 | a |
| b | 0 | 0 | c | a | e | 0 | c | 0 | 0 | b |
| c | 0 | 0 | a | 0 | 0 | 0 | a | 0 | 0 | c |
| d | 0 | 0 | e | 0 | 0 | 0 | a | 0 | 0 | d |
| e | 0 | 0 | 0 | 0 | 0 | 0 | 0 | 0 | 0 | e |
| f | 0 | 0 | c | a | a | 0 | c | 0 | 0 | f |
| g | 0 | 0 | 0 | 0 | 0 | 0 | 0 | 0 | a | g |
| h | 0 | 0 | 0 | 0 | 0 | 0 | 0 | a | 0 | h |
| 1 | 0 | a | b | c | d | e | f | g | h | 1 |

and

| $x$ | $x^-$ |
|---|---|
| 0 | 1 |
| a | b |
| b | a |
| c | d |
| d | c |
| e | f |
| f | e |
| g | g |
| h | h |
| 1 | 0 |

,

with

| $\oplus$ | 0 | a | b | c | d | e | f | g | h | 1 |
|---|---|---|---|---|---|---|---|---|---|---|
| 0 | 0 | a | b | c | d | e | f | g | h | 1 |
| a | a | d | 1 | f | b | d | 1 | 1 | 1 | 1 |
| b | b | 1 | 1 | 1 | 1 | 1 | 1 | 1 | 1 | 1 |
| c | c | f | 1 | 1 | 1 | b | 1 | 1 | 1 | 1 |
| d | d | b | 1 | 1 | 1 | b | 1 | 1 | 1 | 1 |
| e | e | d | 1 | b | b | d | 1 | 1 | 1 | 1 |
| f | f | 1 | 1 | 1 | 1 | 1 | 1 | 1 | 1 | 1 |
| g | g | 1 | 1 | 1 | 1 | 1 | 1 | 1 | b | 1 |
| h | h | 1 | 1 | 1 | 1 | 1 | 1 | b | 1 | 1 |
| 1 | 1 | 1 | 1 | 1 | 1 | 1 | 1 | 1 | 1 | 1 |

.

The matrix of $\wedge_m^M$ is:

| $\wedge_m^M$ | 0 | a | b | c | d | e | f | g | h | 1 |
|---|---|---|---|---|---|---|---|---|---|---|
| 0 | 0 | 0 | 0 | 0 | 0 | 0 | 0 | 0 | 0 | 0 |
| a | 0 | a | e | a | a | e | a | g | h | a |
| b | 0 | a | b | c | d | e | f | g | h | b |
| c | 0 | a | c | c | d | e | c | g | h | c |
| d | 0 | a | c | c | d | e | c | g | h | d |
| e | 0 | a | e | a | e | e | a | g | h | e |
| f | 0 | a | b | c | d | e | f | g | h | f |
| g | 0 | a | b | c | d | e | f | g | 0 | g |
| h | 0 | a | b | c | d | e | f | 0 | h | h |
| 1 | 0 | a | b | c | d | e | f | g | h | 1 |

The binary relation $\leq_m^M$ is not transitive (hence (trans) is not verified) for $(a, c, b)$: $a \leq_m^M c$, $c \leq_m^M b$, but $a \not\leq_m^M b$, since $a \wedge_m^M b = e \neq a$.

**Example 12.4.5 Proper involutive m-pre-BCK algebra (not verifying (trans) and ($\Delta_m$)): m-pre-BCK$_{(DN)}$ (= m-tBE$_{(DN)}$)**

By MACE4 program, we found that the algebra $\mathcal{A}^L = (A_8 = \{0, a, b, c, d, e, f, 1\}, \odot, ^-, 1)$, with the following tables of $\odot$ and $^-$ and of the additional operation $\oplus$, is an involutive left-m-BE algebra verifying (m-BB) $\Longleftrightarrow$ (m-Tr), and (prel$_m$), and not verifying (m-An) for $(a, f)$, (m-Pimpl) for $(a, 0)$, (m-Pabs-i) for $(a, 0)$, (G) for $a$, (Pqmv) for $(b, 0, d)$, (Pom) for $(b, d)$, (Pmv) for $(c, a)$, ($\Delta_m$) for $(a, c)$, (WNM$_m$) and (aWNM$_m$) for $(b, b)$, (m-Pdis) for $(a, a, b)$.

| $\odot$ | 0 | a | b | c | d | e | f | 1 |
|---|---|---|---|---|---|---|---|---|
| 0 | 0 | 0 | 0 | 0 | 0 | 0 | 0 | 0 |
| a | 0 | 0 | 0 | 0 | a | 0 | 0 | a |
| b | 0 | 0 | a | 0 | e | a | 0 | b |
| c | 0 | 0 | 0 | 0 | 0 | 0 | 0 | c |
| d | 0 | a | e | 0 | d | e | f | d |
| e | 0 | 0 | a | 0 | e | a | 0 | e |
| f | 0 | 0 | 0 | 0 | f | 0 | 0 | f |
| 1 | 0 | a | b | c | d | e | f | 1 |

with

| $\oplus$ | 0 | a | b | c | d | e | f | 1 |
|---|---|---|---|---|---|---|---|---|
| 0 | 0 | a | b | c | d | e | f | 1 |
| a | a | b | 1 | f | 1 | 1 | b | 1 |
| b | b | 1 | 1 | b | 1 | 1 | 1 | 1 |
| c | c | f | b | c | 1 | e | f | 1 |
| d | d | 1 | 1 | 1 | 1 | 1 | 1 | 1 |
| e | e | 1 | 1 | e | 1 | 1 | 1 | 1 |
| f | f | b | 1 | f | 1 | 1 | b | 1 |
| 1 | 1 | 1 | 1 | 1 | 1 | 1 | 1 | 1 |

and $(0, a, b, c, d, e, f, 1)^- = (1, b, a, d, c, f, e, 0)$.

The matrix of $\wedge_m^M$ is:

| $\wedge_m^M$ | 0 | a | b | c | d | e | f | 1 |
|---|---|---|---|---|---|---|---|---|
| 0 | 0 | 0 | 0 | 0 | 0 | 0 | 0 | 0 |
| a | 0 | a | a | c | f | a | f | a |
| b | 0 | a | b | c | e | e | f | b |
| c | 0 | 0 | 0 | c | 0 | 0 | 0 | c |
| d | 0 | a | b | c | d | e | f | d |
| e | 0 | a | b | c | e | e | f | e |
| f | 0 | a | a | c | f | a | f | f |
| 1 | 0 | a | b | c | d | e | f | 1 |

The binary relation $\leq_m^M$ is not transitive (hence (trans) is not verified) for $(a, e, d)$: $a \leq_m^M e$, $e \leq_m^M d$, but $a \not\leq_m^M d$, since $a \wedge_m^M d = f \neq a$.

**Example 12.4.6 Involutive m-pre-BCK algebra verifying (trans): tTRANS**

By a PASCAL program, we found that the algebra
$\mathcal{A}^L = (A_6 = \{0, a, b, c, d, 1\}, \odot, ^-, 1)$, with the following tables of $\odot$ and $^-$ and of the additional operation $\oplus$, is an involutive left-m-BE algebra verifying (m-Tr) $\Longleftrightarrow \ldots \Longleftrightarrow$ (m-BB), and also (prel$_m$), (WNM$_m$), (aWNM$_m$), and not verifying (m-An) for $(a, d)$, (m-Pimpl) for $(a, 0)$, (Pqmv) for $(b, a, 0)$, (Pom) for $(c, a)$, (Pmv) for $(b, a)$, $(\Delta_m)$ for $(a, b)$, (m-Pabs-i) for $(b, 0)$, (G) for $a$, (m-Pdis) for $(a, a, a)$.

| $\odot$ | 0 | a | b | c | d | 1 |
|---|---|---|---|---|---|---|
| 0 | 0 | 0 | 0 | 0 | 0 | 0 |
| a | 0 | 0 | 0 | a | 0 | a |
| b | 0 | 0 | 0 | 0 | 0 | b |
| c | 0 | a | 0 | c | a | c |
| d | 0 | 0 | 0 | a | 0 | d |
| 1 | 0 | a | b | c | d | 1 |

and

| $x$ | $x^-$ |
|---|---|
| 0 | 1 |
| a | d |
| b | c |
| c | b |
| d | a |
| 1 | 0 |

, with

| $\oplus$ | 0 | a | b | c | d | 1 |
|---|---|---|---|---|---|---|
| 0 | 0 | a | b | c | d | 1 |
| a | a | 1 | d | 1 | 1 | 1 |
| b | b | d | b | 1 | d | 1 |
| c | c | 1 | 1 | 1 | 1 | 1 |
| d | d | 1 | d | 1 | 1 | 1 |
| 1 | 1 | 1 | 1 | 1 | 1 | 1 |

.

Note that $\leq_m^M$ is transitive (hence (trans) is verified), hence $\leq_m^M$ is an order relation, by Corollary 6.4.25, but not a lattice order w.r. to $\wedge_m^M$, $\vee_m^M$, since $\wedge_m^M$ is not commutative.

**Example 12.4.7 Proper transitive PreMV algebra (not verifying (trans)): tPreMV**

By MACE4 program, we found that the algebra
$\mathcal{A}^L = (A_8 = \{0, a, b, c, d, e, f, 1\}, \odot, ^-, 1)$, with the following tables of $\odot$ and $^-$ and of the additional operation $\oplus$, is an involutive left-m-BE algebra verifying (Pmv) (hence $(\Delta_m)$) and (m-Tr) $\Longleftrightarrow \ldots \Longleftrightarrow$ (m-BB), and also (prel$_m$), and not verifying (m-An) for $(a, e)$, (m-Pimpl) for $(a, 0)$, (m-Pabs-i) for $(a, 0)$, (G) for $a$, (Pqmv) for $(b, 0, c)$, (Pom) for $(b, c)$, (WNM$_m$) and (aWNM$_m$) for $(b, b)$, (m-Pdis) for $(a, a, b)$.

| $\odot$ | 0 | a | b | c | d | e | f | 1 |
|---|---|---|---|---|---|---|---|---|
| 0 | 0 | 0 | 0 | 0 | 0 | 0 | 0 | 0 |
| a | 0 | 0 | 0 | 0 | 0 | 0 | 0 | a |
| b | 0 | 0 | c | a | e | 0 | c | b |
| c | 0 | 0 | a | 0 | 0 | 0 | a | c |
| d | 0 | 0 | e | 0 | 0 | 0 | a | d |
| e | 0 | 0 | 0 | 0 | 0 | 0 | 0 | e |
| f | 0 | 0 | c | a | a | 0 | c | f |
| 1 | 0 | a | b | c | d | e | f | 1 |

with

| $\oplus$ | 0 | a | b | c | d | e | f | 1 |
|---|---|---|---|---|---|---|---|---|
| 0 | 0 | a | b | c | d | e | f | 1 |
| a | a | d | 1 | f | b | d | 1 | 1 |
| b | b | 1 | 1 | 1 | 1 | 1 | 1 | 1 |
| c | c | f | 1 | 1 | 1 | b | 1 | 1 |
| d | d | b | 1 | 1 | 1 | b | 1 | 1 |
| e | e | d | 1 | b | b | d | 1 | 1 |
| f | f | 1 | 1 | 1 | 1 | 1 | 1 | 1 |
| 1 | 1 | 1 | 1 | 1 | 1 | 1 | 1 | 1 |

and $(0, a, b, c, d, e, f, 1)^- = (1, b, a, d, c, f, e, 0)$.

Then, the table of $\wedge_m^M$ is the following:

| $\wedge_m^M$ | 0 | a | b | c | d | e | f | 1 |
|---|---|---|---|---|---|---|---|---|
| 0 | 0 | 0 | 0 | 0 | 0 | 0 | 0 | 0 |
| a | 0 | a | e | a | a | e | a | a |
| b | 0 | a | b | c | d | e | f | b |
| c | 0 | a | c | c | d | e | c | c |
| d | 0 | a | c | c | d | e | c | d |
| e | 0 | a | e | a | e | e | a | e |
| f | 0 | a | b | c | d | e | f | f |
| 1 | 0 | a | b | c | d | e | f | 1 |

.

Note that $\leq_m^M$ is not transitive (hence (trans) is not verified) for $(a, c, b)$.

**Example 12.4.8 Proper tMMV algebra (not verifying (Pmv) and (trans)): tMMV** tMMV-trans-Pmv

By MACE4 program, we found that the algebra $\mathcal{A}^L = (A_{12} = \{0, a, b, c, d, e, f, g, h, i, j, 1\}, \odot, ^-, 1)$, with the following tables of $\odot$ and $^-$ and of the additional operation $\oplus$, is an involutive left-m-BE algebra verifying $(\Delta_m)$ and (m-BB) $\Longleftrightarrow$ (m-Tr), and (prel$_m$), and not verifying (m-An) for $(a, g)$, (m-Pimpl) for $(b, 0)$, (m-Pabs-i) for $(a, 0)$, (G) for $a$, (Pqmv) for $(a, 0, a)$, (Pom) for $(a, a)$, (Pmv) for $(a, a)$, (WNM$_m$) and (aWNM$_m$) for $(c, c)$, (m-Pdis) for $(a, a, a)$.

| $\odot$ | 0 | a | b | c | d | e | f | g | h | i | j | 1 |
|---|---|---|---|---|---|---|---|---|---|---|---|---|
| 0 | 0 | 0 | 0 | 0 | 0 | 0 | 0 | 0 | 0 | 0 | 0 | 0 |
| a | 0 | g | 0 | g | 0 | g | 0 | g | 0 | 0 | g | a |
| b | 0 | 0 | b | i | d | d | i | 0 | b | i | d | b |
| c | 0 | g | i | e | 0 | a | d | g | f | d | g | c |
| d | 0 | 0 | d | 0 | 0 | 0 | 0 | 0 | d | 0 | 0 | d |
| e | 0 | g | d | a | 0 | g | 0 | g | d | 0 | g | e |
| f | 0 | 0 | i | d | 0 | 0 | d | 0 | i | d | 0 | f |
| g | 0 | g | 0 | g | 0 | g | 0 | g | 0 | 0 | g | g |
| h | 0 | 0 | b | f | d | d | i | 0 | b | i | d | h |
| i | 0 | 0 | i | d | 0 | 0 | d | 0 | i | d | 0 | i |
| j | 0 | g | d | g | 0 | g | 0 | g | d | 0 | a | j |
| 1 | 0 | a | b | c | d | e | f | g | h | i | j | 1 |

with

| $\oplus$ | 0 | a | b | c | d | e | f | g | h | i | j | 1 |
|---|---|---|---|---|---|---|---|---|---|---|---|---|
| 0 | 0 | a | b | c | d | e | f | g | h | i | j | 1 |
| a | a | a | 1 | c | j | j | c | a | 1 | c | j | 1 |
| b | b | 1 | h | 1 | h | 1 | h | 1 | h | h | 1 | 1 |
| c | c | c | 1 | 1 | 1 | 1 | 1 | c | 1 | 1 | 1 | 1 |
| d | d | j | h | 1 | f | c | b | e | h | h | c | 1 |
| e | e | j | 1 | 1 | c | c | 1 | j | 1 | 1 | c | 1 |
| f | f | c | h | 1 | b | 1 | h | c | h | h | 1 | 1 |
| g | g | a | 1 | c | e | j | c | a | 1 | c | j | 1 |
| h | h | 1 | h | 1 | h | 1 | h | 1 | h | h | 1 | 1 |
| i | i | c | h | 1 | h | 1 | h | c | h | b | 1 | 1 |
| j | j | j | 1 | 1 | c | c | 1 | j | 1 | 1 | c | 1 |
| 1 | 1 | 1 | 1 | 1 | 1 | 1 | 1 | 1 | 1 | 1 | 1 | 1 |

and $(0, a, b, c, d, e, f, g, h, i, j, 1)^- = (1, b, a, d, c, f, e, h, g, j, i, 0)$.

The matrix of $\wedge_m^M$ is:

| $\wedge_m^M$ | 0 | a | b | c | d | e | f | g | h | i | j | 1 |
|---|---|---|---|---|---|---|---|---|---|---|---|---|
| 0 | 0 | 0 | 0 | 0 | 0 | 0 | 0 | 0 | 0 | 0 | 0 | 0 |
| a | 0 | a | 0 | g | 0 | a | 0 | g | 0 | 0 | g | a |
| b | 0 | 0 | b | f | d | d | f | 0 | h | i | d | b |
| c | 0 | a | i | c | d | e | f | g | f | i | j | c |
| d | 0 | 0 | d | d | d | d | d | 0 | d | d | d | d |
| e | 0 | a | d | e | d | e | d | g | d | d | j | e |
| f | 0 | 0 | i | i | d | d | f | 0 | f | i | d | f |
| g | 0 | a | 0 | a | 0 | a | 0 | g | 0 | 0 | g | g |
| h | 0 | 0 | b | f | d | d | f | 0 | h | i | d | h |
| i | 0 | 0 | i | f | d | d | f | 0 | f | i | d | i |
| j | 0 | a | d | e | d | e | d | g | d | d | j | j |
| 1 | 0 | a | b | c | d | e | f | g | h | i | j | 1 |

.

The binary relation $\leq_m^M$ is not transitive (hence (trans) is not verified) for $(a, e, c)$: $a \leq_m^M e$, $e \leq_m^M c$, but $a \not\leq_m^M c$, since $a \wedge_m^M c = g \neq a$.

**Example 12.4.9 Proper orthomodular algebra: OM**

By MACE4 program, we found that the algebra $\mathcal{A}^L = (A_5 = \{0, a, b, c, 1\}, \odot, ^-, 1)$, with the following tables of $\odot$ and $^-$ and of the additional operation $\oplus$, is an involutive left-m-BE algebra verifying (Pom), and also (prel$_m$), (WNM$_m$), (aWNM$_m$), and not verifying (m-B) for $(a, c, a)$, (m-BB) for $(a, a, c)$, (m-*) for $(c, b, a)$, (m-**) for $(a, c, a)$, (m-Tr) for $(a, c, b)$, (m-An) for $(a, c)$, (m-Pimpl) for $(a, 0)$, (m-Pabs-i) for $(b, 0)$, (G) for $b$, (Pqmv) for $(b, b, 0)$, (Pmv) for $(b, b)$, $(\Delta_m)$ for $(a, b)$, (m-Pdis) for $(a, a, a)$.

| $\odot$ | 0 | a | b | c | 1 |
|---|---|---|---|---|---|
| 0 | 0 | 0 | 0 | 0 | 0 |
| a | 0 | a | 0 | 0 | a |
| b | 0 | 0 | 0 | 0 | b |
| c | 0 | 0 | 0 | 0 | c |
| 1 | 0 | a | b | c | 1 |

and

| $x$ | $x^-$ |
|---|---|
| 0 | 1 |
| a | b |
| b | a |
| c | c |
| 1 | 0 |

, with

| $\oplus$ | 0 | a | b | c | 1 |
|---|---|---|---|---|---|
| 0 | 0 | a | b | c | 1 |
| a | a | 1 | 1 | 1 | 1 |
| b | b | 1 | b | 1 | 1 |
| c | c | 1 | 1 | 1 | 1 |
| 1 | 1 | 1 | 1 | 1 | 1 |

.

## 12.4.2 At involutive m-aBE algebras level

**Example 12.4.10 Proper involutive m-aBE algebra (not verifying (trans)): m-aBE$_{(DN)}$**

By MACE4 program, we found that the algebra $\mathcal{A}^L = (A_{10} = \{0, a, b, c, d, e, f, g, h, 1\}, \odot, ^-, 1)$, with the following tables of $\odot$ and $^-$ and of the additional operation $\oplus$, is an involutive left-m-BE algebra verifying (m-An) and not verifying (m-B) for $(a, g, a)$, (m-BB) for $(a, a, g)$, (m-*) for $(a, g, c)$, (m-**) for $(a, g, a)$, (m-Tr) for $(a, g, b)$, (m-Pimpl) for $(a, 0)$, (m-Pabs-i) for $(a, 0)$, (G) for $b$, (Pqmv) for $(a, d, 0)$, (Pom) for $(b, b)$, (Pmv) for $(a, d)$, $(\Delta_m)$ for $(a, b)$, (prel$_m$) for $(a, b)$, (WNM$_m$) and (aWNM$_m$) for $(b, b)$, (m-Pdis) for $(a, a, a)$.

| $\odot$ | 0 | a | b | c | d | e | f | g | h | 1 |
|---|---|---|---|---|---|---|---|---|---|---|
| 0 | 0 | 0 | 0 | 0 | 0 | 0 | 0 | 0 | 0 | 0 |
| a | 0 | a | 0 | a | 0 | a | 0 | 0 | 0 | a |
| b | 0 | 0 | g | g | 0 | d | 0 | d | d | b |
| c | 0 | a | g | e | 0 | a | d | d | f | c |
| d | 0 | 0 | 0 | 0 | 0 | 0 | 0 | 0 | 0 | d |
| e | 0 | a | d | a | 0 | a | 0 | 0 | d | e |
| f | 0 | 0 | 0 | d | 0 | 0 | 0 | 0 | d | f |
| g | 0 | 0 | d | d | 0 | 0 | 0 | 0 | 0 | g |
| h | 0 | 0 | d | f | 0 | d | d | 0 | f | h |
| 1 | 0 | a | b | c | d | e | f | g | h | 1 |

and

| $x$ | $x^-$ |
|---|---|
| 0 | 1 |
| a | b |
| b | a |
| c | d |
| d | c |
| e | f |
| f | e |
| g | h |
| h | g |
| 1 | 0 |

with

| $\oplus$ | 0 | a | b | c | d | e | f | g | h | 1 |
|---|---|---|---|---|---|---|---|---|---|---|
| 0 | 0 | a | b | c | d | e | f | g | h | 1 |
| a | a | h | 1 | 1 | h | 1 | c | c | c | 1 |
| b | b | 1 | b | 1 | b | 1 | b | 1 | 1 | 1 |
| c | c | 1 | 1 | 1 | 1 | 1 | 1 | 1 | 1 | 1 |
| d | d | h | b | 1 | f | c | b | e | c | 1 |
| e | e | 1 | 1 | 1 | c | 1 | 1 | c | 1 | 1 |
| f | f | c | b | 1 | b | 1 | b | c | 1 | 1 |
| g | g | c | 1 | 1 | e | c | c | e | 1 | 1 |
| h | h | c | 1 | 1 | c | 1 | 1 | 1 | 1 | 1 |
| 1 | 1 | 1 | 1 | 1 | 1 | 1 | 1 | 1 | 1 | 1 |

.

The matrix of $\wedge_m^M$ is:

| $\wedge_m^M$ | 0 | a | b | c | d | e | f | g | h | 1 |
|---|---|---|---|---|---|---|---|---|---|---|
| 0 | 0 | 0 | 0 | 0 | 0 | 0 | 0 | 0 | 0 | 0 |
| a | 0 | a | d | f | d | a | f | d | f | a |
| b | 0 | 0 | b | g | d | d | f | g | h | b |
| c | 0 | a | b | c | d | e | f | g | h | c |
| d | 0 | 0 | d | d | d | d | d | d | d | d |
| e | 0 | a | b | e | d | e | f | g | f | e |
| f | 0 | 0 | g | g | d | d | f | g | f | f |
| g | 0 | a | g | a | d | a | d | g | d | g |
| h | 0 | a | g | e | d | e | f | g | h | h |
| 1 | 0 | a | b | c | d | e | f | g | h | 1 |

.

The binary relation $\leq_m^M$ is not transitive (hence (trans) is not verified) for $(a, e, c)$: $a \leq_m^M e$, $e \leq_m^M c$, but $a \not\leq_m^M c$, since $a \wedge_m^M c = f \neq a$.

**Example 12.4.11 Proper aTRANS (involutive m-aBE algebra verifying (trans) and not verifying (m-Tr)): aTRANS**

By MACE4 program, we found that the algebra $\mathcal{A}^L = (A_7 = \{0, a, b, c, d, e, 1\}, \odot, ^-, 1)$, with the following tables of $\odot$ and $^-$ and of the additional operation $\oplus$, is a proper involutive left-m-aBE algebra, since it does not verify (m-B) for $(b, c, b)$, (m-BB) for $(b, b, c)$, (m-*) for $(c, a, b)$, (m-**) for $(b, c, b)$, (m-Tr) for $(b, c, a)$, ($\wedge_m$-comm) for $(a, e)$, (m-Pabs-i) for $(a, 0)$, (G) for $a$, (m-Pimpl) for $(a, 0)$, (Pqmv) for $(a, d, 0)$, (Pom) for $(a, a)$, (Pmv) for $(a, d)$, $(\Delta_m)$ for $(c, b)$ and also $(\text{prel}_m)$ for $(a, b)$, $(\text{WNM}_m)$ and $(\text{aWNM}_m)$ for $(a, a)$, (m-Pdis) for $(a, a, a)$.

| $\odot$ | 0 | a | b | c | d | e | 1 |
|---|---|---|---|---|---|---|---|
| 0 | 0 | 0 | 0 | 0 | 0 | 0 | 0 |
| a | 0 | c | 0 | d | 0 | c | a |
| b | 0 | 0 | d | 0 | 0 | d | b |
| c | 0 | d | 0 | 0 | 0 | d | c |
| d | 0 | 0 | 0 | 0 | 0 | 0 | d |
| e | 0 | c | d | d | 0 | c | e |
| 1 | 0 | a | b | c | d | e | 1 |

,

| $x$ | $x^-$ |
|---|---|
| 0 | 1 |
| a | b |
| b | a |
| c | c |
| d | e |
| e | d |
| 1 | 0 |

, with

| $\oplus$ | 0 | a | b | c | d | e | 1 |
|---|---|---|---|---|---|---|---|
| 0 | 0 | a | b | c | d | e | 1 |
| a | a | e | 1 | 1 | e | 1 | 1 |
| b | b | 1 | c | e | c | 1 | 1 |
| c | c | 1 | e | 1 | e | 1 | 1 |
| d | d | e | c | e | c | 1 | 1 |
| e | e | 1 | 1 | 1 | 1 | 1 | 1 |
| 1 | 1 | 1 | 1 | 1 | 1 | 1 | 1 |

.

Note that $\leq_m^M$ is transitive (hence (trans) is verified), hence $\leq_m^M$ is an order relation, by Corollary 6.4.25, but not a lattice order w.r.t. to $\wedge_m^M$, $\vee_m^M$, since $\wedge_m^M$ is not commutative.

## Examples 12.4.12 Proper transitive, antisymmetric OM algebras: taOM

• **Example 1: taOM is NM**

By a PASCAL program, we found that the algebra $\mathcal{A}^L = (A_4 = \{0, a, b, 1\}, \odot, ^-, 1)$, with the following tables of $\odot$ and $^-$ and of the additional operation $\oplus$, is a proper transitive, antisymmetric left-orthomodular algebra (taOM) (= m-BCK algebra verifying (Pom)), i.e. (PU), (Pcomm), (Pass), (m-La), (m-Re), (m-An), (DN), (m-Tr) $\Longleftrightarrow \ldots \Longleftrightarrow$ (m-BB), (Pom), but also $(\text{prel}_m)$, $(\text{WNM}_m)$, $(\text{aWNM}_m)$ hold and it does not verify (m-Pabs-i) for $(a, 0)$, (G) for $a$, (m-Pimpl) for $(b, 0)$, $(\wedge_m\text{-comm})$ for $(a, b)$, (Pqmv) for $(a, a, 0)$, (Pmv) for $(a, a)$, $(\Delta_m)$ for $(b, a)$ and also (m-Pdis) for $(a, b, a)$.

| $\odot$ | 0 | a | b | 1 |
|---|---|---|---|---|
| 0 | 0 | 0 | 0 | 0 |
| a | 0 | 0 | 0 | a |
| b | 0 | 0 | b | b |
| 1 | 0 | a | b | 1 |

and

| $x$ | $x^-$ |
|---|---|
| 0 | 1 |
| a | b |
| b | a |
| 1 | 0 |

, with

| $\oplus$ | 0 | a | b | 1 |
|---|---|---|---|---|
| 0 | 0 | a | b | 1 |
| a | a | a | 1 | 1 |
| b | b | 1 | 1 | 1 |
| 1 | 1 | 1 | 1 | 1 |

.

Then, the tables of $\wedge_m^M$ and its dual, $\vee_m^M$, are the following:

| $\wedge_m^M$ | 0 | a | b | 1 |
|---|---|---|---|---|
| 0 | 0 | 0 | 0 | 0 |
| a | 0 | a | 0 | a |
| b | 0 | a | b | b |
| 1 | 0 | a | b | 1 |

and

| $\vee_m^M$ | 0 | a | b | 1 |
|---|---|---|---|---|
| 0 | 0 | a | b | 1 |
| a | a | a | b | 1 |
| b | b | 1 | b | 1 |
| 1 | 1 | 1 | 1 | 1 |

.

Note that $\leq_m^M$ is an order relation, by Corollary 12.1.7, but not a lattice order w.r.t. to $\wedge_m^M$, $\vee_m^M$, since $\wedge_m^M$ is not commutative. The bounded poset $(A_4, \leq_m^M, 0, 1)$ is reprezented by the Hasse diagram from the Figure 12.7:

The binary relation $\leq_m$ ($\Longleftrightarrow \leq_m^B$) is an order relation also, since (m-Re), (m-An) and (m-Tr) hold; hence, $\leq_m^B$ is an order relation too. To see if the order relations $\leq_m$ and $\leq_m^B$ are lattice orders, we make the tables of $\wedge_m^B$ and $\vee_m^B$, which

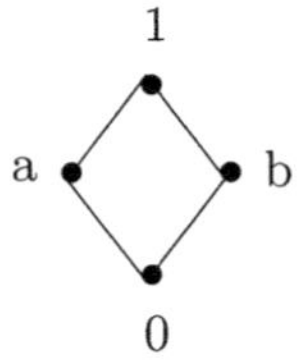


Figure 12.7: The Hasse diagram of the bounded poset $(A_4, \leq_m^M, 0, 1)$

are the transposed of the tables of $\wedge_m^M$ and $\vee_m^M$, respectively:

| $\wedge_m^B$ | 0 | a | b | 1 |
|---|---|---|---|---|
| 0 | 0 | 0 | 0 | 0 |
| a | 0 | a | a | a |
| b | 0 | 0 | b | b |
| 1 | 0 | a | b | 1 |

and

| $\vee_m^B$ | 0 | a | b | 1 |
|---|---|---|---|---|
| 0 | 0 | a | b | 1 |
| a | a | a | 1 | 1 |
| b | b | b | b | 1 |
| 1 | 1 | 1 | 1 | 1 |

.

It follows that the binary relations $\leq_m^B \Longleftrightarrow \leq_m$ $(x \leq_m^B y \stackrel{def.}{\Longleftrightarrow} x \wedge_m^B y = x)$ are linearly ordered (chains) : $0 \leq_m a \leq_m b \leq_m 1$.

Note that the operations $\wedge_m^B$ and $\vee_m^B$ are not commutative, therefore the order relation $\leq_m^B$ is not a lattice order w.r. to $\wedge_m^B$, $\vee_m^B$; but, the order relation $\leq_m$ is a lattice order w.r. to $\wedge = \wedge_m = \inf_m$, $\vee = \vee_m = \sup_m$:

| $\wedge$ | 0 | a | b | 1 |
|---|---|---|---|---|
| 0 | 0 | 0 | 0 | 0 |
| a | 0 | a | a | a |
| b | 0 | a | b | b |
| 1 | 0 | a | b | 1 |

and

| $\vee$ | 0 | a | b | 1 |
|---|---|---|---|---|
| 0 | 0 | a | b | 1 |
| a | a | a | b | 1 |
| b | b | b | b | 1 |
| 1 | 1 | 1 | 1 | 1 |

.

To see if the properties (prel) $((x \to y) \vee (y \to x) = 1)$ and (WNM) $((x \odot y)^- \vee [(x \wedge y) \to (x \odot y)] = 1)$ are verified, we need the table of $\to$ $(x \to y \stackrel{def.}{=} (x \odot y^-)^-)$:

| $\to$ | 0 | a | b | 1 |
|---|---|---|---|---|
| 0 | 1 | 1 | 1 | 1 |
| a | b | 1 | 1 | 1 |
| b | a | a | 1 | 1 |
| 1 | 0 | a | b | 1 |

.

It is easy to see that (prel) is verified. We check by a PASCAL program that (WNM) is also verified. Hence, this taOM algebra is just the NM (Nilpotent Minimum) algebra $\mathcal{F}_4$ (Fodor's) - see Examples 15.3.6. By Examples 15.3.6, we can say that, for any $n \geq 4$, $\mathcal{F}_{n+1}$ is no more a taOM algebra.

• **Example 2: taOM is IMTL**

By a PASCAL program, we found that the algebra $\mathcal{A}^L = (A_6 = \{0, a, b, c, d, 1\}, \odot, ^-, 1)$, with the following tables of $\odot$ and $^-$ and of the additional operation $\oplus$, is a proper transitive, antisymmetric left-orthomodular algebra (taOM) (= m-BCK algebra verifying (Pom)), i.e. (PU), (Pcomm), (Pass), (m-La), (m-Re), (m-An), (DN), (m-Tr) ($\Leftrightarrow$ ... $\Leftrightarrow$ (m-BB)), (Pom), but also (prel$_m$), hold and it does not verify (m-Pabs-i) for $(d, 0)$, (G) for $b$, (m-Pimpl) for $(a, 0)$, (Pqmv) for $(c, d, 0)$, (Pmv) for $(c, d)$, ($\Delta_m$) for $(a, c)$, (WNM$_m$) and (aWNM$_m$) for $(c, c)$.

| $\odot$ | 0 | a | b | c | d | 1 |
|---|---|---|---|---|---|---|
| 0 | 0 | 0 | 0 | 0 | 0 | 0 |
| a | 0 | a | b | b | 0 | a |
| b | 0 | b | 0 | 0 | 0 | b |
| c | 0 | b | 0 | d | 0 | c |
| d | 0 | 0 | 0 | 0 | 0 | d |
| 1 | 0 | a | b | c | d | 1 |

and

| $x$ | $x^-$ |
|---|---|
| 0 | 1 |
| a | d |
| b | c |
| c | b |
| d | a |
| 1 | 0 |

, with

| $\oplus$ | 0 | a | b | c | d | 1 |
|---|---|---|---|---|---|---|
| 0 | 0 | a | b | c | d | 1 |
| a | a | 1 | 1 | 1 | 1 | 1 |
| b | b | 1 | a | 1 | c | 1 |
| c | c | 1 | 1 | 1 | c | 1 |
| d | d | 1 | c | c | d | 1 |
| 1 | 1 | 1 | 1 | 1 | 1 | 1 |

.

Then, the tables of $\wedge_m^M$ and its transposed, $\wedge_m^B$, are the following:

| $\wedge_m^M$ | 0 | a | b | c | d | 1 |
|---|---|---|---|---|---|---|
| 0 | 0 | 0 | 0 | 0 | 0 | 0 |
| a | 0 | a | b | c | d | a |
| b | 0 | b | b | b | d | b |
| c | 0 | b | b | c | d | c |
| d | 0 | 0 | 0 | d | d | d |
| 1 | 0 | a | b | c | d | 1 |

and

| $\wedge_m^B$ | 0 | a | b | c | d | 1 |
|---|---|---|---|---|---|---|
| 0 | 0 | 0 | 0 | 0 | 0 | 0 |
| a | 0 | a | b | b | 0 | a |
| b | 0 | b | b | b | 0 | b |
| c | 0 | c | b | c | d | c |
| d | 0 | d | d | d | d | d |
| 1 | 0 | a | b | c | d | 1 |

.

Note that $\leq_m^M$ is an order relation, by Corollary 12.1.7, but not a lattice order w.r. to $\wedge_m^M$, $\vee_m^M$, since $\wedge_m^M$ is not commutative. From the table of $\wedge_m^M$, we see that $a \leq_m^M 1$; $b \leq_m^M a, c, 1$ ; $c \leq_m^M 1$; $d \leq_m^M c, 1$; hence, the bounded poset $(A_6, \leq_m^M, 0, 1)$ is reprezented by the Hasse diagram from the Figure 12.8.

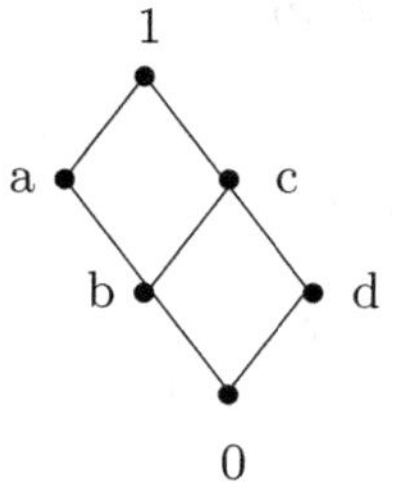


Figure 12.8: The Hasse diagram of the bounded poset $(A_6, \leq_m^M, 0, 1)$

The binary relation $\leq_m$ ($\Longleftrightarrow \leq_m^B$) is an order relation also, since (m-Re), (m-An) and (m-Tr) hold; hence, $\leq_m^B$ is an order relation too. From the table of $\wedge_m^B$, we see that $0 \leq_m d \leq_m b \leq_m c \leq_m a \leq_m 1$, hence the binary relations $\leq_m^B \Longleftrightarrow \leq_m$ ($x \leq_m^B y \overset{def.}{\Longleftrightarrow} x \wedge_m^B y = x$) are linearly ordered. Note that the operation $\wedge_m^B$ is not commutative, therefore the order relation $\leq_m^B$ is not a lattice order w.r. to $\wedge_m^B$, $\vee_m^B$; but, the order relation $\leq_m$ is a lattice order w.r. to $\wedge = \wedge_m = \inf_m$, $\vee = \vee_m = \sup_m$:

| ∧ | 0 | a | b | c | d | 1 |
|---|---|---|---|---|---|---|
| 0 | 0 | 0 | 0 | 0 | 0 | 0 |
| a | 0 | a | b | c | d | a |
| b | 0 | b | b | b | d | b |
| c | 0 | c | b | c | d | c |
| d | 0 | d | d | d | d | d |
| 1 | 0 | a | b | c | d | 1 |

and

| ∨ | 0 | a | b | c | d | 1 |
|---|---|---|---|---|---|---|
| 0 | 0 | a | b | c | d | 1 |
| a | a | a | a | a | a | 1 |
| b | b | a | b | c | b | 1 |
| c | c | a | c | c | c | 1 |
| d | d | a | b | c | d | 1 |
| 1 | 1 | 1 | 1 | 1 | 1 | 1 |

.

To see if the properties (prel) ($(x \to y) \vee (y \to x) = 1$) and (WNM) ($(x \odot y)^- \vee [(x \wedge y) \to (x \odot y)] = 1$) are verified, we need the table of $\to$ ($x \to y \overset{def.}{=} (x \odot y^-)^-$):

| → | 0 | a | b | c | d | 1 |
|---|---|---|---|---|---|---|
| 0 | 1 | 1 | 1 | 1 | 1 | 1 |
| a | d | 1 | c | c | d | 1 |
| b | c | 1 | 1 | 1 | c | 1 |
| c | b | 1 | a | 1 | c | 1 |
| d | a | 1 | 1 | 1 | 1 | 1 |
| 1 | 0 | a | b | c | d | 1 |

.

It is easy to see that (prel) is verified. We check by a PASCAL program that (WNM) is not verified for $(a, b)$, $(c, b)$, $(c, d)$. Hence, this taOM algebra is an IMTL algebra.

Note that, by denoting $(0, 1, 2, 3, 4, 5) = (0, d, b, c, a, 1)$, we obtain that this IMTL algebra is one of the two linearly ordered IMTL algebras with 6 elements verifying (Pom) from ([104], 5.1.1): $\text{IMTL}_6^4$, where $0 \leq_m 1 \leq_m 2 \leq_m 3 \leq_m 4 \leq_m 5$; the other one is $\text{IMTL}_6^3$.

• **Example 3: taOM is a distributive lattice**

By Mace4 program, we found that the algebra $\mathcal{A}^L = (A_6 = \{0, a, b, c, d, 1\}, \odot, ^-, 1)$, with the following tables of $\odot$ and $^-$ and of the additional operation $\oplus$, is a proper transitive, antisymmetric left-orthomodular algebra (taOM) (= m-BCK algebra verifying (Pom)), i.e. (PU), (Pcomm), (Pass), (m-La), (m-Re), (m-An), (DN), (m-Tr) (⇔ ... ⇔ (m-BB)), (Pom) hold and it does not verify (m-Pabs-i) for $(b, 0)$, (G) for $b$, (m-Pimpl) for $(a, 0)$, (Pqmv) for $(b, b, 0)$, (Pmv) for $(b, b)$, ($\Delta_m$) for $(a, b)$, (prel$_m$) for $(b, a)$, (WNM$_m$) and (aWNM$_m$) for $(b, b)$.

| ⊙ | 0 | a | b | c | d | 1 |
|---|---|---|---|---|---|---|
| 0 | 0 | 0 | 0 | 0 | 0 | 0 |
| a | 0 | a | 0 | 0 | a | a |
| b | 0 | 0 | c | 0 | c | b |
| c | 0 | 0 | 0 | 0 | 0 | c |
| d | 0 | a | c | 0 | a | d |
| 1 | 0 | a | b | c | d | 1 |

and

| $x$ | $x^-$ |
|---|---|
| 0 | 1 |
| a | b |
| b | a |
| c | d |
| d | c |
| 1 | 0 |

, with

| ⊕ | 0 | a | b | c | d | 1 |
|---|---|---|---|---|---|---|
| 0 | 0 | a | b | c | d | 1 |
| a | a | d | 1 | d | 1 | 1 |
| b | b | 1 | b | b | 1 | 1 |
| c | c | d | b | b | 1 | 1 |
| d | d | 1 | 1 | 1 | 1 | 1 |
| 1 | 1 | 1 | 1 | 1 | 1 | 1 |

.

Then, the tables of $\wedge_m^M$ and its transposed, $\wedge_m^B$, are the following:

| $\wedge_m^M$ | 0 | a | b | c | d | 1 |
|---|---|---|---|---|---|---|
| 0 | 0 | 0 | 0 | 0 | 0 | 0 |
| a | 0 | a | c | c | a | a |
| b | 0 | 0 | b | c | c | b |
| c | 0 | 0 | c | c | c | c |
| d | 0 | a | b | c | d | d |
| 1 | 0 | a | b | c | d | 1 |

and

| $\wedge_m^B$ | 0 | a | b | c | d | 1 |
|---|---|---|---|---|---|---|
| 0 | 0 | 0 | 0 | 0 | 0 | 0 |
| a | 0 | a | 0 | 0 | a | a |
| b | 0 | c | b | c | b | b |
| c | 0 | c | c | c | c | c |
| d | 0 | a | c | c | d | d |
| 1 | 0 | a | b | c | d | 1 |

.

Note that $\leq_m^M$ is an order relation, by Corollary 12.1.7, but not a lattice order w.r. to $\wedge_m^M$, $\vee_m^M$, since $\wedge_m^M$ is not commutative. From the table of $\wedge_m^M$, we see that $a \leq_m^M d, 1$; $b \leq_m^M 1$; $c \leq_m^M b, d, 1$; $d \leq_m^M 1$; hence, the bounded poset $(A_6, \leq_m^M, 0, 1)$ is reprezented by the Hasse diagram from Figure 12.9:

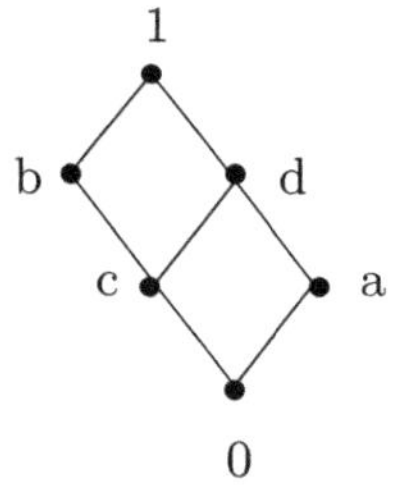


Figure 12.9: The Hasse diagram of the bounded poset $(A_6, \leq_m^M, 0, 1)$

The binary relation $\leq_m$ ($\Longleftrightarrow \leq_m^B$) is an order relation also, since (m-Re), (m-An) and (m-Tr) hold; hence, $\leq_m^B$ is an order relation too. From the table of $\wedge_m^B$, we see that $a \leq_m d, 1$; $b \leq_m d, 1$; $c \leq_m a, b, d, 1$; $d \leq_m 1$; hence, the binary relations $\leq_m^B \Longleftrightarrow \leq_m$ ($x \leq_m^B y \overset{def.}{\Longleftrightarrow} x \wedge_m^B y = x$) are represented by the Hasse diagram from Figure 12.10.

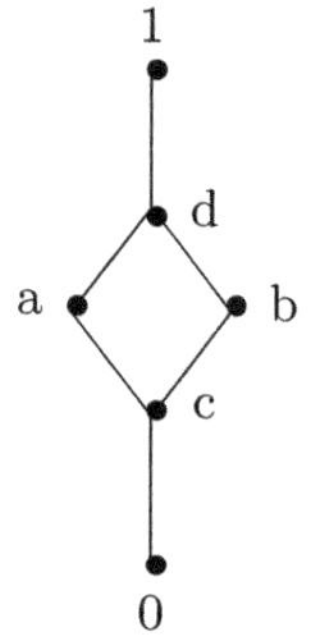


Figure 12.10: The Hasse diagram of the bounded poset $(A_6, \leq_m, 0, 1)$

Note that the operation $\wedge_m^B$ is not commutative, therefore the order relation $\leq_m^B$ is not a lattice order w.r. to $\wedge_m^B$, $\vee_m^B$; but, the order relation $\leq_m$ is a distributive lattice order w.r.t. to $\wedge = \wedge_m = \inf_m$, $\vee = \vee_m = \sup_m$:

| $\wedge$ | 0 | a | b | c | d | 1 |
|---|---|---|---|---|---|---|
| 0 | 0 | 0 | 0 | 0 | 0 | 0 |
| a | 0 | a | c | c | a | a |
| b | 0 | c | b | c | b | b |
| c | 0 | c | c | c | c | c |
| d | 0 | a | b | c | d | d |
| 1 | 0 | a | b | c | d | 1 |

and

| $\vee$ | 0 | a | b | c | d | 1 |
|---|---|---|---|---|---|---|
| 0 | 0 | a | b | c | d | 1 |
| a | a | a | d | a | d | 1 |
| b | b | d | b | b | d | 1 |
| c | c | a | b | c | d | 1 |
| d | d | d | d | d | d | 1 |
| 1 | 1 | 1 | 1 | 1 | 1 | 1 |

.

To see if the property (prel) ($(x \to y) \vee (y \to x) = 1$) is verified, we need the table of $\to$ ($x \to y \overset{def.}{=} (x \odot y^-)^-$):

| $\to$ | 0 | a | b | c | d | 1 |
|---|---|---|---|---|---|---|
| 0 | 1 | 1 | 1 | 1 | 1 | 1 |
| a | b | 1 | b | b | 1 | 1 |
| b | a | d | 1 | d | 1 | 1 |
| c | d | 1 | 1 | 1 | 1 | 1 |
| d | c | d | b | b | 1 | 1 |
| 1 | 0 | a | b | c | d | 1 |

.

It is easy to see that (prel) is not verified for $(a, b)$: $(a \to b) \vee (b \to a) = b \vee d = d \neq 1$. Hence, this taOM algebra is not an IMTL algebra, it is only a distributive lattice.

• **Example 4: taOM is a distributive lattice**

By Mace4 program, we found that the algebra $\mathcal{A}^L = (A_8 = \{0, a, b, c, d, e, f, 1\}, \odot, ^-, 1)$, with the following tables of $\odot$ and $^-$, is a proper transitive, antisymmetric left-orthomodular algebra (taOM) (= m-BCK algebra verifying (Pom)), and also verifying (prel$_m$), (WNM$_m$) and (aWNM$_m$), and not verifying (m-Pabs-i) for $(b, 0)$, (G) for $b$, (m-Pimpl) for $(a, 0)$, (Pqmv) for $(b, b, 0)$, (Pmv) for $(b, b)$, ($\Delta_m$) for $(a, b)$, (m-Pdis) for $(a, a, a)$.

| $\odot$ | 0 | a | b | c | d | e | f | 1 |
|---|---|---|---|---|---|---|---|---|
| 0 | 0 | 0 | 0 | 0 | 0 | 0 | 0 | 0 |
| a | 0 | a | 0 | e | d | e | d | a |
| b | 0 | 0 | 0 | b | 0 | 0 | 0 | b |
| c | 0 | e | b | c | 0 | e | b | c |
| d | 0 | d | 0 | 0 | 0 | 0 | 0 | d |
| e | 0 | e | 0 | e | 0 | e | 0 | e |
| f | 0 | d | 0 | b | 0 | 0 | 0 | f |
| 1 | 0 | a | b | c | d | e | f | 1 |

and

| $x$ | $x^-$ |
|---|---|
| 0 | 1 |
| a | b |
| b | a |
| c | d |
| d | c |
| e | f |
| f | e |
| 1 | 0 |

.

The matrix of $\wedge_m^M$ and $\wedge_M^B$ are:

| $\wedge_m^M$ | 0 | a | b | c | d | e | f | 1 |
|---|---|---|---|---|---|---|---|---|
| 0 | 0 | 0 | 0 | 0 | 0 | 0 | 0 | 0 |
| a | 0 | a | b | e | d | e | f | a |
| b | 0 | 0 | b | b | 0 | 0 | b | b |
| c | 0 | e | b | c | d | e | f | c |
| d | 0 | d | 0 | 0 | d | 0 | d | d |
| e | 0 | e | b | e | d | e | f | e |
| f | 0 | d | b | b | d | 0 | f | f |
| 1 | 0 | a | b | c | d | e | f | 1 |

and

| $\wedge_m^B$ | 0 | a | b | c | d | e | f | 1 |
|---|---|---|---|---|---|---|---|---|
| 0 | 0 | 0 | 0 | 0 | 0 | 0 | 0 | 0 |
| a | 0 | a | 0 | e | d | e | d | a |
| b | 0 | b | b | b | 0 | b | b | b |
| c | 0 | e | b | c | 0 | e | b | c |
| d | 0 | d | 0 | d | d | d | d | d |
| e | 0 | e | 0 | e | 0 | e | 0 | e |
| f | 0 | f | b | f | d | f | f | f |
| 1 | 0 | a | b | c | d | e | f | 1 |

.

From the table of $\wedge_m^B$ we can see easily that: $a \leq_m 1$; $b \leq_m a, c, e, f, 1$; $c \leq_m 1$; $d \leq_m a, c, e, f, 1$; $e \leq_m a, c, 1$; $f \leq_m a, c, e, 1$. It follows that the binary relations $(x \leq_m y \overset{def.}{\Longleftrightarrow} x \odot y^- = 0) \leq_m \Longleftrightarrow \leq_m^B (x \leq_m^B y \overset{def.}{\Longleftrightarrow} x \wedge_m^B y = x)$ are represented by the Hasse diagram from Figure 12.11.

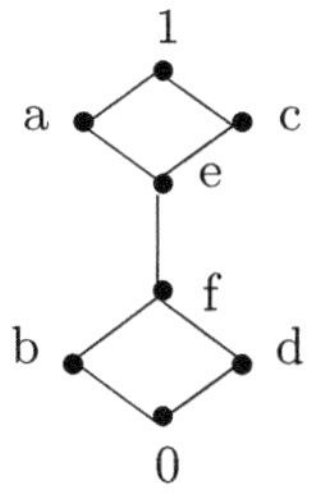


Figure 12.11: The Hasse diagram of the bounded poset (lattice) $(A_8, \leq_m, 0, 1)$

Note that the operation $\wedge_m^B$ is not commutative, therefore the order relation $\leq_m^B$ is not a lattice order w.r. to $\wedge_m^B$, $\vee_m^B$; but, the order relation $\leq_m$ is a distributive lattice order w.r. to $\wedge = \wedge_m = \inf_m$, $\vee = \vee_m = \sup_m$.

**• Example 5: taOM is a non-distributive lattice (Michael Kinyon)**

By Mace4 program (see below), we found that the algebra $\mathcal{A}^L = (A_8 = \{0, a, b, c, d, e, f, 1\}, \odot, ^-, 1)$, with the following tables of $\odot$ and $^-$, is a proper transitive, antisymmetric left-orthomodular algebra (taOM) (= m-BCK algebra verifying (Pom)), verifying (prel$_m$), (WNM$_m$) and (aWNM$_m$), not verifying (m-Pabs-i) for $(b, 0)$, (G) for $b$, (m-Pimpl) for $(a, 0)$, (Pqmv) for $(b, b, 0)$, (Pmv) for $(b, b)$, $(\Delta_m)$ for $(a, b)$, (m-Pdis) for $(a, a, a)$.

| $\odot$ | 0 | a | b | c | d | e | f | 1 |
|---|---|---|---|---|---|---|---|---|
| 0 | 0 | 0 | 0 | 0 | 0 | 0 | 0 | 0 |
| a | 0 | a | 0 | 0 | a | 0 | a | a |
| b | 0 | 0 | c | c | e | 0 | c | b |
| c | 0 | 0 | c | c | 0 | 0 | c | c |
| d | 0 | a | e | 0 | a | 0 | a | d |
| e | 0 | 0 | 0 | 0 | 0 | 0 | 0 | e |
| f | 0 | a | c | c | a | 0 | f | f |
| 1 | 0 | a | b | c | d | e | f | 1 |

and

| $\oplus$ | 0 | a | b | c | d | e | f | 1 |
|---|---|---|---|---|---|---|---|---|
| 0 | 0 | a | b | c | d | e | f | 1 |
| a | a | d | 1 | f | d | d | 1 | 1 |
| b | b | 1 | b | b | 1 | b | 1 | 1 |
| c | c | f | b | b | 1 | b | 1 | 1 |
| d | d | d | 1 | 1 | d | d | 1 | 1 |
| e | e | d | b | b | d | e | 1 | 1 |
| f | f | 1 | 1 | 1 | 1 | 1 | 1 | 1 |
| 1 | 1 | 1 | 1 | 1 | 1 | 1 | 1 | 1 |

,

and $(0, a, b, c, d, e, f, 1)^- = (1, b, a, d, c, f, e, 0)$.

The tables of $\wedge_m^B$ and $\rightarrow$ are:

| $\wedge_m^B$ | 0 | a | b | c | d | e | f | 1 |
|---|---|---|---|---|---|---|---|---|
| 0 | 0 | 0 | 0 | 0 | 0 | 0 | 0 | 0 |
| a | 0 | a | 0 | 0 | a | 0 | a | a |
| b | 0 | e | b | c | e | e | b | b |
| c | 0 | 0 | c | c | 0 | 0 | c | c |
| d | 0 | a | e | e | d | e | d | d |
| e | 0 | e | e | e | e | e | e | e |
| f | 0 | a | c | c | a | 0 | f | f |
| 1 | 0 | a | b | c | d | e | f | 1 |

and

| $\rightarrow$ | 0 | a | b | c | d | e | f | 1 |
|---|---|---|---|---|---|---|---|---|
| 0 | 1 | 1 | 1 | 1 | 1 | 1 | 1 | 1 |
| a | b | 1 | b | b | 1 | b | 1 | 1 |
| b | a | d | 1 | f | d | d | 1 | 1 |
| c | d | d | 1 | 1 | d | d | 1 | 1 |
| d | c | f | b | b | 1 | b | 1 | 1 |
| e | f | 1 | 1 | 1 | 1 | 1 | 1 | 1 |
| f | e | d | b | b | d | e | 1 | 1 |
| 1 | 0 | a | b | c | d | e | f | 1 |

.

From the table of $\wedge_m^B$ we can see easily that: $a \leq_m d, f, 1$; $b \leq_m f, 1$; $c \leq_m b, f, 1$; $d \leq_m f, 1$; $e \leq_m a, b, c, d, f, 1$; $f \leq_m 1$; it follows that the Hasse diagram of the bounded poset $(A_8, \leq_m, 0, 1)$ is that from Figure 12.12.

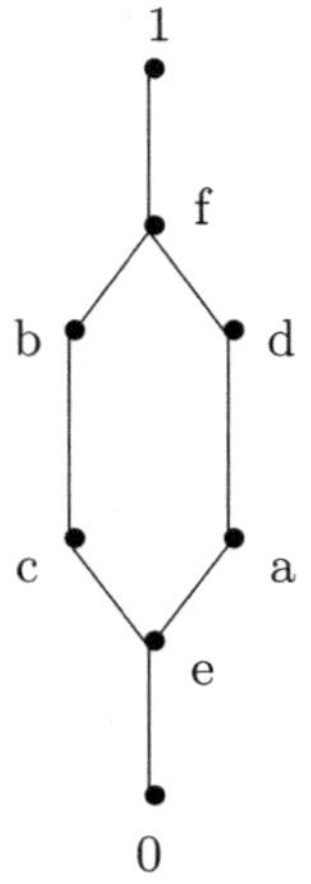


Figure 12.12: The bounded poset $(A_8, \leq_m, 0, 1)$, which is a non-distributive lattice

The lattice is not distributive (it contains as sublattice the pentagon $e, a, c, b, f$) for $(b, a, c)$: $b \wedge_m (a \vee_m c) = b \wedge_m f = b \neq (b \wedge_m a) \vee_m (b \wedge_m c) = e \vee_m c = c$.

From the table of $\rightarrow$ we can see that (prel) is not satisfied for $(b, d)$: $(b \rightarrow$

$d) \vee (d \to b) = d \vee b = f \neq 1$. Hence, this taOM algebra is not an IMTL algebra and it is a non-distributive lattice.

```
% Mace4 input for finding example(s) of non-distributive taOM lattice(s)
1 * x = x                    # label("PU").
x * y = y * x                # label("Pcomm").
(x * y) * z = x * (y * z)    # label("Pass").
x * 0 = 0                    # label("m-L").
x * -x = 0                   # label("m-Re").
-1 = 0                       # label("Neg1-0").
-0 = 1                       # label("Neg0-1").
- -x = x                     # label("DN").
x * -y = 0 & y * -x = 0 -> x = y # label("m-An").
(-(z * x) * (y * x)) * -(y * -z) = 0 # label("m-BB").

x + y = -(-x * -y) # label("plus").
x * y = -(-x + -y) # label("product").
(x * y) + (-(x * y) * x) = x # label("Pom"). % using *, -, +

% we impose condition that order <= m is a lattice
% with meet operation m and join operation j
m(x,y) = x <-> x * -y = 0.
j(x,y) = y <-> x * -y = 0.
m(x,x) = x.
m(x,y) = m(y,x).
m(m(x,y),z) = m(x,m(y,z)).
j(x,x) = x.
j(x,y) = j(y,x).
j(j(x,y),z) = j(x,j(y,z)).
m(x,j(x,y)) = x.
j(x,m(x,y)) = x.

% not distributive
m(a,j(b,c)) != j(m(a,b),m(a,c)).
```

- **Example 6: taOM is a non-dstributive lattice**

By Mace4 program, we found that the algebra
$\mathcal{A}^L = (A_{24} = \{0, a, b, c, d, e, f, g, h, i, j, k, m, n, p, r, s, t, x, y, z, u, v, 1\}, \odot, ^-, 1)$, with the following tables of $\odot$ and $^-$, is a proper transitive, antisymmetric left-orthomodular algebra (taOM) (= m-BCK algebra verifying (Pom)) not verifying (m-Pabs-i) for $(a, 0)$, (G) for $a$, (m-Pimpl) for $(a, 0)$, (Pqmv) for $(a, a, 0)$, (Pmv) for $(a, a)$, $(\Delta_m)$ for $(a, c)$, (m-Pdis) for $(a, a, b)$, $(\text{prel}_m)$ for $(b, d)$, $(\text{WNM}_m)$ and $(\text{aWNM}_m)$ for $(a, d)$.

| $\odot$ | 0 | a | b | c | d | e | f | g | h | i | j | k | m | n | p | r | s | t | x | y | z | u | v | 1 |
|---|---|---|---|---|---|---|---|---|---|---|---|---|---|---|---|---|---|---|---|---|---|---|---|---|
| 0 | 0 | 0 | 0 | 0 | 0 | 0 | 0 | 0 | 0 | 0 | 0 | 0 | 0 | 0 | 0 | 0 | 0 | 0 | 0 | 0 | 0 | 0 | 0 | 0 |
| a | 0 | 0 | 0 | 0 | i | 0 | 0 | 0 | 0 | 0 | i | 0 | 0 | 0 | a | 0 | 0 | a | 0 | 0 | i | 0 | 0 | a |
| b | 0 | 0 | s | r | e | e | c | i | e | 0 | h | n | e | 0 | m | 0 | e | v | y | 0 | s | r | e | b |
| c | 0 | 0 | r | 0 | 0 | 0 | 0 | 0 | 0 | 0 | n | 0 | 0 | 0 | i | 0 | 0 | y | 0 | 0 | r | 0 | 0 | c |
| d | 0 | i | e | 0 | m | e | a | a | e | 0 | m | i | e | 0 | d | 0 | e | d | a | 0 | m | i | e | d |
| e | 0 | 0 | e | 0 | e | e | 0 | 0 | e | 0 | e | 0 | e | 0 | e | 0 | e | e | 0 | 0 | e | 0 | e | e |
| f | 0 | 0 | c | 0 | a | 0 | i | i | n | 0 | k | 0 | i | 0 | g | 0 | r | x | i | 0 | u | 0 | y | f |
| g | 0 | 0 | i | 0 | a | 0 | i | i | 0 | 0 | a | 0 | i | 0 | g | 0 | 0 | g | i | 0 | a | 0 | i | g |
| h | 0 | 0 | e | 0 | e | e | n | 0 | e | 0 | s | r | e | 0 | e | 0 | e | s | r | 0 | e | 0 | e | h |
| i | 0 | 0 | 0 | 0 | 0 | 0 | 0 | 0 | 0 | 0 | 0 | 0 | 0 | 0 | i | 0 | 0 | i | 0 | 0 | 0 | 0 | 0 | i |
| j | 0 | i | h | n | m | e | k | a | s | 0 | b | c | e | r | d | 0 | e | z | u | r | v | y | s | j |
| k | 0 | 0 | n | 0 | i | 0 | 0 | 0 | r | 0 | c | 0 | 0 | 0 | a | 0 | 0 | u | 0 | 0 | y | 0 | r | k |
| m | 0 | 0 | e | 0 | e | e | i | i | e | 0 | e | 0 | e | 0 | m | 0 | e | m | i | 0 | e | 0 | e | m |
| n | 0 | 0 | 0 | 0 | 0 | 0 | 0 | 0 | 0 | 0 | r | 0 | 0 | 0 | 0 | 0 | 0 | r | 0 | 0 | 0 | 0 | 0 | n |
| p | 0 | a | m | i | d | e | g | g | e | i | d | a | m | 0 | p | 0 | e | p | g | i | d | a | m | p |
| r | 0 | 0 | 0 | 0 | 0 | 0 | 0 | 0 | 0 | 0 | 0 | 0 | 0 | 0 | 0 | 0 | 0 | 0 | 0 | 0 | 0 | 0 | 0 | r |
| s | 0 | 0 | e | 0 | e | e | r | 0 | e | 0 | e | 0 | e | 0 | e | 0 | e | e | 0 | 0 | e | 0 | e | s |
| t | 0 | a | v | y | d | e | x | g | s | i | z | u | m | r | p | 0 | e | p | g | i | d | a | m | t |
| x | 0 | 0 | y | 0 | a | 0 | i | i | r | 0 | u | 0 | i | 0 | g | 0 | 0 | g | i | 0 | a | 0 | i | x |
| y | 0 | 0 | 0 | 0 | 0 | 0 | 0 | 0 | 0 | 0 | r | 0 | 0 | 0 | i | 0 | 0 | i | 0 | 0 | 0 | 0 | 0 | y |
| z | 0 | i | s | r | m | e | u | a | e | 0 | v | y | e | 0 | d | 0 | e | d | a | 0 | m | i | e | z |
| u | 0 | 0 | r | 0 | i | 0 | 0 | 0 | 0 | 0 | y | 0 | 0 | 0 | a | 0 | 0 | a | 0 | 0 | i | 0 | 0 | u |
| v | 0 | 0 | e | 0 | e | e | y | i | e | 0 | s | r | e | 0 | m | 0 | e | m | i | 0 | e | 0 | e | v |
| 1 | 0 | a | b | c | d | e | f | g | h | i | j | k | m | n | p | r | s | t | x | y | z | u | v | 1 |

and
$(0, a, b, c, d, e, f, g, h, i, j, k, m, n, p, r, s, t, x, y, z, u, v, 1)^- =$
$(1, b, a, d, c, f, e, h, g, j, i, m, k, p, n, t, x, r, s, z, y, v, u, 0)$.

The table of $\wedge = \wedge_m^M$ is the following:

| $\wedge$ | $0$ | $a$ | $b$ | $c$ | $d$ | $e$ | $f$ | $g$ | $h$ | $i$ | $j$ | $k$ | $m$ | $n$ | $p$ | $r$ | $s$ | $t$ | $x$ | $y$ | $z$ | $u$ | $v$ | $1$ |
|---|---|---|---|---|---|---|---|---|---|---|---|---|---|---|---|---|---|---|---|---|---|---|---|---|
| 0 | $0$ | $0$ | $0$ | $0$ | $0$ | $0$ | $0$ | $0$ | $0$ | $0$ | $0$ | $0$ | $0$ | $0$ | $0$ | $0$ | $0$ | $0$ | $0$ | $0$ | $0$ | $0$ | $0$ | $0$ |
| a | $0$ | $a$ | $a$ | $y$ | $ai$ | $0$ | $a$ | $a$ | $n$ | $i$ | $a$ | $a$ | $i$ | $n$ | $a$ | $r$ | $r$ | $a$ | $a$ | $y$ | $a$ | $a$ | $y=$ | $a$ |
| b | $0$ | $a$ | $b$ | $c$ | $m$ | $e$ | $f$ | $gi$ | $h$ | $i$ | $b$ | $k$ | $m$ | $n$ | $m$ | $r$ | $s$ | $v$ | $x$ | $y$ | $v$ | $u$ | $v$ | $b$ |
| c | $0$ | $i$ | $c$ | $c$ | $i$ | $0$ | $c$ | $i$ | $n$ | $i$ | $c$ | $c$ | $i$ | $n$ | $i$ | $r$ | $r$ | $y$ | $y$ | $y$ | $y$ | $y$ | $y$ | $c$ |
| d | $0$ | $a$ | $v$ | $c$ | $d$ | $e$ | $f$ | $g$ | $h$ | $i$ | $d$ | $ki$ | $m$ | $n$ | $d$ | $r$ | $s$ | $d$ | $x$ | $y$ | $d$ | $u$ | $v$ | $d$ |
| e | $0$ | $a$ | $e$ | $c$ | $e$ | $e$ | $k$ | $a$ | $e$ | $i$ | $e$ | $k$ | $e$ | $n$ | $e$ | $r$ | $e$ | $e$ | $u$ | $y$ | $e$ | $u$ | $e$ | $e$ |
| f | $0$ | $a$ | $c$ | $c$ | $a$ | $0$ | $f$ | $g$ | $n$ | $i$ | $k$ | $k$ | $i$ | $n$ | $g$ | $r$ | $r$ | $x$ | $x$ | $y$ | $u$ | $u$ | $y$ | $f$ |
| g | $0$ | $a$ | $c$ | $c$ | $a$ | $0$ | $g$ | $g$ | $n$ | $i$ | $u$ | $u$ | $i$ | $n$ | $g$ | $r$ | $r$ | $g$ | $g$ | $y$ | $u$ | $u$ | $y$ | $g$ |
| h | $0$ | $a$ | $h$ | $c$ | $e$ | $e$ | $k$ | $a$ | $h$ | $i$ | $h$ | $k$ | $e$ | $n$ | $e$ | $r$ | $s$ | $s$ | $u$ | $y$ | $s$ | $u$ | $s$ | $h$ |
| i | $0$ | $i$ | $i$ | $i$ | $i$ | $0$ | $i$ | $i$ | $r$ | $i$ | $i$ | $i$ | $i$ | $r$ | $i$ | $r$ | $r$ | $i$ | $i$ | $i$ | $i$ | $i$ | $i$ | |
| j | $0$ | $a$ | $b$ | $c$ | $d$ | $e$ | $f$ | $g$ | $h$ | $i$ | $j$ | $k$ | $m$ | $n$ | $d$ | $r$ | $s$ | $z$ | $x$ | $y$ | $z$ | $u$ | $v$ | $j$ |
| k | $0$ | $a$ | $c$ | $c$ | $a$ | $0$ | $k$ | $a$ | $n$ | $i$ | $k$ | $k$ | $i$ | $n$ | $a$ | $r$ | $r$ | $u$ | $u$ | $y$ | $u$ | $u$ | $y$ | $k$ |
| m | $0$ | $a$ | $m$ | $c$ | $m$ | $e$ | $f$ | $g$ | $s$ | $i$ | $m$ | $k$ | $m$ | $n$ | $m$ | $r$ | $s$ | $m$ | $x$ | $y$ | $m$ | $u$ | $m$ | $m$ |
| n | $0$ | $0$ | $n$ | $n$ | $0$ | $0$ | $n$ | $0$ | $n$ | $0$ | $n$ | $n$ | $0$ | $n$ | $0$ | $r$ | $r$ | $r$ | $r$ | $r$ | $r$ | $r$ | $r$ | $n$ |
| p | $0$ | $a$ | $b$ | $c$ | $d$ | $e$ | $f$ | $g$ | $h$ | $i$ | $z$ | $k$ | $m$ | $n$ | $p$ | $r$ | $s$ | $p$ | $x$ | $y$ | $z$ | $u$ | $v$ | $p$ |
| r | $0$ | $0$ | $r$ | $r$ | $0$ | $0$ | $r$ | $0$ | $r$ | $0$ | $r$ | $r$ | $0$ | $r$ | $0$ | $r$ | $r$ | $r$ | $r$ | $r$ | $r$ | $r$ | $r$ | $r$ |
| s | $0$ | $a$ | $s$ | $c$ | $e$ | $e$ | $k$ | $a$ | $s$ | $i$ | $s$ | $k$ | $e$ | $n$ | $e$ | $r$ | $s$ | $s$ | $u$ | $y$ | $s$ | $u$ | $s$ | $s$ |
| t | $0$ | $a$ | $b$ | $c$ | $d$ | $e$ | $f$ | $g$ | $h$ | $i$ | $j$ | $k$ | $m$ | $n$ | $p$ | $r$ | $s$ | $t$ | $x$ | $yi$ | $z$ | $u$ | $v$ | $t$ |
| x | $0$ | $a$ | $c$ | $c$ | $a$ | $0$ | $x$ | $g$ | $n$ | $i$ | $k$ | $k$ | $i$ | $n$ | $g$ | $r$ | $r$ | $x$ | $x$ | $y$ | $u$ | $u$ | $y$ | $x$ |
| y | $0$ | $i$ | $y$ | $y$ | $i$ | $0$ | $y$ | $i$ | $n$ | $i$ | $y$ | $y$ | $i$ | $n$ | $i$ | $r$ | $r$ | $y$ | $y$ | $y$ | $y$ | $y$ | $y$ | $y$ |
| z | $0$ | $a$ | $b$ | $c$ | $d$ | $e$ | $f$ | $g$ | $h$ | $i$ | $z$ | $k$ | $m$ | $n$ | $d$ | $r$ | $s$ | $z$ | $x$ | $y$ | $z$ | $u$ | $v$ | $z$ |
| u | $0$ | $a$ | $c$ | $c$ | $a$ | $0$ | $u$ | $a$ | $n$ | $i$ | $u$ | $u$ | $i$ | $n$ | $a$ | $r$ | $r$ | $u$ | $u$ | $y$ | $u$ | $u$ | $y$ | $u$ |
| v | $0$ | $a$ | $v$ | $c$ | $m$ | $e$ | $f$ | $g$ | $h$ | $i$ | $v$ | $k$ | $m$ | $n$ | $m$ | $r$ | $s$ | $v$ | $x$ | $y$ | $v$ | $u$ | $v$ | $v$ |
| 1 | $0$ | $a$ | $b$ | $c$ | $d$ | $e$ | $f$ | $g$ | $h$ | $i$ | $j$ | $k$ | $m$ | $n$ | $p$ | $r$ | $s$ | $t$ | $x$ | $y$ | $z$ | $u$ | $v$ | $1$ |

Note that, since $x \leq_m y \iff x \odot y^- = 0 \iff x \wedge_m^B y = x \iff y \wedge_m^M x = x$, it follows from the table of $\wedge = \wedge_m^M$ that:

$a \leq_m a, b, d, e, f, g, h, j, k, m, p, s, t, x, z, u, v, 1;$
$b \leq_m b, j, p, t, z, 1;$
$c \leq_m c, b, d, e, f, g, h, j, k, m, p, s, t, x, z, u, v, 1;$
$d \leq_m d, j, p, t, z, 1;$
$e \leq_m e, b, d, h, j, m, p, s, t, z, v, 1;$
$f \leq_m f, b, d, j, m, p, t, z, v, 1;$
$g \leq_m g, b, d, f, j, m, p, t, z, v, 1;$
$h \leq_m h, b, d, j, p, t, z, v, 1;$
$i \leq_m i, a, b, c, d, e, f, g, h, j, k, m, p, s, t, x, y, z, u, v, 1;$
$j \leq_m j, t, 1;$
$k \leq_m k, b, d, e, f, h, j, m, p, s, t, x, z, v, 1;$
$m \leq_m m, b, d, j, p, t, z, v, 1;$
$n \leq_m n, a, b, c, d, e, f, g, h, j, k, m, p, s, t, x, y, z, u, v, 1;$
$p \leq_m p, t, 1;$
$r \leq_m r, a, b, c, d, e, f, g, h, i, j, k, m, n, p, s, t, x, y, z, u, v, 1;$
$s \leq_m s, b, d, h, j, m, p, t, z, v, 1;$
$t \leq_m t, 1;$
$x \leq_m x, b, d, f, j, m, p, t, z, v, 1;$

$y \leq_m y, a, b, c, d, e, f, g, h, j, k, m, p, s, t, x, z, u, v, 1;$
$z \leq_m z, j, p, t, 1;$
$u \leq_m u, b, d, e, f, g, h, j, k, m, p, s, t, x, v, 1;$
$v \leq_m v, b, d, j, p, t, z, 1.$

It follows that the order relation $\leq_m$ can be reprezented by the Hasse diagram from Figure 12.13.

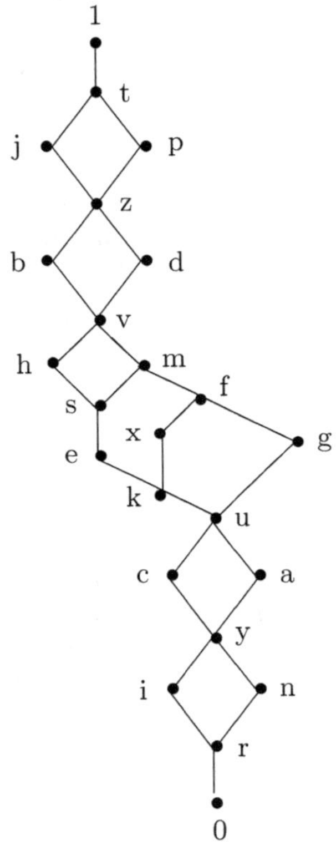


Figure 12.13: The bounded poset $(A_{24}, \leq_m, 0, 1)$

Note that $(A_{24}, \leq_m, 0, 1)$ is a bounded lattice w.r. to $\wedge = \wedge_m = \inf_m$ and $\vee = \vee_m = \sup_m$, but **not a distributive one**, because it contains as sublattice the pentagon lattice $u, k, x, g, f$.

**Example 7: taOM is a distributive lattice**

By Mace4 program, we found that the algebra $\mathcal{A}^L = (A_{12} = \{0, a, b, c, d, e, f, g, h, i, j, 1\}, \odot, ^-, 1)$, with the following tables of $\odot$ and $^-$, is a proper transitive, antisymmetric left-orthomodular algebra (taOM) (= m-BCK algebra verifying (Pom)), verifying (prel$_m$), not verifying (m-Pabs-i) for $(a, 0)$, (G) for $a$, (m-Pimpl) for $(c, 0)$, (Pqmv) for $(a, a, 0)$, (Pmv) for $(a, a)$, $(\Delta_m)$ for $(c, b)$, (m-Pdis) for $(a, d, c)$, (WNM$_m$) and (aWNM$_m$) for $(c, c)$.

| $\odot$ | 0 | a | b | c | d | e | f | g | h | i | j | 1 |
|---|---|---|---|---|---|---|---|---|---|---|---|---|
| 0 | 0 | 0 | 0 | 0 | 0 | 0 | 0 | 0 | 0 | 0 | 0 | 0 |
| a | 0 | 0 | 0 | 0 | 0 | 0 | a | 0 | 0 | 0 | 0 | a |
| b | 0 | 0 | 0 | 0 | 0 | b | 0 | 0 | 0 | 0 | 0 | b |
| c | 0 | 0 | 0 | a | 0 | g | c | 0 | a | g | 0 | c |
| d | 0 | 0 | 0 | 0 | b | d | g | 0 | b | g | 0 | d |
| e | 0 | 0 | b | g | d | e | i | g | d | i | b | e |
| f | 0 | a | 0 | c | g | i | f | g | c | i | a | f |
| g | 0 | 0 | 0 | 0 | 0 | g | g | 0 | 0 | g | 0 | g |
| h | 0 | 0 | 0 | a | b | d | c | 0 | j | g | 0 | h |
| i | 0 | 0 | 0 | g | g | i | i | g | g | i | 0 | i |
| j | 0 | 0 | 0 | 0 | 0 | b | a | 0 | 0 | 0 | 0 | j |
| 1 | 0 | a | b | c | d | e | f | g | h | i | j | 1 |

and

| $x$ | $x^-$ |
|---|---|
| 0 | 1 |
| a | e |
| b | f |
| c | d |
| d | c |
| e | a |
| f | b |
| g | h |
| h | g |
| i | j |
| j | i |
| 1 | 0 |

.

The matrix of $\wedge^B$ is:

| $\wedge_m^B$ | 0 | a | b | c | d | e | f | g | h | i | j | 1 |
|---|---|---|---|---|---|---|---|---|---|---|---|---|
| 0 | 0 | 0 | 0 | 0 | 0 | 0 | 0 | 0 | 0 | 0 | 0 | 0 |
| a | 0 | a | 0 | a | a | a | a | a | a | a | a | a |
| b | 0 | 0 | b | b | b | b | b | b | b | b | b | b |
| c | 0 | a | 0 | c | g | c | c | g | c | c | a | c |
| d | 0 | 0 | b | g | d | d | d | g | d | d | b | d |
| e | 0 | 0 | b | g | d | e | i | g | d | i | b | e |
| f | 0 | a | 0 | c | g | i | f | g | c | i | a | f |
| g | 0 | 0 | 0 | g | g | g | g | g | g | g | 0 | g |
| h | 0 | a | b | c | d | h | h | g | h | h | j | h |
| i | 0 | 0 | 0 | g | g | i | i | g | g | i | 0 | i |
| j | 0 | a | b | j | j | j | j | j | j | j | j | j |
| 1 | 0 | a | b | c | d | e | f | g | h | i | j | 1 |

.

Note, from the table of $\wedge_m^B$, that:
$a \leq_m c, d, e, f, g, h, i, j, 1$;
$b \leq_m c, d, e, f, g, h, i, j, 1$;
$c \leq_m e, f, h, i, 1$;
$d \leq_m e, f, h, i, 1$;
$e \leq_m 1$;
$f \leq_m 1$;
$g \leq_m c, d, e, f, h, i, 1$;
$h \leq_m e, f, i, 1$;
$i \leq_m e, f, 1$;
$j \leq_m c, d, e, f, g, h, i, 1$.

It follows that the order relation $\leq_m$ can be reprezented by the Hasse diagram from Figure 12.14.

**Example 12.4.13 Proper IMTL algebra (not verifying (Pom) and (WNM)): IMTL**

Consider the following linearly ordered IMTL algebra

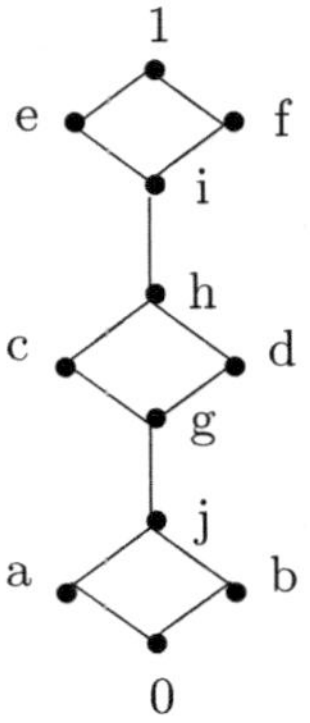


Figure 12.14: The Hasse diagram of the bounded poset (lattice) $(A_{12}, \leq_m, 0, 1)$

$\mathcal{A}^L = (A_5 = \{0, a, b, c, 1\}, \odot, ^-, 1)$ with five elements, IMTL$_5$, from ([104], 5.1.1), where the tables of $\odot$ and $^-$, and of the additional operations $\oplus$ are the following.

| $\odot$ | 0 | a | b | c | 1 |
|---|---|---|---|---|---|
| 0 | 0 | 0 | 0 | 0 | 0 |
| a | 0 | 0 | 0 | 0 | a |
| b | 0 | 0 | 0 | a | b |
| c | 0 | 0 | a | a | c |
| 1 | 0 | a | b | c | 1 |

and

| $x$ | $x^-$ |
|---|---|
| 0 | 1 |
| a | c |
| b | b |
| c | a |
| 1 | 0 |

, with

| $\oplus$ | 0 | a | b | c | 1 |
|---|---|---|---|---|---|
| 0 | 0 | a | b | c | 1 |
| a | a | c | c | 1 | 1 |
| b | b | c | 1 | 1 | 1 |
| c | c | 1 | 1 | 1 | 1 |
| 1 | 1 | 1 | 1 | 1 | 1 |

.

Note that $\mathcal{A}^L$ is a proper left-m-BCK algebra, verifying (prel$_m$), not verifying (m-Pabs-i) for $(a, 0)$, (G) for $a$, (m-Pimpl) for $(a, 0)$, (Pqmv) for $(b, a, 0)$, (Pom) for $(b, c)$, (Pmv) for $(b, a)$, $(\Delta_m)$ for $(c, b)$, (m-Pdis) for $(a, a, b)$, (WNM$_m$) and (aWNM$_m$ for $(b, c)$.

The tables of $\wedge^B$ and $\rightarrow$ are:

| $\wedge_m^B$ | 0 | a | b | c | 1 |
|---|---|---|---|---|---|
| 0 | 0 | 0 | 0 | 0 | 0 |
| a | 0 | a | a | a | a |
| b | 0 | a | b | b | b |
| c | 0 | a | a | c | c |
| 1 | 0 | a | b | c | 1 |

and

| $\rightarrow$ | 0 | a | b | c | 1 |
|---|---|---|---|---|---|
| 0 | 1 | 1 | 1 | 1 | 1 |
| a | c | 1 | 1 | 1 | 1 |
| b | b | c | 1 | 1 | 1 |
| c | a | c | c | 1 | 1 |
| 1 | 0 | a | b | c | 1 |

.

Note from the table of $\wedge_m^B$ $(x \leq_m y \Longleftrightarrow x \wedge_m^B y = x)$ that $a \leq_m b, c, 1$; $b \leq_m c, 1$; $c \leq_m 1$; hence, the chain is $0 \leq_m a \leq_m b \leq_m c \leq_m 1$. Thus, the bounded poset $(A_5, \leq_m, 0, 1)$ is a lattice, with $x \wedge_m y = \inf_m(x, y)$, $x \vee_m y = \sup_m(x, y)$. Hence, the left-m-BCK algebra is a lattice.

Note, from the table of $\rightarrow$, that (prel) is satisfied, but (WNM) is not satisfied for $(b, c)$; $(b \odot c)^- \vee_m ((b \wedge_m c) \rightarrow (b \odot c)) = a^- \vee_m (b \rightarrow a) = c \vee_m c = c \neq 1$; hence, $\mathcal{A}^L$ is a left-IMTL algebra, not verifying (Pom) and (WNM), hence it is a proper IMTL algebra.

**Example 12.4.14 Proper (involutive) m-BCK lattices: m-BCK-L**

By Mace4 program, we found that the algebra $\mathcal{A}^L = (A_9 = \{0, a, b, c, d, e, f, g, 1\}, \odot, ^-, 1)$, with the following tables of $\odot$ and $^-$, is a proper m-BCK algebra, not verifying (m-Pabs-i) for $(a, 0)$, (G) for $a$, (m-Pimpl) for $(a, 0)$, (Pqmv) for $(a, a, 0)$, (Pom) for $(a, d)$, (Pmv) for $(a, a)$, $(\Delta_m)$ for $(c, a)$, (m-Pdis) for $(a, a, a)$, (prel$_m$) for $(a, b)$, (WNM$_m$) and (aWNM$_m$ for $(a, d)$.

| $\odot$ | 0 | a | b | c | d | e | f | g | 1 |
|---|---|---|---|---|---|---|---|---|---|
| 0 | 0 | 0 | 0 | 0 | 0 | 0 | 0 | 0 | 0 |
| a | 0 | 0 | 0 | 0 | e | 0 | e | 0 | a |
| b | 0 | 0 | 0 | e | 0 | 0 | e | 0 | b |
| c | 0 | 0 | e | b | e | 0 | b | e | c |
| d | 0 | e | 0 | e | e | 0 | e | e | d |
| e | 0 | 0 | 0 | 0 | 0 | 0 | 0 | 0 | e |
| f | 0 | e | e | b | e | 0 | b | e | f |
| g | 0 | 0 | 0 | e | e | 0 | e | 0 | g |
| 1 | 0 | a | b | c | d | e | f | g | 1 |

and

| $x$ | $x^-$ |
|---|---|
| 0 | 1 |
| a | c |
| b | d |
| c | a |
| d | b |
| e | f |
| f | e |
| g | g |
| 1 | 0 |

.

The matrix of $\wedge_m^B$ and $\rightarrow$ are:

| $\wedge_M^B$ | 0 | a | b | c | d | e | f | g | 1 |
|---|---|---|---|---|---|---|---|---|---|
| 0 | 0 | 0 | 0 | 0 | 0 | 0 | 0 | 0 | 0 |
| a | 0 | a | e | a | a | e | a | a | a |
| b | 0 | e | b | b | b | e | b | b | b |
| c | 0 | e | b | c | b | e | c | b | c |
| d | 0 | e | e | e | d | e | d | e | d |
| e | 0 | e | e | e | e | e | e | e | e |
| f | 0 | e | b | b | b | e | f | b | f |
| g | 0 | e | e | g | g | e | g | g | g |
| 1 | 0 | a | b | c | d | e | f | g | 1 |

,

| $\rightarrow$ | 0 | a | b | c | d | e | f | g | 1 |
|---|---|---|---|---|---|---|---|---|---|
| 0 | 1 | 1 | 1 | 1 | 1 | 1 | 1 | 1 | 1 |
| a | c | 1 | f | 1 | 1 | f | 1 | 1 | 1 |
| b | d | f | 1 | 1 | 1 | f | 1 | 1 | 1 |
| c | a | d | f | 1 | f | d | 1 | f | 1 |
| d | b | f | f | f | 1 | f | 1 | f | 1 |
| e | f | 1 | 1 | 1 | 1 | 1 | 1 | 1 | 1 |
| f | e | d | f | f | f | d | 1 | f | 1 |
| g | g | f | f | 1 | 1 | f | 1 | 1 | 1 |
| 1 | 0 | a | b | c | d | e | f | g | 1 |

.

Note that $a \leq_m c, d, f, g, 1$; $b \leq_m c, d, f, g, 1$; $c \leq_m f, 1$; $d \leq_m f, 1$; $e \leq_m a, b, c, d, f, g, 1$; $f \leq_m 1$; $g \leq_m c, d, f, 1$. Hence, we have the Hasse diagram of the bounded poset $(A_9, \leq_m, 0, 1)$ in Figure 12.15.

Note that $(a \rightarrow b) \vee (b \rightarrow a) = f \vee f = f \neq 1$, i.e. (prel) is not satisfied, hence $\mathcal{A}^L$ is not an IMTL algebra. It is a proper distributive m-BCK lattice.

**Examples 12.4.15 Proper m-BCK algebras (not lattices): m-BCK**

**Example 1.**

By Mace4 program, we found that the algebra $\mathcal{A}^L = (A_9 = \{0, a, b, c, d, e, f, g, 1\}, \odot, ^-, 1)$, with the following tables of $\odot$ and $^-$, is a proper m-BCK algebra, not verifying (m-Pabs-i) for $(a, 0)$, (G) for $a$, (m-Pimpl) for $(a, 0)$, (Pqmv) for $(a, a, 0)$, (Pom) for $(a, d)$, (Pmv) for $(a, a)$, $(\Delta_m)$ for $(c, a)$, (m-Pdis) for $(a, a, a)$, (prel$_m$) for $(a, b)$, (WNM$_m$) and (aWNM$_m$ for $(a, d)$.

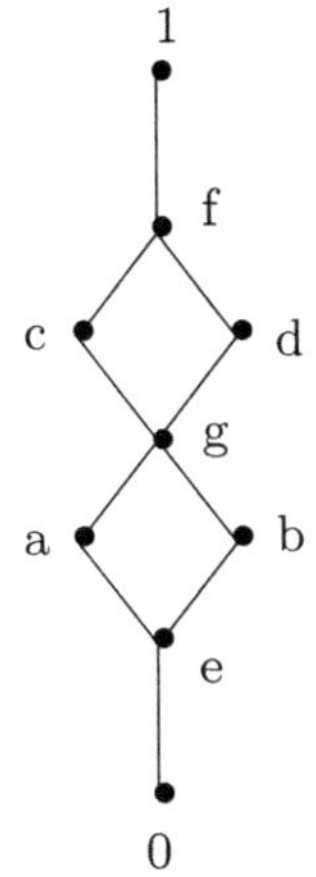


Figure 12.15: The bounded poset $(A_9, \leq_m, 0, 1)$, which is a lattice

| $\odot$ | 0 | a | b | c | d | e | f | g | 1 |
|---|---|---|---|---|---|---|---|---|---|
| 0 | 0 | 0 | 0 | 0 | 0 | 0 | 0 | 0 | 0 |
| a | 0 | 0 | 0 | 0 | e | 0 | e | e | a |
| b | 0 | 0 | 0 | e | 0 | 0 | e | e | b |
| c | 0 | 0 | e | e | e | 0 | e | e | c |
| d | 0 | e | 0 | e | e | 0 | e | e | d |
| e | 0 | 0 | 0 | 0 | 0 | 0 | 0 | 0 | e |
| f | 0 | e | e | e | e | 0 | e | e | f |
| g | 0 | e | e | e | e | 0 | e | 0 | g |
| 1 | 0 | a | b | c | d | e | f | g | 1 |

and

| $x$ | $x^-$ |
|---|---|
| 0 | 1 |
| a | c |
| b | d |
| c | a |
| d | b |
| e | f |
| f | e |
| g | g |
| 1 | 0 |

.

The matrix of $\wedge_m^B$ is:

| $\wedge_m^B$ | 0 | a | b | c | d | e | f | g | 1 |
|---|---|---|---|---|---|---|---|---|---|
| 0 | 0 | 0 | 0 | 0 | 0 | 0 | 0 | 0 | 0 |
| a | 0 | a | e | a | a | e | a | e | a |
| b | 0 | e | b | b | b | e | b | e | b |
| c | 0 | e | e | c | e | e | c | e | c |
| d | 0 | e | e | e | d | e | d | e | d |
| e | 0 | e | e | e | e | e | e | e | e |
| f | 0 | e | e | e | e | e | f | e | f |
| g | 0 | e | e | e | e | e | g | g | g |
| 1 | 0 | a | b | c | d | e | f | g | 1 |

.

Note that $a \leq_m c, d, f, 1$; $b \leq_m c, d, f, 1$; $c \leq_m f, 1$; $d \leq_m f, 1$; $e \leq_m a, b, c, d, f, g, 1$; $f \leq_m 1$; $g \leq_m f, 1$. Hence, the bounded poset $(A_9, \leq_m, 0, 1)$ is represented by the Hasse diagram from Figure 12.16, and it is **not a lattice**.

**Example 2.**

By Mace4 program, we found that the algebra

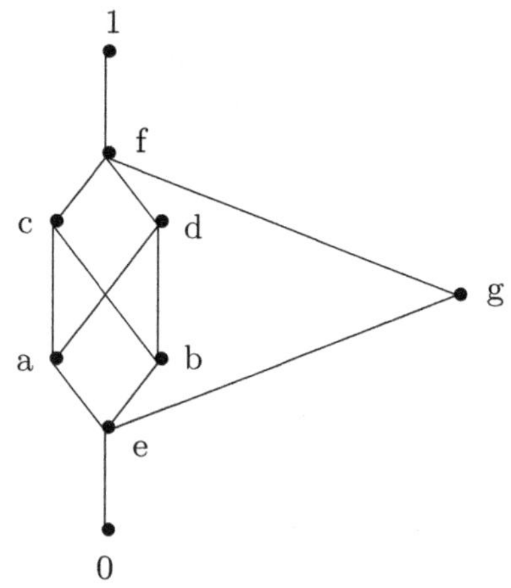


Figure 12.16: The bounded poset $(A_9, \leq_m, 0, 1)$, which is not a lattice

$\mathcal{A}^L = (A_{10} = \{0, a, b, c, d, e, f, g, h, 1\}, \odot, ^-, 1)$, with the following tables of $\odot$ and $^-$, is a proper m-BCK algebra, not verifying (m-Pabs-i) for $(e, 0)$, (G) for $a$, (m-Pimpl) for $(a, 0)$, (Pqmv) for $(a, e, 0)$, (Pom) for $(a, b)$, (Pmv) for $(a, e)$, $(\Delta_m)$ for $(a, g)$, (m-Pdis) for $(a, a, a)$, (prel$_m$) for $(a, b)$, (WNM$_m$) and (aWNM$_m$ for $(a, c)$.

| $\odot$ | 0 | a | b | c | d | e | f | g | h | 1 |
|---|---|---|---|---|---|---|---|---|---|---|
| 0 | 0 | 0 | 0 | 0 | 0 | 0 | 0 | 0 | 0 | 0 |
| a | 0 | 0 | e | e | e | 0 | e | 0 | 0 | a |
| b | 0 | e | 0 | e | e | 0 | e | 0 | 0 | b |
| c | 0 | e | e | e | e | 0 | e | 0 | e | c |
| d | 0 | e | e | e | e | 0 | e | e | 0 | d |
| e | 0 | 0 | 0 | 0 | 0 | 0 | 0 | 0 | 0 | e |
| f | 0 | e | e | e | e | 0 | e | e | e | f |
| g | 0 | 0 | 0 | 0 | e | 0 | e | 0 | 0 | g |
| h | 0 | 0 | 0 | e | 0 | 0 | e | 0 | 0 | h |
| 1 | 0 | a | b | c | d | e | f | g | h | 1 |

and

| $x$ | $x^-$ |
|---|---|
| 0 | 1 |
| a | a |
| b | b |
| c | g |
| d | h |
| e | f |
| f | e |
| g | c |
| h | d |
| 1 | 0 |

.

The matrix of $\wedge_m^B$ is:

| $\wedge_m^B$ | 0 | a | b | c | d | e | f | g | h | 1 |
|---|---|---|---|---|---|---|---|---|---|---|
| 0 | 0 | 0 | 0 | 0 | 0 | 0 | 0 | 0 | 0 | 0 |
| a | 0 | a | e | a | a | e | a | e | e | a |
| b | 0 | e | b | b | b | e | b | e | e | b |
| c | 0 | e | e | c | e | e | c | e | e | c |
| d | 0 | e | e | e | d | e | d | e | e | d |
| e | 0 | e | e | e | e | e | e | e | e | e |
| f | 0 | e | e | e | e | e | f | e | e | f |
| g | 0 | g | g | g | g | e | g | g | e | g |
| h | 0 | h | h | h | h | e | h | e | h | h |
| 1 | 0 | a | b | c | d | e | f | g | h | 1 |

.

Note that $a \leq_m c, d, f, 1$; $b \leq_m c, d, f, 1$; $c \leq_m f, 1$; $d \leq_m f, 1$; $e \leq_m a, b, c, d, f, g, h, 1$; $f \leq_m 1$; $g \leq_m a, b, c, d, f, 1$; $h \leq_m a, b, c, d, f, 1$. Hence, the bounded poset $(A_{10}, \leq_m, 0, 1)$ is reprezented by the Hasse diagram from Figure 12.17, and it is **not a lattice**.

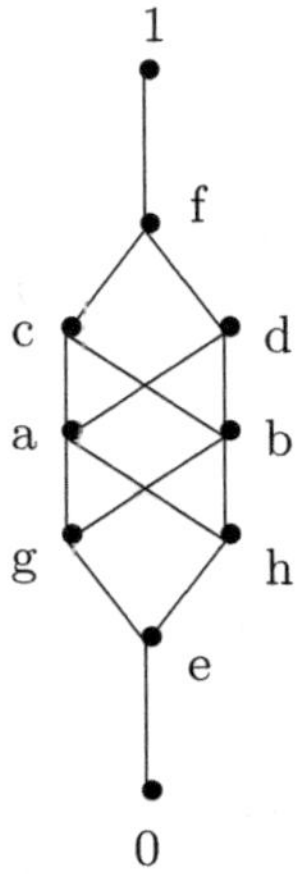


Figure 12.17: The bounded poset $(A_{10}, \leq_m, 0, 1)$, which is not a lattice

# Chapter 13

# Two generalizations of orthomodular lattices

In this chapter, we continue the results from Chapters 10, 11, 12 in the "world" of involutive *m-M algebras*, of the form $(A, \odot, ^-, 1)$, with $1^- = 0$, 1 *being the last element.* In the context of quantum-MV algebras, we study the orthomodular lattices and we introduce and study two generalizations of them: the orthomodular softlattices and the orthomodular widelattices - in connection with the two generalizations of ortholattices: the orthosoftlattices and the orthowidelattices, introduced in Chapter 10.

In Section 13.1, we study in some details the orthomodular lattices (**OML**). We establish connections between OMLs and QMV, OM, pre-MV, metha-MV algebras and ortholattices (**OL**). We prove that any OML is a QMV algebra. We introduce the new notion of *modular algebra* and we prove that modular algebras coincide with modular ortholattices. We prove that transitive OMLs coincide with Boolean algebras.

In Section 13.2, based on the two generalizations of ortholattices (**OL**): **OSL** and **OWL**, introduced in Chapter 10, we introduce and study two corresponding generalizations of **OML**: the orthomodular softlattices (**OMSL**) and the orthomodular widelattices (**OMWL**). We establish connections between **OML**/**OMWL** and **OML** and **QMV**, **OM**, **pre-MV**, **MMV** and **OL**, **OSL**/**OWL**. We prove that the **OMSL** coincide with **OML** and that **OMSL** $\subset$ **OMWL**. The core of the chapter is Proposition 13.2.26, saying that the OMWLs are a (proper) subclass of QMV algebras. Hence, the transitive OMWLs are a proper subclass of transitive QMV algebras.

In Section 13.3, we present 23 examples of the various algebras discussed herein.

The content of this chapter is taken mainly from [120].

## 13.1 Orthomodular lattices: OML

**Definition 13.1.1** An *orthomodular lattice* is an ortholattice $(A, \wedge, \vee, ^-, 0, 1)$ which satisfies the *orthomodular law*: for all $x, y \in A$,
(OML) $\qquad x \leq y \Longrightarrow x \vee (x^- \wedge y) = y.$

Note that property (OML) is not an identity, but there are many identities equivalent to (OML) within the class of ortholattices [166], as for example:

**Proposition 13.1.2** *([166], Corollary 4.10.3) The following identity characterizes orthomodular lattices among ortholattices:*
*(Wom)* $\qquad (x \wedge y) \vee ((x \wedge y)^- \wedge x) = x.$

Note that orthomodular lattices were originally defined as left-algebras.
The dual of (Wom) is:
(Vom) $\qquad (x \vee y) \wedge ((x \vee y)^- \vee x) = x,$
where 'W' comes from 'wedge' (the LaTeX command for the meet, $\wedge$) and 'V' comes from 'vee' (the LaTeX command for the join, $\vee$).

Hence, we consider here the following definition [166].

**Definition 13.1.3** (Definition 1) (The dual one is omitted)
A *left-orthomodular lattice* or an *orthomodular left-lattice*, or a *left-OML* for short, is a left-OL $\mathcal{A}^L = (A^L, \wedge, \vee, ^-, 0, 1)$ verifying: for all $x, y \in A^L$,
(Wom) $\qquad (x \wedge y) \vee ((x \wedge y)^- \wedge x) = x.$

Denote by **OML** the class of all left-OMLs .
Following the equivalent Definition 2 of a left-OL (see Definitions 9.2.17), we obtain immediately the equivalent definition:

**Definition 13.1.4** (Definition 2) (The dual one is omitted)
A *left-orthomodular lattice* (*left-OML*) is an involutive left-m-BE algebra $\mathcal{A}^L = (A^L, \odot, ^-, 1)$ verifying (m-Pimpl) and (Pom), i.e.

$$\mathbf{OML} = \mathbf{m} - \mathbf{BE_{(DN)}} + (m - Pimpl) + (Pom) = \mathbf{OL} \cap \mathbf{OM}. \tag{13.1}$$

Further, we shall work with Definition 2 of left-OMLs. Hence, we have the connections from Figure 13.1.
Recall ([45], Corollary 2.3.13) that:

$$\mathbf{OML} \subset \mathbf{QMV}, \tag{13.2}$$

the inclusion being strict, since there are examples of QMV algebras not verifying (m-Pimpl) - see Example 13.3.4.

**Proposition 13.1.5** *Let* $\mathcal{A}^L = (A^L, \odot, ^-, 1)$ *be a left-OML. We have the equivalence:*

$$(x \odot y = x \overset{def.}{\Longleftrightarrow})\ x \leq_m^P y \Longleftrightarrow x \leq_m^M y\ (\overset{def.}{\Longleftrightarrow} x \wedge_m^M y = x).$$

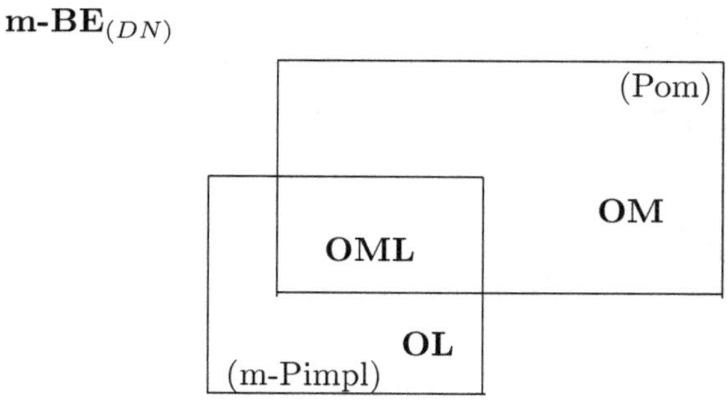


Figure 13.1: Resuming connections between **OL**, **OML** and **OM**

**Proof.** Suppose $x \leq_m^P y$, i.e. $x \odot y = x$. Then,
$x \wedge_m^M y \overset{(6.32)}{=} (x^- \odot y)^- \odot y = ((x \odot y)^- \odot y)^- \odot y$
$\overset{(DN)}{=} (((x^-)^- \odot y)^- \odot y)^- \odot y \overset{(6.32)}{=} (x^- \wedge_m^M y)^- \odot y$
$\overset{(6.39)}{=} ((x^-)^- \vee_m^M y^-)^- \odot y \overset{(m-Wcomm),(DN)}{=} y \odot (x \vee_m^M y^-)$
$\overset{(11.1)}{=} y \odot x = x \odot y = x$, since **OML** $\subset$ **OM**.
Conversely, suppose $x \leq_m^M y$, i.e. $x \wedge_m^M y = x$, i.e. $(x^- \odot y)^- \odot y = x$. Then,
$x \odot y = ((x^- \odot y)^- \odot y) \odot y \overset{(m-Wass)}{=} (x^- \odot y)^- \odot (y \odot y)$
$\overset{(G)}{=} (x^- \odot y)^- \odot y \overset{(6.32)}{=} x \wedge_m^M y = x$, since (m-Pimpl) implies (G), by Proposition 10.1.15. □

### 13.1.1 Connections between OML, PreMV, QMV, MMV, OM, OL

• **OML** + (Pmv) (Connections between **OML** and **PreMV**)

We establish the connections between the OMLs and the pre-MV algebras verifying (m-Pimpl).

**Proposition 13.1.6** *(See Proposition 13.2.3)*
*Let* $\mathcal{A}^L = (A^L, \odot, ^-, 1)$ *be an involutive left-m-BE algebra. Then,*

$$(Pom) + (m - Pimpl) \Longrightarrow (Pmv).$$

**Proof.** Since (m-Pimpl) ($[(x \odot y^-)^- \odot x^-]^- = x$) is equivalent to $(x \odot y^-) \oplus x = x$, hence (by taking $X := x^-$) to $(x^- \odot y^-) \oplus x^- = x^-$, we obtain:
$x \odot (x^- \vee_m^M y) = x \odot ((x^- \odot y^-)^- \odot y^-)^- = (x^- \oplus ((x^- \odot y^-)^- \odot y^-)^=)^-$
$\overset{(DN)}{=} (x^- \oplus ((x^- \odot y^-)^- \odot y^-))^- \overset{(m-Pimpl),(Scomm)}{=} ((x^- \oplus (x^- \odot y^-)) \oplus ((x^- \odot y^-)^- \odot y^-))^-$
$\overset{(Sass),(Pcomm)}{=} (x^- \oplus ((y^- \odot x^-) \oplus ((y^- \odot x^-)^- \odot y^-)))^- \overset{(Pom)}{=} (x^- \oplus y^-)^- = x \odot y$. □

Note that Proposition 13.1.6 says: **OML** $\subset$ **PreMV** - which follows by (13.2).
The following converse of Proposition 13.1.6 also holds:

**Proposition 13.1.7** *Let* $\mathcal{A}^L = (A^L, \odot, ^-, 1)$ *be an involutive m-BE algebra, Then,*

$$(Pmv) \ + \ (m - Pimpl) \ \Longrightarrow \ (Pom).$$

**Proof. (Following a proof by** *Prover9* **of length 25, lasting 0.11 seconds)**
We know that (m-Pimpl) implies (G), and (G) implies:
(a) $x \odot (y \odot x) = y \odot x$;
indeed, $x \odot (y \odot x) \overset{(Pcomm),(Pass)}{=} y \odot (x \odot x) \overset{(G)}{=} y \odot x$.
Then, (m-Pimpl) ($((x \odot y^-)^- \odot x^-)^- = x$) implies, taking $Y := y^-$ and using (DN) and (Pcomm):
(b) $(x^- \odot (y \odot x)^-)^- = x$ and
(b') $x^- \odot (x \odot y)^- = x^-$.
On the other hand, (Pmv) ($x \odot (y^- \odot (x^- \odot y^-)^-)^- = x \odot y$) implies, by (Pcomm):
(c) $x \odot (y^- \odot (y^- \odot x^-)^-)^- = x \odot y$.
Now, by (a), (b), (c), we obtain:
(d) $x \odot (x \odot (y \odot x)^-)^- = y \odot x$;
indeed, in (c), take $Y := y \odot x$ and $X := x$, to obtain:
(x) $x \odot ((y \odot x)^- \odot ((y \odot x)^- \odot x^-)^-)^- = x \odot (y \odot x) \overset{(a)}{=} y \odot x$;
since in (x), $((y \odot x)^- \odot x^-)^- \overset{(Pcomm)}{=} (x^- \odot (y \odot x)^-)^- \overset{(b)}{=} x$,
it follows that (x) becomes:
(x') $x \odot ((y \odot x)^- \odot x)^- = y \odot x$, i.e. (d) holds, by (Pcomm).
Now, by (b'), (d), we obtain:
(e) $(x \odot y)^- \odot (x \odot (x \odot y)^-)^- = x^-$;
indeed, in (d), take $X := (x \odot y)^-$ and $Y := x^-$ to obtain:
(y) $(x \odot y)^- \odot ((x \odot y)^- \odot (x^- \odot (x \odot y)^-)^-)^- = x^- \odot (x \odot y)^-$;
but, in (y), $x^- \odot (x \odot y)^- \overset{(b')}{=} x^-$, hence (y) becomes:
(y') $(x \odot y)^- \odot ((x \odot y)^- \odot x^{=})^- = x^-$, which becomes, by (DN):
(y") $(x \odot y)^- \odot ((x \odot y)^- \odot x)^- = x^-$, which becomes, by (DN) and (Pcomm):
$((x \odot y)^- \odot (x \odot (x \odot y)^-)^-)^- = x$, that is (Pom). □

Note that Proposition 13.1.7 says: **PreMV** $\cap$ **OL** $\subset$ **OM**.
By Propositions 13.1.6 and 13.1.7, we obtain:

**Theorem 13.1.8** *Let* $\mathcal{A}^L = (A^L, \odot, ^-, 1)$ *be an involutive m-BE algebra, Then,*

$$(m - Pimpl) \ \Longrightarrow \ ((Pom) \ \Longleftrightarrow \ (Pmv)) \quad or$$

$$(m - Pimpl) \ + \ (Pom) \ \Longleftrightarrow \ (Pmv) \ + \ (m - Pimpl),$$

*i.e. OMLs coincide with pre-MV algebras verifying (m-Pimpl).*

Hence, Theorem 13.1.8 says:

(13.3) $\quad$ **OML** $=$ **PreMV** $+ \ (m - Pimpl) \ =$ **PreMV** $\cap$ **OL**.

See Example 13.3.10 of involutive m-BE algebra verifying (m-Pimpl) and not verifying (Pom) and (Pmv).

- **OML** + (Pqmv) (Connections between **OML** and **QMV**)

We establish now the connection between the OMLs and the QMV algebras verifying (m-Pimpl).

**Proposition 13.1.9** *Let $\mathcal{A}^L = (A^L, \odot, ^-, 1)$ be a left-OML. Then, $\mathcal{A}^L$ is a left-QMV algebra verifying (m-Pimpl).*
*(i.e. in an involutive m-BE algebra, (Pom) + (m-Pimpl) $\Longrightarrow$ (Pqmv).)*

**Proof.** Since $\mathcal{A}^L$ is a left-OML, it is an involutive m-BE algebra verifying (m-Pimpl) and (Pom) (Definition 2). By Theorem 13.1.6, it verifies (Pmv) also. Hence, $\mathcal{A}^L$ is a left-QMV algebra verifying (m-Pimpl). □

Note that Proposition 13.1.9 says: **OML** $\subset$ **QMV**, which is (13.2). Note also that Proposition 13.1.6 follows from Proposition 13.1.9, since (Pqmv) $\Longrightarrow$ (Pmv).
The following converse of Proposition 13.1.9 also holds.

**Proposition 13.1.10** *Let $\mathcal{A}^L = (A^L, \odot, ^-, 1)$ be a left-QMV algebra verifying (m-Pimpl). Then, $\mathcal{A}^L$ is a left-OML.*
*(i.e. in an involutive m-BE algebra, (Pqmv) + (m-Pimpl) $\Longrightarrow$ (Pom).)*

**Proof.** Since $\mathcal{A}^L$ is a left-QMV algebra verifying (m-Pimpl), it is an involutive m-BE algebra verifying (Pqmv) (hence (Pom), (Pmv) ) and (m-Pimpl). Hence, $\mathcal{A}^L$ is an involutive m-BE algebra verifying (m-Pimpl) and (Pom), i.e. it is a left-OML. □

Note that Proposition 13.1.10 says: **QMV** $\cap$ **OL** $\subset$ **OM**. Note also that Proposition 13.1.10 follows from Proposition 13.1.7.
By Propositions 13.1.9, 13.1.10, we obtain:

**Theorem 13.1.11** *Let $\mathcal{A}^L = (A^L, \odot, ^-, 1)$ be an involutive m-BE algebra. Then,*

$$(m-Pimpl) \Longrightarrow ((Pom) \Leftrightarrow (Pqmv)) \quad or$$

$$(m-Pimpl) + (Pom) \Longleftrightarrow (Pqmv) + (m-Pimpl),$$

*i.e. orthomodular lattices coincide with QMV algebras verifying (m-Pimpl).*

Hence, Theorem 13.1.11 says:

$$\text{(13.4)} \qquad \mathbf{OML} = \mathbf{QMV} + (m-Pimpl) = \mathbf{QMV} \cap \mathbf{OL}.$$

By previous results (13.1), (13.2), (13.3), (13.4), we obtain the connections from Figure 13.2.

- **OML** + ($\Delta_m$) (Connections between **OML** and **MMV**)

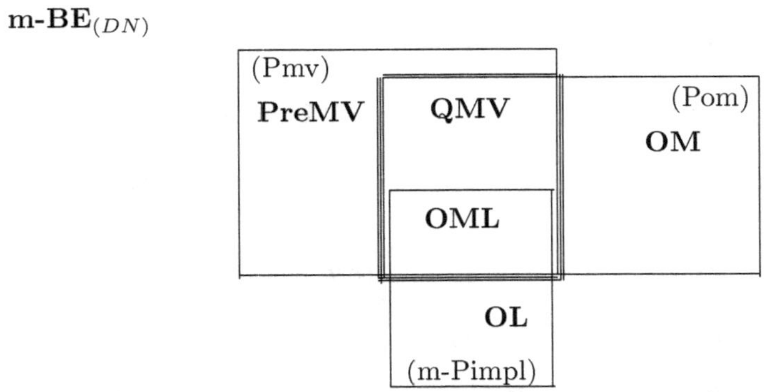


Figure 13.2: Resuming connections between **QMV**, **PreMV**, **OM**, **OL** and **OML**

**Proposition 13.1.12** *Let $\mathcal{A}^L = (A^L, \odot, ^-, 1)$ be an involutive m-BE algebra. Then,*

$$(Pom) \; + \; (m - Pimpl) \; \Longrightarrow \; (\Delta_m).$$

**Proof.** By Proposition 13.1.6, (Pom) + (m-Pimpl) imply (Pmv) and (Pmv) implies $(\Delta_m)$. □

Note that Proposition 13.1.12 says: **OML** $\subset$ **MMV**. which follows also by (13.2). Note also that Proposition 13.1.9 follows also from Proposition 13.1.12, since (Pom) + $(\Delta_m)$ imply (Pqmv) and that Proposition 13.1.12 follows from Proposition 13.1.9, since (Pqmv) implies $(\Delta_m)$.

**Remark 13.1.13** *The following converse of Proposition 13.1.12 ($(\Delta_m)$ + (m-Pimpl) $\Longrightarrow$ (Pom)) does not hold: there are examples of involutive m-BE algebras verifying $(\Delta_m)$ and (m-Pimpl) and not verifying (Pom) - see Example 13.3.10.*

By previous Remark, from the connections from Figure 13.2, we obtain the connections from Figure 13.3.

### 13.1.2 The transitive and/or antisymmetric case

#### 13.1.2.1 Antisymmetric orthomodular lattices: aOML = Boole

Denote by **aOML** the class of all antisymmetric left-OMLs. We prove that **aOML** does not exist as proper class:

**Theorem 13.1.14** *We have:*

$$\mathbf{aOML} = \mathbf{Boole}.$$

**Proof.** **aOML** = **m-BE**$_{(DN)}$ + (m-Pimpl) + (Pom) + (m-An) = **OL** + (Pom) + (m-An) = **Boole** + (Pom) = **Boole**, by Theorem 9.2.19. □

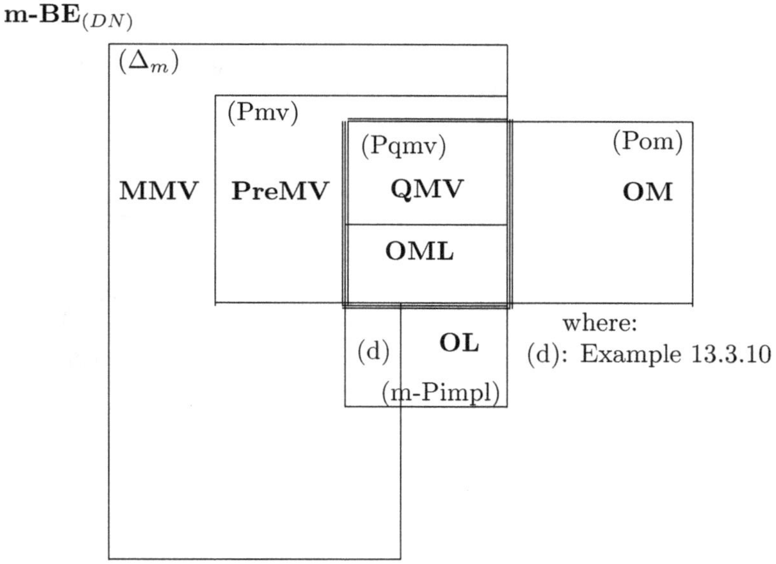


Figure 13.3: Resuming connections between **QMV**, **PreMV**, **MMV**, **OM**, **OL** and **OML**

**Remark 13.1.15** *We have:*

$$\mathbf{OML} \subset \mathbf{QMV} \quad \textit{and} \quad \mathbf{aOML} = \mathbf{Boole} \subset \mathbf{aQMV} = \mathbf{MV}.$$

#### 13.1.2.2 Transitive orthomodular lattices: tOML = Boole

Denote by **tOML** the class of all transitive left-OMLs. We shall prove that **tOML** does not exist as a proper class (**tOML** = **Boole**, by Theorem 13.1.17).

**Theorem 13.1.16** *Let* $\mathcal{A}^L = (A^L, \odot, ^-, 1)$ *be an involutive left-m-BE algebra. Then,*

$$(Pom) + (m - Pimpl) + (m - BB) \Longrightarrow (m - An).$$

**Proof. (By** *Prover9*, **in 0.03 seconds, the length of the proof being 32)**

Suppose: (i) $c_1 \odot c_2^- = 0$ and (j): $c_2 \odot c_1^- = 0$; we have to prove that $c_1 = c_2$.

First, (Pom): $(x \odot y) \oplus ((x \odot y)^- \odot x) = x$ means $[(x \odot y)^- \odot ((x \odot y)^- \odot x)^-]^- = x$, hence by (Pcomm), (DN):

$$(x \odot y)^- \odot (x \odot (x \odot y)^-)^- = x^-. \tag{13.5}$$

Second, (m-BB): $[(x \odot y)^- \odot (z \odot y)] \odot (z \odot x^-)^- = 0$, means, by (Pass):

$$(x \odot y)^- \odot [z \odot (y \odot (z \odot x^-)^-)] = 0. \tag{13.6}$$

Take $x := c_2^-$, $y := c_1$, $z := x$ in (13.6) to obtain:
$(c_2^- \odot c_1)^- \odot [x \odot (c_1 \odot (x \odot c_2)^-)] = 0$, hence by (i), (Neg0-1), (PU):
$x \odot (c_1 \odot (x \odot c_2)^-) = 0$, hence, by (Pass), (Pcomm):

$$c_1 \odot (x \odot (c_2 \odot x)^-) = 0. \tag{13.7}$$

Since (p-Pimpl) implies (G), then (G) ($x \odot x = x$) implies $x \odot y = (x \odot x) \odot y \overset{(Pass)}{=} x \odot (x \odot y)$, hence we have:

$$x \odot (x \odot y) = x \odot y. \tag{13.8}$$

Take now $x := c_1$ in (13.7) to obtain: $c_1 \odot (c_1 \odot (c_2 \odot c_1)^-) = 0$, hence by (13.8) and (Pcomm):

$$c_1 \odot (c_1 \odot c_2)^- = 0. \tag{13.9}$$

Take now $x := c_1$, $y := c_2$ in (13.5) to obtain:
$(c_1 \odot c_2)^- \odot (c_1 \odot (c_1 \odot c_2)^-)^- = c_1^-$; then, by (13.9), (Neg0-1), (PU), we obtain:

$$(c_1 \odot c_2)^- = c_1^-, \quad \textit{hence} \tag{13.10}$$

$$c_1 \odot c_2 = c_1. \tag{13.11}$$

Now, from (m-Pimpl): $[(x \odot y^-)^- \odot x^-]^- = x$, we obtain by (Pcomm) and for $y := y^-$:

$$[x^- \odot (y \odot x)^-]^- = x. \tag{13.12}$$

Take now $x := c_2$, $y := c_1$ in (13.12) to obtain: $[c_2^- \odot (c_1 \odot c_2)^-]^- = c_2$, hence, by (13.10), $[c_2^- \odot c_1^-]^- = c_2$, hence, by (Pcomm):

$$(c_1^- \odot c_2^-)^- = c_2, \quad \textit{hence} \tag{13.13}$$

$$c_1^- \odot c_2^- = c_2^-. \tag{13.14}$$

Finally, take $x := c_1^-$, $y := c_2^-$ in (13.5) to obtain: $(c_1^- \odot c_2^-)^- \odot (c_1^- \odot (c_1^- \odot c_2^-)^-)^- = c_1$; hence, by (13.13), we obtain:
$c_2 \odot (c_1^- \odot c_2)^- = c_1$; hence, by (j), (Pcomm), (Neg0-1) and (PU), we obtain: $c_2 = c_1$. □

Note that Theorem 13.1.16 says: **tOML** $\subset$ **m-aBE**$_{(DN)}$. Hence, **tOML** $\subset$ **taOML**. But **taOML** = **aOML** + (m-Tr) = **Boole** + (m-Tr) = **Boole**, by Theorem 13.1.14. It follows that **tOML** = **Boole**. Thus, we have proved:

**Theorem 13.1.17** *We have:*

$$\mathbf{tOML} = \mathbf{Boole}. \tag{13.15}$$

### 13.1.2.3 The transitive and antisymmetric case

If we make the following table:

| No. | (m-Tr) | (m-Pimpl) | (Pqmv) | Type of m-BE$_{(DN)}$ algebra |
|---|---|---|---|---|
| (1) | 0 | 0 | 0 | proper m-BE$_{(DN)}$ |
| (2) | 0 | 0 | 1 | proper QMV |
| (3) | 0 | 1 | 0 | proper OL |
| (4) | 0 | 1 | 1 | proper OML |
| (5) | 1 | 0 | 0 | proper m-pre-BCK$_{(DN)}$ |
| (6) | 1 | 0 | 1 | proper tQMV |
| (7) | 1 | 1 | 0 | proper tOL |
| (8) | 1 | 1 | 1 | tOML = aOML = Boole |

,

then, we obtain the resuming connections from Figures 13.4 and 13.5.

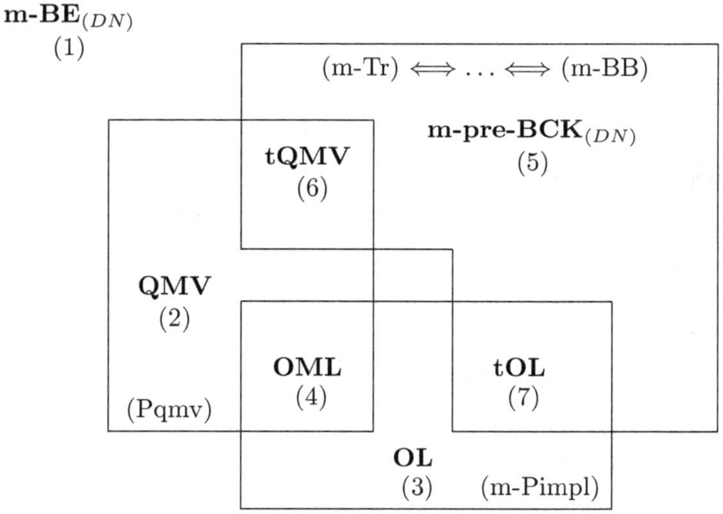


Figure 13.4: Resuming connections in **m-BE**$_{(DN)}$

### 13.1.2.4 The transitive case: tOL $\subset$ tMMV

**Theorem 13.1.18** *Let $\mathcal{A}^L = (A^L, \odot, ^-, 1)$ be an involutive left-m-BE algebra. Then,*

$$(m-Pimpl) + (m-BB) \implies (\Delta_m).$$

**Proof.** Since (m-Pimpl) implies (m-Pabs-i) and since, by Theorem 10.3.13, (m-Pabs-i) + (m-BB) imply ($\Delta_m$), it follows that (m-Pimpl) + (m-BB) imply ($\Delta_m$). □

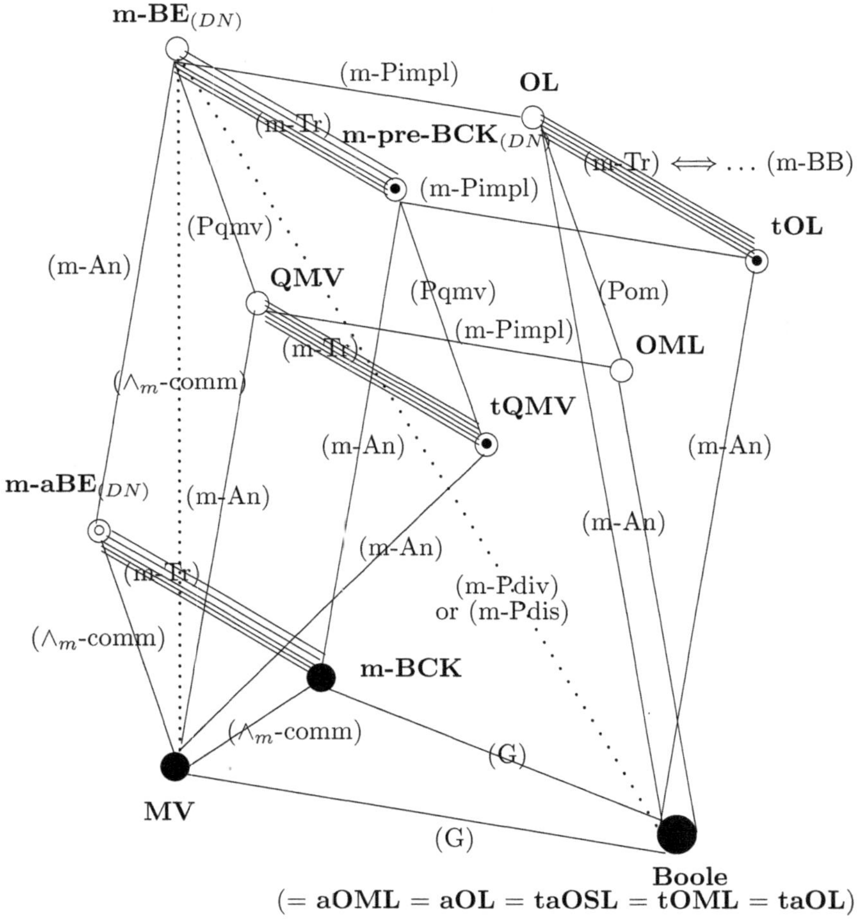


Figure 13.5: Resuming connections in this section

Note that Theorem 13.1.18 says: **tOL** $\subset$ **MMV**, hence **tOL** $\subset$ **tMMV**, since (m-BB) $\Longleftrightarrow$ (m-Tr).

By Theorems 13.1.17 and 13.1.18, from the connections from Figure 13.3, we obtain the connections from Figure 13.6.

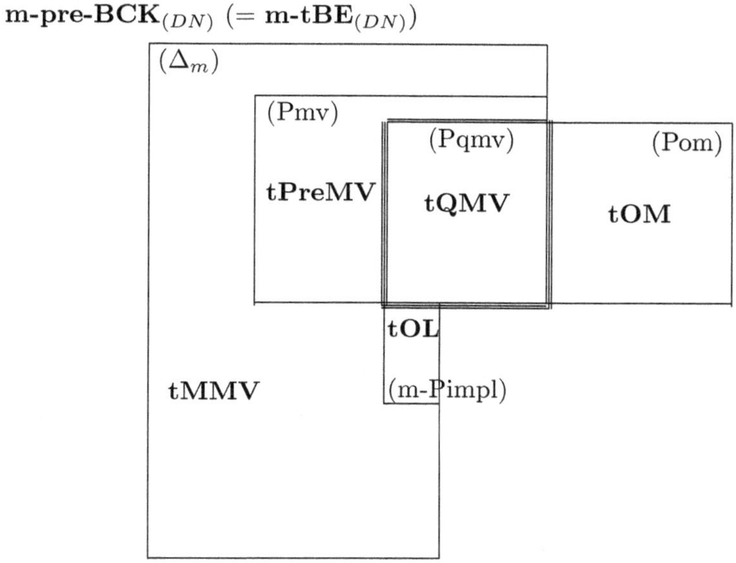


Figure 13.6: Resuming connections between **tQMV**, **tPreMV**, **tMMV**, **tOM** and **tOL**

### 13.1.3 Modular algebras: MOD $\subset$ OML

Recall the following definitions [166]:

(i) A latice $(L, \wedge, \vee)$ is *modular*, if for all $x, y, z \in L$,

(Wmod) $$x \wedge (y \vee (x \wedge z)) = (x \wedge y) \vee (x \wedge z).$$

(i') The dual latice $(L, \vee, \wedge)$ is *modular*, if for all $x, y, z \in L$,

(Vmod) $$x \vee (y \wedge (x \vee z)) = (x \vee y) \wedge (x \vee z).$$

**Definition 13.1.19** (Definition 1) (The dual case is omitted) [166]

A *modular left-ortholattice* is a left-OL $\mathcal{A}^L = (A^L, \wedge, \vee, ^-, 0, 1)$ whose lattice $(A^L, \wedge, \vee)$ is modular.

We shall denote by **MODOL** the class of all modular left-ortholattices.

Recall also [166] that any modular ortholattice is an orthomodular lattice, i.e.

(13.16) $$\mathbf{MODOL} \subset \mathbf{OML}.$$

Following the equivalent definition of OLs, we obtain the following equivalent definition.

**Definition 13.1.20** (Definition 2) (The dual one is omitted)

A *modular left-ortholattice* is an involutive left-m-BE algebra $(A^L, \odot, ^-, 1)$ verifying (m-Pimpl) and (Pmod), where: for all $x, y, z \in A^L$,

(Pmod) $\quad x \odot (y \oplus (x \odot z)) = (x \odot y) \oplus (x \odot z)$, i.e.

**MODOL** = $\mathbf{m\text{-}BE}_{(DN)}$ + (m-Pimpl) + (Pmod) = **OL** + (Pmod).

Then, we introduce the following notion:

**Definitions 13.1.21**

(i) A *left-modular algebra* or a *modular left-algebra*, or a *left-MOD algebra* for short, is an involutive left-m-BE algebra $\mathcal{A}^L = (A^L, \odot, ^- = ^{-^L}, 1)$ verifying: for all $x, y, z \in A^L$,

(Pmod) $\quad x \odot (y \oplus (x \odot z)) = (x \odot y) \oplus (x \odot z)$.

(i') Dually, a *right-modular algebra* or a *modular right-algebra*, or a *right-MOD algebra* for short, is an involutive right-m-BE algebra $\mathcal{A}^R = (A^L, \oplus, ^- = ^{-^R}, 0)$ verifying: for all $x, y, z \in A^R$,

(Smod) $\quad x \oplus (y \odot (x \oplus z)) = (x \oplus y) \odot (x \oplus z)$.

We shall denote by **MOD** the class of all left-MOD algebras and by $\mathbf{MOD}^R$ the class of all right-MOD algebras. Hence, **MOD** = $\mathbf{m\text{-}BE}_{(DN)}$ + (Pmod).

Then,

(13.17) $$\mathbf{MODOL} = \mathbf{OL} + (Pmod) = \mathbf{OL} \cap \mathbf{MOD}.$$

**Proposition 13.1.22** *(The dual one is omitted)*

*Let $\mathcal{A}^L = (A^L, \odot, ^-, 1)$ be an involutive left-m-BE algebra. Then,*

$$(Pmod) \Longrightarrow (Pom).$$

**Proof.** (**Following a proof by Prover9 of length 14, lasting 0,05 seconds**)

(Pmod), i.e. $x \odot (y \oplus (x \odot z)) = (x \odot y) \oplus (x \odot z)$, is equivalent with

(a) $\quad x \odot (y^- \odot (x \odot z)^-)^- = ((x \odot y)^- \odot (x \odot z)^-)^-$, i.e. with:

(a') $\quad ((x \odot y)^- \odot (x \odot z)^-)^- = x \odot (y^- \odot (x \odot z)^-)^-$.

Then,

$((x \odot y)^- \odot (x \odot z)^-)^- \overset{(Pcomm)}{=} ((x \odot z)^- \odot (x \odot y)^-)^- \overset{(a')}{=} x \odot (z^- \odot (x \odot y)^-)^-$,
hence we obtain:

(b) $\quad x \odot (y^- \odot (x \odot z)^-)^- = x \odot (z^- \odot (x \odot y)^-)^-$.

Take now $Z := (x \odot y)^-$ in (b) to obtain:

$x \odot (y^- \odot (x \odot (x \odot y)^-)^-)^- = x \odot ((x \odot y)^= \odot (x \odot y)^-)^-$
$\overset{(DN)}{=} x \odot ((x \odot y) \odot (x \odot y)^-)^- \overset{(m-Re)}{=} x \odot 0^- \overset{(Neg0-1)}{=} x \odot 1 \overset{(PU)}{=} x$;
hence, we have:

(c) $\quad x \odot (y^- \odot (x \odot (x \odot y)^-)^-)^- = x$.

Now, (Pom), i.e. $(x \odot y) \oplus ((x \odot y)^- \odot x) = x$, is equivalent with:

(d) $\quad ((x \odot y)^- \odot ((x \odot y)^- \odot x)^-)^- = x$, which by (Pcomm) means:

(d') $\quad ((x \odot y)^- \odot (x \odot (x \odot y)^-)-)^- = x$;

hence, we must prove that (d') holds.

Indeed, $((x \odot y)^- \odot (x \odot (x \odot y)^-)-)^- \stackrel{(a')}{=} x \odot (y^- \odot (x \odot (x \odot y)^-)^-)^- \stackrel{(c)}{=} x$, hence (d') holds, i.e. (Pom) holds. $\square$

Note that Proposition 13.1.22 says: **MOD** $\subset$ **OM**.

**Proposition 13.1.23** *(The dual one is omitted)*
*Let* $\mathcal{A}^L = (A^L, \odot, ^-, 1)$ *be an involutive left-m-BE algebra. Then,*

$$(Pmod) \implies (m - Pimpl).$$

**Proof. (Following a proof by Prover9 of length 16, lasting 0,00 seconds)**
(Pmod), i.e. $x \odot (y \oplus (x \odot z)) = (x \odot y) \oplus (x \odot z)$, is equivalent with
(a) $x \odot (y^- \odot (x \odot z)^-)^- = ((x \odot y)^- \odot (x \odot z)^-)^-$, i.e. with:
(a') $((x \odot y)^- \odot (x \odot z)^-)^- = x \odot (y^- \odot (x \odot z)^-)^-$.
Now, take in (a') $Y := 1$ and $Z := y$ to obtain, by (PU), (Neg1-0), (Pcomm), (m-La):
$(x^- \odot (x \odot y)^-)^- = ((x \odot 1)^- \odot (x \odot z)^-)^- \stackrel{(a')}{=} x \odot (1^- \odot (x \odot z)^-)^- = x \odot (0 \odot (x \odot z)^-)^- = x \odot 0^- = x \odot 1 = x$, hence:
(b) $(x^- \odot (x \odot y)^-)^- = x$.
Note that (m-Pimpl), i.e. $((x \odot y^-)^- \odot x^-)^- = x$, follows from (b), by (Pcomm). $\square$

Note that Proposition 13.1.23 says: **MOD** $\subset$ **OL**, hence, **MOD** $=$ **OL** $\cap$ **MOD** $\stackrel{(13.17)}{=}$ **MODOL**. Thus, we have:

(13.18) $$\mathbf{MOD} = \mathbf{MODOL}.$$

By Propositions 13.1.22, 13.1.23, we obtain obviously:

**Theorem 13.1.24** *(The dual one is omitted)*
*Let* $\mathcal{A}^L = (A^L, \odot, ^-, 1)$ *be an involutive left-m-BE algebra. Then,*

$$(Pmod) \implies (Pom) + (m - Pimpl).$$

By above Theorem 13.1.24, which says: **MOD** $\subset$ **OM** $\cap$ **OL** $=$ **OML**, by (13.1), we reobtain immediately the recalled known result from (13.16):
**MODOL** ($=$ **MOD**) $\subset$ **OML** ( $\subset$ **OL**).

Recall [166] that inclusion is strict: see Example 13.3.2 of OML that is not MOD algebra and see Example 13.3.1 of proper MOD algebra.

Since **OML** $\subset$ **QMV**, by (13.1), we obtain:

(13.19) $$\mathbf{MOD}\ (= \mathbf{MODOL}) \subset \mathbf{OML} \subset \mathbf{QMV}.$$

Hence, we have:

(13.20) $$\mathbf{aMOD} = \mathbf{aOML} = \mathbf{Boole} \subset \mathbf{aQMV} = \mathbf{MV} \quad and$$

(13.21) $$\mathbf{tMOD} = \mathbf{tOML} = \mathbf{Boole} \subset \mathbf{tQMV}.$$

**Remark 13.1.25** *Recall that any OL that is distributive is a Boolean algebra, by definitions. Consequently, any OML that is distributive is a Boolean algebra and any modular algebra that is distributive is a Boolean algebra.*

**Remark 13.1.26** *In a left-OML $\mathcal{A}^L = (A^L, \odot, ^-, 1)$ (Definition 2):*
*- the initial binary relation, $\leq_m$ ($x \leq_m y \Longleftrightarrow x \odot y^- = 0$), is only* **reflexive**;
*- the binary relation $\leq_m^M$ ($x \leq_m^M y \Longleftrightarrow x \wedge_m^M y = x$) is an* **order**, *by Corollary 12.1.7, but not a lattice order with respect to $\wedge_m^M$, $\vee_m^M$, since $\wedge_m^M$ is not commutative;*
*- the binary relation $\leq_m^P$ ($x \leq_m^P y \Longleftrightarrow x \odot y = x$) is a* **lattice order**, *with respect to $\wedge = \odot$, $\vee = \oplus$, denoted $\leq_m^O$, by Proposition 9.1.5;*
*- we have the equivalence $\leq_m^O \Longleftrightarrow \leq_m^M$, by Proposition 11.2.12; consequently, the tables of $\wedge$ and $\wedge_m^M$ are different, but they coincide for the comparable elements of $A^L$ (with respect to $\leq_m^O$ and $\leq_m^M$, respectively).*

## 13.2 Two generalizations of OML

Starting from the two generalizations of ortholattices (**OL**): the orthosoftlattices (**OSL**) and the orthowidelattices (**OWL**) (Definitions 10.3.1, 10.3.6 and Figure 10.3), we introduce two corresponding generalizations of orthomodular lattices (**OML**): the orthomodular softlattices and the orthomodular widelattices.

### 13.2.1 Orthomodular softlattices: OMSL

We introduce the following notion.

**Definition 13.2.1** (Definition 1) (The dual one is omitted)
A *left-orthomodular softlattice* or an *orthomodular left-softlattice*, or a *left-OMSL* for short, is a left-OSL $\mathcal{A}^L = (A^L, \wedge, \vee, ^-, 0, 1)$ verifying: for all $x, y \in A^L$,
(Wom) $\quad (x \wedge y) \vee ((x \wedge y)^- \wedge x) = x.$

Denote by **OMSL** the class of all left-OMSLs. Following the equivalent Definition 2 of a left-OSL (see Definition 10.3.3), we obtain immediately an equivalent definition:

**Definition 13.2.2** (Definition 2) (The dual one is omitted)
A *left-OMSL* is a left-OSL verifying (Pom), i.e. is an involutive left-m-BE algebra $\mathcal{A}^L = (A^L, \odot, ^-, 1)$ verifying (G) and (Pom), i.e.

$$\textbf{OMSL} = \textbf{m} - \textbf{BE}_{\textbf{(DN)}} + (G) + (Pom) = \textbf{OSL} \cap \textbf{OM}. \quad (13.22)$$

Further, we shall work with Definition 2 of left-OMSLs. Hence, we have the connections from Figure 13.7.

Denote by **tOMSL** the class of all transitive left-OMSLs. We shall prove that **OMSL** and **tOMSL** do not exist (as proper classes) (**OMSL** = **OML**, by (13.26)), hense **tOMSL** = **Boole**, by Theorem 13.1.17).

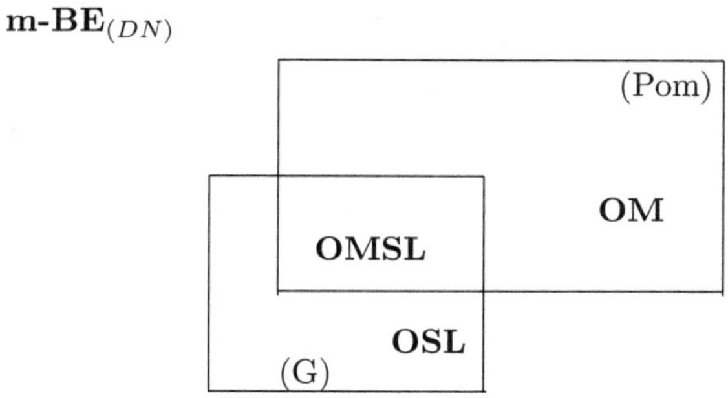


Figure 13.7: Resuming connections between **OSL**, **OMSL** and **OM**

#### 13.2.1.1 Connections between OMSL and PreMV, QMV, MMV, OM, OSL

- **OMSL** + (Pmv) (Connections between **OMSL** and **PreMV**)

We establish the connections between the OMSLs and the pre-MV algebras verifying (G).

**Proposition 13.2.3** *(See Theorem 13.1.6)*
*Let $\mathcal{A}^L = (A^L, \odot, ^-, 1)$ be an involutive left-m-BE algebra, Then,*

$$(Pom) + (G) \Longrightarrow (Pmv).$$

**Proof. (following a proof by *Prover9*, of length 24, lasting 0.36 seconds)**
First, from (Pom) ($((x \odot y)^- \odot (x \odot (x \odot y)^-)^-)^- = x$), by (DN), we obtain:
(a) $\quad (x \odot y)^- \odot (x \odot (x \odot y)^-)^- = x^-$.
Then, (a) implies:
(b) $\quad x^- \odot (x \odot y)^- = x^-$;
indeed, $x^- \odot (x \odot y)^- \overset{(Pcomm)}{=} (x \odot y)^- \odot x^-$
$\overset{(a)}{=} (x \odot y)^- \odot ((x \odot y)^- \odot (x \odot (x \odot y)^-)^-)$
$\overset{(Pass)}{=} ((x \odot y)^- \odot (x \odot y)^-) \odot (x \odot (x \odot y)^-)^-$
$\overset{(G)}{=} (x \odot y)^- \odot (x \odot (x \odot y)^-)^- \overset{(a)}{=} x^-$.
Then, (b) implies (c), by (DN):
(c) $\quad x \odot (x^- \odot y)^- = x$.
On the other hand, (a) implies (d), by interchanging $x$ with $y$:
(d) $\quad (y \odot x)^- \odot (y \odot (y \odot x)^-)^- = y^-$,
and (d) implies (e), by taking $X := x^-$ and by (Pcomm):
(e) $\quad (x^- \odot y)^- \odot (y \odot (x^- \odot y)^-)^- = y^-$.
Finally, (c) and (e) imply:
(f) $\quad x \odot y^- = x \odot (y \odot (x^- \odot y)^-)^-$;
indeed, $x \odot y^- \overset{(e)}{=} x \odot ((x^- \odot y)^- \odot (y \odot (x^- \odot y)^-)^-)$

$\overset{(Pass)}{=} (x \odot (x^- \odot y)^-) \odot (y \odot (x^- \odot y)^-)^-$
$\overset{(c)}{=} x \odot (y \odot (x^- \odot y)^-)^-$; thus, (f) holds.
By taking $Y := y^-$ in (f), we obtain, by (DN):
$x \odot y = x \odot (y^- \odot (x^- \odot y^-)^-)^-$, that is (Pmv). □

Note that Proposition 13.1.6 follows from Proposition 13.2.3, since (m-Pimpl) implies (G). Note also that Proposition 13.2.3 says: **OMSL ⊂ PreMV**.

The following converse of Proposition 13.2.3 also holds:

**Proposition 13.2.4** *(See Proposition 13.1.7)*
*Let $\mathcal{A}^L = (A^L, \odot, ^-, 1)$ be an involutive m-BE algebra. Then,*

$$(Pmv) + (G) \Longrightarrow (Pom).$$

**Proof. (following a proof by *Prover9*, of length 27, lasting 0.12 seconds)**
From (Pmv) ( $x \odot (y^- \odot (x^- \odot y^-)^-)^- = x \odot y$), by taking $Y := y^-$ and by (DN), we obtain:
(a) $x \odot (y \odot (x^- \odot y)^-)^- = x \odot y^-$.
On the other hand, from (G) ($x \odot x = x$), we obtain:
(b) $x \odot (x \odot y) = x \odot y$;
indeed, $x \odot (x \odot y) \overset{(Pass)}{=} (x \odot x) \odot y \overset{(G)}{=} x \odot y$; hence, by (Pcomm), we obtain:
(b') $x \odot (y \odot x) = y \odot x$.
Now, from (b) and (a) we obtain:
(c) $x \odot (x^- \odot y)^- = x$;
indeed, in (a), take $X := x$ and $Y := x^- \odot y$ to obtain:
(x) $x \odot ((x^- \odot y) \odot (x^- \odot (x^- \odot y))^-)^- = x \odot (x^- \odot y)^-$;
but, the part from (x): $x^- \odot (x^- \odot y) \overset{(b)}{=} x^- \odot y$, hence (x) becomes:
(x') $x \odot ((x^- \odot y) \odot (x^- \odot y)^-)^- = x \odot (x^- \odot y)^-$,
which by (m-Re) and (Neg0-1) becomes:
(x") $x \odot 1 = x \odot (x^- \odot y)^-$,
which, by (PU), becomes (c).
Now, from (c), by (Pcomm), we obtain:
(c') $x \odot (y \odot x^-)^- = x$ and
from (c), by taking $X := x^-$ we obtain:
(c") $x^- \odot (x \odot y)^- = x^-$.
Now, from (c') and (a), we obtain:
(d) $x \odot (x \odot (y \odot x)^-)^- = y \odot x$;
indeed, in (a), take $X := x$ and $Y := (y \odot x^=)^-$ to obtain:
(y) $x \odot ((y \odot x^=)^- \odot (x^- \odot (y \odot x^=)^-)^-)^- = x \odot (y \odot x^=)^=$;
but, the part from (y) $x^- \odot (y \odot x^=)^- \overset{(c')}{=} x^-$, hence (y) becomes, by (DN):
(y') $x \odot ((y \odot x)^- \odot x^=)^- = x \odot (y \odot x)$;
but (y'), by (DN) and (b') becomes:
(y") $x \odot ((y \odot x)^- \odot x)^- = y \odot x$;
and (y"), by (Pcomm), becomes (d).

Now, from (c") and (d), we obtain:
(e) $(x \odot y)^- \odot (x \odot (x \odot y)^-)^- = x^-$;
indeed, in (d), take $X := (x \odot y)^-$ and $Y := x^-$ to obtain:
(u) $(x \odot y)^- \odot ((x \odot y)^- \odot (x^- \odot (x \odot y)^-)^-)^- = x^- \odot (x \odot y)^-$;
but, the parts from (u) $x^- \odot (x \odot y)^- \overset{(c'')}{=} x^-$, hence (u) becomes:
(u') $(x \odot y)^- \odot ((x \odot y)^- \odot x^=)^- = x^-$,
which by (DN) and (Pcomm) becomes:
$(x \odot y)^- \odot (x \odot (x \odot y)^-)^- = x^-$, that is (e).

Finally, from (e), by (DN), we obtain:
$((x \odot y)^- \odot (x \odot (x \odot y)^-)^-)^- = x$, that is (Pom). □

Note that Proposition 13.1.7 follows from Proposition 13.2.4, since (m-Pimpl) $\Longrightarrow$ (G). Note also that Proposition 13.2.4 says: **PreMV** $\cap$ **OSL** $\subset$ **OM**.

By Propositions 13.2.3 and 13.2.4, we obtain:

**Theorem 13.2.5** *(See Theorem 13.1.8)*
*Let $\mathcal{A}^L = (A^L, \odot, ^-, 1)$ be an involutive m-BE algebra. Then,*

$$(G) \Longrightarrow ((Pom) \Leftrightarrow (Pmv)) \quad or$$

$$(G) + (Pom) \Longleftrightarrow (Pmv) + (G),$$

*i.e. OMSLs coincide with pre-MV algebras verifying (G).*

See Example 13.3.12 of involutive m-BE algebra verifying (G) and not verifying (Pom) and (Pmv).

Hence, Theorem 13.2.5 says:

(13.23) $$\mathbf{OMSL} = \mathbf{PreMV} + (G) = \mathbf{PreMV} \cap \mathbf{OSL}.$$

Note that Theorem 13.1.8 follows from Theorem 13.2.5, since (m-Pimpl) implies (G).

- **OMSL** + (Pqmv) (Connections between **OMSL** and **QMV**)

We establish now the connection between the OMSLs and the QMV algebras verifying (G).

**Proposition 13.2.6** *(See Proposition 13.1.9)*
*Let $\mathcal{A}^L = (A^L, \odot, ^-, 1)$ be a left-OMSL. Then, $\mathcal{A}^L$ is a left-QMV algebra verifying (G).*
*(i.e. in an involutive m-BE algebra, (Pom) + (G) $\Longrightarrow$ (Pqmv).)*

**Proof.** Since $\mathcal{A}^L$ is a left-OMSL, it is an involutive m-BE algebra verifying (G) and (Pom) (Definition 2). By Proposition 13.2.3, it verifies (Pmv) also. Hence, $\mathcal{A}^L$ is a left-QMV algebra verifying (G). □

Note that Proposition 13.2.6 says:

$$\mathbf{OMSL} \subset \mathbf{QMV}, \tag{13.24}$$

the inclusion being strict, since there are examples of QMV algebras not verifying (G) - see Example 13.3.4. Note also that Proposition 13.1.9 follows from Proposition 13.2.6 and also that Proposition 13.2.3 follows from Proposition 13.2.6, since (Pqmv) implies (Pmv).

The following converse of Proposition 13.2.6 holds.

**Proposition 13.2.7** *(See Proposition 13.1.10)*
*Let $\mathcal{A}^L = (A^L, \odot, ^-, 1)$ be a left-QMV algebra verifying (G). Then, $\mathcal{A}^L$ is a left-OMSL.*
*(i.e. in an involutive m-BE algebra, (Pqmv) + (G) $\Longrightarrow$ (Pom).)*

**Proof.** Since $\mathcal{A}^L$ a left-QMV algebra verifying (G), it is an involutive m-BE algebra verifying (Pmv), (Pom) and (G) (Definition 2). Hence, $\mathcal{A}^L$ is an involutive m-BE algebra verifying (G) and (Pom), i.e. it is a left-OMSL. □

Note that Proposition 13.2.7 says: $\mathbf{QMV} \cap \mathbf{OSL} \subset \mathbf{OM}$. Note also that Proposition 13.1.10 follows from Proposition 13.2.7, since (m-Pimpl) implies (G), and also that Proposition 13.2.7 follows from Proposition 13.2.4, since (Pqmv) implies (Pmv).

By Propositions 13.2.6, 13.2.7, we obtain:

**Theorem 13.2.8** *(See Theorem 13.1.11)*
*Let $\mathcal{A}^L = (A^L, \odot, ^-, 1)$ be an involutive left-m-BE algebra. Then,*

$$(G) \Longrightarrow ((Pom) \Leftrightarrow (Pqmv)) \quad or$$

$$(G) + (Pom) \Longleftrightarrow (Pqmv) + (G),$$

*i.e. orthomodular softlattices coincide with QMV algebras verifying (G).*

Hence, Theorem 13.2.8 says:

$$\mathbf{OMSL} = \mathbf{QMV} + (G) = \mathbf{QMV} \cap \mathbf{OSL}. \tag{13.25}$$

Note that Theorem 13.1.11 follows from Theorem 13.2.8, since (m-Pimpl) implies (G).

By previous results (13.22), (13.23), (13.24), (13.25), we obtain the connections from Figure 13.8.

- **OMSL** $+ (\Delta_m)$ (Connections between **OMSL** and **MMV**)

**Proposition 13.2.9** *Let $\mathcal{A}^L = (A^L, \odot, ^-, 1)$ be an involutive m-BE algebra, Then,*

$$(Pom) + (G) \Longrightarrow (\Delta_m).$$

**m-BE**$_{(DN)}$

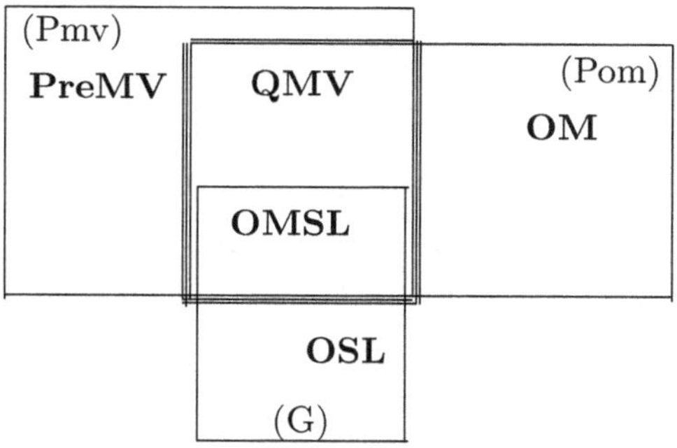


Figure 13.8: Resuming connections between **QMV**, **PreMV**, **OM**, **OSL** and **OMSL**

**Proof.** By Proposition 13.2.3, (Pom) + (G) implies (Pmv) and (Pmv) implies $(\Delta_m)$. □

Note that Proposition 13.2.9 says: **OMSL** $\subset$ **MMV**.

Note also that Proposition 13.1.12 follows from Proposition 13.2.9, since (m-Pimpl) implies (G), that Proposition 13.2.6 follows from Proposition 13.2.9, since (Pom) + $(\Delta_m)$ imply (Pqmv), and that Proposition 13.2.9 follows also from Proposition 13.2.6, since (Pqmv) implies $(\Delta_m)$.

**Remark 13.2.10** *The following converse of Proposition 13.2.9 ($(\Delta_m)$ + (G) $\Longrightarrow$ (Pom)) does not hold: there are examples of involutive m-BE algebras verifying $(\Delta_m)$ and (G) and not verifying (m-Pimpl) and (Pom) - see Example 13.3.12.*

By previous Remark, from the connections from Figure 13.8, we obtain the connections from Figure 13.9.

#### 13.2.1.2 OMSL = OML

**Proposition 13.2.11** *We have:*
*(mPom1) (Pom) + (Pcomm) + (Neg0-1) + (PU) + (DN) $\Longrightarrow$ (m-Re),*
*(mPom2) (Pom) + (G) + (Pass) + (DN) $\Longrightarrow$ (m-Pimpl).*

**Proof.** (mPom2) : **(By** ***Prover9*****, in 0.01 seconds, the length of the proof being 15)**
First, we have:

(a) $\quad x \odot y \overset{(G)}{=} (x \odot x) \odot y \overset{(Pass)}{=} x \odot (x \odot y)$.

Then, in (a), take $X := (x \odot y)^-$ and $Y := ((x \odot y)^- \odot x)^-$ to obtain:

(b) $\quad X \odot Y = (x \odot y)^- \odot ((x \odot y)^- \odot x)^- \overset{(Pom)}{=} x^-$.

Then, $x^- = X \odot Y \overset{(a)}{=} X \odot (X \odot Y) \overset{(b)}{=} X \odot x^- = (x \odot y)^- \odot x^-$; hence, $((x \odot y)^- \odot x^-)^- = (x^-)^- \overset{(DN)}{=} x$, i.e. (m-Pimpl) holds. □

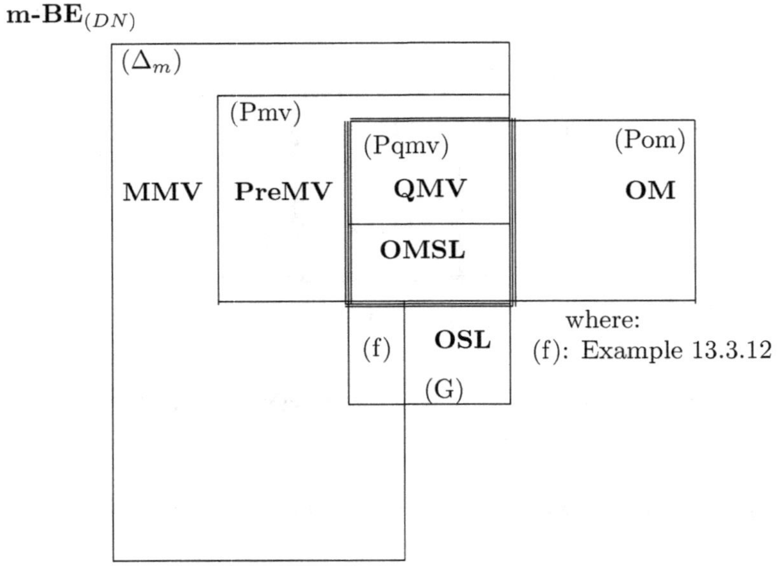


Figure 13.9: Resuming connections between **QMV**, **PreMV**, **MMV**, **OM**, **OSL** and **OMSL**

We know already, by Proposition 10.1.15, that:

**Proposition 13.2.12** *Let $\mathcal{A}^L = (A^L, \odot, ^-, 1)$ be an involutive left-m-BE algebra. Then,*

$$(m-Pimpl) \Longrightarrow (G),$$

*i.e.* **OL** $\subset$ **OSL**.

**Proposition 13.2.13** *Let $\mathcal{A}^L = (A^L, \odot, ^-, 1)$ be an involutive left-m-BE algebra. Then,*

$$(Pom) + (G) \Longrightarrow (m-Pimpl),$$

*i.e.* **OMSL** $\subset$ **OL**.

**Proof.** By (mPom2). □

By Propositions 13.2.12 and 13.2.13, we obtain:

**Theorem 13.2.14** *Let $\mathcal{A}^L = (A^L, \odot, ^-, 1)$ be an involutive left-m-BE algebra. Then,*

$$(Pom) \Longrightarrow ((m-Pimpl) \Leftrightarrow (G))$$

*or*

$$(Pom) + (m-Pimpl) \Longleftrightarrow (Pom) + (G).$$

See Example 13.3.3 which satisfies (Pom) and does not satisfy (m-Pimpl) and (G).

By Theorem 13.2.14 and the equivalent definitions (Definition 2) of left-OMLs and of left-OMSLs, we obtain: **OML**= **OM** + (G) = **OSL** + (Pom) = **OSL** ∩ **OM** = **OMSL**, by (13.22). Hence, we have:

(13.26) $$\mathbf{OMSL} = \mathbf{OML}.$$

By (13.1), (13.22) and (13.26), we obtain the connections from Figure 13.10.

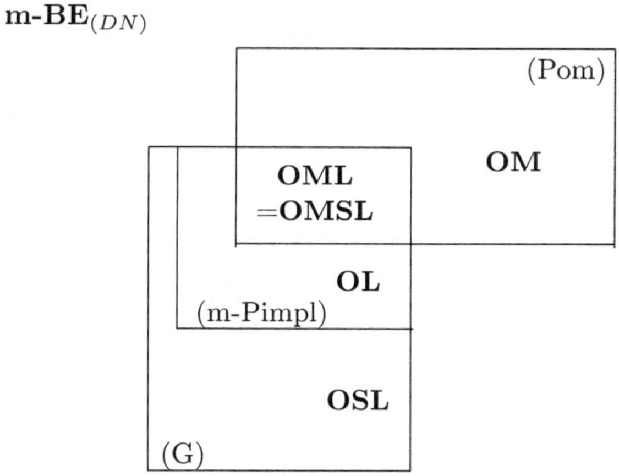


Figure 13.10: Resuming connections between **OML** = **OMSL**, **OL**, **OSL** and **OM**

Finally, since **OML** = **OMSL**, it follows, by Theorems 13.1.11 and 13.2.8:

**Corollary 13.2.15** *We have:*
(13.27)
$$\mathbf{OML} = \mathbf{OMSL} = \mathbf{QMV}+(m-Pimpl) = \mathbf{QMV}\cap\mathbf{OL} = \mathbf{QMV}+(G) = \mathbf{QMV}\cap\mathbf{OSL}.$$

**Corollary 13.2.16** *(See [45], Theorem 2.3.12)*
*Let $\mathcal{A}^L = (A^L, \odot, ^-, 1)$ be a left-QMV algebra. Consider the set of all idempotent elements of $A^L$ (i.e. elements verifying (G)):*

$$Id(A^L) = \{x \in A^L \mid x \odot x = x\}.$$

*Then, $(Id(A^L), \odot, ^-, 1)$ is a left-OML.*

**Proof.** Note that $(Id(A^L), \odot, ^-, 1)$ is a subalgebra of $\mathcal{A}^L$ verifying (G). Then apply above Corollary 13.2.15. □

Moreover,
- there are examples of involutive m-BE algebras verifying (G) and not verifying $(\Delta_m)$, (m-Pimpl) and (Pom) - see Example 13.3.13;

- there are examples of involutive m-BE algebras verifying (m-Pimpl) and not verifying ($\Delta_m$) and (Pom) - see Example 13.3.11.

By the connections from Figures 13.2, 13.8 and 13.10, we obtain the connections from Figure 13.11.

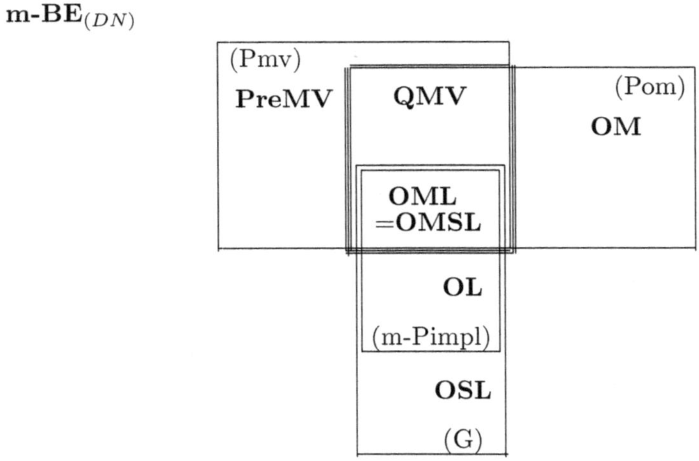


Figure 13.11: Resuming connections between **QMV**, **PreMV**, **OSL**, **OL**, **OM** and **OML** = **OMSL**

By the connections from Figures 13.3, 13.9 and 13.11, we obtain the connections from Figure 13.12.

### 13.2.1.3 The transitive case: tOSL $\subset$ tMMV

**Theorem 13.2.17** *Let* $\mathcal{A}^L = (A^L, \odot, ^-, 1)$ *be an involutive left-m-BE algebra. Then,*

$$(G) + (m - BB) \implies (\Delta_m).$$

**Proof. (following a proof by *Prover9* in 10.75 seconds, the length of the proof being 28)**

First, (G) ($x \odot x = x$) implies:

$$x \odot (x \odot y) = x \odot y. \tag{13.28}$$

Indeed, $x \odot (x \odot y) \overset{(Pass)}{=} (x \odot x) \odot y \overset{(G)}{=} x \odot y$.

Second, (m-BB) ($[(x \odot y)^- \odot (z \odot y)] \odot (z \odot x^-)^- = 0$) implies:

$$x \odot (y \odot ((x \odot z^-)^- \odot (z \odot y)^-)) = 0. \tag{13.29}$$

Indeed, interchange $x$ with $z$ in (m-BB) to obtain:

(x) $\qquad [(z \odot y)^- \odot (x \odot y)] \odot (x \odot z^-)^- = 0;$

then, in (x), apply (Pass) and (Pcomm) to obtain:

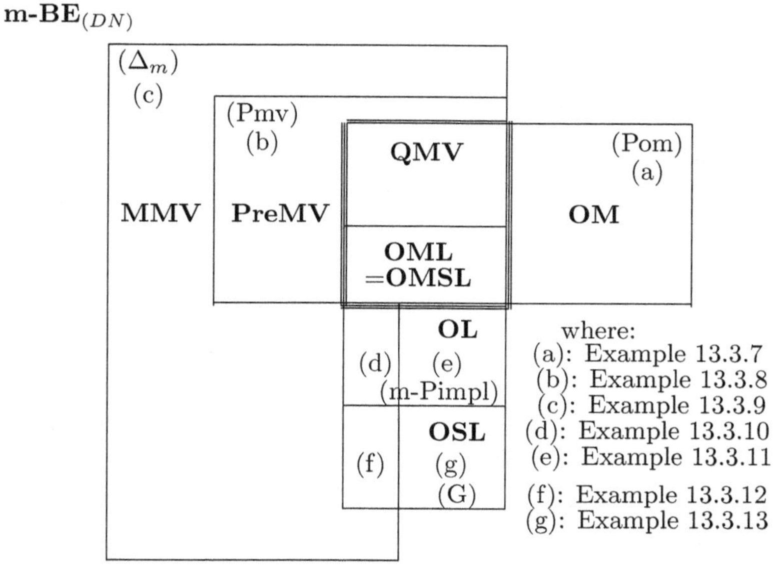


Figure 13.12: Resuming connections between **QMV**, **PreMV**, **MMV**, **OL**, **OSL** and **OML** = **OMSL**

(x') $\qquad [(x \odot y) \odot (x \odot z^-)^-] \odot (z \odot y)^- = 0;$
then apply (Pass) to obtain (13.29).

Also (m-BB) ($[(x \odot y)^- \odot (z \odot y)] \odot (z \odot x^-)^- = 0$) implies:

(13.30) $$(x \odot y)^- \odot (z \odot (x \odot (z \odot y^-)^-)) = 0.$$

Indeed, interchange $x$ with $y$ in (m-BB) to obtain, by (Pcomm):
$[(x \odot y)^- \odot (z \odot x)] \odot (z \odot y^-)^- = 0;$
then apply (Pass) to obtain (13.30).

Now, from (13.30), we obtain:

(13.31) $$x \odot (y \odot (x \odot (y^- \odot z)^-)^-) = 0.$$

Indeed, in (13.30) take $X := x$ and $Y := x^- \odot y$ to obtain:
(y) $\qquad (x \odot (x^- \odot y))^- \odot (z \odot (x \odot (z \odot (x^- \odot y)^-)^-)) = 0;$
but, in (y), $x \odot (x^- \odot y) \overset{(Pass)}{=} (x \odot x^-) \odot y \overset{(m-Re)}{=} 0 \odot y \overset{(Pcomm)}{=} y \odot 0 \overset{(m-La)}{=} 0$,
hence (y) becomes:
(y') $\qquad 0^- \odot (z \odot (x \odot (z \odot (x^- \odot y)^-)^-)) = 0,$
which by (Neg0-1) and (PU) becomes:
(y") $\qquad z \odot (x \odot (z \odot (x^- \odot y)^-)^-) = 0;$
now, in (y") take $X := y$, $Y := z$ and $Z := x$ to obtain:
$x \odot (y \odot (x \odot (y^- \odot z)^-)^-) = 0$, that is (13.31).

Now, from (13.31) and (13.28), we obtain:

(13.32) $$x \odot (x \odot (y \odot x^-)^-)^- = 0.$$

Indeed, in (13.28) take $X := x$ and $Y := (x \odot (x^- \odot y)^-)^-$ to obtain:
(u) $x \odot (x \odot (x \odot (x^- \odot y)^-)^-) = x \odot (x \odot (x^- \odot y)^-)^-$;
also in (13.31) take $X := x$, $Y := x$ and $Z := y$ to obtain:
(v) $x \odot (x \odot (x \odot (x^- \odot y)^-)^-) = 0$;
then, (u) becomes, by (v):
(u') $0 = x \odot (x \odot (x^- \odot y)^-)^-$,
which by (Pcomm) becomes (13.32).

Now, from (m-BB) and (13.32) we obtain:

$$(13.33) \qquad x \odot (y \odot (x \odot ((z \odot x^-)^- \odot y))^-) = 0.$$

Indeed, in (m-BB) ($[(x \odot y)^- \odot (z \odot y)] \odot (z \odot x^-)^- = 0$) take $X := x \odot (y \odot x^-)^-$, $Y := z$ and $Z := x$ to obtain:
(w) $[((x \odot (y \odot x^-)^-) \odot z)^- \odot (x \odot z)] \odot (x \odot (x \odot (y \odot x^-)^-)^-)^- = 0$;
but, in (w), the part $x \odot (x \odot (y \odot x^-)^-)^- = 0$, by (13.32); hence, (w) becomes:
(w') $[((x \odot (y \odot x^-)^-) \odot z)^- \odot (x \odot z)] \odot 0^- = 0$,
which by (Neg0-1) and (PU) becomes:
(w") $((x \odot (y \odot x^-)^-) \odot z)^- \odot (x \odot z) = 0$,
which by (Pcomm), (Pass) becomes:
(w"') $(x \odot z) \odot (x \odot ((y \odot x^-)^- \odot z))^- = 0$,
which by interchanging $y$ with $z$ and by (Pass) becomes:
$x \odot (y \odot (x \odot ((z \odot x^-)^- \odot y))^-) = 0$, that is (13.33).

Now, from (13.29) and (13.33), we obtain:

$$(13.34) \qquad (x \odot (x \odot y^-)^-) \odot (y \odot (y \odot x^-)^-)^- = 0.$$

Indeed, in (13.33), take $X := x$, $Y := (x \odot y^-)^- \odot (y \odot (z \odot x^-)^-)^-$ and $Z := z$ to obtain:
(z) $x \odot (((x \odot y^-)^- \odot (y \odot (z \odot x^-)^-)^-) \odot (x \odot ((z \odot x^-)^- \odot Y))^-) = 0$,
where the part of (z):
$A \overset{notation}{=} x \odot ((z \odot x^-)^- \odot Y) = x \odot ((z \odot x^-)^- \odot ((x \odot y^-)^- \odot (y \odot (z \odot x^-)^-)^-)) = 0$;
indeed, in (13.29) take $X := x$, $Y := (z \odot x^-)^-$ and $Z := y$ to obtain:
$x \odot ((z \odot x^-)^- \odot ((x \odot y^-)^- \odot (y \odot (z \odot x^-)^-)^-)) = 0$, i.e. $A = 0$;
hence, (z) becomes:
(z') $x \odot (((x \odot y^-)^- \odot (y \odot (z \odot x^-)^-)^-) \odot 0^-) = 0$;
then, by (Neg0-1) and (PU), (z') becomes:
(z") $x \odot ((x \odot y^-)^- \odot (y \odot (z \odot x^-)^-)^-) = 0$,
and (z") by (Pass) and by taking $z = y$ becomes:
$(x \odot (x \odot y^-)^-) \odot (y \odot (y \odot x^-)^-)^- = 0$, that is (13.34).

Finally, from (13.34), by interchanging $x$ with $y$, we obtain:
$(y \odot (y \odot x^-)^-) \odot (x \odot (x \odot y^-)^-)^- = 0$, that is $(\Delta_m)$. □

Note that Theorem 13.2.17 says: **tOSL** $\subset$ **MMV**. Hence, **tOSL** $\subset$ **tMMV**.

Note also that Theorem 13.1.18 follows from Theorem 13.2.17, since (m-Pimpl) implies (G).

By Theorems 13.1.18 and 13.2.17 and by the connections from Figures 13.6 and 13.12, we obtain the connections from Figure 13.13.

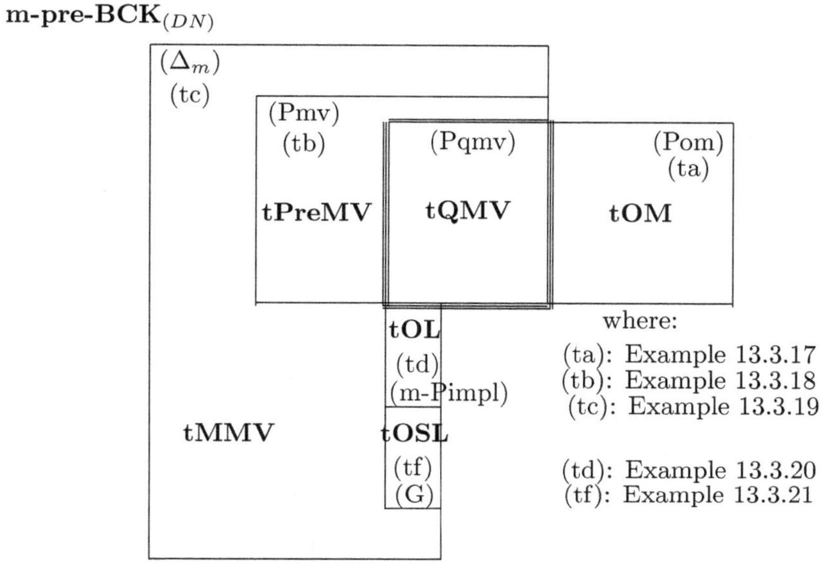


Figure 13.13: Resuming connections between **tQMV**, **tMMV**, **tOSL** and **tOL**

## 13.2.2 Orthomodular widelattices: OMWL

We introduce the following notion.

**Definition 13.2.18** (Definition 1) (The dual one is omitted)
A *left-orthomodular widelattice* or an *orthomodular left-widelattice*, or a *left-OMWL* for short, is a left-OWL verifying: for all $x, y \in A^L$,
(Wom) $\qquad (x \wedge y) \vee ((x \wedge y)^- \wedge x) = x.$

Denote by **OMWL** the class of all left-OMWLs. Following the equivalent Definition 2 of a left-OWL (see Definition 10.3.8), we obtain immediately an equivalent definition:

**Definition 13.2.19** (Definition 2) (The dual one is omitted)
A *left-OMWL* is a left-OWL verifying (Pom), i.e. is an involutive left-m-BE algebra $\mathcal{A}^L = (A^L, \odot, ^-, 1)$ verifying (m-Pabs-i) and (Pom), i.e.

(13.35) $\mathbf{OMWL} \; = \; \mathbf{m-BE_{(DN)}} + (m\!-\!Pabs\!-\!i) + (Pom) \; = \; \mathbf{OWL} \cap \mathbf{OM}.$

Further, we shall work with Definition 2 of OMWLs. Hence, we have the connections from Figure 13.14.

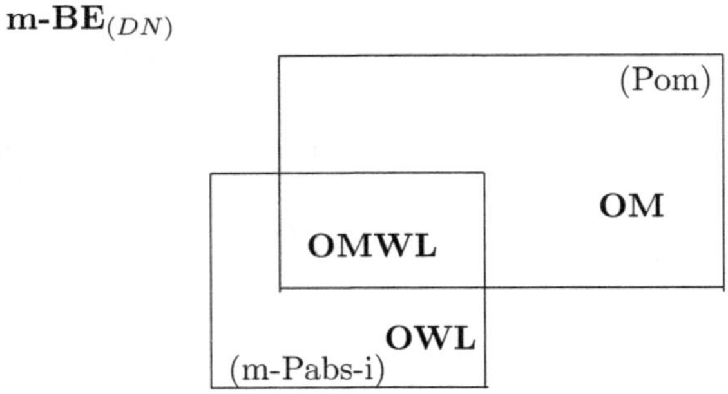


Figure 13.14: Resuming connections between **OWL**, **OMWL** and **OM**

#### 13.2.2.1 Connections between OMWL and PreMV, QMV, MMV, OM, OWL

• **OMWL** + (Pmv) (Connections between **OMWL** and **PreMV**)

The next Proposition 13.2.21 (saying that (Pom) and (m-Pabs-i) imply (Pmv)) **was proved by *Prover9* in 1706 seconds and the proof produced by *Prover9* has the length 23. We divide the proof produced by *Prover9* into the proof of Lemma 13.2.20 and Proposition 13.2.21.**

**Lemma 13.2.20** *Let $\mathcal{A}^L = (A^L, \odot, ^-, 1)$ be an involutive m-BE algebra verifying (Pom) (i.e. an OM algebra). Then, we have:*

(13.36) $$(x \odot y)^- \odot (y \odot (y \odot x)^-)^- = y^-,$$

(13.37) $$(x \odot y)^- \odot [(y \odot (y \odot x)^-)^- \odot z] = y^- \odot z,$$

(13.38) $$(x \odot (y \odot z))^- \odot (x \odot (y \odot (x \odot (y \odot z))^-))^- = (x \odot y)^-,$$

(13.39) $$(x \odot y)^- \odot (z \odot (x \odot (x \odot y)^-)^-) = z \odot x^-,$$

(13.40) $$(x \odot y^-)^- \odot [(y \odot z)^- \odot (x \odot (x \odot y^-)^-)]^- = ((y \odot z)^- \odot x)^-.$$

**Proof.**
(13.36): From (Pom), by interchanging $x$ with $y$ and by (Pcomm).
(13.37): From (13.36), by "multiplying" by $z$.
(13.38): From (Pom), taking $X := x \odot y$ and $Y := z$ and by (Pass).
(13.39): By "multiplying" (Pom) by $z$, and by (Pcomm), (Pass).
(13.40): In (13.38), take $X := (y \odot z)^-$, $Y := x$, $Z := (y \odot (y \odot z)^-)^-$ to obtain:
(13.41)
$[(y\odot z)^- \odot(x\odot(y\odot(y\odot z)^-)^-)]^- \odot[(y\odot z)^- \odot(x\odot[(y\odot z)^- \odot(x\odot(y\odot(y\odot z)^-)^-)]^-)]^-$

$$= ((y \odot z)^- \odot x)^-.$$

On the other hand, in (13.39), take $X := y$, $Y := z$, $Z := x$ to obtain:

$$(y \odot z)^- \odot (x \odot (y \odot (y \odot z)^-)^-) = x \odot y^-. \tag{13.42}$$

Now, from (13.41), by (13.42), we obtain:
$(x \odot y^-)^- \odot ((y \odot z)^- \odot (x \odot (x \odot y^-)^-))^- = ((y \odot z)^- \odot x)^-$, i.e. (13.40) holds. □

**Proposition 13.2.21** *(See Proposition 13.1.6)*
*Let $\mathcal{A}^L = (A^L, \odot, ^-, 1)$ be an involutive m-BE algebra. Then,*

$$(Pom) \ + \ (m - Pabs - i) \implies (Pmv).$$

**Proof.** (**By** *Prover9*)
First, from (m-Pabs-i) $(x \odot (x^- \odot (x^- \odot y^-))^- = x)$, by taking $Y := y^-$, we obtain:

$$x \odot (x^- \odot (x^- \odot y))^- = x. \tag{13.43}$$

Now, we prove:

$$x \odot (y \odot (y^- \odot ((x \odot y)^- \odot z))^-) = x \odot y. \tag{13.44}$$

Indeed, in (13.43), take $X := x \odot y$, Y:= $(y \odot (y \odot x)^-)^- \odot z$ to obtain:

$$(x \odot y) \odot ((x \odot y)^- \odot ((x \odot y)^- \odot [(y \odot (y \odot x)^-)^- \odot z]))^- = x \odot y. \tag{13.45}$$

Now, from (13.45), by (13.37), we obtain:

$$(x \odot y) \odot ((x \odot y)^- \odot (y^- \odot z))^- = x \odot y. \tag{13.46}$$

From (13.46), by (Pass), (Pcomm), we obtain:
$x \odot (y \odot (y^- \odot ((x \odot y)^- \odot z))^-) = x \odot y$, i.e. (13.44) holds.
Now, we prove:

$$x \odot (y^- \odot (y \odot ((y \odot z)^- \odot x)^-)^-) = x \odot y^-. \tag{13.47}$$

Indeed, in (13.44), take $X := x$, $Y := y^-$, $Z := [(y \odot z)^- \odot (x \odot (x \odot y^-)^-)]^-$ to obtain:

$$x \odot (y^- \odot (y \odot ((x \odot y^-)^- \odot [(y \odot z)^- \odot (x \odot (x \odot y^-)^-)]^-))^-) = x \odot y^-. \tag{13.48}$$

From (13.48), by (13.40), we obtain:
$x \odot (y^- \odot (y \odot ((y \odot z)^- \odot x)^-)^-) = x \odot y^-$, i.e. (13.47) holds.
Now, we prove:

$$x^- \odot (y \odot (y \odot x)^-)^- = x^- \odot y^-. \tag{13.49}$$

Indeed, in (13.47), take $X := (x \odot (x \odot y)^-)^-$, $Y := y$, $Z := x$ to obtain:
(13.50)
$(x \odot (x \odot y)^-)^- \odot (y^- \odot [y \odot ((y \odot x)^- \odot (x \odot (x \odot y)^-)^-)^-]^-) = (x \odot (x \odot y)^-)^- \odot y^-.$

In (13.36), take $X := y$, $Y := x$, to obtain:

$$(y \odot x)^- \odot (x \odot (x \odot y)^-)^- = x^-. \tag{13.51}$$

Then, from (13.50), by (13.51), we obtain:

$$(x \odot (x \odot y)^-)^- \odot (y^- \odot (y \odot x^=)^-) = (x \odot (x \odot y)^-)^- \odot y^-. \tag{13.52}$$

From (13.52), by (DN), we obtain:
$(x \odot (x \odot y)^-)^- \odot (y^- \odot (y \odot x)^-) = (x \odot (x \odot y)^-)^- \odot y^-$, hence, by (Pcomm), (Pass), we obtain:
$y^- \odot ((x \odot y)^- \odot ((x \odot y)^- \odot x)^-) = (x \odot (x \odot y)^-)^- \odot y^-$, hence by (Pom), we obtain:
$y^- \odot x^- = (x \odot (x \odot y)^-)^- \odot y^-$, hence, by interchanging $x$, $y$, we obtain:
$x^- \odot y^- = (y \odot (y \odot x)^-)^- \odot x^-$, hence, by (Pcomm), $x^- \odot (y \odot (y \odot x)^-)^- = x^- \odot y^-$, i.e. (13.49) holds.

Now, finally, from (13.49), by $X := x^-$, $Y := y^-$ and (DN), (Pcomm), we obtain:
$x \odot ((x^- \odot y^-)^- \odot y^-)^- = x \odot y$, i.e. (Pmv) holds. □

Note that Proposition 13.1.6 follows from Proposition 13.2.21, since (m-Pimpl) implies (m-Pabs-i).

Note also that Proposition 13.2.21 says: **OMWL** $\subset$ **PreMV**.

**Remark 13.2.22** *The following converse of Proposition 13.2.21 ((Pmv) + (m-Pabs-i) $\Longrightarrow$ (Pom)) does not hold: there are examples of involutive m-BE algebras verifying (Pmv) and (m-Pabs-i) and not verifying (Pom) - see Example 13.3.16.*

- **OMWL** + (Pqmv) (Connections between **OMWL** and **QMV**)

We establish now the connection between the OMWLs and the QMV algebras verifying (m-Pabs-i).

**Proposition 13.2.23** *(See Proposition 13.1.9)*

*Let $\mathcal{A}^L = (A^L, \odot, ^-, 1)$ be a left-OMWL. Then, $\mathcal{A}^L$ is a left-QMV algebra verifying (m-Pabs-i).*
*(i.e. in an involutive m-BE algebras, (Pom) + (m-Pabs-i) $\Longrightarrow$ (Pqmv).)*

**Proof.** Since $\mathcal{A}^L$ is a left-OMWL, it is an involutive m-BE algebra verifying (m-Pabs-i) and (Pom) (Definition 2). By Proposition 13.2.21, it verifies (Pmv) also. Hence, $\mathcal{A}^L$ is a left-QMV algebra verifying (m-Pabs-i). □

Note that Proposition 13.2.23 says:

$$\textbf{OMWL} \subset \textbf{QMV}, \tag{13.53}$$

the inclusion being strict since there are examples of QMV algebras not verifying (m-Pabs-i) - see Example 13.3.4. Note also that Propositions 13.1.9 and 13.2.21 follow from Proposition 13.2.23.

The following converse of Proposition 13.2.23 holds.

**Proposition 13.2.24** *(See Proposition 13.1.10)*
*Let $\mathcal{A}^L = (A^L, \odot, ^-, 1)$ be a left-QMV algebra verifying (m-Pabs-i). Then, $\mathcal{A}^L$ is a left-OMWL.*
*(i.e. in involutive m-BE algebras, (Pqmv) + (m-Pabs-i) $\Longrightarrow$ (Pom).)*

**Proof.** Since $\mathcal{A}^L$ a left-QMV algebra verifying (m-Pabs-i), it is an involutive left-m-BE algebra verifying (Pqmv) (hence (Pmv), (Pom)) and (m-Pabs-i) (Definition 2). Hence, $\mathcal{A}^L$ is an involutive m-BE algebra verifying (m-Pabs-i) and (Pom), i.e. it is a left-orthomodular widelattice. □

Note that Proposition 13.2.24 says: **QMV** $\cap$ **OWL** $\subset$ **OM**. Note also that Proposition 13.1.10 follows from Proposition 13.2.24, since (m-Pimpl) $\Longrightarrow$ (m-Pabs-i).

By Propositions 13.2.23, 13.2.24, we obtain:

**Theorem 13.2.25** *(See Theorem 13.1.11)*
*Let $\mathcal{A}^L = (A^L, \odot, ^-, 1)$ be an involutive m-BE algebra. Then,*

$$(m - Pabs - i) \implies ((Pom) \Leftrightarrow (Pqmv)) \quad or$$

$$(m - Pabs - i) + (Pom) \iff (Pqmv) + (m - Pabs - i),$$

*i.e. orthomodular widelattices coincide with QMV algebras verifying (m-Pabs-i).*

See Example 13.3.16 which verifies (m-Pabs-i) and does not verify (Pom) and (Pqmv).

Note that Theorem 13.2.25 says:

$$\mathbf{OMWL} = \mathbf{QMV} + (m - Pabs - i) = \mathbf{QMV} \cap \mathbf{OWL}. \tag{13.54}$$

Note also that Theorem 13.1.11 follows from Theorem 13.2.25.

By (13.35), (13.54) and Remark 13.2.22, we obtain the connections from Figure 13.15.

- **OMWL** + ($\Delta_m$) (Connections between **OMWL** and **MMV**)

**Proposition 13.2.26** *(See Proposition 13.1.12)*
*Let $\mathcal{A}^L = (A^L, \odot, ^-, 1)$ be an involutive m-BE algebra. Then,*

$$(Pom) + (m - Pabs - i) \implies (\Delta_m).$$

**Proof.** By Proposition 13.2.21, (Pom) + (m-Pabs-i) imply (Pmv) and (Pmv) implies ($\Delta_m$), thus (Pom) + (m-Pabs-i) imply ($\Delta_m$). □

Note that Proposition 13.2.26 says: **OMWL** $\subset$ **MMV**.

Note also that Proposition 13.1.12 follows from Proposition 13.2.26, since (m-Pimpl) implies (m-Pabs-i), that Proposition 13.2.23 follows also from Proposition 13.2.26, since (Pom) + ($\Delta_m$) imply (Pqmv), and that Proposition 13.2.26 follows also from Proposition 13.2.23, since (Pqmv) implies ($\Delta_m$).

**m-BE**$_{(DN)}$

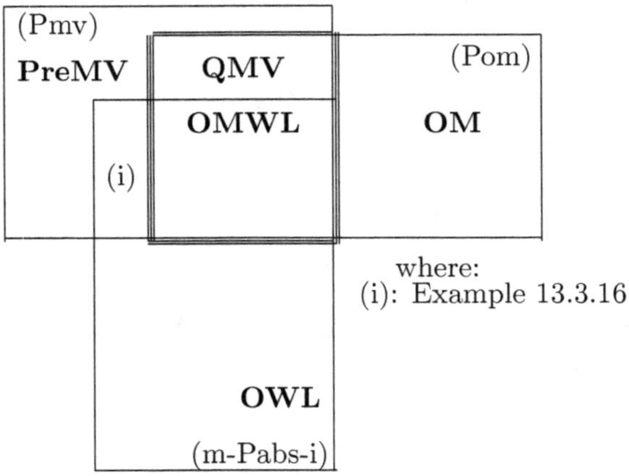


Figure 13.15: Resuming connections between **QMV**, **PreMV**, **OWL**, **OM** and **OMWL**

**Remark 13.2.27** *The following converse of Proposition 13.2.26 (($\Delta_m$) + (m-Pabs-i) $\Longrightarrow$ (Pom)) does not hold: there are examples of involutive m-BE algebras verifying ($\Delta_m$) and (m-Pabs-i) and not verifying (m-Pimpl) and (Pom) - see Example 13.3.14.*

By previous Remark and by the connections from Figure 13.15, we obtain the connections from Figure 13.16.

#### 13.2.2.2 OML $\subset$ OMWL

We know (by Proposition 10.1.15) that:

**Proposition 13.2.28** *Let $\mathcal{A}^L = (A^L, \odot, ^-, 1)$ be an involutive left-m-BE algebra. Then,*

$$(m - Pimpl) \Longrightarrow (m - Pabs - i),$$

*i.e.* **OL** $\subset$ **OWL**.

**Proposition 13.2.29** *Let $\mathcal{A}^L = (A^L, \odot, ^-, 1)$ be an involutive left-m-BE algebra. Then,*

$$(Pom) + (G) \Longrightarrow (m - Pabs - i).$$

**Proof.** By Proposition 13.2.13, (Pom) + (G) imply (m-Pimpl), and by Proposition 13.2.28, (m-Pimpl) implies (m-Pabs-i). □

Note that Proposition 13.2.29 follows from Proposition 13.2.13.

Note also that Proposition 13.2.29 says: **OML** (= **OMSL** ) $\subset$ **OWL**, hence,

(13.55) **OML** (= **OMSL**) $\subset$ **OMWL**,

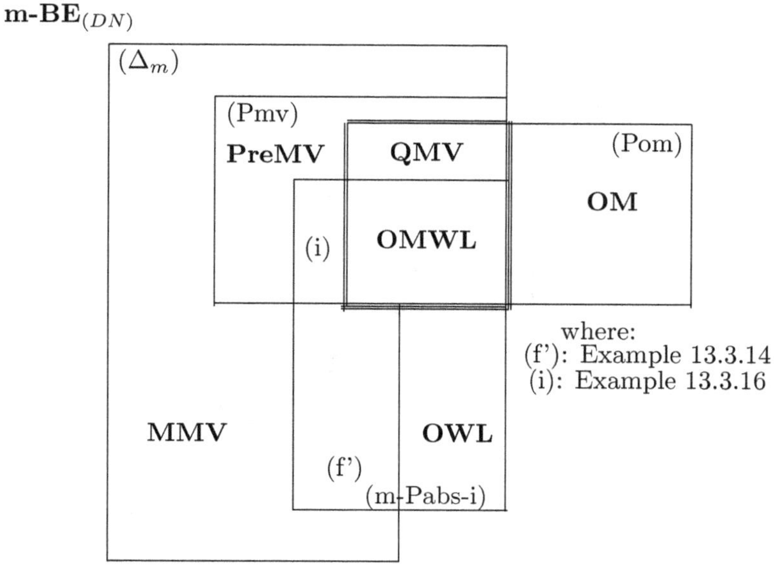


Figure 13.16: Resuming connections between **QMV**, **PreMV**, **MMV**, **OWL** and **OMWL**

the inclusion being strict, since there are examples of OMWLs not verifying (G) - see Example 13.3.3.

Note also that **OML** (= **OMSL**) $\subset$ **OMWL** means (see 10.132):

$$\mathbf{OML} = \mathbf{OMSL} \cap \mathbf{OMWL}.$$

By (13.1), (13.35), (13.55), we obtain the connections from Figure 13.17.

Since **OML** = **OMSL** $\subset$ **OMWL**, by Theorems 13.2.14 and 13.2.29, and **OMWL** $\subset$ **QMV**, by (13.53), we obtain:

$$\mathbf{MOD} \subset \mathbf{OML} = \mathbf{OMSL} \subset \mathbf{OMWL} \subset \mathbf{QMV}.$$

Note that:
- Example 13.3.4 is an example of proper QMV algebra,
- Example 13.3.3 is an example of proper OMWL,
- Example 13.3.2 is an example of proper OML = OMSL.

By the connections from Figures 13.2, 13.15 and 13.17, we obtain the connections from Figure 13.18.

By the connections from Figures 13.3, 13.16 and 13.18, we obtain the connections from Figure 13.19.

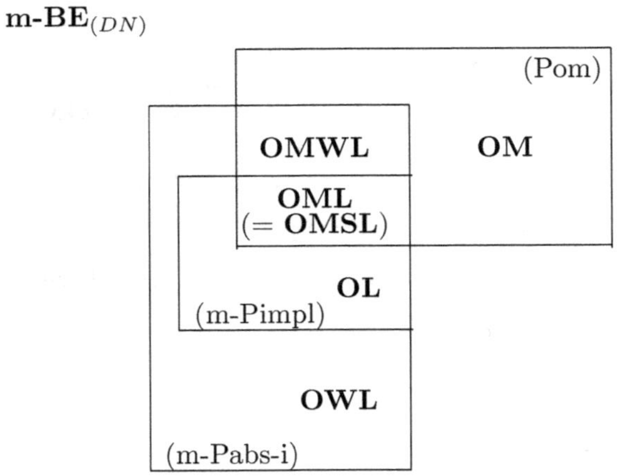


Figure 13.17: Resuming connections between **OMWL**, **OML**, **OL**, **OWL** and **OM**

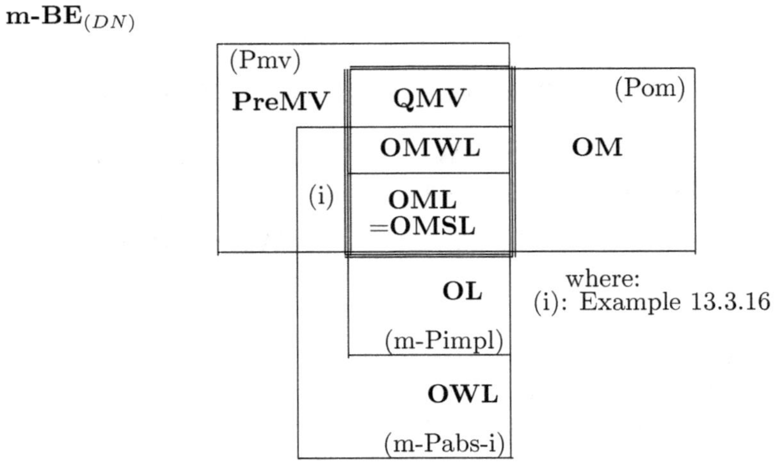


Figure 13.18: Resuming connections between **QMV**, **PreMV**, **OML**, **OWL**, **OL**, **OM** and **OMWL**

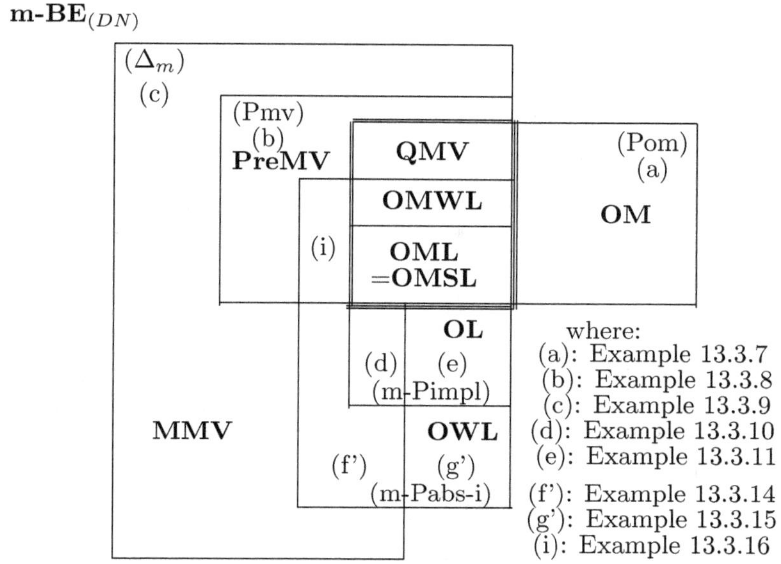


Figure 13.19: Resuming connections between **QMV**, **PreMV**, **MMV**, **OML**, **OL**, **OWL** and **OMWL**

#### 13.2.2.3 The transitive and/or antisymmetric case

- **The transitive case: tOWL $\subset$ tMMV**

Denote by **tOMWL** the class of all transitive left-OMWLs.

**Theorem 13.2.30** *(See Theorem 13.2.17)*
*Let $\mathcal{A}^L = (A^L, \odot, ^-, 1)$ be an involutive m-BE algebra. Then,*

$$(m - Pabs - i) \ + \ (m - BB) \implies (\Delta_m).$$

Note that this theorem is Theorem 10.3.13, proved by *Prover9*. It says that: **tOWL** $\subset$ **MMV**. Hence, **tOWL** $\subset$ **tMMV**.

If, additionally, (Pom) holds, then, as expected: **tOMWL** $\subset$ **tQMV**.

Note that Theorem 13.1.18 follows also from Theorem 13.2.30, since (m-Pimpl) implies (m-Pabs-i).

By (13.15), by Theorems 13.1.18 and 13.2.30 and the connections from Figure 13.19, we obtain the connections from Figure 13.20.

Note that we present, in Section 13.3, Example 13.3.6 of proper tQMV algebra and Example 13.3.5 of proper tOMWL.

- **The transitive and the antisymmetric case**

Denote by **aOMWL** the class of all antisymmetric left-OMWLs.

**m-pre-BCK**$_{(DN)}$

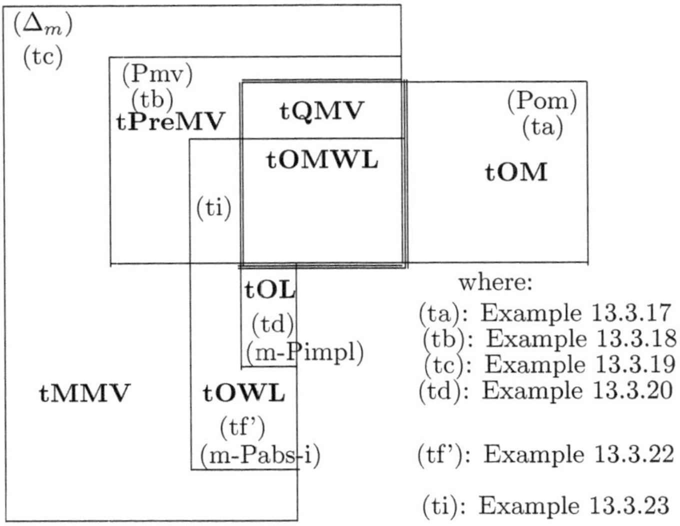


Figure 13.20: Resuming connections between **tQMV**, **tMMV**, **tOWL** and **tOL**

**Theorem 13.2.31** *We have:*

$$\mathbf{aOMWL} = \mathbf{taOMWL}.$$

**Proof.** Since **OMWL** $\subset$ **QMV**, by adding (m-An), we obtain: **aOMWL** $\subset$ **aQMV** $=$ **MV**, by Corollary 11.3.10, and since any MV algebra verifies (m-Tr), it follows that **aOMWL** $=$ **taOMWL**. □

While **tOMWL** $\subset$ **tOWL**, we obtain the following results.

**Theorem 13.2.32** *We have:*
*(i)* **taOWL** $\subset$ **MV**;
*(ii)* **taOMWL** $\subset$ **MV**;
*(iii)* **taOWL** $=$ **taOMWL**.

**Proof.** (i): Since **tOWL** $\subset$ **tMMV**, by applying (m-An), we obtain:
**taOWL** $\subset$ **taMMV** $=$ **MV**, by Corollary 11.3.10 and Remark 11.3.11.

(ii): Since **tOMWL** $\subset$ **tQMV**, by applying (m-An), we obtain:
**taOMWL** $\subset$ **taQMV** $=$ **MV**, by Corollary 11.3.10 and Remark 11.3.11.

(iii): Since any MV algebra verifies (Pom), it follows by (i) that **taOWL** $=$ **taOMWL**. □

**Theorem 13.2.33** *We have:*

$$\mathbf{taOWL} = \mathbf{taOMWL} = \mathbf{aOMWL} \subset \mathbf{MV}.$$

**Proof.** By Theorems 13.2.31, 13.2.32. □

• **Final remarks.** We have:

| | | |
|---|---|---|
| **tOMWL** | $\subset$ | **tQMV** |
| (m-An) $\downarrow$ | | $\downarrow$ (m-An) |
| **taOMWL** = **taOWL** | $\subset$ | **MV**. |

The tOMWLs (inside the tQMV algebras) will be deeply analysed in next chapter, in connection with the taOWLs (inside the MV algebras).

## 13.3 Examples

• **Quantum-MV algebras**

**Example 13.3.1 Proper modular algebra (that is not distributive): MOD**

By MACE4 program, we found that the algebra $\mathcal{A}^L = (A_6 = \{0, a, b, c, d, 1\}, \odot, ^-, 1)$, with the following tables of $\odot$ and $^-$ and of the additional operation $\oplus$, is an involutive left-m-BE algebra verifying (Pqmv) (hence (Pom), (Pmv), hence ($\Delta_m$)), (m-Pimpl) (hence (G) and (m-Pabs-i)) and (Pmod), and (prel$_m$), (WNM$_m$), (aWNM$_m$), and not verifying (m-B) for $(a, c, a)$, (m-BB) for $(a, a, c)$, (m-*) for $(a, c, d)$, (m-**) for $(a, c, a)$, (m-Tr) for $(a, c, b)$, (m-An) for $(a, c)$, (m-Pdis) for $(a, b, c)$.

| $\odot$ | 0 | a | b | c | d | 1 |
|---|---|---|---|---|---|---|
| 0 | 0 | 0 | 0 | 0 | 0 | 0 |
| a | 0 | a | 0 | 0 | 0 | a |
| b | 0 | 0 | b | 0 | 0 | b |
| c | 0 | 0 | 0 | c | 0 | c |
| d | 0 | 0 | 0 | 0 | d | d |
| 1 | 0 | a | b | c | d | 1 |

and

| $x$ | $x^-$ |
|---|---|
| 0 | 1 |
| a | b |
| b | a |
| c | d |
| d | c |
| 1 | 0 |

, with

| $\oplus$ | 0 | a | b | c | d | 1 |
|---|---|---|---|---|---|---|
| 0 | 0 | a | b | c | d | 1 |
| a | a | a | 1 | 1 | 1 | 1 |
| b | b | 1 | b | 1 | 1 | 1 |
| c | c | 1 | 1 | c | 1 | 1 |
| d | d | 1 | 1 | 1 | d | 1 |
| 1 | 1 | 1 | 1 | 1 | 1 | 1 |

.

Consider the lattice order relation w.r. to $\wedge = \odot$, $\vee = \oplus$: $\leq_m^O = \leq_m^P$, defined by $x \leq_m^P y \stackrel{def.}{\Longleftrightarrow} x \odot y = x$. From the table of $\odot$, we see that: $a \leq_m^O 1$; $b \leq_m^O 1$; $c \leq_m^O 1$; $d \leq_m^O 1$; hence, the bounded lattice $(A_6, \leq_m^O, 0, 1)$ has the Hasse diagram from Figure 13.21.

The table of $\wedge_M^M$ is the following:

| $\wedge_m^M$ | 0 | a | b | c | d | 1 |
|---|---|---|---|---|---|---|
| 0 | 0 | 0 | 0 | 0 | 0 | 0 |
| a | 0 | a | 0 | c | d | a |
| b | 0 | 0 | b | c | d | b |
| c | 0 | a | b | c | 0 | c |
| d | 0 | a | b | 0 | d | d |
| 1 | 0 | a | b | c | d | 1 |

.

Note that $\leq_m^M$ ($x \leq_m^M y \overset{def.}{=} x \wedge_m^M y = x$) is an order relation, by Corollary 12.1.7, but not a lattice order w.r. to $\wedge_m^M$, $\vee_m^M$, since $\wedge_m^M$ is not commutative. From the table of $\wedge_m^M$, we see that: $a \leq_m^M 1$; $b \leq_m^M 1$; $c \leq_m^M 1$; $d \leq_m^m O1$; hence, the bounded poset $(A_6, \leq_m^M, 0, 1)$ has the same Hasse diagram from Figure 13.21, since the order relations are equivalent: $x \leq_m^O y \Longleftrightarrow x \leq_m^M y$, by Proposition 11.2.12.

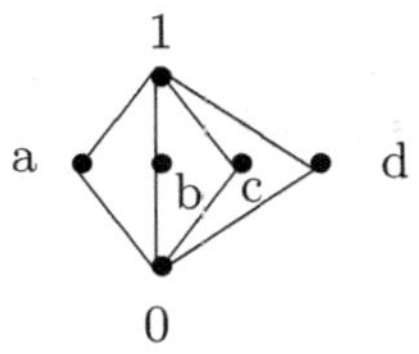


Figure 13.21: The Hasse diagram both for the bounded lattice $(A_6, \leq_m^O, 0, 1)$ and for the bounded poset $(A_6, \leq_m^M, 0, 1)$

**Example 13.3.2 Proper OML (which is not MOD algebra): OML**

By MACE4 program, we found that the algebra $\mathcal{A}^L = (A_{10} = \{0, a, b, c, d, e, f, g, h, 1\}, \odot, ^-, 1)$, with the following tables of $\odot$ and $^-$ and of the additional operation $\oplus$, is an involutive left-m-BE algebra verifying (Pqmv) (hence (Pom), (Pmv), hence ($\Delta_m$)) and (m-Pimpl) (hence (G) and (m-Pabs-i)), and also (prel$_m$), (WNM$_m$), (aWNM$_m$), and not verifying (m-B) for $(a, c, a)$, (m-BB) for $(a, a, c)$, (m-*) for $(a, c, d)$, (m-**) for $(a, c, a)$, (m-Tr) for $(a, c, b)$, (m-An) for $(a, c)$, (Pmod) for $(a, c, e)$, (m-Pdis) for $(a, b, c)$. Thus, $\mathcal{A}^L$ is an orthomodular lattice (OML) which is not a modular ortholattice (MOD).

| $\odot$ | 0 | a | b | c | d | e | f | g | h | 1 |
|---|---|---|---|---|---|---|---|---|---|---|
| 0 | 0 | 0 | 0 | 0 | 0 | 0 | 0 | 0 | 0 | 0 |
| a | 0 | a | 0 | 0 | 0 | e | g | g | e | a |
| b | 0 | 0 | b | 0 | 0 | 0 | b | 0 | b | b |
| c | 0 | 0 | 0 | c | 0 | 0 | 0 | 0 | 0 | c |
| d | 0 | 0 | 0 | 0 | d | 0 | 0 | 0 | 0 | d |
| e | 0 | e | 0 | 0 | 0 | e | 0 | 0 | e | e |
| f | 0 | g | b | 0 | 0 | 0 | f | g | b | f |
| g | 0 | g | 0 | 0 | 0 | 0 | g | g | 0 | g |
| h | 0 | e | b | 0 | 0 | e | b | 0 | h | h |
| 1 | 0 | a | b | c | d | e | f | g | h | 1 |

and

| $x$ | $x^-$ |
|---|---|
| 0 | 1 |
| a | b |
| b | a |
| c | d |
| d | c |
| e | f |
| f | e |
| g | h |
| h | g |
| 1 | 0 |

,

with

| $\oplus$ | 0 | a | b | c | d | e | f | g | h | 1 |
|---|---|---|---|---|---|---|---|---|---|---|
| 0 | 0 | a | b | c | d | e | f | g | h | 1 |
| a | a | a | 1 | 1 | 1 | a | 1 | a | 1 | 1 |
| b | b | 1 | b | 1 | 1 | h | f | f | h | 1 |
| c | c | 1 | 1 | c | 1 | 1 | 1 | 1 | 1 | 1 |
| d | d | 1 | 1 | 1 | d | 1 | 1 | 1 | 1 | 1 |
| e | e | a | h | 1 | 1 | e | 1 | a | h | 1 |
| f | f | 1 | f | 1 | 1 | 1 | f | f | 1 | 1 |
| g | g | a | f | 1 | 1 | a | f | g | 1 | 1 |
| h | h | 1 | h | 1 | 1 | h | 1 | 1 | h | 1 |
| 1 | 1 | 1 | 1 | 1 | 1 | 1 | 1 | 1 | 1 | 1 |

.

Consider the lattice order relation w.r.t. to $\wedge = \odot$, $\vee = \oplus$: $\leq_m^O = \leq_m^P$, defined by $x \leq_m^P y \overset{def.}{\Longleftrightarrow} x \odot y = x$. From the table of $\odot$, we see that: $a \leq_m^O 1$; $b \leq_m^O f, h, 1$; $c \leq_m^O 1$; $d \leq_m^O 1$; $e \leq_m^O a, h, 1$; $f \leq_m^O 1$; $g \leq_m^O a, f, 1$; $h \leq_m^O 1$; hence, the bounded non-distributive lattice $(A_{10}, \leq_m^O, 0, 1)$ has the Hasse diagram from Figure 13.22.

The table of $\wedge_m^M$ is the following:

| $\wedge_m^M$ | 0 | a | b | c | d | e | f | g | h | 1 |
|---|---|---|---|---|---|---|---|---|---|---|
| 0 | 0 | 0 | 0 | 0 | 0 | 0 | 0 | 0 | 0 | 0 |
| a | 0 | a | 0 | c | d | e | g | g | e | a |
| b | 0 | 0 | b | c | d | 0 | b | 0 | b | b |
| c | 0 | a | b | c | 0 | e | f | g | h | c |
| d | 0 | a | b | 0 | d | e | f | g | h | d |
| e | 0 | e | 0 | c | d | e | 0 | 0 | e | e |
| f | 0 | g | b | c | d | 0 | f | g | b | f |
| g | 0 | g | 0 | c | d | 0 | g | g | 0 | g |
| h | 0 | e | b | c | d | e | b | 0 | h | h |
| 1 | 0 | a | b | c | d | e | f | g | h | 1 |

.

Note that $\leq_m^M$ ($x \leq_m^M y \overset{def.}{=} x \wedge_m^M y = x$) is an order relation, by Corollary 12.1.7, but not a lattice order w.r.t. to $\wedge_m^M$, $\vee_m^M$, since $\wedge_m^M$ is not commutative. From the table of $\wedge_m^M$, we see that: $a \leq_m^M 1$; $b \leq_m^M f, h, 1$; $c \leq_m^M 1$; $d \leq_m^M 1$; $e \leq_m^M a, h, 1$; $f \leq_m^M 1$; $g \leq_m^M a, f, 1$; $h \leq_m^M 1$; hence, the bounded poset $(A_{10}, \leq_m^M, 0, 1)$ has the same Hasse diagram from Figure 13.22, since the order relations are equivalent: $x \leq_m^O y \Longleftrightarrow x \leq_m^M y$, by Proposition 11.2.12.

**Example 13.3.3 Proper orthomodular widelattice: OMWL**

By a PASCAL program, we found the algebra $\mathcal{A}^L = (A_6 = \{0, a, b, c, d, 1\}, \odot, ^-, 1)$, with the following tables of $\odot$, $^-$ and of the additional operation $\oplus$:

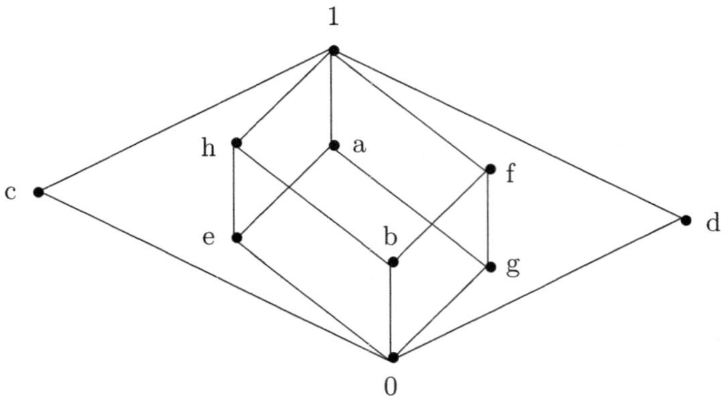


Figure 13.22: The Hasse diagram both for the bounded lattice $(A_{10}, \leq_m^O, 0, 1)$ and for the bounded poset $(A_{10}, \leq_m^M, 0, 1)$

| $\odot$ | 0 | a | b | c | d | 1 |
|---|---|---|---|---|---|---|
| 0 | 0 | 0 | 0 | 0 | 0 | 0 |
| a | 0 | a | 0 | 0 | 0 | a |
| b | 0 | 0 | 0 | 0 | 0 | b |
| c | 0 | 0 | 0 | 0 | 0 | c |
| d | 0 | 0 | 0 | 0 | d | d |
| 1 | 0 | a | b | c | d | 1 |

and

| $x$ | $x^-$ |
|---|---|
| 0 | 1 |
| a | d |
| b | c |
| c | b |
| d | a |
| 1 | 0 |

, with

| $\oplus$ | 0 | a | b | c | d | 1 |
|---|---|---|---|---|---|---|
| 0 | 0 | a | b | c | d | 1 |
| a | a | a | 1 | 1 | 1 | 1 |
| b | b | 1 | 1 | 1 | 1 | 1 |
| c | c | 1 | 1 | 1 | 1 | 1 |
| d | d | 1 | 1 | 1 | d | 1 |
| 1 | 1 | 1 | 1 | 1 | 1 | 1 |

.

$\mathcal{A}^L$ is an involutive left-m-aBE algebra verifying (m-Pabs-i), (Pqmv) (hence (Pom), (Pmv), ($\Delta_m$)), (prel$_m$), (WNM$_m$), (aWNM$_m$) and not verifying (m-B) for $(a, b, a)$, (m-BB) for $(a, a, b)$, (m-*) for $(b, a, d)$, (m-**) for $(a, b, a)$, (m-Tr) for $(a, b, d)$, (m-An) for $(a, b)$, (m-Pimpl) for $(b, 0)$, (G) for $b$, (m-Pdis) for $(a, b, b)$. Hence, it is a proper OMWL (Definition 2).

**Example 13.3.4 Proper quantum-MV algebra: QMV**

By a PASCAL program, we found the algebra $\mathcal{A}^L = (A_6 = \{0, a, b, c, d, 1\}, \odot, ^-, 1)$, with the following tables of $\odot$, $^-$ and of the additional operation $\oplus$:

| $\odot$ | 0 | a | b | c | d | 1 |
|---|---|---|---|---|---|---|
| 0 | 0 | 0 | 0 | 0 | 0 | 0 |
| a | 0 | c | d | 0 | 0 | a |
| b | 0 | d | 0 | 0 | 0 | b |
| c | 0 | 0 | 0 | 0 | 0 | c |
| d | 0 | 0 | 0 | 0 | 0 | d |
| 1 | 0 | a | b | c | d | 1 |

and

| $x$ | $x^-$ |
|---|---|
| 0 | 1 |
| a | d |
| b | c |
| c | b |
| d | a |
| 1 | 0 |

, with

| $\oplus$ | 0 | a | b | c | d | 1 |
|---|---|---|---|---|---|---|
| 0 | 0 | a | b | c | d | 1 |
| a | a | 1 | 1 | 1 | 1 | 1 |
| b | b | 1 | 1 | 1 | 1 | 1 |
| c | c | 1 | 1 | 1 | a | 1 |
| d | d | 1 | 1 | a | b | 1 |
| 1 | 1 | 1 | 1 | 1 | 1 | 1 |

.

$\mathcal{A}^L$ is an involutive left-m-aBE algebra verifying (Pqmv) (hence (Pom), (Pmv), ($\Delta_m$)), (prel$_m$) and not verifying (m-Pabs-i) for $(d, 0)$, (m-B) for $(a, b, b)$, (m-BB) for $(a, b, c)$, (m-*) for $(b, c, a)$, (m-**) for $(a, b, b)$, (m-Tr) for $(a, b, c)$, (m-An) for $(a, b)$, (m-Pimpl) for $(a, 0)$, (m-Pdis) for $(a, a, a)$, (WNM$_m$) and (aWNM$_m$) for $(a, a)$. Hence, it is a proper QMV algebra.

- **Transitive quantum-MV algebras**

**Example 13.3.5 Proper transitive orthomodular widelattice: tOMWL**

By a PASCAL program, we found the algebra $\mathcal{A}^L = (A_6 = \{0, a, b, c, d, 1\}, \odot, ^-, 1)$, with the following tables of $\odot$, $^-$ and of the additional operation $\oplus$:

| $\odot$ | 0 | a | b | c | d | 1 |
|---|---|---|---|---|---|---|
| 0 | 0 | 0 | 0 | 0 | 0 | 0 |
| a | 0 | 0 | 0 | 0 | 0 | a |
| b | 0 | 0 | 0 | 0 | 0 | b |
| c | 0 | 0 | 0 | 0 | 0 | c |
| d | 0 | 0 | 0 | 0 | 0 | d |
| 1 | 0 | a | b | c | d | 1 |

and

| $x$ | $x^-$ |
|---|---|
| 0 | 1 |
| a | d |
| b | c |
| c | b |
| d | a |
| 1 | 0 |

, with

| $\oplus$ | 0 | a | b | c | d | 1 |
|---|---|---|---|---|---|---|
| 0 | 0 | a | b | c | d | 1 |
| a | a | 1 | 1 | 1 | 1 | 1 |
| b | b | 1 | 1 | 1 | 1 | 1 |
| c | c | 1 | 1 | 1 | 1 | 1 |
| d | d | 1 | 1 | 1 | 1 | 1 |
| 1 | 1 | 1 | 1 | 1 | 1 | 1 |

.

$\mathcal{A}^L$ is an involutive left-m-aBE algebra verifying (m-Pabs-i), (Pqmv) (hence (Pom), (Pmv), ($\Delta_m$)), (m-B) $\Longleftrightarrow$ (m-BB) $\Longleftrightarrow$ ... (m-Tr) and (prel$_m$), (WNM$_m$), (aWNM$_m$) and not verifying (m-An) for $(a, b)$, (m-Pimpl) for $(a, 0)$, (m-Pdis) for $(a, a, a)$. Hence, it is a proper tOMWL.

**Example 13.3.6 Proper transitive quantum-MV algebra: tQMV**

By a PASCAL program, we found the algebra $\mathcal{A}^L = (A_6 = \{0, a, b, c, d, 1\}, \odot, ^-, 1)$, with the following tables of $\odot$, $^-$ and of the additional operation $\oplus$:

| $\odot$ | 0 | a | b | c | d | 1 |
|---|---|---|---|---|---|---|
| 0 | 0 | 0 | 0 | 0 | 0 | 0 |
| a | 0 | c | d | 0 | 0 | a |
| b | 0 | d | c | 0 | 0 | b |
| c | 0 | 0 | 0 | 0 | 0 | c |
| d | 0 | 0 | 0 | 0 | 0 | d |
| 1 | 0 | a | b | c | d | 1 |

and

| $x$ | $x^-$ |
|---|---|
| 0 | 1 |
| a | d |
| b | c |
| c | b |
| d | a |
| 1 | 0 |

, with

| $\oplus$ | 0 | a | b | c | d | 1 |
|---|---|---|---|---|---|---|
| 0 | 0 | a | b | c | d | 1 |
| a | a | 1 | 1 | 1 | 1 | 1 |
| b | b | 1 | 1 | 1 | 1 | 1 |
| c | c | 1 | 1 | b | a | 1 |
| d | d | 1 | 1 | a | b | 1 |
| 1 | 1 | 1 | 1 | 1 | 1 | 1 |

.

$\mathcal{A}^L$ is an involutive left-m-aBE algebra verifying (Pqmv) (hence (Pom), (Pmv), ($\Delta_m$)), (m-B) $\Longleftrightarrow$ (m-BB) $\Longleftrightarrow$ ... (m-Tr) and (prel$_m$) and not verifying (m-Pabs-i) for $(c, 0)$, (m-An) for $(a, b)$, (m-Pimpl) for $(a, 0)$, (m-Pdis) for $(a, a, a)$, (WNM$_m$) and (aWNM$_m$) for $(a, a)$. Hence, it is a proper tQMV algebra.

- **Other examples of involutive m-BE algebras**

**Example 13.3.7 Proper orthomodular algebra: OM**

By a PASCAL program, we found the algebra $\mathcal{A}^L = (A_5 = \{0, a, b, c, 1\}, \odot, ^-, 1)$ with the following tables of $\odot$ and $^-$ and of the additional operation $\oplus$:

| $\odot$ | 0 | a | b | c | d | 1 |
|---|---|---|---|---|---|---|
| 0 | 0 | 0 | 0 | 0 | 0 | 0 |
| a | 0 | 0 | 0 | 0 | 0 | a |
| b | 0 | 0 | c | 0 | 0 | b |
| c | 0 | 0 | 0 | 0 | 0 | c |
| d | 0 | 0 | 0 | 0 | d | d |
| 1 | 0 | a | b | c | d | 1 |

and

| $x$ | $x^-$ |
|---|---|
| 0 | 1 |
| a | d |
| b | c |
| c | b |
| d | a |
| 1 | 0 |

, with

| $\oplus$ | 0 | a | b | c | d | 1 |
|---|---|---|---|---|---|---|
| 0 | 0 | a | b | c | d | 1 |
| a | a | a | 1 | 1 | 1 | 1 |
| b | b | 1 | 1 | 1 | 1 | 1 |
| c | c | 1 | 1 | b | 1 | 1 |
| d | d | 1 | 1 | 1 | 1 | 1 |
| 1 | 1 | 1 | 1 | 1 | 1 | 1 |

.

$\mathcal{A}^L$ is an involutive left-m-BE algebra verifying (Pom) and (prel$_m$) and not verifying (m-Pabs-i) for $(a, 0)$, (m-B) for $(b, a, b)$, (m-BB) for $(b, b, a)$, (m-*) for $(a, c, b)$, (m-**) for $(b, a, b)$, (m-Tr) for $(b, a, c)$, (m-An) for $(a, b)$, (m-Pimpl) for $(b, 0)$, (Pqmv) for $(a, a, 0)$, (Pmv) for $(a, a)$, ($\Delta_m$) for $(d, a)$, (WNM$_m$) and (aWNM$_m$) for $(b, b)$, (m-Pdis) for $(a, b, a)$.

**Example 13.3.8 Proper PreMV algebra: PreMV**

By a PASCAL program, we found the algebra $\mathcal{A}^L = (A_5 = \{0, a, b, c, 1\}, \odot, ^-, 1)$ with the following tables of $\odot$ and $^-$ and of the additional operation $\oplus$:

| $\odot$ | 0 | a | b | c | d | 1 |
|---|---|---|---|---|---|---|
| 0 | 0 | 0 | 0 | 0 | 0 | 0 |
| a | 0 | 0 | 0 | 0 | 0 | a |
| b | 0 | 0 | 0 | 0 | 0 | b |
| c | 0 | 0 | 0 | 0 | a | c |
| d | 0 | 0 | 0 | a | a | d |
| 1 | 0 | a | b | c | d | 1 |

and

| $x$ | $x^-$ |
|---|---|
| 0 | 1 |
| a | d |
| b | c |
| c | b |
| d | a |
| 1 | 0 |

, with

| $\oplus$ | 0 | a | b | c | d | 1 |
|---|---|---|---|---|---|---|
| 0 | 0 | a | b | c | d | 1 |
| a | a | d | d | 1 | 1 | 1 |
| b | b | d | 1 | 1 | 1 | 1 |
| c | c | 1 | 1 | 1 | 1 | 1 |
| d | d | 1 | 1 | 1 | 1 | 1 |
| 1 | 1 | 1 | 1 | 1 | 1 | 1 |

.

$\mathcal{A}^L$ is an involutive left-m-BE algebra verifying (Pmv) (and hence ($\Delta_m$)) and (prel$_m$) and not verifying (m-Pabs-i) for $(a, 0)$, (m-B) for $(c, b, d)$, (m-BB) for $(c, d, c)$, (m-*) for $(b, a, c)$, (m-**) for $(c, b, d)$, (m-Tr) for $(c, b, a)$, (m-An) for $(a, b)$, (m-Pimpl) for $(a, 0)$, (Pqmv) for $(c, 1, d)$, (Pom) for $(c, d)$, (WNM$_m$) and (aWNM$_m$) for $(c, d)$, (m-Pdis) for $(a, a, c)$.

**Example 13.3.9 Involutive m-Be algebra verifying ($\Delta_m$) and not verifying (m-Pabs-i) and (Pmv)**

By MACE4, we found the algebra $\mathcal{A}^L = (A_5 = \{0, a, b, c, 1\}, \odot, ^-, 1)$ with the following tables of $\odot$ and $^-$ and of the additional operation $\oplus$:

| $\odot$ | 0 | a | b | c | 1 |
|---|---|---|---|---|---|
| 0 | 0 | 0 | 0 | 0 | 0 |
| a | 0 | a | 0 | 0 | a |
| b | 0 | 0 | c | 0 | b |
| c | 0 | 0 | 0 | 0 | c |
| 1 | 0 | a | b | c | 1 |

and

| $x$ | $x^-$ |
|---|---|
| 0 | 1 |
| a | b |
| b | a |
| c | c |
| 1 | 0 |

, with

| $\oplus$ | 0 | a | b | c | 1 |
|---|---|---|---|---|---|
| 0 | 0 | a | b | c | 1 |
| a | a | c | 1 | 1 | 1 |
| b | b | 1 | b | 1 | 1 |
| c | c | 1 | 1 | 1 | 1 |
| 1 | 1 | 1 | 1 | 1 | 1 |

.

$\mathcal{A}^L$ is an involutive left-m-BE algebra verifying ($\Delta_m$), (WNM$_m$), (aWNM$_m$) and not verifying (m-Pabs-i) for $(a, 0)$, (m-B) for $(a, c, a)$, (m-BB) for $(a, a, c)$, (m-*) for $(c, a, b)$, (m-**) for $(a, c, a)$, (m-Tr) for $(a, c, b)$, (m-An) for $(a, c)$, (m-Pimpl) for $(a, 0)$, (Pqmv) for $(b, 0, b)$, (Pom) for $(b, b)$, (Pmv) for $(b, b)$, (prel$_m$) for $(a, b)$, (m-Pdis) for $(a, a, a)$.

**Example 13.3.10 Involutive m-BE algebra verifying ($\Delta_m$) and (m-Pimpl) and not verifying (Pom)**

By MACE4, we found the algebra $\mathcal{A}^L = (A_8 = \{0, a, b, c, d, e, f, 1\}, \odot, ^-, 1)$ with the following tables of $\odot$ and $^-$ and of the additional operation $\oplus$:

| $\odot$ | 0 | a | b | c | d | e | f | 1 |
|---|---|---|---|---|---|---|---|---|
| 0 | 0 | 0 | 0 | 0 | 0 | 0 | 0 | 0 |
| a | 0 | a | 0 | c | 0 | 0 | 0 | a |
| b | 0 | 0 | b | 0 | b | 0 | 0 | b |
| c | 0 | c | 0 | c | 0 | 0 | 0 | c |
| d | 0 | 0 | b | 0 | d | 0 | 0 | d |
| e | 0 | 0 | 0 | 0 | 0 | e | 0 | e |
| f | 0 | 0 | 0 | 0 | 0 | 0 | f | f |
| 1 | 0 | a | b | c | d | e | f | 1 |

,

| $\oplus$ | 0 | a | b | c | d | e | f | 1 |
|---|---|---|---|---|---|---|---|---|
| 0 | 0 | a | b | c | d | e | f | 1 |
| a | a | a | 1 | a | 1 | 1 | 1 | 1 |
| b | b | 1 | b | 1 | d | 1 | 1 | 1 |
| c | c | a | 1 | c | 1 | 1 | 1 | 1 |
| d | d | 1 | d | 1 | d | 1 | 1 | 1 |
| e | e | 1 | 1 | 1 | 1 | e | 1 | 1 |
| f | f | 1 | 1 | 1 | 1 | 1 | f | 1 |
| 1 | 1 | 1 | 1 | 1 | 1 | 1 | 1 | 1 |

and $(0, a, b, c, d, e, f, 1)^- = (1, b, a, d, c, f, e, 0)$.

$\mathcal{A}^L$ is an involutive left-m-BE algebra verifying $(\Delta_m)$, (m-Pimpl) (hence (G) and (m-Pabs-i)), (prel$_m$), (WNM$_m$), (aWNM$_m$) and not verifying (m-B) for $(a, e, a)$, (m-BB) for $(a, a, e)$, (m-*) for $(a, e, f)$, (m-**) for $(a, e, a)$, (m-Tr) for $(a, e, b)$, (m-An) for $(a, c)$, (Pqmv) for $(a, 0, c)$, (Pom) for $(a, c)$, (Pmv) for $(a, c)$, (m-Pdis) for (a,b,d).

**Example 13.3.11 Involutive m-BE algebra verifying (m-Pimpl) and not verifying $(\Delta_m)$ and (Pom)**

By MACE4, we found the algebra $\mathcal{A}^L = (A_8 = \{0, a, b, c, d, e, f, 1\}, \odot, ^-, 1)$ with the following tables of $\odot$ and $^-$ and of the additional operation $\oplus$:

| $\odot$ | 0 | a | b | c | d | e | f | 1 |
|---|---|---|---|---|---|---|---|---|
| 0 | 0 | 0 | 0 | 0 | 0 | 0 | 0 | 0 |
| a | 0 | a | 0 | c | 0 | 0 | c | a |
| b | 0 | 0 | b | 0 | b | 0 | 0 | b |
| c | 0 | c | 0 | c | 0 | 0 | c | c |
| d | 0 | 0 | b | 0 | d | e | 0 | d |
| e | 0 | 0 | 0 | 0 | e | e | 0 | e |
| f | 0 | c | 0 | c | 0 | 0 | f | f |
| 1 | 0 | a | b | c | d | e | f | 1 |

,

| $\oplus$ | 0 | a | b | c | d | e | f | 1 |
|---|---|---|---|---|---|---|---|---|
| 0 | 0 | a | b | c | d | e | f | 1 |
| a | a | a | 1 | a | 1 | 1 | 1 | 1 |
| b | b | 1 | b | 1 | d | d | 1 | 1 |
| c | c | a | 1 | c | 1 | 1 | f | 1 |
| d | d | 1 | d | 1 | d | d | 1 | 1 |
| e | e | 1 | d | 1 | d | e | 1 | 1 |
| f | f | 1 | 1 | f | 1 | 1 | f | 1 |
| 1 | 1 | 1 | 1 | 1 | 1 | 1 | 1 | 1 |

and $(0, a, b, c, d, e, f, 1)^- = (1, b, a, d, c, f, e, 0)$.

$\mathcal{A}^L$ is an involutive left-m-BE algebra verifying (m-Pabs-i), (m-Pimpl), (prel$_m$), (WNM$_m$), (aWNM$_m$) and not verifying (m-B) for $(b, e, b)$, (m-BB) for $(b, b, e)$, (m-*) for $(a, c, e)$, (m-**) for $(b, e, b)$, (m-Tr) for $(b, e, a)$, (m-An) for $(a, c)$, (Pqmv) for $(a, 0, c)$, (Pom) for $(a, c)$, (Pmv) for $(a, c)$, $(\Delta_m)$ for $(a, e)$, (m-Pdis) for (a,b,d).

**Example 13.3.12 Involutive m-BE algebra verifying $(\Delta_m)$ and (G) and not verifying (m-Pimpl) and (Pom)**

By MACE4, we found the algebra $\mathcal{A}^L = (A_8 = \{0, a, b, c, d, e, f, 1\}, \odot, ^-, 1)$ with the following tables of $\odot$ and $^-$ and of the additional operation $\oplus$:

| $\odot$ | 0 | a | b | c | d | e | f | 1 |
|---|---|---|---|---|---|---|---|---|
| 0 | 0 | 0 | 0 | 0 | 0 | 0 | 0 | 0 |
| a | 0 | a | 0 | c | 0 | 0 | 0 | a |
| b | 0 | 0 | b | 0 | d | 0 | 0 | b |
| c | 0 | c | 0 | c | 0 | 0 | 0 | c |
| d | 0 | 0 | d | 0 | d | 0 | 0 | d |
| e | 0 | 0 | 0 | 0 | 0 | e | 0 | e |
| f | 0 | 0 | 0 | 0 | 0 | 0 | f | f |
| 1 | 0 | a | b | c | d | e | f | 1 |

,

| $\oplus$ | 0 | a | b | c | d | e | f | 1 |
|---|---|---|---|---|---|---|---|---|
| 0 | 0 | a | b | c | d | e | f | 1 |
| a | a | a | 1 | c | 1 | 1 | 1 | 1 |
| b | b | 1 | b | 1 | d | 1 | 1 | 1 |
| c | c | c | 1 | c | 1 | 1 | 1 | 1 |
| d | d | 1 | d | 1 | d | 1 | 1 | 1 |
| e | e | 1 | 1 | 1 | 1 | e | 1 | 1 |
| f | f | 1 | 1 | 1 | 1 | 1 | f | 1 |
| 1 | 1 | 1 | 1 | 1 | 1 | 1 | 1 | 1 |

and $(0, a, b, c, d, e, f, 1)^- = (1, b, a, d, c, f, e, 0)$.

$\mathcal{A}^L$ is an involutive left-m-BE algebra verifying $(\Delta_m)$, (G), (prel$_m$), (WNM$_m$), (aWNM$_m$) and not verifying (m-Pabs-i) for $(a, c)$, (m-B) for $(a, e, a)$, (m-BB) for

$(a, a, e)$, (m-*) for $(a, e, f)$, (m-**) for $(a, e, a)$, (m-Tr) for $(a, e, b)$, (m-An) for $(a, c)$, (m-Pimpl) for $(a, d)$, (Pqmv) for $(a, 0, c)$, (Pom) for $(a, c)$, (Pmv) for $(a, c)$, (m-Pdis) for (a,b,e).

**Example 13.3.13 Involutive m-BE algebra verifying (G) and not verifying ($\Delta_m$), (m-Pimpl) and (Pom) (Chapter 10)**

By a PASCAL program, we found the algebra $\mathcal{A}^L = (A_6 = \{0, a, b, c, d, 1\}, \odot, ^-, 1)$, with the following tables of $\odot$, $^-$ and of the additional operation $\oplus$:

| $\odot$ | 0 | a | b | c | d | 1 |
|---|---|---|---|---|---|---|
| 0 | 0 | 0 | 0 | 0 | 0 | 0 |
| a | 0 | a | 0 | 0 | 0 | a |
| b | 0 | 0 | b | 0 | 0 | b |
| c | 0 | 0 | 0 | c | c | c |
| d | 0 | 0 | 0 | c | d | d |
| 1 | 0 | a | b | c | d | 1 |

and

| $x$ | $x^-$ |
|---|---|
| 0 | 1 |
| a | d |
| b | c |
| c | b |
| d | a |
| 1 | 0 |

, with

| $\oplus$ | 0 | a | b | c | d | 1 |
|---|---|---|---|---|---|---|
| 0 | 0 | a | b | c | d | 1 |
| a | a | a | b | 1 | 1 | 1 |
| b | b | b | b | 1 | 1 | 1 |
| c | c | 1 | 1 | c | 1 | 1 |
| d | d | 1 | 1 | 1 | d | 1 |
| 1 | 1 | 1 | 1 | 1 | 1 | 1 |

.

$\mathcal{A}^L$ is an involutive left-m-BE algebra verifying (G), (prel$_m$), (WNM$_m$), (aWNM$_m$) and not verifying (m-Pabs-i) for $(a, b)$, (m-B) for $(a, b, a)$, (m-BB) for $(a, a, b)$, (m-*) for $(a, c, b)$, (m-**) for $(a, b, a)$, (m-Tr) for $(a, b, d)$, (m-An) for $(a, b)$, (m-Pimpl) for $(d, b)$, (Pqmv) for $(a, b, 0)$, (Pom) for $(d, c)$, (Pmv) for $(a, b)$, ($\Delta_m$) for $(c, a)$, (m-Pdis) for $(a, b, a)$.

**Example 13.3.14 Involutive m-Be algebra verifying (m-Pabs-i) and ($\Delta_m$) and not verifying (Pom) and (m-Pimpl)**

By MACE4, we found the algebra $\mathcal{A}^L = (A_5 = \{0, a, b, c, 1\}, \odot, ^-, 1)$ with the following tables of $\odot$ and $^-$ and of the additional operation $\oplus$:

| $\odot$ | 0 | a | b | c | 1 |
|---|---|---|---|---|---|
| 0 | 0 | 0 | 0 | 0 | 0 |
| a | 0 | 0 | c | 0 | a |
| b | 0 | c | 0 | 0 | b |
| c | 0 | 0 | 0 | 0 | c |
| 1 | 0 | a | b | c | 1 |

and

| $x$ | $x^-$ |
|---|---|
| 0 | 1 |
| a | a |
| b | b |
| c | c |
| 1 | 0 |

, with

| $\oplus$ | 0 | a | b | c | 1 |
|---|---|---|---|---|---|
| 0 | 0 | a | b | c | 1 |
| a | a | 1 | c | 1 | 1 |
| b | b | c | 1 | 1 | 1 |
| c | c | 1 | 1 | 1 | 1 |
| 1 | 1 | 1 | 1 | 1 | 1 |

.

$\mathcal{A}^L$ is an involutive left-m-BE algebra verifying (m-Pabs-i), ($\Delta_m$), (WNM$_m$), (aWNM$_m$) and not verifying (m-B) for $(a, c, b)$, (m-BB) for $(a, b, c)$, (m-*) for $(c, a, b)$, (m-**) for $(a, c, b)$, (m-Tr) for $(a, c, b)$, (m-An) for $(a, c)$, (m-Pimpl) for $(a, 0)$, (Pqmv) for $(a, 0, b)$, (Pom) for $(a, b)$, (Pmv) for $(a, b)$, (prel$_m$) for $(a, b)$, (m-Pdis) for $(a, a, a)$.

**Example 13.3.15 Involutive m-Be algebra verifying (m-Pabs-i) and not verifying (m-Pimpl) and ($\Delta_m$)**

By MACE4, we found the algebra $\mathcal{A}^L = (A_5 = \{0, a, b, c, 1\}, \odot, ^-, 1)$ with the following tables of $\odot$ and $^-$ and of the additional operation $\oplus$:

| $\odot$ | 0 | a | b | c | 1 |
|---|---|---|---|---|---|
| 0 | 0 | 0 | 0 | 0 | 0 |
| a | 0 | 0 | 0 | b | a |
| b | 0 | 0 | 0 | 0 | b |
| c | 0 | b | 0 | 0 | c |
| 1 | 0 | a | b | c | 1 |

and

| $x$ | $x^-$ |
|---|---|
| 0 | 1 |
| a | b |
| b | a |
| c | c |
| 1 | 0 |

, with

| $\oplus$ | 0 | a | b | c | 1 |
|---|---|---|---|---|---|
| 0 | 0 | a | b | c | 1 |
| a | a | 1 | 1 | 1 | 1 |
| b | b | 1 | 1 | a | 1 |
| c | c | 1 | a | 1 | 1 |
| 1 | 1 | 1 | 1 | 1 | 1 |

.

$\mathcal{A}^L$ is an involutive left-m-BE algebra verifying (m-Pabs-i) and (prel$_m$) and not verifying (m-B) for $(a,b,c)$, (m-BB) for $(a,c,a)$, (m-*) for $(a,b,c)$, (m-**) for $(a,b,c)$, (m-Tr) for $(a,b,c)$, (m-An) for $(a,b)$, (m-Pimpl) for $(a,0)$, (Pqmv) for $(a,0,c)$, (Pom) for $(a,c)$, (Pmv) for $(c,b)$, $(\Delta_m)$ for $(a,c)$, (WNM$_m$) and (aWNM$_m$) for $(c,a)$, (m-Pdis) for $(a,a,a)$.

**Example 13.3.16 Involutive m-BE algebra verifying (Pmv) and (m-Pabs-i) and not verifying (Pom)**

By MACE4, we found the algebra $\mathcal{A}^L = (A_9 = \{0,a,b,c,d,e,f,g,1\}, \odot, ^-, 1)$ with the following tables of $\odot$ and $^-$ and of the additional operation $\oplus$:

| $\odot$ | 0 | a | b | c | d | e | f | g | 1 |
|---|---|---|---|---|---|---|---|---|---|
| 0 | 0 | 0 | 0 | 0 | 0 | 0 | 0 | 0 | 0 |
| a | 0 | 0 | c | 0 | 0 | d | 0 | 0 | a |
| b | 0 | c | 0 | 0 | 0 | g | c | 0 | b |
| c | 0 | 0 | 0 | 0 | 0 | 0 | 0 | 0 | c |
| d | 0 | 0 | 0 | 0 | 0 | 0 | 0 | 0 | d |
| e | 0 | d | g | 0 | 0 | 0 | d | 0 | e |
| f | 0 | 0 | c | 0 | 0 | d | 0 | 0 | f |
| g | 0 | 0 | 0 | 0 | 0 | 0 | 0 | 0 | g |
| 1 | 0 | a | b | c | d | e | f | g | 1 |

and

| $x$ | $x^-$ |
|---|---|
| 0 | 1 |
| a | a |
| b | d |
| c | e |
| d | b |
| e | c |
| f | g |
| g | f |
| 1 | 0 |

,

with

| $\oplus$ | 0 | a | b | c | d | e | f | g | 1 |
|---|---|---|---|---|---|---|---|---|---|
| 0 | 0 | a | b | c | d | e | f | g | 1 |
| a | a | 1 | 1 | b | e | 1 | 1 | 1 | 1 |
| b | b | 1 | 1 | 1 | 1 | 1 | 1 | 1 | 1 |
| c | c | b | 1 | 1 | f | 1 | 1 | b | 1 |
| d | d | e | 1 | f | 1 | 1 | 1 | e | 1 |
| e | e | 1 | 1 | 1 | 1 | 1 | 1 | 1 | 1 |
| f | f | 1 | 1 | 1 | 1 | 1 | 1 | 1 | 1 |
| g | g | 1 | 1 | b | e | 1 | 1 | 1 | 1 |
| 1 | 1 | 1 | 1 | 1 | 1 | 1 | 1 | 1 | 1 |

.

$\mathcal{A}^L$ is an involutive left-m-BE algebra verifying (m-Pabs-i), (Pmv) (hence $(\Delta_m)$), (prel$_m$) and not verifying (m-B) for $(a,b,b)$, (m-BB) for $(a,b,d)$, (m-*) for $(b,d,a)$, (m-**) for $(a,b,b)$, (m-Tr) for $(a,b,d)$, (m-An) for $(a,f)$, (m-Pimpl) for $(a,0)$, (Pqmv) for $(a,1,b)$, (Pom) for $(a,b)$, (WNM$_m$) for $(a,b)$, (aWNM$_m$) for $(a,b)$, (m-Pdis) for (a,a,a).

- **Other examples of transitive involutive m-BE algebras**

**Example 13.3.17 Proper transitive OM algebra : tOM**

By MACE4 program, we found that the algebra
$\mathcal{A}^L = (A_8 = \{0,a,b,c,d,e,f,1\}, \odot, ^-, 1)$, with the following tables of $\odot$ and $^-$ and of the additional operation $\oplus$, is an involutive left-m-BE algebra verifying (Pom) and (m-Tr) $\Longleftrightarrow \ldots \Longleftrightarrow$ (m-BB), and also (prel$_m$), and not verifying (m-An) for $(a,e)$, (m-Pimpl) for $(a,0)$, (m-Pabs-i) for $(c,0)$, (G) for $a$, (Pqmv) for $(b,c,0)$, (Pmv) for $(b,c)$, $(\Delta_m)$ for $(a,c)$, (WNM$_m$) and (aWNM$_m$) for $(b,b)$, (m-Pdis) for

$(a, a, a)$, (Pmod) for $(a, 0, d)$.

| $\odot$ | 0 | a | b | c | d | e | f | 1 |
|---|---|---|---|---|---|---|---|---|
| 0 | 0 | 0 | 0 | 0 | 0 | 0 | 0 | 0 |
| a | 0 | 0 | 0 | 0 | a | 0 | 0 | a |
| b | 0 | 0 | c | 0 | a | 0 | c | b |
| c | 0 | 0 | 0 | 0 | 0 | 0 | 0 | c |
| d | 0 | a | a | 0 | d | e | e | d |
| e | 0 | 0 | 0 | 0 | e | 0 | 0 | e |
| f | 0 | 0 | c | 0 | e | 0 | c | f |
| 1 | 0 | a | b | c | d | e | f | 1 |

,

| $\oplus$ | 0 | a | b | c | d | e | f | 1 |
|---|---|---|---|---|---|---|---|---|
| 0 | 0 | a | b | c | d | e | f | 1 |
| a | a | d | 1 | b | 1 | d | 1 | 1 |
| b | b | 1 | 1 | b | 1 | 1 | 1 | 1 |
| c | c | b | b | c | 1 | f | f | 1 |
| d | d | 1 | 1 | 1 | 1 | 1 | 1 | 1 |
| e | e | d | 1 | f | 1 | d | 1 | 1 |
| f | f | 1 | 1 | f | 1 | 1 | 1 | 1 |
| 1 | 1 | 1 | 1 | 1 | 1 | 1 | 1 | 1 |

and $(0, a, b, c, d, e, f, 1)^- = (1, b, a, d, c, f, e, 0)$.

**Example 13.3.18 Involutive m-BE algebra verifying (Pmv) (hence ($\Delta_m$)), (m-B) $\Longleftrightarrow$ (m-BB) $\Longleftrightarrow$ ... (m-Tr), and not verifying the rest: tPreMV**

By a PASCAL program, we found the algebra $\mathcal{A}^L = (A_6 = \{0, a, b, c, d, 1\}, \odot, ^-, 1)$ with the following tables of $\odot$ and $^-$ and of the additional operation $\oplus$:

| $\odot$ | 0 | a | b | c | d | 1 |
|---|---|---|---|---|---|---|
| 0 | 0 | 0 | 0 | 0 | 0 | 0 |
| a | 0 | 0 | 0 | 0 | 0 | a |
| b | 0 | 0 | 0 | 0 | 0 | b |
| c | 0 | 0 | 0 | a | a | c |
| d | 0 | 0 | 0 | a | a | d |
| 1 | 0 | a | b | c | d | 1 |

and

| $x$ | $x^-$ |
|---|---|
| 0 | 1 |
| a | d |
| b | c |
| c | b |
| d | a |
| 1 | 0 |

, with

| $\oplus$ | 0 | a | b | c | d | 1 |
|---|---|---|---|---|---|---|
| 0 | 0 | a | b | c | d | 1 |
| a | a | d | d | 1 | 1 | 1 |
| b | b | d | d | 1 | 1 | 1 |
| c | c | 1 | 1 | 1 | 1 | 1 |
| d | d | 1 | 1 | 1 | 1 | 1 |
| 1 | 1 | 1 | 1 | 1 | 1 | 1 |

.

$\mathcal{A}^L$ is an involutive left-m-BE algebra verifying (Pmv) (and hence ($\Delta_m$)), (m-B) $\Longleftrightarrow$ (m-BB) $\Longleftrightarrow$ ... (m-Tr) and (prel$_m$) and not verifying (m-Pabs-i) for $(a, 0)$, (m-An) for $(a, b)$, (m-Pimpl) for $(a, 0)$, (Pqmv) for $(c, 1, c)$, (Pom) for $(c, c)$, (WNM$_m$) and (aWNM$_m$) for $(c, c)$, (m-Pdis) for $(a, a, c)$.

**Example 13.3.19 Involutive m-BE algebra verifying ($\Delta_m$), (m-B) $\Longleftrightarrow$ (m-BB) $\Longleftrightarrow$ ... (m-Tr), and not verifying the rest: tMMV**

By Mace4 program, we found the algebra $\mathcal{A}^L = (A_6 = \{0, a, b, c, d, 1\}, \odot, ^-, 1)$ with the following tables of $\odot$ and $^-$ and of the additional operation $\oplus$:

| $\odot$ | 0 | a | b | c | d | 1 |
|---|---|---|---|---|---|---|
| 0 | 0 | 0 | 0 | 0 | 0 | 0 |
| a | 0 | a | 0 | c | 0 | a |
| b | 0 | 0 | b | 0 | b | b |
| c | 0 | c | 0 | a | 0 | c |
| d | 0 | 0 | b | 0 | b | d |
| 1 | 0 | a | b | c | d | 1 |

and

| $x$ | $x^-$ |
|---|---|
| 0 | 1 |
| a | b |
| b | a |
| c | d |
| d | c |
| 1 | 0 |

, with

| $\oplus$ | 0 | a | b | c | d | 1 |
|---|---|---|---|---|---|---|
| 0 | 0 | a | b | c | d | 1 |
| a | a | a | 1 | a | 1 | 1 |
| b | b | 1 | b | 1 | d | 1 |
| c | c | a | 1 | a | 1 | 1 |
| d | d | 1 | d | 1 | b | 1 |
| 1 | 1 | 1 | 1 | 1 | 1 | 1 |

.

$\mathcal{A}^L$ is an involutive left-m-BE algebra verifying ($\Delta_m$), (m-B) $\Longleftrightarrow$ (m-BB) $\Longleftrightarrow$ ... (m-Tr) and (prel$_m$), (WNM$_m$), (aWNM$_m$) and not verifying (m-Pabs-i) for $(d, 0)$, (m-An) for $(a, c)$, (m-Pimpl) for $(c, 0)$, (Pqmv) for $(a, 0, c)$, (Pom) for $(a, c)$, (Pmv) for $(a, c)$, (m-Pdis) for $(a, a, c)$.

**Example 13.3.20 Involutive m-BE algebra verifying ($\Delta_m$) and (m-Pimpl), (m-B) $\Longleftrightarrow$ (m-BB) $\Longleftrightarrow$ ... (m-Tr), and not verifying (Pmv) and (Pom)**

By Mace4 program, we found the algebra $\mathcal{A}^L = (A_6 = \{0, a, b, c, d, 1\}, \odot, ^-, 1)$ with the following tables of $\odot$ and $^-$ and of the additional operation $\oplus$:

| $\odot$ | 0 | a | b | c | d | 1 |
|---|---|---|---|---|---|---|
| 0 | 0 | 0 | 0 | 0 | 0 | 0 |
| a | 0 | a | 0 | c | 0 | a |
| b | 0 | 0 | b | 0 | b | b |
| c | 0 | c | 0 | c | 0 | c |
| d | 0 | 0 | b | 0 | d | d |
| 1 | 0 | a | b | c | d | 1 |

and

| $x$ | $x^-$ |
|---|---|
| 0 | 1 |
| a | b |
| b | a |
| c | d |
| d | c |
| 1 | 0 |

, with

| $\oplus$ | 0 | a | b | c | d | 1 |
|---|---|---|---|---|---|---|
| 0 | 0 | a | b | c | d | 1 |
| a | a | a | 1 | a | 1 | 1 |
| b | b | 1 | b | 1 | d | 1 |
| c | c | a | 1 | c | 1 | 1 |
| d | d | 1 | d | 1 | d | 1 |
| 1 | 1 | 1 | 1 | 1 | 1 | 1 |

.

$\mathcal{A}^L$ is an involutive left-m-BE algebra verifying ($\Delta_m$) and (m-Pimpl) (hence (G) and (m-Pabs-i)), (m-B) $\Longleftrightarrow$ (m-BB) $\Longleftrightarrow$ ... (m-Tr) and (prel$_m$), (WNM$_m$) and (aWNM$_m$) and not verifying (m-An) for $(a, c)$, (Pqmv) for $(a, 0, c)$, (Pom) for $(a, c)$, (Pmv) for $(a, c)$, (m-Pdis) for $(a, b, d)$.

**Example 13.3.21 Transitive involutive m-BE algebra verifying ($\Delta_m$) and (G), (m-B) $\Longleftrightarrow$ (m-BB) $\Longleftrightarrow$ ... (m-Tr), and not verifying (m-Pimpl), (Pmv), (Pom)**

By Mace4 program, we found the algebra $\mathcal{A}^L = (A_6 = \{0, a, b, c, d, 1\}, \odot, ^-, 1)$ with the following tables of $\odot$ and $^-$ and of the additional operation $\oplus$:

| $\odot$ | 0 | a | b | c | d | 1 |
|---|---|---|---|---|---|---|
| 0 | 0 | 0 | 0 | 0 | 0 | 0 |
| a | 0 | a | 0 | 0 | d | a |
| b | 0 | 0 | b | c | 0 | b |
| c | 0 | 0 | c | c | 0 | c |
| d | 0 | d | 0 | 0 | d | d |
| 1 | 0 | a | b | c | d | 1 |

and

| $x$ | $x^-$ |
|---|---|
| 0 | 1 |
| a | b |
| b | a |
| c | d |
| d | c |
| 1 | 0 |

, with

| $\oplus$ | 0 | a | b | c | d | 1 |
|---|---|---|---|---|---|---|
| 0 | 0 | a | b | c | d | 1 |
| a | a | a | 1 | 1 | d | 1 |
| b | b | 1 | b | c | 1 | 1 |
| c | c | 1 | c | c | 1 | 1 |
| d | d | d | 1 | 1 | d | 1 |
| 1 | 1 | 1 | 1 | 1 | 1 | 1 |

.

$\mathcal{A}^L$ is an involutive left-m-BE algebra verifying ($\Delta_m$) and (G), (m-B) $\Longleftrightarrow$ (m-BB) $\Longleftrightarrow$ ... (m-Tr) and (prel$_m$), (WNM$_m$), (aWNM$_m$) and not verifying (m-Pabs-i) for $(a, d)$, (m-An) for $(a, d)$, (m-Pimpl) for $(a, c)$, (Pqmv) for $(a, 0, d)$, (Pom) for $(a, d)$, (Pmv) for $(a, d)$, (m-Pdis) for $(a, c, b)$.

**Example 13.3.22 Involutive m-BE algebra verifying ($\Delta_m$), (m-Pabs-i) and (m-B) $\Leftrightarrow$ (m-BB) $\Leftrightarrow$ ... (m-Tr), and not verifying (m-Pimpl) and (Pom)**

By MACE4, we found the algebra $\mathcal{A}^L = (A_8 = \{0, a, b, c, d, e, f, 1\}, \odot, ^-, 1)$ with the following tables of $\odot$ and $^-$ and of the additional operation $\oplus$:

| $\odot$ | 0 | a | b | c | d | e | f | 1 |
|---|---|---|---|---|---|---|---|---|
| 0 | 0 | 0 | 0 | 0 | 0 | 0 | 0 | 0 |
| a | 0 | d | 0 | b | d | e | b | a |
| b | 0 | 0 | 0 | b | 0 | 0 | b | b |
| c | 0 | b | b | c | 0 | 0 | c | c |
| d | 0 | d | 0 | 0 | d | e | 0 | d |
| e | 0 | e | 0 | 0 | e | e | 0 | e |
| f | 0 | b | b | c | 0 | 0 | f | f |
| 1 | 0 | a | b | c | d | e | f | 1 |

with

| $\oplus$ | 0 | a | b | c | d | e | f | 1 |
|---|---|---|---|---|---|---|---|---|
| 0 | 0 | a | b | c | d | e | f | 1 |
| a | a | 1 | 1 | 1 | a | a | 1 | 1 |
| b | b | 1 | c | c | a | a | f | 1 |
| c | c | 1 | c | c | 1 | 1 | f | 1 |
| d | d | a | a | 1 | d | d | 1 | 1 |
| e | e | a | a | 1 | d | e | 1 | 1 |
| f | f | 1 | f | f | 1 | 1 | f | 1 |
| 1 | 1 | 1 | 1 | 1 | 1 | 1 | 1 | 1 |

and $(0, a, b, c, d, e, f, 1)^- = (1, b, a, d, c, f, e, 0)$.

$\mathcal{A}^L$ is an involutive left-m-BE algebra verifying ($\Delta_m$), (m-Pabs-i), (m-B) $\Longleftrightarrow$ (m-BB) $\Longleftrightarrow \ldots$ (m-Tr), (prel$_m$), (WNM$_m$), (aWNM$_m$) and not verifying (m-An) for $(c, f)$, (m-Pimpl) for $(a, 0)$, (Pqmv) for $(a, 0, e)$, (Pom) for $(d, e)$, (Pmv) for $(a, e)$, (m-Pdis) for (a,a,a).

**Example 13.3.23 Involutive m-BE algebra verifying (Pmv), (m-Pabs-i) and (m-Tr), and not verifying (m-Pimpl) and (Pom)**

By MACE4, we found the algebra $\mathcal{A}^L = (A_{10} = \{0, a, b, c, d, e, f, g, h, 1\}, \odot, ^-, 1)$ with the following tables of $\odot$ and $^-$ and of the additional operation $\oplus$:

| $\odot$ | 0 | a | b | c | d | e | f | g | h | 1 |
|---|---|---|---|---|---|---|---|---|---|---|
| 0 | 0 | 0 | 0 | 0 | 0 | 0 | 0 | 0 | 0 | 0 |
| a | 0 | 0 | c | 0 | 0 | d | c | d | 0 | a |
| b | 0 | c | f | c | 0 | h | f | d | c | b |
| c | 0 | 0 | c | 0 | 0 | 0 | c | 0 | 0 | c |
| d | 0 | 0 | 0 | 0 | 0 | d | 0 | d | 0 | d |
| e | 0 | d | h | 0 | d | g | c | g | d | e |
| f | 0 | c | f | c | 0 | c | f | 0 | c | f |
| g | 0 | d | d | 0 | d | g | 0 | g | d | g |
| h | 0 | 0 | c | 0 | 0 | d | c | d | 0 | h |
| 1 | 0 | a | b | c | d | e | f | g | h | 1 |

and

| $x$ | $x^-$ |
|---|---|
| 0 | 1 |
| a | a |
| b | d |
| c | e |
| d | b |
| e | c |
| f | g |
| g | f |
| h | h |
| 1 | 0 |

,

with

| $\oplus$ | 0 | a | b | c | d | e | f | g | h | 1 |
|---|---|---|---|---|---|---|---|---|---|---|
| 0 | 0 | a | b | c | d | e | f | g | h | 1 |
| a | a | 1 | 1 | b | e | 1 | b | e | 1 | 1 |
| b | b | 1 | 1 | b | 1 | 1 | b | 1 | 1 | 1 |
| c | c | b | b | f | h | 1 | f | e | b | 1 |
| d | d | e | 1 | h | g | e | b | g | e | 1 |
| e | e | 1 | 1 | 1 | e | 1 | 1 | e | 1 | 1 |
| f | f | b | b | f | b | 1 | f | 1 | b | 1 |
| g | g | e | 1 | e | g | e | 1 | g | e | 1 |
| h | h | 1 | 1 | b | e | 1 | b | e | 1 | 1 |
| 1 | 1 | 1 | 1 | 1 | 1 | 1 | 1 | 1 | 1 | 1 |

.

$\mathcal{A}^L$ is an involutive left-m-BE algebra verifying (m-Pabs-i), (m-B) $\Longleftrightarrow$ (m-BB) $\Longleftrightarrow \ldots$ (m-Tr), (Pmv) (hence ($\Delta_m$)), (prel$_m$), (WNM$_m$), (aWNM$_m$) and not verifying (m-An) for $(a, h)$, (m-Pimpl) for $(a, 0)$, (Pqmv) for $(a, b, e)$, (Pom) for $(a, b)$, (m-Pdis) for (a,a,a).

# Chapter 14

# The properties (m-Pabs-i) and ($\mathbf{WNM}_m$)

In this chapter, we continue the results from Chapters 11, 12, 13 in the "world" of involutive *m-M algebras*.

In Section 14.1, we introduce the properties ($\mathrm{WNM}_m$) and ($\mathrm{aWNM}_m$) in connection with the property (WNM) introduced in [50] and recalled in Chapter 12 - and we prove Theorem 14.1.10, saying that in a left-MV algebra, we have:

$$(\text{m-Pabs-i}) \iff (\mathrm{aWNM}_m) \iff (\mathrm{WNM}_m)\ (= (\mathrm{WNM})),$$

i.e. $\mathbf{atOWL} = {}_{(WNM)}\mathbf{MV}$.

In Section 14.2, we generalize the results from Section 14.1, proving Theorem 14.2.12, which says that in a left-tQMV algebra, we have:

$$(\text{m-Pabs-i}) \iff (\mathrm{aWNM}_m) \iff (\mathrm{WNM}_m),$$

i.e. $\mathbf{tOMWL} = {}_{(WNM_m)}\mathbf{tQMV}$.

The content of this chapter is taken from [121].

## 14.1 At involutive m-aBE algebras level

### 14.1.1 The properties (WNM) and ($\mathrm{WNM}_m$), ($\mathrm{aWNM}_m$)

Recall the following property from Chapter 12:

(WNM) (Weak Nilpotent Minimum) $\quad (x \odot y)^- \vee [(x \wedge y) \to (x \odot y)] = 1,$

where $\wedge$ and $\vee$ were lattice operations.

Let $\mathcal{A}^L = (A^L, \odot, ^-, 1)$ be an involutive m-BE algebra. Here, we shall consider, formally, that:

$x \wedge y \overset{def.}{=} x \wedge_m^B y = y \wedge_m^M x = (y^- \odot x)^- \odot x \overset{(Pcomm)}{=} x \odot (x \odot y^-)^- = x \odot (x \to y)$

and

$x \vee y \stackrel{def.}{=} x \vee_m^B y = y \vee_m^M x = x \oplus (x \oplus y^-)^- = (y^- \wedge_m x^-)^- = (x^- \wedge y^-)^-$,
hence $x \wedge y = (x^- \vee y^-)^-$, and $x \to y \stackrel{def.}{=} (x \odot y^-)^-$; hence, here, $\wedge$ and $\vee$ are no more lattice operations; in this case, the property (WNM) will be denoted by ($\text{WNM}_m$):

($\text{WNM}_m$) $$(x \odot y)^- \vee_m^B [(x \wedge_m^B y) \to (x \odot y)] = 1.$$

Recall that, in MV algebras, (WNM) and ($\text{WNM}_m$) coincide, because $\wedge = \wedge_m^B = \wedge_m^M$ and $\vee = \vee_m^B = \vee_m^M$.

Hence, ($\text{WNM}_m$) becomes, equivalently:
$(x \odot y)^- \vee_m^B [(x \odot (x \odot y^-)^-) \odot (x \odot y)^-]^- = 1$ or, equivalently,
$(x \odot y) \wedge_m^B [x \odot (x \odot y^-)^- \odot (x \odot y)^-] = 0$ or, equivalently,
$(x \odot y) \odot [(x \odot y) \odot [x \odot (x \odot y^-)^- \odot (x \odot y)^-]^-]^- = 0$
or, equivalently,
(a) $$x \odot y \leq_m (x \odot y) \odot [x \odot (x \odot y^-)^- \odot (x \odot y)^-]^-.$$
By (Chapter 6, (mA14)), (Pcomm) + (Pass) + (m-La) + (m-Re) $\Longrightarrow$ (m-Pleq); hence, (m-Pleq) holds in $\mathcal{A}^L$; thus, we also have:
(b) $$(x \odot y) \odot [x \odot (x \odot y^-)^- \odot (x \odot y)^-]^- \leq_m x \odot y.$$

Consequently, if (m-An) holds, then, from (a) and (b), we obtain the new property ($\text{aWNM}_m$):

($\text{aWNM}_m$) $$(x \odot y) \odot [x \odot (x \odot y^-)^- \odot (x \odot y)^-]^- = x \odot y.$$

Thus, we have proved the following result:

**Lemma 14.1.1** *Let $\mathcal{A}^L = (A^L, \odot, ^-, 1)$ be an involutive m-aBE algebra. Then:*

$$(WNM_m) \Longrightarrow (aWNM_m).$$

We also have:

**Lemma 14.1.2** *Let $\mathcal{A}^L = (A^L, \odot, ^-, 1)$ be an involutive m-BE algebra. Then:*

$$(aWNM_m) \Longrightarrow (WNM_m).$$

**Proof.** If ($\text{aWNM}_m$) holds, then:
$(x \odot y) \odot [(x \odot y) \odot [x \odot (x \odot y^-)^- \odot (x \odot y)^-]^-]^- = (x \odot y) \odot [x \odot y]^- \stackrel{(m-Re)}{=} 0$,
hence ($\text{WNM}_m$) holds. □

By Lemmas 14.1.1 and 14.1.2, we obtain:

**Proposition 14.1.3** *Let $\mathcal{A}^L = (A^L, \odot, ^-, 1)$ be an involutive m-aBE algebra. Then:*

$$(aWNM_m) \Longleftrightarrow (WNM_m).$$

**Remarks 14.1.4** *Note that:*
*- the proper OWL from Example 10.4.8 verifies (a$WNM_m$) and (hence) ($WNM_m$),*
*- the first proper tOWL from Examples 10.4.9 verifies (a$WNM_m$) and (hence) ($WNM_m$), while the second proper tOWL verifies ($WNM_m$), but does not verify (a$WNM_m$) for $(a, e)$;*
*- both taOWLs (MV algebras) from Examples 10.4.10 verify (a$WNM_m$) and (hence) ($WNM_m$);*
*- the proper MV algebra from Example 10.4.11 does not verify (a$WNM_m$) and ($WNM_m$) for $(b, b)$.*

In the sequel, we shall continue the study of the properties ($\mathrm{WNM}_m$), (a$\mathrm{WNM}_m$) in connection with the property (m-Pabs-i).

### 14.1.2 At (involutive) m-BCK algebras level

Consider the following properties:
($\wedge_m$-comm) $(y^- \odot x)^- \odot x = (x^- \odot y)^- \odot y$,
(a$\mathrm{WNM}_m$) $(x \odot y) \odot [x \odot (x \odot y^-)^- \odot (x \odot y)^-]^- = x \odot y$,
(m-Pabs-i) $x \odot (x^- \odot x^- \odot y^-)^- = x$ or $x \odot (x \oplus x \oplus y) = x$.

**The below Theorem 14.1.6 was proved by *Prover9* in 0.03 seconds; the proof by *Prover9* had 29 lines and was divided into the proofs of next Lemma 14.1.5 and Theorem 14.1.6.**

**Lemma 14.1.5** *Let $\mathcal{A}^L = (A^L, \odot, ^-, 1)$ be a left-MV algebra. Then, we have:*

$$x \odot (x \odot (x \odot y)^-)^- = x \odot y, \tag{14.1}$$

$$(x^- \odot y^-)^- \odot (x \odot (y \odot x)^-)^- = y. \tag{14.2}$$

**Proof.** (By *Prover9*.)
(14.1): In ($\wedge_m$-comm), take $Y := x \odot y$ to obtain:
$x \odot (x \odot (x \odot y)^-)^- = (x \odot y) \odot (x^- \odot (x \odot y))^-$
$\overset{(Pass),(Pcomm)}{=} (x \odot y) \odot ((x \odot x^-) \odot y)^-$
$\overset{(m-Re),(m-La),(Neg0-1)}{=} (x \odot y) \odot 1 \overset{(PU)}{=} x \odot y$; thus, (14.1) holds.
(14.2): In ($\wedge_m$-comm), take $X := (x^- \odot y^-)^-$ and $Y := y$ to obtain:
$(x^- \odot y^-)^- \odot (y^- \odot (x^- \odot y^-)^-)^- = y \odot (y \odot (x^- \odot y^-))^-$
$\overset{(Pass),(Pcomm)}{=} y \odot ((y \odot y^-) \odot x^-)^- \overset{(m-Re),(m-La),(Neg0-1)}{=} y \odot 1 \overset{(PU)}{=} y$.
Then, by ($\wedge_m$-comm) again, for $Y := y^-$, we obtain:
$y = (x^- \odot y^-)^- \odot (y^- \odot (x^- \odot y^-)^-)^- = (x^- \odot y^-)^- \odot (x \odot (y \odot x)^-)^-$,
hence $(x^- \odot y^-)^- \odot (x \odot (y \odot x)^-)^- = y$, i.e. (14.2) holds. □

**Theorem 14.1.6** *Let $\mathcal{A}^L = (A^L, \odot, ^-, 1)$ be a left-MV algebra. Then:*

$$(aWNM_m) \Longrightarrow (m - Pabs - i).$$

**Proof.** (By *Prover9.*)

From ($aWNM_m$), by (Pass), (Pcomm), we obtain:

$$x \odot (y \odot [x \odot ((x \odot y)^- \odot (x \odot y^-)^-)]^-) = x \odot y. \tag{14.3}$$

By (14.3), for $y := x$, we obtain:
$x \odot x = x \odot (x \odot [x \odot ((x \odot x)^- \odot (x \odot x^-)^-)]^-)$
$\overset{(m-Re),(Neg0-1)}{=} x \odot (x \odot [x \odot ((x \odot x)^- \odot 1)]^-)$
$\overset{(PU)}{=} x \odot (x \odot [x \odot (x \odot x)^-]^-) \overset{(14.1)}{=} x \odot (x \odot x)$, i.e. we have:

$$x \odot (x \odot x) = x \odot x. \tag{14.4}$$

By (14.4), we obtain:
$(x \odot (x \odot x)) \odot y = (x \odot x) \odot y$;
then, by (Pass), we obtain:

$$x \odot (x \odot (x \odot y)) = x \odot (x \odot y). \tag{14.5}$$

Take now in (14.2) $X := x \odot (x \odot y)$ and $Y := x$ to obtain:
$[(x \odot (x \odot y))^- \odot x^-]^- \odot [(x \odot (x \odot y)) \odot (x \odot (x \odot (x \odot y)))^-]^- = x$,
hence, by (14.5), to obtain:
$[(x \odot (x \odot y))^- \odot x^-]^- \odot [(x \odot (x \odot y)) \odot (x \odot (x \odot y))^-]^- = x$,
hence by (m-Re), (Neg0-1), (PU), to obtain:

$$[(x \odot (x \odot y))^- \odot x^-]^- = x. \tag{14.6}$$

Finally, from (14.6), we obtain, by (DN):
$(x \odot (x \odot y))^- \odot x^- = x^-$,
which for $X := x^-$ and $Y := y^-$ gives, by (Pcomm) and (DN):
$X \odot (X^- \odot (X^- \odot Y^-))^- = X$, i.e. (m-Pabs-i) holds. □

**The converse of Theorem 14.1.6, which is the below Theorem 14.1.9, was proved by *Prover9* in 7344.88 seconds; the proof by *Prover9* had 49 lines and was divided into the proofs of next Lemmas 14.1.7, 14.1.8 and Theorem 14.1.9.**

**Lemma 14.1.7** *Let $\mathcal{A}^L = (A^L, \odot, ^-, 1)$ be an involutive m-BE algebra. We have:*

$$x \odot (x^- \odot y) = 0, \tag{14.7}$$

$$x \odot (y \odot x^-) = 0, \tag{14.8}$$

$$x \odot (y \odot (z \odot (x \odot y)^-)) = 0. \tag{14.9}$$

**Proof.** (By *Prover9.*)

(14.7): By (Pass), (m-Re), (m-La).

(14.8): From (14.7), by (Pcomm).

(14.9): $x \odot (y \odot (z \odot (x \odot y)^-)) \overset{(Pass)}{=} (x \odot y) \odot (z \odot (x \odot y)^-) = 0$, by (14.8). □

**Lemma 14.1.8** *Let $\mathcal{A}^L = (A^L, \odot, ^-, 1)$ be a left-MV algebra. We have:*

(14.10) $$x \odot (y^- \odot x)^- = y \odot (y \odot x^-)^-,$$

(14.11) $$x \odot ((y^- \odot x)^- \odot z) = y \odot ((x^- \odot y)^- \odot z),$$

(14.12) $$(x^- \odot y^-)^- \odot (x \odot (y \odot x)^-)^- = y, \quad \textit{that is } (14.2)$$

(14.13) $$x \odot (y \odot (z^- \odot y)^-) = z \odot (x \odot (y^- \odot z)^-),$$

(14.14) $$x^- \odot (x^- \odot y^-)^- = y \odot (x \odot y)^-,$$

(14.15) $$x^- \odot (y^- \odot x^-)^- = y \odot (y \odot x)^-,$$

(14.16) $$(x^- \odot y^-)^- \odot (y \odot (x \odot y)^-)^- = x,$$

(14.17) $$x^- \odot ((y \odot (x \odot y)^-)^- \odot z) = y^- \odot (x^- \odot z),$$

(14.18) $$x^- \odot (x^- \odot y)^- = y^- \odot (x \odot y^-)^-,$$

(14.19) $$x \odot (y \odot ((x^- \odot y)^- \odot z))^- = x \odot (x \odot ((y^- \odot x)^- \odot z))^-,$$

(14.20) $$(x^- \odot y)^- \odot (y^- \odot (x \odot y^-)^-)^- = x,$$

(14.21) $$(y^- \odot (x \odot y^-)^-)^- \odot [z \odot (y \odot x^-)^-] = z \odot x,$$

(14.22) $$x \odot (x \odot (y \odot x)^-)^- = y \odot x, \quad \textit{see } (14.1)$$

(14.23) $$(x^- \odot y)^- \odot (z \odot (y^- \odot (x \odot y^-)^-)^-) = z \odot x,$$

(14.24) $$x \odot ((x^- \odot y)^- \odot (z \odot (x^- \odot y)^-)^-)^- = z \odot x,$$

(14.25) $$x \odot (y \odot (x^- \odot (z \odot y))^-)^- = y^- \odot x,$$

(14.26) $$x \odot (y \odot (x^- \odot (z \odot (u \odot y)))^-)^- = y^- \odot x.$$

**Proof.** (By *Prover9.*)

(14.10): From ($\wedge_m$-comm), by (Pcomm).

(14.11): From ($\wedge_m$-comm), i.e. from $x \odot (y^- \odot x)^- = y \odot (x^- \odot y)^-$, we obtain: $(x \odot (y^- \odot x)^-) \odot z = (y \odot (x^- \odot y)^-) \odot z$; then, by (Pass), we obtain (14.11).

(14.13): In ($\wedge_m$-comm), take $X := y$ and $Y := z$ to obtain:
$y \odot (z^- \odot y)^- = z \odot (y^- \odot z)^-$;
then, "multiply" both sides by $x$ and use (Pass) and (Pcomm).

(14.14): By (14.10), for $X := x^-$ and by (Pcomm).

(14.15): By (14.14) and (Pcomm).

(14.16): In (14.12), interchange $x$ with $y$ to obtain:
$(y^- \odot x^-)^- \odot (y \odot (x \odot y)^-)^- = x$; then, use (Pcomm).

(14.17): $x^- \odot ((y \odot (x \odot y)^-)^- \odot z)$
$\overset{(14.14)}{=} x^- \odot ((x^- \odot (x^- \odot (x^- \odot y^-)^-)^- \odot z)$
$\overset{(14.11)}{=} (x^- \odot y^-) \odot ((x \odot (x^- \odot y^-))^- \odot z)$
$\overset{(14.7)}{=} (x^- \odot y^-) \odot (0^- \odot z) = (x^- \odot y^-) \odot z \overset{(Pcomm),(Pass)}{=} y^- \odot (x^- \odot z)$,
for $X := x^-$ and $Y := x^- \odot y^-$ in (14.11).

(14.18): By (14.14), for $X := x$ and $Y := y^-$.

(14.19): By (14.11), we obtain:
$(y \odot ((x^- \odot y)^- \odot z))^- = (x \odot ((y^- \odot x)^- \odot z))^-$;
then, "multiply" both sides by $x$.

(14.20): By (14.16), for $X := x$ and $Y := y^-$.

(14.21): In (14.20), "multiply" by $z$ the both sides, to obtain:
$z \odot [(x^- \odot y)^- \odot (y^- \odot (x \odot y^-)^-)^-] = z \odot x$,
hence, by (Pass), (Pcomm), we obtain:
$(y^- \odot (x \odot y^-)^-)^- \odot [z \odot (y \odot x^-)^-] = z \odot x$.

(14.22): By (14.1) and (Pcomm).

(14.23): $(x^- \odot y)^- \odot (z \odot (y^- \odot (x \odot y^-)^-)^-)$
$\overset{(14.18)}{=} (x^- \odot y)^- \odot (z \odot (x^- \odot (x^- \odot y)^-)^-)$
$\overset{(14.13)}{=} z \odot (x \odot ((x^- \odot y) \odot x)^-) = z \odot (x \odot 0^-) = z \odot x$.

(14.24): In (14.22), take $X := (y \odot x^-)^-$ and $Y := z$ to obtain:
(a) $(y \odot x^-)^- \odot ((y \odot x^-)^- \odot (z \odot (y \odot x^-)^-)^-)^- = z \odot (y \odot x^-)^-$.
Now, we replace $z \odot (y \odot x^-)^-$ from (a) in (14.21) to obtain:
(b) $(y^- \odot (x \odot y^-)^-)^- \odot [(y \odot x^-)^- \odot ((y \odot x^-)^- \odot (z \odot (y \odot x^-)^-)^-)^-] = z \odot x$,
then, by (Pcomm), (Pass), we obtain:
(c) $((y \odot x^-)^- \odot (z \odot (y \odot x^-)^-)^-)^- \odot [(y \odot x^-)^- \odot (y^- \odot (x \odot y^-)^-)^-] = z \odot x$,
then, by (Pcomm), we obtain:
(d) $((x^- \odot y)^- \odot (z \odot (x^- \odot y)^-)^-)^- \odot [(x^- \odot y)^- \odot (y^- \odot (x \odot y^-)^-)^-] = z \odot x$.
Then, from (d), by (14.20), we obtain:
$((x^- \odot y)^- \odot (z \odot (x^- \odot y)^-)^-)^- \odot x = z \odot x$,
then, by (Pcomm), we obtain (14.24).

(14.25): In (14.17), take $X := x^- \odot y$, $Y := z$ and $Z := (y^- \odot (x \odot y^-)^-)^-$ to obtain:
(a) $(x^- \odot y)^- \odot [(z \odot ((x^- \odot y) \odot z)^-)^- \odot (y^- \odot (x \odot y^-)^-)^-]$

$= z^- \odot ((x^- \odot y)^- \odot (y^- \odot (x \odot y^-)^-)^-) \overset{(14.20)}{=} z^- \odot x$,
hence, we have:
(b) $(x^- \odot y)^- \odot [(z \odot ((x^- \odot y) \odot z)^-)^- \odot (y^- \odot (x \odot y^-)^-)^-] = z^- \odot x$.
Now, from (b), we obtain, by (Pass):
(c) $(x^- \odot y)^- \odot [(z \odot (x^- \odot (y \odot z))^-)^- \odot (y^- \odot (x \odot y^-)^-)^-] = z^- \odot x$.
Then, in (14.23), take: $X := x$, $Y := y$ and $Z := (z \odot (x^- \odot (y \odot z))^-)^-$ to obtain:
(d) $(x^- \odot y)^- \odot [(z \odot (x^- \odot (y \odot z))^-)^- \odot (y^- \odot (x \odot y^-)^-)^-]$
$= (z \odot (x^- \odot (y \odot z))^-)^- \odot x$.
Then, by (c), from (d) we obtain:
(e) $(z \odot (x^- \odot (y \odot z))^-)^- \odot x = z^- \odot x$.
Finally, from (e), by interchanging $z$ with $y$, we obtain:
(f) $(y \odot (x^- \odot (z \odot y))^-)^- \odot x = y^- \odot x$,
hence, by (Pcomm), we obtain (14.25).

(14.26): In (14.25), take $X := x$, $Y := y$, $Z := z \odot u$ to obtain:
$x \odot (y \odot (x^- \odot ((z \odot u) \odot y))^-)^- = y^- \odot x$;
then, by (Pass), we obtain (14.26). □

**Theorem 14.1.9** *Let $\mathcal{A}^L = (A^L, \odot, ^-, 1)$ be a left-MV algebra. Then:*

$$(m - Pabs - i) \implies (aWNM_m).$$

**Proof.** (By *Prover9*.)

First, we prove:

$$x \odot (x^- \odot (y \odot (y \odot x)^-))^- = x. \tag{14.27}$$

Indeed, $x \odot (x^- \odot (y \odot (y \odot x)^-))^- \overset{(14.15)}{=} x \odot (x^- \odot (x^- \odot (y^- \odot x^-)^-))^- \overset{(m-Pabs-i)}{=} x$,
for $Y := y^- \odot x^-$ in (m-Pabs-i).

Next, we prove:

$$x \odot (y \odot ((z \odot x)^- \odot [z \odot (x \odot (y \odot (z \odot x)^-))^-]^-)) = x \odot (y \odot (z \odot x)^-). \tag{14.28}$$

Indeed, in (14.27), take $X := x \odot (y \odot (z \odot x)^-)$, $Y := z$ to obtain:
(a) $X \odot [X^- \odot (Y \odot (Y \odot X)^-)]^- = X$;
but, in the left side of (a),
$Y \odot (Y \odot X)^- = z \odot (z \odot (x \odot (y \odot (z \odot x)^-)))^- \overset{(14.9)}{=} z \odot 0^- = z$;
hence, (a) becomes:
(b) $X \odot [X^- \odot z]^- = X$,
hence, by (Pcomm), (b) becomes:
(c) $X \odot [z \odot X^-]^- = X$, i.e.
(d) $(x \odot (y \odot (z \odot x)^-)) \odot [z \odot (x \odot (y \odot (z \odot x)^-))^-]^- = x \odot (y \odot (z \odot x)^-)$,
hence, by (Pass),
(e) $x \odot ((y \odot (z \odot x)^-) \odot [z \odot (x \odot (y \odot (z \odot x)^-))^-]^-) = x \odot (y \odot (z \odot x)^-)$,
hence by (Pass),
$x \odot (y \odot ((z \odot x)^- \odot [z \odot (x \odot (y \odot (z \odot x)^-))^-]^-)) = x \odot (y \odot (z \odot x)^-)$, that is (14.28).

Next, we prove:

(14.29) $$x \odot (y \odot (x^- \odot (z \odot (y \odot x^-)^-))^-) = y \odot x.$$

Indeed, first, in (14.26), take $X := x$, $Y := (y \odot (x^- \odot (z \odot (y \odot x^-)^-))^-)^-$, $Z := z$ and $U := (y \odot x^-)^-$ to obtain:
(a) $x \odot (Y \odot (x^- \odot (z \odot (U \odot Y)))^-)^- = Y^- \odot x$;
but, in the left side of (a),
$(x^- \odot (z \odot (U \odot Y)))^-$
$= (x^- \odot (z \odot ((y \odot x^-)^- \odot (y \odot (x^- \odot (z \odot (y \odot x^-)^-))^-)^-)))^-$
$\overset{(14.28)}{=} (x^- \odot (z \odot (y \odot x^-)^-))^-$,
where we took in (14.28) $X := x^-$, $Y := z$ and $Z := y$;
hence, (a) becomes:
(b) $x \odot [(y \odot (x^- \odot (z \odot (y \odot x^-)^-))^-)^- \odot (x^- \odot (z \odot (y \odot x^-)^-))^-]^-$
$= (y \odot (x^- \odot (z \odot (y \odot x^-)^-))^-)^= \odot x$.
Then, from (b), by (Pcomm), we obtain:
(c) $x \odot [(x^- \odot (z \odot (y \odot x^-)^-))^- \odot (y \odot (x^- \odot (z \odot (y \odot x^-)^-))^-)^-]^-$
$= (y \odot (x^- \odot (z \odot (y \odot x^-)^-))^-)^= \odot x$.
Then, in (14.24), take $X := x$, $Y := z \odot (y \odot x^-)^-$ and $Z := y$ to obtain:
(d) $x \odot [(x^- \odot (z \odot (y \odot x^-)^-))^- \odot (y \odot (x^- \odot (z \odot (y \odot x^-)^-))^-)^-]^- = y \odot x$;
then, from (d), by (c), we obtain:
(e) $(y \odot (x^- \odot (z \odot (y \odot x^-)^-))^-)^= \odot x = y \odot x$.
Finally, by (DN) and (Pcomm), from (e), we obtain:
$x \odot (y \odot (x^- \odot (z \odot (y \odot x^-)^-))^-) = y \odot x$, that is (14.29).

Next, we prove:

(14.30) $$x \odot (y \odot [y \odot ((x \odot y)^- \odot (y \odot x^-)^-)]^-) = y \odot x.$$

Indeed, in (14.19), take $X := y$, $Y := x^-$, $Z := z$ to obtain:
(a) $y \odot (x^- \odot ((y^- \odot x^-)^- \odot z))^- = y \odot [y \odot ((x^= \odot y)^- \odot z)]^-$;
now, we "multiply" both sides of (a) with $x$ and take in the result $X := x$, $Y := y$, $Z := (y \odot x^-)^-$ to obtain:
(b) $x \odot (y \odot (x^- \odot ((y^- \odot x^-)^- \odot (y \odot x^-)^-))^-)$
$= x \odot (y \odot [y \odot ((x^= \odot y)^- \odot (y \odot x^-)^-)]^-)$;
the left side of (b) equals $y \odot x$, by (14.29) (where $z := (y^- \odot x^-)^-$);
thus, by (DN), we obtain, from (b):
$y \odot x = x \odot (y \odot [y \odot ((x \odot y)^- \odot (y \odot x^-)^-)]^-)$, that is (14.30).

Finally, from (14.30), by (Pass), (Pcomm), we obtain:
(a) $(x \odot y) \odot [(x \odot y)^- \odot (y \odot (y \odot x^-)^-)]^- = x \odot y$;
from (a), by ($\wedge_m$-comm), we obtain:
(b) $(x \odot y) \odot [(x \odot y)^- \odot (x \odot (x \odot y^-)^-)]^- = x \odot y$,
which by (Pcomm), (Pass) again, gives:
$(x \odot y) \odot [x \odot (x \odot y^-)^- \odot (x \odot y)^-]^- = x \odot y$, that is ($aWNM_m$). □

By Theorems 14.1.6, 14.1.9 and Proposition 14.1.3, we obtain:

**Theorem 14.1.10** *Let* $\mathcal{A}^L = (A^L, \odot, ^-, 1)$ *be a left-MV algebra. Then:*

$$(m - Pabs - i) \iff (aWNM_m) \ (\iff (WNM_m) = (WNM)).$$

Recall that the class of MV algebras verifying (WNM) was denoted by $_{(WNM)}\mathbf{MV}$. Then, Theorem 14.1.10 says that:

$$\mathbf{atOWL} = {}_{(\mathbf{WNM})}\mathbf{MV} \ (\subset \mathbf{MV}).$$

See Examples 15.3.4 and Examples 15.3.8. See Remarks 6.5.9.

## 14.2 At involutive m-BE algebras level

In this section, we shall generalize the results from the previous section.

Consider the following properties:

(Pom) $(x \odot y) \oplus ((x \odot y)^- \odot x) = x$ or, better, $(x \odot y)^- \odot ((x \odot y)^- \odot x)^- = x^-$,

(Pmv) $x \odot ((x^- \odot y^-)^- \odot y^-)^- = x \odot y$,

($\text{WNM}_m$) $(x \odot y) \odot [(x \odot y) \odot [x \odot (x \odot y^-)^- \odot (x \odot y)^-]^-]^- = 0$,

($\text{aWNM}_m$) $(x \odot y) \odot [x \odot (x \odot y^-)^- \odot (x \odot y)^-]^- = x \odot y$,

(m-Pabs-i) $x \odot (x^- \odot x^- \odot y^-)^- = x$ or $x \odot (x \oplus x \oplus y) = x$,

(m-BB) $((x \odot y)^- \odot (z \odot y)) \odot (z \odot x^-)^- = 0$.

**The below Theorem 14.2.4 was proved by *Prover9* in 0.12 seconds; the proof by *Prover9* had 27 lines and was divided into the proofs of next Lemmas 14.2.1, 14.2.2, 14.2.3 and Theorem 14.2.4.**

**Lemma 14.2.1** *Let* $\mathcal{A}^L = (A^L, \odot, ^-, 1)$ *be an involutive left-m-BE algebra verifying (Pom). Then, we have:*

$$(x \odot y)^- \odot (x \odot (x \odot y)^-)^- = x^-, \tag{14.31}$$

$$(x \odot (y \odot z))^- \odot [x \odot (y \odot (x \odot (y \odot z))^-)]^- = (x \odot y)^-, \tag{14.32}$$

$$x \odot (x \odot (x \odot y)^-)^- = x \odot y \ \ (see (14.1)). \tag{14.33}$$

**Proof.** (By *Prover9*.)

(14.31): From (Pom), by (Pcomm).

(14.32): In (14.31), take $X := x \odot y$ and $Y := z$ to obtain:
$((x \odot y) \odot z)^- \odot ((x \odot y) \odot ((x \odot y) \odot z)^-)^- = (x \odot y)^-$;
then, apply (Pass) to obtain (14.32).

(14.33): In (14.31), take $X := (x \odot y)^-$ and $Y := x$ to obtain:
$((x \odot y)^- \odot x)^- \odot ((x \odot y)^- \odot ((x \odot y)^- \odot x)^-)^- = (x \odot y)^=$;
then, by (Pom), we obtain:
$((x \odot y)^- \odot x)^- \odot x^= = (x \odot y)^=$,
hence, by (DN), we obtain:
$((x \odot y)^- \odot x)^- \odot x = x \odot y$,
hence, by (Pcomm), we obtain (14.33). □

**Lemma 14.2.2** *Let $\mathcal{A}^L = (A^L, \odot, ^-, 1)$ be an involutive left-m-BE algebra verifying (Pmv). Then, we have:*

(14.34) $$x \odot (y^- \odot (x^- \odot y^-)^-)^- = x \odot y,$$

(14.35) $$x \odot (y \odot (x^- \odot y)^-)^- = x \odot y^-.$$

**Proof.** (By *Prover9.*)
(14.34): From (Pom), by (Pcomm).
(14.35): From (14.34), taking $Y := y^-$ and by (DN). □

**Lemma 14.2.3** *Let $\mathcal{A}^L = (A^L, \odot, ^-, 1)$ be an involutive left-m-BE algebra verifying ($WNM_m$). Then, we have:*

(14.36) $$x \odot (y \odot [x \odot (y \odot [x \odot ((x \odot y^-)^- \odot (x \odot y)^-)]^-)]^-) = 0.$$

**Proof.** (By *Prover9.*) From ($WNM_m$), by (Pass). □

**Theorem 14.2.4** *(See Theorem 14.1.6)*
*Let $\mathcal{A}^L = (A^L, \odot, ^-, 1)$ be an involutive left-m-BE algebra verifying (Pom), (Pmv) (i.e. $\mathcal{A}^L$ is a left-QMV algebra). Then,*

$$(WNM_m) \implies (m - Pabs - i).$$

**Proof.** (By *Prover9.*)
First, we prove:

(14.37) $$x \odot (x \odot [x \odot (x \odot x)]^-) = 0.$$

Indeed, in (14.36), take $y = x$ to obtain:
$x \odot (x \odot [x \odot (x \odot [x \odot ((x \odot x^-)^- \odot (x \odot x)^-)]^-)]^-) = 0$,
hence, by (m-Re), (m-La), (Neg0-1) and (PU), we obtain:
(a) $x \odot (x \odot [x \odot (x \odot [x \odot (x \odot x)^-]^-)]^-) = 0$;
then, since $x \odot [x \odot (x \odot x)^- = x \odot x$, by (14.33),
we obtain from (a) that:
$x \odot (x \odot [x \odot (x \odot x)]^-) = 0$, that is (14.37).
Next, we prove:

(14.38) $$x \odot (x \odot x) = x \odot x.$$

Indeed, in (14.32), take $x = y = z = x$ to obtain:
(a) $(x \odot (x \odot x))^- \odot [x \odot (x \odot (x \odot (x \odot x))^-)]^- = (x \odot x)^-$
and, since $x \odot (x \odot (x \odot (x \odot x))^-) = 0$, by (14.37),
then (a) becomes, by (Neg0-1) and (PU):
$(x \odot (x \odot x))^- = (x \odot x)^-$, hence, by (DN),
$x \odot (x \odot x) = x \odot x$, that is (14.38).
Next, we prove:

(14.39) $$x \odot (x \odot (x \odot y)) = x \odot (x \odot y).$$

Indeed, $x \odot (x \odot y) \overset{(Pass)}{=} (x \odot x) \odot y \overset{(14.38)}{=} (x \odot (x \odot x)) \odot y$
$\overset{(Pass)}{=} ((x \odot x) \odot x) \odot y \overset{(Pass)}{=} (x \odot x) \odot (x \odot y) \overset{(Pass)}{=} x \odot (x \odot (x \odot y))$, i.e. (14.39) holds.

Finally, we prove that $x \odot [x^- \odot x^- \odot y^-]^- = x$, i.e. (m-Pabs-i) holds. Indeed, in (14.35), take $X := x$ and $Y := x^- \odot (x^- \odot y^-)$ to obtain:
(a) $\quad X \odot (Y \odot (X^- \odot Y)^-)^- = X \odot Y^-$, i.e.
(b) $\quad x \odot [(x^- \odot (x^- \odot y^-)) \odot [x^- \odot (x^- \odot (x^- \odot y^-))]^-]^-$
$= x \odot [x^- \odot (x^- \odot y^-)]^-$;
but, in the left side of (b),
$x^- \odot (x^- \odot (x^- \odot y^-)) = x^- \odot (x^- \odot y^-)$, by (14.39);
hence, (b) becomes:
$x = x \odot 1 = x \odot 0^- \overset{(m-Re)}{=} x \odot [(x^- \odot (x^- \odot y^-)) \odot [x^- \odot (x^- \odot y^-)]^-]^-$
$= x \odot [x^- \odot (x^- \odot y^-)]^-$; thus, (m-Pabs-i) holds. □

The converse of Theorem 14.2.4 does not hold in general; there are examples of QMV algebras verifying (m-Pabs-i) and not verifying ($WNM_m$) - see next Example 14.2.5.

**Example 14.2.5 QMV algebra verifying (m-Pabs-i), not verifying ($WNM_m$), ($aWNM_m$): OMWL**

By MACE4 program, we found the algebra
$\mathcal{A}^L = (A_8 = \{0, a, b, c, d, e, f, 1\}, \odot, ^-, 1)$ with the following tables of $\odot$ and $^-$ and of the additional operation $\oplus$:

| $\odot$ | 0 | a | b | c | d | e | f | 1 |
|---|---|---|---|---|---|---|---|---|
| 0 | 0 | 0 | 0 | 0 | 0 | 0 | 0 | 0 |
| a | 0 | 0 | 0 | e | 0 | 0 | d | a |
| b | 0 | 0 | 0 | 0 | 0 | 0 | 0 | b |
| c | 0 | e | 0 | 0 | 0 | 0 | b | c |
| d | 0 | 0 | 0 | 0 | 0 | 0 | 0 | d |
| e | 0 | 0 | 0 | 0 | 0 | 0 | 0 | e |
| f | 0 | d | 0 | b | 0 | 0 | 0 | f |
| 1 | 0 | a | b | c | d | e | f | 1 |

,

| $\oplus$ | 0 | a | b | c | d | e | f | 1 |
|---|---|---|---|---|---|---|---|---|
| 0 | 0 | a | b | c | d | e | f | 1 |
| a | a | 1 | 1 | 1 | 1 | 1 | 1 | 1 |
| b | b | 1 | 1 | 1 | f | c | 1 | 1 |
| c | c | 1 | 1 | 1 | 1 | 1 | 1 | 1 |
| d | d | 1 | f | 1 | 1 | a | 1 | 1 |
| e | e | 1 | c | 1 | a | 1 | 1 | 1 |
| f | f | 1 | 1 | 1 | 1 | 1 | 1 | 1 |
| 1 | 1 | 1 | 1 | 1 | 1 | 1 | 1 | 1 |

,

and $(0, a, b, c, d, e, f, 1)^- = (1, b, a, d, c, f, e, 0)$.

$\mathcal{A}^L$ is an involutive left-m-BE algebra verifying (Pqmv) (hence (Pom), (Pmv), ($\Delta_m$)) and (m-Pabs-i), ($prel_m$), and not verifying (m-B) for $(a, b, c)$, (m-BB) for $(a, c, a)$, (m-*) for $(a, b, c)$, (m-**) for $(a, b, c)$, (m-Tr) for $(a, b, d)$, (m-An) for $(a, b)$, (m-Pimpl) for $(a, 0)$, ($aWNM_m$) for $(a, c)$, ($WNM_m$) for $(a, c)$, (m-Pdis) for $(a, a, a)$, (G) for $a$. Hence, it is a left-OMWL (Definition 2).

### 14.2.1 At involutive m-pre-BCK algebras level

By Theorem 14.2.4, we obtain immediately the following result, in the transitive case.

**Theorem 14.2.6** *(See Theorem 14.1.6)*
*Let* $\mathcal{A}^L = (A^L, \odot, ^-, 1)$ *be an involutive left-m-BE algebra verifying (Pom), (Pmv) and (m-BB) (⇔ ... ⇔ (m-Tr)) (i.e.* $\mathcal{A}^L$ *is a left-tQMV algebra). Then,*

$$(WNM_m) \implies (m - Pabs - i).$$

**The converse of Theorem 14.2.6, which is the below Theorem 14.2.11, was proved by *Prover9* in 13122 seconds (about 4 hours); the proof by *Prover9* had the length 69 and was divided into the proofs of next Lemmas 14.2.7, 14.2.8, 14.2.9, 14.2.10 and Theorem 14.2.11.**

**Lemma 14.2.7** *Let* $\mathcal{A}^L = (A^L, \odot, ^-, 1)$ *be an involutive left-m-BE algebra verifying (Pom). Then, we have:*

$$(14.40) \qquad (x \odot y)^- \odot [x \odot (x \odot y)^-]^- = x^-, \quad \textit{that is } (14.31)$$

$$(14.41) \qquad (x \odot y)^- \odot [y \odot (y \odot x)^-]^- = y^-,$$

$$(14.42) \qquad (x \odot y)^- \odot ([x \odot (x \odot y)^-]^- \odot z) = x^- \odot z,$$

$$(14.43) \quad [x \odot (y \odot z)]^- \odot [x \odot (y \odot [x \odot (y \odot z)]^-)]^- = (x \odot y)^-, \quad \textit{that is } (14.32)$$

$$(14.44) \qquad x \odot [x \odot (x \odot y)^-]^- = x \odot y, \quad \textit{that is } (14.33)$$

$$(14.45) \qquad x \odot [x \odot (y \odot x)^-]^- = y \odot x,$$

$$(14.46) \qquad x \odot (y \odot [x \odot (x \odot z)^-]^-) = y \odot (x \odot z),$$

$$(14.47) \qquad x \odot (y \odot [x \odot (z \odot x)^-]^-) = y \odot (z \odot x),$$

$$(14.48) \qquad (x \odot y^-)^- \odot [x \odot ((y \odot z)^- \odot (x \odot y^-)^-)]^- = [x \odot (y \odot z)^-]^-.$$

**Proof.** (By *Prover9*.)
(14.41): In (14.40), interchange $x$ with $y$ and apply (Pcomm).
(14.42): In (14.40), "multiply" the both sides by $z$, on the right, and apply (Pass).
(14.45): From (14.44), by (Pcomm).

(14.46): In (14.44), take $y = z$ to obtain:
(a) $$x \odot [x \odot (x \odot z)^-]^- = x \odot z;$$
then, "multiply" the both sides of (a) by $y$, on the left, to obtain:
(b) $$y \odot (x \odot [x \odot (x \odot z)^-]^-) = y \odot (x \odot z);$$
then, by (Pass) and (Pcomm), from (b), we obtain (14.46).

(14.47): From (14.46), by (Pcomm).

(14.48): In (14.43), take $X := x$, $Y := (y \odot z)^-$, $Z := [y \odot (y \odot z)^-]^-$ to obtain:
(a) $$[X \odot (Y \odot Z)]^- \odot [X \odot (Y \odot [X \odot (Y \odot Z)]^-)]^- = (X \odot Y)^-;$$
but, in the left side of (a):
(b) $$X \odot (Y \odot Z) = x \odot ((y \odot z)^- \odot [y \odot (y \odot z)^-]^-)$$
and, in (b), $(y \odot z)^- \odot [y \odot (y \odot z)^-]^- \overset{(14.40)}{=} y^-$,
hence (b) becomes: $X \odot (Y \odot Z) = x \odot y^-$;
then, using (b), (a) becomes:
$(x \odot y^-)^- \odot [x \odot ((y \odot z)^- \odot (x \odot y^-)^-)]^- = [x \odot (y \odot z)^-]^-$, that is (14.48). □

**Lemma 14.2.8** *Let $\mathcal{A}^L = (A^L, \odot, ^-, 1)$ be an involutive left-m-BE algebra verifying (Pmv). Then, we have:*

(14.49) $$x \odot [y^- \odot (x^- \odot y^-)^-]^- = x \odot y, \quad \textit{that is } (14.34)$$

(14.50) $$x \odot [y^- \odot (y^- \odot x^-)^-]^- = x \odot y,$$

(14.51) $$x \odot ([y^- \odot (x^- \odot y^-)^-]^- \odot z) = x \odot (y \odot z).$$

**Proof.** (By *Prover9*.)

(14.50): From (14.49), by (Pcomm).

(14.51): In (14.49), "multiply" both sides by $z$, on the right, then apply (Pass).
□

**Lemma 14.2.9** *Let $\mathcal{A}^L = (A^L, \odot, ^-, 1)$ be an involutive left-m-BE algebra verifying (m-BB). Then, we have:*

(14.52) $$(x \odot y)^- \odot (z \odot (y \odot (z \odot x^-)^-)) = 0,$$

(14.53) $$(x \odot y)^- \odot (z \odot (x \odot (z \odot y^-)^-)) = 0,$$

(14.54) $$(x \odot y)^- \odot (z \odot (y \odot (x^- \odot z)^-)) = 0,$$

(14.55) $$(x \odot y)^- \odot (z \odot (x \odot (y^- \odot z)^-)) = 0,$$

(14.56) $$(x \odot y)^- \odot (z \odot (u \odot (y \odot [z \odot (x^- \odot u)]^-))) = 0,$$

(14.57) $$(x \odot y^-)^- \odot (z \odot (x \odot (y \odot z)^-)) = 0.$$

**Proof.** (By *Prover9.*)

(14.52): From (m-BB), by (Pass).

(14.53): From (14.52), by interchanging $x$ with $y$, and (Pcomm).

(14.54): From (14.52) and (Pcomm).

(14.55): From (14.53) and (Pcomm).

(14.56): In (14.54), take $X := x$, $Y := y$, $Z := z \odot u$ to obtain:

(a) $$(x \odot y)^- \odot ((z \odot u) \odot (y \odot [x^- \odot (z \odot u)]^-)) = 0;$$

then, from (a), by (Pass), (Pcomm), we obtain (14.56).

(14.57): From (14.55), by taking $Y := y^-$ and by (DN). □

**Lemma 14.2.10** *Let $\mathcal{A}^L = (A^L, \odot, ^-, 1)$ be an involutive left-m-BE algebra verifying (m-Pabs-i). Then, we have:*

(14.58) $$x \odot (x^- \odot x^-)^- = x,$$

(14.59) $$x \odot [x^- \odot (x^- \odot y)]^- = x,$$

(14.60) $$x \odot ((x^- \odot x^-)^- \odot y) = x \odot y,$$

(14.61) $$x \odot [x^- \odot (y \odot x^-)]^- = x,$$

(14.62) $$x \odot (y \odot [x^- \odot (x^- \odot z)]^-) = y \odot x,$$

(14.63) $$x^- \odot [x \odot (y \odot x)]^- = x^-.$$

**Proof.** (By *Prover9.*)

(14.58): From (m-Pabs-i), for $y = 0$.

(14.59): From (m-Pabs-i), for $Y := y^-$.

(14.60): In (14.58), "multiply" both sides by $y$, on the right, then, apply (Pass).

(14.61): From (14.59), by (Pcomm).

(14.62): In (14.59): $x \odot [x^- \odot (x^- \odot z)]^- = x$,

"multiply" both sides by $y$, on the left, to obtain:

$y \odot (x \odot [x^- \odot (x^- \odot z)]^-) = y \odot x$; then, apply (Pass), (Pcomm) to obtain (14.62).

(14.63): From (14.61), for $X := x^-$. □

**Theorem 14.2.11** *(See Theorem 14.1.9)*

*Let $\mathcal{A}^L = (A^L, \odot, ^-, 1)$ be an involutive left-m-BE algebra verifying (Pom), (Pmv) and (m-BB) ($\Leftrightarrow \ldots \Leftrightarrow$ (m-Tr)) (i.e. $\mathcal{A}^L$ is a left-tQMV algebra). Then,*

$$(m - Pabs - i) \implies (aWNM_m).$$

**Proof.** (By *Prover9.*)

By using the results from the above lemmas, we prove the following sequence of 25 properties:

(14.64) $$x \odot (x \odot x) = x \odot x.$$

Indeed, in (14.50), take $X := x^= \odot x^=$, $Y := x$ to obtain:
(a) $(x^= \odot x^=) \odot [x^- \odot [x^- \odot (x^= \odot x^=)^-]^-]^- = (x^= \odot x^=) \odot x$;
but, in (a), $x^- \odot (x^= \odot x^=)^- \overset{(14.58)}{=} x^-$, hence (a) becomes:
(b) $(x^= \odot x^=) \odot (x^- \odot x^=)^- = (x^= \odot x^=) \odot x$;
but, in (b), $x^- \odot x^= \overset{(m-Re)}{=} 0$, hence (b) becomes, by (DN), (Neg0-1), (PU): $x \odot x = (x \odot x) \odot x$, hence, by (Pass), (14.64) holds.

(14.65) $$x \odot (y \odot [y^- \odot [y^- \odot (x^- \odot y^-)^-]^-]^-) = x \odot y.$$

Indeed, in (14.51), take $X := x$, $Y := y$, $Z := [y^- \odot [y^- \odot (x^- \odot y^-)^-]^-]^-$ to obtain:
(a) $x \odot ([y^- \odot (x^- \odot y^-)^-]^- \odot [y^- \odot [y^- \odot (x^- \odot y^-)^-]^-]^-) = x \odot (y \odot Z)$;
now, in (14.40), take $X := y^-$, $Y := (x^- \odot y^-)^-$ to obtain:
(b) $(y^- \odot (x^- \odot y^-)^-)^- \odot [y^- \odot [y^- \odot (x^- \odot y^-)^-]^-]^- = y^=$;
then, by (b), (a) becomes: $x \odot y^= = x \odot (y \odot Z)$, which, by (DN), becomes (14.65).

(14.66) $$x \odot (y \odot (x^- \odot y^-)^-) = x \odot y.$$

Indeed, in (14.45), take $X := y^-$, $Y := x^-$ to obtain:
(a) $y^- \odot [y^- \odot (x^- \odot y^-)^-]^- = x^- \odot y^-$;
now, (14.65) becomes, by (a):
(b) $x \odot (y \odot (x^- \odot y^-)^-) = x \odot y$, that is (14.66).

(14.67) $$(x \odot x)^- \odot (x \odot x)^- = (x \odot x)^-.$$

Indeed, in (14.63), take $X := x \odot x$ and $Y := 1$ to obtain:
(a) $(x \odot x)^- \odot [(x \odot x) \odot (1 \odot (x \odot x))]^- = (x \odot x)^-$;
but, in (a),
$(x \odot x) \odot (1 \odot (x \odot x)) \overset{(PU)}{=} (x \odot x) \odot (x \odot x)$
$\overset{(Pass)}{=} x \odot (x \odot (x \odot x)) \overset{(14.64)}{=} x \odot (x \odot x) \overset{(14.64)}{=} x \odot x$;
hence, (a) becomes (14.67).

(14.68) $$x \odot ((x \odot y)^- \odot [y \odot (x \odot y)]^-) = x \odot (x \odot y)^-.$$

Indeed, in (14.61), take $X := x \odot (x \odot y)^-$ and $Y := x$ to obtain:
(a) $(x \odot (x \odot y)^-) \odot [[x \odot (x \odot y)^-]^- \odot (x \odot [x \odot (x \odot y)^-]^-)]^- = x \odot (x \odot y)^-$;
but, in (a), $x \odot [x \odot (x \odot y)^-]^- \overset{(14.44)}{=} x \odot y$,
hence (a) becomes:
(b) $(x \odot (x \odot y)^-) \odot [[x \odot (x \odot y)^-]^- \odot (x \odot y)]^- = x \odot (x \odot y)^-$
which, by (Pcomm), becomes:
(c) $(x \odot (x \odot y)^-) \odot [(x \odot y) \odot [x \odot (x \odot y)^-]^-]^- = x \odot (x \odot y)^-$
which, by (Pass), becomes:
(d) $(x \odot (x \odot y)^-) \odot [x \odot (y \odot [x \odot (x \odot y)^-]^-)]^- = x \odot (x \odot y)^-$;
but, in (d),
$x \odot (y \odot [x \odot (x \odot y)^-]^-) \overset{(14.46)}{=} y \odot (x \odot y)$;
thus, (d) becomes:

(e) $(x \odot (x \odot y)^-) \odot [y \odot (x \odot y)]^- = x \odot (x \odot y)^-$
which, by (Pass), becomes (14.68).

(14.69) $$x \odot ([y \odot (x^- \odot x^-)^-]^- \odot (x \odot y^-)^-) = 0.$$

Indeed, in (14.52), take $X := x$, $Y := (y^- \odot y^-)^-$ and $Z := y$ to obtain:
(a) $[x \odot (y^- \odot y^-)^-]^- \odot (y \odot ((y^- \odot y^-)^- \odot (y \odot x^-)^-)) = 0$
which, by (Pass), becomes:
(b) $[x \odot (y^- \odot y^-)^-]^- \odot ((y \odot (y^- \odot y^-)^-) \odot (y \odot x^-)^-) = 0$;
now, in (14.60), take $X := y$, $Y := 1$ to obtain:
$y \odot ((y^- \odot y^-)^- \odot 1) = y \odot 1$
which, by (PU), becomes:
(c) $y \odot (y^- \odot y^-)^- = y$;
hence, (b), by (c), becomes:
(d) $[x \odot (y^- \odot y^-)^-]^- \odot (y \odot (y \odot x^-)^-) = 0$;
now, in (d), interchange $x$ with $y$ to obtain:
(e) $[y \odot (x^- \odot x^-)^-]^- \odot (x \odot (x \odot y^-)^-) = 0$
which, by (Pass), (Pcomm) becomes (14.69).

(14.70) $$(x \odot x)^- \odot ((x \odot x)^- \odot y) = (x \odot x)^- \odot y,$$

which follows from (14.67), by "multiplying" its both sides by $y$.

(14.71) $$x \odot ((x \odot y)^- \odot [x \odot (y \odot y)]^-) = x \odot (x \odot y)^-,$$

which follows from (14.68), by (Pass).

(14.72) $$[x \odot [y \odot (x^- \odot x^-)^-]^-]^- = (x \odot y^-)^-.$$

Indeed, in (14.48), take $X := x$, $Y := y$ and $Z := (x^- \odot x^-)^-$ to obtain:
(a) $(x \odot y^-)^- \odot [x \odot ([y \odot (x^- \odot x^-)^-]^- \odot (x \odot y^-)^-)]^- = [x \odot [y \odot (x^- \odot x^-)^-]^-]^-$;
but, in (a),
$x \odot ([y \odot (x^- \odot x^-)^-]^- \odot (x \odot y^-)^-) \overset{(14.69)}{=} 0$,
hence (a) becomes:
(b) $(x \odot y^-)^- \odot 0^- = [x \odot [y \odot (x^- \odot x^-)^-]^-]^-$
which, by (Neg0-1), (PU), becomes (14.72).

(14.73) $$x \odot (x \odot (y \odot [(x \odot x)^- \odot z]^-)) = y \odot (x \odot x).$$

Indeed, in (14.62), take $X := x \odot x$, $Y := y$, $Z := z$ to obtain:
(a) $(x \odot x) \odot (y \odot [(x \odot x)^- \odot ((x \odot x)^- \odot z)]^-) = y \odot (x \odot x)$;
but, in (a),
$(x \odot x)^- \odot ((x \odot x)^- \odot z) \overset{(14.70)}{=} (x \odot x)^- \odot z$,
hence (a) becomes:
(b) $(x \odot x) \odot (y \odot [(x \odot x)^- \odot z]^-) = y \odot (x \odot x)$
which, by (Pass), becomes:
$x \odot (x \odot (y \odot [(x \odot x)^- \odot z]^-)) = y \odot (x \odot x)$, that is (14.73).

(14.74) $$x \odot [y \odot (x^- \odot x^-)^-]^- = x \odot y^-,$$

which follows from (14.72), by (DN).

(14.75) $$x^- \odot [y \odot (x \odot x)^-]^- = x^- \odot y^-.$$

Indeed, in (14.72), take $X := x^-$, $Y := y$ to obtain, by (DN):
$[x^- \odot [y \odot (x \odot x)^-]^-]^- = (x^- \odot y^-)^-$,
hence, by (DN) again, (14.75).

(14.76) $$x \odot [(x^- \odot x^-)^- \odot y]^- = x \odot y^-,$$

follows from (14.74), by (Pcomm).

(14.77) $$x^- \odot [(x \odot x)^- \odot y]^- = x^- \odot y^-,$$

which follows from (14.75), by (Pcomm).

(14.78) $$x \odot [y \odot [(x^- \odot x^-)^- \odot y]^-]^- = x.$$

Indeed, in (14.76), take $X := x$, $Y := y \odot [(x^- \odot x^-)^- \odot y]^-$ to obtain:
(a) $x \odot [(x^- \odot x^-)^- \odot (y \odot [(x^- \odot x^-)^- \odot y]^-)]^- = x \odot [y \odot [(x^- \odot x^-)^- \odot y]^-]^-$;
but, in the left side of (a),
$(x^- \odot x^-)^- \odot (y \odot [(x^- \odot x^-)^- \odot y]^-) \overset{(Pass),(m-Re)}{=} 0$,
hence (a) becomes:
$x \overset{(PU)}{=} x \odot 1 = x \odot 0^- = x \odot [y \odot [(x^- \odot x^-)^- \odot y]^-]^-$, that is (14.78).

(14.79) $$x \odot [y \odot (z \odot (x^- \odot [y \odot (x \odot z)]^-))]^- = x.$$

Indeed, in (14.76), take $X := x$, $Y := y \odot (z \odot (x^- \odot [y \odot (x^= \odot z)]^-))$ to obtain:
(a) $x \odot [(x^- \odot x^-)^- \odot (y \odot (z \odot (x^- \odot [y \odot (x^= \odot z)]^-)))]^- = x \odot Y^-$;
now, in (14.56), take $X := x^-$, $Y := x^-$, $Z := y$, $U := z$ to obtain:
(b) $(x^- \odot x^-)^- \odot (y \odot (z \odot (x^- \odot [y \odot (x^= \odot z)]^-))) = 0$;
now, replacing (b) in (a), (a) becomes:
$x \odot 0^- = x \odot Y^-$, i.e.
$x = x \odot [y \odot (z \odot (x^- \odot [y \odot (x^= \odot z)]^-))]^-$, that is (14.79), by (DN).

(14.80) $$x^- \odot [y \odot [(x \odot x)^- \odot y]^-]^- = x^-.$$

Indeed, in (14.77), take $X := x$, $Y := y \odot [(x \odot x)^- \odot y]^-$ to obtain:
(a) $x^- \odot [(x \odot x)^- \odot (y \odot [(x \odot x)^- \odot y]^-)]^- = x^- \odot Y^-$;
but, in the left part of (a),
$(x \odot x)^- \odot (y \odot [(x \odot x)^- \odot y]^-) \overset{(Pass),(m-Re)}{=} 0$;
hence, (a) becomes:
$x^- = x^- \odot 1 = x^- \odot 0^- = x^- \odot [y \odot [(x \odot x)^- \odot y]^-]^-$, that is (14.80).

(14.81) $$x \odot (y \odot [(x \odot x)^- \odot y]^-) = y \odot (x \odot x).$$

Indeed, in (14.66), take $X := x$, $Y := y \odot [(x^= \odot x^=)^- \odot y]^-$ to obtain:
(a) $x \odot ((y \odot [(x^= \odot x^=)^- \odot y]^-) \odot (x^- \odot Y^-)^-) = x \odot Y$;

but, in the left side of (a),
$x^- \odot Y^- = x^- \odot [y \odot [(x^= \odot x^=)^- \odot y]^-]^- \overset{(14.78)}{=} x^-$,
hence (a) becomes:
(b) $$x \odot ((y \odot [(x^= \odot x^=)^- \odot y]^-) \odot x^=) = x \odot (y \odot [(x^= \odot x^=)^- \odot y]^-);$$
now, (b) becomes, by (DN):
(c) $$x \odot ((y \odot [(x \odot x)^- \odot y]^-) \odot x) = x \odot (y \odot [(x \odot x)^- \odot y]^-);$$
then, (c) becomes, by (Pcomm), (Pass):
(d) $$x \odot (x \odot (y \odot [(x \odot x)^- \odot y]^-)) = x \odot (y \odot [(x \odot x)^- \odot y]^-)$$
which, by (14.73), becomes:
(e) $$y \odot (x \odot x) = x \odot (y \odot [(x \odot x)^- \odot y]^-),$$ that is (14.81).

(14.82) $$x \odot [(y \odot y)^- \odot x]^- = x \odot (y \odot y).$$

Indeed, in (14.61), take $X := x \odot [(y \odot y)^- \odot x]^-$ and $Y := y^-$ to obtain:
(a) $$(x \odot [(y \odot y)^- \odot x]^-) \odot [[x \odot [(y \odot y)^- \odot x]^-]^- \odot (Y \odot X^-)]^- = X;$$
but, in the left side of (a),
$Y \odot X^- = y^- \odot [x \odot [(y \odot y)^- \odot x]^-]^- \overset{(14.80)}{=} y^-$;
hence, (a) becomes:
(b) $(x \odot [(y \odot y)^- \odot x]^-) \odot [[x \odot [(y \odot y)^- \odot x]^-]^- \odot y^-]^- = x \odot [(y \odot y)^- \odot x]^-$;
then, (b), by (Pcomm), becomes:
(c) $(x \odot [(y \odot y)^- \odot x]^-) \odot [y^- \odot [x \odot [(y \odot y)^- \odot x]^-]^-]^- = x \odot [(y \odot y)^- \odot x]^-$;
but, in the left side of (c),
$y^- \odot [x \odot [(y \odot y)^- \odot x]^-]^- \overset{(14.80)}{=} y^-$;
hence, (c) becomes:
(d) $$(x \odot [(y \odot y)^- \odot x]^-) \odot y^= = x \odot [(y \odot y)^- \odot x]^-,$$
which by (DN), (Pcomm), (Pass) becomes:
(e) $$y \odot (x \odot [(y \odot y)^- \odot x]^-) = x \odot [(y \odot y)^- \odot x]^-$$
which, by (14.81), becomes
$x \odot (y \odot y) = x \odot [(y \odot y)^- \odot x]^-$, that is (14.82).

(14.83) $$x \odot [x \odot (y \odot y)]^- = (y \odot y)^- \odot x.$$

Indeed, in (14.45), take $X := x$, $Y := (y \odot y)^-$ to obtain:
$x \odot [x \odot [(y \odot y)^- \odot x]^-]^- = (y \odot y)^- \odot x$
which, by (14.82), becomes:
$x \odot [x \odot (y \odot y)]^- = (y \odot y)^- \odot x$, that is (14.83).

(14.84) $$x \odot (y \odot [x \odot (z \odot z)]^-) = y \odot ((z \odot z)^- \odot x).$$

Indeed, in (14.47), take $X := x$, $Y := y$ and $Z := (z \odot z)^-$ to obtain:
$x \odot (y \odot [x \odot [(z \odot z)^- \odot x]^-]^-) = y \odot ((z \odot z)^- \odot x)$
which, by (14.82), becomes:
$x \odot (y \odot [x \odot (z \odot z)]^-) = y \odot ((z \odot z)^- \odot x)$, that is (14.84).

(14.85) $$(x \odot y)^- \odot ((y \odot y)^- \odot x) = x \odot (x \odot y)^-.$$

Indeed, in (14.84), take $X := x$, $Y := (x \odot y)^-$ and $Z := y$ to obtain:
$x \odot ((x \odot y)^- \odot [x \odot (y \odot y)]^-) = (x \odot y)^- \odot ((y \odot y)^- \odot x)$

which, by (14.71), becomes:
$x \odot (x \odot y)^- = (x \odot y)^- \odot ((y \odot y)^- \odot x)$, that is (14.85).

(14.86) $$x \odot [y \odot (x \odot y)]^- = (y \odot y)^- \odot x,$$

which follows by (Pass), (Pcomm) and (14.83).

(14.87) $$x \odot [(y \odot z^-)^- \odot (x^- \odot (y \odot (z \odot x)^-))]^- = x.$$

Indeed, in (14.79), take $X := x$, $Y := (y \odot z^-)^-$ and $Z := y \odot (z \odot x)^-$ to obtain:
(a) $x \odot [(y \odot z^-)^- \odot ((y \odot (z \odot x)^-) \odot (x^- \odot [(y \odot z^-)^- \odot (x \odot (y \odot (z \odot x)^-))]^-))]^- = x$;
now, in (14.57), take $X := y$, $Y := z$, $Z := x$ to obtain:
(b) $(y \odot z^-)^- \odot (x \odot (y \odot [z \odot x]^-)) = 0$;
then, (a) by (b) becomes:
(c) $x \odot [(y \odot z^-)^- \odot ((y \odot (z \odot x)^-) \odot (x^- \odot [0]^-))]^- = x$;
then, (c), by (Neg0-1), (PU), (Pcomm), becomes:
$x \odot [(y \odot z^-)^- \odot (x^- \odot (y \odot (z \odot x)^-))]^- = x$, that is (14.87).

(14.88) $(x \odot y) \odot [x \odot ((x \odot y^-)^- \odot (x \odot y)^-)]^- = x \odot y$, *that is* $(aWNM_m)$.

Indeed, in (14.87), take $X := x \odot y$, $Y := x$, $Z := y$ to obtain:
(a) $(x \odot y) \odot [(x \odot y^-)^- \odot ((x \odot y)^- \odot (x \odot [y \odot (x \odot y)]^-))]^- = x \odot y$;
but, in the left side of (a),
$x \odot [y \odot (x \odot y)]^- \overset{(14.86)}{=} (y \odot y)^- \odot x$,
hence (a) becomes:
(b) $(x \odot y) \odot [(x \odot y^-)^- \odot ((x \odot y)^- \odot ((y \odot y)^- \odot x))]^- = x \odot y$;
but, in the left side of (b),
$(x \odot y)^- \odot ((y \odot y)^- \odot x) \overset{(14.85)}{=} x \odot (x \odot y)^-$,
hence (b) becomes:
(c) $(x \odot y) \odot [(x \odot y^-)^- \odot (x \odot (x \odot y)^-)]^- = x \odot y$
which, by (Pcomm), (Pass), becomes:
$(x \odot y) \odot [x \odot ((x \odot y^-)^- \odot (x \odot y)^-)]^- = x \odot y$, that is (14.88), i.e. (aWNM$_m$). □

By Theorems 14.2.6, 14.2.11 and Lemma 14.1.2, we obtain:

**Theorem 14.2.12** *(See Theorem 14.1.10)*
*Let $\mathcal{A}^L = (A^L, \odot, ^-, 1)$ be an involutive left-m-BE algebra verifying (Pom), (Pmv) and (m-BB) ($\Leftrightarrow \ldots \Leftrightarrow$ (m-Tr)) (i.e. $\mathcal{A}^L$ is a left-tQMV algebra). Then,*

$$(m - Pabs - i) \iff (aWNM_m) \iff (WNM_m).$$

Note that, if we denote the class of tQMV algebras verifying (WNM$_m$) by $_{(WNM_m)}$**tQMV**, then Theorem 14.2.12 says that:

$$\mathbf{tOMWL} = {}_{(\mathbf{WNM_m})}\mathbf{tQMV} \ (\subset \mathbf{tQMV}).$$

See Examples 15.3.14.

**Remarks 14.2.13** *(See Remarks 11.2.13)*

*In a left-tQMV algebra* $\mathcal{A}^L = (A^L, \odot, ^-, 1)$*:*

*- the initial binary relation,* $\leq_m$ $(x \leq_m y \Longleftrightarrow x \odot y^- = 0)$*, is a* **pre-order**, *since it is reflexive and transitive;*

*- the binary relation* $\leq_m^M$ $(x \leq_m^M y \Longleftrightarrow x \wedge_m^M y = x)$ *is an* **order**, *but not a lattice order with respect to* $\wedge_m^M$, $\vee_m^M$, *since* $x \wedge_m^M y \neq y \wedge_m^M x$*;*

*- the binary relation* $\leq_m^P$ $(x \leq_m^P y \Longleftrightarrow x \odot y = x)$ *is only* **antisymmetric and transitive**, *by Proposition 6.4.9.*

The results from this section - mainly Theorem 14.2.12 - show that tQMV algebras are the most important generalizations of MV algebras (among QMV, PreMV, MMV, tQMV, tPreMV, tMMV algebras), being the richest in properties.

Since tQMV algebras are involutive m-pre-BCK algebras (i.e. involutive transitive m-BE algebras), next chapter will be focused on the involutive m-pre-BCK algebras, on their binary relation $\leq_m$ $(\Longleftrightarrow \leq_m^B)$, which is a pre-order, and on their subclasses **tQMV**, **tPreMV**, **tMMV**, **tOMWL**, **tOWL**, **tOL**, **tOSL**, **tOM**.

The results from this section gave us the idea of defining "transitive quantum-IMTL algebras" (**tQIMTL**) and "transitive quantum-NM algebras" (**tQNM**), in the next chapter.

# Chapter 15

# (Involutive) m-BCK algebras and involutive m-pre-BCK algebras

In this final chapter of Part II, we continue the results from the previous chapters culminating with the building up of transitive quantum algebras/structures.

In Section 15.1, we prove that any IMTL algebra verifies ($\text{prel}_m$) and any NM algebra verifies ($\text{prel}_m$) and ($\text{WNM}_m$).

In Section 15.2, we introduce, on any involutive m-pre-BCK algebra, a congruence relation, $\|_Q$, based on the pre-order $\leq_m$, and the corresponding quotient algebra, that is an m-BCK algebra. We prove some results concerning tQMV, tMMV, tPreMV algebras and tOMWLs, tOWLs, tOLs, tOSLs, tOM algebras; we introduce and study two new quantum structures, the tQIMTL and tQNM algebras, as non-lattice generalizations of (transitive) IMTL and NM algebras, respectively; we give a general method, the "method of Q-parallel rows/columns", and a particular method, the "method of identic rows/columns", to obtain involutive m-pre-BCK algebras from a given finite m-BCK algebra.

In Section 15.3, we present 29 types of examples: 8 finite m-BCK algebras and 21 finite involutive m-pre-BCK algebras (10 given, 8 built by the particular method, 3 built by the general method); we build transitive quantum algebras: involutive m-pre-BCK algebras/lattices, tQIMTL, tQNM, tPreMV, tMMV algebras, tOMWLs $= {}_{(WNM_m)}$tQMV algebras, tOWLs, tOSLs, tOLs, tQMV algebras.

In Section 15.4, conclusions end the chapter.

The content of this chapter is entirely new, never published.

## 15.1 (Involutive) m-BCK algebras

A (involutive) left-m-BCK algebra is an involutive left-m-aBE algebra verifying (m-BB) ($\Leftrightarrow \ldots \Leftrightarrow$ (m-Tr)); hence, **m-BCK**, the class of all left-m-BCK algebras, is a variety.

Recall Remark 12.3.4, saying that in a left-m-BCK algebra $\mathcal{A}^L = (A^L, \odot, ^-, 1)$:
- the initial binary relation, $\leq_m$ ($x \leq_m y \overset{def.}{\Longleftrightarrow} x \odot y^- = 0$), is an order (since (m-Re), (m-An), (m-Tr) hold); it can be a lattice order, but we do not know when, in general, excepting the case of MV algebras;
- the binary relation $\leq_m^M$ ($x \leq_m^M y \overset{def.}{\Longleftrightarrow} x \wedge_m^M y = x$) is an order, by Corollary 6.4.25 and since, by Theorem 12.2.3, the binary relation $\leq_m^M$ is transitive (i.e. the property (trans) is verified); but it is not a lattice order, in general, with respect to $\wedge_m^M$, $\vee_m^M$, since $x \wedge_m^M y \neq y \wedge_m^M x$; it is a distributive lattice order if and only if ($\wedge_m$-comm) holds (i.e. $x \wedge_m^M y = y \wedge_m^M x$), i.e. in the case of MV algebras, when $\leq_m \Longleftrightarrow \leq_m^M$;
- the binary relation $\leq_m^P$ ($x \leq_m^P y \overset{def.}{\Longleftrightarrow} x \odot y = x$) is only antisymmetric and transitive, by Proposition 6.4.9; in Boolean algebras, it is a distributive lattice order and $\leq_m^P \Longleftrightarrow \leq_m \Longleftrightarrow \leq_m^M$.

In this section, we shall analyse only the order relation $\leq_m$ ($\Longleftrightarrow \leq_m^B$). To represent the bounded partially ordered set $(A^L, \leq_m, 0, 1)$, we use the Hasse diagram.

**Definition 15.1.1** Let $\mathcal{A}^L = (A^L, \odot, ^-, 1)$ be a left-m-BCK algebra and $\leq_m$ be the order relation.

We say that $\mathcal{A}^L$ is *linearly ordered* or a *chain*, if $x \leq_m y$ or $y \leq_m x$, for any $x, y \in A^L$.

**Lemma 15.1.2** *If $\mathcal{A}^L$ is linearly ordered, then it is a lattice - a left-m-BCK lattice -, where $\wedge = \inf_m$ and $\vee = \sup_m$ w.r. to $\leq_m$.*

Consider the following properties:

(prel) $\quad (x \to y) \vee (y \to x) = 1$,

(WNM) $\quad (x \odot y)^- \vee ((x \wedge y) \to (x \odot y)) = 1$

and

(prel$_m$) $\quad (x \to y) \vee_m^B (y \to x) = 1$ or, equivalently,
$(y \to x) \vee_m^M (x \to y) = 1$ or, equivalently,
$(y \odot x^-)^- \vee_m^M (x \odot y^-)^- = 1$ or, equivalently,
$[(x \odot y^-)^= \odot ((y \odot x^-)^- \odot (x \odot y^-)^=)^-]^- = 1$ or, equivalently,
$[(x \odot y^-) \odot ((y \odot x^-)^- \odot (x \odot y^-))^-]^- = 1$ or, equivalently,
$(x \odot y^-) \odot ((x \odot y^-) \odot (y \odot x^-)^-)^- = 0$;

(WNM$_m$) $\quad (x \odot y)^- \vee_m^B [(x \wedge_m^B y) \to (x \odot y)] = 1$ or, equivalently,
$(x \odot y)^- \vee_m^B [(x \odot (x \odot y^-)^-) \odot (x \odot y)^-]^- = 1$ or, equivalently,
$(x \odot y) \wedge_m^B [x \odot (x \odot y^-)^- \odot (x \odot y)^-] = 0$ or, equivalently,
$(x \odot y) \odot [(x \odot y) \odot [x \odot (x \odot y^-)^- \odot (x \odot y)^-]^-]^- = 0$,

where $x \to y = (x \odot y^-)^-$.

**Proposition 15.1.3** *Let $\mathcal{A}^L = (A^L, \odot, ^-, 1)$ be a linearly ordered left-m-BCK algebra (i.e. a left-m-BCK lattice). Then, (prel) and (prel$_m$) hold (and hence (prel) $\Longleftrightarrow$ (prel$_m$)).*

**Proof.** Suppose that $x \leq_m y$, i.e. $x \odot y^- = 0$. Then, $(x \to y) \vee (y \to x) = (x \odot y^-)^- \vee (y \to x) = 0^- \vee (y \to x) = 1 \vee (y \to x) = 1$, hence (prel) holds.

Similarly, $(x \to y) \vee_m^B (y \to x) = (x \odot y^-)^- \vee_m^B (y \to x) = 1 \vee_m^B (y \to x)$ $= (y \to x) \vee_m^M 1 = 1$, hence (prel$_m$) holds too. □

**Corollary 15.1.4** *Any linearly ordered m-BCK algebra is an IMTL chain.*

**The below Theorem 15.1.6 was proved by *Prover9* in 171,36 seconds; the length of proof by *Prover9* is 51 and was divided into the proofs of next Proposition 15.1.5 and Theorem 15.1.6.**

**Proposition 15.1.5** *Let $\mathcal{A}^L = (A^L, \odot, ^-, 1)$ be a left-m-BCK lattice, with the lattice order $\leq_m$ and the lattice operations $\wedge \overset{notation}{=} \wedge_m = \inf_m$ and $\vee \overset{notation}{=} \vee_m = \sup_m$. Then, we have: for all $x, y \in A^L$,*

$$x \odot (y \vee x^-) = x \odot y. \tag{15.1}$$

**Proof.** (By *Prover9*.)

First, by (Pass), (m-Re), (m-La), (Pcomm), we have:

$$x \odot (x^- \odot y) = 0 \tag{15.2}$$

and also

$$x \odot (y \odot (y \odot x)^-) = 0. \tag{15.3}$$

Second, from (m-BB) $(((z \odot x)^- \odot (y \odot x)) \odot (y \odot z^-)^- = 0)$, by changing $z \mapsto x$, $x \mapsto y$, $y \mapsto z$, we obtain:

(a) $\qquad ((x \odot y)^- \odot (z \odot y)) \odot (z \odot x^-)^- = 0;$

from (a), by (Pass), we obtain:

(b) $\qquad (x \odot y)^- \odot (z \odot (y \odot (z \odot x^-)^-)) = 0;$

from (b), by interchanging $x$ with $y$ and by (Pcomm), we obtain:

$$(x \odot y)^- \odot (z \odot (x \odot (z \odot y^-)^-)) = 0. \tag{15.4}$$

Then, since in logic $p \Longrightarrow q$ means: **non** $p$ **or** $q$ and, hence, $p \Longleftrightarrow q$ means: $p \Longrightarrow q$ **and** $q \Longrightarrow p$, i.e. **(non** $p$ **or** $q$**) and (**$p$ **or non** $q$**)**, it follows that:

$x \wedge y = x \iff x \leq_m y \overset{def.}{\iff} x \odot y^- = 0$ means:

$$x \wedge y \neq x \quad or \quad x \odot y^- = 0 \tag{15.5}$$

and

$$x \wedge y = x \quad or \quad x \odot y^- \neq 0, \tag{15.6}$$

while $x \vee y = y \iff x \leq_m y \overset{def.}{\iff} x \odot y^- = 0$ means:

(15.7) $$x \vee y \neq y \quad or \quad x \odot y^- = 0$$

and

(15.8) $$x \vee y = y \quad or \quad x \odot y^- \neq 0.$$

Now, from (15.6) and (DN), we obtain:

(15.9) $$x \wedge y^- = x \quad or \quad x \odot y \neq 0.$$

From (15.8) and (Pass), we obtain:

(15.10) $$(x \odot y) \vee z = z \quad or \quad x \odot (y \odot z^-) \neq 0.$$

From (15.8) and (DN), we obtain:

(15.11) $$x \vee y^- = y^- \quad or \quad x \odot y \neq 0.$$

We resolve (15.5) by (m-Wabs), and obtain:

(15.12) $$x \odot (x \vee y)^- = 0.$$

From (m-Wcomm) and (m-Vabs), we obtain:

(15.13) $$x \vee (y \wedge x) = x.$$

Now, in (15.12), put $X := x \vee y$ and $Y := z$ to obtain:
$0 = (x \vee y) \odot ((x \vee y) \vee z)^- = (x \vee y) \odot (x \vee (y \vee z))^-$, hence:

(15.14) $$(x \vee y) \odot (x \vee (y \vee z))^- = 0.$$

Now, in (15.4), we put $X := y$, $Y := (y \vee z)^-$ and $Z := x$ to obtain:
$(y \odot (y \vee z)^-)^- \odot (x \odot (y \odot (x \odot (y \vee z)^{=})^-)) = 0$,
which by (15.12) and (DN), becomes:
$0^- \odot (x \odot (y \odot (x \odot (y \vee z))^-)) = 0$,
which by (Neg0-1) and (PU), becomes:

(15.15) $$x \odot (y \odot (x \odot (y \vee z))^-) = 0.$$

Now, we resolve (15.9): in (15.2), take $X := x$ and $Y := x^- \odot y$ to obtain $X \odot Y = 0$, which by (15.9) gives $X \wedge Y^- = X$, hence: $x \wedge (x^- \odot y)^- = x$, which by (DN), taking $X := x^-$, becomes:

(15.16) $$x^- \wedge (x \odot y)^- = x^-.$$

Now, from (15.10), by (Pcomm), we obtain:
$(x \odot y) \vee z = z \quad or \quad (y \odot z^-) \odot x \neq 0$, which by (Pass), becomes:

(15.17) $$(x \odot y) \vee z = z \quad or \quad y \odot (z^- \odot x) \neq 0.$$

Now, we resolve (15.11): in (15.11), put $X := x$ and $Y := y \odot (y \odot x)^-$ to obtain:
$X \vee Y^- = Y^-$ *or* $X \odot Y \neq 0$;
but $X \odot Y = x \odot (y \odot (y \odot x)^-) = 0$, by (15.3); it follows that $X \vee Y^- = Y^-$, i.e.

(15.18) $$x \vee (y \odot (y \odot x)^-)^- = (y \odot (y \odot x)^-)^-.$$

Now, in (15.13), put $X := (x \odot y)^-$, $Y := x^-$ to obtain:
$(x \odot y)^- \vee (x^- \wedge (x \odot y)^-) = (x \odot y)^-$, which by (15.16), becomes:
$(x \odot y)^- \vee x^- = (x \odot y)^-$, which by (m-Vcomm), becomes:

(15.19) $$x^- \vee (x \odot y)^- = (x \odot y)^-.$$

Now, we resolve (15.10): since $X \odot (Y \odot Z^-) = x \odot (y \odot (x \odot (y \vee z))^-) = 0$, by (15.15), it follows from (15.10) that $(X \odot Y) \vee Z = Z$, i.e.

(15.20) $$(x \odot y) \vee (x \odot (y \vee z)) = x \odot (y \vee z).$$

Now, in (15.14), put $X := x$, $Y := y^-$ and $Z := (y \odot z)^-$ to obtain:
$(x \vee y^-) \odot (x \vee (y^- \vee (y \odot z)^-))^- = 0$, which, by (15.19) and (Pcomm), becomes:

(15.21) $$(x \vee y^-) \odot (x \vee (y \odot z)^-)^- = 0.$$

Now, in (15.21), put $X := x$, $Y := y$ and $Z := (y \odot x)^-$ to obtain:
$(x \vee y^-) \odot (x \vee (y \odot (y \odot x)^-)^-)^- = 0$, which by (15.18), becomes:
$(x \vee y^-) \odot (y \odot (y \odot x)^-)^= = 0$, which by (DN), (Pcomm), (Pass), becomes:
$y \odot ((y \odot x)^- \odot (x \vee y^-)) = 0$, which, by interchanging $x$ with $y$, becomes:

(15.22) $$x \odot ((x \odot y)^- \odot (y \vee x^-)) = 0.$$

Now, we resolve (15.17): in (15.17), we put $X := y \vee x^-$, $Y := x$ and $Z := x \odot y$ to obtain:
$((y \vee x^-) \odot x) \vee (x \odot y) = x \odot y$ *or* $x \odot ((x \odot y)^- \odot (y \vee x^-)) \neq 0$;
but $x \odot ((x \odot y)^- \odot (y \vee x^-)) = 0$, by (15.22), hence we obtain:
$((y \vee x^-) \odot x) \vee (x \odot y) = x \odot y$, which, by (Pcomm), (m-Vcomm), becomes:

(15.23) $$(x \odot y) \vee (x \odot (y \vee x^-)) = x \odot y.$$

Finally, (15.23), by (15.20), for $z := x^-$, becomes: $x \odot (y \vee x^-) = x \odot y$, that is (15.1). □

**Theorem 15.1.6** *Let $\mathcal{A}^L = (A^L, \odot, ^-, 1)$ be a left-m-BCK lattice verifying (prel) (i.e. a left-IMTL algebra). Then, (prel$_m$) holds.*

**Proof. (By** *Prover9.***)**
In (15.1), put $X := x \odot y^-$ and $Y := (y \odot x^-)^-$ to obtain:
$(x \odot y^-) \odot ((y \odot x^-)^- \vee (x \odot y^-)^-) = (x \odot y^-) \odot (y \odot x^-)^-$,
which by (prel) $((y \odot x^-)^- \vee (x \odot y^-)^- = 1)$, becomes:
$(x \odot y^-) \odot 1 = (x \odot y^-) \odot (y \odot x^-)^-$, which by (PU) becomes:

(15.24) $$x \odot y^- = (x \odot y^-) \odot (y \odot x^-)^-.$$

Finally, $(x \odot y^-) \odot ((x \odot y^-) \odot (y \odot x^-)^-)^- \overset{(15.24)}{=} (x \odot y^-) \odot (x \odot y^-)^- \overset{(m-Re)}{=} 0$; thus, (prel$_m$) holds. □

The converse of Theorem 15.1.6 does not hold; there are (distributive) m-BCK lattices verifying (prel$_m$) and (aWNM$_m$), (WNM$_m$) and (WNM), and not verifying (prel) - see Examples 15.3.2, 4.

**Proposition 15.1.7** *Let $\mathcal{A}^L = (A^L, \odot, ^-, 1)$ be a linearly ordered left-m-BCK algebra (i.e. a left-m-BCK lattice). If (WNM) holds (i.e. $\mathcal{A}^L$ is a linearly ordered left-NM algebra), then (WNM$_m$) holds (beside (prel$_m$)).*

**Proof.** The proof by *Prover9* takes 822.12 seconds and has the length 194; it will be omitted. See the proof in Examples 15.3.6. □

**Proposition 15.1.8** *Let $\mathcal{A}^L = (A^L, \odot, ^-, 1)$ be a linearly ordered left-m-BCK algebra (i.e. a left-m-BCK lattice). If (WNM$_m$) holds, then (WNM) holds (i.e. $\mathcal{A}^L$ is a linearly ordered left-NM algebra).*

**Proof.** The proof by *Prover9* takes 3959,45 seconds and has the length 237; it will be omitted. □

By Propositions 15.1.3, 15.1.7 and 15.1.8, we obtain:

**Theorem 15.1.9** *Let $\mathcal{A}^L = (A^L, \odot, ^-, 1)$ be a linearly ordered left-m-BCK algebra (i.e. a left-m-BCK lattice). Then,*

$$(prel) \iff (prel_m) \quad and \quad (WNM) \iff (WNM_m).$$

Since any NM algebra is a subdirect product of linearly ordered ones [50], we obtain the following result:

**Theorem 15.1.10** *Let $\mathcal{A}^L = (A^L, \odot, ^-, 1)$ be a left-m-BCK lattice verifying (prel) and (WNM) (i.e. a left-NM algebra, by Definition 2). Then, (prel$_m$) and (WNM$_m$) hold.*

The converse of Theorem 15.1.10 does not hold; there are (distributive) m-BCK lattices verifying (prel$_m$) and (aWNM$_m$), (WNM$_m$) and not verifying (prel) and (WNM) - see Examples 15.3.2, 3.

**Open problem 15.1.11** *We were not able to solve (by Prover9) the following problem: if a distributive m-BCK lattice verifying (prel) and (WNM$_m$), verifies also (WNM); see below the corresponding* **Prover9** *program:*

```
  (Prover9) Assumptions:
 1 * x = x                          # label("PU").
 x * y = y * x                      # label("Pcomm").
 (x * y) * z = x * (y * z)          # label("Pass").
 x * 0 = 0                          # label("m-L").
 x * -x = 0                         # label("m-Re").
 -1 = 0                             # label("Neg1-0").
 -0 = 1                             # label("Neg0-1").
 - -x = x                           # label("DN").

x * -y = 0 & y * -x = 0 -> x = y # label("m-An").
(-(z * x) * (y * x)) * -(y * -z) = 0 # label("m-BB").

% we impose condition that order <= m is a distributive lattice
% with meet operation m and join operation j
m(x,y) = x <-> x * -y = 0.
j(x,y) = y <-> x * -y = 0.
m(x,x) = x.
m(x,y) = m(y,x).
m(m(x,y),z) = m(x,m(y,z)).
j(x,x) = x.
j(x,y) = j(y,x).
j(j(x,y),z) = j(x,j(y,z)).
m(x,j(x,y)) = x.
j(x,m(x,y)) = x.

% distributive
m(a,j(b,c)) = j(m(a,b),m(a,c)).

j(-(x * -y), -(y *-x)) = 1 #label ("prel").
(x * y) * -((x * y)* -((x * -(x * -y)) * -(x * y))) =0 #label ("WNM-m").

  Goals:
j( -(x * y), -(m(x,y) * -(x * y)) ) = 1 #label ("WNM").
```

Recall that, in m-BCK algebras, we have the connections from Figure 12.4.

In Section 15.3 (**Examples**), we recall and analyse in some details some examples of (involutive) m-BCK algebras needed in the sequel: 15.3.1 **m-BCK**, 15.3.2 **m-BCK-L**, 15.3.3 **IMTL**, 15.3.4 **MV**, 15.3.5 non-linearly ordered **MV**, 15.3.6 **NM**, 15.3.7 non-linearly ordered **NM**, 15.3.8 nonlinearly ordered $_{(WNM)}$**MV**.

## 15.2 Involutive m-pre-BCK algebras

An involutive left-m-pre-BCK algebra is an involutive left-m-BE algebra verifying (m-BB) ($\Leftrightarrow \ldots \Leftrightarrow$ (m-Tr)); hence, **m-pre-BCK**$_{(DN)}$ is a variety also and **m-pre-BCK**$_{(DN)}$ + (m-An) = **m-BCK**. It is *proper*, if it is not a left-m-BCK algebra.

**Remark 15.2.1** *In a proper involutive left-m-pre-BCK algebra* $\mathcal{A}^L = (A^L, \odot, ^-, 1)$*:*
*- the initial binary relation,* $\leq_m$ *(*$x \leq_m y \stackrel{def.}{\Longleftrightarrow} x \odot y^- = 0$*), is only a* **pre-order**, *being only reflexive and transitive (i.e. (m-Re) and (m-Tr) hold);*
*- the binary relation* $\leq_m^M$ *(*$x \leq_m^M y \stackrel{def.}{\Longleftrightarrow} x \wedge_m^M y = x$*) is* **reflexive and antisymmetric**, *by Corollary 6.4.25, and in some cases is transitive also (i.e. the property (trans) holds), hence is an order, but not a lattice order with respect to* $\wedge_m^M$, $\vee_m^M$, *since* $x \wedge_m^M y \neq y \wedge_m^M x$;
*- the binary relation* $\leq_m^P$ *(*$x \leq_m^P y \stackrel{def.}{\Longleftrightarrow} x \odot y = x$*) is* **antisymmetric and transitive**, *by Proposition 6.4.9; if (G) holds (i.e. in tOSLs),* $\leq_m^P$ *is reflexive also, hence is an order relation, but not a lattice order by Remark 9.1.4; if (m-Pimpl) holds (i.e. in transitive ortholattices (tOLs)),* $\leq_m^P$ *is a lattice order w.r. to* $\wedge = \odot$, $\vee = \oplus$*, by Proposition 9.1.5.*

In this section, we shall analyse only the pre-order relation $\leq_m$ ($\Longleftrightarrow \leq_m^B$). To represent the bounded partially pre-ordered set $(A^L, \leq_m, 0, 1)$, we shall use the Hasse-type diagram, defined in ([111], Remark 3.11) - see also ([116], Remark 2.1.21) and Chapter 1:

• **Hasse-type diagrams**
Let $\mathcal{A} = (A, \leq)$ be a structure, where $\leq$ is a binary relation on $A$. If $\leq$ is not an order relation (i.e. the binary relation $\leq$ is neither reflexive nor antisymmetric nor transitive, or it is only reflexive, or reflexive and transitive - our case - , or reflexive and antisymmetric), then it is represented by a *Hasse-type diagram*: each element of $A$ is represented by a circ $\circ$, and if $x \leq y$ and $y \leq x$ and $x \neq y$ (i.e. $x$ and $y$ have the *same height*, or are *parallel*), then a horizontal line will connect them.

We introduce the following notions:

**Definitions 15.2.2** Let $\mathcal{A}^L = (A^L, \odot, ^-, 1)$ be an involutive left-m-pre-BCK algebra and let $\leq_m$ be the pre-order relation.

(1) We say that $\mathcal{A}^L$ is *linearly pre-ordered*, if $x \leq_m y$ or $y \leq_m x$, for any $x, y \in A^L$.

(2) We say that $x, y \in A^L$ have the *same Q-height*, or are *Q-parallel* ('Q' comes from 'quantum' ), and we denote this by $x \parallel_Q y$, if $x \leq_m y$ and $y \leq_m x$ (following the similar notion from the theory of quasi-algebras [111], [116]).

**Lemma 15.2.3** *There are no elements* $x \neq 0, 1$ *of* $A^L$ *that are Q-parallel with* 0 *and* 1.

**Proof.** Suppose there exists $x \in A^L$ such that $x \parallel_Q 0$, i.e. $x \leq_m 0$ and $0 \leq_m x$, hence $x \odot 0^- = 0$ and $0 \odot x^- = 0$, hence $x = x \odot 1 = 0$, by (m-La), (Neg0-1), (PU). Similarly, suppose there exists $x \in A^L$ such that $x \parallel_Q 1$, i.e. $x \leq_m 1$ and $1 \leq_m x$, hence $x \odot 1^- = 0$ and $1 \odot x^- = 0$, hence $x^- = 1 \odot x^- = 0$, hence $x = 1$. □

Lemma says, in other words, that if $x \parallel_Q 0$, then $x = 0$, and if $x \parallel_Q 1$, then $x = 1$.

Note that, for example, the smallest bounded linearly pre-ordered set $(A_4 = \{0, a, b, 1\}, \leq_m, 0, 1)$ will be represented by three equivalent Hasse-type diagrams, as can be seen in Figure 15.1.

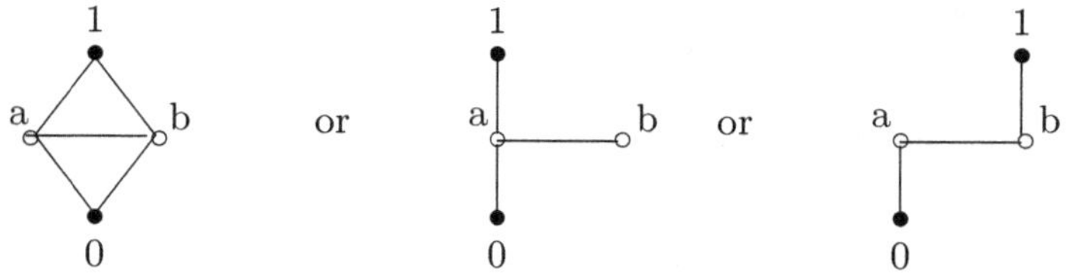


Figure 15.1: Three Hasse-type diagrams for $(A_4, \leq_m, 0, 1)$

**Remark 15.2.4** *If (m-An) holds (i.e. $\mathcal{A}^L$ is a left-m-BCK algebra), then $\leq_m$ is an order relation and, if $x \parallel_Q y$, then $x = y$.*

### 15.2.1 Giving an involutive m-pre-BCK algebra

Let $\mathcal{A}^L = (A^L, \odot, ^-, 1)$ be an involutive left-m-pre-BCK algebra.

**Proposition 15.2.5** *The binary relation $\parallel_Q$ is an equivalence relation.*

**Proof.** - *Reflexivity*: for any $x \in A^L$, $x \parallel_Q x$ means $x \leq_m x$ and $x \leq_m x$, i.e. $x \leq_m x$, that is true by (m-Re).

- *Symmetry*: for any $x, y \in A^L$, if $x \parallel_Q y$, i.e. $x \leq_m y$ and $y \leq_m x$, then $y \leq_m x$ and $x \leq_m y$, i.e. $y \parallel_Q x$.

- *Transitivity*: for any $x, y, z \in A^L$, if $x \parallel_Q y$ and $y \parallel_Q z$, i.e. ($x \leq_m y$ and $y \leq_m x$) and ($y \leq_m z$ and $z \leq_m y$), then ($x \leq_m y$ and $y \leq_m z$) and ($z \leq_m y$ and $y \leq_m x$), hence $x \leq_m z$ and $z \leq_m x$, by (m-Tr), i.e. $x \parallel_Q z$. □

Recall the following properties from Chapter 6:
(m-P- -') $x \leq_m y$ and $a \leq_m b \Longrightarrow x \odot a \leq_m y \odot b$,
(m-Neg2') $x \leq_m y \Longrightarrow y^- \leq_m x^-$.

**Proposition 15.2.6** *The properties (m-P- -') and (m-Neg2') hold.*

**Proof.** By (Chapter 6, (mA12)), (m-**) + (Pcomm) + (m-Tr) $\Longrightarrow$ (m-P- -), and (m-**), (m-Tr) hold, since (m-BB) $\Leftrightarrow \ldots \Leftrightarrow$ (m-Tr); hence, (m-P- -') holds. By (Chapter 6, (mC1)), (m-*) + (PU) $\Longrightarrow$ (m-Neg2), and (m-*) holds, hence (m-Neg2') holds. □

**Proposition 15.2.7** *The binary relation $\parallel_Q$ is a congruence relation, i.e. for any $x, y, a, b \in A^L$,*
*(1) if $x \parallel_Q y$ and $a \parallel_Q b$, then $(x \odot a) \parallel_Q (y \odot b)$ and*
*(2) if $x \parallel_Q y$, then $x^- \parallel_Q y^-$.*

**Proof.** (1): If $x \parallel_Q y$ and $a \parallel_Q b$, i.e. ($x \leq_m y$ and $y \leq_m x$) and ($a \leq_m b$ and $b \leq_m a$), then ($x \leq_m y$ and $a \leq_m b$) and ($y \leq_m x$ and $b \leq_m a$), hence $x \odot a \leq_m y \odot b$ and $y \odot b \leq_m x \odot a$, by (m-P- -'), i.e. $(x \odot a) \parallel_Q (y \odot b)$.

(2): If $x \parallel_Q y$, i.e. $x \leq_m y$ and $y \leq_m x$, then $y^- \leq_m x^-$ and $x^- \leq_m y^-$, by (m-Neg2'), i.e. $y^- \parallel_Q x^-$, hence $x^- \parallel_Q y^-$, by symmetry of $\parallel_Q$. □

Since $\parallel_Q$ is an equivalence relation, we shall denote by $|x|$ the equivalence class of $x$: $|x| = \{y \in A^L \mid y \parallel_Q x\}$ and we shall denote by $A^L/\parallel_Q = \{|x| \mid x \in A^L\}$. We shall then define on $A^L/\parallel_Q$ the operations: $|x| \odot |y| \stackrel{def.}{=} |x \odot y|$, $|x|^- \stackrel{def.}{=} |x^-|$, $|1| = \{x \in A^L \mid x \parallel_Q 1\} = \{1\}$, by Lemma 15.2.3. The operations are well defined, since $\parallel_Q$ is a congruence. Since the involutive m-pre-BCK algebras form an equational class (a variety), then the quotient algebra $\mathcal{A}^L/\parallel_Q = (A^L/\parallel_Q, \odot, ^-, \{1\})$ is also an involutive m-pre-BCK algebra. Then, $|1|^- = |1^-| = |0| = \{0\}$, by Lemma 15.2.3.

Note that for any $|x|, |y| \in A^L/\parallel_Q$, $|x| = |y| \iff$ (*for all* $z \in A^L$, $z \in |x| \Leftrightarrow z \in |y|$) $\iff$ (*for all* $z \in A^L$, $z \parallel_Q x \Leftrightarrow z \parallel_Q y$) $\iff x \parallel_Q y$, hence

$$|x| = |y| \iff x \parallel_Q y. \tag{15.25}$$

Let us define on the quotient algebra $\mathcal{A}^L/\parallel_Q$ a binary relation denoted also by $\leq_m$, defined as follows: for any $|x|, |y| \in A^L/\parallel_Q$, $|x| \leq_m |y| \stackrel{def.}{\iff} |x| \odot |y|^- = |0|$; then, by Lemma 15.2.3, we have:
(15.26)

$$|x| \leq_m |y| \stackrel{def.}{\iff} |x \odot y^-| = |0| \stackrel{(15.25)}{\iff} (x \odot y^-) \parallel_Q 0 \iff x \odot y^- = 0 \stackrel{def.}{\iff} x \leq_m y.$$

**Proposition 15.2.8** *The binary relation* $\leq_m$ *is an order on* $A^L/\parallel_Q$.

**Proof.** - *Reflexivity*: $|x| \leq_m |x| \stackrel{(15.26)}{\iff} x \leq_m x$, that is true, by (m-Re).

- *Transitivity*: $|x| \leq_m |y|$ and $|y| \leq_m |z|$ mean $x \leq_m y$ and $y \leq_m z$, by (15.26), hence $x \leq_m z$, by (m-Tr), hence $|x| \leq_m |z|$.

- *Antisymmetry*: $|x| \leq_m |y|$ and $|y| \leq_m |x|$ mean $x \leq_m y$ and $y \leq_m x$, i.e. $x \parallel_Q y$, hence $|x| = |y|$, by (15.25). □

**Theorem 15.2.9** *Let* $\mathcal{A}^L = (A^L, \odot, ^-, 1)$ *be an involutive left-m-pre-BCK algebra. Then, the quotient algebra* $\mathcal{A}^L/\parallel_Q = (A^L/\parallel_Q, \odot, ^-, \{1\})$ *is a (involutive) left-m-BCK algebra.*

**Proof.** $\mathcal{A}^L/\parallel_Q$ is an involutive left-m-pre-BCK algebra such that $\leq_m$ is antisymmetric (i.e. (m-An) holds); hence, it is a left-m-BCK algebra. □

**Definition 15.2.10** Let $\mathcal{A}^L = (A^L, \odot, ^-, 1)$ be an involutive left-m-pre-BCK algebra. The left-m-BCK algebra $\mathcal{A}^L/\parallel_Q = (A^L/\parallel_Q, \odot, ^-, \{1\})$ will be called the *associated left-m-BCK algebra* of $\mathcal{A}^L$.

Note that all the properties of $\mathcal{A}^L$ are preserved by $\mathcal{A}^L/\parallel_Q$, but $\mathcal{A}^L/\parallel_Q$ has additional properties, including (m-An).

**Proposition 15.2.11** *$\mathcal{A}^L$ is linearly pre-ordered iff $\mathcal{A}^L/\parallel_Q$ is linearly ordered.*

**Proof.** By (15.26). □

Since an m-BCK algebra can be a lattice w.r. to $\leq_m$, we shall introduce the following definition.

**Definition 15.2.12** We shall say that the involutive left-m-pre-BCK algebra $\mathcal{A}^L = (A^L, \odot, ^-, 1)$ is an *involutive left-m-pre-BCK lattice*, if the associated left-m-BCK algebra $\mathcal{A}^L/\parallel_Q = (A^L/\parallel_Q, \odot, ^-, \{1\})$ is a left-m-BCK lattice.

We shall denote by **m-pre-BCK-L**$_{(DN)}$ the class of involutive left-m-pre-BCK lattices.

Note that, if $\mathcal{A}^L$ is a linearly pre-ordered involutive left-m-pre-BCK algebra, then $\mathcal{A}^L$ is an (linearly pre-ordered) involutive left-m-pre-BCK lattice, by Proposition 15.2.11 and Lemma 15.1.2.

**Corollary 15.2.13** *Let $\mathcal{A}^L = (A^L, \odot, ^-, 1)$ be an involutive left-m-pre-BCK algebra.*

*(0) If $\mathcal{A}^L$ is an involutive left-m-pre-BCK lattice, then $\mathcal{A}^L/\parallel_Q$ is a left-m-BCK lattice.*

*(1) If $\mathcal{A}^L$ is a left-tQMV algebra, then $\mathcal{A}^L/\parallel_Q$ is a left-MV algebra.*

*(2) If $\mathcal{A}^L$ is a left-tMMV algebra, then $\mathcal{A}^L/\parallel_Q$ is a left-MV algebra.*

*(3) If $\mathcal{A}^L$ is a left-tPreMV algebra, then $\mathcal{A}^L/\parallel_Q$ is a left-MV algebra.*

*(4) If $\mathcal{A}^L$ is a left-tOMWL $= _{(WNM_m)}$tQMV, then $\mathcal{A}^L/\parallel_Q$ is a left-taOWL $= _{(WNM)}$MV algebra.*

*(5) If $\mathcal{A}^L$ is a left-tOWL, then $\mathcal{A}^L/\parallel_Q$ is a left-taOWL.*

*(6) If $\mathcal{A}^L$ is a left-tOL, then $\mathcal{A}^L/\parallel_Q$ is a left-Boolean algebra.*

*(7) If $\mathcal{A}^L$ is a left-tOSL, then $\mathcal{A}^L/\parallel_Q$ is a left-Boolean algebra.*

*(8) If $\mathcal{A}^L$ is a left-tOM algebra , then $\mathcal{A}^L/\parallel_Q$ is a left-taOM algebra.*

**Proof.** (0): By definition.

(1): By Corollary 11.3.10 and Remark 11.3.11, **taQMV** = **tQMV** + (m-An) = **tMV** = **MV**.

(2): By Corollary 11.3.10 and Remark 11.3.11, **taMMV** = **tMMV** + (m-An) = **MV**.

(3): By Corollary 11.3.10 and Remark 11.3.11, **taPreMV** = **tPreMV** + (m-An) = **MV**.

(4): By Theorem 13.2.32 (iii)), **atOMWL** = **taOMWL** = **taOWL**, hence **atOMWL** = **tOMWL** + (m-An) = $_{(WNM)}$**MV**.

(5): By Theorem 14.1.10, **atOWL** = **taOWL** = **tOWL** + (m-An) = $_{(WNM)}$**MV**.

(6): By Theorem 9.2.19, **aOL** = **OL** + (m-An) = **Boole**, hence **taOL** = **tOL** + (m-An) = **Boole**.

(7): By Chapter 10, **atOSL** = **taOSL** = **tOSL** + (m-An) = **Boole**.

(8): By Chapter 12, **atOM** = **taOM** = **tOM** + (m-An) $\supset$ **MV**. □

See Examples 15.3.11 - 15.3.18.

### 15.2.2 Quantum-IMTL algebras and quantum-NM algebras

We shall introduce the following two new quantum algebras.

**Definitions 15.2.14** (The dual ones are omitted)

(1) A *transitive left-quantum-IMTL algebra*, or a *left-tQIMTL algebra* for short, is an involutive left-m-pre-BCK lattice $\mathcal{A}^L = (A^L, \odot, ^-, 1)$ such that $\mathcal{A}^L/\parallel_Q = (A^L/\parallel_Q, \odot, ^-, \{1\})$ is a left-IMTL algebra (Definition 2).

(2) A *transitive left-quantum-NM algebra*, or a *left-tQNM algebra* for short, is an involutive left-m-pre-BCK lattice $\mathcal{A}^L = (A^L, \odot, ^-, 1)$ such that $\mathcal{A}^L/\parallel_Q = (A^L/\parallel_Q, \odot, ^-, \{1\})$ is a left-NM algebra (Definition 2).

By definitions, it follows that any left-tQNM algebra is a left-tQIMTL algebra.

We shall denote by **tQIMTL** the class of all left-tQIMTL algebras and by **tQNM** the class of all left-tQNM algebras. Hence, we have:

**tQNM** $\subset$ **tQIMTL** $\subset$ **m-pre-BCK-L**$_{(DN)}$,
**tQIMTL** $=$ **m-pre-BCK-L**$_{(DN)}$ $+_a$ (prel),
where $+_a$ means that (prel) is added to the m-BCK lattice level,
**tQNM** $=$ **tQIMTL** $+_a$ (WNM),
where $+_a$ means that (WNM) is added to the m-BCK lattice level.

Consider the following properties:

(prel$_m$) $(x \to y) \vee_m^B (y \to x) = 1$ or, equivalently,
$(x \odot y^-) \odot ((x \odot y^-) \odot (y \odot x^-)^-)^- = 0$;

(WNM$_m$) $(x \odot y)^- \vee_m^B [(x \wedge_m^B y) \to (x \odot y)] = 1$ or, equivalently,
$(x \odot y) \odot [(x \odot y) \odot [x \odot (x \odot y^-)^- \odot (x \odot y)^-]^-]^- = 0$,

where $x \to y = (x \odot y^-)^-$.

Note that any left-tQMV algebra verifies the property (prel$_m$), since any left-QMV algebra verifies the property (prel$_m$), by Corollary 11.2.6.

**Proposition 15.2.15** *Any linearly pre-ordered involutive m-pre-BCK algebra verifies (prel$_m$).*

**Proof.** Since $\leq_m$ is linearly pre-ordered, we have $x \odot y^- = 0$ or $y \odot x^- = 0$, i.e. $(x \odot y^-)^- = 1$ or $(y \odot x^-)^- = 1$, hence $x \to y = 1$ or $y \to x = 1$; hence, $(x \to y) \vee_m^B (y \to x) = (y \to x) \vee_m^M (x \to y) = 1$, so (prel$_m$) holds. □

**Theorem 15.2.16** *Any linearly pre-ordered involutive m-pre-BCK algebra is a (linearly pre-ordered) tQIMTL algebra verifying (prel$_m$).*

**Proof.** Let $\mathcal{A}^L = (A^L, \odot, ^-, 1)$ be a linearly pre-ordered involutive m-pre-BCK algebra. Then, the associated m-BCK algebra $\mathcal{A}^L/_{\parallel_Q}$ is linearly ordered, by Proposition 15.2.11, hence it is a (distributive) lattice, by Lemma 15.1.2; then, (prel) and (prel$_m$) hold in $\mathcal{A}^L/_{\parallel_Q}$, by Proposition 15.1.3; hence, it is an IMTL chain, by Corollary 15.1.4. Then, $\mathcal{A}^L$ is a linearly pre-ordered tQIMTL algebra, by definition; it verifies (prel$_m$) by Proposition 15.2.15. □

**Proposition 15.2.17** *Any linearly pre-ordered tQIMTL algebra verifies (prel$_m$).*

**Proof.** Let $\mathcal{A}^L = (A^L, \odot, ^-, 1)$ be a linearly pre-ordered tQIMTL algebra; by definition, it is a linearly pre-ordered involutive m-pre-BCK algebra and hence verifies (prel$_m$) by Proposition 15.2.15. □

See Example 15.3.9 of linearly pre-ordered tQIMTL algebra (verifying (prel$_m$)).

We know that any IMTL algebra verifies (prel$_m$), by Theorem 15.1.6, but we do not know if any tQIMTL algebra verifies (prel$_m$) and we have no a counterexample (a tQIMTL algebra not verifying (prel$_m$)). Therefore, we shall denote the subclass of those tQIMTL algebras verifying (prel$_m$) by **tQIMTL$_m$**. Hence, we have:

$$\mathbf{tQIMTL_m} \subseteq \mathbf{tQIMTL},$$

$$\mathbf{tQIMTL_m} \;=\; \mathbf{m-pre-BCK-L_{(DN)}} \;+_a\; (prel) + (prel_m),$$

where $+_a$ means that (prel) is added to the m-BCK lattice level.

**Proposition 15.2.18** *Any left-tQMV algebra is a left-tQIMTL algebra verifying (prel$_m$).*

**Proof.** Obviously, since any MV algebra is an IMTL algebra and any tQMV algebra verifies (prel$_m$). □

**The below Theorem 15.2.20 was proved by** ***Prover9*** **in 2028,55 seconds; the length of proof by** ***Prover9*** **is 28 and was divided into the proofs of next Lemma 15.2.19 and Theorem 15.2.20.**

**Lemma 15.2.19** *Let $\mathcal{A}^L = (A^L, \odot, ^-, 1)$ be an involutive left-m-BE algebra verifying (m-BB) (i.e. an involutive left-m-pre-BCK algebra). Then, we have:*

(15.27) $$(x \odot y)^- \odot (z \odot (y \odot (z \odot x^-)^-)) = 0,$$

(15.28) $$(x \odot (y \odot z))^- \odot (u \odot (z \odot (u \odot (x \odot y)^-)^-)) = 0,$$

(15.29) $$(x \odot y)^- \odot (z \odot (u \odot (y \odot (z \odot (u \odot x^-))^-))) = 0.$$

**Proof.** (By *Prover9*) (15.27): In (m-BB) $(((z \odot x)^- \odot (y \odot x)) \odot (y \odot z^-)^- = 0)$, replace $x$ by $y$, $y$ by $z$ and $z$ by $x$ to obtain:
$((x \odot y)^- \odot (z \odot y)) \odot (z \odot x^-)^- = 0$; then, apply (Pass), to obtain:
$(x \odot y)^- \odot (z \odot (y \odot (z \odot x^-)^-)) = 0$, that is (15.27).

(15.28): In (15.27), put $X := x \odot y$, $Y := z$, $Z := u$ to obtain, by (Pass), (15.28).

(15.29): In (15.27) also, put $Z := z \odot u$ to obtain, by (Pass):
$(x \odot y)^- \odot ((z \odot u) \odot (y \odot (z \odot (u \odot x^-))^-)) = 0$; then, by (Pass) again, we obtain (15.29). □

**Theorem 15.2.20** *Let $\mathcal{A}^L = (A^L, \odot, ^-, 1)$ be an involutive left-m-pre-BCK algebra verifying $(\Delta_m)$ (i.e. a tMMV algebra). Then, $(prel_m)$ holds.*

**Proof.** (By *Prover9*.)
First, note that, by (m-Re), (m-La), (Pcomm) and (Pass), we obtain:

(15.30) $$x \odot (y \odot x^-) = 0.$$

Second, note that $(\Delta_m)$ $((x \wedge_m^M y) \odot (y \wedge_m^M x)^- = 0)$ means:
$(y \odot (x^- \odot y)^-) \odot (x \odot (y^- \odot x)^-)^- = 0$;
then, by interchanging $x$ with $y$ and by (Pass), we obtain:

(15.31) $$x \odot ((y^- \odot x)^- \odot (y \odot (x^- \odot y)^-)^-) = 0.$$

Next, in (15.27), put $X := z$, $Y := (y^- \odot z)^- \odot (y \odot (z^- \odot y)^-)^-$ and $Z := x$ to obtain:
(a) $(z \odot Y)^- \odot (x \odot (Y \odot (x \odot z^-)^-)) = 0$;
then note that:
$z \odot Y = z \odot ((y^- \odot z)^- \odot (y \odot (z^- \odot y)^-)^-) = 0$,
by (15.31), for $x := z$; hence, (a) becomes:
(a') $0^- \odot (x \odot (((y^- \odot z)^- \odot (y \odot (z^- \odot y)^-)^-) \odot (x \odot z^-)^-)) = 0$,
which by (Neg0-1), (PU) gives:
(a") $x \odot (((y^- \odot z)^- \odot (y \odot (z^- \odot y)^-)^-) \odot (x \odot z^-)^-) = 0$,
which by (Pass) gives:

(15.32) $$x \odot ((y^- \odot z)^- \odot ((y \odot (z^- \odot y)^-)^- \odot (x \odot z^-)^-)) = 0.$$

Next, in (15.28), put $X := z$, $Y := (y^- \odot z)^-$, $Z := (y \odot (z^- \odot y)^-)^-$ and $U := x$ to obtain:
(b) $(X \odot (Y \odot Z))^- \odot (x \odot ((y \odot (z^- \odot y)^-)^- \odot (x \odot (z \odot (y^- \odot z)^-)^-)^-)) = 0$;
but $X \odot (Y \odot Z) = z \odot ((y^- \odot z)^- \odot (y \odot (z^- \odot y)^-)^-) = 0$ by (15.31) for $x := z$;
from (b), by (Neg0-1), (PU), we obtain:

(15.33) $$x \odot ((y \odot (z^- \odot y)^-)^- \odot (x \odot (z \odot (y^- \odot z)^-)^-)^-) = 0.$$

Next, in (15.29), put $X := x \odot (y^- \odot x)^-$, $Y := z$, $Z := y$ and $U := (x^- \odot y)^-$ to obtain:
(c) $((x \odot (y^- \odot x)^-) \odot z)^- \odot (y \odot ((x^- \odot y)^- \odot (z \odot (Z \odot (U \odot X^-))^-))) = 0$;
but $Z \odot (U \odot X^-) = y \odot ((x^- \odot y)^- \odot (x \odot (y^- \odot x)^-)^-) = 0$ by (15.31) for $x \leftrightarrow y$;
from (c), by (Neg0-1), (PU), we obtain:
(c') $((x \odot (y^- \odot x)^-) \odot z)^- \odot (y \odot ((x^- \odot y)^- \odot z)) = 0$;
from (c'), by (Pcomm) and (Pass), we obtain:
(c") $y \odot (((x^- \odot y)^- \odot z) \odot (x \odot ((y^- \odot x)^- \odot z))^-) = 0$,
which, by interchanging $x$ with $y$ and (Pass), becomes:

(15.34) $$x \odot ((y^- \odot x)^- \odot (z \odot (y \odot ((x^- \odot y)^- \odot z))^-)) = 0.$$

Next, in (15.28), put $X := u$, $Y := (y^- \odot z)^-$, $Z := (y \odot (z^- \odot y)^-)^- \odot (u \odot z^-)^-$ and $U := x$ to obtain:

(d) $(X \odot (Y \odot Z))^- \odot (x \odot (((y \odot (z^- \odot y)^-)^- \odot (u \odot z^-)^-) \odot (x \odot (u \odot (y^- \odot z)^-)^-)^-)) = 0$;
but $X \odot (Y \odot Z) = u \odot ((y^- \odot z)^- \odot ((y \odot (z^- \odot y)^-)^- \odot (u \odot z^-)^-)) = 0$ by (15.32) for $x := u$;
from (d), by (Neg0-1), (PU) and (Pass), we obtain:

$$x \odot ((y \odot (z^- \odot y)^-)^- \odot ((u \odot z^-)^- \odot (x \odot (u \odot (y^- \odot z)^-)^-)^-)) = 0. \tag{15.35}$$

Next, in (15.34), put $X := x$, $Y := (y^- \odot x^-)^-$ and $Z := ((y^- \odot x^-)^- \odot (y \odot (x^= \odot y)^-)^-)^-$ to obtain:
(e) $x \odot (((y^- \odot x^-)^= \odot x)^- \odot (((y^- \odot x^-)^- \odot (y \odot (x^= \odot y)^-)^-)^- \odot A^-)) = 0$,
where $A \overset{notation}{=} Y \odot ((X^- \odot Y)^- \odot Z) =$
$= (y^- \odot x^-)^- \odot ((x^- \odot (y^- \odot x^-)^-)^- \odot ((y^- \odot x^-)^- \odot (y \odot (x^= \odot y)^-)^-)^-) = 0$ by (15.33);
indeed, in (15.33), put $X := (y^- \odot x^-)^-$, $Y := x^-$ and $Z := y$ to obtain $A$;
hence, (e) becomes, by (Neg0-1), (PU), (DN):
(e') $\qquad x \odot (((y^- \odot x^-) \odot x)^- \odot ((y^- \odot x^-)^- \odot (y \odot (x \odot y)^-)^-)^- = 0$;
but, in (e'), $(y^- \odot x^-) \odot x = 0$, by (15.30), hence (e') becomes, by (Neg0-1), (PU):

$$x \odot ((y^- \odot x^-)^- \odot (y \odot (x \odot y)^-)^-)^- = 0. \tag{15.36}$$

Finally, in (15.35), put $X := x$, $Y := y$, $Z := (x \odot y^-)^-$ and $U := (y^= \odot x^-)^-$ to obtain:
(f) $\qquad x \odot ((y \odot ((x \odot y^-)^= \odot y)^-)^- \odot (((y^= \odot x^-)^- \odot (x \odot y^-)^=)^- \odot B^-)) = 0$,
where $B \overset{notation}{=} X \odot (U \odot (Y^- \odot Z)^-)^- =$
$= x \odot ((y \odot x^-)^- \odot (y^- \odot (x \odot y^-)^-)^-)^- = 0$ by (15.36) for $y := y^-$;
hence, (f) becomes, by (Neg0-1), (PU), (DN):
(f') $\qquad x \odot ((y \odot ((x \odot y^-) \odot y)^-)^- \odot ((y \odot x^-)^- \odot (x \odot y^-))^-) = 0$;
but, in (f'), $(x \odot y^-) \odot y = 0$, by (15.30); hence, (f') becomes, by (Neg0-1), (PU):
(f") $\qquad x \odot (y^- \odot ((y \odot x^-)^- \odot (x \odot y^-))^-) = 0$,
which by (Pass), (Pcomm) becomes:
$(x \odot y^-) \odot ((x \odot y^-) \odot (y \odot x^-)^-)^- = 0$, that is equivalent to ($prel_m$). □

Note that, by Theorem 15.2.20 also, any left-tQMV algebra verifies ($prel_m$), since it is a left-tMMV algebra.

**Corollary 15.2.21** *Let $\mathcal{A}^L = (A^L, \odot, ^-, 1)$ be an involutive left-m-pre-BCK algebra verifying (G) (i.e. a tOSL). Then, ($prel_m$) holds.*

**Proof.** By Theorem 13.2.17, **tOSL** $\subset$ **tMMV**. Then, apply Theorem 15.2.20. □

**Corollary 15.2.22** *Let $\mathcal{A}^L = (A^L, \odot, ^-, 1)$ be an involutive left-m-pre-BCK algebra verifying (m-Pabs-i) (i.e. a tOWL). Then, ($prel_m$) holds.*

**Proof.** By Theorem 13.2.30 (Theorem 10.3.13), **tOWL** $\subset$ **tMMV**. Then, apply Theorem 15.2.20. □

**Corollary 15.2.23** *Let $\mathcal{A}^L = (A^L, \odot, ^-, 1)$ be an involutive left-m-pre-BCK algebra verifying (m-Pimpl) (i.e. a tOL). Then, (prel$_m$) holds.*

**Proof.** By Proposition 10.1.15, (m-Pimpl) $\Longleftrightarrow$ (G) + (m-Pabs-i), hence **tOL** = **tOSL** $\cap$ **tOWL**. Then, apply Corollary 15.2.21 or 15.2.22. □

**Remark 15.2.24** *Some proper tOM algebras (i.e. not being tQMV algebras - see Figure 13.13) verify (prel$_m$), others do not verify - see Examples 15.3.18.*

**Proposition 15.2.25** *Let $\mathcal{A}^L = (A^L, \odot, ^-, 1)$ is an involutive left-m-pre-BCK algebra verifying (G) (i.e. a tOSL). Then, (WNM$_m$) holds.*

**Proof. By Prover9 (in 0.05 seconds, length of proof is 19)**

In (m-BB) $(((z \odot x)^- \odot (y \odot x)) \odot (y \odot z^-)^- = 0)$, replace $x$ by $y$, $y$ by $z$ and $z$ by $x$ to obtain:

(a) $\quad ((x \odot y)^- \odot (z \odot y)) \odot (z \odot x^-)^- = 0;$

then, apply (Pass), to obtain from (a):

(b) $\quad (x \odot y)^- \odot (z \odot (y \odot (z \odot x^-)^-)) = 0;$

then, in (b), interchange $x$ with $y$, to obtain, by (Pcomm):

(c) $\quad (x \odot y)^- \odot (z \odot (x \odot (z \odot y^-)^-)) = 0;$

from (c), take $z = x$ to obtain:

(d) $\quad (x \odot y)^- \odot (x \odot (x \odot (x \odot y^-)^-)) = 0.$

Also, by (G) and (Pass), we obtain:

(e) $\quad x \odot (x \odot y) = x \odot y.$

Now, in (d) apply (e) to obtain:

(d') $\quad (x \odot y)^- \odot (x \odot (x \odot y^-)^-) = 0.$

We must prove that (WNM$_m$) holds, i.e.

$(x \odot y) \odot [(x \odot y) \odot [x \odot (x \odot y^-)^- \odot (x \odot y)^-]^-]^- = 0$ or, equivalently, by (Pass):

(f) $\quad (x \odot y) \odot [(x \odot y) \odot [(x \odot y)^- \odot (x \odot (x \odot y^-)^-)]^-]^- = 0;$

indeed, the left side of (f) becomes, by (d'):

$(x \odot y) \odot [(x \odot y) \odot 0^-]^- = (x \odot y) \odot [(x \odot y) \odot 1]^- = (x \odot y) \odot (x \odot y)^- = 0$, by (Neg0-1), (PU), (m-Re); thus, (WNM$_m$) holds. □

**Corollary 15.2.26** *Let $\mathcal{A}^L = (A^L, \odot, ^-, 1)$ be an involutive left-m-pre-BCK algebra verifying (m-Pimpl) (i.e. a tOL). Then, (WNM$_m$) holds.*

**Proof.** By Proposition 10.1.15, (m-Pimpl) $\Longleftrightarrow$ (G) + (m-Pabs-i), hence **tOL** = **tOSL** $\cap$ **tOWL**. Then, apply Proposition 15.2.25 □

Concerning the tOWLs, note that all the examples we have verify (WNM$_m$) - see Examples 15.3.15. But, we have not been able to resolve the following problem.

**Open problem 15.2.27** *Find an example of tOWL which does not verify (WNM$_m$) (using Mace4, we have searched exhaustively for an example up through and including size 21), or prove that any tOWL satisfies (WNM$_m$) (we have also tried to find a proof using Prover9, but despite letting it run for several days, it was unable to find one).*

**Theorem 15.2.28**
*(1) Any tOMWL is a particular case of tQNM algebra verifying ($prel_m$) and ($WNM_m$).*
*(2) Any tOL is a particular case of tQNM algebra verifying ($prel_m$) and ($WNM_m$).*
*(3) Any tOSL is a particular case of tQNM algebra verifying ($prel_m$) and ($WNM_m$).*
*(4) Any tOWL is a particular case of tQIMTL algebra (see the Open problem) verifying ($prel_m$).*
*(5) Some tMMV algebras are particular cases of tQIMTL algebras verifying ($prel_m$).*

**Proof.** (1): By Theorems 14.2.12 and 15.2.20.
(2): By Corollaries 15.2.13 (6), 15.2.23, 15.2.26.
(3): By Corollaries 15.2.13 (7) and 15.2.21 and by Proposition 15.2.25.
(4): By Corollaries 15.2.13 (5), 15.2.22.
(5): By Corollaries 15.2.13 (2). □

See Examples 15.3.10 of tQNM algebras, all verifying ($\text{prel}_m$) and ($\text{WNM}_m$).
Note that we have examples of involutive m-pre-BCK lattices verifying ($\text{prel}_m$) and ($\text{WNM}_m$) and not being tQNM algebras - see Examples 15.3.20, 2.

We know that any NM algebra verifies ($\text{prel}_m$) and ($\text{WNM}_m$), by Theorem 15.1.10, but we do not know if any tQNM algebra verifies ($\text{prel}_m$) and ($\text{WNM}_m$) and we have no a counterexample (a tQNM algebra not verifying ($\text{prel}_m$) and ($\text{WNM}_m$)). Therefore, we shall denote the subclass of those tQNM algebras verifying ($\text{prel}_m$) and ($\text{WNM}_m$) by $\mathbf{tQNM}_m$. Hence, we have:

$$\mathbf{tQNM_m} \subseteq \mathbf{tQNM};$$

$$\mathbf{tQNM_m} \subset \mathbf{tQIMTL_m} \subset \mathbf{m-pre-BCK-L_{(DN)}},$$

$$\mathbf{tQIMTL_m} = \mathbf{m-pre-BCK-L_{(DN)}} + (prel_m) +_a (prel),$$

where $+_a$ means that (prel) is added to the m-BCK lattice level,

$$\mathbf{tQNM_m} = \mathbf{tQIMTL_m} +_a (WNM),$$

where $+_a$ means that (WNM) is added to the m-BCK lattice level.

Resuming, we have the connections from Figure 15.2.

### 15.2.3 Building involutive m-pre-BCK algebras from m-BCK algebras by two methods

**Proposition 15.2.29** *Let $\mathcal{A}^L = (A^L, \odot, ^-, 1)$ be an involutive left-m-pre-BCK algebra and $a, b \in A^L$. If $b \parallel_Q a$, then, in the table of $\odot$, we have:*
*(i) the row of b is Q-parallel with the row of a, i.e. for any $y \in A^L$, $(b\odot y) \parallel_Q (a\odot y)$ and*
*(ii) the column of b is Q-parallel with the column of a, i.e. for any $x \in A^L$, $(x \odot b) \parallel_Q (x \odot a)$.*

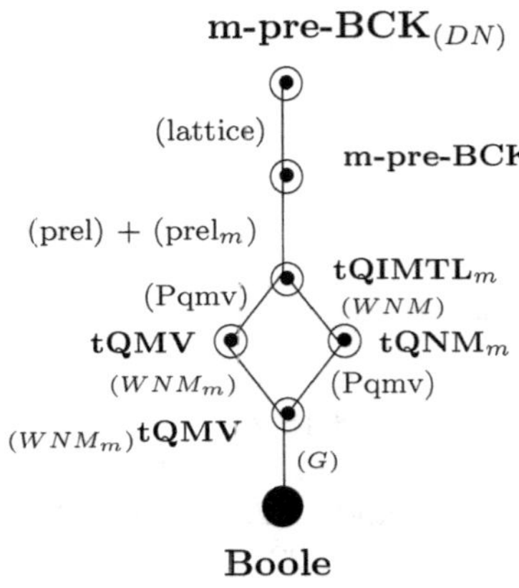


Figure 15.2: The connections between **m-pre-BCK**$_{(DN)}$, **m-pre-BCK-L**$_{(DN)}$, **tQIMTL**$_m$, **tQMV**, **tQNM**$_m$, $_{(WNM_m)}$**tQMV** and **Boole**

**Proof.** Obviously, by Proposition 15.2.7 (1), since $y \parallel_Q y$ and $x \parallel_Q x$, by reflexivity of $\parallel_Q$ and by (Pcomm). See the table of $\odot$ below.

| | | | | | | columns | | | | |
|---|---|---|---|---|---|---|---|---|---|---|
| | $\odot$ | 0 | ... | a | ... | y | ... | b | ... | 1 |
| | 0 | 0 | ... | 0 | ... | 0 | ... | 0 | ... | 0 |
| | ⋮ | ⋮ | | ⋮ | | ⋮ | | ⋮ | | ⋮ |
| | a | 0 | ... | $a \odot a$ | ... | $\mathbf{a \odot y}$ | ... | $a \odot b$ | ... | a |
| rows | ⋮ | ⋮ | | ⋮ | | ⋮ | | ⋮ | | ⋮ |
| | x | 0 | ... | $\mathbf{x \odot a}$ | ... | $x \odot y$ | ... | $\mathbf{x \odot b}$ | ... | x |
| | ⋮ | ⋮ | | ⋮ | | ⋮ | | ⋮ | | ⋮ |
| | b | 0 | ... | $b \odot a$ | ... | $\mathbf{b \odot y}$ | ... | $b \odot b$ | ... | b |
| | ⋮ | ⋮ | | ⋮ | | ⋮ | | ⋮ | | ⋮ |
| | 1 | 0 | ... | a | ... | y | ... | b | ... | 1 |

Consequently, we can build an involutive left-m-pre-BCK algebra from a given **finite left-m-BCK algebra** $\mathcal{A}^L = (A^L, \odot, ^-, 1)$, by adding to $A^L$ new elements (i.e. not belonging to $A^L$) that are Q-parallel with old ones (from $A^L$) and extending the table of $\odot$ by using Proposition 15.2.29 and extending the operation $^-$ such that the property (DN) be preserved. Thus, we obtain a method that will be called the "*method of Q-parallel rows/columns*": if we add to $A^L$ the new element $u \notin A^L$ such that it is Q-parallel with the old element $a \in A^L$, $u \parallel_Q a$, then, in the extended table of $\odot$ on $A^L \cup \{u, \dots\}$, we will have:

for any $y \in A^L$, $(u \odot y) \parallel_Q (a \odot y)$ and for any $x \in A^L$, $(x \odot u) \parallel_Q (x \odot a)$.

But there are multiple choices for the elements $u \odot y$ and $x \odot u$ in applying the "method of Q-parallel rows/columns", so we have to choose those elements $u \odot y$ and $x \odot u$ that verify a desired property - see Examples 15.3.27 - 15.3.29.

Note that a particular case of the "method of Q-parallel rows/columns" is what we shall call the "*method of identic rows/columns*" (based on the reflexivity of $\|_Q$): the row/column of $u$ is identic with the row/column of $a$, excepting the last elements ($u = u \odot 1 = 1 \odot u$ and $a = a \odot 1 = 1 \odot u$), which are Q-parallel, i.e. if we add $u$ such that $u \|_Q a$, then, in the extended table of $\odot$, we will have:

for any $y \in A^L \setminus \{1\}$, $u \odot y = a \odot y$ and for any $x \in A^L \setminus \{1\}$, $x \odot u = x \odot a$.

By this simple method, we can build the majority of quantum structures - see Examples 15.3.19 - 15.3.26 (15.3.19 **m-pre-BCK**$_{(DN)}$, 15.3.20 **m-pre-BCK-L**$_{(DN)}$, 15.3.21 **tQIMTL**, 15.3.22 **tQNM**, 15.3.23 **tPreMV**, 15.3.24 **tMMV**, 15.3.25 **tOMWL**, 15.3.26 **tOWL**).

It remains a problem to be solved in the future: to find which are the properties of the finite m-BCK algebra $\mathcal{A}^L$ that are preserved by the "method of identic rows/columns".

We shall prove next that the propery (Pom) (in some cases) and the property (G) are not preserved by the "method of identic rows/columns", so we can not build tQMV algebras and tOLs, tOSLs by this method; in this cases, we must use the general "method of Q-parallel rows/columns" (see Examples 15.3.27 - 15.3.29).

**Proposition 15.2.30** *Let $\mathcal{A}^L = (A^L, \odot, ^-, 1)$ be a left-m-BCK algebra verifying the property (Pom) and such that there exists $a \in A^L \setminus \{0, 1\}$, $a \odot a \neq 0$. If we add to $A^L$ an element $u \notin A^L$ such that $u \|_Q a$ and we extend the table of $\odot$ by the "method of identic rows/columns", then the resulting algebra does not verify (Pom) anymore.*

**Proof.** Suppose that $a \odot a = b \neq 0 \in A^L$ as in the extended table of $\odot$ given below.

| | | | | | | columns | | | | | |
|---|---|---|---|---|---|---|---|---|---|---|---|
| | $\odot$ | 0 | … | a | … | y | … | u | … | 1 |
| | 0 | 0 | … | 0 | … | 0 | … | 0 | … | 0 |
| | ⋮ | ⋮ | | ⋮ | | ⋮ | | ⋮ | | ⋮ |
| | a | 0 | … | $b$ | … | **a** $\odot$ **y** | … | $b$ | … | a |
| rows | ⋮ | ⋮ | | ⋮ | | ⋮ | | ⋮ | | ⋮ |
| | x | 0 | … | **x** $\odot$ **a** | … | $x \odot y$ | … | **x** $\odot$ **a** | … | x |
| | ⋮ | ⋮ | | ⋮ | | ⋮ | | ⋮ | | ⋮ |
| | u | 0 | … | $b$ | … | **a** $\odot$ **y** | … | $b$ | … | u |
| | ⋮ | ⋮ | | ⋮ | | ⋮ | | ⋮ | | ⋮ |
| | 1 | 0 | … | a | … | y | … | u | … | 1 |

Since $\mathcal{A}^L$ verifies (Pom) $((x \odot y) \oplus ((x \odot y)^- \odot x) = x)$ for any $x, y \in A^L$, it follows that we have: $(a \odot a) \oplus ((a \odot a)^- \odot a) = a$, i.e. $b \oplus (b^- \odot a) = a$. Then, the property (Pom) is not verified for $(u, a)$: indeed,
$(u \odot a) \oplus ((u \odot a)^- \odot u) = (a \odot a) \oplus ((a \odot a)^- \odot u) = b \oplus (b^- \odot u) = b \oplus (b^- \odot a) = a \neq u$. $\Box$

Note that any MV algebra verifies (Pom), but only the linearly ordered proper MV algebras (i.e. not being $_{(WNM)}$MV algebras) (see $\mathcal{L}_4^L$, $\mathcal{L}_5^L$ from Examples 15.3.4) and the direct products of linearly ordered $_{(WNM)}$MV algebras (see $\mathcal{L}_{2\times 2}^L$, $\mathcal{L}_{3\times 2}^L$, $\mathcal{L}_{3\times 3}^L$ from Examples 15.3.8) have elements $a \neq 0, 1$ such that $a \odot a \neq 0$; hence, building involutive m-pre-BCK algebras starting from such algebras, they will not verify (Pom), therefore they will be only tPreMV algebras or tMMV algebras, or even tOWLs, and not tQMV algebras - see Example 15.3.23, Examples 15.3.24 and Examples 15.3.26. To build tQMV algebras, we must use the general "method of Q-parallel rows/columns" - see Examples 15.3.27.

**Proposition 15.2.31** *Let $\mathcal{A}^L = (A^L, \odot, ^-, 1)$ be a linearly ordered left-m-BCK algebra verifying the property (Pom) and such that for any $a \in A^L \setminus \{0, 1\}$, $a \odot a = 0$. If we add to $A^L$ an element $u \notin A^L$ such that $u \parallel_Q a$ and we extend the table of $\odot$ by the "method of identic rows/columns", then the resulting algebra verifies (Pom).*

**Proof.** We have to prove that $u, a$ verify (Pom); indeed,
$(u \odot a) \oplus ((u \odot a)^- \odot u) = (a \odot a) \oplus ((a \odot a)^- \odot u) = 0 \oplus (0^- \odot u) = 0 \oplus (1 \odot u) = 0 \oplus u = u$.
$\Box$

Note that any MV algebra verifies (Pom), but only the linearly ordered $_{(WNM)}$MV algebras ($\mathcal{L}_3^L$ from Examples 15.3.4) have all the elements $a \neq 0, 1$ such that $a \odot a = 0$; hence, building involutive m-pre-BCK algebras starting from such algebras, they will verify (Pom), hence they will be tOMWL = $_{(WNM_m)}$ tQMV algebras - see Examples 15.3.25.

**Proposition 15.2.32** *Let $\mathcal{A}^L = (A^L, \odot, ^-, 1)$ be a left-Boolean algebra (hence (G) holds) and $a \in A^L$, $a \neq 0, 1$. If we add an element $u \notin A^L$ such that $u \parallel_Q a$ and we extend the table of $\odot$ by the "method of identic rows/columns", then the resulting algebra does not verify (G) anymore.*

**Proof.** $\mathcal{A}^L$ verifies (G) $(x \odot x = x)$, hence $a \odot a = a$ and we have the situation from the extended table of $\odot$ presented below.

| | | | | | | columns | | | | |
|---|---|---|---|---|---|---|---|---|---|---|
| | ⊙ | 0 | ... | a | ... | y | ... | u | ... | 1 |
| | 0 | 0 | ... | 0 | ... | 0 | ... | 0 | ... | 0 |
| | ⋮ | ⋮ | | ⋮ | | ⋮ | | ⋮ | | ⋮ |
| | a | 0 | ... | $a$ | ... | **a ⊙ y** | ... | $a$ | ... | a |
| rows | ⋮ | ⋮ | | ⋮ | | ⋮ | | ⋮ | | ⋮ |
| | x | 0 | ... | **x ⊙ a** | ... | $x \odot y$ | ... | **x ⊙ a** | ... | x |
| | ⋮ | ⋮ | | ⋮ | | ⋮ | | ⋮ | | ⋮ |
| | u | 0 | ... | $a$ | ... | **a ⊙ y** | ... | $a$ | ... | u |
| | ⋮ | ⋮ | | ⋮ | | ⋮ | | ⋮ | | ⋮ |
| | 1 | 0 | ... | a | ... | y | ... | u | ... | 1 |

Then, the property (G) is not verified for $u$, since $u \odot u = a \neq u$. □

Note that, following Proposition 15.2.32, we can not build a tOL or a tOSL from Boolean algebras by using the "method of identic rows/columns", because the property (G) is not preserved. Therefore, we must use the general "method of Q-parallel rows/columns" - see Examples 15.3.28 and 15.3.29.

**Remarks 15.2.33**

*(i) The only finite m-BCK algebra, from which we cannot build any involutive m-pre-BCK algebra by any of the above two methods, is the only Boolean algebra that is linearly ordered, namely* $\mathcal{L}_2^L = (L_2 = \{0, 1\}, \odot, ^-, 1)$ *(see Examples 15.3.4). But,* $\mathcal{L}_2^L$ *can be used for obtaining new involutive m-pre-BCK algebras as direct products between a given involutive m-pre-BCK algebra and* $\mathcal{L}_2^L$ *(see Examples 15.3.10, 3 and 15.3.14, 4).*

*(ii) Given a finite m-BCK algebra* $(A, \odot, ^-, 1)$ *with* $|A| \geq 3$, *we can obtain an infinity of finite involutive m-pre-BCK algebras:* $(A^1, \odot, ^-, 1)$, $(A^2, \odot, ^-, 1)$, *... such that their associated m-BCK algebras* $(A^1/\|_Q, \odot, ^-, \{1\})$, $(A^2/\|_Q, \odot, ^-, \{1\})$, *... be (isomorphic to)* $(A, \odot, ^-, 1)$, *by adding one or more new elements (i.e. not belonging to A) Q-parallel with some (old) elements of A by using one of the above two methods (see Examples 15.3.19 - 15.3.29).*

It remains a problem to be solved in the future: to find if the properties (prel$_m$) and (WNM$_m$) of a finite m-BCK lattice are preserved by the "method of Q-parallel rows/columns". If the answer is YES, then **tQIMTL**$_m$ = **tQIMTL** and **tQNM**$_m$ = **tQNM** ?

## 15.3 Examples

The section has two subsections.

In Subsection 15.3.1 (**(Involutive) m-BCK algebras**), we recall and analyse in some details some examples of (involutive) m-BCK algebras (Examples 15.3.1 - 15.3.8) needed in the second subsection:

15.3.1 **m-BCK**, 15.3.2 **m-BCK-L**, 15.3.3 **IMTL**, 15.3.4 **MV**, 15.3.5 non-linearly ordered **MV**, 15.3.6 **NM**, 15.3.7 non-linearly ordered **NM**, 15.3.8 nonlinearly ordered $_{(WNM)}$**MV**.

In Subsection 15.3.2 (**Involutive m-pre-BCK algebras**), we give or we build examples of transitive quantum algebras (as examples of involutive m-pre-BCK algebras):
- in Subsubsection 15.3.2.1 (**Giving transitive quantum algebras**), we analyse in some details mainly the examples of transitive quantum structures presented in our previous chapters 9, 10, 11, 12, 13 (Examples 15.3.9 - 15.3.18), in order to find which is their associated m-BCK algebra:
15.3.9 **tQIMTL**, 15.3.10 **tQNM**, 15.3.11 **tPreMV**, 15.3.12 **tMMV**, 15.3.13 **tQMV**, 15.3.14 **tOMWL**, 15.3.15 **tOWL**, 15.3.16 **tOL**, 15.3.17 **tOSL**, 15.3.18 **tOM**;
- in Subsubsection 15.3.2.2 (**Building transitive quantum algebras by the "method of identic rows/ columns"**), starting from particular m-BCK algebras presented in subsection 15.3.1, we build transitive quantum algebras by the "method of identic rows/ columns" (Examples 15.3.19 - 15.3.26):
15.3.19 **m-pre-BCK**$_{(DN)}$, 15.3.20 **m-pre-BCK-L**$_{(DN)}$, 15.3.21 **tQIMTL**, 15.3.22 **tQNM**, 15.3.23 **tPreMV**, 15.3.24 **tMMV**, 15.3.25 **tOMWL**, 15.3.26 **tOWL**;
- in Subsubsection 15.3.2.3 (**Building transitive quantum algebras by the "method of Q-parallel rows/ columns"**), starting from particular m-BCK algebras presented in subsection 15.3.1, we build transitive quantum algebras by the "method of Q-parallel rows/columns" (Examples 15.3.27 - 15.3.29):
15.3.27 **tQMV**, 15.3.28 **tOL**, 15.3.29 **tOSL**.

For each example, we analyse at least $\leq_m$ ($\Longleftrightarrow \leq_m^B$) among the three binary relations $\leq_m$, $\leq_m^M$, $\leq_m^P$ that exist.

We have written the following PASCAL programs that were used in this section (they can be sent at the request by e-mail):

**ALL-VM0.PAS** which, given the tables of the product $\odot$ and of the involutive negation $^-$, verifies if the properties (m-Re), (m-Pabs-i), (Pass), (m-B), (m-BB), (m-*), (m-**), (m-Tr), (m-An), (m-Pimpl), (Pqmv), (Pom), (Pmv), ($\Delta_m$), (prel$_m$), (WNM$_m$), (aWNM$_m$), (m-Pdis) are satisfied;

**P-SUM.PAS** which, given the tables of the product $\odot$ and of the involutive negation $^-$, provides the table of the sum $\oplus$;

**WEDGES.PAS** which, given the tables of the product $\odot$ and of the involutive negation $^-$, provides the tables of $\wedge_m^M$ and $\wedge_m^B$ and verifies if the binary relations $\leq_m^M$ and $\leq_m^B$ are transitive;

**PSI.PAS** which, given the tables of the product $\odot$ and of the involutive negation $^-$, provides the table of the implication $\rightarrow$;

**VER-WNM.PAS** which, given the tables of $\rightarrow$, $\odot$, $\vee$, $\wedge$, verifies if the property (WNM) is verified.

### 15.3.1 (Involutive) m-BCK algebras

Recall that (involutive) m-BCK algebras mean transitive m-aBE algebras.

**Example 15.3.1 Proper m-BCK algebra (not lattice): m-BCK**

Consider the algebra $\mathcal{A}^L = (A_8 = \{0, m, a, b, c, d, n, 1\}, \odot, ^-, 1)$, with the following tables of $\odot$ , $^-$, $\oplus$ from ([104], 7.2.1):

| $\odot$ | 0 | m | a | b | c | d | n | 1 |
|---|---|---|---|---|---|---|---|---|
| 0 | 0 | 0 | 0 | 0 | 0 | 0 | 0 | 0 |
| m | 0 | 0 | 0 | 0 | 0 | 0 | 0 | m |
| a | 0 | 0 | 0 | 0 | 0 | m | m | a |
| b | 0 | 0 | 0 | 0 | m | 0 | m | b |
| c | 0 | 0 | 0 | m | m | m | m | c |
| d | 0 | 0 | m | 0 | m | m | m | d |
| n | 0 | 0 | m | m | m | m | m | n |
| 1 | 0 | m | a | b | c | d | n | 1 |

,

| $\oplus$ | 0 | m | a | b | c | d | n | 1 |
|---|---|---|---|---|---|---|---|---|
| 0 | 0 | m | a | b | c | d | n | 1 |
| m | m | n | n | n | n | n | 1 | 1 |
| a | a | n | n | n | 1 | n | 1 | 1 |
| b | b | n | n | n | n | 1 | 1 | 1 |
| c | c | n | 1 | n | 1 | 1 | 1 | 1 |
| d | d | n | n | 1 | 1 | 1 | 1 | 1 |
| n | n | 1 | 1 | 1 | 1 | 1 | 1 | 1 |
| 1 | 1 | 1 | 1 | 1 | 1 | 1 | 1 | 1 |

,

and $(0, m, a, b, c, d, n, 1)^- = (1, n, c, d, a, b, m, 0)$.

It is an involutive left-m-BE algebra verifying (m-An) and (m-BB) ($\Leftrightarrow \ldots \Leftrightarrow$ (m-Tr)), and not verifying (m-Pimpl) for $(m, 0)$, (m-Pabs-i) for $(m, 0)$, (G) for $m$, (Pqmv) for $(a, m, 0)$, (Pom) for $(a, 0)$, (Pmv) for $(a, m)$, ($\Delta_m$) for $(c, a)$, (prel$_m$) for $(a, b)$, (WNM$_m$) and (aWNM$_m$) for $(a, d)$. Hence, $\mathcal{A}^L$ is a left-m-BCK algebra.

To see if the m-BCK algebra is lattice or not w.r. to the order relation $\leq_m$ ($\Longleftrightarrow \leq_m^B$), we make the table of $\wedge_m^B$:

| $\wedge_m^B$ | 0 | m | a | b | c | d | n | 1 |
|---|---|---|---|---|---|---|---|---|
| 0 | 0 | 0 | 0 | 0 | 0 | 0 | 0 | 0 |
| m | 0 | m | m | m | m | m | m | m |
| a | 0 | m | a | m | a | a | a | a |
| b | 0 | m | m | b | b | b | b | b |
| c | 0 | m | m | m | c | m | c | c |
| d | 0 | m | m | m | m | d | d | d |
| n | 0 | m | m | m | m | m | n | n |
| 1 | 0 | m | a | b | c | d | n | 1 |

.

From the table of $\wedge_m^B$, we see that $m \leq_m a, b, c, d, n, 1$; $a \leq_m c, d, n, 1$; $b \leq_m c, d, n, 1$; $c \leq_m n, 1$; $d \leq_m n, 1$; $n \leq_m 1$. Hence, the Hasse diagram of the bounded poset $(A_8, \leq_m, 0, 1)$ is that from Figure 15.3.

Note that the poset $(A_8, \leq_m, 0, 1)$ is not a lattice. Hence, $\mathcal{A}^L$ is a proper m-BCK algebra, which is not a lattice.

**Examples 15.3.2 (Involutive) m-BCK lattices: m-BCK-L**

**Example 1: proper m-BCK lattice**

By a PASCAL program, we found that the algebra $\mathcal{A}^L = (A_6 = \{0, a, b, c, d, 1\}, \odot, ^-, 1)$, with the following tables of $\odot$ and $^-$ and of the additional operation $\oplus$, is a proper left-m-BCK algebra, i.e. (PU), (Pcomm), (Pass), (m-La), (m-Re), (m-An), (DN) and (m-BB) ($\Leftrightarrow \ldots \Leftrightarrow$ (m-Tr)) hold and it does not verify (m-Pabs-i) for $(b, 0)$, (G) for $a$, (m-Pimpl) for $(a, 0)$, ($\wedge_m$-comm) for $(a, b)$, (Pqmv) for $(b, d, 0)$, (Pom) for $(b, a)$, (Pmv) for $(b, d)$, ($\Delta_m$) for $(a, b)$ and also (prel$_m$) for $(b, c)$, (WNM$_m$) and (aWNM$_m$) for $(a, a)$, (m-Pdis) for $(a, a, b)$.

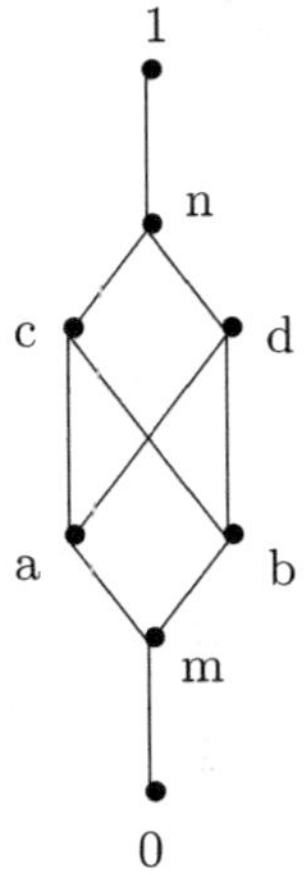


Figure 15.3: The Hasse diagram of the bounded poset $(A_8, \leq_m, 0, 1)$, which is not a lattice

| $\odot$ | 0 | a | b | c | d | 1 |
|---|---|---|---|---|---|---|
| 0 | 0 | 0 | 0 | 0 | 0 | 0 |
| a | 0 | d | d | d | 0 | a |
| b | 0 | d | d | 0 | 0 | b |
| c | 0 | d | 0 | d | 0 | c |
| d | 0 | 0 | 0 | 0 | 0 | d |
| 1 | 0 | a | b | c | d | 1 |

and

| $x$ | $x^-$ |
|---|---|
| 0 | 1 |
| a | d |
| b | c |
| c | b |
| d | a |
| 1 | 0 |

, with

| $\oplus$ | 0 | a | b | c | d | 1 |
|---|---|---|---|---|---|---|
| 0 | 0 | a | b | c | d | 1 |
| a | a | 1 | 1 | 1 | 1 | 1 |
| b | b | 1 | a | 1 | a | 1 |
| c | c | 1 | 1 | a | a | 1 |
| d | d | 1 | a | a | a | 1 |
| 1 | 1 | 1 | 1 | 1 | 1 | 1 |

.

The tables of $\wedge_m^M$, $\wedge_m^B$ and $\rightarrow$ are:

| $\wedge_m^M$ | 0 | a | b | c | d | 1 |
|---|---|---|---|---|---|---|
| 0 | 0 | 0 | 0 | 0 | 0 | 0 |
| a | 0 | a | b | c | d | a |
| b | 0 | d | b | d | d | b |
| c | 0 | d | d | c | d | c |
| d | 0 | d | d | d | d | d |
| 1 | 0 | a | b | c | d | 1 |

,

| $\wedge_m^B$ | 0 | a | b | c | d | 1 |
|---|---|---|---|---|---|---|
| 0 | 0 | 0 | 0 | 0 | 0 | 0 |
| a | 0 | a | **d** | **d** | d | a |
| b | 0 | b | b | d | d | b |
| c | 0 | c | d | c | d | c |
| d | 0 | d | d | d | d | d |
| 1 | 0 | a | b | c | d | 1 |

and

| $\rightarrow$ | 0 | a | b | c | d | 1 |
|---|---|---|---|---|---|---|
| 0 | 1 | 1 | 1 | 1 | 1 | 1 |
| a | d | 1 | a | a | a | 1 |
| b | c | 1 | 1 | a | a | 1 |
| c | b | 1 | a | 1 | a | 1 |
| d | a | 1 | 1 | 1 | 1 | 1 |
| 1 | 0 | a | b | c | d | 1 |

.

• Note that $\leq_m^M$ is an order relation, but not a lattice order w.r. to $\wedge_m^M$, $\vee_m^M$, since $\wedge_m^M$ is not commutative.

• The binary relation $\leq_m$ ($\Longleftrightarrow \leq_m^B$) ($x \leq_m^B y \stackrel{def.}{\Longleftrightarrow} x \wedge_m^B y = x$) is an order relation also, since (m-Re), (m-An) and (m-Tr) hold; hence, $\leq_m^B$ is an order relation too. From the table of $\wedge_m^B$, we see that: $a \leq_m 1$; $b \leq_m a, 1$; $c \leq_m a, 1$; $d \leq_m a, b, c, 1$. It follows that the Hasse diagram of the bounded poset $(A_6, \leq_m, 0, 1)$ is that from Figure 15.4. The operation $\wedge_m^B$ is not commutative, therefore the order relation $\leq_m^B$ is not a lattice order w.r. to $\wedge_m^B$, $\vee_m^B$; but, the order relation $\leq_m$ is a distributive lattice order w.r. to $\wedge = \wedge_m = \inf_m$, $\vee = \vee_m = \sup_m$.

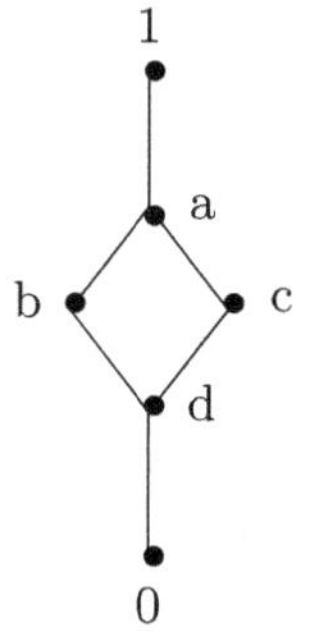


Figure 15.4: The Hasse diagram of the bounded poset $(A_6, \leq_m, 0, 1)$, that is a distributive lattice

From the table of $\rightarrow$, note that $(b \rightarrow c) \vee (c \rightarrow b) = a \vee a = a \neq 1$, hence (prel) is not verified for $(b, c)$; (WNM) is not verified for $(a, a)$. Hence, the algebra $\mathcal{A}^L$ is a proper (involutive) distributive left-m-BCK lattice.

**Example 2: m-BCK (distributive) lattice verifying (aWNM$_m$), (WNM$_m$) and not verifying (prel$_m$)**

By *Mace4* program, we found the algebra $\mathcal{A}^L = (A_{12} = \{0, a, b, c, d, e, f, g, h, i, j, 1\}, \odot, ^-, 1)$ with the following tables of $\odot$, $^-$, $\oplus$:

| $\odot$ | 0 | a | b | c | d | e | f | g | h | i | j | 1 |
|---|---|---|---|---|---|---|---|---|---|---|---|---|
| 0 | 0 | 0 | 0 | 0 | 0 | 0 | 0 | 0 | 0 | 0 | 0 | 0 |
| a | 0 | c | 0 | c | g | 0 | c | 0 | c | c | g | a |
| b | 0 | 0 | e | 0 | e | e | g | 0 | e | g | e | b |
| c | 0 | c | 0 | c | 0 | 0 | c | 0 | c | c | 0 | c |
| d | 0 | g | e | 0 | e | e | g | 0 | e | g | e | d |
| e | 0 | 0 | e | 0 | e | e | 0 | 0 | e | 0 | e | e |
| f | 0 | c | g | c | g | 0 | c | 0 | c | c | g | f |
| g | 0 | 0 | 0 | 0 | 0 | 0 | 0 | 0 | 0 | 0 | 0 | g |
| h | 0 | c | e | c | e | e | c | 0 | h | c | e | h |
| i | 0 | c | g | c | g | 0 | c | 0 | c | c | 0 | i |
| j | 0 | g | e | 0 | e | e | g | 0 | e | 0 | e | j |
| 1 | 0 | a | b | c | d | e | f | g | h | i | j | 1 |

,

| $\oplus$ | 0 | a | b | c | d | e | f | g | h | i | j | 1 |
|---|---|---|---|---|---|---|---|---|---|---|---|---|
| 0 | 0 | a | b | c | d | e | f | g | h | i | j | 1 |
| a | a | f | 1 | f | 1 | h | f | f | 1 | f | h | 1 |
| b | b | 1 | d | h | d | d | 1 | d | 1 | h | d | 1 |
| c | c | f | h | f | 1 | h | f | f | 1 | f | h | 1 |
| d | d | 1 | d | 1 | d | d | 1 | d | 1 | 1 | d | 1 |
| e | e | h | d | h | d | d | 1 | d | 1 | h | d | 1 |
| f | f | f | 1 | f | 1 | 1 | f | f | 1 | f | 1 | 1 |
| g | g | f | d | f | d | d | f | g | 1 | f | d | 1 |
| h | h | 1 | 1 | 1 | 1 | 1 | 1 | 1 | 1 | 1 | 1 | 1 |
| i | i | f | h | f | 1 | h | f | f | 1 | f | 1 | 1 |
| j | j | h | d | h | d | d | 1 | d | 1 | 1 | d | 1 |
| 1 | 1 | 1 | 1 | 1 | 1 | 1 | 1 | 1 | 1 | 1 | 1 | 1 |

and $(0, a, b, c, d, e, f, g, h, i, j, 1)^- = (1, b, a, d, c, f, e, h, g, j, i, 0)$.

It is an involutive left-m-aBE algebra verifying (m-Tr) ($\Leftrightarrow \ldots \Leftrightarrow$ (m-BB)) and (aWNM$_m$), (WNM$_m$), and not verifying (m-Pimpl) for $(a, 0)$, (m-Pabs-i) for $(a, 0)$, (G) for $a$, (Pqmv) for $(a, a, d)$, (Pom) for $(a, a)$, (Pmv) for $(a, e)$, ($\Delta_m$) for $(c, b)$, (prel$_m$) for $(a, i)$, (m-Pdis) for $(a, a, a)$. Hence, $\mathcal{A}^L$ is a left-m-BCK algebra verifying (aWNM$_m$), (WNM$_m$), and not verifying (prel$_m$).

The table of $\wedge_m^B$ is:

| $\wedge_m^B$ | 0 | a | b | c | d | e | f | g | h | i | j | 1 |
|---|---|---|---|---|---|---|---|---|---|---|---|---|
| 0 | 0 | 0 | 0 | 0 | 0 | 0 | 0 | 0 | 0 | 0 | 0 | 0 |
| a | 0 | a | g | c | g | g | a | g | a | c | g | a |
| b | 0 | g | b | g | b | e | g | g | b | g | e | b |
| c | 0 | c | 0 | c | 0 | 0 | c | 0 | c | c | 0 | c |
| d | 0 | g | e | g | d | e | g | g | d | g | e | d |
| e | 0 | 0 | e | 0 | e | e | 0 | 0 | e | 0 | e | e |
| f | 0 | c | g | c | g | g | f | g | f | c | g | f |
| g | 0 | g | g | g | g | g | g | g | g | g | g | g |
| h | 0 | c | e | c | e | e | c | 0 | h | c | e | h |
| i | 0 | c | g | c | g | g | i | g | i | i | g | i |
| j | 0 | g | e | g | j | e | g | g | j | g | j | j |
| 1 | 0 | a | b | c | d | e | f | g | h | i | j | 1 |

.

• The binary relation $\leq_m$ is an order, since it is reflexive, antisymmetric and transitive. From the table of $\wedge_m^B$, we obtain: $a \leq_m a, f, h, 1$; $b \leq_m b, d, h, 1$; $c \leq_m c, a, f, h, i, 1$; $d \leq_m d, h, 1$; $e \leq_m e, b, d, h, j, 1$; $g \leq_m g, a, b, c, d, e, f, h, i, j, 1$; $f \leq_m f, h, 1$; $h \leq_m h, 1$; $i \leq_m i, f, h, 1$; $j \leq_m j, d, h, 1$; hence, the bounded poset $(A_{12}, \leq, 0, 1)$ is represented by the Hasse diagram from Figure 15.5.

**Example 3: m-BCK (distributive) lattice verifying (prel$_m$), (aWNM$_m$), (WNM$_m$) and not verifying (prel), (WNM)**

By *Mace4* program, we found the algebra $\mathcal{A}^L = (A_8 = \{0, a, b, c, d, e, f, 1\}, \odot, ^-, 1)$ with the following tables of $\odot$ , $^-$, $\oplus$:

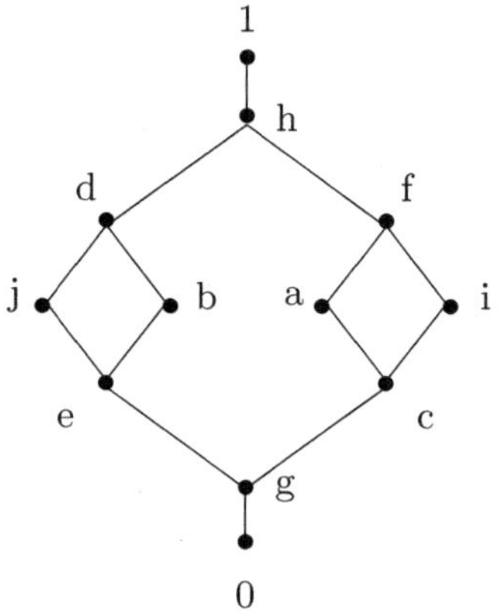


Figure 15.5: The distributive lattice $(A_{12}, \leq_m, 0, 1)$

| $\odot$ | 0 | a | b | c | d | e | f | 1 |
|---|---|---|---|---|---|---|---|---|
| 0 | 0 | 0 | 0 | 0 | 0 | 0 | 0 | 0 |
| a | 0 | c | 0 | c | b | 0 | a | a |
| b | 0 | 0 | 0 | 0 | b | 0 | b | b |
| c | 0 | c | 0 | c | 0 | 0 | c | c |
| d | 0 | b | b | 0 | d | 0 | d | d |
| e | 0 | 0 | 0 | 0 | 0 | 0 | 0 | e |
| f | 0 | a | b | c | d | 0 | f | f |
| 1 | 0 | a | b | c | d | e | f | 1 |

with

| $\oplus$ | 0 | a | b | c | d | e | f | 1 |
|---|---|---|---|---|---|---|---|---|
| 0 | 0 | a | b | c | d | e | f | 1 |
| a | a | 1 | 1 | a | 1 | a | 1 | 1 |
| b | b | 1 | d | a | d | b | 1 | 1 |
| c | c | a | a | c | 1 | c | 1 | 1 |
| d | d | 1 | d | 1 | d | d | 1 | 1 |
| e | e | a | b | c | d | e | 1 | 1 |
| f | f | 1 | 1 | 1 | 1 | 1 | 1 | 1 |
| 1 | 1 | 1 | 1 | 1 | 1 | 1 | 1 | 1 |

and $(0, a, b, c, d, e, f, 1)^- = (1, b, a, d, c, f, e, 0)$.

It is an involutive left-m-BE algebra verifying (m-An) and (m-BB) ($\Leftrightarrow \ldots \Leftrightarrow$ (m-Tr)) and (prel$_m$), (aWNM$_m$), (WNM$_m$), and not verifying (m-Pabs-i) for $(e, 0)$, (G) for $a$, (m-Pimpl) for $(a, 0)$, (Pqmv) for $(e, a, 0)$, (Pom) for $(f, a)$, (Pmv) for $(e, a)$, ($\Delta_m$) for $(a, e)$, (m-Pdis) for $(a, a, a)$. Hence, $\mathcal{A}^L$ is a left-m-BCK algebra verifying (prel$_m$), (aWNM$_m$), (WNM$_m$).

The table of $\wedge_m^M$ and $\wedge_m^B$ are:

| $\wedge_m^M$ | 0 | a | b | c | d | e | f | 1 |
|---|---|---|---|---|---|---|---|---|
| 0 | 0 | 0 | 0 | 0 | 0 | 0 | 0 | 0 |
| a | 0 | a | b | c | b | e | a | a |
| b | 0 | b | b | 0 | b | e | b | b |
| c | 0 | c | 0 | c | 0 | e | c | c |
| d | 0 | b | b | 0 | d | e | d | d |
| e | 0 | 0 | 0 | 0 | 0 | e | 0 | e |
| f | 0 | a | b | c | d | e | f | f |
| 1 | 0 | a | b | c | d | e | f | 1 |

and

| $\wedge_m^B$ | 0 | a | b | c | d | e | f | 1 |
|---|---|---|---|---|---|---|---|---|
| 0 | 0 | 0 | 0 | 0 | 0 | 0 | 0 | 0 |
| a | 0 | a | b | c | b | 0 | a | a |
| b | 0 | b | b | 0 | b | 0 | b | b |
| c | 0 | c | 0 | c | 0 | 0 | c | c |
| d | 0 | b | b | 0 | d | 0 | d | d |
| e | 0 | e | e | e | e | e | e | e |
| f | 0 | a | b | c | d | 0 | f | f |
| 1 | 0 | a | b | c | d | e | f | 1 |

.

• The binary relation $\leq_m^M$ is an order, but not a lattice order, since $\wedge_m^M$ is not commutative.

• The binary relation $\leq_m$ is an order, since it is reflexive, transitive, antisymmetric. We shall see if it is a lattice order. From the table of $\wedge_m^B$, we obtain: $a \leq_m a, f, 1$; $b \leq_m b, a, d, f, 1$; $c \leq_m c, a, f, 1$; $d \leq_m d, f, 1$; $e \leq_m e, a, b, c, d, f, 1$; $f \leq_m f, 1$; hence, the Hasse diagram of the bounded poset $(A_8, \leq_m, 0, 1)$ is that from Figure 15.6, and we see that it is a distributive lattice.

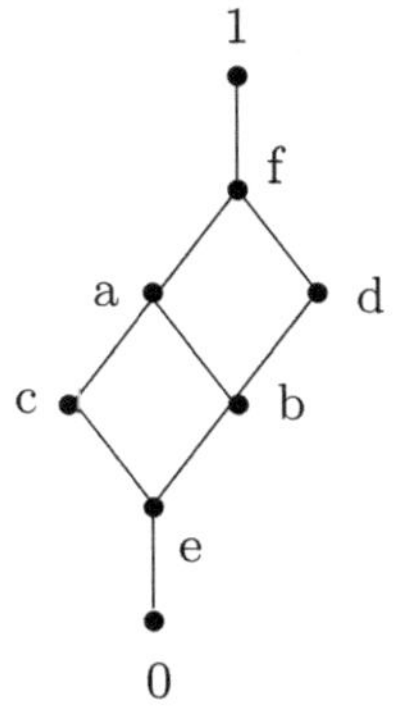


Figure 15.6: The bounded distributive lattice $(A_8, \leq_m, 0, 1)$

We omit the tables of $\wedge = \inf_m$ and $\vee = \sup_m$. The table of $\rightarrow$ $(x \rightarrow y = (x \odot y^-)^-)$ is the following:

| $\rightarrow$ | 0 | a | b | c | d | e | f | 1 |
|---|---|---|---|---|---|---|---|---|
| 0 | 1 | 1 | 1 | 1 | 1 | 1 | 1 | 1 |
| a | b | 1 | d | a | d | b | 1 | 1 |
| b | a | 1 | 1 | a | 1 | a | 1 | 1 |
| c | d | 1 | d | 1 | d | d | 1 | 1 |
| d | c | a | a | c | 1 | c | 1 | 1 |
| e | f | 1 | 1 | 1 | 1 | 1 | 1 | 1 |
| f | e | a | b | c | d | e | 1 | 1 |
| 1 | 0 | a | b | c | d | e | f | 1 |

Note that (prel) $((x \rightarrow y) \vee (y \rightarrow x) = 1)$ is not verified for $(c, b)$: $(c \rightarrow b) \vee (b \rightarrow c) = d \vee a = f \neq 1$. Note also that (WNM) $((x \odot y)^- \vee [(x \wedge y) \rightarrow (x \odot y)] = 1)$ is not verified for $(a, a)$: $(a \odot a)^- \vee [(a \wedge a) \rightarrow (a \odot a)] = c^- \vee [a \rightarrow c] = d \vee a = f \neq 1$. Thus, $\mathcal{A}^L$ is not an IMTL algebra, it is just a m-BCK (distributive) lattice verifying (prel$_m$), (aWNM$_m$), (WNM$_m$) and not verifying (prel), (WNM), hence not a proper m-BCK lattice.

**Example 4: m-BCK (distributive) lattice verifying (prel$_m$), (aWNM$_m$), (WNM$_m$), and (WNM), and not verifying (prel)**

By *Mace4* program, we found the algebra $\mathcal{A}^L = (A_6 = \{0, a, b, c, d, 1\}, \odot, ^-, 1)$ with the following tables of $\odot$ , $^-$, $\oplus$:

| $\odot$ | 0 | a | b | c | d | 1 |
|---|---|---|---|---|---|---|
| 0 | 0 | 0 | 0 | 0 | 0 | 0 |
| a | 0 | a | 0 | a | 0 | a |
| b | 0 | 0 | b | b | 0 | b |
| c | 0 | a | b | c | 0 | c |
| d | 0 | 0 | 0 | 0 | 0 | d |
| 1 | 0 | a | b | c | d | 1 |

and

| $x$ | $x^-$ |
|---|---|
| 0 | 1 |
| a | b |
| b | a |
| c | d |
| d | c |
| 1 | 0 |

, with

| $\oplus$ | 0 | a | b | c | d | 1 |
|---|---|---|---|---|---|---|
| 0 | 0 | a | b | c | d | 1 |
| a | a | a | 1 | 1 | a | 1 |
| b | b | 1 | b | 1 | b | 1 |
| c | c | 1 | 1 | 1 | 1 | 1 |
| d | d | a | b | 1 | d | 1 |
| 1 | 1 | 1 | 1 | 1 | 1 | 1 |

.

It is an involutive left-m-BE algebra verifying (m-An) and (m-BB) ($\Leftrightarrow \ldots \Leftrightarrow$ (m-Tr)) and (prel$_m$), (aWNM$_m$), (WNM$_m$), and not verifying (m-Pabs-i) for $(d, 0)$, (G) for $d$, (m-Pimpl) for $(c, 0)$, (Pqmv) for $(c, a, b)$, (Pom) for $(c, a)$, (Pmv) for $(d, a)$, ($\Delta_m$) for $(a, d)$, (m-Pdis) for $(a, b, c)$. Hence, $\mathcal{A}^L$ is a left-m-BCK algebra verifying (prel$_m$), (aWNM$_m$), (WNM$_m$).

The tables of $\wedge_m^B$ and $\rightarrow$ are:

| $\wedge_m^B$ | 0 | a | b | c | d | 1 |
|---|---|---|---|---|---|---|
| 0 | 0 | 0 | 0 | 0 | 0 | 0 |
| a | 0 | a | 0 | a | 0 | a |
| b | 0 | 0 | b | b | 0 | b |
| c | 0 | a | b | c | 0 | c |
| d | 0 | d | d | d | d | d |
| 1 | 0 | a | b | c | d | 1 |

and

| $\rightarrow$ | 0 | a | b | c | d | 1 |
|---|---|---|---|---|---|---|
| 0 | 1 | 1 | 1 | 1 | 1 | 1 |
| a | b | 1 | b | 1 | b | 1 |
| b | a | a | 1 | 1 | a | 1 |
| c | d | a | b | 1 | d | 1 |
| d | c | 1 | 1 | 1 | 1 | 1 |
| 1 | 0 | a | b | c | d | 1 |

.

• The binary relation $\leq_m$ is an order. From the table of $\wedge_m^B$, we obtain: $a \leq_m a, c, 1$; $b \leq_m b, c, 1$; $c \leq_m c, 1$; $d \leq_m a, b, c, d, 1$; hence, the bounded poset $(A_6, \leq_m, 0, 1)$ is represented by the Hasse diagram from Figure 15.7.

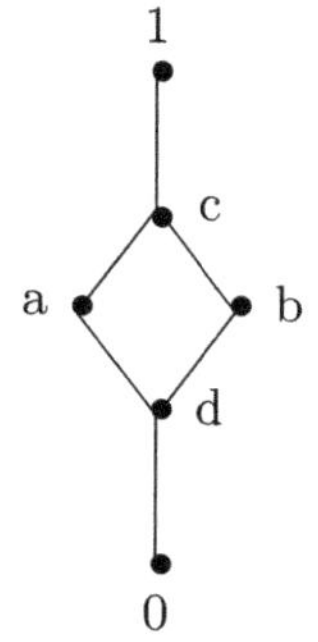


Figure 15.7: The Hasse diagram of the bounded poset $(A_6, \leq_m, 0, 1)$, that is a distributive lattice

The tables of $\vee$ and $\wedge$ are:

| $\vee$ | 0 | a | b | c | d | 1 |
|---|---|---|---|---|---|---|
| 0 | 0 | a | b | c | d | 1 |
| a | a | a | c | c | a | 1 |
| b | b | c | b | c | b | 1 |
| c | c | c | c | c | c | 1 |
| d | d | a | b | c | d | 1 |
| 1 | 1 | 1 | 1 | 1 | 1 | 1 |

and

| $\wedge$ | 0 | a | b | c | d | 1 |
|---|---|---|---|---|---|---|
| 0 | 0 | 0 | 0 | 0 | 0 | 0 |
| a | 0 | a | d | a | d | a |
| b | 0 | d | b | b | d | b |
| c | 0 | a | b | c | d | c |
| d | 0 | d | d | d | d | d |
| 1 | 0 | a | b | c | d | 1 |

.

Note that $(a \to b) \vee (b \to a) = b \vee a = c \neq 1$, hence (prel) is not verified. Condition (WNM) is verified.

**Examples 15.3.3 Linearly ordered IMTL algebras: IMTL**

**Example 1: IMTL$_5$**

Let us consider from ([104], page 211) (see [75]) the chain $L_5 = \{0, 1, 2, 3, 4\}$ (w.r.to $\leq = \leq_m$, which is a lattice w.r.to $\wedge, \vee$), organized as a left-IRL $IMTL_5 = (L_5, \wedge, \vee, \odot, \to, 0, 4)$, with the operations $\to$ and $x \odot y \overset{notation}{=} \min\{z \mid x \leq y \to z\} = (x \to y^-)^-$ as in the following tables:

$IMTL_5$

| $\to$ | 0 | 1 | 2 | 3 | 4 |
|---|---|---|---|---|---|
| 0 | 4 | 4 | 4 | 4 | 4 |
| 1 | 3 | 4 | 4 | 4 | 4 |
| 2 | 2 | **3** | 4 | 4 | 4 |
| 3 | 1 | **3** | **3** | 4 | 4 |
| 4 | 0 | 1 | 2 | 3 | 4 |

| $\odot$ | 0 | 1 | 2 | 3 | 4 |
|---|---|---|---|---|---|
| 0 | 0 | 0 | 0 | 0 | 0 |
| 1 | 0 | 0 | 0 | 0 | 1 |
| 2 | 0 | 0 | 0 | **1** | 2 |
| 3 | 0 | 0 | **1** | **1** | 3 |
| 4 | 0 | 1 | 2 | 3 | 4 |

Note that $IMTL_5$ satisfies (prel) and does not satisfy condition (WNM) for $(2, 3)$. Consequently, $IMTL_5$ is a linearly ordered proper IMTL algebra (you have the values of $x^- = x \to 0$ in the table of $\to$, column of 0).

Equivalently, by Theorem 12.3.3, consider the algebra $IMTL_5 = (A_5 = \{0, a, b, c, 1\}, \odot, ^-, 1)$ with the following operations:

$IMTL_5$

| $\odot$ | 0 | a | b | c | 1 |
|---|---|---|---|---|---|
| 0 | 0 | 0 | 0 | 0 | 0 |
| a | 0 | 0 | 0 | 0 | a |
| b | 0 | 0 | 0 | a | b |
| c | 0 | 0 | a | a | c |
| 1 | 0 | a | b | c | 1 |

and

| $x$ | $x^-$ |
|---|---|
| 0 | 1 |
| a | c |
| b | b |
| c | a |
| 1 | 0 |

, with

| $\oplus$ | 0 | a | b | c | 1 |
|---|---|---|---|---|---|
| 0 | 0 | a | b | c | 1 |
| a | a | c | c | 1 | 1 |
| b | b | c | 1 | 1 | 1 |
| c | c | 1 | 1 | 1 | 1 |
| 1 | 1 | 1 | 1 | 1 | 1 |

.

Then, $IMTL_5$ is an involutive m-aBE algebra verifying (m-BB) ($\Leftrightarrow \ldots \Leftrightarrow$ (m-Tr)), and (prel$_m$), and not verifying (m-Pabs-i) for $(a, 0)$, (m-Pimpl) for $(a, 0)$, (Pqmv) for $(b, a, 0)$, (Pom) for $(b, c)$, (Pmv) for $(b, a)$, ($\Delta_m$) for $(c, b)$, (WNM$_m$) and (aWNM$_m$) for $(b, c)$, (m-Pdis) for $(a, a, b)$.

The tables of $\wedge_m^M$, $\wedge_m^B$ and $\to$ are:

| $\wedge_m^M$ | 0 | a | b | c | 1 |
|---|---|---|---|---|---|
| 0 | 0 | 0 | 0 | 0 | 0 |
| a | 0 | a | a | a | a |
| b | 0 | a | b | a | b |
| c | 0 | a | b | c | c |
| 1 | 0 | a | b | c | 1 |

,

| $\wedge_m^B$ | 0 | a | b | c | 1 |
|---|---|---|---|---|---|
| 0 | 0 | 0 | 0 | 0 | 0 |
| a | 0 | a | a | a | a |
| b | 0 | a | b | b | b |
| c | 0 | a | a | c | c |
| 1 | 0 | a | b | c | 1 |

,

| $\to$ | 0 | a | b | c | 1 |
|---|---|---|---|---|---|
| 0 | 1 | 1 | 1 | 1 | 1 |
| a | c | 1 | 1 | 1 | 1 |
| b | b | c | 1 | 1 | 1 |
| c | a | c | c | 1 | 1 |
| 1 | 0 | a | b | c | 1 |

.

• The binary relation $\leq_m^M$ is an order, but not a lattice order, since $\wedge_m^M$ is not commutative. From the table of $\wedge_m^M$, we obtain: $a \leq_m^M a, b, c, 1$; $b \leq_m^M b, 1$; $c \leq_m^M c, 1$; hence, the bounded poset $(A_5, \leq_m^M, 0, 1)$ is represented by the Hasse diagram from Figure 15.8.

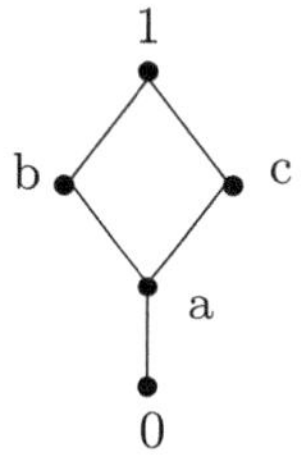


Figure 15.8: The bounded poset $(A_5, \leq_m^M, 0, 1)$ of the smallest IMTL algebra

• The binary relation $\leq_m$ is an order too, since it is reflexive, antisymmetric and transitive. From the table of $\wedge_m^B$, we obtain: $a \leq_m a, b, c, 1$; $b \leq_m b, c, 1$; $c \leq_m c, 1$; hence, the bounded poset $(A_5, \leq_m, 0, 1)$ is the **chain** $0 \leq_m a \leq_m b \leq_m c \leq_m 1$, which is a distributive lattice w.r. to $\wedge = \inf_m$ and $\vee = \sup_m$. Being a chain, the property (prel) is verified. The property (WNM) is not verified for $(b, c)$. Hence, indeed, $IMTL_5$ is the linearly ordered IMTL algebra with the minimum number of elements, five.

**Example 2: IMTL$_6^4$**

Let us consider from ([104], page 212) (see [75]) the chain $L_6 = \{0, 1, 2, 3, 4, 5\}$ (w.r.to $\leq = \leq_m$), organized as a lattice by $x \wedge y = \min\{x, y\}$, $x \vee y = \max\{x, y\}$ and as a left-IRL $IMTL_6^4 = (L_6, \wedge, \vee, \odot, \to, 0, 5)$, with the operations $\to$ and $x \odot y \overset{notation}{=} \min\{z \mid x \leq y \to z\} = (x \to y^-)^-$ as in the following tables:

$IMTL_6^4$

| $\to$ | 0 | 1 | 2 | 3 | 4 | 5 |
|---|---|---|---|---|---|---|
| 0 | 5 | 5 | 5 | 5 | 5 | 5 |
| 1 | 4 | 5 | 5 | 5 | 5 | 5 |
| 2 | 3 | **3** | 5 | 5 | 5 | 5 |
| 3 | 2 | **3** | **4** | 5 | 5 | 5 |
| 4 | 1 | **1** | **3** | **3** | 5 | 5 |
| 5 | 0 | 1 | 2 | 3 | 4 | 5 |

| $\odot$ | 0 | 1 | 2 | 3 | 4 | 5 |
|---|---|---|---|---|---|---|
| 0 | 0 | 0 | 0 | 0 | 0 | 0 |
| 1 | 0 | 0 | 0 | 0 | 0 | 1 |
| 2 | 0 | 0 | 0 | 0 | **2** | 2 |
| 3 | 0 | 0 | 0 | **1** | **2** | 3 |
| 4 | 0 | 0 | **2** | **2** | **4** | 4 |
| 5 | 0 | 1 | 2 | 3 | 4 | 5 |

.

Equivalently, by Theorem 12.3.3, consider the algebra $IMTL_6^4 = (A_6 = \{0, a, b, c, d, 1\}, \odot, ^-, 1)$ with the following operations:

$IMTL_6^4$

| $\odot$ | 0 | a | b | c | d | 1 |
|---|---|---|---|---|---|---|
| 0 | 0 | 0 | 0 | 0 | 0 | 0 |
| a | 0 | 0 | 0 | 0 | 0 | a |
| b | 0 | 0 | 0 | 0 | b | b |
| c | 0 | 0 | 0 | a | b | c |
| d | 0 | 0 | b | b | d | d |
| 1 | 0 | a | b | c | d | 1 |

,

| $x$ | $x^-$ |
|---|---|
| 0 | 1 |
| a | d |
| b | c |
| c | b |
| d | a |
| 1 | 0 |

,

| $\oplus$ | 0 | a | b | c | d | 1 |
|---|---|---|---|---|---|---|
| 0 | 0 | a | b | c | d | 1 |
| a | a | a | c | c | 1 | 1 |
| b | b | c | d | 1 | 1 | 1 |
| c | c | c | 1 | 1 | 1 | 1 |
| d | d | 1 | 1 | 1 | 1 | 1 |
| 1 | 1 | 1 | 1 | 1 | 1 | 1 |

.

Then, $IMTL_6^4$ is an involutive m-aBE algebra verifying (m-BB) ($\Leftrightarrow \ldots \Leftrightarrow$ (m-Tr)) and **(Pom)**, and (prel$_m$), and not verifying (m-Pabs-i) for $(a, 0)$, (m-Pimpl) for $(b, 0)$, (Pqmv) for $(a, a, 0)$, (Pmv) for $(a, a)$, $(\Delta_m)$ for $(b, a)$, (WNM$_m$) and (aWNM$_m$) for $(c, c)$, (m-Pdis) for $(a, b, c)$.

Similarly, we obtain that $A_6$ is the **chain** $0 \leq_m a \leq_m b \leq_m c \leq_m d \leq_m 1$, hence $IMTL_6^4$ satisfies condition (prel), but it does not satisfy condition (WNM) for $(c, c)$. Consequently, $IMTL_6^4$ is, indeed, a linearly ordered proper IMTL algebra, that is a tOM algebra.

**Example 3: IMTL$_6^5$**

Let us consider ([104], page 212) (see [75]) the chain $L_6 = \{0, 1, 2, 3, 4, 5\}$ (w.r.to $\leq = \leq_m$), organized as a lattice by $x \wedge y = \min\{x, y\}$, $x \vee y = \max\{x, y\}$ and as a left-IRL $IMTL_6^5 = (L_6, \wedge, \vee, \odot, \rightarrow, 0, 5)$, with the operations $\rightarrow$ and $x \odot y \overset{notation}{=} \min\{z \mid x \leq y \rightarrow z\} = (x \rightarrow y^-)^-$ as in the following tables:

$IMTL_6^5$

| $\rightarrow$ | 0 | 1 | 2 | 3 | 4 | 5 |
|---|---|---|---|---|---|---|
| 0 | 5 | 5 | 5 | 5 | 5 | 5 |
| 1 | 4 | 5 | 5 | 5 | 5 | 5 |
| 2 | 3 | **3** | 5 | 5 | 5 | 5 |
| 3 | 2 | **2** | **3** | 5 | 5 | 5 |
| 4 | 1 | **1** | **2** | **3** | 5 | 5 |
| 5 | 0 | 1 | 2 | 3 | 4 | 5 |

| $\odot$ | 0 | 1 | 2 | 3 | 4 | 5 |
|---|---|---|---|---|---|---|
| 0 | 0 | 0 | 0 | 0 | 0 | 0 |
| 1 | 0 | 0 | 0 | 0 | 0 | 1 |
| 2 | 0 | 0 | 0 | 0 | **2** | 2 |
| 3 | 0 | 0 | 0 | **2** | **3** | 3 |
| 4 | 0 | 0 | **2** | **3** | **4** | 4 |
| 5 | 0 | 1 | 2 | 3 | 4 | 5 |

.

Equivalently, by Theorem 12.3.3, consider the algebra
$IMTL_6^5 = (A_6 = \{0, a, b, c, d, 1\}, \odot, ^-, 1)$ with the following operations:

$IMTL_6^5$

| $\odot$ | 0 | a | b | c | d | 1 |
|---|---|---|---|---|---|---|
| 0 | 0 | 0 | 0 | 0 | 0 | 0 |
| a | 0 | 0 | 0 | 0 | 0 | a |
| b | 0 | 0 | 0 | 0 | b | b |
| c | 0 | 0 | 0 | b | c | c |
| d | 0 | 0 | b | c | d | d |
| 1 | 0 | a | b | c | d | 1 |

,

| $x$ | $x^-$ |
|---|---|
| 0 | 1 |
| a | d |
| b | c |
| c | b |
| d | a |
| 1 | 0 |

,

| $\oplus$ | 0 | a | b | c | d | 1 |
|---|---|---|---|---|---|---|
| 0 | 0 | a | b | c | d | 1 |
| a | a | a | b | c | 1 | 1 |
| b | b | b | c | 1 | 1 | 1 |
| c | c | c | 1 | 1 | 1 | 1 |
| d | d | 1 | 1 | 1 | 1 | 1 |
| 1 | 1 | 1 | 1 | 1 | 1 | 1 |

.

Then, $IMTL_6^5$ is an involutive m-aBE algebra verifying (m-BB) ($\Leftrightarrow \ldots \Leftrightarrow$ (m-Tr)), and (prel$_m$), and not verifying (m-Pabs-i) for $(a, 0)$, (m-Pimpl) for $(b, 0)$, (Pqmv) for $(a, a, 0)$, (Pom) for $(d, b)$, (Pmv) for $(a, a)$, $(\Delta_m)$ for $(b, a)$, (WNM$_m$) and (aWNM$_m$) for $(c, c)$, (m-Pdis) for $(a, d, a)$.

Similarly, we obtain that $A_6$ is the **chain** $0 \leq_m a \leq_m b \leq_m c \leq_m d \leq_m 1$, hence $IMTL_6^5$ satisfies condition (prel), but it does not satisfy condition (WNM) for $(c, c)$. Consequently, $IMTL_6^5$ is, indeed, a linearly ordered proper IMTL algebra.

**Examples 15.3.4 Linearly ordered MV and $_{(WNM)}$MV algebras: MV** and **$_{(WNM)}$MV**

Recall from ([104], 4.1.1) the following classes of examples of finite linearly ordered MV algebras and $_{(WNM)}$MV algebras.

Consider the linearly ordered (where $\leq \overset{notation}{=} \leq_m$) set (chain) $L_{n+1} = \{0, 1, 2, \ldots, n\}$, $(n \geq 1)$, organized as a lattice with $\wedge = \min$ and $\vee = \max$, and organized as left-MV algebra: $\mathcal{L}_{n+1}^L = (L_{n+1}, \odot, ^-, n)$ with:

$$x \odot y = \max\{0, x + y - n\},\ x^- = n - x = x \to 0, \quad 0 = n^-.$$

Hence, for $n = 1, 2, 3, 4$, we have the linearly ordered left-MV algebras $\mathcal{L}_2^L$, $\mathcal{L}_3^L$, $\mathcal{L}_4^L$, $\mathcal{L}_5^L$, whose tables are the following:

$\mathcal{L}_2^L$

| $\odot$ | 0 | 1 |
|---|---|---|
| 0 | 0 | 0 |
| 1 | 0 | 1 |

and

| $x$ | $x^-$ |
|---|---|
| 0 | 1 |
| 1 | 0 |

, with

| $\oplus$ | 0 | 1 |
|---|---|---|
| 0 | 0 | 1 |
| 1 | 1 | 1 |

;

$\mathcal{L}_3^L$

| $\odot$ | 0 | 1 | 2 |
|---|---|---|---|
| 0 | 0 | 0 | 0 |
| 1 | 0 | 0 | 1 |
| 2 | 0 | 1 | 2 |

and

| $x$ | $x^-$ |
|---|---|
| 0 | 2 |
| 1 | 1 |
| 2 | 0 |

, with

| $\oplus$ | 0 | 1 | 2 |
|---|---|---|---|
| 0 | 0 | 1 | 2 |
| 1 | 1 | 2 | 2 |
| 2 | 2 | 2 | 2 |

or, equivalently,

$\mathcal{L}_3^L$

| $\odot$ | 0 | a | 1 |
|---|---|---|---|
| 0 | 0 | 0 | 0 |
| a | 0 | 0 | a |
| 1 | 0 | a | 1 |

and

| $x$ | $x^-$ |
|---|---|
| 0 | 1 |
| a | a |
| 1 | 0 |

, with

| $\oplus$ | 0 | a | 1 |
|---|---|---|---|
| 0 | 0 | a | 1 |
| a | a | 1 | 1 |
| 1 | 1 | 1 | 1 |

;

$\mathcal{L}_4^L$

| $\odot$ | 0 | 1 | 2 | 3 |
|---|---|---|---|---|
| 0 | 0 | 0 | 0 | 0 |
| 1 | 0 | 0 | 0 | 1 |
| 2 | 0 | 0 | 1 | 2 |
| 3 | 0 | 1 | 2 | 3 |

and

| $x$ | $x^-$ |
|---|---|
| 0 | 3 |
| 1 | 2 |
| 2 | 1 |
| 3 | 0 |

, with

| $\oplus$ | 0 | 1 | 2 | 3 |
|---|---|---|---|---|
| 0 | 0 | 1 | 2 | 3 |
| 1 | 1 | 2 | 3 | 3 |
| 2 | 2 | 3 | 3 | 3 |
| 3 | 3 | 3 | 3 | 3 |

or, equivalently,

$\mathcal{L}_4^L$

| $\odot$ | 0 | a | b | 1 |
|---|---|---|---|---|
| 0 | 0 | 0 | 0 | 0 |
| a | 0 | 0 | 0 | a |
| b | 0 | 0 | a | b |
| 1 | 0 | a | b | 1 |

and

| $x$ | $x^-$ |
|---|---|
| 0 | 1 |
| a | b |
| b | a |
| 1 | 0 |

, with

| $\oplus$ | 0 | a | b | 1 |
|---|---|---|---|---|
| 0 | 0 | a | b | 1 |
| a | a | b | 1 | 1 |
| b | b | 1 | 1 | 1 |
| 1 | 1 | 1 | 1 | 1 |

;

$\mathcal{L}_5^L$

| $\odot$ | 0 | 1 | 2 | 3 | 4 |
|---|---|---|---|---|---|
| 0 | 0 | 0 | 0 | 0 | 0 |
| 1 | 0 | 0 | 0 | 0 | 1 |
| 2 | 0 | 0 | 0 | 1 | 2 |
| 3 | 0 | 0 | 1 | 2 | 3 |
| 4 | 0 | 1 | 2 | 3 | 4 |

and

| $x$ | $x^-$ |
|---|---|
| 0 | 4 |
| 1 | 3 |
| 2 | 2 |
| 3 | 1 |
| 4 | 0 |

, with

| $\oplus$ | 0 | 1 | 2 | 3 | 4 |
|---|---|---|---|---|---|
| 0 | 0 | 1 | 2 | 3 | 4 |
| 1 | 1 | 2 | 3 | 4 | 4 |
| 2 | 2 | 3 | 4 | 4 | 4 |
| 3 | 3 | 4 | 4 | 4 | 4 |
| 4 | 4 | 4 | 4 | 4 | 4 |

or, equivalently,

$\mathcal{L}_5^L$

| $\odot$ | 0 | a | b | c | 1 |
|---|---|---|---|---|---|
| 0 | 0 | 0 | 0 | 0 | 0 |
| a | 0 | 0 | 0 | 0 | a |
| b | 0 | 0 | 0 | a | b |
| c | 0 | 0 | a | b | c |
| 1 | 0 | a | b | c | 1 |

and

| $x$ | $x^-$ |
|---|---|
| 0 | 1 |
| a | c |
| b | b |
| c | a |
| 1 | 0 |

, with

| $\oplus$ | 0 | a | b | c | 1 |
|---|---|---|---|---|---|
| 0 | 0 | a | b | c | 1 |
| a | a | b | c | 1 | 1 |
| b | b | c | 1 | 1 | 1 |
| c | c | 1 | 1 | 1 | 1 |
| 1 | 1 | 1 | 1 | 1 | 1 |

.

Note that:

(1) For $n = 1, 2$, the MV algebras $\mathcal{L}_2^L$ and $\mathcal{L}_3^L$ verify condition (WNM) ($(x \odot y)^- \vee [(x \wedge y) \to (x \odot y)] = 1$), hence they are examples of $_{(WNM)}$MV algebras. Note that $\mathcal{L}_2^L$ is just the Boolean algebra with two elements.

(2) For $n = 3$, the MV algebra $\mathcal{L}_4^L$ does not verify condition (WNM) for $(2, 2)$. Hence, $\mathcal{L}_4^L$ is a proper MV algebra.

(3) For $n \geq 4$, the MV algebra $\mathcal{L}_{n+1}^L$ does not verify condition (WNM) for $(n-2,\ n-1)$; indeed,
$[(n-2) \odot (n-1)]^- \vee [(n-2) \wedge (n-1) \to (n-2) \odot (n-1)]$
$= (n-3)^- \vee [(n-2) \to (n-3)] = 3 \vee (n-1) = n-1 \neq n$, because:
$(n-2) \odot (n-1) = \max(0, (n-2) + (n-1) - n) = \max(0, n-3) = n-3$, since $n - 3 \geq 4 - 3 = 1$,
$(n-3)^- = n - (n-3) = 3$,
$(n-2) \to (n-3) = \min(n, (n-3) - (n-2) + n) = \min(n, n-1) = n-1$ and $n - 1 \geq 4 - 1 = 3$.
Hence, $\mathcal{L}_{n+1}^L$ ($n \geq 3$) is a proper MV algebra.

**Remark 1.** *We shall analyse in some details the linearly ordered Boolean algebra $\mathcal{L}_2^L = (L_2 = \{0, 1\}, \odot, ^-, 1)$. Note that $\mathcal{L}_2^L$ is an involutive m-aBE algebra verifying (m-BB) ($\Leftrightarrow \ldots \Leftrightarrow$ (m-Tr)), (m-Pimpl), (m-Pabs-i), (Pqmv), (Pom), (Pmv), ($\Delta_m$), (prel$_m$), (WNM$_m$), (aWNM$_m$), (m-Pdis). Hence, it is an $_{(WNM)}$MV algebra, since (m-Pabs-i) $\Longleftrightarrow$ (aWNM$_m$) ($\Longleftrightarrow$ (WNM$_m$) = (WNM)), by Theorem 14.1.10. We have $\wedge_m^M = \wedge_m^B = \odot$, hence $\leq_m^M \Longleftrightarrow \leq_m^B \Longleftrightarrow \leq_m \Longleftrightarrow \leq_m^P$ and they are lattice orders.*

**Remark 2.** *We shall analyse in some details the algebra $\mathcal{L}_3^L = (A_3 = \{0, a, 1\}, \odot, ^-, 1)$. Note that $\mathcal{L}_3^L$ is an involutive m-aBE algebra verifying (m-BB) ($\Leftrightarrow \ldots \Leftrightarrow$ (m-Tr)), (m-Pabs-i), (Pqmv), (Pom), (Pmv), ($\Delta_m$), (prel$_m$), (WNM$_m$), (aWNM$_m$) and not verifying (m-Pimpl) for $(a, 0)$, (m-Pdis) for $(a, a, a)$. Hence, it is an $_{(WNM)}$MV algebra, since (m-Pabs-i) $\Longleftrightarrow$ (aWNM$_m$) ($\Longleftrightarrow$ (WNM$_m$) = (WNM)), by Theorem 14.1.10. We have $\wedge_m^M = \wedge_m^B \neq \odot$, hence $\leq_m^M \Longleftrightarrow \leq_m^B \Longleftrightarrow \leq_m$ and they are lattice orders.*

**Remark 3.** *We shall analyse now in some details the algebra* $\mathcal{L}_4^L = (A_4 = \{0,a,b,1\}, \odot, ^-, 1)$. *Note that* $\mathcal{L}_4^L$ *is an involutive m-aBE algebra verifying (m-BB) (⇔ ... ⇔ (m-Tr)) and (Pqmv) (hence (Pom), (Pmv),* $(\Delta_m)$*) and* $(prel_m)$*, and not verifying (m-Pabs-i) for* $(a,0)$*, (m-Pimpl) for* $(a,0)$*,* $(aWNM_m)$ *and* $(WNM_m)$ *for* $(b,b)$*, (m-Pdis) for* $(a,a,b)$*; hence, it is a left-MV algebra.*

*The tables of* $\wedge_m^M$*,* $\wedge_m^B$ *and* $\to$ *are:*

| $\wedge_m^M$ | 0 | a | b | 1 |
|---|---|---|---|---|
| 0 | 0 | 0 | 0 | 0 |
| a | 0 | a | a | a |
| b | 0 | a | b | b |
| 1 | 0 | a | b | 1 |

and

| $\wedge_m^B$ | 0 | a | b | 1 |
|---|---|---|---|---|
| 0 | 0 | 0 | 0 | 0 |
| a | 0 | a | a | a |
| b | 0 | a | b | b |
| 1 | 0 | a | b | 1 |

and

| $\to$ | 0 | a | b | 1 |
|---|---|---|---|---|
| 0 | 1 | 1 | 1 | 1 |
| a | b | 1 | 1 | 1 |
| b | a | **b** | 1 | 1 |
| 1 | 0 | a | b | 1 |

.

*Note that, as expected, the tables of* $\wedge_m^M$ *and* $\wedge_m^B$ *coincide and they are commutative. Hence, we have* $\leq_m^M \Longleftrightarrow \leq_m$ $(\Longleftrightarrow \leq_m^B)$ *and they are lattice orders. We have:* $a \leq_m a, b, 1$ *and* $b \leq_m b, 1$*, hence the bounded lattice* $(A_4, \leq_m, 0, 1)$ *is the* **chain**: $0 \leq_m a \leq_m b \leq_m 1$*, which is a lattice w.r. to* $\wedge = \inf_m$ *and* $\vee = \sup_m$. *Note that* $\wedge = \wedge_m^M = \wedge_m^B$. *Being a chain, the property (prel), which coincides with* $(prel_m)$*, is then verified. The property (WNM)* $((x \odot y)^- \vee [(x \wedge y) \to (x \odot y)] = 1)$ *coincides with* $(WNM_m)$ *and, hence, it is not verified also for* $(b,b)$. *Hence,* $\mathcal{L}_4^L$ *is, indeed, the linearly ordered proper MV algebra with the minimum number of elements, four.*

### Example 15.3.5 Non-linearly ordered proper MV algebra: MV

The set $L_{4\times 2} = \{0,a,b,c,d,e,f,1\} \cong L_4 \times L_2 = \{0,1,2,3\} \times \{0,1\}$
$= \{(0,0),(0,1),(1,0),(1,1),(2,0),(2,1),(3,0),(3,1)\}$,
organized as a lattice as in Figure 15.9 and as an involutive m-BE algebra $(L_{4\times 2}, \odot, ^-, 1)$, with the operations $\odot$, $^-$ and $\oplus$ obtained component-wise as in the following tables, is a non-linearly ordered proper MV algebra (since it does not verify condition (WNM) for $(d,d)$), denoted by $\mathcal{L}_{4\times 2}^L$ ([104], page 167).

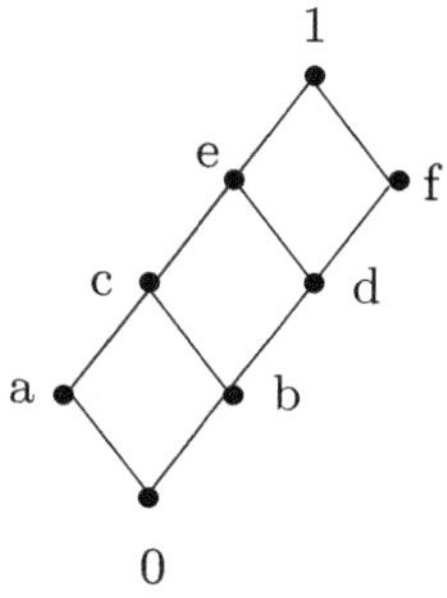


Figure 15.9: The non-linearly ordered proper MV algebra $\mathcal{L}_{4\times 2}^L$ and NM algebra $\mathcal{F}_{4\times 2}^m$

| $\odot$ | 0 | a | b | c | d | e | f | 1 |
|---|---|---|---|---|---|---|---|---|
| 0 | 0 | 0 | 0 | 0 | 0 | 0 | 0 | 0 |
| a | 0 | a | 0 | a | 0 | a | 0 | a |
| b | 0 | 0 | 0 | 0 | 0 | 0 | b | b |
| c | 0 | a | 0 | a | 0 | a | b | c |
| d | 0 | 0 | 0 | 0 | **b** | **b** | d | d |
| e | 0 | a | 0 | a | **b** | **c** | d | e |
| f | 0 | 0 | b | b | d | d | f | f |
| 1 | 0 | a | b | c | d | e | f | 1 |

,

| $\oplus$ | 0 | a | b | c | d | e | f | 1 |
|---|---|---|---|---|---|---|---|---|
| 0 | 0 | a | b | c | d | e | f | 1 |
| a | a | a | c | c | e | e | 1 | 1 |
| b | b | c | d | e | f | 1 | f | 1 |
| c | c | c | e | e | 1 | 1 | 1 | 1 |
| d | d | e | f | 1 | f | 1 | f | 1 |
| e | e | e | 1 | 1 | 1 | 1 | 1 | 1 |
| f | f | 1 | f | 1 | f | 1 | f | 1 |
| 1 | 1 | 1 | 1 | 1 | 1 | 1 | 1 | 1 |

and $(0, a, b, c, d, e, f, 1)^- = (1, f, e, d, c, b, a, 0)$.

Indeed, $\mathcal{L}^L_{4\times 2}$ is an involutive m-aBE algebra verifying (m-BB) ($\Leftrightarrow \ldots \Leftrightarrow$ (m-Tr)) and (Pqmv) (hence (Pom), (Pmv), ($\Delta_m$)) and (prel$_m$), and not verifying (m-Pabs-i) for $(b, 0)$, (m-Pimpl) for $(b, 0)$, (aWNM$_m$) and (WNM$_m$) for $(d, d)$, (m-Pdis) for $(b, b, d)$. We have $\wedge = \inf_m = \wedge^M_m = \wedge^B_m$. Consequently, $\leq^M_m \Longleftrightarrow \leq_m$ ($\Longleftrightarrow \leq^B_m$) and they are lattice orders, and the property (WNM) ($(x \odot y)^- \vee [(x \wedge y) \rightarrow (x \odot y)] = 1$) coincides with (WNM$_m$) and, hence, it is not verified also for $(d, d)$.

**Examples 15.3.6 Linearly ordered NM and $_{(WNM)}$MV algebras: NM** and **$_{(WNM)}$MV** Recall from ([104], 5.1.1) the following classes of examples of linearly ordered (where $\leq = \leq_m$) NM algebras and $_{(WNM)}$MV algebras.

For each $n \geq 1$, let us consider the chain $L_{n+1} = \{0, 1, 2, \ldots, n\}$, organized as a lattice w.r. to the lattice order $\leq \overset{notation}{=} \leq_m$ and the lattice operations $\wedge = \min$ and $\vee = \max$, and organized as:

- a linearly ordered left-IRL $\mathcal{F}^L_{n+1} = (L_{n+1}, \wedge, \vee, \odot_F, \rightarrow_F, 0, n)$ in the following way: we take the strong negation $^-$, defined on $L_{n+1}$ by $x^- = n - x$, and Fodor's implication $\rightarrow_F$ with the corresponding Fodor's t-norm $\odot_F$ [55], [50], defined by:

$$x \rightarrow_F y = \begin{cases} n, & \text{if } x \leq y \\ \max(n - x, y), & \text{if } x > y, \end{cases}$$

$$(15.37) \qquad x \odot_F y = (x \rightarrow_F y^-)^- = \begin{cases} 0, & \text{if } x \leq n - y = y^- \\ \min(x, y), & \text{if } x > n - y = y^-. \end{cases}$$

or, equivalently, by Theorem 12.3.3, as

- a linearly ordered left-m-BCK lattice $\mathcal{F}^m_{n+1} = (L_{n+1}, \odot_F, ^-, n)$ (Definition 2).

Hence, for $n = 1, 2, 3, 4$, we have the left-m-BCK lattices $\mathcal{F}^m_2$, $\mathcal{F}^m_3$, $\mathcal{F}^m_4$, $\mathcal{F}^m_5$, whose tables are the following:

$\mathcal{F}^m_2$

| $\odot_F$ | 0 | 1 |
|---|---|---|
| 0 | 0 | 0 |
| 1 | 0 | 1 |

and

| $x$ | $x^-$ |
|---|---|
| 0 | 1 |
| 1 | 0 |

, with

| $\oplus_F$ | 0 | 1 |
|---|---|---|
| 0 | 0 | 1 |
| 1 | 1 | 1 |

;

$\mathcal{F}^m_3$

| $\odot_F$ | 0 | 1 | 2 |
|---|---|---|---|
| 0 | 0 | 0 | 0 |
| 1 | 0 | 0 | 1 |
| 2 | 0 | 1 | 2 |

and

| $x$ | $x^-$ |
|---|---|
| 0 | 2 |
| 1 | 1 |
| 2 | 0 |

, with

| $\oplus_F$ | 0 | 1 | 2 |
|---|---|---|---|
| 0 | 0 | 1 | 2 |
| 1 | 1 | 2 | 2 |
| 2 | 2 | 2 | 2 |

or, equivalently,

$\mathcal{F}_3^m$

| $\odot_F$ | 0 | a | 1 |
|---|---|---|---|
| 0 | 0 | 0 | 0 |
| a | 0 | 0 | a |
| 1 | 0 | a | 1 |

and

| $x$ | $x^-$ |
|---|---|
| 0 | 1 |
| a | a |
| 1 | 0 |

, with

| $\oplus_F$ | 0 | a | 1 |
|---|---|---|---|
| 0 | 0 | a | 1 |
| a | a | 1 | 1 |
| 1 | 1 | 1 | 1 |

;

$\mathcal{F}_4^m$

| $\odot_F$ | 0 | 1 | 2 | 3 |
|---|---|---|---|---|
| 0 | 0 | 0 | 0 | 0 |
| 1 | 0 | 0 | 0 | 1 |
| 2 | 0 | 0 | **2** | 2 |
| 3 | 0 | 1 | 2 | 3 |

and

| $x$ | $x^-$ |
|---|---|
| 0 | 3 |
| 1 | 2 |
| 2 | 1 |
| 3 | 0 |

, with

| $\oplus_F$ | 0 | 1 | 2 | 3 |
|---|---|---|---|---|
| 0 | 0 | 1 | 2 | 3 |
| 1 | 1 | 1 | 3 | 3 |
| 2 | 2 | 3 | 3 | 3 |
| 3 | 3 | 3 | 3 | 3 |

or,

equivalently,

$\mathcal{F}_4^m$

| $\odot_F$ | 0 | a | b | 1 |
|---|---|---|---|---|
| 0 | 0 | 0 | 0 | 0 |
| a | 0 | 0 | 0 | a |
| b | 0 | 0 | b | b |
| 1 | 0 | a | b | 1 |

and

| $x$ | $x^-$ |
|---|---|
| 0 | 1 |
| a | b |
| b | a |
| 1 | 0 |

, with

| $\oplus_F$ | 0 | a | b | 1 |
|---|---|---|---|---|
| 0 | 0 | a | b | 1 |
| a | a | a | 1 | 1 |
| b | b | 1 | 1 | 1 |
| 1 | 1 | 1 | 1 | 1 |

;

$\mathcal{F}_5^m$

| $\odot_F$ | 0 | 1 | 2 | 3 | 4 |
|---|---|---|---|---|---|
| 0 | 0 | 0 | 0 | 0 | 0 |
| 1 | 0 | 0 | 0 | 0 | 1 |
| 2 | 0 | 0 | 0 | **2** | 2 |
| 3 | 0 | 0 | **2** | **3** | 3 |
| 4 | 0 | 1 | 2 | 3 | 4 |

and

| $x$ | $x^-$ |
|---|---|
| 0 | 4 |
| 1 | 3 |
| 2 | 2 |
| 3 | 1 |
| 4 | 0 |

, with

| $\oplus_F$ | 0 | 1 | 2 | 3 | 4 |
|---|---|---|---|---|---|
| 0 | 0 | 1 | 2 | 3 | 4 |
| 1 | 1 | 1 | 2 | 4 | 4 |
| 2 | 2 | 2 | 4 | 4 | 4 |
| 3 | 3 | 4 | 4 | 4 | 4 |
| 4 | 4 | 4 | 4 | 4 | 4 |

or, equivalently,

$\mathcal{F}_5^m$

| $\odot_F$ | 0 | a | b | c | 1 |
|---|---|---|---|---|---|
| 0 | 0 | 0 | 0 | 0 | 0 |
| a | 0 | 0 | 0 | 0 | 1 |
| b | 0 | 0 | 0 | **b** | b |
| c | 0 | 0 | **b** | **c** | c |
| 1 | 0 | a | b | c | 1 |

and

| $x$ | $x^-$ |
|---|---|
| 0 | 1 |
| a | c |
| b | b |
| c | a |
| 1 | 0 |

, with

| $\oplus_F$ | 0 | a | b | c | 1 |
|---|---|---|---|---|---|
| 0 | 0 | a | b | c | 1 |
| a | a | a | b | 1 | 1 |
| b | b | b | 1 | 1 | 1 |
| c | c | 1 | 1 | 1 | 1 |
| 1 | 1 | 1 | 1 | 1 | 1 |

.

For each $n \geq 1$, $\mathcal{F}_{n+1}^L = (L_{n+1}, \wedge, \vee, \odot_F, \rightarrow_F, 0, 1)$ is an involutive residuated left-lattice that is linearly ordered, hence it satisfies condition (prel) ($(x \rightarrow_F y) \vee (y \rightarrow_F x) = 1$), but also (prel$_m$) ($(x \rightarrow_F y) \vee_m^B (y \rightarrow_F x) = 1$); it satisfies also condition (WNM) ($(x \odot y)^- \vee ((x \wedge y) \rightarrow (x \odot y)) = 1$), thus, $\mathcal{F}_{n+1}^L$ is a NM algebra (Definition 1).

We shall prove now that, for each $n \geq 1$, $\mathcal{F}_{n+1}^L$ **verifies also (aWNM$_m$) and (WNM$_m$)**, where:
(aWNM$_m$) $(x \odot y) \odot [x \odot (x \odot y^-)^- \odot (x \odot y)^-]^- = x \odot y$,
(WNM$_m$) $(x \odot_F y)^- \vee_m^B [(x \wedge_m^B y) \rightarrow_F (x \odot_F y)] = 1$.
But, by Proposition 14.1.3, we have (aWNM$_m$) $\Longleftrightarrow$ (WNM$_m$), hence it is sufficient to prove, for example, that (aWNM$_m$) holds.
Indeed, if we put $A \overset{notation}{=} x \odot_F y$, $B \overset{notation}{=} x \odot_F (x \odot_F y)^-$ and $C \overset{notation}{=}$

$(x \odot_F y^-)^- \odot_F (x \odot_F y)^-$, then note that: if $A = 0$, then (aWNM$_m$) holds, by (m-La); if $B = 0$ or $C = 0$, then (aWNM$_m$) holds, by (m-La), (Neg0-1), (PU). So, we shall prove that, for any $x, y \in L_{n+1}$, either $A = 0$ or $B = 0$ or $C = 0$:

• Suppose $x \leq y$; then,

- if $y^- < y$, then: if $x \leq y^- < y$, then $A = x \odot_F y = 0$, by (15.37); if $y^- < x \leq y$, then $x \odot_F y = \min(x, y) = x$, by (15.37), hence $B = x \odot_F (x \odot_F y)^- = x \odot_F x^- = 0$, by (m-Re);

- if $y \leq y^-$, then $x \leq y \leq y^-$ and hence $A = x \odot_F y = 0$, by (15.37).

• Suppose $x > y$; then,

- for $n = 2k$, there are three cases: (1) $y = k$, (2) $y > k$, (3) $y < k$;

- (1) $y = k$ implies $y^- = 2k - k = k$, hence $y = y^-$; then, we have $y = y^- < x$, hence $x \odot_F y = \min(x, y) = y$ and $x \odot_F y^- = \min(x, y^-) = y^-$, by (15.37), hence $C = (x \odot_F y^-)^- \odot_F (x \odot_F y)^- = y^= \odot_F y^- = y \odot_F y^- = 0$, by (DN) and (m-Re);

- (2) $y > k$ implies $y \geq k + 1$, hence $y^- \leq (k+1)^- = 2k - (k+1) = k - 1$; hence, $y^- < y$ and hence we have $y^- < y < x$; then, by (15.37), $x \odot_F y = \min(x, y) = y$ and $x \odot_F y^- = \min(x, y^-) = y^-$, and, hence, $C = y^= \odot_F y^- = 0$.

- (3) $y < k$ implies $y \leq k - 1$, hence $y^- \geq (k-1)^- = 2k - (k-1) = k + 1$, hence $y < y^-$; then, if $y < x \leq y^-$, then $A = x \odot_F y = 0$, and if $y < y^- < x$, then $x \odot_F y = \min(x, y) = y$ and $x \odot_F y^- = \min(x, y^-) = y^-$, by (15.37), hence, $C = y^= \odot_F y^- = 0$;

- for $n = 2k + 1$, there are two cases: (1') $y \leq k$, (2') $y > k$;

- (1') $y \leq k$ implies $y^- \geq k^- = 2k + 1 - k = k + 1$, hence $y < y^-$; then, if $y < x \leq y^-$, then $A = x \odot_F y = 0$, and if $y < y^- < x$, then $x \odot_F y = \min(x, y) = y$ and $x \odot_F y^- = \min(x, y^-) = y^-$, by (15.37), hence $C = y^= \odot_F y^- = 0$;

- (2') $y > k$ means $y \geq k + 1$, hence $y^- \leq (k+1)^- = 2k + 1 - (k+1) = k$; hence, $y^- < y$; then, we have $y^- < y < x$, hence $x \odot_F y = \min(x, y) = y$ and $x \odot_F y^- = \min(x, y^-) = y^-$, by (15.37), hence $C = y^= \odot_F y^- = 0$. The proof that $\mathcal{F}^L_{n+1}$ verifies (aWNM$_m$) is now complete.

Note that:

(1) For $n = 1, 2$, $\mathcal{F}^m_2 = \mathcal{L}^L_2$ and $\mathcal{F}^m_3 = \mathcal{L}^L_3$, i.e. $\mathcal{F}^m_2$ and $\mathcal{F}^m_3$ are examples of linearly ordered $_{(WNM)}$MV algebras. Note that $\mathcal{F}^m_2 = \mathcal{L}^L_2$ is the Boolean algebra with two elements.

(2) For each $n \geq 3$, $\mathcal{F}^L_{n+1}$ is a linearly ordered proper NM algebra (i.e. not being MV algebra).

Consequently, for each $n \geq 1$, $\mathcal{F}^m_{n+1} = (L_{n+1}, \odot_F, ^-, n)$ is a NM algebra (Definition 2).

Note that:

(1') For $n = 1, 2$, since $\mathcal{F}^m_2 = \mathcal{L}^L_2$ and $\mathcal{F}^m_3 = \mathcal{L}^L_3$, it follows that $\mathcal{F}^m_2$ and $\mathcal{F}^m_3$ are NM algebras verifying (Pom) (since they are also MV algebras).

(2') For $n = 3$, $\mathcal{F}^m_4 = (L_4, \odot_F, ^-, 3)$ is a NM algebra verifying (Pom), hence **it is a proper taOM algebra** (i.e. not being MV algebra).

(3') For $n = 4$, $\mathcal{F}^m_5 = (L_5, \odot_F, ^-, 4)$ is a NM algebra which does not verify (Pom) for $(x, y) = (3, 2)$.

(4') For any $n \geq 4$, there exists $(x, y) = (n - 1, 2)$ such that the NM algebra $\mathcal{F}^m_{n+1} = (L_{n+1}, \odot_F, ^-, n)$ does not verify (Pom), where:

(Pom) $(x \odot_F y) \oplus_F ((x \odot_F y)^- \odot_F x) = x$ or, equivalently,
$(x \odot_F y)^- \odot_F ((x \odot_F y)^- \odot_F x)^- = x^-$.

Indeed, for any $n \geq 4$, $(x \odot_F y)^- \odot_F ((x \odot_F y)^- \odot_F x)^-$
$= ((n-1) \odot_F 2)^- \odot_F (((n-1) \odot_F 2)^- \odot_F (n-1))^-$
$= 2^- \odot_F (2^- \odot_F (n-1))^-$
$= (n-2) \odot_F ((n-2) \odot_F (n-1))^-$
$= (n-2) \odot_F (n-2)^- = (n-2) \odot_F 2 = 0$,
while $x^- = (n-1)^- = 1$;
since $0 \neq 1$, it follows that $(x \odot_F y)^- \odot_F ((x \odot_F y)^- \odot_F x)^- \neq x^-$, i.e. (Pom) does not hold for $(x, y) = (n-1, 2)$.

Hence, for $n \geq 4$, $\mathcal{F}^m_{n+1} = (L_{n+1}, \odot_F, ^-, n)$ is a proper NM algebra (i.e. not being taOM algebra).

**Remark.** *We shall analyse now in some details the algebra* $\mathcal{F}^m_4 = (A_4 = \{0, a, b, 1\}, \odot_F, ^-, 1)$*. Note that* $\mathcal{F}^m_4$ *is an involutive m-aBE algebra verifying (m-BB) (⇔ ... ⇔ (m-Tr)) and (Pom) and (prel$_m$), (WNM$_m$), (aWNM$_m$), and not verifying (m-Pabs-i) for* $(a, 0)$*, (m-Pimpl) for* $(b, 0)$*, (Pqmv) for* $(a, a, 0)$*, (Pmv) for* $(a, a)$*, ($\Delta_m$) for* $(b, a)$*, (m-Pdis) for* $(a, b, a)$

*The tables of* $\wedge^M_m$ *and* $\wedge^B_m$ *are:*

| $\wedge^M_m$ | 0 | a | b | 1 |
|---|---|---|---|---|
| 0 | 0 | 0 | 0 | 0 |
| a | 0 | a | 0 | a |
| b | 0 | a | b | b |
| 1 | 0 | a | b | 1 |

*and*

| $\wedge^B_m$ | 0 | a | b | 1 |
|---|---|---|---|---|
| 0 | 0 | 0 | 0 | 0 |
| a | 0 | a | a | a |
| b | 0 | 0 | b | b |
| 1 | 0 | a | b | 1 |

.

• *The binary relation* $\leq^M_m$ *is an order, but not a lattice order, since the table of* $\wedge^M_m$ *is not commutative. From the table of* $\wedge^M_m$*, we obtain:* $a \leq^M_m a, 1$ *and* $b \leq^M_m b, 1$*; hence, the bounded poset* $(A_4, \leq^M_m, 0, 1)$ *is represented by the Hasse diagram from Figure 15.10.*

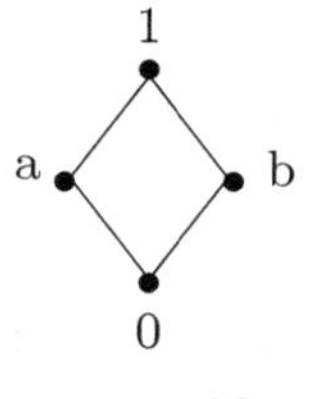


Figure 15.10: The bounded poset $(L_4, \leq^M_m, 0, 1)$ of the smallest NM algebra

• *The binary relation* $\leq_m$ *is an order also, since it is reflexive, antisymmetric and transitive. From the table of* $\wedge^B_m$*, we obtain:* $a \leq_m a, b, 1$ *and* $b \leq_m b, 1$*, hence the bounded poset* $(A_4, \leq_m, 0, 1$ *is the* **chain***:* $0 \leq_m a \leq_m b \leq_m 1$*, which is a lattice w.r.t. to* $\wedge = \inf_m$ *and* $\vee = \sup_m$*. Being a chain, then the property (prel) is verified. The tables of* $\wedge$*,* $\vee$ *and* $\rightarrow$ *are the following:*

| $\wedge$ | 0 | a | b | 1 |
|---|---|---|---|---|
| 0 | 0 | 0 | 0 | 0 |
| a | 0 | a | a | a |
| b | 0 | a | b | b |
| 1 | 0 | a | b | 1 |

,

| $\vee$ | 0 | a | b | 1 |
|---|---|---|---|---|
| 0 | 0 | a | b | 1 |
| a | a | a | b | 1 |
| b | b | b | b | 1 |
| 1 | 1 | 1 | 1 | 1 |

and

| $\rightarrow$ | 0 | a | b | 1 |
|---|---|---|---|---|
| 0 | 1 | 1 | 1 | 1 |
| a | b | 1 | 1 | 1 |
| b | a | **a** | 1 | 1 |
| 1 | 0 | a | b | 1 |

.

*Note that $\wedge \neq \wedge_m^M \neq \wedge_m^B$. The property (WNM) is verified. Hence, $\mathcal{F}_4^m$ is, indeed, the linearly ordered proper NM algebra with the minimum number of elements, four, verifying (Pom) (hence, is a proper left-taOM algebra).*

**Example 15.3.7 Non-linearly ordered proper NM algebra: NM**

The set $F_{4\times 2} = \{0, a, b, c, d, e, f, 1\} \cong F_4 \times F_2 = \{0, 1, 2, 3\} \times \{0, 1\}$
$= \{(0, 0), (0, 1), (1, 0), (1, 1), (2, 0), (2, 1), (3, 0), (3, 1)\}$,
organized as a distributive lattice as in the previous Figure 15.9 and as an involutive m-BE algebra $(F_{4\times 2}, \odot, ^-, 1)$, with the operations $\odot$, $^-$ and $\oplus$ obtained component-wise as in the following tables, is a non-linearly ordered proper NM algebra denoted by $\mathcal{F}_{4\times 2}^m$ ([104], page 163).

| $\odot$ | 0 | a | b | c | d | e | f | 1 |
|---|---|---|---|---|---|---|---|---|
| 0 | 0 | 0 | 0 | 0 | 0 | 0 | 0 | 0 |
| a | 0 | a | 0 | a | 0 | a | 0 | a |
| b | 0 | 0 | 0 | 0 | 0 | 0 | b | b |
| c | 0 | a | 0 | a | 0 | a | b | c |
| d | 0 | 0 | 0 | 0 | **d** | **d** | d | d |
| e | 0 | a | 0 | a | **d** | **e** | d | e |
| f | 0 | 0 | b | b | d | d | f | f |
| 1 | 0 | a | b | c | d | e | f | 1 |

,

| $\oplus$ | 0 | a | b | c | d | e | f | 1 |
|---|---|---|---|---|---|---|---|---|
| 0 | 0 | a | b | c | d | e | f | 1 |
| a | a | a | c | c | e | e | 1 | 1 |
| b | b | c | b | c | f | 1 | f | 1 |
| c | c | c | c | c | 1 | 1 | 1 | 1 |
| d | d | e | f | 1 | f | 1 | f | 1 |
| e | e | e | 1 | 1 | 1 | 1 | 1 | 1 |
| f | f | 1 | f | 1 | f | 1 | f | 1 |
| 1 | 1 | 1 | 1 | 1 | 1 | 1 | 1 | 1 |

and $(0, a, b, c, d, e, f, 1)^- = (1, f, e, d, c, b, a, 0)$.

Indeed, $\mathcal{F}_{4\times 2}^m$ is an involutive m-aBE algebra verifying (m-BB) ($\Leftrightarrow \ldots \Leftrightarrow$ (m-Tr)) and (Pom) and (prel$_m$), (aWNM$_m$), (WNM$_m$), and not verifying (m-Pabs-i) for $(b, 0)$, (m-Pimpl) for $(d, 0)$, (Pqmv) for $(b, b, 0)$, (Pmv) for $(b, b)$, $(\Delta_m)$ for $(d, b)$, (m-Pdis) for $(b, d, b)$. The binary relation $\leq_m$ is a lattice order, $\wedge = \inf_m \neq \wedge_m^M \neq \wedge_m^B$ and the properties (prel) and (WNM) are satisfied.

**Examples 15.3.8 Non-linearly ordered MV algebras verifying (WNM): atOWL $=_{(WNM)}$ MV**

**Example 1: non-linearly ordered Boolean algebra $\mathcal{L}_{2\times 2}^L$**

The set $L_{2\times 2} = \{0, a, b, 1\} \cong L_2 \times L_2 = \{0, 1\} \times \{0, 1\} = \{(0, 0), (0, 1), (1, 0), (1, 1)\}$, organized as a lattice as in Figure 15.11 and organized as an involutive m-BE algebra $(L_{2\times 2}, \odot, ^-, 1)$ with the following operations $\odot$, $^-$, $\oplus$, is a non-linearly ordered Boolean algebra, denoted by $\mathcal{L}_{2\times 2}^L$ ([104], page 163).

$\mathcal{L}_{2\times 2}^L$

| $\odot$ | 0 | a | b | 1 |
|---|---|---|---|---|
| 0 | 0 | 0 | 0 | 0 |
| a | 0 | a | 0 | a |
| b | 0 | 0 | b | b |
| 1 | 0 | a | b | 1 |

and

| $x$ | $x^-$ |
|---|---|
| 0 | 1 |
| a | b |
| b | a |
| 1 | 0 |

, with

| $\oplus$ | 0 | a | b | 1 |
|---|---|---|---|---|
| 0 | 0 | a | b | 1 |
| a | a | a | 1 | 1 |
| b | b | 1 | b | 1 |
| 1 | 1 | 1 | 1 | 1 |

.

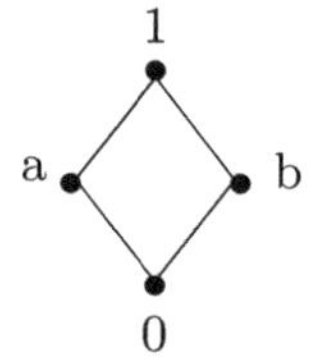


Figure 15.11: The bounded lattice $(L_{2\times 2}, \leq_m, 0, 1)$ of the Boolean algebra

Indeed, $\mathcal{L}^L_{2\times 2}$ verifies (m-B), (m-BB), (m-*), (m-**), (m-Tr), (m-An), (m-Pimpl), (m-Pabs-i)), (Pqmv), (Pom), (Pmv), $(\Delta_m)$, $(\text{prel}_m)$, $(\text{WNM}_m)$, $(\text{aWNM}_m)$, (m-Pdis). We have $\wedge^M_m = \wedge^B_m = \odot$, hence $\leq^M_m \Longleftrightarrow \leq^B_m \Longleftrightarrow \leq_m \Longleftrightarrow \leq^P_m$ and they are lattice orders.

**Example 2: non-linearly ordered MV/NM algebra $\mathcal{L}^L_{3\times 2}$**

The set $L_{3\times 2} = \{0, a, b, c, d, 1\} \cong L_3 \times L_2 = \{0, 1, 2\} \times \{0, 1\} = \{(0,0), (0,1), (1,0), (1,1), (2,0), (2,1)\}$,
organized as a lattice as in Figure 15.12 and as an involutive m-BE algebra with the operations $\odot$, $^-$ and $\oplus$ as in the following tables, is a non-linearly ordered left-MV algebra, denoted by $\mathcal{L}^L_{3\times 2}$ ([104], page 165).

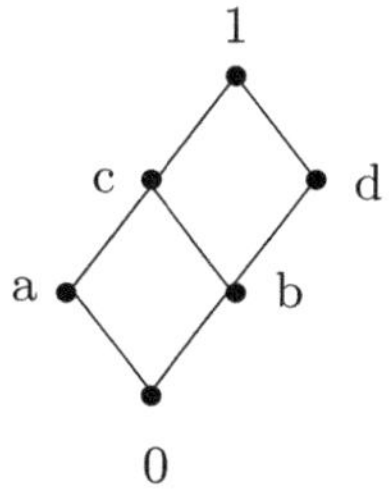


Figure 15.12: The non-linearly ordered $_{(WNM)}$ MV algebra $\mathcal{L}^L_{3\times 2}$

$\mathcal{L}^L_{3\times 2}$

| $\odot$ | 0 | a | b | c | d | 1 |
|---|---|---|---|---|---|---|
| 0 | 0 | 0 | 0 | 0 | 0 | 0 |
| a | 0 | a | 0 | a | 0 | a |
| b | 0 | 0 | 0 | 0 | b | b |
| c | 0 | a | 0 | a | b | c |
| d | 0 | 0 | b | b | d | d |
| 1 | 0 | a | b | c | d | 1 |

and

| $x$ | $x^-$ |
|---|---|
| 0 | 1 |
| a | d |
| b | c |
| c | b |
| d | a |
| 1 | 0 |

,

| $\oplus$ | 0 | a | b | c | d | 1 |
|---|---|---|---|---|---|---|
| 0 | 0 | a | b | c | d | 1 |
| a | a | a | c | c | 1 | 1 |
| b | b | c | d | 1 | d | 1 |
| c | c | c | 1 | 1 | 1 | 1 |
| d | d | 1 | d | 1 | d | 1 |
| 1 | 1 | 1 | 1 | 1 | 1 | 1 |

.

It verifies condition (WNM), hence it is a proper $_{(WNM)}$MV algebra.

**Example 3: non-linearly ordered MV/NM algebra $\mathcal{L}^L_{3\times 3}$**

The set $L_{3\times 3} = \{0, a, b, c, d, e, f, g, 1\} \cong L_3 \times L_3 = \{0, 1, 2\} \times \{0, 1, 2\} =$
$= \{(0,0), (0,1), (0,2), (1,0), (1,1), (1,2), (2,0), (2,1), (2,2)\}$,

organized as a lattice as in Figure 15.13 and as an involutive m-BE algebra $(L_{3\times 3}, \odot, ^-, 1)$ with the operations $\odot$, $^-$ and $\oplus$ as in the following tables, is a non-linearly ordered left-MV algebra, denoted by $\mathcal{L}^L_{3\times 3}$ ([104], page 166).

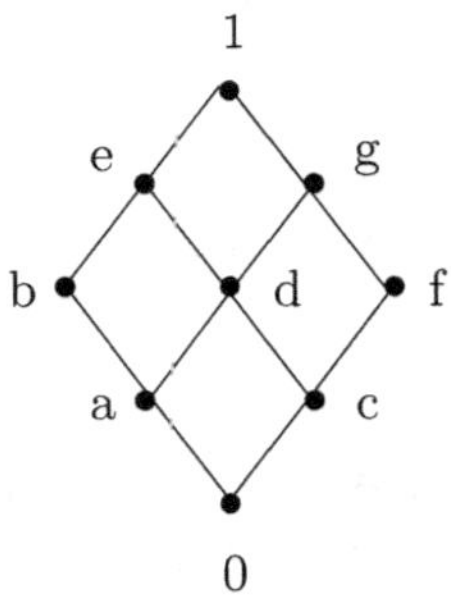


Figure 15.13: The non-linearly ordered $_{(WNM)}$MV algebra $\mathcal{L}^L_{3\times 3}$

$\mathcal{L}^L_{3\times 3}$

| $\odot$ | 0 | a | b | c | d | e | f | g | 1 |
|---|---|---|---|---|---|---|---|---|---|
| 0 | 0 | 0 | 0 | 0 | 0 | 0 | 0 | 0 | 0 |
| a | 0 | 0 | a | 0 | 0 | a | 0 | 0 | a |
| b | 0 | a | b | 0 | a | b | 0 | a | b |
| c | 0 | 0 | 0 | 0 | 0 | 0 | c | c | c |
| d | 0 | 0 | a | 0 | 0 | a | c | c | d |
| e | 0 | a | b | 0 | a | b | c | d | e |
| f | 0 | 0 | 0 | c | c | c | f | f | f |
| g | 0 | 0 | a | c | c | d | f | f | g |
| 1 | 0 | a | b | c | d | e | f | g | 1 |

and

| $x$ | $x^-$ |
|---|---|
| 0 | 1 |
| a | g |
| b | f |
| c | e |
| d | d |
| e | c |
| f | b |
| g | a |
| 1 | 0 |

, with

| $\oplus$ | 0 | a | b | c | d | e | f | g | 1 |
|---|---|---|---|---|---|---|---|---|---|
| 0 | 0 | a | b | c | d | e | f | g | 1 |
| a | a | b | b | d | e | e | g | 1 | 1 |
| b | b | b | b | e | e | e | 1 | 1 | 1 |
| c | c | d | e | f | g | 1 | f | g | 1 |
| d | d | e | e | g | 1 | 1 | g | 1 | 1 |
| e | e | e | e | 1 | 1 | 1 | 1 | 1 | 1 |
| f | f | g | 1 | f | g | 1 | f | g | 1 |
| g | g | 1 | 1 | g | 1 | 1 | g | 1 | 1 |
| 1 | 1 | 1 | 1 | 1 | 1 | 1 | 1 | 1 | 1 |

.

It verifies condition (WNM), hence it is a proper non-linearly ordered $_{(WNM)}$MV algebra.

### 15.3.2 Involutive m-pre-BCK algebras

Recall that involutive m-pre-BCK algebras mean transitive m-BE algebras.

### 15.3.2.1 Giving transitive quantum algebras

**Example 15.3.9 Transitive quantum-IMTL algebra: tQIMTL**

By MACE4 program, we found (see Example 12.4.5) that the algebra $\mathcal{A}^L = (A_8 = \{0, a, b, c, d, e, f, 1\}, \odot, ^-, 1)$, with the following tables of $\odot$ and $^-$ and of the additional operation $\oplus$, is an involutive left-m-BE algebra verifying (m-BB) ($\Leftrightarrow \ldots \Leftrightarrow$ (m-Tr)), and (prel$_m$), and not verifying (m-An) for $(a, f)$, (m-Pimpl) for $(a, 0)$, (m-Pabs-i) for $(a, 0)$, (G) for $a$, (Pqmv) for $(b, 0, d)$, (Pom) for $(b, d)$, (Pmv) for $(c, a)$, ($\Delta_m$) for $(a, c)$, (WNM$_m$) and (aWNM$_m$) for $(b, b)$, (m-Pdis) for $(a, a, b)$, i.e. it is an **involutive m-pre-BCK algebra verifying (prel$_m$)** and not verifying (WNM$_m$) and (aWNM$_m$) for $(b, b)$.

| $\odot$ | 0 | a | b | c | d | e | f | 1 |
|---|---|---|---|---|---|---|---|---|
| 0 | 0 | 0 | 0 | 0 | 0 | 0 | 0 | 0 |
| a | 0 | 0 | 0 | 0 | a | 0 | 0 | a |
| b | 0 | 0 | a | 0 | e | a | 0 | b |
| c | 0 | 0 | 0 | 0 | 0 | 0 | 0 | c |
| d | 0 | a | e | 0 | d | e | f | d |
| e | 0 | 0 | a | 0 | e | a | 0 | e |
| f | 0 | 0 | 0 | 0 | f | 0 | 0 | f |
| 1 | 0 | a | b | c | d | e | f | 1 |

,

| $\oplus$ | 0 | a | b | c | d | e | f | 1 |
|---|---|---|---|---|---|---|---|---|
| 0 | 0 | a | b | c | d | e | f | 1 |
| a | a | b | 1 | f | 1 | 1 | b | 1 |
| b | b | 1 | 1 | b | 1 | 1 | 1 | 1 |
| c | c | f | b | c | 1 | e | f | 1 |
| d | d | 1 | 1 | 1 | 1 | 1 | 1 | 1 |
| e | e | 1 | 1 | e | 1 | 1 | 1 | 1 |
| f | f | b | 1 | f | 1 | 1 | b | 1 |
| 1 | 1 | 1 | 1 | 1 | 1 | 1 | 1 | 1 |

and $(0, a, b, c, d, e, f, 1)^- = (1, b, a, d, c, f, e, 0)$.

The tables of $\wedge_m^M$ and $\wedge_m^B$ are:

| $\wedge_m^M$ | 0 | a | b | c | d | e | f | 1 |
|---|---|---|---|---|---|---|---|---|
| 0 | 0 | 0 | 0 | 0 | 0 | 0 | 0 | 0 |
| a | 0 | a | a | c | f | a | f | a |
| b | 0 | a | b | c | e | e | f | b |
| c | 0 | 0 | 0 | c | 0 | 0 | 0 | c |
| d | 0 | a | b | c | d | e | f | d |
| e | 0 | a | b | c | e | e | f | e |
| f | 0 | a | a | c | f | a | f | f |
| 1 | 0 | a | b | c | d | e | f | 1 |

and

| $\wedge_m^B$ | 0 | a | b | c | d | e | f | 1 |
|---|---|---|---|---|---|---|---|---|
| 0 | 0 | 0 | 0 | 0 | 0 | 0 | 0 | 0 |
| a | 0 | a | a | 0 | a | a | a | a |
| b | 0 | a | b | 0 | b | b | a | b |
| c | 0 | c | c | c | c | c | c | c |
| d | 0 | f | e | 0 | d | e | f | d |
| e | 0 | a | e | 0 | e | e | a | e |
| f | 0 | f | f | 0 | f | f | f | f |
| 1 | 0 | a | b | c | d | e | f | 1 |

.

• The binary relation $\leq_m^M$ is not transitive (hence (trans) is not verified) for $(a, e, d)$: $a \leq_m^M e$, $e \leq_m^M d$, but $a \not\leq_m^M d$, since $a \wedge_m^M d = f \neq a$; hence, $\leq_m^M$ is not an order relation.

• The binary relation $\leq_m$ ($\Longleftrightarrow \leq_m^B$) is a pre-order. From the table of $\wedge_m^B$ we obtain: $a \leq_m a, b, d, e, f, 1$; $b \leq_m b, d, e, 1$; $c \leq_m c, a, b, d, e, f, 1$; $d \leq_m d, 1$; $e \leq_m e, b, d, 1$; $f \leq_m f, a, b, d, e, 1$; the bounded partially pre-ordered set $(A_8, \leq_m, 0, 1)$ is represented by the Hasse-type diagram from Figure 15.14, left side.

Note that $a \parallel_Q f$ and $b \parallel_Q e$. Hence, the quotient set is $A_8/_{\parallel_Q} = \{\{0\}, \{c\}, \{a, f\}, \{b, e\}, \{d\}, \{1\}\}$, denoted by $A_6 = \{0, c, a, b, d, 1\}$, respectively, in the sequel. Then, the associated m-BCK algebra is $(A_6 = \{0, a, b, c, d, 1\}, \odot, ^-, 1)$, where the tables of $\odot$, $^-$, $\oplus$ are:

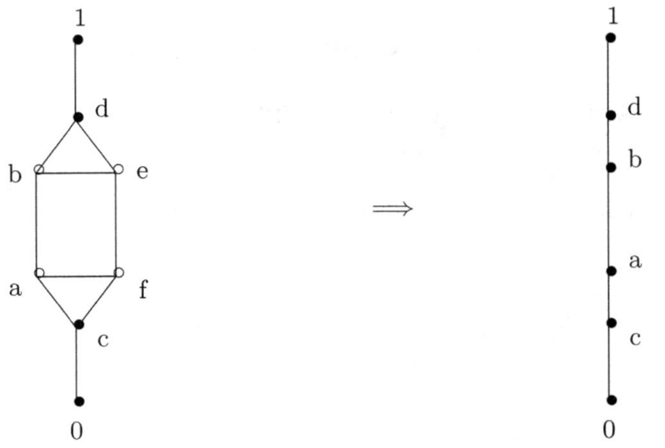


The bounded preordered set $(A_8, \leq_m, 0, 1)$ The bounded chain $(A_6, \leq_m, 0, 1)$

Figure 15.14: $(A_8, \leq_m, 0, 1)$ and $(A_6, \leq_m, 0, 1)$

| $\odot$ | 0 | a | b | c | d | 1 |
|---|---|---|---|---|---|---|
| 0 | 0 | 0 | 0 | 0 | 0 | 0 |
| a | 0 | 0 | 0 | 0 | a | a |
| b | 0 | 0 | a | 0 | b | b |
| c | 0 | 0 | 0 | 0 | 0 | c |
| d | 0 | a | b | 0 | d | d |
| 1 | 0 | a | b | c | d | 1 |

and

| $x$ | $x^-$ |
|---|---|
| 0 | 1 |
| a | b |
| b | a |
| c | d |
| d | c |
| 1 | 0 |

, with

| $\oplus$ | 0 | a | b | c | d | 1 |
|---|---|---|---|---|---|---|
| 0 | 0 | a | b | c | d | 1 |
| a | a | b | 1 | f | 1 | 1 |
| b | b | 1 | 1 | b | 1 | 1 |
| c | c | f | b | c | 1 | 1 |
| d | d | 1 | 1 | 1 | 1 | 1 |
| 1 | 1 | 1 | 1 | 1 | 1 | 1 |

.

Then, $(A_6, \odot, ^-, 1)$ is an (involutive) m-BCK algebra verifyng (prel$_m$) and not verifying (m-Pabs-i) for $(a, 0)$, (m-Pimpl) for $(a, 0)$, (Pqmv) for $(c, a, 0)$, (Pom) for $(d, a)$, (Pmv) for $(c, a)$, $(\Delta_m)$ for $(d, c)$, (WNM$_m$) and (aWNM$_m$) for $(b, b)$, (m-Pdis) for $(a, a, 0)$.

The tables of $\wedge_m^M$ and $\wedge_m^B$ are:

| $\wedge_m^M$ | 0 | a | b | c | d | 1 |
|---|---|---|---|---|---|---|
| 0 | 0 | 0 | 0 | 0 | 0 | 0 |
| 0 | 0 | 0 | 0 | 0 | 0 | 0 |
| a | 0 | a | a | c | a | a |
| b | 0 | a | b | c | b | b |
| c | 0 | 0 | 0 | c | 0 | c |
| d | 0 | a | b | c | d | d |
| 1 | 0 | a | b | c | d | 1 |

and

| $\wedge_m^B$ | 0 | a | b | c | d | 1 |
|---|---|---|---|---|---|---|
| 0 | 0 | 0 | 0 | 0 | 0 | 0 |
| a | 0 | a | a | 0 | a | a |
| b | 0 | a | b | 0 | b | b |
| c | 0 | c | c | c | c | c |
| d | 0 | a | b | 0 | d | d |
| 1 | 0 | a | b | c | d | 1 |

.

• Note that $\leq_m^M$ is transitive, as expected, hence it is an order relation, but not a lattice order w.r. to $\wedge_m^M$, $\vee_m^M$.

• Note, from the table of $\wedge_m^B$, that $a \leq_m a, b, d, 1$; $b \leq_m b, d, 1$; $c \leq_m c, a, b, d, 1$; $d \leq_m d, 1$; hence, the bounded poset $(A_6, \leq_m, 0, 1)$ is the **chain**: $0 \leq_m c \leq_m a \leq_m$

$b \leq_m d \leq_m 1$, represented, as expected, by the Hasse diagram from the same Figure 15.14, right side. Being a chain, it is a lattice w.r. to the following lattice operations $\vee$ and $\wedge$. The tables of union $\vee$, meet $\wedge$ and $\to$ are:

| $\vee$ | 0 | a | b | c | d | 1 |
|---|---|---|---|---|---|---|
| 0 | 0 | a | b | c | d | 1 |
| a | a | a | b | a | d | 1 |
| b | b | b | b | b | d | 1 |
| c | c | a | b | c | d | 1 |
| d | d | d | d | d | d | 1 |
| 1 | 1 | 1 | 1 | 1 | 1 | 1 |

, and

| $\wedge$ | 0 | a | b | c | d | 1 |
|---|---|---|---|---|---|---|
| 0 | 0 | 0 | 0 | 0 | 0 | 0 |
| a | 0 | a | a | c | a | a |
| b | 0 | a | b | c | b | b |
| c | 0 | c | c | c | c | c |
| d | 0 | a | b | c | d | d |
| 1 | 0 | a | b | c | d | 1 |

and

| $\to$ | 0 | a | b | c | d | 1 |
|---|---|---|---|---|---|---|
| 0 | 1 | 1 | 1 | 1 | 1 | 1 |
| a | b | 1 | 1 | b | 1 | 1 |
| b | a | b | 1 | a | 1 | 1 |
| c | d | 1 | 1 | 1 | 1 | 1 |
| d | c | a | b | c | 1 | 1 |
| 1 | 0 | a | b | c | d | 1 |

.

Note that (prel) $((x \to y) \vee (y \to x) = 1)$ is verified, while (WNM) is not verified for $(b, b)$. Consequently, $(A_6, \odot, ^-, 1)$ is an **IMTL chain** (Definition 2), isomorphic to $IMTL_6^5$ ($c \mapsto a$, $a \mapsto b$, $b \mapsto c$) from Examples 15.3.3, 3, and, therefor, $\mathcal{A}^L = (A_8, \odot, ^-, 1)$ is a **linearly pre-ordered transitive quantum-IMTL algebra**.

**Examples 15.3.10 Proper transitive quantum-NM algebras: tQNM**

A tQNM algebra is *proper*, if it is not a tQMV, i.e. if it does not verify (Pqmv) - see Figure 15.2.

**Example 1: linearly pre-ordered tQNM algebra**

By a PASCAL program, we found (see Example 12.4.6), that the algebra $\mathcal{A}^L = (A_6 = \{0, a, b, c, d, 1\}, \odot, ^-, 1)$, with the following tables of $\odot$ and $^-$ and of the additional operation $\oplus$, is an involutive left-m-BE algebra verifying (m-BB) ($\Leftrightarrow \ldots \Leftrightarrow$ (m-Tr)) and also (prel$_m$), (WNM$_m$), (aWNM$_m$), and not verifying (m-An) for $(a, d)$, (m-Pimpl) for $(a, 0)$, (Pqmv) for $(b, a, 0)$, (Pom) for $(c, a)$, (Pmv) for $(b, a)$, ($\Delta_m$) for $(a, b)$, (m-Pabs-i) for $(b, 0)$, (G) for $a$, (m-Pdis) for $(a, a, a)$, i.e. it is an **involutive m-pre-BCK algebra verifying (prel$_m$) and (WNM$_m$), (aWNM$_m$)**.

| $\odot$ | 0 | a | b | c | d | 1 |
|---|---|---|---|---|---|---|
| 0 | 0 | 0 | 0 | 0 | 0 | 0 |
| a | 0 | 0 | 0 | a | 0 | a |
| b | 0 | 0 | 0 | 0 | 0 | b |
| c | 0 | a | 0 | c | a | c |
| d | 0 | 0 | 0 | a | 0 | d |
| 1 | 0 | a | b | c | d | 1 |

and

| $x$ | $x^-$ |
|---|---|
| 0 | 1 |
| a | d |
| b | c |
| c | b |
| d | a |
| 1 | 0 |

, with

| $\oplus$ | 0 | a | b | c | d | 1 |
|---|---|---|---|---|---|---|
| 0 | 0 | a | b | c | d | 1 |
| a | a | 1 | d | 1 | 1 | 1 |
| b | b | d | b | 1 | d | 1 |
| c | c | 1 | 1 | 1 | 1 | 1 |
| d | d | 1 | d | 1 | 1 | 1 |
| 1 | 1 | 1 | 1 | 1 | 1 | 1 |

.

The tables of $\wedge_m^M$ and $\wedge_m^B$ are:

| $\wedge_m^M$ | 0 | a | b | c | d | 1 |
|---|---|---|---|---|---|---|
| 0 | 0 | 0 | 0 | 0 | 0 | 0 |
| a | 0 | a | b | a | d | a |
| b | 0 | 0 | b | 0 | 0 | b |
| c | 0 | a | b | c | d | c |
| d | 0 | a | b | a | d | d |
| 1 | 0 | a | b | c | d | 1 |

and

| $\wedge_m^B$ | 0 | a | b | c | d | 1 |
|---|---|---|---|---|---|---|
| 0 | 0 | 0 | 0 | 0 | 0 | 0 |
| a | 0 | a | 0 | a | a | a |
| b | 0 | b | b | b | b | b |
| c | 0 | a | 0 | c | a | c |
| d | 0 | d | 0 | d | d | d |
| 1 | 0 | a | b | c | d | 1 |

.

• Note that $\leq_m^M$ is transitive (hence (trans) is verified), hence $\leq_m^M$ is an order relation, but not a lattice order w.r.t. to $\wedge_m^M$, $\vee_m^M$, since $\wedge_m^M$ is not commutative. From the table of $\wedge_m^M$ we obtain: $a \leq_m^M a, c, 1$; $b \leq_m^M b, 1$; $c \leq_m^M c, 1$; $d \leq_m^M d, 1$. We omit the representation.

• The binary relation $\leq_m$ ($\Longleftrightarrow \leq_m^B$) is a pre-order. From the table of $\wedge_m^B$, we obtain: $a \leq_m a, c, d, 1$; $b \leq_m b, a, c, d, 1$; $c \leq_m c, 1$; $d \leq_m d, a, c, 1$; hence, the partially pre-ordered set $(A_6, \leq_m, 0, 1)$ is represented by the Hasse-type diagram from Figure 15.15, left side.

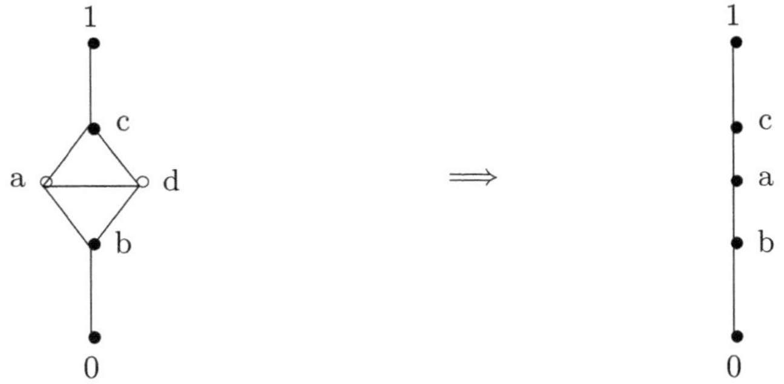


The bounded pre-ordered set $(A_6, \leq_m, 0, 1)$ The bounded chain $(A_5, \leq_m, 0, 1)$

Figure 15.15: $(A_6, \leq_m, 0, 1)$ and $(A_5, \leq_m, 0, 1)$

Note that $a \parallel_Q d$. Hence, the quotient set is $A_6/_{\parallel_Q} = \{\{0\}, \{b\}, \{a, d\}, \{c\}, \{1\}\}$, denoted by $A_5 = \{0, b, a, c, 1\}$, respectively, in the sequel. Then, the associated m-BCK algebra is $(A_5 = \{0, a, b, c, 1\}, \odot, ^-, 1)$, where the tables of $\odot$, $^-$, $\oplus$ are:

| $\odot$ | 0 | a | b | c | 1 |
|---|---|---|---|---|---|
| 0 | 0 | 0 | 0 | 0 | 0 |
| a | 0 | 0 | 0 | a | a |
| b | 0 | 0 | 0 | 0 | b |
| c | 0 | a | 0 | c | c |
| 1 | 0 | a | b | c | 1 |

and

| $x$ | $x^-$ |
|---|---|
| 0 | 1 |
| a | a |
| b | c |
| c | b |
| 1 | 0 |

, with

| $\oplus$ | 0 | a | b | c | 1 |
|---|---|---|---|---|---|
| 0 | 0 | a | b | c | 1 |
| a | a | 1 | a | 1 | 1 |
| b | b | a | b | 1 | 1 |
| c | c | 1 | 1 | 1 | 1 |
| 1 | 1 | 1 | 1 | 1 | 1 |

.

$(A_5, \odot, ^-, 1)$ is an (involutive) m-BCK algebra verifying (prel$_m$) and (WNM$_m$), (aWNM$_m$), and not verifying (m-Pabs-i) for $(b, 0)$, (m-Pimpl) for $(a, 0)$, (Pqmv) for $(b, a, 0)$, (Pom) for $(c, a)$, (Pmv) for $(b, a)$, ($\Delta_m$) for $(a, b)$, (m-Pdis) for $(a, a, a)$.

The tables of $\wedge_m^M$ and $\wedge_m^B$ are:

| $\wedge_m^M$ | 0 | a | b | c | 1 |
|---|---|---|---|---|---|
| 0 | 0 | 0 | 0 | 0 | 0 |
| a | 0 | a | b | a | a |
| b | 0 | 0 | b | 0 | b |
| c | 0 | a | b | c | c |
| 1 | 0 | a | b | c | 1 |

and

| $\wedge_m^B$ | 0 | a | b | c | 1 |
|---|---|---|---|---|---|
| 0 | 0 | 0 | 0 | 0 | 0 |
| a | 0 | a | 0 | a | a |
| b | 0 | b | b | b | b |
| c | 0 | a | 0 | c | c |
| 1 | 0 | a | b | c | 1 |

.

• Note that $\leq_m^M$ is transitive, as expected, hence it is an order relation, but not a lattice order w.r. to $\wedge_m^M$, $\vee_m^M$.

• From the table of $\wedge_m^B$ we obtain: $a \leq_m a, c, 1$; $b \leq_m b, a, c, 1$; $c \leq_m c, 1$; hence, the bounded poset $(A_5, \leq_m, 0, 1)$ is the **chain**: $0 \leq_m b \leq_m a \leq_m c \leq_m 1$, represented, as expected, by the Hasse diagram from the same Figure 15.15, right side. Being a chain, it is a lattice w.r. to the following lattice operations $\vee$ and $\wedge$. The tables of union $\vee$, meet $\wedge$ and $\rightarrow$ are:

| $\vee$ | 0 | a | b | c | 1 |
|---|---|---|---|---|---|
| 0 | 0 | a | b | c | 1 |
| a | a | a | a | c | 1 |
| b | b | a | b | c | 1 |
| c | c | c | c | c | 1 |
| 1 | 1 | 1 | 1 | 1 | 1 |

and

| $\wedge$ | 0 | a | b | c | 1 |
|---|---|---|---|---|---|
| 0 | 0 | 0 | 0 | 0 | 0 |
| a | 0 | a | b | a | a |
| b | 0 | b | b | b | b |
| c | 0 | a | b | c | c |
| 1 | 0 | a | b | c | 1 |

and

| $\rightarrow$ | 0 | a | b | c | 1 |
|---|---|---|---|---|---|
| 0 | 1 | 1 | 1 | 1 | 1 |
| a | a | 1 | a | 1 | 1 |
| b | c | 1 | 1 | 1 | 1 |
| c | b | a | b | 1 | 1 |
| 1 | 0 | a | b | c | 1 |

.

Note that (prel) ($(x \rightarrow y) \vee (y \rightarrow x) = 1$) is satisfied and also (WNM) is satisfied. Consequently, $(A_5, \odot, ^-, 1)$ is a **linearly ordered NM algebra** (Definition 2), isomorphic with $\mathcal{F}_5^m$ ($b \mapsto a$, $a \mapsto b$) from Examples 15.3.6, therefor, $\mathcal{A}^L$ is a **linearly pre-ordered proper transitive quantum-NM algebra**.

**Example 2: linearly pre-ordered tQNM algebra**

Consider the algebra $\mathcal{A}^L = (A_6 = \{0, a, b, c, d, 1\}, \odot, ^-, 1)$ with the following tables of $\odot$, $^-$ and $\oplus$ (see Examples 15.3.22,1):

| $\odot_F$ | 0 | a | b | c | d | 1 |
|---|---|---|---|---|---|---|
| 0 | 0 | 0 | 0 | 0 | 0 | 0 |
| a | 0 | 0 | 0 | 0 | 0 | a |
| b | 0 | 0 | b | 0 | b | b |
| c | 0 | 0 | 0 | 0 | 0 | c |
| d | 0 | 0 | b | 0 | b | d |
| 1 | 0 | a | b | c | d | 1 |

and

| $x$ | $x^-$ |
|---|---|
| 0 | 1 |
| a | b |
| b | a |
| c | d |
| d | c |
| 1 | 0 |

, with

| $\oplus_F$ | 0 | a | b | c | d | 1 |
|---|---|---|---|---|---|---|
| 0 | 0 | a | b | c | d | 1 |
| a | a | a | 1 | a | 1 | 1 |
| b | b | 1 | 1 | 1 | 1 | 1 |
| c | c | a | 1 | a | 1 | 1 |
| d | d | 1 | 1 | 1 | 1 | 1 |
| 1 | 1 | 1 | 1 | 1 | 1 | 1 |

.

Then, $\mathcal{A}^L$ is an involutive m-BE algebra verifying (m-BB) ($\Leftrightarrow \ldots \Leftrightarrow$ (m-Tr)) and (prel$_m$), (WNM$_m$), (aWNM$_m$), and not verifying (m-Pabs-i) for $(a, 0)$, (m-An) for $(a, c)$, (m-Pimpl) for $(b, 0)$, (Pqmv) for $(a, a, 0)$, (Pom) for $(d, b)$, (Pmv) for $(a, a)$, ($\Delta_m$) for $(b, a)$, (m-Pdis) for $(a, b, a)$; hence, it is an **involutive left-m-pre-BCK algebra verifying (prel$_m$), (aWNM$_m$), (WNM$_m$)**.

The tables of $\wedge_m^M$ and $\wedge_m^B$ are:

| $\wedge_m^M$ | 0 | a | b | c | d | 1 |
|---|---|---|---|---|---|---|
| 0 | 0 | 0 | 0 | 0 | 0 | 0 |
| a | 0 | a | 0 | c | 0 | a |
| b | 0 | a | b | c | d | b |
| c | 0 | a | 0 | c | 0 | c |
| d | 0 | a | b | c | d | d |
| 1 | 0 | a | b | c | d | 1 |

and

| $\wedge_m^B$ | 0 | a | b | c | d | 1 |
|---|---|---|---|---|---|---|
| 0 | 0 | 0 | 0 | 0 | 0 | 0 |
| a | 0 | a | a | a | a | a |
| b | 0 | 0 | b | 0 | b | b |
| c | 0 | c | c | c | c | c |
| d | 0 | 0 | d | 0 | d | d |
| 1 | 0 | a | b | c | d | 1 |

.

• The binary relation $\leq_m^M$ is transitive, hence is an order, but not a lattice order, since $\wedge_m^M$ is not commutative.

• The binary relation $\leq_m$ is a pre-order. From the table of $\wedge_m^B$, we obtain: $a \leq_m a, b, c, d, 1$; $b \leq_m b, d, 1$; $c \leq_m a, b, c, d, 1$; $d \leq_m d, b, 1$; hence, the bounded pre-ordered set $(A_6, \leq_m, 0, 1)$ is represented by the Hasse-type diagram from Figure 15.16, left side.

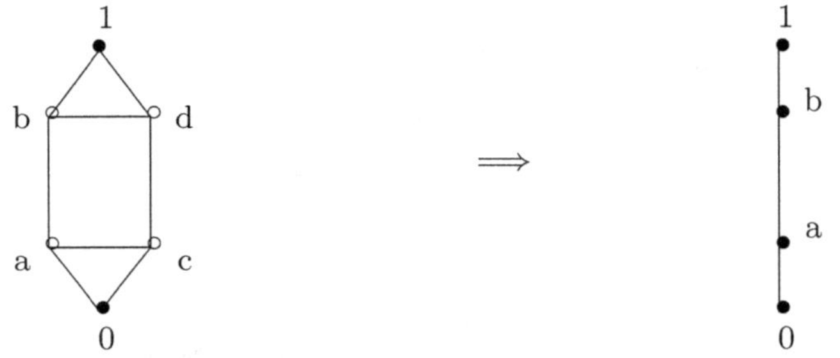


The bounded pre-ordered set $(A_6, \leq_m, 0, 1)$ The bounded chain $(A_4, \leq_m, 0, 1)$

Figure 15.16: $(A_6, \leq_m, 0, 1)$ and $(A_4, \leq_m, 0, 1)$

Note that $a \parallel_Q c$ and $b \parallel_Q d$, hence the quotient set is $A_6/_{\parallel_Q} = \{\{0\}, \{a, c\}, \{b, d\}, \{1\}\}$, denoted by $A_4 = \{0, a, b, 1\}$, respectively, in the sequel. Then, the associated m-BCK algebra is $(A_4 = \{0, a, b, 1\}, \odot, ^-, 1)$, where the tables of $\odot$, $^-$, $\oplus$ are:

| $\odot$ | 0 | a | b | 1 |
|---|---|---|---|---|
| 0 | 0 | 0 | 0 | 0 |
| a | 0 | 0 | 0 | a |
| b | 0 | 0 | b | b |
| 1 | 0 | a | b | 1 |

and

| $x$ | $x^-$ |
|---|---|
| 0 | 1 |
| a | b |
| b | a |
| 1 | 0 |

, with

| $\oplus$ | 0 | a | b | 1 |
|---|---|---|---|---|
| 0 | 0 | a | b | 1 |
| a | a | a | 1 | 1 |
| b | b | 1 | 1 | 1 |
| 1 | 1 | 1 | 1 | 1 |

.

Note that the associated m-BCK algebra $(A_4 = \{0, a, b, 1\}, \odot, ^-, 1)$ is the **linearly ordered proper NM algebra** $\mathcal{F}_4^m$ from Examples 15.3.6, represented by the Hasse diagram from the same Figure 15.16, right side. Hence, $\mathcal{A}^L$ is a **linearly pre-ordered proper tQNM algebra**.

**Example 3: non-linearly pre-ordered tQNM algebra**

Consider the above linearly pre-ordered tQNM algebra $\mathcal{A}^L = (A'_6 = \{0, a', b', c', d', 1\}, \odot, ^-, 1)$ and consider also the canonical Boolean algebra $\mathcal{L}_2^L = (L_2 = \{0, 1\}, \odot, ^-, 1)$ from Examples 15.3.4.

Consider the direct product $A_{6' \times 2} \cong A'_6 \times L_2 = \{0, a', b', c', d', 1\} \times \{0, 1\}$ $= \{(0,0), (0,1), (a',0), (a',1), (b',0), (b',1), (c',0), (c',1), (d',0), (d',1), (1,0), (1,1)\}$ that wiill be denoted, equivalently, with $A_{12} = \{0, a, b, c, d, e, f, g, h, i, j, 1\}$, respectively. Then, the tables of $\odot$ and $^-$, calculated component-wise, are the following:

| $\odot$ | 0 | a | b | c | d | e | f | g | h | i | j | 1 |
|---|---|---|---|---|---|---|---|---|---|---|---|---|
| 0 | 0 | 0 | 0 | 0 | 0 | 0 | 0 | 0 | 0 | 0 | 0 | 0 |
| a | 0 | a | 0 | a | 0 | a | 0 | a | 0 | a | 0 | a |
| b | 0 | 0 | 0 | 0 | 0 | 0 | 0 | 0 | 0 | 0 | b | b |
| c | 0 | a | 0 | a | 0 | a | 0 | a | 0 | a | b | c |
| d | 0 | 0 | 0 | 0 | d | d | 0 | 0 | d | d | d | d |
| e | 0 | a | 0 | a | d | e | 0 | a | d | e | d | e |
| f | 0 | 0 | 0 | 0 | 0 | 0 | 0 | 0 | 0 | 0 | f | f |
| g | 0 | a | 0 | a | 0 | a | 0 | a | 0 | a | f | g |
| h | 0 | 0 | 0 | 0 | d | d | 0 | 0 | d | d | h | h |
| i | 0 | a | 0 | a | d | e | 0 | a | d | e | h | i |
| j | 0 | 0 | b | b | d | d | f | f | h | h | j | j |
| 1 | 0 | a | b | c | d | e | f | g | h | i | j | 1 |

and

| $x$ | $x^-$ |
|---|---|
| 0 | 1 |
| a | j |
| b | e |
| c | d |
| d | c |
| e | b |
| f | i |
| g | h |
| h | g |
| i | f |
| j | a |
| 1 | 0 |

.

Then, $(A_{12}, \odot, ^-, 1)$ is an involutive m-BE algebra verifying (m-BB) ($\Leftrightarrow \ldots \Leftrightarrow$ (m-Tr)) and (prel$_m$), (aWNM$_m$), (WNM$_m$), and not verifying (m-An) for $(b, f)$, (m-Pimpl) for $(d, 0)$, (G) for $b$, (m-Pabs-i) for $(b, 0)$, (Pqmv) for $(b, b, 0)$, (Pom) for $(h, d)$, (Pmv) for $(b, b)$, ($\Delta_m$) for $(d, b)$, (m-Pdis) for $(b, d, b)$.

The table of $\wedge_m^B$ is the following:

| $\wedge_m^B$ | 0 | a | b | c | d | e | f | g | h | i | j | 1 |
|---|---|---|---|---|---|---|---|---|---|---|---|---|
| 0 | 0 | 0 | 0 | 0 | 0 | 0 | 0 | 0 | 0 | 0 | 0 | 0 |
| a | 0 | a | 0 | a | 0 | a | 0 | a | 0 | a | 0 | a |
| b | 0 | 0 | b | b | b | b | b | b | b | b | b | b |
| c | 0 | a | b | c | b | c | b | c | b | c | b | c |
| d | 0 | 0 | 0 | 0 | d | d | 0 | 0 | d | d | d | d |
| e | 0 | a | 0 | a | d | e | 0 | a | d | e | d | e |
| f | 0 | 0 | f | f | f | f | f | f | f | f | f | f |
| g | 0 | a | f | g | f | g | f | g | f | g | f | g |
| h | 0 | 0 | 0 | 0 | h | h | 0 | 0 | h | h | h | h |
| i | 0 | a | 0 | a | h | i | 0 | a | h | i | h | i |
| j | 0 | 0 | b | b | d | d | f | f | h | h | j | j |
| 1 | 0 | a | b | c | d | e | f | g | h | i | j | 1 |

.

• The binary relation $\leq_m$ is a pre-order. From the table of $\wedge_m^B$, we obtain: $a \leq_m a, c, e, g, i, 1$; $b \leq_m b, c, d, e, f, g, h, i, j, 1$; $c \leq_m c, e, g, i, 1$; $d \leq_m d, e, h, i, j, 1$; $e \leq_m e, i, 1$; $f \leq_m b, c, d, e, f, g, h, i, j, 1$; $g \leq_m g, c, e, i, 1$; $h \leq_m h, d, e, i, j, 1$; $i \leq_m i, e, 1$; $j \leq_m j, 1$: hence, the bounded pre-ordered set $(A_{12}, \leq_m, 0, 1)$ is represented by the Hasse-type diagram from Figure 15.17, left side.

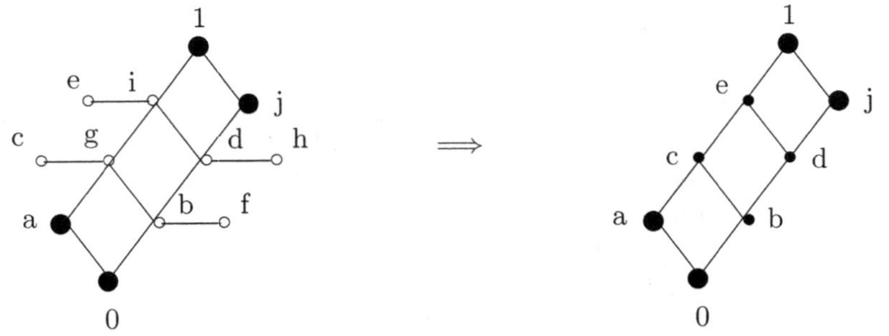


The bounded pre-ordered set $(A_{12}, \leq_m, 0, 1)$    The bounded poset $(A_8, \leq_m, 0, 1)$

Figure 15.17: $(A_{12}, \leq_m, 0, 1)$ and $(A_8, \leq_m, 0, 1)$

Note that $c \parallel_Q g$, $e \parallel_Q i$, $b \parallel_Q f$ and $d \parallel_Q h$, hence the quotient set is $A_{12}/_{\parallel_Q} = \{\{0\}, \{a\}, \{b, f\}, \{c, g\}, \{d, h\}, \{e, i\}, \{j\}, \{1\}\}$, denoted by $A_8 = \{0, a, b, c, d, e, j, 1\}$, respectively, in the sequel. Then, the associated m-BCK algebra is $(A_8 = \{0, a, b, c, d, e, j, 1\}, \odot, ^-, 1)$, where the tables of $\odot$, $^-$ are:

| $\odot$ | 0 | a | b | c | d | e | j | 1 |
|---|---|---|---|---|---|---|---|---|
| 0 | 0 | 0 | 0 | 0 | 0 | 0 | 0 | 0 |
| a | 0 | a | 0 | a | 0 | a | 0 | a |
| b | 0 | 0 | 0 | 0 | 0 | 0 | b | b |
| c | 0 | a | 0 | a | 0 | a | b | c |
| d | 0 | 0 | 0 | 0 | d | d | d | d |
| e | 0 | a | 0 | a | d | e | d | e |
| j | 0 | 0 | b | b | d | d | j | j |
| 1 | 0 | a | b | c | d | e | j | 1 |

and

| $x$ | $x^-$ |
|---|---|
| 0 | 1 |
| a | j |
| b | e |
| c | d |
| d | c |
| e | b |
| j | a |
| 1 | 0 |

.

Then, the m-BCK algebra $(A_8 = \{0, a, b, c, d, e, j, 1\}, \odot, ^-, 1)$ verifies (Pom) and (prel$_m$), (aWNM$_m$), (WNM$_m$) and does not verify (m-Pimpl) for $(d, 0)$, (G) for $b$, (m-Pabs-i) for $(b, 0)$, (Pqmv) for $(b, b, 0)$, (Pmv) for $(b, b)$, $(\Delta_m)$ for $(d, b)$, (m-Pdis) for $(b, d, b)$; note that it is isomorphic with the **non-linearly ordered proper NM algebra** $\mathcal{F}^m_{4\times 2}$ $(j \mapsto f)$ from Examples 15.3.7, represented by the Hasse diagram from the same Figure 15.17, right side. Hence, $(A_{12}, \odot, ^-, 1)$ is a **non-linearly pre-ordered proper tQNM algebra**.

**Examples 15.3.11 Transitive PreMV algebras: tPreMV**

**Example 1: transitive PreMV algebra not verifying (trans)**

By MACE4 program, we found (see Example 12.4.7) that the algebra $\mathcal{A}^L = (A_8 = \{0, a, b, c, d, e, f, 1\}, \odot, ^-, 1)$, with the following tables of $\odot$ and $^-$ and of the additional operation $\oplus$, is an involutive left-m-BE algebra verifying (Pmv) (hence $(\Delta_m)$) and (m-BB) ($\Leftrightarrow \ldots \Leftrightarrow$ (m-Tr)), and also (prel$_m$), and not verifying (m-An) for $(a, e)$, (m-Pimpl) for $(a, 0)$, (m-Pabs-i) for $(a, 0)$, (G) for $a$, (Pqmv) for $(b, 0, c)$, (Pom) for $(b, c)$, (WNM$_m$) and (aWNM$_m$) for $(b, b)$, (m-Pdis) for $(a, a, b)$, i.e. it is an **involutive m-pre-BCK algebra verifying (Pmv) and (prel$_m$).**

| $\odot$ | 0 | a | b | c | d | e | f | 1 |
|---|---|---|---|---|---|---|---|---|
| 0 | 0 | 0 | 0 | 0 | 0 | 0 | 0 | 0 |
| a | 0 | 0 | 0 | 0 | 0 | 0 | 0 | a |
| b | 0 | 0 | c | a | e | 0 | c | b |
| c | 0 | 0 | a | 0 | 0 | 0 | a | c |
| d | 0 | 0 | e | 0 | 0 | 0 | a | d |
| e | 0 | 0 | 0 | 0 | 0 | 0 | 0 | e |
| f | 0 | 0 | c | a | a | 0 | c | f |
| 1 | 0 | a | b | c | d | e | f | 1 |

,

| $\oplus$ | 0 | a | b | c | d | e | f | 1 |
|---|---|---|---|---|---|---|---|---|
| 0 | 0 | a | b | c | d | e | f | 1 |
| a | a | d | 1 | f | b | d | 1 | 1 |
| b | b | 1 | 1 | 1 | 1 | 1 | 1 | 1 |
| c | c | f | 1 | 1 | 1 | b | 1 | 1 |
| d | d | b | 1 | 1 | 1 | b | 1 | 1 |
| e | e | d | 1 | b | b | d | 1 | 1 |
| f | f | 1 | 1 | 1 | 1 | 1 | 1 | 1 |
| 1 | 1 | 1 | 1 | 1 | 1 | 1 | 1 | 1 |

and $(0, a, b, c, d, e, f, 1)^- = (1, b, a, d, c, f, e, 0)$.

Then, the tables of $\wedge_m^M$ and $\wedge_m^B$ are the following:

| $\wedge_m^M$ | 0 | a | b | c | d | e | f | 1 |
|---|---|---|---|---|---|---|---|---|
| 0 | 0 | 0 | 0 | 0 | 0 | 0 | 0 | 0 |
| a | 0 | a | e | a | a | e | a | a |
| b | 0 | a | b | c | d | e | f | b |
| c | 0 | a | c | c | d | e | c | c |
| d | 0 | a | c | c | d | e | c | d |
| e | 0 | a | e | a | e | e | a | e |
| f | 0 | a | b | c | d | e | f | f |
| 1 | 0 | a | b | c | d | e | f | 1 |

and

| $\wedge_m^B$ | 0 | a | b | c | d | e | f | 1 |
|---|---|---|---|---|---|---|---|---|
| 0 | 0 | 0 | 0 | 0 | 0 | 0 | 0 | 0 |
| a | 0 | a | a | a | a | a | a | a |
| b | 0 | e | b | c | c | e | b | b |
| c | 0 | a | c | c | c | a | c | c |
| d | 0 | a | d | d | d | e | d | d |
| e | 0 | e | e | e | e | e | e | e |
| f | 0 | a | f | c | c | a | f | f |
| 1 | 0 | a | b | c | d | e | f | 1 |

.

• Note that $\leq_m^M$ is not transitive (hence (trans) is not verified) for $(a, c, b)$, hence it is not an order.

• The binary relation $\leq_m$ is a pre-order. From the table of $\wedge_m^B$, we obtain: $a \leq_m a, b, c, d, e, f, 1$; $b \leq_m b, f, 1$; $c \leq_m c, b, d, f, 1$; $d \leq_m d, b, c, f, 1$; $e \leq_m e, a, b, c, d, f, 1$; $f \leq_m f, b, 1$; hence, the bounded pre-ordered set $(A_8, \leq_m, 0, 1)$ is represented by the Hasse-type diagram from Figure 15.18, left side.

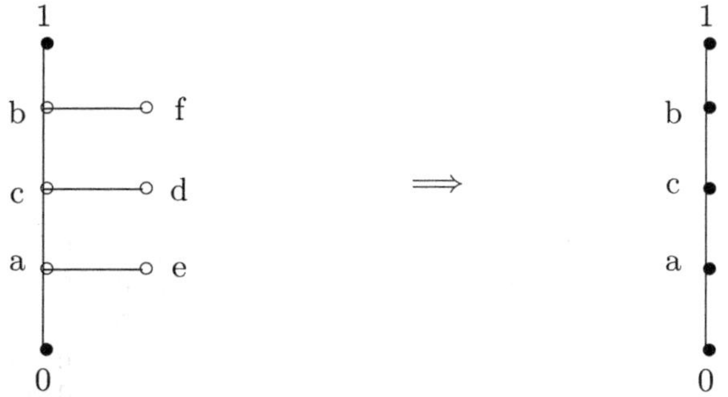


The bounded pre-ordered set $(A_8, \leq_m, 0, 1)$ The MV chain $(A_5, \leq_m, 0, 1)$

Figure 15.18: $(A_8, \leq_m, 0, 1)$ and $(A_5, \leq_m, 0, 1)$

Note that $a \parallel_Q e$, $c \parallel_Q d$, $b \parallel_Q f$, hence the quotient set is $A_8/_{\parallel_Q} = \{\{0\}, \{a, e\}, \{c, d\}, \{b, f\}, \{1\}\}$, denoted by $A_5 = \{0, a, c, b, 1\}$, respectively, in the sequel. Then, the associated m-BCK algebra is $(A_5 = \{0, a, c, b, 1\}, \odot, ^-, 1)$, where the tables of $\odot$, $^-$, $\oplus$ are:

| $\odot$ | 0 | a | b | c | 1 |
|---|---|---|---|---|---|
| 0 | 0 | 0 | 0 | 0 | 0 |
| a | 0 | 0 | 0 | 0 | a |
| b | 0 | 0 | c | a | b |
| c | 0 | 0 | a | 0 | c |
| 1 | 0 | a | b | c | 1 |

and

| $x$ | $x^-$ |
|---|---|
| 0 | 1 |
| a | b |
| b | a |
| c | c |
| 1 | 0 |

, with

| $\oplus$ | 0 | a | b | c | 1 |
|---|---|---|---|---|---|
| 0 | 0 | a | b | c | 1 |
| a | a | d | 1 | f | 1 |
| b | b | 1 | 1 | 1 | 1 |
| c | c | f | 1 | 1 | 1 |
| 1 | 1 | 1 | 1 | 1 | 1 |

.

The m-BCK algebra $(A_5 = \{0, a, b, c, 1\}, \odot, ^-, 1)$ verifies (Pqmv), (Pom), (Pmv), $(\Delta_m)$ and (prel$_m$), and does not verify (m-Pabs-i) for $(a, 0)$, (m-Pimpl) for $(a, 0)$, (m-Pabs-i) for $(a, 0)$, (G) for $a$, (WNM$_m$) and (aWNM$_m$) for $(b, b)$, (m-Pdis) for $(a, a, b)$; hence, it is a **linearly ordered MV algebra with 5 elements**, isomorphic with $\mathcal{L}_5^L$ $(c \mapsto b,\ b \mapsto c)$ from Example 15.3.4, represented in the same Figure 15.18, right-side.

**Example 2: transitive PreMV algebra verifying (trans)**

By a PASCAL program, we found (see Example 11.4.6) that the algebra $\mathcal{A}^L = (A_6 = \{0, a, b, c, d, 1\}, \odot, ^-, 1)$, with the following tables of $\odot$ and $^-$ and of the additional operation $\oplus$, is an involutive left-m-BE algebra verifying (Pmv) (hence $(\Delta_m)$) and (m-BB) ($\Leftrightarrow \ldots \Leftrightarrow$ (m-Tr)), (prel$_m$) and not verifying (m-An) for $(b, c)$, (m-Pabs-i) for $(a, 0)$, (m-Pimpl) for $(a, 0)$, (G) for $a$, (Pqmv) for $(b, 1, d)$, (Pom) for $(b, d)$, (WNM$_m$) and (aWNM$_m$) for $(b, d)$, (m-Pdis) for $(a, a, d)$. i.e. it is an **involutive m-pre-BCK algebra verifying (Pmv) and (prel$_m$).**

| $\odot$ | 0 | a | b | c | d | 1 |
|---|---|---|---|---|---|---|
| 0 | 0 | 0 | 0 | 0 | 0 | 0 |
| a | 0 | 0 | 0 | 0 | 0 | a |
| b | 0 | 0 | 0 | 0 | a | b |
| c | 0 | 0 | 0 | 0 | a | c |
| d | 0 | 0 | a | a | b | d |
| 1 | 0 | a | b | c | d | 1 |

and

| $x$ | $x^-$ |
|---|---|
| 0 | 1 |
| a | d |
| b | c |
| c | b |
| d | a |
| 1 | 0 |

, with

| $\oplus$ | 0 | a | b | c | d | 1 |
|---|---|---|---|---|---|---|
| 0 | 0 | a | b | c | d | 1 |
| a | a | c | d | d | 1 | 1 |
| b | b | d | 1 | 1 | 1 | 1 |
| c | c | d | 1 | 1 | 1 | 1 |
| d | d | 1 | 1 | 1 | 1 | 1 |
| 1 | 1 | 1 | 1 | 1 | 1 | 1 |

.

The tables of $\wedge_m^M$ and $\wedge_m^B$ are:

| $\wedge_m^M$ | 0 | a | b | c | d | 1 |
|---|---|---|---|---|---|---|
| 0 | 0 | 0 | 0 | 0 | 0 | 0 |
| a | 0 | a | a | a | a | a |
| b | 0 | a | b | c | b | b |
| c | 0 | a | b | c | b | c |
| d | 0 | a | b | c | d | d |
| 1 | 0 | a | b | c | d | 1 |

and

| $\wedge_m^B$ | 0 | a | b | c | d | 1 |
|---|---|---|---|---|---|---|
| 0 | 0 | 0 | 0 | 0 | 0 | 0 |
| a | 0 | a | a | a | a | a |
| b | 0 | a | b | b | b | b |
| c | 0 | a | c | c | c | c |
| d | 0 | a | b | b | d | d |
| 1 | 0 | a | b | c | d | 1 |

.

- Note that $\leq_m^M$ is transitive, hence $\leq_m^M$ is an order relation, but not a lattice order w.r.t. to $\wedge_m^M$, $\vee_m^M$, since $\wedge_m^M$ is not commutative.

• The binary relation $\leq_m$ is a pre-order. From the table of $\wedge_m^B$, we obtain: $a \leq_m a, b, c, d, 1$; $b \leq_m b, c, d, 1$; $c \leq_m c, b, d, 1$; $d \leq_m d, 1$; hence, the bounded pre-ordered set $(A_6, \leq_m, 0, 1)$ is represented by the Hasse-type diagram from Figure 15.19, left side.

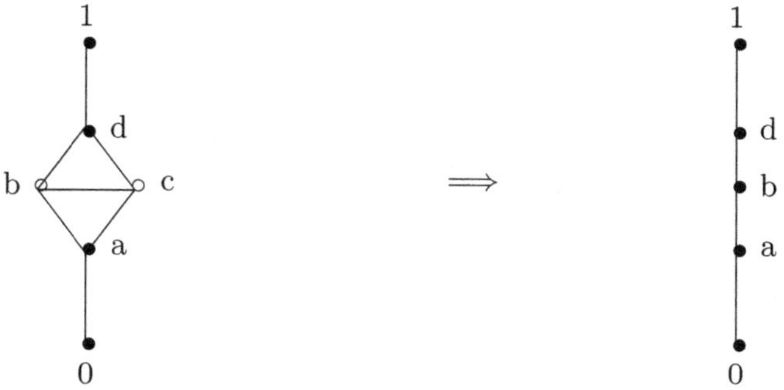


The bounded pre-ordered set $(A_6, \leq_m, 0, 1)$ The bounded MV chain $(A_5, \leq_m, 0, 1)$

Figure 15.19: $(A_6, \leq_m, 0, 1)$ and $(A_5, \leq_m, 0, 1)$

Note that $b \parallel_Q c$, hence the quotient set is $A_6/_{\parallel_Q} = \{\{0\}, \{a\}, \{b, c\}, \{d\}, \{1\}\}$, denoted by $A_5 = \{0, a, b, d, 1\}$, respectively, in the sequel. Then, the associated m-BCK algebra is $(A_5 = \{0, a, b, d, 1\}, \odot, ^-, 1)$, where the tables of $\odot$, $^-$, $\oplus$ are:

| $\odot$ | 0 | a | b | d | 1 |
|---|---|---|---|---|---|
| 0 | 0 | 0 | 0 | 0 | 0 |
| a | 0 | 0 | 0 | 0 | a |
| b | 0 | 0 | 0 | a | b |
| d | 0 | 0 | a | b | d |
| 1 | 0 | a | b | d | 1 |

and

| $x$ | $x^-$ |
|---|---|
| 0 | 1 |
| a | d |
| b | b |
| d | a |
| 1 | 0 |

, with

| $\oplus$ | 0 | a | b | d | 1 |
|---|---|---|---|---|---|
| 0 | 0 | a | b | d | 1 |
| a | a | c | d | 1 | 1 |
| b | b | d | 1 | 1 | 1 |
| d | d | 1 | 1 | 1 | 1 |
| 1 | 1 | 1 | 1 | 1 | 1 |

.

Note that the associated m-BCK algebra $(A_5 = \{0, a, b, d, 1\}, \odot, ^-, 1)$ is the **linearly ordered left-MV algebra $\mathcal{L}_5^L$** from Examples 15.3.4 (with $d$ instead of $c$), represented in the same Figure 15.19, right side.

**Examples 15.3.12 Transitive MMV algebras: tMMV**

**Example 1: transitive MMV algebra not verifying (trans)**

Consider the algebra $\mathcal{A}^L = (A_{12} = \{0, a, b, c, d, e, f, g, h, i, j, 1\}, \odot, ^-, 1)$ from Example 12.4.8. It is an involutive left-m-BE algebra verifying $(\Delta_m)$ and (m-BB) ($\Leftrightarrow \ldots \Leftrightarrow$ (m-Tr)), and (prel$_m$), and not verifying (m-An) for $(a, g)$, (m-Pimpl) for $(b, 0)$, (m-Pabs-i) for $(a, 0)$, (G) for $a$, (Pqmv) for $(a, 0, a)$, (Pom) for $(a, a)$, (Pmv) for $(a, a)$, (WNM$_m$) and (aWNM$_m$) for $(c, c)$, (m-Pdis) for $(a, a, a)$, i.e. it is an **involutive m-pre-BCK algebra verifying $(\Delta_m)$ and (prel$_m$).**

• The binary relation $\leq_m^M$ is not transitive (hence (trans) is not verified) for $(a, e, c)$: $a \leq_m^M e$, $e \leq_m^M c$, but $a \not\leq_m^M c$, since $a \wedge_m^M c = g \neq a$. Hence, $\leq_m^M$ is not an order.

• The binary relation $\leq_m$ is a pre-order. From the table of $\wedge_m^B$ - the transposed of $\wedge_m^M$ - we obtain: $a \leq_m a, c, e, g, j, 1$; $b \leq_m b, h, 1$; $c \leq_m c, 1$; $d \leq_m d, b, c, e, f, h, i, j, 1$; $e \leq_m e, c, j, 1$; $f \leq_m f, b, c, h, i, 1$; $g \leq_m g, a, c, e, j, 1$; $h \leq_m h, b, 1$; $i \leq_m i, b, c, f, h, 1$; $j \leq_m j, c, e, 1$; hence, the bounded pre-ordered set $(A_{12}, \leq_m, 0, 1)$ is represented by the Hasse-type diagram from Figure 15.20, left side.

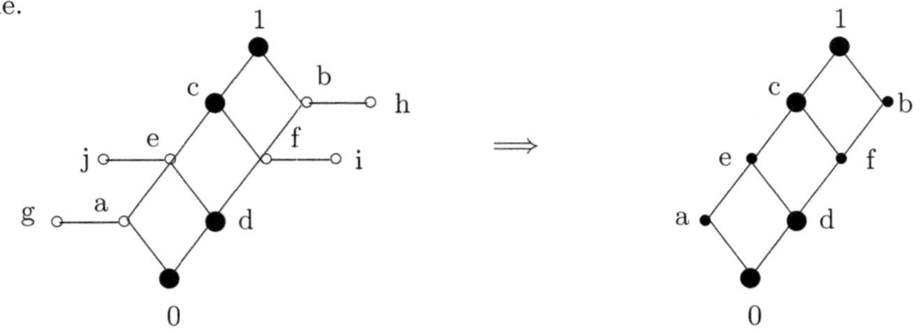


The bounded pre-ordered set $(A_{12}, \leq_m, 0, 1)$ The bounded poset $(A_8, \leq_m, 0, 1)$

Figure 15.20: $(A_{12}, \leq_m, 0, 1)$ and $(A_8, \leq_m, 0, 1)$

Note that $a \parallel_Q g$, $b \parallel_Q h$, $e \parallel_Q j$ and $f \parallel_Q i$; hence the quotient set is $A_{12}/_{\parallel_Q} = \{\{0\}, \{a, g\}, \{b, h\}, \{c\}, \{d\}, \{e, j\}\{f, i\}, \{1\}\}$, denoted by $A_8 = \{0, a, b, c, d, e, f, 1\}$, respectively, in the sequel. Then, the associated m-BCK algebra is $(A_8 = \{0, a, b, c, d, e, f, 1\}, \odot, ^-, 1)$, where the tables of $\odot$, $^-$, $\oplus$ are:

| $\odot$ | 0 | a | b | c | d | e | f | 1 |
|---|---|---|---|---|---|---|---|---|
| 0 | 0 | 0 | 0 | 0 | 0 | 0 | 0 | 0 |
| a | 0 | a | 0 | a | 0 | a | 0 | a |
| b | 0 | 0 | b | f | d | d | f | b |
| c | 0 | a | f | e | 0 | a | d | c |
| d | 0 | 0 | d | 0 | 0 | 0 | 0 | d |
| e | 0 | a | d | a | 0 | a | 0 | e |
| f | 0 | 0 | f | d | 0 | 0 | d | f |
| 1 | 0 | a | b | c | d | e | f | 1 |

and

| $\oplus$ | 0 | a | b | c | d | e | f | 1 |
|---|---|---|---|---|---|---|---|---|
| 0 | 0 | a | b | c | d | e | f | 1 |
| a | a | a | 1 | c | e | e | c | 1 |
| b | b | 1 | b | 1 | b | 1 | b | 1 |
| c | c | c | 1 | 1 | 1 | 1 | 1 | 1 |
| d | d | e | b | 1 | f | c | b | 1 |
| e | e | e | 1 | 1 | c | c | 1 | 1 |
| f | f | c | b | 1 | b | 1 | b | 1 |
| 1 | 1 | 1 | 1 | 1 | 1 | 1 | 1 | 1 |

and $(0, a, b, c, d, e, f, 1)^- = (1, b, a, d, c, f, e, 0)$.

Note that the associated m-BCK algebra $(A_8, \odot, ^-, 1)$ is isomorphic (interchange $e$ with $c$ and $f \mapsto b$, $b \mapsto d$, $d \mapsto f$) with the **non-linearly ordered left-MV algebra** $\mathcal{L}^L_{4\times 2}$ from Example 15.3.5, and is represented in the same Figure 15.20, right side.

**Remark.** *This tMMV algebra is a particular case of tQIMTL algebra.*

**Example 2: transitive MMV algebra verifying (trans)**

By a PASCAL program, we found (see Example 11.4.7) that the algebra $\mathcal{A}^L = (A_6 = \{0, a, b, c, d, 1\}, \odot, ^-, 1)$, with the following tables of $\odot$ and $^-$ and of the additional operation $\oplus$, is an involutive left-m-BE algebra verifying $(\Delta_m)$ and (m-BB) ($\Leftrightarrow \ldots \Leftrightarrow$ (m-Tr)), and also (prel$_m$), (aWNM$_m$), (WNM$_m$), and not verifying (m-An) for $(a,b)$, (m-Pabs-i) for $(b,0)$, (G) for $b$, (Pqmv) for $(b,0,a)$, (Pom) for $(b,a)$, (Pmv) for $(b,a)$, (m-Pdis) for $(a,c,b)$, i.e. it is an **involutive m-pre-BCK algebra verifying ($\Delta_m$) and (prel$_m$), (aWNM$_m$), (WNM$_m$).**

| $\odot$ | 0 | a | b | c | d | 1 |
|---|---|---|---|---|---|---|
| 0 | 0 | 0 | 0 | 0 | 0 | 0 |
| a | 0 | a | a | 0 | 0 | a |
| b | 0 | a | a | 0 | 0 | b |
| c | 0 | 0 | 0 | c | c | c |
| d | 0 | 0 | 0 | c | d | d |
| 1 | 0 | a | b | c | d | 1 |

and

| $x$ | $x^-$ |
|---|---|
| 0 | 1 |
| a | d |
| b | c |
| c | b |
| d | a |
| 1 | 0 |

, with

| $\oplus$ | 0 | a | b | c | d | 1 |
|---|---|---|---|---|---|---|
| 0 | 0 | a | b | c | d | 1 |
| a | a | a | b | 1 | 1 | 1 |
| b | b | b | b | 1 | 1 | 1 |
| c | c | 1 | 1 | d | d | 1 |
| d | d | 1 | 1 | d | d | 1 |
| 1 | 1 | 1 | 1 | 1 | 1 | 1 |

.

The tables of $\wedge_m^M$ and $\wedge_m^B$ are:

| $\wedge_m^M$ | 0 | a | b | c | d | 1 |
|---|---|---|---|---|---|---|
| 0 | 0 | 0 | 0 | 0 | 0 | 0 |
| a | 0 | a | b | 0 | 0 | a |
| b | 0 | a | b | 0 | 0 | b |
| c | 0 | 0 | 0 | c | d | c |
| d | 0 | 0 | 0 | c | d | d |
| 1 | 0 | a | b | c | d | 1 |

and

| $\wedge_m^B$ | 0 | a | b | c | d | 1 |
|---|---|---|---|---|---|---|
| 0 | 0 | 0 | 0 | 0 | 0 | 0 |
| a | 0 | a | a | 0 | 0 | a |
| b | 0 | b | b | 0 | 0 | b |
| c | 0 | 0 | 0 | c | c | c |
| d | 0 | 0 | 0 | d | d | d |
| 1 | 0 | a | b | c | d | 1 |

.

• Note that $\leq_m^M$ is transitive, hence $\leq_m^M$ is an order relation, but not a lattice order w.r. to $\wedge_m^M$, $\vee_m^M$, since $\wedge_m^M$ is not commutative.

• The binary relation $\leq_m$ is a pre-order. From the table of $\wedge_m^B$, we obtain: $a \leq_m a, b, 1$; $b \leq_m b, a, 1$; $c \leq_m c, d, 1$; $d \leq_m d, c, 1$; hence, the bounded pre-ordered set $(A_6, \leq_m, 0, 1)$ is represented by the Hasse-type diagram from Figure 15.21, left side.

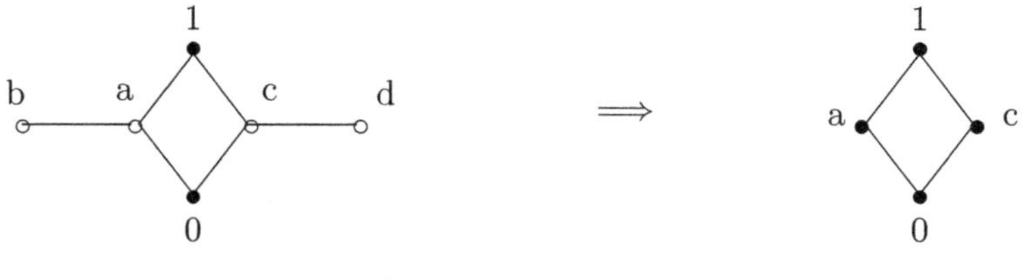


The bounded pre-ordered set $(A_6, \leq_m, 0, 1)$ The bounded lattice $(A_4, \leq_m, 0, 1)$

Figure 15.21: $(A_6, \leq_m, 0, 1)$ and $(A_4, \leq_m, 0, 1)$

Note that $a \parallel_Q b$ and $c \parallel_Q d$, hence the quotient set is $A_6/_{\parallel_Q} = \{\{0\}, \{a,b\}, \{c,d\}, \{1\}\}$, denoted by $A_4 = \{0, a, c, 1\}$, respectively, in the sequel. Then, the associated m-BCK algebra is $(A_4 = \{0, a, c, 1\}, \odot, ^-, 1)$, where the tables of $\odot$, $^-$, $\oplus$ are:

| $\odot$ | 0 | a | c | 1 |
|---|---|---|---|---|
| 0 | 0 | 0 | 0 | 0 |
| a | 0 | a | 0 | a |
| c | 0 | 0 | c | c |
| 1 | 0 | a | c | 1 |

and

| $x$ | $x^-$ |
|---|---|
| 0 | 1 |
| a | c |
| c | a |
| 1 | 0 |

, with

| $\oplus$ | 0 | a | c | 1 |
|---|---|---|---|---|
| 0 | 0 | a | c | 1 |
| a | a | a | 1 | 1 |
| c | c | 1 | c | 1 |
| 1 | 1 | 1 | 1 | 1 |

.

Note that the associated m-BCK algebra $(A_4, \odot, ^-, 1)$ is the **non-linearly ordered Boolean algebra - hence MV algebra -** $\mathcal{L}^L_{2\times 2}$ from Examples 15.3.8, 1, represented by the Hasse diagram from the same Figure 15.21, right side.

**Remark.** *This tMMV algebra is a particular case of tQNM algebra.*

**Examples 15.3.13 Proper transitive QMV algebras: tQMV**

A tQMV algebra is *proper*, if it is not a tOMWL, i.e. if it does not verify (m-Pabs-i) - see Figure 13.20.

**Example 1: linearly pre-ordered tQMV algebra**

By a PASCAL program, we found (see Example 11.4.8) that the algebra $\mathcal{A}^L = (A_6 = \{0, a, b, c, d, 1\}, \odot, ^-, 1)$, with the following tables of $\odot$ and $^-$ and of the additional operation $\oplus$, is a **proper left-tQMV algebra**, i.e. (PU), (Pcomm), (Pass), (m-La), (m-Re), (Pqmv) (hence (Pom), (Pmv), $(\Delta_m)$, $(\text{prel}_m)$), (DN), (m-BB) ($\Leftrightarrow \ldots \Leftrightarrow$ (m-Tr)) hold and it does not verify (m-An) for $(a, b)$. Additionally, it does not verify (m-Pabs-i) for $(a, 0)$, (m-Pimpl) for $(a, 0)$, $(\text{aWNM}_m)$ and $(\text{aWNM}_m)$ for $(c, c)$, (m-Pdis) for $(a, a, c)$.

| $\odot$ | 0 | a | b | c | d | 1 |
|---|---|---|---|---|---|---|
| 0 | 0 | 0 | 0 | 0 | 0 | 0 |
| a | 0 | 0 | 0 | 0 | 0 | a |
| b | 0 | 0 | 0 | 0 | 0 | b |
| c | 0 | 0 | 0 | a | b | c |
| d | 0 | 0 | 0 | b | a | d |
| 1 | 0 | a | b | c | d | 1 |

and

| $x$ | $x^-$ |
|---|---|
| 0 | 1 |
| a | d |
| b | c |
| c | b |
| d | a |
| 1 | 0 |

, with

| $\oplus$ | 0 | a | b | c | d | 1 |
|---|---|---|---|---|---|---|
| 0 | 0 | a | b | c | d | 1 |
| a | a | d | c | 1 | 1 | 1 |
| b | b | c | d | 1 | 1 | 1 |
| c | c | 1 | 1 | 1 | 1 | 1 |
| d | d | 1 | 1 | 1 | 1 | 1 |
| 1 | 1 | 1 | 1 | 1 | 1 | 1 |

.

The tables of $\wedge^M_m$ and $\wedge^B_m$ are:

| $\wedge^M_m$ | 0 | a | b | c | d | 1 |
|---|---|---|---|---|---|---|
| 0 | 0 | 0 | 0 | 0 | 0 | 0 |
| a | 0 | a | b | a | a | a |
| b | 0 | a | b | b | b | b |
| c | 0 | a | b | c | d | c |
| d | 0 | a | b | c | d | d |
| 1 | 0 | a | b | c | d | 1 |

and

| $\wedge^B_m$ | 0 | a | b | c | d | 1 |
|---|---|---|---|---|---|---|
| 0 | 0 | 0 | 0 | 0 | 0 | 0 |
| a | 0 | a | a | a | a | a |
| b | 0 | b | b | b | b | b |
| c | 0 | a | b | c | c | c |
| d | 0 | a | b | d | d | d |
| 1 | 0 | a | b | c | d | 1 |

.

• The binary relation $\leq^M_m$ is transitive, hence is an order, but not a lattice order, since $\wedge^M_m$ is not commutative. From the table of $\wedge^M_m$, we obtain: $a \leq^M_m a, c, d, 1$; $b \leq^M_m b, c, d, 1$; $c \leq^M_m c, 1$; $d \leq^M_m d, 1$; hence, the bounded poset $(A_6, \leq^M_m, 0, 1)$ is represented by the Hasse diagram from Figure 15.22.

• The binary relation $\leq_m$ is a pre-order. From the table of $\wedge^B_m$, we obtain:

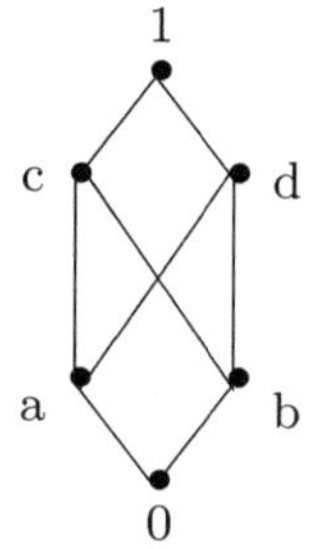


Figure 15.22: The bounded poset $(A_6, \leq_m^M, 0, 1)$

$a \leq_m a, b, c, d, 1$; $b \leq_m b, a, c, d, 1$; $c \leq_m c, d, 1$; $d \leq_m d, c, 1$; hence, the bounded pre-ordered set $(A_6, \leq_m, 0, 1)$ is represented by the Hasse-type diagram from Figure 15.23, left side.

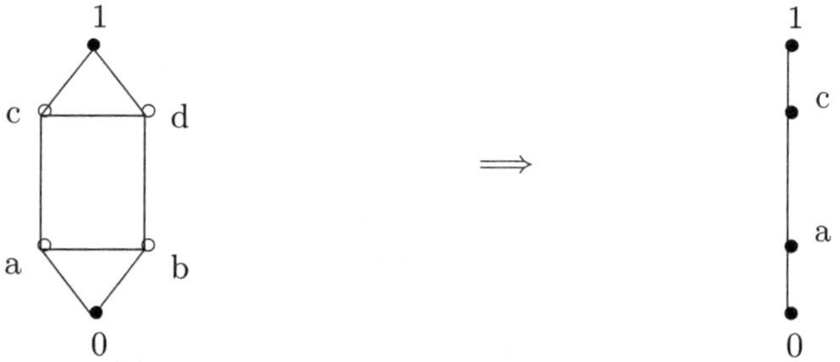


The bounded pre-ordered set $(A_6, \leq_m, 0, 1)$ The bounded chain $(A_4, \leq_m, 0, 1)$

Figure 15.23: $(A_6, \leq_m, 0, 1)$ and $(A_4, \leq_m, 0, 1)$

Note that $a \parallel_Q b$ and $c \parallel_Q d$, hence the quotient set is $A_6/_{\parallel_Q} = \{\{0\}, \{a, b\}, \{c, d\}, \{1\}\}$, denoted by $A_4 = \{0, a, c, 1\}$, respectively, in the sequel. Then, the associated m-BCK algebra is $(A_4 = \{0, a, c, 1\}, \odot, ^-, 1)$, where the tables of $\odot$, $^-$, $\oplus$ are:

| $\odot$ | 0 | a | c | 1 |
|---|---|---|---|---|
| 0 | 0 | 0 | 0 | 0 |
| a | 0 | 0 | 0 | a |
| c | 0 | 0 | a | c |
| 1 | 0 | a | c | 1 |

and

| $x$ | $x^-$ |
|---|---|
| 0 | 1 |
| a | c |
| c | a |
| 1 | 0 |

, with

| $\oplus$ | 0 | a | c | 1 |
|---|---|---|---|---|
| 0 | 0 | a | c | 1 |
| a | a | c | 1 | 1 |
| c | c | 1 | 1 | 1 |
| 1 | 1 | 1 | 1 | 1 |

.

Note that the associated m-BCK algebra $(A_4 = \{0, a, c, 1\}, \odot, ^-, 1)$ is the **linearly ordered proper MV algebra** $\mathcal{L}_4^L$ from Examples 15.3.4, represented by the Hasse diagram from the same Figure 15.23, right side. Hence, $\mathcal{A}^L$ is a **linearly pre-ordered tQMV algebra**.

**Example 2: non-linearly pre-ordered tQMV algebra**
Consider the above linearly pre-ordered tQMV algebra

$\mathcal{A}^L = (A'_6 = \{0, a', b', c', d', 1\}, \odot, ^-, 1)$ and consider also the canonical Boolean algebra $\mathcal{L}_2^L = (L_2 = \{0, 1\}, \odot, ^-, 1)$ from Examples 15.3.4.

Consider the direct product $A_{6' \times 2} \cong A'_6 \times L_2 = \{0, a', b', c', d', 1\} \times \{0, 1\}$ $= \{(0,0), (0,1), (a',0), (a',1), (b',0), (b',1), (c',0), (c',1), (d',0), (d',1), (1,0), (1,1)\}$ that wiill be denoted, equivalently, with $A_{12} = \{0, a, b, c, d, e, f, g, h, i, j, 1\}$, respectively. Then, the tables of $\odot$ and $^-$, calculated component-wise, are the following:

| $\odot$ | 0 | a | b | c | d | e | f | g | h | i | j | 1 |
|---|---|---|---|---|---|---|---|---|---|---|---|---|
| 0 | 0 | 0 | 0 | 0 | 0 | 0 | 0 | 0 | 0 | 0 | 0 | 0 |
| a | 0 | a | 0 | a | 0 | a | 0 | a | 0 | a | 0 | a |
| b | 0 | 0 | 0 | 0 | 0 | 0 | 0 | 0 | 0 | 0 | b | b |
| c | 0 | a | 0 | a | 0 | a | 0 | a | 0 | a | b | c |
| d | 0 | 0 | 0 | 0 | 0 | 0 | 0 | 0 | 0 | 0 | d | d |
| e | 0 | a | 0 | a | 0 | a | 0 | a | 0 | a | d | e |
| f | 0 | 0 | 0 | 0 | 0 | 0 | b | b | d | d | f | f |
| g | 0 | a | 0 | a | 0 | a | b | c | d | e | f | g |
| h | 0 | 0 | 0 | 0 | 0 | 0 | d | d | b | b | h | h |
| i | 0 | a | 0 | a | 0 | a | d | e | b | c | h | i |
| j | 0 | 0 | b | b | d | d | f | f | h | h | j | j |
| 1 | 0 | a | b | c | d | e | f | g | h | i | j | 1 |

and

| $x$ | $x^-$ |
|---|---|
| 0 | 1 |
| a | j |
| b | i |
| c | h |
| d | g |
| e | f |
| f | e |
| g | d |
| h | c |
| i | b |
| j | a |
| 1 | 0 |

.

Then, $(A_{12}, \odot, ^-, 1)$ is an involutive m-BE algebra verifying (m-BB) ($\Leftrightarrow \ldots \Leftrightarrow$ (m-Tr)), (Pqmv) (hence (Pom), (Pmv), ($\Delta_m$)) and (prel$_m$), and not verifying (m-An) for $(b, d)$, (m-Pimpl) for $(b, 0)$, (G) for $b$, (m-Pabs-i) for $(b, 0)$, (aWNM$_m$) and (WNM$_m$) for $(f, f)$, (m-Pdis) for $(b, b, f)$, hence, is a proper tQMV algebra.

The table of $\wedge_m^B$ is the following:

| $\wedge_m^B$ | 0 | a | b | c | d | e | f | g | h | i | j | 1 |
|---|---|---|---|---|---|---|---|---|---|---|---|---|
| 0 | 0 | 0 | 0 | 0 | 0 | 0 | 0 | 0 | 0 | 0 | 0 | 0 |
| a | 0 | a | 0 | a | 0 | a | 0 | a | 0 | a | 0 | a |
| b | 0 | 0 | b | b | b | b | b | b | b | b | b | b |
| c | 0 | a | b | c | b | c | b | c | b | c | b | c |
| d | 0 | 0 | d | d | d | d | d | d | d | d | d | d |
| e | 0 | a | d | e | d | e | d | e | d | e | d | e |
| f | 0 | 0 | b | b | d | d | f | f | f | f | f | f |
| g | 0 | a | b | c | d | e | f | g | f | g | f | g |
| h | 0 | 0 | b | b | d | d | h | h | h | h | h | h |
| i | 0 | a | b | c | d | e | h | i | h | i | h | i |
| j | 0 | 0 | b | b | d | d | f | f | h | h | j | j |
| 1 | 0 | a | b | c | d | e | f | g | h | i | j | 1 |

.

• The binary relation $\leq_m$ is a pre-order. From the table of $\wedge_m^B$, we obtain: $a \leq_m a, c, e, g, i, 1$; $b \leq_m b, c, d, e, f, g, h, i, j, 1$; $c \leq_m c, e, g, i, 1$; $d \leq_m b, c, d, e, f, g, h, i, j, 1$; $e \leq_m e, c, g, i, 1$; $f \leq_m f, g, h, i, j, 1$; $g \leq_m g, i, 1$; $h \leq_m h, f, g, i, j, 1$; $i \leq_m i, g, 1$; $j \leq_m j, 1$: hence, the bounded pre-ordered set $(A_{12}, \leq_m, 0, 1)$ is represented by the Hasse-type diagram from Figure 15.24, left side.

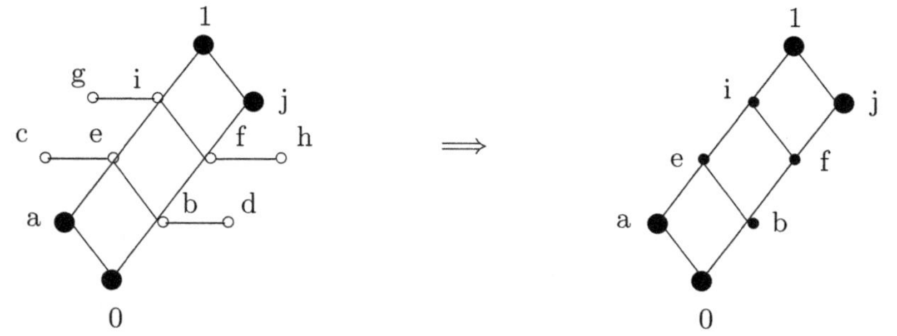


The bounded pre-ordered set $(A_{12}, \leq_m, 0, 1)$ The bounded poset $(A_8, \leq_m, 0, 1)$

Figure 15.24: $(A_{12}, \leq_m, 0, 1)$ and $(A_8, \leq_m, 0, 1)$

Note that $c \parallel_Q e$, $g \parallel_Q i$, $b \parallel_Q d$ and $f \parallel_Q h$, hence the quotient set is $A_{12}/_{\parallel_Q} = \{\{0\}, \{a\}, \{b, d\}, \{e, c\}, \{f, h\}, \{g, i\}, \{j\}, \{1\}\}$, denoted by $A_8 = \{0, a, b, e, f, i, j, 1\}$, respectively, in the sequel. Then, the associated m-BCK algebra is $(A_8 = \{0, a, b, e, f, i, j, 1\}, \odot, ^-, 1)$, where the tables of $\odot$, $^-$ are:

| $\odot$ | 0 | a | b | e | f | i | j | 1 |
|---|---|---|---|---|---|---|---|---|
| 0 | 0 | 0 | 0 | 0 | 0 | 0 | 0 | 0 |
| a | 0 | a | 0 | a | 0 | a | 0 | a |
| b | 0 | 0 | 0 | 0 | 0 | 0 | b | b |
| e | 0 | a | 0 | a | 0 | a | b | e |
| f | 0 | 0 | 0 | 0 | b | b | f | f |
| i | 0 | a | 0 | a | b | e | f | i |
| j | 0 | 0 | b | b | f | f | j | j |
| 1 | 0 | a | b | e | f | i | j | 1 |

and

| $x$ | $x^-$ |
|---|---|
| 0 | 1 |
| a | j |
| b | i |
| e | f |
| f | e |
| i | b |
| j | a |
| 1 | 0 |

.

Then, the m-BCK algebra $(A_8 = \{0, a, b, e, f, i, j, 1\}, \odot, ^-, 1)$ verifies (Pqmv) (hence (Pom), (Pmv), $(\Delta_m)$) and $(\text{prel}_m)$, and does not verify (m-Pimpl) for $(b, 0)$, (G) for $b$, (m-Pabs-i) for $(b, 0)$, $(\text{aWNM}_m)$ and $(\text{WNM}_m)$ for $(f, f)$, (m-Pdis) for $(b, b, f)$, hence, is a proper MV algebra; note that it is isomorphic with the **non-linearly ordered proper MV algebra** $\mathcal{L}^L_{4\times 2}$ $(e \mapsto c,\ f \mapsto d,\ i \mapsto e,\ j \mapsto f)$ from Examples 15.3.5, represented by the Hasse diagram from the same Figure 15.24, right side. Hence, $(A_{12}, \odot, ^-, 1)$ is a **non-linearly pre-ordered tQMV algebra**.

**Examples 15.3.14 tQMV algebras verifying (m-Pabs-i):**
**tOMWL** $=$ $_{(WNM_m)}$**tQMV**

**Example 1:** $\mathcal{A}^L_4$ (see [45], Example 2.3.8)

Let $\mathcal{A}^L_4 = (A_4 = \{0, a, b, 1\}, \odot, ^-, 1)$ be an algebra, where:

| $\odot$ | 0 | a | b | 1 |
|---|---|---|---|---|
| 0 | 0 | 0 | 0 | 0 |
| a | 0 | 0 | 0 | a |
| b | 0 | 0 | 0 | b |
| 1 | 0 | a | b | 1 |

and

| $x$ | $x^-$ |
|---|---|
| 0 | 1 |
| a | a |
| b | b |
| 1 | 0 |

, with

| $\oplus$ | 0 | a | b | 1 |
|---|---|---|---|---|
| 0 | 0 | a | b | 1 |
| a | a | 1 | 1 | 1 |
| b | b | 1 | 1 | 1 |
| 1 | 1 | 1 | 1 | 1 |

.

Then, $\mathcal{A}^L$ is a transitive left-QMV algebra (i.e. (PU), (Pcomm), (Pass), (m-La), (m-Re), (m-BB) ($\Leftrightarrow \ldots \Leftrightarrow$ (m-Tr)), (DN), (Pqmv) (hence (Pom), (Pmv), ($\Delta_m$)) hold) verifying (prel$_m$) and (m-Pabs-i) $\Longleftrightarrow$ (aWNM$_m$) $\Longleftrightarrow$ (WNM$_m$), and not verifying (m-An) for $(a, b)$, (m-Pimpl) for $(a, 0)$, (G) for $a$, (m-Pdis) for $(a, a, a)$, i.e. it is a **left-tOMWL** (see Figure 13.20).

The tables of $\wedge_m^M$ and $\wedge_m^B$ are the following:

| $\wedge_m^M$ | 0 | a | b | 1 |
|---|---|---|---|---|
| 0 | 0 | 0 | 0 | 0 |
| a | 0 | a | b | a |
| b | 0 | a | b | b |
| 1 | 0 | a | b | 1 |

and

| $\wedge_m^B$ | 0 | a | b | 1 |
|---|---|---|---|---|
| 0 | 0 | 0 | 0 | 1 |
| a | 0 | a | a | a |
| b | 0 | b | b | b |
| 1 | 0 | a | b | 1 |

.

• The binary relation $\leq_m^M$ is an order, but not a lattice order w.r. to $\wedge_m^M$ and $\vee_m^M$, sincet $\wedge_m^M$ is not commutative (for $(a, b)$, for example). Note that $a \leq_m^M a, 1$ and $b \leq_m^M b, 1$, hence the bounded poset $(A_4, \leq_m^M, 0, 1)$ is represented by the Hasse diagram from Figure 15.25.

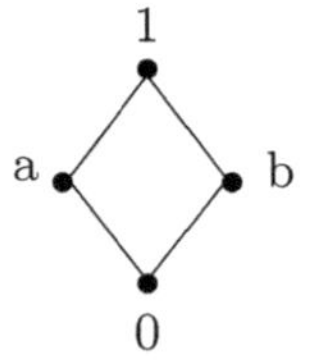


Figure 15.25: The bounded poset $(A_4, \leq_m^M, 0, 1)$ of the smallest $_{(WNM_m)}$tQMV

• The binary relation $\leq_m$ is a pre-order. From the table of $\wedge_m^B$ we obtain: $a \leq_m a, b, 1$ and $b \leq_m b, a, 1$; hence, the bounded pre-ordered set $(A_4, \leq_m, 0, 1)$ is represented by the Hasse-type diagram from Figure 15.26, left side.

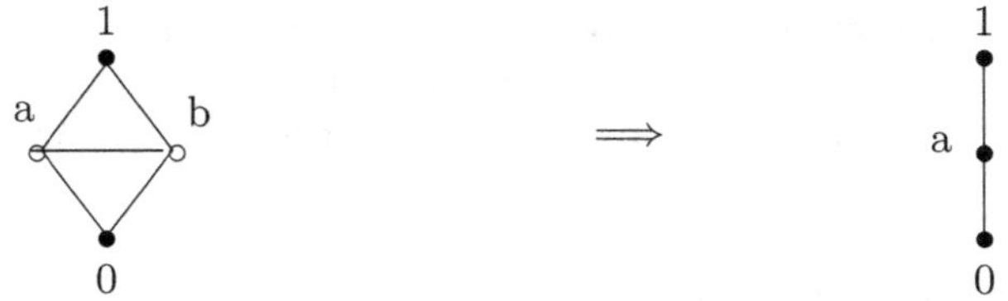


The bounded pre-ordered set $(A_4, \leq_m, 0, 1)$ The bounded chain $(A_3, \leq_m, 0, 1)$

Figure 15.26: $(A_4, \leq_m, 0, 1)$ and $(A_3, \leq_m, 0, 1)$

Note that $a \parallel_Q b$. Hence, the quotient set is $A_4/_{\parallel_Q} = \{\{0\}, \{a, b\}, \{1\}\}$, denoted by $A_3 = \{0, a, 1\}$, respectively, in the sequel. Then, the associated m-BCK algebra is $(A_3 = \{0, a, 1\}, \odot, ^-, 1)$, where the tables of $\odot$, $^-$ and $\oplus$ are:

| $\odot$ | 0 | a | 1 |
|---|---|---|---|
| 0 | 0 | 0 | 0 |
| a | 0 | 0 | a |
| 1 | 0 | a | 1 |

and

| $x$ | $x^-$ |
|---|---|
| 0 | 1 |
| a | a |
| 1 | 0 |

, with

| $\oplus$ | 0 | a | 1 |
|---|---|---|---|
| 0 | 0 | a | 1 |
| a | a | 1 | 1 |
| 1 | 1 | 1 | 1 |

.

Note that this m-BCK algebra is the **linearly ordered $_{(WNM)}$MV algebra** $\mathcal{L}_3^L$ from Examples 15.3.4, and is represented by the Hasse diagram from the same Figure 15.26, right side.

**Example 2: $\mathcal{A}_5^L$**
Let $\mathcal{A}_5^L = (A_5 = \{0, a, b, c, 1\}, \odot, ^-, 1)$ be an algebra, where:

| $\odot$ | 0 | a | b | c | 1 |
|---|---|---|---|---|---|
| 0 | 0 | 0 | 0 | 0 | 0 |
| a | 0 | 0 | 0 | 0 | a |
| b | 0 | 0 | 0 | 0 | b |
| c | 0 | 0 | 0 | 0 | c |
| 1 | 0 | a | b | c | 1 |

and

| $x$ | $x^-$ |
|---|---|
| 0 | 1 |
| a | a |
| b | b |
| c | c |
| 1 | 0 |

, with

| $\oplus$ | 0 | a | b | c | 1 |
|---|---|---|---|---|---|
| 0 | 0 | a | b | c | 1 |
| a | a | 1 | 1 | 1 | 1 |
| b | b | 1 | 1 | 1 | 1 |
| c | c | 1 | 1 | 1 | 1 |
| 1 | 1 | 1 | 1 | 1 | 1 |

.

Then, $\mathcal{A}^L$ is a transitive left-QMV algebra (i.e. (PU), (Pcomm), (Pass), (m-La), (m-Re), (m-BB) ($\Leftrightarrow \ldots \Leftrightarrow$ (m-Tr)), (DN), (Pqmv) (hence (Pom), (Pmv), ($\Delta_m$)) hold) verifying (prel$_m$) and (m-Pabs-i) $\Longleftrightarrow$ (aWNM$_m$) $\Longleftrightarrow$ (WNM$_m$), and not verifying (m-An) for $(a, b)$, (m-Pimpl) for $(a, 0)$, (G) for $a$, (m-Pdis) for $(a, a, a)$, hence it is a **left-tOMWL**.
The tables of $\wedge_m^M$ and $\wedge_m^B$ are the following:

| $\wedge_m^M$ | 0 | a | b | c | 1 |
|---|---|---|---|---|---|
| 0 | 0 | 0 | 0 | 0 | 0 |
| a | 0 | a | b | c | a |
| b | 0 | a | b | c | b |
| c | 0 | a | b | c | c |
| 1 | 0 | a | b | c | 1 |

and

| $\wedge_m^B$ | 0 | a | b | c | 1 |
|---|---|---|---|---|---|
| 0 | 0 | 0 | 0 | 0 | 1 |
| a | 0 | a | a | a | a |
| b | 0 | b | b | b | b |
| c | 0 | c | c | c | c |
| 1 | 0 | a | b | c | 1 |

.

• The binary relation $\leq_m^M$ is an order, but not a lattice order w.r. to $\wedge_m^M$ and $\vee_m^M$, since $\wedge_m^M$ is not commutative (for $(a, b)$, for example). Note that $a \leq_m^M a, 1$; $b \leq_m^M b, 1$ and $c \leq_m^M c, 1$, hence the bounded poset $(A_5, \leq_m^M, 0, 1)$ is represented by the Hasse diagram from Figure 15.27.

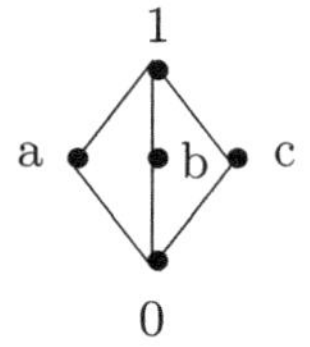


Figure 15.27: The bounded poset $(A_5, \leq_m^M, 0, 1)$ of $_{(WNM_m)}$tQMV algebra with 5 elements

• The binary relation $\leq_m$ is a pre-order. From the table of $\wedge_m^B$ we obtain: $a \leq_m a, b, c, 1$; $b \leq_m b, a, c, 1$; $c \leq_m c, a, b, 1$; hence, the bounded pre-ordered set $(A_5, \leq_m, 0, 1)$ is represented by the Hasse-type diagram from Figure 15.28, left side.

The bounded pre-ordered set $(A_5, \leq_m, 0, 1)$ The bounded chain $(A_3, \leq_m, 0, 1)$

Figure 15.28: $(A_5, \leq_m, 0, 1)$ and $(A_3, \leq_m, 0, 1)$

Note that $a \parallel_Q b \parallel_Q c$. Hence, the quotient set is $A_5/_{\parallel_Q} = \{\{0\}, \{a, b, c\}, \{1\}\}$, denoted by $A_3 = \{0, a, 1\}$, respectively, in the sequel. Then, the associated m-BCK algebra is $(A_3 = \{0, a, 1\}, \odot, ^-, 1)$, where the tables of $\odot$, $^-$ and $\oplus$ are:

| $\odot$ | 0 | a | 1 |
|---|---|---|---|
| 0 | 0 | 0 | 0 |
| a | 0 | 0 | a |
| 1 | 0 | a | 1 |

and

| $x$ | $x^-$ |
|---|---|
| 0 | 1 |
| a | a |
| 1 | 0 |

, with

| $\oplus$ | 0 | a | 1 |
|---|---|---|---|
| 0 | 0 | a | 1 |
| a | a | 1 | 1 |
| 1 | 1 | 1 | 1 |

.

Note that this associated m-BCK algebra is the **linearly ordered** $_{(WNM)}$**MV algebra** $\mathcal{L}_3^L$ from Examples 15.3.4, and is represented by the Hasse diagram from the same Figure 15.28, right side.

**Example 3:** $\mathcal{A}_{n+2}^L$, $n \geq 2$

Note that, for $n = 1$, $\mathcal{A}_3^L$ is the linearly ordered left-MV algebra $\mathcal{L}_3^L$ from Examples 15.3.4.

Let $\mathcal{A}_{n+2}^L = (A_{n+2} = \{0, a_1, a_2, \ldots, a_n, 1\}, \odot, ^-, 1)$ be an algebra, $n \geq 2$, where:

| $\odot$ | 0 | $a_1$ | $a_2$ | $\ldots$ | $a_n$ | 1 |
|---|---|---|---|---|---|---|
| 0 | 0 | 0 | 0 | $\ldots$ | 0 | 0 |
| $a_1$ | 0 | 0 | 0 | $\ldots$ | 0 | $a_1$ |
| $a_2$ | 0 | 0 | 0 | $\ldots$ | 0 | $a_2$ |
| $\vdots$ | $\vdots$ | $\vdots$ | $\vdots$ | $\ldots$ | $\vdots$ | $\vdots$ |
| $a_n$ | 0 | 0 | 0 | $\ldots$ | 0 | $a_n$ |
| 1 | 0 | $a_1$ | $a_2$ | $\ldots$ | $a_n$ | 1 |

,

| $x$ | $x^-$ |
|---|---|
| 0 | 1 |
| $a_1$ | $a_1$ |
| $a_2$ | $a_2$ |
| $\vdots$ | $\vdots$ |
| $a_n$ | $a_n$ |
| 1 | 0 |

,

| $\oplus$ | 0 | $a_1$ | $a_2$ | $\ldots$ | $a_n$ | 1 |
|---|---|---|---|---|---|---|
| 0 | 0 | $a_1$ | $a_2$ | $\ldots$ | $a_n$ | 1 |
| $a_1$ | $a_1$ | 1 | 1 | $\ldots$ | 1 | 1 |
| $a_2$ | $a_2$ | 1 | 1 | $\ldots$ | 1 | 1 |
| $\vdots$ | $\vdots$ | $\vdots$ | $\vdots$ | $\ldots$ | $\vdots$ | $\vdots$ |
| $a_n$ | $a_n$ | 1 | 1 | $\ldots$ | 1 | 1 |
| 1 | 1 | 1 | 1 | $\ldots$ | 1 | 1 |

.

Then, $\mathcal{A}^L$ is a transitive left-QMV algebra (i.e. (PU), (Pcomm), (Pass), (m-La), (m-Re), (m-BB) ($\Leftrightarrow \ldots \Leftrightarrow$ (m-Tr)), (DN), (Pqmv) (hence (Pom), (Pmv), $(\Delta_m)$) hold) verifying (prel$_m$) and (m-Pabs-i) $\Longleftrightarrow$ (aWNM$_m$) $\Longleftrightarrow$ (WNM$_m$), and not verifying (m-An) for ($a_1$, $a_2$), (m-Pimpl) for ($a_1$,0), (G) for $a_1$, (m-Pdis) for $(a_1, a_1, a_1)$. Hence, it is a **left-tOMWL**.

The tables of $\wedge_m^M$ and $\wedge_m^B$ are the following:

| $\wedge_m^M$ | 0 | $a_1$ | $a_2$ | $\dots$ | $a_n$ | 1 |
|---|---|---|---|---|---|---|
| 0 | 0 | 0 | 0 | $\dots$ | 0 | 0 |
| $a_1$ | 0 | $a_1$ | $a_2$ | $\dots$ | $a_n$ | $a_1$ |
| $a_2$ | 0 | $a_1$ | $a_2$ | $\dots$ | $a_n$ | $a_2$ |
| $\vdots$ | $\vdots$ | $\vdots$ | $\vdots$ | $\dots$ | $\vdots$ | $\vdots$ |
| $a_n$ | 0 | $a_1$ | $a_2$ | $\dots$ | $a_n$ | $a_n$ |
| 1 | 0 | $a_1$ | $a_2$ | $\dots$ | $a_n$ | 1 |

and

| $\wedge_m^B$ | 0 | $a_1$ | $a_2$ | $\dots$ | $a_n$ | 1 |
|---|---|---|---|---|---|---|
| 0 | 0 | 0 | 0 | $\dots$ | 0 | 0 |
| $a_1$ | 0 | $a_1$ | $a_1$ | $\dots$ | $a_1$ | $a_1$ |
| $a_2$ | 0 | $a_2$ | $a_2$ | $\dots$ | $a_2$ | $a_2$ |
| $\vdots$ | $\vdots$ | $\vdots$ | $\vdots$ | $\dots$ | $\vdots$ | $\vdots$ |
| $a_n$ | 0 | $a_n$ | $a_n$ | $\dots$ | $a_n$ | $a_n$ |
| 1 | 0 | $a_1$ | $a_2$ | $\dots$ | $a_n$ | 1 |

.

• The binary relation $\leq_m^M$ is an order, but not a lattice order w.r. to $\wedge_m^M$ and $\vee_m^M$, since $\wedge_m^M$ is not commutative. Note that $a_1 \leq_m^M a_1, 1$; $a_2 \leq_m^M a_2, 1$; ...; $a_n \leq_m a_n, 1$, hence the bounded poset $(A_{n+2}, \leq_m^M, 0, 1)$ is represented by the Hasse diagram from Figure 15.29.

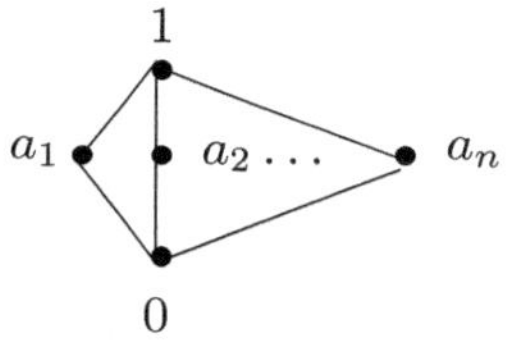


Figure 15.29: The bounded poset $(A_{n+2}, \leq_m^M, 0, 1)$ of ${}_{(WNM_m)}$tQMV algebra with $n+2$ elements

• The binary relation $\leq_m$ is a pre-order. From the table of $\wedge_m^B$ we obtain: $a_1 \leq_m a_1, a_2, \dots, a_n, 1$; $a_2 \leq_m a_2, a_1, \dots, a_n, 1$; $\dots$ $a_n \leq_m a_n, a_1, a_2, \dots, 1$; hence, the bounded pre-ordered set $(A_{n+2}, \leq_m, 0, 1)$ is represented by the Hasse-type diagram from the Figure 15.30, left side.

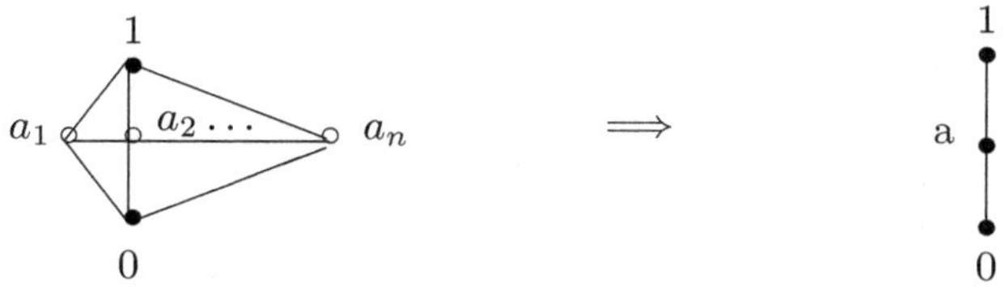


The bounded pre-ordered set $(A_{n+2}, \leq_m, 0, 1)$ The bounded chain $(A_3, \leq_m, 0, 1)$

Figure 15.30: $(A_{n+2}, \leq_m, 0, 1)$ and $(A_3, \leq_m, 0, 1)$

Note that $a_1 \parallel_Q a_2 \parallel_Q \dots \parallel_Q a_n$. Hence, the quotient set is $A_{n+2}/_{\parallel_Q} = \{\{0\}, \{a_1, a_2, \dots, a_n\}, \{1\}\}$, denoted by $A_3 = \{0, a, 1\}$, respectively, in the sequel. Then, the associated m-BCK algebra is $(A_3 = \{0, a, 1\}, \odot, ^-, 1)$, where the tables of $\odot$, $^-$ and $\oplus$ are:

| $\odot$ | 0 | a | 1 |
|---|---|---|---|
| 0 | 0 | 0 | 0 |
| a | 0 | 0 | a |
| 1 | 0 | a | 1 |

and

| $x$ | $x^-$ |
|---|---|
| 0 | 1 |
| a | a |
| 1 | 0 |

, with

| $\oplus$ | 0 | a | 1 |
|---|---|---|---|
| 0 | 0 | a | 1 |
| a | a | 1 | 1 |
| 1 | 1 | 1 | 1 |

.

Note that this m-BCK algebra is the **linearly ordered** $_{(WNM)}$**MV algebra** $\mathcal{L}_3^L$ from Examples 15.3.4, and is represented by the Hasse diagram from the same Figure 15.30, right side.

**Example 4: non-linearly ordered tOMWL as a direct product**

Consider the canonical Boolean algebra $\mathcal{L}_2^L = (L_2 = \{0,1\}, \odot, ^-, 1)$ from Examples 15.3.4, with the tables of $\odot$, $^-$ and $\oplus$ as follows:

| $\odot$ | 0 | 1 |
|---|---|---|
| 0 | 0 | 0 |
| 1 | 0 | 1 |

and

| $x$ | $x^-$ |
|---|---|
| 0 | 1 |
| 1 | 0 |

, with

| $\oplus$ | 0 | 1 |
|---|---|---|
| 0 | 0 | 1 |
| 1 | 1 | 1 |

.

Consider also the linearly ordered tOMWL $\mathcal{A}_4^L = (A_4' = \{0, a', b', 1\}, \odot, ^-, 1)$ from above Example 1, with the tables of $\odot$, $^-$ and $\oplus$ as follows:

| $\odot$ | 0 | a' | b' | 1 |
|---|---|---|---|---|
| 0 | 0 | 0 | 0 | 0 |
| a' | 0 | 0 | 0 | a' |
| b' | 0 | 0 | 0 | b' |
| 1 | 0 | a' | b' | 1 |

and

| $x$ | $x^-$ |
|---|---|
| 0 | 1 |
| a' | a' |
| b' | b' |
| 1 | 0 |

, with

| $\oplus$ | 0 | a' | b' | 1 |
|---|---|---|---|---|
| 0 | 0 | a' | b' | 1 |
| a' | a' | 1 | 1 | 1 |
| b' | b' | 1 | 1 | 1 |
| 1 | 1 | 1 | 1 | 1 |

.

Consider the direct product $A_{4' \times 2} \cong A_4' \times L_2 = \{0, a', b', 1\} \times \{0, 1\}$ $= \{(0,0), (0,1), (a',0), (a',1), (b',0), (b',1), (1,0), (1,1)\}$ that will be denoted, equivalently, with $A_8 = \{0, a, b, c, d, e, f, 1\}$, respectively. Then, the tables of $\odot$ and $^-$, calculated component-wise ($(x,a) \odot (y,b) \stackrel{def.}{=} (x \odot y, a \odot b)$ and $(x,a)^- \stackrel{def.}{=} (x^-, a^-)$ for $x, y \in A_4'$ and $a, b \in L_2$) and the table of $\oplus$ are the following:

| $\odot$ | 0 | a | b | c | d | e | f | 1 |
|---|---|---|---|---|---|---|---|---|
| 0 | 0 | 0 | 0 | 0 | 0 | 0 | 0 | 0 |
| a | 0 | a | 0 | a | 0 | a | 0 | a |
| b | 0 | 0 | 0 | 0 | 0 | 0 | b | b |
| c | 0 | a | 0 | a | 0 | a | b | c |
| d | 0 | 0 | 0 | 0 | 0 | 0 | d | d |
| e | 0 | a | 0 | a | 0 | a | d | e |
| f | 0 | 0 | b | b | d | d | f | f |
| 1 | 0 | a | b | c | d | e | f | 1 |

with

| $\oplus$ | 0 | a | b | c | d | e | f | 1 |
|---|---|---|---|---|---|---|---|---|
| 0 | 0 | a | b | c | d | e | f | 1 |
| a | a | a | c | c | e | e | 1 | 1 |
| b | b | c | f | 1 | f | 1 | f | 1 |
| c | c | c | 1 | 1 | 1 | 1 | 1 | 1 |
| d | d | e | f | 1 | f | 1 | f | 1 |
| e | e | e | 1 | 1 | 1 | 1 | 1 | 1 |
| f | f | 1 | f | 1 | f | 1 | f | 1 |
| 1 | 1 | 1 | 1 | 1 | 1 | 1 | 1 | 1 |

and $(0, a, b, c, d, e, f, 1)^- = (1, f, e, d, c, b, a, 0)$.

Then, the direct product algebra $\mathcal{A}_8^L = (A_8, \odot, ^-, 1)$ is an involutive m-BE algebra verifying (m-BB) $\Leftrightarrow \ldots \Leftrightarrow$ (m-Tr), (m-Pabs-i), (Pqmv) (hence (Pom), (Pmv), ($\Delta_m$)) and (prel$_m$), (WNM$_m$), (aWNM$_m$), and not verifying (m-An) for $(b, d)$, (m-Pimpl) for $(b, 0)$, (m-Pdis) for $(b, b, b)$, i.e. it is a **non-linearly ordered tOMWL** $=$ $_{(WNM_m)}$**tQMV**. Hence, (m-Pabs-i) $\Longleftrightarrow$ (aWNM$_m$) $\Longleftrightarrow$ (WNM$_m$).

The tables of $\wedge_m^M$ and $\wedge_m^B$ are:

| $\wedge_m^M$ | 0 | a | b | c | d | e | f | 1 |
|---|---|---|---|---|---|---|---|---|
| 0 | 0 | 0 | 0 | 0 | 0 | 0 | 0 | 0 |
| a | 0 | a | 0 | a | 0 | a | 0 | a |
| b | 0 | 0 | b | b | d | d | b | b |
| c | 0 | a | b | c | d | e | b | c |
| d | 0 | 0 | b | b | d | d | d | d |
| e | 0 | a | b | c | d | e | d | e |
| f | 0 | 0 | b | b | d | d | f | f |
| 1 | 0 | a | b | c | d | e | f | 1 |

and

| $\wedge_m^B$ | 0 | a | b | c | d | e | f | 1 |
|---|---|---|---|---|---|---|---|---|
| 0 | 0 | 0 | 0 | 0 | 0 | 0 | 0 | 0 |
| a | 0 | a | 0 | a | 0 | a | 0 | a |
| b | 0 | 0 | b | b | b | b | b | b |
| c | 0 | a | b | c | b | c | b | c |
| d | 0 | 0 | d | d | d | d | d | d |
| e | 0 | a | d | e | d | e | d | e |
| f | 0 | 0 | b | b | d | d | f | f |
| 1 | 0 | a | b | c | d | e | f | 1 |

.

• The binary relation $\leq_m^M$ is an order relation. From the table of $\wedge_m^M$, we obtain: $a \leq_m^M a, c, e, 1$; $b \leq_m^M b, c, f, 1$; $c \leq_m^M c, 1$; $d \leq_m^M d, e, f, 1$; $e \leq_m^M e, 1$; $f \leq_m^M f, 1$; hence, the bounded poset $(A_8, \leq_m^M, 0, 1)$ is represented by the Hasse diagram from Figure 15.31.

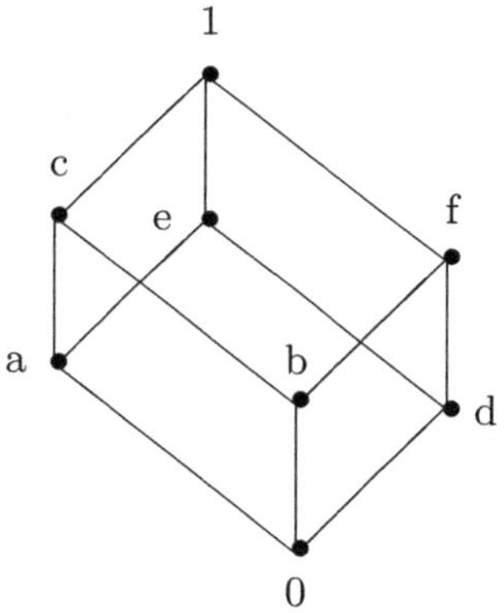


Figure 15.31: The Hasse diagram of the bounded poset $(A_8, \leq_m^M, 0, 1)$

• The binary relation $\leq_m$ ($\Longleftrightarrow \leq_m^B$) is a pre-order. From the table of $\wedge_m^B$, we obtain: $a \leq_m a, c, e, 1$; $b \leq_m b, c, d, e, f, 1$; $c \leq_m c, e, 1$; $d \leq_m d, b, c, e, f, 1$; $e \leq_m e, c, 1$; $f \leq_m f, 1$; hence, the bounded pre-ordered set $(A_8, \leq_m, 0, 1)$ is represented by the Hasse-type diagram from the Figure 15.32, left side.

Note that $b \parallel_Q d$ and $c \parallel_Q e$. Then, the quotient set is $A_8/_{\parallel_Q} = \{\{0\}, \{a\}, \{b, d\}, \{c, e\}, \{f\}, \{1\}\}$ and will be denoted by $A_6 = \{0, a, b, c, f, 1\}$ in the sequel.

Hence, the associated m-BCK algebra $(A_6, \odot, ^-, 1)$ has the following tables of $\odot$, $^-$, $\oplus$:

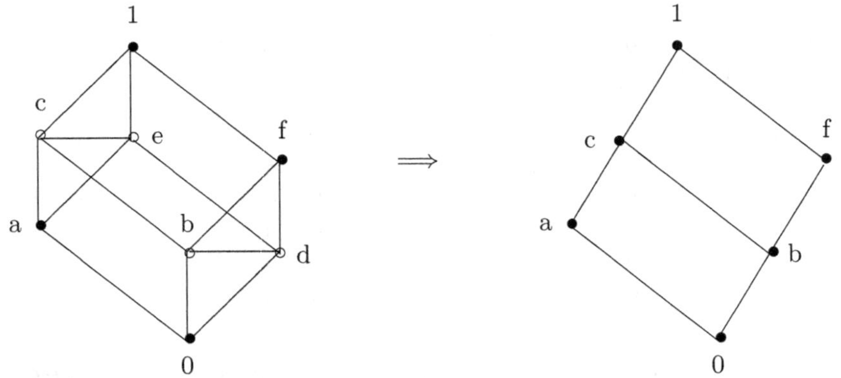


The bounded pre-ordered set $(A_8, \leq_m, 0, 1)$ The bounded lattice $(A_6, \leq_m, 0, 1)$

Figure 15.32: $(A_8, \leq_m, 0, 1)$ and $(A_6, \leq_m, 0, 1)$

| $\odot$ | 0 | a | b | c | f | 1 |
|---|---|---|---|---|---|---|
| 0 | 0 | 0 | 0 | 0 | 0 | 0 |
| a | 0 | a | 0 | a | 0 | a |
| b | 0 | 0 | 0 | 0 | b | b |
| c | 0 | a | 0 | a | b | c |
| f | 0 | 0 | b | b | f | f |
| 1 | 0 | a | b | c | f | 1 |

and

| $x$ | $x^-$ |
|---|---|
| 0 | 1 |
| a | f |
| b | c |
| c | d |
| f | a |
| 1 | 0 |

, with

| $\oplus$ | 0 | a | b | c | f | 1 |
|---|---|---|---|---|---|---|
| 0 | 0 | a | b | c | f | 1 |
| a | a | a | c | c | 1 | 1 |
| b | b | c | f | 1 | f | 1 |
| c | c | c | 1 | 1 | 1 | 1 |
| f | f | 1 | f | 1 | f | 1 |
| 1 | 1 | 1 | 1 | 1 | 1 | 1 |

.

Note that the associated m-BCK algebra $(A_6, \odot, ^-, 1)$ is the **non-linearly ordered atOWL =** $_{(WNM)}$**MV algebra** $\mathcal{L}^L_{3\times 2}$ from Examples 15.3.8, 2, represented in the same Figure 15.32, right side.

**Examples 15.3.15 Proper transitive OWLs: tOWL**

A tOWL is *proper*, if it is not a tOMWL or a tOL - see Figure 13.20.

**Example 1: tOWL verifying (aWNM$_m$) (hence (WNM$_m$)), and not verifying (Pmv)**

By MACE4 program, we found (see Examples 10.4.9, 1) the algebra $\mathcal{A}^L = (A_8 = \{0, a, b, c, d, e, f, 1\}, \odot, ^-, 1)$ with the following tables of $\odot$ and $^-$ and of the additional operation $\oplus$:

| $\odot$ | 0 | a | b | c | d | e | f | 1 |
|---|---|---|---|---|---|---|---|---|
| 0 | 0 | 0 | 0 | 0 | 0 | 0 | 0 | 0 |
| a | 0 | a | 0 | 0 | a | a | 0 | a |
| b | 0 | 0 | b | c | 0 | 0 | f | b |
| c | 0 | 0 | c | f | 0 | 0 | f | c |
| d | 0 | a | 0 | 0 | a | d | 0 | d |
| e | 0 | a | 0 | 0 | d | e | 0 | e |
| f | 0 | 0 | f | f | 0 | 0 | f | f |
| 1 | 0 | a | b | c | d | e | f | 1 |

with

| $\oplus$ | 0 | a | b | c | d | e | f | 1 |
|---|---|---|---|---|---|---|---|---|
| 0 | 0 | a | b | c | d | e | f | 1 |
| a | a | a | 1 | 1 | d | e | 1 | 1 |
| b | b | 1 | b | b | 1 | 1 | b | 1 |
| c | c | 1 | b | b | 1 | 1 | c | 1 |
| d | d | d | 1 | 1 | e | e | 1 | 1 |
| e | e | e | 1 | 1 | e | e | 1 | 1 |
| f | f | 1 | b | c | 1 | 1 | f | 1 |
| 1 | 1 | 1 | 1 | 1 | 1 | 1 | 1 | 1 |

and $(0, a, b, c, d, e, f, 1)^- = (1, b, a, d, c, f, e, 0)$.

$\mathcal{A}^L$ is an involutive left-m-BE algebra verifying (m-Pabs-i), ($\Delta_m$), (m-BB) ($\Leftrightarrow \ldots \Leftrightarrow$ (m-Tr)), and (prel$_m$), (aWNM$_m$), hence (WNM$_m$), and not verifying (m-An) for $(a, d)$, (m-Pimpl) for $(c, 0)$, (Pqmv) for $(b, 0, c)$, (Pom) for $(b, c)$, (Pmv) for $(b, c)$, (m-Pdis) for $(a, b, d)$. Hence, it is a **proper left-tOWL** (Definition 2).

The tables of $\wedge_m^M$ and $\wedge_m^B$ are:

| $\wedge_m^M$ | 0 | a | b | c | d | e | f | 1 |
|---|---|---|---|---|---|---|---|---|
| 0 | 0 | 0 | 0 | 0 | 0 | 0 | 0 | 0 |
| a | 0 | a | 0 | 0 | d | e | 0 | a |
| b | 0 | 0 | b | c | 0 | 0 | f | b |
| c | 0 | 0 | b | c | 0 | 0 | f | c |
| d | 0 | a | 0 | 0 | d | e | 0 | d |
| e | 0 | a | 0 | 0 | d | e | 0 | e |
| f | 0 | 0 | b | c | 0 | 0 | f | f |
| 1 | 0 | a | b | c | d | e | f | 1 |

and

| $\wedge_m^B$ | 0 | a | b | c | d | e | f | 1 |
|---|---|---|---|---|---|---|---|---|
| 0 | 0 | 0 | 0 | 0 | 0 | 0 | 0 | 0 |
| a | 0 | a | 0 | 0 | a | a | 0 | a |
| b | 0 | 0 | b | b | 0 | 0 | b | b |
| c | 0 | 0 | c | c | 0 | 0 | c | c |
| d | 0 | d | 0 | 0 | d | d | 0 | d |
| e | 0 | e | 0 | 0 | e | e | 0 | e |
| f | 0 | 0 | f | f | 0 | 0 | f | f |
| 1 | 0 | a | b | c | d | e | f | 1 |

.

• The binary relation $\leq_m^M$ is transitive, hence it is an order. From the table of $\wedge_m^M$, we obtain: $a \leq_m^M a, 1$; $b \leq_m^M b, 1$; $c \leq_m^M c, 1$; $d \leq_m^M d, 1$; $e \leq_m^M e, 1$; $f \leq_m^M f, 1$; hence, the bounded poset $(A_8, \leq_m^M, 0, 1)$ is represented by the Hasse diagram from Figure 15.33.

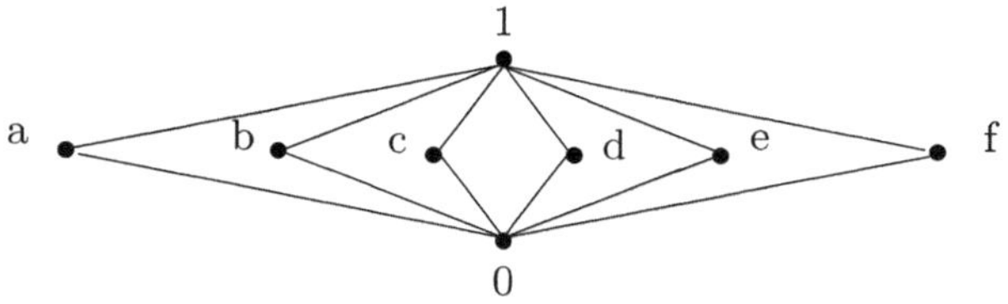


Figure 15.33: The bounded poset $(A_8, \leq_m^M, 0, 1)$

• The binary relation $\leq_m$ is a pre-order. From the table of $\wedge_m^B$, we obtain: $a \leq_m a, d, e, 1$; $b \leq_m b, c, f, 1$; $c \leq_m c, b, f, 1$; $d \leq_m d, a, e, 1$; $e \leq_m e, a, d, 1$; $f \leq_m f, b, c, 1$; hence, the bounded pre-ordered set $(A_8, \leq_m, 0, 1)$ is represented by the Hasse-type diagram from Figure 15.34, left side.

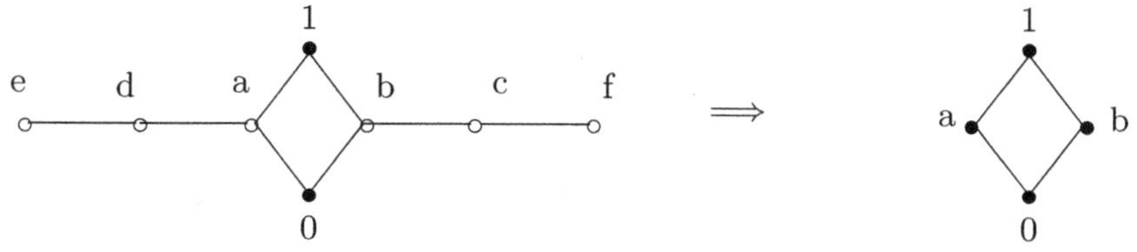


The bounded pre-ordered set $(A_8, \leq_m, 0, 1)$ The bounded lattice $(A_4, \leq_m, 0, 1)$

Figure 15.34: $(A_8, \leq_m, 0, 1)$ and $(A_4, \leq_m, 0, 1)$ of the tOWL

Note that $a \parallel_Q d \parallel_Q e$ and $b \parallel_Q c \parallel_Q f$. Then, the quotient set is

$A_8/_{\|_Q} = \{\{0\}, \{a, d, e\}, \{b, c, f\}, \{1\}\}$ and will be denoted by $A_4 = \{0, a, b, 1\}$ in the sequel.

Hence, the associated m-BCK algebra $(A_4, \odot, ^-, 1)$ has the following tables of $\odot$, $^-$, $\oplus$:

| $\odot$ | 0 | a | b | 1 |
|---|---|---|---|---|
| 0 | 0 | 0 | 0 | 0 |
| a | 0 | a | 0 | a |
| b | 0 | 0 | b | b |
| 1 | 0 | a | b | 1 |

and

| $x$ | $x^-$ |
|---|---|
| 0 | 1 |
| a | b |
| b | a |
| 1 | 0 |

, with

| $\oplus$ | 0 | a | b | 1 |
|---|---|---|---|---|
| 0 | 0 | a | b | 1 |
| a | a | a | 1 | 1 |
| b | b | 1 | b | 1 |
| 1 | 1 | 1 | 1 | 1 |

.

Note that the associated m-BCK algebra is the **non-linearly ordered Boolean algebra** $\mathcal{L}^L_{2\times 2}$ from Examples 15.3.8, 1, and its bounded lattice $(A_4, \leq_m, 0, 1)$ is represented by the Hasse diagram from the same Figure 15.34, right side.

**Example 2: tOWL verifying (WNM$_m$) and not verifying (aWNM$_m$), (Pmv)**

By MACE4, we found (see Examples 10.4.9, 2) the algebra $\mathcal{A}^L = (A_{14} = \{0, a, b, c, d, e, f, g, h, i, j, k, m, 1\}, \odot, ^-, 1)$ with the following tables of $\odot$ and $^-$:

| $\odot$ | 0 | a | b | c | d | e | f | g | h | i | j | k | m | 1 |
|---|---|---|---|---|---|---|---|---|---|---|---|---|---|---|
| 0 | 0 | 0 | 0 | 0 | 0 | 0 | 0 | 0 | 0 | 0 | 0 | 0 | 0 | 0 |
| a | 0 | c | 0 | c | b | g | 0 | c | k | g | f | 0 | c | a |
| b | 0 | 0 | 0 | 0 | b | 0 | 0 | 0 | k | 0 | f | 0 | 0 | b |
| c | 0 | c | 0 | c | 0 | c | 0 | c | 0 | c | 0 | 0 | c | c |
| d | 0 | b | b | 0 | d | b | f | 0 | h | 0 | j | k | b | d |
| e | 0 | g | 0 | c | b | i | 0 | g | k | i | f | 0 | c | e |
| f | 0 | 0 | 0 | 0 | f | 0 | 0 | 0 | f | 0 | f | 0 | 0 | f |
| g | 0 | c | 0 | c | 0 | g | 0 | c | 0 | g | 0 | 0 | c | g |
| h | 0 | k | k | 0 | h | k | f | 0 | j | 0 | j | f | k | h |
| i | 0 | g | 0 | c | 0 | i | 0 | g | 0 | i | 0 | 0 | c | i |
| j | 0 | f | f | 0 | j | f | f | 0 | j | 0 | j | f | f | j |
| k | 0 | 0 | 0 | 0 | k | 0 | 0 | 0 | f | 0 | f | 0 | 0 | k |
| m | 0 | c | 0 | c | b | c | 0 | c | k | c | f | 0 | c | m |
| 1 | 0 | a | b | c | d | e | f | g | h | i | j | k | m | 1 |

and

| $x$ | $x^-$ |
|---|---|
| 0 | 1 |
| a | b |
| b | a |
| c | d |
| d | c |
| e | f |
| f | e |
| g | h |
| h | g |
| i | j |
| j | i |
| k | m |
| m | k |
| 1 | 0 |

.

$\mathcal{A}^L$ is an involutive left-m-BE algebra verifying (m-Pabs-i), ($\Delta_m$) and (m-BB) ($\Leftrightarrow \ldots \Leftrightarrow$ (m-Tr)), and (prel$_m$), (WNM$_m$), and not verifying (m-An) for $(a, e)$, (m-Pimpl) for $(a, 0)$, (Pqmv) for $(a, 0, e)$, (Pom) for $(a, e)$, (Pmv) for $(b, h)$, (m-Pdis) for $(a, a, a)$, (aWNM$_m$) for $(a, e)$. Hence, it is a **proper left-tOWL** (Definition 2).

The table of $\wedge^M_m$ is:

| $\wedge_m^M$ | 0 | a | b | c | d | e | f | g | h | i | j | k | m | 1 |
|---|---|---|---|---|---|---|---|---|---|---|---|---|---|---|
| 0 | 0 | 0 | 0 | 0 | 0 | 0 | 0 | 0 | 0 | 0 | 0 | 0 | 0 | 0 |
| a | 0 | a | b | c | b | e | f | g | k | i | f | k | m | a |
| b | 0 | b | b | 0 | b | k | f | 0 | k | 0 | f | k | b | b |
| c | 0 | c | 0 | c | 0 | g | 0 | g | 0 | i | 0 | 0 | c | c |
| d | 0 | b | b | 0 | d | b | f | 0 | h | 0 | j | k | b | d |
| e | 0 | a | b | c | b | e | f | g | k | i | f | k | m | e |
| f | 0 | k | b | 0 | b | f | f | 0 | k | 0 | f | k | b | f |
| g | 0 | c | 0 | c | 0 | c | 0 | g | 0 | i | 0 | 0 | c | g |
| h | 0 | b | b | 0 | d | k | f | 0 | h | 0 | j | k | b | h |
| i | 0 | g | 0 | c | 0 | i | 0 | g | 0 | i | 0 | 0 | c | i |
| j | 0 | k | b | 0 | d | f | f | 0 | h | 0 | j | k | b | j |
| k | 0 | b | b | 0 | b | b | f | 0 | k | 0 | f | k | b | k |
| m | 0 | a | b | c | b | e | f | g | k | i | f | k | m | m |
| 1 | 0 | a | b | c | d | e | f | g | h | i | j | k | m | 1 |

.

• The binary relation $\leq_m^M$ is transitive, hence it is an order, but not a lattice order, since $\wedge_m^M$ is not commutative. From the table of $\wedge_m^M$, we obtain: $a \leq_m^M a, 1$; $b \leq_m^M b, a, d, m, 1$; $c \leq_m^M c, a, m, 1$; $d \leq_m^M d, 1$; $e \leq_m^M e, 1$; $f \leq_m^M f, e, j, 1$; $g \leq_m^M g, 1$; $h \leq_m^M h, 1$; $i \leq_m^M i, e, 1$, $j \leq_m^M j, 1$; $k \leq_m^M k, h, 1$; $m \leq_m^M m, 1$; we omit the representation of the bounded poset $(A_{14}, \leq_m^M, 0, 1)$.

The table of $\wedge_m^B$ is:

| $\wedge_m^B$ | 0 | a | b | c | d | e | f | g | h | i | j | k | m | 1 |
|---|---|---|---|---|---|---|---|---|---|---|---|---|---|---|
| 0 | 0 | 0 | 0 | 0 | 0 | 0 | 0 | 0 | 0 | 0 | 0 | 0 | 0 | 0 |
| a | 0 | a | b | c | b | a | k | c | b | g | k | b | a | a |
| b | 0 | b | b | 0 | b | b | b | 0 | b | 0 | b | b | b | b |
| c | 0 | c | 0 | c | 0 | c | 0 | c | 0 | c | 0 | 0 | c | c |
| d | 0 | b | b | 0 | d | b | b | 0 | d | 0 | d | b | b | d |
| e | 0 | e | k | g | b | e | f | c | k | i | f | b | e | e |
| f | 0 | f | f | 0 | f | f | f | 0 | f | 0 | f | f | f | f |
| g | 0 | g | 0 | g | 0 | g | 0 | g | 0 | g | 0 | 0 | g | g |
| h | 0 | k | k | 0 | h | k | k | 0 | h | 0 | h | k | k | h |
| i | 0 | i | 0 | i | 0 | i | 0 | i | 0 | i | 0 | 0 | i | i |
| j | 0 | f | f | 0 | j | f | f | 0 | j | 0 | j | f | f | j |
| k | 0 | k | k | 0 | k | k | k | 0 | k | 0 | k | k | k | k |
| m | 0 | m | b | c | b | m | b | c | b | c | b | b | m | m |
| 1 | 0 | a | b | c | d | e | f | g | h | i | j | k | m | 1 |

.

• The binary relation $\leq_m$ is a pre-order. From the table of $\wedge_m^B$, we obtain: $a \leq_m a, e, m, 1$; $b \leq_m b, a, d, e, f, h, j, k, m, 1$; $c \leq_m c, a, e, g, i, m, 1$; $d \leq_m d, h, j, 1$; $e \leq_m e, a, m, 1$; $f \leq_m f, a, b, d, e, h, j, k, m, 1$; $g \leq_m g, a, c, e, i, m, 1$; $h \leq_m h, d, j, 1$; $i \leq_m i, a, c, e, g, m, 1$; $j \leq_m j, d, h, 1$; $k \leq_m k, a, b, d, e, f, h, j, m, 1$; $m \leq_m m, a, e, 1$; hence, the bounded pre-ordered set $(A_{14}, \leq_m, 0, 1)$ is represented by the Hasse-type diagram from Figure 15.35, left side.

Note that $a \parallel_Q e \parallel_Q m$; $b \parallel_Q f \parallel_Q k$; $c \parallel_Q g \parallel_Q i$; $d \parallel_Q h \parallel_Q j$; hence, the

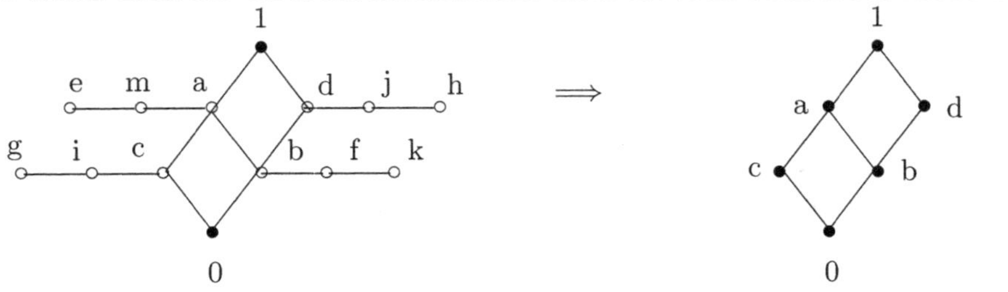


The bounded pre-ordered set $(A_{14}, \leq_m, 0, 1)$ The bounded lattice $(A_6, \leq_m, 0, 1)$

Figure 15.35: $(A_{14}, \leq_m, 0, 1)$ and $(A_6, \leq_m, 0, 1)$

quotient set is $A_{14}/_{\|_Q} = \{\{0\}, \{a, e, m\}, \{b, f, k\}, \{c, g, i\}, \{d, h, j\}, \{1\}\}$ and will be denoted by $A_6 = \{0, a, b, c, d, 1\}$ in the sequel. Hence, the associated m-BCK algebra $(A_6, \odot, ^-, 1)$ has the following tables of $\odot$, $^-$ and $\oplus$:

| $\odot$ | 0 | a | b | c | d | 1 |
|---|---|---|---|---|---|---|
| 0 | 0 | 0 | 0 | 0 | 0 | 0 |
| a | 0 | c | 0 | c | b | a |
| b | 0 | 0 | 0 | 0 | b | b |
| c | 0 | c | 0 | c | 0 | c |
| d | 0 | b | b | 0 | d | d |
| 1 | 0 | a | b | c | d | 1 |

and

| $x$ | $x^-$ |
|---|---|
| 0 | 1 |
| a | b |
| b | a |
| c | d |
| d | c |
| 1 | 0 |

, with

| $\oplus$ | 0 | a | b | c | d | 1 |
|---|---|---|---|---|---|---|
| 0 | 0 | a | b | c | d | 1 |
| a | a | 1 | 1 | a | 1 | 1 |
| b | b | 1 | d | a | d | 1 |
| c | c | a | a | c | 1 | 1 |
| d | d | 1 | d | 1 | d | 1 |
| 1 | 1 | 1 | 1 | 1 | 1 | 1 |

.

Note that this associated m-BCK algebra $(A_6, \odot, ^-, 1)$ is isomorphic with the **non-linearly ordered** $_{(WNM)}$**MV algebra** $\mathcal{L}^L_{3\times 2}$ ($a \mapsto c$, $c \mapsto a$) from Examples 15.3.8, 2, and is represented by the Hasse diagram from the same Figure 15.35, right side.

**Example 3: tOWL verifying (aWNM$_m$) (hence (WNM$_m$)), and not verifying (Pmv)**

By MACE4, we found (see Example 13.3.22) the algebra $\mathcal{A}^L = (A_8 = \{0, a, b, c, d, e, f, 1\}, \odot, ^-, 1)$ with the following tables of $\odot$ and $^-$ and of the additional operation $\oplus$:

| $\odot$ | 0 | a | b | c | d | e | f | 1 |
|---|---|---|---|---|---|---|---|---|
| 0 | 0 | 0 | 0 | 0 | 0 | 0 | 0 | 0 |
| a | 0 | d | 0 | b | d | e | b | a |
| b | 0 | 0 | 0 | b | 0 | 0 | b | b |
| c | 0 | b | b | c | 0 | 0 | c | c |
| d | 0 | d | 0 | 0 | d | e | 0 | d |
| e | 0 | e | 0 | 0 | e | e | 0 | e |
| f | 0 | b | b | c | 0 | 0 | f | f |
| 1 | 0 | a | b | c | d | e | f | 1 |

with

| $\oplus$ | 0 | a | b | c | d | e | f | 1 |
|---|---|---|---|---|---|---|---|---|
| 0 | 0 | a | b | c | d | e | f | 1 |
| a | a | 1 | 1 | 1 | a | a | 1 | 1 |
| b | b | 1 | c | c | a | a | f | 1 |
| c | c | 1 | c | c | 1 | 1 | f | 1 |
| d | d | a | a | 1 | d | d | 1 | 1 |
| e | e | a | a | 1 | d | e | 1 | 1 |
| f | f | 1 | f | f | 1 | 1 | f | 1 |
| 1 | 1 | 1 | 1 | 1 | 1 | 1 | 1 | 1 |

and $(0, a, b, c, d, e, f, 1)^- = (1, b, a, d, c, f, e, 0)$.

$\mathcal{A}^L$ is an involutive left-m-BE algebra verifying $(\Delta_m)$, (m-Pabs-i), (m-BB) ($\Leftrightarrow$

... ⇔ (m-Tr)), and ($\text{prel}_m$), ($\text{aWNM}_m$), hence ($\text{WNM}_m$), and not verifying (m-An) for $(c, f)$, (m-Pimpl) for $(a, 0)$, (Pqmv) for $(a, 0, e)$, (Pom) for $(d, e)$, (Pmv) for $(a, e)$, (m-Pdis) for (a,a,a). Hence, it is a **proper left-tOWL**.

The tables of $\wedge_m^M$ and $\wedge_m^B$ are:

| $\wedge_m^M$ | 0 | a | b | c | d | e | f | 1 |
|---|---|---|---|---|---|---|---|---|
| 0 | 0 | 0 | 0 | 0 | 0 | 0 | 0 | 0 |
| a | 0 | a | b | b | d | e | b | a |
| b | 0 | b | b | b | 0 | 0 | b | b |
| c | 0 | b | b | c | 0 | 0 | f | c |
| d | 0 | d | 0 | 0 | d | e | 0 | d |
| e | 0 | d | 0 | 0 | d | e | 0 | e |
| f | 0 | b | b | c | 0 | 0 | f | f |
| 1 | 0 | a | b | c | d | e | f | 1 |

and

| $\wedge_m^B$ | 0 | a | b | c | d | e | f | 1 |
|---|---|---|---|---|---|---|---|---|
| 0 | 0 | 0 | 0 | 0 | 0 | 0 | 0 | 0 |
| a | 0 | a | b | b | d | d | b | a |
| b | 0 | b | b | b | 0 | 0 | b | b |
| c | 0 | b | b | c | 0 | 0 | c | c |
| d | 0 | d | 0 | 0 | d | d | 0 | d |
| e | 0 | e | 0 | 0 | e | e | 0 | e |
| f | 0 | b | b | f | 0 | 0 | f | f |
| 1 | 0 | a | b | c | d | e | f | 1 |

.

• The binary relation $\leq_m^M$ is transitive, hence is an order, but not a lattice order, since $\wedge_m^M$ is not commutative.

• The binary relation $\leq_m$ is a pre-order, since (m-Re) and (m-Tr) hold. From the table of $\wedge_m^B$, we obtain that: $a \leq_m a, 1$; $b \leq_m b, a, c, f, 1$; $c \leq_m c, f, 1$; $d \leq_m d, a, e, 1$; $e \leq_m e, a, d, 1$; $f \leq_m f, c, 1$; hence, the bounded pre-ordered set $(A_8, \leq_m, 0, 1)$ is represented by the Hasse-type diagram from Figure 15.36, left side.

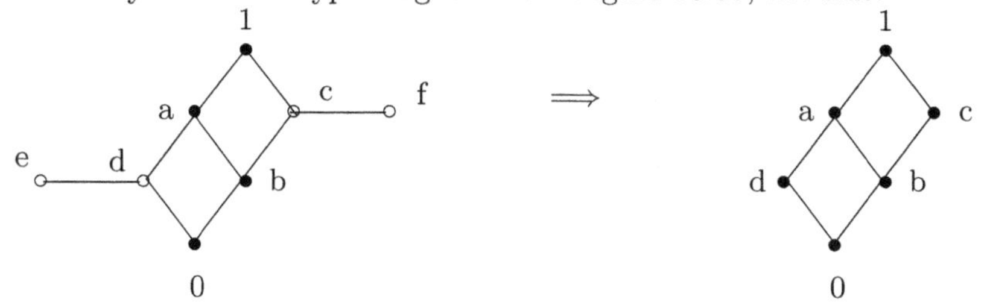


The bounded pre-ordered set $(A_8, \leq_m, 0, 1)$ The bounded lattice $(A_6, \leq_m, 0, 1)$

Figure 15.36: $(A_8, \leq_m, 0, 1)$ and $(A_6, \leq_m, 0, 1)$

Note that $c \parallel_Q f$ and $d \parallel_Q e$; hence, the quotient set is $A_8/_{\parallel_Q} = \{\{0\}, \{d, e\}, \{b\}, \{a\}, \{c, f\}, \{1\}\}$ and will be denoted by $A_6 = \{0, d, b, a, c, 1\}$ in the sequel. Hence, the associated m-BCK algebra $(A_6, \odot, ^-, 1)$ has the following tables of $\odot$, $^-$ and $\oplus$:

| $\odot$ | 0 | a | b | c | d | 1 |
|---|---|---|---|---|---|---|
| 0 | 0 | 0 | 0 | 0 | 0 | 0 |
| a | 0 | d | 0 | b | d | a |
| b | 0 | 0 | 0 | b | 0 | b |
| c | 0 | b | b | c | 0 | c |
| d | 0 | d | 0 | 0 | d | d |
| 1 | 0 | a | b | c | d | 1 |

and

| $x$ | $x^-$ |
|---|---|
| 0 | 1 |
| a | b |
| b | a |
| c | d |
| d | c |
| 1 | 0 |

, with

| $\oplus$ | 0 | a | b | c | d | 1 |
|---|---|---|---|---|---|---|
| 0 | 0 | a | b | c | d | 1 |
| a | a | 1 | 1 | 1 | a | 1 |
| b | b | 1 | c | c | a | 1 |
| c | c | 1 | c | c | 1 | 1 |
| d | d | a | a | 1 | d | 1 |
| 1 | 1 | 1 | 1 | 1 | 1 | 1 |

.

Note that this associated m-BCK algebra $(A_6, \odot, ^-, 1)$ is isomorphic with the

**non-linearly ordered** $_{(WNM)}$**MV algebra** $\mathcal{L}^L_{3\times 2}$ from Examples 15.3.8, 2, and is represented by the Hasse diagram from the same Figure 15.36, right side.

**Example 4: tOWL verifying (aWNM$_m$) (hence (WNM$_m$)) and (Pmv)**

Consider the algebra $\mathcal{A}^L = (A_{10} = \{0, a, b, c, d, e, f, g, h, 1\}, \odot, ^-, 1)$ from Example 13.3.23.

$\mathcal{A}^L$ is an involutive left-m-BE algebra verifying (m-Pabs-i), (m-BB) ($\Leftrightarrow \ldots \Leftrightarrow$ (m-Tr)), (Pmv) (hence ($\Delta_m$)), and (prel$_m$), (aWNM$_m$), (WNM$_m$), and not verifying (m-An) for $(a, h)$, (m-Pimpl) for $(a, 0)$, (Pqmv) for $(a, b, e)$, (Pom) for $(a, b)$, (m-Pdis) for (a,a,a). Hence, it is a **proper left-tOWL verifying (Pmv)**.

The tables of $\wedge^M_m$ and $\wedge^B_m$ are:

| $\wedge^M_m$ | 0 | a | b | c | d | e | f | g | h | 1 |
|---|---|---|---|---|---|---|---|---|---|---|
| 0 | 0 | 0 | 0 | 0 | 0 | 0 | 0 | 0 | 0 | 0 |
| a | 0 | a | h | c | d | h | c | d | h | a |
| b | 0 | a | b | c | d | h | f | d | h | b |
| c | 0 | c | c | c | 0 | c | c | 0 | c | c |
| d | 0 | d | d | 0 | d | d | 0 | d | d | d |
| e | 0 | a | h | c | d | e | c | g | h | e |
| f | 0 | c | f | c | 0 | c | f | 0 | c | f |
| g | 0 | d | d | 0 | d | g | 0 | g | d | g |
| h | 0 | a | h | c | d | h | c | d | h | h |
| 1 | 0 | a | b | c | d | e | f | g | h | 1 |

and

| $\wedge^B_m$ | 0 | a | b | c | d | e | f | g | h | 1 |
|---|---|---|---|---|---|---|---|---|---|---|
| 0 | 0 | 0 | 0 | 0 | 0 | 0 | 0 | 0 | 0 | 0 |
| a | 0 | a | a | c | d | a | c | d | a | a |
| b | 0 | h | b | c | d | h | f | d | h | b |
| c | 0 | c | c | c | 0 | c | c | 0 | c | c |
| d | 0 | d | d | 0 | d | d | 0 | d | d | d |
| e | 0 | h | h | c | d | e | c | g | h | e |
| f | 0 | c | f | c | 0 | c | f | 0 | c | f |
| g | 0 | d | d | 0 | d | g | 0 | g | d | g |
| h | 0 | h | h | c | d | h | c | d | h | h |
| 1 | 0 | a | b | c | d | e | f | g | h | 1 |

.

• The binary relation $\leq^M_m$ is transitive, hence is an order, but not a lattice order, since $\wedge^M_m$ is not commutative.

• The binary relation $\leq_m$ is a pre-order, since (m-Re) and (m-Tr) hold. From the table of $\wedge^B_m$, we obtain that: $a \leq_m a, b, e, h, 1$; $b \leq_m b, 1$; $c \leq_m c, a, b, e, f, h, 1$; $d \leq_m d, a, b, e, g, h, 1$; $e \leq_m e, 1$; $f \leq_m f, b, 1$; $g \leq_m g, e, 1$; $h \leq_m h, a, b, e, 1$; hence, the bounded pre-ordered set $(A_{10}, \leq_m, 0, 1)$ is represented by the Hasse-type diagram from Figure 15.37, left side.

Note that $a \parallel_Q h$, hence, the quotient set is
$A_{10}/_{\parallel_Q} = \{\{0\}, \{a, h\}, \{b\}, \{c\}, \{d\}, \{e\}, \{f\}, \{g\}, \{1\}\}$ and will be denoted by $A_9 = \{0, a, b, c, d, e, f, g, 1\}$ in the sequel. Hence, the associated m-BCK algebra $(A_9, \odot, ^-, 1)$ has the following tables of $\odot$, $^-$ and $\oplus$:

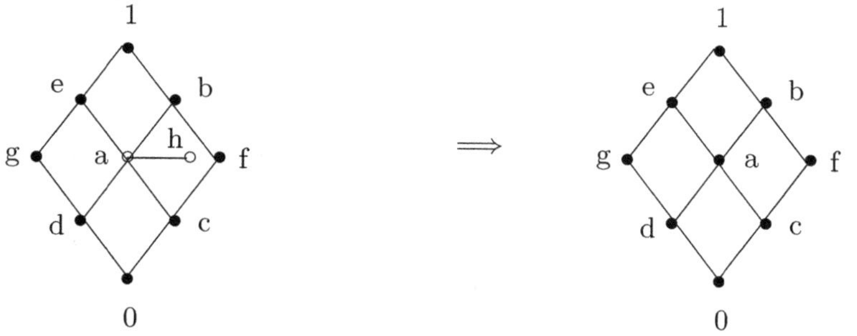


The bounded pre-ordered set $(A_{10}, \leq_m, 0, 1)$ The bounded lattice $(A_9, \leq_m, 0, 1)$

Figure 15.37: $(A_{10}, \leq_m, 0, 1)$ and $(A_9, \leq_m, 0, 1)$

| $\odot$ | 0 | a | b | c | d | e | f | g | 1 |
|---|---|---|---|---|---|---|---|---|---|
| 0 | 0 | 0 | 0 | 0 | 0 | 0 | 0 | 0 | 0 |
| a | 0 | 0 | c | 0 | 0 | d | c | d | a |
| b | 0 | c | f | c | 0 | a | f | d | b |
| c | 0 | 0 | c | 0 | 0 | 0 | c | 0 | c |
| d | 0 | 0 | 0 | 0 | 0 | d | 0 | d | d |
| e | 0 | d | a | 0 | d | g | c | g | e |
| f | 0 | c | f | c | 0 | c | f | 0 | f |
| g | 0 | d | d | 0 | d | g | 0 | g | g |
| 1 | 0 | a | b | c | d | e | f | g | 1 |

,

| $\oplus$ | 0 | a | b | c | d | e | f | g | 1 |
|---|---|---|---|---|---|---|---|---|---|
| 0 | 0 | a | b | c | d | e | f | g | 1 |
| a | a | 1 | 1 | b | e | 1 | b | e | 1 |
| b | b | 1 | 1 | b | 1 | 1 | b | 1 | 1 |
| c | c | b | b | f | a | 1 | f | e | 1 |
| d | d | e | 1 | a | g | e | b | g | 1 |
| e | e | 1 | 1 | 1 | e | 1 | 1 | e | 1 |
| f | f | b | b | f | b | 1 | f | 1 | 1 |
| g | g | e | 1 | e | g | e | 1 | g | 1 |
| 1 | 1 | 1 | 1 | 1 | 1 | 1 | 1 | 1 | 1 |

and $(0, a, b, c, d, e, f, g, 1)^- = (1, a, d, e, b, c, g, f, 0)$.

Note that this associated m-BCK algebra $(A_9, \odot, ^-, 1)$ is isomorphic (interchange $a$ with $d$ and $b$ with $g$) with the **non-linearly ordered** $_{(WNM)}$**MV algebra** $\mathcal{L}^L_{3\times 3}$ from Examples 15.3.8, 3, and is represented by the Hasse diagram from the same Figure 15.37, right side.

**Remark.** *These proper tOWLs are particular cases of tQNM algebras.*

**Example 15.3.16 Transitive ortholattice: tOL**

By MACE4, we found (see Example 9.3.17) that the algebra $\mathcal{A}^L = (A_6 = \{0, a, b, c, d, 1\}, \odot, ^-, 1)$, with the following tables of $\odot$ and $^-$ and of the additional operation $\oplus$, is an involutive left-m-BE algebra verifying (m-Pimpl) (hence (G) and (m-Pabs-i)) and (m-BB) ($\Leftrightarrow \ldots \Leftrightarrow$ (m-Tr)) and ($\Delta_m$), and (prel$_m$), (WNM$_m$), (aWNM$_m$), and not verifying (m-An) for $(a, c)$, (Pqmv) for $(b, 0, d)$, (Pom) for $(b, d)$, (Pmv) for $(b, d)$, (m-Pdis) for $(a, b, c)$. Hence, $\mathcal{A}^L$ is a **transitive left-ortholattice** (Definition 2) - see Figure 13.13.

| $\odot$ | 0 | a | b | c | d | 1 |
|---|---|---|---|---|---|---|
| 0 | 0 | 0 | 0 | 0 | 0 | 0 |
| a | 0 | a | 0 | a | 0 | a |
| b | 0 | 0 | b | 0 | d | b |
| c | 0 | a | 0 | c | 0 | c |
| d | 0 | 0 | d | 0 | d | d |
| 1 | 0 | a | b | c | d | 1 |

and

| $x$ | $x^-$ |
|---|---|
| 0 | 1 |
| a | b |
| b | a |
| c | d |
| d | c |
| 1 | 0 |

, with

| $\oplus$ | 0 | a | b | c | d | 1 |
|---|---|---|---|---|---|---|
| 0 | 0 | a | b | c | d | 1 |
| a | a | a | 1 | c | 1 | 1 |
| b | b | 1 | b | 1 | b | 1 |
| c | c | c | 1 | c | 1 | 1 |
| d | d | 1 | b | 1 | d | 1 |
| 1 | 1 | 1 | 1 | 1 | 1 | 1 |

.

• Note that the binary relation $\leq_m^P$ is a lattice order w.r. to $\wedge = \odot$ and $\vee = \oplus$, since (G) holds. From the table of $\odot$, we obtain: $a \leq_m^P a, c, 1$; $b \leq_m^P b, 1$; $c \leq_m^P c, 1$; $d \leq_m^P d, b, 1$; hence, the bounded lattice $(A_6, \leq_m^P, 0, 1)$ is represented by the Hasse diagram from Figure 15.38.

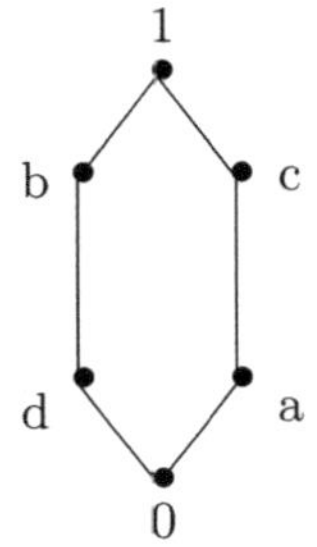


Figure 15.38: The bounded lattice $(A_6, \leq_m^P, 0, 1)$

The tables of $\wedge_m^M$ and $\wedge_m^B$ are:

| $\wedge_m^M$ | 0 | a | b | c | d | 1 |
|---|---|---|---|---|---|---|
| 0 | 0 | 0 | 0 | 0 | 0 | 0 |
| a | 0 | a | 0 | c | 0 | a |
| b | 0 | 0 | b | 0 | d | b |
| c | 0 | a | 0 | c | 0 | c |
| d | 0 | 0 | b | 0 | d | d |
| 1 | 0 | a | b | c | d | 1 |

and

| $\wedge_m^B$ | 0 | a | b | c | d | 1 |
|---|---|---|---|---|---|---|
| 0 | 0 | 0 | 0 | 0 | 0 | 0 |
| a | 0 | a | 0 | a | 0 | a |
| b | 0 | 0 | b | 0 | b | b |
| c | 0 | c | 0 | c | 0 | c |
| d | 0 | 0 | d | 0 | d | d |
| 1 | 0 | a | b | c | d | 1 |

.

• Note that the binary relation $\leq_m^M$ is transitive, hence it is an order, but not a lattice order w.r.to $\wedge_m^M$, $\vee_m^M$. From the table of $\wedge_m^M$ we obtain: $a \leq_m^M a, 1$; $b \leq_m^M b, 1$; $c \leq_m^M c, 1$; $d \leq_m^M d, 1$; hence, the bounded poset $(A_6, \leq_m^M, 0, 1)$ is represented by the Hasse diagram from Figure 15.39.

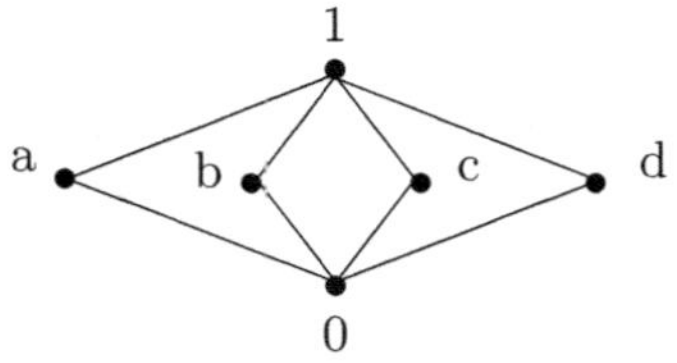


Figure 15.39: The bounded poset $(A_6, \leq_m^M, 0, 1)$

• The binary relation $\leq_m$ is a pre-order. From the table of $\wedge_m^B$, we obtain:

$a \leq_m a, c, 1$; $b \leq_m b, d, 1$; $c \leq_m c, a, 1$; $d \leq_m d, b, 1$; hence, the bounded partially pre-ordered set $(A_6, \leq_m, 0, 1)$ is represented by the Hasse-type diagram from Figure 15.40, left side.

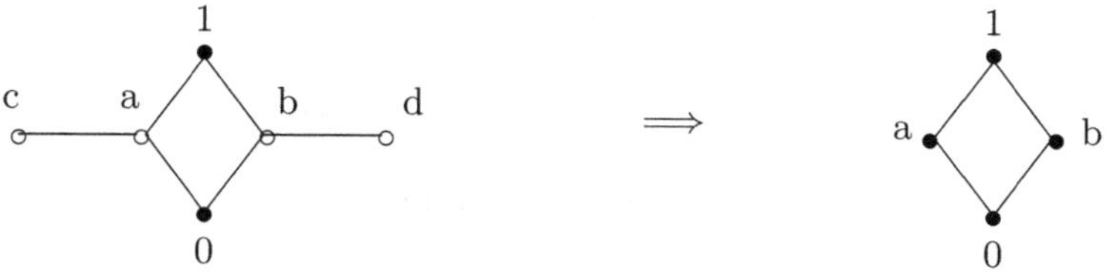


The bounded pre-ordered set $(A_6, \leq_m, 0, 1)$ The bounded lattice $(A_4, \leq_m, 0, 1)$

Figure 15.40: $(A_6, \leq_m, 0, 1)$ and $(A_4, \leq_m, 0, 1)$ of the tOL

Note that $a \parallel_Q c$ and $b \parallel_Q d$. Then, the quotient set is $A_6/_{\parallel_Q} = \{\{0\}, \{a, c\}, \{b, d\}, \{1\}\}$ and will be denoted by $A_4 = \{0, a, b, 1\}$ in the sequel. Hence, the associated m-BCK algebra $(A_4, \odot, ^-, 1)$ has the following tables of $\odot$, $^-$, $\oplus$:

| $\odot$ | 0 | a | b | 1 |
|---|---|---|---|---|
| 0 | 0 | 0 | 0 | 0 |
| a | 0 | a | 0 | a |
| b | 0 | 0 | b | b |
| 1 | 0 | a | b | 1 |

and

| $x$ | $x^-$ |
|---|---|
| 0 | 1 |
| a | b |
| b | a |
| 1 | 0 |

, with

| $\oplus$ | 0 | a | b | 1 |
|---|---|---|---|---|
| 0 | 0 | a | b | 1 |
| a | a | a | 1 | 1 |
| b | b | 1 | b | 1 |
| 1 | 1 | 1 | 1 | 1 |

.

Note that the associated m-BCK algebra is the **Boolean algebra** $\mathcal{L}^L_{2\times 2}$ from Examples 15.3.8, 1, and its bounded lattice $(A_4, \leq_m, 0, 1)$ is represented by the Hasse diagram from the same Figure 15.40, right side.

**Remark.** *The tOLs are particular cases of tQNM algebras.*

**Example 15.3.17 Proper transitive OSL: tOSL**

A tOSL is *proper*, if it is not a tOL - see Figure 13.13.

The algebra $\mathcal{A}^L = (A_6 = \{0, a, b, c, d, 1\}, \odot, ^-, 1)$ from Example 10.4.6 is an involutive left-m-BE algebra verifying (m-BB) ($\Leftrightarrow \ldots \Leftrightarrow$ (m-Tr)), (G), ($\Delta_m$), and (prel$_m$), (aWNM$_m$), hence (WNM$_m$), and not verifying (m-An) for $(a, c)$, (m-Pimpl) for $(b, a)$, (m-Pabs-i) for $(b, d)$, (Pqmv) for $(b, 0, d)$, (Pom) for $(b, d)$, (Pmv) for $(b, d)$, (m-Pdis) for $(a, b, c)$. Hence, it is a **proper transitive left-orthosoftlattice** (Definition 2).

• The binary relation $\leq^P_m$ is reflexive, because (G) holds, hence is an order, but not a lattice order w.r.t. to $\wedge = \odot$, $\vee = \oplus$, since the absorbtion laws do not hold.

The tables of $\wedge^M_m$ and $\wedge^B_m$ are:

| $\wedge_m^M$ | 0 | a | b | c | d | 1 |
|---|---|---|---|---|---|---|
| 0 | 0 | 0 | 0 | 0 | 0 | 0 |
| a | 0 | a | 0 | c | 0 | a |
| b | 0 | 0 | b | 0 | d | b |
| c | 0 | a | 0 | c | 0 | c |
| d | 0 | 0 | b | 0 | d | d |
| 1 | 0 | a | b | c | d | 1 |

and

| $\wedge_m^B$ | 0 | a | b | c | d | 1 |
|---|---|---|---|---|---|---|
| 0 | 0 | 0 | 0 | 0 | 0 | 0 |
| a | 0 | a | 0 | a | 0 | a |
| b | 0 | 0 | b | 0 | b | b |
| c | 0 | c | 0 | c | 0 | c |
| d | 0 | 0 | d | 0 | d | d |
| 1 | 0 | a | b | c | d | 1 |

.

• The binary relation $\leq_m^M$ is transitive, hence is an order, but not a lattice order since $\wedge_m^M$ is not commutative.

• The binary relation $\leq_m$ is a pre-order. From the table of $\wedge_m^B$, we obtain: $a \leq_m a, c, 1$; $b \leq_m b, d, 1$; $c \leq_m c, a, 1$; $d \leq_m d, b, 1$; hence, the bounded partially pre-ordered set $(A_6, \leq_m, 0, 1)$ is represented by the same Hasse-type diagram from the above Figure 15.40, left side.

Note that $a \parallel_Q c$ and $b \parallel_Q d$. Then, the quotient set is $A_6/_{\parallel_Q} = \{\{0\}, \{a, c\}, \{b, d\}, \{1\}\}$ and will be denoted by $A_4 = \{0, a, b, 1\}$ in the sequel. Hence, the associated m-BCK algebra $(A_4, \odot, ^-, 1)$ has the same tables of $\odot$, $^-$, $\oplus$ as in the Example 9.3.17. Consequently, the associated m-BCK algebra is also the **Boolean algebra** $\mathcal{L}^L_{2\times 2}$ from Examples 15.3.8, 1, and its bounded lattice $(A_4, \leq_m, 0, 1)$ is represented by the Hasse diagram from the same Figure 15.40, right side.

**Remark.** *The proper tOSLs are particular cases of tQNM algebras.*

**Examples 15.3.18 Proper transitive OM algebras: tOM**

A tOM is *proper*, if it is not a tQMV algebra - see Figure 13.20.

**Example 1: tOM verifying (prel$_m$)**

By MACE4 program, we found that the algebra $\mathcal{A}^L = (A_8 = \{0, a, b, c, d, e, f, 1\}, \odot, ^-, 1)$, with the following tables of $\odot$ and $^-$ and of the additional operation $\oplus$, is an involutive left-m-BE algebra verifying (Pom) and (m-BB) ($\Leftrightarrow \ldots \Leftrightarrow$ (m-Tr)), and also (prel$_m$), and not verifying (m-An) for $(c, e)$, (m-Pimpl) for $(b, 0)$, (m-Pabs-i) for $(a, 0)$, (G) for $a$, (Pqmv) for $(a, a, 0)$, (Pmv) for $(a, a)$, ($\Delta_m$) for $(b, a)$, (WNM$_m$) and (aWNM$_m$) for $(d, d)$, (m-Pdis) for $(a, b, a)$. Hence, it is an **involutive m-pre-BCK algebra verifying (Pom) and (prel$_m$)**.

| $\odot$ | 0 | a | b | c | d | e | f | 1 |
|---|---|---|---|---|---|---|---|---|
| 0 | 0 | 0 | 0 | 0 | 0 | 0 | 0 | 0 |
| a | 0 | 0 | 0 | 0 | 0 | 0 | 0 | a |
| b | 0 | 0 | b | c | c | e | e | b |
| c | 0 | 0 | c | 0 | 0 | 0 | 0 | c |
| d | 0 | 0 | c | 0 | a | 0 | a | d |
| e | 0 | 0 | e | 0 | 0 | 0 | 0 | e |
| f | 0 | 0 | e | 0 | a | 0 | a | f |
| 1 | 0 | a | b | c | d | e | f | 1 |

with

| $\oplus$ | 0 | a | b | c | d | e | f | 1 |
|---|---|---|---|---|---|---|---|---|
| 0 | 0 | a | b | c | d | e | f | 1 |
| a | a | a | 1 | d | d | f | f | 1 |
| b | b | 1 | 1 | 1 | 1 | 1 | 1 | 1 |
| c | c | d | 1 | b | 1 | b | 1 | 1 |
| d | d | d | 1 | 1 | 1 | 1 | 1 | 1 |
| e | e | f | 1 | b | 1 | b | 1 | 1 |
| f | f | f | 1 | 1 | 1 | 1 | 1 | 1 |
| 1 | 1 | 1 | 1 | 1 | 1 | 1 | 1 | 1 |

and $(0, a, b, c, d, e, f, 1)^- = (1, b, a, d, c, f, e, 0)$.

Then, the tables of $\wedge_m^M$ and its transposed, $\wedge_m^B$, are the following:

| $\wedge_m^M$ | 0 | a | b | c | d | e | f | 1 |
|---|---|---|---|---|---|---|---|---|
| 0 | 0 | 0 | 0 | 0 | 0 | 0 | 0 | 0 |
| a | 0 | a | 0 | 0 | a | 0 | a | a |
| b | 0 | a | b | c | d | e | f | b |
| c | 0 | a | c | c | c | e | e | c |
| d | 0 | a | c | c | d | e | f | d |
| e | 0 | a | e | c | c | e | e | e |
| f | 0 | a | e | c | d | e | f | f |
| 1 | 0 | a | b | c | d | e | f | 1 |

and

| $\wedge_m^B$ | 0 | a | b | c | d | e | f | 1 |
|---|---|---|---|---|---|---|---|---|
| 0 | 0 | 0 | 0 | 0 | 0 | 0 | 0 | 0 |
| a | 0 | a | a | a | a | a | a | a |
| b | 0 | 0 | b | c | c | e | e | b |
| c | 0 | 0 | c | c | c | c | c | c |
| d | 0 | a | d | c | d | c | d | d |
| e | 0 | 0 | e | e | e | e | e | e |
| f | 0 | a | f | e | f | e | f | f |
| 1 | 0 | a | b | c | d | e | f | 1 |

.

• The binary relation $\leq_m^M$ is an order relation, but not a lattice order w.r. to $\wedge_m^M$, $\vee_m^M$, since $\wedge_m^M$ is not commutative. From the table of $\wedge_m^M$, we see that $a \leq_m^M d, f, 1$; $b \leq_m^M 1$; $c \leq_m^M b, d, 1$; $d \leq_m^M 1$; $e \leq_m^M b, f, 1$; $f \leq_m^M 1$; then, the bounded poset $(A_8, \leq_m^M, 0, 1)$ is reprezented by the Hasse diagram from Figure 15.41.

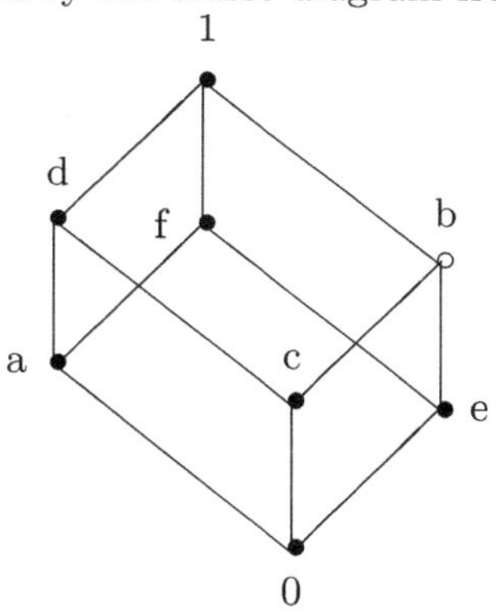


Figure 15.41: The Hasse diagram of the bounded poset $(A_8, \leq_m^M, 0, 1)$

• The binary relation $\leq_m$ ($\Longleftrightarrow \leq_m^B$) is a pre-order relation, since (m-Re) and (m-Tr) hold; hence, $\leq_m^B$ is a pre-order relation too. From the table of $\wedge_m^B$, we see that $a \leq_m b, c, d, e, f, 1$; $b \leq_m 1$; $c \leq_m b, d, e, f, 1$; $d \leq_m b, f, 1$; $e \leq_m b, c, d, f, 1$; $f \leq_m b, d, 1$; hence, the bounded pre-ordered set $(A_8, \leq_m, 0, 1)$ is represented by the Hasse-type diagram from Figure 15.42, left side. Note that $c \parallel_Q e$ and $d \parallel_Q f$. Hence, the quotient set is $A_8/_{\parallel_Q} = \{\{0\}, \{a\}, \{c, e\}, \{d, f\}, \{b\}, \{1\}\}$ and it will be represented by $A_6 = \{0, a, c, d, b, 1\}$ respectively in the sequel. Hence, the associated m-BCK algebra $(A_6, \odot, ^-, 1)$ has the following tables of $\odot$, $^-$ and $\oplus$:

| $\odot$ | 0 | a | b | c | d | 1 |
|---|---|---|---|---|---|---|
| 0 | 0 | 0 | 0 | 0 | 0 | 0 |
| a | 0 | 0 | 0 | 0 | 0 | a |
| b | 0 | 0 | b | c | c | b |
| c | 0 | 0 | c | 0 | 0 | c |
| d | 0 | 0 | c | 0 | a | d |
| 1 | 0 | a | b | c | d | 1 |

and

| $x$ | $x^-$ |
|---|---|
| 0 | 1 |
| a | b |
| b | a |
| c | d |
| d | c |
| 1 | 0 |

, with

| $\oplus$ | 0 | a | b | c | d | 1 |
|---|---|---|---|---|---|---|
| 0 | 0 | a | b | c | d | 1 |
| a | a | a | 1 | d | d | 1 |
| b | b | 1 | 1 | 1 | 1 | 1 |
| c | c | d | 1 | b | 1 | 1 |
| d | d | d | 1 | 1 | 1 | 1 |
| 1 | 1 | 1 | 1 | 1 | 1 | 1 |

.

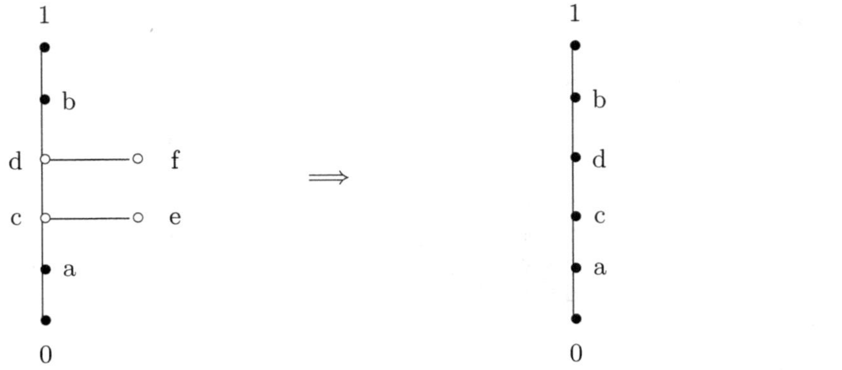


The bounded preordered set $(A_8, \leq_m, 0, 1)$ The bounded chain $(A_6, \leq_m, 0, 1)$

Figure 15.42: $(A_8, \leq_m, 0, 1)$ and $(A_6, \leq_m, 0, 1)$

Its bounded poset $(A_6, \leq_m, 0, 1)$ is the **chain**: $0 \leq_m a \leq_m c \leq_m d \leq_m b \leq_m 1$ (hence a lattice), represented by the Hasse diagram from the same Figure 15.42, right side.

The tables of $\vee$, $\wedge$ and $\rightarrow$ are:

| $\vee$ | 0 | a | b | c | d | 1 |
|---|---|---|---|---|---|---|
| 0 | 0 | a | b | c | d | 1 |
| a | a | a | b | c | d | 1 |
| b | b | b | b | b | b | 1 |
| c | c | c | b | c | d | 1 |
| d | d | d | b | d | d | 1 |
| 1 | 1 | 1 | 1 | 1 | 1 | 1 |

and

| $\wedge$ | 0 | a | b | c | d | 1 |
|---|---|---|---|---|---|---|
| 0 | 0 | 0 | 0 | 0 | 0 | 0 |
| a | 0 | a | a | a | a | a |
| b | 0 | a | b | c | d | b |
| c | 0 | a | c | c | c | c |
| d | 0 | a | d | c | d | d |
| 1 | 0 | a | b | c | d | 1 |

and

| $\rightarrow$ | 0 | a | b | c | d | 1 |
|---|---|---|---|---|---|---|
| 0 | 1 | 1 | 1 | 1 | 1 | 1 |
| a | b | 1 | 1 | 1 | 1 | 1 |
| b | a | a | 1 | d | d | 1 |
| c | d | d | 1 | 1 | 1 | 1 |
| d | c | d | 1 | b | 1 | 1 |
| 1 | 0 | a | b | c | d | 1 |

.

Note that (prel) is satisfied, while (WNM) is not satisfied for (d,d). Hence, $(A_6, \odot, ^-, 1)$ is a **linearly ordered IMTL algebra** (Definition 2), that is isomorphic with $IMTL_6^4$ $(c \mapsto b,\ d \mapsto c,\ b \mapsto d)$ from Examples 15.3.3, 2.

**Remark.** *This tOM $\mathcal{A}^L$ is a particular case of linearly pre-ordered tQIMTL algebra.*

**Example 2: tOM not verifying (prel$_m$)**

By MACE4 program, we found that the algebra
$\mathcal{A}^L = (A_8 = \{0, a, b, c, d, e, f, 1\}, \odot, ^-, 1)$, with the following tables of $\odot$ and $^-$ and

of the additional operation $\oplus$, is an involutive left-m-BE algebra verifying (Pom) and (m-BB) ($\Leftrightarrow \ldots \Leftrightarrow$ (m-Tr)), and not verifying (m-An) for $(a, c)$, (m-Pimpl) for $(a, 0)$, (m-Pabs-i) for $(a, 0)$, (G) for $a$, (Pqmv) for $(a, f, 0)$, (Pmv) for $(a, f)$, ($\Delta_m$) for $(b, f)$, (prel$_m$) for $(f, e)$, (WNM$_m$) and (aWNM$_m$) for $(f, b)$, (m-Pdis) for $(a, a, b)$. Hence, it is an **involutive m-pre-BCK algebra verifying (Pom) (tOM algebra)**.

| $\odot$ | 0 | a | b | c | d | e | f | 1 |
|---|---|---|---|---|---|---|---|---|
| 0 | 0 | 0 | 0 | 0 | 0 | 0 | 0 | 0 |
| a | 0 | 0 | 0 | 0 | 0 | 0 | 0 | a |
| b | 0 | 0 | e | 0 | e | e | a | b |
| c | 0 | 0 | 0 | 0 | 0 | 0 | 0 | c |
| d | 0 | 0 | e | 0 | e | e | c | d |
| e | 0 | 0 | e | 0 | e | e | 0 | e |
| f | 0 | 0 | a | 0 | c | 0 | a | f |
| 1 | 0 | a | b | c | d | e | f | 1 |

with

| $\oplus$ | 0 | a | b | c | d | e | f | 1 |
|---|---|---|---|---|---|---|---|---|
| 0 | 0 | a | b | c | d | e | f | 1 |
| a | a | f | 1 | f | 1 | b | f | 1 |
| b | b | 1 | 1 | 1 | 1 | 1 | 1 | 1 |
| c | c | f | 1 | f | 1 | d | f | 1 |
| d | d | 1 | 1 | 1 | 1 | 1 | 1 | 1 |
| e | e | b | 1 | d | 1 | b | 1 | 1 |
| f | f | f | 1 | f | 1 | 1 | f | 1 |
| 1 | 1 | 1 | 1 | 1 | 1 | 1 | 1 | 1 |

and $(0, a, b, c, d, e, f, 1)^- = (1, b, a, d, c, f, e, 0)$.

Then, the tables of $\wedge_m^M$ and its transposed, $\wedge_m^B$, are the following:

| $\wedge_m^M$ | 0 | a | b | c | d | e | f | 1 |
|---|---|---|---|---|---|---|---|---|
| 0 | 0 | 0 | 0 | 0 | 0 | 0 | 0 | 0 |
| a | 0 | a | a | c | c | 0 | a | a |
| b | 0 | a | b | c | d | e | f | b |
| c | 0 | a | a | c | c | 0 | c | c |
| d | 0 | a | b | c | d | e | f | d |
| e | 0 | a | e | c | e | e | a | e |
| f | 0 | a | a | c | c | 0 | f | f |
| 1 | 0 | a | b | c | d | e | f | 1 |

and

| $\wedge_m^B$ | 0 | a | b | c | d | e | f | 1 |
|---|---|---|---|---|---|---|---|---|
| 0 | 0 | 0 | 0 | 0 | 0 | 0 | 0 | 0 |
| a | 0 | a | a | a | a | a | a | a |
| b | 0 | a | b | a | b | e | a | b |
| c | 0 | c | c | c | c | c | c | c |
| d | 0 | c | d | c | d | e | c | d |
| e | 0 | 0 | e | 0 | e | e | 0 | e |
| f | 0 | a | f | c | f | a | f | f |
| 1 | 0 | a | b | c | d | e | f | 1 |

.

• The binary relation $\leq_m^M$ is an order relation, but not a lattice order w.r. to $\wedge_m^M$, $\vee_m^M$, since $\wedge_m^M$ is not commutative.

• The binary relation $\leq_m$ ($\Longleftrightarrow \leq_m^B$) is a pre-order relation, since (m-Re) and (m-Tr) hold. From the table of $\wedge_m^B$, we see that $a \leq_m b, c, d, e, f, 1$; $b \leq_m 1$; $c \leq_m b, d, e, f, 1$; $d \leq_m b, f, 1$; $e \leq_m b, c, d, f, 1$; $f \leq_m b, d, 1$; hence, the bounded pre-ordered set $(A_8, \leq_m, 0, 1)$ is represented by the Hasse-type diagram from Figure 15.43, left side.

Note that $c \parallel_Q a$ and $d \parallel_Q b$. Hence, the quotient set is $A_8/_{\parallel_Q} = \{\{0\}, \{a, c\}, \{b, d\}, \{e\}, \{f\}, \{1\}\}$ and it will be represented by $A_6 = \{0, a, b, e, f, 1\}$ respectively in the sequel. Hence, the associated m-BCK algebra $(A_6, \odot, ^-, 1)$ has the following tables of $\odot$, $^-$ and $\oplus$:

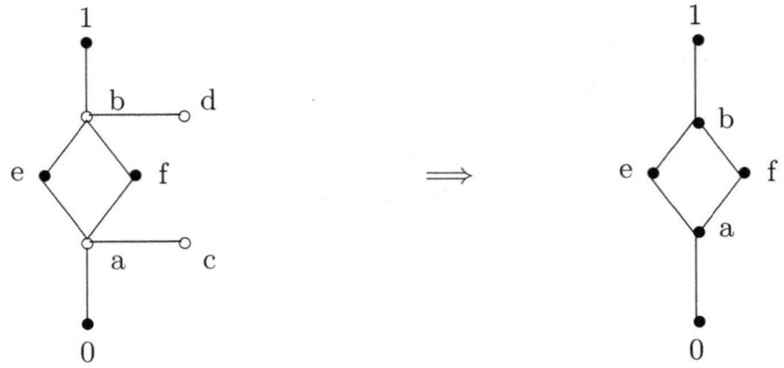


The bounded pre-ordered set $(A_8, \leq_m, 0, 1)$ The bounded lattice $(A_6, \leq_m, 0, 1)$

Figure 15.43: $(A_8, \leq_m, 0, 1)$ and $(A_6, \leq_m, 0, 1)$

| $\odot$ | 0 | a | b | e | f | 1 |
|---|---|---|---|---|---|---|
| 0 | 0 | 0 | 0 | 0 | 0 | 0 |
| a | 0 | 0 | 0 | 0 | 0 | a |
| b | 0 | 0 | e | e | a | b |
| e | 0 | 0 | e | e | 0 | e |
| f | 0 | 0 | a | 0 | a | f |
| 1 | 0 | a | b | e | f | 1 |

and

| $x$ | $x^-$ |
|---|---|
| 0 | 1 |
| a | b |
| b | a |
| e | f |
| f | e |
| 1 | 0 |

, with

| $\oplus$ | 0 | a | b | e | f | 1 |
|---|---|---|---|---|---|---|
| 0 | 0 | a | b | e | f | 1 |
| a | a | f | 1 | b | f | 1 |
| b | b | 1 | 1 | 1 | 1 | 1 |
| e | e | b | 1 | b | 1 | 1 |
| f | f | f | 1 | 1 | f | 1 |
| 1 | 1 | 1 | 1 | 1 | 1 | 1 |

.

Then, $(A_6, \odot, ^-, 1)$ is an involutive left-m-aBE algebra verifying (Pom) and (m-BB) ($\Leftrightarrow \ldots \Leftrightarrow$ (m-Tr)), and not verifying (m-Pimpl) for $(a, 0)$, (m-Pabs-i) for $(a, 0)$, (G) for $a$, (Pqmv) for $(a, f, 0)$, (Pmv) for $(a, f)$, $(\Delta_m)$ for $(b, f)$, (prel$_m$) for $(f, e)$, (WNM$_m$) and (aWNM$_m$) for $(f, b)$, (m-Pdis) for $(a, a, b)$. Its bounded non-linearly ordered poset $(A_6, \leq_m, 0, 1)$ is the **distributive lattice** represented by the Hasse diagram from the same Figure 15.43, right side.

The table of $\rightarrow$ is:

| $\rightarrow$ | 0 | a | b | e | f | 1 |
|---|---|---|---|---|---|---|
| 0 | 1 | 1 | 1 | 1 | 1 | 1 |
| a | b | 1 | 1 | 1 | 1 | 1 |
| b | a | f | 1 | b | f | 1 |
| e | f | f | 1 | 1 | f | 1 |
| f | e | b | 1 | b | 1 | 1 |
| 1 | 0 | a | b | e | f | 1 |

.

Note that (prel) is not satisfied for $(e, f)$: $(e \rightarrow f) \vee (f \rightarrow e) = f \vee b = b \neq 1$. Hence, $(A_6, \odot, ^-, 1)$ is only an **involutive left-m-BCK lattice verifying (Pom)** (see Example 15.3.2).

**Remark.** *This tOM $\mathcal{A}^L$ is only a particular case of non-linearly pre-ordered m-pre-BCK lattice, verifying (Pom).*

### 15.3.2.2 Building transitive quantum algebras by the "method of identic rows/columns"

**Example 15.3.19 Building an involutive m-pre-BCK algebra: m-pre-BCK$_{(DN)}$**

Consider the algebra $\mathcal{A}^L = (A_8 = \{0, m, a, b, c, d, n, 1\}, \odot, ^-, 1)$ from Example 15.3.1.

**Example 1: we add two elements, $x \parallel_Q m$ and $y \parallel_Q n$**

We add to the bounded poset $(A_8, \leq_m, 0, 1)$ from Figure 15.3 (see also Figure 15.44, right side) two elements: $x \parallel_Q m$ and $y \parallel_Q n$; thus, we obtain the set $A_{10} = \{0, m, a, b, c, d, n, x, y, 1\}$. We then extend the operations $\odot$ and $\oplus$ on this set $A_{10}$ by the "method of identic rows/columns" (the row/column of $x$ is identic with the row/column of $m$, excepting the last elements, which are Q-parallel). The operation $^-$ is extended such that the property (DN) be preserved. We obtain the following extended tables:

| $\odot$ | 0 | m | a | b | c | d | n | x | y | 1 |
|---|---|---|---|---|---|---|---|---|---|---|
| 0 | 0 | 0 | 0 | 0 | 0 | 0 | 0 | 0 | 0 | 0 |
| m | 0 | 0 | 0 | 0 | 0 | 0 | 0 | 0 | 0 | m |
| a | 0 | 0 | 0 | 0 | 0 | m | m | 0 | m | a |
| b | 0 | 0 | 0 | 0 | m | 0 | m | 0 | m | b |
| c | 0 | 0 | 0 | m | m | m | m | 0 | m | c |
| d | 0 | 0 | m | 0 | m | m | m | 0 | m | d |
| n | 0 | 0 | m | m | m | m | m | 0 | m | n |
| x | 0 | 0 | 0 | 0 | 0 | 0 | 0 | 0 | 0 | x |
| y | 0 | 0 | m | m | m | m | m | 0 | m | y |
| 1 | 0 | m | a | b | c | d | n | x | y | 1 |

and

| $x$ | $x^-$ |
|---|---|
| 0 | 1 |
| m | n |
| a | c |
| b | d |
| c | a |
| d | b |
| n | m |
| x | y |
| y | x |
| 1 | 0 |

, with

| $\oplus$ | 0 | m | a | b | c | d | n | x | y | 1 |
|---|---|---|---|---|---|---|---|---|---|---|
| 0 | 0 | m | a | b | c | d | n | x | y | 1 |
| m | m | n | n | n | n | n | 1 | n | 1 | 1 |
| a | a | n | n | n | 1 | n | 1 | n | 1 | 1 |
| b | b | n | n | n | n | 1 | 1 | n | 1 | 1 |
| c | c | n | 1 | n | 1 | 1 | 1 | n | 1 | 1 |
| d | d | n | n | 1 | 1 | 1 | 1 | n | 1 | 1 |
| n | n | 1 | 1 | 1 | 1 | 1 | 1 | 1 | 1 | 1 |
| x | x | n | n | n | n | n | 1 | n | 1 | 1 |
| y | y | 1 | 1 | 1 | 1 | 1 | 1 | 1 | 1 | 1 |
| 1 | 1 | 1 | 1 | 1 | 1 | 1 | 1 | 1 | 1 | 1 |

.

Then, the resulting algebra $(A_{10}, \odot, ^-, 1)$ is an involutive m-BE algebra verifying (m-BB) ($\Leftrightarrow \ldots \Leftrightarrow$ (m-Tr)) and not verifying (m-An) for $(m, x)$, (m-Pabs-i) for $(m, 0)$, (G) for $a$, (m-Pimpl) for $(m, 0)$, (Pqmv) for $(a, m, 0)$, (Pom) for $(a, d)$, (Pmv) for $(a, m)$, ($\Delta_m$) for $(c, a)$ and also (prel$_m$) for $(a, b)$, (WNM$_m$) and (aWNM$_m$) for $(a, d)$, (m-Pdis) for $(m, m, a)$. Hence, it is a **proper involutive left-m-pre-BCK algebra**.

Next, we follow the converse way, as in Subsubsection 15.3.2.1.
The tables of $\wedge_m^M$ and $\wedge_m^B$ are:

| $\wedge_m^M$ | 0 | m | a | b | c | d | n | x | y | 1 |
|---|---|---|---|---|---|---|---|---|---|---|
| 0 | 0 | 0 | 0 | 0 | 0 | 0 | 0 | 0 | 0 | 0 |
| m | 0 | m | m | m | m | m | m | x | m | m |
| a | 0 | m | a | m | m | m | m | x | m | a |
| b | 0 | m | m | b | m | m | m | x | m | b |
| c | 0 | m | a | b | c | m | m | x | m | c |
| d | 0 | m | a | b | m | d | m | x | m | d |
| n | 0 | m | a | b | c | d | n | x | y | n |
| x | 0 | m | m | m | m | m | m | x | m | x |
| y | 0 | m | a | b | c | d | n | x | y | y |
| 1 | 0 | m | a | b | c | d | n | x | y | 1 |

and

| $\wedge_m^B$ | 0 | m | a | b | c | d | n | x | y | 1 |
|---|---|---|---|---|---|---|---|---|---|---|
| 0 | 0 | 0 | 0 | 0 | 0 | 0 | 0 | 0 | 0 | 0 |
| m | 0 | m | m | m | m | m | m | m | m | m |
| a | 0 | m | a | m | a | a | a | m | a | a |
| b | 0 | m | m | b | b | b | b | m | b | b |
| c | 0 | m | m | m | c | m | c | m | c | c |
| d | 0 | m | m | m | m | d | d | m | d | d |
| n | 0 | m | m | m | m | m | n | m | n | n |
| x | 0 | x | x | x | x | x | x | x | x | x |
| y | 0 | m | m | m | m | m | y | m | y | y |
| 1 | 0 | m | a | b | c | d | n | x | y | 1 |

.

• The binary relation $\leq_m^M$ is transitive, hence it is an order, but not a lattice order, since $\wedge_m^M$ is not commutative.

• The binary relation $\leq_m$ is a pre-order. From the table of $\wedge_m^B$, we obtain: $m \leq_m m, a, b, c, d, n, x, y, 1$; $a \leq_m a, c, d, n, y, 1$; $b \leq_m b, c, d, n, y, 1$; $c \leq_m c, n, y, 1$; $d \leq_m d, n, y, 1$; $n \leq_m n, y, 1$; $x \leq_m x, m, a, b, c, d, n, y, 1$; $y \leq_m y, n, 1$; hence, the bounded pre-ordered set $(A_{10}, \leq_m, 0, 1)$ is represented by the Hasse-type diagram from Figure 15.44, left side. We have $x \parallel_Q m$ and $y \parallel_Q n$, hence the quotient set is $A_{10}/_{\parallel_Q} = \{\{0\}, \{m, x\}, \{a\}, \{b\}, \{c\}, \{d\}, \{n, y\}, \{1\}\}$, denoted by $A_8 = \{0, m, a, b, c, d, n, 1\}$ in the sequel. The bounded poset $(A_8, \leq_m, 0, 1)$ is represented by the Hasse diagram from the same Figure 15.44, right side. Hence, the **associated m-BCK algebra** $(A_8, \odot, ^-, 1)$ **is just the starting algebra** $\mathcal{A}^L$.

**Examples 15.3.20 Building involutive m-pre-BCK lattices: m-pre-BCK-L$_{(DN)}$**

**Example 1**

Consider the **(involutive) m-BCK lattice** $\mathcal{A}^L = (A_6 = \{0, a, b, c, d, 1\}, \odot, ^-, 1)$ from Example 15.3.2, 1.

**Example 1.1: we add two elements,** $x \parallel_Q a$ **and** $y \parallel_Q d$

We add to the bounded (distributive) lattice $(A_6, \leq_m, 0, 1)$ from Figure 15.4

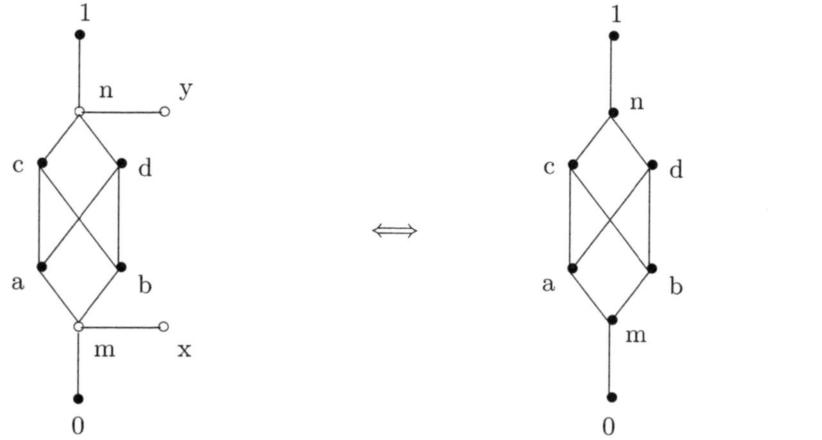


The bounded pre-ordered set $(A_{10}, \leq_m, 0, 1)$ The bounded poset $(A_8, \leq_m, 0, 1)$

Figure 15.44: $(A_{10}, \leq_m, 0, 1)$ and $(A_8, \leq_m, 0, 1)$

(see also Figure 15.45, right side) two elements: $x \parallel_Q a$ and $y \parallel_Q d$; thus, we obtain the set $A_8 = \{0, a, b, c, d, x, y, 1\}$. We then extend the operations $\odot$ and $\oplus$ on this set $A_8$ by the "method of identic rows/columns" (the row/column of $x$ is identic with the row/column of $a$, excepting the last elements, which are Q-parallel). The operation $^-$ is extended such that the property (DN) be preserved. We obtain the following extended tables:

| $\odot$ | 0 | a | b | c | d | x | y | 1 |
|---|---|---|---|---|---|---|---|---|
| 0 | 0 | 0 | 0 | 0 | 0 | 0 | 0 | 0 |
| a | 0 | d | d | d | 0 | d | 0 | a |
| b | 0 | d | d | 0 | 0 | d | 0 | b |
| c | 0 | d | 0 | d | 0 | d | 0 | c |
| d | 0 | 0 | 0 | 0 | 0 | 0 | 0 | d |
| x | 0 | d | d | d | 0 | d | 0 | x |
| y | 0 | 0 | 0 | 0 | 0 | 0 | 0 | y |
| 1 | 0 | a | b | c | d | x | y | 1 |

with

| $\oplus$ | 0 | a | b | c | d | x | y | 1 |
|---|---|---|---|---|---|---|---|---|
| 0 | 0 | a | b | c | d | x | y | 1 |
| a | a | 1 | 1 | 1 | 1 | 1 | 1 | 1 |
| b | b | 1 | a | 1 | a | 1 | a | 1 |
| c | c | 1 | 1 | a | a | 1 | a | 1 |
| d | d | 1 | a | a | a | 1 | a | 1 |
| x | x | 1 | 1 | 1 | 1 | 1 | 1 | 1 |
| y | y | 1 | a | a | a | 1 | a | 1 |
| 1 | 1 | 1 | 1 | 1 | 1 | 1 | 1 | 1 |

and $(0, a, b, c, d, x, y, 1)^- = (1, d, c, b, a, y, x, 0)$.

Then, the resulting algebra $(A_8, \odot, ^-, 1)$ is an involutive m-BE algebra verifying (m-BB) ($\Leftrightarrow \ldots \Leftrightarrow$ (m-Tr)) and not verifying (m-An) for $(a, e)$, (m-Pabs-i) for $(b, 0)$, (G) for $a$, (m-Pimpl) for $(a, 0)$, (Pqmv) for $(b, d, 0)$, (Pom) for $(b, a)$, (Pmv) for $(b, d)$, ($\Delta_m$) for $(a, b)$ and also (prel$_m$) for $(b, c)$, (WNM$_m$) and (aWNM$_m$) for $(a, a)$, (m-Pdis) for $(a, a, b)$. Hence, it is a **proper involutive left-m-pre-BCK lattice**.

Next, we follow the converse way, as in Subsubsection 15.3.2.1.

The tables of $\wedge_m^M$ and $\wedge_m^B$ are:

| $\wedge_m^M$ | 0 | a | b | c | d | x | y | 1 |
|---|---|---|---|---|---|---|---|---|
| 0 | 0 | 0 | 0 | 0 | 0 | 0 | 0 | 0 |
| a | 0 | a | b | c | d | x | y | a |
| b | 0 | d | b | d | d | d | y | b |
| c | 0 | d | d | c | d | d | y | c |
| d | 0 | d | d | d | d | d | y | d |
| x | 0 | a | b | c | d | x | y | x |
| y | 0 | d | d | d | d | d | y | y |
| 1 | 0 | a | b | c | d | x | y | 1 |

and

| $\wedge_m^B$ | 0 | a | b | c | d | x | y | 1 |
|---|---|---|---|---|---|---|---|---|
| 0 | 0 | 0 | 0 | 0 | 0 | 0 | 0 | 0 |
| a | 0 | a | d | d | d | a | d | a |
| b | 0 | b | b | d | d | b | d | b |
| c | 0 | c | d | c | d | c | d | c |
| d | 0 | d | d | d | d | d | d | d |
| x | 0 | x | d | d | d | x | d | x |
| y | 0 | y | y | y | y | y | y | y |
| 1 | 0 | a | b | c | d | x | y | 1 |

.

• The binary relation $\leq_m^M$ is transitive, hence it is an order, but not a lattice order, since $\wedge_m^M$ is not commutative.

• The binary relation $\leq_m$ is a pre-order. From the table of $\wedge_m^B$, we obtain: $a \leq_m a, x, 1$; $b \leq_m b, a, x, 1$; $c \leq_m c, a, x, 1$; $d \leq_m d, a, b, c, x, y, 1$; $x \leq_m x, a, 1$; $y \leq_m y, a, b, c, d, x, 1$; hence, the bounded pre-ordered set $(A_8, \leq_m, 0, 1)$ is represented by the Hasse-type diagram from Figure 15.45, left side. ... Hence, the **associated m-BCK algebra** $(A_6, \odot, ^-, 1)$ **is just the starting algebra** $\mathcal{A}^L$, represented by the Hasse diagram from the same Figure 15.45, right side.

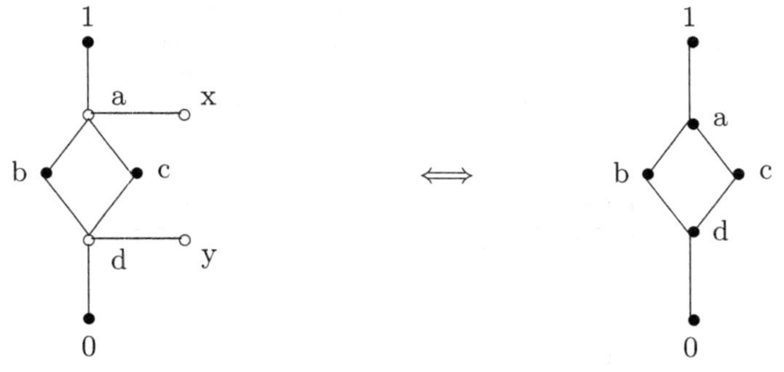


The bounded pre-ordered set $(A_8, \leq_m, 0, 1)$ The bounded lattice $(A_6, \leq_m, 0, 1)$

Figure 15.45: $(A_8, \leq_m, 0, 1)$ and $(A_6, \leq_m, 0, 1)$

### Example 1.2: we add two elements, $u \parallel_Q b$ and $v \parallel_Q c$

We add to the bounded (distributive) lattice $(A_6, \leq_m, 0, 1)$ from Figure 15.4 (see also Figure 15.46, right side) two elements: $u \parallel_Q b$ and $v \parallel_Q c$; thus, we obtain the set $A_8 = \{0, a, b, c, d, u, v, 1\}$. We then extend the operations $\odot$ and $\oplus$ on this set $A_8$ by the "method of identic rows/columns" (the row/column of $u$ is identic with the row/column of $b$, excepting the last elements, which are Q-parallel). The operation $^-$ is extended such that the property (DN) be preserved. We obtain the following extended tables:

| $\odot$ | 0 | a | b | c | d | u | v | 1 |
|---|---|---|---|---|---|---|---|---|
| 0 | 0 | 0 | 0 | 0 | 0 | 0 | 0 | 0 |
| a | 0 | d | d | d | 0 | d | d | a |
| b | 0 | d | d | 0 | 0 | d | 0 | b |
| c | 0 | d | 0 | d | 0 | 0 | d | c |
| d | 0 | 0 | 0 | 0 | 0 | 0 | 0 | d |
| u | 0 | d | d | 0 | 0 | d | 0 | u |
| v | 0 | d | 0 | d | 0 | 0 | d | v |
| 1 | 0 | a | b | c | d | u | v | 1 |

with

| $\oplus$ | 0 | a | b | c | d | u | v | 1 |
|---|---|---|---|---|---|---|---|---|
| 0 | 0 | a | b | c | d | u | v | 1 |
| a | a | 1 | 1 | 1 | 1 | 1 | 1 | 1 |
| b | b | 1 | a | 1 | a | a | 1 | 1 |
| c | c | 1 | 1 | a | a | 1 | a | 1 |
| d | d | 1 | a | a | a | a | a | 1 |
| u | u | 1 | a | 1 | a | a | 1 | 1 |
| v | v | 1 | 1 | a | a | 1 | a | 1 |
| 1 | 1 | 1 | 1 | 1 | 1 | 1 | 1 | 1 |

and $(0, a, b, c, d, u, v, 1)^- = (1, d, c, b, a, v, u, 0)$.

Then, the resulting algebra $(A_8, \odot, ^-, 1)$ is an involutive m-BE algebra verifying (m-BB) ($\Leftrightarrow \ldots \Leftrightarrow$ (m-Tr)) and not verifying (m-An) for $(b, u)$, (m-Pabs-i) for $(b, 0)$, (G) for $a$, (m-Pimpl) for $(a, 0)$, (Pqmv) for $(b, d, 0)$, (Pom) for $(b, a)$, (Pmv) for $(b, d)$, $(\Delta_m)$ for $(a, b)$ and also (prel$_m$) for $(b, c)$, (WNM$_m$) and (aWNM$_m$) for $(a, a)$, (m-Pdis) for $(a, a, b)$. Hence, it is a **proper involutive left-m-pre-BCK lattice.**

Next, we follow the converse way, as in Subsubsection 15.3.2.1.

The tables of $\wedge_m^M$ and $\wedge_m^B$ are:

| $\wedge_m^M$ | 0 | a | b | c | d | u | v | 1 |
|---|---|---|---|---|---|---|---|---|
| 0 | 0 | 0 | 0 | 0 | 0 | 0 | 0 | 0 |
| a | 0 | a | b | c | d | u | v | a |
| b | 0 | d | b | d | d | u | d | b |
| c | 0 | d | d | c | d | d | v | c |
| d | 0 | d | d | d | d | d | d | d |
| u | 0 | d | b | d | d | u | d | u |
| v | 0 | d | d | c | d | d | v | v |
| 1 | 0 | a | b | c | d | u | v | 1 |

and

| $\wedge_m^B$ | 0 | a | b | c | d | u | v | 1 |
|---|---|---|---|---|---|---|---|---|
| 0 | 0 | 0 | 0 | 0 | 0 | 0 | 0 | 0 |
| a | 0 | a | d | d | d | d | d | a |
| b | 0 | b | b | d | d | b | d | b |
| c | 0 | c | d | c | d | d | c | c |
| d | 0 | d | d | d | d | d | d | d |
| u | 0 | u | u | d | d | u | d | u |
| v | 0 | v | d | v | d | d | v | v |
| 1 | 0 | a | b | c | d | u | v | 1 |

.

• The binary relation $\leq_m^M$ is transitive, hence it is an order, but not a lattice order, since $\wedge_m^M$ is not commutative.

• The binary relation $\leq_m$ is a pre-order. From the table of $\wedge_m^B$, we obtain: $a \leq_m a, 1$; $b \leq_m b, a, u, 1$; $c \leq_m c, a, v, 1$; $d \leq_m d, a, b, c, u, v, 1$; $u \leq_m u, a, b, 1$; $v \leq_m v, a, c, 1$; hence, the bounded pre-ordered set $(A_8, \leq_m, 0, 1)$ is represented by the Hasse-type diagram from Figure 15.46, left side. ... Hence, the **associated m-BCK algebra** $(A_6, \odot, ^-, 1)$ **is just the starting algebra** $\mathcal{A}^L$, represented by the Hasse diagram from the same Figure 15.46, right side.

**Example 1.3: we add four elements,** $x \parallel_Q a$, $y \parallel_Q d$, $u \parallel_Q b$ **and** $v \parallel_Q c$

We add to the bounded (distributive) lattice $(A_6, \leq_m, 0, 1)$ from Figure 15.4 (see also Figure 15.47, right side) four elements: $x \parallel_Q a$, $y \parallel_Q d$, $u \parallel_Q b$, $v \parallel_Q c$; thus, we obtain the set $A_{10} = \{0, a, b, c, d, x, y, u, v, 1\}$. We then extend the operations $\odot$ and $\oplus$ on this set $A_{10}$ by the "method of identic rows/columns" (the row/column of $x$ is identic with the row/column of $a$, excepting the last elements, which are Q-parallel). The operation $^-$ is extended such that the property (DN) be preserved. We obtain the following extended tables:

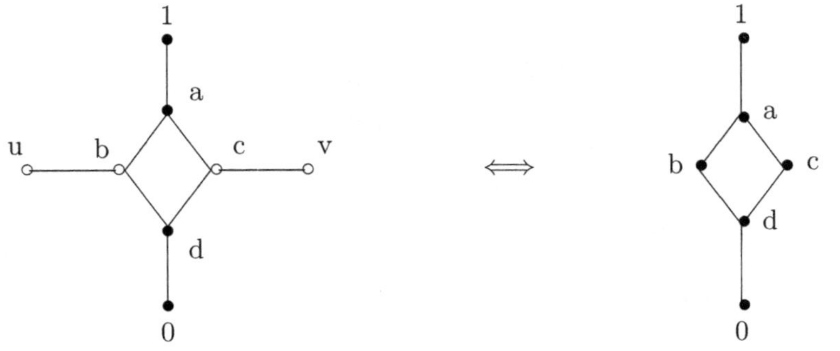


The bounded pre-ordered set $(A_8, \leq_m, 0, 1)$ The bounded lattice $(A_6, \leq_m, 0, 1)$

Figure 15.46: $(A_8, \leq_m, 0, 1)$ and $(A_6, \leq_m, 0, 1)$

| $\odot$ | 0 | a | b | c | d | x | y | u | v | 1 |
|---|---|---|---|---|---|---|---|---|---|---|
| 0 | 0 | 0 | 0 | 0 | 0 | 0 | 0 | 0 | 0 | 0 |
| a | 0 | d | d | d | 0 | d | 0 | d | d | a |
| b | 0 | d | d | 0 | 0 | d | 0 | d | 0 | b |
| c | 0 | d | 0 | d | 0 | d | 0 | 0 | d | c |
| d | 0 | 0 | 0 | 0 | 0 | 0 | 0 | 0 | 0 | d |
| x | 0 | d | d | d | 0 | d | 0 | d | d | x |
| y | 0 | 0 | 0 | 0 | 0 | 0 | 0 | 0 | 0 | y |
| u | 0 | d | d | 0 | 0 | d | 0 | d | 0 | u |
| v | 0 | d | 0 | d | 0 | d | 0 | 0 | d | v |
| 1 | 0 | a | b | c | d | x | y | u | v | 1 |

and

| $x$ | $x^-$ |
|---|---|
| 0 | 1 |
| a | d |
| b | c |
| c | b |
| d | a |
| x | y |
| y | x |
| u | v |
| v | u |
| 1 | 0 |

, with

| $\oplus$ | 0 | a | b | c | d | x | y | u | v | 1 |
|---|---|---|---|---|---|---|---|---|---|---|
| 0 | 0 | a | b | c | d | x | y | u | v | 1 |
| a | a | 1 | 1 | 1 | 1 | 1 | 1 | 1 | 1 | 1 |
| b | b | 1 | a | 1 | a | 1 | a | a | 1 | 1 |
| c | c | 1 | 1 | a | a | 1 | a | 1 | a | 1 |
| d | d | 1 | a | a | a | 1 | a | a | a | 1 |
| x | x | 1 | 1 | 1 | 1 | 1 | 1 | 1 | 1 | 1 |
| y | y | 1 | a | a | a | 1 | a | a | a | 1 |
| u | u | 1 | a | 1 | a | 1 | a | a | 1 | 1 |
| v | v | 1 | 1 | a | a | 1 | a | 1 | a | 1 |
| 1 | 1 | 1 | 1 | 1 | 1 | 1 | 1 | 1 | 1 | 1 |

.

Then, the resulting algebra $(A_{10}, \odot, ^-, 1)$ is an involutive m-BE algebra verifying (m-BB) ($\Leftrightarrow$ ... $\Leftrightarrow$ (m-Tr)) and not verifying (m-An) for $(a, x)$, (m-Pabs-i) for $(b, 0)$, (G) for $a$, (m-Pimpl) for $(a, 0)$, (Pqmv) for $(b, d, 0)$, (Pom) for $(b, a)$, (Pmv) for $(b, d)$, $(\Delta_m)$ for $(a, b)$ and also ($\text{prel}_m$) for $(b, c)$, ($\text{WNM}_m$) and ($\text{aWNM}_m$) for $(a, a)$, (m-Pdis) for $(a, a, b)$. Hence, it is a **proper involutive left-m-pre-BCK lattice.**

Next, we follow the converse way, as in Subsubsection 15.3.2.1.

The tables of $\wedge_m^M$ and $\wedge_m^B$ are:

| $\wedge_m^M$ | 0 | a | b | c | d | x | y | u | v | 1 |
|---|---|---|---|---|---|---|---|---|---|---|
| 0 | 0 | 0 | 0 | 0 | 0 | 0 | 0 | 0 | 0 | 0 |
| a | 0 | a | b | c | d | x | y | u | v | a |
| b | 0 | d | b | d | d | d | y | u | d | b |
| c | 0 | d | d | c | d | d | y | d | v | c |
| d | 0 | d | d | d | d | d | y | d | d | d |
| x | 0 | a | b | c | d | x | y | u | v | x |
| y | 0 | d | d | d | d | d | y | d | d | y |
| u | 0 | d | b | d | d | d | y | u | d | u |
| v | 0 | d | d | c | d | d | y | d | v | v |
| 1 | 0 | a | b | c | d | x | y | u | v | 1 |

and

| $\wedge_m^B$ | 0 | a | b | c | d | x | y | u | v | 1 |
|---|---|---|---|---|---|---|---|---|---|---|
| 0 | 0 | 0 | 0 | 0 | 0 | 0 | 0 | 0 | 0 | 0 |
| a | 0 | a | d | d | d | a | d | d | d | a |
| b | 0 | b | b | d | d | b | d | b | d | b |
| c | 0 | c | d | c | d | c | d | d | c | c |
| d | 0 | d | d | d | d | d | d | d | d | d |
| x | 0 | x | d | d | d | x | d | d | d | x |
| y | 0 | y | y | y | y | y | y | y | y | y |
| u | 0 | u | u | d | d | u | d | u | d | u |
| v | 0 | v | d | v | d | v | d | d | v | v |
| 1 | 0 | a | b | c | d | x | y | u | v | 1 |

.

• The binary relation $\leq_m^M$ is transitive, hence it is an order, but not a lattice order, since $\wedge_m^M$ is not commutative.

• The binary relation $\leq_m$ is a pre-order. From the table of $\wedge_m^B$, we obtain: $a \leq_m a, x, 1$; $b \leq_m b, a, x, u, 1$; $c \leq_m c, a, x, u, 1$; $d \leq_m d, a, b, c, x, y, u, v, 1$; $x \leq_m x, a, 1$; $y \leq_m y, a, b, c, d, x, u, v, 1$; $u \leq_m u, a, b, x, 1$; $v \leq_m v, a, c, x, 1$; hence, the bounded pre-ordered set $(A_{10}, \leq_m, 0, 1)$ is represented by the Hasse-type diagram from Figure 15.47, left side. ... Hence, the **associated m-BCK algebra** $(A_6, \odot, ^-, 1)$ **is just the starting algebra** $\mathcal{A}^L$, represented by the Hasse diagram from the same Figure 15.47, right side.

**Example 2**

Consider the **(involutive) m-BCK (distributive) lattice** $\mathcal{A}^L = (A_8 = \{0, a, b, c, d, e, f, 1\}, \odot, ^-, 1)$ from Example 15.3.2, 3, which verifies (prel$_m$), (aWNM$_m$), (aWNM$_m$) and does not verify (prel), (WNM) and the rest (hence, it is not an NM algebra).

**Example 2.1: we add two elements,** $x \parallel_Q c$ **and** $y \parallel_Q d$

We add to the bounded (distributive) lattice $(A_8, \leq_m, 0, 1)$ from Figure 15.6 (see also Figure 15.48, right side) two elements: $x \parallel_Q c$ and $y \parallel_Q d$; thus, we obtain the set $A_{10} = \{0, a, b, c, d, e, f, x, y, 1\}$. We then extend the operation $\odot$ on this set $A_{10}$ by the "method of identic rows/columns" (the row/column of $x$ is identic

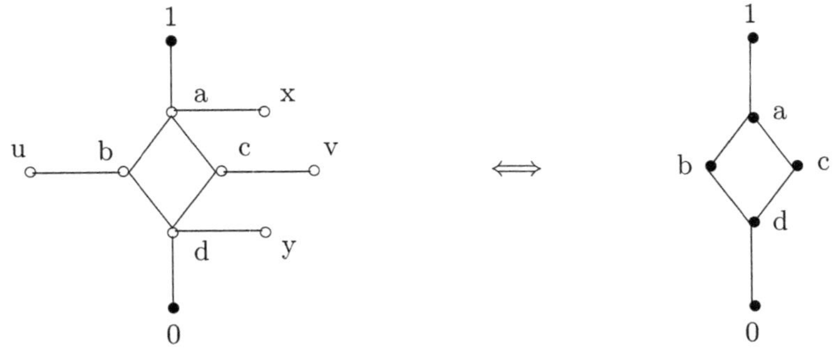


The bounded pre-ordered set $(A_{10}, \leq_m, 0, 1)$ The bounded lattice $(A_6, \leq_m, 0, 1)$

Figure 15.47: $(A_{10}, \leq_m, 0, 1)$ and $(A_6, \leq_m, 0, 1)$

with the row/column of $c$, excepting the last elements, which are Q-parallel). The operation $^-$ is extended such that the property (DN) be preserved. We obtain the following extended tables:

| $\odot$ | 0 | a | b | c | d | e | f | x | y | 1 |
|---|---|---|---|---|---|---|---|---|---|---|
| 0 | 0 | 0 | 0 | 0 | 0 | 0 | 0 | 0 | 0 | 0 |
| a | 0 | c | 0 | c | b | 0 | a | c | b | a |
| b | 0 | 0 | 0 | 0 | b | 0 | b | 0 | b | b |
| c | 0 | c | 0 | c | 0 | 0 | c | c | 0 | c |
| d | 0 | b | b | 0 | d | 0 | d | 0 | d | d |
| e | 0 | 0 | 0 | 0 | 0 | 0 | 0 | 0 | 0 | e |
| f | 0 | a | b | c | d | 0 | f | c | d | f |
| x | 0 | c | 0 | c | 0 | 0 | c | c | 0 | x |
| y | 0 | b | b | 0 | d | 0 | d | 0 | d | y |
| 1 | 0 | a | b | c | d | e | f | x | y | 1 |

and

| $x$ | $x^-$ |
|---|---|
| 0 | 1 |
| a | b |
| b | a |
| c | d |
| d | c |
| e | f |
| f | e |
| x | y |
| y | x |
| 1 | 0 |

.

Then, the resulting algebra $(A_{10}, \odot, ^-, 1)$ is an involutive m-BE algebra verifying (m-BB) ($\Leftrightarrow \ldots \Leftrightarrow$ (m-Tr)) and (prel$_m$), (aWNM$_m$), (WNM$_m$), and not verifying (m-An) for $(c, x)$, (m-Pabs-i) for $(e, 0)$, (G) for $a$, (m-Pimpl) for $(a, 0)$, (Pqmv) for $(e, a, 0)$, (Pom) for $(f, a)$, (Pmv) for $(e, a)$, ($\Delta_m$) for $(a, e)$ and also , (m-Pdis) for $(a, a, a)$. Hence, it is an **involutive left-m-pre-BCK lattice** verifying (prel$_m$), (aWNM$_m$), (WNM$_m$).

Next, we follow the converse way, as in Subsubsection 15.3.2.1. The bounded pre-ordered set $(A_{10}, \leq_m, 0, 1)$ is represented by the Hasse-type diagram from Figure 15.48, left side, and the **associated m-BCK algebra** $(A_8, \odot, ^-, 1)$ **is just the starting algebra** $\mathcal{A}^L$, represented by the Hasse diagram from the same Figure 15.48, right side.

**Examples 15.3.21 Building transitive QIMTL algebras: tQIMTL**

Consider the **linearly ordered IMTL algebra**
$IMTL_5 = (A_5 = \{0, a, b, c, 1\}, \odot, ^-, 1)$ from Examples 15.3.3, 1.

**Example 1: we add one element,** $x \parallel_Q b$

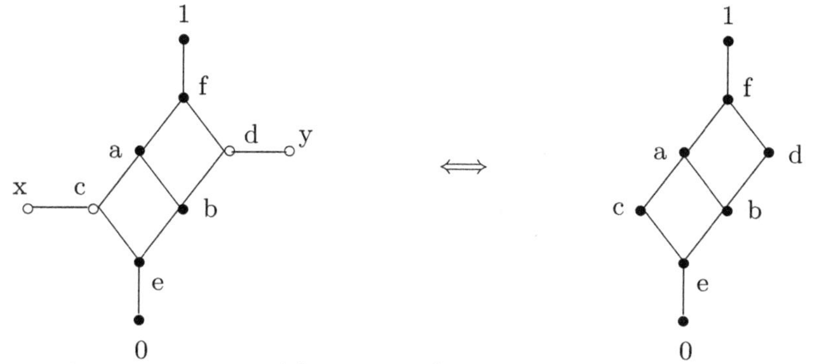


The bounded pre-ordered set $(A_{10}, \leq_m, 0, 1)$ The bounded lattice $(A_8, \leq_m, 0, 1)$

Figure 15.48: $(A_{10}, \leq_m, 0, 1)$ and $(A_8, \leq_m, 0, 1)$

We add to the IMTL chain $0 \leq_m a \leq_m b \leq_m c \leq_m 1$ (see Figure 15.49, right side) one element $x \notin A_5$, $x \parallel_Q b$; thus, we obtain the set $A_6 = \{0, a, b, c, x, 1\}$. We then extend the operations $\odot$ and $\oplus$ on this set $A_6$ by the "method of identic rows/columns" (the row/column of $x$ is identic with the row/column of $b$, excepting the last elements, which are Q-parallel). The operation $^-$ is extended such that the property (DN) be preserved. We obtain the following extended tables:

| $\odot$ | 0 | a | b | c | x | 1 |
|---|---|---|---|---|---|---|
| 0 | 0 | 0 | 0 | 0 | 0 | 0 |
| a | 0 | 0 | 0 | 0 | 0 | a |
| b | 0 | 0 | 0 | a | 0 | b |
| c | 0 | 0 | a | a | a | c |
| x | 0 | 0 | 0 | a | 0 | x |
| 1 | 0 | a | b | c | x | 1 |

and

| $x$ | $x^-$ |
|---|---|
| 0 | 1 |
| a | c |
| b | x |
| c | a |
| x | b |
| 1 | 0 |

, with

| $\oplus$ | 0 | a | b | c | x | 1 |
|---|---|---|---|---|---|---|
| 0 | 0 | a | b | c | b | 1 |
| a | a | c | c | 1 | c | 1 |
| b | b | c | 1 | 1 | 1 | 1 |
| c | c | 1 | 1 | 1 | 1 | 1 |
| x | b | c | 1 | 1 | 1 | 1 |
| 1 | 1 | 1 | 1 | 1 | 1 | 1 |

.

Then, the resulting algebra $(A_6, \odot, ^-, 1)$ is an involutive m-BE algebra verifying (m-BB) ($\Leftrightarrow \ldots \Leftrightarrow$ (m-Tr)) and (prel$_m$), and not verifying (m-An) for $b$ and also (m-Pabs-i) for $(a, 0)$, (m-Pimpl) for $(a, 0)$, (Pqmv) for $(b, a, 0)$, (Pom) for $(b, c)$, (Pmv) for $(b, a)$, $(\Delta_m)$ for $(c, b)$, (WNM$_m$) and (aWNM$_m$) for $(b, c)$, (m-Pdis) for $a, a, b)$. Hence, it is a **linearly pre-ordered left-tQIMTL algebra verifying (prel$_m$)**.

Next, we follow the converse way, as in Subsubsection 15.3.2.1.

The tables of $\wedge_m^M$ and $\wedge_m^B$ are:

| $\wedge_m^M$ | 0 | a | b | c | x | 1 |
|---|---|---|---|---|---|---|
| 0 | 0 | 0 | 0 | 0 | 0 | 0 |
| a | 0 | a | a | a | a | a |
| b | 0 | a | b | a | x | b |
| c | 0 | a | b | c | x | c |
| x | 0 | a | b | a | x | x |
| 1 | 0 | a | b | c | x | 1 |

and

| $\wedge_m^B$ | 0 | a | b | c | x | 1 |
|---|---|---|---|---|---|---|
| 0 | 0 | 0 | 0 | 0 | 0 | 0 |
| a | 0 | a | a | a | a | a |
| b | 0 | a | b | b | b | b |
| c | 0 | a | a | c | a | c |
| x | 0 | a | x | x | x | x |
| 1 | 0 | a | b | c | x | 1 |

.

• The binary relation $\leq_m^M$ is transitive, hence is an order, but not a lattice order, since $\wedge_m^M$ is not commutative.

• The binary relation $\leq_m$ is a pre-order. From the table of $\wedge_m^B$, we obtain: $a \leq_m a, b, c, x, 1$; $b \leq_m b, c, x, 1$; $c \leq_m c, 1$; $x \leq_m x, b, c, 1$: hence, the bounded pre-ordered set $(A_6, \leq_m, 0, 1)$ is represented by the Hasse-type diagram from Figure 15.49, left side. ... Hence, the **associated m-BCK algebra** $(A_5, \odot, ^-, 1)$ **is just the starting algebra** $IMTL_5$, represented by the Hasse diagram from the same Figure 15.49, right side.

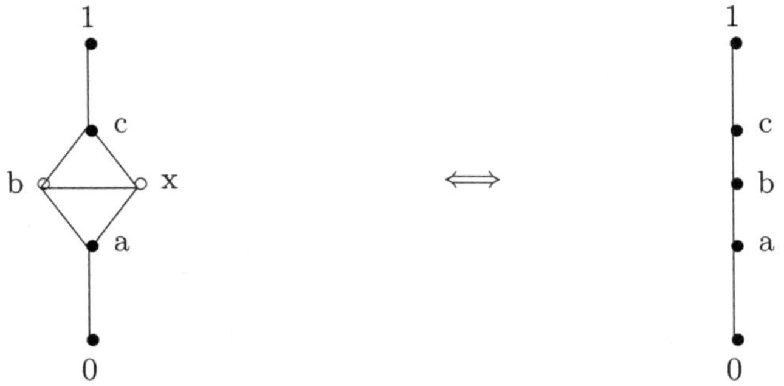


The bounded pre-ordered set $(A_6, \leq_m, 0, 1)$ The IMTL chain $(A_5, \leq_m, 0, 1)$

Figure 15.49: $(A_6, \leq_m, 0, 1)$ and $(A_5, \leq_m, 0, 1)$

**Example 2: we add two elements,** $y \parallel_Q x \parallel_Q b$

We add to the IMTL chain $0 \leq_m a \leq_m b \leq_m c \leq_m 1$ (see also Figure 15.50, right side) two elements $x, y \notin A_5$: $y \parallel_Q x \parallel_Q b$; thus, we obtain the set $A_7 = \{0, a, b, c, x, y, 1\}$. We then extend the operations $\odot$ and $\oplus$ on this set $A_7$ by the "method of identic rows/columns" (the row/column of $x$ is identic with the row/column of $b$, excepting the last elements, which are Q-parallel). The operation $^-$ is extended such that the property (DN) be preserved. We obtain the following extended tables:

| $\odot$ | 0 | a | b | c | x | y | 1 |
|---|---|---|---|---|---|---|---|
| 0 | 0 | 0 | 0 | 0 | 0 | 0 | 0 |
| a | 0 | 0 | 0 | 0 | 0 | 0 | a |
| b | 0 | 0 | 0 | a | 0 | 0 | b |
| c | 0 | 0 | a | a | a | a | c |
| x | 0 | 0 | 0 | a | 0 | 0 | x |
| y | 0 | 0 | 0 | a | 0 | 0 | y |
| 1 | 0 | a | b | c | x | y | 1 |

and

| $x$ | $x^-$ |
|---|---|
| 0 | 1 |
| a | c |
| b | b |
| c | a |
| x | y |
| y | x |
| 1 | 0 |

,

| $\oplus$ | 0 | a | b | c | x | y | 1 |
|---|---|---|---|---|---|---|---|
| 0 | 0 | a | b | c | x | y | 1 |
| a | a | c | c | 1 | c | c | 1 |
| b | b | c | 1 | 1 | 1 | 1 | 1 |
| c | c | 1 | 1 | 1 | 1 | 1 | 1 |
| x | x | c | 1 | 1 | 1 | 1 | 1 |
| y | y | c | 1 | 1 | 1 | 1 | 1 |
| 1 | 1 | 1 | 1 | 1 | 1 | 1 | 1 |

.

Then, the resulting algebra $(A_7, \odot, ^-, 1)$ is an involutive m-BE algebra verifying (m-BB) ($\Leftrightarrow \ldots \Leftrightarrow$ (m-Tr)) and (prel$_m$), and not verifying (m-An) for $b$ and also (m-Pabs-i) for $(a, 0)$, (m-Pimpl) for $(a, 0)$, (Pqmv) for $(b, a, 0)$, (Pom) for $(b, c)$, (Pmv) for $(b, a)$, ($\Delta_m$) for $(c, b)$, (WNM$_m$) and (aWNM$_m$) for $(b, c)$, (m-Pdis) for $a, a, b)$. Hence, it is a **linearly pre-ordered left-tQIMTL algebra verifying**

**(prel$_m$)**.

Next, we follow the converse way, as in Subsubsection 15.3.2.1.

The tables of $\wedge_m^M$ and $\wedge_m^B$ are:

| $\wedge_m^M$ | 0 | a | b | c | x | y | 1 |
|---|---|---|---|---|---|---|---|
| 0 | 0 | 0 | 0 | 0 | 0 | 0 | 0 |
| a | 0 | a | a | a | a | a | a |
| b | 0 | a | b | a | x | y | b |
| c | 0 | a | b | c | x | y | c |
| x | 0 | a | b | a | x | y | x |
| y | 0 | a | b | a | x | y | y |
| 1 | 0 | a | b | c | x | y | 1 |

and

| $\wedge_m^B$ | 0 | a | b | c | x | y | 1 |
|---|---|---|---|---|---|---|---|
| 0 | 0 | 0 | 0 | 0 | 0 | 0 | 0 |
| a | 0 | a | a | a | a | a | a |
| b | 0 | a | b | b | b | b | b |
| c | 0 | a | a | c | a | a | c |
| x | 0 | a | x | x | x | x | x |
| y | 0 | a | y | y | y | y | y |
| 1 | 0 | a | b | c | x | y | 1 |

.

• The binary relation $\leq_m^M$ is transitive, hence is an order, but not a lattice order, since $\wedge_m^M$ is not commutative.

• The binary relation $\leq_m$ is a pre-order. From the table of $\wedge_m^B$, we obtain: $a \leq_m a, b, c, x, y, 1$; $b \leq_m b, c, x, y, 1$; $c \leq_m c, 1$; $x \leq_m x, b, c, y, 1$; $y \leq_m y, b, c, x, 1$; hence, the bounded pre-ordered set $(A_7, \leq_m, 0, 1)$ is represented by the Hasse-type diagram from Figure 15.50, left side. ... Hence, the **associated m-BCK algebra** $(A_5, \odot, ^-, 1)$ **is just the starting algebra** $IMTL_5$, represented by the Hasse diagram from the same Figure 15.50, right side.

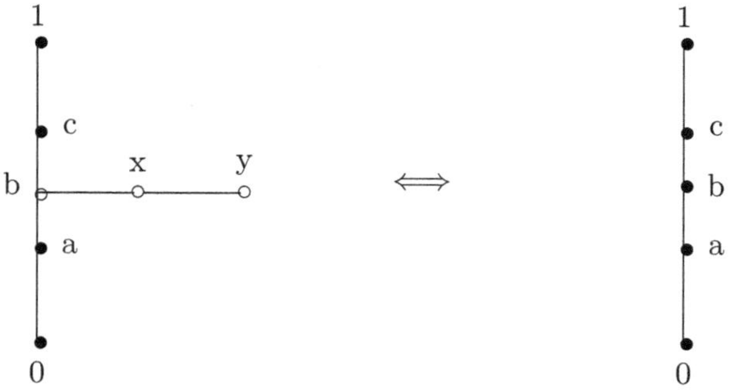


The bounded pre-ordered set $(A_7, \leq_m, 0, 1)$ The IMTL chain $(A_5, \leq_m, 0, 1)$

Figure 15.50: $(A_7, \leq_m, 0, 1)$ and $(A_5, \leq_m, 0, 1)$

**Examples 15.3.22 Building transitive QNM algebras: tQNM**

Consider the **linearly ordered left-NM algebra**
$\mathcal{F}_4^m = (A_4 = \{0, a, b, 1\}, \odot_F, ^-, 1)$ from Examples 15.3.6.

**Example 1: we add two elements,** $x \parallel_Q a$ **and** $y \parallel_Q b$

We add to the NM chain $0 \leq_m a \leq_m b \leq_m 1$ (see Figure 15.52, right side) two elements $x, y \notin A_4$: $x \parallel_Q a$ and $y \parallel_Q b$; thus, we obtain the set $A_6 = \{0, a, b, x, y, 1\}$. We then extend the operations $\odot_F$ and $\oplus_F$ on this set $A_6$ by the "method of identic rows/columns" (the row/column of $x$ is identic with the row/column of $a$, excepting the last elements, which are Q-parallel). The operation $^-$ is extended such that the property (DN) be preserved. We obtain the following extended tables:

| $\odot_F$ | 0 | a | b | x | y | 1 |
|---|---|---|---|---|---|---|
| 0 | 0 | 0 | 0 | 0 | 0 | 0 |
| a | 0 | 0 | 0 | 0 | 0 | a |
| b | 0 | 0 | b | 0 | b | b |
| x | 0 | 0 | 0 | 0 | 0 | x |
| y | 0 | 0 | b | 0 | b | y |
| 1 | 0 | a | b | x | y | 1 |

and

| $x$ | $x^-$ |
|---|---|
| 0 | 1 |
| a | b |
| b | a |
| x | y |
| y | x |
| 1 | 0 |

, with

| $\oplus_F$ | 0 | a | b | x | y | 1 |
|---|---|---|---|---|---|---|
| 0 | 0 | a | b | x | y | 1 |
| a | a | a | 1 | a | 1 | 1 |
| b | b | 1 | 1 | 1 | 1 | 1 |
| x | x | a | 1 | a | 1 | 1 |
| y | y | 1 | 1 | 1 | 1 | 1 |
| 1 | 1 | 1 | 1 | 1 | 1 | 1 |

.

Then, the resulting algebra $(A_6, \odot_F, ^-, 1)$ is an involutive m-BE algebra verifying (m-BB) ($\Leftrightarrow \ldots \Leftrightarrow$ (m-Tr)) and (prel$_m$), (WNM$_m$), (aWNM$_m$), and not verifying (m-Pabs-i) for $(a, 0)$, (m-An) for $(a, x)$, (m-Pimpl) for $(b, 0)$, (Pqmv) for $(a, a, 0)$, (Pom) for $(y, b)$, (Pmv) for $(a, a)$, ($\Delta_m$) for $(b, a)$, (m-Pdis) for $(a, b, a)$; hence, it is an **involutive left-m-pre-BCK algebra verifying (prel$_m$), (aWNM$_m$), (WNM$_m$).**

Next, we follow the converse way, as in Subsubsection 15.3.2.1.

The tables of $\wedge_m^M$ and $\wedge_m^B$ are:

| $\wedge_m^M$ | 0 | a | b | x | y | 1 |
|---|---|---|---|---|---|---|
| 0 | 0 | 0 | 0 | 0 | 0 | 0 |
| a | 0 | a | 0 | x | 0 | a |
| b | 0 | a | b | x | y | b |
| x | 0 | a | 0 | x | 0 | x |
| y | 0 | a | b | x | y | y |
| 1 | 0 | a | b | x | y | 1 |

and

| $\wedge_m^B$ | 0 | a | b | x | y | 1 |
|---|---|---|---|---|---|---|
| 0 | 0 | 0 | 0 | 0 | 0 | 0 |
| a | 0 | a | a | a | a | a |
| b | 0 | 0 | b | 0 | b | b |
| x | 0 | x | x | x | x | x |
| y | 0 | 0 | y | 0 | y | y |
| 1 | 0 | a | b | x | y | 1 |

.

• The binary relation $\leq_m^M$ is transitive, hence it is an order, but not a lattice order, since $\wedge_m^M$ is not commutative. From the table of $\wedge_m^M$, we obtain: $a \leq_m^M a, 1$; $b \leq_m^M b, 1$; $x \leq_m^M x, 1$; $y \leq_m^M y, 1$; hence, the bounded poset $(A_6, \leq_m^M, 0, 1)$ is represented by the Hasse diagram from Figure 15.51.

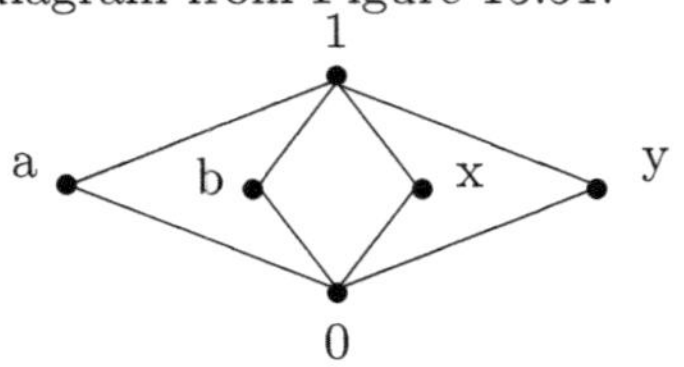


Figure 15.51: The bounded poset $(A_6, \leq_m^M, 0, 1)$

• The binary relation $\leq_m$ is a pre-order. From the table of $\wedge_m^B$, we obtain: $a \leq_m a, b, x, y, 1$; $b \leq_m b, y, 1$; $x \leq_m x, a, b, y, 1$; $y \leq_m y, b, 1$; hence, the bounded pre-ordered set $(A_6, \leq_m, 0, 1)$ is represented by the Hasse-type diagram from Figure 15.52, left side. ... Hence, the **associated m-BCK algebra** $(A_4, \odot_F, ^-, 1)$ **is just the starting algebra** $\mathcal{F}_4^m$, represented by the Hasse diagram from the same Figure 15.52, right side. Hence, $(A_6, \odot, ^-, 1)$ **is a linearly pre-ordered tQNM algebra.**

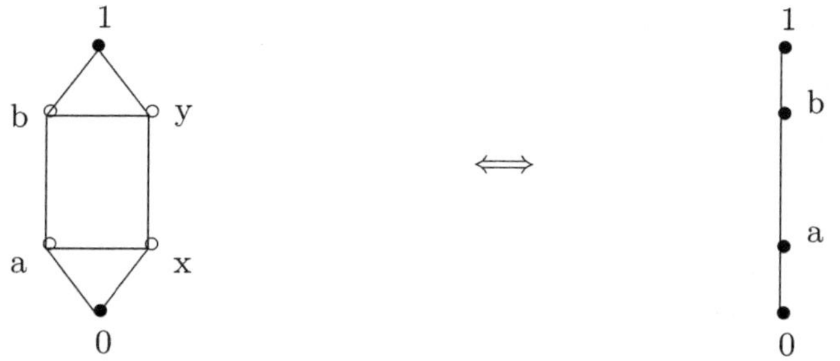


The bounded pre-ordered set $(A_6, \leq_m, 0, 1)$ The bounded chain $(A_4, \leq_m, 0, 1)$

Figure 15.52: $(A_6, \leq_m, 0, 1)$ and $(A_4, \leq_m, 0, 1)$

**Remark.** *If we extend the operations $\odot_F$ and $\oplus_F$ on this set $A_6$ by the "method of Q-parallel rows/columns" (the row/column of $x$ is Q-parallel with the row/column of $a$), then we obtain, for example, the following extended tables:*

| $\odot_F$ | 0 | a | b | x | y | 1 |
|---|---|---|---|---|---|---|
| 0 | 0 | 0 | 0 | 0 | 0 | 0 |
| a | 0 | 0 | 0 | 0 | 0 | a |
| b | 0 | 0 | b | 0 | y | b |
| x | 0 | 0 | 0 | 0 | 0 | x |
| y | 0 | 0 | y | 0 | b | y |
| 1 | 0 | a | b | x | y | 1 |

and

| $x$ | $x^-$ |
|---|---|
| 0 | 1 |
| a | b |
| b | a |
| x | y |
| y | x |
| 1 | 0 |

, with

| $\oplus_F$ | 0 | a | b | x | y | 1 |
|---|---|---|---|---|---|---|
| 0 | 0 | a | b | x | y | 1 |
| a | a | a | 1 | x | 1 | 1 |
| b | b | 1 | 1 | 1 | 1 | 1 |
| x | x | x | 1 | a | 1 | 1 |
| y | y | 1 | 1 | 1 | 1 | 1 |
| 1 | 1 | 1 | 1 | 1 | 1 | 1 |

.

*Then, the resulting algebra $(A_6, \odot_F, ^-, 1)$ is an involutive m-BE algebra verifying (m-BB) ($\Leftrightarrow \ldots \Leftrightarrow$ (m-Tr)) and (prel$_m$), (WNM$_m$), (aWNM$_m$), and not verifying (m-Pabs-i) for $(a, 0)$, (m-An) for $(a, x)$, (m-Pimpl) for $(b, 0)$, (Pqmv) for $(a, a, 0)$,* **(Pom) for** *$(b, y)$, (Pmv) for $(a, a)$, ($\Delta_m$) for $(b, a)$, (m-Pdis) for $(a, b, a)$; hence, it is also an involutive left-m-pre-BCK algebra verifying (prel$_m$), (WNM$_m$), (aWNM$_m$) and not verifying the rest.*

**Example 2: we add four elements, $u \parallel_Q x \parallel_Q a$ and $v \parallel_Q y \parallel_Q b$**

We add to the NM chain $0 \leq_m a \leq_m b \leq_m 1$ (see Figure 15.54, right side) four elements $x, y, u, v \notin A_4$: $u \parallel_Q x \parallel_Q a$ and $v \parallel_Q y \parallel_Q b$; thus, we obtain the set $A_8 = \{0, a, b, x, y, u, v, 1\}$. We then extend the operations $\odot_F$ and $\oplus_F$ on this set $A_8$ by the "method of identic rows/columns" (the row/column of $x$ is identic with the row/column of $a$, excepting the last elements, which are Q-parallel). The operation $^-$ is extended such that the property (DN) be preserved. We obtain the following extended tables:

| $\odot_F$ | 0 | a | b | x | y | u | v | 1 |
|---|---|---|---|---|---|---|---|---|
| 0 | 0 | 0 | 0 | 0 | 0 | 0 | 0 | 0 |
| a | 0 | 0 | 0 | 0 | 0 | 0 | 0 | a |
| b | 0 | 0 | b | 0 | b | 0 | b | b |
| x | 0 | 0 | 0 | 0 | 0 | 0 | 0 | x |
| y | 0 | 0 | b | 0 | b | 0 | b | y |
| u | 0 | 0 | 0 | 0 | 0 | 0 | 0 | u |
| v | 0 | 0 | b | 0 | b | 0 | b | v |
| 1 | 0 | a | b | x | y | u | v | 1 |

with

| $\oplus_F$ | 0 | a | b | x | y | u | v | 1 |
|---|---|---|---|---|---|---|---|---|
| 0 | 0 | a | b | x | y | u | v | 1 |
| a | a | a | 1 | a | 1 | a | 1 | 1 |
| b | b | 1 | 1 | 1 | 1 | 1 | 1 | 1 |
| x | x | a | 1 | a | 1 | a | 1 | 1 |
| y | y | 1 | 1 | 1 | 1 | 1 | 1 | 1 |
| u | u | a | 1 | a | 1 | a | 1 | 1 |
| v | v | 1 | 1 | 1 | 1 | 1 | 1 | 1 |
| 1 | 1 | 1 | 1 | 1 | 1 | 1 | 1 | 1 |

and $(0, a, b, x, y, u, v, 1)^- = (1, b, a, y, x, v, u, 0)$.

Then, the resulting algebra $(A_8, \odot_F, -, 1)$ is an involutive m-BE algebra verifying (m-BB) ($\Leftrightarrow$ ... $\Leftrightarrow$ (m-Tr)) and (prel$_m$), (WNM$_m$), (aWNM$_m$), and not verifying (m-Pabs-i) for $(a, 0)$, (m-An) for $(a, x)$, (m-Pimpl) for $(b, 0)$, (Pqmv) for $(a, a, 0)$, (Pom) for $(y, b)$, (Pmv) for $(a, a)$, $(\Delta_m)$ for $(b, a)$, (m-Pdis) for $(a, b, a)$; hence, it is an **involutive m-pre-BCK algebra verifying (prel$_m$), (aWNM$_m$), (WNM$_m$)**.

Next, we follow the converse way, as in Subsubsection 15.3.2.1.

The tables of $\wedge_m^M$ and $\wedge_m^B$ are:

| $\wedge_m^M$ | 0 | a | b | x | y | u | v | 1 |
|---|---|---|---|---|---|---|---|---|
| 0 | 0 | 0 | 0 | 0 | 0 | 0 | 0 | 0 |
| a | 0 | a | 0 | x | 0 | u | 0 | a |
| b | 0 | a | b | x | y | u | v | b |
| x | 0 | a | 0 | x | 0 | u | 0 | x |
| y | 0 | a | b | x | y | u | v | y |
| u | 0 | a | 0 | x | 0 | u | 0 | u |
| v | 0 | a | b | x | y | u | v | v |
| 1 | 0 | a | b | x | y | u | v | 1 |

and

| $\wedge_m^B$ | 0 | a | b | x | y | u | v | 1 |
|---|---|---|---|---|---|---|---|---|
| 0 | 0 | 0 | 0 | 0 | 0 | 0 | 0 | 0 |
| a | 0 | a | a | a | a | a | a | a |
| b | 0 | 0 | b | 0 | b | 0 | b | b |
| x | 0 | x | x | x | x | x | x | x |
| y | 0 | 0 | y | 0 | y | 0 | y | y |
| u | 0 | u | u | u | u | u | u | u |
| v | 0 | 0 | v | 0 | v | 0 | v | v |
| 1 | 0 | a | b | x | y | u | v | 1 |

.

• The binary relation $\leq_m^M$ is transitive, hence it is an order, but not a lattice order, since $\wedge_m^M$ is not commutative. From the table of $\wedge_m^M$, we obtain: $a \leq_m^M a, 1$; $b \leq_m^M b, 1$; $x \leq_m^M x, 1$; $y \leq_m^M y, 1$; $u \leq_m^M u, 1$; $v \leq_m^M v, 1$; hence, the bounded poset $(A_8, \leq_m^M, 0, 1)$ is represented by the Hasse diagram from Figure 15.53.

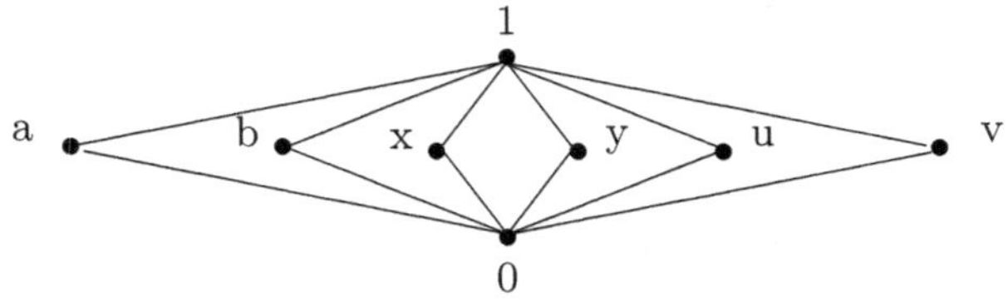


Figure 15.53: The bounded poset $(A_8, \leq_m^M, 0, 1)$

• The binary relation $\leq_m$ is a pre-order. From the table of $\wedge_m^B$, we obtain: $a \leq_m a, b, x, y, u, v, 1$; $b \leq_m b, y, v, 1$; $x \leq_m x, a, b, y, u, v, 1$; $y \leq_m y, b, v, 1$; $u \leq_m u, a, b, x, y, v, 1$; $v \leq_m v, b, y, 1$; hence, the bounded pre-ordered set $(A_8, \leq_m, 0, 1)$ is represented by the Hasse-type diagram from Figure 15.54, left side. ... Hence, the

**associated m-BCK algebra** $(A_4, \odot_F, ^-, 1)$ **is just the starting algebra** $\mathcal{F}_4^m$, represented by the Hasse diagram from the same Figure 15.54, right side. Hence, $(A_8, \odot, ^-, 1)$ **is a linearly pre-ordered tQNM algebra.**

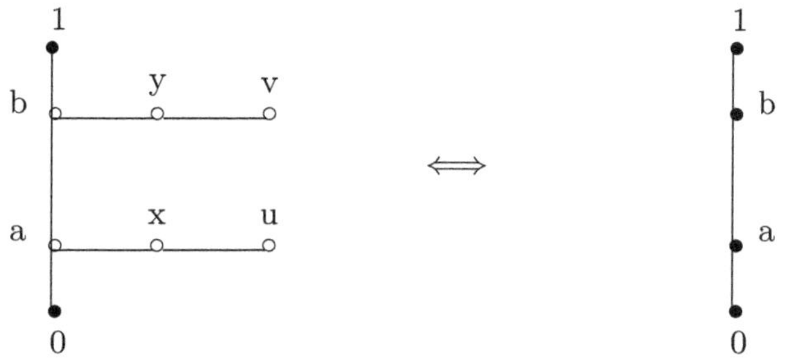


The bounded pre-ordered set $(A_8, \leq_m, 0, 1)$ The bounded chain $(A_4, \leq_m, 0, 1)$

Figure 15.54: $(A_8, \leq_m, 0, 1)$ and $(A_4, \leq_m, 0, 1)$

**Example 15.3.23 Building transitive PreMV algebra: tPreMV**

Consider the proper **linearly ordered left-MV algebra** with the minimum number of elements, $\mathcal{L}_4^L = (L_4 = \{0, a, b, 1\}, \odot, ^-, 1)$, from Examples 15.3.4.

**Example 1: we add** $x \parallel_Q a$ **and** $y \parallel_Q b$

We add to the MV chain $0 \leq_m a \leq_m b \leq_m 1$ (see the previous Figure 15.52, right side) two elements $x, y \notin L_4$ such that $x \parallel_Q a$ and $y \parallel_Q b$; thus, we obtain the set $A_6 = \{0, a, b, x, y, 1\}$. We then extend the operations $\odot$ and $\oplus$ on this set $A_6$ by the "method of identic rows/columns" (the row/column of $x$ is identic with the row/column of $a$, excepting the last elements, which are Q-parallel); the operation $^-$ is extended such that the property (DN) be preserved. Thus, the extended operations are:

| $\odot$ | 0 | a | b | x | y | 1 |
|---|---|---|---|---|---|---|
| 0 | 0 | 0 | 0 | 0 | 0 | 0 |
| a | 0 | 0 | 0 | 0 | 0 | a |
| b | 0 | 0 | a | 0 | a | b |
| x | 0 | 0 | 0 | 0 | 0 | x |
| y | 0 | 0 | a | 0 | a | y |
| 1 | 0 | a | b | x | y | 1 |

and

| $x$ | $x^-$ |
|---|---|
| 0 | 1 |
| a | b |
| b | a |
| x | y |
| y | x |
| 1 | 0 |

, with

| $\oplus$ | 0 | a | b | x | y | 1 |
|---|---|---|---|---|---|---|
| 0 | 0 | a | b | x | y | 1 |
| a | a | b | 1 | b | 1 | 1 |
| b | b | 1 | 1 | 1 | 1 | 1 |
| x | x | b | 1 | b | 1 | 1 |
| y | y | 1 | 1 | 1 | 1 | 1 |
| 1 | 1 | 1 | 1 | 1 | 1 | 1 |

.

Then, the resulting algebra $(A_6, \odot, ^-, 1)$ is an involutive m-BE algebra verifying (m-BB) ($\Leftrightarrow \ldots \Leftrightarrow$ (m-Tr)) and (Pmv) (hence $(\Delta_m)$) and (prel$_m$), and not verifying (m-An) for $(a, x)$, (m-Pabs-i) for $(a, 0)$, (m-Pimpl) for $(a, 0)$, (Pqmv) for $(y, 1, b)$, (Pom) for $(y, b)$, (WNM$_m$) and (aWNM$_m$) for $(b, b)$, (m-Pdis) for $(a, a, b)$; hence, it is a **linearly ordered transitive left-PreMV algebra.**

Next, we follow the converse way, as in Subsubsection 15.3.2.1.

The tables of $\wedge_m^M$ and $\wedge_m^B$ are:

| $\wedge_m^M$ | 0 | a | b | x | y | 1 |
|---|---|---|---|---|---|---|
| 0 | 0 | 0 | 0 | 0 | 0 | 0 |
| a | 0 | a | a | x | a | a |
| b | 0 | a | b | x | y | b |
| x | 0 | a | a | x | a | x |
| y | 0 | a | b | x | y | y |
| 1 | 0 | a | b | x | y | 1 |

and

| $\wedge_m^B$ | 0 | a | b | x | y | 1 |
|---|---|---|---|---|---|---|
| 0 | 0 | 0 | 0 | 0 | 0 | 0 |
| a | 0 | a | a | a | a | a |
| b | 0 | a | b | a | b | b |
| x | 0 | x | x | x | x | x |
| y | 0 | a | y | a | y | y |
| 1 | 0 | a | b | x | y | 1 |

.

• The binary relation $\leq_m$ is a pre-order. From the table of $\wedge_m^B$, we obtain: $a \leq_m a, b, x, y, 1$; $b \leq_m b, y, 1$; $x \leq_m x, a, b, y, 1$; $y \leq_m y, b, 1$; hence, the bounded pre-ordered set $(A_6, \leq_m, 0, 1)$ is represented by the same Hasse-type diagram from the previous Figure 15.52, left side. ... Hence, the associated m-BCK algebra $(A_4, \odot, ^-, 1)$ is just the starting algebra $\mathcal{L}_4^L$, represented by the Hasse diagram from the same previous Figure 15.52, right side.

**Examples 15.3.24 Building transitive MMV algebras: tMMV**

**Example 1: building tMMV from $\mathcal{L}_{2\times 2}^L$**

Consider the **non-linearly ordered Boolean algebra** $\mathcal{L}_{2\times 2}^L = (L_{2\times 2} = \{0, a, b, 1\}, \odot, ^-, 1)$ from Examples 15.3.8, 1.

**Example 1.1: we add two elements,** $x \parallel_Q a$ **and** $y \parallel_Q b$

We add to the bounded lattice $(L_{2\times 2}, \leq_m, 0, 1)$ from Figure 15.11 (see also Figure 15.55, right side) two elements $x, y \notin L_{2\times 2}$, $x \parallel_Q a$ and $y \parallel_Q b$; thus, we obtain the set $A_6 = \{0, a, b, x, y, 1\}$. We then extend the operations $\odot$ and $\oplus$ on this set $A_6$ by the "method of identic rows/columns" (the row/column of $x$ is identic with the row/column of $a$, excepting the last elements, which are Q-parallel). The operation $^-$ is extended such that the property (DN) be preserved. We obtain the following extended tables:

| $\odot$ | 0 | a | b | x | y | 1 |
|---|---|---|---|---|---|---|
| 0 | 0 | 0 | 0 | 0 | 0 | 0 |
| a | 0 | a | 0 | a | 0 | a |
| b | 0 | 0 | b | 0 | b | b |
| x | 0 | a | 0 | a | 0 | x |
| y | 0 | 0 | b | 0 | b | y |
| 1 | 0 | a | b | x | y | 1 |

and

| $x$ | $x^-$ |
|---|---|
| 0 | 1 |
| a | b |
| b | a |
| x | y |
| y | x |
| 1 | 0 |

, with

| $\oplus$ | 0 | a | b | x | y | 1 |
|---|---|---|---|---|---|---|
| 0 | 0 | a | b | x | y | 1 |
| a | a | a | 1 | a | 1 | 1 |
| b | b | 1 | b | 1 | b | 1 |
| x | x | a | 1 | a | 1 | 1 |
| y | y | 1 | b | 1 | b | 1 |
| 1 | 1 | 1 | 1 | 1 | 1 | 1 |

.

Then, the resulting algebra $(A_6, \odot, ^-, 1)$ is an involutive m-BE algebra verifying (m-BB) ($\Leftrightarrow \ldots \Leftrightarrow$ (m-Tr)) and ($\Delta_m$), and (prel$_m$), (WNM$_m$), (aWNM$_m$), and not verifying (m-Pabs-i) for $(x, 0)$, (m-An) for $(a, x)$, (m-Pimpl) for $(x, 0)$, (Pqmv) for $(x, 0, a)$, (Pom) for $(x, a)$, (Pmv) for $(x, a)$, (m-Pdis) for $(a, b, x)$; hence, it is a **transitive MMV algebra verifying (prel$_m$), (aWNM$_m$), (WNM$_m$).**

Next, we follow the converse way, as in Subsubsection 15.3.2.1.

The tables of $\wedge_m^M$ and $\wedge_m^B$ are:

| $\wedge_m^M$ | 0 | a | b | x | y | 1 |
|---|---|---|---|---|---|---|
| 0 | 0 | 0 | 0 | 0 | 0 | 0 |
| a | 0 | a | 0 | x | 0 | a |
| b | 0 | 0 | b | 0 | d | b |
| x | 0 | a | 0 | x | 0 | x |
| y | 0 | 0 | b | 0 | y | y |
| 1 | 0 | a | b | x | y | 1 |

and

| $\wedge_m^B$ | 0 | a | b | x | y | 1 |
|---|---|---|---|---|---|---|
| 0 | 0 | 0 | 0 | 0 | 0 | 0 |
| a | 0 | a | 0 | a | 0 | a |
| b | 0 | 0 | b | 0 | b | b |
| x | 0 | x | 0 | x | 0 | x |
| y | 0 | 0 | y | 0 | y | y |
| 1 | 0 | a | b | x | y | 1 |

.

• The binary relation $\leq_m^M$ is transitive, hence it is an order, but not a lattice order, since $\wedge_m^M$ is not commutative.

• The binary relation $\leq_m$ is a pre-order. From the table of $\wedge_m^B$, we obtain: $a \leq_m a, x, 1$; $b \leq_m b, y, 1$; $x \leq_m x, a, 1$; $y \leq_m y, b, 1$; hence, the bounded pre-ordered set $(A_6, \leq_m, 0, 1)$ is represented by the Hasse-type diagram from Figure 15.55, left side. ... Hence, the **associated m-BCK algebra** $(A_4, \leq_m, 0, 1)$ **is just the starting algebra** $\mathcal{L}_{2\times 2}^L$, represented by the Hasse diagram from Figure 15.55, right side.

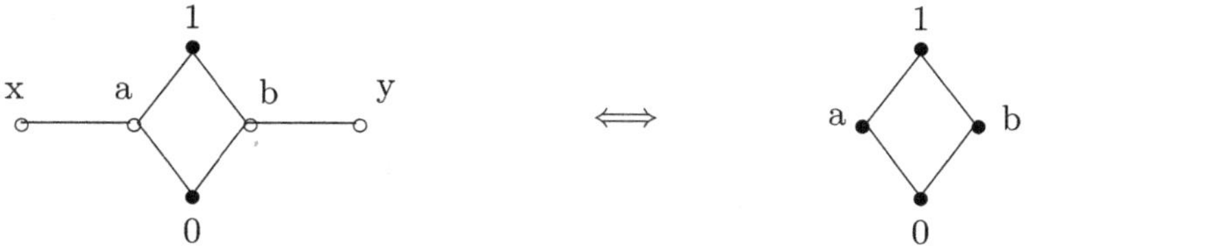


The bounded pre-ordered set $(A_6, \leq_m, 0, 1)$ The bounded lattice $(A_4, \leq_m, 0, 1)$

Figure 15.55: $(A_6, \leq_m, 0, 1)$ and $(A_4 = L_{2\times 2}, \leq_m, 0, 1)$

**Example 2: building tMMV from $\mathcal{L}_{3\times 2}^L$**

Consider the $_{(WNM)}$**MV algebra** $\mathcal{L}_{3\times 2}^L = (L_{3\times 2} = \{0, a, b, c, d, 1\}, \odot, ^-, 1)$ from Examples 15.3.8, 2.

**Example 2.1: we add two elements, $x \parallel_Q a$ and $y \parallel_Q d$**

We add to the bounded lattice $(L_{3\times 2}, \leq_m, 0, 1)$ from Figure 15.12 (see also Figure 15.56, right side) two elements $x, y \notin L_{3\times 2}$, $x \parallel_Q a$ and $y \parallel_Q d$; thus, we obtain the set $A_8 = \{0, a, b, c, d, x, y, 1\}$. We then extend the operations $\odot$ and $\oplus$ on this set $A_8$ by the "method of identic rows/columns" (the row/column of $x$ is identic with the row/column of $a$, excepting the last elements, which are Q-parallel). The operation $^-$ is extended such that the property (DN) be preserved. We obtain the following extended tables:

| $\odot$ | 0 | a | b | c | d | x | y | 1 |
|---|---|---|---|---|---|---|---|---|
| 0 | 0 | 0 | 0 | 0 | 0 | 0 | 0 | 0 |
| a | 0 | a | 0 | a | 0 | a | 0 | a |
| b | 0 | 0 | 0 | 0 | b | 0 | b | b |
| c | 0 | a | 0 | a | b | a | b | c |
| d | 0 | 0 | b | b | d | 0 | d | d |
| x | 0 | a | 0 | a | 0 | a | 0 | x |
| y | 0 | 0 | b | b | d | 0 | d | y |
| 1 | 0 | a | b | c | d | x | y | 1 |

with

| $\oplus$ | 0 | a | b | c | d | x | y | 1 |
|---|---|---|---|---|---|---|---|---|
| 0 | 0 | a | b | c | d | x | y | 1 |
| a | a | a | c | c | 1 | a | 1 | 1 |
| b | b | c | d | 1 | d | c | d | 1 |
| c | c | c | 1 | 1 | 1 | c | 1 | 1 |
| d | d | 1 | d | 1 | d | 1 | d | 1 |
| x | x | a | c | c | 1 | a | 1 | 1 |
| y | y | 1 | d | 1 | d | 1 | d | 1 |
| 1 | 1 | 1 | 1 | 1 | 1 | 1 | 1 | 1 |

and $(0, a, b, c, d, x, y, 1)^- = (1, d, c, b, a, y, x, 0)$.

Then, the resulting algebra $(A_8, \odot, ^-, 1)$ is an involutive m-BE algebra verifying (m-BB) ($\Leftrightarrow \ldots \Leftrightarrow$ (m-Tr)) and ($\Delta_m$), and (prel$_m$), (WNM$_m$), (aWNM$_m$), and not verifying (m-Pabs-i) for $(x, 0)$, (m-An) for $(a, x)$, (m-Pimpl) for $(b, 0)$, (G) for $b$, (Pqmv) for $(x, 0, a)$, (Pom) for $(x, a)$, (Pmv) for $(x, a)$, (m-Pdis) for $(a, d, x)$; hence, it is a **transitive MMV algebra verifying (prel$_m$), (aWNM$_m$), (WNM$_m$).**

Next, we follow the converse way, as in Subsubsection 15.3.2.1.

The tables of $\wedge_m^M$ and $\wedge_m^B$ are:

| $\wedge_m^M$ | 0 | a | b | c | d | x | y | 1 |
|---|---|---|---|---|---|---|---|---|
| 0 | 0 | 0 | 0 | 0 | 0 | 0 | 0 | 0 |
| a | 0 | a | 0 | a | 0 | x | 0 | a |
| b | 0 | 0 | b | b | b | 0 | b | b |
| c | 0 | a | b | c | b | x | b | c |
| d | 0 | 0 | b | b | d | 0 | y | d |
| x | 0 | a | 0 | a | 0 | x | 0 | x |
| y | 0 | 0 | b | b | d | 0 | y | y |
| 1 | 0 | a | b | c | d | x | y | 1 |

and

| $\wedge_m^B$ | 0 | a | b | c | d | x | y | 1 |
|---|---|---|---|---|---|---|---|---|
| 0 | 0 | 0 | 0 | 0 | 0 | 0 | 0 | 0 |
| a | 0 | a | 0 | a | 0 | a | 0 | a |
| b | 0 | 0 | b | b | b | 0 | b | b |
| c | 0 | a | b | c | b | a | b | c |
| d | 0 | 0 | b | b | d | 0 | d | d |
| x | 0 | x | 0 | x | 0 | x | 0 | x |
| y | 0 | 0 | b | b | y | 0 | y | y |
| 1 | 0 | a | b | c | d | x | y | 1 |

.

• The binary relation $\leq_m^M$ is transitive, hence it is an order, but not a lattice order, since $\wedge_m^M$ is not commutative.

• The binary relation $\leq_m$ is a pre-order. From the table of $\wedge_m^B$, we obtain: $a \leq_m a, c, x, 1$; $b \leq_m b, c, d, y, 1$; $c \leq_m c, 1$; $d \leq_m d, y, 1$; $x \leq_m x, a, c, 1$; $y \leq_m y, d, 1$; hence, the bounded pre-ordered set $(A_8, \leq_m, 0, 1)$ is represented by the Hasse-type diagram from Figure 15.56, left side. ... Hence, the **associated m-BCK algebra** $(A_6, \leq_m, 0, 1)$ **is just the starting algebra** $\mathcal{L}^L_{3\times 2}$, represented by the Hasse diagram from the same Figure 15.56, right side.

**Examples 15.3.25 Building tQMV algebras verifying (m-Pabs-i): tOMWL $=$ $_{(WNM_m)}$tQMV**

Consider the **linearly ordered** $_{(WNM)}$**MV algebra** $\mathcal{L}^L_3 = (A_3 = \{0, a, 1\}, \odot, ^-, 1)$ from Examples 15.3.4.

**Example 1: we add one element,** $x \parallel_Q a$

We add to the bounded lattice $(A_3, \leq_m, 0, 1)$ from Figure 15.26, right side, one element $x \notin A_3$, $x \parallel_Q a$; thus, we obtain the set $A_4 = \{0, a, x, 1\}$. We then extend

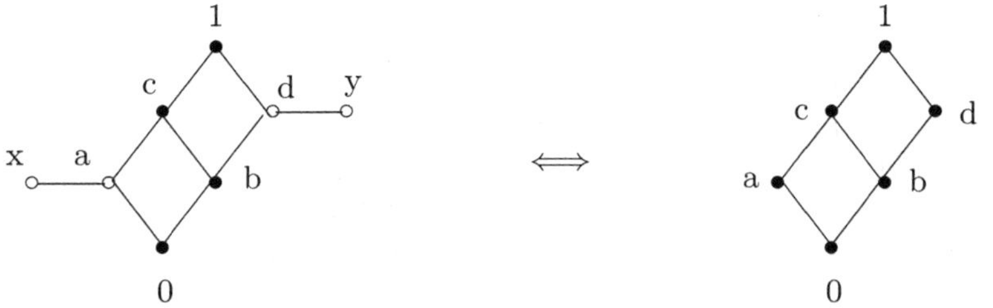


The bounded pre-ordered set $(A_8, \leq_m, 0, 1)$ The bounded lattice $(A_6, \leq_m, 0, 1)$

Figure 15.56: $(A_8, \leq_m, 0, 1)$ and $(A_6 = L_{3\times 2}, \leq_m, 0, 1)$

the operations $\odot$ and $\oplus$ on this set $A_4$ by the "method of identic rows/columns" (the row/column of $x$ is identic with the row/column of $a$, excepting the last elements, which are Q-parallel) - note that, given the table of $\odot$, this is the only choice we can have. The operation $^-$ is extended such that the property (DN) be preserved. We obtain the following extended tables:

| $\odot$ | 0 | a | x | 1 |
|---|---|---|---|---|
| 0 | 0 | 0 | 0 | 0 |
| a | 0 | 0 | 0 | a |
| x | 0 | 0 | 0 | x |
| 1 | 0 | a | x | 1 |

and

| $x$ | $x^-$ |
|---|---|
| 0 | 1 |
| a | a |
| x | x |
| 1 | 0 |

, with

| $\oplus$ | 0 | a | x | 1 |
|---|---|---|---|---|
| 0 | 0 | a | x | 1 |
| a | a | 1 | 1 | 1 |
| x | x | 1 | 1 | 1 |
| 1 | 1 | 1 | 1 | 1 |

.

Then, the resulting algebra $(A_4, \odot, ^-, 1)$ verifies (PU), (Pcomm), (Pass), (m-La), (m-Re), (m-BB) ($\Leftrightarrow \ldots \Leftrightarrow$ (m-Tr)), (DN), (m-Pabs-i), (Pqmv) (hence (Pom), (Pmv), $(\Delta_m)$), and (prel$_m$), (aWNM$_m$), (WNM$_m$), and not verifying (m-An) for $(a, x)$, (m-Pimpl) for $(a, 0)$, (G) for $a$, (m-Pdis) for $(a, a, a)$; hence, it is a transitive left-QMV algebra and, hence, (m-Pabs-i) $\Longleftrightarrow$ (aWNM$_m$) $\Longleftrightarrow$ (WNM$_m$), by Theorem 14.2.12, i.e. $(A_4, \odot, ^-, 1)$ is a **left-tOMWL**.

**Remark.** *This tOMWL is isomorphic to the tOMWL $\mathcal{A}_4^L$ $(x \mapsto b)$ from Examples 15.3.14, 1.*

**Example 2: we add two elements,** $y \parallel_Q x \parallel_Q a$

We add to the bounded lattice $(A_3, \leq_m, 0, 1)$ from Figure 15.28, right side, two elements $x, y \notin A_3$, $y \parallel_Q x \parallel_Q a$; thus, we obtain the set $A_5 = \{0, a, x, y, 1\}$. We then extend the operations $\odot$ and $\oplus$ on this set $A_5$ by the "method of identic rows/columns" (the row/column of $x$ is identic with the row/column of $a$, excepting the last elements, which are Q-parallel) - note that, given the table of $\odot$, this is the only choice we can have. The operation $^-$ is extended such that the property (DN) be preserved. We obtain the following extended tables:

| $\odot$ | 0 | a | x | y | 1 |
|---|---|---|---|---|---|
| 0 | 0 | 0 | 0 | 0 | 0 |
| a | 0 | 0 | 0 | 0 | a |
| x | 0 | 0 | 0 | 0 | x |
| y | 0 | 0 | 0 | 0 | y |
| 1 | 0 | a | b | c | 1 |

and

| $x$ | $x^-$ |
|---|---|
| 0 | 1 |
| a | a |
| x | x |
| y | y |
| 1 | 0 |

, with

| $\oplus$ | 0 | a | x | y | 1 |
|---|---|---|---|---|---|
| 0 | 0 | a | x | y | 1 |
| a | a | 1 | 1 | 1 | 1 |
| x | x | 1 | 1 | 1 | 1 |
| y | y | 1 | 1 | 1 | 1 |
| 1 | 1 | 1 | 1 | 1 | 1 |

.

Then, the resulting algebra $(A_5, \odot, ^-, 1)$ verifies (PU), (Pcomm), (Pass), (m-La), (m-Re), (m-BB) ($\Leftrightarrow \ldots \Leftrightarrow$ (m-Tr)), (DN), (m-Pabs-i), (Pqmv) (hence (Pom), (Pmv), ($\Delta_m$)), and (prel$_m$), (aWNM$_m$), (WNM$_m$), and not verifying (m-An) for $(a, x)$, (m-Pimpl) for $(a, 0)$, (G) for $a$, (m-Pdis) for $(a, a, a)$. Hence, it is a transitive left-QMV algebra and, hence, (m-Pabs-i) $\Longleftrightarrow$ (aWNM$_m$) $\Longleftrightarrow$ (WNM$_m$), by Theorem 14.2.12.

**Remark.** *This tOMWL is isomorphic to the tOMWL $\mathcal{A}_5^L$ ($x \mapsto b$, $y \mapsto c$) from Examples 15.3.14, 2.*

**Examples 15.3.26 Building transitive OWLs: tOWL**

Consider the **non-linearly ordered** $_{(WNM)}$**MV algebra** $\mathcal{L}^L_{3\times 3} = (L_{3\times 3} = \{0, a, b, c, d, e, f, g, 1\}, \odot, ^-, 1)$ from Examples 15.3.8, 3.

**Example 1: we add one element,** $x \parallel_Q d$

We add to the bounded lattice $(L_{3\times 3}, \leq_m, 0, 1)$ from Figure 15.13 (see also Figure 15.37, right side) one element $x \notin L_{3\times 3}$, $x \parallel_Q d$; thus, we obtain the set $A_{10} = \{0, a, b, c, d, e, f, g, x, 1\}$. We then extend the operations $\odot$ and $\oplus$ on this set $A_{10}$ by the "method of identic rows/columns" (the row/column of $x$ is identic with the row/column of $d$, excepting the last elements, which are Q-parallel). The operation $^-$ is extended such that the property (DN) be preserved. We obtain the following extended tables:

| $\odot$ | 0 | a | b | c | d | e | f | g | x | 1 |
|---|---|---|---|---|---|---|---|---|---|---|
| 0 | 0 | 0 | 0 | 0 | 0 | 0 | 0 | 0 | 0 | 0 |
| a | 0 | 0 | a | 0 | 0 | a | 0 | 0 | 0 | a |
| b | 0 | a | b | 0 | a | b | 0 | a | a | b |
| c | 0 | 0 | 0 | 0 | 0 | 0 | c | c | 0 | c |
| d | 0 | 0 | a | 0 | 0 | a | c | c | 0 | d |
| e | 0 | a | b | 0 | a | b | c | d | a | e |
| f | 0 | 0 | 0 | c | c | c | f | f | c | f |
| g | 0 | 0 | a | c | c | d | f | f | c | g |
| x | 0 | 0 | a | 0 | 0 | a | c | c | 0 | x |
| 1 | 0 | a | b | c | d | e | f | g | x | 1 |

and

| $x$ | $x^-$ |
|---|---|
| 0 | 1 |
| a | g |
| b | f |
| c | e |
| d | d |
| e | c |
| f | b |
| g | a |
| x | x |
| 1 | 0 |

with

| ⊕ | 0 | a | b | c | d | e | f | g | x | 1 |
|---|---|---|---|---|---|---|---|---|---|---|
| 0 | 0 | a | b | c | d | e | f | g | x | 1 |
| a | a | b | b | d | e | e | g | 1 | e | 1 |
| b | b | b | b | e | e | e | 1 | 1 | e | 1 |
| c | c | d | e | f | g | 1 | f | g | g | 1 |
| d | d | e | e | g | 1 | 1 | g | 1 | 1 | 1 |
| e | e | e | e | 1 | 1 | 1 | 1 | 1 | 1 | 1 |
| f | f | g | 1 | f | g | 1 | f | g | g | 1 |
| g | g | 1 | 1 | g | 1 | 1 | g | 1 | 1 | 1 |
| x | x | e | e | g | 1 | 1 | g | 1 | 1 | 1 |
| 1 | 1 | 1 | 1 | 1 | 1 | 1 | 1 | 1 | 1 | 1 |

.

Then, the resulting algebra $(A_{10}, \odot, ^-, 1)$ is an involutive left-m-BE algebra verifying (m-Pabs-i), (m-BB) ($\Leftrightarrow \ldots \Leftrightarrow$ (m-Tr)), (Pmv) (hence ($\Delta_m$)), and (prel$_m$), (aWNM$_m$), (WNM$_m$), and not verifying (m-An) for $(d, x)$, (m-Pimpl) for $(a, 0)$, (Pqmv) for $(x, b, f)$, (Pom) for $(x, b)$, (m-Pdis) for (a,a,a). Hence, it is a **proper left-tOWL verifying (Pmv)**.

**Remark.** *This tOWL is isomorphic to the tOWL from Examples 10.4.9, 4.*

**Example 2: we add two elements,** $y \parallel_Q x \parallel_Q d$

We add to the bounded lattice $(L_{3\times3}, \leq_m, 0, 1)$ from Figure 15.13 (see also Figure 15.57, right side) two elements $x, y \notin L_{3\times3}$, $y \parallel_Q x \parallel_Q d$; thus, we obtain the set $A_{11} = \{0, a, b, c, d, e, f, g, x, y, 1\}$. We then extend the operations $\odot$ and $\oplus$ on this set $A_{11}$ by the "method of identic rows/columns" (the row/column of $x$ is identic with the row/column of $d$, excepting the last elements, which are Q-parallel). The operation $^-$ is extended such that the property (DN) be preserved. We obtain the following extended tables:

| ⊙ | 0 | a | b | c | d | e | f | g | x | y | 1 |
|---|---|---|---|---|---|---|---|---|---|---|---|
| 0 | 0 | 0 | 0 | 0 | 0 | 0 | 0 | 0 | 0 | 0 | 0 |
| a | 0 | 0 | a | 0 | 0 | a | 0 | 0 | 0 | 0 | a |
| b | 0 | a | b | 0 | a | b | 0 | a | a | a | b |
| c | 0 | 0 | 0 | 0 | 0 | 0 | c | c | 0 | 0 | c |
| d | 0 | 0 | a | 0 | 0 | a | c | c | 0 | 0 | d |
| e | 0 | a | b | 0 | a | b | c | d | a | a | e |
| f | 0 | 0 | 0 | c | c | c | f | f | c | c | f |
| g | 0 | 0 | a | c | c | d | f | f | c | c | g |
| x | 0 | 0 | a | 0 | 0 | a | c | c | 0 | 0 | x |
| y | 0 | 0 | a | 0 | 0 | a | c | c | 0 | 0 | y |
| 1 | 0 | a | b | c | d | e | f | g | x | y | 1 |

with

| ⊕ | 0 | a | b | c | d | e | f | g | x | y | 1 |
|---|---|---|---|---|---|---|---|---|---|---|---|
| 0 | 0 | a | b | c | d | e | f | g | x | y | 1 |
| a | a | b | b | d | e | e | g | 1 | e | e | 1 |
| b | b | b | b | e | e | e | 1 | 1 | e | e | 1 |
| c | c | d | e | f | g | 1 | f | g | g | g | 1 |
| d | d | e | e | g | 1 | 1 | g | 1 | 1 | 1 | 1 |
| e | e | e | e | 1 | 1 | 1 | 1 | 1 | 1 | 1 | 1 |
| f | f | g | 1 | f | g | 1 | f | g | g | g | 1 |
| g | g | 1 | 1 | g | 1 | 1 | g | 1 | 1 | 1 | 1 |
| x | x | e | e | g | 1 | 1 | g | 1 | 1 | 1 | 1 |
| y | y | e | e | g | 1 | 1 | g | 1 | 1 | 1 | 1 |
| 1 | 1 | 1 | 1 | 1 | 1 | 1 | 1 | 1 | 1 | 1 | 1 |

and $(0, a, b, c, d, e, f, g, x, y, 1)^- = (1, g, f, e, d, c, b, a, y, x, 0)$.

Then, the resulting algebra $(A_{11}, \odot, ^-, 1)$ is an involutive left-m-BE algebra verifying (m-Pabs-i), (m-BB) ($\Leftrightarrow \ldots \Leftrightarrow$ (m-Tr)), (Pmv) (hence $(\Delta_m)$), and (prel$_m$), (aWNM$_m$), (WNM$_m$), and not verifying (m-An) for $(d, x)$, (m-Pimpl) for $(a, 0)$, (Pqmv) for $(x, b, f)$, (Pom) for $(x, b)$, (m-Pdis) for (a,a,a). Hence, it is a **proper left-tOWL verifying (Pmv)**.

Next, we follow the converse way, as in Subsubsection 15.3.2.1, Examples 15.3.15, 4.

The tables of $\wedge_m^M$ and $\wedge_m^B$ are:

| $\wedge_m^M$ | 0 | a | b | c | d | e | f | g | x | y | 1 |
|---|---|---|---|---|---|---|---|---|---|---|---|
| 0 | 0 | 0 | 0 | 0 | 0 | 0 | 0 | 0 | 0 | 0 | 0 |
| a | 0 | a | a | 0 | a | a | 0 | a | a | a | a |
| b | 0 | a | b | 0 | a | b | 0 | a | a | a | b |
| c | 0 | 0 | 0 | c | c | c | c | c | c | c | c |
| d | 0 | a | a | c | d | d | c | d | x | y | d |
| e | 0 | a | b | c | d | e | c | d | x | y | e |
| f | 0 | 0 | 0 | c | c | c | f | f | c | c | f |
| g | 0 | a | a | c | d | d | f | g | x | y | g |
| x | 0 | a | a | c | d | d | c | d | x | y | x |
| y | 0 | a | a | c | d | d | c | d | x | y | y |
| 1 | 0 | a | b | c | d | e | f | g | x | y | 1 |

and

| $\wedge_m^B$ | 0 | a | b | c | d | e | f | g | x | y | 1 |
|---|---|---|---|---|---|---|---|---|---|---|---|
| 0 | 0 | 0 | 0 | 0 | 0 | 0 | 0 | 0 | 0 | 0 | 0 |
| a | 0 | a | a | 0 | a | a | 0 | a | a | a | a |
| b | 0 | a | b | 0 | a | b | 0 | a | a | a | b |
| c | 0 | 0 | 0 | c | c | c | c | c | c | c | c |
| d | 0 | a | a | c | d | d | c | d | d | d | d |
| e | 0 | a | b | c | d | e | c | d | d | d | e |
| f | 0 | 0 | 0 | c | c | c | f | f | c | c | f |
| g | 0 | a | a | c | d | d | f | g | d | d | g |
| x | 0 | a | a | c | x | x | c | x | x | x | x |
| y | 0 | a | a | c | y | y | c | y | y | y | y |
| 1 | 0 | a | b | c | d | e | f | g | x | y | 1 |

• The binary relation $\leq_m^M$ is transitive, hence is an order, but not a lattice order, since $\wedge_m^M$ is not commutative.

• The binary relation $\leq_m$ is a pre-order, since (m-Re) and (m-Tr) hold. From the table of $\wedge_m^B$, we obtain that: $a \leq_m a, b, d, e, g, x, y, 1$; $b \leq_m b, e, 1$; $c \leq_m c, d, e, f, g, x, y, 1$; $d \leq_m d, e, g, x, y, 1$; $e \leq_m e, 1$; $f \leq_m f, g, 1$; $g \leq_m g, 1$; $x \leq_m x, d, e, g, y, 1$; $y \leq_m y, d, e, g, x, 1$; hence, the bounded pre-ordered set $(A_{11}, \leq_m, 0, 1)$ is represented by the Hasse-type diagram from Figure 15.57, left side.

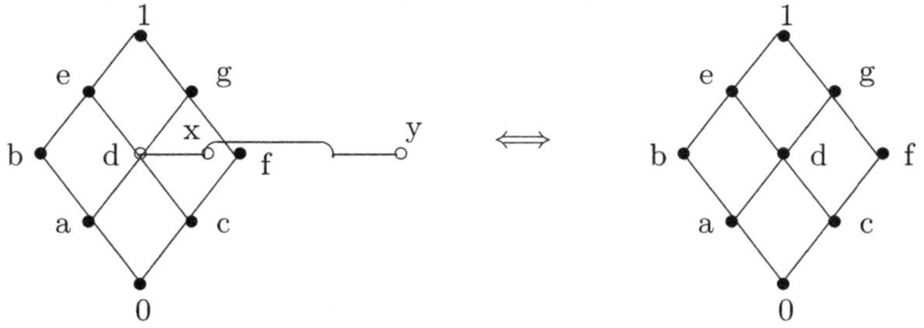


The bounded pre-ordered set $(A_{11}, \leq_m, 0, 1)$ The bounded lattice $(A_9, \leq_m, 0, 1)$

Figure 15.57: $(A_{11}, \leq_m, 0, 1)$ and $(A_9 = L_{3\times3}, \leq_m, 0, 1)$

Note that $d \parallel_Q x \parallel_Q y$, hence, the quotient set is $A_{11}/_{\parallel_Q} = \{\{0\}, \{a\}, \{b\}, \{c\}, \{d, x, y\}, \{e\}, \{f\}, \{g\}, \{1\}\}$ and will be denoted by $A_9 = \{0, a, b, c, d, e, f, g, 1\}$ in the sequel. Hence, the **associated m-BCK algebra** $(A_9, \odot, ^-, 1)$ **is just the starting algebra** $\mathcal{L}_{3\times3}^L$, represented by the Hasse diagram from the same Figure 15.57, right side.

### 15.3.2.3 Building transitive quantum algebras by the "method of Q-parallel rows/columns"

We cannot build tQMV algebras, tOLs and tOSLs by the "method of identic rows/columns", so we shall use the "method of Q-parallel rows/columns" in these cases. But, by this general method, there are multiple choices for the elements, so we must check each case by our PASCAL program ALL-VM0.PAS.

**Examples 15.3.27 Building transitive QMV algebras: tQMV**

Consider the **proper linearly ordered MV algebra** with the minimum number of elements, $\mathcal{L}_4^L = (A_4 = \{0, a, b, 1\}, \odot, ^-, 1)$, from Examples 15.3.4.

**Example 1: we add $x \parallel_Q a$ and $y \parallel_Q b$**

We add to the MV chain $0 \leq_m a \leq_m b \leq_m 1$ (see the previous Figure 15.52, right side) two elements $x, y \notin L_4$ such that $x \parallel_Q a$ and $y \parallel_Q b$; thus, we obtain the set $A_6 = \{0, a, b, x, y, 1\}$. We then extend the operations $\odot$ and $\oplus$ on this set $A_6$ by the "method of Q-parallel rows/columns" (the row/column of $x$ is Q-parallel with the row/column of $a$) in the following way (note that any other way leads to a transitive PreMV algebra); the operation $^-$ is extended such that the property (DN) be preserved:

| $\odot$ | 0 | a | b | x | y | 1 |
|---|---|---|---|---|---|---|
| 0 | 0 | 0 | 0 | 0 | 0 | 0 |
| a | 0 | 0 | 0 | 0 | 0 | a |
| b | 0 | 0 | a | 0 | x | b |
| x | 0 | 0 | 0 | 0 | 0 | x |
| y | 0 | 0 | x | 0 | a | y |
| 1 | 0 | a | b | x | y | 1 |

and

| $x$ | $x^-$ |
|---|---|
| 0 | 1 |
| a | b |
| b | a |
| x | y |
| y | x |
| 1 | 0 |

, with

| $\oplus$ | 0 | a | b | x | y | 1 |
|---|---|---|---|---|---|---|
| 0 | 0 | a | b | x | y | 1 |
| a | a | b | 1 | y | 1 | 1 |
| b | b | 1 | 1 | 1 | 1 | 1 |
| x | x | y | 1 | b | 1 | 1 |
| y | y | 1 | 1 | 1 | 1 | 1 |
| 1 | 1 | 1 | 1 | 1 | 1 | 1 |

.

Then, the resulting algebra $(A_6, \odot, ^-, 1)$ is an involutive m-BE algebra verifying (m-BB) ($\Leftrightarrow \ldots \Leftrightarrow$ (m-Tr)) and (Pqmv) (hence (Pom), (Pmv), ($\Delta_m$)) and (prel$_m$), and not verifying (m-An) for $(a, x)$, (m-Pabs-i) for $(a, 0)$, (m-Pimpl) for $(a, 0)$, (WNM$_m$) and (aWNM$_m$) for $(b, b)$, (m-Pdis) for $(a, a, b)$; hence, it is a **linearly pre-ordered transitive left-QMV algebra**.

**Remark.** *This tQMV algebra is isomorphic to the tQMV algebra from Example 15.3.13 ($x \mapsto b$, $b \mapsto c$, $y \mapsto d$).*

**Example 2: we add $u \parallel_Q x \parallel_Q a$ and $v \parallel_Q y \parallel_Q b$**

We add to the MV chain $0 \leq_m a \leq_m b \leq_m 1$ (see the previous Figure 15.54, right side) four elements $x, y, u, v \notin L_4$ such that $u \parallel_Q x \parallel_Q a$ and $v \parallel_Q y \parallel_Q b$; thus, we obtain the set $A_8 = \{0, a, b, x, y, u, v, 1\}$. We then extend the operations $\odot$ and $\oplus$ on this set $A_8$ by the "method of Q-parallel rows/columns" (the row/column of $x$ is Q-parallel with the row/column of $a$) in the following way (note that any other way leads to a transitive PreMV algebra); the operation $^-$ is extended such that the property (DN) be preserved:

| $\odot$ | 0 | a | b | x | y | u | v | 1 |
|---|---|---|---|---|---|---|---|---|
| 0 | 0 | 0 | 0 | 0 | 0 | 0 | 0 | 0 |
| a | 0 | 0 | 0 | 0 | 0 | 0 | 0 | a |
| b | 0 | 0 | a | 0 | x | 0 | u | b |
| x | 0 | 0 | 0 | 0 | 0 | 0 | 0 | x |
| y | 0 | 0 | x | 0 | a | 0 | a | y |
| u | 0 | 0 | 0 | 0 | 0 | 0 | 0 | u |
| v | 0 | 0 | u | 0 | a | 0 | a | v |
| 1 | 0 | a | b | x | y | u | v | 1 |

with

| $\oplus$ | 0 | a | b | x | y | u | v | 1 |
|---|---|---|---|---|---|---|---|---|
| 0 | 0 | a | b | x | y | u | v | 1 |
| a | a | b | 1 | y | 1 | v | 1 | 1 |
| b | b | 1 | 1 | 1 | 1 | 1 | 1 | 1 |
| x | x | y | 1 | b | 1 | b | 1 | 1 |
| y | y | 1 | 1 | 1 | 1 | 1 | 1 | 1 |
| u | u | v | 1 | b | 1 | b | 1 | 1 |
| v | v | 1 | 1 | 1 | 1 | 1 | 1 | 1 |
| 1 | 1 | 1 | 1 | 1 | 1 | 1 | 1 | 1 |

and $(0, a, b, x, y, u, v, 1)^- = (1, b, a, y, x, v, u, 0)$.

Then, the resulting algebra $(A_8, \odot, ^-, 1)$ is an involutive m-BE algebra verifying (m-BB) ($\Leftrightarrow \ldots \Leftrightarrow$ (m-Tr)) and (Pqmv) (hence (Pom), (Pmv), ($\Delta_m$)) and (prel$_m$), and not verifying (m-An) for $(a, x)$, (m-Pabs-i) for $(a, 0)$, (m-Pimpl) for $(a, 0)$, (WNM$_m$) and (aWNM$_m$) for $(b, b)$, (m-Pdis) for $(a, a, b)$; hence, it is a **linearly pre-ordered transitive left-QMV algebra**.

Next, we follow the converse way, as in Subsubsection 15.3.2.1.

The tables of $\wedge_m^M$ and $\wedge_m^B$ are:

| $\wedge_m^M$ | 0 | a | b | x | y | u | v | 1 |
|---|---|---|---|---|---|---|---|---|
| 0 | 0 | 0 | 0 | 0 | 0 | 0 | 0 | 0 |
| a | 0 | a | a | x | a | u | a | a |
| b | 0 | a | b | x | y | u | v | b |
| x | 0 | a | x | x | x | u | u | x |
| y | 0 | a | b | x | y | u | v | y |
| u | 0 | a | u | x | x | u | u | u |
| v | 0 | a | b | x | y | u | v | v |
| 1 | 0 | a | b | x | y | u | v | 1 |

and

| $\wedge_m^B$ | 0 | a | b | x | y | u | v | 1 |
|---|---|---|---|---|---|---|---|---|
| 0 | 0 | 0 | 0 | 0 | 0 | 0 | 0 | 0 |
| a | 0 | a | a | a | a | a | a | a |
| b | 0 | a | b | x | b | u | b | b |
| x | 0 | x | x | x | x | x | x | x |
| y | 0 | a | y | x | y | x | y | y |
| u | 0 | u | u | u | u | u | u | u |
| v | 0 | a | v | u | v | u | v | v |
| 1 | 0 | a | b | x | y | u | v | 1 |

.

• The binary relation $\leq_m^M$ is transitive, hence it is an order, but not a lattice order, since $\wedge_m^M$ is not commutative.

• The binary relation $\leq_m$ is a pre-order. From the table of $\wedge_m^B$, we obtain: $a \leq_m a, b, x, y, u, v, 1$; $b \leq_m b, y, v, 1$; $x \leq_m x, a, b, y, u, v, 1$; $y \leq_m y, b, v, 1$; $u \leq_m u, a, b, x, y, v, 1$; $v \leq_m v, b, y, 1$; hence, the bounded pre-ordered set $(A_8, \leq_m, 0, 1)$ is represented by the Hasse-type diagram from the previous Figure 15.54, left side. ... Hence, the **associated m-BCK algebra** $(A_4, \leq_m, 0, 1)$ **is just the starting algebra** $\mathcal{L}_4^L$, represented by the Hasse diagram from the same Figure 15.54, right side.

**Example 3: we add** $z \parallel_Q u \parallel_Q x \parallel_Q a$ **and** $w \parallel_Q v \parallel_Q y \parallel_Q b$

We add to the chain $0 \leq_m a \leq_m b \leq_m 1$ (see Figure 15.58, right side) six elements $x, y, u, v, z, w \notin L_4$ such that $z \parallel_Q u \parallel_Q x \parallel_Q a$ and $w \parallel_Q v \parallel_Q y \parallel_Q b$; thus, we obtain the set $A_{10} = \{0, a, b, x, y, u, v, z, w, 1\}$. We then extend the operations $\odot$ and $\oplus$ on this set $A_{10}$ by the "method of Q-parallel rows/columns" (the row/column of $x$ is Q-parallel with the row/column of $a$) in the following way (note that any other way leads to a transitive PreMV algebra); the operation $^-$ is extended such that the property (DN) be preserved:

| $\odot$ | 0 | a | b | x | y | u | v | z | w | 1 |
|---|---|---|---|---|---|---|---|---|---|---|
| 0 | 0 | 0 | 0 | 0 | 0 | 0 | 0 | 0 | 0 | 0 |
| a | 0 | 0 | 0 | 0 | 0 | 0 | 0 | 0 | 0 | a |
| b | 0 | 0 | a | 0 | x | 0 | u | 0 | z | b |
| x | 0 | 0 | 0 | 0 | 0 | 0 | 0 | 0 | 0 | x |
| y | 0 | 0 | x | 0 | a | 0 | a | 0 | a | y |
| u | 0 | 0 | 0 | 0 | 0 | 0 | 0 | 0 | 0 | u |
| v | 0 | 0 | u | 0 | a | 0 | a | 0 | a | v |
| z | 0 | 0 | 0 | 0 | 0 | 0 | 0 | 0 | 0 | z |
| w | 0 | 0 | z | 0 | a | 0 | a | 0 | a | w |
| 1 | 0 | a | b | x | y | u | v | z | w | 1 |

and

| $x$ | $x^-$ |
|---|---|
| 0 | 1 |
| a | b |
| b | a |
| x | y |
| y | x |
| u | v |
| v | u |
| z | w |
| w | z |
| 1 | 0 |

, with

| $\oplus$ | 0 | a | b | x | y | u | v | z | w | 1 |
|---|---|---|---|---|---|---|---|---|---|---|
| 0 | 0 | a | b | x | y | u | v | z | w | 1 |
| a | a | b | 1 | y | 1 | v | 1 | w | 1 | 1 |
| b | b | 1 | 1 | 1 | 1 | 1 | 1 | 1 | 1 | 1 |
| x | x | y | 1 | b | 1 | b | 1 | b | 1 | 1 |
| y | y | 1 | 1 | 1 | 1 | 1 | 1 | 1 | 1 | 1 |
| u | u | v | 1 | b | 1 | b | 1 | b | 1 | 1 |
| v | v | 1 | 1 | 1 | 1 | 1 | 1 | 1 | 1 | 1 |
| z | z | w | 1 | b | 1 | b | 1 | b | 1 | 1 |
| w | w | 1 | 1 | 1 | 1 | 1 | 1 | 1 | 1 | 1 |
| 1 | 1 | 1 | 1 | 1 | 1 | 1 | 1 | 1 | 1 | 1 |

.

Then, the resulting algebra $(A_{10}, \odot, ^-, 1)$ is an involutive m-BE algebra verifying (m-BB) ($\Leftrightarrow \ldots \Leftrightarrow$ (m-Tr)) and (Pqmv) (hence (Pom), (Pmv), $(\Delta_m)$) and $(\text{prel}_m)$, and not verifying (m-An) for $(a, x)$, (m-Pabs-i) for $(a, 0)$, (m-Pimpl) for $(a, 0)$, $(\text{WNM}_m)$ and $(\text{aWNM}_m)$ for $(b, b)$, (m-Pdis) for $(a, a, b)$; hence, it is a **linearly pre-ordered transitive left-QMV algebra**.

Next, we follow the converse way, as in Subsubsection 15.3.2.1.

The table of $\wedge_m^B$ is:

| $\wedge_m^B$ | 0 | a | b | x | y | u | v | z | w | 1 |
|---|---|---|---|---|---|---|---|---|---|---|
| 0 | 0 | 0 | 0 | 0 | 0 | 0 | 0 | 0 | 0 | 0 |
| a | 0 | a | a | a | a | a | a | a | a | a |
| b | 0 | a | b | x | b | u | b | z | b | b |
| x | 0 | x | x | x | x | x | x | x | x | x |
| y | 0 | a | y | x | y | x | y | x | y | y |
| u | 0 | u | u | u | u | u | u | u | u | u |
| v | 0 | a | v | u | v | u | v | u | v | v |
| z | 0 | z | z | z | z | z | z | z | z | z |
| w | 0 | a | w | z | w | z | w | z | w | w |
| 1 | 0 | a | b | x | y | u | v | z | w | 1 |

.

• The binary relation $\leq_m^M$ is transitive, hence it is an order, but not a lattice order, since $\wedge_m^M$ is not commutative.

• The binary relation $\leq_m$ is a pre-order. From the table of $\wedge_m^B$, we obtain: $a \leq_m a, b, x, y, u, v, z, w, 1$; $b \leq_m b, y, v, w, 1$; $x \leq_m x, a, b, y, u, v, z, w, 1$; $y \leq_m y, b, v, w, 1$; $u \leq_m u, a, b, x, y, v, z, w, 1$; $v \leq_m v, b, y, w, 1$; $z \leq_m z, a, b, x, y, u, v, w, 1$; $w \leq_m w, b, y, v, 1$; hence, the bounded pre-ordered set $(A_{10}, \leq_m, 0, 1)$ is represented by the Hasse-type diagram from Figure 15.58, left side. ... Hence, the **associated m-BCK algebra** $(A_4, \leq_m, 0, 1)$ **is just the starting algebra** $\mathcal{L}_4^L$, represented by the Hasse diagram from the same Figure 15.58, right side.

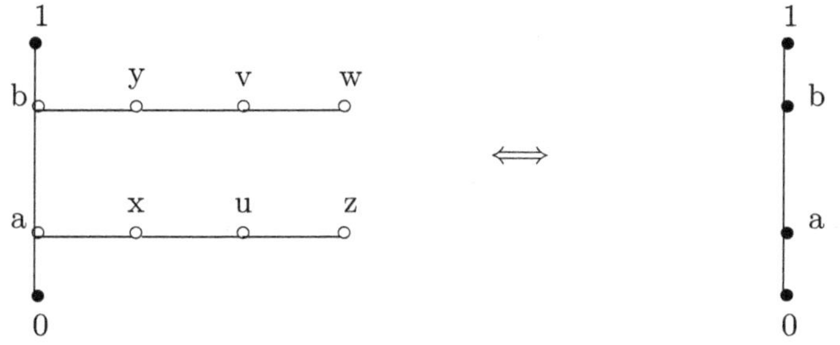


The bounded pre-ordered set $(A_{10}, \leq_m, 0, 1)$ The bounded chain $(A_4, \leq_m, 0, 1)$

Figure 15.58: $(A_{10}, \leq_m, 0, 1)$ and $(A_4, \leq_m, 0, 1)$

**Remark** *By adding Q-parallel elements and by extending the tables of* $\odot$ *and* $\oplus$ *of* $\mathcal{L}_4^L$ *by the "method of identic rows/columns", we obtain only linearly pre-ordered transitive PreMV algebras - see Example 15.3.23.*

**Examples 15.3.28 Building transitive OLs: tOL**

Consider the **non-linearly ordered Boolean algebra** with the minimum number of elements, $\mathcal{L}_{2\times 2}^L = (L_{2\times 2} = \{0, a, b, 1\}, \odot, ^-, 1)$, from Examples 15.3.8, 1.

**Example 1: we add two elements,** $x \parallel_Q a$ **and** $y \parallel_Q b$

We add to the bounded lattice $(L_{2\times 2}, \leq_m, 0, 1)$ from Figure 15.11 (see also the previous Figure 15.55, right side) two elements $x, y \notin L_{2\times 2}$ such that $x \parallel_Q a$ and $y \parallel_Q b$; thus, we obtain the set $A_6 = \{0, a, b, x, y, 1\}$. We then extend the operations $\odot$ and $\oplus$ on this set $A_6$ by the "method of Q-parallel rows/columns" (the row/column of $x$ is Q-parallel with the row/column of $a$) in the following way such that (G) be verified; the operation $^-$ is extended such that the property (DN) be preserved:

| $\odot$ | 0 | a | b | x | y | 1 |
|---|---|---|---|---|---|---|
| 0 | 0 | 0 | 0 | 0 | 0 | 0 |
| a | 0 | a | 0 | **a** | 0 | a |
| b | 0 | 0 | b | 0 | **y** | b |
| x | 0 | **a** | 0 | x | 0 | x |
| y | 0 | 0 | **y** | 0 | y | y |
| 1 | 0 | a | b | x | y | 1 |

and

| $x$ | $x^-$ |
|---|---|
| 0 | 1 |
| a | b |
| b | a |
| x | y |
| y | x |
| 1 | 0 |

, with

| $\oplus$ | 0 | a | b | x | y | 1 |
|---|---|---|---|---|---|---|
| 0 | 0 | a | b | x | y | 1 |
| a | a | a | 1 | x | 1 | 1 |
| b | b | 1 | b | 1 | b | 1 |
| x | x | x | 1 | x | 1 | 1 |
| y | y | 1 | b | 1 | y | 1 |
| 1 | 1 | 1 | 1 | 1 | 1 | 1 |

.

Then, the resulting algebra $(A_6, \odot, ^-, 1)$ is an involutive m-BE algebra verifying (m-BB) ($\Leftrightarrow \ldots \Leftrightarrow$ (m-Tr)) and (m-Pimpl) (hence (G) and (m-Pabs-i)) and ($\Delta_m$) and (prel$_m$), (WNM$_m$), (aWNM$_m$), and not verifying (m-An) for $(a, x)$, (Pqmv) for $(b, 0, y)$, (Pom) for $(b, y)$, (Pmv) for $(b, y)$, (m-Pdis) for $(a, b, x)$; hence, it is a **non-linearly pre-ordered transitive left-OL**.

**Remark.** *This tOL is just the tOL from Example 9.3.17.*

**Example 1': we add $x \parallel_Q a$ and $y \parallel_Q b$**

We add to the bounded lattice $(L_{2\times 2}, \leq_m, 0, 1)$ from Figure 15.11 (see also the previous Figure 15.55, right side) two elements $x, y \notin L_{2\times 2}$ such that $x \parallel_Q a$ and $y \parallel_Q b$; thus, we obtain the set $A_6 = \{0, a, b, x, y, 1\}$. We then extend the operations $\odot$ and $\oplus$ on this set $A_6$ by the "method of Q-parallel lines and columns" (the row/column of $x$ is Q-parallel with the row/column of $a$) in the following **different** way such that (G) be verified; the operation $^-$ is extended such that the property (DN) be preserved:

| $\odot$ | 0 | a | b | x | y | 1 |
|---|---|---|---|---|---|---|
| 0 | 0 | 0 | 0 | 0 | 0 | 0 |
| a | 0 | a | 0 | **x** | 0 | a |
| b | 0 | 0 | b | 0 | **b** | b |
| x | 0 | **x** | 0 | x | 0 | x |
| y | 0 | 0 | **b** | 0 | y | y |
| 1 | 0 | a | b | x | y | 1 |

and

| $x$ | $x^-$ |
|---|---|
| 0 | 1 |
| a | b |
| b | a |
| x | y |
| y | x |
| 1 | 0 |

, with

| $\oplus$ | 0 | a | b | x | y | 1 |
|---|---|---|---|---|---|---|
| 0 | 0 | a | b | x | y | 1 |
| a | a | a | 1 | a | 1 | 1 |
| b | b | 1 | b | 1 | y | 1 |
| x | x | a | 1 | x | 1 | 1 |
| y | y | 1 | y | 1 | y | 1 |
| 1 | 1 | 1 | 1 | 1 | 1 | 1 |

.

Then, the resulting algebraa $(A_6, \odot, ^-, 1)$ is an involutive m-BE algebra verifying (m-BB) ($\Leftrightarrow \ldots \Leftrightarrow$ (m-Tr)) and (m-Pimpl) (hence (G) and (m-Pabs-i)) and ($\Delta_m$) and (prel$_m$), (WNM$_m$), (aWNM$_m$), and not verifying (m-An) for $(a, x)$, (Pqmv) for $(a, 0, x)$, (Pom) for $(a, x)$, (Pmv) for $(a, x)$, (m-Pdis) for $(a, b, y)$; hence, it is also a **non-linearly pre-ordered transitive left-OL**.

Next, we follow the converse way, as in Subsubsection 15.3.2.1, Example 9.3.17.

• Note that the binary relation $\leq_m^P$ is a lattice order w.r. to $\wedge = \odot$ and $\vee = \oplus$, since (G) holds. From the table of $\odot$, we obtain: $a \leq_m^P a, 1$; $b \leq_m^P b, y, 1$; $x \leq_m^P x, a, 1$; $y \leq_m^P y, 1$; hence, the bounded lattice $(A_6, \leq_m^P, 0, 1)$ is represented by the Hasse diagram from Figure 15.59.

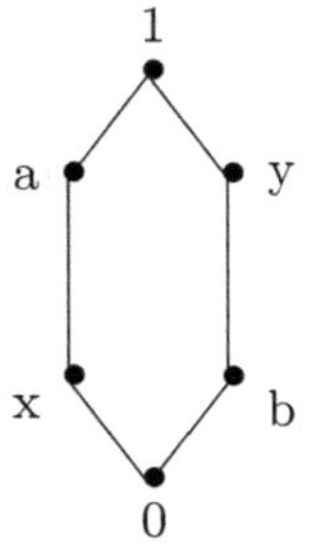


Figure 15.59: $(A_6, \leq_m^P, 0, 1)$

The tables of $\wedge_m^M$ and $\wedge_m^B$ are:

| $\wedge_m^M$ | 0 | a | b | x | y | 1 |
|---|---|---|---|---|---|---|
| 0 | 0 | 0 | 0 | 0 | 0 | 0 |
| a | 0 | a | 0 | x | 0 | a |
| b | 0 | 0 | b | 0 | y | b |
| x | 0 | a | 0 | x | 0 | x |
| y | 0 | 0 | b | 0 | y | y |
| 1 | 0 | a | b | x | y | 1 |

and

| $\wedge_m^B$ | 0 | a | b | x | y | 1 |
|---|---|---|---|---|---|---|
| 0 | 0 | 0 | 0 | 0 | 0 | 0 |
| a | 0 | a | 0 | a | 0 | a |
| b | 0 | 0 | b | 0 | b | b |
| x | 0 | x | 0 | x | 0 | x |
| y | 0 | 0 | y | 0 | y | y |
| 1 | 0 | a | b | x | y | 1 |

.

• The binary relation $\leq_m^M$ is transitive, hence it is an order, but not a lattice order, since $\wedge_m^M$ is not commutative.

• The binary relation $\leq_m$ is a pre-order. From the table of $\wedge_m^B$, we obtain: $a \leq_m a, x, 1$; $b \leq_m b, y, 1$; $x \leq_m x, a, 1$; $y \leq_m y, b, 1$; hence, the bounded partially pre-ordered set $(A_6, \leq_m, 0, 1)$ is represented by the same Hasse-type diagram from the previous Figure 15.55, left side. ... Hence, the **associated m-BCK algebra** $(A_4, \leq_m, 0, 1)$ **is just the starting algebra** $\mathcal{L}_{2\times 2}^L$, represented by the Hasse diagram from the same Figure 15.55, right side.

**Examples 15.3.29 Building transitive OSLs: tOSL**

Consider the **non-linearly ordered Boolean algebra** $\mathcal{L}_{2\times 2}^L = (L_{2\times 2} = \{0, a, b, 1\}, \odot, ^-, 1)$, from Examples 15.3.8, 1.

**Example 1: we add two elements,** $x \parallel_Q a$ **and** $y \parallel_Q b$

We add to the bounded lattice $(L_{2\times 2}, \leq_m, 0, 1)$ from Figure 15.11 (see also the previous Figure 15.55, right side) two elements $x, y \notin L_{2\times 2}$ such that $x \parallel_Q a$ and $y \parallel_Q b$; thus, we obtain the set $A_6 = \{0, a, b, x, y, 1\}$. We then extend the operations $\odot$ and $\oplus$ on this set $A_6$ by the "method of Q-parallel rows/columns" (the row/column of $x$ is Q-parallel with the row/column of $a$) in the following **different** way such that (G) be verified; the operation $^-$ is extended such that the property (DN) be preserved:

| $\odot$ | 0 | a | b | x | y | 1 |
|---|---|---|---|---|---|---|
| 0 | 0 | 0 | 0 | 0 | 0 | 0 |
| a | 0 | a | 0 | **a** | 0 | a |
| b | 0 | 0 | b | 0 | **b** | b |
| x | 0 | **a** | 0 | x | 0 | x |
| y | 0 | 0 | **b** | 0 | y | y |
| 1 | 0 | a | b | x | y | 1 |

and

| $x$ | $x^-$ |
|---|---|
| 0 | 1 |
| a | b |
| b | a |
| x | y |
| y | x |
| 1 | 0 |

, with

| $\oplus$ | 0 | a | b | x | y | 1 |
|---|---|---|---|---|---|---|
| 0 | 0 | a | b | x | y | 1 |
| a | a | a | 1 | a | 1 | 1 |
| b | b | 1 | b | 1 | b | 1 |
| x | x | a | 1 | x | 1 | 1 |
| y | y | 1 | b | 1 | y | 1 |
| 1 | 1 | 1 | 1 | 1 | 1 | 1 |

.

Then, the resulting algebra $(A_6, \odot, ^-, 1)$ is an involutive m-BE algebra verifying (m-BB) ($\Leftrightarrow \ldots \Leftrightarrow$ (m-Tr)) and (G) and ($\Delta_m$), and (prel$_m$), (WNM$_m$), (aWNM$_m$), and not verifying (m-Pabs-i) for $(x, a)$, (m-An) for $(a, x)$, (m-Pimpl) for $(x, b)$, (Pqmv) for $(x, 0, a)$, (Pom) for $(x, a)$, (Pmv) for $(x, a)$, (m-Pdis) for $(a, b, x)$; hence, it is also a **non-linearly pre-ordered transitive left-OSL**.

**Remark.** *This tOSL is isomorphic to the tOSL from Example 10.4.6 ($y \mapsto b$, $x \mapsto c$, $b \mapsto d$).*

**Example 1': we add two elements, $x \parallel_Q a$ and $y \parallel_Q b$**

We add to the bounded lattice $(L_{2\times 2}, \leq_m, 0, 1)$ from Figure 15.11 (see also the previous Figure 15.55, right side) two elements $x, y \notin L_{2\times 2}$ such that $x \parallel_Q a$ and $y \parallel_Q b$; thus, we obtain the set $A_6 = \{0, a, b, x, y, 1\}$. We then extend the operations $\odot$ and $\oplus$ on this set $A_6$ by the "method of Q-parallel rows/columns" (the row/column of $x$ is Q-parallel with the row/column of $a$) in the following way such that (G) be verified; the operation $^-$ is extended such that the property (DN) be preserved:

| $\odot$ | 0 | a | b | x | y | 1 |
|---|---|---|---|---|---|---|
| 0 | 0 | 0 | 0 | 0 | 0 | 0 |
| a | 0 | a | 0 | **x** | 0 | a |
| b | 0 | 0 | b | 0 | **y** | b |
| x | 0 | **x** | 0 | x | 0 | x |
| y | 0 | 0 | **y** | 0 | y | y |
| 1 | 0 | a | b | x | y | 1 |

and

| $x$ | $x^-$ |
|---|---|
| 0 | 1 |
| a | b |
| b | a |
| x | y |
| y | x |
| 1 | 0 |

, with

| $\oplus$ | 0 | a | b | x | y | 1 |
|---|---|---|---|---|---|---|
| 0 | 0 | a | b | x | y | 1 |
| a | a | a | 1 | x | 1 | 1 |
| b | b | 1 | b | 1 | y | 1 |
| x | x | x | 1 | x | 1 | 1 |
| y | y | 1 | y | 1 | y | 1 |
| 1 | 1 | 1 | 1 | 1 | 1 | 1 |

.

Then, the resulting algebra $(A_6, \odot, ^-, 1)$ is an involutive m-BE algebra verifying (m-BB) ($\Leftrightarrow \ldots \Leftrightarrow$ (m-Tr)) and (G) and ($\Delta_m$), and (prel$_m$), (WNM$_m$), (aWNM$_m$), and not verifying (m-Pabs-i) for $(a, x)$, (m-An) for $(a, x)$, (m-Pimpl) for $(a, y)$, (Pqmv) for $(a, 0, x)$, (Pom) for $(a, x)$, (Pmv) for $(a, x)$, (m-Pdis) for $(a, y, b)$; hence, it is a **non-linearly pre-ordered transitive left-OSL**.

Next, we follow the converse way, as in Subsubsection 15.3.2.1.

The tables of $\wedge_m^M$ and $\wedge_m^B$ are:

| $\wedge_m^M$ | 0 | a | b | x | y | 1 |
|---|---|---|---|---|---|---|
| 0 | 0 | 0 | 0 | 0 | 0 | 0 |
| a | 0 | a | 0 | x | 0 | a |
| b | 0 | 0 | b | 0 | y | b |
| x | 0 | a | 0 | x | 0 | x |
| y | 0 | 0 | b | 0 | y | y |
| 1 | 0 | a | b | x | y | 1 |

and

| $\wedge_m^B$ | 0 | a | b | x | y | 1 |
|---|---|---|---|---|---|---|
| 0 | 0 | 0 | 0 | 0 | 0 | 0 |
| a | 0 | a | 0 | a | 0 | a |
| b | 0 | 0 | b | 0 | b | b |
| x | 0 | x | 0 | x | 0 | x |
| y | 0 | 0 | y | 0 | y | y |
| 1 | 0 | a | b | x | y | 1 |

.

• The binary relation $\leq_m^M$ is transitive, hence it is an order, but not a lattice order, since $\wedge_m^M$ is not commutative.

• The binary relation $\leq_m$ is a pre-order. From the table of $\wedge_m^B$, we obtain: $a \leq_m a, x, 1$; $b \leq_m b, y, 1$; $x \leq_m x, a, 1$; $y \leq_m y, b, 1$; hence, the bounded partially pre-ordered set $(A_6, \leq_m, 0, 1)$ is represented by the same Hasse-type diagram from the previous Figure 15.55, left side. ... Hence, the **associated m-BCK algebra** $(A_4, \leq_m, 0, 1)$ **is just the starting algebra** $\mathcal{L}_{2\times 2}^L$, represented by the Hasse diagram from the same Figure 15.55, right side.

## 15.4 Conclusions and future work

This chapter ends the series of five chapters 11-15 on QMV algebras, but also ends the larger series of ten chapters 6-15 which connects all quantum structures, old ones and new ones: BILs, De Morgan algebras, OLs (tOLs), OSLs (tOSLs), OWLs (tOWLs), QMV (tQMV) algebras, PreMV (tPreMV) and MMV (tMMV) algebras, OM (tOM) algebras, OMLs (tOMLs = Boolean algebras), OMWLs (tOMWLs), tQIMTL and tQNM algebras.

Almost all the results from these chapters were obtained by the powerful computer program *Prover9/Mace4* (version dec. 2007) created by William W. McCune (1953-2011) [155]. Most probably, they could not be obtained otherwise. For the examples presented in Section 15.3, the mentioned PASCAL programs were crucial in finding their properties.

We have proved in all these chapters the deep connections existing between the algebraic structures connected to the classical and non-classical logics and the algebraic structures connected to the quantum logics: they exist on the same "map" ("m-map"), but at different levels (parallels).

The 'story' of the algebras involved in these chapters is connected to the 'story' of the three/four binary relations that can be defined in such algebras:

$x \leq_m y \overset{def.}{\Longleftrightarrow} x \odot y^- = 0$ (with $x \leq_m y \iff x \leq_m^B y$, $\quad x \leq_m^B y \overset{def}{\Longleftrightarrow} x \wedge_m^B y = x$),

$x \leq_m^M y \overset{def}{\Longleftrightarrow} x \wedge_m^M y = x$ and

$x \leq_m^P y \overset{def}{\Longleftrightarrow} x \odot y = x$.

Note that the central role is played by the binary relation $\leq_m$, that determines the "parallels" and the "meridians" of the "map". The three/four binary relations $\leq_m$ ($\Longleftrightarrow \leq_m^B$), $\leq_m^M$ and $\leq_m^P$ are all equivalent in Boolean algebras, while in MV algebras only $\leq_m$ ( $\Longleftrightarrow \leq_m^B$) and $\leq_m^M$ are equivalent.

This chapter can be (must be) also a beginning for vast researches into at least three directions:

- a first direction is to continue the results from Section 15.2 on involutive m-pre-BCK algebras versus (involutive) m-BCK algebras, finding characterizations of tQIMTL and tQNM algebras, studying the filter theory w.r. to $\leq_m$ (but also w.r. to $\leq_m^M$, $\leq_m^P$), etc.

- a second direction is to find and study the analogous quantum structures from the "world of involutive algebras of logic", described in Chapters 2 - 4, and then to find and study the corresponding quantum logics. By the inverse mappings $\Phi$ ($x \odot y \overset{def.}{=} (x \to y^-)^-$) and $\Psi$ ($x \to y \overset{def.}{=} (x \odot y^-)^-$) (Theorem 17.1.1) that connect the "world" of *algebras of logic* of the form $(A, \to, ^-, 1)$ and the "world" of *algebras* of the form $(A, \odot, ^-, 1)$, in the involutive case, one can obtain simply, by choosing the appropriate definitions of the algebras, the definitionally equivalent *involutive algebras of logic* corresponding to the *involutive algebras* from these Chapters 6 - 15 and the corresponding results and examples. Note that, in Chapter 3, Definition 3.4.3, the *implicative-ortholattices* were already introduced as involutive *BE algebras* verifying (impl) ($(x \to y) \to x = x$) and their d.e. with the ortholattices was proved by Corollary 17.1.6. Similarly, one can introduce now the *quantum-Wajsberg algebras*, as algebras of logic (involutive *BE algebras* verifying, say, (qw)) d. e. with the quantum-MV algebras, etc. Note that, by Theorem 17.1.1, the involutive MEL algebras are d.e. with the involutive m-MEL algebras (Theorem 17.1.2), the involutive BE algebras are d.e. with the involutive m-BE algebras (Corollay 17.1.3), the involutive pre-BCK algebras are d.e. with the involutive m-pre-BCK algebras, the involutive BCK algebras are d.e. with the (involutive) m-BCK algebras (Corollary 17.1.4), etc.

- a third direction is to generalize to the non-commutative case algebras from Chapters 6 - 8, 11 - 15 in order to obtain, for example, that the involutive pseudo-BCK algebras are d.e. with the involutive m-pseudo-BCK algebras.

# Part III

# Bridge theorems between the two 'worlds'

# Chapter 16

# Connections in the non-commutative case

In this chapter, we present bridge theorems in the non-commutative case.

In Section 16.1, we present bridge theorems connecting po($l$)-pi-magmas (Section 1.4) and po($l$)-magmas (Section 5.3).

In Section 16.2, we present bridge theorems connecting i-groups (Section 1.5) and groups (Section 5.4).

## 16.1 Connections between po($l$)-pi-magmas and po($l$)-magmas

Most of the results from this section are taken from [116].

### 16.1.1 Connections between structures with (pRP') and (pPR')

The connections between the Galois dual *structures with (pRP')* (from Section 1.4) and *structures with (pPR')* (from Section 5.3) are the following.

**Theorem 16.1.1** *(The dual case is omitted)*

*(1) Let $\mathcal{A}^L = (A^L, \leq = \leq^e, \to, \rightsquigarrow, \odot)$ be a po-pe(pRP').*
*Define $\Phi'(\mathcal{A}^L) = (A^L, \leq, \odot, \to, \rightsquigarrow)$.*
*Then, $\Phi'(\mathcal{A}^L)$ is a po-s(pPR').*

*(1') Conversely, let $\mathcal{A}^L = (A^L, \leq = \leq^e, \odot, \to, \rightsquigarrow)$ be a po-s(pPR').*
*Define $\Psi'(\mathcal{A}^L) = (A^L, \leq, \to, \rightsquigarrow, \odot)$.*
*Then, $\Psi'(\mathcal{A}^L)$ is a po-pe(pRP').*

*(2) The above mappings $\Phi'$ and $\Psi'$ are mutually inverse.*

**Proof.** (1): Let $\mathcal{A}^L = (A^L, \leq = \leq^e, \to, \rightsquigarrow, \odot)$ be a po-pe(pRP'), i.e. $(A^L, \leq)$ is a poset and (pEx), (pRP') hold. By Proposition 1.4.36, $(A^L, \odot)$ is a semigroup (i.e. (Pass) holds). Hence, $\Phi'(\mathcal{A}^L)$ is a po-s(pPR').

(1'): Let $\mathcal{A}^L = (A^L, \leq = \leq^e, \odot, \to, \rightsquigarrow)$ be a po-s(pPR'), i.e. $(A^L, \leq)$ is a poset and (Pass), (pPR') hold. By Proposition 5.3.27, $(A^L, \to, \rightsquigarrow)$ is an exchange (i.e. (pEx) holds). Hence, $\Psi'(\mathcal{A}^L)$ is a po-pe(pRP').

(2): Obviously. □

Hence, we have the following equivalences beteen the two 'worlds':

$$\begin{array}{lccc} & \textbf{po-pe(pRP')} & \Longleftrightarrow & \textbf{po-s(pPR')}, \\ \textbf{QB(pRP')-L} = & l\textbf{-pe(pRP')} & \Longleftrightarrow & l\textbf{-s(pPR')}. \end{array}$$

**Theorem 16.1.2** *(The dual case is omitted)*

*(1) Let $\mathcal{A}^L = (A^L, \leq, \to, \rightsquigarrow, \odot, 1)$ be a po-pr(pRP').*
*Define $\Phi'(\mathcal{A}^L) = (A^L, \leq, \odot, \to, \rightsquigarrow, 1)$.*
*Then, $\Phi'(\mathcal{A}^L)$ is a po-m(pPR').*

*(1') Conversely, let $\mathcal{A}^L = (A^L, \leq, \odot, \to, \rightsquigarrow, 1)$ be a po-m(pPR').*
*Define $\Psi'(\mathcal{A}^L) = (A^L, \leq, \to, \rightsquigarrow, \odot, 1)$.*
*Then, $\Psi'(\mathcal{A}^L)$ is a po-pr(pRP').*

*(2) The above mappings $\Phi'$ and $\Psi'$ are mutually inverse.*

**Proof.** (1): Let $\mathcal{A}^L = (A^L, \leq, \to, \rightsquigarrow, \odot, 1)$ be a po-pr(pRP'), i.e. $(A^L, \leq)$ is a poset and (pEx), (pM), (pRP') hold. By Proposition 1.4.37, $(A^L, \odot, 1)$ is a monoid (i.e. (Pass), (PU) hold). Hence, $\Phi'(\mathcal{A}^L)$ is a po-m(pPR').

(1'): Let $\mathcal{A}^L = (A^L, \leq, \odot, \to, \rightsquigarrow, 1)$ be a po-m(pPR'), i.e. $(A^L, \leq)$ is a poset and (Pass), (PU), (pPR') hold. By Proposition 5.3.28, $(A^L, \to, \rightsquigarrow, 1)$ is a pseudo-residoid (i.e. (pEx), (pM) hold). Hence, $\Psi'(\mathcal{A}^L)$ is a po-pr(pRP').

(2): Obviously. □

Hence, we have the following equivalences between the two 'worlds':

$$\begin{array}{lccc} & \textbf{po-pr(pRP')} & \Longleftrightarrow & \textbf{po-m(pPR')}, \\ \textbf{uQB(pRP')-L} = & l\textbf{-pr(pRP')} & \Longleftrightarrow & l\textbf{-m(pPR')}. \end{array}$$

**Theorem 16.1.3** *(The dual case is omitted)*

*(1) Let $\mathcal{A}^L = (A^L, \leq, \to, \rightsquigarrow, \odot, 1)$ be a po-mpr(pRP').*
*Define $\Phi'(\mathcal{A}^L) = (A^L, \leq, \odot, \to, \rightsquigarrow, 1)$.*
*Then, $\Phi'(\mathcal{A}^L)$ is a po-mm(pPR').*

*(1') Conversely, let $\mathcal{A}^L = (A^L, \leq, \odot, \to, \rightsquigarrow, 1)$ be a po-mm(pPR').*
*Define $\Psi'(\mathcal{A}^L) = (A^L, \leq, \to, \rightsquigarrow, \odot, 1)$.*
*Then, $\Psi'(\mathcal{A}^L)$ is a po-mpr(pRP').*

*(2) The above mappings $\Phi'$ and $\Psi'$ are mutually inverse.*

**Proof.** (1): Let $\mathcal{A}^L = (A^L, \leq, \to, \rightsquigarrow, \odot, 1)$ be a po-mpr(pRP'); hence, $\mathcal{A}^L$ is a po-pr(pRP') verifying additionally (N'). By above Theorem 16.1.2 (1), $\Phi'(\mathcal{A}^L)$ is a po-m(pPR') verifying additionally (N'), hence $\Phi'(\mathcal{A}^L)$ is a po-mm(pPR').

(1'): Let $\mathcal{A}^L = (A^L, \leq, \odot, \to, \rightsquigarrow, 1)$ be a po-mm(pPR'); hence, $\mathcal{A}^L$ is a po-m(pPR') verifying additionally (N'). By above Theorem 16.1.2 (1'), $\Psi'(\mathcal{A}^L)$ is a po-pr(pRP') verifying additionally (N'), hence $\Psi'(\mathcal{A}^L)$ is a po-mpr(pRP').

(2): Obviously. □

Hence, we have the following equivalences between the two 'worlds':

$$\begin{array}{llll} & \textbf{po-mpr(pRP')} & \Longleftrightarrow & \textbf{po-mm(pPR')}, \\ \textbf{pBCI(pRP')-L} = & l\textbf{-mpr(pRP')} & \Longleftrightarrow & l\textbf{-mm(pPR')}. \end{array}$$

**Theorem 16.1.4** *(The dual case is omitted)*
*(1) Let $\mathcal{A}^L = (A^L, \leq, \to, \rightsquigarrow, \odot, 1)$ be a po-ipr(pRP').*
*Define $\Phi'(\mathcal{A}^L) = (A^L, \leq, \odot, \to, \rightsquigarrow, 1)$.*
*Then, $\Phi'(\mathcal{A}^L)$ is a po-im(pPR').*
*(1') Conversely, let $\mathcal{A}^L = (A^L, \leq, \odot, \to, \rightsquigarrow, 1)$ be a po-im(pPR').*
*Define $\Psi'(\mathcal{A}^L) = (A^L, \leq, \to, \rightsquigarrow, \odot, 1)$.*
*Then, $\Psi'(\mathcal{A}^L)$ is a po-ipr(pRP').*
*(2) The above mappings $\Phi'$ and $\Psi'$ are mutually inverse.*

**Proof.** (1): Let $\mathcal{A}^L = (A^L, \leq, \to, \rightsquigarrow, \odot, 1)$ be a po-ipr(pRP'); hence, $\mathcal{A}^L$ is a po-pr(pRP') verifying additionally (La'). By above Theorem 16.1.2 (1), $\Phi'(\mathcal{A}^L)$ is a po-m(pPR') verifying additionally (La'), hence $\Phi'(\mathcal{A}^L)$ is a po-im(pPR').

(1'): Let $\mathcal{A}^L = (A^L, \leq, \odot, \to, \rightsquigarrow, 1)$ be a po-im(pPR'); hence, $\mathcal{A}^L$ is a po-m(pPR') verifying additionally (La'). By above Theorem 16.1.2 (1'), $\Psi'(\mathcal{A}^L)$ is a po-pr(pRP') verifying additionally (La'), hence $\Psi'(\mathcal{A}^L)$ is a po-ipr(pRP').

(2): Obviously. □

Hence, we have the following equivalences between the two 'worlds':

$$\begin{array}{lllll} & \textbf{po-ipr(pRP')} & \Longleftrightarrow & \textbf{po-im(pPR')}, & \\ \textbf{pBCK(pRP')-L} = & l\textbf{-ipr(pRP')} & \Longleftrightarrow & l\textbf{-im(pPR')} & = \textbf{pR-L}. \end{array}$$

### 16.1.2 Connections (pR') - (pRP') - (pPR') - (pP')

By Theorems 1.4.33, 5.3.24 and 16.1.1, 16.1.2, 16.1.3, 16.1.4, we obtain, finally, the equivalences:

$$\begin{array}{lllllll} \textbf{po-pe(pP')} & \Leftrightarrow & \textbf{po-pe(pRP')} & \Leftrightarrow & \textbf{po-s(pPR')} & \Leftrightarrow & \textbf{po-s(pR')}, \\ \textbf{po-pr(pP')} & \Leftrightarrow & \textbf{po-pr(pRP')} & \Leftrightarrow & \textbf{po-m(pPR')} & \Leftrightarrow & \textbf{po-m(pR')}, \\ \textbf{po-mpr(pP')} & \Leftrightarrow & \textbf{po-mpr(pRP')} & \Leftrightarrow & \textbf{po-mm(pPR')} & \Leftrightarrow & \textbf{po-mm(pR')}, \\ \textbf{po-ipr(pP')} & \Leftrightarrow & \textbf{po-ipr(pRP')} & \Leftrightarrow & \textbf{po-im(pPR')} & \Leftrightarrow & \textbf{po-im(pR')} \end{array}$$

and

$$\begin{array}{lllllll} l\textbf{-pe(pP')} & \Leftrightarrow & l\textbf{-pe(pRP')} & \Leftrightarrow & l\textbf{-s(pPR')} & \Leftrightarrow & l\textbf{-s(pR')}, \\ l\textbf{-pr(pP')} & \Leftrightarrow & l\textbf{-pr(pRP')} & \Leftrightarrow & l\textbf{-m(pPR')} & \Leftrightarrow & l\textbf{-m(pR')}, \\ l\textbf{-mpr(pP')} & \Leftrightarrow & l\textbf{-mpr(pRP')} & \Leftrightarrow & l\textbf{-mm(pPR')} & \Leftrightarrow & l\textbf{-mm(pR')}, \\ l\textbf{-ipr(pP')} & \Leftrightarrow & l\textbf{-ipr(pRP')} & \Leftrightarrow & l\textbf{-im(pPR')} & \Leftrightarrow & l\textbf{-im(pR')}. \end{array}$$

## 16.2 Connections between i-groups and groups

In this section, we present from [107], [110], [116] the equivalences (EIG1), (EIG2), (EIG3) between the intermediary notions, in Subsection 16.2.1, and the equivalences (Eq1), (Eq2), (Eq3) between the corresponding four notions, in Subsection 16.2.2.

The section has 2 subsections.

### 16.2.1 Connections between the two intermediary notions

#### 16.2.1.1 The basic equivalence (EIG1)

**Theorem 16.2.1**

*(1) Let* $\mathcal{G} = (G, \to, \rightsquigarrow, \cdot, 1)$ *be an X-i-group.*
*Define* $\Phi'(\mathcal{G}) \stackrel{def.}{=} (G, \cdot, \to, \rightsquigarrow, 1)$.
*Then,* $\Phi'(\mathcal{G})$ *is an X-group.*
*(1') Let* $\mathcal{G} = (G, \cdot, \to, \rightsquigarrow, 1)$ *be an X-group.*
*Define* $\Psi'(\mathcal{G}) \stackrel{def.}{=} (G, \to, \rightsquigarrow, \cdot, 1)$.
*Then,* $\Psi'(\mathcal{G})$ *is an X-i-group.*
*(2) The mappings* $\Phi'$ *and* $\Psi'$ *are mutually inverse.*

**Proof.** (1): By Proposition 1.5.41.
(1'): By Proposition 5.4.22.
(2): Obviously. □

Hence, we have the equivalence:

(EIG1) **X-i-group** $\Longleftrightarrow$ **X-group**,

i.e. the X-implicative-groups and the X-groups are termwise equivalent.

#### 16.2.1.2 The equivalences (EIG2)

**Theorem 16.2.2**

*(1) Let* $\mathcal{G} = (G, \leq, \to, \rightsquigarrow, \cdot, 1)$ *be an X-po-i-group.*
*Define* $\Phi'(\mathcal{G}) \stackrel{def.}{=} (G, \leq, \cdot, \to, \rightsquigarrow, 1)$.
*Then,* $\Phi'(\mathcal{G})$ *is an X-po-group.*
*(1') Let* $\mathcal{G} = (G, \leq, \cdot, \to, \rightsquigarrow, 1)$ *be an X-po-group.*
*Define* $\Psi'(\mathcal{G}) \stackrel{def.}{=} (G, \leq, \to, \rightsquigarrow, \cdot, 1)$.
*Then,* $\Psi'(\mathcal{G})$ *is an X-po-i-group.*
*(2) The mappings* $\Phi'$ *and* $\Psi'$ *are mutually inverse.*

**Proof.** (1): By Proposition 1.5.61.
(1'): By Proposition 5.4.40.
(2): Obviously. □

Hence, we have the equivalences (EIG2) (the analogous of (iELM5)):

(EIG2) **X-po-i-group** $\Longleftrightarrow$ **X-po-group**, i.e. the X-po-implicative-groups and the X-po-groups are termwise equivalent.

#### 16.2.1.3 The equivalences (EIG3)

Here we present the equivalences (EIG3) between the X-$l$-i-groups and the X-$l$-groups. The announced result follows immediately by Theorem 16.2.2.

Hence we have the dual equivalences (EIG3) (the analogous of (iELM6)):

(EIG3) **X-$l$-i-group $\Longleftrightarrow$ X-$l$-group**,

i.e. the X-$l$-implicative-groups are termwise equivalent to the X-$l$-groups.

### 16.2.2 Connections between the four notions

#### 16.2.2.1 The basic equivalences (Eq1), at group level

By Theorems 5.4.23, 1.5.47 and 16.2.1, we obtain:

**Corollary 16.2.3** *The groups, the X-groups, the i-groups and the X-i-groups are all termwise equivalent.*

Hence, by (EG1), (EI1) and (EIG1), we obtain the following basic equivalences:

(Eq1) **i-group $\Longleftrightarrow$ X-i-group $\Longleftrightarrow$ X-group $\Longleftrightarrow$ group**,

that are illustrated in Figure 16.1 (we use Definition 5 of i-groups).

| **i-group** | $\Longleftrightarrow$ | **X-i-group** | $\Longleftrightarrow$ | **X-group** | $\Longleftrightarrow$ | **group** |
|---|---|---|---|---|---|---|
| $(G, \to, \rightsquigarrow, 1)$<br>(pEx),<br>(IdEq),<br>(pEqrelR=) | | $(G, \to, \rightsquigarrow, \cdot, 1)$<br>(pM),<br>(pEx),<br><br>(pGa$^=$) | | $(G, \cdot, \to, \rightsquigarrow, 1)$<br>(PU),<br>(Pass),<br><br>(pGa$^=$) | | $(G, \cdot, ^{-1}, 1)$<br>(PU),<br>(Pass),<br>(pIv) |
| $x^{-1} = x \to 1,$<br>$= x \rightsquigarrow 1$<br>$x \cdot y$<br>$= (x \to y^{-1})^{-1}$<br>$= (y \rightsquigarrow x^{-1})^{-1}$ | | $x^{-1} = x \to 1$<br>$= x \rightsquigarrow 1,$ | | $x^{-1} = x \to 1$<br>$= x \rightsquigarrow 1$ | | $x \to y =$<br>$(x \cdot y^{-1})^{-1}$<br>$= y \cdot x^{-1},$<br>$x \rightsquigarrow y =$<br>$(y^{-1} \cdot x)^{-1}$<br>$= x^{-1} \cdot y$ |
| (I.1) | | (I.2) | | (II.2) | | (II.1) |

Figure 16.1: The equivalences (Eq1)

**Remark 16.2.4** *(See Remark 5.4.7) For all $x \in G$, we have:*
$1 = x \to x = x \rightsquigarrow x$, $x^{-1} = x \to 1 = x \rightsquigarrow 1$ *and* $(x^{-1})^{-1} = x$
*(there is only one* **involutive** *negation).*

Hence, we have that groups are termwise equivalent to i-groups, i.e. we have the following theorem, for which we present a direct proof.

**Theorem 16.2.5**
*(1) Let $\mathcal{G} = (G, \cdot, ^{-1}, 1)$ be a group.*
*Define $\Psi(\mathcal{G}) \stackrel{def.}{=} (G, \to, \rightsquigarrow, 1)$ by: for all $x, y \in G$,*

$$x \to y \stackrel{def.}{=} ((x \cdot y^{-1})^{-1} \stackrel{(pIvG)}{=}) \; y \cdot x^{-1},$$

$$x \rightsquigarrow y \stackrel{def.}{=} ((y^{-1} \cdot x)^{-1} \stackrel{(pIvG)}{=}) \; x^{-1} \cdot y.$$

*Then $\Psi(\mathcal{G})$ is an i-group.*
*(1') Conversely, let $\mathcal{G} = (G, \to, \rightsquigarrow, 1)$ be an i-group.*
*Define $\Phi(\mathcal{G}) \stackrel{def.}{=} (G, \cdot, ^{-1}, 1)$ by: for all $x, y \in G$,*

$$x^{-1} \stackrel{def.}{=} x \to 1 \stackrel{(Id)}{=} x \rightsquigarrow 1,$$

$$x \cdot y \stackrel{def.}{=} ((x \to y^{-1})^{-1} \stackrel{(1.38)}{=}) \; y^{-1} \to x \stackrel{(pDNeg2)}{=} ((y \rightsquigarrow x^{-1})^{-1} \stackrel{(1.38)}{=}) \; x^{-1} \rightsquigarrow y.$$

*Then $\Phi(\mathcal{G})$ is a group.*
*(2) The mappings $\Phi$ and $\Psi$ are mutually inverse.*

**Proof.** (1): By Corollary 5.4.13.
(1'): By Corollary 1.5.36.
(2): Let $(G, \cdot, ^{-1}, 1) \xrightarrow{\Psi} (G, \to, \rightsquigarrow, 1) \xrightarrow{\Phi} (G, \odot, ^*, 1)$. Then, for all $x, y \in G$,
$x^* = x \to 1 = (x \cdot 1^{-1})^{-1} = (x \cdot 1)^{-1} = x^{-1}$ and
$x \odot y = (x \to y^*)^* = (x \to y^{-1})^{-1} = ((x \cdot ((y^{-1})^{-1}))^{-1})^{-1} = x \cdot y$, by (DN).
Let now $(G, \to, \rightsquigarrow, 1) \xrightarrow{\Phi} (G, \cdot, ^{-1}, 1) \xrightarrow{\Psi} (G, \Rightarrow, \approx>, 1)$. Then, for all $x, y \in G$,
$x \Rightarrow y = (x \cdot y^{-1})^{-1} = ((x \to ((y^{-1})^{-1}))^{-1})^{-1} = x \to y$ and
$x \approx> y = (y^{-1} \cdot x)^{-1} = ((x \rightsquigarrow ((y^{-1})^{-1}))^{-1})^{-1} = x \rightsquigarrow y$. □

**Corollary 16.2.6** *The i-group is commutative iff the termwise equivalent group is commutative, i.e. $x \to y = x \rightsquigarrow y$ for all $x, y$ if and only if $x \cdot y = y \cdot x$ for all $x, y$.*

**Proof.** $x \to y = x \rightsquigarrow y$ for all $x, y$ implies $x^{-1} \to y = x^{-1} \rightsquigarrow y \Leftrightarrow y \cdot (x^{-1})^{-1} = (x^{-1})^{-1} \cdot y$, i.e. $y \cdot x = x \cdot y$, by (DN). Conversely, $x \cdot y = y \cdot x$ for all $x, y$ implies $x^{-1} \cdot y = y \cdot x^{-1}$, i.e. $x \rightsquigarrow y = x \to y$. □

### 16.2.2.2 The equivalences (Eq2), at po-group level

By Theorems 5.4.41, 1.5.62 and 16.2.2, we obtain:

**Corollary 16.2.7** *The po-groups, the X-po-groups, the po-i-groups and the X-po-i-groups are all termwise equivalent.*

Hence, by (EG2), (EI2), (EIG2), we obtain the dual equivalences:

(Eq2) **po-i-group** $\Longleftrightarrow$ **X-po-i-group** $\Longleftrightarrow$ **X-po-group** $\Longleftrightarrow$ **po-group**,

that are illustrated in Figure 16.2 (note that (pBB) + (pM) + (pEqrelR) $\Longrightarrow$ (p*)).

| **po-i-group** | $\Longleftrightarrow$ | **X-po-i-group** | $\Longleftrightarrow$ | **X-po-g.** | $\Longleftrightarrow$ | **po-group** |
|---|---|---|---|---|---|---|
| $(G, \leq,$<br>$\rightarrow, \rightsquigarrow, 1)$ | | $(G, \leq,$<br>$\rightarrow, \rightsquigarrow, \cdot, 1)$ | | $(G, \leq,$<br>$\cdot, \rightarrow, \rightsquigarrow, 1)$ | | $(G, \leq,$<br>$\cdot, ^{-1}, 1)$ |
| poset<br>(pEx),<br>(IdEq),<br>(pEqrelR=),<br>**(p*)** | | poset<br>(pM),<br>(pEx),<br><br>(pGa$^{\leq}$),<br>(pGa$^{\geq}$) | | poset<br>(PU),<br>(Pass),<br><br>(pGa$^{\leq}$),<br>(pGa$^{\geq}$) | | poset<br>(PU),<br>(Pass),<br>(pIv),<br>**(pCp)** |
| $x^{-1} = x \rightarrow 1$<br>$= x \rightsquigarrow 1,$<br>$x \cdot y =$<br>$(x \rightarrow y^{-1})^{-1}$<br>$= (y \rightsquigarrow x^{-1})^{-1}$ | | $x^{-1} = x \rightarrow 1$<br>$= x \rightsquigarrow 1$ | | $x^{-1}$<br>$= x \rightarrow 1$<br>$= x \rightsquigarrow 1$ | | $x \rightarrow y =$<br>$(x \cdot y^{-1})^{-1},$<br>$x \rightsquigarrow y =$<br>$(y^{-1} \cdot x)^{-1}$ |
| (I.1) | | (I.2) | | (II.2) | | (II.1) |

Figure 16.2: The equivalences (Eq2)

Hence, we have that po-groups are termwise equivalent to po-implicative-groups, i.e. we have the following theorem, for which we present a direct proof.

**Theorem 16.2.8** *(See Theorem 16.2.5)*

*(1) Let $\mathcal{G} = (G, \leq, \cdot, ^{-1}, 1)$ be a po-group.*

*Define $\Psi'(\mathcal{G}) \stackrel{def.}{=} (G, \leq, \rightarrow, \rightsquigarrow, 1)$, where $(G, \rightarrow, \rightsquigarrow, 1) = \Psi(G, \cdot, ^{-1}, 1)$, with $\Psi$ from Theorem 16.2.5(1).*

*Then $\Psi'(\mathcal{G})$ is a po-i-group.*

*(1') Conversely, let $\mathcal{G} = (G, \leq, \rightarrow, \rightsquigarrow, 1)$ be a po-i-group.*

*Define $\Phi'(\mathcal{G}) \stackrel{def.}{=} (G, \leq, \cdot, ^{-1}, 1)$, where $(G, \cdot, ^{-1}, 1) = \Phi(G, \rightarrow, \rightsquigarrow, 1)$, with $\Phi$ from Theorem 16.2.5(1').*

*Then $\Phi'(\mathcal{G})$ is a po-group.*

*(2) The mappings $\Phi'$ and $\Psi'$ are mutually inverse.*

**Proof.** To prove (1), by Theorem 16.2.5 (1), it remains to prove ($p^{*L}$), which holds by Proposition 5.4.34.

(1') follows by Corollary 12.4.1.

(2) follows by Theorem 16.2.5 (2). □

#### 16.2.2.3 The equivalences (Eq3), at *l*-group level

By the analagous of Theorems 5.4.41, 1.5.62 and 16.2.2 in lattice-ordered case, we obtain:

**Corollary 16.2.9** *The l-groups, the X-l-groups, the l-i-groups and the X-l-i-groups are all termwise equivalent.*

Hence (by (EG3), (EI3), (EIG3)), we have the equivalences:

(Eq3) $\quad$ ***l*-i-group** $\Longleftrightarrow$ **X-*l*-i-group** $\Longleftrightarrow$ **X-*l*-group** $\Longleftrightarrow$ ***l*-group**,

that are represented in Figure 16.3 (see Figures 16.1, 16.2).

| ***l*-i-group** | $\Longleftrightarrow$ | **X-*l*-i-group** | $\Longleftrightarrow$ | **X-*l*-group** | $\Longleftrightarrow$ | ***l*-group** |
|---|---|---|---|---|---|---|
| $(G,\wedge,\vee,$<br>$\to,\rightsquigarrow,1)$<br>(pEx),<br>(IdEq),<br>(pEqrelR=),<br>**(p*)** | | $(G,\wedge,\vee,$<br>$\to,\rightsquigarrow,\cdot,1)$<br>(pM),<br>(pEx),<br><br>(pGa$^{\leq}$),<br>(pGa$^{\geq}$) | | $(G,\wedge,\vee,$<br>$\cdot,\to,\rightsquigarrow,1)$<br>(PU),<br>(Pass),<br><br>(pGa$^{\leq}$),<br>(pGa$^{\geq}$) | | $(G,\wedge,\vee,$<br>$\cdot,^{-1},1)$<br>(PU),<br>(Pass),<br>(pIv),<br>**(pCp)** |
| $x^{-1}=x\to 1$<br>$=x\rightsquigarrow 1,$<br>$x\cdot y$<br>$=(x\to y^{-1})^{-1}$<br>$=(y\rightsquigarrow x^{-1})^{-1}$ | | $x^{-1}=x\to 1$<br>$=x\rightsquigarrow 1$ | | $x^{-1}$<br>$=x\to 1$<br>$=x\rightsquigarrow 1$ | | $x\to y=$<br>$(x\cdot y^{-1})^{-1},$<br>$x\rightsquigarrow y=$<br>$(y^{-1}\cdot x)^{-1}$ |
| (I.1) | | (I.2) | | (II.2) | | (II.1) |

Figure 16.3: The equivalences (Eq3)

# Chapter 17

# Connections between M and m-M, $m_1$-M algebras in the involutive case

In this chapter, we present bridge theorems in the commutative case. We establish two general theorems connecting the two 'worlds' in the involutive case, one for algebras *with last element*, the other for algebras *without last element.* More precisely,
- the *'world' of M algebras* (of logic) with additional operation (the 'world' of $\rightarrow, ^-, 1$) (from Chapters 2, 3) and the *'world' of m-M algebras* (i.e. the 'world' of $\odot, ^-, 1$) (from Chapters 6, 8) are connected, *in the involutive case*, by two mutually inverse transformations, $\Phi$ and $\Psi$, in Section 17.1, and
- the *'world' of M algebras* (of logic) with additional operation (the 'world' of $\rightarrow, ^{-1}, 1$ ) (from Chapters 2, 4) and the *'world' of $m_1$-M algebras* (i.e. the 'world' of $\cdot, ^{-1}, 1$) (from Chapters 7, 8) are connected, *in the involutive case*, by two mutually inverse transformations, $\Phi_1$ and $\Psi_1$, in Section 17.2.

These theorems can be used to prove the *definitionally equivalence* (d.e.) between the analogous involutive algebras from the two 'worlds' simply by choosing appropriate definitions of the algebras.

The content of this chapter is taken mainly from [118].

## 17.1 Connections in the case $1^- = 0$

The bounded M algebras from the hierarchies 2, 2'$^b$ (from Sections 2, 3) and the corresponding m-M algebras from the hierarchies m-2, m-2' (from Sections 6, 8) are connected, *in the involutive case*, by two mutually inverse transformations, $\Phi$ and $\Psi$, as we shall see in next Theorem 17.1.1.

Recall the properties (Ex0): $x \rightarrow (y \rightarrow 0) = y \rightarrow (x \rightarrow 0)$ and (Neg3): $x \rightarrow y^- = y \rightarrow x^-$ and that (Ex) implies (Ex0). We then have the following general result:

**Theorem 17.1.1**

*(1) Let $\mathcal{A}^L = (A^L, \to, 0, 1)$ be an algebra of type $(2,0,0)$ verifying (M), (Ex0) and (DN), where a negation $^-$ is defined by (dfneg) ($x^- \stackrel{def.}{=} x \to 0$). (Equivalently, let $\mathcal{A}^L = (A^L, \to, ^-, 1)$ be an algebra of type $(2,1,0)$ verifying (M), (Neg3) and (DN), such that $0 \stackrel{def.}{=} 1^-$).*

*Define an internal binary relation $\leq$ by (dfrelR) ($x \leq y \stackrel{def.}{\Longleftrightarrow} x \to y = 1$).*

*Define $\Phi(\mathcal{A}^L) \stackrel{def.}{=} (A^L, \odot, ^-, 1)$ by: for all $x, y \in A^L$,*

$$x \odot y \stackrel{def.}{=} (x \to y^-)^-.$$

*Then, $\Phi(\mathcal{A}^L)$ satisfies (PU), (Pcomm) and (DN).*

*(1') Conversely, let $\mathcal{A}^L_m = (A^L, \odot, ^-, 1)$ be an algebra of type $(2,1,0)$. Define $0 \stackrel{def.}{=} 1^-$ and suppose that (PU), (Pcomm) and (DN) hold.*

*Define an internal binary relation $\leq_m$ by (m-dfrelP) ($x \leq_m y \stackrel{def}{\Longleftrightarrow} x \odot y^- = 0$).*

*Define $\Psi(\mathcal{A}^L_m) \stackrel{def.}{=} (A, \to, 0, 1)$ (or, equivalently, $\Psi(\mathcal{A}^L_m) \stackrel{def.}{=} (A, \to, ^-, 1)$) by: for all $x, y \in A^L$,*

$$x \to y \stackrel{def.}{=} (x \odot y^-)^-.$$

*Then, $\Psi(\mathcal{A}^L_m)$ satisfies (M), (Ex0) and (DN) (or (M), (Neg3), (DN), respectively).*

*(2) The mappings $\Phi$ and $\Psi$ are mutually inverse.*

*(3) $x \leq y \Longleftrightarrow x \leq_m y$, for all $x, y \in A^L$.*

*(3') $x \leq^W y \Longleftrightarrow y^- \leq^M_m x^-$, for all $x, y \in A^L$.*

*(3") $x \leq^P y \Longleftrightarrow x \leq^P_m y$, for all $x, y \in A^L$.*

*(4) The following properties are equivalent:*

*(Ex) $\Longleftrightarrow$ (Pass), (La) $\Longleftrightarrow$ (m-La), (Re) $\Longleftrightarrow$ (m-Re),*
*(An) $\Longleftrightarrow$ (m-An), (DN4) $\Longleftrightarrow$ (m-DN4),*
*(**) $\Longleftrightarrow$ (m-**), (*) $\Longleftrightarrow$ (m-*), (BB) $\Longleftrightarrow$ (m-BB), (B) $\Longleftrightarrow$ (m-B),*
*(N) $\Longleftrightarrow$ (m-N) and (V) $\Longleftrightarrow$ (m-V),*
*(C) $\Longleftrightarrow$ (m-C), (Tr) $\Longleftrightarrow$ (m-Tr), (D) $\Longleftrightarrow$ (m-D),*
*($\vee$-comm) $\Longleftrightarrow$ ($\wedge_m$-comm), (impl) $\Longleftrightarrow$ (m-Pimpl),*
*(DN6) = (i-G) $\Longleftrightarrow$ (G), (@) $\Longleftrightarrow$ (m-@)=(m-Wdiv).*

**Proof.** First note that the equivalence of the conditions from the hypothesis was proved in Theorems 3.1.4 and 3.1.5 from Chapter 3.

(1): Let $\mathcal{A}^L = (A^L, \to, 0, 1)$ be an algebra such that (M), (Ex0) and (DN) hold, where $x^- \stackrel{def.}{=} x \to 0$, by (dfneg). Then, (Neg1-0) holds, by (M): $1^- = 1 \to 0 = 0$; (Neg0-1) holds too, by (DN): $0^- = (1^-)^- = 1$; (Neg3) holds, by (Ex0): $x \to y^- = x \to (y \to 0) = y \to (x \to 0) = y \to x^-$. By (C3), (DN) + (Neg3) imply (DN3); by (C2'), (DN) + (DN3) imply (DN2); by (C4), (DN2) implies (DN4), hence (DN4') holds too.

Equivalently, let $\mathcal{A}^L = (A^L, \to, ^-, 1)$ be an algebra such (M), (Neg3) and (DN) hold, where $0 \stackrel{def.}{=} 1^-$, hence (Neg1-0) holds; then, (Neg0-1) holds too, by (DN), and (DN4') holds, by (C3), (C2'), (C4).

Define $x \odot y \stackrel{def.}{=} (x \to y^-)^-$. We must prove that (Pcomm) and (PU) hold. Indeed,
$x \odot y \stackrel{def.}{=} (x \to y^-)^- \stackrel{(Neg3)}{=} (y \to x^-)^- = y \odot x$; thus, (Pcomm) holds.
$1 \odot x \stackrel{def.}{=} (1 \to x^-)^- \stackrel{(M)}{=} (x^-)^- \stackrel{(DN)}{=} x$; thus, (PU) holds.

(1'): Let $\mathcal{A}^L_m = (A^L, \odot, ^-, 1)$ be an algebra such that (Pcomm), (PU) and (DN) hold; let $0 \stackrel{def.}{=} 1^-$, hence (Neg1-0) holds; then, (Neg0-1) holds too, by (DN). By (mCDN2), (Pcomm) + (DN) imply (m-DN4'), hence (m-DN4') holds too.
Define $x \to y \stackrel{def.}{=} (x \odot y^-)^-$. We must prove that (M) and (Ex0) hold. Indeed,
$1 \to x \stackrel{def.}{=} (1 \odot x^-)^- \stackrel{(PU)}{=} (x^-)^- \stackrel{(DN)}{=} x$; thus, (M) holds.
$x \to (y \to 0) \stackrel{def.}{=} (x \odot ((y \odot 0^-)^-)^-)^- \stackrel{(DN),(Neg0-1)}{=} (x \odot (y \odot 1))^-$
$\stackrel{(PU)}{=} (x \odot y)^- \stackrel{(Pcomm)}{=} (y \odot x)^- \stackrel{(PU)}{=} (y \odot (x \odot 1))^-$
$\stackrel{(Neg0-1),(DN)}{=} (y \odot ((x \odot 0^-)^-)^-)^- \stackrel{def.}{=} y \to (x \to 0)$; thus, (Ex0) holds.

(2): Let
$$(A^L, \to, 0, 1) \quad \xrightarrow{\Phi} \quad (A^L, \odot, ^-, 1) \quad \xrightarrow{\Psi} \quad (A^L, \Rightarrow, \mathbf{0}, 1).$$
Then, for all $x, y \in A^L$, we have:
$x \Rightarrow y = (x \odot y^-)^- = ((x \to (y^-)^-)^-)^- \stackrel{(DN)}{=} x \to y$ and
$\mathbf{0} = 1^- = 1 \to 0 \stackrel{(M)}{=} 0$.

Conversely, let
$$(A^L, \odot, ^-, 1) \quad \xrightarrow{\Psi} \quad (A^L, \to, 0, 1) \quad \xrightarrow{\Phi} \quad (A^L, \otimes, ', 1).$$
Then, for all $x, y \in A^L$, we have:
$x' = x \to 0 = (x \odot 0^-)^- \stackrel{(Neg0-1)}{=} (x \odot 1)^- \stackrel{(PU)}{=} x^-$ and
$x \otimes y = (x \to y')' = (x \to y^-)^- = (x \odot (y^-)^-)^- \stackrel{(DN)}{=} x \odot y$.

(3): Suppose $x \leq y$, i.e. $x \to y = 1$, by (dfrelR); hence, $(x \odot y^-)^- = 1$, by $\Phi$; then, $x \odot y^- = 0$, by (DN) and (Neg1-0), hence $x \leq_m y$, by (m-dfrelP).
Suppose now $x \leq_m y$, i.e. $x \odot y^- = 0$, by (m-dfrelP); then, $(x \odot y^-)^- = 1$, by (Neg0-1); hence $x \to y = 1$, by $\Psi$; hence $x \leq y$, by (dfrelR).

(3'): $x \leq^W y \Longleftrightarrow x \vee^W y = y \Longleftrightarrow (y \to x) \to x = y \Longleftrightarrow (y \odot x^-)^- \to x = y \Longleftrightarrow ((y \odot x^-)^- \odot x^-)^- = y \Longleftrightarrow (y \odot x^-)^- \odot x^- = y^- \Longleftrightarrow (y^= \odot x^-)^- \odot x^- = y^- \Longleftrightarrow y^- \wedge^M_m x^- = y^- \Longleftrightarrow y^- \leq^M_m x^-$.

(3"): $x \leq^P y \Longleftrightarrow (x \to y^-)^- = x \Longleftrightarrow x \odot y = x \Longleftrightarrow x \leq^P_m y$.

(4): (Ex) $\Longrightarrow$ (Pass): $x \odot (y \odot z) \stackrel{(Pcomm)}{=} x \odot (z \odot y)$
$= (x \to ((z \to y^-)^-)^-)^- \stackrel{(DN)}{=} (x \to (z \to y^-))^- \stackrel{(Ex)}{=} (z \to (x \to y^-))^-$
$\stackrel{(DN)}{=} (z \to ((x \to y^-)^-)^-)^- = z \odot (x \odot y) \stackrel{(Pcomm)}{=} (x \odot y) \odot z$.
(Pass) $\Longrightarrow$ (Ex): $x \to (y \to z) = (x \odot ((y \odot z^-)^-)^-)^- \stackrel{(DN)}{=} (x \odot (y \odot z^-))^-$
$\stackrel{(Pass)}{=} ((x \odot y) \odot z^-)^- \stackrel{(Pcomm)}{=} ((y \odot x) \odot z^-)^- \stackrel{(Pass)}{=} (y \odot (x \odot z^-)^-$
$\stackrel{(DN)}{=} (y \odot ((x \odot z^-)^-)^-)^- = y \to (x \to z)$.

(La) $\Longleftrightarrow$ (m-La): (La') $\Longleftrightarrow x \leq 1 \stackrel{(3)}{\Longleftrightarrow} x \leq_m 1 \Longleftrightarrow$ (m-La').

(Re) $\Longleftrightarrow$ (m-Re): (Re') $\Longleftrightarrow x \leq x \stackrel{(3)}{\Longleftrightarrow} x \leq_m x \Longleftrightarrow$ (m-Re').

(An) $\Longleftrightarrow$ (m-An): Suppose that (An') holds and take $x \leq_m y$ and $y \leq_m x$; then, $x \leq y$ and $y \leq x$, by above (3); then, $x = y$, by (An'); thus, (m-An') holds too.
Conversely, suppose that (m-An') holds and take $x \leq y$ and $y \leq x$; then, $x \leq_m y$ and $y \leq_m x$, by above (3); then, $x = y$, by (m-An'); thus, (An') holds too.

(DN4) $\Longleftrightarrow$ (m-DN4): (DN4'), i.e. $x \leq y \Longleftrightarrow y^- \leq x^-$, is equivalent to (m-DN4'), i.e. $x \leq_m y \Longleftrightarrow y^- \leq_m x^-$, by above (3).

(**) $\Longleftrightarrow$ (m-**): Suppose that (**') holds and take $x \leq_m y$; then, $x \leq y$, by above (3); then, $y \to z^- \leq x \to z^-$, by (**'); hence, $(y \odot (z^-)^-)^- \leq_m (x \odot (z^-)^-)^-$, by (3) again; then, $(y \odot z)^- \leq_m (x \odot z)^-$, by (DN); then $x \odot z \leq_m y \odot z$, by (m-DN4'); thus, (m-**') holds too.
Conversely, suppose that (m-**') holds and take $x \leq y$; then, $x \leq_m y$, by above (3); then, $x \odot z^- \leq_m y \odot z^-$, by (m-**'); hence, $x \odot z^- \leq y \odot z^-$, by (3) again, hence $(x \to (z^-)^-)^- \leq (y \to (z^-)^-)^-$, by $\Psi$; then, $(x \to z)^- \leq (y \to z)^-$, by (DN); then $y \to z \leq x \to z$, by (DN4'); thus, (**') holds too.

(*) $\Longleftrightarrow$ (m-*): Suppose that (*') holds and take $x \leq_m y$; then, $x \leq y$, by above (3); then, $z \to x \leq z \to y$, by (*'); then $(z \to y)^- \leq (z \to x)^-$, by (DN4') and $(z \to (y^-)^-)^- \leq (z \to (x^-)^-)^-$, by (DN); hence, $z \odot y^- \leq_m z \odot x^-$, by (3) again and by $\Phi$; thus, (m-*') holds too.
Conversely, suppose that (m-*') holds and take $x \leq y$; then, $x \leq_m y$, by above (3); then, $z \odot y^- \leq_m z \odot x^-$, by (m-*'); hence, $(z \to (y^-)^-)^- \leq (z \to (x^-)^-)^-$, by (3) again and by $\Psi$; then, $(z \to y)^- \leq (z \to x)^-$, by (DN) and $z \to x \leq z \to y$, by (DN4'); thus, (*') holds too.

(BB) $\Longleftrightarrow$ (m-BB): Suppose that (BB) holds; then,
$[(z \odot x)^- \odot (y \odot x)] \odot (y \odot z^-)^- \overset{(Pcomm)}{=} (y \odot z^-)^- \odot [(z \odot x)^- \odot (y \odot x)]$
$\overset{(DN)}{=} (y \odot z^-)^- \odot [[(z \odot (x^-)^-)^- \odot ((y \odot (x^-)^-)^-)^-]^-]^-$
$= (y \to z) \odot [[(z \to x^-) \odot (y \to x^-)^-]^-]^-$
$\overset{(DN)}{=} ((y \to z) \to [(z \to x^-) \to (y \to x^-)])^- \overset{(BB)}{=} 1^- \overset{(Neg1-0)}{=} 0$; thus, (m-BB) holds too.
Conversely, suppose that (m-BB) holds; then,
$(y \to z) \to [(z \to x) \to (y \to x)] = ((y \odot z^-)^- \odot [[(z \odot x^-)^- \odot ((y \odot x^-)^-)^-]^-]^-)^-$
$\overset{(DN)}{=} ((y \odot z^-)^- \odot [(z \odot x^-)^- \odot (y \odot x^-)])^-$
$\overset{(Pcomm)}{=} ([(z \odot x^-)^- \odot (y \odot x^-)] \odot (y \odot z^-)^-)^- \overset{(m-BB)}{=} 0^- \overset{(Neg0-1)}{=} 1$; thus, (BB) holds.

(B) $\Longleftrightarrow$ (m-B): Suppose that (B) holds; then,
$[(x \odot y^-)^- \odot (x \odot z)] \odot (y \odot z)^- \overset{(Pcomm)}{=} (y \odot z)^- \odot [(x \odot y^-)^- \odot (x \odot z)]$
$\overset{(DN)}{=} (y \odot (z^-)^-)^- \odot [[(x \odot y^-)^- \odot (x \odot (z^-)^-)]^-]^-$
$= (y \to z^-) \odot [(x \to y) \to (x \to z^-)]^-$
$\overset{(DN)}{=} ((y \to z^-) \to [(x \to y) \to (x \to z^-)])^- \overset{(B)}{=} 1^- \overset{(Neg1-0)}{=} 0$; thus, (m-B) holds too.
Conversely, suppose that (m-B) holds; then,
$(y \to z) \to [(x \to y) \to (x \to z)] = ((y \odot z^-)^- \odot [[(x \odot y^-)^- \odot ((x \odot z^-)^-)^-]^-]^-)^-$

$\overset{(DN)}{=} ((y \odot z^-)^- \odot [(x \odot y^-)^- \odot (x \odot z^-)])^-$
$\overset{(Pcomm)}{=} ([(x \odot y^-)^- \odot (x \odot z^-)] \odot (y \odot z^-)^-)^- \overset{(m-B)}{=} 0^- \overset{(Neg0-1)}{=} 1$; thus, (B) holds.

(N) $\Longleftrightarrow$ (m-N): $(1 \leq x \Longrightarrow x = 1) \overset{(3)}{\Longleftrightarrow} (1 \leq_m x \Longrightarrow x = 1) \Longleftrightarrow$ (m-N');
(V) $\Longleftrightarrow$ (m-V): $(x \leq 0 \Longrightarrow x = 0) \overset{(3)}{\Longleftrightarrow} (x \leq_m 0 \Longrightarrow x = 0) \Longleftrightarrow$ (m-V').

(C) $\Longleftrightarrow$ (m-C): Suppose that (C') holds; then,
$y \odot (x \odot z) \leq_m x \odot (y \odot z) \overset{(DN)}{\Longleftrightarrow} y \odot (x \odot (z^-)^-) \leq_m x \odot (y \odot (z^-)^-)$
$\overset{(m-DN4')}{\Longleftrightarrow} (x \odot (y \odot (z^-)^-))^- \leq_m (y \odot (x \odot (z^-)^-))^-$
$\overset{(DN)}{\Longleftrightarrow} (x \odot ((y \odot (z^-)^-)^-)^-)^- \leq_m (y \odot ((x \odot (z^-)^-)^-)^-)^-$
$\overset{(3)}{\Longleftrightarrow} x \to (y \to z^-) \leq y \to (x \to z^-)$ that is true by (C'); thus, (m-C') holds too.
Conversely, suppose that (m-C') holds; then,
$x \to (y \to z) \leq y \to (x \to z) \overset{(3)}{\Longleftrightarrow} (x \odot ((y \odot z^-)^-)^-)^- \leq_m (y \odot ((x \odot z^-)^-)^-)^-$
$\overset{(DN)}{\Longleftrightarrow} (x \odot (y \odot z^-))^- \leq_m (y \odot (x \odot z^-))^-$
$\overset{(DN4')}{\Longleftrightarrow} y \odot (x \odot z^-) \leq_m x \odot (y \odot z^-)$ that is true by (m-C'); thus, (C') holds too.

(Tr) $\Longleftrightarrow$ (m-Tr): (Tr') $\Longleftrightarrow (x \leq y$ and $y \leq z \Longrightarrow x \leq z)$
$\overset{(3)}{\Longleftrightarrow} (x \leq_m y$ and $y \leq_m z \Longrightarrow x \leq_m z) \Longleftrightarrow$ (m-Tr').

(D) $\Longleftrightarrow$ (m-D): Suppose that (D') holds; then,
$(y^- \odot x)^- \odot x \leq_m y \overset{(m-DN4')}{\Longleftrightarrow} y^- \leq_m [(y^- \odot x)^- \odot x]^-$
$\overset{(DN)}{\Longleftrightarrow} y^- \leq_m [(y^- \odot (x^-)^-)^- \odot (x^-)^-]^-$
$\overset{(3)}{\Longleftrightarrow} y^- \leq (y^- \to x^-) \to x^-$ that is true by (D'); thus, (m-D') holds too.
Conversely, suppose that (m-D') holds; then,
$y \leq (y \to x) \to x \overset{(DN4')}{\Longleftrightarrow} [(y \to x) \to x]^- \leq y^-$
$\overset{(3)}{\Longleftrightarrow} [[(y \odot x^-)^- \odot x^-]^-]^- \leq_m y^- \overset{(DN)}{\Longleftrightarrow} (y \odot x^-)^- \odot x^- \leq_m y^-$
$\overset{(DN)}{\Longleftrightarrow} ((y^-)^- \odot x^-)^- \odot x^- \leq_m y^-$ that is true by (m-D'); thus, (D') holds too.

($\vee$-comm) $\Longleftrightarrow$ ($\wedge_m$-comm): Suppose that ($\vee$-comm) holds; then,
$(x^- \odot y)^- \odot y = (y^- \odot x)^- \odot x$
$\overset{(DN)}{\Longleftrightarrow} [[(x^- \odot (y^-)^-)^- \odot (y^-)^-]^-]^- = [[(y^- \odot (x^-)^-)^- \odot (x^-)^-]^-]^-$
$\Longleftrightarrow [(x^- \to y^-) \to y^-]^- = [(y^- \to x^-) \to x^-]^-$
$\Longleftrightarrow (x^- \to y^-) \to y^- = (y^- \to x^-) \to x^-$ that is true by ($\vee$-comm); thus, ($\wedge_m$-comm) holds too.
Conversely, suppose that ($\wedge_m$-comm) holds; then,
$(x \to y) \to y = (y \to x) \to x \Longleftrightarrow ((x \odot y^-)^- \odot y^-)^- = ((y \odot x^-)^- \odot x^-)^-$
$\overset{(DN)}{\Longleftrightarrow} (((x^-)^- \odot y^-)^- \odot y^-)^- = (((y^-)^- \odot x^-)^- \odot x^-)^-$
$\overset{(DN)}{\Longleftrightarrow} ((x^-)^- \odot y^-)^- \odot y^- = ((y^-)^- \odot x^-)^- \odot x^-$
that is true by ($\wedge_m$-comm); thus, ($\vee$-comm) holds too.

(impl) $\Longleftrightarrow$ (m-Pimpl): (impl) $\Longleftrightarrow (x \to y) \to x = x$
$\Longleftrightarrow [(x \odot y^-)^- \odot x^-]^- = x \Longleftrightarrow$ (m-Pimpl).

(DN6) $\Longleftrightarrow$ (G): Suppose that (DN6) holds; then, $x \odot x = (x \to x^-)^- \overset{(DN)}{=}$

$((x^-)^- \to x^-)^- \overset{(DN6)}{=} (x^-)^- \overset{(DN)}{=} x$; thus, (G) holds.
Conversely, suppose that (G) holds; then, $x^- \to x = (x^- \odot x^-)^- \overset{(G)}{=} (x^-)^- = x$; thus, (DN6) holds.

(@) $\Longleftrightarrow$ (m-@): Suppose that (@) holds; then,
$(y \odot x^-)^- \odot y = x \odot y \overset{(DN)}{\Longleftrightarrow} [[((y^-)^- \odot x^-)^- \odot (y^-)^-]^-]^- = [[x \odot (y^-)^-]^-]^-$
$\Longleftrightarrow [((y^-)^- \to x) \to y^-]^- = [x \to y^-]^-$ that is true by (@); thus, (m-@) holds.
Conversely, suppose that (m-@) holds; then,
$(y^- \to x) \to y = x \to y \Longleftrightarrow ((y^- \odot x^-)^- \odot y^-)^- = (x \odot y^-)^-$ that is true by (m-@); thus, (@) holds too. □

By this Theorem 17.1.1, choosing appropriate definitions, one can prove almost all definitionally equivalences (d.e.) between the corresponding involutive algebras from the two 'worlds'. As examples, we shall prove next the d.e. between involutive MEL and m-MEL algebras, between involutive BE and m-BE algebras, between involutive BCK and (involutive) m-BCK algebras, between Wajsberg and MV algebras, between i-ortholattices and ortholattices and between i-Boolean and Boolean algebras.

**Theorem 17.1.2**

*(mel1) Let $\mathcal{A}^L = (A^L, \to, 0, 1)$ be an involutive left-MEL algebra, where the negation $^-$ is defined by (dfneg) ($x^- \overset{def.}{=} x \to 0$). Define $\Phi(\mathcal{A}^L) \overset{def.}{=} (A^L, \odot, ^-, 1)$ as follows: for all $x, y \in A^L$,*

$$x \odot y \overset{def.}{=} (x \to y^-)^-.$$

*Then, $\Phi(\mathcal{A}^L)$ is an involutive left-m-MEL algebra.*

*(mel1') Conversely, let $\mathcal{A}^L_m = (A^L, \odot, ^-, 1)$ be an involutive left-m-MEL algebra and $0 \overset{def.}{=} 1^-$. Define $\Psi(\mathcal{A}^L_m) \overset{def.}{=} (A^L, \to, 0, 1)$ as follows: for all $x, y \in A^L$,*

$$x \to y \overset{def.}{=} (x \odot y^-)^-.$$

*Then, $\Psi(\mathcal{A}^L_m)$ is an involutive left-MEL algebra.*

*(mel2) The two mappings, $\Phi$ and $\Psi$ are mutually inverse.*

**Proof.** Let $\mathcal{A}^L = (A^L, \to, 0, 1)$ be an involutive left-MEL algebra, i.e. (M), (Ex), (La), (Fi), (DN) hold, where $x^- \overset{def.}{=} x \to 0$, by (dfneg). Since (Ex) holds, it follows that (Ex0) holds. Let also $\mathcal{A}^L_m = (A^L, \odot, ^-, 1)$ be an involutive left-m-MEL algebra and $0 \overset{def.}{=} 1^-$, i.e. (Neg1-0), (Neg0-1), (PU), (Pcomm), (Pass), (m-La), (DN) hold.

Since $\mathcal{A}^L$ verifies (M), (Ex0), (DN) and $\mathcal{A}^L_m$ verifies (PU), (Pcomm), (DN), we can apply Theorem 17.1.1 and we have that (1), (1') and (2) (hence (mel2)) hold. Then, to finish the proof of (mel1), since (Ex) $\Longleftrightarrow$ (Pass) and (La) $\Longleftrightarrow$ (m-La), it follows that (Pass) and (m-La) hold; (Neg1-0) holds by (M) and (Neg0-1) holds by (DN); thus, (mel1) holds. To finish the proof of (mel1'), since (Pass) $\Longleftrightarrow$ (Ex) and (m-La) $\Longleftrightarrow$ (La), it follows that (Ex) and (La) hold; it remains to prove that

(Fi) holds; indeed, $0 \to x = (0 \odot x^-)^- \overset{(Pcomm)}{=} (x^- \odot 0)^- \overset{(m-La)}{=} 0^- \overset{(Neg0-1)}{=} 1$, i.e. (Fi) holds; thus, (mel1') holds. □

**Corollary 17.1.3** *Involutive left-BE algebras are d.e. with involutive left-m-BE algebras.*

**Proof.** An involutive left-BE algebra is an involutive left-MEL algebra verifying (Re). An involutive left-m-BE algebra is an involutive left-m-MEL algebra verifying (m-Re). By Theorem 17.1.2, involutive left-MEL algebras are d.e. with involutive left-m-MEL algebras and, by Theorem 17.1.1, (Re) $\Longleftrightarrow$ (m-Re). The proof is complete. □

**Corollary 17.1.4** *Involutive left-BCK algebras are d.e. with (involutive) left-m-BCK algebras.*

**Proof.** An involutive left-BCK algebra is an involutive left-BE algebra verifying (An) and (BB). An (involutive) left-m-BCK algebra is an involutive left-m-BE algebra verifying (m-An) and (m-BB). By Corollary 17.1.3, involutive left-BE algebras are d.e. with involutive left-m-BE algebras and, by Theorem 17.1.1, (An) $\Longleftrightarrow$ (m-An) and (BB) $\Longleftrightarrow$ (m-BB). The proof is complete. □

**Corollary 17.1.5** *Wajsberg algebras are d.e. with MV algebras.*

**Proof.** A left-Wajsberg algebra (Definition 2) is a bounded left-MEL algebra verifying ($\vee$-comm), hence an involutive left-MEL algebra verifying ($\vee$-comm). A left-MV algebra (Remark 6.5.2) is an involutive left-m-MEL algebra verifying ($\wedge_m$-comm). By Theorem 17.1.2, involutive left-MEL algebras are d.e. with involutive left-m-MEL algebras and, by Theorem 17.1.1, ($\vee$-comm) $\Longleftrightarrow$ ($\wedge_m$-comm). The proof is complete. □

**Corollary 17.1.6** *Implicative-ortholattices are d.e. with ortholattices.*

**Proof.** A left-i-ortholattice (Definition 2) is an involutive left-BE algebra verifying (impl). A left-ortholattice (Definition 2) is a (involutive) left-m-BE algebra verifying (m-Pimpl). By Corollary 17.1.3, involutive left-BE algebras are d.e. with involutive left-m-BE algebras and, by Theorem 17.1.1, (impl) $\Longleftrightarrow$ (m-Pimpl). The proof is complete. □

**Corollary 17.1.7** *Implicative-Boolean algebras are d.e. with Boolean algebras.*

**Proof.** I-Boolean algebras are d.e. with i-ortholattices verifying (@), by Theorem 3.4.11. Boolean algebras are d.e. with ortholattices verifying (m-@) = (m-Wdiv), by Theorem 9.2.25. By Corollary 17.1.6, implicative-ortholattices are d.e. with ortholattices and, by Theorem 17.1.1, (@) $\Longleftrightarrow$ (m-@). The proof is complete. □

## 17.2 Connections in the case $1^{-1} = 1$

The M algebras without last element from the hierarchies 1, 1' (from Chapters 2, 4) and the corresponding $m_1$-M algebras from the hierarchies m-1, m-1' (from Chapters 7, 8) are connected, *in the involutive case*, by two mutually inverse transformations, $\Phi_1$ and $\Psi_1$, as we shall see in next Theorem 17.2.1.

Recall the properties (Ex1): $x \to (y \to 1) = y \to (x \to 1)$ and (Neg3): $x \to y^{-1} = y \to x^{-1}$ and that (Ex) implies (Ex1). We then have the following general result:

**Theorem 17.2.1**

*(1) Let $\mathcal{A}^L = (A^L, \to, 1)$ be an algebra of type $(2,0)$ verifying (M), (Ex1) and (DN), where a negation $^{-1}$ is defined by (dfneg1) ($x^{-1} \stackrel{def.}{=} x \to 1$). (Equivalently, let $\mathcal{A}^L = (A^L, \to, ^{-1}, 1)$ be an algebra of type $(2,1,0)$ such that $1^{-1} = 1$ and verifying (M), (Neg3) and (DN)).*

*Define an internal binary relation $\leq$ by (dfrelR) ($x \leq y \stackrel{def.}{\Longleftrightarrow} x \to y = 1$).*

*Define $\Phi_1(\mathcal{A}^L) \stackrel{def.}{=} (A^L, \cdot, ^{-1}, 1)$ by: for all $x, y \in A^L$,*

$$x \cdot y \stackrel{def.}{=} (x \to y^{-1})^{-1}.$$

*Then, $\Phi_1(\mathcal{A}^L)$ satisfies (PU), (Pcomm) and (DN).*

*(1') Conversely, let $\mathcal{A}^L_{m_1} = (A^L, \cdot, ^{-1}, 1)$ be an algebra of type $(2,1,0)$ such that $1^{-1} = 1$ and suppose that (PU), (Pcomm) and (DN) hold.*

*Define an internal binary relation $\leq_{m_1}$ by (m-dfrelP$_1$) ($x \leq_{m_1} y \stackrel{def}{\Longleftrightarrow} x \cdot y^{-1} = 1$).*

*Define $\Psi_1(\mathcal{A}^L_{m_1}) \stackrel{def.}{=} (A^L, \to, 1)$ (or, equivalently, $\Psi_1(\mathcal{A}^L_{m_1}) \stackrel{def.}{=} (A^L, \to, ^{-1}, 1)$) by: for all $x, y \in A^L$,*

$$x \to y \stackrel{def.}{=} (x \cdot y^{-1})^{-1}.$$

*Then, $\Psi_1(\mathcal{A}^L_{m_1})$ satisfies (M), (Ex1) and (DN) (or (M), (Neg3) and (DN), respectively).*

*(2) The mappings $\Phi_1$ and $\Psi_1$ are mutually inverse.*

*(3) $x \leq y \Longleftrightarrow x \leq_{m_1} y$, for all $x, y \in A^L$.*

*(4) The following properties are equivalent:*
*(Ex) $\Longleftrightarrow$ (Pass), (Re) $\Longleftrightarrow$ ($m_1$-Re), (An) $\Longleftrightarrow$ ($m_1$-An), (DN4) $\Longleftrightarrow$ ($m_1$-DN4), (**) $\Longleftrightarrow$ ($m_1$-**), (*) $\Longleftrightarrow$ ($m_1$-*), (BB) $\Longleftrightarrow$ ($m_1$-BB), (BB=) $\Longleftrightarrow$ ($m_1$-BB=), (B) $\Longleftrightarrow$ ($m_1$-B), (B=) $\Longleftrightarrow$ ($m_1$-B=), (C) $\Longleftrightarrow$ ($m_1$-C), (Tr) $\Longleftrightarrow$ ($m_1$-Tr), (D) $\Longleftrightarrow$ ($m_1$-D), (D=) $\Longleftrightarrow$ ($m_1$-D=), ($\vee$-comm) $\Longleftrightarrow$ ($\wedge_m$-comm), (impl) $\Longleftrightarrow$ (m-Pimpl), (p-s) $\Longleftrightarrow$ (m-p-s).*

**Proof.** First note that the equivalence of the conditions from the hypothesis was proved in Theorem 4.1.4 from Chapter 4.

(1): Let $\mathcal{A}^L = (A^L, \to, 1)$ be an algebra such that (M), (Ex1) and (DN) hold, where $x^{-1} \stackrel{def.}{=} x \to 1$, by (dfneg1). Then, (N1) holds, by (M): $1^{-1} = 1 \to 1 = 1$; (Neg3) holds, by (Ex1): $x \to y^{-1} = x \to (y \to 1) \stackrel{(Ex1)}{=} y \to (x \to 1) = y \to x^{-1}$.

By (C3), (DN) + (Neg3) imply (DN3); by (C2'), (DN) + (DN3) imply (DN2); by (C4), (DN2) implies (DN4), hence (DN4') holds too.

Define $x \cdot y \stackrel{def.}{=} (x \to y^{-1})^{-1}$. We must prove that (Pcomm) and (PU) hold. Indeed,

$x \cdot y \stackrel{def.}{=} (x \to y^{-1})^{-1} \stackrel{(Neg3)}{=} (y \to x^{-1})^{-1} = y \cdot x$; thus, (Pcomm) holds.

$1 \cdot x \stackrel{def.}{=} (1 \to x^{-1})^{-1} \stackrel{(M)}{=} (x^{-1})^{-1} \stackrel{(DN)}{=} x$; thus, (PU) holds.

(1'): Let $\mathcal{A}^L_{m_1} = (A^L, \cdot, ^{-1}, 1)$ be an algebra such that (Pcomm), (PU) and (DN) hold and $1^{-1} = 1$, hence (N1) holds; then, by (m1CDN2), (Pcomm) + (DN) imply ($\text{m}_1$-DN4'), hence ($\text{m}_1$-DN4') holds too.

Define $x \to y \stackrel{def.}{=} (x \cdot y^{-1})^{-1}$. We must prove that (M) and (Ex1) hold. Indeed,

$1 \to x \stackrel{def.}{=} (1 \cdot x^{-1})^{-1} \stackrel{(PU)}{=} (x^{-1})^{-1} \stackrel{(DN)}{=} x$; thus, (M) holds.

$x \to (y \to 1) \stackrel{def.}{=} (x \cdot ((y \cdot 1^{-1})^{-1})^{-1})^{-1} \stackrel{(DN),(N1)}{=} (x \cdot (y \cdot 1))^{-1}$

$\stackrel{(PU)}{=} (x \cdot y)^{-1} \stackrel{(Pcomm)}{=} (y \cdot x)^{-1} \stackrel{(PU)}{=} (y \cdot (x \cdot 1))^{-1}$

$\stackrel{(N1),(DN)}{=} (y \cdot ((x \cdot 1^{-1})^{-1})^{-1})^{-1} \stackrel{def.}{=} y \to (x \to 1)$; thus, (Ex1) holds.

(2): Let

$$(A^L, \to, 1) \quad \stackrel{\Phi_1}{\longrightarrow} \quad (A^L, \cdot, ^{-1}, 1) \quad \stackrel{\Psi_1}{\longrightarrow} \quad (A^L, \Rightarrow, 1).$$

Then, for all $x, y \in A^L$, we have:

$x \Rightarrow y = (x \cdot y^{-1})^{-1} = ((x \to (y^{-1})^{-1})^{-1})^{-1} \stackrel{(DN)}{=} x \to y$.

Conversely, let

$$(A^L, \cdot, ^{-1}, 1) \quad \stackrel{\Psi_1}{\longrightarrow} \quad (A^L, \to, 1) \quad \stackrel{\Phi_1}{\longrightarrow} \quad (A^L, \otimes, ', 1).$$

Then, for all $x, y \in A^L$, we have:

$x' = x \to 1 = (x \cdot 1^{-1})^{-1} \stackrel{(N1)}{=} (x \cdot 1)^{-1} \stackrel{(PU)}{=} x^{-1}$ and

$x \otimes y = (x \to y')' = (x \to y^{-1})^{-1} = (x \cdot (y^{-1})^{-1})^{-1} \stackrel{(DN)}{=} x \cdot y$.

(3): Suppose $x \leq y$, i.e. $x \to y = 1$, by (dfrelR); hence, $(x \cdot y^{-1})^{-1} = 1$, by $\Phi_1$; then, $x \cdot y^{-1} = 1$, by (DN) and (N1), hence $x \leq_{m_1} y$, by (m-dfrelP$_1$).

Conversely, suppose $x \leq_{m_1} y$, i.e. $x \cdot y^{-1} = 1$, by (m-dfrelP$_1$); then, $(x \cdot y^{-1})^{-1} = 1$, by (N1); hence $x \to y = 1$, by $\Psi_1$; hence $x \leq y$, by (dfrelR).

The proof of (4) is similar to the proof of (4) from Theorem 17.1.1. We prove only the following:

(BB=) $\Longleftrightarrow$ ($\text{m}_1$-BB=): Suppose that (BB=) holds; then,

$(z \cdot x)^{-1} \cdot (y \cdot x) = y \cdot z^{-1} \Longleftrightarrow (y \cdot z^{-1})^{-1} = [(z \cdot x)^{-1} \cdot (y \cdot x)]^{-1}$

$\stackrel{(DN)}{\Longleftrightarrow} (y \cdot z^{-1})^{-1} = [(z \cdot (x^{-1})^{-1})^{-1} \cdot ((y \cdot (x^{-1})^{-1})^{-1})^{-1}]^{-1}$

$\Longleftrightarrow y \to z = [(z \to x^{-1}) \cdot (y \to x^{-1})^{-1}]^{-1}$

$\Longleftrightarrow y \to z = (z \to x^{-1}) \to (y \to x^{-1})$ that is true by (BB=); thus, ($\text{m}_1$-BB=) holds too.

Conversely, suppose that ($\text{m}_1$-BB=) holds; then,

$y \to z = (z \to x) \to (y \to x) \Longleftrightarrow (y \cdot z^{-1})^{-1} = [(z \cdot x^{-1})^{-1} \cdot ((y \cdot x^{-1})^{-1})^{-1}]^{-1}$

$\stackrel{(DN)}{\Longleftrightarrow} (y \cdot z^{-1})^{-1} = [(z \cdot x^{-1})^{-1} \cdot (y \cdot x^{-1})]^{-1}$

$\stackrel{(DN)}{\Longleftrightarrow} (z \cdot x^{-1})^{-1} \cdot (y \cdot x^{-1}) = y \cdot z^{-1}$ that is true by ($\text{m}_1$-BB=); thus, (BB=) holds.

(B=) $\Longleftrightarrow$ ($m_1$-B=): Suppose that (B=) holds; then,
$(x \cdot y^{-1})^{-1} \cdot (x \cdot z) = y \cdot z \Longleftrightarrow (y \cdot z)^{-1} = [(x \cdot y^{-1})^{-1} \cdot (x \cdot z)]^{-1}$
$\overset{(DN)}{\Longleftrightarrow} (y \cdot (z^{-1})^{-1})^{-1} = [(x \cdot y^{-1})^{-1} \cdot (x \cdot (z^{-1})^{-1})^{-1})^{-1}]^{-1}$
$\Longleftrightarrow y \to z^{-1} = (x \to y) \to (x \to z^{-1})$ that is true by (B=); thus, ($m_1$-B=) holds.
Conversely, suppose that ($m_1$-B=) holds; then,
$y \to z = (x \to y) \to (x \to z) \Longleftrightarrow (y \cdot z^{-1})^{-1} = [(x \cdot y^{-1})^{-1} \cdot ((x \cdot z^{-1})^{-1})^{-1}]^{-1}$
$\overset{(DN)}{\Longleftrightarrow} (y \cdot z^{-1})^{-1} = [(x \cdot y^{-1})^{-1} \cdot (x \cdot z^{-1})]^{-1}$
$\Longleftrightarrow (x \cdot y^{-1})^{-1} \cdot (x \cdot z^{-1}) = y \cdot z^{-1}$ that is true by ($m_1$-B=); thus, (B=) holds.

(D=) $\Longleftrightarrow$ ($m_1$-D=): Suppose that (D=) holds; then,
$(y^{-1} \cdot x)^{-1} \cdot x = y \Longleftrightarrow y^{-1} = [(y^{-1} \cdot x)^{-1} \cdot x]^{-1}$
$\overset{(DN)}{\Longleftrightarrow} y^{-1} = [(y^{-1} \cdot (x^{-1})^{-1})^{-1} \cdot (x^{-1})^{-1}]^{-1}$
$\Longleftrightarrow y^{-1} = (y^{-1} \to x^{-1}) \to x^{-1}$ that is true by (D=); thus, ($m_1$-D=) holds too.
Conversely, suppose that ($m_1$-D=) holds; then,
$y = (y \to x) \to x \Longleftrightarrow [(y \to x) \to x]^{-1} = y^{-1}$
$\Longleftrightarrow [[(y \cdot x^{-1})^{-1} \cdot x^{-1}]^{-1}]^{-1} = y^{-1} \overset{(DN)}{\Longleftrightarrow} (y \cdot x^{-1})^{-1} \cdot x^{-1} = y^{-1}$
$\overset{(DN)}{\Longleftrightarrow} ((y^{-1})^{-1} \cdot x^{-1})^{-1} \cdot x^{-1} = y^{-1}$ that is true by ($m_1$-D=); thus, (D=) holds too.

(p-s) $\Longleftrightarrow$ (m-p-s):
(p-s) $\Longleftrightarrow (x \leq y \Longleftrightarrow x = y) \overset{(3)}{\Longleftrightarrow} (x \leq_{m_1} y \Longleftrightarrow x = y) \Longleftrightarrow$ (m-p-s). □

By this Theorem 17.2.1, choosing appropriate definitions, one can prove all the definitionally equivalences between the corresponding involutive algebras from the two 'worlds', as for examples between commutative i-goops and commutative goops, between commutative i-groups and commutative groups, etc.

# Bibliography


[1] Marlow Anderson, and Todd Feil, Lattice-Ordered Groups, *An Introduction*, D. Reidel Publishing Company, Dordrecht, etc. 1988.

[2] Emil Artin, *Theory of braids*, Ann. Math., 48, 1947, pp. 101–126.

[3] Raymond Balbes, and Philip Dwinger, Distributive lattices, Univ. Missouri Press, 1974.

[4] Garrett Birkhoff, Lattice Theory, American Mathematical Society, Providence RI 1967 (3rd edition) [*Colloquium Publications 25*].

[5] Garrett Birkhoff, and John von Neumann, *The logic of quantum mechanics*, Ann. Math. 37, 1936, 823=843= J. von Neumann, Collected Papers, Pergamon Press, 1961, vol. IV, 105–125.

[6] Willem J. Blok, and Don Pigozzi, Algebraizable logics, Memoirs of the American Mathematical Society, 396, Amer. Math. Soc., Providence, 1989.

[7] Willem J. Blok, and James G. Raftery, *Varieties of Commutative Residuated Integral Pomonoids and Their Residuation Subreducts*, Journal of Algebra 190, 1997, 280–328.

[8] Willem J. Blok, and Jordi Rebagliato, *Algebraic Semantics for Deductive Systems*, Studia Logica, 74, 2003, pp. 153–180.

[9] George Boole, The Mathematical Analysis of Logic, Macmillan, Barclay & Macmillan, Cambridge 1847.

[10] — An Investigation into the Laws of Thought, Macmillan, London 1854 [reprint: Open Court Publ. Co., Chicago IL 1940].

[11] Rajab Ali Borzooei, Arsham Borumand Saeid, Akbar Rezaei, Akefe Radfar, and Reza Ameri, *On pseudo-BE algebras,* Discussiones Mathematicae, General Algebra and Applications, 33, 2013, 95–108.

[12] Bruno Bosbach, *Rechtskomplementäre Halbgruppen*, Math. Z., 124, 1972, pp. 273–288.

[13] — *Concerning cone algebras*, Algebra univers., 15 , 1982, pp. 58–66.

[14] Felix Bou, Francesco Paoli, Antonio Ledda, and Hector Freytes, *On some properties of quasi-MV algebras and* $\sqrt{'}$*quasi-MV algebras. Part II*, Soft Comput., 12 (4), 2008, pp. 341–352.

[15] Felix Bou, Francesco Paoli, Aantonio Ledda, Matthew Spinks, and Roberto Giuntini, *The Logic of Quasi-MV algebras*, Journal of Logic and Computation, 20, 2, 2010, pp. 619–643.

[16] Egbert Brieskorn, and Kyoji Saito, *Artin-Gruppen und Coxeter-Gruppen*, Invent. Math. 17 , 1972, pp. 245–271.

[17] Martin W. Bunder, *Simpler axioms for* BCK*-algebras and the connection between the axioms and the combinators* B, C *and* K, Mathematica Japonica 26, 1981, pp. 415–418.

[18] Dumitru Buşneag, Contributions to the study of Hilbert algebras [Romanian], PhD Dissertation, University of Bucharest, 1985.

[19] Dumitru Buşneag, and Sergiu Rudeanu *A glimpse of deductive systems in algebra*, Cent. Eur. J. Math. 8 (4), 2010, pp. 688–705.

[20] Ivan Chajda, Radomir Halas, and Jan Kühr, *Distributive lattices with sectionally antitone involutions*, Acta Sci. Math. (Szeged) 71, 2005, 19–33.

[21] — Semilattice Structures, Research and Exposition in Mathematics 30, Heldermann, Verlag, 2007.

[22] — *Many-valued quantum algebras*, Algebra Universalis 60, 2009, 63–90.

[23] Chen Chung Chang, *Algebraic analysis of many valued logics*, Trans. Amer. Math. Soc. 88, 1958, pp. 467–490.

[24] Roberto L. O. Cignoli, Itala M. L. D'Ottaviano, and Daniele Mundici, Algebraic Foundations of Many-valued Reasoning, Kluwer Academic Publishers & Springer Science, Dordrecht 2000 [*Trends in Logic - Studia Logica Library 7*].

[25] Jänis Cirulis, *Implication in sectionally pseudocomplemented posets*, Acta Sci. Math. (Szeged) 74, 2008, pp. 477–491.

[26] — *Subtraction-like operations in nearsemilattices*, Dem. Math. 43, 2010, pp. 725–738.

[27] — *Quasi-orthomodular posets and weak* BCK-*algebras*, Order, DOI 10.1007/s11083-013-9309-1.

[28] Lavinia Corina Ciungu, Non-commutative Multiple-Valued Logic Algebras, Springer, Cham, Heidelberg, New York, Dordrecht, London, 2014.

[29] — *Monadic classes of quantum B-algebras*, Soft Computing 25, 2021, 1–14.

[30] — *Quantum B-algebras with involutions*, Journal of algebra and its applications 20(12), 2021, Article ID 2150233

[31] — *Valued quantum B-algebras*, Fuzzy Sets and Systems, https://doi.org/10.1016/j.fss.2022.05.002

[32] — *Deductive systems in unital quantum B-algebras*, Bulletin of the Belgian Mathematical Society, to appear.

[33] — *Results in L-algebras*, Algebra Universalis 82, 2021, No. 1, Paper No 7

[34] — *Quantifiers on L-algebras*, Math. Slovaca, to appear

[35] — *The category of L-algebras*, Transactions on Fuzzy Sets and Systems, to appear.

[36] Paul F. Conrad, Lattice-ordered Groups, Tulane Lecture Notes, New Orleans, 1970.

[37] Haskell B. Curry, Robert Feys, and William Craig, Combinatory logic, Volume 1, North Holland, Amsterdam, 1958.

[38] Maria Luisa Dalla Chiara, Roberto Giuntini, and Richard Greechie, Reasoning in Quantum Theory, *Sharp and Unsharp Quantum Logics*, Springer 2004 [*Trends in Logic – Studia Logica Library 22*].

[39] Richard Dedekind, *Über Zerlegung von Zahlen durch ihre grössten gemeinsamen Teiler*, Festchrift Tech. Hochschule Braunschweig, 1-40. = Gesammelte Math. Werke, vol. II, 1897, pp. 103–147.

[40] Patrick Dehornoy, *Groupes de Garside*, Ann. Sci. Éc. Norm. Supér. 35(2), 2002, pp. 267–306.

[41] Pierre Deligne, *Les immeubles des groupes de tresses généralisés*, Invent. Math. 17, 1972, pp. 273–302.

[42] Antonio Diego, Sur les algèbres de Hilbert, PhD Disserrtation, Gauthier-Villars, Paris 1966. [*Collection de Logique mathématique, Serie A, XXI*]. (A translation of the PhD Dissertation Sobre álgebras de Hilbert, Instituto de Matemática, Universidad Nacional del Sur, Bahia Blanca 1965 [*Nótas de Logica Matemática 12*].)

[43] Robert P. Dilworth, *Non-commutative residuated lattices*, Trans. of the American Math. Soc. 46, 1939, pp. 426–444.

[44] Wieslaw A. Dudek, and Young-Bae Jun, *Pseudo-*BCI *algebras*, East Asian Math. J., 24, 2008, pp. 187–190.

[45] Anatolij Dvurečenskij, and Sylvia Pulmanová, New Trends in Quantum Algebras, Kluwer Academic Publishers, Dordrecht & Ister Science, Bratislava, 2000.

[46] Grzegorz Dymek, *On two classes of pseudo* BCI-*algebras*, Discuss. Math., Gen. Algebra Appl. 31, 2011, pp. 217–229.

[47] — *p-semisimple pseudo*-BCI-*algebras*, J. of Multiple-Valued Logic and Soft Computing, 19, 2012, pp. 461–474.

[48] — *On compatible deductive systems of pseudo*-BCI-*algebras*, J. of Multiple-Valued Logic and Soft Computing, 22, 2014, pp. 167–187.

[49] Grzegorz Dymek, and Anna Kozanecka-Dymek, *Pseudo*-BCI-*logic*, Bull. Sect. Logic. 42, 2013, pp. 33–41.

[50] Francesc Esteva, and Lluis Godo, *Monoidal t-norm based logic: towards a logic for left-continuous t-norms*, Fuzzy Sets and Systems 124 (3), 2001, pp. 271–288.

[51] Francesc Esteva, Lluis Godo, and Carles Noguera, *On rational weak nilpotent minimum logics*, J. Mult.-Valued Logic Soft Comput. 12, 2006, pp. 9–32.

[52] Pavel Etingof, Travis Schedler, and Alexandre Soloviev, *Set-theoretical solutions to the quantum Yang-Baxter equation*, Duke Math. J. 100, 1999, pp. 169–209.

[53] Davide Fazio, Antonio Ledda, and Francesco Paoli, *Residuated Structures and Orthomodular Lattices*, Studia Logica, https://doi.org/10.1007/s11225-021-09946-1, Published online 17 April 2021.

[54] Paul Flondor, George Georgescu, and Afrodita Iorgulescu, *Pseudo-t-norms and pseudo*-BL *algebras*, Soft Computing 5 (5), 2001, pp. 355–371.

[55] János C. Fodor, *Contrapositive symmetry of fuzzy implications*, Fuzzy Sets and Systems, 69, 1995, pp. 141–156.

[56] Josep Maria Font, Antonio J. Rodríguez, and Antoni Torrens, *Wajsberg algebras*, Stochastica 8 (1), 1984, pp. 5–31.

[57] Harry Furstenberg, *The inverse operation in groups*, Proc. Amer. Math. Soc., 6, 1955, pp. 991–997.

[58] Nick Galatos, Peter Jipsen, Tomasz Kowalski, and Hiroakira Ono, *Residuated lattices: An Algebraic Glimpse at Substrucural Logics*, Studies in Logic and the Foundation of Mathematics, Vol. 151, Elsevier, 2007.

[59] George Georgescu, and Afrodita Iorgulescu, *Pseudo-MV Algebras: a non-commutative Extension of MV Algebras*, The Proceedings of the Fourth International Symposium on Economic Informatics, INFOREC Printing House, Bucharest, Romania, May 1999, pp. 961–968.

[60] — *Pseudo-BL algebras: A non-commutative extension of BL algebras*, Abstracts of The Fifth International Conference FSTA 2000, Slovakia, February 2000, pp. 90–92.

[61] — *Pseudo-MV algebras*, Mult. Val. Logic (A special issue dedicated to the memory of Gr. C. Moisil), Vol. 6, Nr. 1-2, 2001, pp. 95–135.

[62] — *Pseudo*-BCK *algebras: An extension of* BCK *algebras*, Proceedings of DMTCS'01: Combinatorics, Computability and Logic (C.S. Calude, M.J. Dinneen, S. Sburlan (Eds)), Springer, London, 2001, pp. 97–114.

[63] — Logică matematică [Mathematical Logic] [Romanian], Editura A.S.E. [Academy of Economic Studies Publishing House], Bucharest 2010.

[64] George Georgescu, Laurenţiu Leuştean, and Viorel Preoteasa, *Pseudo-hoops*, J. Mult.-Valued Logic Soft Comput. 11, 2005, pp. 153–184.

[65] José Gil-Férez, Antonio Ledda, Francesco Paoli, and Constantine Tsinakis, *Projectable l-groups and algebras of logic: Categorical and algebraic connections,* Journal of Pure and Applied Algebra 220, 2016, pp. 3514–3532.

[66] Roberto Giuntini, Quasilinear QMV algebras, *Inter. J. Theor. Phys. 34*, 1995, pp. 1397–1407.

[67] — *Quantum MV algebras*, Studia Logica 56, 1996, pp. 393–417.

[68] — *Quantum Logic and Hidden Variables*, **8**, B.I. Wissenschaftfsverlag, Mannheim, Wien, Zürich, 1991.

[69] — *Unsharp orthoalgebras and quantum MV algebras*, In: The Foundations of Quantum Mechanics, (C. Garola, A. Rossi, eds.), Kluwer Acad. Publ., Dordrecht, 1996, pp. 325–337.

[70] — *Quantum MV-algebras and commutativity*, Inter. J. Theor. Phys. 37, 1998, pp. 65–74.

[71] — *Weakly linear quantum MV-algebras*, Algebra Universalis 53, 2005, pp. 45–72. (doi: https://doi.org/10.1007/s00012-005-1907-310.1007/s00012-005-1907-3

[72] — *An independent axiomatization of quantum MV algebras*. In: C. Carola, A. Rossi (eds.), The Foundations of Quantum Mechanics, World Scientific, Singapore, 2000, pp. 233–249.

[73] Robert I. Goldblatt, *The Stone space of an ortholattice*, Bulletin of the London Mathematical Society 7, 1, 1975, pp. 45–48.

[74] George Grätzer, Universal Algebra, D. Van Nostrand Company, Inc., 1968.

[75] Andrzej Grzaślewicz, *On some problem on* BCK-*algebras*, Math. Japonica 25, No. 4, 1980, pp. 497–500.

[76] Stanley Gudder, *Total extension of effect algebras*, Found. Phys. Letters 8, 1995, pp. 243–252.

[77] Petr Hájek, *Metamathematics of fuzzy logic*, Technical Report 682, Institute of Computer Science, Academy of Science of the Czech Republic, 1996.

[78] — *Basic Fuzzy Logic and* BL-*Algebras*, Soft Computing 2, 1998, pp. 124–128.

[79] — Metamathematics of Fuzzy Logic, Kluwer Academic Publishers, Dordrecht 1998 [*Trends in Logic 4*].

[80] — *Observations on non-commutative logic*, Soft Comput., 8, 2003, pp. 38–43.

[81] Shengwei Han, Xiaoting Xu, and Feng Qin, *The unitality of quantum* B-*algebras*, Int. J. Theor. Phys. 57, 2018, pp. 1582–1590.

[82] Shengwei Han, Rongrong Wang, and Xiaoting Xu, *On the injective hulls of quantum* B-*algebras*, Fuzzy Sets Syst. **369**, 2019, pp. 114–121.

[83] Leon Henkin *An algebraic characterization of quantifiers*, Fundamenta Mathematicae 37, 1950, pp. 63–74.

[84] L. Herman, E.L. Marsden, and Robert Piziak, *Implication connectives in orthomodular lattices*, Notre Dame J. Form. Log. **16**, 1975, pp. 305–328.

[85] Graham Higman, and Bernhard Hermann Neumann, *Groups as grupoids with one law*, Publicationes Mathematicae, Debrezin, 2, 1952, pp. 215–221.

[86] David Hilbert, *Die logischen Grundlagen der Mathematik*, Mathematischen Annalen 88, 1923, pp. 151–165.

[87] David Hilbert, and Paul Bernays, Grundlagen der Mathematik 1, Springer Verlag, Berlin 1934.

[88] Ulrich Höhle, *Commutative, residuated l-monoids*. in: Ulrich Höhle, and Erich Peter Klement (eds.), Non-Classical Logics and Their Applications to Fuzzy Subsets, Kluwer Acad. Publ., Dordrecht 1995, pp. 53–106.

[89] Qingping Hu, and Xin Li, *On* BCH-*algebras*, *Math. Sem. Notes 11*, 1983, pp. 313–320.

[90] — *On proper* BCH-*algebras*, Mathematica Japonica 30, 1985, pp. 659–661.

[91] Yisheng Huang, BCI-algebras, Science Press, Beijing 2006. (Distributed by Elsevier Science.)

[92] Pawel M. Idziak, *Lattice operations in BCK-algebras,* Mathematica Japonica 29, 1984, pp. 839–846.

[93] Yasuyuki Imai, and Kioshi Iséki, *On axiom systems of propositional calculi* XIV, Proc. Japan Academy 42, 1966, pp. 19–22.

[94] Afrodita Iorgulescu, *$\mathcal{S}$-prealgebras*, Discrete Mathematics 126, 1994, pp. 415–419.

[95] — *Some direct ascendents of Wajsberg and* MV *algebras*, Scientiae Mathematicae Japonicae, Vol. 57, No. 3, 2003, pp. 583–647.

[96] — *On pseudo-*BCK *algebras and porims*, Scientiae Mathematicae Japonicae, Vol. 10, Nr. 16, 2004, pp. 293–305.

[97] — *Iséki algebras. Connection with BL algebras*, Soft Computing 8, No. 7, 2004, pp. 449–463.

[98] — *Pseudo-Iséki algebras. Connection with pseudo-*BL *algebras*, J. of Multiple-Valued Logic and Soft Computing, Vol. 11 (3-4), 2005, pp. 263–308.

[99] — *Classes of pseudo-*BCK *algebras - Part I*, J. of Multiple-Valued Logic and Soft Computing, Vol. 12, No. 1-2, 2006, pp. 71–130.

[100] — *Classes of pseudo-*BCK *algebras - Part II*, J. of Multiple-Valued Logic and Soft Computing, Vol. 12, No. 5-6, 2006, pp. 575–629.

[101] — *On* BCK *algebras - Part I.a: An attempt to treat unitarily the algebras of logic. New algebras*, J. of Universal Computer Science, Vol. 13, no. 11, 2007, pp. 1628–1654.

[102] — *On* BCK *algebras - Part I.b: An attempt to treat unitarily the algebras of logic. New algebras*, J. of Universal Computer Science, Vol. 14, no. 22, 2008, pp. 3686–3715.

[103] — *Monadic Involutive Pseudo-*BCK *algebras*, Acta Universitatis Apulensis, No. 15, 2008, pp. 159–178.

[104] — *Algebras of logic as* BCK *algebras*, Academy of Economic Studies Press, Bucharest 2008.

[105] — *Asupra algebrelor Booleene* [Romanian], Revista de logică, `http://egovbus.net/rdl`, 25.01.2009, pp. 1–25.

[106] — *Classes of examples of pseudo-*MV *algebras, pseudo-BL algebras and divisible bounded non-commutative residuated lattices*, Soft Computing, Vol. 14, no. 4, 2010, pp. 313–327.

[107] — *The implicative-group - a term equivalent definition of the group coming from algebras of logic*, Part I, Preprint nr. 11/2011, Institutul de Matematică "Simion Stoilow" al Academiei Române, Bucharest 2011.

[108] — *The implicative-group - a term equivalent definition of the group coming from algebras of logic*, Part II, Preprint nr. 12/2011, Institutul de Matematică "Simion Stoilow" al Academiei Române, Bucharest 2011.

[109] — *On l-implicative-groups and associated algebras of logic*, presented at Congmatro 2011-Braşov, Bulletin of the Transilvania University of Braşov [# Series III: Mathematics, Informatics, Physics], 5 (54) 2012, pp. 179–194. (Special Issue: Proceedings of the Seventh Congress of Romanian Mathematicians, published by Transilvania University Press, Braşov, and the Publishing House of the Romanian Academy, Bucharest.)

[110] — *New algebras and new connections between/in the algebras of logic and the monoidal algebras*, Libertas Mathematica [Arligtom TX], 35 (1), 2015, pp. 13–55. (Dedicated to Nicolae Dinculeanu and Solomon Marcus in celebration of their 90th Birthday.)

[111] — *Quasi-algebras versus regular algebras*, Part I (Dedicated to Sergiu Rudeanu), Scientific Annals of Computer Science 25 (1), 2015, pp.1–43 (doi: 10.7561/SACS.2015.1.ppp).

[112] — *Quasi-algebras vs. regular algebras*, Part II, 2015, Preprint nr. 3/2017, `http://imar.ro/~increst/2017/3_2017.pdf`.

[113] — *Quasi-algebras vs. regular algebras*, Part III, 2015, Preprint nr. 4/2017, `http://imar.ro/~increst/2017/4_2017.pdf`.

[114] — *New generalizations of* BCI, BCK *and Hilbert algebras*, Parts I, II (Dedicated to Dragoş Vaida), J. of Multiple-Valued Logic and Soft Computing 27 (4), 2016, pp. 353–406, and 407–456. (A previous version available from December 6, 2013, at `http://arxiv.org/abs/1312.2494`.)

[115] — *Implicative-groups*, Colocviu de matematică-informatică, Curtea de Argeş, România, 10-11 octombrie 2016.

[116] — Implicative-groups vs. groups and generalizations, Matrix Rom, Bucharest 2018.

[117] — *Generalizations of MV algebras, ortholattices and Boolean algebras*, ManyVal 2019, Bucharest, Romania, November 1-3, 2019.

[118] — *Algebras of logic vs. Algebras*, Landscapes in Logic, Vol. 1 Contemporary Logic and Computing, Editor Adrian Rezuş, College Publications, 2020, pp. 157–258.

[119] — *On Quantum-MV algebras - Part I: Orthomodular algebras*, Scientific Annals of Computer Science. Vol. 31(2), 2021, pp. 163–221.

[120] — *On Quantum-MV algebras - Part II: Orthomodular lattices, softlattices and widelattices*, Trans. Fuzzy Sets Syst., Vol. 1, No. 1, 2022, pp. 1–41.

[121] — *On quantum-MV algebras - Part III: The properties (m-Pabs-i) and ($WNM_m$)*, Scientiae Mathematicae Japonicae, online version at https://www.jams.or.jp/notice/scmjol/2022.html, 2022, pp. 1–33.

[122] Afrodita Iorgulescu, and Michael Kinyon, *Putting bounded involutive lattices, De Morgan algebras, ortholattices and Boolean algebras on the "map"*, Journal of Applied Logics (JALs-ifCoLog), Special Issue on Multiple-Valued Logic and Applications, Vol. 8, No. 5, 2021, pp. 1169 - 1213.

[123] — *Two generalizations of bounded involutive lattices and of ortholattices*, Journal of Applied Logics (JALs-ifCoLog), Vol. 8, No. 7, 2021, pp. 2173 - 2218.

[124] — *Putting Quantum MV algebras on the "map"*, Scientiae Mathematicae Japonicae, online version at https://www.jams.or.jp/notice/scmjol/2021.html, 2021, pp. 1–27.

[125] Kiyoshi Iséki, *Algebraic formulations of propositional calculi*, Proc. Japan Acad. 41, 1965, pp. 803–807.

[126] — *An algebra related with a propositional calculus*, Proc. Japan Acad. 42, 1966, pp. 26–29.

[127] Kiyoshi Iséki, and Shôtarô Tanaka, *An introduction to the theory of* BCK-*algebras*, Math. Japonica 23 (1), 1978, pp. 1–26.

[128] Luisa Iturrioz, *A representation theory for orthomodular lattices by means of closure spaces*, Acta Mathematica Hungarica 47, 1986, pp. 145–151.

[129] Peter Jipsen, and Constantine Tsinakis, A Survey of Residuated Lattices, www.math.vanderbilt.edu/ pjipsen/

[130] Young Bae Jun, Eun Hwan Roh, and Hee Sik Kim, *On* BH-*algebras*, Scientiae Mathematicae Japonicae 1, 1998, pp. 347–354.

[131] Young Bae Jun, Hee Sik Kim, and Joseph Neggers, *Pseudo d-algebras*, Information Sciences 179, 2009, 1751–1759.

[132] Jacek K. Kabziński, BCI-*algebras from the point of view of logic*, Bull. Sect. Logic, Polish Acad. Sci., Inst. Philos. and Socio., 12, 1983, pp. 126–129.

[133] Gudrun Kalmbach, Orthomodular Lattices, Academic Press, London, New York, etc. 1983. [*London Mathematical Society Monographs 18*]

[134] Hee Sik Kim, and Young Hee Kim, *On* BE-*algebras*, *Sci. Math. Jpn.*, online e-2006 (2006), pp. 1192–1202.

[135] Jae-Doek Kim, Young-Mi Kim, and Eun-Hwan Roh, *A note on* GT-*algebras*, J. Korea Soc. Math. Educ. Ser. B Pure Appl. Math. 16, 2009, pp. 59–68.

[136] Yuichi Komori, *The separation theorem of the $\aleph_0$-valued Łukasiewicz propositional logic*, Reports of Faculty of Science, Shizuoka University 12, 1978, pp. 1–5.

[137] — *Super-Łukasiewicz implicational logics*, Nagoya Math. J. 72, 1978, pp. 127–133.

[138] — *Super-Łukasiewicz propositional logics*, Nagoya Math. J. 84, 1981, pp. 119–133.

[139] — *The variety generated by* BCC-*algebras is finitely based*, Reports Fac. Sci. Shizuoka Univ. 17, 1983, pp. 13–16.

[140] — *The class of BCC-algebras is not a variety*, Math. Japonica 29, 1984, pp. 391–394.

[141] Michiko Kondo, and Young Bae Jun *The class of B-algebras coincides with the class of groups*, Sci. Math. Jpn. 57, 2003, 197–199.

[142] Tomasz Kowalski, and Hiroakira Ono, Residuated lattices: An algebraic glimpse at logics without contraction, monograph, 2001

[143] Tomasz Kowalski, and Francesco Paoli, *Joins and subdirect products of varieties*, Algebra Universalis 65, 2011, pp. 371–391.

[144] Wolfgang Krull, *Axiomatische Begründung der allgemeinen Idealtheorie*, Sitzungsberichte der physikalisch-medizinischen Societät zu Erlangen 56, 1924, pp. 47–63.

[145] Jan Kühr, Pseudo-BCK algebras and related structures, Univerzita Palackého v Olomouci, 2007.

[146] Tiande D. Lei, and ChangChang Xi, *p-radical in BCI-algebras*, Math. Japonica 30 (4), 1985, pp. 511–517.

[147] Laurenţiu Leuştean, Representations of Many-valued Algebras, Editura Universitară, Bucharest, 2010.

[148] Yuan-Hong Lin, *Some properties on commutative BCK-algebras*, Scientiae Mathematicae Japonicae 55, No. 1, 2002, pp. 129–133, :e5, 235–239.

[149] Jiang-Hua Lu, Min Yan, and Yong-Chang Zhu, *On the set-theoretical Yang-Baxter equation*, Duke Math. J. 104, 2000, pp. 1–18.

[150] Jan Łukasiewicz, Elements of Mathematical Logic, Oxford, 1963.

[151] Bao Long Meng, CI-*algebras*, Sci. Math. Jpn. 71 (1), 2010, pp. 11–17.

[152] Jie Meng, *BCI-algebras and abelian groups*, Math. Japonica 32, 1987, pp. 693–696.

[153] Jie Meng, and Young Bae Jun, BCK-algebras, Kyung Moon Sa Co., Seoul 1994.

[154] Carew A. Meredith, and Arthur Norman Prior, *Notes on the axiomatics of the propositional calculus*, Notre Dame J. Formal Logic 4, 1963, pp. 171–187.

[155] William W. McCune, *Prover9 and Mace4*, available at http://www.cs.unm.edu/ mccune/Prover9.

[156] Grigore C. Moisil, *Recherches sur l'algèbre de la logique*, Ann. Sci. Univ. Jassy, 22, 1935, pp. 1–117.

[157] Antonio Monteiro, *Sur la definition des algébres de Łukasiewicz trivalentes*, Notas de Logica Matematica 11, 1974.

[158] — *Construction des algèbres de Łukasiewicz trivalentes dans les algèbres de Boole Monadiques-I*, Notas de Logica Mat. 11, 1974.

[159] Daniele Mundici, *MV-algebras are categorically equivalent to bounded commutative* BCK-*algebras*, Math. Japonica 31 (6), 1986, pp. 889–894.

[160] Joseph Neggers, and Hee Sik Kim, *On d-algebras*, Math. Slovaca 49, 1999, 19–26.

[161] — *On B-algebras*, Mat. Vesnik. 54, 2002, 21–29.

[162] Carles Noguera, Francesc Esteva, and Joan Gispert, *On triangular norm based axiomatic extensions of the Weak Nilpotent Minimum Logic*, Mathematical Logical Quaterly, 54, 2008, pp. 387-409.

[163] Mitsuhiro Okada, and Kazushige Terui, *The finite model property for various fragments of intuitionistic logic*, Journal of Symbolic Logic 64 , 1999, pp. 790–802.

[164] Hiroakira Ono, and Yuichi Komori, *Logics without the contraction rule*, Journal of Symbolic Logic 50, 1985, pp. 169–201.

[165] Óystein Ore, *On the foundations of abstract algebra, I.* Ann. Math. 36, 1935, pp. 406–437.

[166] Ranganathan Padmanabhan, and Sergiu Rudeanu, Axioms for Lattices and Boolean Algebras, World Scientific, Singapore, etc. 2008.

[167] Jan Pavelka, *On fuzzy logic II. Enriched residuated lattices and semantics of propositional calculi*, Zeitschrift für mathematische Logik und Grundlagen der Mathematik 25, 1979, pp. 119–134.

[168] DAOWU PEI, On equivalent forms of fuzzy logic systems NM and IMTL, *Fuzzy Sets and Systems* 138, 2003, pp. 187-195.

[169] Dana Piciu, Algebras of Fuzzy Logic, Editura Universitaria Craiova, Craiova 2007.

[170] Don Pigozzi, *Abstract algebraic logic: past, present and future. A personal view*, Workshop on Abstract Algebraic Logic, July 1–5, 1997, Centre de Recerca Matemàtica, Bellaterra (Spain), Quaderns núm. 10, 1998, pp. 122–138.

[171] — *Abstract algebraic logic*, in: M. Hazewinkel (Ed.), Encyclopaedia of Mathematics, Supplement III, Kluwer, Dordrecht, 2001, pp. 2–13.

[172] — *Partially ordered varieties and quasivarieties*, manuscript, 2004; See http://orion.math.iastate.edu/dpigozzi/.

[173] Daniel Ponasse, *Problémes d'universalité s'introduissant dans l'algebrisation de la logique mathématique*, Thése présenté á la Faculté des Sciences de l'Université de Clermont Ferrand, 1961.

[174] — Logique mathématique, O.C.D.L., Paris 1967.

[175] Daniel Ponasse, and Jean Claude Carrega, Algébre et topologie booléennes, Masson, Paris 1979.

[176] Arthur N. Prior, Formal logic, Second Edition. Clarendon Press, Oxford, 1962.

[177] James G. Raftery, *A perspective on the algebra of logic*, Quaest. Math. 34, 2011, pp. 275–325.

[178] — *Order algebraizable logics*, Annals of Pure and Applied Logic, 164, 2013, pp. 251–283.

[179] Helena Rasiowa An algebraic Approach to Non-Classical Logics, North-Holland, Amsterdam, 1974.

[180] Akbar Rezaei, Arsham Borumand Saeid, and K. Yousefi Sikari Saber, *On pseudo-CI algebras*, Soft Computing 23, 2019, pp. 4643–4654.

[181] Wolfgang Rump, *A decomposition theorem for square-free unitary solutions of the quantum Yang-Baxter equation*, Adv. Math. 193, 2005, pp. 40–55.

[182] — *L-algebras, self-similarity, and l-groups*, J. Algebra 320(6), 2008, pp. 2328–2348.

[183] — *Semidirect products in algebraic logic and solutions of the quantum Yang-Baxter equation*, J. Algebra Appl. 7(4), 2008, pp. 471–490.

[184] — *A general Glivenko theorem*, Algebra Univers. 61(3–4), 2009, pp. 455–473.

[185] — *Right l-groups, geometric Garside groups, and solutions of the quantum Yang-Baxter equation*, J. Algebra 439, 2015, pp. 470–510.

[186] — *Decomposition of Garside groups and self-similar L-algebras*, J. Algebra 485, 2017, pp. 118–141.

[187] — *Von Neumann algebras, L-algebras, Baer-monoids, and Garside groups*, Forum Math. 30(4), 2018, pp. 973–995.

[188] — *Symmetric quantum sets and L-algebras*, Int. Math. Res. Not., 2022 (February 2022) pp. 1770–1810, https://doi .org /10 .1093 /imrn /rnaa135.

[189] — *L-algebras with duality and the structure group of a set-theoretic solution to the Yang-Baxter equation*, J. Pure Appl. Algebra 224(8), 2020, pp. 106–314.

[190] — *Commutative L-algebras and measure theory*, Forum Math. 33(6), 2021, pp. 1527–1548.

[191] — *L-algebras and three main non-classical logics*, Annals of Pure and Applied Logic 173, 2022, pp. 103–121.

[192] — *Structure groups of L-algebras and Hurwitz action*, submitted for publication

[193] — *Quantum* B-*algebras*, Cent. Eur. J. Math. 11, 2013, pp. 1881–1899.

[194] — *The completion of a quantum* B-*algebra*, Cah. Topol. Géom. Différ. Catég. 57, 2016, pp. 203–228.

[195] — *Quantum* B-*algebras: their omnipresence in algebraic logic and beyond*, Soft Comput. 21, 2017, pp. 2521–2529.

[196] Wolfgang Rump, and X. Zhang, *L-effect algebras*, Stud. Log. 108(4), 2020, pp. 725–750.

[197] Wolfgang Rump, and Yi Chuan Yang, *Non-commutative logical algebras and algebraic quantales*, Ann. Pure Appl. Logic 165, 2014, pp. 759–785.

[198] — *Hereditary arithmetics*, J. Algebra 468, 2016, pp. 214–252.

[199] Regivan Santiago, Benjamin Bedregal, João Marcos, Carlos Caleiro, and Jocivania Pinheiro, *Semi-BCI Algebras*, J. of Multiple-Valued Logic and Soft Computing, vol. 32, 2019, 87–109.

[200] Marlow Sholander, *Postulates for distributive lattices*, Canadian J. of Math. 3 (1), 1951, pp. 28–30.

[201] Matthew Spinks, Contributions to the theory of Pre-BCK-Algebras, PhD Dissertation, Monash University, 2003.

[202] Kanduri Lakshminarasimha Swamy, *Dually residuated lattice ordered semigroups*, Mathematische Annalen 159, 1965, pp. 105–114.

[203] Shôtarô Tanaka, *On* $\wedge$*-commutative algebras*, Math. Seminar Notes (Kobe University), 3, (7) 1975.

[204] Tadeusz Traczyk, *On the structure of* BCK-*algebras with* $zx \cdot yx = zy \cdot xy$, Math. Japon. 33, 1988, no. 2, pp. 319–324.

[205] Hiroshi Yutani, *The class of commutative BCK-algebra is equationally definable*, Math. Seminar Notes (Kobe University), 5, 1977, pp. 207–210.

[206] — *On a system of axioms of a commutative* BCK-*algebra*, Math. Seminar Notes (Kobe University), 5, 1977, pp. 255–256.

[207] Mordchaj Wajsberg, *Beiträge zum Metaaussagenkalkül*, Monat. Math. Phys. 42, 1935, p. 240.

[208] Andrzej Walendziak, *Some axiomatizations of B-algebras*, Math. Slovaca 56, 2006, 301–306.

[209] — *Pseudo-BCH algebras*, Discussiones Mathematicae, General algebra and Applications 35, 2015, 5–19.

[210] — *The property of commutativity for some generalizations of* BCK-*algebras*, Soft Computing 23, 2019, pp. 7505–7511.

[211] — *The implicative property for some generalizations of* BCK-*algebras*, J. of Multiple-Valued Logic and Soft Computing, Vol. 31, 2018, pp. 591–611.

[212] Guo-Jun Wang, *A formal deductive system for fuzzy propositional calculus,* Chinese Sci. Bull. 42, 1997, pp. 1521–1526.

[213] Morgan Ward, *Residuated distributive lattices*, Duke Math. Journal 6, 1940, pp. 641–651.

[214] Morgan Ward, and Robert P. Dilworth, *Residuated lattices*, Trans. of the American Math. Soc. 45, 1939, pp. 335–354.

[215] Annika Meike Wille, Residuated Structures with Involution, Ph.D. Thesis, Shaker Verlag, Aachen 2006.

[216] Andrzej Wroński, BCK-*algebras do not form a variety*, Math. Japon. 28, 1983, pp. 211–213.

[217] — *An algebraic motivation for* BCK-*algebras*, Math. Japon. 30, 1985, no. 2, pp. 187–193.

[218] Xiaohong Zhang, $\mathsf{BIK}^+$*-logic and non-commutative fuzzy logics*, Fuzzy Systems Math., 21, 2007, pp. 31–36.

[219] Xiaohong Zhang, and Ruifen Ye, *BZ-algebras and groups*, J. Math. Phys. Sci., 29, 1995, pp. 223–233.

# Index

www.ingramcontent.com/pod-product-compliance
Lightning Source LLC
Chambersburg PA
CBHW071057050326
40690CB00008B/1048